AF804869

BD 040163067 0

Foundations of Regenerative Medicine

Foundations of Regenerative Medicine

Anthony Atala, MD
W.B. Boyce Professor and Director, Wake Forest Institute for Regenerative Medicine, Wake Forest School of Medicine,
Professor, Virginia Tech-WakeForest University
School of Biomedical Engineering and Sciences
Winston-Salem, North Carolina, USA

Robert Lanza, MD
Advanced Cell Technology, Worcester, Massachusetts, USA

James A. Thomson, PhD
Wisconsin Regional Primate Research Center,
Department of Anatomy, Madison, Wisconsin, USA

and

Robert M. Nerem, PhD
Georgia Institute of Technology, Atlanta, Georgia, USA

ELSEVIER

AMSTERDAM • BOSTON • HEIDELBERG • LONDON • NEW YORK • OXFORD
PARIS • SAN DIEGO • SAN FRANCISCO • SINGAPORE • SYDNEY • TOKYO
Academic Press is an imprint of Elsevier

Academic Press is an imprint of Elsevier
525 B Street, Suite 1900, San Diego, CA 92101-4495, USA
30 Corporate Drive, Suite 400, Burlington, MA 01803, USA.
32 Jamestown Road, London NWI 7BY, UK

First edition 2010

Copyright © 2010 Elsevier Inc. All rights reserved

No part of this publication may be reproduced, stored in a retrieval system or transmitted in any form or by any means electronic, mechanical, photocopying, recording or otherwise without the prior written permission of the publisher

Permissions may be sought directly from Elsevier's Science & Technology Rights Department in Oxford, UK: phone (+44) (0) 1865 843830; fax (+44) (0) 1865 853333; email: permissions@elsevier.com. Alternatively you can submit your request online by visiting the Elsevier website at http://elsevier.com/locate/permissions, and selecting *Obtaining permission to use Elsevier material*

Notice
No responsibility is assumed by the publisher for any injury and/or damage to persons or property as a matter of products liability, negligence or otherwise, or from any use or operation of any methods, products, instructions or ideas contained in the material herein. Because of rapid advances in the medical sciences, in particular, independent verification of diagnoses and drug dosages should be made

British Library Cataloguing-in-Publication Data
A catalogue record for this book is available from the British Library

Library of Congress Cataloging-in-Publication Data
A catalog record for this book is available from the Library of Congress

ISBN: 978-0-12-375085-3

For information on all Academic Press publications
visit our web site at books.elsevier.com

Typeset by Macmillan Publishing Solutions
(www.macmillansolutions.com)

Printed and bound in Canada

10 11 12 13 14 10 9 8 7 6 5 4 3 2 1

Working together to grow
libraries in developing countries
www.elsevier.com | www.bookaid.org | www.sabre.org
ELSEVIER BOOK AID International Sabre Foundation

I would like to dedicate this textbook to the joys of my life – my wife, Katherine, and my children, Christopher and Zachary

–Anthony Atala

Contents

Preface xi
List of Contributors xxiii

Part I Introduction to Regenerative Medicine 1

1. **Current and Future Perspectives of Regenerative Medicine** 2
   Mark E. Furth and Anthony Atala

2. **Fundamentals of Cell-Based Therapies** 16
   Ross Tubo

3. **Stem Cell Research** 28
   T. Ahsan, A.M. Doyle, and R.M. Nerem

Part II Biologic and Molecular Basis of Regenerative Medicine 49

4. **Molecular Organization of Cells** 50
   Jon D. Ahlstrom and Carol A. Erickson

5. **Cell–ECM Interactions in Repair and Regeneration** 66
   M. Petreaca and M. Martins-Green

6. **Developmental Mechanisms of Regeneration** 100
   David L. Stocum

7. **The Molecular Basis of Pluripotency in Principles of Regenerative Medicine** 126
   Ariel J. Levine and Ali H. Brivanlou

Part III Cells and Tissue Development 137

8. **Embryonic Stem Cells: Derivation and Properties** 138
   Junying Yu and James A. Thomson

9. **Stem Cells Derived from Amniotic Fluid and Placenta** 154
   Paolo De Coppi, Shay Soker, and Anthony Atala

10. **Bone Marrow Stem Cells: Properties and Pluripotency** 166
    Munira Xaymardan, Massimo Cimini, Richard D. Weisel, and Ren-Ke Li

11. **Mesenchymal Stem Cells** 182
Zulma Gazit, Hadi Aslan, Yossi Gafni, Nadav Kimelman, Gadi Pelled, and Dan Gazit

12. **Islet Cell Therapy and Pancreatic Stem Cells** 208
Juan Domínguez-Bendala, Antonello Pileggi, and Camillo Ricordi

13. **Mechanical Determinants of Tissue Development** 228
Jonathan A. Kluge, Gary G. Leisk, and David L. Kaplan

14. **Morphogenesis and Morphogenetic Proteins** 246
A.H. Reddi

15. **Physical Stress as a Factor in Tissue Growth and Remodeling** 260
Robert E. Guldberg, Christopher S. Gemmiti, Yash Kolambkar, and Blaise Porter

16. **Engineering Cellular Microenvironments** 284
Wendy F. Liu, Elliot E. Hui, Sangeeta N. Bhatia, and Christopher S. Chen

17. **Applications of Nanotechnology** 302
Benjamin S. Harrison

Part IV Biomaterials for Regenerative Medicine 311

18. **Design Principles in Biomaterials and Scaffolds** 312
Hyukjin Lee and Tae Gwan Park

19. **Naturally Occurring Scaffold Materials** 326
Stephen F. Badylak

20. **Synthetic Polymers** 336
M.C. Hacker and A.G. Mikos

21. **Surface Modification of Biomaterials** 368
Andrés J. García

22. **Biocompatibility and Bioresponse to Biomaterials** 378
James M. Anderson

Part V Therapeutic Applications: Cell Therapy 399

23. **Islet Cell Transplantation** 400
Juliet A. Emamaullee and A.M. James Shapiro

24. **Cell-Based Repair for Cardiovascular Regeneration and Neovascularization: What, Why, How, and Where Are We Going in the Next 5–10 Years?** 418
Doris A. Taylor and Andrey G. Zenovich

25. **Cell Therapies for Bone Regeneration** 458
Rehan N. Khanzada, Chantal E. Holy, F. Jerry Volenec, and Scott P. Bruder

26. **Cell-Based Therapies for Musculoskeletal Repair** 478
Wan-Ju Li, Kiran Gollapudi, David P. Patterson, George T.-J. Huang, and Rocky S. Tuan

27. **Hepatocyte Transplantation** 502
Stephen C. Strom and Ewa C.S. Ellis

28. **Cell-Based Drug Delivery** 518
Grace J. Lim, Sang Jin Lee, and Anthony Atala

Part VI **Therapeutic Applications: Tissue Therapy** 531

29. **Engineering of Large Diameter Vessels** 532
Saami K. Yazdani and George J. Christ

30. **Cardiac Tissue** 554
Milica Radisic and Michael V. Sefton

31. **Intracorporeal Kidney Support** 576
James J. Yoo, Akira Joraku, and Anthony Atala

32. **Genitourinary System** 584
Anthony Atala

33. **Tissue Engineering of the Reproductive System** 596
Stefano Giuliani, Laura Perin, Sargis Sedrakyan, and Roger De Filippo

34. **Phalanges and Small Joints** 622
Makoto Komura, Daniel Eberli, James J. Yoo, and Anthony Atala

35. **Functional Tissue Engineering of Ligament and Tendon Injuries** 630
Savio L.-Y. Woo, Alejandro J. Almarza, Sinan Karaoglu, and Steven D. Abramowitch

36. **Tissue Therapy: Implications of Regenerative Medicine for Skeletal Muscle** 656
Shen Wei and Johnny Huard

37. **Peripheral Nerve Regeneration** 672
Mahesh C. Dodla and Ravi V. Bellamkonda

38. **Innovative Regenerative Medicine Approaches to Skin Cell-Based Therapy for Patients with Burn Injuries** 688
Jörg C. Gerlach, Steven E. Wolf, Christa Johnen, and Bernd Hartmann

Part VII Regulations and Ethics 713

39. **Ethical Considerations** 714
Louis M. Guenin

40. **To Make is to Know: The Ethical Issues in Human Tissue Engineering** 726
Laurie Zoloth

41. **Overview of FDA Regulatory Process** 734
Celia Witten, Ashok Batra, Charles N. Durfor, Stephen L. Hilbert, David S. Kaplan, Donald Fink, Deborah Lavoie, Ellen Maher, and Richard McFarland

42. **Current Issues in US Patent Law** 754
Patrea L. Pabst

Index 770

Preface

The first edition of the textbook *Principles of Regenerative Medicine* has had a broad appeal, and is currently ranked as the most widely distributed textbook in the field of regenerative medicine. Demand for a compendium that would incorporate the key chapters from the first edition led to the creation of this volume, the first edition of *Foundations of Regenerative Medicine*. This volume presents key topics in the field of regenerative medicine. The chapters picked for this edition were deemed to be those that form the basis for the field and would give the reader an adequate overview. We would like to thank the authors, all experts in the field, who updated their contributions and made this volume possible.

Regenerative medicine seeks to devise new therapies for patients with severe injuries or chronic diseases in which the body's own responses do not suffice to restore functional tissue. A recent publication from the US National Academy of Sciences, *Stem Cells and the Future of Regenerative Medicine* (Committee on the Biological and Biomedical Applications of Stem Cell Research, 2002), identified a wide array of major unmet medical needs which might be addressed by regenerative technologies. These include congestive heart failure (approximately 5 million US patients) (Murray-Thomas and Cowie, 2003), osteoporosis (10 million US patients), Alzheimer's and Parkinson's diseases (5.5 million patients each), severe burns (0.3 million), spinal cord injuries (0.25 million), and birth defects (0.15 million). Another area of critical need is diabetes mellitus (16 million US patients and more than 217 million worldwide) (Smyth and Heron, 2006). Patients with type 1 diabetes lack pancreatic beta-cells, essential for the production of insulin, because of autoimmune destruction and represent from 10% to 20% of the total. Many patients with type 2 diabetes also show insufficient pancreatic beta-cell mass. Thus, patients in both groups potentially might be treated if methods could be developed to promote endogenous regeneration of beta-cells or to provide enough surrogate beta-cells and pancreatic islets for transplantation (Weir, 2004).

Regenerative therapies comprising living cells have entered practice, initially through the widespread adoption of both allogeneic and autologous bone marrow transplantation (Thomas, 1999). The presence of hematopoietic progenitor and stem cells with great replicative capacity *in vivo*, and their ability to reenter the bone marrow niche from the circulation, enabled this major medical advance. Subsequently, the development of methods to expand *ex vivo* and deliver such cell types as keratinocytes and chondrocytes, through advances in cell culture and scaffold technologies, led to successful tissue engineering for wound repair (Johnson, 2000; Lavik and Langer, 2004). Despite significant challenges in development and manufacturing, several bioartificial skin graft and cartilage replacement products have achieved regulatory approval (Lysaght and Reyes, 2001; Naughton, 2002; Lysaght and Hazlehurst, 2004). These therapies validate the potential of cell-based regenerative approaches.

The source of cells for regenerative medicine remains a major topic of interest. A primary issue remains the choice between using a patient's own cells, or those of a closely matched relative, *versus* those from an unrelated allogeneic donor. More broadly, future developments depend heavily on increased understanding and effective utilization of multiple classes of progenitor and stem cells. The first part of this volume addresses some of the therapeutic needs and how stem and progenitor cells can be used for a broad range of diseases.

Copyright © 2010 Elsevier Inc. All rights reserved

The ability to produce enough cells of the necessary types from the skin, cartilage, or bladder for bioengineered products depended on the presence of stem and progenitor cells in the corresponding adult tissues. It also required the development of culture methods that both permit the expansion of the precursor cells and allow enough differentiation for generation of the desired neo-tissue. Implementation of this strategy for regenerative medicine, based on expansion of autologous cells, cannot yet be extended to all tissues and organs. In some cases, it is not clear how to obtain biopsies containing progenitor or stem cells, or even whether such cells exist. In other cases, culture conditions for expansion of the precursor cell population are not yet available.

The future development of cell-based regenerative medicine depends on further translation of basic discoveries regarding the identity and behavior of stem cells into practical clinical applications. Important targets include cells of organs for which orthotopic transplantation already has been established as an important mode of therapy, but for which the supply of donor organs does not meet the current need. Examples include cells of the heart, kidney, liver, and pancreas—specifically insulin-producing beta-cells. In addition, production of neurons and other cells of the nervous system may permit therapy of degenerative diseases for which no effective treatment yet exists.

Mammalian stem cells have been divided into three general categories, embryonic, fetal, and adult. Embryonic stem (ES) cells and the comparable embryonic germ (EG) cells appear to be able to give rise to all specialized cell types, with the exception of a limited set of extra-embryonic cells. Fetal stem cells are usually derived from the amniotic fluid or the placenta. Adult stem cells, which may actually derive from neonatal, or adult tissue, show varying degrees of restriction to particular lineages.

The ES cells can self-renew apparently without limit in culture, although mechanisms underlying this capacity remain incompletely understood (Rao, 2004; Stewart et al., 2006) and established ES lines may display some genomic instability. Furthermore, ES cells are broadly pluripotent (Evans and Kaufman, 1981; Martin, 1981; Shamblott et al., 1998; Amit et al., 2000). This great degree of plasticity represents both the strongest attraction and a significant potential limitation to the use of ES cells for regenerative medicine. A major remaining challenge is to direct the efficient production of pure populations of specific desired cell types from human ES cells (Odorico et al., 2001).

ES cells appear unique among normal stem cells in being tumorigenic, forming teratomas that contain cell types representing all three embryonic germ layers in a disorganized form (Martin, 1981; Thomson et al., 1998; Cowan et al., 2004). For clinical use, it will be important to exclude undifferentiated stem cells from any products derived from ES cells (Lawrenz et al., 2004). Strategies have been envisaged to increase safety by introducing into ES cells a "suicide" gene, for example that encoding the thymidine kinase of herpes simplex virus, which would render any escaping tumor cells sensitive to the drug ganciclovir (Odorico et al., 2001; Schuldiner et al., 2003). However, the genetic manipulation is itself not without risk, and the need to validate the engineered cell system would likely extend and complicate regulatory review of therapeutic products.

A central issue that must be addressed for tissue engineered products derived from ES cells, and also from any non-autologous adult stem cells, is immune rejection based on mismatches at genetic histocompatibility loci. It generally has been assumed that, because human ES cells and their differentiated derivatives can be induced to express high levels of MHC Class I antigens (e.g., HLA-A and HLA-B), any ES cell-based product will be subject to graft rejection (Drukker et al., 2002).

Therapeutic cloning offers a potential means to generate cells with the exact genetic constitution of each individual patient, so that immune rejection of grafts based on mismatched histocompatibility antigens should not occur. The approach entails transferring the nucleus of a somatic cell into an enucleated oocyte (somatic cell nuclear transfer, SCNT), generating a blastocyst, and then culturing the inner cell mass to obtain an ES cell line (Colman and Kind, 2000). If required, genetic manipulation of the cells may be carried out to correct an inherited defect prior to production of the therapeutic graft (Rideout et al., 2002). Despite a published claim (Hwang et al., 2005) later withdrawn, the generation of human ES cells by SCNT has not yet been achieved. However, the concept of therapeutic cloning to provide cells for tissue engineering applications has been clearly validated in a large animal model. Adult bovine fibroblasts were used as nuclear donors and bioengineered tissues were generated from cloned cardiac, skeletal muscle and kidney cells (Lanza et al., 2002). The grafts, including functioning renal units capable of urine production, were successfully transplanted into the corresponding donor animals long term with no evidence for rejection. Although SCNT is the subject of political, ethical, and scientific debate (Hall et al., 2006), intense efforts in both the private sector (Lysaght and Hazlehurst, 2003) and academic institutions are likely to yield cloned human lines in the near future.

The reprogramming of somatic cell nuclei to yield pluripotent cells after introduction into the cytoplasm of enucleated eggs raises the possibility that additional means may be found to create cells with expanded potential to yield desired differentiated cell types. Counter to intuition, it appears that nuclei taken from certain terminally differentiated cells, such as postmitotic neurons, are readily reprogrammed to yield pluripotent cells by SCNT (Eggan et al., 2004). Nuclei from more differentiated cells may actually be superior for this purpose than nuclei of adult stem cells (Inoue et al., 2006; Sung et al., 2006), although the opposite trend was noted in studies using nuclei from neuronal lineage cells (Blelloch et al., 2006). In addition, fusion of somatic cells to ES cells can reprogram the somatic nuclei to an embryonic state (Cowan et al., 2005). Most remarkably, the expression of a small set of genes usually associated with ES cells (e.g., Oct3/4, Sox2, c-Myc, and Klf4) can induce an "embryonic" state, including pluripotency and the capacity to form teratoma tumors, in at least some somatic cells (fibroblasts) (Takahashi and Yamanaka, 2006).

The properties and differentiation potential of a number of human ES cell lines obtained by traditional means from early embryos currently used for research have been reviewed recently (Hoffman and Carpenter, 2005). The clinical application of ES cells for tissue engineering will depend on the development of robust methods to isolate and grow them under conditions consistent with Good Manufacturing Practice and regulatory review for safety. In particular, it is important to eliminate the requirement for murine feeder cells by using human feeders or, better, feeder-free conditions. In addition, development of culture conditions without the requirement for non-human serum would be advantageous. Progress has been made in the derivation and expansion of human ES cells with human feeder cells (Amit et al., 2003; Hovatta et al., 2003; Yoo et al., 2005; Stacey et al., 2006) or entirely without feeders (Amit et al., 2004; Carpenter et al., 2004; Beattie et al., 2005; Hovatta and Skottman, 2005; Klimanskaya et al., 2005; Sjogren-Jansson et al., 2005).

Perhaps, the greater challenge remains in directing the differentiation of human ES cells to a given desired lineage with high efficiency. The underlying difficulty is that ES cells are developmentally many steps removed from adult, differentiated cells, and to date we have no general way to deterministically control the key steps in lineage restriction. Presumably, the same problem would be encountered with ES cells generated by SCNT or other means of reprogramming somatic cell nuclei.

To induce differentiation *in vitro* ES cells are allowed to attach to plastic in monolayer culture or, more frequently, to form aggregates called embryoid bodies (Itskovitz-Eldor et al., 2000). Overtime, cell types of many lineages are generated within these aggregates, including representatives of the three germ layers. The production of embryoid bodies can be enhanced and made more consistent by incubation in bioreactors (Gerecht-Nir et al., 2004). Further, selection of specific lineages generally requires sequential exposure to a series of inducing conditions, either based on known signaling pathways or identified by trial and error. In most cases, lineage-specific markers are expressed by the differentiated cells, but cells often do not progress to a full terminally differentiated phenotype. As summarized in recent reviews, the cell lineages which have been generated *in vitro* include, among others, several classes of neurons, astrocytes, oligodendrocytes, multipotent mesenchymal precursor cells, osteoblasts, cardiomyocytes, keratinocytes, pneumocytes, hematopoietic cells, hepatocytes, and pancreatic beta-cells (Nir et al., 2003; Tian and Kaufman, 2005; Raikwar et al., 2006; Trounson, 2006).

In general, it appears easier to obtain adult cells derived from ectoderm, including neurons, and mesoderm, including cardiomyocytes, than cells derived from endoderm (Trounson, 2006). This may help determine the first areas in which ES-derived cells enter clinical translation, once the barriers discussed above are surmounted. Dopaminergic neurons generated from primate and human ES cells already have been tested with encouraging results in animal models of Parkinson's disease (Perrier et al., 2004; Sanchez-Pernaute et al., 2005). Promising data have also been obtained with ES-derived oligodendrocytes in spinal cord injury models (Keirstead et al., 2005; Mueller et al., 2005). Cardiomyocytes derived from human ES cells, similarly, are candidates for future clinical use (He et al., 2003; Nir et al., 2003; Goh et al., 2005; Lev et al., 2005). However, the functional criteria that must be met to ensure physiological competence will be stringent because of the risk of inducing arrhythmias (Caspi and Gepstein, 2006; Passier et al., 2006).

The robust generation of pancreatic beta-cells and bioengineered islets from human ES cells or other stem cells would represent a particularly important achievement, with potential to treat diabetes (Weir, 2004; Nir and Dor, 2005). Clusters of insulin-positive cells, resembling pancreatic islets and expressing various additional markers of the endocrine pancreatic lineage, have been produced from mouse ES cells (Lumelsky et al., 2001) and also from non-human primate and human ES cells (Assady et al., 2001; Lester et al., 2004; Brolen et al., 2005; Baharvand et al., 2006). The production of beta-like cells can be enhanced by the expression of pancreatic transcription factors (Miyazaki et al., 2004; Shiroi et al., 2005). However, the assessment of differentiation must take into account the uptake of insulin from the growth medium, in addition to *de novo* synthesis (Paek et al., 2005). It seems fair to conclude that the efficient production of functional beta-cells from ES cells remains a difficult objective to achieve. As in other bioengineering applications with ES-derived cells, efforts to reverse diabetes also will depend on the complete removal of non-differentiated cells to avoid the formation of teratoma tumors, which were observed after implantation of ES-derived beta-cells in an animal model (Fujikawa et al., 2005).

Despite the acknowledged promise of ES cells, the challenges of controlling lineage-specific differentiation and eliminating residual stem cells are likely to extend the timeline for a number of tissue engineering applications. In many cases, adult stem cells may provide a more direct route to clinical translation.

Lineage-restricted stem cells have been isolated from both fetal and postnatal tissues based on selective outgrowth in culture and/or immunoselection for surface markers. Examples with significant potential for new applications in regenerative medicine include neural (Baizabal et al., 2003; Goh et al., 2003), cardiac (Beltrami et al., 2003; Oh et al., 2003), muscle-derived (Cao et al., 2005), and hepatic stem cells (Kamiya et al., 2006; Schmelzer et al., 2006). A significant feature of each of these populations is a high capacity for self-renewal in culture. Their ability to expand may be less than that for ES cells, but in some cases the cells have been shown to express telomerase and may not be subject to replicative senescence. These adult stem cells are multipotent. Neural stem cells can yield neurons, astrocytes, and oligodendrocytes. Cardiac stem cells are reported to yield cardiomyocytes, smooth muscle, and endothelial cells. Muscle-derived stem cells yield skeletal muscle and can be induced to produce chondrocytes. Hepatic stem cells yield hepatocytes and bile duct epithelial cells. The lineage-restricted adult stem cells all appear non-tumorigenic. Thus, unlike ES cells, it is likely that they could be used safely for bioengineered products with or without prior differentiation.

It is possible that some lineage-specific adult stem cells are capable of greater plasticity than might be supposed based solely on their tissue of origin. For example, there is evidence that hepatic stem cells may be induced to generate cells of additional endodermal lineages such as the endocrine pancreas (Yang et al., 2002; Nakajima-Nagata et al., 2004; Yamada et al., 2005; Zalzman et al., 2005). This type of switching of fates among related cell lineages may prove easier than inducing a full developmental program from a primitive precursor such as an ES cell.

Another class of adult cells with enormous potential value for regenerative medicine is the mesenchymal stem cells (MSC) initially described in bone marrow (Bruder et al., 1994; Pittenger et al., 1999). These multipotent cells are able to give rise to differentiated cells of connective tissues including bone, cartilage, muscle, tendon, and fat. The MSC have, therefore, generated considerable interest for musculoskeletal and vascular tissue engineering (Gao and Caplan, 2003; Tuan et al., 2003; Barry and Murphy, 2004; Guilak et al., 2004; Risbud and Shapiro, 2005). Cells with similar differentiation potential and marker profiles have been isolated from a number of tissues in addition to the bone marrow. A notable source is adipose tissue in which the cells are abundant and easily obtained by processing of suction-assisted lipectomy (liposuction) specimens (Zuk et al., 2001; Gimble and Guilak, 2003).

In general, it seems better to view MSC as mixed populations of progenitor cells with varying degrees of replicative potential, rather than homogeneous stem cells. However, some classes of MSC, including lines cloned from single cells in skin (Bartsch et al., 2005), have been maintained in culture for extended periods. A very small subset of mesenchymal cells from bone marrow, termed multipotent adult progenitor cell (MAPC), reportedly are capable of extensive self-renewal and of differentiation into cell lineages not observed with typical MSC, including examples from each embryonic germ layer (Jiang et al., 2002).

Cells originating in a developing fetus and isolated from amniotic fluid or chorionic villi are a new source of stem cells of great potential interest for regenerative medicine (De Coppi et al., 2001; Siddiqui and Atala, 2004; Tsai et al., 2006). Fetal-derived cells with apparently similar properties have also been described in the amnion of term placenta (Miki et al., 2005). Amniotic fluid stem (AFS) cells and amniotic epithelial cells can give rise to differentiated cell types representing the three embryonic germ layers (Siddiqui and Atala, 2004; Miki et al., 2005). Formal proof that single AFS cells can yield this full range of progeny cells was obtained using clones marked by retroviral insertion (unpublished data from A. Atala laboratory). The cells can be expanded for well over 200 population doublings with no sign of telomere shortening or replicative senescence, and retain a normal diploid karyotype. They are readily cultured without the need for feeder cells. The AFS cells express some markers in common with ES cells, such as the surface antigen SSEA4 and the transcription factor Oct3/4, while other markers are shared with mesenchymal and neural stem cells. A broadly multipotent cell population obtained from umbilical cord blood may have certain key properties in common with AFS cells, and was termed "unrestricted somatic stem cells" (USSCs) (Kogler et al., 2004). This population may overlap with or be identical to the so-called "umbilical cord matrix stem" (UCMS) cells isolated from Wharton's jelly (Mitchell et al., 2003; Weiss et al., 2006).

The full developmental potential of the various stem cell populations obtained from fetal and adult sources remains to be determined. It is possible that virtually all of the cell types that might be desired for tissue engineering could be obtained from AFS cells, equivalent stem cells from placenta, those from the non-hematopoietic subset in umbilical cord blood, or comparable populations. Similar approaches to those being taken with ES cells, such as genetic modification with expression vectors for lineage-specific transcription factors, may help in the generation of those differentiated cell types for which it proves difficult to develop a straightforward induction protocol using external signals. However, it will remain necessary to show, beyond induction of a set of characteristic markers, that fully functional mature cells can be generated for any given lineage.

MSC from bone marrow and their differentiated derivatives also have been shown both to escape an allogeneic immune response and to possess immunomodulatory activity to block such a response (Bartholomew et al., 2002; Le Blanc, 2003; Potian et al., 2003; Aggarwal and Pittenger, 2005). The effect likewise is observed with MSC isolated from adipose tissue (Puissant et al., 2005). The successful therapeutic use of allogeneic MSC has been confirmed in animal models (Arinzeh et al., 2003; De Kok et al., 2003). Therefore, beyond the application of MSC as regenerative cells, it is possible that they could be employed to induce immune tolerance to grafts of other cell types. The mechanisms underlying the immunomodulatory properties of MSC are under active investigation and understanding them may have profound impact on regenerative medicine (Plumas et al., 2005; Krampera et al., 2006; Sotiropoulou et al., 2006).

Other stem cell populations should be examined for their ability to escape and/or modulate an allogeneic immune response. While it is important to exercise caution in interpreting the laboratory results and in designing clinical trials, there is some reason to hope that the use of allogeneic stem cell-based bioengineered products will not necessarily imply the need for lifelong treatment with immunosuppressive drugs. In the first FDA-approved clinical trial of allogeneic human neural stem cells, in children with a neural ceroid lipofuscinosis disorder known as Batten disease (Taupin, 2006), immunosuppressive therapy will be utilized for the initial year after cell implantation and then reevaluated.

Banking of stem cells for future therapeutic use extends possibilities both for autologous and allogeneic therapy paradigms, even if it turns out that histocompatibility matching is important for stem cell-based therapies. Amniocentesis specimens, placenta, and cord blood represent sources from which highly multipotent adult stem cells can be obtained and typed with minimal invasiveness. Prospective parents could opt for collection and cryopreservation of such cells for future use by their children in the event of medical need. Furthermore, collection and typing of a sufficient number of samples (*ca.* 100,000 for the US population) to permit nearly perfect histocompatibility matching between unrelated donors and recipients would be readily achieved. Similarly, collection and banking of cells from adult adipose tissue appears straightforward. Although it would entail a greater level of effort and could be politically controversial, it might also be feasible to prepare and bank a relatively large set of human ES lines to facilitate histocompatibility matching.

One recent study suggests that a surprisingly modest number of banked lines or specimens could provide substantial ability to match donor cells to recipients (Taylor et al., 2005). Taken together with the low immunogenicity of certain stem cells, these results support the concept that allogeneic bioengineered products may not inevitably demand intensive immunosuppressive treatment, even if it proves impossible to develop general methods to induce selective immunological tolerance.

Recent clinical studies highlight ongoing efforts to develop new autologous cell-based therapies. The recognition that, in addition to hematopoietic stem cells, bone marrow also contains mesenchymal stem cells (MSC) and endothelial progenitor cells (EPC) has spurred ongoing efforts to use autologous marrow cells for blood vessel tissue engineering and for the treatment of myocardial infarction.

In the case of engineering of blood vessels, vascular grafts of autologous bone marrow cells seeded onto biodegradable synthetic conduits or patches have been implanted in children with congenital heart defects (Shin'oka et al., 2005). Safety data on 42 patients with a mean follow-up period of 490 days postsurgery appeared very encouraging, with no major adverse events reported. The grafted engineered vessels remained patent and functional. Moreover, there was evidence that the vessels increased in diameter as the patients grew, thus highlighting a critical potential advantage of regenerative therapies incorporating living cells. Further advances in blood vessel engineering will likely arise from multidisciplinary approaches demanding advances at the interface of biology and engineering (Stitzel et al., 2006).

Similar strategies may be attempted to treat patients with congestive heart failure (Krupnick et al., 2004). Already, a number of clinical studies have been carried out on the injection of autologous bone marrow cells, sometimes unfractionated, sometimes enriched for stem/progenitor cells, into the heart after myocardial infarction (Stamm et al., 2006). The initial rationale for this approach came from experiments in rodents interpreted as demonstrating the production of new cardiomyocytes through the transdifferentiation of hematopoietic stem cells. Evidence for myogenesis of grafted cells, whether from the hematopoietic lineage or, as seems much more plausible, from mesenchymal progenitors, remains sparse. However, some controlled studies do indicate potential clinical benefits from the autologous cell therapy. This may result from the production of angiogenic factors by the injected cells rather than from integration of donor cells into either muscle or new blood vessels. Nonetheless, although still a daunting challenge, the application of regenerative medicine principles to repair damaged cardiac muscle now seems within the possible realm (Dimmeler et al., 2005). The correct choice of cell source, the development and maturation of tissue engineered cardiac patches, and overcoming chronic fibrotic scarring remain hurdles to be overcome.

In another example of regenerative therapy utilizing autologous cells, here following the general paradigm first established for skin, bladder urothelial and smooth muscle cells were expanded in culture from small biopsies and seeded on scaffolds to produce tissue engineered neo-bladders. Such constructs were implanted in seven pediatric patients with high pressure or poorly compliant bladders, some of whom have now been followed for over 5 years (mean 46 months) (Atala et al., 2006). The results are encouraging and should lead to larger scale studies of safety and efficacy, targeting product approval after regulatory review.

Regenerative medicine is a highly interdisciplinary field. Future progress will continue to depend on synergies between advances in biology, chemistry, and engineering. The clinical use of various cell sources, smart biomaterials, and cytokines and growth factors are sure to increase the tools available for therapies in patients. This volume seeks to give the reader an overview of the field in a compact format. We thank the authors, editors, and editorial staff for their contributions to this compendium.

REFERENCES

Aggarwal, S. and Pittenger, M.F. (2005). Human mesenchymal stem cells modulate allogeneic immune cell responses. *Blood* **105**(4): 1815–1822.

Amit, M., Carpenter, M.K., Inokuma, M.S., Chiu, C.P., Harris, C.P., Waknitz, M.A., Itskovitz-Eldor, J. and Thomson, J.A. (2000). Clonally derived human embryonic stem cell lines maintain pluripotency and proliferative potential for prolonged periods of culture. *Dev. Biol.* **227**(2): 271–278.

Amit, M., Margulets, V., Segev, H., Shariki, K., Laevsky, I., Coleman, R. and Itskovitz-Eldor J. (2003). Human feeder layers for human embryonic stem cells. *Biol. Reprod.* **68**(6): 2150–2156.

Amit, M., Shariki, C., Margulets, V. and Itskovitz-Eldor, J. (2004). Feeder layer- and serum-free culture of human embryonic stem cells. *Biol. Reprod.* **70**(3): 837–845.

Arinzeh, T.L., Peter, S.J., Archambault, M.P., van den Bos, C., Gordon, S., Kraus, K., Smith, A. and Kadiyala, S. (2003). Allogeneic mesenchymal stem cells regenerate bone in a critical-sized canine segmental defect. *J. Bone Joint Surg. Am.* **85-A**(10): 1927–1935.

Assady, S., Maor, G., Amit, M., Itskovitz-Eldor, J., Skorecki, K.L. and Tzukerman, M. (2001). Insulin production by human embryonic stem cells. *Diabetes* **50**(8): 1691–1697.

Atala, A., Bauer, S.B., Soker, S., Yoo, J.J. and Retik, A.B. (2006). Tissue-engineered autologous bladders for patients needing cystoplasty. *Lancet* **367**(9518): 1241–1246.

Baharvand, H., Jafary, H. Massumi, M. and Ashtiani, S.K. (2006). Generation of insulin-secreting cells from human embryonic stem cells. *Dev. Growth Differ.* **48**(5): 323–332.

Baizabal, J.M., Furlan-Magaril, M., Santa-Olalla, J. and Covarrubias, L. (2003). Neural stem cells in development and regenerative medicine. *Arch. Med. Res.* **34**(6): 572–588.

Barry, F.P. and Murphy, J.M. (2004). Mesenchymal stem cells: clinical applications and biological characterization. *Int. J. Biochem. Cell Biol.* **36**(4): 568–584.

Bartholomew, A., Sturgeon, C., Siatskas, M., Ferrer, K., McIntosh, K., Patil, S., Hardy, W., Devine, S., Ucker, D., Deans, R., Moseley, A. and Hoffman, R. (2002). Mesenchymal stem cells suppress lymphocyte proliferation *in vitro* and prolong skin graft survival *in vivo*. *Exp. Hematol.* **30**(1): 42–48.

Bartsch, G., Yoo, J.J., De Coppi, P., Siddiqui, M.M., Schuch, G., Pohl, H.G., Fuhr, J., Perin, L., Soker, S. and Atala, A. (2005). Propagation, expansion, and multilineage differentiation of human somatic stem cells from dermal progenitors. *Stem Cells Dev.* **14**(3): 337–348.

Beattie, G.M., Lopez, A.D., Bucay, N., Hinton, A., Firpo, M.T., King, C.C. and Hayek, A. (2005). Activin A maintains pluripotency of human embryonic stem cells in the absence of feeder layers. *Stem Cells* **23**(4): 489–495.

Beltrami, A.P., Barlucchi, L., Torella, D., Baker, M., Limana, F., Chimenti, S., Kasahara, H., Rota, M., Musso, E., Urbanek, K., Leri, A., Kajstura, J., Nadal-Ginard, B. and Anversa, P. (2003). Adult cardiac stem cells are multipotent and support myocardial regeneration. *Cell* **114**(6): 763–776.

Blelloch, R., Wang, Z., Meissner, A., Pollard, S., Smith, A. and Jaenisch, R. (2006). Reprogramming efficiency following somatic cell nuclear transfer is influenced by the differentiation and methylation state of the donor nucleus. *Stem Cells* **24**(9): 2007–2013.

Brolen, G.K., Heins, N., Edsbagge, J. and Semb, H. (2005). Signals from the embryonic mouse pancreas induce differentiation of human embryonic stem cells into insulin-producing beta-cell-like cells. *Diabetes* **54**(10): 2867–2874.

Bruder, S.P., Fink, D.J. and Caplan, A.I. (1994). Mesenchymal stem cells in bone development, bone repair, and skeletal regeneration therapy. *J. Cell Biochem.* **56**(3): 283–294.

Cao, B., Deasy, B.M., Pollett, J. and Huard, J. (2005). Cell therapy for muscle regeneration and repair. *Phys. Med. Rehabil. Clin. N. Am.* **16**(4): 889–907, viii.

Carpenter, M.K., Rosler, E.S., Fisk, G.J., Brandenberger, R., Ares, X., Miura, T., Lucero, M. and Rao, M.S. (2004). Properties of four human embryonic stem cell lines maintained in a feeder-free culture system. *Dev Dyn.* **229**(2): 243–258.

Caspi, O. and Gepstein, L. (2006). Regenerating the heart using human embryonic stem cells—from cell to bedside. *Isr. Med. Assoc. J.* **8**(3): 208–214.

Colman, A. and Kind, A. (2000). Therapeutic cloning: concepts and practicalities. *Trends Biotechnol.* **18**(5): 192–196.

Committee on the Biological and Biomedical Applications of Stem Cell Research (2002). *Stem Cells and the Future of Regenerative Medicine.* Washington, DC: National Academy Press, 94 pp.

Cowan, C.A., Klimanskaya, I., McMahon, J., Atienza, J., Witmyer, J., Zucker, J.P., Wang, S., Morton, C.C., McMahon, A.P., Powers, D. and Melton, D.A. (2004). Derivation of embryonic stem-cell lines from human blastocysts. *N. Engl. J. Med.* **350**(13): 1353–1356.

Cowan, C.A., Atienza, J., Melton, D.A. and Eggan, K. (2005). Nuclear reprogramming of somatic cells after fusion with human embryonic stem cells. *Science* **309**(5739): 1369–1373.

De Coppi, P., Bartsch, G., Dal Cin, P., Yoo, J.J., Soker, S. and Atala, A. (2001). Human fetal stem cell isolation from amniotic fluid. *American Academy of Pediatrics National Conference*, San Francisco, pp. 210–211.

De Kok, I.J., Peter, S.J., Archambault, M., van den Bos, C., Kadiyala, S., Aukhil, I. and Cooper, L.F. (2003). Investigation of allogeneic mesenchymal stem cell-based alveolar bone formation: preliminary findings. *Clin. Oral Implants Res.* **14**(4): 481–489.

Dimmeler, S., Zeiher, A.M. and Schneider, M.D. (2005). Unchain my heart: the scientific foundations of cardiac repair. *J. Clin. Invest.* **115**(3): 572–583.

Drukker, M., Katz, G., Urbach, A., Schuldiner, M., Markel, G., Itskovitz-Eldor, J., Reubinoff, B., Mandelboim, O., Benvenisty, N. (2002). Characterization of the expression of MHC proteins in human embryonic stem cells. *Proc Natl Acad Sci USA* **23**; 99(15): 9864–9869.

Eggan, K., Baldwin, K., Tackett, M., Osborne, J., Gogos, J., Chess, A., Axel, R. and Jaenisch, R. (2004). Mice cloned from olfactory sensory neurons. *Nature* **428**(6978): 44–49.

Evans, M.J. and Kaufman, M.H. (1981). Establishment in culture of pluripotential cells from mouse embryos. *Nature* **292**(5819): 154–156.

Fujikawa, T., Oh, S.H., Pi, L., Hatch, H.M., Shupe, T. and Petersen, B.E. (2005). Teratoma formation leads to failure of treatment for type I diabetes using embryonic stem cell-derived insulin-producing cells. *Am. J. Pathol.* **166**(6): 1781–1791.

Gao, J. and Caplan, A.I (2003). Mesenchymal stem cells and tissue engineering for orthopaedic surgery. *Chir. Organi Mov.* **88**(3): 305–316.

Gerecht-Nir, S., Cohen, S. and Itskovitz-Eldor, J. (2004). Bioreactor cultivation enhances the efficiency of human embryoid body (hEB) formation and differentiation. *Biotechnol. Bioeng.* **86**(5): 493–502.

Gimble, J. and Guilak, F. (2003). Adipose-derived adult stem cells: isolation, characterization, and differentiation potential. *Cytotherapy* **5**(5): 362–369.

Goh, E.L., Ma, D., Ming, G.L. and Song, H. (2003). Adult neural stem cells and repair of the adult central nervous system. *J. Hematother. Stem Cell Res.* **12**(6): 671–679.

Goh, G., Self, T., Barbadillo Munoz, M.D., Hall, I.P., Young, L. and Denning, C. (2005). Molecular and phenotypic analyses of human embryonic stem cell-derived cardiomyocytes: opportunities and challenges for clinical translation. *Thromb. Haemost.* **94**(4): 728–737.

Guilak, F., Awad, H.A., Fermor, B., Leddy, H.A. and Gimble, J.M. (2004). Adipose-derived adult stem cells for cartilage tissue engineering. *Biorheology* **41**(3–4): 389–399.

Hall, V.J., Stojkovic, P. and Stojkovic, M. (2006). Using therapeutic cloning to fight human disease: a conundrum or reality? *Stem Cells* **24**(7): 1628–1637.

He, J.Q., Ma, Y., Lee, Y., Thomson, J.A. and Kamp, T.J. (2003). Human embryonic stem cells develop into multiple types of cardiac myocytes: action potential characterization. *Circ. Res.* **93**(1): 32–39.

Hoffman, L.M. and Carpenter, M.K. (2005). Characterization and culture of human embryonic stem cells. *Nat. Biotechnol.* **23**(6): 699–708.

Hovatta, O. and Skottman, H. (2005). Feeder-free derivation of human embryonic stem-cell lines. *Lancet* **365**(9471): 1601–1603.

Hovatta, O., Mikkola, M., Gertow, K., Stromberg, A.M., Inzunza, J., Hreinsson, J., Rozell, B., Blennow, E., Andang, M. and Ahrlund-Richter, L. (2003). A culture system using human foreskin fibroblasts as feeder cells allows production of human embryonic stem cells. *Hum. Reprod.* **18**(7): 1404–1409.

Hwang, W.S., Roh, S.I., Lee, B.C., Kang, S.K., Kwon, D.K., Kim, S., Kim, S.J., Park, S.W., Kwon, H.S., Lee, C.K., Lee, J.B., Kim, J.M., Ahn, C., Paek, S.H., Chang, S.S., Koo, J.J., Yoon, H.S., Hwang, J.H., Hwang, Y.Y., Park, Y.S., Oh, S.K., Kim, H.S., Park, J.H., Moon, S.Y. and Schatten, G. (2005). Patient-specific embryonic stem cells derived from human SCNT blastocysts. *Science* **308**(5729): 1777–1783.

Inoue, K., Ogonuki, N., Miki, H., Hirose, M., Noda, S., Kim, J.M., Aoki, F., Miyoshi, H. and Ogura, A. (2006). Inefficient reprogramming of the hematopoietic stem cell genome following nuclear transfer. *J. Cell Sci.* **119**(Pt 10): 1985–1991.

IItskovitz-Eldor, J., Schuldiner, M., Karsenti, D., Eden, A., Yanuka, O., Amit, M., Soreq, H. and Benvenisty, N. (2000). Differentiation of human embryonic stem cells into embryoid bodies compromising the three embryonic germ layers. *Mol. Med.* **6**(2): 88–95.

Jiang, Y., Jahagirdar, B.N., Reinhardt, R.L., Schwartz, R.E., Keene, C.D., Ortiz-Gonzalez, X.R., Reyes, M., Lenvik, T., Lund, T., Blackstad, M., Du, J. Aldrich, S. Lisberg, A. Low, W.C. Largaespada, D.A. and Verfaillie, C.M. (2002). Pluripotency of mesenchymal stem cells derived from adult marrow. *Nature* **418**(6893): 41–49.

Johnson, P.C. (2000). The role of tissue engineering. *Adv. Skin Wound Care* **13**(2 Suppl): 12–14.

Kamiya, A., Gonzalez, F.J. and Nakauchi, H. (2006). Identification and differentiation of hepatic stem cells during liver development. *Front Biosci.* **11**: 1302–1310.

Keirstead, H.S., Nistor, G., Bernal, G., Totoiu, M., Cloutier, F., Sharp, K. and Steward, O. (2005). Human embryonic stem cell-derived oligodendrocyte progenitor cell transplants remyelinate and restore locomotion after spinal cord injury. *J. Neurosci.* **25**(19): 4694–4705.

Klimanskaya, I., Chung, Y., Meisner, L., Johnson, J., West, M.D. and Lanza, R. (2005). Human embryonic stem cells derived without feeder cells. *Lancet* **365**(9471): 1636–1641.

Kogler, G., Sensken, S., Airey, J.A., Trapp, T., Muschen, M., Feldhahn, N., Liedtke, S., Sorg, R.V., Fischer, J., Rosenbaum, C., Greschat, S., Knipper, A., Bender, J., Degistirici, O., Gao, J., Caplan, A.I., Colletti, E.J., Almeida-Porada, G., Muller, H.W., Zanjani, E. and Wernet, P. (2004). A new human somatic stem cell from placental cord blood with intrinsic pluripotent differentiation potential. *J. Exp. Med.* **200**(2): 123–135.

Krampera, M., Cosmi, L., Angeli, R., Pasini, A., Liotta, F., Andreini, A., Santarlasci, V., Mazzinghi, B., Pizzolo, G., Vinante, F., Romagnani, P., Maggi, E., Romagnani, S. and Annunziato, F. (2006). Role for interferon-gamma in the immunomodulatory activity of human bone marrow mesenchymal stem cells. *Stem Cells* **24**(2): 386–398.

Krupnick, A.S., Kreisel, D., Riha, M., Balsara, K.R. and Rosengard, B.R. (2004). Myocardial tissue engineering and regeneration as a therapeutic alternative to transplantation. *Curr. Top Microbiol. Immunol.* **280:**139–164.

Lanza, R.P., Chung, H.Y., Yoo, J.J., Wettstein, P.J., Blackwell, C., Borson, N., Hofmeister, E., Schuch, G., Soker, S., Moraes, C.T., West, M.D. and Atala, A. (2002). Generation of histocompatible tissues using nuclear transplantation. *Nat. Biotechnol.* **20**(7): 689–696.

Lavik, E. and Langer, R. (2004). Tissue engineering: current state and perspectives. *Appl. Microbiol. Biotechnol.* **65**(1): 1–8.

Lawrenz, B., Schiller, H., Willbold, E., Ruediger, M., Muhs, A. and Esser, S. (2004). Highly sensitive biosafety model for stem-cell-derived grafts. *Cytotherapy* **6**(3): 212–222.

Le Blanc, K. (2003). Immunomodulatory effects of fetal and adult mesenchymal stem cells. *Cytotherapy***5**(6): 485–489.

Lester, L.B., Kuo, H.C., Andrews, L., Nauert, B. and Wolf, D.P. (2004). Directed differentiation of rhesus monkey ES cells into pancreatic cell phenotypes. *Reprod. Biol. Endocrinol.* **2**: 42.

Lev, S., Kehat, I. and Gepstein, L. (2005). Differentiation pathways in human embryonic stem cell-derived cardiomyocytes. *Ann. NY Acad. Sci.* **1047**: 50–65.

Lumelsky, N., Blondel, O., Laeng, P., Velasco, I., Ravin, R. and McKay, R. (2001). Differentiation of embryonic stem cells to insulin-secreting structures similar to pancreatic islets. *Science* **292**(5520): 1389–1394.

Lysaght, M.J. and Hazlehurst, A.L. (2003). Private sector development of stem cell technology and therapeutic cloning. *Tissue Eng.* **9**(3): 555–561.

Lysaght, M.J. and Hazlehurst, A.L. (2004). Tissue engineering: the end of the beginning. *Tissue Eng.* **10**(1–2): 309–320.

Lysaght, M.J. and Reyes, J. (2001). The growth of tissue engineering. *Tissue Eng.* **7**(5): 485–493.

Martin, G.R. (1981). Isolation of a pluripotent cell line from early mouse embryos cultured in medium conditioned by teratocarcinoma stem cells. *Proc. Natl Acad. Sci. USA* **78**(12): 7634–7638.

Miki, T., Lehmann, T., Cai, H., Stolz, D.B. and Strom, S.C. (2005). Stem cell characteristics of amniotic epithelial cells. *Stem Cells* **23**(10): 1549–1559.

Mitchell, K.E., Weiss, M.L., Mitchell, B.M., Martin, P., Davis, D., Morales, L., Helwig, B., Beerenstrauch, M., Abou-Easa, K., Hildreth, T., Troyer, D. and Medicetty, S. (2003). Matrix cells from Wharton's jelly form neurons and glia. *Stem Cells* **21**(1): 50–60.

Miyazaki, S., Yamato, E. and Miyazaki, J. (2004). Regulated expression of pdx-1 promotes *in vitro* differentiation of insulin-producing cells from embryonic stem cells. *Diabetes* **53**(4): 1030–1037.

Mueller, D., Shamblott, M.J., Fox, H.E., Gearhart, J.D. and Martin, L.J. (2005). Transplanted human embryonic germ cell-derived neural stem cells replace neurons and oligodendrocytes in the forebrain of neonatal mice with excitotoxic brain damage. *J. Neurosci. Res.* **82**(5): 592–608.

Murray-Thomas, T. and Cowie, M.R. (2003). Epidemiology and clinical aspects of congestive heart failure. *J. Renin Angiotensin Aldosterone Syst.* **4**(3): 131–136.

Nakajima-Nagata, N., Sakurai, T., Mitaka, T., Katakai, T., Yamato, E., Miyazaki, J., Tabata, Y., Sugai, M. and Shimizu, A. (2004). *In vitro* induction of adult hepatic progenitor cells into insulin-producing cells. *Biochem. Biophys. Res. Commun.* **318**(3): 625–630.

Naughton, G.K. (2002). From lab bench to market: critical issues in tissue engineering. *Ann. NY Acad. Sci.* **961**: 372–385.

Nir, S.G., David, R., Zaruba, M., Franz, W.M. and Itskovitz-Eldor, J. (2003). Human embryonic stem cells for cardiovascular repair. *Cardiovasc. Res.* **58**(2): 313–323.

Nir, T. and Dor, Y. (2005). How to make pancreatic beta cells—prospects for cell therapy in diabetes. *Curr. Opin. Biotechnol.* **16**(5): 524–529.

Odorico, J.S., Kaufman, D.S. and Thomson, J.A. (2001). Multilineage differentiation from human embryonic stem cell lines. *Stem Cells* **19**(3): 193–204.

Oh, H., Bradfute, S.B., Gallardo, T.D., Nakamura, T., Gaussin, V., Mishina, Y., Pocius, J., Michael, L.H., Behringer, R.R., Garry, D.J., Entman, M.L. and Schneider, M.D. (2003). Cardiac progenitor cells from adult myocardium: homing, differentiation, and fusion after infarction. *Proc. Natl Acad. Sci. USA* **100**(21): 12313–12318.

Paek, H.J., Morgan, J.R. and Lysaght, M.J. (2005). Sequestration and synthesis: the source of insulin in cell clusters differentiated from murine embryonic stem cells. *Stem Cells* **23**(7): 862–867.

Passier, R., Denning, C. and Mummery, C. (2006). Cardiomyocytes from human embryonic stem cells. *Handb. Exp. Pharmacol.* **174**: 101–122.

Perrier, A.L., Tabar, V., Barberi, T., Rubio, M.E., Bruses, J., Topf, N., Harrison, N.L. and Studer, L. (2004). Derivation of midbrain dopamine neurons from human embryonic stem cells. *Proc. Natl Acad. Sci. USA* **101**(34): 12543–12548.

Pittenger, M.F., Mackay, A.M., Beck, S.C., Jaiswal, R.K., Douglas, R., Mosca, J.D., Moorman, M.A., Simonetti, D.W., Craig, S. and Marshak, D.R. (1999). Multilineage potential of adult human mesenchymal stem cells. *Science* **284**(5411): 143–147.

Plumas, J., Chaperot, L., Richard, M.J., Molens, J.P., Bensa, J.C. and Favrot, M.C. (2005). Mesenchymal stem cells induce apoptosis of activated T cells. *Leukemia* **19**(9): 1597–1604.

Potian, J.A., Aviv, H., Ponzio, N.M., Harrison, J.S. and Rameshwar, P. (2003). Veto-like activity of mesenchymal stem cells: functional discrimination between cellular responses to alloantigens and recall antigens. *J. Immunol.* **171**(7): 3426–3434.

Puissant, B., Barreau, C., Bourin, P., Clavel, C., Corre, J., Bousquet, C., Taureau, C., Cousin, B., Abbal, M., Laharrague, P., Penicaud, L., Casteilla, L. and Blancher, A. (2005). Immunomodulatory effect of human adipose tissue-derived adult stem cells: comparison with bone marrow mesenchymal stem cells. *Br. J. Haematol.* **129**(1): 118–129.

Raikwar, S.P., Mueller, T. and Zavazava, N. (2006). Strategies for developing therapeutic application of human embryonic stem cells. *Physiology (Bethesda)* **21**: 19–28.

Rao, M. (2004). Conserved and divergent paths that regulate self-renewal in mouse and human embryonic stem cells. *Dev. Biol.* **275**(2): 269–286.

Rideout III, W.M., Hochedlinger, K., Kyba, M., Daley, G.Q. and Jaenisch, R. (2002). Correction of a genetic defect by nuclear transplantation and combined cell and gene therapy. *Cell* **109**(1): 17–27.

Risbud, M.V. and Shapiro, I.M. (2005). Stem cells in craniofacial and dental tissue engineering. *Orthod. Craniofac. Res.* **8**(2): 54–59.

Sanchez-Pernaute, R., Studer, L., Ferrari, D., Perrier, A., Lee, H., Vinuela, A. and Isacson, O. (2005). Long-term survival of dopamine neurons derived from parthenogenetic primate embryonic stem cells (cyno-1) after transplantation. *Stem Cells* **23**(7): 914–922.

Schmelzer, E., Wauthier, E. and Reid, L.M. (2006). The phenotypes of pluripotent human hepatic progenitors. *Stem Cells* **24**(8): 1852–1858.

Schuldiner, M., Itskovitz-Eldor, J. and Benvenisty, N. (2003). Selective ablation of human embryonic stem cells expressing a "suicide" gene. *Stem Cells* **21**(3): 257–265.

Shamblott, M.J., Axelman, J., Wang, S., Bugg, E.M., Littlefield, J.W., Donovan, P.J., Blumenthal, P.D., Huggins, G.R. and Gearhart, J.D. (1998). Derivation of pluripotent stem cells from cultured human primordial germ cells. *Proc. Natl Acad. Sci. USA* **95**(23): 13726–13731.

Shin'oka, T., Matsumura, G., Hibino, N., Naito, Y., Watanabe, M., Konuma, T., Sakamoto, T., Nagatsu, M. and Kurosawa, H. (2005). Midterm clinical result of tissue-engineered vascular autografts seeded with autologous bone marrow cells. *J. Thorac. Cardiovasc. Surg.* **129**(6): 1330–1338.

Shiroi, A., Ueda, S., Ouji, Y., Saito, K., Moriya, K., Sugie, Y., Fukui, H., Ishizaka, S. and Yoshikawa, M. (2005). Differentiation of embryonic stem cells into insulin-producing cells promoted by Nkx2.2 gene transfer. *World J. Gastroenterol.* **11**(27): 4161–4166.

Siddiqui, M.M. and Atala, A. (2004). Amniotic fluid-derived pluripotential cells: adult and fetal. In: Lanza, R., Blau, H., Melton, D., Moore, M., Thomas, E.D., Verfaillie, C., Weissman, I. and West, M. (eds.). *Handbook of Stem Cells*, Vol. 2. Amsterdam: Elsevier Academic Press, pp. 175–180.

Sjogren-Jansson, E., Zetterstrom, M., Moya, K., Lindqvist, J., Strehl, R. and Eriksson, P.S. (2005). Large-scale propagation of four undifferentiated human embryonic stem cell lines in a feeder-free culture system. *Dev. Dyn.* **233**(4): 1304–1314.

Smyth, S., and Heron, A. (2006). Diabetes and obesity: the twin epidemics. *Nat. Med.* **12**(1): 75–80.

Sotiropoulou, P.A., Perez, S.A., Gritzapis, A.D., Baxevanis, C.N. and Papamichail, M. (2006). Interactions between human mesenchymal stem cells and natural killer cells. *Stem Cells* **24**(1): 74–85.

Stacey, G.N., Cobo, F., Nieto, A., Talavera, P., Healy, L. and Concha, A. (2006). The development of "feeder" cells for the preparation of clinical grade hES cell lines: challenges and solutions. *J. Biotechnol.* **125**(4): 583–588.

Stamm, C., Liebold, A., Steinhoff, G. and Strunk, D. (2006). Stem cell therapy for ischemic heart disease: beginning or end of the road? *Cell Transplant.* **15**(Suppl 1): S47–S56.

Stewart, R., Stojkovic, M. and Lako, M. (2006). Mechanisms of self-renewal in human embryonic stem cells. *Eur. J. Cancer* **42**(9): 1257–1272.

Stitzel, J., Liu, J., Lee, S.J., Komura, M., Berry, J., Soker, S., Lim, G., Van Dyke, M., Czerw, R., Yoo, J.J. and Atala, A. (2006). Controlled fabrication of a biological vascular substitute. *Biomaterials* **27**(7): 1088–1094.

Sung, L.Y., Gao, S., Shen, H., Yu, H., Song, Y., Smith, S.L., Chang, C.C., Inoue, K., Kuo, L., Lian, J., Li, A., Tian, X.C., Tuck, D.P., Weissman, S.M., Yang, X. and Cheng, T. (2006). Differentiated cells are more efficient than adult stem cells for cloning by somatic cell nuclear transfer. *Nat. Genet.* **38**(11): 1323–1328.

Takahashi, K. and Yamanaka, S. (2006). Induction of pluripotent stem cells from mouse embryonic and adult fibroblast cultures by defined factors. *Cell* **126**(4): 663–676.

Taupin, P. (2006). HuCNS-SC (stem cells). *Curr. Opin. Mol. Ther.* **8**(2): 156–163.

Taylor, C.J., Bolton, E.M., Pocock, S., Sharples, L.D., Pedersen, R.A. and Bradley, J.A. (2005). Banking on human embryonic stem cells: estimating the number of donor cell lines needed for HLA matching. *Lancet* **366**(9502): 2019–2025.

Thomas, E.D. (1999). Bone marrow transplantation: a review. *Semin. Hematol.* **36**(4 Suppl 7): 95–103.

Thomson, J.A., Itskovitz-Eldor, J., Shapiro, S.S., Waknitz, M.A., Swiergiel, J.J., Marshall, V.S. and Jones, J.M. (1998). Embryonic stem cell lines derived from human blastocysts. *Science* **282**(5391): 1145–1147.

Tian, X. and Kaufman, D.S. (2005). Hematopoietic development of human embryonic stem cells in culture. *Methods Mol. Med.* **105**: 425–436.

Trounson, A. (2006). The production and directed differentiation of human embryonic stem cells. *Endocr. Rev.* **27**(2): 208–219.

Tsai, M.S., Hwang, S.M., Tsai, Y.L., Cheng, F.C., Lee, J.L. and Chang, Y.J. (2006). Clonal amniotic fluid-derived stem cells express characteristics of both mesenchymal and neural stem cells. *Biol. Reprod.* **74**(3): 545–551.

Tuan, R.S., Boland, G. and Tuli, R. (2003). Adult mesenchymal stem cells and cell-based tissue engineering. *Arthritis Res. Ther.* **5**(1): 32–45.

Weir, G.C. (2004). Can we make surrogate beta-cells better than the original? *Semin. Cell Dev. Biol.* **15**(3): 347–357.

Weiss, M.L., Medicetty, S., Bledsoe, A.R., Rachakatla, R.S., Choi, M., Merchav, S., Luo,Y., Rao, M.S., Velagaleti, G. and Troyer, D. (2006). Human umbilical cord matrix stem cells: preliminary characterization and effect of transplantation in a rodent model of Parkinson's disease. *Stem Cells* **24**(3): 781–792.

Yamada, S., Terada, K., Ueno, Y., Sugiyama, T., Seno, M. and Kojima, I. (2005). Differentiation of adult hepatic stem-like cells into pancreatic endocrine cells. *Cell Transplant.* **14**(9): 647–653.

Yang, L., Li, S., Hatch, H., Ahrens, K., Cornelius, J.G., Petersen, B.E. and Peck, A.B. (2002). *In vitro* trans-differentiation of adult hepatic stem cells into pancreatic endocrine hormone-producing cells. *Proc. Natl Acad. Sci. USA* **99**(12): 8078–8083.

Yoo, S.J., Yoon, B.S., Kim, J.M., Song, J.M., Roh, S., You, S. and Yoon, H.S. (2005). Efficient culture system for human embryonic stem cells using autologous human embryonic stem cell-derived feeder cells. *Exp. Mol. Med.* **37**(5): 399–407.

Zalzman, M., Anker-Kitai, L. and Efrat, S. (2005). Differentiation of human liver-derived, insulin-producing cells toward the beta-cell phenotype. *Diabetes* **54**(9): 2568–2575.

Zuk, P.A., Zhu, M., Mizuno, H., Huang, J., Futrell, J.W., Katz, A.J., Benhaim, P., Lorenz, H.P. and Hedrick, M.H. (2001). Multilineage cells from human adipose tissue: implications for cell-based therapies. *Tissue Eng.* **7**(2): 211–228.

List of Contributors

Steven D. Abramowitch
Musculoskeletal Research Center
University of Pittsburgh

Jon D. Ahlstrom
Department of Molecular and Cellular Biology
University of California at Davis

Taby Ahsan
Parker H. Petit Institute for Bioengineering and Bioscience
Georgia Institute of Technology

Alejandro J. Almarza
Research Assistant Professor
Musculoskeletal Research Center
Department of Bioengineering
University of Pittsburgh

James M. Anderson*
Department of Pathology
University Hospitals of Cleveland

Hadi Aslan
Hebrew University – Hadassah Medical Center
Hebrew University
Center for Converging Sciences & Technologies
Skeletal Biotechnology Lab

Anthony Atala*
W.B. Boyce Professor and Director,
Wake Forest Institute for Regenerative Medicine,
Wake Forest School of Medicine, Winston-Salem,
North Carolina, USA

Stephen F. Badylak*
McGowan Institute for Regenerative Medicine
University of Pittsburgh

Ashok Batra
Office of Cellular, Tissue and Gene Therapies
Center for Biologics Evaluations and Research

Ravi V. Bellamkonda*
Professor of Biomedical Engineering
Neurological Biomaterials and Therapeutics
Wallace H Coulter Department of Biomedical Engineering
Georgia Institute of Technology/Emory University
Atlanta

Sangeeta N. Bhatia
Division of Medicine
Brigham & Women's Hospital

Ali H. Brivanlou*
Professor and Head of the Laboratory of Molecular Vertebrate Embryology, Rockerfeller University

Scott P. Bruder
DePuy Spine, Inc., a Johnson & Johnson Company

Christopher S. Chen*
University of Pennsylvania
Translational Research Labs

George J. Christ*
Professor of Regenerative Medicine,
Urology and Physiology & Member,
Molecular Medicine Program and Virginia
Tech-Wake Forest University
School of Biomedical Engineering and Sciences
Head of the Cell, Tissue & Organ
Physiology Program
Wake Forest Institute for Regenerative Medicine
Wake Forest University Baptist Medical Center
NC, USA

Massimo Cimini
MaRS Center
Toronto Medical Discover Tower

Paolo De Coppi*
Department of General Paediatric Surgery
Great Ormond Street Hospital and Institute of Child Health

Mahesh C. Dodla
Georgia Institute of Technology

Juan Domínguez-Bendala
Pancreatic Development & Stem Cell Laboratory
Diabetes Research Institute
University of Miami Leonard M. Miller School of Medicine

A.M. Doyle
Georgia Institute of Technology
Parker H. Petit Institute for Bioengineering and Bioscience

Charles N. Durfor
Center for Devices and Radiological Health

Daniel Eberli
University Hospital Zurich
Switzerland

Ewa C.S. Ellis
Department of Pathology
University of Pittsburgh

Juliet A. Emamaullee
University of Alberta

Carol A. Erickson*
Department of Molecular and Cellular Biology
University of California at Davis

Roger De Filippo*
Childrens Hospital Los Angeles
Division of Urology

Donald Fink
Office of Cellular, Tissue and Gene Therapies
Center for Biologics Evaluations and Research

Mark E. Furth
Department of Urology
Wake Forest University

Yossi Gafni
Hebrew University – Hadassah Medical Center
Hebrew University
Center for Converging Sciences & Technologies
Skeletal Biotechnology Lab

Andrés García*
Associate Professor
Woodruff Faculty Fellow
Woodruff School of Mechanical Engineering
Petit Institute for Bioengineering and Bioscience
Georgia Institute of Technology

Daniel Gazit*
Hebrew University – Hadassah Medical Center
Hebrew University
Center for Converging Sciences & Technologies
Skeletal Biotechnology Lab
International Stem Cell Institute – Department of Surgery
Cedars Sinai Medical Center
Los Angeles, CA

Zulma Gazit
Hebrew University – Hadassah Medical Center
Hebrew University
Center for Converging Sciences & Technologies
Skeletal Biotechnology Lab

Christopher S. Gemmiti
Georgia Institute of Technology

Jörg C. Gerlach*
McGowan Institute for Regenerative Medicine,
Bridgeside Point Boulevard
University of Pittsburgh

Kiran Gollapudi
Cartilage Biology and Orthopaedics Branch
National Institute of Arthritis

Louis M. Guenin
Department of Microbiology and Molecular Genetics
Harvard Medical School

Stefano Giuliani
Division of Urology, Childrens Hospital Los Angeles, Saban Research Institute, Keck School of Medicine,
University of Southern California

Robert E. Guldberg*
Professor, Associate Director, Institute for Bioengineering and Bioscience
School of Mechanical Engineering
Georgia Institute of Technology

M.C. Hacker
Rice University
Laboratory of Biomedical Engineering

Benjamin S. Harrison*
Wake Forest University School of Medicine

Bernd Hartmann
Burn Center
Unfalkrankenhaus

Stephen H. Hilbert
Center for Devices and Radiological Health

Chantal E. Holy*
Director of Scientific Affairs DePuy Spine 325
Paramount Drive Raynham, MA, USA

George T.-J. Huang
Division of Endodontics
Baltimore College of Dental Surgery
University of Marlyand

Johnny Huard*
The Growth and Development Lab,
Childrens Hospital of Pittsburgh

Elliot E. Hui
Harvard – M.I.T. Division of Health Sciences and Biology
Electrical Engineering and Computer Science

Christa Johnen
Charite-Campus Virchow
Humboldt University

Akira Joraku
Department of Regenerative Medicine,
Wake Forest University Health Sciences,
Winston Salem

David L. Kaplan*
Department of Biomedical Engineering
Tufts University

David S. Kaplan
Center for Devices and Radiological Health

Sinan Karaoglu
Musculoskeletal Research Center
University of Pittsburgh

Rehan N. Khanzada
Sr. Process Development Engineer
Johnson & Johnson Regenerative Therapeutics, LLC

Nadav Kimelman
Hebrew University – Hadassah Medical Center
Hebrew University
Center for Converging Sciences & Technologies
Skeletal Biotechnology Lab

Jonathan A. Kluge
Department of Biomedical Engineering
Tufts University

Yash Kolambkar
Graduate Research Assistant
Department of Biomedical Engineering
Georgia Tech/Emory Center for the Engineering of Living Tissues

Makoto Komura
The Department of Pediatric Surgery
Tokyo University Hospital

Deborah Lavoie
Office of Cellular, Tissue and Gene Therapies
Center for Biologics Evaluations and Research

Hyukjin Lee
Department of Biological Sciences
Korea Advances Institute of Science and Technology

Gary G. Leisk
Department of Biomedical Engineering
Tufts University

Ariel J. Levine
Rockerfeller University

Ren-Ke Li*
MaRS Center
Toronto Medical Discover Tower

Wan-Ju Li
Cartilage Biology and Orthopaedics Branch
National Institute of Arthritis

Grace J. Lim*
Medical Research Institute
Department of Medical and Biological Engineering
Kyungpook National University School of Medicine, South Korea

Wendy F. Liu
University of Pennsylvania
Translational Research Labs

Ellen Maher
Office of Cellular, Tissue and Gene Therapies
Center for Biologics Evaluations and Research

Manuela Martins-Green*
Department of Cell Biology and Neuroscience
University of California

Richard McFarland
Office of Cellular, Tissue and Gene Therapies
Center for Biologics Evaluations and Research

A.G. Mikos*
Rice University
Laboratory of Biomedical Engineering

Robert M. Nerem*
Georgia Institute of Technology
Parker H. Petit Institute for Bioengineering and Bioscience

Patrea L. Pabst*
Pabst Patent Group LLP

Tae Gwan Park*
Department of Biological Sciences
Korea Advances Institute of Science and Technology

David P. Patterson
Cartilage Biology and Orthopaedics Branch
National Institute of Arthritis

Gadi Pelled
Hebrew University – Hadassah Medical Center
Hebrew University
Center for Converging Sciences & Technologies
Skeletal Biotechnology Lab

Laura Perin
Wake Forest Institute for Regenerative Medicine
Wake Forest University School of Medicine
Medical Center Boulevard
Winston-Salem

M. Petreaca
Department of Cell Biology and Neuroscience
University of California

Antonello Pileggi
Research Associate, Professor of Surgery
Cell Transplant Center
Diabetes Research Institute
University of Miami Miller School of Medicine

Blaise Porter
Georgia Institute of Technology

Milica Radisic
Assistant Professor
Institute of Biomaterials and Biomedical Engineering
Department of Chemical Engineering and Applied Chemistry Heart & Stroke/Richard Lewar Centre of Excellence
University of Toronto

A. Hari Reddi*
Professor and Lawrence J. Ellison Chair
University of California, Davis Sacramento, CA, USA

Camillo Ricordi*
Stacy-Joy Goodman Professor of Surgery
Distinguished Professor of Medicine
Professor of Biomedical Engineering, Microbiology and Immunology
Director, Diabetes Research Institute and Cell Transplant Center
University of Miami Miller School of Medicine

Sargis Sedrakyan
Department of Urology Children's Hospital Los Angeles
Keck School of Medicine
University of Southern California

Michael V. Sefton*
University Professor
Institute of Biomaterials and Biomedical Engineering,
Michael E. Charles Professor, Department of Chemical Engineering and Applied Chemistry
University of Toronto

A.M. James Shapiro
University of Alberta

Shay Soker*
Associate Professor of Regenerative Medicine and Surgical Sciences
Head, Molecular and Cell Biology
Wake Forest Institute for Regenerative Medicine
Wake Forest University School of Medicine

David L. Stocum*
Center for Regenerative Biology and Medicine
Indiana University-Purdue University Indianapolis

Stephen C. Strom*
Department of Pathology
University of Pittsburgh

Doris A. Taylor*
Bakken Professor
Director, Center for Cardiovascular Repair
University of Minnesota

James A. Thomson
Department of Anatomy
Wisconsin Regional Primate Research Center

Rocky S. Tuan*
Chief, Cartilage Biology and Orthopaedics Branch
National Institute of Arthritis, and Musculoskeletal & Skin Diseases
National Institute of Health
MD, USA

Ross S. Tubo*
Senior Director,
Stem Cell Biology, Genzyme Corp.

F. Jerry Volenec
DuPuy Spine

Shen Wei
The Growth and Development Lab,
Children's Hospital of Pittsburgh

Richard D. Weisel
MaRS Center
Toronto Medical Discover Tower

Celia Witten*
Office of Cellular, Tissue and Gene Therapies
Center for Biologics Evaluations and Research

Steven E. Wolf
Burn Center
United State Army Institute of Surgical Research

Savio L.-Y. Woo*
Musculoskeletal Research Center (MSRC)
University of Pittsburgh

Munira Xaymardan
MaRS Center
Toronto Medical Discover Tower

Saami K. Yazdani
Wake Forest University Baptist Medical Center
Medical Center Boulevard

James J. Yoo*
Wake Forest University School of Medicine
Institute for Regenerative Medicine
Medical Center Boulevard

Junying Yu*
The Genetics and Biotechnology Building
University of Wisconsin-Madison

Andrey G. Zenovich
Center for Cardiovascular Repair

Laurie Zoloth*
Feinberg School of Medicine
Northwestern University

*Corresponding Authors

Part I

Introduction to Regenerative Medicine

1

Current and Future Perspectives of Regenerative Medicine

Mark E. Furth and Anthony Atala

REGENERATIVE MEDICINE: CURRENT AND FUTURE PERSPECTIVES

Progress and Challenges for Cell-Based Regenerative Medicine

Regenerative medicine seeks to devise new therapies for patients with severe injuries or chronic diseases in which the body's own responses do not suffice to restore functional tissue. A recent publication from the US National Academy of Sciences, *Stem Cells and the Future of Regenerative Medicine* (Committee, 2002), identified a wide array of major unmet medical needs which might be addressed by regenerative technologies. These include congestive heart failure (approximately 5 million patients in the United States) (Murray-Thomas and Cowie, 2003), osteoporosis (10 million US patients), Alzheimer's and Parkinson's diseases (5.5 million patients each), severe burns (0.3 million), spinal cord injuries (0.25 million), and birth defects (0.15 million). Another area of critical need is diabetes mellitus (16 million US patients and more than 217 million worldwide) (Smyth and Heron, 2006). Patients with type 1 diabetes lack pancreatic beta-cells, essential for the production of insulin, because of autoimmune destruction and represent from 10% to 20% of the total. Many patients with type 2 diabetes also show insufficient pancreatic beta-cell mass. Thus, patients in both groups potentially might be treated if methods could be developed to promote endogenous regeneration of beta-cells or to provide enough surrogate beta-cells and pancreatic islets for transplantation (Weir, 2004).

The therapeutic use of growth factors and cytokines to stimulate the production and/or function of endogenous cells represents the area of regenerative medicine that, arguably, has shown the greatest clinical impact to date (Ioannidou, 2006). Regenerative therapies comprising living cells also have entered into practice, initially through the widespread adoption of both allogeneic and autologous bone marrow transplantation (Thomas, 1999). The presence of hematopoietic progenitor and stem cells with great replicative capacity *in vivo*, and their ability to reenter the bone marrow niche from the circulation, enabled this major medical advance. Subsequently, the development of methods to expand *ex vivo* and deliver such cell types as keratinocytes and chondrocytes, through advances in cell culture and scaffold technologies, led to successful tissue engineering for wound repair (Johnson, 2000; Lavik and Langer, 2004). Despite significant challenges in development and manufacturing, several bioartificial skin graft and cartilage replacement products have achieved regulatory approval (Lysaght and

Copyright © 2008 Elsevier Inc. All rights reserved

Reyes, 2001; Naughton, 2002; Lysaght and Hazlehurst, 2004). These therapies validate the potential of cell-based regenerative approaches.

The extension to new therapeutic areas, especially the development of neo-organs with complex three-dimensional structure, will depend on complementary advances in biology, materials science, and engineering. A major limitation remains the ability to provide oxygen and nutrients to neo-tissues both *in vitro* and after implantation. Advances in scaffold composition and design, in bioreactor technology, and in the use of pro-angiogenic factors may all help to overcome this barrier and are discussed in depth in other chapters of this book.

Here we will focus mainly on sources of cells for regenerative medicines. A primary issue remains the choice between using a patient's own cells, or those of a closely matched relative, *versus* those from an unrelated allogeneic donor. More broadly, future developments depend heavily on increased understanding and effective utilization of multiple classes of progenitor and stem cells.

When populations that include precursor cells (i.e. cells not yet fully differentiated and capable of significant proliferation) can be obtained from a small biopsy of a patient's tissue, and these cells are able to expand and differentiate in culture and/or after implantation back into the patient, autologous therapies are feasible. These have the great advantage of avoiding the risk of immune rejection based on differences in histocompatibility antigens, so that the use of immunosuppressive drugs is not required. However, there is a substantial practical appeal to "off the shelf" products that do not require the cost and time associated with customized manufacturer of an individual product for each individual recipient (Lysaght and Hazlehurst, 2004).

Among the approved bioengineered skin products, Dermagraft (Smith & Nephew) and Apligraf (Organogenesis) utilize allogeneic cells expanded from donated human foreskins to treat many unrelated patients. Despite the genetic mismatch between donor and recipient, the skin cells in Dermagraft and Apligraf do not induce acute immune rejection, possibly because of the absence of antigen-presenting cells in the grafts (Briscoe et al., 1999; Horch et al., 2005). Thus, these products can be utilized without immunosuppressive drug therapy (Moller et al., 1999). Eventually, the donated skin cells may be rejected, but after sufficient time has passed for the patient's endogenous skin cells to recover and take their place.

Products based on autologous cells also have achieved regulatory approval and reached the market. In particular, Genzyme Biosurgery has developed Epicel, a permanent skin replacement product for patients with life-threatening burns, and Carticel, a chondrocyte-based treatment for large articular cartilage lesions. In each case seed cells are obtained from a small biopsy of the patient's tissue. These cells are expanded in culture, processed, and returned to the patient.

New Therapies Using Autologous Cells

Recent clinical studies highlight ongoing efforts to develop new autologous cell-based therapies. The recognition that, in addition to hematopoietic stem cells, bone marrow also contains mesenchymal stem cells (MSC) and endothelial progenitor cells (EPC), has spurred ongoing efforts to use autologous marrow cells for blood vessel tissue engineering and for treatment of myocardial infarction.

In the case of engineering of blood vessels, vascular grafts of autologous bone marrow cells seeded onto biodegradable synthetic conduits or patches have been implanted in children with congenital heart defects (Shin'oka et al., 2005). Safety data on 42 patients with a mean follow-up period of 490 days post-surgery appeared very encouraging, with no major adverse events reported. The grafted engineered vessels remained patent and functional. Moreover, there was evidence that the vessels increased in diameter as the patients grew, thus highlighting a critical potential advantage of regenerative therapies incorporating living cells.

Further advances in blood vessel engineering will likely arise from multidisciplinary approaches demanding advances at the interface of biology and engineering. In recent preclinical studies scaffolds for neo-vessels blending collagen type I and elastin with polylactic-co-glycolic acid (PLGA) were fabricated by electrospinning and showed compliance, burst pressure, and mechanical properties comparable to native vessels (Stitzel et al., 2006). The electrospun vessels also displayed good biocompatibility both *in vitro* and after implantation *in vivo*. When seeded with endothelial and smooth muscle cells, or progenitor MSCs and/or EPCs, these constructs may provide a basis to produce functional vascular grafts suitable for clinical applications such as cardiac bypass procedures. The seeding process itself may demand future advances, since it will be difficult for cells to penetrate a nanofibrillar structure in which pore spaces are considerably smaller than the diameter of a cell (Lutolf and Hubbell, 2005). Electrospinning actually may be used to incorporate living cells into a fibrous matrix. A recent proof of concept study documented that smooth muscle cells could be concurrently electrospun with an elastomeric poly(ester urethane)urea, leading to "microintegration" of the cells in strong, flexible fibers with mechanical properties not greatly inferior to those of the synthetic polymer alone (Stankus et al., 2006). The cell population retained high viability and, when maintained in a perfusion bioreactor, the cellular density in the electrospun fibers doubled over 4 days in culture. One can imagine that in the future, progenitors of vessel cells may be harvested from a patient, incorporated into an electrospun matrix and incubated in a bioreactor, first to drive expansion and differentiation and then, via pulsed flow, to promote vessel maturation (Niklason et al., 1999).

Similar strategies may be attempted to treat patients with congestive heart failure (Krupnick et al., 2004). Already, a number of clinical studies have been carried out on the injection of autologous bone marrow cells, sometimes unfractionated sometimes enriched for stem/progenitor cells, into the heart after myocardial infarction (Stamm et al., 2006). The initial rationale for this approach came from experiments in rodents interpreted as demonstrating the production of new cardiomyocytes through the transdifferentiation of hematopoietic stem cells. Evidence for myogenesis of grafted cells, whether from the hematopoietic lineage or, as seems much more plausible, from mesenchymal progenitors, remains sparse. However, some controlled studies do indicate potential clinical benefits from the autologous cell therapy. This may result from the production of angiogenic factors by the injected cells rather than from integration of donor cells into either muscle or new blood vessels. Nonetheless, although still a daunting challenge, the application of regenerative medicine principles to repair damaged cardiac muscle now seems within the possible realm (Dimmeler et al., 2005). The correct choice of cell source, the development and maturation of tissue engineered cardiac patches, and overcoming chronic fibrotic scarring remain hurdles to be overcome.

In another example of regenerative therapy utilizing autologous cells, here following the general paradigm first established for skin, bladder urothelial and smooth muscle cells were expanded in culture from small biopsies and seeded on scaffolds to produce tissue engineered neo-bladders. Such constructs were implanted in seven pediatric patients with high-pressure or poorly compliant bladders, some of whom have now been followed for over 5 years (mean 46 months) (Atala et al., 2006). The results are strongly encouraging and should lead to larger scale studies of safety and efficacy, targeting product approval after regulatory review.

Cell Sources

The ability to produce enough cells of the necessary types from the skin, cartilage, or bladder for bioengineered products depended on the presence of stem and progenitor cells in the corresponding adult tissues. It also required the development of culture methods that both permit the expansion of the precursor cells and allow enough differentiation for generation of the desired neo-tissue. Implementation of this strategy for regenerative medicine, based on expansion of autologous cells, cannot yet be extended to all tissues and organs. In some cases it is not clear how to obtain biopsies containing progenitor or stem cells, or even whether such cells exist. In other cases, culture conditions for expansion of the precursor cell population are not yet available.

The future development of cell-based regenerative medicine depends on further translation of basic discoveries regarding the identity and behavior of stem cells into practical clinical applications. Important targets include cells of organs for which orthotopic transplantation already has been established as an important mode of therapy, but for which the supply of donor organs does not meet the current need. Examples include cells of the heart, kidney, liver, and pancreas, specifically insulin-producing beta-cells. In addition production of neurons and other cells of the nervous system may permit therapy of degenerative diseases for which no effective treatment yet exists.

Mammalian stem cells have been divided into two general categories: embryonic and adult. Embryonic stem (ES) cells and the comparable embryonic germ (EG) cells appear to give rise to all specialized cell types, with the exception of a limited set of extra-embryonic cells. Adult stem cells, which may actually derive from fetal, neonatal, or truly adult tissue, show varying degrees of restriction to particular lineages.

ES Cells

ES cells and EG cells appear very similar (we will use "ES" to refer to both) and will likely have comparable medical applications. In fact, a recent report indicates that ES cells, which are derived from the inner cell mass of early embryos, most closely resemble early germ cells (Zwaka and Thomson, 2005). The ES cells can self-renew apparently without limit in culture, although mechanisms underlying this capacity remain incompletely understood (Rao, 2004; Stewart et al., 2006) and established ES lines may display some genomic instability. Furthermore, ES cells are broadly pluripotent (Evans and Kaufman, 1981; Martin, 1981; Shamblott et al., 1998; Amit et al., 2000). This great degree of plasticity represents both the strongest attraction and a significant potential limitation to the use of ES cells for regenerative medicine. A major remaining challenge is to direct the efficient production of pure populations of specific desired cell types from human ES cells (Odorico et al., 2001).

ES cells appear unique among normal stem cells in being tumorigenic, forming teratomas that contain cell types representing all three EG layers in a disorganized form (Martin, 1981; Thomson et al., 1998; Cowan et al., 2004). For clinical use it will be important to exclude undifferentiated stem cells from any products derived from ES cells (Lawrenz et al., 2004). Strategies have been envisaged to increase safety by introducing into ES cells a "suicide" gene, for example that encoding the thymidine kinase of Herpes simplex virus, which would render any escaping tumor cells sensitive to the drug ganciclovir (Odorico et al., 2001; Schuldiner et al., 2003). However, the genetic manipulation is itself not without risk, and the need to validate the engineered cell system would likely extend and complicate regulatory review of therapeutic products.

A central issue that must be addressed for tissue engineered products derived from ES cells, and also from any non-autologous adult stem cells, is immune rejection based on mismatches at genetic histocompatibility loci. It generally has been assumed that, because human ES cells and their differentiated derivatives can be induced to express high levels of major histocompatibility complex (MHC) Class I antigens (e.g. HLA-A and HLA-B), any ES cell-based product will be subjected to graft rejection (Drukker et al., 2002).

Therapeutic cloning offers a potential means to generate cells with the exact genetic constitution of each individual patient, so that immune rejection of grafts based on mismatched histocompatibility antigens should not occur. The approach entails transferring the nucleus of a somatic cell into an enucleated oocyte (SCNT), generating a blastocyst, and then culturing the inner cell mass to obtain an ES cell line (Colman and Kind, 2000). If required, genetic manipulation of the cells may be carried out to correct an inherited defect prior to production of the therapeutic graft (Rideout III et al., 2002). Despite a published claim (Hwang et al., 2005) later withdrawn, the generation of human ES cells by SCNT has not yet been achieved. However, the concept of therapeutic cloning to provide cells for tissue engineering applications has been clearly validated in a large animal model. Adult bovine fibroblasts were used as nuclear donors and bioengineered tissues were generated from cloned cardiac, skeletal muscle, and kidney cells (Lanza et al., 2002). The grafts, including functioning renal units capable of urine production, were successfully transplanted into the corresponding donor animals long term with no evidence for rejection. Although SCNT is the subject of political, ethical, and scientific debate (Hall et al., 2006), intense efforts in both the private sector (Lysaght and Hazlehurst, 2003) and academic institutions are likely to yield cloned human lines in the near future.

The reprogramming of somatic cell nuclei to yield pluripotent cells after introduction into the cytoplasm of enucleated eggs raises the possibility that additional means may be found to create cells with expanded potential to yield desired differentiated cell types. Counter to intuition, it appears that nuclei taken from certain terminally differentiated cells, such as postmitotic neurons, are readily reprogrammed to yield pluripotent cells by SCNT (Eggan et al., 2004). Nuclei from more differentiated cells may actually be superior for this purpose than nuclei of adult stem cells (Inoue et al., 2006; Sung et al., 2006), although the opposite trend was noted in studies using nuclei from neuronal lineage cells (Blelloch et al., 2006). In addition, fusion of somatic cells to ES cells can reprogram the somatic nuclei to an embryonic state (Cowan et al., 2005). Most remarkably, the expression of a small set of genes usually associated with ES cells (e.g. Oct3/4, Sox2, c-Myc, and Klf4) can induce an "embryonic" state, including pluripotency and the capacity to form teratoma tumors, in at least some somatic cells (fibroblasts) (Takahashi and Yamanaka, 2006).

The properties and differentiation potential of a number of human ES cell lines obtained by traditional means from early embryos currently used for research have been reviewed recently (Hoffman and Carpenter, 2005). The clinical application of ES cells for tissue engineering will depend on the development of robust methods to isolate and grow them under conditions consistent with Good Manufacturing Practice and regulatory review for safety. In particular, it is important to eliminate the requirement for murine feeder cells by using human feeders or, better, feeder-free conditions. In addition, development of culture conditions without the requirement for non-human serum would be advantageous. Progress has been made in the derivation and expansion of human ES cells with human feeder cells (Amit et al., 2003; Hovatta et al., 2003; Yoo et al., 2005; Stacey et al., 2006) or entirely without feeders (Amit et al., 2004; Carpenter et al., 2004; Beattie et al., 2005; Hovatta and Skottman, 2005; Klimanskaya et al., 2005; Sjogren-Jansson et al., 2005).

Perhaps the greater challenge remains in directing the differentiation of human ES cells to a given desired lineage with high efficiency. The underlying difficulty is that ES cells are developmentally many steps removed from adult, differentiated cells, and to date we have no general way to deterministically control the key steps in lineage restriction. Presumably, the same problem would be encountered with ES cells generated by SCNT or other means of reprogramming somatic cell nuclei.

To induce differentiation *in vitro* ES cells are allowed to attach to plastic in monolayer culture or, more frequently, to form aggregates called embryoid bodies (Itskovitz-Eldor et al., 2000). Over time within these aggregates cell types of many lineages are generated, including representatives of the three germ layers. The production of embryoid bodies can be enhanced and made more consistent by incubation in bioreactors (Gerecht-Nir et al., 2004). Further selection of specific lineages generally requires sequential exposure to a series of inducing conditions, either based on known signaling pathways or identified by trial and error. In most cases lineage-specific markers are expressed by the differentiated cells, but cells often do not progress to a full terminally differentiated phenotype. As summarized in recent reviews, the cell lineages which have been generated *in vitro* include, among others, several classes of neurons, astrocytes, oligodendrocytes, multipotent mesenchymal precursor cells, osteoblasts, cardiomyocytes, keratinocytes, pneumocytes, hematopoietic cells, hepatocytes, and pancreatic beta-cells (Nir et al., 2003; Tian and Kaufman, 2005; Raikwar et al., 2006; Trounson, 2006).

In general, it appears easier to obtain adult cells derived from ectoderm, including neurons, and mesoderm, including cardiomyocytes, than cells derived from endoderm (Trounson, 2006). This may help determine the first areas in which ES-derived cells enter clinical translation, once the barriers discussed above are surmounted. Dopaminergic neurons generated from primate and human ES cells already have been tested with encouraging results in animal models of Parkinson's disease (Perrier et al., 2004; Sanchez-Pernaute et al., 2005). Promising data also have been obtained with ES-derived oligodendrocytes in spinal cord injury models (Keirstead et al., 2005; Mueller et al., 2005). Cardiomyocytes derived from human ES cells, similarly, are candidates for future clinical use (He et al., 2003; Nir et al., 2003; Goh et al., 2005; Lev et al., 2005). However, the functional criteria that must be met to ensure physiological competence will be stringent because of the risk of inducing arrhythmias (Caspi and Gepstein, 2006; Passier et al., 2006).

The robust generation of pancreatic beta-cells and bioengineered islets from human ES cells or other stem cells would represent a particularly important achievement, with potential to treat diabetes (Weir, 2004; Nir and Dor, 2005). Clusters of insulin-positive cells, resembling pancreatic islets and expressing various additional markers of the endocrine pancreatic lineage, have been produced from mouse ES cells (Lumelsky et al., 2001) and also from non-human primate and human ES cells (Assady et al., 2001; Lester et al., 2004; Brolen et al., 2005; Baharvand et al., 2006). The production of beta-like cells can be enhanced by the expression of pancreatic transcription factors (Miyazaki et al., 2004; Shiroi et al., 2005). However, the assessment of differentiation must take into account the uptake of insulin from the growth medium, in addition to *de novo* synthesis (Paek et al., 2005). It seems fair to conclude that the efficient production of functional beta-cells from ES cells remains a difficult objective to achieve. As in other bioengineering applications with ES-derived cells, efforts to reverse diabetes also will depend on the complete removal of non-differentiated cells to avoid the formation of teratoma tumors, which were observed after implantation of ES-derived beta-cells in an animal model (Fujikawa et al., 2005).

Adult Stem Cells

Despite the acknowledged promise of ES cells, the challenges of controlling lineage-specific differentiation and eliminating residual stem cells are likely to extend the timeline for a number of tissue engineering applications. In many cases adult stem cells may provide a more direct route to clinical translation.

Lineage-restricted stem cells have been isolated from both fetal and postnatal tissues based on selective outgrowth in culture and/or immunoselection for surface markers. Examples with significant potential for new applications in regenerative medicine include neural (Baizabal et al., 2003; Goh et al., 2003), cardiac (Beltrami et al., 2003; Oh et al., 2003), muscle-derived (Cao et al., 2005), and hepatic stem cells (Kamiya et al., 2006; Schmelzer et al., 2006). A significant feature of each of these populations is a high capacity for self-renewal in culture. Their ability to expand may be less than that for ES cells, but in some cases the cells have been shown to express telomerase and may not be subjected to replicative senescence. These adult stem cells are multipotent. Neural stem cells can yield neurons, astrocytes, and oligodendrocytes. Cardiac stem cells are reported to yield cardiomyocytes, smooth muscle, and endothelial cells. Muscle-derived stem cells yield skeletal muscle and can be induced to produce chondrocytes. Hepatic stem cells yield hepatocytes and bile duct epithelial cells. The lineage-restricted adult stem cells all appear non-tumorigenic. Thus, unlike ES cells, it is likely that they could be used safely for bioengineered products with or without prior differentiation.

It is possible that some lineage-specific adult stem cells are capable of greater plasticity than might be supposed based solely on their tissue of origin. For example, there is evidence that hepatic stem cells may be induced to generate cells of additional endodermal lineages such as the endocrine pancreas (Yang et al., 2002; Nakajima-Nagata et al., 2004; Yamada et al., 2005; Zalzman et al., 2005). This type of switching of fates among related cell lineages may prove easier than inducing a full developmental program from a primitive precursor such as an ES cell.

Another class of adult cells with enormous potential value for regenerative medicine is the MSC, initially described in bone marrow (Bruder et al., 1994; Pittenger et al., 1999). These multipotent cells are able to give rise to differentiated cells of connective tissues including bone, cartilage, muscle, tendon, and fat. The MSC have, therefore, generated considerable interest for musculoskeletal and vascular tissue engineering (Gao and Caplan, 2003; Tuan et al., 2003; Barry and Murphy, 2004; Guilak et al., 2004; Risbud and Shapiro, 2005). Cells with similar differentiation potential and marker profiles have been isolated from a number of tissues in addition to the bone marrow. A notable source is the adipose tissue in which the cells are abundant and easily obtained by processing of suction-assisted lipectomy (liposuction) specimens (Zuk et al., 2001; Gimble and Guilak, 2003).

In general it seems better to view MSC as mixed populations of progenitor cells with varying degrees of replicative potential, rather than homogeneous stem cells. However, some classes of MSC, including lines cloned from single cells in skin (Bartsch et al., 2005), have been maintained in culture for extended periods. A very small subset of mesenchymal cells from bone marrow, termed multipotent adult progenitor cell (MAPC), reportedly are capable of extensive self-renewal and of differentiation into cell lineages not observed with typical MSC, including examples from each EG layer (Jiang et al., 2002).

Cells originating in a developing fetus and isolated from amniotic fluid or chorionic villi are a new source of stem cells of great potential interest for regenerative medicine (De Coppi et al., 2001; Siddiqui and Atala, 2004; Tsai et al., 2006). Fetal-derived cells with apparently similar properties also have been described in the amnion of term placenta (Miki et al., 2005). Amniotic fluid stem (AFS) cells and amniotic epithelial cells can give rise to differentiated cell types representing the three EG layers (Siddiqui and Atala, 2004; Miki et al., 2005). Formal proof that single AFS cells can yield this full range of progeny cells was obtained using clones marked by retroviral insertion (unpublished data from A. Atala laboratory). The cells can be expanded for well over 200 population doublings with no sign of telomere shortening or replicative senescence, and retain a normal diploid karyotype. They are readily cultured without need for feeder cells. The AFS cells express some markers in common with ES cells, such as the surface antigen SSEA4 and the transcription factor Oct3/4, while other markers are shared with mesenchymal and neural stem cells. A broadly multipotent cell population obtained from umbilical cord blood may have certain key properties in common with AFS cells, and was termed "unrestricted somatic stem cells" (USSCs) (Kogler et al., 2004). This population may overlap with or be identical to the so-called "umbilical cord matrix stem" (UCMS) cells isolated from Wharton's jelly (Mitchell et al., 2003; Weiss et al., 2006).

The full developmental potential of the various stem cell populations obtained from fetal and adult sources remains to be determined. It is possible that virtually all of the cell types that might be desired for tissue engineering could be obtained from AFS cells, equivalent stem cells from placenta, those from the non-hematopoietic subset in umbilical cord blood, or comparable populations. Similar approaches to those being taken with ES cells, such as genetic modification with expression vectors for lineage-specific transcription factors, may help in the generation of those differentiated cell types for which it proves difficult to develop a straightforward induction protocol using external signals. However, it will remain necessary to show, beyond induction of a set of characteristic markers, that fully functional mature cells can be generated for any given lineage.

Immune Compatibility

The growing number of choices of cell sources for bioengineered tissues opens up a range of strategies to obtain the desired differentiated cell populations. The issue of immune compatibility remains central. Although life-long immunosuppression can be successful, as in conjunction with orthotopic organ transplantation, it would be preferable to design bioengineering-based products that will be tolerated by recipients without the need for immunosuppressive drugs. The only cell-based therapies guaranteed to be histocompatible would contain autologous cells or those derived by therapeutic cloning (assuming mitochondrial differences are not critical) (Lanza et al., 2002). When a perfectly matched, personalized therapeutic product is not available, there still should be ways to limit the requirement for immunosuppression. First, there may be a strong intrinsic advantage to developing cell-based products from certain stem cells because there is evidence that they, and possibly differentiated cells derived from them, are immune privileged. Second, it may be possible to develop banks of cells that can be used to permit histocompatibility matching with recipient patients.

Human ES cells express low levels of MHC Class I antigens (HLA-A, HLA-B) and are negative for MHC Class II (HLA-DR) (Drukker et al., 2002). Differentiated derivatives of the ES cells remain negative for MHC II but show some increase in MHC Class I that is further up-regulated by exposure to interferon. These observations gave rise to the natural assumption that ES cells and their differentiated progeny would be subjected to rejection based on MHC mismatches, and led to a search for strategies to induce immunological tolerance in recipients of transplanted cells derived from ES lines (Drukker, 2004). However, it was observed that ES cells in the mouse and comparable stem cells from the inner cell mass of the embryo in the rat could be transplanted successfully in immune competent animals despite mismatches at the MHC loci. Furthermore, rodent ES cells may be able to induce immune tolerance in the recipient animals (Fandrich et al., 2002). Even more remarkably, human ES cells and differentiated derivatives were not rejected by immune competent mice *in vivo*, nor did they stimulate an immune response *in vitro* by human T-lymphocytes specific for mismatched MHC. Rather, the human cells appeared to inhibit the T-cell response (Li et al., 2004). An independent study using mice with a "humanized" immune system confirmed a very low T-cell response to human ES cells and differentiated derivatives (Drukker et al., 2006).

MSC from bone marrow and their differentiated derivatives also have been shown both to escape an allogeneic immune response and to possess immunomodulatory activity to block such a response (Bartholomew et al., 2002; Le Blanc, 2003; Potian et al., 2003; Aggarwal and Pittenger, 2005). The effect likewise is observed with MSC isolated from adipose tissue (Puissant et al., 2005). The successful therapeutic use of allogeneic MSC has been confirmed in animal models (Arinzeh et al., 2003; De Kok et al., 2003). Therefore, beyond the application of MSC as regenerative cells, it is possible that they could be employed to induce immune tolerance to grafts of other cell types. The mechanisms underlying the immunodulatory properties of MSC are under active investigation and understanding them may have profound impact on regenerative medicine (Plumas et al., 2005; Krampera et al., 2006; Sotiropoulou et al., 2006).

Other stem cell populations should be examined for their ability to escape and/or modulate an allogeneic immune response. While it is important to exercise caution in interpreting the laboratory results and in designing clinical trials, there is some reason to hope that the use of allogeneic stem cell-based bioengineered products will not necessarily imply the need for life-long treatment with immunosuppressive drugs. In the first FDA-approved clinical trial of allogeneic human neural stem cells, in children with a Neural Ceroid Lipofuscinosis disorder known as Batten disease (Taupin, 2006), immunosuppressive therapy will be utilized for the initial year after cell implantation and then reevaluated.

Banking of stem cells for future therapeutic use extends possibilities both for autologous and allogeneic therapy paradigms, even if it turns out that histocompatibility matching is important for stem cell-based therapies. Amniocentesis specimens, placenta, and cord blood represent sources from which highly multipotent adult stem cells can be obtained and typed with minimal invasiveness. Prospective parents could opt for collection and cryopreservation of such cells for future use by their children in the event of medical need. Furthermore, collection and typing of a sufficient number of samples (*ca.* 100,000 for the US population) to permit nearly perfect histocompatibility matching between unrelated donors and recipients would be readily achieved. Similarly, collection and banking of cells from adult adipose tissue appears straightforward. Although it would entail a greater level of effort and could be politically controversial, it also might be feasible to prepare and bank a relatively large set of human ES lines to facilitate histocompatibility matching.

One recent study suggests that a surprisingly modest number of banked lines or specimens could provide substantial ability to match donor cells to recipients (Taylor et al., 2005). Taken together with the low immunogenicity of certain stem cells, these results support the concept that allogeneic bioengineered products may not inevitably demand intensive immunosuppressive treatment, even if it proves impossible to develop general methods to induce selective immunological tolerance.

CONCLUSIONS

Regenerative medicine is a highly interdisciplinary field. Future progress will continue to depend on synergies between advances in biology, chemistry, and engineering. Yet the development of new therapies may be rate limited by the need to identify and obtain stem and progenitor cells capable of yielding desired specialized cell types safely and efficiently. Exciting new work indicates unexpected paths that may provide novel solutions to two critical problems: sourcing of progenitors for a potentially unlimited range of specialized cell types and overcoming the need for life-long immunotherapy associated with allogeneic therapies.

REFERENCES

Aggarwal, S. and Pittenger, M.F. (2005). Human mesenchymal stem cells modulate allogeneic immune cell responses. *Blood* **105**(4): 1815–1822.

Amit, M., Carpenter, M.K., Inokuma, M.S., Chiu, C.P., Harris, C.P., Waknitz, M.A., Itskovitz-Eldor, J. and Thomson, J.A. (2000). Clonally derived human embryonic stem cell lines maintain pluripotency and proliferative potential for prolonged periods of culture. *Dev. Biol.* **227**(2): 271–278.

Amit, M., Margulets, V., Segev, H., Shariki, K., Laevsky, I., Coleman, R. and Itskovitz-Eldor, J. (2003). Human feeder layers for human embryonic stem cells. *Biol. Reprod.* **68**(6): 2150–2156.

Amit, M., Shariki, C., Margulets, V. and Itskovitz-Eldor, J. (2004). Feeder layer- and serum-free culture of human embryonic stem cells. *Biol. Reprod.* **70**(3): 837–845.

Arinzeh, T.L., Peter, S.J., Archambault, M.P., van den Bos, C., Gordon, S., Kraus, K., Smith, A. and Kadiyala, S. (2003). Allogeneic mesenchymal stem cells regenerate bone in a critical-sized canine segmental defect. *J. Bone Joint Surg. Am.* **85-A**(10): 1927–1935.

Assady, S., Maor, G., Amit, M., Itskovitz-Eldor, J., Skorecki, K.L. and Tzukerman, M. (2001). Insulin production by human embryonic stem cells. *Diabetes* **50**(8): 1691–1697.

Atala, A., Bauer, S.B., Soker, S., Yoo, J.J. and Retik, A.B. (2006). Tissue-engineered autologous bladders for patients needing cystoplasty. *Lancet* **367**(9518): 1241–1246.

Baharvand, H., Jafary, H., Massumi, M. and Ashtiani, S.K. (2006). Generation of insulin-secreting cells from human embryonic stem cells. *Dev. Growth Differ.* **48**(5): 323–332.

Baizabal, J.M., Furlan-Magaril, M., Santa-Olalla, J. and Covarrubias, L. (2003). Neural stem cells in development and regenerative medicine. *Arch. Med. Res.* **34**(6): 572–588.

Barry, F.P. and Murphy, J.M. (2004). Mesenchymal stem cells: clinical applications and biological characterization. *Int. J. Biochem. Cell. Biol.* **36**(4): 568–584.

Bartholomew, A., Sturgeon, C., Siatskas, M., Ferrer, K., McIntosh, K., Patil, S., Hardy, W., Devine, S., Ucker, D., Deans, R., Moseley, A. and Hoffman, R. (2002). Mesenchymal stem cells suppress lymphocyte proliferation *in vitro* and prolong skin graft survival *in vivo*. *Exp. Hematol.* **30**(1): 42–48.

Bartsch, G., Yoo, J.J., De Coppi, P., Siddiqui, M.M., Schuch, G., Pohl, H.G., Fuhr, J., Perin, L., Soker, S. and Atala, A. (2005). Propagation, expansion, and multilineage differentiation of human somatic stem cells from dermal progenitors. *Stem Cells Dev.* **14**(3): 337–348.

Beattie, G.M., Lopez, A.D., Bucay, N., Hinton, A., Firpo, M.T., King, C.C. and Hayek, A. (2005). Activin A maintains pluripotency of human embryonic stem cells in the absence of feeder layers. *Stem Cells* **23**(4): 489–495.

Beltrami, A.P., Barlucchi, L., Torella, D., Baker, M., Limana, F., Chimenti, S., Kasahara, H., Rota, M., Musso, E., Urbanek, K., Leri, A., Kajstura, J., Nadal-Ginard, B. and Anversa, P. (2003). Adult cardiac stem cells are multipotent and support myocardial regeneration. *Cell* **114**(6): 763–776.

Blelloch, R., Wang, Z., Meissner, A., Pollard, S., Smith, A. and Jaenisch, R. (2006). Reprogramming efficiency following somatic cell nuclear transfer is influenced by the differentiation and methylation state of the donor nucleus. *Stem Cells* **24**(9): 2007–2013.

Briscoe, D.M., Dharnidharka, V.R., Isaacs, C., Downing, G., Prosky, S., Shaw, P., Parenteau, N.L. and Hardin-Young, J. (1999). The allogeneic response to cultured human skin equivalent in the hu-PBL-SCID mouse model of skin rejection. *Transplantation* **67**(12): 1590–1599.

Brolen, G.K., Heins, N., Edsbagge, J. and Semb, H. (2005). Signals from the embryonic mouse pancreas induce differentiation of human embryonic stem cells into insulin-producing beta-cell-like cells. *Diabetes* **54**(10): 2867–2874.

Bruder, S.P., Fink, D.J. and Caplan, A.I. (1994). Mesenchymal stem cells in bone development, bone repair, and skeletal regeneration therapy. *J. Cell Biochem.* **56**(3): 283–294.

Cao, B., Deasy, B.M., Pollett, J. and Huard, J. (2005). Cell therapy for muscle regeneration and repair. *Phys. Med. Rehabil. Clin. N. Am.* **16**(4): 889–907, viii.

Carpenter, M.K., Rosler, E.S., Fisk, G.J., Brandenberger, R., Ares, X., Miura, T., Lucero, M. and Rao, M.S. (2004). Properties of four human embryonic stem cell lines maintained in a feeder-free culture system. *Dev. Dyn.* **229**(2): 243–258.

Caspi, O. and Gepstein, L. (2006). Regenerating the heart using human embryonic stem cells – from cell to bedside. *Isr. Med. Assoc. J.* **8**(3): 208–214.

Colman, A. and Kind, A. (2000). Therapeutic cloning: concepts and practicalities. *Trends Biotechnol.* **18**(5): 192–196.

Committee on the Biological and Biomedical Applications of Stem Cell Research (2002). *Stem Cells and the Future of Regenerative Medicine*. Washington, DC: National Academy Press, 94 pp.

Cowan, C.A., Klimanskaya, I., McMahon, J., Atienza, J., Witmyer, J., Zucker, J.P., Wang, S., Morton, C.C., McMahon, A.P., Powers, D. and Melton, D.A. (2004). Derivation of embryonic stem-cell lines from human blastocysts. *N. Engl. J. Med.* **350**(13): 1353–1356.

Cowan, C.A., Atienza, J., Melton, D.A. and Eggan, K. (2005). Nuclear reprogramming of somatic cells after fusion with human embryonic stem cells. *Science* **309**(5739): 1369–1373.

De Coppi, P., Bartsch, G., Dal Cin, P., Yoo, J.J., Soker, S. and Atala, A. (2001). Human fetal stem cell isolation from amniotic fluid. *American Academy of Pediatrics National Conference*, San Francisco, pp. 210–211.

De Kok, I.J., Peter, S.J., Archambault, M., van den Bos, C., Kadiyala, S., Aukhil, I. and Cooper, L.F. (2003). Investigation of allogeneic mesenchymal stem cell-based alveolar bone formation: preliminary findings. *Clin. Oral Implant. Res.* **14**(4): 481–489.

Dimmeler, S., Zeiher, A.M. and Schneider, M.D. (2005). Unchain my heart: the scientific foundations of cardiac repair. *J. Clin. Invest.* **115**(3): 572–583.

Drukker, M. (2004). Immunogenicity of human embryonic stem cells: can we achieve tolerance? *Springer Semin. Immunopathol.* **26**(1–2): 201–213.

Drukker, M., Katz, G., Urbach, A., Schuldiner, M., Markel, G., Itskovitz-Eldor, J., Reubinoff, B., Mandelboim, O. and Benvenisty, N. (2002). Characterization of the expression of MHC proteins in human embryonic stem cells. *Proc. Natl Acad. Sci. USA* **99**(15): 9864–9869.

Drukker, M., Katchman, H., Katz, G., Even-Tov Friedman, S., Shezen, E., Hornstein, E., Mandelboim, O., Reisner, Y. and Benvenisty, N. (2006). Human embryonic stem cells and their differentiated derivatives are less susceptible to immune rejection than adult cells. *Stem Cells* **24**(2): 221–229.

Eggan, K., Baldwin, K., Tackett, M., Osborne, J., Gogos, J., Chess, A., Axel, R. and Jaenisch, R. (2004). Mice cloned from olfactory sensory neurons. *Nature* **428**(6978): 44–49.

Evans, M.J. and Kaufman, M.H. (1981). Establishment in culture of pluripotential cells from mouse embryos. *Nature* **292**(5819): 154–156.

Fandrich, F., Lin, X., Chai, G.X., Schulze, M., Ganten, D., Bader, M., Holle, J., Huang, D.S., Parwaresch, R., Zavazava, N. and Binas, B. (2002). Preimplantation-stage stem cells induce long-term allogeneic graft acceptance without supplementary host conditioning. *Nat. Med.* **8**(2): 171–178.

Fujikawa, T., Oh, S.H., Pi, L., Hatch, H.M., Shupe, T. and Petersen, B.E. (2005). Teratoma formation leads to failure of treatment for type I diabetes using embryonic stem cell-derived insulin-producing cells. *Am. J. Pathol.* **166**(6): 1781–1791.

Gao, J. and Caplan, A.I. (2003). Mesenchymal stem cells and tissue engineering for orthopaedic surgery. *Chir. Organ. Mov.* **88**(3): 305–316.

Gerecht-Nir, S., Cohen, S. and Itskovitz-Eldor, J. (2004). Bioreactor cultivation enhances the efficiency of human embryoid body (hEB) formation and differentiation. *Biotechnol. Bioeng.* **86**(5): 493–502.

Gimble, J. and Guilak, F. (2003). Adipose-derived adult stem cells: isolation, characterization, and differentiation potential. *Cytotherapy* **5**(5): 362–369.

Goh, E.L., Ma, D., Ming, G.L. and Song, H. (2003). Adult neural stem cells and repair of the adult central nervous system. *J. Hematother. Stem Cell Res.* **12**(6): 671–679.

Goh, G., Self, T., Barbadillo Munoz, M.D., Hall, I.P., Young, L. and Denning, C. (2005). Molecular and phenotypic analyses of human embryonic stem cell-derived cardiomyocytes: opportunities and challenges for clinical translation. *Thromb. Haemost.* **94**(4): 728–737.

Guilak, F., Awad, H.A., Fermor, B., Leddy, H.A. and Gimble, J.M. (2004). Adipose-derived adult stem cells for cartilage tissue engineering. *Biorheology* **41**(3–4): 389–399.

Hall, V.J., Stojkovic, P. and Stojkovic, M. (2006). Using therapeutic cloning to fight human disease: a conundrum or reality? *Stem Cells* **24**(7): 1628–1637.

He, J.Q., Ma, Y., Lee, Y., Thomson, J.A. and Kamp, T.J. (2003). Human embryonic stem cells develop into multiple types of cardiac myocytes: action potential characterization. *Circ. Res.* **93**(1): 32–39.

Hoffman, L.M. and Carpenter, M.K. (2005). Characterization and culture of human embryonic stem cells. *Nat. Biotechnol.* **23**(6): 699–708.

Horch, R.E., Kopp, J., Kneser, U., Beier, J. and Bach, A.D. (2005). Tissue engineering of cultured skin substitutes. *J. Cell Mol. Med.* **9**(3): 592–608.

Hovatta, O. and Skottman, H. (2005). Feeder-free derivation of human embryonic stem-cell lines. *Lancet* **365**(9471): 1601–1603.

Hovatta, O., Mikkola, M., Gertow, K., Stromberg, A.M., Inzunza, J., Hreinsson, J., Rozell, B., Blennow, E., Andang, M. and Ahrlund-Richter, L. (2003). A culture system using human foreskin fibroblasts as feeder cells allows production of human embryonic stem cells. *Hum. Reprod.* **18**(7): 1404–1409.

Hwang, W.S., Roh, S.I., Lee, B.C., Kang, S.K., Kwon, D.K., Kim, S., Kim, S.J., Park, S.W., Kwon, H.S., Lee, C.K., Lee, J.B., Kim, J.M., Ahn, C., Paek, S.H., Chang, S.S., Koo, J.J., Yoon, H.S., Hwang, J.H., Hwang, Y.Y., Park, Y.S., Oh, S.K., Kim, H.S., Park, J.H., Moon, S.Y. and Schatten, G. (2005). Patient-specific embryonic stem cells derived from human SCNT blastocysts. *Science* **308**(5729): 1777–1783.

Inoue, K., Ogonuki, N., Miki, H., Hirose, M., Noda, S., Kim, J.M., Aoki, F., Miyoshi, H. and Ogura, A. (2006). Inefficient reprogramming of the hematopoietic stem cell genome following nuclear transfer. *J. Cell Sci.* **119**(Pt 10): 1985–1991.

Ioannidou, E. (2006). Therapeutic modulation of growth factors and cytokines in regenerative medicine. *Curr. Pharm. Des.* **12**(19): 2397–2408.

Itskovitz-Eldor, J., Schuldiner, M., Karsenti, D., Eden, A., Yanuka, O., Amit, M., Soreq, H. and Benvenisty, N. (2000). Differentiation of human embryonic stem cells into embryoid bodies compromising the three embryonic germ layers. *Mol. Med.* **6**(2): 88–95.

Jiang, Y., Jahagirdar, B.N., Reinhardt, R.L., Schwartz, R.E., Keene, C.D., Ortiz-Gonzalez, X.R., Reyes, M., Lenvik, T., Lund, T., Blackstad, M., Du, J., Aldrich, S., Lisberg, A., Low, W.C., Largaespada, D.A. and Verfaillie, C.M. (2002). Pluripotency of mesenchymal stem cells derived from adult marrow. *Nature* **418**(6893): 41–49.

Johnson, P.C. (2000). The role of tissue engineering. *Adv. Skin Wound Care* **13**(2 Suppl): 12–14.

Kamiya, A., Gonzalez, F.J. and Nakauchi, H. (2006). Identification and differentiation of hepatic stem cells during liver development. *Front Biosci.* **11**: 1302–1310.

Keirstead, H.S., Nistor, G., Bernal, G., Totoiu, M., Cloutier, F., Sharp, K. and Steward, O. (2005). Human embryonic stem cell-derived oligodendrocyte progenitor cell transplants remyelinate and restore locomotion after spinal cord injury. *J. Neurosci.* **25**(19): 4694–4705.

Klimanskaya, I., Chung, Y., Meisner, L., Johnson, J., West, M.D. and Lanza, R. (2005). Human embryonic stem cells derived without feeder cells. *Lancet* **365**(9471): 1636–1641.

Kogler, G., Sensken, S., Airey, J.A., Trapp, T., Muschen, M., Feldhahn, N., Liedtke, S., Sorg, R.V., Fischer, J., Rosenbaum, C., Greschat, S., Knipper, A., Bender, J., Degistirici, O., Gao, J., Caplan, A.I., Colletti, E.J., Almeida-Porada, G., Muller, H.W., Zanjani, E. and Wernet, P. (2004). A new human somatic stem cell from placental cord blood with intrinsic pluripotent differentiation potential. *J. Exp. Med.* **200**(2): 123–135.

Krampera, M., Cosmi, L., Angeli, R., Pasini, A., Liotta, F., Andreini, A., Santarlasci, V., Mazzinghi, B., Pizzolo, G., Vinante, F., Romagnani, P., Maggi, E., Romagnani, S. and Annunziato, F. (2006). Role for interferon-gamma in the immunomodulatory activity of human bone marrow mesenchymal stem cells. *Stem Cells* **24**(2): 386–398.

Krupnick, A.S., Kreisel, D., Riha, M., Balsara, K.R. and Rosengard, B.R. (2004). Myocardial tissue engineering and regeneration as a therapeutic alternative to transplantation. *Curr. Top. Microbiol. Immunol.* **280**: 139–164.

Lanza, R.P., Chung, H.Y., Yoo, J.J., Wettstein, P.J., Blackwell, C., Borson, N., Hofmeister, E., Schuch, G., Soker, S., Moraes, C.T., West, M.D. and Atala, A. (2002). Generation of histocompatible tissues using nuclear transplantation. *Nat. Biotechnol.* **20**(7): 689–696.

Lavik, E. and Langer, R. (2004). Tissue engineering: current state and perspectives. *Appl. Microbiol. Biotechnol.* **65**(1): 1–8.

Lawrenz, B., Schiller, H., Willbold, E., Ruediger, M., Muhs, A. and Esser, S. (2004). Highly sensitive biosafety model for stem-cell-derived grafts. *Cytotherapy* **6**(3): 212–222.

Le Blanc, K. (2003). Immunomodulatory effects of fetal and adult mesenchymal stem cells. *Cytotherapy* **5**(6): 485–489.

Lester, L.B., Kuo, H.C., Andrews, L., Nauert, B. and Wolf, D.P. (2004). Directed differentiation of rhesus monkey ES cells into pancreatic cell phenotypes. *Reprod. Biol. Endocrinol.* **2**: 42.

Lev, S., Kehat, I. and Gepstein, L. (2005). Differentiation pathways in human embryonic stem cell-derived cardiomyocytes. *Ann. NY Acad. Sci.* **1047**: 50–65.

Li, L., Baroja, M.L., Majumdar, A., Chadwick, K., Rouleau, A., Gallacher, L., Ferber, I., Lebkowski, J., Martin, T., Madrenas, J. and Bhatia, M. (2004). Human embryonic stem cells possess immune-privileged properties. *Stem Cells* **22**(4): 448–456.

Lumelsky, N., Blondel, O., Laeng, P., Velasco, I., Ravin, R. and McKay, R. (2001). Differentiation of embryonic stem cells to insulin-secreting structures similar to pancreatic islets. *Science* **292**(5520): 1389–1394.

Lutolf, M.P. and Hubbell, J.A. (2005). Synthetic biomaterials as instructive extracellular microenvironments for morphogenesis in tissue engineering. *Nat. Biotechnol.* **23**(1): 47–55.

Lysaght, M.J. and Hazlehurst, A.L. (2003). Private sector development of stem cell technology and therapeutic cloning. *Tissue Eng.* **9**(3): 555–561.

Lysaght, M.J. and Hazlehurst, A.L. (2004). Tissue engineering: the end of the beginning. *Tissue Eng.* **10**(1–2): 309–320.

Lysaght, M.J. and Reyes, J. (2001). The growth of tissue engineering. *Tissue Eng.* **7**(5): 485–493.

Martin, G.R. (1981). Isolation of a pluripotent cell line from early mouse embryos cultured in medium conditioned by teratocarcinoma stem cells. *Proc. Natl Acad. Sci. USA* **78**(12): 7634–7638.

Miki, T., Lehmann, T., Cai, H., Stolz, D.B. and Strom, S.C. (2005). Stem cell characteristics of amniotic epithelial cells. *Stem Cells* **23**(10): 1549–1559.

Mitchell, K.E., Weiss, M.L., Mitchell, B.M., Martin, P., Davis, D., Morales, L., Helwig, B., Beerenstrauch, M., Abou-Easa, K., Hildreth, T., Troyer, D. and Medicetty, S. (2003). Matrix cells from Wharton's jelly form neurons and glia. *Stem Cells* **21**(1): 50–60.

Miyazaki, S., Yamato, E. and Miyazaki, J. (2004). Regulated expression of pdx-1 promotes *in vitro* differentiation of insulin-producing cells from embryonic stem cells. *Diabetes* **53**(4): 1030–1037.

Moller, E., Soderberg-Naucler, C. and Sumitran-Karuppan, S. (1999). Role of alloimmunity in clinical transplantation. *Rev. Immunogenet.* **1**(3): 309–322.

Mueller, D., Shamblott, M.J., Fox, H.E., Gearhart, J.D. and Martin, L.J. (2005). Transplanted human embryonic germ cell-derived neural stem cells replace neurons and oligodendrocytes in the forebrain of neonatal mice with excitotoxic brain damage. *J. Neurosci. Res.* **82**(5): 592–608.

Murray-Thomas, T. and Cowie, M.R. (2003). Epidemiology and clinical aspects of congestive heart failure. *J. Renin Angiotensin Aldosterone Syst.* **4**(3): 131–136.

Nakajima-Nagata, N., Sakurai, T., Mitaka, T., Katakai, T., Yamato, E., Miyazaki, J., Tabata, Y., Sugai, M. and Shimizu, A. (2004). *In vitro* induction of adult hepatic progenitor cells into insulin-producing cells. *Biochem. Biophys. Res. Commun.* **318**(3): 625–630.

Naughton, G.K. (2002). From lab bench to market: critical issues in tissue engineering. *Ann. NY Acad. Sci.* **961**: 372–385.

Niklason, L.E., Gao, J., Abbott, W.M., Hirschi, K.K., Houser, S., Marini, R. and Langer, R. (1999). Functional arteries grown *in vitro*. *Science* **284**(5413): 489–493.

Nir, S.G., David, R., Zaruba, M., Franz, W.M. and Itskovitz-Eldor, J. (2003). Human embryonic stem cells for cardiovascular repair. *Cardiovasc. Res.* **58**(2): 313–323.

Nir, T. and Dor, Y. (2005). How to make pancreatic beta cells – prospects for cell therapy in diabetes. *Curr. Opin. Biotechnol.* **16**(5): 524–529.

Odorico, J.S., Kaufman, D.S. and Thomson, J.A. (2001). Multilineage differentiation from human embryonic stem cell lines. *Stem Cells* **19**(3): 193–204.

Oh, H., Bradfute, S.B., Gallardo, T.D., Nakamura, T., Gaussin, V., Mishina, Y., Pocius, J., Michael, L.H., Behringer, R.R., Garry, D.J., Entman, M.L. and Schneider, M.D. (2003). Cardiac progenitor cells from adult myocardium: homing, differentiation, and fusion after infarction. *Proc. Natl Acad. Sci. USA* **100**(21): 12313–12318.

Paek, H.J., Morgan, J.R. and Lysaght, M.J. (2005). Sequestration and synthesis: the source of insulin in cell clusters differentiated from murine embryonic stem cells. *Stem Cells* **23**(7): 862–867.

Passier, R., Denning, C. and Mummery, C. (2006). Cardiomyocytes from human embryonic stem cells. *Handb. Exp. Pharmacol.* **174**: 101–122.

Perrier, A.L., Tabar, V., Barberi, T., Rubio, M.E., Bruses, J., Topf, N., Harrison, N.L. and Studer, L. (2004). Derivation of midbrain dopamine neurons from human embryonic stem cells. *Proc. Natl Acad. Sci. USA* **101**(34): 12543–12548.

Pittenger, M.F., Mackay, A.M., Beck, S.C., Jaiswal, R.K., Douglas, R., Mosca, J.D., Moorman, M.A., Simonetti, D.W., Craig, S. and Marshak, D.R. (1999). Multilineage potential of adult human mesenchymal stem cells. *Science* **284**(5411): 143–147.

Plumas, J., Chaperot, L., Richard, M.J., Molens, J.P., Bensa, J.C. and Favrot, M.C. (2005). Mesenchymal stem cells induce apoptosis of activated T cells. *Leukemia* **19**(9): 1597–1604.

Potian, J.A., Aviv, H., Ponzio, N.M., Harrison, J.S. and Rameshwar, P. (2003). Veto-like activity of mesenchymal stem cells: functional discrimination between cellular responses to alloantigens and recall antigens. *J. Immunol.* **171**(7): 3426–3434.

Puissant, B., Barreau, C., Bourin, P., Clavel, C., Corre, J., Bousquet, C., Taureau, C., Cousin, B., Abbal, M., Laharrague, P., Penicaud, L., Casteilla, L. and Blancher, A. (2005). Immunomodulatory effect of human adipose tissue-derived adult stem cells: comparison with bone marrow mesenchymal stem cells. *Br. J. Haematol.* **129**(1): 118–129.

Raikwar, S.P., Mueller, T. and Zavazava, N. (2006). Strategies for developing therapeutic application of human embryonic stem cells. *Physiology (Bethesda)* **21**: 19–28.

Rao, M. (2004). Conserved and divergent paths that regulate self-renewal in mouse and human embryonic stem cells. *Dev. Biol.* **275**(2): 269–286.

Rideout III, W.M., Hochedlinger, K., Kyba, M., Daley, G.Q. and Jaenisch, R. (2002). Correction of a genetic defect by nuclear transplantation and combined cell and gene therapy. *Cell* **109**(1): 17–27.

Risbud, M.V. and Shapiro, I.M. (2005). Stem cells in craniofacial and dental tissue engineering. *Orthod. Craniofac. Res.* **8**(2): 54–59.

Sanchez-Pernaute, R., Studer, L., Ferrari, D., Perrier, A., Lee, H., Vinuela, A. and Isacson, O. (2005). Long-term survival of dopamine neurons derived from parthenogenetic primate embryonic stem cells (cyno-1) after transplantation. *Stem Cells* **23**(7): 914–922.

Schmelzer, E., Wauthier, E. and Reid, L.M. (2006). The phenotypes of pluripotent human hepatic progenitors. *Stem Cells* **24**(8): 1852–1858.

Schuldiner, M., Itskovitz-Eldor, J. and Benvenisty, N. (2003). Selective ablation of human embryonic stem cells expressing a "suicide" gene. *Stem Cells* **21**(3): 257–265.

Shamblott, M.J., Axelman, J., Wang, S., Bugg, E.M., Littlefield, J.W., Donovan, P.J., Blumenthal, P.D., Huggins, G.R. and Gearhart, J.D. (1998). Derivation of pluripotent stem cells from cultured human primordial germ cells. *Proc. Natl Acad. Sci. USA* **95**(23): 13726–13731.

Shin'oka, T., Matsumura, G., Hibino, N., Naito, Y., Watanabe, M., Konuma, T., Sakamoto, T., Nagatsu, M. and Kurosawa, H. (2005). Midterm clinical result of tissue-engineered vascular autografts seeded with autologous bone marrow cells. *J. Thorac. Cardiovasc. Surg.* **129**(6): 1330–1338.

Shiroi, A., Ueda, S., Ouji, Y., Saito, K., Moriya, K., Sugie, Y., Fukui, H., Ishizaka, S. and Yoshikawa, M. (2005). Differentiation of embryonic stem cells into insulin-producing cells promoted by Nkx2.2 gene transfer. *World J. Gastroenterol.* **11**(27): 4161–4166.

Siddiqui, M.M. and Atala, A. (2004). Amniotic fluid-derived pluripotential cells: adult and fetal. In: Lanza, R., Blau, H., Melton, D., Moore, M., Thomas, E.D., Verfaillie, C., Weissman, I. and West, M. (eds.). *Handbook of Stem Cells*, Vol. 2. Amsterdam: Elsevier Academic Press, pp. 175–180.

Sjogren-Jansson, E., Zetterstrom, M., Moya, K., Lindqvist, J., Strehl, R. and Eriksson, P.S. (2005). Large-scale propagation of four undifferentiated human embryonic stem cell lines in a feeder-free culture system. *Dev. Dyn.* **233**(4): 1304–1314.

Smyth, S. and Heron, A. (2006). Diabetes and obesity: the twin epidemics. *Nat. Med.* **12**(1): 75–80.

Sotiropoulou, P.A., Perez, S.A., Gritzapis, A.D., Baxevanis, C.N. and Papamichail, M. (2006). Interactions between human mesenchymal stem cells and natural killer cells. *Stem Cells* **24**(1): 74–85.

Stacey, G.N., Cobo, F., Nieto, A., Talavera, P., Healy, L. and Concha, A. (2006). The development of "feeder" cells for the preparation of clinical grade hES cell lines: challenges and solutions. *J. Biotechnol.* **125**(4): 583–588.

Stamm, C., Liebold, A., Steinhoff, G. and Strunk, D. (2006). Stem cell therapy for ischemic heart disease: beginning or end of the road? *Cell Transplant.* **15** (Suppl 1): S47–S56.

Stankus, J.J., Guan, J., Fujimoto, K. and Wagner, W.R. (2006). Microintegrating smooth muscle cells into a biodegradable, elastomeric fiber matrix. *Biomaterials* **27**(5): 735–744.

Stewart, R., Stojkovic, M. and Lako, M. (2006). Mechanisms of self-renewal in human embryonic stem cells. *Eur. J. Cancer* **42**(9): 1257–1272.

Stitzel, J., Liu, J., Lee, S.J., Komura, M., Berry, J., Soker, S., Lim, G., Van Dyke, M., Czerw, R., Yoo, J.J. and Atala, A. (2006). Controlled fabrication of a biological vascular substitute. *Biomaterials* **27**(7): 1088–1094.

Sung, L.Y., Gao, S., Shen, H., Yu, H., Song, Y., Smith, S.L., Chang, C.C., Inoue, K., Kuo, L., Lian, J., Li, A., Tian, X.C., Tuck, D.P., Weissman, S.M., Yang, X. and Cheng, T. (2006). Differentiated cells are more efficient than adult stem cells for cloning by somatic cell nuclear transfer. *Nat. Genet.*

Takahashi, K. and Yamanaka, S. (2006). Induction of pluripotent stem cells from mouse embryonic and adult fibroblast cultures by defined factors. *Cell* **126**(4): 663–676.

Taupin, P. (2006). HuCNS-SC (StemCells). *Curr. Opin. Mol. Ther.* **8**(2): 156–163.

Taylor, C.J., Bolton, E.M., Pocock, S., Sharples, L.D., Pedersen, R.A. and Bradley, J.A. (2005). Banking on human embryonic stem cells: estimating the number of donor cell lines needed for HLA matching. *Lancet* **366**(9502): 2019–2025.

Thomas, E.D. (1999). Bone marrow transplantation: a review. *Semin. Hematol.* **36**(4 Suppl 7): 95–103.

Embryonic stem cell lines derived from human blastocysts. *Science* **282**(5391): 1145–1147.

Tian, X. and Kaufman, D.S. (2005). Hematopoietic development of human embryonic stem cells in culture. *Method. Mol. Med.* **105**: 425–436.

Trounson, A. (2006). The production and directed differentiation of human embryonic stem cells. *Endocr. Rev.* **27**(2): 208–219.

Tsai, M.S., Hwang, S.M., Tsai, Y.L., Cheng, F.C., Lee, J.L. and Chang, Y.J. (2006). Clonal amniotic fluid-derived stem cells express characteristics of both mesenchymal and neural stem cells. *Biol. Reprod.* **74**(3): 545–551.

Tuan, R.S., Boland, G. and Tuli, R. (2003). Adult mesenchymal stem cells and cell-based tissue engineering. *Arthritis Res. Ther.* **5**(1): 32–45.

Weir, G.C. (2004). Can we make surrogate beta-cells better than the original? *Semin. Cell Dev. Biol.* **15**(3): 347–357.

Weiss, M.L., Medicetty, S., Bledsoe, A.R., Rachakatla, R.S., Choi, M., Merchav, S., Luo, Y., Rao, M.S., Velagaleti, G. and Troyer, D. (2006). Human umbilical cord matrix stem cells: preliminary characterization and effect of transplantation in a rodent model of Parkinson's disease. *Stem Cells* **24**(3): 781–792.

Yamada, S., Terada, K., Ueno, Y., Sugiyama, T., Seno, M. and Kojima, I. (2005). Differentiation of adult hepatic stem-like cells into pancreatic endocrine cells. *Cell Transplant.* **14**(9): 647–653.

Yang, L., Li, S., Hatch, H., Ahrens, K., Cornelius, J.G., Petersen, B.E. and Peck, A.B. (2002). *In vitro* trans-differentiation of adult hepatic stem cells into pancreatic endocrine hormone-producing cells. *Proc. Natl Acad. Sci. USA* **99**(12): 8078–8083.

Yoo, S.J., Yoon, B.S., Kim, J.M., Song, J.M., Roh, S., You, S. and Yoon, H.S. (2005). Efficient culture system for human embryonic stem cells using autologous human embryonic stem cell-derived feeder cells. *Exp. Mol. Med.* **37**(5): 399–407.

Zalzman, M., Anker-Kitai, L. and Efrat, S. (2005). Differentiation of human liver-derived, insulin-producing cells toward the beta-cell phenotype. *Diabetes* **54**(9): 2568–2575.

Zuk, P.A., Zhu, M., Mizuno, H., Huang, J., Futrell, J.W., Katz, A.J., Benhaim, P., Lorenz, H.P. and Hedrick, M.H. (2001). Multilineage cells from human adipose tissue: implications for cell-based therapies. *Tissue Eng.* **7**(2): 211–228.

Zwaka, T.P. and Thomson, J.A. (2005). A germ cell origin of embryonic stem cells? *Development* **132**(2): 227–233.

2

Fundamentals of Cell-Based Therapies

Ross Tubo

INTRODUCTION

Cell-based therapies have been proposed as a solution for a multitude of clinical problems ranging from structural repair of localized tissue damage to physiological restoration of systemic defects (Green et al., 1979; Caplan et al., 1998; Li et al., 1998). The successful treatment of such varied unmet medical needs ultimately depends upon the ability of cells to respond to their environment and function in a clinically relevant manner. This represents one of the most simple, and yet most complex principles for cell-based therapies.

Many factors contribute to deciding on the most appropriate cell-based therapy for any given patient. The clinical problem and type of the tissue repair desired are primary factors. Whether the repair tissue is to be permanent or temporary, structural or biological are important considerations. For instance, replacement of permanent structure may require an autologous cell therapy, while temporary restoration of biology may be better suited for allogeneic cells.

Autologous cell-based therapies represent our best clinical success in terms of permanent structural repair, harnessing the intrinsic capabilities of patient-derived cells to repair their own damaged tissues (Peterson et al., 2000). Studies examining the potential for allogeneic somatic cells for restoration of biology have also been successfully completed, resulting in Food and Drug Administration (FDA) approval for use of three allogeneic tissue-engineered products (Lysaght and Hazlehurst, 2004). The potential for use of allogeneic stem cells for structural repair of biological correction remains the subject of vigorous debate and research (Rao and Civin, 2006).

Our knowledge of cells and their interaction with extracellular matrices and biological factors have continued to grow during the past 20–25 years, with significant progress being made in the *in vitro* generation of three-dimensional tissue-engineered constructs of skin, cartilage, and blood vessels. We have learned the importance of providing proper physical and biological context in order to elicit the desired cellular response. Understanding these interactions will continue to guide the future development of clinically useful engineered tissues or organs in the practice of regenerative medicine.

RATIONALE FOR CELL-BASED THERAPIES

The inability of most adult tissues to regenerate themselves following injury has led to the development of cell-based strategies for structural repair or restoration of tissue physiology. Moreover, our ability to culture just about any somatic cell type has made it possible to consider the development of cell-specific culture systems for rapid proliferative expansion of such cells to treat previously unmet medical needs.

Copyright © 2008 Elsevier Inc. All rights reserved

Cell-based therapies generally fall into two main categories: (1) autologous cells for permanent structural repair and (2) allogeneic cells for short-term structural repair or restoration of physiological function. Autologous cells are derived from the patient to be treated, while allogeneic cells are derived from a donor. Allogeneic cell therapies developed in the past include cultured dermal fibroblasts and keratinocytes as dermal/epidermal constructs for the repair of cutaneous wounds (Parenteau et al., 2000), cultured kidney epithelia for renal assist (Humes and Szczypka, 2004), hepatocytes for liver function (Chan et al., 2004), pancreatic islets for diabetes (Ryan et al., 2002), and hematopoietic stem cells for bone marrow transplantation and immune reconstitution in leukemia and other cancers.

Structural repair using autologous cells seems to be the most straightforward type of cell-based therapy, where the role of the cells is to produce a permanent repair tissue having the structural characteristics of the tissue from which they were derived. Allogeneic cells are expected to elicit a physiological response from the host by the transient production of tissue stimulatory molecules, which alters host disease biology resulting in restoration of physiological function. Use of allogeneic cells for short-term physiological restoration or stimulation of host repair is slightly more complicated, given the potential for immunological rejection of donor cells. Lastly, long-term correction of physiology, as is necessary for replacing organ function, is clearly the most sophisticated and problematic therapy. Careful attention needs to be given to physical structure, biological function, and the immunological component for a successful cell therapy.

Autologous Cell-Based Therapies (Unmet Medical Need)

The two earliest examples of successful cell-based therapies for structural repair are cultured autologous epidermal keratinocytes (Epicel) for permanent skin replacement in severe burns (Gallico et al., 1984) and cultured autologous articular chondrocytes (Carticel) for repair of a patient's own damaged articular cartilage (Brittberg et al., 1994). These products represent the first, the second, and the only autologous cell-based therapies ever commercialized.

Epicel, the First Autologous Cell Therapy

Human epidermal keratinocytes (HEKs) or skin cells can be proliferatively expanded by culture on a mouse 3T3-fibroblast feeder layer under very specific conditions. Single cell suspensions of HEKs are prepared by enzymatic digestion of host skin tissue and placed in monolayer culture on the feeder layer. The feeder layer provides the appropriate physical niche and biological milieu for rapid expansion (Rollins et al., 1989). HEKs change their morphology and characteristic *in vivo* gene expression pattern when placed *in vitro*. This phenomenon, generically called dedifferentiation, is a process quite characteristic of any cell type subjected to proliferative expansion *in vitro* (Haudenschild et al., 2001). For HEKs, dedifferentiation is characterized by rapid change in cellular morphology, increased cell proliferation, and decreased expression of keratins normally found in epidermis with increased expression of keratins found in proliferating cells (Lersch et al., 1989). When propagated HEKs are subsequently applied to the host, they sense their environment and respond by "redifferentiating," expressing genes and proteins characteristic of HEKs found in skin.

When applied to the patient, Epicel grafts have the appearance of a patchwork quilt. The grafts are quite fragile, being only 2–3 HEK cell layers thick (Figure 2.1). As such, they are very sensitive to microbial infection and physical manipulation. The nascent epithelial tissue attaches to the wound bed and further redifferentiates, having three to four differentiated layers of epidermis within about 7–10 days. Over time the epithelium develops into a fully functional epidermis and modulates the development of a neo-dermis or new dermis having all the histological hallmarks of a fully functional dermal–epidermal junction with rete ridges within a year (Compton et al., 1993).

Given that cells respond differently to different environments, whether grown on or in materials such as collagen, polylactide/polygalactide (PLA/PGA), hyaluronic acid, or other polymers, in the presence of varied growth factors and morphogens, understanding the cellular response to these agents at the molecular level is critical to the development of successful cell-based products (Lutolf and Hubbell, 2005; Lisignoli et al., 2006).

Restoration of Physiology Using Tissue-Engineered Organ Equivalents

While our understanding of the interactions of cells, extracellular matrices, and biological factors in some ways seems to be in its infancy, tremendous progress has been made in the development of *ex vivo* organ constructs for renal assist (Humes and Szczypka, 2004) and bladder function (Atala, 2004). The renal assist device makes use of the intrinsic ability of cells from freshly digested kidney tissue to assemble spontaneously in a three-dimensional cartridge, through which the patient blood is then passed. This *ex vivo* dialysis system is currently in a clinical trial. Clinical studies have also been conducted using a more traditional tissue-engineering approach for bladder replacement. Small biopsies of bladder tissue are obtained from the patient requiring bladder replacement. The epithelial and smooth muscle cells are cultivated separately, loaded sequentially onto a three-dimensional biodegradable construct, cultured for a brief period of time and transplanted. The results are truly astonishing and represent the first successful functional replacement of an organ using an "organ" engineered *in vitro* (Atala, 2004).

Allogeneic "Ready-to-Use" Cell Therapies

Two primary reasons have led investigators away from autologous cells toward allogeneic. First, some clinical indications do not require permanent survival of the applied cells, but rather temporary production of a biological agent that will restore host tissue function. Second, providing a patient with their own cells is inherently expensive due to logistical and manufacturing issues.

Autologous cells require several weeks for the isolation, propagation, and return of cells to the patient. Delivery of functional cells providing more immediate clinical benefit is the goal of allogeneic or donor-derived cell therapies. Since allogeneic cells are likely to be rejected immunologically, they are likely better suited for indications not requiring permanent survival of the applied cells, but rather a temporary production of a biological agent to restore host tissue function.

Dermal ulcers are small non-healing cutaneous wounds (10–50 cm^2) which can be induced to heal by covering them with allogeneic skin wound dressings (Parenteau et al., 2000; Metcalfe and Ferguson, 2005). Dermal ulcers have been treated with a temporary epithelial dressing (allogeneic HEK, Acticel) or living-skin equivalents (Dermagraft, Apligraft). Each of these wound-healing dressings was derived from cells isolated from the donor tissue. These allogeneic cells can be propagated under the same conditions as autologous cells, but the expectation for their clinical use is for the temporary covering of cutaneous wounds to facilitate their healing.

Allogeneic cells are intended to be "ready to use" by definition.

Somatic cells – The logistical difficulties of providing a patient with their own cells and the inherent expense of the procedure would be significantly reduced by using allogeneic donor cells. Since each autologous sample is treated as its own manufacturing lot, it must be subjected to individualized culture, quality control testing, and preparation for delivery to a patient. Allogeneic cell preparation would allow for bulk quality control and manufacturing of one batch of cells to treat multiple patients, thus reducing expense. Furthermore, one batch of cells may be used to treat more than one clinical indication.

Stem cells, the new frontier – Perhaps the cell type which has captivated the most attention from both scientists and lay people are stem cells. Stem cells fall broadly into two categories: embryonic or adult tissue derived. Embryonic stem (ES) cells are derived from the inner cell mass of developing embryos, whereas adult stem cells have been derived from a variety of adult tissue sources including bone marrow, dermis, adipose

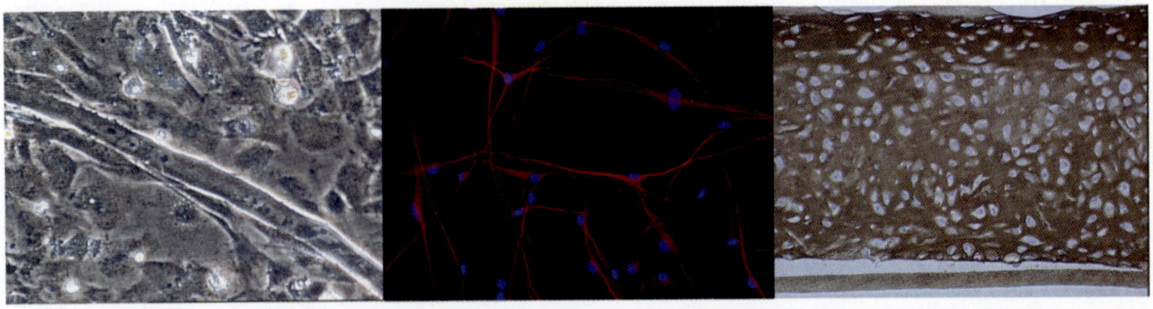

Figure 2.3 **Adult bone-marrow-derived MSCs were cultured under conditions to promote differentiation to the muscle (left), neural (middle), and cartilage (right) lineages. MSCs cultured in the presence of low serum formed myotubes, while those cultured in the presence of forskolin or TGF-beta differentiated into neural (nestin positive) or cartilage cells (type II collagen positive), respectively.**

tissue, and others (Pittenger et al., 1999; Jiang et al., 2002; Gimble and Guilak, 2003; Verfaillie et al., 2003; Bartsch et al., 2005).

The bone marrow provides an attractive source of easily accessible adult pluripotent stem cells. The specialized microenvironment within the connective tissue framework of adult bone marrow supports the existence of at least two distinct populations of stem cells: one hematopoietic and the other mesenchymal. Hematopoietic stem cells (HSCs) in the adult ultimately give rise to all components of the immune and blood systems, while MSCs have the potential to give rise to cells of varied lineages, including bone, cartilage, and adipose tissues. The MSC population can be isolated from the bone marrow and expanded in culture in the absence of differentiation for at least 30–40 population doublings (Lodie et al., 2002). Even after expansion, MSCs can still differentiate to cells of multiple lineages (Bruder et al., 1997). Because MSCs have been shown to give rise to adipocytes, osteoblasts, chondrocytes, myoblasts, neurons, and other cell types (Figure 2.3), they are an intriguing alternative source of cells for cellular replacement therapies.

ES cells have also been shown to exhibit pluripotent differentiation potential *in vitro* and *in vivo* (Schuldiner et al., 2000; Stojkovic et al., 2004). ES cells can spontaneously differentiate in culture into a layer of beating myocardium (He et al., 2003). These kinds of studies demonstrate the tremendous potential for ES cells; however, the exact culture conditions required to reproducibly induce ES cell differentiation in a controllable fashion remains the subject of intense study. Similarly, undifferentiated ES cells spontaneously form teratomas when injected subcutaneously in immune compromised mice (Przyborski, 2005). Histological analysis of ES cell implants reveals that tissue of cardiac, neural, and other tissue lineages spontaneously originate from the same population transplanted ES cells, again illustrating the tremendous differentiation potential of ES cells, and highlighting the need for further study to determine precise control of differentiation. Some investigators are engineering their ES cells to express conditional suicide genes to reduce the risk of inappropriate ES cell differentiation.

Several recent papers suggest that the utility of adult stem cells may not be limited to *in vitro* differentiation for direct cell replacement of damaged tissues. Recent evidence suggests that the adult murine bone marrow cells possess the intrinsic capability to differentiate into β-cells after total bone marrow transplantation in non-diabetic animals (Ianus et al., 2003). In addition, transplantation of bone marrow (Zorina et al., 2003) and bone-marrow-derived stem cells was shown to activate endogenous tissue regeneration, specifically β-cell regeneration in the pancreas (Hess et al., 2003). There is also an emerging evidence which points to the transplant of bone-marrow-derived stem cells as having additional benefits including recruitment of endogenous stem cells (Kocher et al., 2001), vascularization of damaged tissue (Rafii and Lyden, 2003), and, as discussed further below, immune transplant tolerance (Bartholomew et al., 2002).

Intramyocardial injection of bone-marrow-derived MSCs overexpressing Akt (Akt-MSCs) inhibits ventricular remodeling and restores cardiac function measured 2 weeks after myocardial infarction (Gnecchi et al., 2006). When injected into infarcted hearts, the Akt-MSC-conditioned medium significantly limits infarct size and improves ventricular function relative to controls. Support to the paracrine hypothesis is provided by data showing that several genes, coding for factors (VEGF, FGF-2, HGF, IGF-I, and TB4) that are potential mediators of the effects exerted by the Akt-MSC-conditioned medium, are significantly up-regulated in the Akt-MSCs, particularly in response to hypoxia. Taken together, our data support Akt-MSC-mediated paracrine mechanisms of myocardial protection and functional improvement.

Immunosuppressive Properties of MSCs

MSCs have been reported to be immunomodulatory both *in vitro* and *in vivo*. They express very low levels of co-stimulatory molecules and HLA Class I on their cell surface and lack HLA Class II expression (Devine and Hoffamn, 2000; Di Nicola et al., 2002). Class II expression did increase upon MSC differentiation. The immunophenotype of MSCs suggested that they may play a role in modulating T-cell proliferation and the immune response. MSCs have been shown to inhibit a mixed lymphocyte reaction (MLR) using purified CD3 T-cells and third party dendritic cells as antigen presenting cells (Tse et al., 2003). Both autologous and allogeneic MSCs suppress lymphocyte proliferation (Di Nicola et al., 2002). It has been postulated that MSCs may suppress T-cell proliferation by several mechanisms: secretion of growth factors, such as TGF-beta or HGF, suppression of pro-inflammatory (TH-1) cytokines, stimulation of anti-inflammatory (TH-2-type) cytokines, and up-regulation of pro-apoptotic cell surface molecules (Bartholomew et al., 2002; Di Nicola et al., 2002). Studies are ongoing to further elucidate the underlying mechanisms for MSC-mediated suppression of T-lymphocyte proliferation.

It has been reported that intravenously administered allogeneic MSCs are not rejected in a baboon model due to lack of immune recognition. Furthermore, bone marrow transplantation of baboon MSCs into MHC-mismatched recipients prior to a third party skin graft led to prolonged graft survival (Bartholomew et al., 2002). Taken together, these data suggest that MSCs not only possess immunosuppressive properties that inhibit T-cell proliferation *in vitro*, but they also have immunomodulating properties which may enhance graft survival *in vivo*. MSCs also reduced the incidence and severity of graft-versus-host disease (GVHD) during allogeneic transplantation and although the mechanisms remain to be elucidated, the data offer insight into the potential use of MSCs for induction of tolerance for reduction of GVHD, rejection, and modulation of inflammation (Aggarwal and Pittenger, 2005).

Adult bone-marrow-derived MSCs appear to offer several advantages over autologous cell therapies and even ES cells. First, MSCs exhibit multi-potential differentiation in a well-controlled, predictable fashion, in contrast to ES cells. Second, the fact that they appear to down-modulate the host (recipient) immune response (GVHD) may permit their persistence for the longer term, similar to autologous cells. Although this is not necessarily a functional advantage of MSCs over autologous cells, the production of multiple treatment doses from a single donor source is quite attractive from a manufacturing and quality control perspective, thereby reducing costs associated with personalized medicine.

Commercialization of a Cell Therapy

Commercialization of cellular therapies is not easy. Autologous cells, while clinically successful, may not be commercially successful, due in part to the fact that they are logistically difficult and inherently expensive to produce. Clinical evaluation of such therapies is complicated and time consuming. Moreover, the regulatory and reimbursement issues can be very challenging. That having been said, cell therapies can significantly enhance the quality of human healthcare for serious unmet medical needs.

The appeal of allogeneic stem cells is obvious: one cell source for multiple indications; potential for an off-the-shelf product; improved quality control; and reduced cost of goods. However, before we get too carried away with the "promise" of stem cells, we need to do a reality check. We need to apply the same fundamental principles to stem cells that were applied to autologous cell therapies. Many questions remain to be addressed before the potential of stem cell therapy can be realized, such as: Can the cells be routinely isolated and propagated? Can the cells terminally differentiate into the cell type of interest? *In vitro* and *in vivo*? What is their potency and purity? How long do the cells persist *in vivo*? Can the purity of the expanded cells be established prior to shipment? Having the answers to these questions is critical for the successful commercialization of stem cell therapy.

Ensuring Production of the Best Quality Cell Therapy Products

Measurement of identity and functionality of cells following proliferative expansion are the two key features of ensuring the best possible quality of cell-based products, autologous or allogeneic. Cell surface makers can be used to assess identity and purity of the expanded cell population. Differentiation assays can be used to assess functionality of the cells *in vitro*. *In vivo* studies are required to determine differentiation and persistence *in vivo*.

The principle that cultured cells can dedifferentiate and undergo proliferative expansion *in vitro*, and then redifferentiate when placed *in vivo* is central to the success of cell-based therapies. Stem cells may propagate in a multi-potent state and then differentiate *in vivo*. Our ability to assay for this activity *in vitro*, as a matter of "quality control," is critical to the ultimate success of a cell-based therapy for tissue replacement or repair *in vivo*. Given that cells in culture respond differently to varied culture conditions and environmental cues (Haudenschild et al., 2001; Lodie et al., 2002; Solchaga et al., 2005), it is important to confirm that the cell types being propagated are indeed the desired cell type and that they are capable of the intended function.

CONCLUSIONS

Inadequate therapies to repair injured tissue, replace failing organs, and restore structural and metabolic functions remain a driving force behind the demand for cell-based therapies. Cells represent a "lowest common denominator" of sorts for cell-based therapies; their numbers are expandable, they are programmed by the environment within which they find themselves to respond and produce a biological response. The challenge is to harness the tremendous potential within this tiny unit, ultimately providing the proper structure and biological function necessary for successful treatment of the clinical problem at hand.

Generally speaking cell-based therapies fall into two broad categories of use: (1) cells for permanent structural repair or replacement (e.g. cultured keratinocytes as skin replacement, chondrocytes for repair of cartilage or visco-uretal reflux) and (2) cells for correction of a physiological or metabolic problem. Understanding the nature of the problem you are trying to treat and role that the cell may play in solving the problem is critical to developing a successful cell therapy. Issues to be considered include whether the cells are for structural replacement or restoration of metabolism. If structural replacement, autologous cells are likely the cell of choice. If the goal is the correction of metabolism, the length of time required to see physiological benefit and subsequent immunological status may influence the source of cells to be used.

Scientists in regenerative medicine have strived to understand the interaction of cells, extracellular matrices, and biological factors as they have endeavored to develop tissue-engineered constructs for repair and replacement of damaged tissue. Understanding how to produce a "simple" functional tissue *in vitro*, by harnessing our knowledge of these building blocks remains a very complex and yet exciting problem for us to solve. While our understanding of the mechanisms underlying the interactions of cells, extracellular matrices, and biological factors continues to grow, we continue to take advantage of the "intrinsic knowledge" that the cell retains to accomplish the goal of tissue repair.

ACKNOWLEDGMENTS

I would like to thank Drs. Tracey Lodie, Ajeeta Dash, and Michelle Youd for their insightful review of this document, and Ms. Maureen Swartz for her administrative contributions.

REFERENCES

Aggarwal, S. and Pittenger, M.F. (2005). Human mesenchymal stem cells modulate allogeneic immune cell responses. *Blood* **105**: 1815–1822.

Atala, A. (2004). Tissue engineering for the replacement of organ function in the genitourinary system. *Am. J. Transplant.* (Suppl 6), 58–73.

Bartholomew, A., Sturgeon, C., Siatskas, M., Ferrer, K., McIntosh, K., Patil, S., Hardy, W., Devine, S., Ucker, D., Deans, R., Moseley, A. and Hoffman, R. (2002). Mesenchymal stem cells suppress lymphocyte proliferation *in vitro* and prolong skin graft survival *in vivo*. *Exp. Hematol.* **30**: 42–48.

Bartsch, G., Yoo, J.J., De Coppi, P., Siddiqui, M.M., Schuch, G., Pohl, H.G., Fuhr, J., Perin, L., Soker, S. and Atala, A. (2005). Propagation, expansion, and multilineage differentiation of human somatic stem cells from dermal progenitors. *Stem Cells Dev.* **14**: 337–348.

Ben-Yishay, A., Grande, D.A., Menche, D. and Pitman, M. (1992). *Repair of Articular Cartilage Defects Using Collagen–Chondrocyte Allografts*. 38th Annual Meeting, Orthopaedic Research Society, Washington, DC, p. 174.

Brittberg, M., Lindahl, A., Nilsson, A., Ohlsson, C., Isaksson, O. and Peterson, L. (1994). Treatment of deep cartilage defects in the knee with autologous chondrocyte transplantation. *N. Engl. J. Med.* **331**: 889–895.

Caplan, A.I., Reuben, D. and Haynesworth, S.E. (1998). Cell-based tissue engineering therapies: the influence of whole body physiology. *Adv. Drug. Deliv. Rev.* **33**: 3–14.

Chan, C., Berthiaume, F., Nath, B.D., Tilles, A.W., Toner, M. and Yarmush, M.L. (2004). Hepatic tissue engineering for adjunct and temporary liver support: critical technologies. *Liver Transpl.* **10**: 1331–1342.

Compton, C.C., Gill, J.M., Bradford, D.A., Regauer, S., Gallico, G.G. and O'Connor, N.E. (1989). Skin regenerated from cultured epithelial autografts on full-thickness burn wounds from 6 days to 5 years after grafting. A light, electron microscopic and immunohistochemical study. *Lab. Invest.* **60**: 600–612.

Compton, C.C., Hickerson, W., Nadire, K. and Press, W. (1993). Acceleration of skin regeneration from cultured epithelial autografts by transplantation to homograft dermis. *J. Burn Care Rehabil.* **14**: 653–662.

Devine, S.M. and Hoffman, R. (2000). Role of mesenchymal stem cells in hematopoietic stem cell transplantation *Curr. Opin. Hematol.* **7**: 358–363.

Di Nicola, M., Carlo-Stella, C., Magni, M., Milanesi, M., Longoni, P.D., Matteucci, P., Grisanti, S. and Gianni, A.M. (2002). Human bone marrow stromal cells suppress T-lymphocyte proliferation induced by cellular or nonspecific mitogenic stimuli. *Blood* **99**: 3838–3843.

Gallico, G.G., O'Connor, N.E., Compton, C.C., Kehinde, O. and Green, H. (1984). Permanent coverage of large burn wounds with autologous cultured human epithelium. *N. Engl. J. Med.* **311**: 448–451.

Gimble, J. and Guilak, F. (2003). Adipose-derived adult stem cells: isolation, characterization, and differentiation potential. *Cytotherapy* **5**: 362–369.

Gnecchi, M., He, H., Noiseux, N., Liang, O.D., Zhang, L., Morello, F., Mu, H., Melo, L.G., Pratt, R.E., Ingwall, J.S. and Dzau, V.J. (2006). Evidence supporting paracrine hypothesis for Akt-modified mesenchymal stem cell-mediated cardiac protection and functional improvement. *FASEB J.* **20**: 661–669.

Green, H., Kehinde, O. and Thomas, J. (1979). Growth of cultured human epidermal cells into multiple spithellia suitable for grafting. *Proc. Natl Acad. Sci. US Am.* **76**: 5665–5668.

Haudenschild, D.R., McPherson, J.M., Tubo, R. and Binette, F. (2001). Differential expression of multiple genes during articular chondrocyte redifferentiation. *Anat. Rec.* **263**: 91–98.

He, J.Q., Ma, Y., Lee, Y., Thomson, J.A. and Kamp, T.J. (2003). Human embryonic stem cells develop into multiple types of cardiac myocytes: action potential characterization. *Circ. Res.* **93**: 32–39.

Hess, D., Li, L., Martin, M., Sakano, S., Hill, D., Strutt, B., Thyssen, S., Gray, D.A. and Bhatia, M. (2003). Bone marrow-derived stem cells initiate pancreatic regeneration. *Nat. Biotechnol.* **21**: 763–770.

Humes, H.D. and Szczypka, M.S. (2004). Advances in cell therapy for renal failure. *Transpl. Immunol.* **12**: 219–227.

Ianus, A., Holz, G.G., Theise, N.D. and Hussain, M.A. (2003). *In vivo* derivation of glucose-competent pancreatic endocrine cells from bone marrow without evidence of cell fusion. *J. Clin. Invest.* **111**: 843–850.

Jiang, Y., Vaessen, B., Lenvik, T., Blackstad, M., Reyes, M. and Verfaillie, C.M. (2002). Multipotent progenitor cells can be isolated from postnatal murine bone marrow, muscle, and brain. *Exp. Hematol.* **30**: 896–904.

Kocher, A.A., Schuster, M.D., Szabolcs, M.J., Takuma, S., Burkhoff, D., Wang, J., Homma, S., Edwards, N.M. and Itescu, S. (2001). Neovascularization of ischemic myocardium by human bone-marrow-derived angioblasts prevents cardiomyocyte apoptosis, reduces remodeling and improves cardiac function. *Nat. Med.* **7**: 430–436.

Lersch, R., Stellmach, V., Stocks, C., Giudice, G. and Fuchs, E. (1989). Isolation, sequence, and expression of a human keratin K5 gene: transcriptional regulation of keratins and insights into pairwise control. *Mol. Cell. Biol.* **9**: 3685–3697.

Li, R.K., Yau, T.M., Sakai, T., Mickle, D.A. and Weisel, R.D. (1998). Cell therapy to repair broken hearts. *Can. J. Card.* **14**: 735–744.

Lisignoli, G., Cristino, S., Piacentini, A., Cavallo, C., Caplan, A.I. and Facchini, A. (2006). Hyaluronan-based polymer scaffold modulates the expression of inflammatory and degradative factors in mesenchymal stem cells: involvement of Cd44 and Cd54. *J. Cell. Physiol.* **207**: 364–373.

Lodie, T.A., Blickarz, C.E., Devarakonda, T.J., He, C., Dash, A.B., Clarke, J., Gleneck, K., Shihabuddin, L. and Tubo, R. (2002). Systematic analysis of reportedly distinct populations of multipotent bone marrow-derived stem cells reveals a lack of distinction. *Tissue Eng.* **8**: 739–751.

Lutolf, M.P. and Hubbell, J.A. (2005). Synthetic biomaterials as instructive extracellular microenvironments for morphogenesis in tissue engineering. *Nat. Biotechnol.* **23**: 47–55.

Lysaght, M.J. and Hazlehurst, A.L. (2004). Tissue engineering: the end of the beginning. *Tissue Eng.* **10**: 309–320.

Metcalfe, A.D. and Ferguson, M.W. (2005). Harnessing wound healing and regeneration for tissue engineering. *Biochem. Soc. Trans.* **33**: 413–417.

Miyahara, Y., Nagaya, N., Kataoka, M., Yanagawa, B., Tanaka, K., Hao, H., Ishino, K., Ishida, H., Shimizu, T., Kangawa, K., Sano, S., Okano, T., Kitamura, S. and Mori, H. (2006). Monolayered mesenchymal stem cells repair scarred myocardium after myocardial infarction. *Nat. Med.* **12**: 459–465.

Parenteau, N.L., Hardin-Young, J. and Ross, R.N. (2000). Skin. In: Lanza, R., Langer, R. and Vacanti, J. (eds.), *Principles of Tissue Engineering*. San Diego: Academic Press, pp. 879–890.

Peel, S.A.F., Chen, H., Renlund, R., Badylak, S. F. and Kandel, R.A. (1998). Formation of a SIS-cartilage composite graft *in vitro* and its use in the repair of articular cartilage defects. *Tissue Eng.* 143–155.

Peterson, L., Minas, T., Brittberg, M., Nilsson, A., Sjogren-Jansson, E. and Lindahl, A. (2000). Two- to 9-year outcome after autologous chondrocyte transplantation of the knee. *Clin. Orthop. Relat. Res.* **374**: 212–234.

Pittenger, M.F., Mackay, A.M., Beck, S.C., Jaiswal, R.K., Douglas, R., Mosca, J.D., Moorman, M.A., Simonetti, D.W., Craig, S. and Marshak, D.R. (1999). Multilineage potential of adult human mesenchymal stem cells. *Science* **284**: 143–147.

Przyborski, S.A. (2005). Differentiation of human embryonic stem cells after transplantation in immune-deficient mice. *Stem Cells* **23**: 1242–1250.

Rafii, S. and Lyden, D. (2003). Therapeutic stem and progenitor cell transplantation for organ vascularization and regeneration. *Nat. Med.* **9**: 702–712.

Rao, M.S. and Civin, C.I. (2006). How many human embryonic stem cell lines are sufficient? A US perspective. *Stem Cells* March 16 (Epub ahead of print).

Rollins, B.J., O'Connell, T.M., Bennett, G., Burton, L.E., Stiles, C.D. and Rheinwald, J.G. (1989). Environment-dependent growth inhibition of human epidermal keratinocytes by recombinant human transforming growth factor-beta. *J. Cell. Physiol.* **139**: 455–462.

Ryan, E.A., Lakey, J.R., Paty, B.W., Imes, S., Korbutt, G.S., Kneteman, N.M., Bigam, D., Rajotte, R.V. and Shapiro, A.M. (2002). Successful islet transplantation: continued insulin reserve provides long-term glycemic control. *Diabetes* **51**: 2148–2157.

Schuldiner, M., Yanuka, O., Itskovitz-Eldor, J., Melton, D.A. and Benvenisty, N. (2000). Effects of eight growth factors on the differentiation of cells derived from human embryonic stem cells. *Proc. Natl Acad. Sci.* **97**: 11307–11312.

Shortkroff, S., Barone, L., Hsu, H.-P., Wrenn, C., Gagne, T., Chi, T., Breinan, H., Minas, T., Sledge, C.B., Tubo, R. and Spector, M. (1996). Healing of chondral and osteochondral defects in a canine model: the role of cultured chondrocytes in regeneration of articular cartilage. *Biomaterials* **17**: 147–154.

Solchaga, L.A., Temenoff, J.S., Gao, J., Mikos, A.G., Caplan, A.I. and Goldberg, V.M. (2005). Repair of osteochondral defects with hyaluronan- and polyester-based scaffolds. *Osteoarthritis Cartilage* **13**: 297–309.

Stojkovic, M., Lako, M., Strachan, T. and Murdoch, A. (2004). Derivation, growth and applications of human embryonic stem cells. *Reproduction* **128**: 259–267.

Tse, W.T., Pendleton, J.D., Beyer, W.M., Egalka, M.C. and Guinan, E.C. (2003). Suppression of allogeneic T-cell proliferation by human marrow stromal cells: implications in transplantation *Transplantation* **75**: 389–397.

Verfaillie, C.M., Schwartz, R., Reyes, M. and Jiang, Y. (2003). Unexpected potential of adult stem cells. *Ann. NY Acad. Sci.* **996**: 231–234.

Wakitani, S., Kimura, T., Hirooka, A., Ochi, T., Yoneda, M., Owaki, H., Ono, K. and Yasui, N. (1989). Repair of rabbit articular surfaces with allografts of chondrocytes embedded in collagen gels. *J. Jpn Ortho. Assoc.* **63**: 529–538.

Zorina, T.D., Subbotin, V.M., Bertera, S., Alexander, A.M., Haluszczak, C., Gambrell, B., Bottino, R., Styche, A.J. and Trucco, M. (2003). Recovery of the endogenous beta cell function in the NOD model of autoimmune diabetes. *Stem Cells* **21**: 377–388.

3

Stem Cell Research

T. Ahsan, A.M. Doyle, and R.M. Nerem

INTRODUCTION

Regenerative medicine is an emerging branch of medicine whose goal is to restore organ and/or tissue function using a biological approach. A growing crisis in organ transplantation and an aging population have driven a search for new and alternative therapies. There currently are approximately 90,000 patients on the US transplant waiting list. Despite growing numbers of donors, the availability of suitable organs is still insufficient. This discrepancy is only likely to increase during the next 25 years, given that the population of those 65 years and older is projected by the US Census Bureau to more than double.

Recent advances in stem cell technology have shown great promise and propelled regenerative medicine to the forefront of both scientific research and public consciousness. While some believe the therapeutic potential of stem cells has been overstated in the media, an analysis of the potential benefits of stem cell-based therapies indicates that 128 million people in the United States alone may benefit, with the largest impact on patients with cardiovascular disorders, autoimmune diseases, and diabetes (Figure 3.1) (Perry, 2000). The enthusiasm surrounding stem cells is related in part to their potential to treat a broad range of clinical pathologies. Some identified stem cell targets, such as neurological diseases, spinal cord injuries, diabetes, and cardiovascular diseases, currently have few accepted treatments or no cures. In other conditions, such as bone fracture healing or cartilage repair, stem cells may improve upon therapies currently in use. Stem cells may change the very nature of medicine: they have the potential to address the cell sourcing issue of tissue

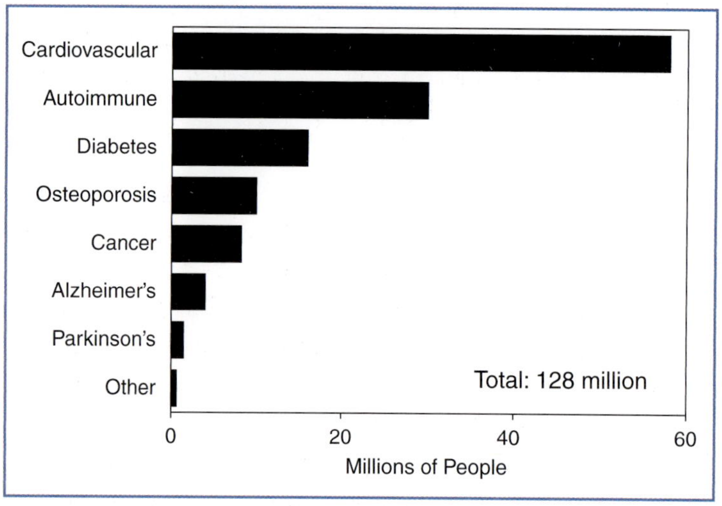

Figure 3.1 Persons in the United States affected by diseases or injuries that may be helped by stem cell research (Perry, 2000).

Copyright © 2008 Elsevier Inc. All rights reserved

engineering and to customize therapies for individual patients. As a self-renewing source of allogeneic cells, stem cells make off-the-shelf products a more probable and closer reality. Through either autologous adult stem cells or somatic nuclear transfer, cell-based therapies can be genetically matched to each patient, addressing immunocompatibility and disease transfer concerns.

Stem cells in regenerative medicine can serve not only as cellular machinery, but also as gene delivery vehicles and systems to promote further understanding of development. Due to their self-renewing, proliferating, differentiating, and distribution potential *in vivo*, stem cells are a natural choice as gene delivery vehicles (Lemoine, 2002). Stem cells grown *in vitro* or implanted in animals can serve as model systems to study many basic science questions. Our understanding of the mechanisms that govern development is expanded using the spontaneous differentiation of embryonic stem cells (ESCs) in embryoid bodies, the directed differentiation of adult stem cells in response to chemical and/or physical cues, and the homing and engraftment of stem cells in animals. Additionally, *in vitro* models of development provide a unique opportunity to study mutations that would otherwise be lethal *in vivo*. As a result of increased understanding from the use of these various stem cell model systems, strategies may be developed that focus on preventative medicine.

The potential of stem cells and regenerative medicine is too vast to cover in its entirety in a single chapter. As a result, we have largely focused this discussion on an overview of genetically unmodified human stem cells and their current status in clinical applications of regenerative medicine. It is important to note, however, that the extensive amount of work done with animal stem cells and in animal models is not only a basis for human applications, but also indicates the long range potential of stem cells in regenerative medicine. Ultimately, the intent of this introductory chapter is to address the range of stem cell technology and leave to subsequent chapters the more exhaustive and in-depth analyses of specific stem cells and their applications. This chapter gives an overview of the different types of stem cells, the modes of stem cell modulation *in vitro*, the general strategies of regenerative medicine, and the role of stem cells in various clinical applications.

STEM CELLS

A stem cell is an unspecialized cell that can both self-renew (reproduce itself) and differentiate into functional phenotypes. Stem cells can originate from embryonic, fetal, or adult tissue and are broadly categorized accordingly. ESCs are commonly derived from the inner cell mass of a blastocyst, an early (4–5 days) stage of the embryo. Embryonic germ cells (EGCs) are isolated from the gonadal ridge of a 5–10 week fetus. In particular, EGCs are derived from the primordial germ cells, which ultimately give rise to eggs or sperms in the adult. Adult stem cells differ from ESCs and EGCs in that they are found in tissues after birth, and to date, have been found to differentiate into a narrower range of cell types, primarily those phenotypes found in the originating tissue.

A major value of stem cells in regenerative medicine is their potential to become different cell types. Our current understanding of differentiation is based on a hierarchical tree structure in which a few unspecialized stem cells branch to ultimately yield a larger number of mature cellular phenotypes (Figure 3.2). Stem cells divide to generate at least one daughter cell that retains the stem cell identity, resulting in a perpetuating population (Ho, 2005). They can also give rise to progenitors, or precursor cells, which typically differentiate into tissue-specific cell types and are only capable of symmetric division. Yet these progenitors play a major role *in vivo* that may be beneficial for cell-based therapies: symmetric division of rapidly proliferating progenitors allows exponential yield of terminally differentiated cells. Thus, as a system, this hierarchical structure allows for a small perpetual population of stem cells to give rise as needed to large numbers of differentiated cells.

The hierarchical tree structure of differentiation is based on observations from developmental biology. Differentiation during embryogenesis begins with gastrulation, when cells separate into three structural

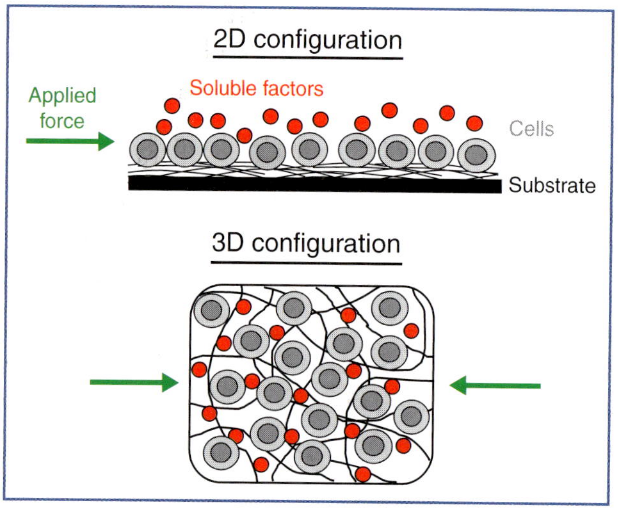

Figure 3.4 The different configurations and environmental cues during cell culture. In two dimensions, cells (gray circles) may be (a) adhered to a surface via a protein substrate (black strands), (b) exposed to soluble factors (red circles) in the medium, and (c) subjected to applied forces (green arrows) via surface distention or fluid motion. In three dimensions, cells may be (a) seeded onto or embedded within a scaffold with matrix molecules, (b) exposed to soluble factors in the medium, and (c) subjected to applied forces via scaffold deformation, fluid motion, or fluid pressurization.

studies on human stem cells is limited and these are based on prior work in animal cells, this discussion will review *in vitro* mechanical modulation of cells from multiple species.

The field of biomaterials has developed a wide range of 3D scaffolds that customize the physical microenvironment within which cells reside. Some studies have found that a 3D culture environment can provide cues that are otherwise missing from 2D stem cell cultures (Tun et al., 2002). In particular, scaffolds can enhance ESC self-renewal (Nur-E-Kamal et al., 2005) and allow the propagation of hematopoietic progenitor cells without the normally requisite growth factors or stromal cells (Bagley et al., 1999; Ehring et al., 2003). Other studies indicate the capacity of stem cells to spontaneously differentiate to cells of all three lineages in 3D (Levenberg et al., 2003). Subsequent work along that line has focused on using 3D biomaterials to direct differentiation of stem cells toward a variety of phenotypes, including hematopoietic (Liu et al., 2003), neural (Ma et al., 2004), and orthopedic (Chaudhry et al., 2004; Hwang et al., 2005).

The use of 3D environments has utility beyond simply promoting stem cell self-renewal and differentiation. The very nature of a 3D environment allows an organization of matrix (Grayson et al., 2004) and the formation of structures (Levenberg et al., 2003) that are not otherwise possible on flat surfaces. This may be particularly useful in tissue engineering, in which initial studies have used a 3D scaffold to serve both as a physical cue for stem cell differentiation, as well as the basis of a tissue construct (Awad et al., 2004; Betre et al., 2006). Additionally, a 3D environment may physically entrap or be bound to chemical cues to provide controlled spatial and temporal gradients (Batorsky et al., 2005).

Biomaterials themselves can provide both chemical and physical cues that influence stem cell fate. It is thought that biological matrix components, such as collagen, fibrin, and laminin, provide bioactive cues typically seen *in vivo* that are difficult to replicate using synthesized polymers. Biological components have thus become the basis of gels (Chen et al., 2003) or have served as coatings (Levenberg et al., 2003) in 3D scaffolds. The influence of biological factors on stem cell differentiation was elucidated in one study in which ESCs differentiated toward the tissue-specific lineages when seeded onto extracts from cartilage versus basement membrane (Philp et al., 2005). The effect of these biologically based scaffolds is more than compositional, as chemically similar collagen in macroscopically different 3D configurations (sponge versus gel) results in different differentiation patterns (Chen et al., 2003). The importance of macroscopic 3D architecture was corroborated in one study where a porous polymeric scaffold without any matrix molecules (Gerecht-Nir et al., 2004b) promoted differentiation of seeded ESCs. It is not only scaffold architecture, however, but also mechanical properties, that play a role in differentiation. Scaffolds that are too stiff have been shown to inhibit embryoid body growth, cavitation, and differentiation (Battista et al., 2005). This suggests the need for more research using engineered scaffolds in which protein presentation, macroscopic architecture, mechanical stiffness, and degradation rates can all be tailored. In general, the engineering of scaffolds has already become quite sophisticated, at times even using biologically derived and synthetic components together. In relation to stem cell research, one example of an innovative-engineered scaffold uses silk, a natural polymer, that can be customized in terms of mechanical and degradation characteristics in 3D configurations (Wang et al., 2005b).

Physical forces, such as compression, tension, and shear, have long been applied to cells via bioreactors, a term commonly used for systems with controlled culture conditions. Some bioreactor systems are used to study the modulation of cells and tissues by well-defined cues. Once the appropriate cues for a given application are determined, bioreactors can be appropriately designed to scale up modulation to large numbers of cells and tissue samples. With the recent commercial availability of a few systems, studies that utilize bioreactors and are designed to understand the importance of environmental cues have become more numerous, with some now focusing on stem cells.

Initial use of bioreactors with stem cells revolved around non-adherent cells, namely hematopoietic and neural progenitors, in suspension cultures to accelerate and augment expansion kinetics and capabilities, respectively. Stir-based and perfusion bioreactors have been used with hematopoietic progenitor cells, in which the increase in cellular yield is attributed to frequent medium changes, as well as controlled oxygen and cytokine concentration gradients (reviewed by others: Nielsen, 1999; Cabrita et al., 2003). Similar stir-based bioreactors have also been used with neural progenitors, where the main objective is to provide fluid motion to regulate neurosphere diameter, a characteristic correlated to proliferation rates and differentiation potential (Kallos and Behie, 1999; Kallos et al., 1999; Sen et al., 2002; Alam et al., 2004). The approach of allowing limited cell aggregation (cells come together to form a cluster), without sphere agglomeration (clusters come together to form larger bodies), is now being applied to ESC studies. The embryoid body model of differentiation is being studied in some fluid shear stress-based bioreactors that control sphere morphology (Dang et al., 2004; Gerecht-Nir et al., 2004a; Bauwens et al., 2005) and promote differentiation toward a particular phenotype (Schroeder et al., 2005). Although these bioreactors often generate a poorly controlled microenvironment (Konstantinov et al., 2004), they are easily operated and can be scaled up for clinical or manufacturing purposes.

The fundamental mechanisms that regulate stem cell responses to applied forces are commonly investigated using well-characterized bioreactor systems for 2D cultures. In systems where cyclic tensile strain ($\sim$10%, $\sim$0.5 Hz) has been applied to silastic membranes seeded with cells, it has been found that both ESCs (Saha et al., 2006) and MSCs (Lee et al., 2005) proliferate and retain their original differentiation potential. Based on the assumption that cells functionally adapt to their microenvironment, many investigators have chosen to mimic certain aspects of *in vivo* mechanical environments in their studies on differentiation. Endothelial cells that line vascular vessels *in situ* experience varying levels of fluid shear stress as blood flows past. Similar fluid shear stresses applied *in vitro* to ESCs (Ahsan and Nerem, 2005; Yamamoto et al., 2005), circulating EPCs (Yamamoto et al., 2003), and mesenchymal progenitor cells (Wang et al., 2005a) have indeed resulted in an increase in protein expression typical of the endothelial phenotype. While bioreactors designed for 2D cell cultures cannot truly mimic *in vivo* conditions or create 3D tissues, they provide simplified mechanical environments that allow for careful study of stem cell responses.

Bioreactors have previously been used for tissue engineering with differentiated cells seeded onto scaffolds to create 3D constructs. As stem cell research progresses, this approach is now incorporating the use of undifferentiated cells. Most studies in this area have until now focused on MSCs and orthopedic applications. MSCs seeded onto a fibrous 3D construct and then subjected to fluid perfusion resulted in an increased cell density and a more uniform cellular distribution (Zhao and Ma, 2005), whereas MSCs embedded in agarose and subjected to compression resulted in cells with a chondrogenic phenotype (Huang, C.Y., et al., 2004). These studies applied a single homogenous input to an entire population of cells. A recent study used a more complicated system that differentially applied complex inputs to subsets of the original cell population. In a layered microporous tubular scaffold, Flk-1+ cells were seeded onto the lumen and submitted to fluid shear stress and circumferential strain. Due to the complex geometry and multiple inputs, the stimuli sensed by each cell were dependent on the spatial location of the cell within the scaffold. Analogous to the organization within a blood vessel, cells lining the lumen assumed an endothelial morphology, while cells in deeper layers took on smooth muscle cell-like characteristics (Huang et al., 2005). Although limited in scope, this study and others support the concept that the *in vitro* microenvironment can be used to differentially regulate stem/progenitor cell self-renewal and differentiation processes, potentially within a single tissue-engineered construct.

REGENERATIVE MEDICINE

Regenerative medicine focuses on strategies to repair, regenerate, and/or replace tissues and organs. The goal in each of these cases is to restore tissue and organ function through the delivery of cells, signaling molecules, and/or support structures. Disease can be thought of as a failure of the normal biological repair mechanisms that are present in the body. If one can detect disease at an early stage, even at a pre-clinical stage, and intervene by enhancing/inducing biological repair, then it may be possible to restore normal biological function without creating new tissue. In this case repair is at a local, cellular level. Once disease progresses to a more advanced stage (i.e. a clinical stage), then it may be necessary to regenerate or create new tissue in order to restore full function. Finally, when such an approach is not sufficient, then the strategy may actually require replacement of the tissue so as to restore the structure and full biological function. This includes the mechanical/electrical/chemical aspects of function. It should be noted, however, that although one can attempt to define each of these three mechanisms so as to distinguish between them, the fact of the matter is that many regenerative medicine therapies incorporate multiple elements. Thus, as an example, inducing repair may in the long term lead to the creation (i.e. regeneration) of new tissue. Another example is the introduction of a replacement that acts as a trigger for a repair and/or regenerative response that ultimately restores function.

Cells are the machinery that promote tissue regeneration and, specifically, stem cells are a useful source for transplantation or tissue engineering. The cells, however, may originate from a variety of locations and be at varying levels of commitment. Certain regenerative medicine approaches may rely on autologous adult stem cells being recruited from the host, such as an osteoinductive graft for critical bone defects, into which stem cells and osteoblasts from the recipient's own tissues migrate to the site of repair. On the other hand, with the capacity to self-renew and differentiate *in vitro*, stem cells could be a means by which to generate large homogenous populations of normal cells, either undifferentiated or committed, for tissue engineering or transplantation. In tissue engineering, cells are used to grow 3D constructs *in vitro* for implantation. Transplant examples include bone marrow (which contains marrow-derived stem cells) to treat various blood disorders and chondrocytes for articular cartilage repair (Brittberg et al., 1994). Overall, stem cells in regenerative medicine may be allogeneic or autologous, added exogenously or recruited from the host, and potentially expanded or differentiated *in vitro*. Complex strategies may eventually be developed to combine approaches, perhaps exploiting the effects of co-culture by implanting donor cells of a particular phenotype, that together with host cells, result in a desired regenerative response.

As regenerative medicine covers a wide spectrum of clinical applications and approaches, the field and the research that supports it include a range of disciplines and professions (i.e. basic scientists, engineers, and clinicians). The intersection of stem cell technology and regenerative medicine can be categorized by various criteria, such as stem cell type, technology, or approach. In the end, however, regenerative medicine is a subset of medical treatments, and so here the discussion is organized based on clinical application.

Neural Applications

Neural applications in regenerative medicine include trauma and diseases, such as spinal cord injuries and Parkinson's, respectively. Spinal cord injury therapies may require multiple cell types, including neurons and oligodendrocytes, to help regenerate transected tissues. Parkinson's is a degenerative condition in which dopaminergic cells are lost, resulting in motor dysfunction such as bradykinesia, rigidity, and tremors. Clinical studies related to neural applications have focused on Parkinson's and aim to restore the presence of dopaminergic neurons. Transplantation of cadaveric and adrenal dopaminergic neurons has been shown to have little or short-lived effects (Quinn, 1990). Beyond survival, integration of transplanted neurons with the host tissue is thought to be pivotal for long-term success. It is thought that stem cells for transplantation may restore normal neural function by either integrating and forming working neurons or acting as a trigger to promote neurogenesis by host cells.

Implantation of undifferentiated cells are feared to form teratomas with undesirable cell types, so a favored strategy for Parkinson's therapy is to use stem or progenitor cells committed to the neurogenic pathway prior to transplantation. Fetal mesencephalic tissue containing dopamine-producing neural progenitors has been transplanted in multiple clinical studies. An "open label" clinical study during the mid-1980s transplanted fetal tissue and found mixed, but promising, results. This was followed by two independent National Institutes of Health (NIH)-funded double-blinded clinical trials: one led by Freed et al. (2001) and the other by Olanow et al. (2003), with each study using slightly different sample preparation and surgical procedures for fetal tissue transplantation. Neither study showed a significant improvement when comparing entire patient populations and in a few cases, unfortunately, side effects actually included periods of increased Parkinson's symptoms (Hagell et al., 2002). The study results indicated, however, that increasing the number of transplanted cells may be beneficial for patients with milder cases of Parkinson's. Due to the logistical and technical issues related to fetal tissue harvest, including the lack of tissue standardization and low cellular yield, alternate cell sources are required. Xenotransplantations, using cells from fetal pigs, were found to be safe, but failed to promote significant improvement in patients (Schumacher et al., 2000). The potential to generate large numbers of stem cell-derived dopaminergic neurons *in vitro* could have a meaningful effect on therapies for Parkinson's. While clinical studies have mostly focused on Parkinson's, there are many opportunities for stem cells to impact treatment of both neurodegenerative diseases and neural injuries.

Cardiovascular Applications

Cardiovascular applications of regenerative medicine include myocardial repair, blood vessel substitutes, and valvular replacements. Each application has unique challenges. In myocardial repair, the ideal repair response includes revascularization of ischemic tissue and electrical synchrony with host cardiomyocytes; blood vessel substitutes need to remain patent, and preferably are vasoactive as well; and valvular replacements must persist in a mechanically severe environment. In all three applications, the use of either allogeneic or autologous stem cells may be beneficial.

A myocardial infarct starts a cascade of events that can lead to congestive heart failure. Initial events include ischemia-induced myocardial necrosis and dysfunction. Necrotic cells are removed through an immunological response and eventually a scar tissue is formed. As a result of this process, heart muscle contractility and remodeling are adversely affected, resulting in the loss of cardiac function. Clinical trials have investigated the use of stem cells as a post-infarction therapy using a myriad of approaches. Some studies used freshly isolated autologous bone marrow-derived mononuclear cells, delivered to the myocardium anytime from less than 3-day post-infarct to late stages of congestive heart failure, and showed improvements in common indicators of cardiac function (Assmus et al., 2002; Strauer et al., 2002; Perin et al., 2003; Tse et al., 2003). Other studies have used *ex vivo* expanded autologous blood-derived EPCs (Assmus et al., 2002) or selected (CD133+) autologous marrow-derived cells and also found promising results. Future possible cell sources may include ESC-derived cardiomyocytes (Caspi and Capstein, 2004) and endothelial cells (Levenberg et al., 2002), as well as muscle- (Qu-Petersen et al., 2002), adipose- (Planat-Benard et al., 2004a, b), and umbilical cord- (Murohara et al., 2000) derived stem cells. Nonetheless, these initial clinical studies using bone marrow- and blood-derived stem cells are among the most advanced applications of stem cells in cell-based therapies to date.

Another application of stem cells in cardiovascular regenerative medicine is as a cell source for engineered tissues. Cardiovascular tissues synthesized *in vitro* include substitute blood vessels, myocardial patches, and valvular replacements. In substitute blood vessels, large (>6 mm) synthetic vessels have been found to be somewhat successful and remain patent. Yet small diameter synthetic vessels, as would be used for coronary bypass, quickly become occluded. It is thought that an endothelial cell lining, as found in native vessels, would provide an anti-thrombogenic inner layer that would prevent clot formation. Future generations of blood vessel substitutes are likely to be capable of vasoactivity and long-term remodeling, to which smooth muscle cells in the medial layer are critical. Similarly, valvular replacements or myocardial patches expected to remodel over years will likely need endothelial and interstitial cells or cardiomyocytes, respectively. Stem cells, either allogeneic or autologous, may provide means to generate these diverse vascular phenotypes *in vitro*.

Orthopedic Applications

One current stem cell-based orthopedic therapy includes bone marrow-derived MSC transplantation for osteogenesis imperfecta, a genetic disorder in which osteoblasts synthesize defective collagen type I, which leads to a variety of skeletal pathologies. In limited clinical studies in children, it has been found that allogeneic bone marrow-derived mesenchymal cells engraft in multiple skeletal sites and improve bone growth velocity (Horwitz et al., 2002). In other applications, stem cells are recruited from the host to help regenerate tissues. Cartilage repair techniques, such as microfracture (Steadman et al., 2001), expose vascularized bone that then forms a conduit by which marrow cells, including MSCs, can access the defect site.

One well-established cellular therapy in orthopedics, autologous chondrocyte transplantation for articular cartilage, may be improved through the use of stem cells. In this procedure, originally published by Brittberg et al. (1994) and subsequently commercialized by Genzyme Biosurgery under the name Carticel®, chondrocytes are harvested from a non-load bearing region of the knee, expanded *in vitro*, injected into an articular cartilage defect, and covered with a periosteal flap. Donor site morbidity is an undesirable consequence of this procedure. A stem cell-derived chondrocyte may provide a marked improvement on this already well-established orthopedic therapy.

Future applications of stem cells in orthopedic regenerative medicine include tissue engineering. The scope of *in vitro* engineered tissues currently being studied in orthopedics includes bone, articular cartilage, temporal mandibular cartilage, meniscus, muscle, tendon, and ligament. Similar to many tissue engineering applications, cell sourcing of terminally differentiated or appropriate progenitor cells is problematic and stem cells are an option. One preliminary study already used stem cells for orthopedic tissue engineering. In just a few patients, marrow-derived osteoprogenitor cells were grown on porous hydroxyapatite scaffolds that were then implanted into critical length defects in long bones. In three patients, radiographs indicated callus formation along the implants and good integration with the adjacent host bone (Quarto et al., 2001). This study, *albeit* very limited, shows the promise of stem cells in orthopedic tissue engineering.

Metabolic and Secretory Applications

The cell types in metabolic and secretory organs are among those in the body that have the most complex functional properties. The Edmonton protocol has shown the value of islet transplantation in addressing insulin regulation in patients with type I diabetes (Shapiro et al., 2000). Islets are a collection of endocrine cells in the pancreas responsible for insulin secretion used in metabolizing glucose. Beta cells, which constitute 80–85% of the islets, sense blood sugar levels and secrete appropriate amounts of insulin in response. These cells are destroyed by an abnormal immunological response in individuals with type I diabetes. Among the key aspects that led to successful insulin-independence in the Edmonton protocol was the large number of islets transplanted into the patients. Collecting those large numbers is problematic due to the paucity of available donor organs and the difficulty in islet isolation (Kobayashi et al., 2004). As a result, there is great interest in generating insulin-secreting cells from stem cells. The literature in this area, however, is conflicting and controversial. While some studies reported the derivation of beta cells from pancreatic adult stem cells, lineage tracing and evidence of clonal expansion to support those claims were lacking. Embryoid bodies, used to differentiate human ESCs, include a small number of insulin producing cells. Better characterization is needed to determine whether those cells are beta cell precursors, neural cells, or extra-embryonic endodermal cells (Otonkoski et al., 2005). While developments in this area are significant, further basic science studies are required before stem cells provide an alternative therapy for diabetes.

Hematopoietic and Autoimmune Applications

Bone marrow transplantation, which originated in the 1950s, is now known to include the transfer of multiple stem cell types, including hematopoietic and MSCs. It is the capacity of HSCs to yield blood components or even to restore the entire immune system that is the basis of marrow-derived cell therapies. Hematological malignancies, such as leukemia, sickle cell, and aplastic anemia, arise as a result of abnormalities in marrow-derived cells. Transplantation of allogeneic bone marrow- (Arcese et al., 1999) and umbilical cord- (Benito et al., 2004) derived stem cells treats pathologies in hematopoiesis and the immune system by providing a new source of blood and immune cells.

Autoimmune diseases are another application of stem cells in regenerative medicine. Such diseases, which can affect either specific organs or the entire system, include multiple sclerosis, rheumatoid arthritis, and systemic lupus erythematosus. Conventional treatments for these conditions include immunosuppression, which can be effective but not curative. Recently, refractory cases of autoimmune diseases are being treated with severe immunosuppression, to the extent of immunoablation, followed by allogeneic or autologous stem cell transplantation (Jantunen and Luosujarvi, 2005). Subsequent treatment with stem cell mobilizers is meant to allow the transplanted cells to rebuild the entire immune system. In most cases, transplanted cells originate from the bone marrow, but now cells obtained from peripheral blood are also being used. In one recent study, a trial of 85 patients with progressive multiple sclerosis found that greater than 60% of the patients benefited from this procedure (Fassas et al., 2002). Another ongoing study focuses on using this approach to arrest the progression of the disease in patients that are less effected (Havrdova, 2005). Treatment regimens that are better tolerated will need to be developed before this approach becomes a widely accepted therapy. Once the risks associated with this therapy are sufficiently low, extensions of this approach may be applied to other uses, such as boosting immune systems after aggressive chemotherapy during cancer treatment.

CONCLUSION

Stem cell technology shows potential in contributing to regenerative medicine. Self-renewing with the potential to differentiate into specialized phenotypes, stem cells may be derived from embryonic, fetal, or adult cells or tissues. These cells are allogeneic or autologous, added exogenously or recruited from the host, and potentially expanded and/or differentiated *in vitro*. In regenerative medicine, stem cells can serve as the machinery to repair, regenerate, and/or replace tissues and organs. The ethical, regulatory, and scientific hurdles will need to be overcome for each stem cell type before clinical use. Applications of stem cells in regenerative medicine will help to confront the organ transplantation crisis and allow customization of therapies for each patient.

ACKNOWLEDGMENTS

The authors thank the Georgia Tech/Emory Center for the Engineering of Living Tissues (National Science Foundation Engineering Research Center: NSF EEC-9731643), an NIH Biotechnology Training Program (T32GM08433), and the Ruth L. Kirschstein National Research Service Award (1F32HL076978-01A1) for financial support.

INTERNET RESOURCES

National Institute for Health (2005). Stem Cell Information. stemcells.nih.gov

National Marrow Donor Program (2005). www.marrow.org

United Network for Organ Sharing (2005). Transplant Living. www.transplantliving.org

United States Census Bureau (2005). www.census.gov

REFERENCES

Adams, A.B., Williams, M.A., Jones, T.R., Shirasugi, N., Durham, M.M., Kaech, S.M., Wherry, E.J., Onami, T., Lanier, J.G., Kokko, K.E., et al. (2003). Heterologous immunity provides a potent barrier to transplantation tolerance. *J. Clin. Invest.* **111**(12): 1887–1895.

Aflatoonian, B. and Moore, H. (2005). Human primordial germ cells and embryonic germ cells, and their use in cell therapy. *Curr. Opin. Biotechnol.* **16**(5): 530–535.

Ahsan, T. and Nerem, R.M. (2005). Effect of fluid shear stress on embryonic stem-derived cells (abstract). *Tissue Engineering Society International*, Shanghai, China (October 22–25).

Alam, S., Sen, A., Behie, L.A. and Kallos, M.S. (2004). Cell cycle kinetics of expanding populations of neural stem and progenitor cells *in vitro*. *Biotechnol. Bioeng.* **88**(3): 332–347.

Alessandri, G., Pagano, S., Bez, A., Benetti, A., Pozzi, S., Iannolo, G., Baronio, M., Invernici, G., Caruso, A., Muneretto, C., et al. (2004). Isolation and culture of human muscle-derived stem cells able to differentiate into myogenic and neurogenic cell lineages. *Lancet* **364**(9448): 1872–1883.

Amit, M., Margulets, V., Segev, H., Shariki, K., Laevsky, I., Coleman, R. and Itskovitz-Eldor, J. (2003). Human feeder layers for human embryonic stem cells. *Biol. Reprod.* **68**(6): 2150–2156.

Amit, M., Shariki, C., Margulets, V. and Itskovitz-Eldor, J. (2004). Feeder layer- and serum-free culture of human embryonic stem cells. *Biol. Reprod.* **70**(3): 837–845.

Arcese, W., Guglielmi, C., Iori, A.P., Screnci, M., Carmini, D., Testi, A.M., Moleti, M.L., Mengarelli, A., Del Giudice, I., Cimino, G., et al. (1999). Umbilical cord blood transplant from unrelated hla-mismatched donors in children with high risk leukemia. *Bone Marrow Transplant.* **23**(6): 549–554.

Asahara, T., Murohara, T., Sullivan, A., Silver, M., van der Zee, R., Li, T., Witzenbichler, B., Schatteman, G. and Isner, J.M. (1997). Isolation of putative progenitor endothelial cells for angiogenesis. *Science* **275**(5302): 964–967.

Asakura, A. and Rudnicki, M.A. (2002). Side population cells from diverse adult tissues are capable of *in vitro* hematopoietic differentiation. *Exp. Hematol.* **30**(11): 1339–1345.

Ashjian, P.H., Elbarbary, A.S., Edmonds, B., DeUgarte, D., Zhu, M., Zuk, P.A., Lorenz, H.P., Benhaim, P. and Hedrick, M.H. (2003). *In vitro* differentiation of human processed lipoaspirate cells into early neural progenitors. *Plast. Reconstr. Surg.* **111**(6): 1922–1931.

Assmus, B., Schachinger, V., Teupe, C., Britten, M., Lehmann, R., Dobert, N., Grunwald, F., Aicher, A., Urbich, C., Martin, H., et al. (2002). Transplantation of progenitor cells and regeneration enhancement in acute myocardial infarction (topcare-ami). *Circulation* **106**(24): 3009–3017.

Awad, H.A., Wickham, M.Q., Leddy, H.A., Gimble, J.M. and Guilak, F. (2004). Chondrogenic differentiation of adipose-derived adult stem cells in agarose, alginate, and gelatin scaffolds. *Biomaterials* **25**(16): 3211–3222.

Bagley, J., Rosenzweig, M., Marks, D.F. and Pykett, M.J. (1999). Extended culture of multipotent hematopoietic progenitors without cytokine augmentation in a novel three-dimensional device. *Exp. Hematol.* **27**(3): 496–504.

Batorsky, A., Liao, J., Lund, A.W., Plopper, G.E. and Stegemann, J.P. (2005). Encapsulation of adult human mesenchymal stem cells within collagen–agarose microenvironments. *Biotechnol. Bioeng.* **92**(4): 492–500.

Battista, S., Guarnieri, D., Borselli, C., Zeppetelli, S., Borzacchiello, A., Mayol, L., Gerbasio, D., Keene, D.R., Ambrosio, L. and Netti, P.A. (2005). The effect of matrix composition of 3d constructs on embryonic stem cell differentiation. *Biomaterials* **26**(31): 6194–6207.

Bauwens, C., Yin, T., Dang, S., Peerani, R. and Zandstra, P.W. (2005). Development of a perfusion fed bioreactor for embryonic stem cell-derived cardiomyocyte generation: oxygen-mediated enhancement of cardiomyocyte output. *Biotechnol. Bioeng.* **90**(4): 452–461.

Benito, A.I., Diaz, M.A., Gonzalez-Vicent, M., Sevilla, J. and Madero, L. (2004). Hematopoietic stem cell transplantation using umbilical cord blood progenitors: review of current clinical results. *Bone Marrow Transplant.* **33**(7): 675–690.

Betre, H., Ong, S.R., Guilak, F., Chilkoti, A., Fermor, B. and Setton, L.A. (2006). Chondrocytic differentiation of human adipose-derived adult stem cells in elastin-like polypeptide. *Biomaterials* **27**(1): 91–99.

Brittberg, M., Lindahl, A., Nilsson, A., Ohlsson, C., Isaksson, O. and Peterson, L. (1994). Treatment of deep cartilage defects in the knee with autologous chondrocyte transplantation. *N. Engl. J. Med.* **331**(14): 889–895.

Broxmeyer, H.E., Douglas, G.W., Hangoc, G., Cooper, S.,Bard, J., English, D., Arny, M., Thomas, L. and Boyse, E.A. (1989). Human umbilical cord blood as a potential source of transplantable hematopoietic stem/progenitor cells. *Proc. Natl Acad. Sci. USA* **86**(10): 3828–3832.

Cabrita, G.J., Ferreira, B.S., da Silva, C.L., Goncalves, R., Almeida-Porada, G. and Cabral, J.M. (2003). Hematopoietic stem cells: from the bone to the bioreactor. *Trends Biotechnol.* **21**(5): 233–240.

Caspi, O. and Gepstein, L. (2004). Potential applications of human embryonic stem cell-derived cardiomyocytes. *Ann. NY Acad. Sci.* **1015**: 285–298.

Chaudhry, G.R., Yao, D., Smith, A. and Hussain, A. (2004). Osteogenic cells derived from embryonic stem cells produced bone nodules in three-dimensional scaffolds. *J. Biomed. Biotechnol.* **2004**(4): 203–210.

Chen, S.S., Revoltella, R.P., Papini, S., Michelini, M., Fitzgerald, W., Zimmerberg, J. and Margolis, L. (2003). Multilineage differentiation of rhesus monkey embryonic stem cells in three-dimensional culture systems. *Stem Cells* **21**(3): 281–295.

Chiu, B., Wan, J.Z., Abley, D. and Akabutu, J. (2005). Induction of vascular endothelial phenotype and cellular proliferation from human cord blood stem cells cultured in simulated microgravity. *Acta Astronaut.* **56**(9–12): 918–922.

Chung, Y., Klimanskaya, I., Becker, S., Marh, J., Lu, S.J., Johnson, J., Meisner, L. and Lanza, R. (2006). Embryonic and extraembryonic stem cell lines derived from single mouse blastomeres. *Nature* **439**(7073): 216–219.

Dang, S.M., Gerecht-Nir, S., Chen, J., Itskovitz-Eldor, J. and Zandstra, P.W. (2004). Controlled, scalable embryonic stem cell differentiation culture. *Stem Cells* **22**(3): 275–282.

Edwards, R.G. (2004). Stem cells today: B1. Bone marrow stem cells. *Reprod. Biomed. Online* **9**(5): 541–583.

Ehring, B., Biber, K., Upton, T.M., Plosky, D., Pykett, M. and Rosenzweig, M. (2003). Expansion of hpcs from cord blood in a novel 3d matrix. *Cytotherapy* **5**(6): 490–499.

Eisenberg, L.M. and Eisenberg, C.A. (2003). Stem cell plasticity, cell fusion, and transdifferentiation. *Birth Defects Res. C Embryo Today* **69**(3): 209–218.

Erices, A., Conget, P. and Minguell, J.J. (2000). Mesenchymal progenitor cells in human umbilical cord blood. *Br. J. Haematol.* **109**(1): 235–242.

Evans, M.J. and Kaufman, M.H. (1981). Establishment in culture of pluripotential cells from mouse embryos. *Nature* **292**(5819): 154–156.

Fassas, A., Passweg, J.R., Anagnostopoulos, A., Kazis, A., Kozak, T., Havrdova, E., Carreras, E., Graus, F., Kashyap, A., Openshaw, H., et al. (2002). Hematopoietic stem cell transplantation for multiple sclerosis. A retrospective multicenter study. *J. Neurol.* **249**(8): 1088–1097.

Freed, C.R., Greene, P.E., Breeze, R.E., Tsai, W.Y., DuMouchel, W., Kao, R., Dillon, S., Winfield, H., Culver, S., Trojanowski, J.Q., et al. (2001). Transplantation of embryonic dopamine neurons for severe Parkinson's disease. *N. Engl. J. Med.* **344**(10): 710–719.

Gerecht-Nir, S., Cohen, S. and Itskovitz-Eldor, J. (2004a). Bioreactor cultivation enhances the efficiency of human embryoid body (heb) formation and differentiation. *Biotechnol. Bioeng.* **86**(5): 493–502.

Gerecht-Nir, S., Cohen, S., Ziskind, A. and Itskovitz-Eldor, J. (2004b). Three-dimensional porous alginate scaffolds provide a conducive environment for generation of well-vascularized embryoid bodies from human embryonic stem cells. *Biotechnol. Bioeng.* **88**(3): 313–320.

Goodell, M.A., Brose, K., Paradis, G., Conner, A.S. and Mulligan, R.C. (1996). Isolation and functional properties of murine hematopoietic stem cells that are replicating *in vivo*. *J. Exp. Med.* **183**(4): 1797–1806.

Grayson, W.L., Ma, T. and Bunnell, B. (2004). Human mesenchymal stem cells tissue development in 3d pet matrices. *Biotechnol. Prog.* **20**(3): 905–912.

Hagell, P., Piccini, P., Bjorklund, A., Brundin, P., Rehncrona, S., Widner, H., Crabb, L., Pavese, N., Oertel, W.H., Quinn, N., et al. (2002). Dyskinesias following neural transplantation in Parkinson's disease. *Nat. Neurosci.* **5**(7): 627–628.

Havrdova, E. (2005). Aggressive multiple sclerosis – is there a role for stem cell transplantation? *J. Neurol.* **252**(Suppl 3): iii34–iii37.

Ho, A.D. (2005). Kinetics and symmetry of divisions of hematopoietic stem cells. *Exp. Hematol.* **33**(1): 1–8.

Hoffman, L.M. and Carpenter, M.K. (2005). Characterization and culture of human embryonic stem cells. *Nat. Biotechnol.* **23**(6): 699–708.

Horwitz, E.M., Gordon, P.L., Koo, W.K., Marx, J.C., Neel, M.D., McNall, R.Y., Muul, L. and Hofmann, T. (2002). Isolated allogeneic bone marrow-derived mesenchymal cells engraft and stimulate growth in children with osteogenesis imperfecta: implications for cell therapy of bone. *Proc. Natl Acad. Sci. USA* **99**(13): 8932–8937.

Hovatta, O., Mikkola, M., Gertow, K., Stromberg, A.M., Inzunza, J., Hreinsson, J., Rozell, B., Blennow, E., Andang, M. and Ahrlund-Richter, L. (2003). A culture system using human foreskin fibroblasts as feeder cells allows production of human embryonic stem cells. *Hum. Reprod.* **18**(7): 1404–1409.

Huang, C.Y., Hagar, K.L., Frost, L.E., Sun, Y. and Cheung, H.S. (2004). Effects of cyclic compressive loading on chondrogenesis of rabbit bone-marrow derived mesenchymal stem cells. *Stem Cells* **22**(3): 313–323.

Huang, H., Nakayama, Y., Qin, K., Yamamoto, K., Ando, J., Yamashita, J., Itoh, H., Kanda, K., Yaku, H., Okamoto, Y., et al. (2005). Differentiation from embryonic stem cells to vascular wall cells under *in vitro* pulsatile flow loading. *J. Artif. Organs* **8**(2): 110–118.

Huang, J.I., Zuk, P.A., Jones, N.F., Zhu, M., Lorenz, H.P., Hedrick, M.H. and Benhaim, P. (2004). Chondrogenic potential of multipotential cells from human adipose tissue. *Plast. Reconstr. Surg.* **113**(2): 585–594.

Hwang, N.S., Kim, M.S., Sampattavanich, S., Baek, J.H., Zhang, Z. and Elisseeff, J. (2006). The effects of three dimensional culture and growth factors on the chondrogenic differentiation of murine embryonic stem cells. *Stem Cells* **24**(2): 284–291.

Hwang, W.S., Ryu, Y.J., Park, J.H., Park, E.S., Lee, E.G., Koo, J.M., Jeon, H.Y., Lee, B.C., Kang, S.K., Kim, S.J., et al. (2004). Evidence of a pluripotent human embryonic stem cell line derived from a cloned blastocyst. *Science* **303**(5664): 1669–1674.

Jantunen, E. and Luosujarvi, R. (2005). Stem cell transplantation in autoimmune diseases: an update. *Ann. Med.* **37**(7): 533–541.

Jiang, Y., Jahagirdar, B.N., Reinhardt, R.L., Schwartz, R.E., Keene, C.D., Ortiz-Gonzalez, X.R., Reyes, M., Lenvik, T.,. Lund, T., Blackstad, M., et al. (2002). Pluripotency of mesenchymal stem cells derived from adult marrow. *Nature* **418**(6893): 41–49.

Kallos, M.S. and Behie, L.A. (1999). Inoculation and growth conditions for high-cell-density expansion of mammalian neural stem cells in suspension bioreactors. *Biotechnol. Bioeng.* **63**(4): 473–483.

Kallos, M.S., Behie, L.A. and Vescovi, A.L. (1999). Extended serial passaging of mammalian neural stem cells in suspension bioreactors. *Biotechnol. Bioeng.* **65**(5): 589–599.

Kobayashi, N., Okitsu, T., Lakey, J.R. and Tanaka, N. (2004). The current situation in human pancreatic islet transplantation: problems and prospects. *J. Artif. Organs* **7**(1): 1–8.

Kocher, A.A., Schuster, M.D., Szabolcs, M.J., Takuma, S.,Burkhoff, D., Wang, J., Homma, S., Edwards, N.M. and Itescu, S. (2001). Neovascularization of ischemic myocardium by human bone-marrow-derived angioblasts prevents cardiomyocyte apoptosis, reduces remodeling and improves cardiac function. *Nat. Med.* **7**(4): 430–436.

Konstantinov, S.M., Mindova, M.M., Gospodinov, P.T. and Genova, P.I. (2004). Three-dimensional bioreactor cultures: a useful dynamic model for the study of cellular interactions. *Ann. NY Acad. Sci.* **1030**: 103–115.

Lakshmipathy, U. and Verfaillie, C. (2005). Stem cell plasticity. *Blood Rev.* **19**(1): 29–38.

Lee, J.B., Lee, J.E., Park, J.H., Kim, S.J., Kim, M.K., Roh, S.I. and Yoon, H.S. (2005). Establishment and maintenance of human embryonic stem cell lines on human feeder cells derived from uterine endometrium under serum-free condition. *Biol. Reprod.* **72**(1): 42–49.

Lemoine, N.R. (2002). The power to deliver: stem cells in gene therapy. *Gene Ther.* **9**(10): 603–605.

Levenberg, S., Golub, J.S., Amit, M., Itskovitz-Eldor, J. and Langer, R. (2002). Endothelial cells derived from human embryonic stem cells. *Proc. Natl Acad. Sci. USA* **99**(7): 4391–4396.

Levenberg, S., Huang, N.F., Lavik, E., Rogers, A.B., Itskovitz-Eldor, J. and Langer, R. (2003). Differentiation of human embryonic stem cells on three-dimensional polymer scaffolds. *Proc. Natl Acad. Sci. USA* **100**(22): 12741–12746.

Levenstein, M.E., Ludwig, T.E., Xu, R.H., Llanas, R.A., Vandenheuvel-Kramer, K., Manning, D. and Thomson, J.A. (2006). Basic fgf support of human embryonic stem cell self-renewal. *Stem Cells* **24**(3): 568–574.

Li, Y., Firpo, M. and Healy, K.E. (2005a). Controlling stem cell fate via biomimetic polymers (abstract). *Pacific Polymer Conference IX*, Maui, Hawaii (December 11–14).

Li, Y., Powell, S., Brunette, E., Lebkowski, J. and Mandalam, R. (2005b). Expansion of human embryonic stem cells in defined serum-free medium devoid of animal-derived products. *Biotechnol. Bioeng.* **91**(6): 688–698.

Liu, L., Zhang, L., Ren, B., Wang, F. and Zhang, Q. (2003). Preparation and characterization of collagen–hydroxyapatite composite used for bone tissue engineering scaffold. *Artif. Cells Blood Substit. Immobil. Biotechnol.* **31**(4): 435–448.

Ma, W., Fitzgerald, W., Liu, Q.Y.,. O'Shaughnessy, T.J., Maric, D., Lin, H.J., Alkon, D.L. and Barker, J.L. (2004). Cns stem and progenitor cell differentiation into functional neuronal circuits in three-dimensional collagen gels. *Exp. Neurol* **190**(2): 276–288.

Martin, G.R. (1981). Isolation of a pluripotent cell line from early mouse embryos cultured in medium conditioned by teratocarcinoma stem cells. *Proc. Natl Acad. Sci. USA* **78**(12): 7634–7638.

Meissner, A. and Jaenisch, R. (2006). Generation of nuclear transfer-derived pluripotent es cells from cloned cdx2-deficient blastocysts. *Nature* **439**(7073): 212–215.

Murohara, T., Ikeda, H., Duan, J., Shintani, S., Sasaki, K., Eguchi, H., Onitsuka, I., Matsui, K. and Imaizumi, T. (2000). Transplanted cord blood-derived endothelial precursor cells augment postnatal neovascularization. *J. Clin. Invest.* **105**(11): 1527–1536.

Nielsen, L.K. (1999). Bioreactors for hematopoietic cell culture. *Annu. Rev. Biomed. Eng.* **1**: 129–152.

Nur-E-Kamal, A., Ahmed, I., Kamal, J.M., Schindler, M. and Meiners, S. (2006). Three-dimensional nanofibrillar surfaces promote self-renewal in mouse embryonic stem cells. *Stem Cells* **24**(2): 426–433.

Ogawa, M. (1993). Differentiation and proliferation of hematopoietic stem cells. *Blood* **81**(11): 2844–2853.

Olanow, C.W., Goetz, C.G., Kordower, J.H., Stoessl, A.J., Sossi, V., Brin, M.F., Shannon, K.M., Nauert, G.M., Perl, D.P., Godbold, J., et al. (2003). A double-blind controlled trial of bilateral fetal nigral transplantation in Parkinson's disease. *Ann. Neurol.* **54**(3): 403–414.

Otonkoski, T., Gao, R. and Lundin, K. (2005). Stem cells in the treatment of diabetes. *Ann. Med.* **37**(7): 513–520.

Perin, E.C., Dohmann, H.F., Borojevic, R., Silva, S.A., Sousa, A.L., Mesquita, C.T., Rossi, M.I., Carvalho, A.C., Dutra, H.S., Dohmann, H. J., et al. (2003). Transendocardial, autologous bone marrow cell transplantation for severe, chronic ischemic heart failure. *Circulation* **107**(18): 2294–2302.

Perry, D. (2000). Patients' voices: the powerful sound in the stem cell debate. *Science* **287**(5457): 1423.

Philp, D., Chen, S.S., Fitzgerald, W., Orenstein, J., Margolis, L. and Kleinman, H.K. (2005). Complex extracellular matrices promote tissue-specific stem cell differentiation. *Stem Cells* **23**(2): 288–296.

Pittenger, M.F., Mackay, A.M., Beck, S.C., Jaiswal, R.K., Douglas, R., Mosca, J.D., Moorman, M.A., Simonetti, D.W., Craig, S. and Marshak, D.R. (1999). Multilineage potential of adult human mesenchymal stem cells. *Science* **284**(5411): 143–147.

Planat-Benard, V., Menard, C., Andre, M., Puceat, M., Perez, A., Garcia-Verdugo, J.M., Penicaud, L. and Casteilla, L. (2004a). Spontaneous cardiomyocyte differentiation from adipose tissue stroma cells. *Circ. Res.* **94**(2): 223–229.

Planat-Benard, V., Silvestre, J.S., Cousin, B., Andre, M.,Nibbelink, M., Tamarat, R., Clergue, M., Manneville, C., Saillan-Barreau, C., Duriez, M., et al. (2004b). Plasticity of human adipose lineage cells toward endothelial cells: physiological and therapeutic perspectives. *Circulation* **109**(5): 656–663.

Qu-Petersen, Z., Deasy, B., Jankowski, R., Ikezawa, M., Cummins, J., Pruchnic, R., Mytinger, J., Cao, B., Gates, C., Wernig, A., et al. (2002). Identification of a novel population of muscle stem cells in mice: potential for muscle regeneration. *J. Cell Biol.* **157**(5): 851–864.

Quarto, R., Mastrogiacomo, M., Cancedda, R., Kutepov, S.M., Mukhachev, V., Lavroukov, A., Kon, E. and Marcacci, M. (2001). Repair of large bone defects with the use of autologous bone marrow stromal cells. *N. Engl. J. Med.* **344**(5): 385–386.

Quesenberry, P.J., Abedi, M., Aliotta, J., Colvin, G., Demers, D., Dooner, M., Greer, D., Hebert, H., Menon, M.K., Pimentel, J., et al. (2004). Stem cell plasticity: an overview. *Blood Cells Mol. Dis.* **32**(1): 1–4.

Quinn, N.P. (1990). The clinical application of cell grafting techniques in patients with Parkinson's disease. *Prog. Brain. Res.* **82**: 619–625.

Rao, R.R., Calhoun, J.D., Qin, X., Rekaya, R., Clark, J.K. and Stice, S.L. (2004). Comparative transcriptional profiling of two human embryonic stem cell lines. *Biotechnol. Bioeng.* **88**(3): 273–286.

Reubinoff, B.E., Pera, M.F., Fong, C.Y., Trounson, A. and Bongso, A. (2000). Embryonic stem cell lines from human blastocysts: somatic differentiation *in vitro*. *Nat. Biotechnol.* **18**(4): 399–404.

Richards, M., Fong, C.Y., Chan, W.K., Wong, P.C. and Bongso, A. (2002). Human feeders support prolonged undifferentiated growth of human inner cell masses and embryonic stem cells. *Nat. Biotechnol.* **20**(9): 933–936.

Saha, S., Ji, L., de Pablo, J.J. and Palecek, S.P. (2006). Inhibition of human embryonic stem cell differentiation by mechanical strain. *J. Cell. Physiol.* **206**(1): 126–137.

Scharenberg, C.W., Harkey, M.A. and Torok-Storb, B. (2002). The abcg2 transporter is an efficient hoechst 33342 efflux pump and is preferentially expressed by immature human hematopoietic progenitors. *Blood* **99**(2): 507–512.

Schroeder, M., Niebruegge, S., Werner, A., Willbold, E., Burg, M., Ruediger, M., Field, L.J., Lehmann, J. and Zweigerdt, R. (2005). Differentiation and lineage selection of mouse embryonic stem cells in a stirred bench scale bioreactor with automated process control. *Biotechnol. Bioeng.* **92**(7): 920–933.

Schumacher, J.M., Ellias, S.A., Palmer, E.P., Kott, H.S., Dinsmore, J., Dempsey, P.K., Fischman, A.J., Thomas, C., Feldman, R.G., Kassissieh, S., et al. (2000). Transplantation of embryonic porcine mesencephalic tissue in patients with pd. *Neurology* **54**(5): 1042–1050.

Sen, A., Kallos, M.S. and Behie, L.A. (2002). Passaging protocols for mammalian neural stem cells in suspension bioreactors. *Biotechnol. Prog.* **18**(2): 337–345.

Shamblott, M.J., Axelman, J., Wang, S., Bugg, E.M., Littlefield, J.W., Donovan, P.J., Blumenthal, P.D., Huggins, G.R. and Gearhart, J.D. (1998). Derivation of pluripotent stem cells from cultured human primordial germ cells. *Proc. Natl Acad. Sci. USA* **95**(23): 13726–13731.

Shamblott, M.J., Axelman, J., Littlefield, J.W., Blumenthal, P.D., Huggins, G.R., Cui, Y., Cheng, L. and Gearhart, J.D. (2001). Human embryonic germ cell derivatives express a broad range of developmentally distinct markers and proliferate extensively *in vitro*. *Proc. Natl Acad. Sci. USA* **98**(1): 113–118.

Shapiro, A.M., Lakey, J.R., Ryan, E.A., Korbutt, G.S., Toth, E., Warnock, G.L., Kneteman, N.M. and Rajotte, R.V. (2000). Islet transplantation in seven patients with type 1 diabetes mellitus using a glucocorticoid-free immunosuppressive regimen. *N. Engl. J. Med.* **343**(4): 230–238.

Steadman, J.R., Rodkey, W.G. and Rodrigo, J.J. (2001). Microfracture: surgical technique and rehabilitation to treat chondral defects. *Clin. Orthop. Relat. Res.* (391 Suppl): S362–S369.

Steindler, D.A. and Pincus, D.W. (2002). Stem cells and neuropoiesis in the adult human brain. *Lancet* **359**(9311): 1047–1054.

Strauer, B.E., Brehm, M., Zeus, T., Kostering, M., Hernandez, A., Sorg, R.V., Kogler, G. and Wernet, P. (2002). Repair of infarcted myocardium by autologous intracoronary mononuclear bone marrow cell transplantation in humans. *Circulation* **106**(15): 1913–1918.

Tao, H. and Ma, D.D. (2003). Evidence for transdifferentiation of human bone marrow-derived stem cells: recent progress and controversies. *Pathology* **35**(1): 6–13.

Thomson, J.A., Itskovitz-Eldor, J., Shapiro, S.S., Waknitz, M.A., Swiergiel, J.J., Marshall, V.S. and Jones, J.M. (1998). Embryonic stem cell lines derived from human blastocysts. *Science* **282**(5391): 1145–1147.

Tosh, D. and Strain, A. (2005). Liver stem cells – prospects for clinical use. *J. Hepatol.* **42**(Suppl1): S75–S84.

Tse, H.F., Kwong, Y.L., Chan, J.K., Lo, G., Ho, C.L. and Lau, C.P. (2003). Angiogenesis in ischaemic myocardium by intramyocardial autologous bone marrow mononuclear cell implantation. *Lancet* **361**(9351): 47–49.

Tun, T., Miyoshi, H., Aung, T., Takahashi, S., Shimizu, R., Kuroha, T., Yamamoto, M. and Ohshima, N. (2002). Effect of growth factors on *ex vivo* bone marrow cell expansion using three-dimensional matrix support. *Artif. Organs* **26**(4): 333–339.

Turnpenny, L., Brickwood, S., Spalluto, C.M., Piper, K., Cameron, I.T., Wilson, D.I. and Hanley, N.A. (2003). Derivation of human embryonic germ cells: an alternative source of pluripotent stem cells. *Stem Cells* **21**(5): 598–609.

Wagers, A.J. and Weissman, I.L. (2004). Plasticity of adult stem cells. *Cell* **116**(5): 639–648.

Wang, H., Riha, G.M., Yan, S., Li, M., Chai, H., Yang, H., Yao, Q. and Chen, C. (2005a). Shear stress induces endothelial differentiation from a murine embryonic mesenchymal progenitor cell line. *Arterioscler. Thromb. Vasc. Biol.* **25**(9): 1817–1823.

Wang, Y., Kim, U.J., Blasioli, D.J., Kim, H.J. and Kaplan, D.L. (2005b). *In vitro* cartilage tissue engineering with 3d porous aqueous-derived silk scaffolds and mesenchymal stem cells. *Biomaterials* **26**(34): 7082–7094.

Yamamoto, K., Sokabe, T., Watabe, T., Miyazono, K., Yamashita, J.K., Obi, S., Ohura, N., Matsushita, A., Kamiya, A. and Ando, J. (2005). Fluid shear stress induces differentiation of flk-1-positive embryonic stem cells into vascular endothelial cells *in vitro*. *Am. J. Physiol. Heart Circ. Physiol.* **288**(4): H1915–H1924.

Yamamoto, K., Takahashi, T., Asahara, T., Ohura, N., Sokabe, T., Kamiya, A. and Ando, J. (2003). Proliferation, differentiation, and tube formation by endothelial progenitor cells in response to shear stress. *J. Appl. Physiol.* **95**(5): 2081–2088.

Zhao, F. and Ma, T. (2005). Perfusion bioreactor system for human mesenchymal stem cell tissue engineering: dynamic cell seeding and construct development. *Biotechnol. Bioeng.* **91**(4): 482–493.

Zuk, P.A., Zhu, M., Ashjian, P., De Ugarte, D.A., Huang, J.I., Mizuno, H. Alfonso, Z.C., Fraser, J.K., Benhaim, P. and Hedrick, M.H. (2002). Human adipose tissue is a source of multipotent stem cells. *Mol. Biol. Cell* **13**(12): 4279–4295.

Zulewski, H., Abraham, E.J., Gerlach, M.J., Daniel, P.B., Moritz, W., Muller, B., Vallejo, M., Thomas M.K. and Habener, J.F. (2001). Multipotential nestin-positive stem cells isolated from adult pancreatic islets differentiate *ex vivo* into pancreatic endocrine, exocrine, and hepatic phenotypes. *Diabetes* **50**(3): 521–533.

Part II

Biologic and Molecular Basis for Regenerative Medicine

E-cadherin (reviewed in De Craene, 2005). E-cadherin protein activity can also be regulated by trafficking and turnover (Bryant and Stow, 2004). The precise endocytic pathways for E-cadherin are still unclear, and there is evidence for both caveolae-dependent endocytosis (Lu et al., 2003) and clathrin-dependent endocytosis of E-cadherin (Ivanov et al., 2004). E-cadherin can also be ubiquitinated in cultured cells by the E3-ligase, Hakai, which targets E-cadherin to the proteasome (Fujita et al., 2002). Another mechanism to disrupt E-cadherin function is through extracellular proteases such as matrix metalloproteases (MMPs), which degrade the extracellular domain of E-cadherin and consequently reduce cadherin-mediated cell adhesion (Egeblad and Werb, 2002). Some or all of these mechanisms may occur simultaneously during an EMT to disrupt cell–cell adhesion.

In addition to their role in cell–cell adhesion, cadherins also function as cell-signaling molecules. Intracellularly, classical cadherins interact with α- and β-catenin, which in turn link with the actin cytoskeleton (Tepass et al., 2000). Hence, β-catenin is an important structural component of the cytoskeleton. β-Catenin can also function in cell signaling when it translocates to the nucleus and acts as a co-activator of the lymphoid enhancer-binding factor/T-cell factor (LEF/TCF) transcription factor family (Sharpe et al., 2001). β-Catenin is pivotal for regulating most EMTs. In vertebrates, β-catenin is required for gastrulation, and misexpression of β-catenin results in ectopic gastrulation events (Moon and Kimelman, 1998). β-Catenin is also necessary for the EMT during cardiac cushion development (Liebner et al., 2004). In breast cancer, β-catenin expression is highly correlated with metastasis and poor survival (Cowin et al., 2005), and blocking β-catenin function in tumor cells inhibits their invasion *in vitro* (Wong and Gumbiner, 2003). It is unclear whether β-catenin overexpression alone is sufficient for all EMTs. If β-catenin is misexpressed in cultured cells, it causes apoptosis (Kim et al., 2000); however, misexpressing a stabilized form of β-catenin in mouse epithelial cells *in vivo* causes metastatic skin tumors (Gat et al., 1998). Therefore, the central role of cadherins in an EMT may not be solely due to their cell–cell adhesive function, but also to cadherin regulation of the β-catenin signaling pathway. In support of this view, ectopic cadherin expression in *Xenopus* embryos sequesters β-catenin to adhesion junctions and consequently inhibits β-catenin migration to the nucleus (Fagotto et al., 1996). In E-cadherin misexpression studies in metastatic cancer cells, the suppression of cancer cell invasion does not require cell–cell adhesion, as only the cytoplasmic β-catenin-binding domain of E-cadherin and not the extracellular adhesion domain is required (Wong and Gumbiner, 2003). In summary, cell–cell adhesions depend on cadherins, and cadherins can regulate additional EMT events through β-catenin signaling.

Cell–ECM Adhesion Changes

Changes in the way that cells interact with the ECM are also important for EMTs and METs. During sea urchin gastrulation, PMCs lose cell–cell adhesions but simultaneously acquire adhesion to the basal lamina through which they invade (Fink and McClay, 1985). Cell–ECM adhesion is mediated principally by integrins (reviewed in Hynes, 2002). Integrins are transmembrane proteins composed of two non-covalently linked subunits, α and β, and require Ca^{2+} or Mg^{2+} for binding to ECM components such as fibronectin, laminin, and collagen. The cytoplasmic domain of integrins links to the cytoskeleton and interacts with other signaling molecules. Changes in integrin function are required for many EMTs. For example, in neural crest delamination, β1 integrin is necessary for neural crest adhesion to fibronectin and becomes functional just a few hours before the EMT (Delannet and Duband, 1992). Likewise, while epiblast cells undergo an EMT to form mesoderm during mouse gastrulation, the cells exhibit increased adhesion to ECM molecules (Burdsal et al., 1993). In both of these cases, blocking integrin function with function-blocking antibodies prevents cell migration. Integrin changes are also associated with increased metastasis in certain cancers (reviewed in Hood and Cheresh, 2002).

One molecule that coordinates the loss of cell–cell adhesion with the gain of cell–ECM adhesion during EMT is the GTPase Rap1. In several cultured cell lines, the endocytosis of E-cadherin activates the Ras family member Rap1. Activated Rap1 is required to form integrin-mediated adhesions, as overexpression of the

Rap1-inactivating enzyme, Rap1GAPv, blocks integrin-ECM adhesion formation (Balzac et al., 2005). The molecules with which Rap1 interacts to activate integrin function are not yet known. Hence, cell–ECM adhesions are maintained by integrins, and changes in cell–ECM interactions are also important for EMTs.

Stimulation of Cell Motility

In order for epithelial cells to undergo an EMT they must become migratory. The gain of cell motility is distinct from simply losing cell–cell adhesions. For example, in EpH4 cells that undergo an EMT after activation of the transcription factor Jun, there is a complete loss of epithelial polarity, but cell migration is not activated (Fialka et al., 1996). Similarly there are two steps during the EMT that generates the cardiac cushion cells: first, the cardiac endothelium is "activated," whereby the cells lose their adhesions to each other, become hypertrophic, and polarize the Golgi toward one end of the cell. Second, these activated cells become motile and invasive. Curiously it is estimated that only 7% of activated endothelial cells ever invade a collagen gel in *in vitro* invasion assays (Boyer et al., 1999). The activation and dispersion steps in the EMT are separable and are regulated by different signaling pathways (Markwald et al., 1977; Krug et al., 1985; Runyan et al., 1990). The cellular changes that are responsible for activating cell motility are not understood. However, in many EMTs, there is an upregulation of integrins (e.g. in the cardiac cushion precursors, integrin $\alpha6$ is upregulated; Boyer et al., 1999). Potentially the ability to adhere to the ECM is sufficient to stimulate motility. Additionally, activation of members of the Rho family of GTPases is required for organizing actin to generate filopodia, lamellipodia, and focal contacts (reviewed in Burridge and Wennerberg, 2004). In many EMTs the loss of Rho family members inhibits the EMT (e.g. RhoB (Liu and Jessell, 1998) and rac (our unpublished data) are required for the neural crest EMT). The extent to which activation of cell motility is needed for the EMT and how it is regulated will be the subject of future research.

Invasion of the Basal Lamina

In most EMTs epithelial cells penetrate the underlying basal lamina. The basal lamina stabilizes epithelial integrity and generally acts as a barrier to migratory cells (Erickson, 1987). One mechanism that cells use to breach the basal lamina is to produce enzymes that degrade it, including plasminogen activator and MMPs. Plasminogen activator is associated with a number of EMTs, including neural crest delamination and the formation of cardiac cushion cells during heart morphogenesis. Experimentally, blocking plasminogen activity will reduce the number of migratory neural crest cells (Erickson and Isseroff, 1989) or migratory cardiac cells (McGuire and Alexander, 1993). MMPs are also important to a number of EMTs. MMP-2 is necessary for the EMT that generates neural crest cells, because when inhibitors of MMP-2 are added to chicken embryos *in vivo*, or if MMP-2 translation is blocked with MMP-2 antisense oligonucleotides, neural crest delamination – but not neural crest migration – is inhibited (Duong and Erickson, 2004). In mouse mammary cells, MMP-3 is sufficient for an EMT *in vitro* and *in vivo* (Sternlicht et al., 1999). MMP-3 induces an alternatively spliced form of Rac1 (Rac1b), which then causes an increase in reactive oxygen species (ROS) intracellularly. Either Rac1b activity or ROS is necessary and sufficient for an MMP-3-induced EMT. Rac1b or ROS can also induce the expression of the transcription factor Snail (Radisky et al., 2005). The role of Rac1b or ROS in controlling other EMT events during development or disease is not yet known.

MOLECULAR CONTROL OF THE EMT

The initiation of an EMT or an MET is a tightly regulated event during development and tissue repair, because deregulation of epithelial organization is disastrous to the organism. A variety of external and internal signaling mechanisms coordinate the complex events of the EMT, and can be disrupted or reactivated during disease processes. Many of the molecules that trigger EMTs or METs have been identified, and in some cases the downstream effectors are known. Yet, in general, complete signaling cascades have not been elucidated. EMT can be induced by either ECM components or diffusible signaling molecules and these inductive signals act either directly on cell adhesion molecules or by regulating the EMT transcriptional cascades. Next we will discuss the role of signaling molecules and ECM in triggering the EMT, and then describe the transcriptional programs that are activated.

Signaling Molecules

During development, five main cellular signaling pathways are employed – the TGF-β, Wnt, receptor tyrosine kinase (RTK), Notch, and Hedgehog pathways (Gerhart, 1999). These pathways and the soluble ligands that activate them have a role in triggering EMTs. Although the activation of a single signaling pathway can be sufficient for an EMT, in most cases the EMT is coordinated by a combination of signaling molecules.

TGF-β Pathway

The TGF-β superfamily includes TGF-β, activin, and bone morphogenetic protein (BMP) families. These ligands signal through receptor serine/threonine kinases to activate a variety of signaling molecules including Smads, mitogen-activated protein kinase (MAPK), and PI3K (Derynck and Zhang, 2003). Most EMTs studied are induced in part, or solely, by TGF-β superfamily members (Zavadil and Bottinger, 2005). TGF-β2 and TGF-β3 have sequential and necessary roles in signaling the formation of heart valves from cardiac endothelium (Camenisch et al., 2002a), and TGF-β3 triggers an EMT in the fusing palate (Nawshad et al., 2004). In experimental models, TGF-β has context-dependent effects, acting as a growth suppressor on normal tissue, but as an EMT inducer in later stages of cancer progression. For example, transgenic mice expressing TGF-β1 in keratinocytes are more resistant to the development of chemically induced skin tumors than controls, suggesting a growth-inhibiting effect of TGF-β1. However, a greater portion of the tumors that do form in the keratinocyte-TGF-β1 transgenic mice are highly invasive spindle-cell carcinomas, indicating that TGF-β1 also promotes an EMT (Cui et al., 1996). Similar effects of TGF-β are observed in breast cancer progression, where the TGF-β pathway inhibits initial tumor growth, but promotes metastasis to the lung (Siegel et al., 2003). Expression of dominant-negative TGF-βR II in cancer cells transplanted into nude mice blocks TGF-β-induced metastasis (Portella et al., 1998). In cultured breast cancer cells, TGF-β in combination with activated Ras induces an irreversible EMT (Janda et al., 2002), and in cultured pig cells TGF-β and epidermal growth factor (EGF) synergistically stimulate the EMT (Grande et al., 2002).

Some of the downstream effectors of TGF-β signaling in EMTs have been determined. One mode of TGF-β action is to cause the dissociation of cell–cell adhesions. For example, in TGF-β-induced EMTs of mammary epithelial cells, TGF-βR II directly phosphorylates the polarity protein, Par6, and phosphorylated Par6 causes the E3 ubiquitin ligase, Smurf1, to target the GTPase, RhoA, for degradation. RhoA is required for the stability of tight junctions and loss of RhoA leads to their dissolution (Ozdamar et al., 2005). The loss of tight junctions causes changes in cell polarity. Exactly how the ubiquitination of RhoA leads to the loss of tight junctions is not yet known.

Besides the action of TGF-β signaling on cell–cell adhesion, the TGF-β pathway also regulates EMT genes. TGF-β signaling through serine/threonine kinases results in the phosphorylation and activation of several Smads that regulate gene expression (reviewed in Shi and Massague, 2003). Smad3 may be the molecule that signals the TGF-β-induced EMT. The deletion of Smad3 in a mouse model leads to the inhibition of injury-induced lens and kidney tissue EMT (Roberts et al., 2005). The precise role and mechanism of Smads in the EMT remain to be elucidated.

Wnt Pathway

The Wnt family of ligands also has a central role in many EMTs. Wnt ligands signal through seven-pass transmembrane proteins of the Frizzled family, and activate G-proteins, PI3K, and β-catenin (Huelsken and Behrens, 2002). Wnt6 is sufficient for the induction of Slug transcription in the neural crest and perturbation of the Wnt pathway reduces neural crest formation (Garcia-Castro et al., 2002). Wnts can also signal an MET; for example, Wnt4 is necessary to induce the coalescence of nephrogenic mesenchyme into epithelial tubules during murine kidney formation (Stark et al., 1994), and Wnt6 is necessary and sufficient for the MET that forms somites (Schmidt et al., 2004). As with the TGF-β superfamily, Wnt signals both adhesion molecules and transcription factors. One mode of Wnt11 activity, which regulates zebrafish gastrulation, is to stimulate the GTPase Rab5c, which results in the endocytosis of E-cadherin and consequently the loss of cell–cell adhesion (Ulrich et al., 2005). Wnt signaling also activates transcription of genes that coordinate the EMT, often through the stabilization of β-catenin and the subsequent nuclear β-catenin co-activation of LEF/TCF transcription factors.

Signaling by RTK Ligands

The RTK family of receptors and the growth factors that activate them also regulate EMTs or METs. RTKs are activated by their respective ligands, which causes receptor dimerization and results in the autophosphorylation of tyrosine residues intracellularly. These cytoplasmic phosphotyrosines act as docking sites for intracellular signaling molecules or adapter proteins, which in turn activate signaling components such as Ras/MAPK, Rac, PI3K, and JAK/STAT (reviewed in Schlessinger, 2000). Next we cite a few examples.

Hepatocyte growth factor (HGF), also known as scatter factor, acts through the RTK c-met. HGF is important for the MET in the developing kidney, since HGF/SF function-blocking antibodies inhibit the assembly of metanephric mesenchymal cells into kidney epithelium in organ culture (Woolf et al., 1995). HGF signaling is required for the EMT that produces myoblasts (limb muscle precursors) from somite tissue in the mouse, because in knockout mice for c-met, myoblasts fail to migrate into the limb bud (Bladt et al., 1995).

Fibroblast growth factor (FGF) signaling regulates the EMT during mouse gastrulation. In FGFR1 mouse mutants, E-cadherin is not downregulated, β-catenin does not move into the nucleus, *snail* is not expressed in gastrulating cells, and gastrulation does not occur. Interestingly, if E-cadherin function is also inhibited in FGFR1 mutants by the addition of function-blocking E-cadherin antibodies, the EMT proceeds normally. The suggested mechanism is that failure to remove E-cadherin allows E-cadherin to sequester free β-catenin and therefore attenuate later Wnt signaling required to complete gastrulation events (Ciruna and Rossant, 2001). FGF signaling also stimulates cell motility and MMP activation. In studies using cultured cancer cells, sustained FGF2 signaling results in cell motility, MMP-9 activation, and the ability to invade ECM (Suyama et al., 2002).

Insulin growth factor (IGF) signaling can also induce an EMT. In cultured epithelial cells, IGFR1 complexes with E-cadherin and β-catenin, and the ligand IGF-II causes nuclear translocation of β-catenin, activation of the transcription factor TCF-3, degradation of E-cadherin, and subsequent EMT (Morali et al., 2001).

Another RTK receptor known for its role in EMTs is the ErbB2/HER-2/Neu receptor, whose ligand is heregulin/neuregulin. Overexpression of HER-2 occurs in 25% of human breast cancers, and misexpression of HER-2 in mouse mammary tissue *in vivo* is sufficient to cause metastatic breast cancer (Muller et al., 1988). Herceptin® (antibody against the anti-HER-2 receptor) treatment is effective in reducing the recurrence of HER-2-positive metastatic breast cancers (Goldenberg, 1999). HER-2 signaling activates *snail* expression in breast cancer (Moody et al., 2005).

Another example of the importance of RTKs in the EMT is the mechanism used by the bacterium *Helicobacter pylori* to promote the breakdown of gastric epithelium that causes peptic ulcers and gastric adenocarcinoma.

This disease process requires that the bacterial protein CagA be transferred into gastric epithelial cells. Once in epithelial cells, CagA is phosphorylated at tyrosine residues located at its C terminus, and results in the activation of RTK signaling pathways. When CagA is expressed in MDCK cells, the cells lose cell–cell adhesions and epithelial polarity, exhibit cell migration, and gain the ability to invade ECM (Bagnoli et al., 2005). Whether or not this CagA-mediated EMT also occurs *in vivo* is not yet known.

Notch Pathway

The Notch signaling family is well known for its role in cell specification, and is now emerging as an important regulator of EMTs (Huber et al., 2005). When the Notch receptor is activated by its ligand Delta, the intracellular portion of the Notch receptor ligand is cleaved and transported to the nucleus where it regulates gene expression (Mumm and Kopan, 2000). In zebrafish Notch1 mutants, cardiac endothelium expresses very little *snail* and does not undergo the EMT required to make the cardiac cushions (Timmerman et al., 2004). In this study, similar results are obtained by treating embryonic heart explants with inhibitors of the Notch pathway. Conversely, misexpression of activated Notch1 is sufficient to activate *snail* expression and promote an EMT in cultured endothelial cells. Notch signaling is also important for TGF-β-induced EMT. Upon TGF-β treatment of cultured kidney, mammary, and epidermal epithelial cells, the transcription of the transcriptional repressor Hey1 and the Notch-ligand Jagged1 is stimulated in a Smad3-dependent process. The use of antisense oligonucleotides against *hey1* mRNA, siRNA against *jagged1* mRNA, or γ-secretase inhibitor (GSI) treatment (to block Notch receptor activation) all inhibit TGF-β-induced EMT in these cell lines (Zavadil et al., 2004).

ECM Signaling

In addition to diffusible signaling molecules, the extracellular environment also regulates EMTs or METs. When lens or thyroid epithelium is embedded in collagen, these tissues undergo an EMT (reviewed in Hay, 2005). Integrin signaling appears to be important in this transition, because if function-blocking antibodies against integrins are added in the collagen gels, the EMT is inhibited (Zuk and Hay, 1994). Hyaluronan is another ECM component that regulates EMTs. In the hyaluronan synthase-2 knockout mouse (Has2 –/–, results in defects in hyaluronan synthesis and secretion), the cardiac endothelium fails to undergo an EMT and produces the migratory mesenchymal cells critical for heart valve formation (Camenisch et al., 2000). The role of hyaluronan in this EMT may be to activate the RTK ErbB2/HER-2/Neu, because treating cultured Has2 –/– heart explants with heregulin (ligand for ErbB2) rescues the EMT. Consistent with this hypothesis, treating cardiac explants with hyaluronan activates ErbB2, and blocking ErbB2 signaling with the drug herstatin reproduces the Has2 knockout phenotype (Camenisch et al., 2002b). A third ECM component that is important for EMTs is the gamma-2 chain of laminin 5, which is cleaved from laminin 5 by MMP-2. The gamma-2 chain causes the scattering and migration of epithelial cancer cells (Koshikawa et al., 2000), and may be a marker of epithelial tumor cell invasion (Katayama et al., 2003).

Integrins are the major mediators of cell interactions with the ECM, but integrins are also involved in cell signaling. Integrins play important roles in regulating cell survival, proliferation, cytoskeletal rearrangements, cell polarity, and cell motility (reviewed in Hood and Cheresh, 2002). One of the intracellular mediators of integrin signaling is integrin-linked kinase (ILK). ILK interacts with the cytoplasmic domains of the β1 and β3 integrin subunits, and ILK can be activated by integrin, TGF-β, Wnt, or RTK signaling (for a review, see Oloumi et al., 2004). Overexpression of ILK in cultured breast or colon cancer cells leads to translocation of β-catenin to the nucleus, activation of Lef-1/β-catenin as transcription factors, and downregulation of E-cadherin (Novak et al., 1998). Inhibition of ILK in cultured colon cancer cells leads to the stabilization of GSK-3β activity, decreased nuclear β-catenin localization, and results in the suppression of *lef-1* and *snail* transcription (Tan et al., 2001).

The EMT Transcriptional Program

All of the molecules that regulate cell–cell adhesion, cell–ECM interactions, cell motility, and basal lamina invasion are encoded by DNA. Therefore, at the heart of an EMT are the transcription factors that control the expression of genes that are required for an EMT. Although many of the transcription factors that regulate an EMT have been identified, these complex transcriptional networks are still being defined. Here we review the transcription factors that control EMTs, and then review how the transcriptional activity and protein function of these transcription factors are regulated.

Transcription Factors that Regulate EMTs

The Snail family of zinc-finger transcription factors, including Snail and Slug, is emerging as the central regulator of adhesion and cell movement during EMTs (for recent reviews, see Barrallo-Gimeno and Nieto, 2005; De Craene et al., 2005). Snail and Slug are transcriptional repressors that are evolutionarily conserved in vertebrates and invertebrates, and are expressed singly or in combination during every EMT yet examined. Snail was first described in *Drosophila*, and *snail* mutants fail to express mesodermal markers or undergo the epithelial invagination that produces mesoderm (Alberga et al., 1991). In the Snail knockout mouse, migratory cells with mesodermal markers form a type of mesoderm; however, these presumptive mesenchymal cells still retain apical/basal polarity, adherens junctions, and express E-cadherin mRNA (Carver et al., 2001). Hence, Snail is only necessary for a part of the process that generates mesoderm. One of the known roles of Snail and Slug in an EMT is to repress the transcription of E-cadherin and thus promote the loss of cell–cell adhesion (reviewed in De Craene et al., 2005). Snail represses the *e-cadherin* promoter by recruiting the mSin3A co-repressor complex and histone deacetylases (Peinado et al., 2004a). Snail is also a transcriptional repressor of the tight junction proteins, Claudin and Occludin (Ikenouchi et al., 2003). The misexpression of Snail and Slug also leads to the transcription of genes important for cell motility. In MDCK, the misexpression of Snail indirectly leads to the expression of fibronectin and vimentin, which are important for mesenchymal cell motility (Cano et al., 2000), and Slug induces *RhoB* expression, a GTPase involved in motility, in avian neural crest cells (Del Barrio and Nieto, 2002). In MDCK cells, the misexpression of Snail also promotes *mmp-9* transcription and basal lamina invasion through a yet unknown pathway (Jorda et al., 2005). Although Snail and Slug are transcriptional repressors, they somehow activate other EMT genes, and the process has not yet been elucidated.

Two other zinc-finger transcription factors regulate EMTs. Delta-crystallin enhancer-binding factor 1 (δEF1), also known as ZEB1, is necessary and sufficient for an EMT in mammary cells transformed by the transcription factor c-Fos (Eger et al., 2005). Smad-interacting protein-1 (Sip1), also known as ZEB2, is structurally similar to δEF1, and Sip1 overexpression is sufficient to downregulate E-cadherin, dissociate adherens junctions, and increase motility in MDCK cells (Comijn et al., 2001). Both δEF1 and Sip1 can bind to the E-cadherin promoter and repress transcription (reviewed in De Craene et al., 2005).

The basic helix-loop-helix (bHLH) transcription factors Twist and E2A also play roles in EMTs. *Twist* is expressed during *Drosophila* gastrulation, and the double *twist* and *snail* mutant has a more severe gastrulation phenotype than either mutant alone, suggesting that *snail* and *twist* have distinct functions. Twist1 is not necessary for mouse gastrulation, yet Twist1 mouse mutants do have neural tube fusion, limb, and somite defects (Chen and Behringer, 1995). Twist is also necessary for the EMT that generates the mouse neural crest (Soo et al., 2002). E2A is not necessary for many EMTs, since mouse mutants for E2A survive and are only defective in B cell production (Zhuang et al., 1994). However, overexpression of E2A in MDCK cells promotes tumor invasion (Perez-Moreno et al., 2001). In MDCK cells, Snail is more efficient at promoting the initial invasion of ECM, whereas E2A is better at inducing later angiogenesis (Peinado et al., 2004b). Twist and E2A can also both repress E-cadherin transcription (De Craene et al., 2005).

Another important family of EMT transcription factors is the LEF/TCF transcription factor family. The limiting factor for LEF/TCF activation is the availability of β-catenin. β-Catenin levels are negatively regulated by GSK-3β or antigen-presenting cell (APC), and a surplus of β-catenin becomes available after being freed from disassembled adherens junctions (Stockinger et al., 2001). Forced expression of Lef-1 in the presence of stabilized β-catenin causes the downregulation of E-cadherin and promotes EMT in cultured colon cancer cells. Inhibition of Lef-1 misexpression (by removing Lef-1 retrovirus from the culture medium) causes cultured cells to revert back to an epithelium (Kim et al., 2002). LEF/TCF transcription factors directly activate genes that regulate cell motility. The LEF/TCF pathway activates the promoter of the L1 adhesion molecule, and L1 is associated with increased motility and invasive behavior of colon cancer cells (Gavert et al., 2005). β-Catenin and LEF/TCF also activate the fibronectin gene (Gradl et al., 1999). Finally, LEF/TCF transcription factors activate genes that stimulate basal lamina invasion, including *mmp-3* and *mmp-7* (Crawford et al., 1999; Gustavson et al., 2004).

The Regulation of Transcription Factors that Control EMT

To fully understand the transcriptional network that regulates EMTs, we should also know how EMT-inducing transcription factors are regulated. Transcription factor activity can be controlled both at the level of transcription as well as at the protein level by nuclear import/export or protein degradation.

The activation of *snail* transcription in *Drosophila* requires the transcription factors Dorsal (NF-κB) and Twist, and the Snail promoter includes both Dorsal and Twist binding sites (Ip et al., 1992). The human Snail promoter also has functional NF-κB sites (Barbera et al., 2004). In cultured human cells transformed by Ras and induced by TGF-β, NF-κB is essential for EMT initiation and maintenance (Huber et al., 2004). A Snail transcriptional repressor has also been identified. In breast cancer cell lines, metastasis-associated protein 3 (MTA3) binds directly to and represses the transcription of Snail in combination with the Mi-2/NuRD complex (Fujita et al., 2003). MTA3 is induced by the estrogen receptor (ER, nuclear hormone) pathway, and the absence of ER signaling or MTA3 leads to the activation of Snail. This suggests a mechanism whereby loss of the ER in breast cancer contributes to metastasis. The role of MTA3 in other EMTs is not known.

Slug transcriptional regulators have also been identified. In *Xenopus*, the Slug promoter has functional LEF/TCF binding sites (Vallin et al., 2001), and in the mouse, MyoD (transcription factor central to muscle cell development) binds to the Slug promoter and activates Slug transcription (Zhao et al., 2002). In humans, the oncogene E2A-HLF (Inukai et al., 1999), and the pigment cell regulator, microthalamia-associated transcription factor (MITF) (Sanchez-Martin et al., 2002), also bind to the Slug promoter and activate transcription.

Lef-1 transcription is directly activated by Smad 2/4 (TGF-β signaling), and the phosphorylated complex of Smad 2/4 in the nucleus can promote Lef-1 transcription in the absence of nuclear β-catenin during fusion of the mouse palate (Nawshad and Hay, 2003). The misexpression of Snail also activates the transcription of δEF-1 and Lef-1 through a yet unknown mechanism (Guaita et al., 2002). The complete transcriptional networks that orchestrate an EMT remain to be elucidated.

In addition to controlling gene expression, another way to regulate the activity of transcription factors is at the protein level, including protein stability (targeting to the proteasome) and nuclear localization. GSK-3β, the same protein kinase that phosphorylates β-catenin and targets it for destruction, also phosphorylates Snail. The human Snail protein contains two GSK-3β phosphorylation consensus sites between amino acids 97 and 123. Blocking GSK-3β stabilizes Snail expression and results in the loss of E-cadherin in cultured epithelial cells (Zhou et al., 2004; Yook et al., 2005). Hence, Wnt signaling stabilizes (and therefore activates) both β-catenin and Snail by inhibiting GSK-3β. Lysyl-oxidase-like proteins, LOXL2 and LOXL3, are two molecules that prevent GSK-3β-mediated phosphorylation of Snail, and thus stabilize Snail activity. LOXL2 and LOXL3 form a complex with Snail near the GSK-3β phosphorylation sites, thus preventing GSK-3β from phosphorylating Snail. Expression of LOXL2 or LOXL3 prevents Snail protein destruction and induces an EMT in culture (Peinado et al., 2005).

In addition to targeting Snail to the proteasome, the activity of Snail as a transcriptional repressor also depends on nuclear localization. Snail contains a nuclear export sequence (NES) at amino acids 132–143 that is sufficient and necessary for the export of Snail from the nucleus to the cytoplasm, and depends on the calreticulin nuclear export pathway (Dominguez et al., 2003). This NES sequence is activated by phosphorylation of the same lysine residues that GSK-3β acts upon, suggesting a mechanism whereby phosphorylation of Snail by GSK-3β leads to the export of Snail from the nucleus, although this has not yet been shown directly. While GSK-3β can cause the export of Snail from the nucleus, the phosphorylation of human Snail by p21-activated kinase 1 (Pak1) at Ser^{246} promotes the nuclear localization of Snail (and therefore Snail activation) in breast cancer cells. Knocking down Pak1 by siRNA blocks Pak1-mediated Snail phosphorylation, increases the cytoplasmic accumulation of Snail, and reduces the invasive behavior of breast cancer cells (Yang et al., 2005). The protein that imports Snail into the nucleus in human cells is not yet known, although a Snail importer has already been described in zebrafish. The zinc-finger transporting protein LIV1 is required for Snail to localize to the nucleus during zebrafish gastrulation, and LIV1 is activated by STAT3 signaling (Yamashita et al., 2004). In zebrafish, the protein kinase that phosphorylates Snail to activate the translocation of Snail to the nucleus has not yet been identified. Therefore, both the stability and the subcellular localization of snail are important for snail function in the EMT.

CONCLUSION

Over the past 20+ years since the term "EMT" was coined (Greenburg and Hay, 1982), great strides have been made in this rapidly expanding field of research. EMT and MET events occur during development and disease, and many of the molecules that regulate the EMT or MET have been characterized, thanks in large part to the advent of cell culture models. Despite this progress, our picture of the EMT is still not complete and there are major gaps in our knowledge of the EMT regulatory networks. Mounting evidence suggests that disease processes such as the metastasis of epithelial-derived cancers and kidney fibrosis are regulated by the same molecules that create migratory and invasive cells from an epithelium during development. A clearer understanding of EMT and MET pathways in the future will no doubt lead to more effective strategies for tissue engineering and novel therapeutic targets.

REFERENCES

Alberga, A., Boulay, J.L., Kempe, E., Dennefeld, C. and Haenlin, M. (1991). The snail gene required for mesoderm formation in Drosophila is expressed dynamically in derivatives of all three germ layers. *Development* **111**: 983–992.

Bagnoli, F., Buti, L., Tompkins, L., Covacci, A. and Amieva, M.R. (2005). *Helicobacter pylori* CagA induces a transition from polarized to invasive phenotypes in MDCK cells. *PNAS* **102**: 16339–16344.

Balzac, F., Avolio, M., Degani, S., Kaverina, I., Torti, M., Silengo, L., Small, J.V. and Retta, S.F. (2005). E-cadherin endocytosis regulates the activity of Rap1: a traffic light GTPase at the crossroads between cadherin and integrin function. *J. Cell Sci.* **118**: 4765–4783.

Barasch, J. (2001). Genes and proteins involved in mesenchymal to epithelial transition. *Curr. Opin. Nephrol. Hypertens.* **10**: 429–436.

Barbera, M.J., Puig, I., Dominguez, D., Julien-Grille, S., Guaita-Esteruelas, S., Peiro, S., Baulida, J., Franci, C., Dedhar, S., Larue, L., et al. (2004). Regulation of Snail transcription during epithelial to mesenchymal transition of tumor cells. *Oncogene* **23**: 7345–7354.

Barrallo-Gimeno, A. and Nieto, M.A. (2005). The Snail genes as inducers of cell movement and survival: implications in development and cancer. *Development* **132**: 3151–3161.

Bladt, F., Riethmacher, D., Isenmann, S., Aguzzi, A. and Birchmeier, C. (1995). Essential role for the c-met receptor in the migration of myogenic precursor cells into the limb bud. *Nature* **376**: 768–771.

Boyer, A.S., Ayerinskas, I.I., Vincent, E.B., McKinney, L.A., Weeks, D.L. and Runyan, R.B. (1999). TGFβ2 and TGFβ3 have separate and sequential activities during epithelial–mesenchymal cell transformation in the embryonic heart. *Dev. Biol.* **208**: 530–545.

Bronner-Fraser, M., Wolf, J.J. and Murray, B.A. (1992). Effects of antibodies against N-cadherin and N-CAM on the cranial neural crest and neural tube. *Dev. Biol.* **153**: 291–301.

Bryant, D.M. and Stow, J.L. (2004). The ins and outs of E-cadherin trafficking. *Trends Cell Biol.* **14**: 427–434.

Burdsal, C., Damsky, C. and Pedersen, R. (1993). The role of E-cadherin and integrins in mesoderm differentiation and migration at the mammalian primitive streak. *Development* **118**: 829–844.

Burridge, K. and Wennerberg, K. (2004). Rho and Rac take center stage. *Cell* **116**: 167–179.

Camenisch, T.D., Molin, D.G.M., Person, A., Runyan, R.B., Gittenberger-De Groot, A.C., McDonald, J.A. and Klewer, S.E. (2002a). Temporal and distinct TGFβ ligand requirements during mouse and avian endocardial cushion morphogenesis. *Dev. Biol.* **248**: 170–181.

Camenisch, T.D., Schroeder, J.A., Bradley, J., Klewer, S.E. and McDonald, J.A. (2002b). Heart-valve mesenchyme formation is dependent on hyaluronan-augmented activation of ErbB2–ErbB3 receptors. *Nat. Med.* **8**: 850–855.

Camenisch, T.D., Spicer, A.P., Brehm-Gibson, T., Biesterfeldt, J., Augustine, M.L., Calabro Jr., A., Kubalak, S., Klewer, S.E. and McDonald, J.A. (2000). Disruption of hyaluronan synthase-2 abrogates normal cardiac morphogenesis and hyaluronan-mediated transformation of epithelium to mesenchyme. *J. Clin. Invest.* **106**: 349–360.

Cano, A., Perez-Moreno, M.A., Rodrigo, I., Locascio, A., Blanco, M.J., Del Barrio, M.G., Portillo, F. and Nieto, M.A. (2000). The transcription factor Snail controls epithelial–mesenchymal transitions by repressing E-cadherin expression. *Nat. Cell Biol.* **2**: 76–83.

Carver, E.A., Jiang, R., Lan, Y., Oram, K.F. and Gridley, T. (2001). The mouse snail gene encodes a key regulator of the epithelial–mesenchymal transition. *Mol. Cell Biol.* **21**: 8184–8188.

Chen, Z.F. and Behringer, R.R. (1995). *Twist* is required in head mesenchyme for cranial neural tube morphogenesis. *Gene Dev.* **9**: 686–699.

Ciruna, B. and Rossant, J. (2001). FGF signaling regulates mesoderm cell fate specification and morphogenetic movement at the primitive streak. *Dev. Cell* **1**: 37–49.

Comijn, J., Berx, G., Vermassen, P., Verschueren, K., Van Grunsven, L., Bruyneel, E., Mareel, M., Huylebroeck, D. and Van Roy, F. (2001). The two-handed E box binding zinc finger protein SIP1 downregulates E-cadherin and induces invasion. *Mol. Cell* **7**: 1267–1278.

Cowin, P., Rowlands, T.M. and Hatsell, S.J. (2005). Cadherins and catenins in breast cancer. *Curr. Opin. Cell Biol.* **17**: 499–508.

Crawford, H.C., Fingleton, B.M., Rudolph-Owen, L.A., Goss, K.J., Rubinfeld, B., Polakis, P. and Matrisian, L.M. (1999). The metalloproteinase matrilysin is a target of beta-catenin transactivation in intestinal tumors. *Oncogene* **18**: 2883–2891.

Cui, W., Fowlis, D.J., Bryson, S., Duffie, E., Ireland, H., Balmain, A. and Akhurst, R.J. (1996). TGFβ1 inhibits the formation of benign skin tumors, but enhances progression to invasive spindle carcinomas in transgenic mice. *Cell* **86**: 531–542.

De Craene, B., Van Roy, F. and Berx, G. (2005). Unraveling signalling cascades for the snail family of transcription factors. *Cell. Signal.* **17**: 535–547.

Del Barrio, M.G. and Nieto, M.A. (2002). Overexpression of snail family members highlights their ability to promote chick neural crest formation. *Development* **129**: 1583–1593.

Delannet, M. and Duband, J.L. (1992). Transforming growth factor-beta control of cell–substratum adhesion during avian neural crest cell emigration *in vitro*. *Development* **116**: 275–287.

Derynck, R. and Zhang, Y.E. (2003). Smad-dependent and Smad-independent pathways in TGF-β family signalling. *Nature* **425**: 577–584.

Dominguez, D., Montserrat-Sentis, B., Virgos-Soler, A., Guaita, S., Grueso, J., Porta, M., Puig, I., Baulida, J., Franci, C. and Garcia De Herreros, A. (2003). Phosphorylation regulates the subcellular location and activity of the snail transcriptional repressor. *Mol. Cell Biol.* **23**: 5078–5089.

Duong, T.D. and Erickson, C.A. (2004). MMP-2 plays an essential role in producing epithelial–mesenchymal transformations in the avian embryo. *Dev. Dyn.* **229**: 42–53.

Ebnet, K., Suzuki, A., Ohno, S. and Vestweber, D. (2004). Junctional adhesion molecules (JAMs): more molecules with dual functions? *J. Cell Sci.* **117**: 19–29.

Egeblad, M. and Werb, Z. (2002). New functions for the matrix metalloproteinases in cancer progression. *Nat. Rev. Cancer* **2**: 161–174.

Eger, A., Aigner, K., Sonderegger, S., Dampier, B., Oehler, S., Schreiber, M., Berx, G., Cano, A., Beug, H. and Foisner, R. (2005). DeltaEF1 is a transcriptional repressor of E-cadherin and regulates epithelial plasticity in breast cancer cells. *Oncogene* **24**: 2375–2385.

Eisenberg, L.M. and Markwald, R.R. (1995). Molecular regulation of atrioventricular valvuloseptal morphogenesis. *Circ. Res.* **77**: 1–6.

Erickson, C.A. (1987). Behavior of neural crest cells on embryonic basal laminae. *Dev. Biol.* **120**: 38–49.

Erickson, C.A. and Isseroff, R.R. (1989). Plasminogen activator activity is associated with neural crest cell motility in tissue culture. *J. Exp. Zool.* **251**: 123–133.

Fagotto, F., Funayama, N., Gluck, U. and Gumbiner, B. (1996). Binding to cadherins antagonizes the signaling activity of beta-catenin during axis formation in *Xenopus*. *J. Cell Biol.* **132**: 1105–1114.

Fialka, I., Schwarz, H., Reichmann, E., Oft, M., Busslinger, M. and Beug, H. (1996). The estrogen-dependent c-JunER protein causes a reversible loss of mammary epithelial cell polarity involving a destabilization of adherens junctions. *J. Cell Biol.* **132**: 1115–1132.

Fink, R.D. and McClay, D.R. (1985). Three cell recognition changes accompany the ingression of sea urchin primary mesenchyme cells. *Dev. Biol.* **107**: 66–74.

Frixen, U., Behrens, J., Sachs, M., Eberle, G., Voss, B., Warda, A., Lochner, D. and Birchmeier, W. (1991). E-cadherin-mediated cell–cell adhesion prevents invasiveness of human carcinoma cells. *J. Cell Biol.* **113**: 173–185.

Fujita, N., Jaye, D.L., Kajita, M., Geigerman, C., Moreno, C.S. and Wade, P.A. (2003). MTA3, a Mi-2/NuRD complex subunit, regulates an invasive growth pathway in breast cancer. *Cell* **113**: 207–219.

Fujita, Y., Krause, G., Scheffner, M., Zechner, D., Leddy, H.E.M., Behrens, J., Sommer, T. and Birchmeier, W. (2002). Hakai, a c-Cbl-like protein, ubiquitinates and induces endocytosis of the E-cadherin complex. *Nat. Cell Biol.* **4**: 222–231.

Garcia-Castro, M.I., Marcelle, C. and Bronner-Fraser, M. (2002). Ectodermal Wnt function as a neural crest inducer. *Science* **297**: 848–851.

Gat, U., Dasgupta, R., Degenstein, L. and Fuchs, E. (1998). De Novo hair follicle morphogenesis and hair tumors in mice expressing a truncated β-catenin in skin. *Cell* **95**: 605–614.

Gavert, N., Conacci-Sorrell, M., Gast, D., Schneider, A., Altevogt, P., Brabletz, T. and Ben-Ze'ev, A. (2005). L1, a novel target of β-catenin signaling, transforms cells and is expressed at the invasive front of colon cancers. *J. Cell Biol.* **168**: 633–642.

Gerhart, J. (1999). 1998 warkany lecture: signaling pathways in development. *Teratology* **60**: 226–239.

Getsios, S., Huen, A.C. and Green, K.J. (2004). Working out the strength and flexibility of desmosomes. *Nat. Rev. Mol. Cell Biol.* **5**: 271–281.

Goldenberg, M.M. (1999). Trastuzumab, a recombinant DNA-derived humanized monoclonal antibody, a novel agent for the treatment of metastatic breast cancer. *Clin. Ther.* **21**: 309–318.

Gradl, D., Kuhl, M. and Wedlich, D. (1999). The Wnt/Wg signal transducer beta-catenin controls fibronectin expression. *Mol. Cell Biol.* **19**: 5576–5587.

Grande, M., Franzen, A., Karlsson, J.O., Ericson, L.E., Heldin, N.-E. and Nilsson, M. (2002). Transforming growth factor-β and epidermal growth factor synergistically stimulate epithelial to mesenchymal transition (EMT) through a MEK-dependent mechanism in primary cultured pig thyrocytes. *J. Cell Sci.* **115**: 4227–4236.

Greenburg, G. and Hay, E. (1982). Epithelia suspended in collagen gels can lose polarity and express characteristics of migrating mesenchymal cells. *J. Cell Biol.* **95**: 333–339.

Guaita, S., Puig, I., Franci, C., Garrido, M., Dominguez, D., Batlle, E., Sancho, E., Dedhar, S., De Herreros, A.G. and Baulida, J. (2002). Snail induction of epithelial to mesenchymal transition in tumor cells is accompanied by MUC1 repression and ZEB1 expression. *J. Biol. Chem.* **277**: 39209–39216.

Gumbiner, B.M. (2005). Regulation of cadherin-mediated adhesion in morphogenesis. *Nat. Rev. Mol. Cell Biol.* **6**: 622–634.

Gustavson, M.D., Crawford, H.C., Fingleton, B. and Matrisian, L.M. (2004). Tcf binding sequence and position determines beta-catenin and Lef-1 responsiveness of MMP-7 promoters. *Mol. Carcinog.* **41**: 125–139.

Hay, E.D. (2005). The mesenchymal cell, its role in the embryo, and the remarkable signaling mechanisms that create it. *Dev. Dyn.* **233**: 706–720.

Hood, J.D. and Cheresh, D.A. (2002). Role of integrins in cell invasion and migration. *Nat. Rev. Cancer* **2**: 91–100.

Huber, M.A., Azoitei, N., Baumann, B., Grunert, S., Sommer, A., Pehamberger, H., Kraut, N., Beug, H. and Wirth, T. (2004). NF-KB is essential for epithelial–mesenchymal transition and metastasis in a model of breast cancer progression. *J. Clin. Invest.* **114**: 569–581.

Huber, M.A., Kraut, N. and Beug, H. (2005). Molecular requirements for epithelial–mesenchymal transition during tumor progression. *Curr. Opin. Cell Biol.* **17**: 548–558.

Huelsken, J. and Behrens, J. (2002). The Wnt signalling pathway. *J. Cell Sci.* **115**: 3977–3978.

Hynes, R.O. (2002). Integrins: bidirectional, allosteric signaling machines. *Cell* **110**: 673–687.

Ikenouchi, J., Matsuda, M., Furuse, M. and Tsukita, S. (2003). Regulation of tight junctions during the epithelium–mesenchyme transition: direct repression of the gene expression of claudins/occludin by snail. *J. Cell Sci.* **116**: 1959–1967.

Imhof, B.A., Vollmers, H.P., Goodman, S.L. and Birchmeier, W. (1983). Cell–cell interaction and polarity of epithelial cells: specific perturbation using a monoclonal antibody. *Cell* **35**: 667–675.

Inukai, T., Inoue, A., Kurosawa, H., Goi, K., Shinjyo, T., Ozawa, K., Mao, M., Inaba, T. and Look, A.T. (1999). SLUG, a ces-1-related zinc finger transcription factor gene with antiapoptotic activity, is a downstream target of the E2A-HLF oncoprotein. *Mol. Cell* **4**: 343–352.

Ip, Y.T., Park, R.E., Kosman, D., Yazdanbakhsh, K. and Levine, M. (1992). Dorsal–twist interactions establish snail expression in the presumptive mesoderm of the Drosophila embryo. *Gene Dev.* **6**: 1518–1530.

Ivanov, A.I., Nusrat, A. and Parkos, C.A. (2004). Endocytosis of epithelial apical junctional proteins by a clathrin-mediated pathway into a unique storage compartment. *Mol. Biol. Cell* **15**: 176–188.

Iwano, M., Plieth, D., Danoff, T.M., Xue, C., Okada, H. and Neilson, E.G. (2002). Evidence that fibroblasts derive from epithelium during tissue fibrosis. *J. Clin. Invest.* **110**: 341–350.

Janda, E., Lehmann, K., Killisch, I., Jechlinger, M., Herzig, M., Downward, J., Beug, H. and Grunert, S. (2002). Ras and TGFβ cooperatively regulate epithelial cell plasticity and metastasis: dissection of Ras signaling pathways. *J. Cell Biol.* **156**: 299–314.

Jorda, M., Olmeda, D., Vinyals, A., Valero, E., Cubillo, E., Llorens, A., Cano, A. and Fabra, A. (2005). Upregulation of MMP-9 in MDCK epithelial cell line in response to expression of the Snail transcription factor. *J. Cell Sci.* **118**: 3371–3385.

Kalluri, R. and Neilson, E.G. (2003). Epithelial–mesenchymal transition and its implications for fibrosis. *J. Clin. Invest.* **112**: 1776–1784.

Katayama, M., Sanzen, N., Funakoshi, A. and Sekiguchi, K. (2003). Laminin gamma 2-chain fragment in the circulation: a prognostic indicator of epithelial tumor invasion. *Cancer Res.* **63**: 222–229.

Katow, H. and Solursh, M. (1980). Ultrastructure of primary mesenchyme cell ingression in the sea urchin *Lytechinus pictus*. *J. Exp. Zool.* **213**: 231–246.

Kim, K., Lu, Z. and Hay, E.D. (2002). Direct evidence for a role of β-catenin/LEF-1 signalling pathway in induction of EMT. *Cell Biol. Int.* **26**: 463–476.

Kim, K., Pang, K.M., Evans, M. and Hay, E.D. (2000). Overexpression of β-catenin induces apoptosis independent of its transactivation function with LEF-1 or the involvement of major G1 cell cycle regulators. *Mol. Biol. Cell* **11**: 3509–3523.

Koshikawa, N., Giannelli, G., Cirulli, V., Miyazaki, K. and Quaranta, V. (2000). Role of cell surface metalloprotease MT1-MMP in epithelial cell migration over laminin-5. *J. Cell Biol.* **148**: 615–624.

Krug, E.L., Runyan, R.B. and Markwald, R.R. (1985). Protein extracts from early embryonic hearts initiate cardiac endothelial cytodifferentiation. *Dev. Biol.* **112**: 414–426.

Kuure, S., Vuolteenaho, R. and Vainio, S. (2000). Kidney morphogenesis: cellular and molecular regulation. *Mech. Dev.* **92**: 31–45.

Leptin, M. (2005). Gastrulation movements: the logic and the nuts and bolts. *Dev. Cell* **8**: 305–320.

Levine, E., Lee, C.H., Kintner, C. and Gumbiner, B.M. (1994). Selective disruption of E-cadherin function in early *Xenopus* embryos by a dominant negative mutant. *Development* **120**: 901–909.

Liebner, S., Cattelino, A., Gallini, R., Rudini, N., Iurlaro, M., Piccolo, S. and Dejana, E. (2004). β-Catenin is required for endothelial–mesenchymal transformation during heart cushion development in the mouse. *J. Cell Biol.* **166**: 359–367.

Liu, J. and Jessell, T. (1998). A role for rhoB in the delamination of neural crest cells from the dorsal neural tube. *Development* **125**: 5055–5067.

Lu, Z., Ghosh, S., Wang, Z. and Hunter, T. (2003). Downregulation of caveolin-1 function by EGF leads to the loss of E-cadherin, increased transcriptional activity of β-catenin, and enhanced tumor cell invasion. *Cancer Cell* **4**: 499–515.

Maeda, M., Johnson, K.R. and Wheelock, M.J. (2005). Cadherin switching: essential for behavioral but not morphological changes during an epithelium-to-mesenchyme transition. *J. Cell Sci.* **118**: 873–887.

Markwald, R.R., Fitzharris, T.P. and Manasek, F.J. (1977). Structural development of endocardial cushions. *Am. J. Anat.* **148**: 85–119.

Matter, K. and Balda, M.S. (2003). Signalling to and from tight junctions. *Nat. Rev. Mol. Cell Biol.* **4**: 225–237.

McGuire, P.G. and Alexander, S.M. (1993). Inhibition of urokinase synthesis and cell surface binding alters the motile behavior of embryonic endocardial-derived mesenchymal cells *in vitro*. *Development* **118**: 931–939.

Miller, J.R. and McClay, D.R. (1997). Characterization of the role of cadherin in regulating cell adhesion during sea urchin development. *Dev. Biol.* **192**: 323–339.

Moody, S.E., Perez, D., Pan, T.-C., Sarkisian, C.J., Portocarrero, C.P., Sterner, C.J., Notorfrancesco, K.L., Cardiff, R.D. and Chodosh, L.A. (2005). The transcriptional repressor Snail promotes mammary tumor recurrence. *Cancer Cell* **8**: 197–209.

Moon, R.T. and Kimelman, D. (1998). From cortical rotation to organizer gene expression: toward a molecular explanation of axis specification in *Xenopus*. *BioEssays* **20**: 536–546.

Morali, O.G., Delmas, V., Moore, R., Jeanney, C., Thiery, J.P. and Larue, L. (2001). IGF-II induces rapid beta-catenin relocation to the nucleus during epithelium to mesenchyme transition. *Oncogene* **20**: 4942–4950.

Muller, W.J., Sinn, E., Pattengale, P.K., Wallace, R. and Leder, P. (1988). Single-step induction of mammary adenocarcinoma in transgenic mice bearing the activated c-neu oncogene. *Cell* **54**: 105–115.

Mumm, J.S. and Kopan, R. (2000). Notch signaling: from the outside in. *Dev. Biol.* **228**: 151–165.

Nagafuchi, A., Shirayoshi, Y., Okazaki, K., Yasuda, K. and Takeichi, M. (1987). Transformation of cell adhesion properties by exogenously introduced E-cadherin cDNA. *Nature* **329**: 341–343.

Nakagawa, S. and Takeichi, M. (1995). Neural crest cell–cell adhesion controlled by sequential and subpopulation-specific expression of novel cadherins. *Development* **121**: 1321–1332.

Nakagawa, S. and Takeichi, M. (1998). Neural crest emigration from the neural tube depends on regulated cadherin expression. *Development* **125**: 2963–2971.

Navarro, P., Gomez, M., Pizarro, A., Gamallo, C., Quintanilla, M. and Cano, A. (1991). A role for the E-cadherin cell–cell adhesion molecule during tumor progression of mouse epidermal carcinogenesis. *J. Cell Biol.* **115**: 517–533.

Nawshad, A. and Hay, E.D. (2003). TGFβ3 signaling activates transcription of the LEF1 gene to induce epithelial mesenchymal transformation during mouse palate development. *J. Cell Biol.* **163**: 1291–1301.

Nawshad, A., Lagamba, D. and Hay, E.D. (2004). Transforming growth factor β (TGFβ) signalling in palatal growth, apoptosis and epithelial mesenchymal transformation (EMT). *Arch. Oral Biol.* **49**: 675–689.

Novak, A., Hsu, S.-C., Leung-Hagesteijn, C., Radeva, G., Papkoff, J., Montesano, R., Roskelley, C., Grosschedl, R. and Dedhar, S. (1998). Cell adhesion and the integrin-linked kinase regulate the LEF-1 and beta-catenin signaling pathways. *PNAS* **95**: 4374–4379.

Oloumi, A., McPhee, T. and Dedhar, S. (2004). Regulation of E-cadherin expression and β-catenin/Tcf transcriptional activity by the integrin-linked kinase. *Biochim. Biophys. Acta* **1691**: 1–15.

Ozdamar, B., Bose, R., Barrios-Rodiles, M., Wang, H.-R., Zhang, Y. and Wrana, J.L. (2005). Regulation of the polarity protein Par6 by TGFβ receptors controls epithelial cell plasticity. *Science* **307**: 1603–1609.

Peinado, H., Ballestar, E., Esteller, M. and Cano, A. (2004a). Snail mediates E-cadherin repression by the recruitment of the Sin3A/histone deacetylase 1 (HDAC1)/HDAC2 complex. *Mol. Cell. Biol.* **24**: 306–319.

Peinado, H., Marin, F., Cubillo, E., Stark, H.-J., Fusenig, N., Nieto, M.A. and Cano, A. (2004b). Snail and E47 repressors of E-cadherin induce distinct invasive and angiogenic properties *in vivo*. *J. Cell Sci.* **117**: 2827–2839.

Peinado, H., Del Carmen Iglesias-De La Cruz, M., Olmeda, D., Csiszar, K., Fong, K.S., Vega, S., Nieto, M.A., Cano, A. and Portillo, F. (2005). A molecular role for lysyl oxidase-like 2 enzyme in Snail regulation and tumor progression. *EMBO J* **24**: 34–46.

Perez-Moreno, M., Jamora, C. and Fuchs, E. (2003). Sticky business: orchestrating cellular signals at adherens junctions. *Cell* **112**: 535–548.

Perez-Moreno, M.A., Locascio, A., Rodrigo, I., Dhondt, G., Portillo, F., Nieto, M.A. and Cano, A. (2001). A new role for E12/E47 in the repression of E-cadherin expression and epithelial–mesenchymal transitions. *J. Biol. Chem.* **276**: 27424–27431.

Perl, A.-K., Wilgenbus, P., Dahl, U., Semb, H. and Christofori, G. (1998). A causal role for E-cadherin in the transition from adenoma to carcinoma. *Nature* **392**: 190–193.

Pla, P., Moore, R., Morali, O.G., Grille, S., Martinozzi, S., Delmas, V. and Larue, L. (2001). Cadherins in neural crest cell development and transformation. *J. Cell Physiol.* **189**: 121–132.

Portella, G., Cumming, S., Liddell, J., Cui, W., Ireland, H., Akhurst, R. and Balmain, A. (1998). Transforming growth factor beta is essential for spindle cell conversion of mouse skin carcinoma *in vivo*: implications for tumor invasion. *Cell Growth Differ.* **9**: 393–404.

Radice, G.L., Rayburn, H., Matsunami, H., Knudsen, K.A., Takeichi, M. and Hynes, R.O. (1997). Developmental defects in mouse embryos lacking N-cadherin. *Dev. Biol.* **181**: 64–78.

Radisky, D.C., Levy, D.D., Littlepage, L.E., Liu, H., Nelson, C.M., Fata, J.E., Leake, D., Godden, E.L., Albertson, D.G., Nieto, M.A., et al. (2005). Rac1b and reactive oxygen species mediate MMP-3-induced EMT and genomic instability. *Nature* **436**: 123–127.

Roberts, A.B., Tian, F., Byfield, S.D., Stuelten, C., Ooshima, A., Saika, S. and Flanders, K.C. (2005). Smad3 is key to TGF-[beta]-mediated epithelial-to-mesenchymal transition, fibrosis, tumor suppression and metastasis. *Cytokine Growth Factor Rev.* 17: 19–27.

Runyan, R.B., Potts, J.D., Sharma, R.V., Loeber, C.P., Chiang, J.J. and Bhalla, R.C. (1990). Signal transduction of a tissue interaction during embryonic heart development. *Cell Regul.* **1**: 301–313.

Sanchez-Martin, M., Rodriguez-Garcia, A., Perez-Losada, J., Sagrera, A., Read, A.P. and Sanchez-Garcia, I. (2002). SLUG (SNAI2) deletions in patients with Waardenburg disease. *Hum. Mol. Genet.* **11**: 3231–3236.

Schlessinger, J. (2000). Cell signaling by receptor tyrosine kinases. *Cell* **103**: 211–225.

Schmidt, C., Stoeckelhuber, M., Mckinnell, I., Putz, R., Christ, B. and Patel, K. (2004). Wnt 6 regulates the epithelialisation process of the segmental plate mesoderm leading to somite formation. *Dev. Biol.* **271**: 198–209.

Sharpe, C., Lawrence, N. and Martinez Arias, A. (2001). Wnt signalling: a theme with nuclear variations. *BioEssays* **23**: 311–318.

Shi, Y. and Massague, J. (2003). Mechanisms of TGF-β signaling from cell membrane to the nucleus. *Cell* **113**: 685–700.

Shook, D. and Keller, R. (2003). Mechanisms, mechanics and function of epithelial–mesenchymal transitions in early development. *Mech. Dev.* **120**: 1351–1383.

Siegel, P.M., Shu, W., Cardiff, R.D., Muller, W.J. and Massague, J. (2003). Transforming growth factor {beta} signaling impairs Neu-induced mammary tumorigenesis while promoting pulmonary metastasis. *PNAS* **100**: 8430–8435.

Soo, K., O'Rourke, M.P., Khoo, P.-L., Steiner, K.A., Wong, N., Behringer, R.R. and Tam, P.P.L. (2002). Twist function is required for the morphogenesis of the cephalic neural tube and the differentiation of the cranial neural crest cells in the mouse embryo. *Dev. Biol.* **247**: 251–270.

Stark, K., Vainio, S., Vassileva, G. and McMahon, A.P. (1994). Epithelial transformation of metanephric mesenchyme in the developing kidney regulated by Wnt-4. *Nature* **372**: 679–683.

Sternlicht, M.D., Lochter, A., Sympson, C.J., Huey, B., Rougier, J.P., Gray, J.W., Pinkel, D., Bissell, M.J. and Werb, Z. (1999). The stromal proteinase MMP3/stromelysin-1 promotes mammary carcinogenesis. *Cell* **98**: 137–146.

Stockinger, A., Eger, A., Wolf, J., Beug, H. and Foisner, R. (2001). E-cadherin regulates cell growth by modulating proliferation-dependent β-catenin transcriptional activity. *J. Cell Biol.* **154**: 1185–1196.

Suyama, K., Shapiro, I., Guttman, M. and Hazan, R.B. (2002). A signaling pathway leading to metastasis is controlled by N-cadherin and the FGF receptor. *Cancer Cell* **2**: 301–314.

Tan, C., Costello, P., Sanghera, J., Dominguez, D., Baulida, J., De Herreros, A.G. and Dedhar, S. (2001). Inhibition of integrin linked kinase (ILK) suppresses beta-catenin-Lef/Tcf-dependent transcription and expression of the E-cadherin repressor, snail, in APC-/- human colon carcinoma cells. *Oncogene* **20**: 133–140.

Tepass, U., Truong, K., Godt, D., Ikura, M. and Peifer, M. (2000). Cadherins in embryonic and neural morphogenesis. *Nat. Rev. Mol. Cell Biol.* **1**: 91–100.

Thiery, J.P. (2002). Epithelial–mesenchymal transitions in tumour progression. *Nat. Rev. Cancer* **2**: 442–454.

Timmerman, L.A., Grego-Bessa, J., Raya, A., Bertran, E., Perez-Pomares, J.M., Diez, J., Aranda, S., Palomo, S., McCormick, F., Izpisua-Belmonte, J.C., et al. (2004). Notch promotes epithelial–mesenchymal transition during cardiac development and oncogenic transformation. *Gene Dev.* **18**: 99–115.

Ulrich, F., Krieg, M., Schotz, E.-M., Link, V., Castanon, I., Schnabel, V., Taubenberger, A., Mueller, D., Puech, P.-H. and Heisenberg, C.-P. (2005). Wnt11 functions in gastrulation by controlling cell cohesion through Rab5c and E-cadherin. *Dev. Cell* **9**: 555–564.

Vallin, J., Thuret, R., Giacomello, E., Faraldo, M.M., Thiery, J.P. and Broders, F. (2001). Cloning and characterization of three *Xenopus* Slug promoters reveal direct regulation by Lef/beta-catenin signaling. *J. Biol. Chem.* **276**: 30350–30358.

Wheelock, M.J. and Johnson, K.R. (2003). Cadherins as modulators of cellular phenotype. *Annu. Rev. Cell Dev. Biol.* **19**: 207–235.

Wheelock, M.J., Soler, A.P. and Knudsen, K.A. (2001). Cadherin junctions in mammary tumors. *J. Mammary Gland Biol. Neoplasia* **6**: 275–285.

Wong, A.S.T. and Gumbiner, B.M. (2003). Adhesion-independent mechanism for suppression of tumor cell invasion by E-cadherin. *J. Cell Biol.* **161**: 1191–1203.

Woolf, A.S., Kolatsi-Joannou, M., Hardman, P., Andermarcher, E., Moorby, C., Fine, L.G., Jat, P.S., Noble, M.D. and Gherardi, E. (1995). Roles of hepatocyte growth factor/scatter factor and the met receptor in the early development of the metanephros. *J. Cell Biol.* **128**: 171–184.

Yamashita, S., Miyagi, C., Fukada, T., Kagara, N., Che, Y.-S. and Hirano, T. (2004). Zinc transporter LIVI controls epithelial– mesenchymal transition in zebrafish gastrula organizer. *Nature* **429**: 298–302.

Yang, Z., Rayala, S., Nguyen, D., Vadlamudi, R.K., Chen, S. and Kumar, R. (2005). Pak1 phosphorylation of Snail, a master regulator of epithelial-to-mesenchyme transition, modulates Snail's subcellular localization and functions. *Cancer Res.* **65**: 3179–3184.

Yook, J.I., Li, X.-Y., Ota, I., Fearon, E.R. and Weiss, S.J. (2005). Wnt-dependent regulation of the E-cadherin repressor Snail. *J. Biol. Chem.* **280**: 11740–11748.

Zavadil, J. and Bottinger, E.P. (2005). TGF-β and epithelial-to-mesenchymal transitions. *Oncogene* **24**: 5764–5774.

Zavadil, J., Cermak, L., Soto-Nieves, N. and Bottinger, E.P. (2004). Integration of TGF-beta/Smad and Jagged1/Notch signalling in epithelial-to-mesenchymal transition. *EMBO J.* **23**: 1155–1165.

Zhao, P., Iezzi, S., Carver, E., Dressman, D., Gridley, T., Sartorelli, V. and Hoffman, E.P. (2002). Slug is a novel downstream target of MyoD. Temporal profiling in muscle regeneration. *J. Biol. Chem.* **277**: 30091–30101.

Zhou, B.P., Deng, J., Xia, W., Xu, J., Li, Y.M., Gunduz, M. and Hung, M.-C. (2004). Dual regulation of Snail by GSK-3β-mediated phosphorylation in control of epithelial–mesenchymal transition. *Nat. Cell Biol.* **6**: 931–940.

Zhuang, Y., Soriano, P. and Weintraub, H. (1994). The helix-loop-helix gene E2A is required for B cell formation. *Cell* **79**: 875–884.

Zuk, A. and Hay, E.D. (1994). Expression of beta 1 integrins changes during transformation of avian lens epithelium to mesenchyme in collagen gels. *Dev. Dyn.* **201**: 378–393.

5

Cell–ECM Interactions in Repair and Regeneration

M. Petreaca and M. Martins-Green

INTRODUCTION

For many years, the extracellular matrix (ECM) was thought to serve only as a structural support for tissues. However, as early as 1966, Hauschka and Konigsberg showed that interstitial collagen promoted the conversion of myoblasts to myotubes, and, shortly thereafter, it was shown that both collagen (Wessells and Cohen, 1968) and glycosaminoglycans (Bernfield et al., 1972) play a crucial role in salivary gland morphogenesis. Based upon these findings as well as other pieces of indirect evidence, Hay (1977) put forth the idea that the ECM is an important component in embryonic inductions, a concept which implicated the presence of binding sites (receptors) for specific matrix molecules on the surface of cells. The stage was then set to begin to investigate in detail the mechanisms by which ECM molecules influence cell behavior. Bissell et al. proposed the model of "dynamic reciprocity." In this model, ECM molecules interact with receptors on the surface of cells which then transmit signals across the cell membrane to molecules in the cytoplasm; these signals initiate a cascade of events through the cytoskeleton into the nucleus, resulting in the expression of specific genes, whose products, in turn, affect the ECM in various ways (Bissell et al., 1982). It has become clear that this concept is essentially correct (Ingber, 1991; Boudreau et al., 1995); cell–ECM interactions participate directly in promoting cell adhesion, migration, growth, differentiation, and programmed cell death (also called apoptosis), as well as in modulation of the activities of cytokines and growth factors, and in directly activating intracellular signaling.

Most of what we know about the molecular basis of cell–ECM interactions in these events comes from studies that have used induced mutations, experimental perturbations *in vivo*, and cell/organ cultures. Below, we will first briefly discuss the composition and diversity of some of the better known ECM molecules and their receptors, then discuss selected examples that illustrate the dynamics of cell–ECM interactions during wound healing and regeneration, as well as the potential mechanisms involved in the signal transduction pathways initiated by these interactions. Finally, we will discuss the implications of cell–ECM interactions in regenerative medicine.

COMPOSITION AND DIVERSITY OF THE ECM

The ECM is a molecular complex that consists of collagens and other glycoproteins, hyaluronic acid, proteoglycans, glycosaminoglycans and elastins, and that harbors molecules such as growth factors, cytokines, and matrix-degrading enzymes and their inhibitors. The distribution and organization of these molecules are not static, but rather vary from tissue to tissue and during development from stage to stage (Ffrench-Constant and Hynes, 1989; Laurie et al., 1989; Sanes et al., 1990; Martins-Green and Bissell, 1995; Tsuda et al., 1998; Werb and Chin, 1998; Zhu et al., 2001), which has significant implications for tissue function (Sechler et al., 1998; Xu et al.,

Copyright © 2008 Elsevier Inc. All rights reserved

1998; So et al., 2001). Mesenchymal cells are immersed in an interstitial matrix that confers specific biomechanical and functional properties to connective tissue (Culav et al., 1999; Suki et al., 2005). In contrast, epithelial and endothelial cells contact a specialized matrix, the basement membrane, via their basal surfaces only, conferring mechanical strength and specific physiological properties to the epithelia (Edwards and Streuli, 1995; Fuchs et al., 1997; Dockery et al., 1998). This diversity of composition, organization, and distribution of ECM results not only from differential gene expression of the various molecules in specific tissues, but also from the existence of differential splicing and post-translational modifications of those molecules. For example, alternative splicing may change the binding potential of proteins to other matrix molecules (Ffrench-Constant and Hynes, 1989; Chiquet-Ehrismann et al., 1991; Wallner et al., 1998; Ghert et al., 2001; Mostafavi-Pour et al., 2001) or to their receptors (Aota et al., 1994; Mould et al., 1994; Akiyama et al., 1995; Cox and Huttenlocher, 1998), and variations in glycosylation can lead to changes in cell adhesion (Dean et al., 1990; Anderson et al., 1994; Vlodavsky et al., 1996; Schamhart and Kurth, 1997; Cotman et al., 1999). In addition, the presence of divalent cations such as Ca^{2+} (Paulsson, 1988; Ekblom et al., 1994; Wess et al., 1998) can affect matrix organization and influence molecular interactions that are important in the way ECM molecules interact with cells (Sjaastad and Nelson, 1997; Kielty et al., 2002).

Growth factors and cytokines interact with the ECM in a variety of ways which allows them to mutually affect each other (Nathan and Sporn, 1991; Adams and Watt, 1993); they can stimulate cells to alter the production of ECM molecules, their inhibitors and/or their receptors (Streuli et al., 1993; Schuppan et al., 1998; Verrecchia and Mauviel, 2002; Gratchev et al., 2005). TGFβ for example, upregulates the expression of matrix molecules and of inhibitors of enzymes that degrade ECM molecules, the combination of which increases ECM levels (Wikner et al., 1990; Bonewald, 1999; Kutz et al., 2001). The ECM can also influence the local concentration and biological activity of growth factors and cytokines by serving as a reservoir that binds them and protects them from being degraded, by presenting them more efficiently to their receptors, or by affecting their synthesis (Roberts et al., 1988; Chiquet-Ehrismann et al., 1991; Flaumenhaft and Rifkin, 1992; Lamszus et al., 1996; Miao et al., 1996; Kagami et al., 1998; Banwell et al., 2000; Schonherr and Hausser, 2000; Miralem et al., 2001; Rahman et al., 2005). Examples of this include the increased production of TNFα by neutrophils after binding to fibronectin (Nathan and Sporn, 1991), the dependence of HGF (hepatocyte growth factor)-mediated hepatocyte proliferation on heparan sulfate proteoglycans (Sakakura et al., 1999), and the increased ability of VEGF (vascular endothelial growth actor) to induce breast cancer cell proliferation and migration in the presence of fibronectin or heparin (Miralem et al., 2001). Growth factor binding to ECM molecules may also exert an inhibitory effect; SPARC (secreted protein acidic and rich in cysteine)/osteonectin binds multiple growth factors, preventing receptor binding and/or downstream signaling events (Lane and Sage, 1994; Kupprion et al., 1998; Francki et al., 2003). In some cases, only particular forms of these growth factors and cytokines bind to specific ECM molecules, for example, PDGF (platelet derived growth factor) (LaRochelle et al., 1991; Pollock and Richardson, 1992), VEGF (Poltorak et al., 1997), and the chemokine cIL-8 (previously called cCAF (chicken chemotactic and angiogenic factor)). cIL-8 is a small cytokine that is overexpressed during wound repair and in the stroma of tumors (Martins-Green and Bissell, 1990; Martins-Green et al., 1992), and is secreted as a 9 kDa protein, although it can be processed by plasmin to yield a 7 kDa protein. Both forms of the protein are found in association with interstitial collagen, but only the smaller form binds to laminin or tenascin, while neither form binds to fibronectin, collagen IV, or heparin (Martins-Green and Bissell, 1995; Martins-Green et al., 1996). Importantly, binding of specific forms of these factors to specific ECM molecules can lead to their localization to particular areas of tissues and affect their biological activities.

Another feature of ECM/growth factor interactions that has been more recently characterized involves the ability of specific domains of various ECM molecules, including laminin-5, tenascin-C, and decorin, to bind and activate growth factor receptors (Tran et al., 2004). The epidermal growth factor (EGF)-like repeats of laminin (Panayotou et al., 1989; Schenk et al., 2003; Koshikawa et al., 2005) and tenascin-C bind and activate the EGF receptor (EGFR) (Swindle et al., 2001). In the case of laminin, the EGF-like repeats interact with EGFR following their release by matrix metalloproteinase (MMP)-mediated proteolysis (Schenk et al., 2003; Koshikawa et al., 2005), whereas tenascin-C repeats are thought to bind EGFR in the context of the full-length protein (Swindle et al., 2001). Decorin also binds and activates EGFR, although this binding occurs via leucine-rich repeats rather than EGF-like repeats (Iozzo et al., 1999; Santra et al., 2002). The ability of ECM molecules to serve as ligands for growth factor receptors may facilitate a stable signaling environment for the associated cells due to the inability for the ligand to either diffuse or be internalized, thus serving as a long-term pro-migratory and/or pro-proliferative signal (Tran et al., 2004, 2005).

RECEPTORS FOR ECM MOLECULES

In order to establish that ECM molecules themselves directly affect cellular behavior, it was important to identify transmembrane receptors for the specific sequences present on these molecules. As early as 1973, it was observed that during salivary gland morphogenesis near the sites of glycosaminoglycan deposition, the intracellular microfilaments contracted (Bernfield et al., 1973). These investigators proposed that the ECM could "be involved in regulating microfilament function," suggesting that these molecules can specifically interact with cell-surface receptors. It was subsequently shown that various ECM molecules contain specific amino acid motifs that allow them to bind directly to cell-surface receptors (Humphries et al., 1991; Hynes, 1992; Gullberg and Ekblom, 1995). The best characterized motif is the tripeptide RGD, first found in fibronectin (Pierschbacher and Ruoslahti, 1984; Yamada and Kennedy, 1984). Peptides containing this amino acid sequence promote adhesion of cells and inhibit the adhesive properties of fibronectin. This and other amino acid adhesive motifs have been found in laminin, entactin, thrombin, tenascin, fibrinogen, vitronectin, collagens I and VI, bone sialoprotein, and osteopondin (Humphries et al., 1991).

Integrins, a family of heterodimeric transmembrane proteins composed of α and β subunits were the first ECM receptors to be identified (Hynes, 1987). At least 18α and 8β subunits have been identified so far; they pair with each other in a variety of combinations, giving rise to a large family which recognizes specific sequences on the ECM molecules (Figure 5.1). Some integrin receptors are very specific, whereas others bind several different epitopes, which may be on the same or different ECM molecules (Figure 5.1), thus facilitating plasticity and redundancy in specific systems (Hynes, 1992; Cotman et al., 1998; Dedhar, 1999; Hynes, 1999). Although the α and β subunits of integrins are unrelated, there is 40–50% homology within each subunit with the highest divergence in the intracellular domain of the α subunit. All but one of these subunits (β_4) have large extracellular domains and very small intracellular domains (Briesewitz et al., 1995; Fornaro and Languino, 1997). The extracellular domain of the α subunits contains four regions that serve as binding sites for divalent cations, which appear to augment ligand binding and increase the strength of the ligand–integrin interactions (Gailit and Ruoslahti, 1988; Loftus et al., 1990; Dickeson et al., 1997; Pujades et al., 1997; Leitinger et al., 2000).

Although not as extensively studied as the integrins, it has been found that transmembrane proteoglycans can also serve as receptors for ECM molecules (Rapraeger et al., 1987; Jalkehen et al., 1991; Couchman and Woods, 1996; McFall and Rapraeger, 1998). Several proteoglycan receptors that bind to ECM molecules have been isolated and characterized: syndecan, CD44, RHAMM (receptor for hyaluronate-mediated motility), and phosphacan (Grumet et al., 1994; Couchman and Woods, 1996; Entwistle et al., 1996; Liu et al., 1998). Syndecan binds cells to matrix via chondroitin- and heparan-sulfate glycosaminoglycans, whose composition varies based upon the type of tissue in which syndecan is expressed; the differential glycosaminoglycan modifications alter the binding capacity of particular ligands (Kim et al., 1994; Salmivirta and Jalkanen, 1995). Syndecan also associates with the cytoskeleton, promoting intracellular signaling events and cytoskeletal reorganization through

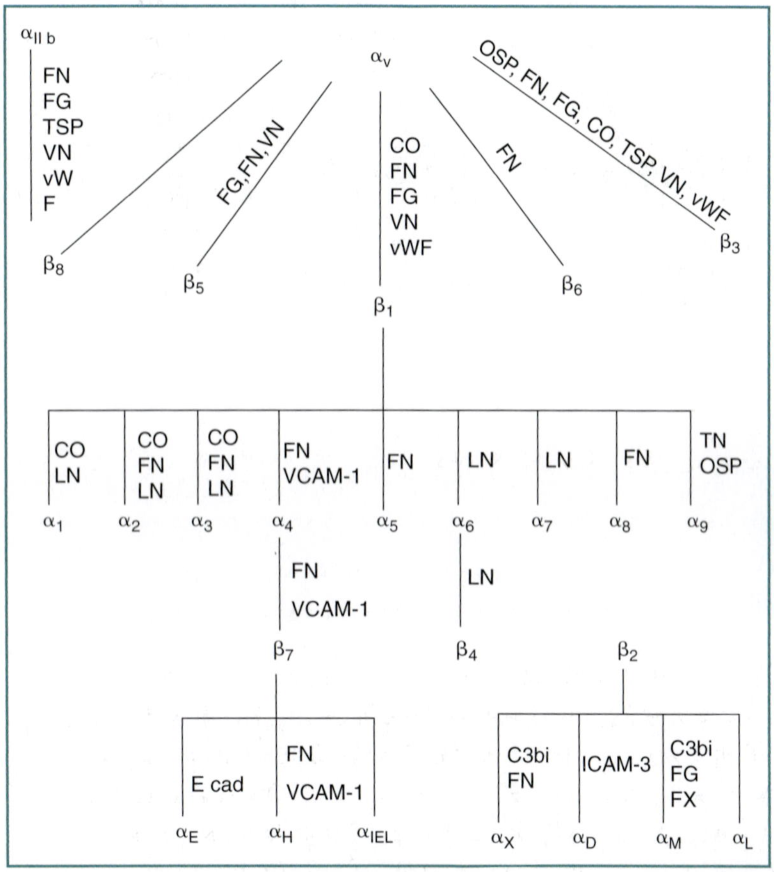

Figure 5.1 Representative members of the integrin family of ECM receptors and their respective ligands. These heterodimeric receptors are composed of one α and one β subunit, and are capable of binding a variety of ligands, including Ig superfamily cell adhesion molecules, complement factors, and clotting factors in addition to ECM molecules. Cell–cell adhesion is largely mediated through integrin heterodimers containing the β_2 subunits, while cell–matrix adhesion is mediated primarily via integrin heterodimers containing the β_1 and β_3 subunits. In general, the β_1 integrins interact with ligands found in the connective tissue matrix, including laminin, fibronectin, and collagen, whereas the β_3 integrins interact with vascular ligands, including thrombospondin, vitronectin, fibrinogen, and von Willebrand factor. Abbreviations: CO, collagens; C3bi, complement component; FG, fibrinogen; FN, fibronectin; FX, Factor X; ICAM-1, intercellular adhesion molecule-1; ICAM-2, intercellular adhesion molecule-2; ICAM-3, intercellular adhesion molecule-3; LN, laminin; OSP, osteopontin; TN, tenascin; TSP, thrombospondin; VCAM-1, vascular cell adhesion molecule-1; VN, vitronectin; vWF, von Willebrand factor.

activation of Rho GTPases (Carey, 1997; Granes et al., 1999; Saoncella et al., 1999; Bass and Humphries, 2002; Yoneda and Couchman, 2003). The CD44 receptor also carries chondroitin sulfate and heparan sulfate chains on its extracellular domain (Milstone et al., 1994), and undergoes tissue-specific splicing and glycosylation to yield multiple isoforms; these may play roles in cell adhesion as well as in ligand binding (Brown et al., 1991; Ehnis et al., 1996; Tuhkanen et al., 1997). One of the extracellular domains of CD44 is structurally similar to the hyaluronan-binding domain of the cartilage link protein and aggrecan, which suggested that CD44 could serve as a hyaluronan receptor. Using a variety of techniques involving antibody binding and mutagenesis, it has been shown that this domain of CD44 as well as an additional domain outside this region can interact directly with hyaluronan (Miyake et al., 1990; Peach et al., 1993; Bajorath et al., 1998); these regions can also mediate CD44 binding to other proteoglycans, although hyaluronic acid is its primary ligand (Marhaba and Zoller, 2004). In addition, studies have shown that CD44 can also interact with collagen, laminin, and fibronectin (Jalkanen and

Jalkanen, 1992; Ishii et al., 1993, 1994; Ehnis et al., 1996), although the exact binding sites of these molecules to CD44, as well as the functional significance of such interactions *in vivo* are not well understood (Ponta et al., 2003). RHAMM has been identified as an additional hyaluronic acid receptor (Hardwick et al., 1992), which is responsible for hyaluronic-acid-mediated cell motility in a number of cell types, and also appears to be important in trafficking of hematopoietic cells (Hall et al., 1994; Masellis-Smith et al., 1996; Pilarski et al., 1999; Savani et al., 2001).

Cell-surface receptors other than integrins or proteoglycans have also been identified as receptors for ECM molecules. A non-integrin 67 kDa protein known as the elastin-laminin receptor (ELR) recognizes the YIGSR sequence of laminin and the VGVAPG sequence of elastin, sequences not recognized by integrins; the ELR co-localizes with cytoskeleton-associated and signaling proteins upon laminin ligation, suggesting a role in laminin-mediated signaling (Grant et al., 1989; Massia et al., 1993; Bushkin-Harav and Littauer, 1998), and has more recently been implicated in the signaling downstream of elastin and laminin during mechanotransduction (Spofford and Chilian, 2003). A second receptor, CD36, functions as a scavenger receptor for long chain fatty acids and oxidized LDL, but also binds collagens I and IV, thrombospondin, and malaria-infected erythrocytes to endothelial cells and some types of epithelial cells (Febbraio et al., 2001). Each of these ligands has a separate binding site, but all are located in the same external loop of CD36 (Asch et al., 1993), and the intracellular signals occurring after ligand binding lead to activation of a variety of signal transduction molecules (Huang et al., 1991; Lipsky et al., 1997). Indeed, the anti-angiogenic effects of thrombospondin are dependent upon signaling downstream of CD36 (Jimenez et al., 2000, 2001; Isenberg et al., 2005). Furthermore, alternative splice variants of tenascin-C interact with cell-surface annexin II, which may mediate the cellular responses to this particular form (Chung and Erickson, 1994). In addition, ECM molecules have been shown to bind and activate tyrosine kinase receptors, including the EGFR via EGF-like domains (see above) as well as the discoidin domain receptors DDR1 and DDR2. DDR1 and DDR2 function as receptors for various collagens and mediate cell adhesion and signaling events (Vogel et al., 1997). The DDR receptors have also been implicated in ECM remodeling, as their overexpression decreases the expression of multiple matrix molecules and their receptors, including collagen, syndecan-1, and integrin α_3, while simultaneously increasing MMP activity (Faraci et al., 2003; Ferri et al., 2004).

SIGNAL TRANSDUCTION EVENTS DURING CELL–ECM INTERACTIONS

The interactions between ECM molecules and their receptors as described above can transmit signals directly or indirectly to signaling molecules within the cell, leading to a cascade of events and the coordinated expression of a variety of genes involved in cell adhesion, migration, proliferation, differentiation, and death (Figure 5.2). There is increasing evidence that cell–ECM interactions, especially through integrins, activate a variety of signaling pathways that can be linked to those specific functions. Some of the signaling events important in these cellular processes are discussed below.

Adhesion and Migration

It is now well established that, upon ligand binding, integrins can directly induce biochemical signals inside cells (Kumar, 1998; Dedhar, 1999). The cytoplasmic domain of integrins interacts with the cytoskeleton, suggesting that ECM signaling through integrins is transduced via the cytoskeletal elements and can induce cell shape changes which, in turn, may lead to growth, migration, and/or differentiation (van der Flier and Sonnenberg, 2001; Hynes, 2002). For example, cell migration is promoted when fibronectin binds simultaneously to integrins through its cell-binding domain and to proteoglycan receptors through its heparin-binding domain (Bernfield et al., 1992; Hardingham and Fosang, 1992; Hynes, 1992; Giancotti, 1997; Schlaepfer and Hunter,

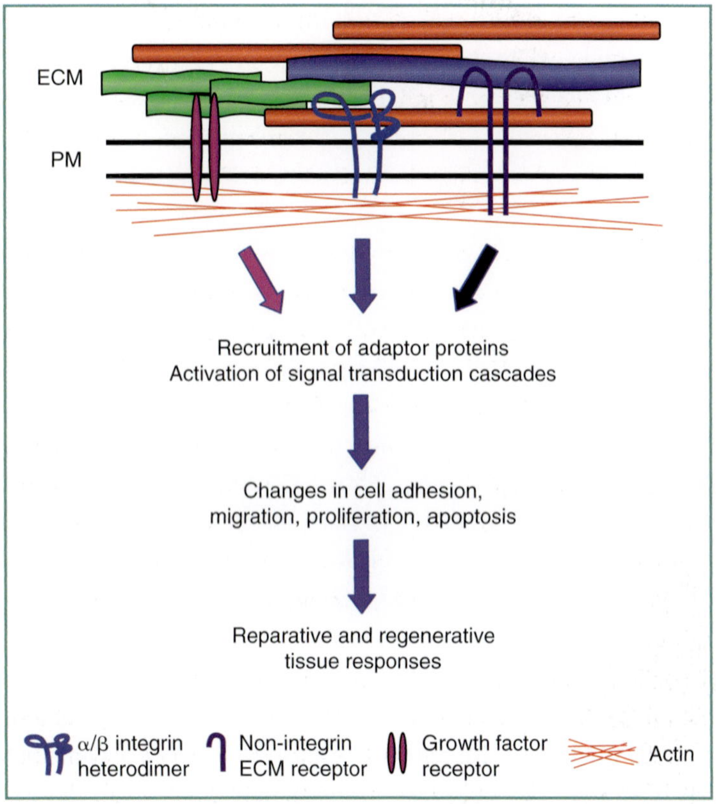

Figure 5.2 Schematic diagram of cell–ECM interactions present during the healing and regenerative responses. Such interactions between the ECM receptors and their respective ligands initiate signal transduction cascades culminating in a variety of cellular events important in repair and regeneration, including changes in cellular adhesion and migration and altered rates of proliferation and apoptosis. The presence and/or extent of such changes may influence the balance of repair and regenerative responses to favor one outcome over another; thus, interventions that alter ECM signaling events may shift this balance to favor tissue regeneration and thus decrease scarring.

1998; Dedhar, 1999; Mercurius and Morla, 2001). These receptors interact and colocalize in areas of adhesion where microfilaments associate with the β_1 subunit of the integrin receptor via structural proteins such as talin and α-actinin present in the actin cytoskeleton of the focal adhesions. The cytoplasmic domain of the β_1 subunit also interacts directly with the focal adhesion tyrosine kinase $pp125^{FAK}$ which, when activated, undergoes autophosphorylation on tyrosine 397 (Hildebrand et al., 1995); this phosphotyrosine residue subsequently serves as the binding site for the SH2 domain of the non-receptor tyrosine kinase c-Src. In turn, c-Src phosphorylates many components of the focal adhesion plaques, including paxillin, tensin, vinculin, and the protein p130cas. Paxillin has been implicated in the regulation of integrin-mediated signaling events and motility; paxillin-deficient fibroblasts exhibit reduced phosphorylation of signaling molecules downstream of integrin ligation, with a concomitant reduction in cell motility (Hagel et al., 2002). The specific role of tensin in the process of adhesion/de-adhesion during migration is not known; however, it interacts with both the cytoskeleton and with other phosphorylated signaling molecules via its SH2 domain, and may thus mediate signals between the plasma membrane and the cytoskeleton and/or facilitate signaling events (Lo, 2004). p130cas activation promotes its interaction with the adaptor molecules Crk and Nck, which appear to form a scaffold for localized activation of Rac-GTPase and the MAP/JNK kinase pathways, thus facilitating migration (Dolfi et al., 1998; Kiyokawa et al., 1998; Klemke et al., 1998; Cho and Klemke, 2002). In addition, it has also been shown that c-Src phosphorylates focal adhesion kinase (FAK) on tyrosine 925 which serves as a site for binding of Grb2/Sos complex with subsequent activation of Ras and the MAP kinase cascade (Schlaepfer and Hunter, 1996, 1997, 1998; Schlaepfer et al., 1997), which may also be involved in adhesion/de-adhesion and migration (Giancotti, 1997; Schlaepfer and Hunter, 1998; Dedhar, 1999; Ly and Corbett, 2005).

Proliferation and Survival

ECM interaction with its receptors can promote cell proliferation and survival, often in conjunction with growth factors or cytokine receptors. Such cooperative effects may occur in a direct manner, as in situations in which the EGF-like repeats of ECM molecules bind and activate growth factor receptors, leading to cell proliferation (Swindle et al., 2001; Tran et al., 2004). However, more is known regarding the importance of indirect cooperative effects, particularly those involved in the anchorage dependence of cell growth. Anchorage is required for cells to enter S phase; even in the presence of growth factors, cells will not enter the DNA synthesis phase without being anchored to a substrate (Zhu and Assoian, 1995). Thus, adhesion of cells to ECM molecules plays a very important role in regulating cell survival and proliferation (Giancotti, 1997; Mainiero et al., 1997; Murgia et al., 1998). Integrin–ECM binding leads to the activation of Fyn and its subsequent interaction with the adaptor protein Shc, which recruits Grb2 and thus activates the Ras/ERK cascade, leading to the phosphorylation of the Elk-1 transcription factor and the expression of early response genes involved in cell cycle progression (Wary et al., 1998; Aplin et al., 2001); integrin ligation is also important for the efficient and prolonged activation of MAPK by growth factors, which may explain, in part, the anchorage dependence of growth factor-mediated proliferation (Aplin and Juliano, 1999; Roovers et al., 1999). It has also been shown that cooperation between integrins and growth factors involves the activation of phosphatidylinositol phosphate kinases, thus increasing the levels of phosphatidylinositol bis-phosphate (PIP2). PIP2 then serves as substrate for phospholipase C_γ (PLC_γ), which is activated by growth factors as well as by integrin ligation, ultimately leading to the activation of protein kinase C (PKC) and the promotion of cell proliferation (Housey et al., 1988; Schwartz, 1992; Cybulsky et al., 1993). Furthermore, PI-3 kinase activated downstream of Ras can rescue cells in suspension from undergoing apoptosis via the activation of the Akt serine/threonine kinase (Khwaja et al., 1997). Signaling downstream of cell–ECM binding may also promote degradation of cell cycle inhibitors, thus facilitating cell proliferation; indeed, fibronectin-mediated adhesion leads to the degradation of p21 in a Rac1- and Cdc42-dependent manner (Bao et al., 2002). The importance of the Rac/JNK pathway in integrin-mediated proliferation is underscored by studies involving a β_1 integrin cytoplasmic domain mutant, which decreased the activation of the Rac/JNK pathway and also negatively affected fibroblast proliferation and survival; these effects were rescued by the expression of constitutively active Rac1 (Hirsch et al., 2002). Likewise, other studies involving integrin inhibition or knockout yield similar negative effects on cell proliferation due to changes in signaling. For example, studies of mice lacking the $\alpha_1\beta_1$ integrin, which is a primary collagen receptor, showed that the fibroblasts of these mice have reduced proliferation even though they attach normally (Pozzi et al., 1998). In addition, mammary epithelial cells over-expressing a dominant negative β_1 integrin subunit exhibit reduced proliferation due to a combination of decreased MAPK and Akt activation (Faraldo et al., 2001); Akt activation is also diminished in cells over-expressing the β_1 integrin mutant mentioned above (Hirsch et al., 2002).

Differentiation

Interaction of cells with ECM molecules, hormones, and growth factors is required to activate genes that are specific for differentiation. Interestingly, the latter studies have shown that the cell–ECM interactions that result in the differentiated phenotype are those that fail to activate Shc and the MAP kinase cascade, at least in some cases. This has been shown for endothelial cells in which the interaction of $\alpha_2\beta_1$ with laminin, which does not activate the Shc pathway, leads to formation of capillary-type structures (Kubota et al., 1988), whereas the interaction of $\alpha_5\beta_1$ in the same cells with fibronectin results in proliferation (Wary et al., 1998). Similar observations have been made with primary bronchial epithelial cells when they are cultured on collagen matrices (Moghal and Neel, 1998). The formation of endothelial capillary-like tubes also relies upon additional signaling pathways, such as occur upon activation of integrin-linked kinase (ILK); over-expression of this kinase can rescue tube formation in the absence of ECM molecules (Cho et al., 2005), while expression of dominant negative ILK prevents tube formation in the presence of ECM and VEGF (Watanabe et al., 2005). Other differentiated phenotypes likewise require integrin-mediated signaling events. Indeed, TGF-β_1-mediated myofibroblast differentiation, an event important in both wound healing and liver regeneration, requires the ligation of specific integrins as well as the activation of FAK and its associated signaling pathways (Thannickal et al., 2003; Lygoe et al., 2004).

Apoptosis

Signal transduction pathways that lead to apoptosis have been delineated for endothelial cells and leukocytes and appear to involve primarily tyrosine kinase activity (Fukai et al., 1998; Ilan et al., 1998; Kettritz et al., 1999; Avdi et al., 2001). For example, the neutrophil apoptosis stimulated by TNF-α is dependent upon β_2 integrin-mediated signaling events involving the activation of the Pyk2 and Syk tyrosine kinases as well as JNK1 (Avdi et al., 2001). In other cell types, alterations in the ligand presentation by ECM can also regulate apoptosis. Studies have suggested that integrin ligation by soluble, rather than intact, ligands can function as integrin antagonists and promote apoptosis rather than survival or proliferation (Brooks et al., 1994; Vogel et al., 2001; Stupack and Cheresh, 2002); such soluble ligands may be created by matrix degradation during tissue remodeling, and thus promote apoptosis. The apoptosis stimulated by soluble ligands or other antagonists appears to occur via the recruitment and activation of caspase 8 by the clustered integrins, without any requirement for death receptors (Stupack et al., 2001). However, the recruitment process itself is not well understood.

CELL–ECM INTERACTIONS DURING HEALING OF SKIN WOUNDS

Interactions of cells with ECM molecules play a crucial role during wound healing and regeneration. It is the continuous crosstalk between cells and the surrounding matrix environment that contribute to the processes of clot formation, inflammation, granulation tissue development, and remodeling, and during regeneration, the matrix interactions are important in restoration of the damaged tissue. As we will see, many different lines of experimental evidence have shown that the basic cellular mechanisms that result in these events involve cell adhesion/de-adhesion, migration, proliferation, differentiation, and apoptosis (Figure 5.2).

Adhesion and Migration

Shortly after tissue damage and during the early stages of wound healing, there is a release of blood contents and tissue factors into the area of the wound, leading to platelet activation and adhesion, and the formation of a vascular plug containing primarily platelets, plasma fibronectin, and fibrin (crosslinked by factor XIII), but also including small amounts of tenascin, thrombospondin, and SPARC. During this process, activated mast cells degranulate, releasing vasodilating and chemotactic factors that will bring polymorphonucleocytes to the wound site. These events constitute the early stages of the inflammatory response. The fibrin–fibronectin meshwork provides a provisional matrix which serves as substrate for the subsequent migration of leukocytes and keratinocytes during the very early stages of healing when inflammation and wound closure are occurring. Leukocyte interactions with ECM molecules via integrin receptors affect many of the functions of these cells, in particular those that lead to cell adhesion and migration or to production of inflammatory mediators (Rosales and Juliano, 1995; Romanic et al., 1997; Wei et al., 1997; Vaday and Lider, 2000). An example of the latter involves the stimulation of pro-inflammatory cytokine release by tissue macrophages upon binding to low molecular weight hyaluronic acid via CD44 (Hodge-Dufour et al., 1997). Because some inflammatory molecules can be damaging to tissues when produced in excess, the course of inflammation can be affected significantly by the types of ECM encountered by these leukocytes (Wei et al., 1997; Vaday and Lider, 2000). ECM molecules can also facilitate leukocyte chemotaxis into the inflamed area by binding chemokines, thus creating a stable chemotactic gradient to promote a specific directional migration (Patel et al., 2001); mutant chemokines unable to bind glycosaminoglycans were unable to promote chemotaxis *in vivo*, underscoring the importance of ECM binding in leukocyte recruitment (Proudfoot et al., 2003).

During re-epithelialization of cutaneous wounds, keratinocytes migrate over the provisional matrix primarily composed of fibrin/fibrinogen, fibronectin, vitronectin, tenascin, and collagen type III. These cells express $\alpha_2\beta_1$, $\alpha_3\beta_1$, $\alpha_5\beta_1$, $\alpha_6\beta_1$, $\alpha_5\beta_4$, and α_v integrin receptors for these ECM molecules, which, in conjunction with MMPs, facilitate their migration to close the wound (Cavani et al., 1993; Juhasz et al., 1993; Gailit et al., 1994; O'Toole, 2001; Li et al., 2004). The importance of individual matrix components in re-epithelialization is underscored by studies done in mice lacking these molecules; for example, fibrinogen-deficient mice experienced disordered re-epithelialization (Drew et al., 2001). This keratinocyte migration may also require new laminin deposition, as an antibody against laminin inhibited keratinocyte migration on fibronectin or collagen (Decline and Rousselle, 2001). Cell–ECM interactions are equally important in the closure of other epithelial wounds. Studies examining the sequential deposition of ECM molecules after wounding of retinal pigment epithelial cells showed "*de novo*" fibronectin deposition 24 h after wounding, which is followed by deposition of collagen IV and laminin. This sequence of matrix deposition is tightly linked to adhesion and migration of cells to close the wound (Kamei et al., 1998), and inhibition of integrin-matrix binding using antibodies or cyclic peptides can prevent both cell adhesion and migration, implicating cell–ECM interactions in the observed epithelial closure (Hergott et al., 1993; Hoffmann et al., 2005). A similar sequence of events is observed during the repair of airway epithelial cells after mechanical injury (Pilewski et al., 1997; White et al., 1999; Sacco et al., 2004); functional inhibition of fibronectin or various expressed integrins likewise diminished cell migration and healing of this epithelium (Herard et al., 1996; White et al., 1999).

As healing progresses, embryonic-type cellular fibronectin produced by macrophages and fibroblasts in the wound bed contributes to formation of the granulation tissue, a provisional connective tissue containing nascent blood vessels and multiple types of ECM molecules (Li et al., 2003). This fibronectin serves as substrate for the migration of the endothelial cells that form the vasculature of the wound bed, myofibroblasts, and lymphocytes that are chemoattracted to the wound site by a variety of small cytokines (chemokines) secreted by both macrophages and fibroblasts (Greiling and Clark, 1997; Feugate et al., 2002b). These chemokines belong to a large superfamily, and have been characterized in humans, other mammals, and in avians (Rossi and Zlotnik, 2000; Gillitzer and Goebeler, 2001). Chemokine-mediated chemoattraction of cells involved in granulation tissue formation, in conjunction with the interaction of these cells with ECM via cell-surface receptors, results in processes that lead to cell adhesion and migration into the area of the wound to form the granulation tissue (Lukacs and Kunkel, 1998; Martins-Green and Feugate, 1998; Feugate et al., 2002b).

One of the most extensively studied chemokines with functions important in wound healing is IL-8 (Martins-Green and Bissell, 1990; Martins-Green et al., 1992; Martins-Green and Hanafusa, 1997; Martins-Green and Feugate, 1998; Martins-Green 2001; Feugate et al., 2002a, 2002b). This has been well illustrated in studies performed using cIL-8/cCAF and chicks as model system. cIL-8 is stimulated to high levels shortly after wounding in the fibroblasts of the wounded tissue (Martins-Green and Bissell, 1990; Martins-Green et al., 1992), and thrombin, an enzyme involved in coagulation that is activated upon wounding, stimulates these cells to overexpress cIL-8 (Vaingankar and Martins-Green, 1998; Li et al., 2000). This chemokine then chemoattracts monocyte/macrophages and lymphocytes (Martins-Green and Feugate, 1998). We have shown that thrombin can promote further increases in hIL-8 levels by stimulation of hIL-8 expression in THP-1 differentiated macrophages (Zheng et al., 2007). Expression of cIL-8 remains elevated during granulation tissue formation due to its secretion by fibroblasts, the endothelial cells of the microvasculature of the wound, and macrophages, as well as from its binding to the interstitial collagens, tenascin, and laminin present in the granulation tissue (Martins-Green and Bissell, 1990; Martins-Green et al., 1992; Martins-Green et al., 1996). Furthermore, both hIL-8 and cIL-8 are angiogenic *in vivo*, and, in the case of cIL-8, the angiogenic portion of the molecule is localized in the C-terminus of the molecule (Martins-Green and Feugate, 1998; Martins-Green and Kelly, 1998). Based on the pattern of expression and functions of IL-8, it appears that this chemokine participates both in inflammation; via chemotaxis for specific leukocytes, and in the formation of the granulation tissue via stimulation of angiogenesis and ECM deposition (Martins-Green and Hanafusa, 1997; Martins-Green, 2001; Feugate et al., 2002b).

ECM interactions with endothelial cells are crucial in the cell migration and in the development of blood vessels during granulation tissue formation (Cockerill et al., 1995; Baldwin, 1996; Hanahan, 1997; Kumar et al., 1998; Li et al., 2003). Human umbilical vein endothelial cells migrate and arrange themselves in tubular structures when cultured for 12 h on a matrix isolated from Engelbreth-Holm-Swarm (EHS) tumors (a basement membrane-like matrix consisting primarily of laminin but also containing collagen IV, proteoglycans, and entactin/nidogen) (Kubota et al., 1988; Grant et al., 1989; Lawley and Kubota, 1989). When these cells are cultured on collagen I, however, tubular structures do not form in this period of time (Kubota et al., 1988), but if they are grown for a week inside collagen gels, giving the endothelial cells time to deposit their own basement membrane, tubes do develop (Montesano et al., 1983; Madri et al., 1988; Bell et al., 2001). The much more rapid tubulogenesis that occurs on EHS suggests that one or more components of the basement membrane plays an important role in the development of the capillary-like structures, a speculation confirmed both in culture and *in vivo* (Sakamoto et al., 1991; Grant et al., 1992). Indeed, preincubation of these endothelial cells with antibodies to laminin, the major component of basement membrane, prevents the formation of tubules *in vitro* (Kubota et al., 1988). Furthermore, synthetic peptides containing the sequence SIKVAV derived from the A chain of laminin induce endothelial cell adhesion and elongation and promote angiogenesis (Grant et al., 1992), while peptides containing the sequence YIGSR derived from the laminin B1 chain promote endothelial tube formation (Grant et al., 1989), although YIGSR peptides block angiogenesis *in vivo* (Sakamoto et al., 1991; Grant et al., 1992) and inhibit endothelial cell migration *in vitro* (Sakamoto et al., 1991). The mechanisms behind the ability of the YIGSR synthetic peptide to yield such different results *in vivo* may result from competition of this peptide with laminin for receptor binding, as this YIGSR peptide is known to block laminin binding to cells and block migration. If such competition does occur, the binding of the soluble YIGSR peptide to this receptor rather than YIGSR in the normal context of the complete laminin protein may alter downstream signaling events due to changes in the mechanical resistance and ligand presentation afforded by soluble, rather than intact, ligand, as has been suggested for integrin signaling (Vogel et al., 2001; Stupack and Cheresh, 2002). Regardless of the actual mechanism of action, the fact that soluble receptor-binding regions of ECM molecules may yield results different from those of the intact molecule may be of particular importance during matrix degradation, which releases ECM fragments. For example, matrix-degrading enzymes are activated during angiogenesis to facilitate the migration and invasion of endothelial cells into adjacent tissues and matrix; this matrix degradation may provide angiogenic or anti-angiogenic factors via release from the matrix or by appropriate cleavage of ECM molecules such as laminin (Werb et al., 1999; Rundhaug, 2005). *In vivo*, angiogenic sequences or factors could be provided locally, and when they have served their purpose, inhibition of further action could similarly be initiated by suitable cleavage to create CDPGYIGSR-NH2 or some other comparable factor present in the ECM (Sakamoto et al., 1991). Therefore, the way matrix molecules are locally cleaved and/or factors are locally released could have important consequences for the formation of the granulation tissue.

Proliferation

Immediately after wounding, the epithelium undergoes changes that lead to wound closure. During this re-epithelialization period, the keratinocytes trailing behind those at the front edge of migration replicate to provide a source of cells to cover the wound. Basement membrane-type ECM still present on the basal surface of these keratinocytes may be important in maintaining this proliferative state. In support of this possibility is the finding that during normal skin remodeling, fibronectin associated with the basal lamina of epithelia is crucial for maintaining the basal keratinocyte layer in a proliferative state for constant replenishment of the suprabasal layers (Nicholson and Watt, 1991). It has also been shown using a dermal wound model, that basement membrane matrices are able to sustain the proliferation of keratinocytes for several days (Dawson et al., 1996). The component of the basement membrane involved in this proliferation may be laminin, as laminin 10/11 can promote keratinocyte proliferation *in vitro* (Pouliot et al., 2002). In contrast, the fibrin-containing provisional matrix may prevent excessive keratinocyte proliferation, as the keratinocytes of fibrinogen-deficient mice do proliferate abnormally during re-epithelialization (Drew et al., 2001).

As re-epithelialization is occurring, the granulation tissue begins to form. This latter tissue is composed of fibroblasts, myofibroblasts, monocytes/macrophages, lymphocytes, endothelial cells of the microvasculature, and ECM molecules, including embryonic fibronectin, hyaluronic acid, type III collagen, and small amounts of type I collagen (Clark, 1996). These ECM molecules, in conjunction with growth factors released by the platelets and secreted by the cells present in the granulation tissue, provide signals to the cells which lead to their proliferation (Tuan et al., 1996; Bissell, 1998). However, ECM molecules themselves such as fibronectin, as well as specific fragments of fibronectin, laminin, collagen VI, and SPARC/osteonectin, have been shown to stimulate fibroblast and endothelial cell proliferation (Bitterman et al., 1983; Panayotou et al., 1989; Atkinson et al., 1996; Grant et al., 1998; Kapila et al., 1998; Ruhl et al., 1999; Sage et al., 2003). In the case of laminin, this proliferative activity appears to be mediated by its EGF-like domains (Panayotou et al., 1989), suggesting a potential dependence upon the activation of EGFR (Schenk et al., 2003; Koshikawa et al., 2005). In contrast, ECM molecules and/or peptides derived from their proteolysis can have inhibitory effects on cell proliferation; intact decorin (Sulochana et al., 2005) and SPARC (Funk and Sage, 1991; Chlenski et al., 2005), as well as peptides derived from decorin (Sulochana et al., 2005), SPARC (Sage et al., 2003), collagens XVIII and XV (endostatin) (O'Reilly et al., 1997; Sasaki et al., 2000), and collagen IV (tumstatin) (Hamano et al., 2003) have anti-angiogenic effects due to their inhibition of endothelial cell proliferation. ECM molecules may also cooperate with growth factors in the proliferation of fibroblasts and the development of new blood vessels in the granulation tissue. During this angiogenic process, growth factors such as VEGFs and fibroblast growth factors (FGFs) associate with ECM molecules and stimulate proliferation of endothelial cells which then migrate to form the new microvessels (Miao et al., 1996; Ikuta et al., 2000, 2001; Sottile, 2004); indeed, recent studies suggest that some anti-angiogenic molecules, including thrombospondin and endostatin, may inhibit angiogenesis by competition with these growth factors for ECM binding (Gupta et al., 1999; Reis et al., 2005). Conversely, ECM–growth factor interactions can be inhibitory, for example, VEGF binding of SPARC can inhibit VEGF-induced proliferation (Kupprion et al., 1998). In addition, the proliferation stimulated by growth factors may be dependent upon the presence of specific ECM molecules; for example, TGF-β_1 stimulation of fibroblast proliferation is dependent upon fibronectin (Clark et al., 1997).

Differentiation

As healing progresses during the formation of granulation tissue, some of the fibroblasts differentiate into myofibroblasts; they acquire the morphological and biochemical characteristics of smooth muscle cells by expressing α-smooth muscle actin (Desmouliere and Gabbiani, 1994; Desmouliere et al., 2005). Matrix molecules are important in this differentiation process. For example, heparin decreases the proliferation of fibroblasts in culture and induces the expression of α-smooth muscle actin in these cells. *In vivo*, the local application of tumor necrosis factor α leads to the development of granulation tissue, but the presence of cells expressing α-smooth muscle actin was only observed when heparin was also applied (Desmouliere et al., 1992). These results suggest that some of the properties of heparin not related to its anticoagulant effects are important in the induction of α-smooth muscle actin. This function may be related to the ability of heparin and heparin sulfate proteoglycans to bind cytokines and/or growth factors, such as TGFβ that regulate myofibroblast differentiation (Kim and Mooney, 1998; Kirkland et al., 1998; Menart et al., 2002; Li, J. et al., 2004).

Specific interactions with the ECM are also important for myofibroblast differentiation; inhibition of the ED-A-containing form of fibronectin or α_v or β_1 integrins can block TGF-β_1-mediated myofibroblast differentiation (Serini et al., 1998; Kato et al., 2001; Lygoe et al., 2004). In addition, cardiac fibroblasts undergo myofibroblast differentiation when plated on collagen VI (Naugle et al., 2005). Interstitial collagens have also been shown to play a role in the acquisition of the myofibroblastic phenotype. When fibroblasts are cultured on relaxed collagen gels or collagen-coated plates, they do not differentiate (Tomasek et al., 1992; Naugle et al., 2005); however, if they are grown on anchored collagen matrices where the collagen fibers are aligned (much like in the granulation tissue) they show myofibroblast characteristics (Bell et al., 1979; Arora et al., 1999). These observations led to the hypothesis that myofibroblast differentiation is regulated by mechanical tension; more recent studies *in vivo*, during wound healing, and *in vitro* have suggested that this hypothesis is, in fact, correct (Hinz et al., 2001; Wang et al., 2003).

Apoptosis

Apoptosis also plays a role during normal wound healing as the granulation tissue evolves into scar tissue. As the wound heals, the number of fibroblasts, myofibroblasts, endothelial cells, and pericytes decreases dramatically, matrix molecules, especially interstitial collagen, accumulate, and a scar forms (Clark, 1996). In this remodeling phase of healing, cell death by apoptosis leads to elimination of many cells of various types at once without causing tissue damage (Clark, 1996). For example, studies using transmission electron microscopy and *in situ* end-labeling of DNA fragments have shown that many myofibroblasts and endothelial cells undergo apoptosis during the remodeling process. Morphometric analysis of the granulation tissue showed that the number of cells undergoing apoptosis increases around days 20–25 after injury and this results in a dramatic reduction in cellularity after day 25 (Desmouliere et al., 1995); similar results were noted in cardiac granulation tissue following infarction (Takemura et al., 1998). Moreover, using model systems that mimic regression of granulation tissue, it has been shown that release of mechanical tension triggers apoptosis of human fibroblasts and myofibroblasts (Fluck et al., 1998; Grinnell et al., 1999; Bride et al., 2004). In these models, apoptotic cell death was regulated by interstitial-type collagens in combination with growth factors and mechanical tension and did not require differentiation of the fibroblasts into myofibroblasts, strongly suggesting that contractile collagens determine the susceptibility of fibroblasts of the wound tissue to undergo apoptotic cell death (Fluck et al., 1998; Grinnell et al., 1999). Further studies have also implicated the interactions between thrombospondin-1 and the $\alpha_v\beta_3$ integrin-CD47 complex in the mechanical tension-mediated stimulation of fibroblast apoptosis (Graf et al., 2002). Such apoptosis may be required for resolution of wound healing and the prevention of scarring. Indeed, fibroblast/myofibroblast apoptosis is reduced in keloid and hypertrophic scars, resulting in the excessive matrix accumulation and scarring (Ladin et al., 1998; Saed et al., 1998; Ishihara et al., 2000). In keloid scars, this decreased apoptosis may be due to p53 mutations and/or growth factor receptor over-expression (Ladin et al., 1998; Saed et al., 1998; Messadi et al., 1999; Ishihara et al., 2000; Moulin et al., 2004); in contrast, it is thought that apoptotic failure in hypertrophic scars results from an over-expression of tissue transglutaminase, leading to increased matrix breakdown and decreased collagen contraction (Linge et al., 2005). In addition to cell death by apoptosis, it has also been shown that bronchoalveolar lavage fluid collected during lung remodeling after injury can promote fibroblast cell death by a process that is distinct from that of necrosis or apoptosis (Polunovsky et al., 1993). Although this process of cell death has not been extensively studied, it suggests that there are other processes of programmed cell death which are distinct from apoptosis and occur preferentially in association with wound repair.

CELL–ECM INTERACTIONS DURING REGENERATION

True tissue regeneration following injury rarely occurs in vertebrate species, but it does occur in specific instances, such as during in fetal cutaneous wound healing, liver regeneration, and urodele amphibian limb regeneration. Unlike wound healing in normal adult animals, which is characterized by scarring, fetal cutaneous wounds heal without fibrosis and scar formation, leading to regeneration of the injured area. Similarly, after injury, injured liver very effectively restores both normal function and normal organ size by proliferation and differentiation of pre-existing cell types. The contribution of cell–ECM interactions to regeneration in fetal healing and liver regeneration are discussed below (Figure 5.3).

Fetal Wound Healing

Adhesion and Migration

Scarless fetal wounds have significant differences in cell–ECM interactions in the injured area when compared with scarring adult wounds; these changes occur due to alterations in the composition of the ECM molecules, the rate of their appearance after wounding, and their duration in the wound area. One crucial ECM molecule in fetal wound healing is hyaluronic acid, which appears to be necessary for the regenerative response; its removal from fetal wounds promotes a healing response more similar to that of adults (Mast et al., 1992), and treatment of normally scarring wounds or wound organ cultures with hyaluronic acid decreases scarring (Iocono et al., 1998a, b; Hu et al., 2003). Hyaluronic acid is present at higher levels (Krummel et al., 1987; Sawai et al., 1997) and for a longer duration in fetal skin wounds compared with adult wounds; the latter may result, in part, from the reduced activity of hyaluronidase in fetal wounds (West et al., 1997). Fetal fibroblasts also express higher levels of the hyaluronic acid receptor CD44 (Adolph et al., 1993), thus increasing receptor–ligand interactions which promote

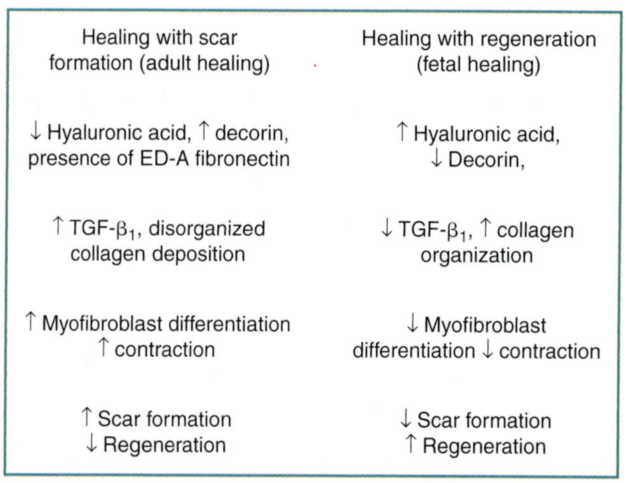

Figure 5.3 **A comparison of particular cell–ECM interactions occurring in scar-forming adult healing versus those occurring during regenerative fetal healing. As shown in this diagram, unique subsets of ECM molecules are associated with scarring versus regenerative healing. As such, therapeutic alteration of ECM composition may allow physicians to modulate healing to promote tissue regeneration. Additional therapeutic approaches may be generated upon further investigation into the importance of additional cell–ECM interactions in scarring and regenerative responses.**

fibroblast migration (Huang-Lee et al., 1994). Increased fetal hyaluronic acid may also facilitate fibroblast migration by decreasing or preventing expression of TGF-β_1, a factor that increases collagen I deposition (Ignotz and Massague, 1986) and inhibits fibroblast migration (Ellis et al., 1992; Hu et al., 2003).

Tenascin C is induced more rapidly and to a greater extent in fetal wounds, thus modulating cell adhesion to fibronectin (Whitby and Ferguson, 1991; Whitby et al., 1991). Fibronectin levels also increase more quickly in fetal wounds than adult wounds (Longaker et al., 1989). This increased expression of tenascin and fibronectin is associated with concomitant increases in the expression of integrins that serve as their receptors. In particular, the α_5 subunit, $\alpha_v\beta_3$, and $\alpha_v\beta_6$ integrins, which bind fibronectin and/or tenascin, are upregulated in the wounded fetal epithelium (Cass et al., 1998). The combined rapid increases in fibronectin and tenascin, coupled with increased expression of their respective integrin receptors in epithelial cells, are likely important in facilitating cell migration and re-epithelialization in fetal wounds.

In addition, fetal fibroblasts produce more collagen (Adzick et al., 1985; Longaker et al., 1990; Lovvorn et al., 1999; Gosiewska et al., 2001), particularly collagen type III (Hallock et al., 1988), than adult cells, and the organization of the fibrils in the fetal wound appears normal, while that of the adult wound exhibits an organization indicative of scarring (Whitby and Ferguson, 1991). The changes in the collagen levels and organization in fetal wounds may result from the increased expression in fetal fibroblasts of the collagen receptor DDR1, which is important in collagen expression and organization (Chin et al., 2001). Furthermore, hyaluronic acid increases collagen synthesis *in vitro*, and may thus contribute to increased collagen deposition in fetal wounds (Mast et al., 1993). In spite of the increased collagen production by fetal fibroblasts, the fetal wounds do not exhibit excessive collagen deposition and fibrosis; this may be due to rapid turnover of these ECM components by protease-mediated degradation. For example, levels of urokinase plasminogen activator (uPA) and MMPs are increased while the levels of their endogenous inhibitors, PAI-1 and tissue inhibitor of metalloproteinases (TIMPs), are decreased in the fetal wounds, ultimately promoting matrix degradation and turnover (Huang et al., 2002; Peled et al., 2002; Dang et al., 2003). Not only does this prevent fibrosis, it also likely facilitates cell migration by reducing matrix density and increases the generation of proteolytic matrix fragments that modulate various stages of wound repair, as mentioned above for laminin and collagen fragments that can alter angiogenesis during granulation tissue formation.

Proliferation

As mentioned above, during fetal wound healing, increased levels of hyaluronic acid are present and *in vitro* studies indicate that hyaluronic acid decreases fetal fibroblast proliferation (Mast et al., 1993). However, early studies comparing fetal wounds with those of newborns and adults showed an increase in fibroblast number in the wounded area in the fetal wounds, and fetal fibroblasts proliferate more rapidly than adult cells (Adzick et al., 1985; Khorramizadeh et al., 1999). It is unclear how these findings may be reconciled; however, it is possible that hyaluronic acid prevents excessive fibroblast proliferation in fetal wounds.

Another critical event in wound healing is re-epithelialization, which requires both keratinocyte migration and proliferation. Keratinocyte proliferation is decreased in mice lacking CD44 expression in keratinocytes (Kaya et al., 1997), suggesting that interactions between hyaluronic acid and CD44 may be important for keratinocyte proliferation during healing, and thus more effective re-epithelialization. This finding may explain, in part, the enhanced rate of healing seen in wounds treated with hyaluronic acid.

Differentiation

Fetal wounds have a decreased number of myofibroblasts, which appear in the wounded site earlier and remain a shorter time than in adult wounds; in fact, one study showed a lack of α-smooth muscle actin-expressing myofibroblasts in the wounds of early-stage fetuses (Estes et al., 1994). This is associated with a general lack of contraction in the fetal wounds themselves (Krummel et al., 1987). Increased levels of hyaluronic acid present during fetal wound healing may alter the differentiation and/or contractility of myofibroblasts in the wound site; studies *in vitro* have shown that addition of hyaluronic acid decreases fibroblast contraction of collagen matrices (Huang-Lee et al., 1994). This may be due, in part, to reduced expression of TGF-β_1, a major inducer of myofibroblast differentiation and fibrosis. Indeed, incisional adult wounds treated with hyaluronic acid healed more rapidly with a significant decrease in TGF-β_1 levels (Hu et al., 2003). The large amounts of hyaluronic acid in fetal wounds may thus explain the greatly reduced levels of TGF-β_1 in fetal wounds (Nath et al., 1994; Chen et al., 2005). Downregulation of TGF-β_1 in adult wounds produces a decrease in scarring similar to that observed with hyaluronic acid treatment (Choi et al., 1996). Conversely, studies have shown that the addition of TGF-β_1 to normally scarless fetal wounds induces a more scarring phenotype, with myofibroblast differentiation, wound contraction, and fibrosis (Lin et al., 1995; Lanning et al., 1999). Thus, hyaluronic acid-mediated inhibition of TGF-β_1 expression may be critical in scarless fetal healing.

If any TGF-β_1 is present during fetal wound healing, it may be regulated by inhibitory ECM molecules present in the injured area. One such inhibitor is fibromodulin, which is capable of binding TGF-β_1 and preventing receptor binding and is expressed to a greater extent in fetal wounds relative to adult wounds (Hildebrand et al., 1994; Soo et al., 2000). Another molecule that may alter TGF-β_1 activity is decorin, although the function of decorin in modulating TGF-β_1 activity is somewhat controversial; some studies indicate that decorin binding decreases TGF-β_1 activity (Noble et al., 1992), while others suggest that this interaction either has no effect on TGF-β_1 or even actually increases activity (Hausser et al., 1994; Takeuchi et al., 1994). The outcome of decorin– TGF-β_1 binding may depend upon the microenvironment, and this has not been extensively studied in fetal wounds. Regardless, decorin levels are decreased in scarless wounds, resulting in decreased decorin–TGF-β_1 interactions and altered TGF-β_1 activity (Beanes et al., 2001). Decreased activity of this growth factor, combined with extremely low levels of expression in fetal wounds, results in decreased fibrosis, myofibroblast differentiation, and wound contraction, leading to regeneration rather than scarring.

Apoptosis

Little is known regarding the apoptotic process in fetal wounds, and whether this differs from that of adult wounds. However, as in adult healing, multiple cell types present within the fetal granulation tissue likely disappear via apoptosis. It is also apparent that any myofibroblasts that do differentiate during granulation tissue formation disappear rapidly (Estes et al., 1994), perhaps due to an altered rate of apoptosis in these wounds. If changes in apoptotic efficiency do indeed occur, they may result from the decreased contraction, and thus decreased mechanical tension, in fetal wounds (Krummel et al., 1987), as well as altered collagen levels within the collagen matrix (Adzick et al., 1985; Longaker et al., 1990; Lovvorn et al., 1999; Gosiewska et al., 2001). It is also possible that apoptosis is not as critical in the healing of fetal wounds as in adult wounds; leukocyte influx and myofibroblast differentiation appear to be minimal in fetal wounds, and thus may not require large numbers of cells to undergo apoptosis for regeneration to occur (Estes et al., 1994; Harty et al., 2003).

Liver Regeneration

Adhesion and Migration

ECM–cell interactions are also altered during mammalian liver regeneration, leading to changes in adhesion and migration. One major molecule upregulated after liver injury is laminin (Martinez-Hernandez et al., 1991; Kato et al., 1992). Hepatocytes isolated soon after liver injury and plated on laminin attach more efficiently than non-injured hepatocytes suggesting a concomitant increase in laminin-binding integrins (Carlsson et al., 1981; Kato et al., 1992). Collagens I, III, IV, and V increase in regenerating liver several days after injury. Hepatocytes isolated from this stage of regenerating liver show increased adhesion to collagen, which may indicate increased expression of collagen adhesion receptors (Kato et al., 1992). The increased levels of laminin and collagen IV during regeneration may also promote hepatocyte migration, as both the basal and stimulated migration of hepatocytes is enhanced on laminin and collagen IV relative to other types of ECM (Ma et al., 1999).

Proliferation

In response to liver injury, hepatocytes proliferate to restore normal liver function and size. *In vitro* studies show that laminin enhances hepatocyte proliferation in general and in response to EGF; thus, the increased laminin present in regenerating tissue may facilitate proliferation (Hirata et al., 1983; Kato et al., 1992). Both the mRNA and the protein levels of plasma fibronectin and its receptor $\alpha_5\beta_1$ integrin increase in regenerating liver following injury (Gluck et al., 1992; Kato et al., 1992; Pujades et al., 1992), which may also increase proliferation. Indeed, intraperitoneal injection of plasma fibronectin further stimulates proliferation in the regenerating liver (Kwon et al., 1990b).

The primary growth factor responsible for hepatocyte proliferation is HGF; thus, processes that stimulate HGF production and/or release from matrix components will also increase hepatocyte numbers in regenerating liver. Heparan sulfate proteoglycans that are upregulated after injury bind HGF and promote its mitogenic activity (Matsumoto et al., 1993; Kato et al., 1994; Lai et al., 2004). Various proteoglycans are also upregulated after injury, potentially increasing HGF activity in the regenerating liver (Otsu et al., 1992; Gallai et al., 1996). Other ECM molecules are known to bind HGF with low affinity, possibly sequestering HGF in the ECM and preventing its activity (Schuppan et al., 1998). In fact, increased MMP expression during regeneration stimulates ECM degradation and hepatocyte proliferation. This increased proliferation is likely due to the proteolytic processing and release of matrix-bound HGF (Nishio et al., 2003; Mohammed et al., 2005). Increases in MMP production are followed by increased TIMP expression, which may prevent excessive hepatocyte proliferation (Rudolph et al., 1999; Mohammed et al., 2005). HGF, and thus hepatocyte proliferation, can also be activated by plasmin, suggesting a role for plasminogen activators in liver regeneration (Shimizu et al., 2001). Indeed, rapid increases in uPA activity after injury is followed by increases in plasmin activation and fibrinogen cleavage and a rapid loss of fibronectin, laminin, and entactin via proteolysis, although the levels of these latter proteins increase at later stages of healing (Kim et al., 1997). The importance of plasmin activation is underscored by studies in which the livers of uPA and tissue plasminogen activator (tPA) single and double knockout mice or plasminogen knockout mice were injured chemically (Bezerra et al., 1999, 2001). It was found that the plasminogen and uPA single knockouts, as well as the uPA/tPA double knockouts experienced significant liver regenerative problems accompanied by excessive fibrin and fibronectin, with a lesser effect seen in the tPA knockout. The observed disruption of regeneration may be due to a reduction of hepatocyte proliferation resulting from decreased HGF activity.

Differentiation

Myofibroblast differentiation can also occur from the stellate cells of the liver, which can then stimulate excessive ECM deposition, leading to fibrosis and cirrhosis rather than regeneration. Thus, myofibroblast differentiation must be very limited to allow appropriate liver regeneration. Plasma fibronectin levels are increased in the liver regenerating tissue, but are reduced in cirrhotic tissue (Kwon et al., 1990a; Chijiiwa et al., 1994). In addition, myofibroblast differentiation appears to require the ED-A domain of fibronectin (Serini et al., 1998; Kato et al., 2001), which is lacking in plasma fibronectin. These results, when taken together, suggest the possibility that plasma fibronectin may limit myofibroblast differentiation and fibrosis in the liver. This may be particularly important, given the increased quantity and activation of TGF-β_1, TGF-β_2, and TGF-β_3 in the regenerating liver, which would otherwise promote differentiation and fibrosis (Jakowlew et al., 1991). In contrast, the stellate cell differentiation state may be maintained by the basement membrane, which appears to both maintain the differentiation state of stellate cells and, *in vitro*, promote myofibroblast de-differentiation back to stellate cells (Friedman et al., 1989; Sohara et al., 2002).

Apoptosis

In liver regeneration, prevention of hepatocyte apoptosis is critical for regeneration, while increased apoptotic rates are associated with impaired regeneration. Indeed, extensive cell death following a large liver resection leads to liver failure rather than regeneration (Panis et al., 1997). Liver ischemia–reperfusion injury can also promote apoptosis and liver failure rather than regeneration (Takeda et al., 2002). In the latter case of ischemia–reperfusion injury, prevention of apoptosis can significantly reduce the incidence of liver failure, underscoring the relationship between apoptosis and impaired regeneration or failure (Vilatoba et al., 2005b). The lack of regeneration in such cases is associated with the upregulation of pro-apoptotic gene expression and the downregulation of pro-survival genes (Morita et al., 2002), and may thus be related to the inability of hepatocytes to proliferate under such pro-apoptotic conditions (Iimuro et al., 1998). This hypothesis is supported by studies indicating that apoptosis and liver failure resulting from extensive liver resection or ischemia–reperfusion injury can be largely prevented by treatment conditions that promote cell proliferation (Longo et al., 2005; Vilatoba et al., 2005a). The prevention of apoptosis may thus require ECM molecules that are important in promoting hepatocyte proliferation, including laminin (Hirata et al., 1983; Kato et al., 1992), plasma fibronectin (Kwon et al., 1990b), and HGF-binding proteoglycans (Matsumoto et al., 1993; Kato et al., 1994; Lai et al., 2004). Different MMPs are activated after ischemia–reperfusion injury when compared with forms of injury that regenerate (Cursio et al., 2002), perhaps leading to the degradation of a different profile of ECM proteins; the activation of specific MMPs is thought to promote hepatocyte proliferation by releasing matrix-sequestered HGF (Nishio et al., 2003; Mohammed et al., 2005). The activation of different MMPs and cleavage of different substrates may alter HGF release and subsequent proliferation, leaving these cells more susceptible to apoptosis. This idea is supported by a study in which liver with ischemia–reperfusion injury was treated with an MMP inhibitor, which decreased apoptosis and necrosis in the injured liver (Cursio et al., 2002).

Although apoptosis of hepatocytes disrupts the regenerative process, apoptosis of myofibroblastic hepatic stellate cells may be critical in preventing fibrosis and scarring during regeneration (Issa et al., 2001). These myofibroblastic hepatic stellate cells disappear via apoptosis (Saile et al., 1997; Issa et al., 2001), and also potentially by de-differentiation back to stellate cells (Friedman et al., 1989; Sohara et al., 2002). The apoptosis of these myofibroblastic cells seems to be dependent upon the activation of specific proteases and the subsequent degradation of matrix components. Mice expressing a collagen I gene that is resistant to proteolysis had decreased stellate cell myofibroblast apoptosis and increased fibrosis, and thus impaired regeneration, relative to wild type (Issa et al., 2003). These myofibroblasts also persist in plasminogen-deficient mice, and are associated with a general accumulation of non-degraded matrix components (Ng et al., 2001), further supporting a role for matrix degradation in the observed apoptosis. The matrix degradation important in apoptosis also likely involves the activation of MMPs, as inhibition of MMP activity using synthetic inhibitors or TIMP-1 (Murphy et al., 2002; Zhou et al., 2004) prevents apoptosis of myofibroblastic stellate cells *in vitro*, whereas MMP-9 activity promotes apoptosis of these cells (Zhou et al., 2004). In *in vitro* models of cutaneous wound healing, a release of mechanical tension within the collagen matrix (Fluck et al., 1998; Grinnell et al., 1999; Bride et al., 2004) can promote myofibroblast apoptosis. It is possible that a similar release of mechanical tension, perhaps via cleavage of collagen I, is critical for myofibroblast apoptosis in the liver. Proteolysis of ECM components may also contribute to stellate cell apoptosis by abolishing integrin signaling downstream of binding to these components. Experimental disruption of ECM–integrin binding via an RGD-containing peptide (Iwamoto et al., 1999) or various $\alpha_v\beta_3$ antagonists (Zhou et al., 2004) induce stellate cell apoptosis *in vitro*, further supporting a role for integrin-mediated signaling in this apoptotic event.

IMPLICATIONS FOR REGENERATIVE MEDICINE

One primary goal of studies comparing differences in cell–ECM interactions, and thus changes in signaling, that accompany regenerative and non-regenerative healing is to determine what types of interactions promote and which inhibit tissue regeneration (for an example, see Figure 5.3). After elucidating the functions of particular interactions, it may be possible to increase the regenerative response through (1) the induction of pro-regenerative ECM molecules or signaling events in the wounded area combined with (2) the antagonism of anti-regenerative/scarring interactions or signaling events using specific inhibitors. This discussion of regenerative medicine will focus upon possible strategies to promote regeneration in adult scarring wounds, thus causing adult wounds to more closely resemble fetal scarless wounds. Such an increased regenerative response would be particularly useful in the treatment of wounds that heal abnormally with increased scar formation, such as keloids and hypertrophic scars, ischemic reperfusion injury, and chronic inflammatory responses.

Different types of approaches may be used to increase pro-regenerative ECM levels in the wounded area, including the introduction of these molecules via direct application of the molecules themselves, through the addition of agents that increase their expression, or through the addition into the wound sites of cells producing these types of ECM that have been prepared to minimize immunogenicity. Several different ECM molecules are present at higher levels in fetal wounds than in adult wounds, including hyaluronic acid, tenascin, fibronectin, and collagen III (Krummel et al., 1987; Hallock et al., 1988; Longaker et al., 1989; Whitby and Ferguson, 1991; Whitby et al., 1991; Sawai et al., 1997), and may play important roles in the regeneration process. Thus, altering the levels of these molecules in a scarring wound may improve regeneration. Indeed, preliminary experiments in rat wounds suggest that hyaluronic acid treatment decreases both the time required for healing and the amount of scar formation (Hu et al., 2003), underscoring the potential for this molecule in therapeutics. It is possible that treatment with tenascin, fibronectin, or collagen III in addition to hyaluronic acid could yield even more favorable outcomes.

When attempting to promote regeneration, it is also imperative to inhibit events associated with scarring, including excessive ECM deposition, fibrosis, and contraction. During the adult healing process, these scar-associated processes are primarily controlled by the myofibroblast, a differentiated cell type that arises during the adult healing process but that is largely absent throughout fetal wound healing. As such, inhibition of myofibroblast differentiation or function along with the addition of pro-regenerative molecules may facilitate a stronger regenerative response. Inhibition of differentiation could be accomplished by blocking the factors that normally stimulate this process, such as TGF-β_1 (Lin et al., 1995; Lanning et al., 1999) and IL-8 (Feugate et al., 2002a), or by preventing fibroblast–ECM interactions that facilitate myofibroblast differentiation, such as ED-A-containing fibronectin (Serini et al., 1998; Kato et al., 2001). Hyaluronic acid and fibromodulin appear to decrease TGF-β_1 levels and activity, respectively, thus treatment of normally scarring wounds with these matrix components may thus decrease TGF-β_1-mediated scarring (Hildebrand et al., 1994; Soo et al., 2000; Hu et al., 2003). IL-8, on the other hand, is a chemokine that activates G-protein linked receptors, which are highly amenable to inhibition by small molecules, which could be used to reduce the effects of this chemokine on myofibroblast differentiation.

In summary, the recent surge in research regarding the ECM molecules themselves and their interactions with particular cells and cell-surface receptors has led to the realization that such interactions are many and complex, and that they are of the utmost importance in determining cell behavior during such events as wound repair and tissue regeneration. As such, the manipulation of specific cell–ECM interactions has the potential to modulate particular aspects of the repair process in order to promote a regenerative response.

REFERENCES

Adams, J.C. and Watt, F.M. (1993). Regulation of development and differentiation by the extracellular matrix. *Development* **117**: 1183–1198.

Adolph, V.R., Bleacher, J.C., Dillon, P.W. and Krummel, T.M. (1993). Hyaluronate uptake and CD-44 activity in fetal and adult fibroblasts. *J. Surg. Res.* **54**: 328–330.

Adzick, N.S., Harrison, M.R., Glick, P.L., Beckstead, J.H., Villa, R.L., Scheuenstuhl, H. and Goodson III, W.H. (1985). Comparison of fetal, newborn, and adult wound healing by histologic, enzyme-histochemical, and hydroxyproline determinations. *J. Pediatr. Surg.* **20**: 315–319.

Akiyama, S.K., Aota, S. and Yamada, K.M. (1995). Function and receptor specificity of a minimal 20 kilodalton cell adhesive fragment of fibronectin. *Cell Adhes. Commun.* **3**: 13–25.

Anderson, S.S., Kim, Y. and Tsilibary, E.C. (1994). Effects of matrix glycation on mesangial cell adhesion, spreading and proliferation. *Kidney Int.* **46**: 1359–1367.

Aota, S., Nomizu, M. and Yamada, K.M. (1994). The short amino acid sequence Pro-His-Ser-Arg-Asn in human fibronectin enhances cell-adhesive function. *J. Biol. Chem.* **269**: 24756–24761.

Aplin, A.E. and Juliano, R.L. (1999). Integrin and cytoskeletal regulation of growth factor signaling to the MAP kinase pathway. *J. Cell. Sci.* **112**(Pt 5): 695–706.

Aplin, A.E., Stewart, S.A., Assoian, R.K. and Juliano, R.L. (2001). Integrin-mediated adhesion regulates ERK nuclear translocation and phosphorylation of Elk-1. *J. Cell Biol.* **153**: 273–282.

Arora, P.D., Narani, N. and McCulloch, C.A. (1999). The compliance of collagen gels regulates transforming growth factor-beta induction of alpha-smooth muscle actin in fibroblasts. *Am. J. Pathol.* **154**: 871–882.

Asch, A.S., Liu, I., Briccetti, F.M., Barnwell, J.W., Kwakye-Berko, F., Dokun, A., Goldberger, J. and Pernambuco, M. (1993). Analysis of CD36 binding domains: ligand specificity controlled by dephosphorylation of an ectodomain. *Science* **262**: 1436–1440.

Atkinson, J.C., Ruhl, M., Becker, J., Ackermann, R. and Schuppan, D. (1996). Collagen VI regulates normal and transformed mesenchymal cell proliferation *in vitro*. *Exp. Cell Res.* **228**: 283–291.

Avdi, N.J., Nick, J.A., Whitlock, B.B., Billstrom, M.A., Henson, P.M., Johnson, G.L. and Worthen, G.S. (2001). Tumor necrosis factor-alpha activation of the c-Jun N-terminal kinase pathway in human neutrophils. Integrin involvement in a pathway leading from cytoplasmic tyrosine kinases apoptosis. *J. Biol. Chem.*, **276**: 2189–2199.

Bajorath, J., Greenfield, B., Munro, S.B., Day, A.J. and Aruffo, A. (1998). Identification of CD44 residues important for hyaluronan binding and delineation of the binding site. *J. Biol. Chem.* **273**: 338–343.

Baldwin, H.S. (1996). Early embryonic vascular development. *Cardiovasc. Res.* **31 Spec No:** E34–E45.

Banwell, C.M., Partington, K.M., Jenkinson, E.J. and Anderson, G. (2000). Studies on the role of IL-7 presentation by mesenchymal fibroblasts during early thymocyte development. *Eur. J. Immunol.* **30**: 2125–2129.

Bao, W., Thullberg, M., Zhang, H., Onischenko, A. and Stromblad, S. (2002). Cell attachment to the extracellular matrix induces proteasomal degradation of p21(CIP1) via Cdc42/Rac1 signaling. *Mol. Cell. Biol.* **22**: 4587–4597.

Bass, M.D. and Humphries, M.J. (2002). Cytoplasmic interactions of syndecan-4 orchestrate adhesion receptor and growth factor receptor signalling. *Biochem. J.* **368**: 1–15.

Beanes, S.R., Dang, C., Soo, C., Wang, Y., Urata, M., Ting, K., Fonkalsrud, E.W., Benhaim, P., Hedrick, M.H., Atkinson, J.B., et al. (2001). Down-regulation of decorin, a transforming growth factor-beta modulator, is associated with scarless fetal wound healing. *J. Pediatr. Surg.* **36**, 1666–1671.

Bell, E., Ivarsson, B. and Merrill, C. (1979). Production of a tissue-like structure by contraction of collagen lattices by human fibroblasts of different proliferative potential *in vitro*. *Proc. Natl Acad. Sci. USA* **76**: 1274–1278.

Bell, S.E., Mavila, A., Salazar, R., Bayless, K.J., Kanagala, S., Maxwell, S.A. and Davis, G.E. (2001). Differential gene expression during capillary morphogenesis in 3D collagen matrices: regulated expression of genes involved in basement membrane matrix assembly, cell cycle progression, cellular differentiation and G-protein signaling. *J. Cell. Sci.* **114**: 2755–2773.

Bernfield, M., Kokenyesi, R., Kato, M., Hinkes, M.T., Spring, J., Gallo, R.L. and Lose, E.J. (1992). Biology of the syndecans: a family of transmembrane heparan sulfate proteoglycans. *Annu. Rev. Cell. Biol.* **8**: 365–393.

Bernfield, M.R., Banerjee, S.D. and Cohn, R.H. (1972). Dependence of salivary epithelial morphology and branching morphogenesis upon acid mucopolysaccharide-protein (proteoglycan) at the epithelial surface. *J. Cell Biol.* **52**: 674–689.

Bernfield, M.R., Cohn, R.H. and Banerjee, S.D. (1973). Glycosaminoglycans and epithelial organ formation. *Am. Zool.* **13**: 1067–1083.

Bezerra, J.A., Bugge, T.H., Melin-Aldana, H., Sabla, G., Kombrinck, K.W., Witte, D.P. and Degen, J.L. (1999). Plasminogen deficiency leads to impaired remodeling after a toxic injury to the liver. *Proc. Natl Acad. Sci. USA* **96**: 15143–15148.

Bezerra, J.A., Currier, A.R., Melin-Aldana, H., Sabla, G., Bugge, T.H., Kombrinck, K.W. and Degen, J.L. (2001). Plasminogen activators direct reorganization of the liver lobule after acute injury. *Am. J. Pathol.* **158**: 921–929.

Bissell, D.M. (1998). Hepatic fibrosis as wound repair: a progress report. *J. Gastroenterol.* **33**: 295–302.

Bissell, M.J., Hall, H.G. and Parry, G. (1982). How does the extracellular matrix direct gene expression? *J. Theor. Biol.* **99**: 31–68.

Bitterman, P.B., Rennard, S.I., Adelberg, S. and Crystal, R.G. (1983). Role of fibronectin as a growth factor for fibroblasts. *J. Cell Biol.* **97**: 1925–1932.

Bonewald, L.F. (1999). Regulation and regulatory activities of transforming growth factor beta. *Crit. Rev. Eukaryot. Gene Expr.* **9**: 33–44.

Boudreau, N., Myers, C. and Bissell, M.J. (1995). From laminin to lamin: regulation of tissue-specific gene expression by the ECM. *Trends Cell Biol.* **5**: 1–4.

Bride, J., Viennet, C., Lucarz-Bietry, A. and Humbert, P. (2004). Indication of fibroblast apoptosis during the maturation of disc-shaped mechanically stressed collagen lattices. *Arch. Dermatol. Res.* **295**: 312–317.

Briesewitz, R., Kern, A. and Marcantonio, E.E. (1995). Assembly and function of integrin receptors is dependent on opposing alpha and beta cytoplasmic domains. *Mol. Biol. Cell* **6**: 997–1010.

Brooks, P.C., Montgomery, A.M., Rosenfeld, M., Reisfeld, R.A., Hu, T., Klier, G. and Cheresh, D.A. (1994). Integrin alpha v beta 3 antagonists promote tumor regression by inducing apoptosis of angiogenic blood vessels. *Cell* **79**: 1157–1164.

Brown, T.A., Bouchard, T., St John, T., Wayner, E. and Carter, W.G. (1991). Human keratinocytes express a new CD44 core protein (CD44E) as a heparan-sulfate intrinsic membrane proteoglycan with additional exons. *J. Cell Biol.* **113**: 207–221.

Bushkin-Harav, I. and Littauer, U.Z. (1998). Involvement of the YIGSR sequence of laminin in protein tyrosine phosphorylation. *FEBS Lett.* **424**: 243–247.

Carey, D.J. (1997). Syndecans: multifunctional cell-surface co-receptors. *Biochem. J.* **327**(Pt 1): 1–16.

Carlsson, R., Engvall, E., Freeman, A. and Ruoslahti, E. (1981). Laminin and fibronectin in cell adhesion: enhanced adhesion of cells from regenerating liver to laminin. *Proc. Natl Acad Sci. USA* **78**: 2403–2406.

Cass, D.L., Bullard, K.M., Sylvester, K.G., Yang, E.Y., Sheppard, D., Herlyn, M. and Adzick, N.S. (1998). Epidermal integrin expression is upregulated rapidly in human fetal wound repair. *J. Pediatr. Surg.* **33**: 312–316.

Cavani, A., Zambruno, G., Marconi, A., Manca, V., Marchetti, M. and Giannetti, A. (1993). Distinctive integrin expression in the newly forming epidermis during wound healing in humans. *J. Invest. Dermatol.* **101**: 600–604.

Chen, W., Fu, X., Ge, S., Sun, T., Zhou, G., Jiang, D. and Sheng, Z. (2005). Ontogeny of expression of transforming growth factor-beta and its receptors and their possible relationship with scarless healing in human fetal skin. *Wound Repair Regen.* **13**: 68–75.

Chijiiwa, K., Nakano, K., Kameoka, N., Nagai, E. and Tanaka, M. (1994). Proliferating cell nuclear antigen, plasma fibronectin, and liver regeneration rate after seventy percent hepatectomy in normal and cirrhotic rats. *Surgery* **116**: 544–549.

Chin, G.S., Lee, S., Hsu, M., Liu, W., Kim, W.J., Levinson, H. and Longaker, M.T. (2001). Discoidin domain receptors and their ligand, collagen, are temporally regulated in fetal rat fibroblasts *in vitro*. *Plast. Reconstr. Surg.* **107**: 769–776.

Chiquet-Ehrismann, R., Matsuoka, Y., Hofer, U., Spring, J., Bernasconi, C. and Chiquet, M. (1991). Tenascin variants: differential binding to fibronectin and distinct distribution in cell cultures and tissues. *Cell Regul.* **2**: 927–938.

Chlenski, A., Liu, S., Guerrero, L.J., Yang, Q., Tian, Y., Salwen, H.R., Zage, P. and Cohn, S.L. (2006). SPARC expression is associated with impaired tumor growth, inhibited angiogenesis and changes in the extracellular matrix. *Int. J. Cancer.* **118**: 310–316.

Cho, H.J., Youn, S.W., Cheon, S.I., Kim, T.Y., Hur, J., Zhang, S.Y., Lee, S.P., Park, K.W., Lee, M.M., Choi, Y.S., et al. (2005). Regulation of endothelial cell and endothelial progenitor cell survival and vasculogenesis by integrin-linked kinase. *Arterioscler. Thromb. Vasc. Biol.* **25**: 1154–1160.

Cho, S.Y. and Klemke, R.L. (2002). Purification of pseudopodia from polarized cells reveals redistribution and activation of Rac through assembly of a CAS/Crk scaffold. *J. Cell Biol.* **156**: 725–736.

Choi, B.M., Kwak, H.J., Jun, C.D., Park, S.D., Kim, K.Y., Kim, H.R. and Chung, H.T. (1996). Control of scarring in adult wounds using antisense transforming growth factor-beta 1 oligodeoxynucleotides. *Immunol. Cell Biol.* **74**: 144–150.

Chung, C.Y. and Erickson, H.P. (1994). Cell surface annexin II is a high affinity receptor for the alternatively spliced segment of tenascin-C. *J. Cell Biol.* **126**: 539–548.

Clark, R.A., McCoy, G.A., Folkvord, J.M. and McPherson, J.M. (1997). TGF-beta 1 stimulates cultured human fibroblasts to proliferate and produce tissue-like fibroplasia: a fibronectin matrix-dependent event. *J. Cell Physiol.* **170**: 69–80.

Clark, R.A.F., ed. (1996). *The Molecular and Cellular Biology of Wound Repair*. New York: Plenum Press.

Cockerill, G.W., Gamble, J.R. and Vadas, M.A. (1995). Angiogenesis: models and modulators. *Int. Rev. Cytol.* **159**: 113–160.

Cotman, C.W., Hailer, N.P., Pfister, K.K., Soltesz, I. and Schachner, M. (1998). Cell adhesion molecules in neural plasticity and pathology: similar mechanisms, distinct organizations? *Prog. Neurobiol.* **55**: 659–669.

Cotman, S.L., Halfter, W. and Cole, G.J. (1999). Identification of extracellular matrix ligands for the heparan sulfate proteoglycan agrin. *Exp. Cell Res.* **249**: 54–64.

Couchman, J.R. and Woods, A. (1996). Syndecans, signaling, and cell adhesion. *J. Cell Biochem.* **61**: 578–584.

Cox, E.A. and Huttenlocher, A. (1998). Regulation of integrin-mediated adhesion during cell migration. *Microsc. Res. Tech.* **43**: 412–419.

Culav, E.M., Clark, C.H. and Merrilees, M.J. (1999). Connective tissues: matrix composition and its relevance to physical therapy. *Phys. Ther.* **79**: 308–319.

Cursio, R., Mari, B., Louis, K., Rostagno, P., Saint-Paul, M.C., Giudicelli, J., Bottero, V., Anglard, P., Yiotakis, A., Dive, V., et al. (2002). Rat liver injury after normothermic ischemia is prevented by a phosphinic matrix metalloproteinase inhibitor. *FASEB J.* **16**: 93–95.

Cybulsky, A.V., Carbonetto, S., Cyr, M.D., McTavish, A.J. and Huang, Q. (1993). Extracellular matrix-stimulated phospholipase activation is mediated by beta 1-integrin. *Am. J. Physiol.* **264**: C323–C332.

Dang, C.M., Beanes, S.R., Lee, H., Zhang, X., Soo, C. and Ting, K. (2003). Scarless fetal wounds are associated with an increased matrix metalloproteinase-to-tissue-derived inhibitor of metalloproteinase ratio. *Plast. Reconstr. Surg.* **111**: 2273–2285.

Dawson, R.A., Goberdhan, N.J., Freedlander, E. and MacNeil, S. (1996). Influence of extracellular matrix proteins on human keratinocyte attachment, proliferation and transfer to a dermal wound model. *Burns* **22**: 93–100.

Dean, J.W. III, Chandrasekaran, S. and Tanzer, M.L. (1990). A biological role of the carbohydrate moieties of laminin. *J. Biol. Chem.* **265**: 12553–12562.

Decline, F. and Rousselle, P. (2001). Keratinocyte migration requires alpha2beta1 integrin-mediated interaction with the laminin 5 gamma2 chain. *J. Cell Sci.* **114**: 811–823.

Dedhar, S. (1999). Integrins and signal transduction. *Curr. Opin. Hematol.* **6**: 37–43.

Desmouliere, A. and Gabbiani, G. (1994). Modulation of fibroblastic cytoskeletal features during pathological situations: the role of extracellular matrix and cytokines. *Cell Motil. Cytoskel.* **29**: 195–203.

Desmouliere, A., Rubbia-Brandt, L., Grau, G. and Gabbiani, G. (1992). Heparin induces alpha-smooth muscle actin expression in cultured fibroblasts and in granulation tissue myofibroblasts. *Lab. Invest.* **67**: 716–726.

Desmouliere, A., Redard, M., Darby, I. and Gabbiani, G. (1995). Apoptosis mediates the decrease in cellularity during the transition between granulation tissue and scar. *Am. J. Pathol.* **146**: 56–66.

Desmouliere, A., Chaponnier, C. and Gabbiani, G. (2005). Tissue repair, contraction, and the myofibroblast. *Wound Repair Regen.* **13**: 7–12.

Dickeson, S.K., Walsh, J.J. and Santoro, S.A. (1997). Contributions of the I and EF hand domains to the divalent cation-dependent collagen binding activity of the alpha2beta1 integrin. *J. Biol. Chem.* **272**: 7661–7668.

Dockery, P., Khalid, J., Sarani, S.A., Bulut, H.E., Warren, M.A., Li, T.C. and Cooke, I.D. (1998). Changes in basement membrane thickness in the human endometrium during the luteal phase of the menstrual cycle. *Hum. Reprod. Update* **4**: 486–495.

Dolfi, F., Garcia-Guzman, M., Ojaniemi, M., Nakamura, H., Matsuda, M. and Vuori, K. (1998). The adaptor protein Crk connects multiple cellular stimuli to the JNK signaling pathway. *Proc. Natl Acad. Sci. USA* **95**: 15394–15399.

Drew, A.F., Liu, H., Davidson, J.M., Daugherty, C.C. and Degen, J.L. (2001). Wound-healing defects in mice lacking fibrinogen. *Blood* **97**: 3691–3698.

Edwards, G. and Streuli, C. (1995). Signalling in extracellular-matrix-mediated control of epithelial cell phenotype. *Biochem. Soc. Trans.* **23**: 464–468.

Ehnis, T., Dieterich, W., Bauer, M., Lampe, B. and Schuppan, D. (1996). A chondroitin/dermatan sulfate form of CD44 is a receptor for collagen XIV (undulin). *Exp. Cell Res.* **229**: 388–397.

Ekblom, P., Ekblom, M., Fecker, L., Klein, G., Zhang, H.Y., Kadoya, Y., Chu, M.L., Mayer, U. and Timpl, R. (1994). Role of mesenchymal nidogen for epithelial morphogenesis *in vitro*. *Development* **120**: 2003–2014.

Ellis, I., Grey, A.M., Schor, A.M. and Schor, S.L. (1992). Antagonistic effects of TGF-beta 1 and MSF on fibroblast migration and hyaluronic acid synthesis. Possible implications for dermal wound healing. *J. Cell Sci.* **102**(Pt 3): 447–456.

Entwistle, J., Hall, C.L. and Turley, E.A. (1996). HA receptors: regulators of signalling to the cytoskeleton. *J. Cell Biochem.* **61**: 569–577.

Estes, J.M., Vande Berg, J.S., Adzick, N.S., MacGillivray, T.E., Desmouliere, A. and Gabbiani, G. (1994). Phenotypic and functional features of myofibroblasts in sheep fetal wounds. *Differentiation* **56**: 173–181.

Faraci, E., Eck, M., Gerstmayer, B., Bosio, A. and Vogel, W.F. (2003). An extracellular matrix-specific microarray allowed the identification of target genes downstream of discoidin domain receptors. *Matrix Biol.* **22**: 373–381.

Faraldo, M.M., Deugnier, M.A., Thiery, J.P. and Glukhova, M.A. (2001). Growth defects induced by perturbation of beta1-integrin function in the mammary gland epithelium result from a lack of MAPK activation via the Shc and Akt pathways. *EMBO Rep.* **2**: 431–437.

Febbraio, M., Hajjar, D.P. and Silverstein, R.L. (2001). CD36: a class B scavenger receptor involved in angiogenesis, atherosclerosis, inflammation, and lipid metabolism. *J. Clin. Invest.* **108**: 785–791.

Ferri, N., Carragher, N.O. and Raines, E.W. (2004). Role of discoidin domain receptors 1 and 2 in human smooth muscle cell-mediated collagen remodeling: potential implications in atherosclerosis and lymphangioleiomyomatosis. *Am. J. Pathol.* **164**: 1575–1585.

Feugate, J.E., Li, Q., Wong, L. and Martins-Green, M. (2002a). The cxc chemokine cCAF stimulates differentiation of fibroblasts into myofibroblasts and accelerates wound closure. *J. Cell Biol.* **156**: 161–172.

Feugate, J.E., Wong, L., Li, Q.J. and Martins-Green, M. (2002b). The CXC chemokine cCAF stimulates precocious deposition of ECM molecules by wound fibroblasts, accelerating development of granulation tissue. *BMC Cell Biol.* **3**: 13.

Ffrench-Constant, C. and Hynes, R.O. (1989). Alternative splicing of fibronectin is temporally and spatially regulated in the chicken embryo. *Development* **106**: 375–388.

Flaumenhaft, R. and Rifkin, D.B. (1992). The extracellular regulation of growth factor action. *Mol. Biol. Cell* **3**: 1057–1065.

Fluck, J., Querfeld, C., Cremer, A., Niland, S., Krieg, T. and Sollberg, S. (1998). Normal human primary fibroblasts undergo apoptosis in three-dimensional contractile collagen gels. *J. Invest. Dermatol.* **110**: 153–157.

Fornaro, M. and Languino, L.R. (1997). Alternatively spliced variants: a new view of the integrin cytoplasmic domain. *Matrix Biol.* **16**: 185–193.

Francki, A., Motamed, K., McClure, T.D., Kaya, M., Murri, C., Blake, D.J., Carbon, J.G. and Sage, E.H. (2003). SPARC regulates cell cycle progression in mesangial cells via its inhibition of IGF-dependent signaling. *J. Cell Biochem.* **88**: 802–811.

Friedman, S.L., Roll, F.J., Boyles, J., Arenson, D.M. and Bissell, D.M. (1989). Maintenance of differentiated phenotype of cultured rat hepatic lipocytes by basement membrane matrix. *J. Biol. Chem.* **264**: 10756–10762.

Fuchs, E., Dowling, J., Segre, J., Lo, S.H. and Yu, Q.C. (1997). Integrators of epidermal growth and differentiation: distinct functions for beta 1 and beta 4 integrins. *Curr. Opin. Genet. Dev.* **7**: 672–682.

Fukai, F., Mashimo, M., Akiyama, K., Goto, T., Tanuma, S. and Katayama, T. (1998). Modulation of apoptotic cell death by extracellular matrix proteins and a fibronectin-derived antiadhesive peptide. *Exp. Cell Res.* **242**: 92–99.

Funk, S.E., and Sage, E.H. (1991). The Ca2(+)-binding glycoprotein SPARC modulates cell cycle progression in bovine aortic endothelial cells. *Proc. Natl Acad. Sci. USA* **88**: 2648–2652.

Gailit, J. and Ruoslahti, E. (1988). Regulation of the fibronectin receptor affinity by divalent cations. *J. Biol. Chem.* **263**: 12927–12932.

Gailit, J., Welch, M.P. and Clark, R.A. (1994). TGF-beta 1 stimulates expression of keratinocyte integrins during re-epithelialization of cutaneous wounds. *J. Invest. Dermatol.* **103**: 221–227.

Gallai, M., Sebestyen, A., Nagy, P., Kovalszky, I., Onody, T. and Thorgeirsson, S.S. (1996). Proteoglycan gene expression in rat liver after partial hepatectomy. *Biochem. Biophys. Res. Commun.* **228**: 690–694.

Ghert, M.A., Qi, W.N., Erickson, H.P., Block, J.A. and Scully, S.P. (2001). Tenascin-C splice variant adhesive/anti-adhesive effects on chondrosarcoma cell attachment to fibronectin. *Cell Struct. Funct.* **26**: 179–187.

Giancotti, F.G. (1997). Integrin signaling: specificity and control of cell survival and cell cycle progression. *Curr. Opin. Cell Biol.* **9**: 691–700.

Gillitzer, R. and Goebeler, M. (2001). Chemokines in cutaneous wound healing. *J. Leukoc. Biol.* **69**: 513–521.

Gluck, U., Rodriguez Fernandez, J.L., Pankov, R. and Ben-Ze'ev, A. (1992). Regulation of adherens junction protein expression in growth-activated 3T3 cells and in regenerating liver. *Exp. Cell Res.* **202**: 477–486.

Gosiewska, A., Yi, C.F., Brown, L.J., Cullen, B., Silcock, D. and Geesin, J.C. (2001). Differential expression and regulation of extracellular matrix-associated genes in fetal and neonatal fibroblasts. *Wound Repair Regen.* **9**: 213–222.

Graf, R., Freyberg, M., Kaiser, D. and Friedl, P. (2002). Mechanosensitive induction of apoptosis in fibroblasts is regulated by thrombospondin-1 and integrin associated protein (CD47). *Apoptosis* **7**: 493–498.

Granes, F., Garcia, R., Casaroli-Marano, R.P., Castel, S., Rocamora, N., Reina, M., Urena, J.M. and Vilaro, S. (1999). Syndecan-2 induces filopodia by active cdc42Hs. *Exp. Cell Res.* **248**: 439–456.

Grant, D.S., Tashiro, K., Segui-Real, B., Yamada, Y., Martin, G.R. and Kleinman, H.K. (1989). Two different laminin domains mediate the differentiation of human endothelial cells into capillary-like structures *in vitro*. *Cell* **58**: 933–943.

Grant, D.S., Kinsella, J.L., Fridman, R., Auerbach, R., Piasecki, B.A., Yamada, Y., Zain, M. and Kleinman, H.K. (1992). Interaction of endothelial cells with a laminin A chain peptide (SIKVAV) *in vitro* and induction of angiogenic behavior *in vivo*. *J. Cell Physiol.* **153**: 614–625.

Grant, M.B., Caballero, S., Bush, D.M. and Spoerri, P.E. (1998). Fibronectin fragments modulate human retinal capillary cell proliferation and migration. *Diabetes* **47**: 1335–1340.

Gratchev, A., Kzhyshkowska, J., Utikal, J. and Goerdt, S. (2005). Interleukin-4 and dexamethasone counterregulate extracellular matrix remodelling and phagocytosis in type-2 macrophages. *Scand. J. Immunol.* **61**: 10–17.

Greiling, D. and Clark, R.A. (1997). Fibronectin provides a conduit for fibroblast transmigration from collagenous stroma into fibrin clot provisional matrix. *J. Cell Sci.* **110**(Pt 7): 861–870.

Grinnell, F., Zhu, M., Carlson, M.A. and Abrams, J.M. (1999). Release of mechanical tension triggers apoptosis of human fibroblasts in a model of regressing granulation tissue. *Exp. Cell Res.* **248**: 608–619.

Grumet, M., Milev, P., Sakurai, T., Karthikeyan, L., Bourdon, M., Margolis, R.K. and Margolis, R.U. (1994). Interactions with tenascin and differential effects on cell adhesion of neurocan and phosphacan, two major chondroitin sulfate proteoglycans of nervous tissue. *J. Biol. Chem.* **269**: 12142–12146.

Gullberg, D. and Ekblom, P. (1995). Extracellular matrix and its receptors during development. *Int. J. Dev. Biol.* **39**: 845–854.

Gupta, K., Gupta, P., Wild, R., Ramakrishnan, S. and Hebbel, R.P. (1999). Binding and displacement of vascular endothelial growth factor (VEGF) by thrombospondin: effect on human microvascular endothelial cell proliferation and angiogenesis. *Angiogenesis* **3**: 147–158.

Hagel, M., George, E.L., Kim, A., Tamimi, R., Opitz, S.L., Turner, C.E., Imamoto, A. and Thomas, S.M. (2002). The adaptor protein paxillin is essential for normal development in the mouse and is a critical transducer of fibronectin signaling. *Mol. Cell. Biol.* **22**: 901–915.

Hall, C.L., Wang, C., Lange, L.A. and Turley, E.A. (1994). Hyaluronan and the hyaluronan receptor RHAMM promote focal adhesion turnover and transient tyrosine kinase activity. *J. Cell Biol.* **126**: 575–588.

Hallock, G.G., Rice, D.C., Merkel, J.R. and DiPaolo, B.R. (1988). Analysis of collagen content in the fetal wound. *Ann. Plast. Surg.* **21**: 310–315.

Hamano, Y., Zeisberg, M., Sugimoto, H., Lively, J.C., Maeshima, Y., Yang, C., Hynes, R.O., Werb, Z., Sudhakar, A. and Kalluri, R. (2003). Physiological levels of tumstatin, a fragment of collagen IV alpha3 chain, are generated by MMP-9 proteolysis and suppress angiogenesis via alphaV beta3 integrin. *Cancer Cell* **3**: 589–601.

Hanahan, D. (1997). Signaling vascular morphogenesis and maintenance. *Science* **277**: 48–50.

Hardingham, T.E. and Fosang, A.J. (1992). Proteoglycans: many forms and many functions. *FASEB J.* **6**: 861–870.

Hardwick, C., Hoare, K., Owens, R., Hohn, H.P., Hook, M., Moore, D., Cripps, V., Austen, L., Nance, D.M. and Turley, E.A. (1992). Molecular cloning of a novel hyaluronan receptor that mediates tumor cell motility. *J. Cell Biol.* **117**: 1343–1350.

Harty, M., Neff, A.W., King, M.W. and Mescher, A.L. (2003). Regeneration or scarring: an immunologic perspective. *Dev. Dyn.* **226**: 268–279.

Hausser, H., Groning, A., Hasilik, A., Schonherr, E. and Kresse, H. (1994). Selective inactivity of TGF-beta/decorin complexes. *FEBS Lett.* **353**: 243–245.

Hay, E.D. (1977). Interaction between the cell surface and extracellular matrix in corneal development. In: Lash, J.W. and Burger, M.M. (eds.), *Cell and Tissue Interactions*. New York: Raven Press, pp. 115–137.

Herard, A.L., Pierrot, D., Hinnrasky, J., Kaplan, H., Sheppard, D., Puchelle, E. and Zahm, J.M. (1996). Fibronectin and its alpha 5 beta 1-integrin receptor are involved in the wound-repair process of airway epithelium. *Am. J. Physiol.* **271**: L726–L733.

Hergott, G.J., Nagai, H. and Kalnins, V.I. (1993). Inhibition of retinal pigment epithelial cell migration and proliferation with monoclonal antibodies against the beta 1 integrin subunit during wound healing in organ culture. *Invest. Ophthalmol. Vis. Sci.* **34**: 2761–2768.

Hildebrand, A., Romaris, M., Rasmussen, L.M., Heinegard, D., Twardzik, D.R., Border, W.A. and Ruoslahti, E. (1994). Interaction of the small interstitial proteoglycans biglycan, decorin and fibromodulin with transforming growth factor beta. *Biochem. J.* **302**(Pt 2): 527–534.

Hildebrand, J.D., Schaller, M.D. and Parsons, J.T. (1995). Paxillin, a tyrosine phosphorylated focal adhesion-associated protein binds to the carboxyl terminal domain of focal adhesion kinase. *Mol. Biol. Cell* **6**: 637–647.

Hinz, B., Mastrangelo, D., Iselin, C.E., Chaponnier, C. and Gabbiani, G. (2001). Mechanical tension controls granulation tissue contractile activity and myofibroblast differentiation. *Am. J. Pathol.* **159**: 1009–1020.

Hirata, K., Usui, T., Koshiba, H., Maruyama, Y., Oikawa, I., Freeman, A.E., Shiramatsu, K. and Hayasaka, H. (1983). Effects of basement membrane matrix on the culture of fetal mouse hepatocytes. *Gann* **74**: 687–692.

Hirsch, E., Barberis, L., Brancaccio, M., Azzolino, O., Xu, D., Kyriakis, J.M., Silengo, L., Giancotti, F.G., Tarone, G., Fassler, R., et al. (2002). Defective Rac-mediated proliferation and survival after targeted mutation of the beta1 integrin cytodomain. *J. Cell Biol.* **157**: 481–492.

Hodge-Dufour, J., Noble, P.W., Horton, M.R., Bao, C., Wysoka, M., Burdick, M.D., Strieter, R.M., Trinchieri, G. and Pure, E. (1997). Induction of IL-12 and chemokines by hyaluronan requires adhesion-dependent priming of resident but not elicited macrophages. *J. Immunol.* **159**: 2492–2500.

Hoffmann, S., He, S., Jin, M., Ehren, M., Wiedemann, P., Ryan, S.J. and Hinton, D.R. (2005). A selective cyclic integrin antagonist blocks the integrin receptors alphavbeta3 and alphavbeta5 and inhibits retinal pigment epithelium cell attachment, migration and invasion. *BMC Ophthalmol.* **5**: 16.

Housey, G.M., Johnson, M.D., Hsiao, W.L., O'Brian, C.A., Murphy, J.P., Kirschmeier, P. and Weinstein, I.B. (1988). Overproduction of protein kinase C causes disordered growth control in rat fibroblasts. *Cell* **52**: 343–354.

Hu, M., Sabelman, E.E., Cao, Y., Chang, J. and Hentz, V.R. (2003). Three-dimensional hyaluronic acid grafts promote healing and reduce scar formation in skin incision wounds. *J. Biomed. Mater. Res. B Appl. Biomater.* **67**: 586–592.

Huang, E.Y., Wu, H., Island, E.R., Chong, S.S., Warburton, D., Anderson, K.D. and Tuan, T.L. (2002). Differential expression of urokinase-type plasminogen activator and plasminogen activator inhibitor-1 in early and late gestational mouse skin and skin wounds. *Wound Repair Regen.* **10**: 387–396.

Huang, M.M., Bolen, J.B., Barnwell, J.W., Shattil, S.J. and Brugge, J.S. (1991). Membrane glycoprotein IV (CD36) is physically associated with the Fyn, Lyn, and Yes protein-tyrosine kinases in human platelets. *Proc. Natl Acad. Sci. USA* **88**: 7844–7848.

Huang-Lee, L.L., Wu, J.H. and Nimni, M.E. (1994). Effects of hyaluronan on collagen fibrillar matrix contraction by fibroblasts. *J. Biomed. Mater. Res.* **28**: 123–132.

Humphries, M.J., Mould, A.P. and Yamada, K.M. (1991). Matrix receptors in cell migration. In McDonald, J.A. and Mecham, R.P. (eds.), *Receptors for Extracellular Matrix*. San Diego: Academic Press, 195–253.

Hynes, R.O. (1987). Integrins: a family of cell surface receptors. *Cell* **48**: 549–554.

Hynes, R.O. (1992). Integrins: versatility, modulation, and signaling in cell adhesion. *Cell* **69**: 11–25.

Hynes, R.O. (1999). The dynamic dialogue between cells and matrices: implications of fibronectin's elasticity. *Proc. Natl Acad. Sci. USA* **96**: 2588–2590.

Hynes, R.O. (2002). Integrins: bidirectional, allosteric signaling machines. *Cell* **110**: 673–687.

Ignotz, R.A. and Massague, J. (1986). Transforming growth factor-beta stimulates the expression of fibronectin and collagen and their incorporation into the extracellular matrix. *J. Biol. Chem.* **261**: 4337–4345.

Iimuro, Y., Nishiura, T., Hellerbrand, C., Behrns, K.E., Schoonhoven, R., Grisham, J.W. and Brenner, D.A. (1998). NFkappaB prevents apoptosis and liver dysfunction during liver regeneration. *J. Clin. Invest.* **101**: 802–811.

Ikuta, T., Ariga, H. and Matsumoto, K. (2000). Extracellular matrix tenascin-X in combination with vascular endothelial growth factor B enhances endothelial cell proliferation. *Genes Cells* **5**: 913–927.

Ikuta, T., Ariga, H. and Matsumoto, K.I. (2001). Effect of tenascin-X together with vascular endothelial growth factor A on cell proliferation in cultured embryonic hearts. *Biol. Pharm. Bull.* **24**: 1320–1323.

Ilan, N., Mahooti, S. and Madri, J.A. (1998). Distinct signal transduction pathways are utilized during the tube formation and survival phases of *in vitro* angiogenesis. *J. Cell Sci.* **111**(Pt 24): 3621–3631.

Ingber, D. (1991). Extracellular matrix and cell shape: potential control points for inhibition of angiogenesis. *J. Cell Biochem.* **47**: 236–241.

Iocono, J.A., Ehrlich, H.P., Keefer, K.A. and Krummel, T.M. (1998a). Hyaluronan induces scarless repair in mouse limb organ culture. *J. Pediatr. Surg.* **33**: 564–567.

Iocono, J.A., Krummel, T.M., Keefer, K.A., Allison, G.M. and Paul, H. (1998b). Repeated additions of hyaluronan alters granulation tissue deposition in sponge implants in mice. *Wound Repair Regen.* **6**: 442–448.

Iozzo, R.V., Moscatello, D.K., McQuillan, D.J. and Eichstetter, I. (1999). Decorin is a biological ligand for the epidermal growth factor receptor. *J. Biol. Chem.* **274**: 4489–4492.

Isenberg, J.S., Ridnour, L.A., Perruccio, E.M., Espey, M.G., Wink, D.A. and Roberts, D.D. (2005). Thrombospondin-1 inhibits endothelial cell responses to nitric oxide in a cGMP-dependent manner. *Proc. Natl Acad. Sci. USA* **102**, 13141–13146.

Ishihara, H., Yoshimoto, H., Fujioka, M., Murakami, R., Hirano, A., Fujii, T., Ohtsuru, A., Namba, H. and Yamashita, S. (2000). Keloid fibroblasts resist ceramide-induced apoptosis by overexpression of insulin-like growth factor I receptor. *J. Invest. Dermatol.* **115**: 1065–1071.

Ishii, S., Ford, R., Thomas, P., Nachman, A., Steele Jr., G. and Jessup, J.M. (1993). CD44 participates in the adhesion of human colorectal carcinoma cells to laminin and type IV collagen. *Surg. Oncol.* **2**: 255–264.

Ishii, S., Steele Jr., G., Ford, R., Paliotti, G., Thomas, P., Andrews, C., Hansen, H.J., Goldenberg, D.M. and Jessup, J.M. (1994). Normal colonic epithelium adheres to carcinoembryonic antigen and type IV collagen. *Gastroenterology* **106**: 1242–1250.

Issa, R., Williams, E., Trim, N., Kendall, T., Arthur, M.J., Reichen, J., Benyon, R.C. and Iredale, J.P. (2001). Apoptosis of hepatic stellate cells: involvement in resolution of biliary fibrosis and regulation by soluble growth factors. *Gut* **48**: 548–557.

Issa, R., Zhou, X., Trim, N., Millward-Sadler, H., Krane, S., Benyon, C. and Iredale, J. (2003). Mutation in collagen-1 that confers resistance to the action of collagenase results in failure of recovery from CCl4-induced liver fibrosis, persistence of activated hepatic stellate cells, and diminished hepatocyte regeneration. *FASEB J.* **17**: 47–49.

Iwamoto, H., Sakai, H., Tada, S., Nakamuta, M. and Nawata, H. (1999). Induction of apoptosis in rat hepatic stellate cells by disruption of integrin-mediated cell adhesion. *J. Lab. Clin. Med.* **134**: 83–89.

Jakowlew, S.B., Mead, J.E., Danielpour, D., Wu, J., Roberts, A.B. and Fausto, N. (1991). Transforming growth factor-beta (TGF-beta) isoforms in rat liver regeneration: messenger RNA expression and activation of latent TGF-beta. *Cell Regul.* **2**: 535–548.

Jalkanen, S. and Jalkanen, M. (1992). Lymphocyte CD44 binds the COOH-terminal heparin-binding domain of fibronectin. *J. Cell Biol.* **116**: 817–825.

Jalkehen, M., Jalkanen, S. and Bernfield, M. (1991). Binding of extracellular effector molecules by cell surface proteoglycans. In: McDonald, J.A. and Mecham, R.P. (eds.), *Receptors for Extracellular Matrix*. San Diego: Academic Press, 1–38.

Jimenez, B., Volpert, O.V., Crawford, S.E., Febbraio, M., Silverstein, R.L. and Bouck, N. (2000). Signals leading to apoptosis-dependent inhibition of neovascularization by thrombospondin-1. *Nat. Med.* **6**: 41–48.

Jimenez, B., Volpert, O.V., Reiher, F., Chang, L., Munoz, A., Karin, M. and Bouck, N. (2001). c-Jun N-terminal kinase activation is required for the inhibition of neovascularization by thrombospondin-1. *Oncogene* **20**: 3443–3448.

Juhasz, I., Murphy, G.F., Yan, H.C., Herlyn, M. and Albelda, S.M. (1993). Regulation of extracellular matrix proteins and integrin cell substratum adhesion receptors on epithelium during cutaneous human wound healing *in vivo*. *Am. J. Pathol.* **143**: 1458–1469.

Kagami, S., Kondo, S., Loster, K., Reutter, W., Urushihara, M., Kitamura, A., Kobayashi, S. and Kuroda, Y. (1998). Collagen type I modulates the platelet-derived growth factor (PDGF) regulation of the growth and expression of beta1 integrins by rat mesangial cells. *Biochem. Biophys. Res. Commun.* **252**: 728–732.

Kamei, M., Kawasaki, A. and Tano, Y. (1998). Analysis of extracellular matrix synthesis during wound healing of retinal pigment epithelial cells. *Microsc. Res. Tech.* **42**: 311–316.

Kapila, Y.L., Lancero, H. and Johnson, P.W. (1998). The response of periodontal ligament cells to fibronectin. *J. Periodontol.* **69**: 1008–1019.

Kato, R., Kamiya, S., Ueki, M., Yajima, H., Ishii, T., Nakamura, H., Katayama, T. and Fukai, F. (2001). The fibronectin-derived antiadhesive peptides suppress the myofibroblastic conversion of rat hepatic stellate cells. *Exp. Cell Res.* **265**: 54–63.

Kato, S., Ishii, T., Hara, H., Sugiura, N., Kimata, K. and Akamatsu, N. (1994). Hepatocyte growth factor immobilized onto culture substrates through heparin and matrigel enhances DNA synthesis in primary rat hepatocytes. *Exp. Cell Res.* **211**: 53–58.

Kato, S., Otsu, K., Ohtake, K., Kimura, Y., Yashiro, T., Suzuki, T. and Akamatsu, N. (1992). Concurrent changes in sinusoidal expression of laminin and affinity of hepatocytes to laminin during rat liver regeneration. *Exp. Cell Res.* **198**: 59–68.

Kaya, G., Rodriguez, I., Jorcano, J.L., Vassalli, P. and Stamenkovic, I. (1997). Selective suppression of CD44 in keratinocytes of mice bearing an antisense CD44 transgene driven by a tissue-specific promoter disrupts hyaluronate metabolism in the skin and impairs keratinocyte proliferation. *Genes Dev.* **11**: 996–1007.

Kettritz, R., Xu, Y.X., Kerren, T., Quass, P., Klein, J.B., Luft, F.C. and Haller, H. (1999). Extracellular matrix regulates apoptosis in human neutrophils. *Kidney Int.* **55**: 562–571.

Khorramizadeh, M.R., Tredget, E.E., Telasky, C., Shen, Q. and Ghahary, A. (1999). Aging differentially modulates the expression of collagen and collagenase in dermal fibroblasts. *Mol. Cell. Biochem.* **194**: 99–108.

Khwaja, A., Rodriguez-Viciana, P., Wennstrom, S., Warne, P.H. and Downward, J. (1997). Matrix adhesion and Ras transformation both activate a phosphoinositide 3-OH kinase and protein kinase B/Akt cellular survival pathway. *EMBO J.* **16**: 2783–2793.

Kielty, C.M., Wess, T.J., Haston, L., Ashworth, J.L., Sherratt, M.J. and Shuttleworth, C.A. (2002). Fibrillin-rich microfibrils: elastic biopolymers of the extracellular matrix. *J. Muscle Res. Cell Motil.* **23**: 581–596.

Kim, B.S. and Mooney, D.J. (1998). Engineering smooth muscle tissue with a predefined structure. *J. Biomed. Mater. Res.* **41**: 322–332.

Kim, C.W., Goldberger, O.A., Gallo, R.L. and Bernfield, M. (1994). Members of the syndecan family of heparan sulfate proteoglycans are expressed in distinct cell-, tissue-, and development-specific patterns. *Mol. Biol. Cell* **5**: 797–805.

Kim, T.H., Mars, W.M., Stolz, D.B., Petersen, B.E. and Michalopoulos, G.K. (1997). Extracellular matrix remodeling at the early stages of liver regeneration in the rat. *Hepatology* **26**: 896–904.

Kirkland, G., Paizis, K., Wu, L.L., Katerelos, M. and Power, D.A. (1998). Heparin-binding EGF-like growth factor mRNA is upregulated in the peri-infarct region of the remnant kidney model: *in vitro* evidence suggests a regulatory role in myofibroblast transformation. *J. Am. Soc. Nephrol.* **9**: 1464–1473.

Kiyokawa, E., Hashimoto, Y., Kobayashi, S., Sugimura, H., Kurata, T. and Matsuda, M. (1998). Activation of Rac1 by a Crk SH3-binding protein, DOCK180. *Genes Dev.* **12**: 3331–3336.

Klemke, R.L., Leng, J., Molander, R., Brooks, P.C., Vuori, K. and Cheresh, D.A. (1998). CAS/Crk coupling serves as a "molecular switch" for induction of cell migration. *J. Cell Biol.* **140**: 961–972.

Koshikawa, N., Minegishi, T., Sharabi, A., Quaranta, V. and Seiki, M. (2005). Membrane-type matrix metalloproteinase-1 (MT1-MMP) is a processing enzyme for human laminin gamma 2 chain. *J. Biol. Chem.* **280**: 88–93.

Krummel, T.M., Nelson, J.M., Diegelmann, R.F., Lindblad, W.J., Salzberg, A.M., Greenfield, L.J. and Cohen, I.K. (1987). Fetal response to injury in the rabbit. *J. Pediatr. Surg.* **22**: 640–644.

Kubota, Y., Kleinman, H.K., Martin, G.R. and Lawley, T.J. (1988). Role of laminin and basement membrane in the morphological differentiation of human endothelial cells into capillary-like structures. *J. Cell Biol.* **107**: 1589–1598.

Kumar, C.C. (1998). Signaling by integrin receptors. *Oncogene* **17**: 1365–1373.

Kumar, R., Yoneda, J., Bucana, C.D. and Fidler, I.J. (1998). Regulation of distinct steps of angiogenesis by different angiogenic molecules. *Int. J. Oncol.* **12**: 749–757.

Kupprion, C., Motamed, K. and Sage, E.H. (1998). SPARC (BM-40, osteonectin) inhibits the mitogenic effect of vascular endothelial growth factor on microvascular endothelial cells. *J. Biol. Chem.* **273**: 29635–29640.

Kutz, S.M., Hordines, J., McKeown-Longo, P.J. and Higgins, P.J. (2001). TGF-beta1-induced PAI-1 gene expression requires MEK activity and cell-to-substrate adhesion. *J. Cell Sci.* **114**: 3905–3914.

Kwon, A.H., Inada, Y., Uetsuji, S., Yamamura, M., Hioki, K. and Yamamoto, M. (1990a). Response of fibronectin to liver regeneration after hepatectomy. *Hepatology* **11**: 593–598.

Kwon, A.H., Uetsuji, S., Yamamura, M., Hioki, K. and Yamamoto, M. (1990b). Effect of administration of fibronectin or aprotinin on liver regeneration after experimental hepatectomy. *Ann. Surg.* **211**: 295–300.

Ladin, D.A., Hou, Z., Patel, D., McPhail, M., Olson, J.C., Saed, G.M. and Fivenson, D.P. (1998). p53 and apoptosis alterations in keloids and keloid fibroblasts. *Wound Repair Regen.* **6**: 28–37.

Lai, J.P., Chien, J.R., Moser, D.R., Staub, J.K., Aderca, I., Montoya, D.P., Matthews, T.A., Nagorney, D.M., Cunningham, J.M., Smith, D.I., et al. (2004). hSulf1 Sulfatase promotes apoptosis of hepatocellular cancer cells by decreasing heparin-binding growth factor signaling. *Gastroenterology* **126**: 231–248.

Lamszus, K., Joseph, A., Jin, L., Yao, Y., Chowdhury, S., Fuchs, A., Polverini, P.J., Goldberg, I. D. and Rosen, E.M. (1996). Scatter factor binds to thrombospondin and other extracellular matrix components. *Am. J. Pathol.* **149**: 805–819.

Lane, T.F. and Sage, E.H. (1994). The biology of SPARC, a protein that modulates cell–matrix interactions. *FASEB J.* **8**: 163–173.

Lanning, D.A., Nwomeh, B.C., Montante, S.J., Yager, D.R., Diegelmann, R.F. and Haynes, J.H. (1999). TGF-beta1 alters the healing of cutaneous fetal excisional wounds. *J. Pediatr. Surg.* **34**: 695–700.

LaRochelle, W.J., May-Siroff, M., Robbins, K.C. and Aaronson, S.A. (1991). A novel mechanism regulating growth factor association with the cell surface: identification of a PDGF retention domain. *Genes Dev.* **5**: 1191–1199.

Laurie, G.W., Horikoshi, S., Killen, P.D., Segui-Real, B. and Yamada, Y. (1989). *In situ* hybridization reveals temporal and spatial changes in cellular expression of mRNA for a laminin receptor, laminin, and basement membrane (type IV) collagen in the developing kidney. *J. Cell Biol.* **109**: 1351–1362.

Lawley, T.J. and Kubota, Y. (1989). Induction of morphologic differentiation of endothelial cells in culture. *J. Invest. Dermatol.* **93**: 59S–61S.

Lei Zheng and Martins-Green, M. (2007). Molecular Mechanism of Thrombin-Induced Interleukin-8 Expression in THP-1-derived and Primary Human Macrophages. JLB 82(3):619–629.

Leitinger, B., McDowall, A., Stanley, P. and Hogg, N. (2000). The regulation of integrin function by Ca(2+). *Biochim. Biophys. Acta.* **1498**: 91–98.

Li, J., Zhang, Y.P. and Kirsner, R.S. (2003). Angiogenesis in wound repair: angiogenic growth factors and the extracellular matrix. *Microsc. Res. Tech.* **60**: 107–114.

Li, J., Kleeff, J., Kayed, H., Felix, K., Penzel, R., Buchler, M.W., Korc, M. and Friess, H. (2004). Glypican-1 antisense transfection modulates TGF-beta-dependent signaling in Colo-357 pancreatic cancer cells. *Biochem. Biophys. Res. Commun.* **320**: 1148–1155.

Li, Q.J., Vaingankar, S., Sladek, F.M. and Martins-Green, M. (2000). Novel nuclear target for thrombin: activation of the Elk1 transcription factor leads to chemokine gene expression. *Blood* **96**: 3696–3706.

Li, W., Henry, G., Fan, J., Bandyopadhyay, B., Pang, K., Garner, W., Chen, M. and Woodley, D.T. (2004). Signals that initiate, augment, and provide directionality for human keratinocyte motility. *J. Invest. Dermatol.* **123**: 622–633.

Lin, R.Y., Sullivan, K.M., Argenta, P.A., Meuli, M., Lorenz, H.P. and Adzick, N.S. (1995). Exogenous transforming growth factor-beta amplifies its own expression and induces scar formation in a model of human fetal skin repair. *Ann. Surg.* **222**: 146–154.

Linge, C., Richardson, J., Vigor, C., Clayton, E., Hardas, B. and Rolfe, K. (2005). Hypertrophic scar cells fail to undergo a form of apoptosis specific to contractile collagen – the role of tissue transglutaminase. *J. Invest. Dermatol.* **125**: 72–82.

Lipsky, R.H., Eckert, D.M., Tang, Y. and Ockenhouse, C.F. (1997). The carboxyl-terminal cytoplasmic domain of CD36 is required for oxidized low-density lipoprotein modulation of NF-kappaB activity by tumor necrosis factor-alpha. *Recept. Signal Transduct.* **7**: 1–11.

Liu, D., Liu, T., Li, R. and Sy, M.S. (1998). Mechanisms regulating the binding activity of CD44 to hyaluronic acid. *Front. Biosci.* **3**: d631–d636.

Lo, S.H. (2004). Tensin. *Int. J. Biochem. Cell Biol.* **36**: 31–34.

Loftus, J.C., O'Toole, T.E., Plow, E.F., Glass, A., Frelinger, A.L III. and Ginsberg, M.H. (1990). A beta 3 integrin mutation abolishes ligand binding and alters divalent cation-dependent conformation. *Science* **249**: 915–918.

Longaker, M.T., Whitby, D.J., Ferguson, M.W., Harrison, M.R., Crombleholme, T.M., Langer, J.C., Cochrum, K.C., Verrier, E.D. and Stern, R. (1989). Studies in fetal wound healing: III. Early deposition of fibronectin distinguishes fetal from adult wound healing. *J. Pediatr. Surg.* **24**: 799–805.

Longaker, M.T., Whitby, D.J., Adzick, N.S., Crombleholme, T.M., Langer, J.C., Duncan, B.W., Bradley, S.M., Stern, R., Ferguson, M.W. and Harrison, M.R. (1990). Studies in fetal wound healing, VI. Second and early third trimester fetal wounds demonstrate rapid collagen deposition without scar formation. *J. Pediatr. Surg.* **25**: 63–68; discussion 68–69.

Longo, C.R., Patel, V.I., Shrikhande, G.V., Scali, S.T., Csizmadia, E., Daniel, S., Sun, D.W., Grey, S.T., Arvelo, M.B. and Ferran, C. (2005). A20 protects mice from lethal radical hepatectomy by promoting hepatocyte proliferation via a p21waf1-dependent mechanism. *Hepatology* **42**: 156–164.

Lovvorn, H.N III, Cheung, D.T., Nimni, M.E., Perelman, N., Estes, J.M. and Adzick, N.S. (1999). Relative distribution and crosslinking of collagen distinguish fetal from adult sheep wound repair. *J. Pediatr. Surg.* **34**: 218–223.

Lukacs, N.W. and Kunkel, S.L. (1998). Chemokines and their role in disease. *Int. J. Clin. Lab. Res.* **28**: 91–95.

Ly, D.P. and Corbett, S.A. (2005). The integrin alpha5beta1 regulates alphavbeta3-mediated extracellular signal-regulated kinase activation. *J. Surg. Res.* **123**: 200–205.

Lygoe, K.A., Norman, J.T., Marshall, J.F. and Lewis, M.P. (2004). AlphaV integrins play an important role in myofibroblast differentiation. *Wound Repair Regen.* **12**: 461–470.

Ma, T.Y., Kikuchi, M., Sarfeh, I.J., Shimada, H., Hoa, N.T. and Tarnawski, A.S. (1999). Basic fibroblast growth factor stimulates repair of wounded hepatocyte monolayer: modulatory role of protein kinase A and extracellular matrix. *J. Lab. Clin. Med.* **134**: 363–371.

Madri, J.A., Pratt, B.M. and Tucker, A.M. (1988). Phenotypic modulation of endothelial cells by transforming growth factor-beta depends upon the composition and organization of the extracellular matrix. *J. Cell Biol.* **106**: 1375–1384.

Mainiero, F., Murgia, C., Wary, K.K., Curatola, A.M., Pepe, A., Blumemberg, M., Westwick, J.K., Der, C.J. and Giancotti, F.G. (1997). The coupling of alpha6beta4 integrin to Ras-MAP kinase pathways mediated by Shc controls keratinocyte proliferation. *EMBO J.* **16**: 2365–2375.

Marhaba, R. and Zoller, M. (2004). CD44 in cancer progression: adhesion, migration and growth regulation. *J. Mol. Histol.* **35**: 211–231.

Martinez-Hernandez, A., Delgado, F.M. and Amenta, P.S. (1991). The extracellular matrix in hepatic regeneration. Localization of collagen types I, III, IV, laminin, and fibronectin. *Lab. Invest.* **64**: 157–166.

Martins-Green, M. (2001). The chicken chemotactic and angiogenic factor (cCAF), a CXC chemokine. *Int. J. Biochem. Cell Biol.* **33**: 427–432.

Martins-Green, M. and Bissell, M.J. (1990). Localization of 9E3/CEF-4 in avian tissues: expression is absent in Rous sarcoma virus-induced tumors but is stimulated by injury. *J. Cell Biol.* **110**: 581–595.

Martins-Green, M. and Bissell, M.J. (1995). Cell–extracellular matrix interactions in development. *Sems. Dev. Biol.* **6**: 149–159.

Martins-Green, M. and Feugate, J.E. (1998). The 9E3/CEF4 gene product is a chemotactic and angiogenic factor that can initiate the wound-healing cascade *in vivo*. *Cytokine* **10**: 522–535.

Martins-Green, M. and Hanafusa, H. (1997). The 9E3/CEF4 gene and its product the chicken chemotactic and angiogenic factor (cCAF): potential roles in wound healing and tumor development. *Cytokine Growth Factor Rev.* **8**: 221–232.

Martins-Green, M. and Kelly, T. (1998). The chicken chemotactic and angiogenic factor (9E3 gene product): its angiogenic properties reside in the C-terminus of the molecule. *Cytokine* **10**: 819–830.

Martins-Green, M., Aotaki-Keen, A., Hjelmeland, L.M. and Bissell, M.J. (1992). The 9E3 protein: immunolocalization *in vivo* and evidence for multiple forms in culture. *J. Cell Sci.* **101**(Pt 3), 701–707.

Martins-Green, M., Stoeckle, M., Hampe, A., Wimberly, S. and Hanafusa, H. (1996). The 9E3/CEF4 cytokine: kinetics of secretion, processing by plasmin, and interaction with extracellular matrix. *Cytokine* **8**: 448–459.

Masellis-Smith, A., Belch, A.R., Mant, M.J., Turley, E.A. and Pilarski, L.M. (1996). Hyaluronan-dependent motility of B cells and leukemic plasma cells in blood, but not of bone marrow plasma cells, in multiple myeloma: alternate use of receptor for hyaluronan-mediated motility (RHAMM) and CD44. *Blood* **87**: 1891–1899.

Massia, S.P., Rao, S.S. and Hubbell, J.A. (1993). Covalently immobilized laminin peptide Tyr-Ile-Gly-Ser-Arg (YIGSR) supports cell spreading and co-localization of the 67-kilodalton laminin receptor with alpha-actinin and vinculin. *J. Biol. Chem.* **268**: 8053–8059.

Mast, B.A., Haynes, J.H., Krummel, T.M., Diegelmann, R.F. and Cohen, I.K. (1992). *In vivo* degradation of fetal wound hyaluronic acid results in increased fibroplasia, collagen deposition, and neovascularization. *Plast. Reconstr. Surg.* **89**: 503–509.

Mast, B.A., Diegelmann, R.F., Krummel, T.M. and Cohen, I.K. (1993). Hyaluronic acid modulates proliferation, collagen and protein synthesis of cultured fetal fibroblasts. *Matrix* **13**: 441–446.

Matsumoto, K., Tajima, H., Okazaki, H. and Nakamura, T. (1993). Heparin as an inducer of hepatocyte growth factor. *J. Biochem. (Tokyo)* **114**: 820–826.

McFall, A.J. and Rapraeger, A.C. (1998). Characterization of the high affinity cell-binding domain in the cell surface proteoglycan syndecan-4. *J. Biol. Chem.* **273**: 28270–28276.

Menart, V., Fonda, I., Kenig, M. and Porekar, V.G. (2002). Increased *in vitro* cytotoxicity of TNF-alpha analog LK-805 is based on the interaction with cell surface heparan sulfate proteoglycan. *Ann. NY Acad. Sci.* **973**: 194–206.

Mercurius, K.O. and Morla, A.O. (2001). Cell adhesion and signaling on the fibronectin 1st type III repeat: requisite roles for cell surface proteoglycans and integrins. *BMC Cell Biol.* **2**: 18.

Messadi, D.V., Le, A., Berg, S., Jewett, A., Wen, Z., Kelly, P. and Bertolami, C.N. (1999). Expression of apoptosis-associated genes by human dermal scar fibroblasts. *Wound Repair Regen.* **7**: 511–517.

Miao, H.Q., Ishai-Michaeli, R., Atzmon, R., Peretz, T. and Vlodavsky, I. (1996). Sulfate moieties in the subendothelial extracellular matrix are involved in basic fibroblast growth factor sequestration, dimerization, and stimulation of cell proliferation. *J. Biol. Chem.* **271**: 4879–4886.

Milstone, L.M., Hough-Monroe, L., Kugelman, L.C., Bender, J.R. and Haggerty, J.G. (1994). Epican, a heparan/chondroitin sulfate proteoglycan form of CD44, mediates cell–cell adhesion. *J. Cell Sci.* **107**(Pt 11): 3183–3190.

Miralem, T., Steinberg, R., Price, D. and Avraham, H. (2001). VEGF(165) requires extracellular matrix components to induce mitogenic effects and migratory response in breast cancer cells. *Oncogene* **20**: 5511–5524.

Miyake, K., Underhill, C.B., Lesley, J. and Kincade, P.W. (1990). Hyaluronate can function as a cell adhesion molecule and CD44 participates in hyaluronate recognition. *J. Exp. Med.* **172**: 69–75.

Moghal, N. and Neel, B.G. (1998). Integration of growth factor, extracellular matrix, and retinoid signals during bronchial epithelial cell differentiation. *Mol. Cell. Biol.* **18**: 6666–6678.

Mohammed, F.F., Pennington, C.J., Kassiri, Z., Rubin, J.S., Soloway, P.D., Ruther, U., Edwards, D.R. and Khokha, R. (2005). Metalloproteinase inhibitor TIMP-1 affects hepatocyte cell cycle via HGF activation in murine liver regeneration. *Hepatology* **41**: 857–867.

Montesano, R., Orci, L. and Vassalli, P. (1983). *In vitro* rapid organization of endothelial cells into capillary-like networks is promoted by collagen matrices. *J. Cell Biol.* **97**: 1648–1652.

Morita, T., Togo, S., Kubota, T., Kamimukai, N., Nishizuka, I., Kobayashi, T., Ichikawa, Y., Ishikawa, T., Takahashi, S., Matsuo, K., et al. (2002). Mechanism of postoperative liver failure after excessive hepatectomy investigated using a cDNA microarray. *J. Hepatobiliary Pancreat. Surg.* **9**: 352–359.

Mostafavi-Pour, Z., Askari, J.A., Whittard, J.D. and Humphries, M.J. (2001). Identification of a novel heparin-binding site in the alternatively spliced IIICS region of fibronectin: roles of integrins and proteoglycans in cell adhesion to fibronectin splice variants. *Matrix Biol.* **20**: 63–73.

Mould, A.P., Askari, J.A., Craig, S.E., Garratt, A.N., Clements, J. and Humphries, M.J. (1994). Integrin alpha 4 beta 1-mediated melanoma cell adhesion and migration on vascular cell adhesion molecule-1 (VCAM-1) and the alternatively spliced IIICS region of fibronectin. *J. Biol. Chem.* **269**: 27224–27230.

Moulin, V., Larochelle, S., Langlois, C., Thibault, I., Lopez-Valle, C.A. and Roy, M. (2004). Normal skin wound and hypertrophic scar myofibroblasts have differential responses to apoptotic inductors. *J. Cell Physiol.* **198**: 350–358.

Murgia, C., Blaikie, P., Kim, N., Dans, M., Petrie, H.T. and Giancotti, F.G. (1998). Cell cycle and adhesion defects in mice carrying a targeted deletion of the integrin beta4 cytoplasmic domain. *EMBO J.* **17**: 3940–3951.

Murphy, F.R., Issa, R., Zhou, X., Ratnarajah, S., Nagase, H., Arthur, M.J., Benyon, C. and Iredale, J.P. (2002). Inhibition of apoptosis of activated hepatic stellate cells by tissue inhibitor of metalloproteinase-1 is mediated via effects on matrix metalloproteinase inhibition: implications for reversibility of liver fibrosis. *J. Biol. Chem.* **277**: 11069–11076.

Nath, R.K., LaRegina, M., Markham, H., Ksander, G.A. and Weeks, P.M. (1994). The expression of transforming growth factor type beta in fetal and adult rabbit skin wounds. *J. Pediatr. Surg.* **29**: 416–421.

Nathan, C. and Sporn, M. (1991). Cytokines in context. *J. Cell Biol.* **113**: 981–986.

Naugle, J.E., Olson, E.R., Zhang, X., Mase, S.E., Pilati, C.F., Maron, M.B., Folkesson, H.G., Horne, W.I., Doane, K.J. and Meszaros, J.G. (2005). Type VI collagen induces cardiac myofibroblast differentiation: implications for post-infarction remodeling. *Am. J. Physiol. Heart. Circ. Physiol.* **290**: H323–H330.

Ng, V.L., Sabla, G.E., Melin-Aldana, H., Kelley-Loughnane, N., Degen, J.L. and Bezerra, J.A. (2001). Plasminogen deficiency results in poor clearance of non-fibrin matrix and persistent activation of hepatic stellate cells after an acute injury. *J. Hepatol.* **35**: 781–789.

Nicholson, L.J. and Watt, F.M. (1991). Decreased expression of fibronectin and the alpha 5 beta 1 integrin during terminal differentiation of human keratinocytes. *J. Cell Sci.* 98(Pt 2): 225–232.

Nishio, T., Iimuro, Y., Nitta, T., Harada, N., Yoshida, M., Hirose, T., Yamamoto, N., Morimoto, T., Brenner, D.A. and Yamaoka, Y. (2003). Increased expression of collagenase in the liver induces hepatocyte proliferation with cytoplasmic accumulation of beta-catenin in the rat. *J. Hepatol.* **38**: 468–475.

Noble, N.A., Harper, J.R. and Border, W.A. (1992). *In vivo* interactions of TGF-beta and extracellular matrix. *Prog. Growth Factor Res.* **4**: 369–382.

O'Reilly, M.S., Boehm, T., Shing, Y., Fukai, N., Vasios, G., Lane, W.S., Flynn, E., Birkhead, J.R., Olsen, B.R. and Folkman, J. (1997). Endostatin: an endogenous inhibitor of angiogenesis and tumor growth. *Cell* **88**: 277–285.

O'Toole, E.A. (2001). Extracellular matrix and keratinocyte migration. *Clin. Exp. Dermatol.* **26**: 525–530.

Otsu, K., Kato, S., Ohtake, K. and Akamatsu, N. (1992). Alteration of rat liver proteoglycans during regeneration. *Arch. Biochem. Biophys.* **294**: 544–549.

Panayotou, G., End, P., Aumailley, M., Timpl, R. and Engel, J. (1989). Domains of laminin with growth-factor activity. *Cell* **56**: 93–101.

Panis, Y., McMullan, D.M. and Emond, J.C. (1997). Progressive necrosis after hepatectomy and the pathophysiology of liver failure after massive resection. *Surgery* **121**: 142–149.

Patel, D.D., Koopmann, W., Imai, T., Whichard, L.P., Yoshie, O. and Krangel, M.S. (2001). Chemokines have diverse abilities to form solid phase gradients. *Clin. Immunol.* **99**: 43–52.

Paulsson, M. (1988). The role of Ca2+ binding in the self-aggregation of laminin–nidogen complexes. *J. Biol. Chem.* **263**: 5425–5430.

Peach, R.J., Hollenbaugh, D., Stamenkovic, I. and Aruffo, A. (1993). Identification of hyaluronic acid binding sites in the extracellular domain of CD44. *J. Cell Biol.* **122**: 257–264.

Peled, Z.M., Phelps, E.D., Updike, D.L., Chang, J., Krummel, T.M., Howard, E.W. and Longaker, M.T. (2002). Matrix metalloproteinases and the ontogeny of scarless repair: the other side of the wound healing balance. *Plast. Reconstr. Surg.* **110**: 801–811.

Pierschbacher, M.D. and Ruoslahti, E. (1984). Cell attachment activity of fibronectin can be duplicated by small synthetic fragments of the molecule. *Nature* **309**: 30–33.

Pilarski, L.M., Pruski, E., Wizniak, J., Paine, D., Seeberger, K., Mant, M.J., Brown, C.B. and Belch, A.R. (1999). Potential role for hyaluronan and the hyaluronan receptor RHAMM in mobilization and trafficking of hematopoietic progenitor cells. *Blood* **93**: 2918–2927.

Pilewski, J.M., Latoche, J.D., Arcasoy, S.M. and Albelda, S.M. (1997). Expression of integrin cell adhesion receptors during human airway epithelial repair *in vivo*. *Am. J. Physiol.* **273**: L256–L263.

Pollock, R.A. and Richardson, W.D. (1992). The alternative-splice isoforms of the PDGF A-chain differ in their ability to associate with the extracellular matrix and to bind heparin *in vitro*. *Growth Factors* **7**: 267–277.

Poltorak, Z., Cohen, T., Sivan, R., Kandelis, Y., Spira, G., Vlodavsky, I., Keshet, E. and Neufeld, G. (1997). VEGF145, a secreted vascular endothelial growth factor isoform that binds to extracellular matrix. *J. Biol. Chem.* **272**: 7151–7158.

Polunovsky, V.A., Chen, B., Henke, C., Snover, D., Wendt, C., Ingbar, D.H. and Bitterman, P.B. (1993). Role of mesenchymal cell death in lung remodeling after injury. *J. Clin. Invest.* **92**: 388–397.

Ponta, H., Sherman, L. and Herrlich, P.A. (2003). CD44: from adhesion molecules to signalling regulators. *Nat. Rev. Mol. Cell Biol.* **4**: 33–45.

Pouliot, N., Saunders, N.A. and Kaur, P. (2002). Laminin 10/11: an alternative adhesive ligand for epidermal keratinocytes with a functional role in promoting proliferation and migration. *Exp. Dermatol.* **11**: 387–397.

Pozzi, A., Wary, K.K., Giancotti, F.G. and Gardner, H.A. (1998). Integrin alpha1beta1 mediates a unique collagen-dependent proliferation pathway *in vivo*. *J. Cell Biol.* **142**: 587–594.

Proudfoot, A.E., Handel, T.M., Johnson, Z., Lau, E.K., LiWang, P., Clark-Lewis, I., Borlat, F., Wells, T.N. and Kosco-Vilbois, M.H. (2003). Glycosaminoglycan binding and oligomerization are essential for the *in vivo* activity of certain chemokines. *Proc. Natl Acad. Sci. USA* **100**: 1885–1890.

Pujades, C., Alon, R., Yauch, R.L., Masumoto, A., Burkly, L.C., Chen, C., Springer, T.A., Lobb, R.R. and Hemler, M.E. (1997). Defining extracellular integrin alpha-chain sites that affect cell adhesion and adhesion strengthening without altering soluble ligand binding. *Mol. Biol. Cell* **8**: 2647–2657.

Pujades, C., Forsberg, E., Enrich, C. and Johansson, S. (1992). Changes in cell surface expression of fibronectin and fibronectin receptor during liver regeneration. *J. Cell Sci.* **102**(Pt 4): 815–820.

Rahman, S., Patel, Y., Murray, J., Patel, K.V., Sumathipala, R., Sobel, M. and Wijelath, E.S. (2005). Novel hepatocyte growth factor (HGF) binding domains on fibronectin and vitronectin coordinate a distinct and amplified Met-integrin induced signalling pathway in endothelial cells. *BMC Cell Biol.* **6**: 8.

Rapraeger, A., Jalkanen, M., Endo, E. and Bernfield, M. (1987). Integral membrane proteoglycans as matrix receptors: role in cytoskeleton and matrix assembly at the epithelial cell surface. In Wight, T. and Mecham, R. (eds.), *Biology of Extracellular Matrix: A Series. Biology of Proteoglycans*. Orlando, FL: Academic Press, pp. 129–154.

Reis, R.C., Schuppan, D., Barreto, A.C., Bauer, M., Bork, J.P., Hassler, G. and Coelho-Sampaio, T. (2005). Endostatin competes with bFGF for binding to heparin-like glycosaminoglycans. *Biochem. Biophys. Res. Commun.* **333**: 976–983.

Roberts, R., Gallagher, J., Spooncer, E., Allen, T.D., Bloomfield, F. and Dexter, T.M. (1988). Heparan sulphate bound growth factors: a mechanism for stromal cell mediated haemopoiesis. *Nature* **332**: 376–378.

Romanic, A.M., Graesser, D., Baron, J.L., Visintin, I., Janeway Jr., C.A. and Madri, J.A. (1997). T cell adhesion to endothelial cells and extracellular matrix is modulated upon transendothelial cell migration. *Lab. Invest.* **76**: 11–23.

Roovers, K., Davey, G., Zhu, X., Bottazzi, M.E. and Assoian, R.K. (1999). Alpha5beta1 integrin controls cyclin D1 expression by sustaining mitogen-activated protein kinase activity in growth factor-treated cells. *Mol. Biol. Cell* **10**: 3197–3204.

Rosales, C. and Juliano, R.L. (1995). Signal transduction by cell adhesion receptors in leukocytes. *J. Leukoc. Biol.* **57**: 189–198.

Rossi, D. and Zlotnik, A. (2000). The biology of chemokines and their receptors. *Annu. Rev. Immunol.* **18**: 217–242.

Rudolph, K.L., Trautwein, C., Kubicka, S., Rakemann, T., Bahr, M.J., Sedlaczek, N., Schuppan, D. and Manns, M.P. (1999). Differential regulation of extracellular matrix synthesis during liver regeneration after partial hepatectomy in rats. *Hepatology* **30**: 1159–1166.

Ruhl, M., Sahin, E., Johannsen, M., Somasundaram, R., Manski, D., Riecken, E.O. and Schuppan, D. (1999). Soluble collagen VI drives serum-starved fibroblasts through S phase and prevents apoptosis via down-regulation of Bax. *J. Biol. Chem.* **274**: 34361–34368.

Rundhaug, J.E. (2005). Matrix metalloproteinases and angiogenesis. *J. Cell. Mol. Med.* **9**: 267–285.

Sacco, O., Silvestri, M., Sabatini, F., Sale, R., Defilippi, A.C. and Rossi, G.A. (2004). Epithelial cells and fibroblasts: structural repair and remodelling in the airways. *Paediatr. Respir. Rev.* **5**(Suppl A), S35–S40.

Saed, G.M., Ladin, D., Olson, J., Han, X., Hou, Z. and Fivenson, D. (1998). Analysis of p53 gene mutations in keloids using polymerase chain reaction-based single-strand conformational polymorphism and DNA sequencing. *Arch. Dermatol.* **134**: 963–967.

Sage, E.H., Reed, M., Funk, S.E., Truong, T., Steadele, M., Puolakkainen, P., Maurice, D.H. and Bassuk, J.A. (2003). Cleavage of the matricellular protein SPARC by matrix metalloproteinase 3 produces polypeptides that influence angiogenesis. *J. Biol. Chem.* **278**: 37849–37857.

Saile, B., Knittel, T., Matthes, N., Schott, P. and Ramadori, G. (1997). CD95/CD95L-mediated apoptosis of the hepatic stellate cell. A mechanism terminating uncontrolled hepatic stellate cell proliferation during hepatic tissue repair. *Am. J. Pathol.* **151**: 1265–1272.

Sakakura, S., Saito, S. and Morikawa, H. (1999). Stimulation of DNA synthesis in trophoblasts and human umbilical vein endothelial cells by hepatocyte growth factor bound to extracellular matrix. *Placenta* **20**: 683–693.

Sakamoto, N., Iwahana, M., Tanaka, N.G. and Osada, Y. (1991). Inhibition of angiogenesis and tumor growth by a synthetic laminin peptide, CDPGYIGSR-NH2. *Cancer Res.* **51**: 903–906.

Salmivirta, M. and Jalkanen, M. (1995). Syndecan family of cell surface proteoglycans: developmentally regulated receptors for extracellular effector molecules. *Experientia* **51**: 863–872.

Sanes, J.R., Engvall, E., Butkowski, R. and Hunter, D.D. (1990). Molecular heterogeneity of basal laminae: isoforms of laminin and collagen IV at the neuromuscular junction and elsewhere. *J. Cell Biol.* **111**: 1685–1699.

Santra, M., Reed, C.C. and Iozzo, R.V. (2002). Decorin binds to a narrow region of the epidermal growth factor (EGF) receptor, partially overlapping but distinct from the EGF-binding epitope. *J. Biol. Chem.* **277**: 35671–35681.

Saoncella, S., Echtermeyer, F., Denhez, F., Nowlen, J.K., Mosher, D.F., Robinson, S.D., Hynes, R.O. and Goetinck, P.F. (1999). Syndecan-4 signals cooperatively with integrins in a Rho-dependent manner in the assembly of focal adhesions and actin stress fibers. *Proc. Natl Acad. Sci. U.S.A.* **96**: 2805–2810.

Sasaki, T., Larsson, H., Tisi, D., Claesson-Welsh, L., Hohenester, E. and Timpl, R. (2000). Endostatins derived from collagens XV and XVIII differ in structural and binding properties, tissue distribution and anti-angiogenic activity. *J. Mol. Biol.* **301**: 1179–1190.

Savani, R.C., Cao, G., Pooler, P.M., Zaman, A., Zhou, Z. and DeLisser, H.M. (2001). Differential involvement of the hyaluronan (HA) receptors CD44 and receptor for HA-mediated motility in endothelial cell function and angiogenesis. *J. Biol. Chem.* **276**: 36770–36778.

Sawai, T., Usui, N., Sando, K., Fukui, Y., Kamata, S., Okada, A., Taniguchi, N., Itano, N. and Kimata, K. (1997). Hyaluronic acid of wound fluid in adult and fetal rabbits. *J. Pediatr. Surg.* **32**: 41–43.

Schamhart, D.H. and Kurth, K.H. (1997). Role of proteoglycans in cell adhesion of prostate cancer cells: from review to experiment. *Urol. Res.* **25**(Suppl 2): S89–S96.

Schenk, S., Hintermann, E., Bilban, M., Koshikawa, N., Hojilla, C., Khokha, R. and Quaranta, V. (2003). Binding to EGF receptor of a laminin-5 EGF-like fragment liberated during MMP-dependent mammary gland involution. *J. Cell Biol.* **161**: 197–209.

Schlaepfer, D.D. and Hunter, T. (1996). Signal transduction from the extracellular matrix – a role for the focal adhesion protein-tyrosine kinase FAK. *Cell Struct. Funct.* **21**: 445–450.

Schlaepfer, D.D. and Hunter, T. (1997). Focal adhesion kinase overexpression enhances ras-dependent integrin signaling to ERK2/mitogen-activated protein kinase through interactions with and activation of c-Src. *J. Biol. Chem.* **272**: 13189–13195.

Schlaepfer, D.D. and Hunter, T. (1998). Integrin signalling and tyrosine phosphorylation: just the FAKs? *Trends Cell Biol.* **8**: 151–157.

Schlaepfer, D.D., Broome, M.A. and Hunter, T. (1997). Fibronectin-stimulated signaling from a focal adhesion kinase–c-Src complex: involvement of the Grb2, p130cas, and Nck adaptor proteins. *Mol. Cell. Biol.* **17**: 1702–1713.

Schonherr, E. and Hausser, H.J. (2000). Extracellular matrix and cytokines: a functional unit. *Dev. Immunol.* **7**: 89–101.

Schuppan, D., Schmid, M., Somasundaram, R., Ackermann, R., Ruehl, M., Nakamura, T. and Riecken, E.O. (1998). Collagens in the liver extracellular matrix bind hepatocyte growth factor. *Gastroenterology* **114**: 139–152.

Schwartz, M.A. (1992). Transmembrane signalling by integrins. *Trends Cell Biol* **2**: 304–308.

Sechler, J.L., Corbett, S.A., Wenk, M.B. and Schwarzbauer, J.E. (1998). Modulation of cell–extracellular matrix interactions. *Ann. NY Acad. Sci.* **857**: 143–154.

Serini, G., Bochaton-Piallat, M.L., Ropraz, P., Geinoz, A., Borsi, L., Zardi, L. and Gabbiani, G. (1998). The fibronectin domain ED-A is crucial for myofibroblastic phenotype induction by transforming growth factor-beta1. *J. Cell Biol.* **142**: 873–881.

Shimizu, M., Hara, A., Okuno, M., Matsuno, H., Okada, K., Ueshima, S., Matsuo, O., Niwa, M., Akita, K., Yamada, Y., et al. (2001). Mechanism of retarded liver regeneration in plasminogen activator-deficient mice: impaired activation of hepatocyte growth factor after Fas-mediated massive hepatic apoptosis. *Hepatology* **33**: 569–576.

Sjaastad, M.D. and Nelson, W.J. (1997). Integrin-mediated calcium signaling and regulation of cell adhesion by intracellular calcium. *Bioessays* **19**: 47–55.

So, C.L., Kaluarachchi, K., Tam, P.P. and Cheah, K.S. (2001). Impact of mutations of cartilage matrix genes on matrix structure, gene activity and chondrogenesis. *Osteoarthritis Cartilage* **9**(Suppl A): S160–S173.

Sohara, N., Znoyko, I., Levy, M.T., Trojanowska, M. and Reuben, A. (2002). Reversal of activation of human myofibroblast-like cells by culture on a basement membrane-like substrate. *J. Hepatol.* **37**: 214–221.

Soo, C., Hu, F.Y., Zhang, X., Wang, Y., Beanes, S.R., Lorenz, H.P., Hedrick, M.H., Mackool, R.J., Plaas, A., Kim, S. J., et al. (2000). Differential expression of fibromodulin, a transforming growth factor-beta modulator, in fetal skin development and scarless repair. *Am. J. Pathol.* **157**: 423–433.

Sottile, J. (2004). Regulation of angiogenesis by extracellular matrix. *Biochim. Biophys. Acta* **1654**: 13–22.

Spofford, C.M. and Chilian, W.M. (2003). Mechanotransduction via the elastin-laminin receptor (ELR) in resistance arteries. *J. Biomech.* **36**: 645–652.

Streuli, C.H., Schmidhauser, C., Kobrin, M., Bissell, M.J. and Derynck, R. (1993). Extracellular matrix regulates expression of the TGF-beta 1 gene. *J. Cell Biol.* **120**: 253–260.

Stupack, D.G. and Cheresh, D.A. (2002). Get a ligand, get a life: integrins, signaling and cell survival. *J. Cell Sci.* **115**: 3729–3738.

Stupack, D.G., Puente, X.S., Boutsaboualoy, S., Storgard, C.M. and Cheresh, D.A. (2001). Apoptosis of adherent cells by recruitment of caspase-8 to unligated integrins. *J. Cell Biol.* **155**: 459–470.

Suki, B., Ito, S., Stamenovic, D., Lutchen, K.R. and Ingenito, E.P. (2005). Biomechanics of the lung parenchyma: critical roles of collagen and mechanical forces. *J. Appl. Physiol.* **98**: 1892–1899.

Sulochana, K.N., Fan, H., Jois, S., Subramanian, V., Sun, F., Kini, R.M. and Ge, R. (2005). Peptides derived from human decorin leucine-rich repeat 5 inhibit angiogenesis. *J. Biol. Chem.* **280**: 27935–27948.

Swindle, C.S., Tran, K.T., Johnson, T.D., Banerjee, P., Mayes, A.M., Griffith, L. and Wells, A. (2001). Epidermal growth factor (EGF)-like repeats of human tenascin-C as ligands for EGF receptor. *J. Cell Biol.* **154**: 459–468.

Takeda, K., Togo, S., Kunihiro, O., Fujii, Y., Kurosawa, H., Tanaka, K., Endo, I., Takimoto, A., Sekido, H., Hara, M., et al. (2002). Clinicohistological features of liver failure after excessive hepatectomy. *Hepatogastroenterology* **49**: 354–358.

Takemura, G., Ohno, M., Hayakawa, Y., Misao, J., Kanoh, M., Ohno, A., Uno, Y., Minatoguchi, S., Fujiwara, T. and Fujiwara, H. (1998). Role of apoptosis in the disappearance of infiltrated and proliferated interstitial cells after myocardial infarction. *Circ. Res.* **82**: 1130–1138.

Takeuchi, Y., Kodama, Y. and Matsumoto, T. (1994). Bone matrix decorin binds transforming growth factor-beta and enhances its bioactivity. *J. Biol. Chem.* **269**: 32634–32638.

Thannickal, V.J., Lee, D.Y., White, E.S., Cui, Z., Larios, J.M., Chacon, R., Horowitz, J.C., Day, R.M. and Thomas, P.E. (2003). Myofibroblast differentiation by transforming growth factor-beta1 is dependent on cell adhesion and integrin signaling via focal adhesion kinase. *J. Biol. Chem.* **278**: 12384–12389.

Tomasek, J.J., Haaksma, C.J., Eddy, R.J. and Vaughan, M.B. (1992). Fibroblast contraction occurs on release of tension in attached collagen lattices: dependency on an organized actin cytoskeleton and serum. *Anat. Rec.* **232**: 359–368.

Tran, K.T., Griffith, L. and Wells, A. (2004). Extracellular matrix signaling through growth factor receptors during wound healing. *Wound Repair Regen.* **12**: 262–268.

Tran, K.T., Lamb, P. and Deng, J.S. (2005). Matrikines and matricryptins: implications for cutaneous cancers and skin repair. *J. Dermatol. Sci.* **40**: 11–20.

Tsuda, T., Majumder, K. and Linask, K.K. (1998). Differential expression of flectin in the extracellular matrix and left-right asymmetry in mouse embryonic heart during looping stages. *Dev. Genet.* **23**: 203–214.

Tuan, T.L., Song, A., Chang, S., Younai, S. and Nimni, M.E. (1996). *In vitro* fibroplasia: matrix contraction, cell growth, and collagen production of fibroblasts cultured in fibrin gels. *Exp. Cell Res.* **223**: 127–134.

Tuhkanen, A.L., Tammi, M. and Tammi, R. (1997). CD44 substituted with heparan sulfate and endo-beta-galactosidase-sensitive oligosaccharides: a major proteoglycan in adult human epidermis. *J. Invest. Dermatol.* **109**: 213–218.

Vaday, G.G. and Lider, O. (2000). Extracellular matrix moieties, cytokines, and enzymes: dynamic effects on immune cell behavior and inflammation. *J. Leukoc. Biol.* **67**: 149–159.

Vaingankar, S.M. and Martins-Green, M. (1998). Thrombin aivation of the 9E3/CEF4 chemokine involves tyrosine kinases including c-src and the epidermal growth factor receptor. *J. Biol. Chem.* **273**: 5226–5234.

van der Flier, A. and Sonnenberg, A. (2001). Function and interactions of integrins. *Cell Tissue Res.* **305**: 285–298.

Verrecchia, F. and Mauviel, A. (2002). Transforming growth factor-beta signaling through the Smad pathway: role in extracellular matrix gene expression and regulation. *J. Invest. Dermatol.* **118**: 211–215.

Vilatoba, M., Eckstein, C., Bilbao, G., Frennete, L., Eckhoff, D.E. and Contreras, J.L. (2005a). 17beta-estradiol differentially activates mitogen-activated protein-kinases and improves survival following reperfusion injury of reduced-size liver in mice. *Transplant Proc.* **37**: 399–403.

Vilatoba, M., Eckstein, C., Bilbao, G., Smyth, C.A., Jenkins, S., Thompson, J.A., Eckhoff, D.E. and Contreras, J.L. (2005b). Sodium 4-phenylbutyrate protects against liver ischemia reperfusion injury by inhibition of endoplasmic reticulum-stress mediated apoptosis. *Surgery* **138**: 342–351.

Vlodavsky, I., Miao, H.Q., Medalion, B., Danagher, P. and Ron, D. (1996). Involvement of heparan sulfate and related molecules in sequestration and growth promoting activity of fibroblast growth factor. *Cancer Metastasis Rev.* **15**: 177–186.

Vogel, V. Thomas, W.E., Craig, D.W., Krammer, A. and Baneyx, G. (2001). Structural insights into the mechanical regulation of molecular recognition sites. *Trends Biotechnol.* **19**: 416–423.

Vogel, W., Gish, G.D., Alves, F. and Pawson, T. (1997). The discoidin domain receptor tyrosine kinases are activated by collagen. *Mol. Cell* **1**: 13–23.

Wallner, E.I., Yang, Q., Peterson, D.R., Wada, J. and Kanwar, Y.S. (1998). Relevance of extracellular matrix, its receptors, and cell adhesion molecules in mammalian nephrogenesis. *Am. J. Physiol.* **275**: F467–F477.

Wang, J., Chen, H., Seth, A. and McCulloch, C.A. (2003). Mechanical force regulation of myofibroblast differentiation in cardiac fibroblasts. *Am. J. Physiol. Heart Circ. Physiol.* **285**: H1871–H1881.

Wary, K.K., Mariotti, A., Zurzolo, C. and Giancotti, F.G. (1998). A requirement for caveolin-1 and associated kinase Fyn in integrin signaling and anchorage-dependent cell growth. *Cell* **94**: 625–634.

Watanabe, M., Fujioka-Kaneko, Y., Kobayashi, H., Kiniwa, M., Kuwano, M. and Basaki, Y. (2005). Involvement of integrin-linked kinase in capillary/tube-like network formation of human vascular endothelial cells. *Biol. Proced. Online* **7**: 41–47.

Wei, J., Shaw, L.M. and Mercurio, A.M. (1997). Integrin signaling in leukocytes: lessons from the alpha6beta1 integrin. *J. Leukoc. Biol.* **61**: 397–407.

Werb, Z. and Chin, J.R. (1998). Extracellular matrix remodeling during morphogenesis. *Ann. NY Acad. Sci.* **857**: 110–118.

Werb, Z., Vu, T.H., Rinkenberger, J.L. and Coussens, L.M. (1999). Matrix-degrading proteases and angiogenesis during development and tumor formation. *Apmis* **107**: 11–18.

Wess, T.J., Purslow, P.P., Sherratt, M.J., Ashworth, J., Shuttleworth, C.A. and Kielty, C.M. (1998). Calcium determines the supramolecular organization of fibrillin-rich microfibrils. *J. Cell Biol.* **141**: 829–837.

Wessells, N.K. and Cohen, J.H. (1968). Effects of collagenase on developing epithelia *in vitro*: lung, ureteric bud, and pancreas. *Dev. Biol.* **18**: 294–309.

West, D.C., Shaw, D.M., Lorenz, P., Adzick, N.S. and Longaker, M.T. (1997). Fibrotic healing of adult and late gestation fetal wounds correlates with increased hyaluronidase activity and removal of hyaluronan. *Int. J. Biochem. Cell Biol.* **29**: 201–210.

Whitby, D.J. and Ferguson, M.W. (1991). The extracellular matrix of lip wounds in fetal, neonatal and adult mice. *Development* **112**: 651–668.

Whitby, D.J., Longaker, M.T., Harrison, M.R., Adzick, N.S. and Ferguson, M.W. (1991). Rapid epithelialisation of fetal wounds is associated with the early deposition of tenascin. *J. Cell Sci.* **99**(Pt 3): 583–586.

White, S.R., Dorscheid, D.R., Rabe, K.F., Wojcik, K.R. and Hamann, K.J. (1999). Role of very late adhesion integrins in mediating repair of human airway epithelial cell monolayers after mechanical injury. *Am. J. Respir. Cell Mol. Biol.* **20**: 787–796.

Wikner, N.E., Elder, J.T., Persichitte, K.A., Mink, P. and Clark, R.A. (1990). Transforming growth factor-beta modulates plasminogen activator activity and plasminogen activator inhibitor type-1 expression in human keratinocytes *in vitro*. *J. Invest. Dermatol.* **95**: 607–613.

Xu, T., Bianco, P., Fisher, L.W., Longenecker, G., Smith, E., Goldstein, S., Bonadio, J., Boskey, A., Heegaard, A.M., Sommer, B., et al. (1998). Targeted disruption of the biglycan gene leads to an osteoporosis-like phenotype in mice. *Nat. Genet.* **20**: 78–82.

Yamada, K.M. and Kennedy, D.W. (1984). Dualistic nature of adhesive protein function: fibronectin and its biologically active peptide fragments can autoinhibit fibronectin function. *J. Cell Biol.* **99**: 29–36.

Yoneda, A. and Couchman, J.R. (2003). Regulation of cytoskeletal organization by syndecan transmembrane proteoglycans. *Matrix Biol.* **22**: 25–33.

Zhou, X., Murphy, F.R., Gehdu, N., Zhang, J., Iredale, J.P. and Benyon, R.C. (2004). Engagement of alphavbeta3 integrin regulates proliferation and apoptosis of hepatic stellate cells. *J. Biol. Chem.* **279**: 23996–24006.

Zhu, X. and Assoian, R.K. (1995). Integrin-dependent activation of MAP kinase: a link to shape-dependent cell proliferation. *Mol. Biol. Cell* **6**: 273–282.

Zhu, Y., McAlinden, A. and Sandell, L.J. (2001). Type IIA procollagen in development of the human intervertebral disc: regulated expression of the NH(2)-propeptide by enzymic processing reveals a unique developmental pathway. *Dev. Dyn.* **220**: 350–362.

6

Developmental Mechanisms of Regeneration

David L. Stocum

INTRODUCTION

All the cell types of the adult vertebrate body are derived from stem cells of the early embryo. In mammals, these embryonic stem cells (ESCs) constitute the inner cell mass of the pre-implantation blastocyst (Rossant, 2001). ESCs are pluripotent, as demonstrated *in vivo* by their ability to make contributions to all tissues after injection into host blastocysts, their ability to form teratomas containing ectodermal, mesodermal and endodermal derivatives when implanted into immunodeficient mice, and their ability to differentiate spontaneously, or as directed, into multiple cell types *in vitro*. They express species and stage-specific embryonic antigens (SSEAs), alkaline phosphatase, and high levels of telomerase (Smith, 2001; Rippon and Bishop, 2004). ES cell lines have been established from a variety of vertebrate early embryos, including fish, birds, mice (Smith, 2001; Rippon and Bishop, 2004) and humans (Thomson et al., 1998; Shamblott et al., 1998). Several transcription factors have been implicated in the acquisition and maintenance of mouse and human ESC self-renewal and pluripotency. These are OCT4 (Smith, 2001), SOX2 (Avilion et al., 2003), Fox D3 (Hanna et al., 2002), all activated by LIF (mouse) or FGF-2 (human) through STAT-3, and the LIF/STAT3-independent transcription factors Nanog (Mitsui et al., 2003; Chambers et al., 2003), Tbx3, Esrrb, and Tcl1 (Ivanova et al., 2006). BMPs also play a role, by inducing the expression of inhibitor of differentiation (Id) genes via Smad transcription factors (Ying et al., 2003).

ESCs give rise during cleavage to prospective ectoderm, endoderm and mesoderm cells. Once these cell types are established, they undergo the morphogenetic movements of gastrulation to position the mesoderm between the ectoderm and endoderm. Nanog is down-regulated as prospective mesoderm cells exit from the primitive streak during gastrulation, while OCT4 is still expressed. Subsequently all the pluripotency genes are down-regulated except in the germ cells (Hart et al., 2004). Cell interactions among these three embryonic tissue layers determine patterns of gene activity that establish the boundaries of organ and appendage fields. The distinct patterns of growth, tissue differentiation and morphogenesis that characterize the organs and appendages emerge as a result of further cell interactions within these fields. The cell and tissue interactions of development take place via autocrine, paracrine and juxtacrine signaling molecules that bind to receptors and activate intracellular signal transduction pathways leading to specific patterns of gene activity. Seven major signaling pathways have been identified: Notch, Wnt, hedgehog, JAK-SAT, RTK (receptor tyrosine kinase), TGF-β and the apoptotic, or cell death, pathway, which is important for eliminating excess cells in developing tissues and for morphogenesis (Gilbert, 2006).

Copyright © 2008 Elsevier Inc. All rights reserved

Regeneration is a developmental process that maintains (in the face of normal cell turnover) and restores (after damage) tissue integrity in the fetus, juvenile and adult. In most cases, it involves a partial recapitulation of tissue embryogenesis. The same signal transduction pathways and transcription factors are used in regeneration as in embryonic development, although they may not be deployed in exactly the same way. This chapter examines the mechanisms of regeneration, examples of each mechanism, and the strategies of regenerative medicine that are being developed from our knowledge of these mechanisms.

MECHANISMS OF REGENERATION

There are three mechanisms of regeneration: compensatory hyperplasia, activation of resident adult stem cells (ASCs), and production of stem cells by the dedifferentiation of mature cells (Figure 6.1, Table 6.1). In all of these mechanisms, the regeneration-competent cells of adult tissues reside in three-dimensional environmental "niches" consisting of specific combinations and concentrations of soluble factors and extracellular matrix (ECM) that promote their survival, precisely regulate their proliferation, and determine the phenotypic direction and histological pattern of their differentiation (Scadden, 2006; Engler et al., 2006). Comprehending the elements and interlocking pathways of this "molecular ecology" (Powell, 2005) is one of the most important tasks of regeneration research today.

Compensatory Hyperplasia

Compensatory hyperplasia is defined as the mitosis of differentiated cells to maintain or restore tissue mass. New cells thus are derived solely from pre-existing differentiated cells. This is the only mechanism of regeneration, that does not recapitulate part of the embryonic developmental program.

The classic example of regeneration by compensatory hyperplasia is the mammalian liver (Michalopoulos and De Francis, 1997; Fausto, 2004). Individual hepatocytes have an enormous capacity for replication, up to at least 70 times. They are maintained in a non-proliferative state by C/EBPα inhibition of cyclin-dependent kinases (cdks). Partial hepatectomy triggers the appearance of TNF-α, IL-6, HGF and EGF, mitogenic signals that prime the hepatocytes for entry into the cell cycle by activation of the transcription factors STAT3, PHF/

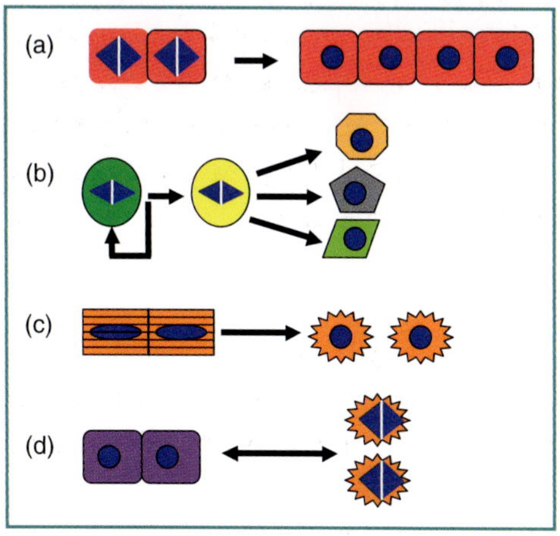

Figure 6.1 Mechanisms of regeneration. (A) Compensatory hyperplasia, the division of differentiated cells to restore tissue mass. (B) Activation and proliferation of adult stem cells. The mother stem cell self-renews while also giving rise to a transit amplifying cell that proliferates and gives rise to single or multiple types of terminally differentiated cells. (C) Dedifferentiation of muscle (left) by cellularization and loss of contractile apparatus to produce mesenchymal-like stem cells (right). (D) Epithelial (left) to mesenchymal (right) transformation and mesenchymal to epithelial transformation.

Table 6.1 Mechanisms of regeneration and types of cells involved

Compensatory Hyperplasia

Hepatocytes
- Liver

Beta cells
- Pancreas

Activation of Adult Stem Cells

Epithelial Stem Cells
- Digestive tract (including canals of Hering and pancreatic ductules)
- Respiratory tract
- Interfollicular epidermis
- Hair follicle (bulge)
- Neural (olfactory epithelium, lateral ventricles of brain, hippocampus, hair cells of cochlea)
- Kidney proximal tubules

Endothelial Stem Cells
- Bone marrow
- Capillaries and venules
- Epicardium?

Hematopoietic Stem Cells
- Bone marrow

Mesenchymal Stem Cells
- Bone marrow
- Periosteum, endosteum
- Dental pulp, periodontal ligament
- Adipose tissue
- Connective tissue compartments

Muscle Stem Cells
- Skeletal muscle (satellite cells)
- Myocardium (cardiac stem cells)

Dedifferentiation

Amphibian
- Tail
- Limb
- Jaws
- Lens, retina
- Myocardium
- Intestine, fish fins

Fish
- Fins
- Retina
- Myocardium

Lizard
- Tail

Epithelial < > Mesenchymal Transformation
- Amphibian spinal cord
- Capillaries and venules
- Kidney proximal tubules

NF-κB, AP-1 and C/EBPβ. These transcription factors induce the activity of sets of "early immediate" and "delayed immediate" genes that encode proteins involved in entering and progressing through the G_1 phase of the cell cycle. HGF appears to play a central role in this process. Pro-HGF is released by liver matrix degradation and its synthesis by sinusoidal endothelial cells is promoted by VEGF (Le Couter et al., 2003). Pro-HGF is activated by urokinase plasminogen activator (uPA) and triggers entry into the cell cycle by binding to its receptor, c-met. Once the original mass of the liver is attained, proliferation ceases and the original histological architecture of the liver is restored.

Beta cells and acinar cells of the pancreas also appear to regenerate *in vivo* by compensatory hyperplasia. Genetic marking experiments have revealed that during growth of the mouse pancreas or during its injury-induced regeneration, new β-cells and acinar cells are derived from pre-existing β and acinar cells (Dor et al., 2004; Desai et al., 2007). Beta cell regeneration can be initiated by a number of proteins: β-cell regeneration protein (Reg), islet neogenesis associated protein (INGAP, a 15 amino acid fragment of Reg), betacellulin (a member of the EGF family), and GLP-1 (Risbud and Bhonde, 2002; Bonner-Weir and Weir, 2005).

Activation of Adult Stem Cells

Adult stem cells (ASCs) are arrested in a pre-terminal differentiation phase of their developmental program, within their tissues of residence. Differentiation of the tissue of residence results in the creation of niche conditions that balance quiescence and activation of ASCs. The mechanisms that sequester small subpopulations of stem cells as other cells differentiate around them are not well understood. When activated, ASCs divide asymmetrically so that one daughter remains a stem cell (self-renewal) and the other gives rise to a transit amplifying population that undergoes terminal differentiation. Epithelia, dental tissues, tissues of the nervous system, musculoskeletal tissues, and hematopoietic cells regenerate via ASCs (Table 6.1). The ASCs of different tissues are commonly maintained in a quiescent state either directly by the Notch signal transduction pathway, or indirectly by absence or inhibition of signaling molecules for other signaling transduction pathways.

Epithelial Stem Cells

Nearly all of the epithelial tissues of the body undergo continual self-renewal throughout life and have a high capacity for injury-induced regeneration. Interfollicular epidermis and hair follicles are among the best studied. The stem cells that regenerate interfollicular skin epidermis are integrin-expressing cells located in the stratum basale. During maintenance regeneration, they give rise to transit amplifying cells that detach from the basement membrane and differentiate into keratinocytes as they move upward to the stratum corneum (Jones et al., 1995; Jensen et al., 1999). Gaps in the epidermis of an excisional wound are filled in by the division of stem cells at the edges of the wound whose progeny migrate laterally through the provisional fibrin matrix of the wound. Migration is initiated by TGF-α and EGF produced by macrophages. Cell division at the wound edges is promoted by KGF and GM-CSF secreted by fibroblasts of the granulation tissue (Werner et al., 1994; Woodley, 1996). Once the wound is re-covered, these same factors promote vertical division to thicken the epidermis. The trigger for stem cell proliferation after wounding may be the binding of EGF family ligands on the apical cell surface to EGF receptors on the basolateral surface (Vermeer et al., 2003). Normally, tight junctions separate the apical and basolateral domains, but the cell separation that occurs upon wounding allows ligands and receptors of the two domains to interact.

The basal epidermal cells are continuous with the basal cells of the outer root sheath of the hair follicle. Hair follicle stem cells are located in a special region of the outer root sheath called the bulge. Marking studies have shown that these stem cells divide asymmetrically to self-renew and produce transit-amplifying cells that feed upward toward the surface of the epidermis where they differentiate into epidermal keratinocytes, and downward to the matrix of the hair follicle where they proliferate and differentiate as the hair shaft (Morris et al., 2004; Tumbar et al., 2004). Hairs go through a three-stage maintenance cycle of catagen (follicle regression), telogen (follicle rest) and anagen (regeneration of the follicle and new hair growth) that is regulated by growth factor signals from the dermal papilla at the base of the hair follicle (Hardy, 1992; Messenger, 1993). Transcriptional repression by Lef-1/Tcf-3 maintains the stem cells of the epidermis and hair follicles in a quiescent state. The cells are activated to proliferate by Wnt signaling, which stabilizes β-catenin, allowing it to translocate to the nucleus and complex with Lef1 to form a transcriptional activating complex leading to proliferation.

Corneal epithelium is regenerated continuously or after injury by epithelial stem cells located in the limbus, the region where the cornea undergoes a transition into the sclera of the eye (Cotsarelis et al., 1989). Limbal stem cells divide asymmetrically to produce transit amplifying cells that migrate centripetally to replace corneal epithelial cells lost by turnover or injury.

The epithelia of the digestive, respiratory and urogenital systems have extensive capacity for regeneration that is regulated by overlapping sets of growth factors (Stocum, 2006, for review). Small intestinal epithelial stem cells are located in the crypts of Lieberkuhn (Potten, 1997; Brittan and Wright, 2004). Liver stem cells are located in the epithelium of the canals of Hering and are activated when the ability of hepatocytes to proliferate by compensatory hyperplasia is compromised (Dabeva and Shafritz, 2003). The ductules of the pancreas may also harbor stem cells that can differentiate into β-cells, and non-β-cells of the islets have been reported to transform *in vitro* into epithelial cells that differentiate into β-cells (Bonner Weir and Weir, 2005; Jamal et al., 2005). Overexpression of the *Arx* gene (which determines the embryonic differentiation of α and PP cells of the islets) in β-cells converts them to α and PP cells (Collombat et al., 2007). A subset of type I pneumocytes in the alveolar epithelium regenerates injured type I pneumocytes (Reddy et al., 2004). The ciliated epithelium of the trachea and bronchial tree is constantly renewed by stem cells in the basal epithelium, as is the epithelium of the bladder, ureters and urethra (Ham and Cormack, 1979).

Some types of neurons regenerate from epithelial stem cells. The sound and motion-sensing auditory hair cells of the sensory epithelium of the cochlea and vestibular apparatus in birds are regenerated by support stem cells interspersed with these sensory neurons (Stone and Rubel, 2000; Stone et al., 2004). Mammalian hippocampal NSCs exhibit a low level of maintenance regeneration that in mice is enhanced by environmental enrichment (Kempermann et al., 1998; Gage, 2000, for review). Neurons of the olfactory nerve and olfactory bulb turn over on a regular schedule. The olfactory nerve neurons are regenerated by NSCs in the nasal epithelium, while NSCs in the walls of the lateral ventricle replenish olfactory bulb neurons (Schwob, 2002, for review). Killing thalamic projection neurons in the cortex of the mammalian brain or granule neurons in the dentate gyrus of the hippocampus by focal or global ischemia results in a low level of NSC proliferation in the lateral ventricle walls and hippocampal ventricle walls. Intraventricular injection of a combination of FGF-2 and EGF elevates the number of regenerated hippocampal neurons to 40% of the number lost; these neurons are functionally integrated into the hippocampal circuitry. The small number of regenerated neurons in the absence of these growth factors may be due to inadequate output of growth factors by astrocytes, since neonatal hippocampal astrocytes induce hippocampal NSCs to differentiate into neurons *in vitro*, whereas adult astrocytes have only half the effect (Nakatomi et al., 2002; Song et al., 2002).

Larval salamanders regenerate the spinal cord after amputation of the tail. Muscle, cartilage and connective tissue dedifferentiate to form a blastema. A tube of dividing ependymal cells extends from the cut end of the spinal cord into the blastema. As the ependymal tube grows distally, the cells closest to the amputation plane extend end feet. The endfeet form channels that promote the regeneration of axons from above the level of transection, while other ependymal cells differentiate into new motor neurons, interneurons and glia (Chernoff et al., 2003). FGF-2, Wnt, BMP and Notch signaling pathways all appear to be involved in regulating this regeneration.

In order to migrate during regeneration, many epithelial stem cells, as well as the endothelium of blood vessels, undergo an epithelial to mesenchymal transformation (EMT), followed by the reverse mesenchymal to epithelial transformation (MET) to reconstitute the epithelium or endothelium (Fig. 6.1, Table 6.1). Well-studied examples of these transformations are the regeneration of wounded epidermis, the ependyma of the transected thoracic or lumbar spinal cord of urodele amphibians (Chernoff et al., 2003) and proximal tubule epithelial cells of the mammalian kidney (Bonventre, 2003). The cells of wounded epidermis are induced to migrate by macrophage-produced EGF and TGF-α. Kidney tubule epithelial cells are induced by TGF-β1 to undergo EMT to cover denuded areas of the basement membrane. Once having filled the gap, the mesenchymal cells are induced to undergo MET by BMP-7 (Zeisberg et al., 2003). In the case of urodele spinal cord regeneration, EMT produces a mass of cells that bridge the gap, followed by MET to restore the ependyma. The ependymal cells form endfeet that project to the glia limitans and form channels that support axon regeneration. In all cases, intermediate filament expression alternates between epithelial markers (cytokeratins) and mesenchymal markers (actin locomotory filaments, vimentin).

The bone marrow harbors endothelial stem cells that circulate in the blood. Circulating EnScs have the phenotype $[CD133\ VEGFR2]^+$ and express the receptor for the chemoattractant, stromal cell derived factor 1 (SDF-1). These cells are recruited to sites of injury by SDF-1 and angiogenic factors such as VEGF-A and placental growth factor (PLGF). There, they are incorporated into regenerating blood vessels. EnSCs make only a minor contribution to the construction of the new vessels, which takes place primarily by sprouting from existing vessels (Stocum, 2006, for review). Endothelial cells in the injured vessel wall are induced by FGF-2, TGF-β1, IL-8 and TNF-α to undergo EMT. The activated cells may be a subpopulation in the vessel wall similar to EnSCs (Ingram et al., 2004). They lose their intercellular junctions and express proteases that break down their basement membrane. In the presence of VEGF synthesized by the epidermis of a healing wound, the cells proliferate and migrate out as mesenchymal cords into the fibrin matrix of the wound. PD-ECGF and TNF-α are chemotactic for endothelial cells; TGF-α, TGF-β, FGF 1 and 2, and PDGF-B stimulate proliferation (Madri et al., 1996; Tomanek and Shatteman, 2000). The proliferating cells then undergo MET and rearrange themselves into endothelial tubes, a process mediated by laminin and a fibroblast-secreted protein, Egfl7 (Parker et al., 2004).

Mesenchymal Stem Cells

Mesenchymal stem cells were first isolated from the bone marrow as an adherent cell capable of differentiating into fibroblasts, chondrocytes, osteoblasts and adipocytes. They are responsible primarily for the regeneration of bone, tendon and ligament, but also for a limited regenerative capacity of dentin in adult teeth.

Bones regenerate after fracture by the proliferation of MSCs residing in the bone marrow, endosteum and periosteum. In fractured membrane (flat) bones, the MSCs differentiate directly to osteoblasts that secrete the bone matrix. The MSCs of fractured endochondral (long) bones first differentiate into a chondrocyte template that is subsequently replaced by osteoblasts (Ham and Cormack, 1979). The molecular mediators of fracture repair appear to be identical to those involved in embryonic bone development. BMPs, TGF-β, FGF-1 and 2, PDGF and IGF-1 expressed by the MSCs regulate chondrocyte differentiation (Bostrom, 1998; Einhorn, 1998; Trippel, 1998). The transcription factor Sox-9 activates the expression of type I, IX and collagen genes and the gene for aggrecan protein. *Ihh* signaling pathway components are expressed in a population of cells on the periphery of the soft callus that will reform the periosteum, indicating that the same mechanism used to regulate the rate at which chondrocytes mature during the embryonic development of long bones is operative during fracture healing. As the cartilage template is replaced by osteoblast invasion, the expression of genes involved in osteoblast differentiation, such as Runx2 and osteocalcin, is detected (Ferguson et al., 1998).

Teeth contain two types of stem cells in the pulp (Gronthos et al., 2000; Shi et al., 2001; Miura et al., 2002). One type has been isolated from adult teeth. It is similar to the bone marrow mesenchymal stem cell and differentiates into odontoblasts that make new dentin to counter the loss of odontoblasts destroyed by trauma or bacterial invasion (Murray and Garcia-Godoy, 2004). The other has been isolated from normally exfoliated deciduous incisors and is associated with capillaries. Mesenchymal stem cells also reside in the periodontal ligament (Seo et al., 2000). These cells continually maintain the ligament, which is under constant stress, but they can also regenerate injured alveolar bone. Adult mammals cannot regenerate lost teeth, but adult urodele amphibians, sharks and crocodilians can do so, and may thus be valuable research models for learning how to regenerate human teeth.

Mesenchymal stem cells are also found in adipose tissue (Zuk et al., 2002) and in most of the connective tissue compartments of the body (Young and Black, 2004). Like MSCs of the bone marrow, these cells can differentiate into fibroblasts, chondrocytes, osteoblasts and adipocytes. Whether they actually have a regenerative function *in vivo* is unknown, and their nature is unclear. It is possible that these cells, as well as the stem cells isolated from deciduous incisors, are pericytes, which are ubiquitous as cells that stabilize capillaries and venules, and have long been known to histologists as multipotential cells.

Hematopoietic and Endothelial Stem Cells

Blood cells are regenerated by hematopoietic stem cells (HSCs) in the bone marrow and blood vessels are regenerated both by endothelial cells in the walls of venules and circulating endothelial stem cells (EnSCs) from the bone marrow. HSCs are dependent on associated stromal cells of the marrow for their survival, proliferation and differentiation. They are small cells with the surface phenotype [CD34 c-Kit Sca-1 VEGFR2]$^{+}$ Thy-1^{lo} Lin^{-} and express the transcription factor Runx-1 (Spangrude et al., 1998; North et al., 2002). They divide asymmetrically to self-renew, while spawning a common erythroid/myeloid progenitor that gives rise to the blood cell lineages and a common lymphoid progenitor that gives rise to the cells of the immune system (Adolfsson et al., 2005). HSCs are maintained in a quiescent state by interaction with a subset of N-cadherin$^+$ CD45$^-$ stromal cells in the marrow, mediated by the Tie-2 receptor on HSCs and its ligand Ang-1 on the stromal cells. Wnt3a and Notch signaling are necessary for proliferation and self-renewal, as is Bmi, a protein that represses the expression of the p16 and p19 genes, which suppress proliferation and promote apoptosis, respectively (Calvi et al., 2003; Park et al., 2003; Arai et al., 2004).

Stem Cells of Skeletal and Cardiac Muscle

Skeletal muscle is regenerated by satellite cells (SCs) expressing the surface phenotype [CXR β1-integrin CD34 c-met]$^+$ [CD45 Sca-1 Mac-1]$^-$ (Sherwood et al., 2004). These cells reside between the sarcolemma and the basement membrane of the myofibers. SCs are held in a quiescent state by Notch signaling and do not express muscle regulatory factors (MRFs). Free-grafted muscle degenerates, followed by a typical inflammatory response. Quiescent SCs are activated, detach from their basement membranes, and proliferate within them, using the anaerobic pentose phosphate metabolic pathway (Hansen-Smith and Carlson, 1979; Carlson, 2003). The proliferating SCs strongly up-regulate Pax7 and MRFs and subsequently fuse and differentiate to form new myofibers. HGF released from muscle ECM and the growth hormone (GH)-stimulated upregulation of the IGF-IEc isoform (mechano growth factor, MGF) by myofibers are the major growth factors that stimulate SC proliferation, augmented by PDGF, FGF-2, LIF and TGF-β (Allen et al., 1995; Tatsumi et al., 1998; Pastoret and Partridge, 1998; Hill and Goldspink, 2003; Goldspink, 2005).

Mammalian cardiac muscle initiates a regenerative response to ischemic injury that is not sustained. Stem cells in the myocardium proliferate, but fibroblast proliferation is faster, suppressing the regenerative response and creating a scar. There are three distinct cardiac stem cell phenotypes that can differentiate *in vivo* and *in vitro* into cardiomyocytes: [c-Kit Sca-1]$^+$, Sca-1$^+$ [c-Kit Lin]$^-$, and Isl-1$^+$ [Sca-1 c-Kit]$^-$ (Beltrami et al., 2003; Oh et al., 2003; Laugwitz et al., 2005). The Isl-1$^+$ cells are found only in those parts of the heart that have an embryonic contribution from the secondary heart field. The relationship between these three subpopulations with regard to cardiac regenerative potential is not clear. Heart muscle regenerates in the MRL/MpJ mouse after cryogenic infarction (Leferovich et al., 2001). The frequency of mitosis in the injured MRL hearts is 10–20%, compared to 1–3% in wild-type animals. This animal model offers the opportunity to investigate how stem cell populations and/or injury environments differ in regenerating vs. non-regenerating mammalian heart tissue.

Dedifferentiation

Dedifferentiation is a mechanism for making mature cells into mesenchymal-like stem cells by the loss of phenotypic specialization. Dedifferentiation is not observed during embryogenesis, but the cells derived by dedifferentiation of adult cells do recapitulate part of the embryonic developmental program. The divas of dedifferentiation are the larval and adult urodeles (salamanders and newts) and anuran (frog and toad) tadpoles. These animals can regenerate many complex structures by dedifferentiation, including lens and neural retina of the eye, spinal cord, intestine, heart muscle, upper and lower jaws and limbs and tails. The major difference between dedifferentiated cells and ASCs is that dedifferentiated cells do not self-renew in the conventional sense. One could argue, however, that because these structures can regenerate repeatedly, each cycle of regeneration represents a self-renewal. Amphibian structures known to regenerate by dedifferentiation are the lens, neural retina, intestine, upper and lower jaws, heart muscle, limbs and tails. We know the most about dedifferentiation from *in vivo* and *in vitro* studies on muscle of regenerating amphibian limbs (Brockes and Kumar, 2005; Stocum, 2006, for reviews).

Limb regeneration in amphibians is achieved by the formation of a blastema derived from the satellite cells of muscle (Morrison et al., 2006) and by the dedifferentiation of dermis, muscle, skeletal and Schwann cells local to the amputation surface (Brockes and Kumar, 2005; Figure 6.2). Dedifferentiation is accomplished

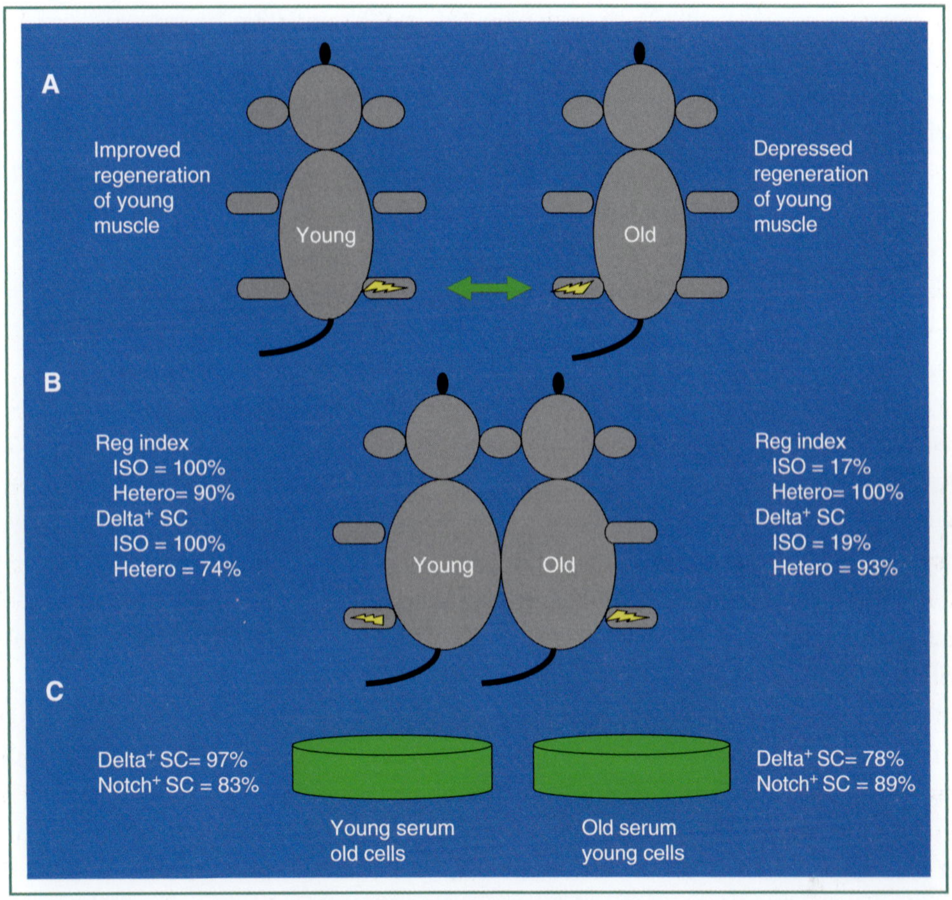

Figure 6.2 Experiments demonstrating that the regenerative capacity of old rat muscle is restored by providing the muscle with a young environment. (A) Reciprocal transplantation of leg muscle between young and old rats. Old muscle regains regenerative capacity, while the capacity of young muscle for regeneration is reduced. (B) Parabiosis of young and old rats followed by cryoinjury to leg muscle. ISO = same age parabionts; HETERO = old/young parabionts. The regeneration index (RI) is the number of regenerated myofibers and Delta$^+$ and Notch$^+$ indicate the number of activated satellite cells. All HETERO values and the ISO value for old/old parabints are measured against the ISO value for young/young parabionts. Left, values for regeneration of young muscle in HETERO parabionts are depressed to 90%, and 74%, respectively of the control (ISO) value. Right, values for regeneration of old muscle are improved from 17% and 19% of the young control (ISO) value to 100% and 93%. (C) Left, old satellite cells in young serum. Delta and Notch are expressed at 97% and 83% of the control value for young cells in young serum. Right, young satellite cells in old serum. Delta and Notch are expressed at 78% and 89% of the control value for young cells in young serum.

by the proteolytic degradation of ECM and the loss of phenotypic specialization by the liberated cells. In the case of muscle, this also involves cellularization. Re-entry of blastema cells into the cell cycle is induced by an as yet unidentified thrombin-activated protein (Tanaka et al., 1997, 1999). We do not yet have a clear picture of the molecular mechanism of dedifferentiation. Destabilization of microtubules is involved, but does not lead to the complete program of dedifferentiation and re-entry into the cell cycle (Duckmanton et al., 2005). Elements of the Notch signal transduction pathway are expressed in blastema cells, but other pathways may be involved as well. The blastema cells require growth and trophic factors from the wound epidermis and regenerating limb nerves for their survival and proliferation. Both the wound epithelium and nerves provide FGFs for this purpose. In addition, the nerves sustain blastema cells by glial growth factor-2, substance P, and transferrin (Stocum, 2006, for review).

The blastema is a self-organizing entity from its inception (Stocum, 2006, for review). The developmental fate and handedness of the blastema cannot be altered by grafting it to an ectopic location, even under conditions that force it to repeat the earliest stages of blastema formation. The mechanism of self-organization can be explained by local cell interactions that intercalate missing limb structures within boundaries established in the early blastema. The limb can be viewed as a three-dimensional "normal neighbor" map in which each cell knows its position relative to all other cells (Mittenthal, 1981). When a limb is amputated, dedifferentiated cells inherit a memory of their position on the circumference and radius of the limb. How the proximodistal positional identities are restored is not yet clear, but probably involves intercalary regeneration via local cell interactions between boundary positional identities established during early blastema formation.

In vitro and *in vivo* adhesion assays, in conjunction with RA treatment, have shown that positional identity is encoded in the cell surface (Nardi and Stocum, 1983; Crawford and Stocum, 1988; Escheverri and Tanaka, 2005). One molecule that has been implicated in PD positional identity is Prod-1, a molecule related to mammalian CD59, whose expression is differentially regulated by RA and whose overexpression causes distal blastema cells to translocate proximally (Morais da Silva et al., 2002; Echeverri and Tanaka, 2005). Patterning genes that are activated by local cell interactions during self-organization are similar to those that have been identified in the developing embryonic limb bud. In the PD axis, *Hoxa9, Hoxd10* and *Meis1* and *2* are involved in specification of the stylopodium and zeugopodium, and *Hoxa13* is involved in specification of the autopodium (Simon and Tabin, 1993; Gardiner and Bryant, 1996; Mercader et al., 2005). In the AP axis, *Shh* plays a role in establishing digit number and identity, and *Lmx-1* in the development of dorsal tissue pattern (Imokawa and Yoshizato, 1997; Endo et al., 1997).

Effects of Aging on Regenerative Capacity

Aging clearly reduces the capacity of tissues for regeneration. A major controversy is whether this decline is due to a decline in number or quality of regeneration-competent cells, a deteriorating niche environment (local and/or systemic), or both. A good example is skeletal muscle. The gastrocnemius muscle of young rats regenerates well, but regenerates poorly in old rats (Carlson and Faulkner, 1989; Carlson et al., 2001). Verdijk et al. (2007) reported that the absolute number of satellite cells per type I myofiber and the cross-sectional area of these myofibers is similar in the vastus lateralis muscle of young and elderly humans, but that the cross-sectional area, absolute number of SCs, percentage of SC myonuclei per myofiber, and the number of SCs per myofiber area is significantly lower in the type II myofibers of elderly muscle. Collins et al. (2007) reported a significantly lower number of SCs in aged rat muscle, but identified a subset of aged SCs *in vitro* that regenerate myofibers as efficiently as SCs from young muscle.

Reciprocal exchange of the gastrocnemius muscle between young and old rats, or parabiosis of old and young rats improves the regeneration of old muscle, while depressing the regenerative capacity of young muscle (Carlson and Faulkner, 1989; Carlson et al., 2001; Conboy et al., 2005). Parabiotic studies indicated that the decline in regenerative capacity of old muscle was associated with a lower percentage of proliferating SCs, not with a decrease in the number or quality of satellite cells (Figure 6.3). The serum of old rats appears to be deficient in factors that in young rats promote the proliferation of satellite cells by increasing the expression levels of Notch and Delta. These factors have not been specifically identified, but one of them may be growth hormone, since strength training in elderly humans significantly retards sarcopenia, and increases MGF production, particularly in combination with administration of GH (Goldspink, 2004, 2005). Similar results have been obtained with young vs. old liver (Conboy et al., 2005). Other ASCs exhibit age-related declines in regenerative capacity as well, but no age-reversal experiments of the type performed on liver and skeletal muscle have been done on these tissues (Stocum, 2006, for review).

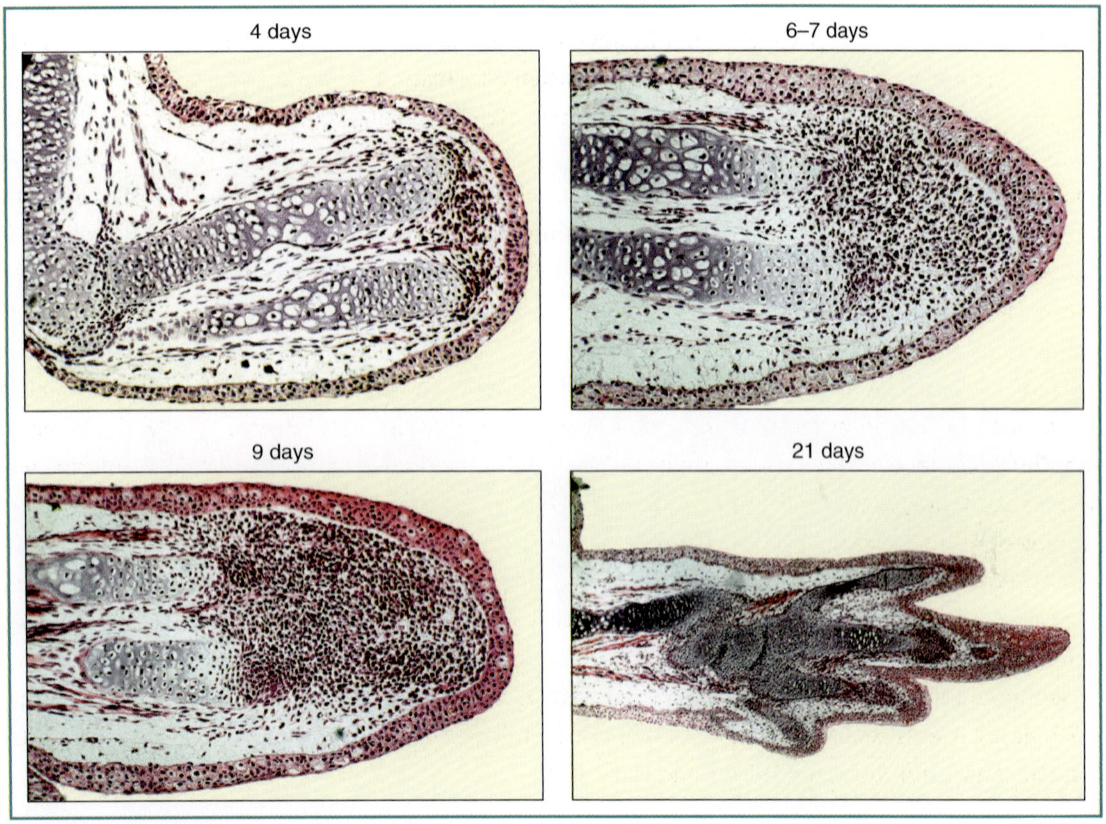

Figure 6.3 **Longitudinal sections of a regenerating axolotl limb amputated through the distal radius and ulna, 4–21 days post-amputation. By 4 days, dedifferentiation has created an accumulation of mesenchymal stem cell-like cells under the wound epidermis, which becomes a cone by 6–7 days due to mitosis. The first signs of differentiation emerge at 9 days, and by 21 days, a replica of the missing wrist and hand has been regenerated.**

STRATEGIES OF REGENERATIVE MEDICINE

Regenerative medicine uses three strategies based on the regenerative biology of regeneration-competent cells: cell transplants, implantation of bioartificial tissues and the chemical induction of regeneration (Figure 6.4). These strategies seek to reconstruct damaged tissues, organs and appendages by cell transplants or bioartificial tissues, or by inducing resident cells to reconstruct them *in situ*.

Cell Transplants and Bioartificial Tissues

Cell Transplants

Fetal cells

Fetal cells have been used primarily to treat Parkinson's and Huntington's disease. Mesencephalic cells from 6–8 week old fetuses appeared to differentiate into dopaminergic neurons, increase dopamine output, and make synaptic connections with host neurons (Bjorklund and Lindvall, 2000, for review; Bjorklund et al., 2003). The results are highly variable, however, due to the differential survival of the transplanted cells, and double-blind studies suggest that there is a large placebo effect of the treatment (Lazic and Barker, 2003). Fetal striatal tissue grafted to the striatum of marmoset or macaque monkeys with NPA-induced Huntington's was reported to reverse the symptoms of the disease (Kendall et al., 1998; Palfi et al., 1998). Immunohistochemical studies indicated good survival and differentiation of the grafted neurons, with establishment of functional connections with host tissue. Preliminary clinical trials in human patients given grafts of human fetal

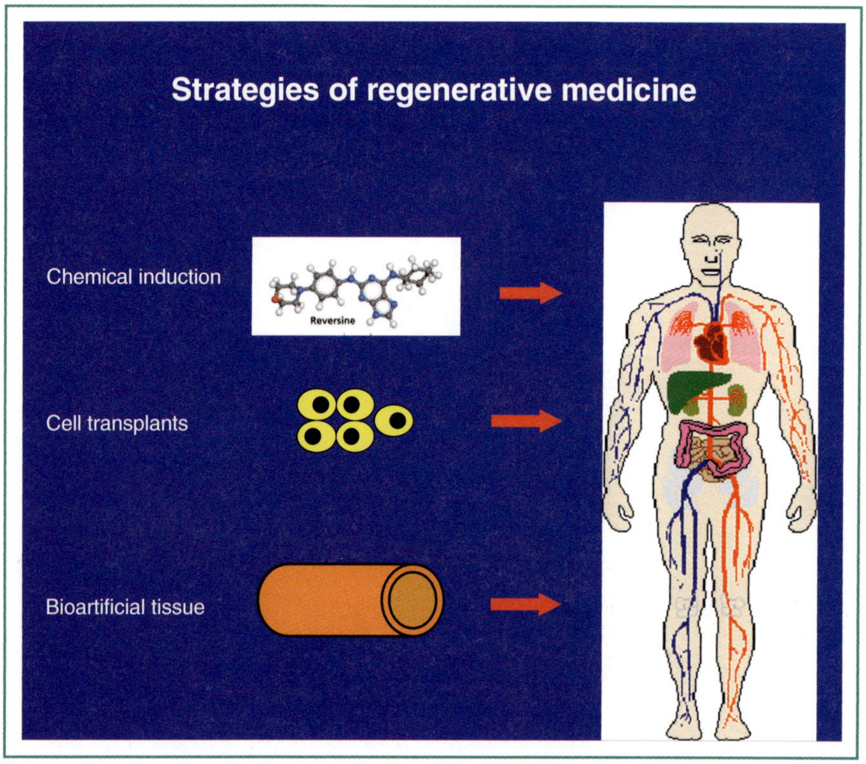

Figure 6.4 The three strategies of regenerative medicine. Chemical induction can involve administration of combinations of cytokines, growth factors, natural or artificial ECM templates, or small natural or synthetic molecules, such as reversine (see text). Cell transplants can be used as bioreactors to provide host tissues with paracrine factors, to rebuild tissue, or to construct bioartificial tissues, such as artificial blood vessels.

striatal tissue indicated that the tissue survived and that the symptoms of the disease were alleviated to some extent, with persistent benefits to some patients three years post-grafting (Rosser et al., 2002).

Adult Stem Cells

The most sophisticated and successful clinical adult stem cell transplants, begun in 1968, are those of bone marrow for hematopoietic malignancies or genetic disorders. Variable success has been had with other types of ASCs. Cultured keratinocytes have been applied to acute and chronic wounds (Liu et al., 2004) and cultured autogeneic limbal or oral epithelial stem cells have been used to replace the cornea in patients who have suffered corneal damage (Tsai et al., 2000; Nishida et al., 2004). Transplantation of cultured satellite cells for human Duchenne muscular dystrophy has not been successful, but in *mdx* mice, fresh satellite cells or satellite cells derived from cultured wild-type MSCs have successfully regenerated normal muscle (Montarras et al., 2005). Human MSCs converted to satellite cells by transfection with the DNA sequence for the Notch intracellular domain (NICD), followed by treatment with satellite cell conditioned medium, regenerated muscle after transplantation to dystrophic *mdx* mice (Dezawa et al., 2005). In a similar experiment, cultured rat and human MSCs were reported to become dopaminergic neurons (41% frequency) when transfected with the NICD sequence and treated with glial derived neurotrophic factor (GDNF) (Dezawa et al., 2004). Transplantation of these cells into Parkinsonian rats significantly increase dopamine production and decreased symptoms.

Over the past decade, one of the great hopes of regenerative medicine has been that adult stem cells will prove to have a plasticity that allows them to be reprogrammed by foreign injury environments *in vivo* or defined chemical factors *in vitro* to cell types of other lineages (lineage conversion) for transplantation or

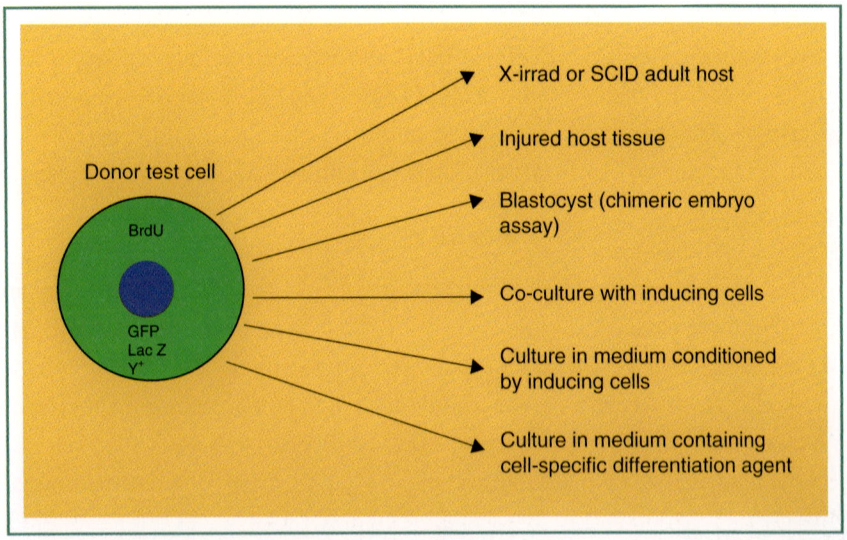

Figure 6.5 Assays for lineage conversion of adult stem cells. Labels to identify the donor cells include the Y chromosome, transgenes for green fluorescent protein (GFP) or β-galactosidase, or some combination thereof. BrdU is added to detect DNA synthesis. The cells are then injected or implanted into a variety of host *in vivo* environments, or cultured *in vitro* with inducing cells or chemical agents.

bioartificial tissue construction. Several types of lineage conversion assays (Figure 6.5) have been used to test the developmental plasticity of various types of ASCs. Bone marrow stem cells have been of the most interest, because they are easy to harvest and expand as autogeneic cells. The results of such assays have been inconsistent and in many cases difficult to repeat, because of differences in ever-evolving experimental protocols, fusion with host cells, and artifacts such as contamination of the donor cell population with other differentiated cell types, incorporation of host leukocytes into donor tissues, or incorporation of donor cells into host tissues without long-term survival or differentiation into authentic cell phenotypes of that tissue. Cells with high putative plasticity have also been isolated from long-term cultures of bone marrow cells and from connective tissue compartments. These cells share some characteristics with ESCs and differentiate *in vivo* and *in vitro* into a wide variety of cell types at frequencies of 5–90% (Jiang et al., 2002; Young and Black, 2004). However, these results have also been difficult to repeat. The overall evaluation to date is that the lineage conversion of adult stem cells is possible and a goal worth pursuing, but requires more consistent and rigorous proof (Wagers et al., 2002; Murry et al., 2004; Balsam et al., 2004; Laflamme and Murry, 2005).

Embryonic Stem Cells

Most ASCs are difficult to harvest and expand. Furthermore, any allogeneic cell transplant or bioartificial tissue will be immunorejected, unless the cells are encapsulated. ESCs are viewed as a prime source of cells for transplant or bioartificial tissue construction because they can be expanded indefinitely in culture to provide the large numbers of cells required to produce derivatives, while retaining their pluripotency. Transplanted neural and glial precursors differentiated from ESCs have successfully reversed the lesions and symptoms of Parkinson's disease and deymelinating disorders in rodents (Kim et al., 2002; Barberi et al., 2003; Brustle et al., 1999). Beyond this, the use of ESC derivatives for tissue regeneration is still in a nascent state.

Like other allogeneic cells, ESC derivatives are subject to immune rejection (Rippon and Bishop, 2004). Immunorejection of ESC derivatives theoretically can be avoided by using autogeneic ESC lines from blastocysts derived by somatic cell nuclear transplant (SCNT). Proof of principle has already been demonstrated in experimental animals (Munsie et al., 2000; Wakayama et al., 2001), but autogeneic human blastocysts have not yet been created, and their derivation is bioethically controversial. This has sparked a search for ways to derive autogeneic ESCs in other ways. Adult mouse fibroblasts have been successfully reprogrammed to express the transcriptional profile of ESC by fusing them with ESCs (Cowan et al., 2005) or transfecting them with the pluripotency transcription factors Oct-4, Sox-2, c-Myc and Klf-4 (Takahashi and Yamanaka, 2006). NSCs have been reprogrammed in the same way by fusing them with F9 embryonal carcinoma cells (Do et al., 2007).

Other cells

Chondrocytes cultured from biopsies of healthy cartilage have been successful in repairing articular cartilage damaged by trauma (Brittberg et al., 1994), and β-cell transplants from cadavers have temporarily reversed the symptoms of diabetes, though such transplants are severely limited by donor shortage (Shapiro et al., 2000). Umbilical cord blood cells, which are easily harvested and preserved, show great promise for hematopoietic regeneration (Takahashi et al., 2004). Stem cells that express both embryonic and adult stem cell markers have been isolated from amniotic fluid. These cells were induced *in vitro* and *in vivo* to differentiate into neuronal, hepatic, and osteogenic phenotypes (De Coppi et al., 2007) and may represent the best of both ASC and ESC worlds.

Bioartificial Tissues

Cell transplants are primarily useful for replacing small areas of tissue. Bioartificial homologues are necessary to replace larger tissue areas or whole organs. Tissue homologues have been successfully created in experimental animals for long bone segments by seeding ceramic scaffolds ("bone blanks") with MSCs that differentiate into osteoblasts (Dennis et al., 2001; Cowan et al., 2004), for intestine, trachea, and urinary bladder by seeding biodegradable polymer meshes with epithelial and smooth muscle cells and for blood vessels by culturing endothelial, smooth muscle and fibroblast cells around a mandrel (Stocum, 2006, for review). Work is ongoing to bioengineer whole organs such as the liver, but success in this endeavor has so far been limited because of the difficulty in providing the tissue with vascular channels *in vitro*. The most spectacular bioartificial tissue made so far is a human mandible constructed of a titanium mesh cage filled with blocks of bone matrix, bone marrow cells (for MSCs) and BMPs (Warnke et al., 2004). This construct was prevascularized and differentiated by growing it for seven weeks in a pocket made in the latissmus dorsi muscle of a patient who had lost his mandible to cancer. The construct was then removed and transplanted successfully into the position of the original mandible.

A major issue for bioartificial tissue construction (or for regeneration templates, see ahead), aside from vascularization, is mimicking the properties of the ECM. The ECM is a complex, three-dimensional assembly of macromolecules synthesized by cells as an adjacent acellular basement membrane, and/or as an interstitial tissue matrix surrounding the cells. Interstitial ECM is composed of fibrous proteins (primarily collagens) embedded in a highly hydrated gel of GAGs and proteoglycans that is also a repository for signaling molecules such as growth factors, proteases and their inhibitors (Voytik-Harbin, 2001). The natural matrix releases the appropriate biological signaling information at the right times and places to promote and maintain cell adhesion, proliferation, differentiation and tissue organization. Thus, processed natural biomaterials such as cadaver dermis and pig SIS have been a logical choice for use as regeneration templates and scaffolds for bioartificial tissues.

The use of synthetic biomaterials is advantageous because they can be manufactured in virtually unlimited quantities to specified standards, with additional shape-shifting features built in, such as liquidity and small volume at room temperature, changing to gelation, expansion, and space-filling at body temperature within a tissue gap. The goal of biomaterials science is to make synthetic scaffolds that mimic the ECM *in vivo*, providing not only the geometry and physical/chemical properties to maximize the migration of cells throughout the scaffold, but also the capability to sequester and release biological signals essential for cell

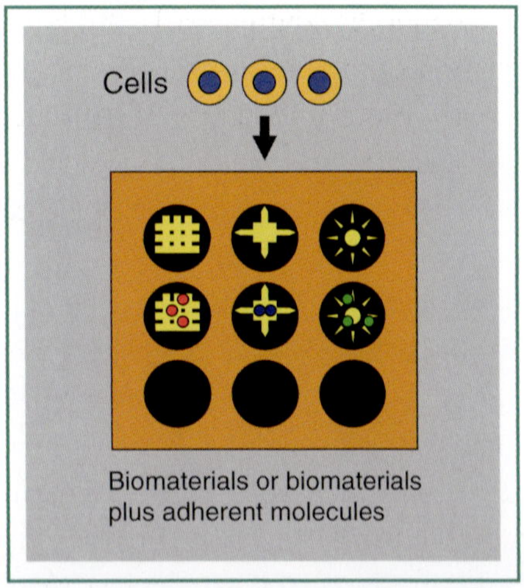

Figure 6.6 **High throughput screening of biomaterials (yellow) or biomaterials with adherent molecules (red, blue and green dots) for their effects on cells layered on top of the biomaterials.**

proliferation and differentiation (Langer and Tirell, 2004). Hench and Polak (2002) have described the evolution of biomaterials from a first generation in the 1960s and 70s that mimicked the physical properties of replaced tissue with minimal toxicity, to a second generation in the 1980s and 90s that was bioactive and biodegradable as well. This second generation of biomaterials is the set from which most scaffolds are currently made (for example, polyglycolic acid and polylactic acid).

Third generation biomaterials focus on micro- and nanofibrillar biomaterial gels, including self-assembling peptide and non-biological amphiphiles, and non-fibrillar synthetic hydrophilic polymer hydrogels that have the physical and chemical properties of natural ECM (Lutolf and Hubbell, 2005). A number of biologically important signaling and enzyme-sensitive entities can be incorporated into these hydrogels, including recognition sequences for cell adhesion proteins, soluble growth factors, and protease-sensitive oligopeptide or protein elements. Derivatized amino reactive polyethylene glycols (PEG) containing both peptide substrates for proteases and binding peptides for soluble factors or cell adhesion molecules appear to be particularly promising for creating mimics of ECM-cell interactive processes (West and Hubbell, 1999; Zisch et al., 2003; Tessmar et al., 2004). There are significant technical hurdles yet to be overcome in making interactive synthetic biomaterials that mimic the specific microniche environments of regeneration-competent cells (Lutolf and Hubbell, 2005). However, the development of new generations of biomaterials with interactive effects on cell behavior is being aided by high-throughput screening of biomaterials (Anderson et al., 2004) (Figure 6.6). Hubbell (2004) has pointed out that polymer biomaterials could be used as tethering platforms to screen combinatorial libraries of molecules that bind to the polymers for their effects on cell activity. If the activity of such molecules is dependent on their association with the polymers, they would not show an effect when presented to cells by themselves, but would reveal their effects if bound to a polymer.

Chemical/Physical Induction of Repair and Regeneration

Topical Agents for Skin Repair

Various topical agents have been tested for their efficacy in accelerating repair of acute wounds and chronic skin wounds (Fu et al., 2005; Stocum, 2006, for reviews). The growth factors TGF-β1 and 2, FGF-2, EGF, and IGF-1, and growth hormone (GH) have been reported to accelerate the repair of acute wounds in experimental animals and FGF-2 and GH have this effect in human patients. Other agents reported to accelerate the repair of acute skin wounds are extract of the *Celosia argentea* leaf, vanadate, oxandrolone, the opoid fentanyl, ketanserin, oleic fatty acids, pig enamel matrix, and the peptide HB-107. These agents increase the rate and extent of re-epithelialization, angiogenesis, or cellularity of granulation tissue. Still other topical agents act by reducing scarring. Extract of *Allium cepia* (onion), chitosan, the COX-2 inhibitor celicoxib, HGF, and anti-TGF-β1 and 2 antibodies all promote healing with less scarring, as do hydrogels composed of cross-linked hyaluronic acid and chondroitin sulfate. Combinations of topically applied PDGF-B plus IGF-1, EGF plus insulin, and TGF-β plus PDGF-B enhance chronic wound repair in experimental animals. PDGF-B, FGF-2, EGF, TGF-β, and rhKGF-2 all accelerate the closure of chronic wounds in patients. Currently, FGF-2, PDGF-B and rhKGF-2 are approved for clinical use. Other topically applied agents that accelerate the repair of chronic wounds are angiotensin (1–7), thymosin β4, L-arginine, and pentoxifylline. These agents exert their effect through anti-inflammatory and angiogenesis-promoting activities.

Regeneration Templates

Natural or bioartificial scaffolds have been used as templates to encourage immigration of resident cells bordering lesions to repair dermis and other connective tissues, peripheral nerves, urinary conduit tissue, digestive tract, and bone (Yannas, 2001; Stocum, 2006, for reviews). Cadaver dermal matrix (Alloderm®) and fetal bovine dermal matrix (Primatrix™) promote repair of burns; porcine dermal matrix (Permacol®) and porcine small intestine submucosa (SIS, Surgisis™) are approved for hernia repair. Primatrix™ and another form of SIS, Oasis™ accelerate the healing of diabetic ulcers. The most widely used bioartificial dermal matrix is Integra®, which consists of bovine dermal collagen and chondroitin 6-sulfate. Clinical assessments of Integra® have reported results superior to those of other constructs for excisional wounds, including burns (Heitland et al., 2004). Epidermal coverings do not take well on dermal regeneration templates when the dermis is badly damaged, due to slow vascularization. Thus they are often applied in a two-step procedure in which the dermal template is put on the wound first and allowed to revascularize, after which keratinocytes or meshed split thickness skin grafts are added.

Collagen tubes filled with a copolymer of type I collagen and chondroitin 6-sulfate with longitudinally oriented pores promoted the regeneration of transected peripheral nerve axons, while a collagen/laminin matrix, alginate gel and intercostal nerve sheath embedded in fibrin matrix have been reported to foster the regeneration of spinal cord axons (Yannas, 2001; Goldsmith and de la Torre, 1992; Cheng et al., 1996; Ramer et al., 2005).

Urinary bladder matrix has proved effective as a template to promote the regeneration of bladder wall tissue in pigs and urethral wall in human patients (Reddy et al., 2000; El-Kassaby et al., 2003). SIS matrix promoted regeneration of small defects in the esophagus, intestine, bile duct, trachea, bladder and ureter in experimental animals, and polyester mesh has been used as a template to regenerate small defects in the trachea and bladder (Stocum, 2006, for review).

A wide variety of scaffolds, including ceramics, polymer combinations, and bioactive glass, encourage the regeneration of small segments of bone by MSCs that migrate into the biomaterial (Seeherman et al., 2002). Plasmid or retroviral growth factor constructs (primarily BMPs) have been incorporated into polymers to promote the commitment of MSCs to osteoblasts (Goldstein and Bonadio, 1998; Bonadio, 2002). Extra bone for transplant has been made in rabbits by filling a subperiosteal space in the tibia with alginate. MSCs from the periosteum migrated into the alginate and formed new bone that was then transplanted to fill a defect made in the contralateral tibia (Stevens et al., 2005).

Soluble Factors

A number of soluble agents have been found to protect neurons of the damaged mammalian spinal cord and to neutralize or remove molecules inhibitory to regeneration (Ramer et al., 2005). Neuroprotectives include molecules such as methylprednisolone and gacyclidine. Myelin proteins inhibitory to axon regeneration in the cord can be neutralized with antibodies, and chondroitinase ABC has been shown to promote spinal cord axon regeneration by cleaving off the chondroitin sulfate side chains from proteoglycans of glial scar (Filbin, 2000; Bradbury et al., 2002).

A significant discovery is that transplanted cells are the source of paracrine factors that promote the survival of host cells in the injured region, reduce scarring and even promote regeneration of host tissue. For example, bioartificial skin equivalents, comprised of allogeneic neonatal foreskin fibroblasts in a collagen or polyester scaffold, act as living wound dressings to enhance the healing of chronic wounds by providing growth factors to host cells of the wound (Ehrlich, 2004; Jimenez and Jimenez, 2004). The fibroblasts are eventually rejected and replaced with host fibroblasts. Transplanted NSCs or NSCs transfected with a lentiviral GDNF construct or injection of the construct itself into the striatum promoted the survival of host dopaminergic neurons in Parkinsonian rats (Kordower et al., 2000). Regeneration of spinal cord axons is promoted by the incorporation of Schwann cells and olfactory ensheathing cells into regeneration templates (Ramer et al., 2005). These cells provide soluble factors and adhesion molecules to cord axons that are used normally in the regeneration of spinal nerve and olfactory nerve axons, respectively. Improvements in the symptoms of ALS patients have also been reported in China after injection of autogeneic olfactory ensheathing cells into the forebrain, presumably by paracrine action, although these results have been criticized because they are uncontrolled (Watts, 2005; Curt and Dietz, 2005).

MSCs improved cardiomyocyte survival when injected into the infarcted hearts of mice. The effect of these cells is due to paracrine action that activates the cell survival gene Akt, as shown by the fact that conditioned medium of hypoxic MSCs activates this gene and reduces infarct size by reducing apoptosis of cardiomyocytes when injected into the infarct region. Thymosin β-4, which plays a role in regulating the assembly of G-actin into F-actin filaments, also enhances cardiomyocyte survival and cardiac function by activation of *Akt* (Mangi et al., 2003; Gnecchi et al., 2005; Bock-Marquette et al., 2004), but also by an effect on the migration of epicardial cells and their differentiation into endothelial cells (Smart et al., 2007). Modest improvement in cardiac function was reported in random, double-blinded clinical trials of bone marrow cells injected into the infarct region of patients (Wollert et al., 2004; Lovell and Mathur, 2004; Mathur and Martin, 2004). However, in another controlled, random double-blind study, G-CSF induced mobilization of bone marrow stem cells had no effect on cardiac function in patients who had suffered myocardial infarct (Zohlnhofer et al., 2006).

Identifying Constellations of Natural Regeneration Promoting and Inhibitory Molecules

Tissues that normally undergo maintenance or injury-induced regeneration clearly possess the niche factors requisite for regeneration. Regeneration-permissive signals must also be present in the injury environments of tissues that fail to regenerate, because such tissues (for example, spinal cord and heart) have regeneration-competent cells that often initiate a regenerative response, which is then suppressed by fibrosis. To regenerate these tissues, it might only be necessary to neutralize molecules that promote scarring. For other tissues that do not initiate a regenerative response, it may prove essential to provide additional regeneration-permissive or inductive signals to the injury site, particularly if the tissue does not contain regeneration-competent cells. Two strategies can be used to identify regeneration-permissive/inductive and inhibitory molecules (Figure 6.7). One is to identify the molecules secreted by cells known to enhance host cell survival and inhibit scarring after transplantation, and determine which combinations are active in these processes. The other is to compare fibrosis and regeneration in three types of *in vivo* models.

The first model compares wild-type tissues to genetic variations that confer a gain or loss of regenerative capacity. Several strains of MRL mouse can regenerate ear and heart tissue (Heber-Katz et al., 2004).

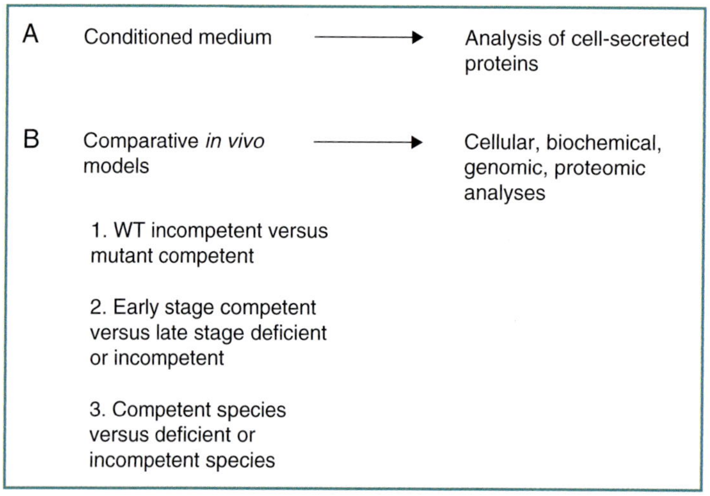

Figure 6.7 Approaches to the identification of natural molecules that constitute the molecular difference between regeneration-competence and regeneration-deficiency or incompetence.

Molecular comparisons can be made between these and regeneration-incompetent wild-type strains to reveal molecules permissive and inhibitory to regeneration.

The second model compares tissues at developmental stages when they are capable of regeneration versus stages when they are not. For example, fetal skin in many mammalian species regenerates perfectly, but late in gestation the injury response switches to scar tissue formation characteristic of the adult (McCallion and Ferguson, 1996), whereas the skin of the neonatal PU.1 null mouse retains the fetal capacity for regeneration (Redd et al., 2004). The frog limb bud regenerates perfectly at early tadpole stages, but becomes regeneration-deficient at late tadpole stages. The loss of regenerative capacity in fetal skin and the frog limb bud may be related to maturation of the immune system and the resultant greater inflammatory response after wounding, while the retention of regenerative capacity in the PU.1 null mouse may be due to failure of the immune system to mature (Mescher and Neff, 2005; Godwin and Brockes, 2006, for reviews).

The third model compares tissues in regenerating vs. non-regenerating species, such as the regenerating axolotl or newt limb vs. the non-regenerating frog or mouse limb. For example, it has been shown that newt dorsal iris cells and myofibers have the ability to respond to a thrombin-activated protein (as yet unidentified) by entering the cell cycle, whereas axolotl lens and mouse myofibers do not (Tanaka et al., 1999; Imokawa and Brockes, 2003).

Comparative genomic analyses using these models have revealed differences in the gene activity of regeneration-competent vs. deficient tissues. For example, subtractive hybridization analysis of regeneration-competent vs. deficient limbs in the frog *Xenopus laevis* has revealed not only the upregulation and downregulation of many known genes, but also many novel genes (King et al., 2003). Proteomic analyses should prove even more revealing. Coupled with bioinformatics and systems biology approaches, such data will be invaluable in providing complete molecular descriptions of regeneration competence vs. deficiency, allowing us to potentially promote regeneration in regeneration-deficient or incompetent tissues by manipulating the environment and/ or cellular responses at the injury site. Proof of concept has already been shown by experiments in which antibodies to TGF-β1 and 2, or application of TGF-β3 to adult skin wounds reduce scarring (Ferguson and O'Kane, 2004), by modest improvements in the regeneration of late frog tadpole limb buds by administration of FGF-8 and 10, BMP-4, HGF (Suzuki et al., 2006, for review), and by experiments showing that mouse muscle cells, which do not normally cellularize and dedifferentiate in response to injury, are induced to do so when treated *in vitro* with a protein extract of regenerating newt limb (McGann et al., 2001).

Small Molecules

A new approach to the chemical induction of regeneration is the use of small molecules to initiate regenerative responses. One such molecule is retinoic acid (RA, MW 300, acid derivative of vitamin A). RA is a key player in tissue embryogenesis, particularly the nervous system (Maden, 2002). It induces alveolar regeneration in the lung (Maden, 2004) and a lentiviral construct of the β2 retinoic acid receptor promotes functional recovery of injured rat spinal cord (Yip et al., 2006). RA has profound effects on the positional identity of blastema cells in regenerating urodele limbs, causing their proximalization, posteriorization and ventralization (Niazi, 1996; Maden, 1998; Stocum, 2006, for reviews).

The emerging field of chemical biology has developed methods to systematically identify synthetic small molecules that have developmental or regeneration-related effects on cells. Combinatorial libraries of molecules are generated from starting molecules, and are screened on cells for specific effects. Two such molecules have been identified that effect dedifferentiation of C2C12 mouse myofibers *in vitro*. Myoseverin (a tri-substituted purine) depolymerizes microtubules and upregulates growth factor, immunomodulatory and stress-response genes. Reversine (a di-substituted purine) interacts with protein kinases and initiates a full dedifferentiation program in C2C12 myofibers to create mesenchymal stem cell-like cells that can differentiate into muscle, osteoblasts and adipocytes (Rosania, 2004; Chen et al., 2004). Another synthetic purine derivative is puromorphine, which induces osteogenesis via the hedgehog signaling pathway (Wu et al., 2004). Neuropathiazol is a synthetic 4-aminothiazole that selectively induces neuronal differentiation of hippocampal neural stem cells (Warashina et al., 2006). Molecules generated in this way clearly have the potential to be useful for initiating regenerative responses and/or suppressing fibrosis in injured regeneration-deficient or incompetent tissues.

CONCLUSION

Developing the potential of regenerative medicine will require wide multidisciplinary efforts in the biological, chemical, physical, engineering and information sciences. The first wave of regenerative medicine was the transplantation of adult stem cells, begun in 1968 with the first bone marrow transplants. Current research aims to expand this success to other kinds of adult stem cells and derivatives of embryonic stem cells. The back part of this wave will be the chemical induction of regeneration using transplanted cells as bioreactors to provide survival and regeneration-permissive factors to host tissues and/or suppress fibrosis. These efforts have not seen much success as yet. The second wave will be the chemical induction of regeneration by cell-interactive regeneration templates, the direct delivery of regeneration-promoting and/or fibrosis inhibiting molecules or genes encoding these molecules to a lesion site, or some combination of these. These types of treatments will not only be relatively simple to administer clinically, but will also be much less expensive than cell transplantation therapies. To make the chemical induction of regeneration feasible, we must understand the biology of regeneration and how it differs from fibrosis to a much greater depth than is currently available. Only then will we know the appropriate places, times and at what concentrations to intervene in the pathways of repair to choose regeneration over fibrosis. The third wave will be the *in vitro* construction of bioartificial tissues and organs that can be implanted in place of the originals.

A single type of regenerative therapy is unlikely to fit all degrees of tissue damage. For example, it may not be possible to regenerate tissues much beyond a critical size defect using a cell transplant, chemical cocktail, or regeneration template. Larger defects may require a regeneration template seeded with cells to make a bioartificial tissue. Nor will success in understanding the biology of regeneration be achieved by a singular focus on mammalian research models. Non-mammalian models that are more powerful regenerators than mammals, such as amphibians, planarians and coelenterates can teach us much about the mechanisms of regeneration that we need to know in order to stimulate the latent regenerative powers of, or even confer such powers on, non-regenerating mammalian tissues.

ACKNOWLEDGMENT

Supported in part by a grant from the W.M. Keck Foundation

REFERENCES

Adolfsson, J., Mansson, R., Buza-Vidas, N., Hultquist, A., Liuba, K., Jensen, C.T., Bryder, D., Yang, L., Borge, O-J., Thoren, L.A.M., Anderson, K., Sitnicka, E., Sasaki, Y., Sigvardsson, M. and Jacobsen, S.E.W. (2005). Identification of $Flt3^+$ lympho-myeloid stem cells lacking erythro-megakaryocytic potential: A revised road map for adult blood lineage commitment. *Cell* **121**: 295–306.

Allen, R.E, Sheehan, S.M, Taylor, R.G, Kendall, T.K. and Rice, G.M. (1995). Hepatocyte growth factor activates quiescent skeletal muscle satellite cells *in vitro*. *J. Cell. Physiol.* **165**: 307–312.

Anderson, D., Levenberg, S. and Langer, R. (2004). Nanoliter-scale synthesis of arrayed biomaterials and application to human embryonic stem cells. *Nature Biotech.* **22**: 863–866.

Arai, F., Hirao, A., Ohmura, M., Sato, H., Matsuoka, S., Takubo, K., Ito, K., Koh, G.Y. and Suda, T. (2004). Tie2.angiopoietin-1 signaling regulates hematopoietic stem cell quiescence in the bone marrow niche. *Cell* **118**: 149–161.

Avilion, A.A., Nicolis, S., Pevny, L.H., Perez, L., Vivian, N. and Lovell-Badge, R. (2003). Multipotent cell lineages in early mouse development depend on SOX2 function. *Genes Dev* **17**: 126–140.

Balsam, L.B., Wagers, A.J., Christensen, J.L., Kofidis, T., Weissman, I.L. and Robbis, R.C. (2004). Haematopoietic stem cells adopt mature haematopoietic fates in ischaemic myocardium. *Nature* **428**: 668–673.

Beltrami, A.P., Barlucchi, L., Torella, D., Baker, M., Limana, F., Chimenti, S., Kasahara, H., Rota, M., Musso, E., Urbanek, K., Leri, A., Kajstura, J., Nidal-Ginard, B. and Anversa, P. (2003). Adult cardiac stem cells are multipotent and support myocardial regeneration. *Cell* **114**: 763–776.

Bjorklund, A. and Lindvall, O. (2000). Cell replacement therapies for central nervous system disorders. *Nature Neurosci.* **3**: 537–544.

Bjorklund, A., Dunnett, S.B., Brundin, P., Stoessl, A.J., Freed, C.R., Breeze, R.E., Levivier, M., Peschanski, M., Studer, L. and Barker, R. (2003). Neural transplantation for the treatment of Parkinson's disease. *Lancet Neurol.* **2**: 437–445.

Bock-Marquette, I., Saxena, A., White, M.D., DiMaio, J.M., Srivastava, D. (2004). Thymosin β4 activates integrin-linked kinase and promotes cardiac cell migration, survival and cardiac repair. *Nature* **432**: 466–472.

Bonadio, J. (2002). Genetic approaches to tissue repair. *Ann NY Acad Sci.* **961**: 58–60.

Bonner-Weir, S. and Weir, G.C. (2005). New sources of pancreatic β-cells. *Nat. Biotech.* **23**: 857–861.

Bonventre, J.V. (2003). Dedifferentiation and proliferation of surviving epithelial cells in acute renal failure. *J. Am. Soc. Nephrol.* **14**: S55–S61.

Bostrom, M.P.G. and Camacho, N.P. (1998). Potential role of bone morphogenetic proteins in fracture healing. *Clin. Orthopaed. Rel. Res.* **355S**: S274–S282.

Bradbury, E.J., Moon, L.D.F., Popat, R.J., King, V.R., Bennet, G.S., Patel, P.N., Fawcett, J.W. and McMahon, S.B. (2002). Chondroitinase ABC promotes functional recovery after spinal cord injury. *Nature* **416**: 636–640.

Brittan, M. and Wright, N.A. (2004). The gastrointestinal stem cell. *Cell Proliferat.* **37**: 35–53.

Brittberg, M., Lindahl, A., Nilsson, A.A., Ohlsson, C., Isaksson, O. and Peterson, L. (1994). Treatment of deep cartilage defects in the knee with autologous chondrocyte transplantation. *N. Engl. J. Med.* **331**: 889–895.

Brockes, J.P. and Kumar, A. (2005). Appendage regeneration in adult vertebrates and implications for regenerative medicine. *Science* **310**: 1919–1923.

Brustle, O., Jones, N., Learish, R., Karram, K., Choudhary, K., Wiestler, O.D., Duncan, I.D. and McKay, R.D.G. (1999). Embryonic stem cell-derived glial precursors. A source of myelinating transplants. *Science* **285**: 754–756.

Calvi, L.M., Adams, G.B., Weibrecht, K.W., Weber, J.M., Olson, D.P., Knight, M.C., Martin, R.P., Schipani, E., Diveti, P., Bringhurst, F.R., Milner, L.A., Kronenberg, H.M. and Scadden, D.T. (2003). Osteoblastic cells regulate the hematopoietic stem cell miche. *Nature* **425**: 841–846.

Carlson, B.M. and Faulkner, J.A. (1989). Muscle transplantation between young and old rats: age of host determines recovery. *Am. J. Physiol.* **256**: C1262–C1266.

Carlson, B.M. (2003). Muscle regeneration in amphibians and mammals: passing the torch. *Dev. Dynam.* **226**: 167–181.

Carlson, B.M., Dedkov, E.I., Borisv, A.B. and Faulkner, J.A. (2001). Skeletal muscle regeneration in very old rats. *J Gerontol.* **A56**: B224–B233.

Chambers, I., Colby, D., Robertson, M., Nichols, J., Lee, S., Tweedie, S. and Smith, A. (2003). Functional expression cloning of Nanog, a pluripotency sustaining factor in embryonic stem cells. *Cell* **113**: 643–655.

Cheng, H., Cao, Y. and Olson, L. (1996). Spinal cord repair in adult paraplegic rats: partial restoration of hind limb function. *Science* **273**: 510–513.

Chernoff, E.A.G., Stocum, D.L., Nye, H.L.D. and Cameron, J.A. (2003). Urodele spinal cord regeneration and related processes. *Dev. Dynam.* **226**: 280–294.

Chen, S., hang, Q., Wu, X., Schultz, P.G. and Ding, S. (2004). Differentiation of lineage-committed cells by a small molecule. *J. Am. Chem. Soc.* **126**: 410–411.

Collins, C.A., Zammit, P.S., Ruiz, A.P., Morgan, J.E. and Partridge, T.A. (2007). A population of myogenic stem cells that survives skeletal muscle aging. *Stem Cells* **25**: 885–894.

Collombat, P., Hecksher-Sorensen, J., Krull, J., Berger, J., Riedel, D., Herrera, P.L., Serup, P. and Mansouri, A. (2007). Embryonic endocrine pancreas and mature β cells acquire α and PP cell phenotypes upon *Arx* misexpression. *J. Clin. Invest.* **117**: 961–970.

Cotsarelis, G., Cheng, S.Z., Dong, G., Sun, T.T. and Lavker, R.M. (1989). Existence of slow-cycling limbal epithelial basal cells that can be preferentially stimulated to proliferate: implications on epithelial stem cells. *Cell* **57**: 201–209.

Conboy, I.M., Conboy, M.J., Wagers, A.J., Girma, E.R., Weissman, I.L. and Rando, T.A. (2005). Rejuvenation of aged progenitor cells by exposure to a young systemic environment. *Nature* **433**: 760–764.

Cowan, C.A., Atienza, J., Melton, D.A. and Eggan, K. (2005). Nuclear reprogramming of somatic cells after fusion with human embryonic stem cells. *Science* **309**: 1369–1373.

Cowan, C.M., Shi, Y.-Y., Aalami, O.O., Chou, Y.-F., Mari, C., Thomas, R., Quarto, N., Contag, C., Wu, B. and Longaker, M.T. (2004). Adipose-drived adult stromal cells heal critical-size mouse calvarial defects. *Nature Biotech.* **22**: 560–567.

Crawford, K. and Stocum, D.L. (1988). Retinoic acid coordinately proximalizes regenerate pattern and blastema differential affinity in axolotl limbs. *Development* **102**: 687–698.

Curt, A. and Dietz, V. (2005). Controversial treatments for spinal-cord injuries. *Lancet* **365**: 841.

Dabeva, M.D. and Shafritz, D.A. (2003). Hepatic stem cells and liver repopulation. In: Berk, P.D., Thorgeirsson, S. and Grisham, J.W. (eds.), *Seminars in Liver Disease*. New York: Thieme Medical Pub Inc, pp. 349–361.

De Coppi, P., Bartsch Jr., G., Siddiqui, M.M., Xu, T., Santos, C.C., Perin, L., Mostoslavsky, G., Serre, A.C., Snyder, E.Y., Yoo, J.J., Furth, M.E., Soker, S. and Atala, A. (2007). Isolation of amniotic stem cell lines with potential for therapy. *Nature Biotech.* **25**: 100–106.

Dennis, J.E., Solchaga, L.A. and Caplan, A.I. (2001). Mesenchymal stem cells for musculoskeletal tissue engineering. *Landes. Biosci.* **1**: 112–115.

Desai, B.M., Oliver-Krasinski, J., De Leon, D.D., Farzad, C., Hong, N., Leach, S.D. and Stoffers, D.A. (2007). Preexisting pancreatic acinar cells contribute to acinar cell, but not islet β cell, regeneration. *J. Clin. Invest.* **117**: 971–977.

Dezawa, M., Kanno, H., Hoshino, M., Cho, H., Matsumoto, N., Itokazu, Y., Tajima, N., Yamada, H., Sawada, H., Ishikawa, H., Mimura, T., Kitada, M., Suzuki, Y. and Ide, C. (2004). Specific induction of neuronal cells from bone marrow stromal cells and application for autologous transplantation. *J. Clin. Invest.* **113**: 1701–1710.

Dezawa, M., Ishikawa, H., Itokazu, Y., Yoshihara, T., Hoshino, M., Takeda, S-i., Ide, C. and Nabeshima, Y-i. (2005). Bone marrow stromal cells generate muscle cells and repair muscle degeneration. *Science* **309**: 314–317.

Do, J.T., Han, D.W., Gentile, L., Sobek-Klocke, I., Stehling, M., Lee, H.T. and Scholer, H.R. (2007). Erasure of cellular memory by fusion with pluripotent cells. *Stem Cells* **25**: 1013–1020.

Dor, Y., Brown, J., Martinez, O.I. and Melton, D.A. (2004). Adult pancreatic β-cells are formed by self-duplication rather than stem cell differentiation. *Nature* **429**: 41–46.

Duckmanton, A., Kumar, A., Chang, Y-T. and Brockes, J. (2005). A single cell analysis of myogenic dedifferentiation induced by small molecules. *Chem. Biol.* **12**: 1117–1126.

Echeverri, K. and Tanaka, E.M. (2005). Proximodistal patterning during limb regeneration. *Dev. Biol.* **279**: 391–401.

Ehrlich, H. (2004). Understanding experimental biology of skin equivalent: from laboratory to clinical use in patients with burns and chronic wounds. *Am. J. Surg.* **187S**: 29S–33S.

Einhorn, T.A. (1998). The cell and molecular biology of fracture healing. *Clin. Orthopaed. Related. Res.* **355S**: 7–21.

El-Kassaby, A.W., Retik, A.B., Yoo, J.J. and Atala, A. (2003). Urethral stricture repair with an off-the-shelf collagen matrix. *Urology* **169**: 170–173.

Engler, A.J., Sen, S., Sweeney, H.L. and Discher, D.E. (2006). Matrix elasticity directs stem cell lineage specification. *Cell* **126**: 677–689.

Endo, T., Yokoyama, H., Tamura, and Ide, H. (1997). *Shh* expression in in developing and regenerating limb buds of *Xenopus laevis. Dev. Dyn.* **209**: 227–232.

Fausto, N. (2004). Liver regeneration and repair: Hepatocytes, progenitor cells and stem cells. *Hepatology* **39**: 1477–1487.

Ferguson, C.M., Miclau, T., Hu, D., Alpern, E. and Helms, J.A. (1998). Common molecular pathways in skeletal morphogenesis and repair. *Ann. NY Acad. Sci.* **857**: 33–42.

Ferguson, M.W.J. and O'Kane, S. (2004). Scar-free healing: From embryonic mechanisms to adult therapeutic intervention. *Phil. Trans. R. Soc. Lond B* **359**: 839–850.

Filbin, M.T. (2000). Axon regeneration: Vaccinating against spinal cord injury. *Curr. Biol.* **10**: R100–R103.

Fu, X., Li, X., Chen, W. and Sheng, Z. (2005). Engineered growth factors and cutaneous wound healing: Success and possible questions in the past 10 years. *Wound Rep. Reg.* **13**: 122–130.

Gage, F.H. (2000). Mammalian neural stem cells. *Science* **287**: 1433–1438.

Gardiner, D.M. and Bryant, S.V. (1996). Molecular mechanisms in the control of limb regeneration: The role of homeobox genes. *Int. J. Dev. Biol.* **40**: 797–805.

Gilbert, S.F. (2006). *Developmental Biology*, 8th edn. Sunderland: Sinauer Associates Inc, pp. 147–164.

Gnecchi, M., He, H., Liang, O.D., Melo, L.G., Morello, F., Mu, H., Noiseux, N., Zhang, L., Pratt, R.E., Ingwall, J.S. and Dzau, V.J. (2005). Paracrine action accounts for marked protection of ischemic heart by Akt-modified mesenchymal stem cells. *Nat. Med.* **11**: 367–368.

Godwin, J.W. and Brockes, J.P. (2006). Regeneration, tissue injury and the immune response. *J. Anat.* **209**: 423–432.

Goldsmith, H.S. and de la Torre, J.C. (1992). Axonal regeneration after spinal cord transection and reconstruction. *Brain Res.* **589**: 217–224.

Goldspink, G. (2004). Age-related loss of skeletal muscle function: impairment of gene expression. *J. Musculoskel. Neuron Interact.* **4**: 143–147.

Goldspink, G. (2005). Mechanical signals, IGF-I gene splicing, and muscle adaptation. *Physiology* **20**: 232–238.

Goldstein, S.A. and Bonadio, J. (1998). Potential role for direct gene transfer in the enhancement of fracture healing. *Clin. Orthopaed. Rel. Res.* **355S**: S154–S162.

Gronthos, S., Mankani, M., Brahim, J., Robey, P. and Shi, S. (2000). Postnatal human dental pulp stem cells (DPSCs) *in vitro* and *in vivo*. *Proc. Natl Acad. Sci. USA* **97**: 13625–13630.

Ham, A.W. and Cormack, D.H. (1979). *Histology*, 8th edn. Philadelphia: JB Lippincott, pp. 377–462.

Hanna, L.A., Foreman, R., Tarasenko, I.A., Kessler, D.S. and Labosky, P.A. (2002). Requirement for Foxd3 in maintaining pluripotent cells of the early mouse embryo. *Genes Dev.* **16**: 2650–2661.

Hansen-Smith, F.M. and Carlson, B.M. (1979). Cellular responses to free grafting of the extensor digitorum longus muscle of the rat. *J. Neurol. Sci.* **41**: 149–173.

Hardy, M.H. (1992). The secret life of the hair follicle. *Trends Genet.* **8**: 55–61.

Hart, A.H., Hartley, L., Ibrahaim, M. and Robb, L. (2004). Identification, cloning and expression analysis of the pluripotency promoting nanog genes in mouse and human. *Dev. Dynam.* **230**: 187–198.

Heber-Katz, E. (2004). Spallanzani's mouse: A model of restoration and regeneration. In: Heber-Katz, E. (ed), Regeneration: Stem Cells and Beyond, *Curr. Topics Microbiol. Immunol.* **280**: 165–189.

Heitland, A., Piatkowski, A., Noah, E.M. and Pallua, N. (2004). Update on the use of collagen/glycosaminoglycan skin substitute – six years of experiences with artificial skin in 15 German burn centers. *Burns* **30**: 471–475.

Hench, L.L. and Polak, J.M. (2002). Third-generation biomedical materials. *Science* **295**: 1014–1017.

Hill, M. and Goldspink, G. (2003). Expression and splicing of the insulin-like growth factor gene in rodent muscle is associated with muscle satellite (stem) cell activation following local tissue damage. *J. Physiol.* **549**: 409–418.

Hubbell, J.A. (2004). Biomaterials science and high-throughput screening. *Nat. Biotech.* **22**: 828–829.

Imokawa, Y. and Yoshizato, K. (1997). Expression of *Sonic hedgehog* gene in regenerating newt limb blastemas recapitulates that in developing limb buds. *Proc. Natl. Acad. Sci. USA* **94**: 9159–9164.

Imokawa, Y. and Brockes, J.P. (2003). Selective activation of thrombin is a critical determinant for vertebrate lens regeneration. *Curr. Biol.* **13**: 877–881.

Ingham, D.A., Mead, L.E., Moore, D.B., Woodard, W., Fenoglio, A. and Yoder, M.C. (2004). Vessel wall derived endothelial cells rapidly proliferate because they contain a complete hierarchy of endothelial progenitor cells. *Blood First Edition Paper*, published online: DOI 10.1182/blood-2004-08-3057.

Ivanova, N., Dobrin, R., Lu, R., Kotenko, I., Levorse, J., DeCoste, C., Schafer, X., Lun, Y. and Lemischka, I.R. (2006). Dissecting self-renewal in stem cells with RNA interferencer. *Nature* **442**: 533–538.

Jamal, A.-M., Lipsett, M., Sladek, R., Laganiere, S., Hanley, S. and Rosenberg, L. (2005). Morphogenetic plasticity of adult human pancreatic islets of Langerhans. *Cell Death Diff* **12**: 702–712.

Jensen, U.B., Lowell, S. and Watt, F. (1999). The spatial relationship between stem cells and their progeny in the basal layer of human epidermis: A new view based on whole mount labeling and lineage analysis. *Development* **126**: 2409–2418.

Jiang, Y., Jahagirdar, B.N., Reinhardt, R., Schwarts, R.E., Keene, C.D., Ortiz-Gonzalez, X.R., Reyes, M., Lenvik, T., Lund, T., Blackstad, M., Du, J., Aldrich, S., Lisberg, A., Low, W.C., Largaespada, D.A. and Verfaille, C.M. (2002). Pluripotency of mesenchymal stem cells derived from adult marrow. *Nature* **418**: 41–49.

Jimenez, P.A. and Jimenez, S.E. (2004). Tissue and cellular approaches to wound repair. *Am. J. Surg.* **187S**: 56S–64S.

Jones, P.H., Harper, S. and Watt, F.M. (1995). Stem cell patterning and fate in human epidermis. *Cell* **80**: 83–93.

Kempermann, G., Kuhn, H.G. and Gage, F.H. (1998). Experience-induced neurogenesis in the senescent dentate gyrus. *J. Neurosci.* **18**: 3206–3212.

Kendall, A.L., Rayment, F.D., Torres, E.M., Baker, H.F., Ridley, R.M. and Dunnett, S.B. (1998). Functional integration of striatal allografts in a primate model of Huntington's disease. *Nature Med.* **4**: 727–729.

King, M.W., Nguyen, T., Calley, J., Harty, M.W., Muzinich, M.C., Mescher, A.L., Chalfant, C., N'Cho, M., McLeaster, K., McEntire, J., Stocum, D.L., Smith, R.C. and Neff, A.W. (2003). Identification of genes expressed during *Xenopus laevis* limb regeneration by using subtractive hybridization. *Dev. Dynam.* **226**: 398–409.

Kordower, J.H., Emborg, M.E., Bloch, J., Ma, S.Y., Chu, Y., Leventhal, L., McBride, J., Chen, E-Y., Palfi, S., Roitberg, B.Z., Brown, W.D., Holden, J.E., Pyzalski, R., Taylor, M.D., Carvey, P., Ling, Z.D., Trono, D., Hantraye, P., Deglon, N. and Aebischer, P. (2000). Neurodegeneration prevented by lentiviral vector delivery of GDNF in primate models of Parkinson's disease. *Science* **290**: 767–773.

Laflamme, M.A. and Murry, C.E. (2005). Regenerating the heart. *Nat. Biotech.* **23**: 845–856.

Langer, R. and Tirell, D.A. (2004). Designing materials for biology and medicine. *Nature* **428**: 487–492.

Laugwitz, K-L., Moretti, A., Lam, J., Gruber, P., Chen, Y., Woodard, S., Lin, L-Z., Cai, C-L., Lu, M.M., Reth, M., Platoshyn, O., Yuan, J. X-J., Evans, S. and Chien, K.R. (2005). Postnatal isl1$^+$ cardioblasts enter fully differentiated cardiomyocyte lineages. *Nature* **433**: 647–653.

Lazic, S.E. and Barker, R.A. (2003). The future of cell-based transplantation therapies for neurodegenerative disorders. *J. Hematother Stem Cell Res.* **12**: 635–642.

LeCouter, J., Moritz, D.R., Li, B., Phillips, G., Liang, X.H., Gerber, H-P., Hillan, K.J. and Ferrara, N. (2003). Angiogenesis-independent endothelial protection of the liver: Role of VEGFR-1. *Science* **299**: 890–893.

Leferovich, J., Bedelbaeva, K., Samulewicz, S., Zhang, X.-M., Zwas, D., Lankford, E.B. and Heber-Katz, E. (2001). Heart regeneration in adult MRL mice. *Proc. Natl Acad. Sci. USA* **98**: 9830–9835.

Liu, J.Y., Hafner, J., Dragieva, G., Seifert, B. and Burg, G. (2004). Autologous cultured keratinocytes on porcine gelatin microbeads effectively heal chronic venous leg ulcers. *Wound Rep. Reg.* **12**: 148–156.

Lovell, M.J. and Mathur, A. (2004). The role of stem cells for treatment of cardiovascular disease. *Cell Proliferat* **37**: 67–87.

Lutolf, M.P. and Hubbell, J.A. (2005). Synthetic biomaterials as instructive extracellular microenvironments for morphogenesis in tissue engineering. *Nat. Biotech.* **23**: 47–55.

Maden, M. (1997). Retinoic acid and its receptors in limb regeneration. *Sem. Cell Dev. Biol.* **8**: 445–453.

Maden, M. (1998). Retinoids as endogenous components of the regenerating limb and tail. *Wound Rep. Reg.* **6**: 358–365.

Maden, M. (2002). Positional information: knowing where you are in a limb. *Current Biol.* **12**: R773–R775.

Maden, M. (2004). Retinoids in lung development and regeneration. *Current Topics in Dev. Biol.* **61**: 154–189.

Madri, J.A., Asankar, S. and Romanic, A.M. (1996). Angiogenesis. In: Clark, R.A.F. (ed.), *The Molecular and Cellular Biology of Wound Repair*, 2nd edn. New York: Plenum Press, pp. 355–372.

Mangi, A.A., Noiseux, N., Kong, D., He, H., Rezvani, M., Ingwall, J.S. and Dzau, V.J. (2003). Mesenchymal stem cells modified with Akt prevent remodeling and restore performance of infarcted hearts. *Nat. Med.* **9**: 1195–1201.

Mathur, A. and Martin, J.F. (2004). Stem cells and repair of the heart. *Lancet* **364**: 183–192.

Mercader, N., Tanaka, E. and Torres, M. (2005). Proximodistal identity during vertebrate limb regeneration is regulated by Meis homeodomain proteins. *Development* **132**: 4131–4142.

Mescher, A.L. and Neff, A.W. (2005). Regenerative capacity and the developing immune system. *Adv. Biochem. Eng/Biotech.* **93**: 39–66.

Mitsui, K., Tokuzawa, Y., Itoh, H., Segawa, K., Murakami, M., Takahashi, K., Maruyama, M., Maeda, M. and Yamanaka. (2003). The homeoprotein Nanog is required for maintenance of pluripotency in mouse epiblast and ES cells. *Cell* **113**: 631–642.

Mittenthal, J.E. (1981). The rule of normal neighbors: A hypothesis for morphogenetic pattern regulation. *Dev. Biol.* **88**: 15–26.

Montarras, D., Morgan, J., Collins, C., Relaix, F., Zaffran, S., Cumano, A., Partridge, T. and Buckingham, M. (2005). Direct isolation of satellite cells for skeletal muscle regeneration. *Science* **309**: 2064–2067.

Morais da Silva, S.M., Gates, P.B. and Brockes, J.P. (2002). The newt orthology of CD59 is implicated in proximodistal identity during amphibian limb regeneration. *Dev. Cell* **3**: 547–555.

Munsie, M., Michalska, A.E., O'Brien, C.M., Trounson, A.O., Pera, M.F. and Mountford, P.S. (2000). Isolation of pluripotent embryonic stem cells from reprogrammed adult mouse somatic cell nuclei. *Curr. Biol.* **10**: 989–992.

Murray, P.E. and Garcia-Godoy, F. (2004). Stem cell responses in tooth regeneration. *Stem Cells Dev.* **13**: 255–262.

Murry, C.E., Soonpa, M.H., Reinecke, H., Nakajima, H., Rubart, M., Pasumarthi, K.B.S., Virag, J.I., Barelmez, S.H., Poppa, V., Bradford, G., Dowell, J.D., Williams, D.A. and Field, L.J. (2004). Haematopoietic stem cells do not transdifferentiate into cardiac myocytes in myocardial infarcts. *Nature* **428**: 664–668.

McCallion, R.L. and Ferguson, M.W.J. (1996). Fetal wound healing and the development of antiscarring therapies for adult wound healing. In: Clark, R.A.F. (ed.), *The Molecular and Cellular Biology of Wound Repair*, 2nd edn. New York: Plenum Press, pp. 561–600.

McGann, C., Odelberg, S.J. and Keating, M.T. (2001). Mammalian myotube dedifferentiation induced by newt regeneration extract. *Proc. Natl Acad. Sci. USA* **98**: 13699–13703.

Messenger, A.G. (1993). The control of hair growth: an overview. *J. Invest. Dermatol.* **101**(Suppl 1): 4S–9S.

Michalopoulos, G.K. and DeFrances, M.C. (1997). Liver regeneration. *Science* **276**: 60–66.

Miura, M., Gronthos, S., Zhao, M., Lu, B., Fisher, L.W., Robey, P.G. and Shi, S. (2002). SHED: Stem cells from human exfoliated deciduous teeth. *Proc. Natl Acad. Sci. USA* **100**: 5807–5812.

Morris, R.J., Liu, Y., Marles, L., Yang, Z., Trempus, C., Li, S., Lin, J.S., Sawicki, J.A. and Cotsarelis, G. (2004). Capturing and profiling adult hair follicle stem cells. *Nat. Biotechnol.* **22**: 411–417.

Morrison, J.I., Loof, S., He, P. and Simon, A. (2006). Salamander limb regeneration involves the activation of a multipotent skeletal muscle satellite cell population. *J. Cell Biol.* **172**: 433–440.

Nakatomi, H., Kuriu, T., Okabe, S., Yamamoto, S., Hatano, O., Kawahara, N., Tamura, A., Kirino, T. and Nakafuku, M. (2002). Regeneration of hippocampal pyramidal neurons after ischemic brain injury by recruitment of endogenous neural progenitors. *Cell* **110**: 429–441.

Nardi, J.B. and Stocum, D.L. (1983). Surface properties of regenerating limb cells: Evidence for gradation along the proximodistal axis. *Differentiation* **25**: 27–31.

Niazi, I.A. (1996). Background to work on retinoids and amphibian limb regeneration: Studies on anuran tadpoles – a retrospect. *J. Biosci.* **21**: 273–297.

Nishida, K., Yamato, M., Hayashida, Y., Watanabe, K., Yamamoto, K., Adachi, E., Nagai, S., Maeda, N., Watanabe, H., Okano, T. and Tano, Y. (2004). Corneal reconstruction with tissue-engineered cell sheets composed of autologous oral mucosal epithelium. *N. Eng. J. Med.* **351**: 1187–1196.

North, T.E., de Bruijin, M.F.T.R., Stacy, T., Talebian, L., Lind, E., Robin, C., Binder, M., Dzierzak, E. and Speck, N. (2002). Runx1 expression marks long-term repopulating hematopoietic cells in the midgestation mouse embryo. *Immunity* **16**: 661–672.

Oh, H., Bradfute, S.B., Gallardo, T.D., Nakamura, T., Gaussin, V., Mishina, Y., Pocius, J., Michael, L.H., Behringer, R.R., Garry, D.J., Entman, M.L. and Schneider, M.D. (2003). Cardiac progenitor cells from adult myocardium: homing, differentiation, and fusion after infarction. *Proc. Natl. Acad. Sci. USA* **100**: 12313–12318.

Palfi, S., Conde, F., Riche, D., Brouillet, E., Dautry, C., Mittoux, V., Chibois, A., Peschanski, M. and Hantraye, P. (1998). Fetal striatal allografts reverse cognitive defects in a primate model of Huntington's disease. *Nat. Med.* **4**: 963–966.

Park, I.-k., Quian, D., Kiel, M., Becker, M.W., Pihalja, M., Weissman, I.L., Morrison, S.J. and Clarke, M.F. (2003). Bmi-1 is required for maintenance of adult self-renewing harmatopoietic stem cells. *Nature* **423**: 302–305.

Parker, L.H., Schmidt, M., Jin, S-W., Gray, A.M., Beis, D., Pham, T., Frantz, G., Palmieri, S., Hillan, K., Stainier, D.Y.R., de Sauvage, F.J. and Ye, W. (2004). The endothelial-cell-derived secreted factor *Egfl7* regulates vascular tube formation. *Nature* **428**: 754–758.

Pastoret, C. and Partridge, T.A. (1998). Muscle regeneration. In: Ferretti, P. and Geraudie, J. (eds.), *Cellular and Molecular Basis of Regeneration*. New York: John Wiley and Sons, pp. 309–334.

Potten, C.S., Booth, C. and Pitchard, D.M. (1997). The intestinal epithelial stem cell: the mucosal governor. *Int. J. Exp. Pathol.* **78**: 219–243.

Powell, K. (2005). It's the ecology, stupid. *Nature* **435**: 268–270.

Ramer, L.M., Ramer, M.S. and Steeves, J.D. (2005). Setting the stage for functional repair of spinal cord injuries: a cast of thousands. *Spinal Cord* **43**: 134–161.

Redd, M.J., Cooper, L., Wood, W., Stramer, B. and Martin, P. (2004). Wound healing and inflammation: embryos reveal the way to perfect repair. *Phil. Trans. R. Soc. Lond. B* **359**: 777–784.

Reddy, P.P., Barrieras, D.J., Wilson, G., Bagli, D.J., McLorie, G.A., Khoury, A.E. and Merguerian, P.A. (2000). Regeneration of functional bladder substitutes using large segment acellular matrix allografts in a porcine model. *J. Urol.* **164**: 936–941.

Reddy, R., Buckley, S., Doerken, M., Barsky, L., Weinberg, K., Anderson, K.D., Warburton, D. and Driscoll, B. (2004). Isolation of a putative progenitor subpopulation of alveolar epithelial type 2 cells. *Am. J. Physiol. Lung Cell Mol. Physiol.* **286**: L658–L667.

Rippon, H.J. and Bishop, A.E. (2004). Embryonic stem cells. *Cell Prolif.* **37**: 23–34.

Risbud, M.V. and Bhonde, R.R. (2002). Models of pancreatic regeneration in diabetes. *Diabet. Res. Clin. Practice* **58**: 155–165.

Rosania, G.R., Chang, Y.-T., Perez, O., Sutherlin, D., Dong, H., Lockhart, D.J. and Schultz, P.G. (2000). Myoseverin, a microtubule-binding molecule with novel cellular effects. *Nature Biotech* **18**: 304–308.

Rossant, J. (2001). Stem cells from the mammalian blastocyst. *Stem Cells* **19**: 477–482.

Rosser, A.E., Barker, R.A., Harrower, T., Watts, C., Farrington, M., Ho, A.K., Burnstein, R.M., Menon, D.K., Gillard, J.H., Pickard, J. and Dunnett, S.B. (2002). Unilater transplantation of human primary fetal tissue in four patients with Huntington's disease: NEST-UK safety report ISRCTN #36485475. *J Neurol. Neurosurg. Psychiatr.* 73678–73685.

Scadden, D.T. (2006). The stem-cell niche as an entity of action. *Nature* **441**: 1075–1079.

Schwob, J.E. (2002). Neural regeneration and the peripheral olfactory system. *Anat. Rec.* **269**: 33–49.

Seeherman, H., Wozney, J. and Li, R. (2002). Bone morphogenetic protein delivery systems. *Spine* **27**(Suppl 16): S16–S23.

Seo, B-M., Miura, M., Gronthos, S., Bartold, P.M., Batouli, S., Brahim, J., Young, M., Robey, P.G., Wang, C-U. and Shi, S. (2004). Investigation of multipotent postnatal stem cells from human periodontal ligament. *The Lancet* **364**: 149–155.

Shamblott, M.J., Axelman, J., Wang, S., Bugg, E.M., Littlefield, J.W., Donovan, P.J., Blumenthal, D., Huggins, G.R. and Gearhart, J.D. (1998). Derivation of pluripotent stem cells from cultured human primordial germ cells. *Proc. Natl. Acad. Sci. USA* **95**: 13726–13731.

Shapiro, A.M.J., Lakey, J.R.T., Ryan, E.A., Korbutt, G.S., Toth, E., Warnock, G.L., Kneteman, N.M. and Rajotte, R.V. (2000). Islet transplantation in seven patients with type I diabetes mellitus using a glucocorticoid-free immunosuppressive regimen. *N. Eng. J. Med.* **343**: 230–238.

Sherwood, R., Christensen, J.L., Conboy, I.M., Conboy, M.J., Rando, T.A., Weissman, I.L. and Wagers, A.J. (2004). Isolation of adult mouse myogenic progenitors: functional heterogeneity of cells within engrafting skeletal muscle. *Cell* **119**: 543–554.

Shi, S., Robey, P.G. and Grnthos, S. (2001). Comparison of human dental pulp and bone marrow stromal cells by cDN microarray analysis. *Bone* **29**: 532–539.

Simon, H-G. and Tabin, C.J. (1993). Analysis of *Hox- 4.5* and *Hox-3.6* expression during newt limb regeneration: differential regulation of paralogous *Hox* genes suggests different roles for members of different *Hox* clusters. *Development* **111**: 1397–1407.

Smart, N., Risebro, C.A., Melville, A.A.D., Moses, K., Schwartz, R.J., Chien, K.R. and Riley, P.R. (2007). Thymosin β4 induces adult epicardial progenitor mobilization and neovascularization. *Nature* **445**: 177–182.

Smith, A. (2001). Embryonic stem cells. In: Marshak ,D.R., Gardner, R. and Gottlieb, D. (eds.), *Stem Cell Biology*. Cold Spring Harbor: Cold Spring Harbor Laboratory Press, pp. 205–230.

Song, H., Stevens, C.S. and Gage, F.H. (2002). Astroglia induce neurogenesis from adult neural stem cells. *Nature* **417**: 39–44.

Spangrude, G.J., Heimfeld, S. and Weissman I.L. (1988). Purification and characterization of mouse hematopoietic cells. *Science* **241**: 58–62.

Stevens, M.M., Marini, R.P., Schaefer, D., Aronson, J., Langer, R. and Shastri, V.P. (2005). *In vivo* engineering of organs: the bone bioreactor. *Proc. Natl Acad. Sci .USA* **102**: 11450–11455.

Stocum, D.L. (2006). *Regenerative Biology and Medicine*. San Diego/New York: Academic Press/Elsevier, 448 pp.

Stone, J.S. and Rubel, E.W. (2000). Avian auditory hair cell regeneration. *Proc. Natl Acad. Sci. USA* **97**: 11714–11721.

Stone, J.S., Shang, L. and Tomarev, S. (2004). cProx1 immunoreactivity distinguishes progenitor cells and predicts hair cell fate during avian hair cell regeneration. *Dev. Dynam.* **230**: 597–614.

Suzuki, M., Yakushiji, N., Nakada, Y., Satoh, A., Ide, H. and Tamura, K. (2006). Limb regeneration in *Xenopus laevis froglet. TSW Dev. Embryol.* **1**(Suppl 1): 26–37.

Takahashi, S., Iseki, T., Ooi, J., Tomonari, A., Takasugi, K., Shimohakamada, Y., Yamada, T., Uchimaru, K., Tojo, A., Shirafuji, N., Kodo, H., Tani, K., Takahashi, T., Yamaguchi, T. and Asano, S. (2004). Single-institute comparative analysis of unrelated bone marrow transplantation and cord blood transplantation for adult patients with hematologic malignancies. *Blood* **104**: 3813–3820.

Takahashi, K. and Yamanaka, S. (2006). Induction of pluripotent stem cells from mouse embryonic and adult fibroblast cultures by defined factors. *Cell* **126**: 663–676.

Tanaka, E.M., Gann, F., Gates, P.B. and Brockes, J.P. (1997). Newt myotubes re-enter the cell cycle by phosphorylation of the retinoblastoma protein. *J. Cell. Biol.* **136**: 155–165.

Tanaka, E.M., Dreschel, N. and Brockes, J.P. (1999). Thrombin regulates S phase re-entry by cultured newt myoblasts. *Curr. Biol.* **9**: 792–799.

Tatsumi, R., Anderson, E., Nevoret, C.J., Halevy, O. and Allen, R.A. (1998). HGF/SF is present in normal adult skeletal muscle and is capable of activating satellite cells. *Dev. Biol.* **194**: 114–128.

Tessmar, J., Kellner, K., Schultz, M.B., Blunk, T. and Gopferich, A. (2004). Toward the development of biomimetic polymers by protein immobilization: PEGylation of insulin as a model reaction. *Tiss. Eng.* **10**: 441–453.

Thomson, J.A., Itskovitz-Eldor, J., Shapiro, S.S., Waknitz, M.A., Swiergiel, J.J., Marshall, V.S. and Jones, J.M. (1998). Embryonic stem cell lines derived from human blastocysts. *Science* **282**: 1145–1147.

Tomanek, R.J. and Schatteman, G.C. (2000). Angiogenesis: new insights and therapeutic potential. *Anat. Rec. (New Anat.)* **261**: 126–135.

Trippel, S.B. (1998). Potential role of insulinlike growth factors in fracture healing. *Clin. Orthopaed. Rel. Res.* (Suppl 355): S301–S313.

Tsai, R.J., Li, L.M. and Chen, J.K. (2000). Reconstruction of damaged corneas by transplantation of autologous limbal epithelial cells. *N. Eng. J. Med.* **343**: 86–93.

Tumbar, T., Guasch, G., Greco, V., Blanpain, C., Lowry, W.E., Rendl, M. and Fuchs, E. (2004). Defining the epithelial stem cell niche in skin. *Science* **303**: 359–363.

Verdijk, L.B., Koopman, R., Schaart, G., Meijer, K., Savalberg, H.H.C.M. and van Loon, L.J.C. (2007). Satellite cell content is specifically reduced in type II skeletal muscle fibers in the elderly. *Am. J. Physiol. Endocrinol. Metabol.* **292**: E151–E157.

Vermeer, P.D., Einwalter, L.A., Moninger, T.O., Rokhlina, T., Kern, J.A., Zabner, J. and Welsh, M.J. (2003). Segregation of receptor and ligand regulates activation of epithelial growth factor receptor. *Nature* **422**: 322–326.

Voytik-Harbin, S.L. (2001). Three-dimensional extracellular matrix substrates for cell culture. *Methods Cell. Biol.* **63**: 561–581.

Wagers, A.J., Sherwood, R.I., Christensen, J.L. and Weissman, I.L. (2002). Little evidence for developmental plasticity of adult hematopoietic stem cells. *Science* **297**: 2256–2259.

Wakayama, T., Tabar, V., Rodriguez, I., Perry, A.F.C., Studer, L. and Mombaerts, P. (2001). Differentiation of embryonic stem cell lines generated from adult somatic cells by nuclear transfer. *Science* **292**: 740–742.

Warashina, M., Min, K.H., Kuwabara, T., Huynh, A., Gage, F.H., Schultz, P.G. and Ding, S. (2006). A synthetic small molecule that induces neuronal differentiation of adult hippocampal neural progenitor cells. *Angew. Chem. Int. Ed.* **45**: 591–593.

Warnke, P.H., Springer, I.N.G., Wiltfanf, J., Acil, Y., Eufinger, H., Wehmoller, M., Russo, P.A.J., Bolte, H., Sherry, E., Behrens, E. and Terheyden, H. (2004) Growth and transplantation of a custom vascularized bone graft in a man. *Lancet* **364**: 766–770.

Watts, J. (2005). Controversy in China. *Lancet* **365**: 109–110.

Werner, S., Smola, H., Liao, X., Longaker, M.T., Krieg, T., Hofschneider, P.H. and Williams, L.T. (1994). The function of KGF in morphogenesis of epithelium and reepithelialization of wounds. *Science* **266**: 819–822.

West, J.L. and Hubbell, J.A. (1999). Polymeric biomaterials with degradation sites for proteases involved in cell migration. *Macromolecules* **32**: 241–244.

Wollert, K.C., Meyer, G.P., Lotz, J., Ringes-Lichtenberg, S., Lippolt, P., Breidenbach, C., Fichtner, S., Korte, T., Hornig, B., Messinger, D., Arseniev, L., Hertenstein, B., Ganser, A. and Drexler, H. (2004). Intracoronary autologous bone-marrow cell transfer after myocardial infarction: the BOOST randomized controlled clinical trial. *Lancet* **364**: 141–148.

Woodley, D.T. (1996). Reepithelialization. In: Clark, R.A.F. (ed.), *The Molecular and Cellular Biology of Wound Repair*, 2nd edn. New York: Plenum Press, pp. 339–354.

Wu, X., Walker, J., Zhang, J., Ding, S. and Schultz, P.G. (2004). Puromorphamine induces osteogenesis by activation of the hedgehog signaling pathway. *Chem. & Biol.* **11**: 1229–1238.

Yannas, I.V. (2001). Tissue and Organ Regeneration in Adults. New York: Springer.

Ying, Q-L., Nichols, J., Chambers, I. and Smith, A. (2003). BMP induction of Id proteins suppresses differentiation and sustains embryonic stem cell self-renewal in collaboration with STAT3. *Cell* **115**: 281–292.

Yip, P.K., Wong, L-F., Pattinson, D., Battaglia, A., Grist, J., Bradbury, E.J., Maden, M., McMahon, S.B. and Mazarakis, N.D. (2006). Lentiviral vector expressing retinoic acid receptor b2 promotes recovery of function after corticospinal injury in the adult rat spinal cord. *Hum. Mol. Genet.* **15**: 1–12.

Young, H. and Black Jr., A.C. (2004). Adult stem cells. *Anat. Rec. Part A* **276A**: 75–102.

Zeisberg, M., Hanai, J.-I., Sugimoto, H., Mammato, T., Charytan, D., Strutz, F. and Kalluri, R. (2003). BMP-7 counteracts TGF-β1-induced epithelial-to-mesenchymal transition and reverses chronic renal injury. *Nature Med.* **9**: 964–968.

Zisch, A.H., Lutolf, M.P., Ehrbar, M., Raeber, G.P., Rizzi, S.C., Davies, N., Schmokel, H., Bezuidenhout, D., Djonov, V., Zilla, P. and Hubbell, J.A. (2003). Cell-demanded release of VEGF from synthetic, biointeractive cell ingrowth matrices for vascularized tissue growth. *FASEB J.* **17**: 2260–2262.

Zohlnhofer, D., Ott, I., Mehilli, J., Schomig, K., Michalk, F., Ibrahim, T., Meisetschlager, G., von Wedel, J., Bollwein, H., Seyfarth, M., Dirschinger, J., Schmitt, C., Schwaiger, M., Kastrati, A. and Schomig, A. (2006). Stem cell mobilization by granulocyte colony-stimulating factor in patients with acute myocardial infarction. *J. Am. Med. Assoc.* **295**: 1003–1010.

Zuk, P.A., Zhu, M., Ashjian, P., De Ugarte, D.A., Huang, J.I., Mizuno, H., Alfonso, Z.C., Fraser, J.K., Benhaim, P. and Hedrick, M.H. (2002). Human adipose tissue is a source of multipotent stem cells. *Mol. Biol. Cell.* **13**: 4279–4295.

7

The Molecular Basis of Pluripotency in Principles of Regenerative Medicine

Ariel J. Levine and Ali H. Brivanlou

INTRODUCTION TO PLURIPOTENCY

The union of sperm and egg, two highly differentiated cell types, gives rise to the zygote – the totipotent cell. The zygote has the potential to form every cell type of the embryo and the adult organism through a series of sequential cell fate decisions that successively limit its range of potency. For example, the cells of the very early mammalian embryo divide, maintaining their totipotency until they reach 16–32 cells, at which point outer cells will give rise to extra-embryonic tissues such as the placenta, and inner cells are fated to give rise to the embryo proper. This, the choice between the outer trophoblast and the "inner cell mass," represents the first restriction in cell fate potential and therefore the end of *toti*potency.

The inner cell mass will give rise to the reproductive germ lineage and all three germ layers of the embryo *in vivo* while, *in vitro*, the inner cell mass of the mouse embryo can give rise to embryonic stem cells (ESC) that share this *pluri*potency (Martin, 1981). Of note, human ESC that are also derived from the inner cell mass can form extra-embryonic derivatives (Xu et al., 2002) and may also spontaneously form primordial germ cells (Clark et al., 2004) and thus may be totipotent.

The molecular basis of pluripotency has been best studied in ESC, about which this review will focus. In addition, other pluripotent cells types include "multipotent" adult progenitor cells (Reyes and Verfaillie, 2001) derived after prolonged culture of bone marrow cells, primordial germ cells cultured as "embryonic germ cells" (Matsui et al., 1992; Shamblott et al., 1998), embryonic carcinoma cells derived from teratomas (Finch and Ephrussi, 1967), and "multipotent" adult male germline stem cells (Guan et al., 2006). It is interesting how many of the pluripotent cell types are related to germ cells, highlighting the developmental proximity between the gametes and the totipotent zygote and potentially between the embryonic epiblast and the origin of primordial germ cells.

Potency, or cell fate potential, is a functional characterization of cell types and does not necessarily describe the range of genes expressed in these cells, their origin, or whether they represent an endogenous cell type in the organism. The hallmark of pluripotent cells is the potential to give rise to germ cells, endoderm (gut, liver, pancreas), mesoderm (muscle, blood, bone), and ectoderm (neurons, glia, skin). This potential is determined using cell type specific molecular markers (such as insulin, cardiac actin, and neurofilament heavy chain), morphological criteria (such as typical histology, beating foci of cardiomyocytes, and branched axons of neurons), and functionally (through secretion of appropriate hormones or neurotransmitters in response to stimuli). The potency of cells may be revealed experimentally *in vitro*, using "embryoid bodies" in

Copyright © 2008 Elsevier Inc. All rights reserved

culture, within a teratoma upon injection into immunocompromised mice, and ultimately, *in vivo* within an embryo upon injection of plurpotent cells into a blastocyst (Brivanlou and Darnell, 2002).

In addition to the ability to differentiate into all of the cell types of the organism, these pluripotent cells possess the ability to self-renew. These two characteristics together endow these cells with "stemness." This quality is also measured by molecular markers (such as Oct4 and nanog), typical morphological criteria, and functionally by the ability to self-renew indefinitely and to differentiate into the broad spectrum of cell fates, *in vitro* and *in vivo*.

Other stem cell types, such as adult hematopoietic stem cells (HSC), intestinal stem cells (ISC), hair follicle stem cells (HFSC), and neural stem cells (NSC) are *multi*potent, meaning that they can give rise to a range of cell types restricted to a particular tissue type. For example, HSC, found in the bone marrow, can give rise to all of the cell types of blood including macrophages, erythrocytes, and leukocytes. Many reports in the past several years have claimed transdifferentiation of multipotent cells into other tissues but these findings are contested and the molecular basis for them is not well understood so they are not considered in this review.

Based on the above definitions of pluripotency and multipotency, it is clear that cells could exist with intermediate potencies. For example, recent data has suggested the existence of mesoangioblasts: cells derived from the embryonic aorta that can self-renew indefinitely in culture and give rise to many mesodermal cell types such as blood, bone, and muscle (Minasi et al., 2002). Further, these terms are defined along a unidirectional undifferentiated-to-differentiated vector, within a given window of time for a cell and barring major changes to a cell's state. A fully differentiated "unipotent" cell type may be used to support totipotent development through nuclear transplantation and cloning. And in tumors, a fully differentiated cell may "dedifferentiate" by losing its markers of differentiation while gaining factors that support self-renewal. While these are somewhat semantic matters, they raise the point that our current abilities are limited for describing, and therefore fully characterizing and understanding the multiple states of potency and stemness.

EXTRACELLULAR SIGNALING FACTORS AND SIGNAL TRANSDUCTION

Pluripotent cells *in vivo* exist in communication with other cell types, or in a "niche" that help to regulate their cell fate through extracellular signaling factors that activate signal transduction cascades within the stem cells. *In vitro*, the first pluripotent cell types, embryonic carcinoma and ESC, were cultured on feeder cells or in media conditioned by these cells (Martin, 1981). However, the factors secreted by these feeder cells were not known, and only a few have been characterized to date, and the media used for maintaining pluripotency included serum, which itself is replete with many known and unknown growth and signaling factors. While mouse embryonic stem cells may be grown on defined substrates such as gelatin, human ESC are still grown on either feeder cells or on a complex, and not defined, tumor cell extracellular matrix.

Despite these many unknown inputs on pluripotent cells, several major signal transduction pathways have been shown to be sufficient and/or required for pluripotency. These pathways are the coded information that pluripotent and support cells exchange with pluripotent cells. The first of these was leukemia inhibitory factor (LIF). In addition, the Wnt pathway, the fibroblast growth factor (FGF) pathway, TGF-β/activin/nodal pathway, and the bone morphogenetic protein (BMP) pathway have all been shown to regulate pluripotency. Importantly, all of these latter pathways are initiated by morphogens – proteins that can produce different cell fates at different doses so it is imperative to consider the dose range of each pathway. Complete inhibition of a pathway is different than low levels of signaling which is qualitatively different than moderate or high levels of signaling (Figure 7.1).

LIF

LIF was the first factor that was demonstrated to maintain mouse ESC in the pluripotent state. It was identified as a pluripotency factor secreted by feeder cells (Smith et al., 1988) and can now be added to cells in

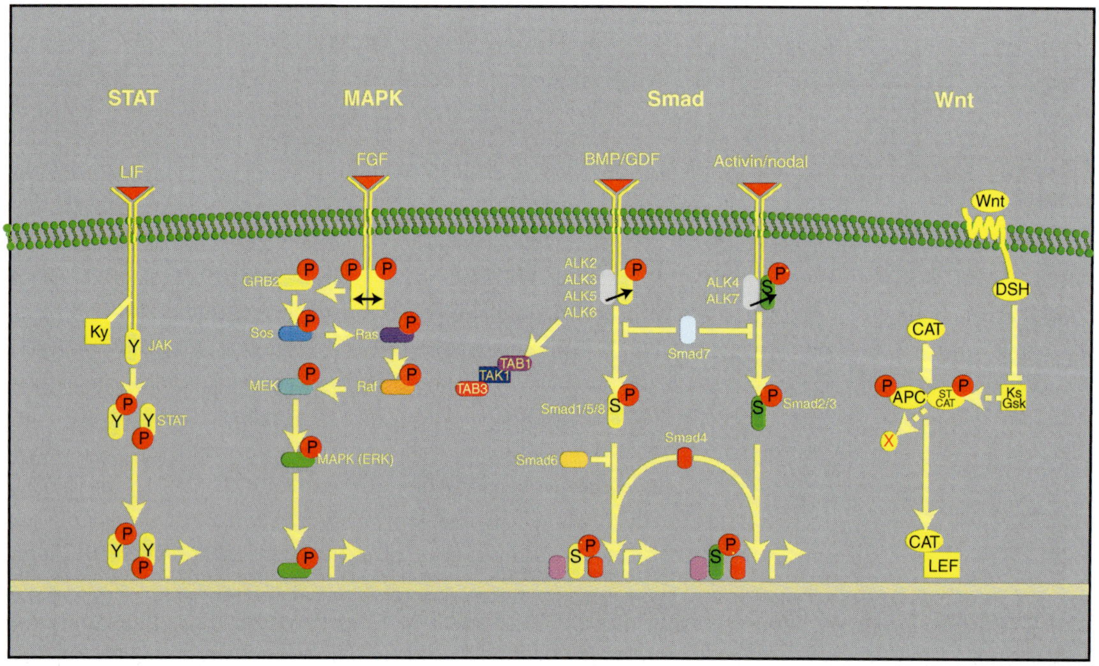

+	−	+	−	+	MESCs
−	+	−	+	+	HESCs

Figure 7.1 Signal transduction pathways implicated in the molecular basis of pluripotency in ESC. The STAT pathway is activated in mouse ESC by LIF, but is not required for pluripotency and is not active in human ESC. The MAPK pathway can be activated by FGF signaling in human ESC to maintain stemness but promotes differentiation in mouse ESC. BMP/GDFs promote differentiation in human ESC but can support pluripotency in mouse ESC (in the presence of LIF) through Smad1/5/8 or through signaling to the MAPK pathway. Activin/nodal ligands activate signaling through Smad2/3 to maintain pluripotency in human ESC; this pathway is active in mouse ESC but not required for stemness. Wnt signaling through the canonical pathway maintains pluripotency in both human and mouse ESC.

a recombinant form. LIF binds to the LIF receptor (LIFR) and these proteins then form a complex with gp130 that activates STAT3 through tyrosine phosphorylation (Heinrich et al., 2003). While LIF can also activate other signal transduction pathways, such as ERK–MAPK, STAT3 is the major factor that mediates the affects of LIF on pluripotency. STAT3 activation alone is sufficient to maintain pluripotency in the presence of serum (Matsuda et al., 1999), bypassing a requirement for LIF, while STAT3 inhibition forces differentiation of mouse ESC (Niwa et al., 1998). STAT activation (of STAT5) also plays a role in the multipotency of HSC (Bradley et al., 2002; Schuringa et al., 2004).

Surprisingly, though, neither LIF nor STAT3 is sufficient to maintain human ESC in a pluripotent state and STAT3 is not even activated in human ESC (Thomson et al., 1998; Humphrey et al., 2004; Sato et al., 2004). Further, LIF signaling is not required *in vivo* within the embryo for either pluripotency or viability of the organism (Stewart et al., 1992).

BMP/GDF

Recently, it has been shown in mouse ESC, BMPs can substitute for serum in cooperating with LIF to support the undifferentiated state (Ying et al., 2003; Qi et al., 2004). In contrast, it has also been demonstrated that BMP inhibition can synergize with FGF signaling to support pluripotency in human ESC (Xu et al., 2005).

BMPs are TGF-β superfamily members that bind to Type 1 TGF-β receptors Alk1, Alk2, Alk3, or Alk6 together with Type 2 receptors to activate phosphorylation and subsequent nuclear localization of Smad1/5/8 (Shi and Massague, 2003). In addition, BMPs signal through Smad-independent means to regulate other signal transduction pathways, such as mitogen-activated protein kinase (MAPK) (von Bubnoff and Cho, 2001).

There are two proposed mechanisms for BMPs effects on mouse ESC. The first suggests that BMPs mediate their effect on mouse ESC through Smad1 induction of Id proteins (Ying et al., 2003), while the other proposes that BMPs cooperate with LIF to maintain pluripotency by inhibiting MAPK signaling (Qi et al., 2004). The observation that BMPs can support pluripotency through Smad activation is surprising because endogenously, ESC and early mammalian embryos do not have active BMP signaling through Smad1/5/8 (James et al., 2005) and even in this work, the authors found that high levels of BMP signaling promoted differentiation of mouse ESC, even in the presence of LIF (Ying et al., 2003). However, as BMPs are morphogens (Wilson et al., 1997), it is possible that very low levels of BMP signaling support pluripotency while higher levels push the cells to differentiate. In support of this model, reduction of levels of GDF-3, a stem cell-associated BMP inhibitor, precludes normal differentiation of mouse ESC (Levine and Brivanlou, 2006).

In human ESC, BMPs promote rapid differentiation to extra-embryonic cell fates even when these cells are cultured in feeder conditioned media that normally maintains their pluripotent state (Xu et al., 2002), despite the fact that stem cells express both the BMP inhibitor GDF-3 and the inhibitor Lefty (Sato et al., 2003) and that feeder cells secrete a BMP inhibitor as well (Xu et al., 2005). However, human ESC can be maintained without conditioned media by an exogenous combination of FGF activation and BMP inhibition (Xu et al., 2005). These findings suggest that the normal inhibition of BMP signaling in stem cells and early embryos are required to suppress differentiation to extra-embryonic fates.

A role for BMP inhibition in maintaining the potency of stem cell types is conserved in adult, multipotent cells such as ISC (where it limits self-renewal of stem cells and antagonizes Wnt signaling (Haramis et al., 2004; He et al., 2004)), HFSC (where it antagonizes the ability of Wnt signaling to maintain the stem cells (Jamora et al., 2003)) and hematopoietic stem cells (where BMP signaling through Alk3 regulates the stem cell niche) (Zhang et al., 2003).

TGF-β/Activin/Nodal

The other branch of TGF-β signaling, the classic TGF-βs, activins, and nodal, support the pluripotent state and are required for stemness in human ESC (James et al., 2005). The members of this branch of the TGF-β pathway bind to Type 1 receptors Alk4, Alk5, or Alk7 together with a Type 2 receptor to activate signal transduction through Smad2/3 (Shi and Massague, 2003).

Activin/nodal signaling is active in early mouse embryos and in both mouse and human ESC, as revealed by phosphorylation and nuclear localization of Smad2/3 (James et al., 2005). This activation is significant for the pluripotent state as exogenous activin or nodal promote pluripotency in human ESC (Vallier et al., 2004; Beattie et al., 2005; James et al., 2005). Further, activin/nodal signaling is required for the maintenance of stemness in human ESC, such that abrogation of signaling through a small molecule inhibitor of Alk4/5/7 (SB431542) or through excess extracellular domains of the receptors forces differentiation of human ESC even in conditioned media, or downstream of Wnt or FGF activation that maintain pluripotency, as described below (James et al., 2005; Vallier et al., 2005).

While inhibition of Alk4/5/7 signaling in mouse ESC does not affect pluripotency (Dunn et al., 2004; James et al., 2005; Vallier et al., 2005), mice lacking both Smads2 and 3 are deficient in maintaining the epiblast, the immediate derivative of the inner cell mass, and have significantly reduced levels of Oct4 (Dunn et al., 2004). These findings show that activin/nodal signaling is important for pluripotency in human ESC and in mouse embryos, and suggest that the signaling events in mouse ESC may not represent the *in vivo* scenario.

FGF

In many cell types, FGF signaling promotes survival and proliferation through its activation of ras and ERK/MAPK. These downstream mediators of FGF signaling have been shown to play roles in pluripotent stem cells, but, as with many signaling pathways in these cells, the results are somewhat contradictory. In mouse ESC, eRas has been shown to promote the proliferation, self-renewal, and tumorigenicity of stem cells (Takahashi et al., 2003); however, active ERK/MAPK has also been shown to promote differentiation to extra-embryonic endoderm cell fates (Yoshida-Koide et al., 2004). In human ESC, non-physiological levels of FGF signaling can act independently (Levenstein et al., 2006) or combine with BMP inhibition (Xu et al., 2005) or nodal activation (Vallier et al., 2005) to support pluripotency of the cells without conditioned media. In fact, FGF signaling itself can act as a BMP inhibitor in these cells, perhaps through an inhibitory phosphorylation by MAPK on the linker region of Smad1/5/8 (Pera et al., 2003; Xu et al., 2005).

Wnt

All of the above pathways are required or sufficient in either mouse or human ESC, but not both, reflecting a curious degree of species-dependent differences in the molecular basis of pluripotency. However, Wnt signaling has been shown to support pluripotency or multipotency in mouse ESC, human ESC, HSC, HFSC, and ISC, suggesting that it plays a core role in the molecular basis of pluripotency. Wnt ligands signal to cells through multiple pathways including the "canonical" pathway in which Wnt binds to frizzled receptors, which signal through disheveled to relieve GSK3β inhibition of β-catenin (Reya and Clevers, 2005).

In mouse and human ESC, Wnt signaling can be activated with a small molecule BIO, an inhibitor of the Wnt inhibitor GSK3β and thereby maintain the pluripotent state, as determined by marker gene analysis and chimera formation (Sato et al., 2004). This is in agreement with the observation that Wnt signaling is normally active in mouse ESC and is decreased upon differentiation (Sato et al., 2004). BIO is only able to sustain pluripotency in human ESC for a limited number of passages. A possible explantation for this phenomenon is that the primary input of Wnt activation is on the self-renewal aspect of stemness rather than the maintenance of pluripotency. In this case, a fraction of each passage would differentiate spontaneously and be lost upon further passage. Interestingly, Wnt signaling requires intact activin/nodal signaling as inhibition of Alk4/5/7 abrogates the ability of BIO to maintain pluripotency (James et al., 2005). While these data relied on inhibition of GSK3β (which has Wnt-independent targets), the role of Wnt ligands in supporting stemness has been demonstrated in mouse ESC, in experiments that show that Wnts secreted by feeder cells or Wnt-conditioned media maintain stemness in mouse ESC (Hao et al., 2006; Ogawa et al., 2006). Further, constitutive activation of β-catenin synergizes with LIF to maintain pluripotency in mouse ESC (Ogawa et al., 2006).

Wnt signaling also plays important roles in maintaining the mulitpotency of adult intestinal and HFSC. Loss of the Wnt-responsive transcription factor Tcf4 allows normal development of the gut but results in complete loss of the stem cells such that instead of a normal arrangement of differentiated villi and crypts that contain progenitors, only differentiated cells are present (van de Wetering et al., 2002). Forced activation of Wnt signaling in skin cells allows formation of new hair follicles and, eventually, skin tumors (Gat et al., 1998); indeed, mutations in a Wnt transcription factor are found in many human skin tumors (Chan et al., 1999).

In HSC, Wnt3a has been shown to promote expansion of stem cells and activated β-catenin promotes self-renewal and maintains the undifferentiated state of HSC. These cells normally have active Wnt signaling and blocking Wnt activity inhibits self-renewal and the ability of HSC to reconstitute bone marrow. Importantly, in these experiments, the authors examined potential targets of Wnt signaling that could act to mediate the effects of Wnts on stemness and found that HoxB4 and Notch1 were upregulated by Wnt signaling and could play this role (Reya et al., 2003).

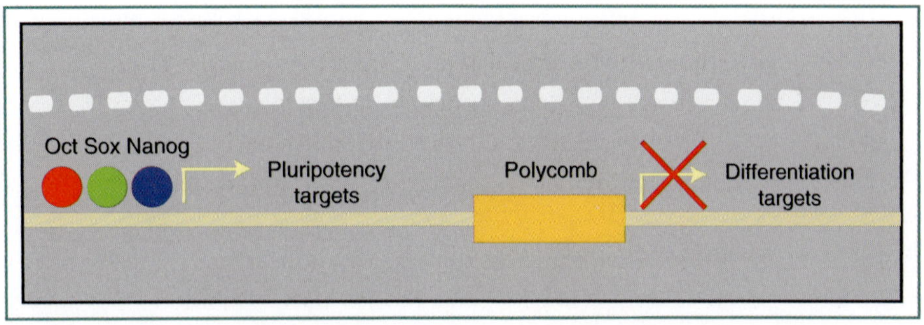

Figure 7.2 **Intrinsic factors implicated in the molecular basis of pluripotency. Oct4, Sox2, and nanog coordinately regulate transcription of pluripotency targets. Differentiation targets are repressed in ESC by polycomb group epigenetic mechanisms.**

TRANSCRIPTIONAL NETWORKS

The nuclear factors that regulate pluripotency and convert extrinsic signals into intrinsic cellular responses have been the subject of intense scrutiny. Three principal transcription factors coordinately regulate the pluripotency program: Oct4, Sox2, and nanog. Each of these genes is expressed in the early mammalian embryo and within the blastocyst stage, they are localized to the inner cell mass (Rosner et al., 1990; Avilion et al., 2003; Chambers et al., 2003; Mitsui et al., 2003) (Figure 7.2).

Mutants for these factors cannot maintain the pluripotent epiblast but, interestingly, different outcomes result from this common deficiency. In Oct4 knock-out embryos and stem cells, the cells differentiate into extra-embryonic trophectoderm. Reduction of Oct4 levels in human ESC confirms these findings, as these cells upregulate markers of trophoblast (Zaehres et al, 2005). Sox2 mutant embryos have a similar phenotype but fail slightly later in development and Sox2 mutant outgrowths of blastocyst embryos divert to trophectoderm (Avilion et al., 2003). In contrast, nanog mutant embryos form extra-embryonic endoderm (Mitsui et al., 2003), a fate that is shared upon nanog reduction in human ESC (Hyslop et al., 2005), although these cells also express a marker of trophoblast (Zaehres et al., 2005).

Mouse ESC that overexpress Oct4 become primitive endoderm (Niwa et al., 2000), suggesting a possible morphogen effect mediated by Oct4. In contrast, nanog overexpressing stem cells retain pluripotency cell-autonomously and do not require LIF or other factors (Chambers et al., 2003; Mitsui et al., 2003). Nanog overexpression similarly frees human ESC of exogenous factors to support pluripotency but converts these cells into a type that more closely resembles epiblast rather than inner cell mass (Darr et al., 2006). These results highlight the need for a critical balance of these stem cells factors to achieve pluripotency.

Recent work has analyzed the targets of Oct4, Sox2, and nanog on a genome-wide scale and has found that these three factors coordinately regulate the stem cell program through both positive and negative regulation of target genes. Of the promoters bound by Oct4, more than half are bound by all three factors and the binding sites for these proteins are often very close together (Boyer et al., 2005). Further, synergistic co-regulation of the FGF4 promoter by Oct4 and Sox2 has been well established (Yuan et al., 1995).

EPIGENETIC AND ENVIRONMENTAL REGULATION

When the differentiated sperm and egg are converted into the totipotent zygote, a process known as "nuclear reprogramming" plays a critical role. This process reprograms the chromatin structure characteristic of differentiated cells into a new conformation typical of the pre-implantation mammalian embryo. Nuclear reprogramming is also a critical step in animal cloning and is required for the nucleus of a differentiated cell to support complete embryonic differentiation when placed into a host egg.

Recently, two groups have shown that in pluripotent ESC, many genes whose expression is associated with differentiation are kept in either a suppressed state (Boyer et al., 2006) or a particular "bivalent" state in which the genes are expressed at very low levels but are easy to activate (Bernstein et al., 2006). One group of chromatin regulators that may be important for these epigenetic modifications is the polycomb group proteins. Specifically, ESC mutant for the member Eed aberrantly express many genes typical of differentiated tissues. These findings imply that unless these genes are silenced, the pluripotent cells will differentiate, meaning that differentiation is a default phenotype in ESC and maintaining cells in an undifferentiated, pluripotent state is an active process.

A reciprocal relationship between pluripotent cells and differentiated cells has also been demonstrated, showing that DNA methylation is required for differentiation. For instance, ESC mutant for the DNA methyl transferase proteins Dnmt3a and Dnmt3b (Chen et al., 2003) or for the CpG binding protein CGBP (Carlone et al., 2005) do not differentiate normally *in vitro* or upon formation of teratomas; instead, these cells maintain expression of the pluripotency markers Oct4 and alkaline phosphatase (another stemness marker).

In this case, targets for suppression by DNA methylation are the proteins that mediate pluripotency. The Oct4 locus is DNA methylated very early in development (Gidekel and Bergman, 2002). Another potential target is a region of human chromosome 12p13 that contains several genes involved in stemness and early germ cells including nanog, GDF-3, and Stella. Interestingly, this cluster of genes is overexpressed in almost all male germ cell tumors and nanog and GDF-3 are specifically overexpressed within pluripotent embryonic carcinomas relative to seminomas and their expression is decreased upon differentiation of embryonic carcinomas (Korkola et al., 2006). A local concentration of pluripotency genes would allow their coordinate regulation by epigenetic mechanisms such that they could be silenced after early development to avoid undue proliferation or inhibition of normal differentiation.

Another non-classical type of molecular regulation of pluripotency includes environmental factors such as oxygen concentration. Low oxygen levels, or hypoxia, have been shown to promote more pluripotent and multipotent cell types at the expense of their differentiated progeny. For instance, it has been shown that low oxygen decreases the differentiation of human ESC, enhances the multipotency of NSC, and expands hematopoietic stem cells (Morrison et al., 2000; Danet et al., 2003; Ezashi et al., 2005).

A possible mechanism for these observations is the fact that HIF2α, a key regulator of the cellular response to hypoxia, directly activates Oct4 (Covello et al., 2006). Accordingly, HIF2α knock-in ESC form tera-tomas with an increased percentage of undifferentiated cells (Covello et al., 2005) and knock-in embryos die shortly after implantation and often contain an expanded epiblast (Covello et al., 2006). In adult tissues, damage may be sensed by hypoxia, triggering local stem cells to self-renew and differentiate to repair the damaged tissue. In the embryonic environment the inner cell mass, from which ESC are derived, could be located further from a source of oxygen so that low levels of oxygen support that internal, pluripotent fate.

SUMMARY AND PERSPECTIVES

The molecular basis of pluripotency is a complex coordination of extracellular and environmental factors, intracellular signal transduction and transcriptional networks, and global regulation of transcription through epigenetic mechanisms. The output of all of these factors is "stemness:" the ability of these special cells to self-renew and to differentiate into the cell types of the embryo proper.

Several themes emerge from this review of our understanding of pluripotency. First, a delicate balance of instructive and inhibitory signals maintain pluripotency. Second, Wnt activation and BMP inhibition are shared signaling characteristics of several types of pluripotent and multipotent cells. Third, Oct4, Sox2, and nanog regulate the transcriptional program of both human and mouse ESC.

Despite these important discoveries, many questions remain regarding the molecular basis of pluripotency. Among the priorities is the molecular basis for the differences in human and mouse ESC. These differences may be artifacts of cell culture, may reflect differences in the endogenous cell type that each represents, or may be true differences in the potencies ("default" or otherwise) of human and mouse ESC. Further, it is unclear how the signaling pathways implicated in stemness mediate their effects. What are the targets of these pathways and how do they converge onto the transcriptional regulators of pluripotency? Ultimately, in addition to understanding the basis of stemness for purposes of basic biological knowledge, it is important to determine how these pathways can be manipulated and controlled to provide the material for regenerative medicine.

REFERENCES

Avilion, A.A., Nicolis, S.K., et al. (2003). Multipotent cell lineages in early mouse development depend on SOX2 function. *Gene Dev.* **17**: 126–140.

Beattie, G.M., Lopez, A.D., et al. (2005). Activin a maintains pluripotency of human embryonic stem cells in the absence of feeder layers. *Stem Cells* **23**: 489–495.

Bernstein, B.E., Mikkelsen, T.S., et al. (2006). A bivalent chromatin structure marks key developmental genes in embryonic stem cells. *Cell* **125**: 315–326.

Boyer, L.A., Lee, T.I., et al. (2005). Core transcriptional regulatory circuitry in human embryonic stem cells. *Cell* **122**: 947–956.

Boyer, L.A., Plath, K., et al. (2006). Polycomb complexes repress developmental regulators in murine embryonic stem cells. *Nature.*

Bradley, H.L., Hawley, T.S., et al. (2002). Cell intrinsic defects in cytokine responsiveness of stat5-deficient hematopoietic stem cells. *Blood* **100**: 3983–3989.

Brivanlou, A.H. and Darnell Jr., J.E. (2002). Signal transduction and the control of gene expression. *Science* **295**: 813–818.

Carlone, D.L., Lee, J.H., et al. (2005). Reduced genomic cytosine methylation and defective cellular differentiation in embryonic stem cells lacking CpG binding protein. *Mol. Cell. Biol.* **25**: 4881–4891.

Chambers, I., Colby, D., et al. (2003). Functional expression cloning of nanog, a pluripotency sustaining factor in embryonic stem cells. *Cell* **113**: 643–655.

Chan, E.F., Gat, U., et al. (1999). A common human skin tumour is caused by activating mutations in beta-catenin. *Nat. Genet.* **21**: 410–413.

Chen, T., Ueda, Y., et al. (2003). Establishment and maintenance of genomic methylation patterns in mouse embryonic stem cells by dnmt3a and dnmt3b. *Mol. Cell. Biol.* **23**: 5594–5605.

Clark, A.T., Bodnar, M.S., et al. (2004). Spontaneous differentiation of germ cells from human embryonic stem cells *in vitro*. *Hum. Mol. Genet.* **13**: 727–739.

Covello, K.L., Kehler, J., et al. (2006). Hif-2alpha regulates Oct-4: effects of hypoxia on stem cell function, embryonic development, and tumor growth. *Gene Dev.* **20**: 557–570.

Covello, K.L., Simon, M.C., et al. (2005). Targeted replacement of hypoxia-inducible factor-1alpha by a hypoxia-inducible factor-2alpha knock-in allele promotes tumor growth. *Cancer Res.* **65**: 2277–2286.

Danet, G.H., Pan, Y., et al. (2003). Expansion of human scid-repopulating cells under hypoxic conditions. *J. Clin. Invest.* **112**: 126–135.

Darr, H., Mayshar, Y., et al. (2006). Overexpression of nanog in human ES cells enables feeder-free growth while inducing primitive ectoderm features. *Development* **133**: 1193–1201.

Dunn, N.R., Vincent, S.D., et al. (2004). Combinatorial activities of Smad2 and Smad3 regulate mesoderm formation and patterning in the mouse embryo. *Development* **131**: 1717–1728.

Ezashi, T., Das, P., et al. (2005). Low O_2 tensions and the prevention of differentiation of hES cells. *Proc. Natl. Acad. Sci. USA* **102**: 4783–4788.

Finch, B.W. and Ephrussi, B. (1967). Retention of multiple developmental potentialities by cells of a mouse testicular teratocarcinoma during prolonged culture *in vitro* and their extinction upon hybridization with cells of permanent lines. *Proc. Natl. Acad. Sci. USA* **57**: 615–621.

Gat, U., DasGupta, R., et al. (1998). *De novo* hair follicle morphogenesis and hair tumors in mice expressing a truncated beta-catenin in skin. *Cell* **95**: 605–614.

Gidekel, S. and Bergman, Y. (2002). A unique developmental pattern of Oct-3/4 DNA methylation is controlled by a *cis*-demodification element. *J. Biol. Chem.* **277**: 34521–34530.

Guan, K., Nayernia, K., et al. (2006). Pluripotency of spermatogonial stem cells from adult mouse testis. *Nature* **440**: 1199–1203.

Hao, J., Li, T.G., et al. (2006). Wnt/beta-catenin pathway up-regulates Stat3 and converges on LIF to prevent differentiation of mouse embryonic stem cells. *Dev. Biol.* **290**: 81–91.

Haramis, A.P., Begthel, H., et al. (2004). *De novo* crypt formation and juvenile polyposis on BMP inhibition in mouse intestine. *Science* **303**: 1684–1686.

He, X.C., Zhang, J., et al. (2004). BMP signaling inhibits intestinal stem cell self-renewal through suppression of wnt-beta-catenin signaling. *Nat. Genet.* **36**: 1117–1121.

Heinrich, P.C., Behrmann, I., et al. (2003). Principles of interleukin (il)-6-type cytokine signalling and its regulation. *Biochem. J.* **374**: 1–20.

Humphrey, R.K., Beattie, G.M., et al. (2004). Maintenance of pluripotency in human embryonic stem cells is Stat3 independent. *Stem Cells* **22**: 522–530.

Hyslop, L., Stojkovic, M., et al. (2005). Downregulation of nanog induces differentiation of human embryonic stem cells to extraembryonic lineages. *Stem Cells* **23**: 1035–1043.

James, D., Levine, A.J., et al. (2005). Tgfbeta/activin/nodal signaling is necessary for the maintenance of pluripotency in human embryonic stem cells. *Development* **132**: 1273–1282.

Jamora, C., DasGupta, R., et al. (2003). Links between signal transduction, transcription and adhesion in epithelial bud development. *Nature* **422**: 317–322.

Korkola, J.E., Houldsworth, J., et al. (2006). Down-regulation of stem cell genes, including those in a 200-kb gene cluster at 12p13.31, is associated with *in vivo* differentiation of human male germ cell tumors. *Cancer Res.* **66**: 820–827.

Levenstein, M.E., Ludwig, T.E., et al. (2006). Basic fibroblast growth factor support of human embryonic stem cell self-renewal. *Stem Cells* **24**: 568–574.

Levine, A.J. and Brivanlou, A.H. (2006). Gdf3, a BMP inhibitor, regulates cell fate in stem cells and early embryos. *Development* **133**: 209–216.

Martin, G.R. (1981). Isolation of a pluripotent cell line from early mouse embryos cultured in medium conditioned by teratocarcinoma stem cells. *Proc. Natl. Acad. Sci. USA* **78**: 7634–7638.

Matsuda, T., Nakamura,T., et al. (1999). Stat3 activation is sufficient to maintain an undifferentiated state of mouse embryonic stem cells. *EMBO J.* **18**: 4261–4269.

Matsui, Y., Zsebo, K., et al. (1992). Derivation of pluripotential embryonic stem cells from murine primordial germ cells in culture. *Cell* **70**: 841–847.

Minasi, M.G., Riminucci, M., et al. (2002). The meso-angioblast: a multipotent, self-renewing cell that originates from the dorsal aorta and differentiates into most mesodermal tissues. *Development* **129**: 2773–2783.

Mitsui, K., Tokuzawa, Y., et al. (2003). The homeoprotein nanog is required for maintenance of pluripotency in mouse epiblast and ES cells. *Cell* **113**: 631–642.

Morrison, S.J., Csete, M., et al. (2000). Culture in reduced levels of oxygen promotes clonogenic sympathoadrenal differentiation by isolated neural crest stem cells. *J. Neurosci.* **20**: 7370–7376.

Niwa, H., Burdon, T., et al. (1998). Self-renewal of pluripotent embryonic stem cells is mediated via activation of Stat3. *Gene Dev.* **12**: 2048–2060.

Niwa, H., Miyazaki, J., et al. (2000). Quantitative expression of Oct-3/4 defines differentiation, dedifferentiation or self-renewal of ES cells. *Nat. Genet.* **24**: 372–376.

Ogawa, K., Nishinakamura, R., et al. (2006). Synergistic action of Wnt and LIF in maintaining pluripotency of mouse ES cells. *Biochem. Biophys. Res. Commun.* **343**: 159–166.

Pera, E.M., Ikeda, A., et al. (2003). Integration of IGF, FGF, and anti-BMP signals via Smad1 phosphorylation in neural induction. *Gene Dev.* **17**: 3023–3028.

Qi, X., Li, T.G., et al. (2004). BMP4 supports self-renewal of embryonic stem cells by inhibiting mitogen-activated protein kinase pathways. *Proc. Natl. Acad. Sci. USA* **101**: 6027–6032.

Reya, T. and Clevers, H. (2005). Wnt signalling in stem cells and cancer. *Nature* **434**: 843–850.

Reya, T., Duncan, A.W., et al. (2003). A role for Wnt signalling in self-renewal of haematopoietic stem cells. *Nature* **423**: 409–414.

Reyes, M. and Verfaillie, C.M. (2001). Characterization of multipotent adult progenitor cells, a subpopulation of mesenchymal stem cells. *Ann. NY Acad. Sci.* **938**: 231–233; discussion 233–235.

Rosner, M.H., Vigano, M.A., et al. (1990). A POU-domain transcription factor in early stem cells and germ cells of the mammalian embryo. *Nature* **345**: 686–692.

Sato, N., Meijer, L., et al. (2004). Maintenance of pluripotency in human and mouse embryonic stem cells through activation of Wnt signaling by a pharmacological GSK-3-specific inhibitor. *Nat. Med.* **10**: 55–63.

Sato, N., Sanjuan, I.M., et al. (2003). Molecular signature of human embryonic stem cells and its comparison with the mouse. *Dev. Biol.* **260**: 404–413.

Schuringa, J.J., Chung, K.Y., et al. (2004). Constitutive activation of STAT5A promotes human hematopoietic stem cell self-renewal and erythroid differentiation. *J. Exp. Med.* **200**: 623–635.

Shamblott, M.J., Axelman, J., et al. (1998). Derivation of pluripotent stem cells from cultured human primordial germ cells. *Proc. Natl. Acad. Sci. USA* **95**: 13726–13731.

Shi, Y. and Massague, J. (2003). Mechanisms of TGF-beta signaling from cell membrane to the nucleus. *Cell* **113**: 685–700.

Smith, A.G., Heath, J.K., et al. (1988). Inhibition of pluripotential embryonic stem cell differentiation by purified polypeptides. *Nature* **336**: 688–690.

Stewart, C.L., Kaspar, P., et al. (1992). Blastocyst implantation depends on maternal expression of leukaemia inhibitory factor. *Nature* **359**: 76–79.

Takahashi, K., Mitsui, K., et al. (2003). Role of eras in promoting tumour-like properties in mouse embryonic stem cells. *Nature* **423**: 541–545.

Thomson, J.A., Itskovitz-Eldor, J., et al. (1998). Embryonic stem cell lines derived from human blastocysts. *Science* **282**: 1145–1147.

Vallier, L., Alexander, M. et al. (2005). Activin/nodal and FGF pathways cooperate to maintain pluripotency of human embryonic stem cells. *J. Cell Sci.* **118**: 4495–4509.

Vallier, L., Reynolds, D., et al. (2004). Nodal inhibits differentiation of human embryonic stem cells along the neuroectodermal default pathway. *Dev. Biol.* **275**: 403–421.

van de Wetering, M., Sancho, E., et al. (2002). The beta-catenin/TCF-4 complex imposes a crypt progenitor phenotype on colorectal cancer cells. *Cell* **111**: 241–250.

von Bubnoff, A. and Cho, K.W. (2001). Intracellular BMP signaling regulation in vertebrates: pathway or network? *Dev. Biol.* **239**: 1–14.

Wilson, P.A., Lagna, G., et al. (1997). Concentration-dependent patterning of the xenopus ectoderm by BMP4 and its signal transducer Smad1. *Development* **124**: 3177–3184.

Xu, R.H., Chen, X., et al. (2002). BMP4 initiates human embryonic stem cell differentiation to trophoblast. *Nat. Biotechnol.* **20**: 1261–1264.

Xu, R.H., Peck, R.M., et al. (2005). Basic FGF and suppression of BMP signaling sustain undifferentiated proliferation of human ES cells. *Nat. Meth.* **2**: 185–190.

Ying, Q.L., Nichols, J., et al. (2003). BMP induction of Id proteins suppresses differentiation and sustains embryonic stem cell self-renewal in collaboration with Stat3. *Cell* **115**: 281–292.

Yoshida-Koide, U., Matsuda, T., et al. (2004). Involvement of ras in extraembryonic endoderm differentiation of embryonic stem cells. *Biochem. Biophys. Res. Commun.* **313**: 475–481.

Yuan, H., Corbi, N., et al. (1995). Developmental-specific activity of the FGF-4 enhancer requires the synergistic action of SOX2 and Oct-3. *Gene Dev.* **9**: 2635–2645.

Zaehres, H., Lensch, M.W., et al. (2005). High-efficiency RNA interference in human embryonic stem cells. *Stem Cells* **23**: 299–305.

Zhang, J., Niu, C., et al. (2003). Identification of the haematopoietic stem cell niche and control of the niche size. *Nature* **425**: 836–841.

Part III

Cells and Tissue Development

8

Embryonic Stem Cells: Derivation and Properties

Junying Yu and James A. Thomson

INTRODUCTION

Embryonic stem (ES) cells are derived from early embryos, and are capable of indefinite self-renewal *in vitro* while maintaining the potential to develop into all cell types of the body – they are pluripotent. With these remarkable features, ES cells hold great promise in both regenerative medicine and basic biological research. In this chapter, we will discuss how ES cells are derived and what is known about the mechanisms that allow these cells to maintain their pluripotency while proliferating *in vitro*.

DERIVATION OF ES CELLS

Embryonic Carcinoma Cells

Teratocarcinoma is a form of malignant germ cell tumor that occurs in both animals and humans. These tumors comprise an undifferentiated embryonal carcinoma (EC) component and differentiated derivatives that can include all three germ layers. Although teratocarcinomas had been known as medical curiosities for centuries (Wheeler, 1983), it was the discovery that male mice of strain 129 had a high incidence of testicular teratocarcinomas (Stevens and Little, 1954) that made these tumors more routinely amenable to experimental analysis. Because their growth is sustained by a persistent EC cell component, teratocarcinomas can be serially transplanted between mice. In 1964, Kleinsmith and Pierce demonstrated that a single EC cell was capable of both self-renewal and multilineage differentiation, and this formal demonstration of a pluripotent stem cell provided the intellectual framework for both mouse and human ES cells.

The first mouse EC cell lines were established in the early 1970s (Kahan and Ephrussi, 1970; Evans, 1972). EC cells exhibit similar antigen and protein expression as the cells present in the inner cell mass (ICM) (Klavins et al., 1971; Comoglio et al., 1975; Gachelin et al., 1977; Solter and Knowles, 1978; Calarco and Banka, 1979; Howe et al., 1980; Henderson et al., 2002), and this led to the notion that EC cells are the counterpart of pluripotent cells present in the ICM (Martin, 1980; Rossant and Papaioannou, 1984). When injected into mouse blastocysts, some EC cell lines are able to contribute to various somatic cell types (Brinster, 1974; Mintz and Illmensee, 1975; Papaioannou et al., 1975; Illmensee and Mintz, 1976), but most EC cell lines have limited developmental potential and contribute poorly to chimeric mice, probably reflecting genetic changes acquired during teratocarcinoma formation (Atkin et al., 1974; McBurney, 1976; Bronson et al., 1980; Zeuthen et al., 1980). Mutations that confer growth advantages to EC cells are likely to accumulate during tumorigenesis, and EC cells in chimeras can result in tumor formation (Papaioannou et al., 1978). As a result, there are limitations in the application of EC cells to both regenerative medicine and research in basic developmental biology.

Copyright © 2008 Elsevier Inc. All rights reserved

Following fertilization, as the one-cell embryo migrates down the oviduct, it undergoes a series of cleavage divisions resulting in a morula. During blastocyst formation, the outer cell layer of the morula delaminates from the rest of the embryo to form the trophectoderm. The ICM of the blastocyst gives rise to all the fetal tissues (ectoderm, mesoderm and endoderm) and some extra-embryonic tissues, and the trophectoderm gives rise to the trophoblast. Although the early ICM can contribute to the trophoblast, the late ICM does not (Winkel and Pedersen, 1988), suggesting there is some restriction in developmental potential at this stage. In normal embryos, the pluripotent cells of the embryo have a transient existence, as these cells quickly give rise to other non-pluripotent cells through the normal developmental program. Thus, the pluripotent cells of the intact embryo really function *in vivo* as precursor cells and not as stem cells. However, if early mouse embryos are transferred to extra-uterine sites, such as the kidney or testis capsules of adult mice, they can develop into teratocarcinomas that include pluripotent stem (EC) cells (Solter et al., 1970; Stevens, 1970). These ectopic transplantation experiments result in teratocarcinomas at high frequencies, even in strains that do not spontaneously have elevated incidence of germ cell tumors, suggesting that this process is not the result of rare neoplastic transformation events. These key transplantation experiments led to the search for culture conditions that would allow the *in vitro* derivation of pluripotent stem cells directly from the embryo, without the intermediate need to form teratocarcinomas *in vivo*.

Derivation of ES Cells

In 1981, pluripotent ES cell lines were derived directly from the ICM of mouse blastocysts using culture conditions previously developed for mouse EC cells (Evans and Kaufman, 1981; Martin, 1981). ES cell cultures derived from a single cell could differentiate into a wide variety of cell types, or could form teratocarcinomas when injected into mice (Martin, 1981). Unlike EC cells, however, these karyotypically normal cells contributed at a high frequency to a variety of tissues in chimeras, including germ cells, and thus provided a practical way to introduce modifications to the mouse germ line (Bradley et al., 1984).

The efficiency in mouse ES cell derivation is influenced by genetic background. For example, ES cells can be easily derived from the inbred 129/ter-Sv strain, but less efficiently from C57BL/6 and other mouse strains (Ledermann and Burki, 1991; Kitani et al., 1996), and these strain differences somewhat correspond with the propensity of mice of different strains to develop teratocarcinomas. These observations suggested that genetic and/or epigenetic components play an important role in the derivation of mouse ES cells. On the other hand, the efficiency of teratocarcinoma formation induced through extra-uterine mouse embryo transplantations appears to be somewhat less strain dependent (Damjanov et al., 1983). This indicates that the difference in the efficiency of ES cell derivation from different mouse strains might be due to suboptimal culture conditions. Indeed, mouse ES cells can be derived from some non-permissive strains using modified protocols (McWhir et al., 1996; Brook and Gardner, 1997).

ES cell lines are generally derived from the culture of the ICM, but this does not mean that ES cells are the *in vitro* equivalent to ICM cells, or even that ICM cells are the immediate precursor to ES cells. It is possible that during culture, ICM cells give rise to other cells that serve as the immediate precursors. Some experiments suggest that ES cells more closely resemble cells from the primitive ectoderm, the cell layer derived from the ICM after delamination of the primitive endoderm. Isolated primitive ectoderm from the mouse gives rise to ES cell lines at a high frequency, and allows the isolation of ES cell lines from mouse strains that had previously been refractory to ES cell isolation (Brook and Gardner, 1997). Indeed, single primitive ectoderm cells can give rise to ES cell lines at a reasonable frequency, something not possible with early ICM cells (Brook and Gardner, 1997). Although these experiments do suggest that ES cells are more closely related to primitive ectoderm than to ICM, they do not reveal whether ES cells more closely resemble primitive ectoderm or another cell type (e.g. very early germ cells) derived from it *in vitro* (Zwaka and Thomson, 2005). As no pluripotent cell in the intact embryo undergoes long-term self-renewal, ES cells are in some ways tissue culture artifacts. It is surprising that even more than 20 years after their derivation, the origin of these cells is not completely understood. Given the dramatic improvement in molecular techniques since the initial derivation in the 1980s, there is considerable value in reexamining the origin of ES cells to better understand the control of their proliferative pluripotent state (Zwaka and Thomson, 2005).

In addition to derivation from the ICM and isolated primitive ectoderm, mouse ES cells have also been derived from morula-stage embryos and even from individual blastomeres (Eistetter, 1989; Delhaise et al., 1996; Chung et al., 2005; Tesar, 2005). Again, although the ES cell lines were derived from morula, there may well be a progression of intermediate states during the derivation process. The frequencies of success were lower when starting with morula or blastomeres, but these results do suggest that it might be possible to derive human ES cells without the destruction of an embryo. Such cell lines could prove useful to the child resulting from the transfer of a biopsied embryo, as they would be genetically matched to the child.

Derivation of Human ES Cells

In 1978 the first baby was born from an embryo fertilized *in vitro* (Steptoe and Edwards, 1978), and without this event, the derivation of human ES cells would not have been possible. Although there were attempts to derive human ES cells as early as the 1980s, species-specific differences and suboptimal human embryo culture media delayed their successful isolation until 1998 (Thomson et al., 1998). For example, the culture of isolated ICMs from human blastocysts was reported (Bongso et al., 1994), but stable undifferentiated cell lines were not produced in medium supplemented with leukemia inhibitory factor (LIF) in the presence of feeder layers, conditions that allow the isolation of mouse ES cells. In the mid-1990s, ES cell lines were derived from two non-human primates: the rhesus monkey and the common marmoset (Thomson et al., 1995, 1996). Experience with these ES cell lines and concomitant improvements in culture conditions for human *in vitro* fertilization (IVF) embryos (Gardner et al., 1998) resulted in the successful derivation of human ES cell lines (Thomson et al., 1998). These human ES cells had normal karyotypes, and even after prolonged undifferentiated proliferation, maintained the developmental potential to contribute to advanced derivatives of all three germ layers.

To date, more than 120 human ES cell lines have been established worldwide (Stojkovic et al., 2004b). Although most were derived from isolated ICMs, some were derived from morulae or later blastocyst stage embryos (Stojkovic et al., 2004a; Strelchenko et al., 2004). It is not yet known whether ES cells derived from these different developmental stages have any consistent differences or whether they are developmentally equivalent. Human ES cell lines have also been derived from embryos carrying various disease-associated genetic changes, which provide new *in vitro* models of disease (Verlinsky et al., 2005). Recently, and with a remarkably high efficiency, human ES cell lines have been derived through a process of somatic cell nuclear transfer (SCNT) (Hwang et al., 2004, 2005). By using the nuclear transfer technology, the nuclei of human somatic cells, such as skin cells, were transferred to donated human oocytes that were already stripped of their own genetic material. The oocytes were then activated and cultured *in vitro* to the blastocyst stage for ES cell derivation. Because such ES cells contain the genetic material present in the donor cell, it is hoped that they could provide immune-compatible ES cells for cell replacement therapies.

CULTURE OF ES CELLS

Culture of Mouse ES Cells

Mitotically inactivated feeder layers were first used to support difficult-to-culture epithelial cells (Puck et al., 1956), and were later successfully adapted for the culture of mouse EC cells (Martin and Evans, 1975; Martin et al., 1977) and mouse ES cells (Evans and Kaufman, 1981; Martin, 1981). Medium which is "conditioned" by co-culture with fibroblasts sustains EC cells (Smith and Hooper, 1983). Fractionation of conditioned medium led to the identification of a cytokine, LIF, that sustains ES cells (Smith et al., 1988; Williams et al., 1988). LIF and its related cytokines act via the gp130 receptor (Yoshida et al., 1994). Binding of LIF induces dimerization of LIF/gp130 receptors, which in turn activates the latent transcription factor STAT3 (Lutticken et al., 1994; Wegenka et al., 1994), and ERK mitogen-activated protein kinase (MAPK) cascade (Takahashi-Tezuka et al., 1998). STAT3 activation is sufficient for LIF-mediated self-renewal of mouse ES cells in the presence of serum (Matsuda et al., 1999). In contrast, suppression of the ERK pathway promotes ES cell proliferation (Burdon et al., 1999). In serum-free medium, LIF alone is insufficient to prevent mouse ES cell differentiation, but in combination with BMP (bone morphogenetic protein, a member of the TGFβ superfamily), mouse ES cells are sustained (Ying et al., 2003a). BMPs induce expression of Id (inhibitor of differentiation) proteins and inhibit the ERK and p38 MAPK pathways, thus attenuating the pro-differentiation activation of ERK MAPK pathway by LIF.

Culture of Human ES Cells

Mitotically inactivated fibroblast feeder layers and serum-containing medium were used in the initial derivation of human ES cells, essentially the same conditions used for the derivation of mouse ES cells prior to the identification of LIF (Thomson et al., 1998; Reubinoff et al., 2000). However, it now appears largely to be a lucky coincidence that fibroblast feeder layers support both mouse and human ES cells, as the specific factors identified to date that sustain mouse ES cells do not support human ES cells. LIF and its related cytokines fail to support human or non-human primate ES cells in serum-containing media that supports mouse ES cells (Thomson et al., 1998; Daheron et al., 2004; Humphrey et al., 2004; Sumi et al., 2004), and BMPs, when added to human ES cells, cause rapid differentiation in conditions that would otherwise support their self-renewal (Xu et al., 2002; Pera et al., 2004). Indeed, the LIF/STAT3 pathway has yet to be shown to have any relevance to the self-renewal of human ES cells (Thomson et al., 1998; Daheron et al., 2004; Humphrey et al., 2004).

In contrast to mouse ES cells, fibroblast growth factor (FGF) signaling appears to be of central importance in the self-renewal of human ES cells. Basic FGF (bFGF or FGF2) allows the clonal growth of human ES cells on fibroblasts in the presence of a commercially available serum replacement (Amit et al., 2000; Xu et al., 2001). At higher concentrations, bFGF allows feeder independent growth of human ES cells cultured in the same serum replacement (Wang et al., 2005; Xu, C. et al., 2005; Xu, R.H. et al., 2005). The mechanism through which these high concentrations of bFGF exert their functions is incompletely known, although one of the effects is suppression of BMP signaling (Xu, R.H. et al., 2005). Serum and the serum replacement currently used have significant BMP-like activity, which is sufficient to induce differentiation of human ES cells, and conditioning this medium on fibroblasts reduces this activity (Xu, R.H. et al., 2005). At moderate concentrations of bFGF (40 ng/ml), the addition of noggin or other inhibitors of BMP signaling significantly decreases background differentiation of human ES cells. At higher concentrations (100 ng/ml), bFGF itself suppresses BMP signaling in human ES cells to levels comparable to those observed in fibroblast-conditioned medium, and the addition of noggin is no longer needed for feeder independent growth (Xu, R.H. et al., 2005). As more defined culture conditions are developed for human ES cells that lack serum products containing BMP activity, it is not yet clear how important the suppression of the BMP pathway will be, unless there is significant production of BMPs by the ES cells themselves. Also, the effects of BMP signaling could change depending on context. Even in mouse ES cells, BMPs are inducers of differentiation unless they are presented in combination with LIF, and it is entirely possible that in a different signaling context, the effects of BMPs on human ES cells could change.

Suppression of BMP activity by itself is insufficient to maintain human ES cells (Xu, R.H. et al., 2005), thus bFGF must be serving other signaling functions. Human ES cells themselves produce FGFs, and in high-density cultures either on fibroblasts or in fibroblast-conditioned medium, it is not necessary to add FGFs. However, chemical inhibitors of FGF receptor-mediated phosphorylation cause differentiation of human ES cells under these standard culture conditions (Dvorak et al., 2005). The required downstream events are not yet well worked out, but some evidence implicates activation of the ERK pathway (Kang et al., 2005).

Although FGF signaling appears to have a central role in the self-renewal of human ES cells, other pathways have also been implicated. When combined with low to moderate levels of FGFs, TGFβ/activin/nodal signaling has a positive effect on the undifferentiated proliferation of human ES cells (Amit et al., 2004; Beattie et al., 2005; James et al., 2005; Vallier et al., 2005), and inhibition of this pathway leads to differentiation (James et al., 2005; Vallier et al., 2005). However, one of the effects of inhibiting the TGFβ/activin/nodal pathway is a stimulation of the BMP pathway (James et al., 2005), which in itself would be sufficient to induce differentiation. Thus, it is not yet clear whether TGFβ/activin/nodal signaling has a role in human ES cell self-renewal independent of its effects on BMP signaling. Further studies directly inhibiting the BMP pathway in the context of inhibition or stimulation of the TGFβ/activin/nodal are needed to resolve this issue.

The molecular components of the Wnt pathway are well represented in human ES cells (Sperger et al., 2003). In short-term cultures, activation of Wnt signaling by a pharmacological GSK-3-specfic inhibitor (6-bromoindirubin-3′-oxime (BIO)) has been reported to have a positive effect on human ES cell self-renewal (Sato et al., 2004), but in a different study, inhibition of Wnt signaling or stimulation of Wnt signaling by the addition of recombinant Wnt proteins showed no effect on the maintenance of human ES cells (Dravid et al., 2005). It is possible that the positive observed effect of BIO on human ES cells is mediated through other pathways (James et al., 2005).

For human ES cells to be used in a clinical setting, it would be useful for these cells to be derived and maintained in conditions that are free of animal products. For example, human ES cells derived with mouse embryonic fibroblasts were shown to be contaminated with immunogenic non-human sialic acid, which would cause an immune reaction if the cells were used in human patients (Martin et al., 2005). Toward this goal, protein matrices including laminin and fibronectin, and different types of human feeder cells were developed to sustain human ES cells (Xu et al., 2001; Amit et al., 2003; Richards et al., 2003). New human ES cell lines have been derived in the absence of feeder cells, but in the presence of a mouse-derived matrix and a bovine-derived serum replacement product (Klimanskaya et al., 2005). Existing human ES cell lines have been grown in defined serum-free medium that included sphingosine-1-phosphate (S1P) and platelet-derived growth factor (PDGF) (Pebay et al., 2005), but this medium does not eliminate the need for feeder layers. Existing human ES cells lines have also been adapted to feeder-free conditions in which none of the protein components are animal derived, but it is not yet known whether these specific conditions will allow derivation of new lines (Li et al., 2005). Clearly, however, recent improvements in human ES cell culture suggest that the development of completely defined, feeder-free culture conditions are near at hand, and that such conditions will allow the derivation of new cell lines that will be more directly applicable to therapeutic purposes.

During extended culture, genetic changes can accumulate in human ES cells (Draper et al., 2004; Maitra et al., 2005). The status of imprinted genes appears to be relatively stable in human ES cells, but can also change (Rugg-Gunn et al., 2005). Such genetic and epigenetic alterations present a challenge that must be appropriately managed if human ES cells are to be used in cell replacement therapy. The rates at which these changes accumulate in culture likely depend on the culture system used, and the particular selective pressures applied. For example, in all current culture conditions, the cloning efficiency of human ES cells is poor, typically 1% or less (Amit et al., 2000). If cells are dispersed into a suspension of single cells, there is a tremendous selective pressure for cells that clone at a higher efficiency, and indeed, such an increase in cloning efficiency is observed in karyotypically abnormal cells (Enver et al., 2005). Enzymatic methods of passaging ES cells can allow long-term passage without karyotypic changes if the clump size is carefully controlled (Amit et al., 2000), but if such methods are used to disperse cells to single cell suspensions or small clumps, karyotypic changes are more frequent (Cowan et al., 2004). This is a likely explanation for why mechanical splitting of individual colonies allows such long-term karyotypic stability (Buzzard et al., 2004). Understanding the rates at which genetic changes occur and the selective pressures that allow them to over-grow a culture in different culture conditions will be critical to the large-scale expansion and clinical use of human ES cells.

DEVELOPMENTAL POTENTIAL OF ES CELLS

Differentiation of ES Cells

Since ES cells have the ability to differentiate into clinically relevant cell types such as dopamine neurons, cardiomyocytes, and β cells, there is tremendous interest in using these cells both in basic biological research and in transplantation medicine. Both uses demand a great deal of control over lineage allocation and expansion. There are several experimental approaches to demonstrate the developmental potential of ES cells and to direct their differentiation to specific lineages. These approaches range in complexity and experimental control from allowing the ES cells to respond to normal developmental cues in a chimera within an intact embryo, to the addition of defined growth factors to a monolayer culture.

Mouse ES cells reintroduced into blastocysts participate in normal embryogenesis, even after prolonged culture and extensive manipulation *in vitro*. In such chimeras, the progeny of ES cells contributes to both somatic tissues and germ cells (Bradley et al., 1984). When ES cells are introduced into tetraploid blastocysts, mice entirely derived from ES cells can be produced, as the teraploid component is out-competed in the ICM-derived somatic tissues (Nagy et al., 1993; Ueda et al., 1995). Although mice entirely derived from ES cells can be generated, signals from the ICM of the blastocyst are likely necessary for mouse ES cells to contribute to offspring, as fetal development has not been reported when the ICM is completely replaced with ES cells.

ES cells injected into syngeneic or immunocompromised adult mice form teratomas that contain differentiated derivatives of all three germ layers (ectoderm, mesoderm and endoderm) (Martin, 1981). This property is similar to both early embryos and EC cells, and is an approach now routinely used to demonstrate the pluripotency of human ES cells (Thomson et al., 1998). Very complex structures resembling neural tube, gut, teeth and hair form in these teratomas in a very consistent temporal pattern, and these teratomas do offer an experimental model to study the development of these structures in human material, but the environment of differentiation is complex and difficult to manipulate.

Aggregates of EC cells or ES cells cultured in conditions that prevent their attachment form cystic "embryoid bodies" (Martin and Evans, 1975; Martin et al., 1977) that recapitulate some of the events of early development. Differentiated derivatives of all three germ layers form in these structures, and for ES cells, the temporal events occurring mimic *in vivo* embryogenesis. The formation of embryoid bodies has been used, for example, to produce neural cells (Bain et al., 1995; Zhang et al., 2001), cardiomyocyte (Klug et al., 1996; He et al., 2003), hematopoietic precursors (Keller et al., 1993; Chadwick et al., 2003), β-like cells (Assady et al., 2001; Lumelsky et al., 2001), hepatocytes (Hamazaki et al., 2001; Rambhatla et al., 2003), and germ cells (Hubner et al., 2003; Toyooka et al., 2003; Geijsen et al., 2004). The formation of a three-dimensional structure in embryonic bodies (EBs) is useful to promote certain developmental events, but the complicated cell–cell interactions make it difficult to elucidate the essential signaling pathways involved.

A somewhat more controlled method to differentiate ES cells is to co-culture them with differentiated cells that induce their differentiation to specific lineages. For example, MS5, S2 and PA6 stromal cells have been used to derive dopamine neurons from human ES cells (Perrier et al., 2004; Zeng et al., 2004); bone marrow stromal cell lines S17 and OP9 support efficient hematopoietic differentiation (Kaufman et al., 2001; Vodyanik et al., 2005). The inducing activity provided by such stromal cells, while efficient in directing ES cell differentiation, contains many unknown factors, and such activity can change both between and within cell lines as a function of culture conditions.

An even more controlled method is differentiation in monolayers on defined matrices in the presence of specific growth factors. Both mouse and human ES cells differentiate into neuroectodermal precursors in monolayer culture (Ying et al., 2003b; Gerrard et al., 2005), and human ES cells can be efficiently induced to differentiate into trophoblasts with addition of BMPs (Xu et al., 2002). This method eliminates many unknown factors provided by either EBs or stromal cells, thus allowing precise analysis of specific factors on the differentiation of ES cells into lineages of choice. With improved understanding of regulatory events governing germ layer and cell lineage specifications, more cell types will likely be derived from ES cells in increasingly defined conditions.

Molecular Control of Pluripotency

We remain remarkably ignorant about why one cell is pluripotent and another is not, although some of the key players important to maintaining this remarkable state have been identified. Oct4, a member of the POU family of transcription factors, is essential for both the derivation and maintenance of ES cells (Pesce et al., 1998). The expression of Oct4 in the mouse is restricted to early embryos and germ cells (Scholer et al., 1989; Okamoto et al., 1990), and homozygous deletion of this gene causes a failure in the formation of the ICM (Nichols et al., 1998). For mouse ES cells to remain undifferentiated, the expression of Oct4 must be maintained within a critical range. Overexpression of this protein causes differentiation into endoderm and mesoderm, while decreased expression leads to differentiation into trophoblast (Niwa et al., 2000). The expression of Oct4 is also a hallmark of human ES cells (Hansis et al., 2000), and its down-regulation also leads to differentiation and expression of trophoblast markers (Matin et al., 2004).

Another transcription factor important for the pluripotency of ES cells is Nanog (Chambers et al., 2003; Mitsui et al., 2003). Similar to Oct4, the expression of Nanog decreases rapidly as ES cells differentiate. However, unlike Oct4, overexpression of this protein in mouse ES cells allows their self-renewal to be independent of LIF/STAT3, though Nanog appears not to be a direct downstream target of LIF/STAT3 pathway (Chambers et al., 2003). In both mouse and human ES cells, reduced expression of Nanog causes differentiation into extra-embryonic lineages (Chambers et al., 2003; Mitsui et al., 2003; Hyslop et al., 2005).

The expression of genes enriched in ES cells has been extensively studied by several groups (see for example Rao and Stice, 2004 and references therein), and includes, for example, transcription factors *Sox2*, *FOXD3*, RNA-binding protein *Esg-1* (Dppa5), and *de novo* DNA methyltransferase 3b. Deletion of some of them in mice does demonstrate a critical function in early development (Table 8.1). ES cells also express high levels of genes involved in protein synthesis and mRNA processing (Richards et al., 2004), and non-coding RNAs unique to ES cells (Suh et al., 2004). A surprisingly high percentage of genes enriched in ES cells have unknown functions (Tanaka et al., 2002; Robson, 2004 and references therein).

A recent genome-wide location analysis of human ES cells showed that Oct4 and Nanog, along with *Sox2*, co-occupy the promoters of a high number of genes, many of which are transcription factors such as Oct4, Nanog and *Sox2* (Boyer et al., 2005). These three proteins, in addition to regulating their own transcription as previously shown (Catena et al., 2004; Kuroda et al., 2005; Okumura-Nakanishi et al., 2005; Rodda et al., 2005), could also activate or repress the expression of many other genes. These genome-wide approaches hold great promise in elucidating the networks that control the pluripotent state.

Table 8.1 Examples of genes with enriched expression in ES cells

Genes	Protein features and functions	References
Sox2	HMG-box transcription factor; interacts with Oct4 to regulate transcription; *Sox2–/–* mouse embryos died shortly after implantation with loss of epiblast at ~E6.0.	Avilion et al. (2003)
FOXD3	Forkhead family transcription factor; *FoxD3–/–* mouse embryos died shortly after implantation with loss of epiblast (~E6.5); no *FoxD3–/–* ES cells can be established.	Hanna et al. (2002)
Rex-1(Zfp-42)	Zinc-finger transcription factor; direct target of Oct4; *Rex-1–/–* EC cells failed to differentiate into primitive and visceral endoderm.	(Rosfjord and Rizzino, 1994; Thompson and Gudas, 2002)
Gbx2(Stra7)	Homeobox-containing transcription factor; *Gbx–/–* embryos displayed defects in neural crest cell patterning and pharyngeal arch artery.	(Byrd and Meyers, 2005)
Sall1	Potent zinc-finger transcription repressor; heterozygous mutations in humans cause Townes-Brocks syndrome; *Sall1–/–* mice died perinatally.	Kiefer et al. (2002); Kohlhase et al. (1998); Nishinakamura et al. (2001)
Sall2	Homolog of *Sall1*; *Sall–/–* mice showed no phenotype.	Sato et al. (2003))
Hoxa11	Transcription factor; *Hoxa11–/–* mice showed defects in male and female fertility.	Hsieh-Li et al. (1995)
UTF1	Transcriptional co-activator; stimulate ES cell proliferation.	Nishimoto et al. (2005)
TERT	Reverse transcriptase (catalytic component of telomerase).	Liu et al. (2000)
TERF1	Telomere repeat-binding factor 1; *TERF1–/–* mouse embryos died at E5-6 with severe growth defect in ICM.	Karlseder et al. (2003)
TERF2	Telomere repeat-binding factor 2.	Sakaguchi et al. (1998)
DNMT3b	*De novo* DNA methyltransferase; required for methylation of centrimeric minor satellite repeats; *DNMT3b–/–* embryos died before birth.	Okano et al. (1999)
DNMT3a	*De novo* DNA methyltransferase; *DNMT3a–/–* mice died at age of 4 weeks.	Okano et al. (1999)
Dppa2	Putative DNA binding motif SAP.	Bortvin et al. (2003)
Dppa3 (PGC7, Stella)	Putative DNA binding motif SAP.	Bortvin et al. (2003); Bowles et al. (2003); Saitou et al. (2002); Sato et al., (2002)
Dppa4 (FLJ10713)	Putative DNA binding motif SAP.	Bortvin et al. (2003); Sperger et al. (2003)
Dppa5 (Ph34, Esg-1)	Similar to KH RNA-binding motif.	Astigiano et al. (1991); Tanaka et al. (2002))
ECAT11 (FLJ10884)	Conserved transposase 22 domain.	Sperger et al. (2003)

CONCLUSION

Progress in developmental biology has been dramatic over the last few decades, and one of the legacies of the derivation of human ES cells is that they provide a compelling link between that progress and the understanding and treatment of human disease. The derivation of mouse ES cells in 1981 and subsequent development of

Enver, T., Soneji, S., Joshi, C., Brown, J., Iborra, F., Orntoft, T., Thykjaer, T., Maltby, E., Smith, K., Abu Dawud, R., Jones, M., Matin, M., Gokhale, P., Draper, J. and Andrews, P.W. (2005). Cellular differentiation hierarchies in normal and culture adapted human embryonic stem cells. *Hum. Mol. Genet.*

Evans, M.J. (1972). The isolation and properties of a clonal tissue culture strain of pluripotent mouse teratoma cells. *J. Embryol. Exp. Morphol.* **28**: 163–176.

Evans, M.J. and Kaufman, M.H. (1981). Establishment in culture of pluripotential cells from mouse embryos. *Nature* **292**: 154–156.

Gachelin, G., Kemler, R., Kelly, F. and Jacob, F. (1977). PCC4, a new cell surface antigen common to multipotential embryonal carcinoma cells, spermatozoa, and mouse early embryos. *Dev. Biol.* **57**: 199–209.

Gardner, D.K., Vella, P., Lane, M., Wagley, L., Schlenker, T. and Schoolcraft, W.B. (1998). Culture and transfer of human blastocysts increases implantation rates and reduces the need for multiple embryo transfers. *Fertil. Steril.* **69**: 84–88.

Geijsen, N., Horoschak, M., Kim, K., Gribnau, J., Eggan, K. and Daley, G.Q. (2004). Derivation of embryonic germ cells and male gametes from embryonic stem cells. *Nature* **427**: 148–154.

Gerrard, L., Rodgers, L. and Cui, W. (2005). Differentiation of human embryonic stem cells to neural lineages in adherent culture by blocking BMP signaling. *Stem Cells.*

Hamazaki, T., Iiboshi, Y., Oka, M., Papst, P.J., Meacham, A.M., Zon, L.I. and Terada, N. (2001). Hepatic maturation in differentiating embryonic stem cells *in vitro*. *FEBS Lett.* **497**: 15–19.

Hanna, L.A., Foreman, R.K., Tarasenko, I.A., Kessler, D.S. and Labosky, P.A. (2002). Requirement for Foxd3 in maintaining pluripotent cells of the early mouse embryo. *Genes Dev.* **16**: 2650–2661.

Hansis, C., Grifo, J.A. and Krey, L.C. (2000). Oct-4 expression in inner cell mass and trophectoderm of human blastocysts. *Mol. Hum. Reprod.* **6**: 999–1004.

He, J.Q., Ma, Y., Lee, Y., Thomson, J.A. and Kamp, T.J. (2003). Human embryonic stem cells develop into multiple types of cardiac myocytes: action potential characterization. *Circ. Res.* **93**: 32–39.

Henderson, J.K., Draper, J.S., Baillie, H.S., Fishel, S., Thomson, J.A., Moore, H. and Andrews, P.W. (2002). Preimplantation human embryos and embryonic stem cells show comparable expression of stage-specific embryonic antigens. *Stem Cells* **20**: 329–337.

Howe, C.C., Gmur, R. and Solter, D. (1980). Cytoplasmic and nuclear protein synthesis during *in vitro* differentiation of murine ICM and embryonal carcinoma cells. *Dev. Biol.* **74**: 351–363.

Hsieh-Li, H.M., Witte, D.P., Weinstein, M., Branford, W., Li, H., Small, K. and Potter, S.S. (1995). Hoxa 11 structure, extensive antisense transcription, and function in male and female fertility. *Development* **121**: 1373–1385.

Hubner, K., Fuhrmann, G., Christenson, L.K., Kehler, J., Reinbold, R., De La Fuente, R., Wood, J., Strauss III, J.F., Boiani, M. and Scholer, H.R. (2003). Derivation of oocytes from mouse embryonic stem cells. *Science* **300**: 1251–1256.

Humphrey, R.K., Beattie, G.M., Lopez, A.D., Bucay, N., King, C.C., Firpo, M.T., Rose-John, S. and Hayek, A. (2004). Maintenance of pluripotency in human embryonic stem cells is STAT3 independent. *Stem Cells* **22**: 522–530.

Hwang, W.S., Ryu, Y.J., Park, J.H., Park, E.S., Lee, E.G., Koo, J.M., Jeon, H.Y., Lee, B.C., Kang, S.K., Kim, S.J., Ahn, C., Hwang, J.H., Park, K.Y., Cibelli, J.B. and Moon, S.Y. (2004). Evidence of a pluripotent human embryonic stem cell line derived from a cloned blastocyst. *Science* **303**: 1669–1674.

Hwang, W.S., Roh, S.I., Lee, B.C., Kang, S.K., Kwon, D.K., Kim, S., Kim, S.J., Park, S.W., Kwon, H.S., Lee, C.K., Lee, J.B., Kim, J.M., Ahn, C., Paek, S.H., Chang, S.S., Koo, J.J., Yoon, H.S., Hwang, J.H., Hwang, Y.Y., Park, Y.S., Oh, S.K.,. Kim, H.S., Park, J.H., Moon, S.Y. and Schatten, G. (2005). Patient-specific embryonic stem cells derived from human SCNT blastocysts. *Obstet. Gynecol. Surv.* **60**: 587–589.

Hyslop, L., Stojkovic, M., Armstrong, L., Walter, T., Stojkovic, P., Przyborski, S., Herbert, M., Murdoch, A., Strachan, T. and Lako, M. (2005). Downregulation of NANOG induces differentiation of human embryonic stem cells to extraembryonic lineages. *Stem Cells* **23**: 1035–1043.

Illmensee, K. and Mintz, B. (1976). Totipotency and normal differentiation of single teratocarcinoma cells cloned by injection into blastocysts. *Proc. Natl Acad. Sci. USA* **73**: 549–553.

James, D., Levine, A.J., Besser, D. and Hemmati-Brivanlou, A. (2005). TGFbeta/activin/nodal signaling is necessary for the maintenance of pluripotency in human embryonic stem cells. *Development* **132**: 1273–1282.

Kahan, B.W. and Ephrussi, B. (1970). Developmental potentialities of clonal *in vitro* cultures of mouse testicular teratoma. *J. Natl Cancer Inst.* **44**: 1015–1036.

Kang, H.B., Kim, J.S.,. Kwon, H.J., Nam, K.H., Youn, H.S., Sok, D.E. and Lee, Y. (2005). Basic fibroblast growth factor activates ERK and induces c-fos in human embryonic stem cell line MizhES1. *Stem Cells Dev.* **14**: 395–401.

Karlseder, J., Kachatrian, L., Takai, H., Mercer, K., Hingorani, S., Jacks, T. and de Lange, T. (2003). Targeted deletion reveals an essential function for the telomere length regulator Trf1. *Mol. Cell. Biol.* **23**: 6533–6541.

Kaufman, D.S., Hanson, E.T., Lewis, R.L., Auerbach, R. and Thomson, J.A. (2001). Hematopoietic colony-forming cells derived from human embryonic stem cells. *Proc. Natl Acad. Sci. USA* **98**: 10716–10721.

Keller, G., Kennedy, M., Papayannopoulou, T. and Wiles, M.V. (1993). Hematopoietic commitment during embryonic stem cell differentiation in culture. *Mol. Cell Biol.* **13**: 473–486.

Kiefer, S.M., McDill, B.W., Yang, J. and Rauchman, M. (2002). Murine Sall1 represses transcription by recruiting a histone deacetylase complex. *J. Biol. Chem.* **277**: 14869–14876.

Kitani, H., Takagi, N., Atsumi, T., Kawakura, K., Imamura, K., Goto, S., Kusakabe, M. and Fukuta, K. (1996). Isolation of a germline-transmissible embryonic stem (ES) cell line from C3H/He mice. *Zoolog. Sci.* **13**: 865–871.

Klavins, J.V., Mesa-Tejada, R. and Weiss, M. (1971). Human carcinoma antigens cross reacting with anti-embryonic antibodies. *Nat. New Biol.* **234**: 153–154.

Klimanskaya, I., Chung, Y., Meisner, L., Johnson, J., West, M.D. and Lanza, R. (2005). Human embryonic stem cells derived without feeder cells. *Lancet* **365**: 1636–1641.

Klug, M.G., Soonpaa, M.H., Koh, G.Y. and Field, L.J. (1996). Genetically selected cardiomyocytes from differentiating embronic stem cells form stable intracardiac grafts. *J. Clin. Invest.* **98**: 216–224.

Kohlhase, J., Wischermann, A., Reichenbach, H., Froster, U. and Engel, W. (1998). Mutations in the SALL1 putative transcription factor gene cause Townes-Brocks syndrome. *Nat. Genet.* **18**: 81–83.

Kuroda, T., Tada, M., Kubota, H., Kimura, H., Hatano, S.Y., Suemori, H., Nakatsuji, N. and Tada, T. (2005). Octamer and Sox elements are required for transcriptional cis regulation of Nanog gene expression. *Mol. Cell Biol.* **25**: 2475–2485.

Ledermann, B. and Burki, K. (1991). Establishment of a germ-line competent C57BL/6 embryonic stem cell line. *Exp. Cell Res.* **197**: 254–258.

Li, Y., Powell, S., Brunette, E., Lebkowski, J. and Mandalam, R. (2005). Expansion of human embryonic stem cells in defined serum-free medium devoid of animal-derived products. *Biotechnol. Bioeng.* **91**: 688–698.

Liu, Y., Snow, B.E., Hande, M.P., Yeung, D., Erdmann, N.J., Wakeham, A., Itie, A., Siderovski, D.P., Lansdorp, P.M. Robinson, M.O. and Harrington, L. (2000). The telomerase reverse transcriptase is limiting and necessary for telomerase function *in vivo*. *Curr. Biol.* **10**: 1459–1462.

Lumelsky, N., Blondel, O., Laeng, P., Velasco, I., Ravin, R. and McKay, R. (2001). Differentiation of embryonic stem cells to insulin-secreting structures similar to pancreatic islets. *Science* **292**: 1389–1394.

Lutticken, C., Wegenka, U.M., Yuan, J., Buschmann, J., Schindler, C., Ziemiecki, A., Harpur, A.G., Wilks, A.F., Yasukawa, K., Taga, T., et al. (1994). Association of transcription factor APRF and protein kinase Jak1 with the interleukin-6 signal transducer gp130. *Science* **263**: 89–92.

Maitra, A., Arking, D.E., Shivapurkar, N., Ikeda, M., Stastny, V., Kassauei, K., Sui, G., Cutler, D.J., Liu, Y., Brimble, S.N., Noaksson, K., Hyllner, J., Schulz, T.C., Zeng, X., Freed, W.J., Crook, J., Abraham, S., Colman, A., Sartipy, P., Matsui, S.I., Carpenter, M., Gazdar, A.F., Rao, M. and Chakravarti, A. (2005). Genomic alterations in cultured human embryonic stem cells. *Nat. Genet.*

Martin, G.R. (1980). Teratocarcinomas and mammalian embryogenesis. *Science* **209**: 768–776.

Martin, G.R. (1981). Isolation of a pluripotent cell line from early mouse embryos cultured in medium conditioned by teratocarcinoma stem cells. *Proc. Natl Acad. Sci. USA* **78**: 7634–7638.

Martin, G.R. and Evans, M.J. (1975). Differentiation of clonal lines of teratocarcinoma cells: formation of embryoid bodies *in vitro*. *Proc Natl Acad. Sci. USA* **72**: 1441–1445.

Martin, G.R., Wiley, L.M. and Damjanov, I. (1977). The development of cystic embryoid bodies *in vitro* from clonal teratocarcinoma stem cells. *Dev. Biol.* **61**: 230–244.

Martin, M.J., Muotri, A., Gage, F. and Varki, A. (2005). Human embryonic stem cells express an immunogenic nonhuman sialic acid. *Nat. Med.* **11**: 228–232.

Matin, M.M., Walsh, J.R., Gokhale, P.J., Draper, J.S., Bahrami, A.R., Morton, I., Moore, H.D. and Andrews, P.W. (2004). Specific knockdown of Oct4 and beta2-microglobulin expression by RNA interference in human embryonic stem cells and embryonic carcinoma cells. *Stem Cells* **22**: 659–668.

Matsuda, T., Nakamura, T., Nakao, K., Arai, T., Katsuki, M., Heike, T. and Yokota, T. (1999). STAT3 activation is sufficient to maintain an undifferentiated state of mouse embryonic stem cells. *EMBO J.* **18**: 4261–4269.

McBurney, M.W. (1976). Clonal lines of teratocarcinoma cells *in vitro*: differentiation and cytogenetic characteristics. *J. Cell Physiol.* **89**: 441–455.

McWhir, J., Schnieke, A.E., Ansell, R., Wallace, H., Colman, A., Scott, A.R. and Kind, A.J. (1996). Selective ablation of differentiated cells permits isolation of embryonic stem cell lines from murine embryos with a non-permissive genetic background. *Nat. Genet.* **14**: 223–226.

Mintz, B. and Illmensee, K. (1975). Normal genetically mosaic mice produced from malignant teratocarcinoma cells. *Proc. Natl Acad. Sci. USA* **72**: 3585–3589.

Mitsui, K., Tokuzawa, Y., Itoh, H., Segawa, K., Murakami, M., Takahashi, K., Maruyama, M., Maeda, M. and Yamanaka, S. (2003). The homeoprotein Nanog is required for maintenance of pluripotency in mouse epiblast and ES cells. *Cell* **113**: 631–642.

Nagy, A., Rossant, J., Nagy, R., Abramow-Newerly, W. and Roder, J.C. (1993). Derivation of completely cell culture-derived mice from early-passage embryonic stem cells. *Proc. Natl Acad. Sci. USA* **90**: 8424–8428.

Nichols, J., Zevnik, B., Anastassiadis, K., Niwa, H., Klewe-Nebenius, D., Chambers, I., Scholer, H. and Smith, A. (1998). Formation of pluripotent stem cells in the mammalian embryo depends on the POU transcription factor Oct4. *Cell* **95**: 379–391.

Nishimoto, M., Miyagi, S., Yamagishi, T., Sakaguchi, T., Niwa, H., Muramatsu, M. and Okuda, A. (2005). Oct-3/4 maintains the proliferative embryonic stem cell state via specific binding to a variant octamer sequence in the regulatory region of the UTF1 locus. *Mol. Cell Biol.* **25**: 5084–5094.

Nishinakamura, R., Matsumoto, Y., Nakao, K., Nakamura, K., Sato, A., Copeland, N.G., Gilbert, D.J., Jenkins, N.A., Scully, S., Lacey, D.L., Katsuki, M., Asashima, M. and Yokota, T. (2001). Murine homolog of SALL1 is essential for ureteric bud invasion in kidney development. *Development* **128**: 3105–3115.

Niwa, H., Miyazaki, J. and Smith, A.G. (2000). Quantitative expression of Oct-3/4 defines differentiation, dedifferentiation or self-renewal of ES cells. *Nat. Genet.* **24**: 372–376.

Okamoto, K., Okazawa, H., Okuda, A., Sakai, M., Muramatsu, M. and Hamada, H. (1990). A novel octamer binding transcription factor is differentially expressed in mouse embryonic cells. *Cell* **60**: 461–472.

Okano, M., Bell, D.W., Haber, D.A. and Li, E. (1999). DNA methyltransferases Dnmt3a and Dnmt3b are essential for de novo methylation and mammalian development. *Cell* **99**: 247–257.

Okumura-Nakanishi, S., Saito, M., Niwa, H. and Ishikawa, F. (2005). Oct-3/4 and Sox2 regulate Oct-3/4 gene in embryonic stem cells. *J. Biol. Chem.* **280**: 5307–5317.

Papaioannou, V.E., McBurney, M.W., Gardner, R.L. and Evans, M.J. (1975). Fate of teratocarcinoma cells injected into early mouse embryos. *Nature* **258**: 70–73.

Papaioannou, V.E., Gardner, R.L., McBurney, M.W., Babinet, C. and Evans, M.J. (1978). Participation of cultured teratocarcinoma cells in mouse embryogenesis. *J. Embryol. Exp. Morphol.* **44**: 93–104.

Pebay, A., Wong, R.C., Pitson, S.M., Wolvetang, E., Peh, G.S., Filipczyk, A., Koh, K.L., Tellis, I., Nguyen, L.T. and Pera, M.F. (2005). Essential roles of sphingosine-1-phosphate and platelet-derived growth factor in the maintenance of human embryonic stem cells. *Stem Cells.*

Pera, M.F., Andrade, J., Houssami, S., Reubinoff, B., Trounson, A., Stanley, E.G., Ward-van Oostwaard, D. and Mummery, C. (2004). Regulation of human embryonic stem cell differentiation by BMP-2 and its antagonist noggin. *J. Cell Sci.* **117**: 1269–1280.

Perrier, A.L., Tabar, V., Barberi, T., Rubio, M.E., Bruses, J., Topf, N., Harrison, N.L. and Studer, L. (2004). Derivation of midbrain dopamine neurons from human embryonic stem cells. *Proc. Natl Acad. Sci. USA* **101**: 12543–12548.

Pesce, M., Gross, M.K. and Scholer, H.R. (1998). In line with our ancestors: Oct-4 and the mammalian germ. *Bioessays* **20**: 722–732.

Puck, T.T., Marcus, P.I. and Cieciura, S.J. (1956). Clonal growth of mammalian cells *in vitro*; growth characteristics of colonies from single HeLa cells with and without a feeder layer. *J. Exp. Med.* **103**: 273–283.

Rambhatla, L., Chiu, C.P., Kundu, P., Peng, Y. and Carpenter, M.K. (2003). Generation of hepatocyte-like cells from human embryonic stem cells. *Cell Transplant.* **12**: 1–11.

Rao, R.R. and Stice, S.L. (2004). Gene expression profiling of embryonic stem cells leads to greater understanding of pluripotency and early developmental events. *Biol. Reprod.* **71**: 1772–1778.

Reubinoff, B.E., Pera, M.F., Fong, C.Y., Trounson, A. and Bongso, A. (2000). Embryonic stem cell lines from human blastocysts: somatic differentiation *in vitro*. *Nat. Biotechnol.* **18**: 399–404.

Richards, M., Tan, S., Fong, C.Y., Biswas, A., Chan, W.K. and Bongso, A. (2003). Comparative evaluation of various human feeders for prolonged undifferentiated growth of human embryonic stem cells. *Stem Cells* **21**: 546–556.

Richards, M., Tan, S.P., Tan, J.H., Chan, W.K. and Bongso, A. (2004). The transcriptome profile of human embryonic stem cells as defined by SAGE. *Stem Cells* **22**: 51–64.

Robson, P. (2004). The maturing of the human embryonic stem cell transcriptome profile. *Trends Biotechnol.* **22**: 609–612.

Rodda, D.J., Chew, J.L., Lim, L.H., Loh, Y.H., Wang, B., Ng, H.H. and Robson, P. (2005). Transcriptional regulation of nanog by OCT4 and SOX2. *J. Biol. Chem.* **280**: 24731–24737.

Rosfjord, E. and Rizzino, A. (1994). The octamer motif present in the Rex-1 promoter binds Oct-1 and Oct-3 expressed by EC cells and ES cells. *Biochem. Biophys. Res. Commun.* **203**: 1795–1802.

Rossant, J. and Papaioannou, V.E. (1984). The relationship between embryonic, embryonal carcinoma and embryo-derived stem cells. *Cell Differ.* **15**: 155–161.

Rugg-Gunn, P.J., Ferguson-Smith, A.C. and Pedersen, R.A. (2005). Epigenetic status of human embryonic stem cells. *Nat. Genet.* **37**: 585–587.

Saitou, M., Barton, S.C. and Surani, M.A. (2002). A molecular programme for the specification of germ cell fate in mice. *Nature* **418**: 293–300.

Sakaguchi, A.Y., Padalecki, S.S., Mattern, V., Rodriguez, A., Leach, R.J., McGill, J.R., Chavez, M. and Giambernardi, T.A. (1998). Chromosomal sublocalization of the transcribed human telomere repeat binding factor 2 gene and comparative mapping in the mouse. *Somat. Cell Mol. Genet.* **24**: 157–163.

Sato, A., Matsumoto, Y., Koide, U., Kataoka, Y., Yoshida, N., Yokota, T., Asashima, M. and Nishinakamura, R. (2003). Zinc finger protein sall2 is not essential for embryonic and kidney development. *Mol. Cell Biol.* **23**: 62–69.

Sato, M., Kimura, T., Kurokawa, K., Fujita, Y., Abe, K., Masuhara, M., Yasunaga, T., Ryo, A., Yamamoto, M. and Nakano, T. (2002). Identification of PGC7, a new gene expressed specifically in preimplantation embryos and germ cells. *Mech. Dev.* **113**: 91–94.

Sato, N., Meijer, L., Skaltsounis, L., Greengard, P. and Brivanlou, A.H. (2004). Maintenance of pluripotency in human and mouse embryonic stem cells through activation of Wnt signaling by a pharmacological GSK-3-specific inhibitor. *Nat. Med.* **10**: 55–63.

Scholer, H.R., Balling, R., Hatzopoulos, A.K., Suzuki, N. and Gruss, P. (1989). Octamer binding proteins confer transcriptional activity in early mouse embryogenesis. *EMBO J.* **8**: 2551–2557.

Smith, A.G., Heath, J.K., Donaldson, D.D., Wong, G.G., Moreau, J., Stahl, M. and Rogers, D. (1988). Inhibition of pluripotential embryonic stem cell differentiation by purified polypeptides. *Nature* **336**: 688–690.

Smith, T.A. and Hooper, M.L. (1983). Medium conditioned by feeder cells inhibits the differentiation of embryonal carcinoma cultures. *Exp. Cell Res.* **145**: 458–462.

Solter, D. and Knowles, B.B. (1978). Monoclonal antibody defining a stage-specific mouse embryonic antigen (SSEA-1). *Proc. Natl Acad. Sci. USA* **75**: 5565–5569.

Solter, D., Skreb, N. and Damjanov, I. (1970). Extrauterine growth of mouse egg-cylinders results in malignant teratoma. *Nature* **227**: 503–504.

Sperger, J.M., Chen, X., Draper, J.S., Antosiewicz, J.E., Chon, C.H., Jones, S.B., Brooks, J.D., Andrews, P.W., Brown, P.O. and Thomson, J.A. (2003). Gene expression patterns in human embryonic stem cells and human pluripotent germ cell tumors. *Proc. Natl Acad. Sci. USA.* **100**: 13350–13355.

Steptoe, P.C. and Edwards, R.G. (1978). Birth after the reimplantation of a human embryo. *Lancet* **2**: 366.

Stevens, L.C. (1970). The development of transplantable teratocarcinomas from intratesticular grafts of pre- and postimplantation mouse embryos. *Dev. Biol.* **21**: 364–382.

Stevens, L.C. and Little, C.C. (1954). Spontaneous testicular teratomas in an inbred strain of mice. *Proc. Natl Acad Sci. USA* **40**: 1080–1087.

Stojkovic, M., Lako, M., Stojkovic, P., Stewart, R., Przyborski, S., Armstrong, L., Evans, J., Herbert, M., Hyslop, L., Ahmad, S., Murdoch, A. and Strachan, T. (2004a). Derivation of human embryonic stem cells from day-8 blastocysts recovered after three-step *in vitro* culture. *Stem Cells* **22**: 790–797.

Stojkovic, M., Lako, M., Strachan, T. and Murdoch, A. (2004b). Derivation, growth and applications of human embryonic stem cells. *Reproduction* **128**: 259–267.

Strelchenko, N., Verlinsky, O., Kukharenko, V. and Verlinsky, Y. (2004). Morula-derived human embryonic stem cells. *Reprod. Biomed. Online* **9**: 623–629.

Suh, M.R., Lee, Y., Kim, J.Y., Kim, S.K., Moon, S.H., Lee, J.Y., Cha, K.Y., Chung, H.M., Yoon, H.S., Moon, S.Y., Kim, V.N. and Kim, K.S. (2004). Human embryonic stem cells express a unique set of microRNAs. *Dev. Biol.* **270**: 488–498.

Sumi, T., Fujimoto, Y., Nakatsuji, N. and Suemori, H. (2004). STAT3 is dispensable for maintenance of self-renewal in nonhuman primate embryonic stem cells. *Stem Cells* **22**: 861–872.

Takahashi-Tezuka, M., Yoshida, Y., Fukada, T., Ohtani, T., Yamanaka, Y., Nishida, K., Nakajima, K., Hibi, M. and Hirano, T. (1998). Gab1 acts as an adapter molecule linking the cytokine receptor gp130 to ERK mitogen-activated protein kinase. *Mol. Cell Biol.* **18**: 4109–4117.

Tanaka, T.S., Kunath, T., Kimber, W.L., Jaradat, S.A., Stagg, C.A., Usuda, M., Yokota, T., Niwa, H., Rossant, J. and Ko, M.S. (2002). Gene expression profiling of embryo-derived stem cells reveals candidate genes associated with pluripotency and lineage specificity. *Genome Res.* **12**: 1921–1928.

Tesar, P.J. (2005). Derivation of germ-line-competent embryonic stem cell lines from preblastocyst mouse embryos. *Proc. Natl Acad. Sci. USA* **102**: 8239–8244.

Thompson, J.R. and Gudas, L.J. (2002). Retinoic acid induces parietal endoderm but not primitive endoderm and visceral endoderm differentiation in F9 teratocarcinoma stem cells with a targeted deletion of the Rex-1 (Zfp-42) gene. *Mol. Cell. Endocrinol.* **195**: 119–133.

Thomson, J.A., Kalishman, J., Golos, T.G., Durning, M., Harris, C.P., Becker, R.A. and Hearn, J.P. (1995). Isolation of a primate embryonic stem cell line. *Proc. Natl Acad. Sci. USA* **92**: 7844–7848.

Thomson, J.A., Kalishman, J., Golos, T.G., Durning, M., Harris, C.P. and Hearn, J.P. (1996). Pluripotent cell lines derived from common marmoset (Callithrix jacchus) blastocysts. *Biol. Reprod.* **55**: 254–259.

Thomson, J.A., Itskovitz-Eldor, J., Shapiro, S.S., Waknitz, M.A., Swiergiel, J.J., Marshall, V.S. and Jones, J.M. (1998). Embryonic stem cell lines derived from human blastocysts. *Science* **282**: 1145–1147.

Toyooka, Y., Tsunekawa, N., Akasu, R. and Noce, T. (2003). Embryonic stem cells can form germ cells *in vitro*. *Proc. Natl Acad. Sci. USA* **100**: 11457–11462.

Ueda, O., Jishage, K., Kamada, N., Uchida, S. and Suzuki, H. (1995). Production of mice entirely derived from embryonic stem (ES) cell with many passages by coculture of ES cells with cytochalasin B induced tetraploid embryos. *Exp. Anim.* **44**: 205–210.

Vallier, L., Alexander, M. and Pedersen, R.A. (2005). Activin/Nodal and FGF pathways cooperate to maintain pluripotency of human embryonic stem cells. *J. Cell Sci.* **118**: 4495–4509.

Verlinsky, Y., Strelchenko, N., Kukharenko, V., Rechitsky, S., Verlinsky, O., Galat, V. and Kuliev, A. (2005). Human embryonic stem cell lines with genetic disorders. *Reprod. Biomed. Online.* **10**: 105–110.

Vodyanik, M.A., Bork, J.A. Thomson, J.A. and Slukvin, I.I. (2005). Human embryonic stem cell-derived CD34+ cells: efficient production in the coculture with OP9 stromal cells and analysis of lymphohematopoietic potential. *Blood* **105**: 617–626.

Wang, L., Li, L., Menendez, P., Cerdan, C. and Bhatia, M. (2005). Human embryonic stem cells maintained in the absence of mouse embryonic fibroblasts or conditioned media are capable of hematopoietic development. *Blood* **105**: 4598–4603.

Wegenka, U.M., Lutticken, C., Buschmann, J., Yuan, J., Lottspeich, F., Muller-Esterl, W., Schindler, C., Roeb, E., Heinrich, P.C. and Horn, F. (1994). The interleukin-6-activated acute-phase response factor is antigenically and functionally related to members of the signal transducer and activator of transcription (STAT) family. *Mol. Cell. Biol.* **14**: 3186–3196.

Wheeler, J.E. (1983). *The Human Teratomas: Experimental and Clinical Biology*. Clifton, NJ: Humana Press, pp. 1–22.

Williams, R.L., Hilton, D.J., Pease, S., Willson, T.A., Stewart, C.L., Gearing, D.P., Wagner, E.F., Metcalf, D., Nicola, N.A. and Gough, N.M. (1988). Myeloid leukaemia inhibitory factor maintains the developmental potential of embryonic stem cells. *Nature* **336**: 684–687.

Winkel, G.K. and Pedersen, R.A. (1988). Fate of the inner cell mass in mouse embryos as studied by microinjection of lineage tracers. *Dev. Biol.* **127**: 143–156.

Xu, C., Inokuma, M.S., Denham, J., Golds, K., Kundu, P., Gold, J.D. and Carpenter, M.K. (2001). Feeder-free growth of undifferentiated human embryonic stem cells. *Nat. Biotechnol.* **19**: 971–974.

Xu, C., Rosler, E., Jiang, J., Lebkowski, J.S., Gold, J.D., O'Sullivan, C., Delavan-Boorsma, K., Mok, M., Bronstein, A. and Carpenter, M.K. (2005). Basic fibroblast growth factor supports undifferentiated human embryonic stem cell growth without conditioned medium. *Stem Cells* **23**: 315–323.

Xu, R.H., Chen, X., Li, D.S., Li, R., Addicks, G.C., Glennon, C., Zwaka, T.P. and Thomson, J.A. (2002). BMP4 initiates human embryonic stem cell differentiation to trophoblast. *Nat. Biotechnol.* **20**: 1261–1264.

Xu, R.H., Peck, R.M., Li, D.S., Feng, X., Ludwig, T. and Thomson, J.A. (2005). Basic FGF and suppression of BMP signaling sustain undifferentiated proliferation of human ES cells. *Nat. Methods* **2**: 185–190.

Ying, Q.L., Nichols, J., Chambers, I. and Smith, A. (2003a). BMP induction of Id proteins suppresses differentiation and sustains embryonic stem cell self-renewal in collaboration with STAT3. *Cell* **115**: 281–292.

Ying, Q.L., Stavridis, M., Griffiths, D., Li, M. and Smith, A. (2003b). Conversion of embryonic stem cells into neuroectodermal precursors in adherent monoculture. *Nat. Biotechnol.* **21**: 183–186.

Yoshida, K., Chambers, I., Nichols, J., Smith, A., Saito, M., Yasukawa, K., Shoyab, M., Taga, T. and Kishimoto, T. (1994). Maintenance of the pluripotential phenotype of embryonic stem cells through direct activation of gp130 signalling pathways. *Mech. Dev.* **45**: 163–171.

Zeng, X., Cai, J., Chen, J., Luo, Y., You, Z.B., Fotter, E., Wang, Y., Harvey, B., Miura, T., Backman, C., Chen, G.J., Rao, M.S. and Freed, W.J. (2004). Dopaminergic differentiation of human embryonic stem cells. *Stem Cells* **22**: 925–940.

Zeuthen, J., Norgaard, J.O., Avner, P., Fellous, M., Wartiovaara, J., Vaheri, A., Rosen, A. and Giovanella, B.C. (1980). Characterization of a human ovarian teratocarcinoma-derived cell line. *Int. J. Cancer* **25**: 19–32.

Zhang, S.C., Wernig, M., Duncan, I.D., Brustle, O. and Thomson, J.A. (2001). *In vitro* differentiation of transplantable neural precursors from human embryonic stem cells. *Nat. Biotechnol.* **19**: 1129–1133.

Zwaka, T.P. and Thomson, J.A. (2005). A germ cell origin of embryonic stem cells? *Development* **132**: 227–233.

9

Stem Cells Derived from Amniotic Fluid and Placenta

Paolo De Coppi, Shay Soker, and Anthony Atala

INTRODUCTION

Amniotic fluid and placenta have been recently taken into consideration as potential sources of progenitor cells. Amniocentesis and chorionic villus sampling (CVS) are widely accepted methods for prenatal diagnosis. Minimal or no ethical concerns would be present if embryonic and fetal stem cells would be taken from amniotic fluid and placenta before or at birth. In the last few years our and other groups have described the presence of stem cell with various differentiative and proliferative potential in the amniotic fluid and placenta. We will briefly describe the techniques in use for amniocentesis and CVS, and we will examine the different progenitors that have been described.

CVS AND AMNIOCENTESIS

The first reported amniocentesis took place in 1930 when attempts were being made to correlate the cytologic examination of cell concentration, count and phenotypes in the amniotic fluid to the sex and the health of the baby. Since then, the development technique of karyotype and the discovery of reliable diagnostic markers, such as alpha-fetoprotein, as well as the development of ultrasound-guided amniocentesis, have greatly increased the reliability of the procedure as a valid diagnostic tool as well as the safety of the procedure (Milunsky, 1979; Hoehn and Salk, 1982; Gosden, 1983; Crane and Cheung, 1988). One of the primary uses of amniocentesis is a safe method of isolating cells from the fetus that can be karyotyped and examined for chromosomal abnormalities. In general, the protocol consists of acquiring 10–20 ml of fluid using a transabdominal approach. Amniotic fluid samples are then centrifuged, and the cell supernatant is resuspended in culture medium. Approximately 10^4 cells are seeded on 22×22 mm cover slips. Cultures are grown to confluence for 3–4 weeks in 5% CO_2 at 37°C, and the chromosomes are characterized from mitotic phase cells (Brace and Resnik, 1999). Amniocentesis is performed typically around 16 weeks of gestation, although in some cases it may be performed as early as 14 weeks when the amnion fuses with the chorion and the risk of bursting the amniotic sac by needle puncture is minimized. Amniocentesis can be performed as late as term. The amniotic sac is usually noticed first by ultrasound around the 10-week gestational time point.

With the introduction of CVS in the 1980s, first-trimester diagnosis became a reality. A small sample of chorionic villi (tissue from the developing placenta) is obtained from the mother's uterus under ultrasound guidance, either transvaginally or transabdominally. Sampling of chorionic villi from the fetus is performed from 10 weeks of gestation. The biopsy is usually taken under ultrasound guidance via a transabdominal approach. Alternatively, the cervical approach may be utilized. Each biopsy yields 5–30 mg of tissue that can be used for fetal sexing, karyotyping, biochemical studies and DNA analysis. A direct fetal chromosomal

Copyright © 2008 Elsevier Inc. All rights reserved

analysis on cultured cells is possible within 24 h. However, given the problem with mosaicism in CVS samples, this should always be followed by chromosomal analysis on cultured cells from the sample 2–3 weeks later. An advantage of CVS is that termination can be completed in the first trimester when it is technically easier.

DIFFERENTIATED CELLS FROM AMNIOTIC FLUID AND PLACENTA

Amniotic fluid cell culture consists of a heterogeneous cell population displaying a range of morphologies and behaviors. Studies on these cells have characterized them into many shapes and sizes varying from 6 μm to 50 μmm in diameter and from round to squamous in shape. Most cells in the fluid are terminally differentiated along epithelial lineages and have limited proliferative and differentiation capabilities. Previous studies have noted an interesting composition of the fluid consisting of a heterogeneous cell population expressing markers from all three germ layers (Sarkar et al., 1980; Cousineau et al., 1982; Medina-Gomez and Johnston, 1982; von Koskull et al., 1984). The source of these cells and of the fluid itself underwent a great deal of research. Current theories suggest that the fluid is largely derived from the urine and pulmonary secretion from the fetus as well as from some ultra filtrate from the plasma of the mother entering though the placenta. The cells in the fluid have been shown to be overwhelmingly from the fetus and are thought to be mostly cells sloughed off the epithelium and digestive and urinary tracts of the fetus as well as the amnion (Lotgering and Wallenburg, 1986; Underwood et al., 2005).

MESENCHYMAL CELL FROM PLACENTA AND AMNIOTIC FLUID

Preliminary studies have been published a few years ago describing very simple protocols for the isolation of a non-specific population of cells with "mesenchymal" characteristics from amniotic fluid and placenta (Haigh et al., 1999; Kaviani et al., 2001, 2002, 2003). These cells were able to proliferate *in vitro*, to be engineered in a three-dimensional structure and used *in vivo* to repair a tissue defect (Kaviani et al., 2003). A few years later In't Anker et al. were able to prove for the first time that both amniotic fluid and placenta were abundant sources of fetal mesenchymal stem cells (MSCs) that exhibit a phenotype and multilineage differentiation potential similar to that of postnatal bone marrow (BM)-derived MSCs (In't Anker et al., 2003). They described a simple and repeatable protocol for their isolation and expansion. Briefly, amniotic fluid samples were centrifuged for 10 min at 1,283 rpm. Pellets were resuspended in Iscove's modified Dulbecco's medium containing 2% fetal calf serum (FCS) and antibiotics (defined as washing medium). Similarly, for the placenta, approximately 1 cm^3 was washed in phosphate-buffered saline (PBS) and single-cell suspensions were made by mincing and flushing the tissue parts through a 100 μm nylon filter with washing medium. Single-cell suspensions of amniotic fluid and placenta were plated in six-well plates and cultured in M199 supplemented with 10% FCS, 20 μg/ml endothelial cell growth factor, heparin (8 U/ml), and antibiotics. After 7 days, non-adherent cells were removed and the medium was refreshed. When grown to confluence, adherent cells were detached with trypsin/EDTA and expanded in culture flasks pre-coated with 1% gelatin and kept in a humidified atmosphere at 37°C. The expansion potency of fetal MSCs was higher compared with adult BM-derived MSCs. As a result, they were able to expand amniotic fluid MSCs to about 180×10^6 cells within 4 weeks (three passages). The phenotype of the culture-expanded amniotic fluid-derived cells was similar to that reported for MSCs derived from second-trimester fetal tissues and adult BM. They were able to show that amniotic fluid-derived MSCs showed multilineage differentiation potential into fibroblasts, adipocytes, and osteocytes (In't Anker et al., 2004). Furthermore, amniotic fluid-derived MSCs were successfully isolated, cultured, and enriched without interfering with the routine process of fetal karyotyping. Flow cytometry analyses showed that they were positive for SH2, SH3, SH4, CD29 and CD44, low positive for CD90 and CD105, but negative for CD10, CD11b, CD14, CD34, CD117, and EMA (Tsai et al., 2004). Most importantly, immunophenotypic analyses demonstrated that these cells expressed HLA-ABC, class I major histocompatibility complex (MHC-I), but they did not express HLA-DR, DP, and DQ (MHC-II molecules) (Li et al., 2005a). Li et al. have

Dunn, J.C., Yarmush, M.L., Koebe, H.G. and Tompkins, R.G. (1989). Hepatocyte function and extracellular matrix geometry: long-term culture in a sandwich configuration. *FASEB J.* **3**: 174–177.

Elwan, M.A. and Sakuragawa, N. (1997). Evidence for synthesis and release of catecholamines by human amniotic epithelial cells. *Neuroreport* **8**: 3435–3438.

Ferrari, G., et al. (1998). Muscle regeneration by bone marrow-derived myogenic progenitors. *Science* **279**: 1528–1530.

Fuchs, J.R., Kaviani, A., Oh, J.T., LaVan, D., Udagawa, T., Jennings, R.W., Wilson, J.M. and Fauza, D.O. (2004). Diaphragmatic reconstruction with autologous tendon engineered from mesenchymal amniocytes. *J. Pediatr. Surg.* **39**(6): 834–838; discussion 834–838.

Fukuchi, Y., Nakajima, H., Sugiyama, D., Hirose, I., Kitamura, T. and Tsuji, K. (2004). Human placenta-derived cells have mesenchymal stem/progenitor cell potential. *Stem Cell* **22**(5): 649–658.

Gekas, C., Dieterlen-Lievre, F., Orkin, S.H. and Mikkola, H.K. (2005). The placenta is a niche for hematopoietic stem cells. *Dev. Cell* **8**(3): 365–375.

Gosden, C.M. (1983). Amniotic fluid cell types and culture. *Br. Med. Bull.* **39**: 348–354.

Haigh, T., Chen, C., Jones, C.J. and Aplin, J.D. (1999). Studies of mesenchymal cells from 1st trimester human placenta: expression of cytokeratin outside the trophoblast lineage. *Placenta* **20**: 615–625.

Hamazaki, T., et al. (2001). Hepatic maturation in differentiating embryonic stem cells *in vitro*. *FEBS Lett.* **497**: 15–19.

Hoehn, H. and Salk, D. (1982). Morphological and biochemical heterogeneity of amniotic fluid cells in culture. *Meth. Cell Biol.* **26**: 11–34.

Igura, K., Zhang, X., Takahashi, K., Mitsuru, A., Yamaguchi, S. and Takashi, T.A. (2004). Isolation and characterization of mesenchymal progenitor cells from chorionic villi of human placenta. *Cytotherapy* **6**(6): 543–553.

In't Anker, P.S., Scherjon, S.A., Kleijburg-van der Keur, C., Noort, W.A., Claas, F.H., Willemze, R., Fibbe, W.E. and Kanhai, H.H. (2003). Amniotic fluid as a novel source of mesenchymal stem cells for therapeutic transplantation. *Blood* **102**(4): 1548–1549.

In't Anker, P.S., Scherjon, S.A., Kleijburg-van der Keur, C., de Groot-Swings, G.M., Claas, F.H., Fibbe, W.E. and Kanhai, H.H. (2004). Isolation of mesenchymal stem cells of fetal or maternal origin from human placenta. *Stem Cell* **22**(7): 1338–1345.

Karlmark, K.R., Freilinger, A., Marton, E., Rosner, M., Lubec, G. and Hengstschlager, M. (2005). Activation of ectopic Oct-4 and Rex-1 promoters in human amniotic fluid cells. *Int. J. Mol. Med.* **16**(6): 987–992.

Karsenty, G. (2000). Role of Cbfa1 in osteoblast differentiation and function. *Semin. Cell Dev. Biol.* **11**: 343–346.

Kaviani, A., Perry, T.E., Dzakovic, A., et al. (2001). The amniotic fluid as a source of cells for fetal tissue engineering. *J. Pediatr. Surg.* **36**: 1662–1665.

Kaviani, A., Perry, T.E., Barnes, C.M., et al. (2002). The placenta as a cell source in fetal tissue engineering. *J. Pediatr. Surg.* **37**: 995–999.

Kaviani, A., Guleserian, K., Perry, T.E., et al. (2003). Tissue engineering from amniotic fluid. *J. Am. Coll. Surg.* **196**: 592–597.

Kim, J.B., Wright, H.M., Wright, M. and Spiegelman, B.M. (1998). ADD1/SREBP1 activates PPARgamma through the production of endogenous ligand. *Proc. Natl Acad. Sci. USA* **95**: 4333–4337.

Kogler, G., Sensken, S., Airey, J.A., Trapp, T., Muschen, M., Feldhahn, N., Liedtke, S., Sorg, R.V., Fischer, J., Rosenbaum, C., Greschat, S., Knipper, A., Bender, J., Degistirici, O., Gao, J., Caplan, A.I., Colletti, E.J., Almeida-Porada, G., Muller, H.W., Zanjani, E. and Wernet, P. (2004). A new human somatic stem cell from placental cord blood with intrinsic pluripotent differentiation potential. *J. Exp. Med.* **200**(2): 123–135.

Li, C.D., Zhang, W.Y., Li, H.L., Jiang, X.X., Zhang, Y., Tang, P. and Mao, N. (2005a). Isolation and identification of a multi-lineage potential mesenchymal cell from human placenta. *Placenta.*

Li, C.D., Zhang, W.Y., Li, H.L., Jiang, X.X., Zhang, Y., Tang, P.H. and Mao, N. (2005b). Mesenchymal stem cells derived from human placenta suppress allogeneic umbilical cord blood lymphocyte proliferation. *Cell Res.* **15**(7): 539–547.

Lotgering, F.K. and Wallenburg, H.C. (1986). Mechanisms of production and clearance of amniotic fluid. *Semin. Perinatol.* **10**(2): 94–102.

Medina-Gomez, P. and Johnston, T.H. (1982). Cell morphology in long-term cultures of normal and abnormal amniotic fluids. *Hum. Genet.* **60**(4): 310–313.

Miki, T., Lehmann, T., Cai, H., Stolz, D.B. and Strom, S.C. (2005). Stem cell characteristics of amniotic epithelial cells. *Stem Cell* **23**(10): 1549–1559.

Mikkola, H.K., Gekas, C., Orkin, S.H. and Dieterlen-Lievre, F. (2005). Placenta as a site for hematopoietic stem cell development. *Exp. Hematol.* **33**(9): 1048–1054.

Milunsky, A. (1979). Amniotic fluid cell culture. In: Milunsky, A. (ed.), *Genetic Disorder of the Fetus*. New York: Plenum Press, pp. 75–84.

Mosquera, A., Fernandez, J.L., Campos, A., et al. (1999). Simultaneous decrease of telomerase length and telomerase activity with ageing of human amniotic fluid cells. *J. Med. Genet.* **36**: 494–496.

Okawa, H., Okuda, O., Arai, H., Sakuragawa, N. and Sato, K. (2001). Amniotic epithelial cells transform into neuron-like cells in the ischemic brain. *Neuroreport* **12**(18): 4003–4007.

Olmsted-Davis, E.A., et al. (2003). Primitive adult hematopoietic stem cells can function as osteoblast precursors. *Proc. Natl Acad. Sci. USA* **100**: 15877–15882.

Ottersbach, K. and Dzierzak, E. (2005). The murine placenta contains hematopoietic stem cells within the vascular labyrinth region. *Dev. Cell* **8**(3): 377–387.

Prusa, A.R., Marton, E., Rosner, M., Bernaschek, G. and Hengstschlager, M. (2003). Oct-4-expressing cells in human amniotic fluid: a new source for stem cell research? *Hum. Reprod.* **18**(7): 1489–1493.

Rosenblatt, J.D., Lunt, A.I., Parry, D.J. and Partridge, T.A. (1995). Culturing satellite cells from living single muscle fiber explants. *In Vitro Cell Dev. Biol. Anim.* **31**: 773–779.

Sakuragawa, N., Thangavel, R., Mizuguchi, M., et al. (1996). Expression of markers for both neuronal and glial cells in human amniotic epithelial cells. *Neurosci. Lett.* **209**: 9–12, 23.

Sakuragawa, N., Enosawa, S., Ishii, T., et al. (2000). Human amniotic epithelial cells are promising transgene carriers for allogeneic cell transplantation into liver. *J. Hum. Genet.* **45**: 171–176.

Sankar, V. and Muthusamy, R. (2003). Role of human amniotic epithelial cell transplantation in spinal cord injury repair research. *Neuroscience* **118**(1): 11–17.

Sarkar, S., Chang, H.C., Porreco, R.P. and Jones, O.W. (1980). Neural origin of cells in amniotic fluid. *Am. J. Obstet. Gynecol.* **136**(1): 67–72.

Sartore, S., Lenzi, M., Angelini, A., Chiavegato, A., Gasparotto, L., De Coppi, P., Bianco, R. and Gerosa, G. (2005). Amniotic mesenchymal cells autotransplanted in a porcine model of cardiac ischemia do not differentiate to cardiogenic phenotypes. *Eur. J. Cardiothorac. Surg.* **28**(5): 677–684.

Takahashi, N., Enosawa, S., Mitani, T., et al. (2002). Transplantation of amniotic epithelial cells into fetal rat liver by *in utero* manipulation. *Cell Transplant.* **11**: 443–449.

Takeda, J., Seino, S. and Bell, G.I. (1992). Human Oct-3 gene family: cDNA sequences, alternative splicing, gene organization, chromosomal location, and expression at low levels in adult tissues. *Nucl. Acid Res.* **20**: 4613–4620.

Tamagawa, T., Ishiwata, I. and Saito, S. (2004). Establishment and characterization of a pluripotent stem cell line derived from human amniotic membranes and initiation of germ layers *in vitro*. *Hum. Cell* **17**(3): 125–130.

Tsai, M.S., Lee, J.L., Chang, Y.J. and Hwang, S.M. (2004). Isolation of human multipotent mesenchymal stem cells from second-trimester amniotic fluid using a novel two-stage culture protocol. *Hum. Reprod.* **19**(6): 1450–1456.

Tsai, M.S., Hwang, S.M., Tsai, Y.L., Cheng, F.C., Lee, J.L. and Chang, Y.J. (2005). Clonal amniotic fluid-derived stem cells express characteristics of both mesenchymal and neural stem cells. *Biol. Reprod.*

Underwood, M.A., Gilbert, W.M. and Sherman, M.P. (2005). Amniotic fluid: not just fetal urine anymore. *J. Perinatol.* **25**(5): 341–348.

von Koskull, H., Aula, P., Trejdosiewicz, L.K. and Virtanen, I. (1984). Identification of cells from fetal bladder epithelium in human amniotic fluid. *Hum. Genet.* **65**(3): 262–267.

Wulf, G.G., Viereck, V., Hemmerlein, B., Haase, D., Vehmeyer, K., Pukrop, T., Glass, B., Emons, G. and Trumper, L. (2004). Mesengenic progenitor cells derived from human placenta. *Tissue Eng.* **10**(7–8): 1136–1147.

Yen, B.L., Huang, H.I., Chien, C.C., Jui, H.Y., Ko, B.S., Yao, M., Shun, C.T., Yen, M.L., Lee, M.C. and Chen, Y.C. (2005). Isolation of multipotent cells from human term placenta. *Stem Cell* **23**(1): 3–9.

Zhang, X., Mitsuru, A., Igura, K., Takahashi, K., Ichinose, S., Yamaguchi, S. and Takahashi, T.A. (2005). Mesenchymal progenitor cells derived from chorionic villi of human placenta for cartilage tissue engineering. *Biochem. Biophys. Res. Commun.* **340**(3): 944–952.

Zhao, P., Ise, H., Hongo, M., Ota, M., Konishi, I., Nikaido, T. (2005). Human amniotic mesenchymal cells have some characteristics of cardiomyocytes. *Transplantation* **79**(5): 528–535.

10

Bone Marrow Stem Cells: Properties and Pluripotency

Munira Xaymardan, Massimo Cimini, Richard D. Weisel, and Ren-Ke Li

INTRODUCTION

The formation of new tissue in animals is generally confined to the embryonic and developmental stages. Regeneration of highly complex tissue does occur in some amphibians and reptiles. In mammals, however, healing of damaged tissue essentially results in the replacement of functional cells by highly fibrotic reparative tissue, which leads to diminished or even lost organ function. In the past 5 years, accumulating evidence has shown that multipotential stem cells are in fact present in many adult tissues. Bone marrow and tissue-specific stem cells can be induced to differentiate into adult cell types previously thought terminally differentiated, including cardiomyocytes, skeletal muscle cells, and neurons.

Stem cells, by definition, have two characteristics: (1) the ability to self-renew and generate more stem cells through cell division and (2) under appropriate induction, the ability to give rise to clonal progency that continue to differentiate into one or more specialized cell types. In adults, bone marrow is a major reservoir for stem cells. Unlike totipotent stem cells (such as a fertilized egg), which can give rise to entire organism, bone marrow stem cells (BMSCs) are multipotent cells, which can give rise to most of the adult cell types, but not yet proven to be able to develop into a fetus (Stocum, 2001). Stem- and progenitor-based therapies are currently being developed for the treatment of cardiovascular diseases, which represent the major cause of death in the Western world (NIH, 2000). A number of clinical trials are underway to test the efficacy of local and systemic delivery of bone marrow-derived stem cells for the replacement of cardiomyocytes and vascular endothelial cells (Britten et al., 2003; Perin and Silva, 2004) (Table 10.1). Preliminary results are suggestive, but their widespread application necessitates a thorough understanding of the mechanisms of cellular replacement in order to optimize the efficient use of BMSCs for vascular repair and cardioprotection.

BONE MARROW STEM CELLs

Bone marrow is hematopoietic tissue that lies within the trabecular bone. The trabecular and the bone marrow stroma are the elements that physically support and physiologically maintain the hematopoietic tissue. In adult humans, bone marrow is the site for production of all hematopoietic cells; the supporting stroma consists of reticular cells, osteocytes, adipocytes, vascular endothelium, and extracellular matrix. And together with the blood vessels, the bone marrow forms a hematopoietic inductive microenvironment that controls adult hematopoiesis, where five billion blood cells are produced every day.

The vascular structure of bone marrow consists of sinusoidal vasculature in which the endothelial cells do not have subsequent encapsulation of other types of cells; this is highly permissive for the emigration and immigration of the bone marrow cells.

Copyright © 2008 Elsevier Inc. All rights reserved

Table 10.1 Summary of clinical trials

Author	Trial name	Disease	Trial size	Length of follow-up	Cell source and type	Delivery route	Outcome
Perin and Silva (2004)		Ischemic heart failure	21	4 months	BM, mononuclear cells	Trans-endocardial	Increase LVEF
Kang et al. (2004)	MAGIC; randomized-controlled	>2 days	27	6 months	G-CSF + CPC CPC	Intra-coronary	G-CSF induced restenosis
Schachinger et al. (2004)	TOPCARE-AMI	AMI, 4.7 +/−1.7 days	59	6 months and 1 year	BM, CPC (contain CD133; CD34)	Intra-coronary	Improved global and regional contractility; decreased MI size
Wollert et al. (2004)	BOOST; random-controlled	AMI, 4.5 days	60	6 months	BM, CD34	Intra-coronary	Improved global LVEF
Lunde et al. (2005)	ASTAMI Random-controlled	AMI, 5–8 days	100	1 year	BMC	Intra-coronary	No benefit
Cleland et al. (2006)	REPAIR-AMI; random-controlled	4 days after AMI	204	18 months plus	BMC	Intra-coronary	Benefit up to 18 months

Cx-non-treatment control; AMI-acute myocardial infarction; BM-bone marrow; CPCs-circulating blood progenitor cells; LV-left ventricular; EF-ejection fraction.

Hematopoietic Stem Cells

Hematopoietic stem cells (HSCs) are the stem cells from which all red and white blood cells develop. They are entirely responsible for the development, maintenance, and regeneration of the blood forming tissue for life (Weissman, 2000). Because HSCs can reconstitute and restore the **hematopoietic** system of a myeloablated host, they have been traditionally used for treating hematologic disorders, starting in 1945 (Gengozian and Makinodan, 1956), when donor-derived HSCs were first used to protect a lethally irradiated civilian population.

In adult mouse bone marrow, HSC activity has been shown in a cell population marked by c-kit^{pos}, thy-1^{low}, and sca-1^{pos} (Bradfute et al., 2005). In adult humans, HSCs are marked by c-kit^{pos}, thy-1^{pos}, and CD34^{pos} (Weissman, 2000). HSCs from mice and humans are being isolated, starting with a lineage depletion step in which all the lineage-specific cells (B220, CD3, 4, 8, 11b Mac-1, Gr-1 and Tcr-119 for mice and CD10, 14, 15, 16, 19, and 20 in human) are removed (Figure 10.1). The resultant population, referred to as Lin^{neg}, can be enriched 10–100-fold, and is able to re-populate of bone marrow of a lethally irradiated host. *In vitro* expansion of HSCs can be achieved by co-culturing them with stromal cells from bone marrow.

Researchers have found several subpopulations within Lin^{neg} HSCs. One homogenous population is characterized as side population (SP) cells based on their unique ability to extrude hoechst dye. When examined by fluorescence activated cell sorter (FACS) analysis, SP cells fall within a separate population to the side of the rest of the cells on a dot plot of emission data. SP cells express the ABCG2 transporter, a transmembrane protein, which allows them to actively exclude hoechst dye and fluoresce in this specific manner. These cells are also able

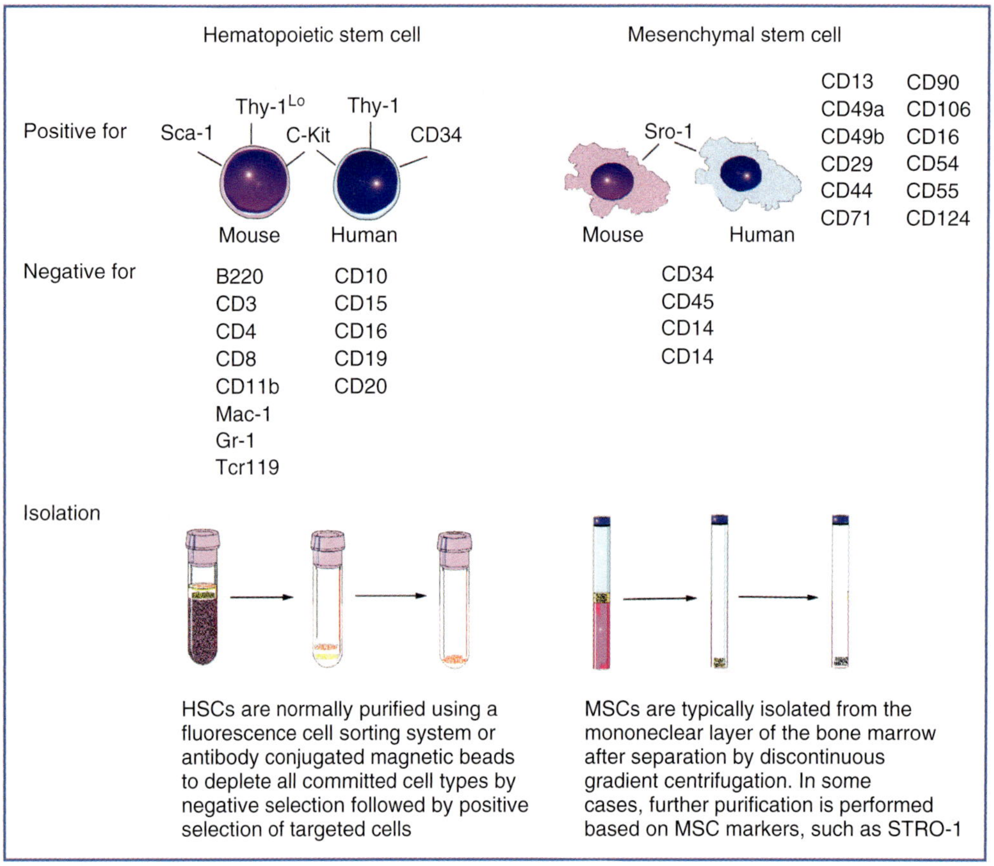

Figure 10.1 Isolation of HSCs and MSCs from bone marrow.

to home rapidly to the bone marrow of a lethally irradiated host (Goodell et al., 1996) and contribute progeny to the lung and liver in irradiated mice, and infiltrate into the infarcted heart (Abe et al., 2003). SP cells are also present in other tissues, including skeletal muscle and skin (Liadaki et al., 2005). Data are conflicting: some suggest that SP cells can be tissue-specific stem cells within these organs, and others suggest that are actually bone marrow-derived SP cells lodged within these tissues.

Another group of highly plastic stem cells isolated from bone marrow are known as **bone marrow-derived stem cells** (Leone et al., 2005). Several recent studies indicate that these cells are highly plastic, exhibiting tremendous differentiation activity in numerous non-hematopoietic organs. It is unclear whether these populations are enriched for pre-hematopoietic cells that maintain greater pluripotentiality than HSCs. An additional possibility is that a differentiated hematopoietic cell, such as a macrophage, may be able to assume the gene expression pattern of a different cell type by fusion (Ozturk et al., 2004).

Mesenchymal Stem Cells

Mesenchymal stem cells (MSCs) are stem cells found in bone marrow, from where they can generate bone, cartilage, fat, and fibrous connective tissue. They are the non-hematopoietic, structural components of bone marrow that support hematopoiesis by providing extracellular matrix components, cytokines, and growth factors. MSCs represent 0.001–0.01% of the bone marrow cell population, and in culture as clonal, plastic adherent cells that assume a spindle cell morphology with a finite life span. Friedenstein first discovered MSCs in 1970s (1978), when his laboratory was able to culture these cells in media and induce them to differentiate into multilineage cell types, including osteoblasts, chondroblasts, and adipocytes, in response to appropriate stimuli. No specific constellation of surface markers has been agreed upon for these cells, but human MSCs are typically isolated from the mononuclear layer of the bone marrow after separation by discontinuous gradient centrifugation. In some cases, further purification is performed based on MSC markers, such as STRO-1 (Gronthos et al., 1994). Other surface antigens reported to exist on MSCs are: CD13 (aminopeptidase-N), CD49a and CD49b (integrins-alpha), CD29 (integrin-beta), CD44 (hyaluronate), CD71 (transferrin), CD90 (thy-1), CD106 (vascular cell adhesion molecule-1), CD166 (activated leukocyte cell adhesion molecule), CD54 (intercellular adhesion molecule-1), CD55 (decay accelerating factor), and CD124 (interleukin-4 (IL-4) receptor). MSCs uniformly lack antigens CD34, CD45, CD14, and CD31 that typically identify hematopoietic cells (Pittenger and Martin, 2004) (Figure 10.1). A wide array of cytokines, including fibroblast growth factor-2 (FGF-2), FGF-4, platelet-derived growth factor-BB (PDGF-BB), and leukemia inhibitory factor (LIF), have been used to expand MSCs (Gregory et al., 2005).

Because MSCs are easily expandable in culture and differentiate into multiple tissue lineages, there has been much interest in their clinical potential for tissue repair and gene therapy. In particular, a population of highly plastic, adult-derived bone marrow cells, referred to as multipotent adult progenitor cells (MAPCs), can be grown *in vitro* from the postnatal marrow (and other organs) of mice, rats, and humans. These cells co-purify initially with MSCs and grow as adherent cells *in vitro* (Reyes et al., 2001). However, unlike MSCs, MAPCs can be cultured indefinitely in a relatively nutrient-poor medium (Jiang et al., 2002). Specific changes in growth factors induce differentiation of MAPCs into myoblasts, hepatocytes, and even neural tissue (Jiang et al., 2002; Schwartz et al., 2002).

Endothelial Progenitor Cells

Endothelial progenitor cells (EPCs) are a group of non-endothelial cells that can give rise to endothelial cells. Stemness of the cells is not clear, but they can be expanded, and increasing evidence shows that EPCs play a major role in postnatal *neovascularization*. Bone marrow HSCs and MSCs, as well as other tissues (fat, cord blood, and circulating blood), are the sources of the EPCs, of which HSC-derived EPCs are perhaps the best characterized. HSC-derived EPCs are maintained in the BMSC niche and are released upon mobilization with cytokines such as vascular endothelial growth factor (VEGF) or stromal cell-derived factor-1 (SDF-1), which are synthesized by ischemic tissue (Leone et al., 2005).

Indeed, Asahara et al. (1999) demonstrated that bone marrow-derived HPCs give rise to endothelial cells and contribute to endothelial recovery and new capillary formation after ischemia. EPCs have been subsequently defined as cells that express HSC markers such as CD34 or CD133, and an endothelial marker protein, VEGF receptor 2 (VEGFR2 or flk-1). Isolated cells express the classic HSC marker protein CD34 or the more immature HSC marker protein CD133. Both cell populations differentiate to endothelial cells *in vitro* under appropriate endothelial differentiation-promoting factors (Gehling et al., 2000). Most importantly, injection of $CD34^{pos}$ or $CD133^{pos}$ cells enhanced neovascularization in animal models after ischemia (Asahara et al., 1999).

Likewise, MSCs can differentiate into endothelial cells (Oswald et al., 2004) and improve neovascularization *in vivo* (Pittenger and Martin, 2004). Because MSCs can release a variety of angiogenic growth factors, this cocktail of growth factors may also act in a paracrine manner to support angiogenesis and arteriogenesis. Verfailliea's group (Reyes et al., 2002) reported that MAPCs that co-purify with MSCs isolated from postnatal human bone marrow can differentiate into cells that express endothelial markers, function *in vitro* as mature endothelial cells, and contribute to *in vivo* neoangiogenesis during tumor angiogenesis and wound healing (Reyes et al., 2002).

Interaction of Bone Marrow Cells and Stem Cell Niches

The stem cells in the bone marrow are not randomly distributed. They reside in specific compartments consisting of support cells known as niche, the microenvironment, which in turn controls the fate of the stem cells.

presence of specific signals such as those given by growth factors, cytokines, and components of the extracellular matrix, a certain population of daughter cells undergoes a sequential cascade of differentiation that alters the cells' original gene expression pattern. Self-renewal and differentiation potential are two criteria that define MSCs as real stem cells; however, these characteristics have only been proved after *in vitro* manipulation, and there is no clear description of the characteristics displayed by unmanipulated MSCs *in vivo* (Javazon et al., 2004). Our limited knowledge of MSCs is due to the fact that MSCs lack a unique marker, in contrast to other stem cells such as HSCs, which are identified by the expression of the CD34 surface marker. The CD105 surface antigen (endoglin) has been recently used to isolate hMSCs from BM and such an approach enabled the characterization of freshly isolated hMSCs before culture. A distinct expression of certain surface antigens such as CD45 and CD31 was demonstrated in freshly isolated hMSCs and the expression of these molecules was lower in culture-expanded hMSCs (Aslan et al., 2006b). These data suggest, again, the alterations that hMSCs may undergo during culture.

In several studies, cultured MSCs have been characterized either by using cell surface antigens or by examining the cells' differentiation potential. The most accepted characteristics for *in vitro* grown MSCs are the following: (1) the ability to form CFU-Fs when plated in plastic tissue culture plates in the presence of an animal serum such as FCS in a basic medium such as Dulbecco modified Eagle medium; (2) the expandability of these cells without losing their differentiation potential; and (3) the high levels of expression of the surface antigens CD105, CD73, CD29, CD44, CD71, CD90, CD106, CD120a, and CD124, and the low levels of expression of CD14, CD34, and the leukocyte common antigen CD45 (reviewed by Deans and Moseley, 2000).

THE STEM CELL NATURE OF MSCs

Stem cells are defined by their ability to self-renew and by their potential to undergo differentiation into functional cells under the right conditions. MSCs exhibit the potential to differentiate into the osteogenic, chondrogenic, adipogenic, tenogenic, myogenic, or stromal lineages (Haynesworth et al., 1992; Mackay et al., 1998; Yoo et al., 1998; Young et al., 1998; reviewed by Caplan and Bruder, 2001). In the presence of certain agents, such as a combination of ascorbic acid, β-glycerophosphate, and dexamethasone, or in the presence of BMPs, MSCs undergo a series of morphological and metabolic changes until they exhibit characteristics of osteogenic cells, which include elevated levels of alkaline phosphatase, osteopontin, and osteocalcin, and accumulation of calcium. Culturing MSCs in a three-dimensional manner (such as a pellet culture) and in the presence of TGFβ1 can induce the formation of collagen II and glucosaminoglycans within these cultures, therefore creating a cartilage-like tissue. Differentiation of MSCs into adipogenic cells has also been achieved *in vitro*, as demonstrated by the accumulation of fatty acid droplets within these cells (Pittenger et al., 1999).

In addition to their *in vitro* differentiation potential, MSCs have been shown to home to and engraft into several organs and tissues when injected systemically. Human BM-derived MSCs transplanted into the peritoneum of lamb fetuses at 65 days of gestation (before the development of the immune system) engrafted and underwent site-specific differentiation into chondrocytes, adipocytes, myocytes, cardiomyocytes, BM stromal cells, and stromal cells of the thymus. Surprisingly, when this transplantation took place at 85 days of gestation (an age at which there is active hematopoiesis and a competent immune system), hMSCs also integrated in a manner similar to cells transplanted at 65 days of gestation. These results suggest that systemically administered hMSCs are widely distributed to many tissues and organs, and that within these organs, specific signals and factors induce tissue-specific differentiation of MSCs.

In local models, as opposed to systemic, hMSCs can induce bone formation *in vivo* following transplantation in ectopic sites and in sites of segmental bone defect (Bruder et al., 1998a; Mankani et al., 2001). Direct injection of hMSCs into the brain tissue of rats resulted in the cells' long-term engraftment and subsequent migration along pathways similar to those used by neural stem cells (Azizi et al., 1998). The results of these studies demonstrate the multilineage differentiation potential of BM-derived adult MSCs and aid in defining them as suitable candidates for the regeneration of several mesenchymal tissues.

SOURCE OF MSCs AND ISOLATION TECHNIQUES

The embryonic origin of MSCs is still unclear, and much of our knowledge of these cells lies in the biological characteristics they display in adult tissues. Nevertheless, some findings indicate a possible origin of MSCs in a supporting layer of the dorsal aorta in the aorto-gonadal–mesonephric region in human and murine fetuses (Cortes et al., 1999; Marshall et al., 1999; Tavian et al., 1999). Consistent with these findings, the presence of circulating MSC-like cells within early human blood suggests remnants of migrating MSCs in the circulation (Campagnoli et al., 2001).

In adults, MSCs are found in the BM compartments of long bones, iliac crest, sternum, and cranial bones. BM has two major compartments: the hematopoietic compartment, in which hematopoiesis occurs, and the stroma-supportive system, which is associated with the former compartment and is composed of MSCs, ECs, and adipocytes (Bianco et al., 2001). Recent reports and our unpublished data have shown the presence of a potent MSC population in the BM of the craniofacial complex (Matsubara et al., 2005; Akintoye et al., 2006).

The lack of a surface marker unique to MSCs poses a further challenge for isolating them as a pure and unmanipulated population. Originally, Friendestein identified "stromal stem cells" by their ability to adhere to standard plastic in the presence of animal serum (Friedenstein et al., 1982, 1987). Pittenger et al. (1999) found that particular lots of fetal bovine serum (FBS) are highly preferable for initial cell adherence and the subsequent survival and proliferation of MSCs isolated from human iliac crest BM (Kadiyala et al., 1997; Pittenger et al., 1999) According to their report, a density gradient should first be used to separate and isolate fractions of mononuclear cells (MNCs) and red blood cells in the BM. The MNCs are then collected and seeded in medium containing 10% FBS at a density of 10–15 $\times$ 10^5 cells/cm^2 growth area. Adherent spindle-shaped cells appear within 48 h after the initial seeding, and the estimated percentage of MNCs ranges from 0.001% to 0.01%. These cells continue to grow, and when they have reached 100% confluence the cells should be detached and replated in fresh culture medium at a density of 5,000–6,000 cells/cm^2 growth area. This MSC isolation approach has been broadly followed by many groups. The major disadvantages of using this method are the presence of adherent cells of hematopoietic origin within the cultures during the first days and the need for a specific lot of FBS.

Based on the expression of surface molecules on MSCs, some techniques have been developed to isolate MSCs at a higher yield and purity and even without the need to seed them in culture. The expression of endoglin (CD105) by MSCs was used to distinguish these cells and isolate them from other BM cells (Majumdar et al., 2000). CD105-immunoisolated MSCs exhibit the same immunophenotype and differentiation potential described for MSCs that have been isolated using the plastic-adherence method. Using antibodies directed against the CD105 molecule, MNCs can be labeled with microbeads that possess magnetic properties and attach to anti-CD105 antibodies. Within the MNCs, CD105+ cells become coated with a magnetic shield and can be separated from the rest of the cells by passing them through a magnetic field (Majumdar et al., 2000; Aslan et al., 2006b).

Similarly, anti-Stro-1 antibodies were also used to isolate MSCs from BM (Gronthos et al., 1995; Gronthos et al., 2003). Stro-1 is an unidentified cell surface antigen expressed by a minor subpopulation of adult human BM. Anti-Stro-1 antibodies can be used to identify all clonogenic CFU-Fs within the BM, but they do not react to cells of hematopoietic origin (Simmons et al., 1991). Stro-1+ cells have been shown to contain an MSC fraction with the capacity to form a supportive microenvironment for hematopoietic cells *in vitro* and to differentiate into stromal cell types including smooth muscle cells, adipocytes, osteoblasts, and chondrocytes (Gronthos et al., 1994; Dennis et al., 2002).

The use of anti-CD49a antibodies to isolate hMSCs has also reported to yield a CFU-Fs-forming population that displays MSC characteristics (Deschaseaux et al., 2003).

WHICH TISSUES CONTAIN MSCs?

We have already described BM as the original and main source of MSCs. However, many studies have demonstrated the presence of MSCs or MSC-like cells within other tissues such as adipose tissue (Zuk et al., 2001, 2002), cord blood (Hong et al., 2005; Jeong et al., 2005; Moon et al., 2005), BM of the craniofacial complex (Matsubara et al., 2005; Akintoye et al., 2006), and peripheral blood, although the latter finding is still controversial (Fernandez et al., 1997; Conrad et al., 2002).

A plastic-adherent, CFU-F-forming cell population can be isolated from adipose tissue after treatment with enzymatic collagenase (Zuk et al., 2001, 2002; Katz et al., 2005). Following this treatment, a stromal vascular fraction is obtained that parallels the MNC fraction in BM. This fraction is collected while the adipocytes-containing fraction is removed during the first steps of centrifugation due to its high content of fatty acids. Plastic-adherent cells within the stromal vascular fraction were originally named processed lipoaspirate (PLA) cells, and were shown to have a high potential for *in vitro* expansion and a high potential for differentiation into several mesodermal lineages including the adipogenic, chondrogenic, myogenic, and osteogenic lineages (Zuk et al., 2001, 2002). PLA cells are quite similar to BM-derived MSCs morphologically and immunophenotypically; however, PLA cells form more CFU-Fs when plated in culture (Kern et al., 2006). Because adipose tissue is usually more available, can be collected with the use of local anesthesia, and its aspiration is associated with minimal discomfort and risks, it has been proposed as an additional or even alternative source for obtaining MSCs for regenerative medical purposes (Mizuno et al., 2003).

Cord blood is a source of MSCs that has been viewed with growing interest. MSCs have been isolated from umbilical cord blood (Hong et al., 2005; Hutson et al., 2005; Jeong et al., 2005; Moon et al., 2005) following gradient centrifugation in a manner similar to that used to obtain them from BM. The success rate of MSC isolation from umbilical cord blood is less than 100% (34% Wagner et al., 2005 and 63% Kern et al., 2006) compared with the 100% rate found in using BM or adipose tissue.

Other sources of MSCs include maxillofacial BM (Matsubara et al., 2005; Akintoye et al., 2006), and dermal tissue (Bartsch et al., 2005). Recent reports have shown isolation of MSCs from BM of craniofacial bones (craniofacial MSCs) and compared them to iliac crest and long bones-derived MSCs. Craniofacial MSCs were shown to have highly osteogenic potential and share the main basic characters as iliac crest-derived MSCs (Matsubara et al., 2005). Akintoye et al. (2006) compared maxillofacial- to iliac crest-derived MSCs from the same individuals and reported higher osteogenic and adipogenic potential of maxillofacial MSCs. Our unpublished data have shown that hMSCs isolated from maxillofacial BM can be genetically engineered using adenoviral vectors and utilized for inducing bone formation. MSCs appear to be "resident" stem cells in many tissues, and they function in the normal turnover of these tissues. When tissue repair is required, these cells can be stimulated to proliferate and differentiate. The use of MSCs for appropriate tissue healing may require isolation of the right stem cells and directing the differentiation of these cells into the appropriate lineage.

SKELETAL TISSUE REGENERATION BY MSCs

Bone

Bone regeneration is required for a number of orthopedic, neurosurgical, and maxillofacial clinical indications. Spinal fusion, treatment of nonunion bone defects in long bones, and treatment of substantial bone loss due to trauma or osteoporosis are only a few examples. Currently, these conditions are treated by using synthetic implants that do not resemble natural bone and tend to fail in the long term. The few biological solutions that exist for restoration of bone loss include autologous bone grafts, which may cause donor-site morbidity (Quarto et al., 2001), and bone-inducing protein-based treatment (BMP-2 therapy, for example), which requires highly expensive megadoses of protein, and they do not always lead to beneficial results. Given that MSCs can differentiate into the osteogenic lineage, they are considered good candidates for tissue-engineered bone replacement. To promote bone regeneration *in vivo* by using cultured MSCs, it is essential to seed the cells onto a ceramic scaffold, which is usually composed of hydroxyapatite and β-tricalcium phosphate. Without the osteoinductive properties of these ceramic scaffolds, implanted MSCs tend to form a nonspecific connective tissue in bone defects, as we have shown in several studies (Moutsatsos et al., 2001; Turgeman et al., 2001). The potential for MSC-loaded ceramic scaffolds to repair bone defects has been shown in a number of animal models by using MSCs isolated from the BM of different species. Bruder et al. have shown that critically sized defects in dog femora can be filled with bone newly formed from autologous MSCs (Bruder et al., 1998b) and that a similar result can be achieved by placing hMSCs in femoral bone defects in athymic rats (Bruder et al., 1998a). Using a similar approach, Arinzeh et al. (2003) were able to regenerate femoral bone defects in adult dogs by using allogeneic MSCs, without any evidence of an immune response targeted to the tissue-engineered graft. Another animal model in which autologous MSCs have been used to repair large bone defects is sheep (Kon et al., 2000). In this instance, the same approach was used to generate a substantial amount of newly formed bone tissue to create bone fusion between adjacent vertebras, a method also known as spinal fusion. Such a fusion can eliminate the need for metal screws, which nowadays are used for spinal fusion. The validity of this approach was demonstrated in both rabbit and rhesus monkey models, in which implantation of autologous MSCs led to much greater bone formation than other experimental grafts devoid of cells (Cinotti et al., 2004). Following the solid experimental proof of principle, Quarto et al. attempted this tissue-engineering method in the treatment of three human patients who suffered a bone loss of 4–7 cm in long bones (Quarto et al., 2001). Autologous MSCs were isolated and expanded *in vitro* for each patient. The cells were seeded onto macroporous hydroxyapatite scaffolds, which had been molded into the shape of the missing piece of the bone, and were implanted in the defect. Two months after implantation, a good integration of implant to bone was evident. Although the patients recovered function in 6–7 months after surgery (one-half to one-third of the time needed for recovery using bone grafts (Quarto et al., 2001) and did not report any problems during a 6-year follow-up period, the ceramic scaffolds were still not absorbed after 5 years (Mastrogiacomo et al., 2005).

Other approaches to MSC-aided bone repair include the use of MSCs that have been osteogenically differentiated *in vitro* prior to implantation *in vivo*. This strategy allows the seeding of cells onto nonosteoinductive scaffolds, which degrade better *in vivo*. However, this method requires prolonged periods of culture.

Because MSCs are relatively easily isolated from BM and fat tissue, it is conceivable to use them as vehicles for the delivery of therapeutic genes *in vivo*, a strategy known as stem cell-based gene therapy (Gazit et al., 1999). The aim of most gene therapy studies directed at bone healing is to induce bone formation either in a model of nonunion bone fractures or as a means to achieve spinal fusion. Indeed, some studies have involved the use of primary MSCs and cell lines for expression and delivery of osteogenic genes, which induce bone formation (Engstrand et al., 2000). These studies have implemented various types of MSCs including cell lines such as C3H10T1/2 and primary marrow-derived stem cells for the delivery of BMP-2. The delivery of growth factors of the BMP family is often used in these studies, because these factors promote osteogenic differentiation and bone formation (Wozney et al., 1988). In particular, BMP-2 has commonly been used because it is a highly osteoinductive agent that has been well studied and is known to induce bone *in vivo* in ectopic and orthotopic sites (Wozney et al., 1988; Wang et al., 1990; Volek-Smith et al., 1996; Yamaguchi et al., 1996; Chaudhari et al., 1997; Hanada et al., 1997; Lecanda et al., 1997; Fromigue et al., 1998; Gori et al., 1999). Other members of the BMP family, such as BMP-4 and BMP-9, have also been used for stem cell-mediated

MSCs can be used for beta-cell replacement therapy by harvesting the MSCs and applying *in vitro* differentiation protocols before delivering the cell back to the target tissue, or by enhancing biological mechanisms of mobilization and the homing of MSCs followed by biologically induced differentiation. Sordi et al. were able to define the chemokine receptor repertoire of hMSCs derived from BM that determines their migratory activity. Using a pancreatic cell coculture, these researchers concluded that modulation of the homing capacity of MSCs may be instrumental for harnessing the therapeutic potential of MSCs derived from BM (Sordi et al., 2005).

Recently it was reported that *in vitro* human BM stem cells are able to differentiate into insulin-expressing cells through a mechanism involving several transcription factors of the beta-cell developmental pathway when cultured in an appropriate microenvironment (Moriscot et al., 2005). Nevertheless, the insulin-producing capacity of BM-derived cells is still controversial. Recently, Choi et al. suggested that there is little evidence of transdifferentiation of BM-derived cells into pancreatic beta cells *in vivo*. However, their studies did not exclude the possibility that BM-derived MSCs could differentiate into beta cells *in vitro* by using the right inducer, for example, RPE, or as recently suggested, that the expression of the Pdx1 gene into various cells can provoke differentiation into cells similar to pancreatic beta cells (Choi et al., 2003).

In vitro models of parenchymal liver cells are of great importance in toxicology and in bioartificial liver research (Azar et al., 1996; Locasciulli et al., 1997), because primary cultures of hepatocytes are hindered by a short life span and a rapid loss of hepatic function under *in vitro* conditions (Kim et al., 2000).

Schwartz and associates reported for the first time that under *in vitro* conditions an adult marrow-derived stem cell, MAPC, can differentiate into functional hepatocyte-like cells (Schwartz et al., 2002) as well as into mesodermal and ectodermal cell lineages (Reyes et al., 2001; Jiang et al., 2002; Verfaillie et al., 2003). Following this study, Lee et al. used MSCs and demonstrated differentiation into cells of the endoderm as well as into those of the mesoderm (Lee et al., 2004). Rat MSCs require specific culture conditions and growth factors to differentiate into hepatocytes. Regarding this, several controversial reports have emerged: some demonstrating that differentiation was achieved only by using fibroblast growth factor-4 and HGF (Wang et al., 2004; Kang et al., 2005) and some showing that rat MSCs must be cultured in supplemented medium and the presence of freshly isolated rat liver cells (Lange et al., 2005b). Under specified culture conditions, only rat MSCs cocultured with liver cells acquired the hepatocytic phenotype.

In vivo transplantation of HGF-induced differentiated rat MSCs into liver-injured rats restored serum albumin levels and significantly suppressed transaminase activity and liver fibrosis (Oyagi et al., 2005). The next generation of experiments involved hMSCs, which were examined by directly xenografting them to allylalcohol-treated rat liver. When BM-derived cells were fractioned into MSCs, CD34+ cells, and non-MSC CD34- cells, and transplanted *in vivo*, hepatocyte-like cells were observed only in the recipient livers that contained MSC fractions (Sato et al., 2005).

The ultimate goal of differentiation studies is the amendment of damaged tissue by cellular transplantation. The recovery of damaged liver may be clearly attained if one uses a syngeneic model on a larger scale of transplantation (Lange et al., 2005a, b).

In summary, BM-derived hMSCs indeed possess great potential as the future treatment of choice for several nonskeletal tissue injuries and diseases.

CONCLUSIONS

MSCs constitute a unique population of adult stem cells that hold great promise for various tissue-engineering applications. These cells can readily be isolated from various sites in the human body, especially from BM and adipose tissues. Established protocols exist for the induction of specific differentiation patterns of MSCs into different committed cells, most notably into osteoblasts, chondrocytes, and adipocytes. So far it has been demonstrated that the use of genetically modified MSCs, overexpression of various therapeutic transgenes, is a powerful tool in the induction of differentiation and in the promotion of tissue regeneration *in vivo*. Novel technologies, which utilize electroporation-based systems, allow for the safe and efficient gene delivery into MSCs and bypass the need for using non-safe viral vectors. It has been shown that the ultrastructural, chemical and nanobiomechanical properties of engineered bone derived from MSCs were similar to that of native origin. Bioinformatics techniques can be applied to genetically modified MSCs in order to find new candidate genes for therapeutic purposes. The conventional method of MSC isolation using plastic adherence has shown to be costly and might reduce the stemness of the cells. Therefore an attractive alternative has been developed and it includes the immediate use of immunoisolated, non-cultured MSCs for *in vivo* implantation. Future challenges require the identification of an optimal scaffold for MSC implantation *in vivo* and, finally, the development of a preservation method for future reuse of autologous cells. Noninvasive imaging will continue to play an important role in analyzing the power of MSCs to regenerate tissues in various defect models. Overcoming these hurdles will no doubt make MSCs the optimal tool for biological tissue replacement in this century.

REFERENCES

Abedin, M., Tintut, Y. and Demer, L.L. (2004). Mesenchymal stem cells and the artery wall. *Circ. Res.* **95**(7): 671–676.

Aggarwal, S. and Pittenger, M.F. (2005). Human mesenchymal stem cells modulate allogeneic immune cell responses. *Blood* **105**(4): 1815–1822.

Akintoye, S.O., Lam, T., Shi, S., Brahim, J., Collins, M.T. and Robey, P.G. (2006). Skeletal site-specific characterization of orofacial and iliac crest human bone marrow stromal cells in same individuals. *Bone* **38**(6): 758–768.

Akiyama, Y., Radtke, C. and Kocsis, J.D. (2002). Remyelination of the rat spinal cord by transplantation of identified bone marrow stromal cells. *J. Neurosci.* **22**(15): 6623–6630.

Altman, G.H., Horan, R.L., Martin, I., Farhadi, J., Stark, P.R., Volloch, V., Richmond, J.C., Vunjak-Novakovic, G. and Kaplan, D.L. (2002). Cell differentiation by mechanical stress. *FASEB J.* **16**(2): 270–272.

Amado, L.C., Saliaris, A.P., Schuleri, K.H., St John, M., Xie, J.S., Cattaneo, S., Durand, D.J., Fitton, T., Kuang, J.Q. Stewart, G., Lehrke, S., Baumgartner, W.W., Martin, B.J., Heldman, A.W. and Hare, J.M. (2005). Cardiac repair with intramyocardial injection of allogeneic mesenchymal stem cells after myocardial infarction. *Proc Natl Acad Sci USA* **102**(32): 11474–11479.

Arinzeh, T.L., Peter, S.J., Archambault, M.P., van den Bos, C., Gordon, S., Kraus, K., Smith, A. and Kadiyala, S. (2003). Allogeneic mesenchymal stem cells regenerate bone in a critical-sized canine segmental defect. *J. Bone Joint Surg. Am.* **85A**(10): 1927–1935.

Aslan, H., Ravid, O., Cohen, A., Tzur, L., Pelled, G., Turgeman, G., Zilberman, Y., Gazit, Z., Clancy, B., Domany, E. and Gazit, D. (2003). Novel bioinformatic analysis of Tet-regulated bone formation *in vivo* indicates novel cross talk between WNT and BMP2 signalling pathways. *J. Bone Miner. Res.* **18**(Suppl 2): S305.

Aslan, H., Zilberman, Y., Arbeli, V., Sheyn, D., Matan, Y., Liebergall, M., Li, J.Z., Helm, G.A., Gazit, D. and Gazit, Z. (2006a). Nucleofection-based *ex vivo* nonviral gene delivery to human stem cells as a platform for tissue regeneration. *Tissue Eng.* **12**(4): 877–889.

Aslan, H., Zilberman, Y., Kendel, A., Liebergall, M., Oskouian, R.J. Gazit, D. and Gazit, Z. (2006b). Osteogenic differentiation of noncultured immunoisolated bone marrow-derived CD105+ Cells. *Stem Cells* **24**(7): 1728–1737.

Awad, H.A., Butler, D.L., Boivin, G.P., Smith, F.N., Malaviya, P., Huibregtse, B. and Caplan, A.I. (1999). Autologous mesenchymal stem cell-mediated repair of tendon. *Tissue Eng.* **5**(3): 267–277.

Azar, N., Valla, D., Abdel-Samad, I., Hoang, C., Fretz, C., Sutton, L., Fournel, J.J., Le Charpentier, Y., Binet, J.L. and Leblond, V. (1996). Liver dysfunction in allogeneic bone marrow transplantation recipients. *Transplantation* **62**(1): 56–61.

Azizi, S.A., Stokes, D., Augelli, B.J., DiGirolamo, C. and Prockop, D.J. (1998). Engraftment and migration of human bone marrow stromal cells implanted in the brains of albino rats–similarities to astrocyte grafts. *Proc. Natl Acad. Sci. USA* **95**(7): 3908–3913.

Bartholomew, A., Sturgeon, C., Siatskas, M., Ferrer, K., McIntosh, K., Patil, S., Hardy, W., Devine, S., Ucker, D., Deans, R., Moseley, A. and Hoffman, R. (2002). Mesenchymal stem cells suppress lymphocyte proliferation *in vitro* and prolong skin graft survival *in vivo*. *Exp. Hematol.* **30**(1): 42–48.

Bartsch, G., Yoo, J.J., De Coppi, P., Siddiqui, M.M., Schuch, G., Pohl, H.G., Fuhr, J., Perin, L., Soker, S. and Atala, A. (2005). Propagation, expansion, and multilineage differentiation of human somatic stem cells from dermal progenitors. *Stem Cells Dev.* **14**(3): 337–348.

Bayes-Genis, A., Roura, S., Soler-Botija, C., Farre, J., Hove-Madsen, L., Llach, A. and Cinca, J. (2005). Identification of cardiomyogenic lineage markers in untreated human bone marrow-derived mesenchymal stem cells. *Transplant Proc.* **37**(9): 4077–4079.

Beltrami, A.P., Urbanek, K., Kajstura, J., Yan, S.M., Finato, N., Bussani, R., Nadal-Ginard, B., Silvestri, F., Leri, A., Beltrami, C.A. and Anversa, P. (2001). Evidence that human cardiac myocytes divide after myocardial infarction. *N. Engl. J. Med.* **344**(23): 1750–1757.

Bertram, H., Kroeber, M., Wang, H., Unglaub, F., Guehring, T., Carstens, C. and Richter, W. (2005). Matrix-assisted cell transfer for intervertebral disc cell therapy. *Biochem. Biophys. Res. Commun.* **331**(4): 1185–1192.

Beyth, S., Borovsky, Z., Mevorach, D., Liebergall, M., Gazit, Z., Aslan, H.,. Galun, E. and Rachmilewitz, J. (2005). Human mesenchymal stem cells alter antigen-presenting cell maturation and induce T-cell unresponsiveness. *Blood* **105**(5): 2214–2219.

Bianco, P., Riminucci, M., Gronthos, S. and Robey, P.G. (2001). Bone marrow stromal stem cells: nature, biology, and potential applications. *Stem Cells* **19**(3): 180–192.

Black, I.B. and Woodbury, D. (2001). Adult rat and human bone marrow stromal stem cells differentiate into neurons. *Blood Cell. Mol. Dis.* **27**(3): 632–636.

Bonafe, F., Muscari, C., Guarnieri, C. and Caldarera, C.M. (2003). [Regeneration of infarcted cardiac tissue: the route of stem cells]. *Ital. Heart J.* Suppl **4**(4): 299–305.

Brittberg, M., Lindahl, A., Nilsson, A., Ohlsson, C., Isaksson, O. and Peterson, L. (1994). Treatment of deep cartilage defects in the knee with autologous chondrocyte transplantation. *N. Engl. J. Med.* **331**(14): 889–895.

Bruder, S.P., Kraus, K.H., Goldberg, V.M. and Kadiyala, S. (1998a). The effect of implants loaded with autologous mesenchymal stem cells on the healing of canine segmental bone defects. *J. Bone Joint Surg. Am.* **80**(7): 985–996.

Bruder, S.P., Kurth, A.A., Shea, M., Hayes, W.C., Jaiswal, N. and Kadiyala, S. (1998b). Bone regeneration by implantation of purified, culture-expanded human mesenchymal stem cells. *J. Orthop. Res.* **16**(2): 155–162.

Burt, R.K., Oyama, Y., Traynor, A. and Kenyon, N.S. (2002). Hematopoietic stem cell therapy for type 1 diabetes: induction of tolerance and islet cell neogenesis. *Autoimmun. Rev.* **1**(3): 133–138.

Campagnoli, C., Roberts, I.A., Kumar, S., Bennett, P.R., Bellantuono, I. and Fisk, N.M. (2001). Identification of mesenchymal stem/progenitor cells in human first-trimester fetal blood, liver, and bone marrow. *Blood* **98**(8): 2396–2402.

Caplan, A.I. and Bruder, S.P. (2001). Mesenchymal stem cells: building blocks for molecular medicine in the 21st century. *Trends Mol. Med.* **7**(6): 259–264.

Caplan, A.I., Elyaderani, M., Mochizuki, Y., Wakitani, S. and Goldberg, V.M. (1997). Principles of cartilage repair and regeneration. *Clin. Orthop. Relat. Res.* (342): 254–269.

Chaudhari, A., Ron, E. and Rethman, M.P (1997). Recombinant human bone morphogenetic protein-2 stimulates differentiation in primary cultures of fetal rat calvarial osteoblasts. *Mol. Cell Biochem.* **167**(1–2): 31–39.

Chen, J., Li, Y., Wang, L., Lu, M., Zhang, X. and Chopp, M. (2001). Therapeutic benefit of intracerebral transplantation of bone marrow stromal cells after cerebral ischemia in rats. *J. Neurol. Sci.* **189**(1–2): 49–57.

Chen, L.B., Jiang, X.B. and Yang, L. (2004). Differentiation of rat marrow mesenchymal stem cells into pancreatic islet beta-cells. *World J. Gastroenterol.* **10**(20): 3016–3020.

Chen, Y., Cheung, K.M., Kung, H.F., Leong, J.C., Lu, W.W. and Luk, K.D. (2002). *In vivo* new bone formation by direct transfer of adenoviral-mediated bone morphogenetic protein-4 gene. *Biochem. Biophys. Res. Commun.* **298**(1): 121–127.

Choi, J.B., Uchino, H., Azuma, K., Iwashita, N., Tanaka, Y., Mochizuki, H., Migita, M., Shimada, T., Kawamori, R. and Watada, H. (2003). Little evidence of transdifferentiation of bone marrow-derived cells into pancreatic beta cells. *Diabetologia* **46**(10): 1366–1374.

Choi, K.S., Shin, J.S., Lee, J.J., Kim, Y.S., Kim, S.B. and Kim, C.W. (2005). *In vitro* trans-differentiation of rat mesenchymal cells into insulin-producing cells by rat pancreatic extract. *Biochem. Biophys. Res. Commun.* **330**(4): 1299–1305.

Chopp, M., Xue Hui, Z., Li, Y., Wang, L., Chen, J., Lu, D., Lu, M. and Rosenblum, M. (2000). Spinal cord injury in rat: Treatment with bone marrow stromal cell transplantation. *NeuroReport* **11**(13): 3001–3005.

Cinotti, G., Patti, A.M., Vulcano, A., Della Rocca, C., Polveroni, G., Giannicola, G. and Postacchini, F. (2004). Experimental posterolateral spinal fusion with porous ceramics and mesenchymal stem cells. *J. Bone Joint Surg. Br.* **86**(1): 135–142.

Conrad, C., Gottgens, B., Kinston, S., Ellwart, J. and Huss, R. (2002). GATA transcription in a small rhodamine 123(low)CD34(+) subpopulation of a peripheral blood-derived CD34(+)CD105(+) mesenchymal cell line. *Exp. Hematol.* **30**(8): 887–895.

Cortes, F., Deschaseaux, F., Uchida, N., Labastie, M.C., Friera, A.M., He, D., Charbord, P. and Peault, B. (1999). HCA, an immunoglobulin-like adhesion molecule present on the earliest human hematopoietic precursor cells, is also expressed by stromal cells in blood-forming tissues. *Blood* **93**(3): 826–837.

Crevensten, G., Walsh, A.J., Ananthakrishnan, D., Page, P., Wahba, G.M., Lotz, J.C. and Berven, S. (2004). Intervertebral disc cell therapy for regeneration: mesenchymal stem cell implantation in rat intervertebral discs. *Ann. Biomed. Eng.* **32**(3): 430–434.

Crigler, L., Robey, R.C., Asawachaicharn, A., Gaupp, D. and Phinney, D.G. (2006). Human mesenchymal stem cell subpopulations express a variety of neuro-regulatory molecules and promote neuronal cell survival and neuritogenesis. *Exp. Neurol.* **198**(1): 54–64.

Cristino, S., Grassi, F., Toneguzzi, S., Piacentini, A., Grigolo, B., Santi, S., Riccio, M., Tognana, E., Facchini, A. and Lisignoli, G. (2005). Analysis of mesenchymal stem cells grown on a three-dimensional HYAFF 11-based prototype ligament scaffold. *J. Biomed. Mater. Res. A* **73**(3): 275–283.

Dabeva, M.D. and Shafritz, D.A. (2003). Hepatic stem cells and liver repopulation. *Semin. Liver Dis.* **23**(4): 349–362.

Davani, S., Marandin, A., Mersin, N., Royer, B., Kantelip, B., Herve, P., Etievent, J.P. and Kantelip, J.P. (2003). Mesenchymal progenitor cells differentiate into an endothelial phenotype, enhance vascular density, and improve heart function in a rat cellular cardiomyoplasty model. *Circulation* **108** (Suppl 1): II253–II258.

Deans, R.J. and Moseley, A.B. (2000). Mesenchymal stem cells: biology and potential clinical uses. *Exp. Hematol.* **28**(8): 875–884.

Dennis, J.E., Carbillet, J.P., Caplan, A.I. and Charbord, P. (2002). The STRO-1+ marrow cell population is multipotential. *Cells Tissues Organs* **170**(2–3): 73–82.

Deschaseaux, F., Gindraux, F., Saadi, R., Obert, L., Chalmers, D. and Herve, P. (2003). Direct selection of human bone marrow mesenchymal stem cells using an anti-CD49a antibody reveals their CD45med, low phenotype. *Br. J. Haematol.* **122**(3): 506–517.

Di Nicola, M., Carlo-Stella, C., Magni, M., Milanesi, M., Longoni, P.D., Matteucci, P., Grisanti, S. and Gianni, A.M. (2002). Human bone marrow stromal cells suppress T-lymphocyte proliferation induced by cellular or nonspecific mitogenic stimuli. *Blood* **99**(10): 3838–3843.

Djouad, F., Plence, P., Bony, C., Tropel, P., Apparailly, F., Sany, J., Noel, D. and Jorgensen, C. (2003). Immunosuppressive effect of mesenchymal stem cells favors tumor growth in allogeneic animals. *Blood* **102**(10): 3837–3844.

Dragoo, J.L., Choi, J.Y., Lieberman, J.R., Huang, J., Zuk, P.A., Zhang, J. Hedrick, M.H. and Benhaim, P. (2003). Bone induction by BMP-2 transduced stem cells derived from human fat. *J. Orthop. Res.* **21**(4): 622–629.

Dressler, M.R., Butler, D.L. and Boivin, G.P. (2005). Effects of age on the repair ability of mesenchymal stem cells in rabbit tendon. *J. Orthop. Res.* **23**(2): 287–293.

Dumont, R.J., Dayoub, H., Li, J.Z., Dumont, A.S., Kallmes, D.F., Hankins, G.R. and Helm, G.A. (2002). *Ex vivo* bone morphogenetic protein-9 gene therapy using human mesenchymal stem cells induces spinal fusion in rodents. *Neurosurgery* **51**(5): 1239–1244; discussion 1244–1245.

Eliopoulos, N., Stagg, J., Lejeune, L., Pommey, S. and Galipeau, J. (2005). Allogeneic marrow stromal cells are immune rejected by MHC class I- and class II-mismatched recipient mice. *Blood* **106**(13): 4057–4065.

Engstrand, T., Daluiski, A., Bahamonde, M.E., Melhus, H. and Lyons, K.M. (2000). Transient production of bone morphogenetic protein 2 by allogeneic transplanted transduced cells induces bone formation. *Hum. Gene Ther.* **11**(1): 205–211.

Fang, L.J., Fu, X.B., Sun, T.Z., Li, J.F., Cheng, B., Yang, Y.H. and Wang, Y.X. (2003). [An experimental study on the differentiation of bone marrow mesenchymal stem cells into vascular endothelial cells]. *Zhonghua Shao Shang Za Zhi* **19**(1): 22–24.

Fazel, S., Chen, L., Weisel, R.D., Angoulvant, D., Seneviratne, C., Fazel, A., Cheung, P., Lam, J., Fedak, P.W., Yau, T.M. and Li, R.K. (2005). Cell transplantation preserves cardiac function after infarction by infarct stabilization: augmentation by stem cell factor. *J. Thorac. Cardiovasc. Surg.* **130**(5): 1310.

Fernandez, M., Simon, V. Herrera, G., Cao, C., Del Favero, H. and Minguell, J.J. (1997). Detection of stromal cells in peripheral blood progenitor cell collections from breast cancer patients. *Bone Marrow Transplant* **20**(4): 265–271.

Friedenstein, A.J., Latzinik, N.W., Grosheva, A.G. and Gorskaya, U.F. (1982). Marrow microenvironment transfer by heterotopic transplantation of freshly isolated and cultured cells in porous sponges. *Exp. Hematol.* **10**(2): 217–227.

Friedenstein, A.J., Chailakhyan, R.K. and Gerasimov, U.V. (1987). Bone marrow osteogenic stem cells: *in vitro* cultivation and transplantation in diffusion chambers. *Cell Tissue Kinet.* **20**(3): 263–272.

Fromigue, O., Marie, P.J. and Lomri, A. (1998). Bone morphogenetic protein-2 and transforming growth factor-beta2 interact to modulate human bone marrow stromal cell proliferation and differentiation. *J. Cell Biochem.* **68**(4): 411–426.

Fukuda, K. (2001). Development of regenerative cardiomyocytes from mesenchymal stem cells for cardiovascular tissue engineering. *Artif. Organs* **25**(3): 187–193.

Fukuda, K. (2002). Molecular characterization of regenerated cardiomyocytes derived from adult mesenchymal stem cells. *Congenit Anom (Kyoto)* **42**(1): 1–9.

Fukuda, K. (2003a). Application of mesenchymal stem cells for the regeneration of cardiomyocyte and its use for cell transplantation therapy. *Hum. Cell* **16**(3): 83–94.

Fukuda, K. (2003b). Use of adult marrow mesenchymal stem cells for regeneration of cardiomyocytes. *Bone Marrow Transplant.* **32**(Suppl 1): S25–S7.

Fukuda, K. (2005). Progress in myocardial regeneration and cell transplantation. *Circ. J.* **69**(12): 1431–1446.

Gafni, Y., Turgeman, G., Liebergal, M., Pelled, G., Gazit, Z. and Gazit, D. (2004). Stem cells as vehicles for orthopedic gene therapy. *Gene Ther.* **11**(4): 417–426.

Gamradt, S.C. and Lieberman J.R. (2004). Genetic modification of stem cells to enhance bone repair. *Ann. Biomed. Eng.* **32**(1): 136–147.

Gan, J.C., Ducheyne, P., Vresilovic, E.J., Swaim, W. and Shapiro, I.M. (2003). Intervertebral disc tissue engineering I: characterization of the nucleus pulposus. *Clin. Orthop. Relat. Res.* 411: 305–314.

Gazit, D., Turgeman, G., Kelley, P., Wang, E., Jalenak, M., Zilberman, Y. and Moutsatsos, I. (1999). Engineered pluripotent mesenchymal cells integrate and differentiate in regenerating bone: a novel cell-mediated gene therapy. *J. Gene Med.* **1**(2): 121–133.

Gille, J., Ehlers, E.M., Okroi, M., Russlies, M. and Behrens, P. (2002). Apoptotic chondrocyte death in cell–matrix biocomposites used in autologous chondrocyte transplantation. *Ann. Anat.* **184**(4): 325–332.

Gojo, S., Gojo, N., Takeda, Y., Mori, T., Abe, H., Kyo, S., Hata, J. and Umezawa, A. (2003). *In vivo* cardiovasculogenesis by direct injection of isolated adult mesenchymal stem cells. *Exp. Cell Res.* **288**(1): 51–59.

Goncalves, M.A., de Vries, A.A., Holkers, M., van de Watering, M.J., van der Velde, I., van Nierop, G.P., Valerio, D. and Knaan-Shanzer, S. (2006). Human mesenchymal stem cells ectopically expressing full-length dystrophin can complement Duchenne muscular dystrophy myotubes by cell fusion. *Hum. Mol. Genet.* **15**(2): 213–221.

Gori, F., Thomas, T., Hicok, K.C., Spelsberg, T.C. and Riggs, B.L. (1999). Differentiation of human marrow stromal precursor cells: bone morphogenetic protein-2 increases OSF2/CBFA1, enhances osteoblast commitment, and inhibits late adipocyte maturation. *J. Bone Miner. Res.* **14**(9): 1522–1535.

Gotherstrom, C., Ringden, O., Tammik, C., Zetterberg, E., Westgren, and Le Blanc, K. (2004). Immunologic properties of human fetal mesenchymal stem cells. *Am. J. Obstet. Gynecol.* **190**(1): 239–245.

Gronthos, S., Graves, S.E., Ohta, S. and Simmons, P.J. (1994). The STRO-1+ fraction of adult human bone marrow contains the osteogenic precursors. *Blood* **84**(12): 4164–4173.

Gronthos, S. and Simmons, P.J. (1995). The growth factor requirements of STRO-1-positive human bone marrow stromal precursors under serum-deprived conditions *in vitro*. *Blood* **85**(4): 929–940.

Gronthos, S., Zannettino, A.C., Hay, S.J., Shi, S., Graves, S.E., Kortesidis, A. and Simmons, P.J. (2003). Molecular and cellular characterisation of highly purified stromal stem cells derived from human bone marrow. *J. Cell Sci.* **116**(Pt 9): 1827–1835.

Gysin, R., Wergedal, J.E., Sheng, M.H., Kasukawa, Y., Miyakoshi, N., Chen, S.T., Peng, H., Lau, K.H., Mohan, S. and Baylink, D.J. (2002). *Ex vivo* gene therapy with stromal cells transduced with a retroviral vector containing the BMP4 gene completely heals critical size calvarial defect in rats. *Gene Ther.* **9**(15): 991–999.

Hanada, K., Dennis, J.E. and Caplan, A.I. (1997). Stimulatory effects of basic fibroblast growth factor and bone morphogenetic protein-2 on osteogenic differentiation of rat bone marrow-derived mesenchymal stem cells. *J. Bone Miner. Res.* **12**(10): 1606–1614.

Harris, M.T., Butler, D.L., Boivin, G.P., Florer, J.B., Schantz, E.J. and Wenstrup, R.J. (2004). Mesenchymal stem cells used for rabbit tendon repair can form ectopic bone and express alkaline phosphatase activity in constructs. *J. Orthop. Res.* **22**(5): 998–1003.

Hasharoni, A., Zilberman, Y., Turgeman, G., Helm, G.A., Liebergall, M. and Gazit, D. (2005). Murine spinal fusion induced by engineered mesenchymal stem cells that conditionally express bone morphogenetic protein-2. *J. Neurosurg. Spine* **3**(1): 47–52.

Haynesworth, S.E., Goshima, J., Goldberg, V.M. and Caplan, A.I. (1992). Characterization of cells with osteogenic potential from human marrow. *Bone* **13**(1): 81–88.

Hoffmann, A., Czichos, S., Kaps, C., Bachner, D., Mayer, H., Kurkalli, B.G., Zilberman, Y., Turgeman, G., Pelled, G., Gross, G. and Gazit, D. (2002). The T-box transcription factor Brachyury mediates cartilage development in mesenchymal stem cell line C3H10T1/2. *J. Cell Sci.* **115**(Pt 4): 769–781.

Hoffmann, A., Pelled, G., Turgeman, G., Eberle, P., Zilberman, Y., Shinar, H., Keinan-Adamsky, K., Winkel, A., Shahab, S., Navon, G., Gross, G. and Gazit, D. (2006). Neotendon formation induced by manipulation of the Smad8 signaling pathway in mesenchymal stem cells. *J. Clin. Invest.* **116**(4): 940–952.

Hofstetter, C.P., Schwarzt, E.J., Hess, D., Widenfalk, J., El Manira, A., Prockop, D.J. and Olson, L. (2002). Marrow stromal cells form guiding strands in the injured spinal cord and promote recovery. *Proc. Natl Acad. Sci. USA* **99**(4): 2199–2204.

Hong, S.H., Gang, E.J., Jeong, J.A., Ahn, C., Hwang, S.H., Yang, I.H., Park, H.K., Han, H. and Kim, H. (2005). *In vitro* differentiation of human umbilical cord blood-derived mesenchymal stem cells into hepatocyte-like cells. *Biochem. Biophys. Res. Commun.* **330**(4): 1153–1161.

Honigman, A., Zeira, E., Ohana, P., Abramovitz, R., Tavor, E., Bar, I., Zilberman, Y., Rabinovsky, R., Gazit, D., Joseph, A., Panet, A., Shai, E., Palmon, A., Laster, M. and Galun, E. (2001). Imaging transgene expression in live animals. *Mol. Ther.* **4**(3): 239–249.

Horwitz, E.M., Prockop, D.J., Fitzpatrick, L.A., Koo, W.W.,. Gordon, P.L., Neel, M., Sussman, M., Orchard, P., Marx, J.C., Pyeritz, R.E. and Brenner, M.K. (1999). Transplantability and therapeutic effects of bone marrow-derived mesenchymal cells in children with osteogenesis imperfecta. *Nat. Med.* **5**(3): 309–313.

Horwitz, E.M., Prockop, D.J., Gordon, P.L., Koo, W.W., Fitzpatrick, L.A., Neel, M.D., McCarville, M.E., Orchard, P.J., Pyeritz, R.E. and Brenner, M.K. (2001). Clinical responses to bone marrow transplantation in children with severe osteogenesis imperfecta. *Blood* **97**(5): 1227–1231.

Hutson, E.L., Boyer, S. and Genever, P.G. (2005). Rapid isolation, expansion, and differentiation of osteoprogenitors from full-term umbilical cord blood. *Tissue Eng.* **11**(9–10): 1407–1420.

Im, G.I., Kim, D.Y., Shin, J.H., Hyun, C.W. and Cho, W.H. (2001). Repair of cartilage defect in the rabbit with cultured mesenchymal stem cells from bone marrow. *J. Bone Joint Surg. Br.* **83**(2): 289–294.

Iris, B., Zilberman, Y., Zeira, E., Galun, E., Honigman, A., Turgeman, G., Clemens, T., Gazit, Z. and Gazit, D. (2003). Molecular imaging of the skeleton: quantitative real-time bioluminescence monitoring gene expression in bone repair and development. *J. Bone Miner. Res.* **18**(3): 570–578.

Itescu, S., Schuster, M.D. and Kocher, A.A. (2003). New directions in strategies using cell therapy for heart disease. *J. Mol. Med.* **81**(5): 288–296.

Jain, M., Pfister, O., Hajjar, R.J. and Liao, R. (2005). Mesenchymal stem cells in the infarcted heart. *Coron. Artery Dis.* **16**(2): 93–97.

Javazon, E.H., Beggs, K.J. and Flake, A.W. (2004). Mesenchymal stem cells: paradoxes of passaging. *Exp. Hematol.* **32**(5): 414–425.

Jeong, J.A., Hong, S.H., Gang, E.J., Ahn, C., Hwang, S.H., Yang, I.H., Han, H. and Kim, H. (2005). Differential gene expression profiling of human umbilical cord blood-derived mesenchymal stem cells by DNA microarray. *Stem Cells* **23**(4): 584–593.

Jiang, Y., Jahagirdar, B.N., Reinhardt, R.L., Schwartz, R.E., Keene, C.D., Ortiz-Gonzalez, X.R., Reyes, M., Lenvik, T., Lund, T., Blackstad, M., Du, J., Aldrich, S., Lisberg, A., Low, W.C., Largaespada, D.A. and Verfaillie, C.M. (2002). Pluripotency of mesenchymal stem cells derived from adult marrow. *Nature* **418**(6893): 41–49.

Juncosa-Melvin, N., Boivin, G.P., Galloway, M.T., Gooch, C., West, J. R., Sklenka, A.M. and Butler, D.L. (2005). Effects of cell-to-collagen ratio in mesenchymal stem cell-seeded implants on tendon repair biomechanics and histology. *Tissue Eng.* **11**(3–4): 448–457.

Kadiyala, S., Young, R.G., Thiede, M.A. and Bruder, S.P. (1997). Culture expanded canine mesenchymal stem cells possess osteochondrogenic potential *in vivo* and *in vitro*. *Cell Transplant.* **6**(2): 125–134.

Kang, X.Q., Zang, W.J., Song, T.S., Xu, X.L., Yu, X.J., Li, D.L., Meng, K.W., Wu, S.L. and Zhao, Z.Y. (2005). Rat bone marrow mesenchymal stem cells differentiate into hepatocytes *in vitro*. *World J. Gastroenterol.* **11**(22): 3479–3484.

Katz, A.J., Tholpady, A., Tholpady, S.S., Shang, H. and Ogle, R.C. (2005). Cell surface and transcriptional characterization of human adipose-derived adherent stromal (hADAS) cells. *Stem Cells* **23**(3): 412–423.

Kawada, H., Fujita, J., Kinjo, K., Matsuzaki, Y., Tsuma, M., Miyatake, H., Muguruma, Y., Tsuboi, K., Itabashi, Y., Ikeda, Y., Ogawa, S., Okano, H., Hotta, T., Ando, K. and Fukuda, K. (2004). Nonhematopoietic mesenchymal stem cells can be mobilized and differentiate into cardiomyocytes after myocardial infarction. *Blood* **104**(12): 3581–3587.

Kawamura, K., Chu, C.R., Sobajima, S., Robbins, P.D., Fu, F.H., Izzo, N.J. and Niyibizi, C. (2005). Adenoviral-mediated transfer of TGF-beta1 but not IGF-1 induces chondrogenic differentiation of human mesenchymal stem cells in pellet cultures. *Exp. Hematol.* **33**(8): 865–872.

Kern, S., Eichler, H., Stoeve, J., Kluter, H. and Bieback, K. (2006). Comparative analysis of mesenchymal stem cells from bone marrow, umbilical cord blood or adipose tissue. *Stem Cells* **24**(5): 1294–1301.

Kim, B.K., Chung, K.W., Sun, H.S., Suh, J.G., Min, W.S., Kang, C.S., Sim, S.I., Shin, W.S. and Kim, C.C. (2000). Liver disease during the first post-transplant year in bone marrow transplantation recipients: retrospective study. *Bone Marrow Transplant.* **26**(2): 193–197.

Kim, S.Y., Lee, S.H., Kim, B.M., Kim, E.H., Min, B.H. Bendayan M. and Park, I.S. (2004). Activation of nestin-positive duct stem (NPDS) cells in pancreas upon neogenic motivation and possible cytodifferentiation into insulin-secreting cells from NPDS cells. *Dev. Dyn.* **230**(1): 1–11.

Kimelman, N., Oskouian, R.J., Zilberman, Y., Pelled,G., Gazit, Z. and Gazit, D. (2006). Genetically engineered adult stem cells and hybrid scaffolds as a platform for intervertebral disc regeneration. *Mol. Ther.* (ASGT abstracts).

Koc, O.N., Day, J., Nieder, M., Gerson, S.L., Lazarus, H.M. and Krivit, W. (2002). Allogeneic mesenchymal stem cell infusion for treatment of metachromatic leukodystrophy (MLD) and Hurler syndrome (MPS-IH). *Bone Marrow Transplant.* **30**(4): 215–22.

Kon, E., Muraglia, A., Corsi, A., Bianco, P., Marcacci, M., Martin, I., Boyde, A., Ruspantini, I., Chistolini, P., Rocca, M., Giardino, R., Cancedda, R. and Quarto, R. (2000). Autologous bone marrow stromal cells loaded onto porous hydroxyapatite ceramic accelerate bone repair in critical-size defects of sheep long bones. *J. Biomed. Mater. Res.* **49**(3): 328–337.

Lange, C., Bassler, P., Lioznov, M.V., Bruns, H., Kluth, D., Zander, A.R. and Fiegel, H.C. (2005a). Hepatocytic gene expression in cultured rat mesenchymal stem cells. *Transplant Proc.* **37**(1): 276–279.

Lange, C., Bassler, P., Lioznov, M.V., Bruns, H., Kluth, D., Zander, A.R. and Fiegel, H.C. (2005b). Liver-specific gene expression in mesenchymal stem cells is induced by liver cells. *World J. Gastroenterol.* **11**(29): 4497–4504.

Lardon, J., Rooman, I. and Bouwens, L. (2002). Nestin expression in pancreatic stellate cells and angiogenic endothelial cells. *Histochem. Cell Biol.* **117**(6): 535–540.

Laurencin, C.T., Attawia, M.A., Lu, L.Q., Borden, M.D., Lu, H.H., Gorum, W.J. and Lieberman, J.R. (2001). Poly(lactide-co-glycolide)/hydroxyapatite delivery of BMP-2-producing cells: a regional gene therapy approach to bone regeneration. *Biomaterials* **22**(11): 1271–1277.

Le Blanc, K., Rasmusson, I., Sundberg, B., Gotherstrom, C., Hassan, M., Uzunel, M. and Ringden, O. (2004). Treatment of severe acute graft-versus-host disease with third party haploidentical mesenchymal stem cells. *Lancet* **363**(9419): 1439–1441.

Le Blanc, K., Gotherstrom, C., Ringden, O., Hassan, M., McMahon, R., Horwitz, E., Anneren,G., Axelsson, O., Nunn, J., Ewald, U., Norden-Lindeberg, S., Jansson, M., Dalton, A., Astrom, E. and Westgren, M. (2005). Fetal mesenchymal stem-cell engraftment in bone after *in utero* transplantation in a patient with severe osteogenesis imperfecta. *Transplantation* **79**(11): 1607–1614.

Lecanda, F., Avioli, L.V. and Cheng, S.L. (1997). Regulation of bone matrix protein expression and induction of differentiation of human osteoblasts and human bone marrow stromal cells by bone morphogenetic protein-2. *J. Cell Biochem.* **67**(3): 386–396.

Lee, K., Majumdar, M.K., Buyaner, D., Hendricks, J.K., Pittenger, M.F. and Mosca, J.D. (2001). Human mesenchymal stem cells maintain transgene expression during expansion and differentiation. *Mol. Ther.* **3**(6): 857–866.

Lee, K.D., Kuo, T.K., Whang-Peng, J., Chung, Y.F., Lin, C.T., Chou, S.H., Chen, J.R., Chen, Y.P. and Lee, O.K. (2004). *In vitro* hepatic differentiation of human mesenchymal stem cells. *Hepatology* **40**(6): 1275–1284.

Li, Y., Chen, J., Wang, L., Zhang, L., Lu, M. and Chopp, M. (2001). Intracerebral transplantation of bone marrow stromal cells in a 1-methyl-4-phenyl-1,2,3,6-tetrahydropyridine mouse model of Parkinson's disease. *Neurosci. Lett.* **316**(2): 67–70.

Liechty, K.W., MacKenzie, T.C., Shaaban, A.F., Radu, A., Moseley, A.M., Deans, R., Marshak, D.R. and Flake, A.W. (2000). Human mesenchymal stem cells engraft and demonstrate site-specific differentiation after *in utero* transplantation in sheep. *Nat. Med.* **6**(11): 1282–1286.

Liu, Y., Chen, F., Liu, W., Cui, L., Shang, Q., Xia, W., Wang, J., Cui, Y., Yang, G., Liu, D., Wu, J., Xu, R., Buonocore, S.D. and Cao, Y. (2002). Repairing large porcine full-thickness defects of articular cartilage using autologous chondrocyte-engineered cartilage. *Tissue Eng.* **8**(4): 709–721.

Locasciulli, A., Testa, M., Valsecchi, M.G., Vecchi, L., Longoni, D., Sparano, P., Rovelli, A., Uderzo, C., Masera, G. and Alberti, A. (1997). Morbidity and mortality due to liver disease in children undergoing allogeneic bone marrow transplantation: a 10-year prospective study. *Blood* **90**(9): 3799–3805.

Lu, D., Mahmood, A., Wang, L., Li, Y., Lu, M. and Chopp, M. (2001). Adult bone marrow stromal cells administered intravenously to rats after traumatic brain injury migrate into brain and improve neurological outcome. *NeuroReport* **12**(3): 559–563.

MacDonald, D.J., Luo, J. Saito, T., Duong, M., Bernier, P.L., Chiu, R.C. and Shum-Tim, D. (2005). Persistence of marrow stromal cells implanted into acutely infarcted myocardium: observations in a xenotransplant model. *J. Thorac. Cardiovasc. Surg.* **130**(4): 1114–1121.

Mackay, A.M., Beck, S.C., Murphy, J.M., Barry, F.P., Chichester, C.O. and Pittenger, M.F. (1998). Chondrogenic differentiation of cultured human mesenchymal stem cells from marrow. *Tissue Eng.* **4**(4): 415–428.

Mahmood, A., Lu, D., Lu, M., Chopp, M., Amar, A.P., Brawanski, A. and Valadka, A.B. (2003). Treatment of traumatic brain injury in adult rats with intravenous administration of human bone marrow stromal cells. *Neurosurgery* **53**(3): 697–703.

Majumdar, M.K., Banks, V., Peluso, D.P. and Morris, E.A. (2000). Isolation, characterization, and chondrogenic potential of human bone marrow-derived multipotential stromal cells. *J. Cell Physiol.* **185**(1): 98–106.

Majumdar, M.K., Keane-Moore, M., Buyaner, D., Hardy, W.B., Moorman, M.A., McIntosh, K.R. and Mosca, J.D. (2003). Characterization and functionality of cell surface molecules on human mesenchymal stem cells. *J. Biomed. Sci.* **10**(2): 228–241.

Makino, S., Fukuda, K., Miyoshi, S., Konishi, F., Kodama, H., Pan, J., Sano, M., Takahashi, T., Hori, S. Abe, H.. Hata, J.-i., Umezawa, A. and Ogawa, S. (1999). Cardiomyocytes can be generated from marrow stromal cells *in vitro*. *J. Clin. Invest.* **103**(5): 697–705.

Mangi, A.A., Noiseux, N., Kong, D., He, H., Rezvani, M. Ingwall, J.S. and Dzau, V.J. (2003). Mesenchymal stem cells modified with Akt prevent remodeling and restore performance of infarcted hearts. *Nat. Med.* **9**(9): 1195–1201.

Mankani, M.H., Kuznetsov, S.A., Fowler, B., Kingman, A. and Robey, P.G. (2001). *In vivo* bone formation by human bone marrow stromal cells: effect of carrier particle size and shape. *Biotechnol. Bioeng.* **72**(1): 96–107.

Marshall, C.J., Moore, R.L., Thorogood, P., Brickell, P.M., Kinnon, C. and Thrasher, A.J. (1999). Detailed characterization of the human aorta-gonad-mesonephros region reveals morphological polarity resembling a hematopoietic stromal layer. *Dev. Dyn.* **215**(2): 139–147.

Mastrogiacomo, M., Muraglia, A., Komlev, V., Peyrin, F., Rustichelli, F., Crovace, A. and Cancedda, R. (2005). Tissue engineering of bone: search for a better scaffold. *Orthod. Craniofac. Res.* **8**(4): 277–284.

Matsubara, T., Suardita, K., Ishii, M., Sugiyama, M., Igarashi, A., Oda, R., Nishimura, M., Saito, M., Nakagawa, K., Yamanaka, K., Miyazaki, K., Shimizu, M., Bhawal, U.K., Tsuji, K., Nakamura, K. and Kato, Y. (2005). Alveolar bone marrow as a cell source for regenerative medicine: differences between alveolar and iliac bone marrow stromal cells. *J. Bone Miner. Res.* **20**(3): 399–409.

Minamino, K., Adachi, Y., Okigaki, M., Ito, H., Togawa, Y., Fujita, K., Tomita, M., Suzuki, Y., Zhang, Y., Iwasaki, M., Nakano, K., Koike, Y., Matsubara, H., Iwasaka, T., Matsumura, M. and Ikehara, S. (2005). Macrophage colony-stimulating factor (M-CSF), as well as granulocyte colony-stimulating factor (G-CSF), accelerates neovascularization. *Stem Cells* **23**(3): 347–354.

Minguell, J.J. and Erices, A. (2006). Mesenchymal stem cells and the treatment of cardiac disease. *Exp. Biol. Med. (Maywood)* **231**(1): 39–49.

Mizuno, H. and Hyakusoku, H. (2003). Mesengenic potential and future clinical perspective of human processed lipoaspirate cells. *J. Nippon Med. Sch.* **70**(4): 300–306.

Moon, Y.J., Lee, M.W., Yang, M.S., Kim, S.K., Park, J.S., Kim, H.C., Kim, H.S., Lee, K.H., Kim, Y.J. and Choi, J. (2005). Expression profile of genes representing varied spectra of cell lineages in human umbilical cord blood-derived mesenchymal stem cells. *Acta Haematol.* **114**(2): 117–120.

Moriscot, C., de Fraipont, F., Richard, M.J., Marchand, M., Savatier, P., Bosco, D., Favrot, M. and Benhamou, P.Y. (2005). Human bone marrow mesenchymal stem cells can express insulin and key transcription factors of the endocrine pancreas developmental pathway upon genetic and/or microenvironmental manipulation *in vitro. Stem Cells* **23**(4): 594–603.

Moutsatsos, I.K., Turgeman, G., Zhou, S., Kurkalli, B.G., Pelled, G., Tzur, L., Kelley, P., Stumm, N., Mi, S., Muller, R., Zilberman, Y. and Gazit, D. (2001). Exogenously regulated stem cell-mediated gene therapy for bone regeneration. *Mol. Ther.* **3**(4): 449–461.

Muschler, G.F., Nitto, H., Matsukura, Y., Boehm, C., Valdevit, A., Kambic, H., Davros, W., Powell, K. and Easley, K. (2003). Spine fusion using cell matrix composites enriched in bone marrow-derived cells. *Clin. Orthop. Relat. Res.* **407**: 102–118.

Nagaya, N., Kangawa, K., Itoh, T., Iwase, T., Murakami, S., Miyahara, Y., Fujii, T., Uematsu, M., Ohgushi, H., Yamagishi, M., Tokudome, T., Mori, H., Miyatake, K. and Kitamura, S. (2005). Transplantation of mesenchymal stem cells improves cardiac function in a rat model of dilated cardiomyopathy. *Circulation* **112**(8): 1128–1135.

Okuyama, R., Yanai, N. and Obinata, M. (1995). Differentiation capacity toward mesenchymal cell lineages of bone marrow stromal cells established from temperature-sensitive SV40 T-antigen gene transgenic mouse. *Exp. Cell Res.* **218**(2): 424–429.

Oreffo, R.O., Virdi, A.S. and Triffitt, J.T. (2001). Retroviral marking of human bone marrow fibroblasts: *in vitro* expansion and localization in calvarial sites after subcutaneous transplantation *in vivo. J. Cell Physiol.* **186**(2): 201–209.

Orlic, D. (2003). Adult bone marrow stem cells regenerate myocardium in ischemic heart disease. *Ann. NY Acad. Sci.* **996**: 152–157.

Oswald, J., Boxberger, S., Jorgensen, B., Feldmann, S., Ehninger, G., Bornhauser, M. and Werner, C. (2004). Mesenchymal stem cells can be differentiated into endothelial cells *in vitro. Stem Cells* **22**(3): 377–384.

Oyagi, S., Hirose, M., Kojima, M., Okuyama, M., Kawase, M., Nakamura, T., Ohgushi, H. and Yagi, K. (2006). Therapeutic effect of transplanting HGF-treated bone marrow mesenchymal cells into CCl(4)-injured rats. *J. Hepatol* **44**(4): 742–748.

Palmer, G.D., Steinert, A., Pascher, A., Gouze, E., Gouze, J.N., Betz, O., Johnstone, B., Evans, C.H. and Ghivizzani, S.C. (2005). Gene-induced chondrogenesis of primary mesenchymal stem cells *in vitro. Mol. Ther.* **12**(2): 219–28.

Pelled, G., Tai, K., Sheyn, D., Zilberman, Y., Kumbar, S.S., Nair, L.S., Laurencin, C.T., Gazit, D. and Ortiz, C. (2006). Structural and nanoindentation studies of stem cell-based tissue-engineered bone. *J. Biomech.* **40**(2): 399–411.

Peng, H., Wright, V., Usas, A., Gearhart, B., Shen, H.C., Cummins, J. and Huard, J. (2002). Synergistic enhancement of bone formation and healing by stem cell-expressed VEGF and bone morphogenetic protein-4. *J. Clin. Invest.* **110**(6): 751–759.

Peterson, B., Zhang, J. Iglesias, R. Kabo, M. Hedrick, M. Benhaim P., and Lieberman, J.R. (2005). Healing of critically sized femoral defects, using genetically modified mesenchymal stem cells from human adipose tissue. *Tissue Eng.* **11**(1–2): 120–129.

Pittenger, M.F., Mackay, A.M., Beck, S.C., Jaiswal, R.K., Douglas, R. Mosca, J.D., Moorman, M.A., Simonetti, D.W., Craig, S. and Marshak, D.R. (1999). Multilineage potential of adult human mesenchymal stem cells. *Science* **284**(5411): 143–147.

Prockop, D.J. (1997). Marrow stromal cells as stem cells for nonhematopoietic tissues. *Science* **276**(5309): 71–74.

Quarto, R., Mastrogiacomo, M., Cancedda, R., Kutepov, S.M., Mukhachev, V., Lavroukov, A., Kon, E. and Marcacci, M. (2001). Repair of large bone defects with the use of autologous bone marrow stromal cells. *N. Engl. J. Med.* **344**(5): 385–386.

Rasmusson, I., Ringden, O., Sundberg, B. and Le Blanc, K. (2005). Mesenchymal stem cells inhibit lymphocyte proliferation by mitogens and alloantigens by different mechanisms. *Exp. Cell Res.* **305**(1): 33–41.

Reyes, M., Lund, T. Lenvik, T. Aguiar, D. Koodie, L. and Verfaillie, C.M. (2001). Purification and ex vivo expansion of postnatal human marrow mesodermal progenitor cells. *Blood* **98**(9): 2615–2625.

Reyes, M., Dudek, A., Jahagirdar, B., Koodie, L., Marker, P.H. and Verfaillie, C.M. (2002). Origin of endothelial progenitors in human postnatal bone marrow. *J. Clin. Invest.* **109**(3): 337–346.

Rombouts, W.J. and Ploemacher, R.E. (2003). Primary murine MSC show highly efficient homing to the bone marrow but lose homing ability following culture. *Leukemia* **17**(1): 160–170.

Roughley, P.J. (2004). Biology of intervertebral disc aging and degeneration: involvement of the extracellular matrix. *Spine* **29**(23): 2691–2699.

Rubio, D., Garcia-Castro, J., Martin, M.C., de la Fuente, R., Cigudosa, J.C., Lloyd, A.C. and Bernad, A. (2005). Spontaneous human adult stem cell transformation. *Cancer Res.* **65**(8): 3035–3039.

Sakai, D., Mochida, J., Iwashina, T., Watanabe, T., Nakai, T., Ando, K. and Hotta, T. (2005). Differentiation of mesenchymal stem cells transplanted to a rabbit degenerative disc model: potential and limitations for stem cell therapy in disc regeneration. *Spine* **30**(21): 2379–2387.

Sakai, D., Mochida, J., Iwashina, T., Hiyama, A., Omi, H., Imai, M., Nakai, T., Ando, K. and Hotta, T. (2006). Regenerative effects of transplanting mesenchymal stem cells embedded in atelocollagen to the degenerated intervertebral disc. *Biomaterials* **27**(3): 335–345.

Sato, Y., Araki, H., Kato, J., Nakamura, K., Kawano, Y., Kobune, M., Sato, T., Miyanishi, K., Takayama, T., Takahashi, M., Takimoto, R. Iyama, S., Matsunaga, T., Ohtani, S., Matsuura, A., Hamada, H. and Niitsu, Y. (2005). Human mesenchymal stem cells xenografted directly to rat liver are differentiated into human hepatocytes without fusion. *Blood* **106**(2): 756–763.

Schwartz, R.E., Reyes, M., Koodie, L., Jiang, Y., Blackstad, M., Lund, T., Lenvik, T., Johnson, S., Hu, W.S. and Verfaillie, C.M. (2002). Multipotent adult progenitor cells from bone marrow differentiate into functional hepatocyte-like cells. *J. Clin. Invest.* **109**(10): 1291–1302.

Siepe, M., Heilmann, C., von Samson, P., Menasche, P. and Beyersdorf, F. (2005). Stem cell research and cell transplantation for myocardial regeneration. *Eur. J. Cardiothorac. Surg.* **28**(2): 318–324.

Silva, G.V., Litovsky, S., Assad, J.A., Sousa, A.L., Martin, B.J., Vela, D., Coulter, S.C., Lin, J., Ober, J., Vaughn, W.K., Branco, R.V., Oliveira, E.M., He, R., Geng, Y.J., Willerson, J.T. and Perin, E.C. (2005). Mesenchymal stem cells differentiate into an endothelial phenotype, enhance vascular density, and improve heart function in a canine chronic ischemia model. *Circulation* **111**(2): 150–156.

Simmons, P.J. and Torok-Storb, B. (1991). Identification of stromal cell precursors in human bone marrow by a novel monoclonal antibody, STRO-1. *Blood* **78**(1): 55–62.

Smits, A.M., van Vliet, P., Hassink, R.J., Goumans, M.J. and Doevendans, P.A. (2005). The role of stem cells in cardiac regeneration. *J. Cell Mol. Med.* **9**(1): 25–36.

Sonoyama, W., Coppe, C., Gronthos, S. and Shi, S. (2005). Skeletal stem cells in regenerative medicine. *Curr. Top. Dev. Biol.* **67**: 305–323.

Sordi, V., Malosio, M.L., Marchesi, F., Mercalli, A., Melzi, R., Giordano, T., Belmonte, N., Ferrari, G., Leone, B.E., Bertuzzi, F., Zerbini, G., Allavena, P., Bonifacio, E. and Piemonti, L. (2005). Bone marrow mesenchymal stem cells express a restricted set of functionally active chemokine receptors capable of promoting migration to pancreatic islets. *Blood* **106**(2): 419–427.

Sotiropoulou, P.A., Perez, S.A., Gritzapis, A.D., Baxevanis, C.N. and Papamichail, M. (2006). Interactions between human mesenchymal stem cells and natural killer cells. *Stem Cells* **24**(1): 74–85.

Takizawa, S. (2003). [Differentiation of adult bone marrow cells into neurons and endothelial cells in rat brain after stroke in the presence of cytokines]. *Rinsho Shinkeigaku* **43**(11): 830–831.

Tatebe, M., Nakamura, R., Kagami, H., Okada, K. and Ueda, M. (2005). Differentiation of transplanted mesenchymal stem cells in a large osteochondral defect in rabbit. *Cytotherapy* **7**(6): 520–530.

Tavian, M., Hallais, M.F. and Peault, B. (1999). Emergence of intraembryonic hematopoietic precursors in the pre-liver human embryo. *Development* **126**(4): 793–803.

Toma, C., Pittenger, M.F., Cahill, K.S., Byrne, B.J. and Kessler, P.D. (2002). Human mesenchymal stem cells differentiate to a cardiomyocyte phenotype in the adult murine heart. *Circulation* **105**(1): 93–98.

Tse, W.T., Pendleton, J.D., Beyer, W.M., Egalka, M.C. and Guinan, E.C. (2003). Suppression of allogeneic T-cell proliferation by human marrow stromal cells: implications in transplantation. *Transplantation* **75**(3): 389–397.

Turgeman, G., Pittman, D.D., Muller, R., Kurkalli, B.G., Zhou, S., Pelled, G., Peyser, A., Zilberman, Y., Moutsatsos, I.K. and Gazit, D. (2001). Engineered human mesenchymal stem cells: a novel platform for skeletal cell mediated gene therapy. *J. Gene Med.* **3**(3): 240–251.

Valentin-Opran, A., Wozney, J., Csimma, C., Lilly, L. and Riedel, G.E. (2002). Clinical evaluation of recombinant human bone morphogenetic protein-2. *Clin. Orthop. Relat. Res.* (395): 110–120.

Verfaillie, C.M., Schwartz, R., Reyes, M. and Jiang, Y. (2003). Unexpected potential of adult stem cells. *Ann. NY. Acad. Sci.* **996**: 231–234.

Volek-Smith, H. and Urist, M.R. (1996). Recombinant human bone morphogenetic protein (rhBMP) induced heterotopic bone development *in vivo* and *in vitro*. *Proc. Soc. Exp. Biol. Med.* **211**(3): 265–272.

12

Islet Cell Therapy and Pancreatic Stem Cells

Juan Domínguez-Bendala, Antonello Pileggi, and Camillo Ricordi

INTRODUCTION

Replacement of insulin-producing cell function represents an appealing approach for the treatment of diabetes mellitus, a condition characterized by loss of β-cell mass and/or function (Ricordi, 2003; Ricordi and Strom, 2004) consequent to autoimmunity (**type 1 diabetes mellitus**, T1DM), **metabolic disorders** (i.e. cystic fibrosis, hemochromatosis, and liver cirrhosis), surgery (i.e. **iatrogenic diabetes** following pancreatectomy for relapsing, chronic pancreatitis) (Ricordi, 2003), or β-cell dysfunction secondary to insulin resistance and hyperinsulinism (**type 2 diabetes mellitus**, T2DM). Exogenous insulin injections have represented a life-saving treatment in T1DM, changing the natural history of diabetes and remarkable progress has been achieved in recent years in the management of glycemic control by combining diet, exercise, and improved exogenous insulin treatment options. However, this approach requires continuous adjustments in insulin administrations with significant challenges in attaining tight glycemic control in the absence of severe hypoglycemic episodes. Tight metabolic control with avoidance of wide glycemic excursions is necessary to decrease the risk of development and/or progression of the chronic complications that can negatively impact the quality of life and life expectancy of patients with diabetes.

Hundreds of thousands endocrine cell clusters, from <50 to >500 μm in diameter (islets of Langerhans) are scattered into the pancreatic tissue, representing approximately 1–2% of the entire organ. The islets are "micro-organs" with a unique cytoarchitecture, composed of heterogeneous cell subsets specialized in the production and secretion of endocrine hormones (α-cells for glucagon; β-cells for insulin; δ-cells for somatostatin; PP-cells for pancreatic polypeptide) that are essential for the regulation of glucose homeostasis in the blood (Brissova et al., 2005; Cabrera et al., 2006). Complex interactions between the cell subsets composing the islets, their innervation and the rich vascular bed result in "real-time" secretion of endocrine hormones that maintain glycemic values within physiologic ranges. Better understanding of pancreatic islet cell ontogeny, biology, and physiology will be of assistance in developing efficient protocols for cellular therapies for the restoration of metabolic control in patients with diabetes.

Considerable progress has been achieved in the last two decades in the field of β-cell replacement therapy, either by transplantation of the pancreas as a vascularized organ or by infusion of islet cell products. The encouraging results of recent clinical trials support the value of this approach, which has been shown to improve both quality of life and metabolic control in patients with T1DM following intrahepatic islet transplantation (Ryan et al., 2001, 2002, 2005; Froud et al., 2005a, b; Pileggi et al., 2005, 2006; Poggioli et al., 2006). Current challenges to the widespread application of β-cell replacement therapies include the shortage of transplantable tissue and the need for more effective and safer immune interventions that favor long-term graft function. Ultimately, successful strategies for immunoisolation, tissue engineering with local

Copyright © 2008 Elsevier Inc. All rights reserved

immunosuppression, or the development of successful strategies for tolerance induction to avoid the need for life-long immunosuppression of the recipients will be necessary for the widespread applicability of islet cell therapy. In fact, the current requirements for life-long immunosuppression of the recipients severely limit the current indications for islet transplantation to the most severe cases of T1DM or in patients already undergoing organ transplantation and therefore already undergoing immunosuppressive treatment (Pileggi et al., 2001; Ricordi and Strom, 2004). When islet transplantation will be possible without chronic recipient immunosuppression, current sources of donor pancreata will clearly be insufficient to meet the demand. That is why it is so critical to continue to work toward the identification of alternative sources of insulin-producing cells. Encouraging data are emerging in the field of islet cell neogenesis and stem cell research that justify a cautious optimism for the years to come (Ricordi et al., 2005; Pileggi et al., 2006).

This chapter will review the current status, challenges, and perspectives in clinical islet transplantation for treatment of diabetes and the progress of selected areas of stem cell and β-cell regeneration.

BENEFITS OF β-CELL REPLACEMENT THERAPY

Transplantation of β-cells is currently performed as vascularized pancreas or isolated islet cell grafts. Both procedures can result in improved glycemic control in patients with diabetes (Pileggi et al., 2006). Transplantation of pancreatic islets offers substantial advantage over whole pancreas transplantation because of the lower risks for procedure-related complications and the possibility of preconditioning the graft *in vitro* prior to implantation (Pileggi et al., 2006).

Islets are isolated from the donor pancreas by a mechanically enhanced, enzymatic digestion process that allows for the physical dissociation of pancreatic tissue into small fragments and liberation of the endocrine cell clusters with preserved integrity (Ricordi et al., 1988). The dissociation phase is followed by purification on density gradients that enriches for fractions with higher endocrine cell clusters (~2% of the whole pancreatic tissue) while minimizing contamination with non-endocrine tissue (Alejandro et al., 1990; Ichii et al., 2005b). After isolation and culture, fractions with different degrees of purity are pooled for transplantation.

Islet transplantation is performed using minimally invasive interventional radiology techniques consisting of percutaneous, transhepatic cannulation of the portal vein, and infusion of the islets by gravity. After intraportal infusion, the islet cell clusters remain trapped at the presinusoidal level (Alejandro and Mintz, 1988; Baidal et al., 2003; Froud et al., 2004; Pileggi et al., 2005). The purification procedure allows to substantially reduce the volume of tissue to be infused, therefore minimizing the previously reported risk of portal thrombosis and portal hypertension consequent to the intrahepatic embolization of the islet grafts (Froud et al., 2004, 2005), which has been described when unpurified islet preparations or inadequate islet infusion techniques were used.

Islet transplantation is indicated in patients who have lost insulin-producing cell function. Recent clinical trials have shown the importance of intensive insulin treatment to obtain tight glycemic control and its ability to prevent or delay the dreadful complications of unstable glycemic control, including neuropathy, vasculopathy, and nephropathy (DCCT, 1993). Unfortunately, intensive insulin treatment cannot maintain glycemic values within normal ranges throughout the day and is associated with an increased risk of severe hypoglycemia, at times fatal. Restoration of islet β-cell function is a highly desirable goal for patients with T1DM as it can provide a more physiological glycemic metabolic control than exogenous insulin.

Transplantation of **autologous islets** (**autotransplantation**) is generally performed to prevent iatrogenic diabetes in patients undergoing pancreatectomy due to severe pain for chronic, relapsing pancreatitis, or

It is conceivable that this approach will not suffice overcoming the increasing demand of islet grafts due to the disproportioned pancreas donor-to-recipient ratio: there will be a large number of patients with diabetes who would benefit of restoration of β-cell function and not sufficient pancreata for processing. Alternative sources of insulin-producing β-cells (from either allogeneic or xenogeneic donors) or induction of self-regeneration of the patient's own β-cells (in combination with adequate immunomodulation to prevent recurrence of autoimmunity) (Ogawa et al., 2004) may help achieving the desired metabolic control in the near future (Ricordi et al., 2005). For β-cell replacement therapies to become the treatment of choice for patients with diabetes, successful restoration of metabolic function should be achieved long term. For this reason, implementing a sequential, integrated approach that combines strategies aiming at improving β-cell mass together with those focusing on the modulation of the immune response (i.e. preventing rejection and recurrence of autoimmunity) could represent an essential element toward definition of successful therapeutic strategies (Ricordi and Strom, 2004; Ricordi et al., 2005). Promising data on the induction of donor-specific unresponsiveness and tolerance to transplanted tissues in experimental models justify optimism for the near future, and may allow achieving long-term function of transplanted insulin-producing cells in the absence of rejection and recurrence of autoimmunity without the need for chronic immunosuppression in the clinical setting (Inverardi and Ricordi, 2001; Inverardi et al., 2004; Ricordi and Strom, 2004).

ALTERNATIVE SOURCES OF INSULIN-PRODUCING CELLS: STEM CELLS AND β-CELL REGENERATION

Stem cells could be defined as **undifferentiated cells that have the ability to proliferate while retaining the potential to fully mature into other cell types**. The extent to which stem cells can be induced to proliferate or differentiate depends on their origin and stage of development. Arguably, the most powerful stem cells available are **embryonic stem (ES) cells**. These cells, which are obtained from the inner cell mass (ICM) of the blastocyst, can be maintained indefinitely in an undifferentiated, proliferative stage *in vitro* (Odorico et al., 2001; Thomson et al., 2005; Thomson et al., 1998). When transplanted into immunodeficient animals or otherwise induced to spontaneously differentiate, they can give rise to cells of all three embryonal layers (endoderm, ectoderm, and mesoderm). The prospects of turning human ES (huES) cells into islet cell types are therefore substantiated, but not exempt of safety and ethical concerns.

Stem cells of fetal origin may still retain some degree of multilineage differentiation, as well as the potential to proliferate *in vitro*. Despite their embryonic origin, these cells should not be confused with the blastocyst-derived ES cells. In fact, in many respects, these cells are more akin to adult cell types than to ES cells. This, together with the controversy surrounding their procurement, makes them unlikely candidates to become an alternative source of islets.

Expansion of fully differentiated, adult β-cells has been reported *in vitro*. However, the induction of β-cell proliferation has been generally associated with loss of mature cell phenotype and of functional competence that is only partially recovered after re-differentiation. Many **adult tissues** have also stem cells involved in their physiologic maintenance and repair mechanisms. Whether the adult pancreas contains endocrine stem cells or not is still the subject of heated debate. In general, tissue-specific stem cells are elusive and difficult to culture *in vitro*, and their differentiation potential is much more restricted than that of ES cells. One possible exception to this rule is the bone marrow (BM)-derived **multipotent adult progenitor cells** (MAPCs) described by Verfaillie and colleagues. These cells have been shown to proliferate extensively *in vitro* and are able to give rise to the three embryonal layers when injected into mouse blastocysts (Jiang et al., 2002). However, the routine isolation and culture of these cells is still far from mainstream, as it has proven more challenging than working with ES cells.

An additional potential approach to obtain insulin-producing cells is **transdifferentiation** of adult cells (i.e. hepatocytes) under selected conditions both *in vivo* or *ex vivo*.

PANCREATIC DEVELOPMENT

Research conducted over the last decade has outlined a basic "roadmap" of the major molecular events that shape islet development (Figure 12.1) (Edlund, 2001; Edlund, 2002). Critical developmental milestones are: (a) generation of endoderm/gut endothelium; (b) pancreatic differentiation; (c) endocrine specification; and (d) β-cell differentiation. Transition between each of these stages of development appears to be catalyzed by a surprisingly manageable number of transcription factors, which are highly conserved between mouse and man.

Generation of Endoderm/Gut Endothelium

ES cells are an artificially frozen snapshot of the ICMcells found at the blastocyst stage (embryonic day e3.5). Expression of genes such as **telomerase**, *Oct3/4*, and *Nanog* make these cells immortal and pluripotent under defined conditions *in vitro*. Subsequent differentiation will be marked by the permanent down-regulation of these genes. **Visceral endoderm** and **epiblast**, respectively, constitute the outer and inner layers of the ICM immediately before gastrulation. The visceral endoderm will become part of the yolk sac, without contribution to the embryo proper. In contrast, the **definitive endoderm** is formed during gastrulation when epiblast cells leave the ICM through the primitive streak. There is an intermediate stage in definitive endoderm formation, called **mesendoderm**. Although visceral and definitive endoderm are similar, mesendoderm-specific genes such as *Gsc* and *Bry* do not appear during visceral endoderm differentiation (Kispert and Herrmann, 1994; Tam et al., 2003; Kubo et al., 2004; Yasunaga et al., 2005), and therefore can be used to identify true definitive endoderm (Yasunaga et al., 2005). The **anterior** part of the definitive endoderm will evolve into the foregut, from which pancreas, liver, and lungs will eventually bud out. The **posterior** definitive endoderm, on the other hand, becomes the midgut and hindgut, which will differentiate into large and small intestine. Graded *Nodal* signaling is responsible for the initial patterning of the primitive gut endothelium. Many genes

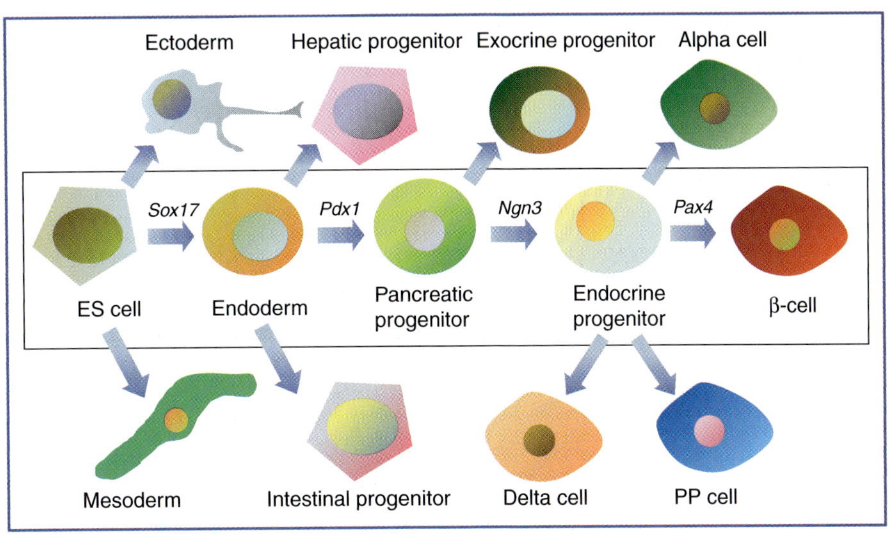

Figure 12.1 ***Schematic representation of the differentiation pathway from ES cells to β-cells.*** **Genes whose expression is necessary for the transition between each step are indicated in italics.**

have been associated with the formation of true endoderm, including *Foxa2, Mixl1, GATA-4,* and several members of the *Sox* family, chiefly *Sox17* (de Santa Barbara et al., 2003).

Pancreatic Differentiation

There is a cross-communication between the gut endoderm and the surrounding mesoderm, mediated by *Shh* signaling. *Shh* is highly expressed throughout the gut endothelium, but is down-regulated in a *Pdx1*-positive region that will later become the pancreas at e8. Both *Shh* repression and *Pdx1* activation are defining events of pancreatic specification. *Pdx1* knockouts are born without pancreas (Jonsson et al., 1994). Chemical inhibition of *Shh* enhances pancreatic differentiation at the expense of intestine (Kim et al., 1997). Conversely, ectopic expression of *Shh* under the control of the *Pdx1* promoter induces intestinal fates at the expense of the pancreas (Apelqvist et al., 1997).

Endocrine Specification

Endocrine differentiation occurs through a **lateral inhibition** process, mediated by *Notch* signaling. Cells in which the *Notch* receptor is activated by the ligands **delta** or **serrate** express high levels of *HES-1*, which in turn represses the pro-endocrine gene *Ngn3*. Lower levels of *Notch* signaling may randomly occur in individual cells, where *HES-1* expression will not be up-regulated. In the absence of its repressor, *Ngn3* will be expressed robustly, and the cell will adopt a pro-endocrine fate (Apelqvist et al., 1999; Gradwohl et al., 2000; Jensen et al., 2000). The differentiation into each of the five endocrine cell types within the islet (α-, β-, δ-, PP- and ε-cells) is preferentially observed at specific time points during embryogenesis, suggesting that *Ngn3*-positive cells adapt their responses to an evolving milieu of signals.

β-Cell Differentiation

Little is known about the extracellular signals that drive β-cell specification from *Ngn3*-positive progenitors. Animals lacking *Nkx6.1* (Sander et al., 2000) and *Nkx2.2* (Sussel et al., 1998) have defects in β-cell formation. However, several observations point to *Pax4* as the main hallmark of β-cell differentiation: (i) the knockout of this gene results in the total absence of β-cells (Sosa-Pineda, 1997), but not α-cells; (ii) its expression peaks between e13.5 and e15.5, which coincides with the period of maximal differentiation of β-cell precursors (Sosa-Pineda et al., 2004); and (iii) shortly after endocrine specification, *Ngn3* co-localizes with *Pax4* (Wang et al., 2004), which suggests that the latter may be one of the targets of the former. Recent evidence indicates that *Pax4* and *Arx* are mutually repressed, and that the balance between the two determines α- (*Arx*) or β-cell (*Pax4*) specification from *Ngn3+* progenitors (Collombat et al., 2003, 2005).

ISLET NEOGENESIS FROM ES CELLS

Ideally, the "education" of human ES (huES) cells along the β-cell lineage would require the exact recapitulation of the differentiation steps described earlier. If we could identify the "instructive" extracellular signals that naturally drive this process, such signals could then be added in the proper sequence to the culture medium in the hope that the cells would respond accordingly (Figure 12.1). However, our understanding of the fine regulation of extracellular signaling is still somewhat limited at the present time. In fact, the combined action of signals such as *FGF, Nodal, Hedgehog, Notch, BMP/TGF-β,* or *Wnt* is responsible for the patterning and development of most organs (Edlund, 2002). During development, cells respond differentially to environmental cues depending on their exact location, their interaction with other developing tissues and time. Fine gradients of *Nodal* (for endoderm/gut endothelium specification), *FGF,* and *Shh* (for pancreatic differentiation), as well as random cell-to-cell interactions in the *Notch* pathway, are examples of the complex differentiation mechanisms that we are just starting to unravel.

Given these limitations, it is not surprising that all attempts at generating β-cells from ES cells have been unsuccessful so far. The observation by Assady and colleagues that spontaneous *in vitro* differentiation of huES cells resulted in the scattered appearance of insulin-producing cells (Assady et al., 2001) merely confirmed the well-known fact that these cells have unlimited differentiation potential. Protocols for the efficient differentiation of β-cells were still necessary, and several groups set up to develop them. Lumelsky and colleagues, for instance, described a five-step method to generate islet-like cells from murine ES cells (Lumelsky et al., 2001) based on the derivation of cells positive for the intermediate filament protein *Nestin*, a known marker of neuroectodermal and mesodermal fates. Further analyses on these cells demonstrated that they were not true pancreatic endocrine cells, but rather neuroectodermal derivatives that absorbed insulin from the culture medium. Further refinements on this method have led to somewhat improved results, although the amount of insulin expressed by these cells is still quite reduced compared to that of mature β-cells (Fujikawa et al., 2005). Using a genetic engineering approach, Soria and colleagues (Soria et al., 2000) generated murine ES cell lines where a selectable marker (neomycin, which confers resistance to the drug G418) was placed under the control of the insulin promoter. Thus, when allowed to spontaneously differentiate, G418 selection yielded insulin-producing clones. Although elegant, this method requires the introduction of foreign genes. Also, it must be taken into account that insulin expression is not a very stringent criterion for the selection of β-cells, as many other tissues do express it. Indeed, the same authors later confirmed the ectodermal identity of some of the selected clones (Roche et al., 2005).

The most exciting developments in the field of ES cell differentiation have been the result of a seemingly less ambitious approach. Instead of attempting the direct differentiation of ES cells into insulin-producing cells, several groups have focused on the key first step of the process, namely endoderm specification. The difficulty of this enterprise is highlighted by the fact that standard culture conditions strongly favor ectoderm and mesoderm over endoderm specification (hence the proliferation of ectoderm-based differentiation protocols). Also, early attempts to generate endoderm could not direct ES cells specifically toward **definitive**, rather than **visceral**, endoderm. Kubo and colleagues were the first to report the generation of definitive endoderm from murine ES cells, *albeit* at a low frequency (Kubo et al., 2004). Far more striking results were successively described by Tada and colleagues (Tada et al., 2005), also in mouse ES cells, and D'Amour and collaborators in huES cells (D'Amour et al., 2005). The latter was based on the addition of high concentrations of **Activin A**, a **TGF-β**-related agonist of *Nodal*, in low-serum conditions.

Since endoderm specification had been widely regarded as the main obstacle toward pancreatic differentiation, we now expect a steady, accelerated progress of these lines of research. Of particular interest are the new differentiation strategies that make use of protein transduction technology (Wadia and Dowdy, 2002, 2003) for the delivery of key transcription factors (*Pdx1*, *Ngn3*, *Pax4*, etc.) to stem cells *in vitro*. This approach would be particularly useful to bypass the enormous challenge of mimicking *in vitro* the complex signaling pattern that is responsible for the sequential activation of such transcription factors *in vivo* (Domínguez-Bendala et al., 2005).

ISLET NEOGENESIS FROM ADULT STEM CELLS

The ability of adult pancreatic islet cells to retain regenerative potential during adulthood has been recognized. Several experimental models such as partial pancreatectomy (Bonner-Weir et al., 1993), cellophane wrapping of the pancreas (Rafaeloff et al., 1992), duct ligation (Wang et al., 1995), or treatment with streptozotocin (Guz et al., 2001), as well as physiological conditions such as pregnancy (Nielsen et al., 1999;

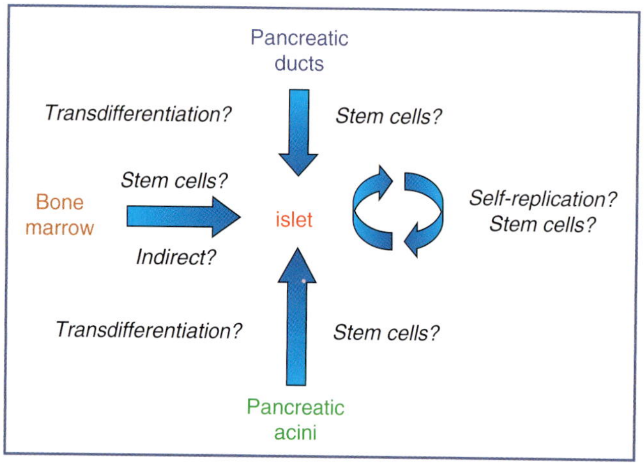

Figure 12.2 ***Islet regeneration during adulthood*** **may occur through several mechanisms. Yet unidentified endocrine stem cells within the islet may be responsible for beta cell turnover, although data obtained in a mouse model suggest that islet self-maintenance is preferentially due to replication of existing beta cells. Other investigators hypothesize that islets can be regenerated from ductal or acinar tissue, although it is not clear yet whether this phenomenon would be mediated by putative stem cells or by transdifferentiation. Finally, the BM has also been proposed as a reservoir of β-cell progenitors. Recent evidence, however, suggest that the regenerative capacity of migrating BM cells might rather be due to their *in situ* differentiation into supporting endothelial cell types.**

Johansson et al., 2006; Sorenson and Brelje, 1997) and perhaps long-standing T1DM (Meier et al., 2005) confirm that insulin-producing cells can regenerate in adult life. However, the quest for endocrine pancreatic stem cells has been an elusive one (Figure 12.2). Numerous observations suggest that these cells may reside in the **ductal epithelium**. Aside from countless microphotographic snapshots showing insulin-positive cells that appear to sprout from the pancreatic ducts (Bonner-Weir et al., 1993; Wang et al., 1995; Meier et al., 2005; Sarvetnick and Gu, 1992), cultured ductal cells respond to various stimuli *in vitro* by expressing several β-cell markers and even secreting low levels of insulin (Bonner-Weir, 2000). Other groups have identified **Nestin-positive cells** within the pancreas with a remarkable ability to expand, although their ability to emulate β-cells upon differentiation was less impressive (Zulewski et al., 2001). More recently, Gershengorn and colleagues demonstrated that **adult islets** can undergo a reversible epithelial-to-mesenchymal transition *in vitro* (Gershengorn et al., 2004). Upon "de-differentiation," these cells could be expanded by a factor of 10^{12}, which is well within the realm of clinical applicability. However, when "re-differentiated," these putative β-cells expressed a mere 0.02% of the amount of insulin found in mature, primary islets. A variation on this protocol resulted in enhanced insulin production (Ouziel-Yahalom et al., 2006), but the ability of these cells to proliferate was much more modest. Finally, it has been proposed that **acinar tissue** may also contain endocrine stem cells (Hao et al., 2006). In this case, transdifferentiation was almost negligible, and no effort was made at characterizing either the neogenic insulin-positive cells or their putative progenitors.

In short, thus far nobody has been able to present conclusive evidence that adult stem cells can generate genuine β-cells *in vitro*. Most of these cellular byproducts are, at best, oddities that co-express markers found in many diverse cell types. Concerns that these cells may just be culture artifacts are justified, and were further fueled when Dor and collaborators (Dor et al., 2004) suggested that adult β-cells regenerate by replication rather than differentiation. Lineage-tracing experiments conducted in rodents convincingly demonstrated that the regeneration and normal turnover of islets occur preferentially by division of existing β-cells.

This report did not rule out the possibility that stem cells may still exist in the pancreas, but their importance was suddenly – and dramatically – reduced.

Although the burden of proof is now on those who defend the existence and biological significance of pancreatic stem cells, the jury is still out. For instance, it has been argued that human β-cells replicate at a much lower rate than their mouse counterparts, which would be inconsistent with the "replication only" hypothesis (Meier et al., 2005; Hao et al., 2006; Tyrberg et al., 2001; Finegood et al., 1995). Also, it is conceivable that the adult human pancreas may have evolved different mechanisms for normal β-cell turnover and damage-induced regeneration. The impossibility of conducting lineage-tracing experiments in humans will keep this controversy alive for years to come.

TRANSDIFFERENTIATION

Adult, differentiated cells from specific tissues can turn into completely different cell types in certain conditions. This phenomenon has been termed **transdifferentiation**. We will examine here two cell substrates (namely bone marrow and liver) that have shown some promise at transdifferentiating into pancreatic cell types.

Stem cells derived from the **bone marrow** (BM) have been associated with numerous examples of tissue repair and regeneration *in vivo*. It has been documented that transplanted BM cells can migrate from their niche to various tissues in response to injury (Goodell, 2001). In some cases, this migration was accompanied by a significant regeneration of the damaged tissue, which led to the hypothesis that some cell types within the BM may have either ES cell-like properties or the ability to transdifferentiate. However, as it was confirmed in a model of liver disease (Grompe, 2003), the regenerative effect can also be due to the fusion of the BM cells with cells of the target tissue. Regarding the pancreas, an early study showed that up to 3% of islet β-cells were of donor origin 1month after BM cell transplantation, without evidence of cell fusion (Ianus et al., 2003). The conclusions of this report, however, were recently contested by Lechner and colleagues (Lechner et al., 2004), who could not find any significant contribution of the BM to islets either in healthy mice or in models of pancreatic injury. Furthermore, Kang and colleagues reported that while BM cell transplantation was enough to prevent diabetes onset in nonobese diabetic (NOD) mice, there was little or no involvement of the BM cells in islet cell regeneration once the disease was overt (Kang et al., 2005). Still, yet another report presented evidence that donor BM cells **do** promote islet regeneration in a mouse model of diabetes (Hess et al., 2003). Interestingly, the authors of this study did not find any evidence of transdifferentiation of BM cells into β-cells: the beneficial effect was rather due to the recruitment of donor-derived endothelial cells to the injured islets, where they induced their regeneration.

There is also a wealth of observations indicating that **liver** and pancreas are especially susceptible of interconversion. Many invertebrates have a single organ that comprises both hepatic and pancreatic functions, which suggests that the separation of these two organs is a relatively late evolutionary event. Indeed, both originate from common endodermal progenitors in the early foregut of vertebrate embryos (Deutsch et al., 2001; Jung et al., 1999). In general, hepatocytes and β-cells share not only many developmental features, but also similar molecular machinery for glucose sensing and secretion (Nordlie et al., 1999; Kim and Ahn, 2004). Many studies confirm that interconversion of liver and pancreas occurs under a variety of experimental conditions (Rao et al., 1988; Rao and Reddy, 1995; Rao et al., 1986; Shen et al., 2000), as well as in certain diseases (Lee et al., 1989; Wolf et al., 1990).

Based on the above evidence, Ferber and colleagues (Ferber et al., 2000) set up to demonstrate that ectopic expression of *Pdx1* in liver cells could induce transdifferentiation into pancreatic cell types. Using an adenovirus vector, a *Pdx1* cassette was delivered to the livers of recipient mice, where normally silent, β-cell-specific

13

Mechanical Determinants of Tissue Development

Jonathan A. Kluge, Gary G. Leisk, and David L. Kaplan

INTRODUCTION

The field of tissue engineering offers promising new solutions for replacement or repair of damaged tissues and organs. The ultimate goal of these strategies is to fully restore normal tissue function. The most common tissue engineering approach toward this goal is to develop viable constructs *in vitro* that can be implanted in the human body. Ideally, the implanted tissue continues to develop, providing the structure, composition, cell signaling, and functions that the native tissue exhibits (Vunjak-Novakovic et al., 2005).

To develop viable constructs *in vitro*, it is believed that the *in vivo* conditions which promote growth and differentiation of target cell types should be replicated as closely as possible. Unfortunately, *in vivo* environmental conditions, native tissue mechanical loading, and the complex signaling critical to cell function and tissue development are all difficult to quantify, let alone replicate. Toward this end, a wide array of bioreactors has been developed by researchers to provide an *in vitro* environment that recapitulates the *in vivo* environment as faithfully as possible. Providing a limited set of loading and environmental conditions, often specific to the type of tissue being produced, modern bioreactor designs are increasingly sophisticated and produce ever-improving tissue quality.

In this chapter, we focus on the effect that a specific epigenetic factor, mechanical stimulation, has on tissue development. To understand the complex relationship between mechanical loading and tissue development, we draw on research from biomechanics, cell biology, and biochemistry, three fields rapidly discovering overlapping themes and unsolved challenges. These challenges will be met through collaborative interdisciplinary efforts, sparked by recent initiatives (Kaplan et al., 2005).

By the end of this chapter, the reader will have an understanding for the role of mechanics on two levels of cell and tissue function: the macro, full tissue level, and the microscopic cell level, including the extracellular matrix (ECM) and intracellular signaling mechanisms. By reviewing principles in both mechanics and biology, and then proceeding from a macroscopic to a microscopic perspective, the interplay between levels, and how they influence tissue engineering, will become clearer.

The challenges presented to the field due to the complexity found in biology require a confluence of modeling, bioreactor design, and biomaterials engineering that best replicate *in vivo* tissue development *in vitro*. This complexity derives in part from the structural hierarchy found in biological materials, which creates difficulty in measuring and applying mechanical forces in a developmentally relevant temporal and spatial manner *in vitro*. Compressed time frames are needed to satisfy potential therapeutic benefits *in vivo*. Further complicating the situation are the diversity of cell types and states of cell function that exist cooperatively in any given tissue type, the presence of gradients of structure and function, and the effects of water and related environmental variables on tissue structure and thus mechanical properties and overall function. The challenges ahead are

Copyright © 2008 Elsevier Inc. All rights reserved

immense, however, scientific progress achieved over the last 5 years suggest future options remain bright to bridge the needs of biomechanics and functional tissue engineering.

MECHANICAL FORCES EXPERIENCED BY TISSUES

Proceeding with a top-down approach to the role of mechanical forces in tissue development, we begin with a brief introduction to mechanics of materials, a review of the gross loading conditions on the body's major connective tissues, and the measurement of tissue response. The reader seeking additional information on mechanics of materials, specifically biomechanics, is referred to full texts on the subject (Fung, 1993; Mow and Huiskes, 2005).

Mechanics of Materials

An understanding of how biological materials respond under mechanical loading requires knowledge of the types of external loads that may be applied, the internal forces and stresses that are generated, and the properties that govern the material response. This area of biomechanics, known as **mechanics of materials**, is briefly introduced here. This information will be important when studying gross mechanical forces on tissues and mechanical forces that act directly on cells and their local environment.

Force and Stress

An object that is externally loaded may move if it is unconstrained and deform if made from a deformable material. For an object that is physically constrained, such as a girder in a building frame or a tendon of the human body, forces and moments (i.e. bending or twisting action) at the constraints counteract the applied loads and may restrict object motion or deformation. This presence of simultaneous external loads and reaction forces and moments can generate a variety of internal forces within an object. At an arbitrary section through the object, one could isolate: normal forces that act perpendicular to the section, pushing or pulling on the object; shear forces that act along the plane of the section; torsional moments (torques) that twist the object about an axis perpendicular to the section; and bending moments that bend the object about an axis within the plane of the section (Hibbeler, 2000).

Stress, which is a quantity representing the intensity of force, is separated into two types: normal and shear stress. Normal stress acts to cause local expansion or contraction within a material, while shear stress causes distortion. The deformation due to normal stress is simply called strain, while the distortion caused by shear stress is called shear strain. In the case of simple normal and shear force application, resulting normal stress and shear stress will cause the material to undergo deformation and distortion, respectively. Torsional moments tend to cause distortion only, while bending moments generate both deformation and distortion (Hibbeler, 2000).

The magnitude of stress experienced by an object is affected by its geometric shape and dimensions and the nature and magnitude of external loading. The nature of loading on human connective tissues, for example, can range from biomechanical loads, imposed on the body through the actions of sitting or engaging in athletic activities, to physiological loads, such as pressure and flow effects of blood and other bodily fluids. Stress is a derived quantity and, therefore, not directly measurable. To quantify the stress in an object of known geometry, indirect measurement techniques are often used, such as measuring strain directly and correlating it to the stress from which it was produced (Hibbeler, 2000).

Material Properties

How a material responds to stress is dictated by its mechanical properties. These properties are derived from destructive or non-destructive testing, usually employing standardized equipment and test procedures.

For example, the uniaxial tension test is a standard destructive test that involves pulling a carefully prepared material specimen with uniform tension, while recording the applied load and resulting material deformation. Material properties, such as Young's modulus, which characterizes material stiffness, as well as strength properties, such as yield and failure strength, can be calculated from a tension test. Many other tests can be applied, such as compression, shear, bending, fatigue, and torsion tests. In material property testing, the closer the test conditions, such as the loading magnitude, rate, and specimen geometry of a material, resemble the actual parameters, the greater the confidence in the derived property.

Forces in Biological Tissues

The types of loading conditions experienced by biological tissues are varied, depending on the specific tissue. Some tissues experience continuous loading and unloading cycles, often in response to the body's movement (e.g. bone and cartilage response to walking); others experience a state of prestress, in which a low load level is constantly applied (e.g. ligament tension and bone compression). In contrast to most commonly used engineering materials, many biological tissues respond to loading regimes with nonlinear, time-dependent deformation (i.e. a viscoelastic response). Such nonlinear response is more challenging to characterize and model.

Mechanical Properties of Tissues

Like most commonly used engineering materials, the tissues of the body do not last forever. Just as every component in an automobile has a finite lifetime, individual structures in a human body eventually fail, whether due to catastrophic events, disease, or normal wear and tear. One approach that automobile manufacturers employ to ensure component longevity is to design relevant assemblies such that stress levels and the number of stress cycles experienced by the component are minimized. Manufacturers can then select constituent materials whose mechanical properties (e.g. breaking strength, cycles to failure) comfortably exceed anticipated stress levels. We can view tissues of the human body in a similar light: to function properly over time and through many cycles, tissue strength properties should comfortably exceed stress levels generated by anticipated loading conditions.

The reader should note that in tissue engineering practice, one does not necessarily exercise growing tissue constructs to the perceived structural potential of functioning tissue *in vivo*. Typically, low-level, continual loading regimes are applied to the constructs in a bioreactor environment. Regardless of the loading magnitudes used, one must ensure that the tissues have sufficient mechanical integrity through the regenerative process. The mechanical properties of specific (engineered) tissues are discussed in other chapters.

Measurement of Force in Tissues

In determining the mechanical properties of tissues, special equipment is often needed to isolate the tissue in question, measure its response to a predetermined loading criterion, and capture and record the data (Fung, 1993). This equipment ranges from implantable strain transducers and data acquisition devices to standard material testing equipment and non-standard video imaging, depending on whether measurements are to be made *in vivo* or *ex vivo*. Since such techniques typically provide individual measurement parameters, such as a load level or an amount of deformation, they are often combined with analytical or empirical information on a tissue's constitutive behavior to derive additional measurement parameters. Especially for more complicated tissue structures, such as intervertebral disks, the *in vitro* or *in vivo* measurements are used to corroborate mathematical or computer-based models, sometimes referred to as *in silico* modeling (Prendergast et al., 2005). With the understanding that researchers in the field of biomechanics are continually discovering new ways to measure the mechanical properties of tissues, only a brief review of those techniques will follow.

Ex Vivo Measurements

Whole sample tissues can be tested *ex vivo*. For example, the distention forces caused by applied pressure can be measured on isolated vascular tissue explant samples. Measurement data generated in this fashion can be valuable for the design of viable tissue constructs (McCulloch et al., 2004). *In vitro* testing involves subjecting samples to a limited set of environmental and loading conditions that mimic *in vivo* conditions. Specific responses can be isolated in this way. Despite the prevalence of *in vitro* testing, some disadvantages exist. For example, *in vitro* testing tends to be short term and cannot exactly mimic natural conditions. In addition, the loading conditions of tissues in the body, such as biaxial loading in pressurized vessels, are difficult to replicate in bench-top testing.

In Vivo Measurements

Another common practice for deriving mechanical properties of tissues is through *in vivo* measurement of native tissue function. Various invasive and non-invasive techniques have been pursued by researchers to acquire data regarding internal tissues. Invasive techniques include the use of a shaped indenter or an aspirator to create a measurable deformation of the tissue, from which the material's elastic response can be deduced. Surgical instruments that have been modified to incorporate force and position sensors have also been used. Each of these techniques is intended to quantify the resistance of the material to deformation (Ottensmeyer and Salisbury, 2001). Several non-invasive techniques, all categorized as "elastography," are also based on tissue deformation. Strain fields (deformations) produced using this technique are measured using magnetic resonance imaging (MRI), optics, ultrasound, or another technology. Another new non-invasive ultrasonic technique involves the use of an ultrasonic pulser to send an ultrasonic wave through a tissue and the use of a second ultrasonic sensor to measure the displacement (Doyley et al., 2005). Unfortunately, there are limitations to many invasive techniques used for *in vivo* measurement, including an inability to isolate tissue response from a single variable, a dearth of appropriate internal force sensors, and ethical concerns.

In vivo measurements on animal subjects are sometimes pursued as an ethical and practical alternative to invasive human procedures. In these cases, attempts should be made to select animal models whose morphologies and relative sizes closely match the human tissue of interest. Strain gauge-based force transducers have been implanted to monitor tissue response in certain animals. In the case of research on rabbit tendons, force response was monitored during various activity levels, such as "in-cage" and vigorous activities (Juncosa et al., 2003). Similarly, strain gauges used to measure microlevel deformations in various animal bone tissues revealed the prevalence of different strain regimes throughout regular daily loading (Fritton et al., 2000). This type of *in vivo* monitoring may be used to develop specific design parameters for tissue engineering.

THE CELL AS A SIGNAL RECEIVER AND PROCESSOR

Shifting the discussion from the macroscopic tissue level to the microscopic cell level, we now focus on the underpinnings of tissues, cells, and their molecular constituents within the framework of mechanics. The cells are the workforce behind tissue-engineered constructs, as they serve to generate the ECM, or the material which gives tissue its mechanical integrity. In addition, the cells contain their own internal structural hierarchies and means of adaptation to external mechanical forces, which may lead to the formation of new tissue (proliferation and metabolism) or establishment of terminal phenotypes (differentiation). The following sections will review the mechanosensing components of the cell, the overall process by which cells integrate mechanical signals to direct tissue-specific growth, and the role that mechanosensation (MS) has on cell proliferation and differentiation.

Cell Receptors and Sensors

Cells within living tissues transduce mechanical force by using a variety of mechanisms. Although the signaling processes of MS are complex, involving many different molecules and pathways, they may all be activated by similar mechanisms in response to a variety of incoming signals (Huang et al., 2004). Among possible mechanical determinants previously discussed, shear forces due to fluid flow, strains imparted through cell/ECM constraints, and high-frequency vibrations are among the most prevalent to which a cell will respond (Hamill and Martinac, 2001). Surface mechanisms, which allow cells to transmit these forces throughout the cell, and the inner supportive cell structures (the cytoskeleton) are the vehicles by which signals can be integrated from the host tissue. In this section, transmembrane matrix molecules will be reviewed in the context of their suspected ability to convert external physical forces to intracellular biochemical signals, followed by a review of the tensegrity model, which offers a correlation between surface-level alterations and widespread cell changes in a global network.

Cell–Matrix Adhesions

The cell makes contact with its surrounding ECM through "adhesions," a term used to describe a wide array of protein-mediated molecular links. The membrane portion of a cell's ECM adhesions contains specific integrins, which are heterodimers of α and β subunits that bind to specific sequences on ECM molecules through a large extracellular domain (Geiger et al., 2001). Intracellularly, these integrins will interact with plaque proteins, which could be bridging proteins that connect the integrins to the cytoskeleton or signaling molecules. Several intracellular multi-molecular proteins serve to link the actin portion of the cell cytoskeleton to membrane integrins; these linkers include α-actinin and talin among others. Signaling molecules, another widely classified group of intracellular plaque proteins, are often activated by integrins or their bridging proteins, and include focal adhesion kinase (FAK) and mitogen-activated protein kinase (MAPK). Extracellularly, ligands can act as part of the ECM adhesion receptors such as fibronectin (α_5 and β_1), vitronectin (α_v and β_3), and various collagens (α_1 and β_1), which are all supplemented by membrane-bound non-integrin proteoglycan components such as syndecan-4 and CD44 (Geiger et al., 2001) (Figure 13.1).

Following the occupation of integrins by their ligands, the initial step in reinforcing adhesions involves the clustering of integrin molecules. Focal complexes, small dot-like structures associated with the cell lamellipodium (thin, flat extensions at the cell periphery), are typically the first structures formed at cell/ECM junctions, and are either transient or evolve into more stable focal adhesions. The creation of a more stable adhesion is thought to be generated internally by responses of the cytoskeleton to applied forces (Geiger et al., 2001). Intracellular activators (such as talin) are thought to interact with either the α or β subunit tail of integrins and induce their separation, thereby causing further conformational changes that open the binding site on the headpiece and allow the integrin to create this enhanced focal adhesion (Giancotti, 2003).

Since most intracellular adhesion components are multi-domain molecules, having the ability to partner with several different molecules, there are innumerable combinations of molecular interactions that could be occurring to create a signal pathway or generate a stimulus response (Geiger et al., 2001). The complexity of such relationships does not diminish the impact that individual integrins have on the development of tissue. Specifically, studies using β_1 integrin-deficient knock-out mice showed basement membrane defects (Stephens et al., 1995). Additionally, the early organization of collagen fibrils *in vitro* depends on fibronectin, whose attachment is determined by integrins (Geiger et al., 2001).

Integrins exist in two major allosteric conformations that are determined by their activity: an inactive (low-affinity state) and active (high-affinity) states. When ECM ligands bind to the molecular headpiece, they induce conformational changes in the integrin that are propagated along its length. Release from the inactive state causes the intracellular tails to move away from each other so that the β subunit is free to engage the underlying cytoskeleton (Giancotti, 2003). Since the cytoplasmic segment undergoes conformational changes via intracellular linkers, and the extracellular segment is controlled by ECM interactions, integrin molecules appear to have two functions: to regulate the extracellular binding activity from inside the cell (inside-out

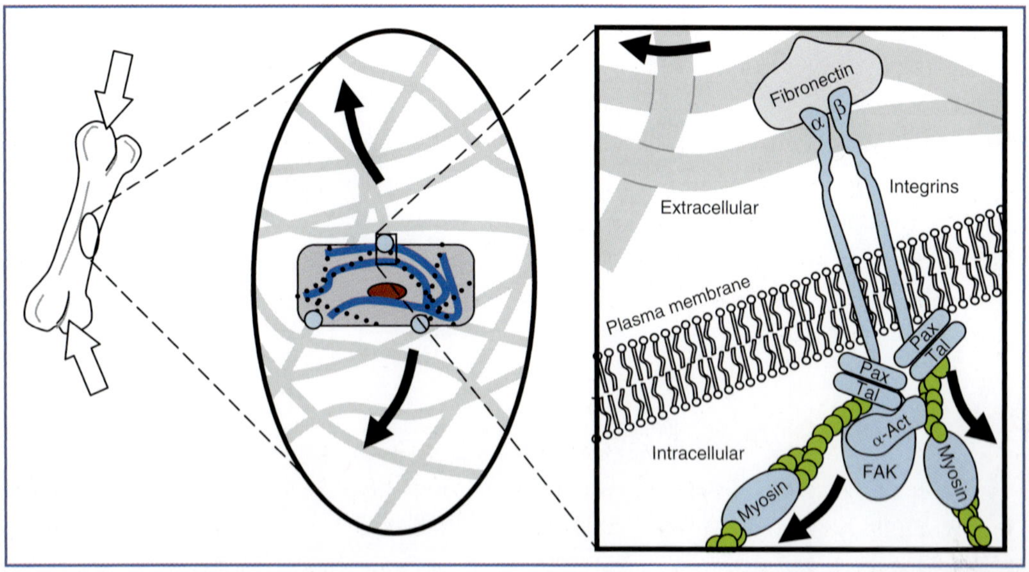

Figure 13.1 **Diagrammatic view of mechanical force propagation (seen as dark arrows) throughout tissue from macroscopic to nanoscopic levels. Human connective tissue, such as bone, is affected by repeated external loads (left). The underlying ECM (mostly collagen, represented by random fibrils) will be subjected to these forces, which are then transferred to cells through insoluble ligands (fibronectin, circles at cell periphery). Surrounded by ECM, the cell contains its own sub-structure: lines, both continuous and dotted, represent the cytoskeletal microtubules and actin microfilaments, respectively (middle). The result of external forces is intracellular signaling: recruitment of bridging proteins, paxillin (Pax), α-actinin (α-Act), and talin (Tal), and signaling molecules, such as FAK, to the site of developing integrin-mediated focal adhesions. The result is a force balance which presumably affects actin-bound signaling proteins (actin helices with connecting myosin) (right).**

signaling) and to elicit intracellular changes through ECM binding (outside-in signaling) (Giancotti and Ruoslahti, 1999).

The signaling molecules are associated with enzymes that can trigger pathways, which ultimately lead to changes in protein production and cell fates. By following the response of a signaling molecule, FAK, one can appreciate the complexity of signaling pathways and the importance of focal adhesion in initiating cell responses. For several years, FAK has been associated with both the growth and the disassembly of integrin-based focal adhesion sites (Geiger et al., 2001). Recently, the relationship of FAK with various GTPase proteins (Rho, Rac, and Cdc42) and indirect association with integrins through bridging proteins have been elucidated, and place FAK at the forefront of several intracellular signaling pathways (Mitra et al., 2005).

The molecular signaling pathways are too numerous and complex to provide a full review in the context of this discussion; instead, the possible role of FAK activation in various signaling channels will be outlined. To briefly illustrate these channels, a simple schematic is provided which links FAK to intracellular activity (Figure 13.2). Formation of new integrin-mediated focal complexes or the transduction of forces through integrins may lead to an activation of FAK signaling mechanisms, mainly recruitment of other focal adhesion proteins. One such mechanism, responsible for the assembly and disassembly of focal complexes, is FAK's ability to control phosphorylation of the bridging protein α-actinin, which can cross-link and tether actin/myosin stress fibers (Mitra et al., 2005). The second mechanism, although not completely understood, may be the activation of the Rho effector diaphanous (mDia) which leads to the stabilization of the cytoskeleton (i.e. microtubules) at the leading edge of migrating cells (Geiger et al., 2001). The third mechanism is the activation and/or inhibition of the various GTPase proteins that lead to regulation of cytoskeletal extensions, such as stress fibers,

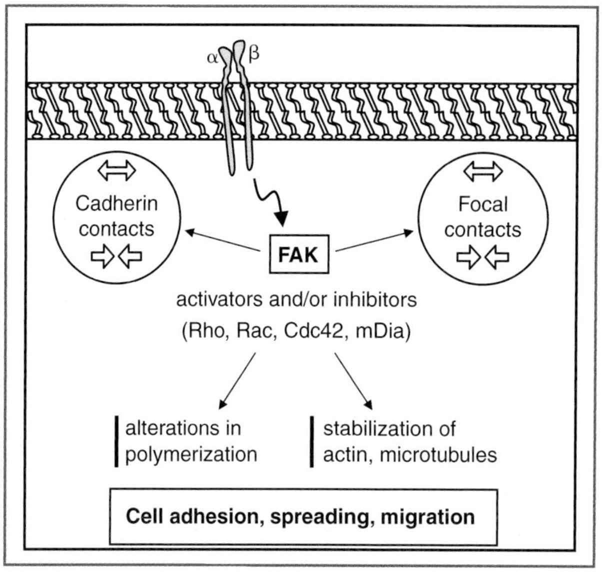

Figure 13.2 **Cell signaling mediated by integrin responses at membrane interfaces, such as due to changes in external mechanical states, leads to intracellular cascades, as shown. FAK plays a central role in these responses. Mediation of cell interactions with the external environment are summarized, including responses such as cell adhesion, spreading, and movement based on changes in focal contacts related to cell–matrix adhesion and cadherins related to cell–cell-mediated interactions. FAK functions to recruit other focal contact proteins or their regulators, leading to changes in the internal structure through polymerization and stabilization of cytoskeletal elements. All of the events illustrated occur in a complex symphony of orchestrated events to modulate internal and external changes in response to changes in external mechanical signaling (Figure is in part patterned after Figure 1 from Mitra et al. (2005)).**

lamellipodia, and filopodia. The final mechanism is the formation and disassembly of cell–cell (cadherin-based) connections, providing an added route for solute exchange and signaling (Mitra et al., 2005). The importance of relationships between the cytoskeleton and molecular signaling pathways will gain further emphasis in the following discussion.

Tensegrity Model of Cell

An understanding of intracellular microstructure and hierarchy is critical for grasping the interactions between incoming signals and their propagation throughout the cell. The cytoskeleton is not merely a randomly configured collection of molecules; instead, it is believed that each different cytoskeletal molecule is integrated in a unique way to maintain the mechanical signaling pathways. Furthermore, the mechanical behavior of the whole cell is driven by both the cytoskeletal elements found just below the surface of the plasma membrane and also the internal cytoskeletal lattice, a component often overlooked because of its misunderstood complexity. The role of the cytoskeleton as a support and shape-retaining structure has long been recognized, but it is now known that it can also provide directed signals to the intracellular elements, and is thus capable of inducing endogenous changes to occur (Ingber, 2003a).

Structural molecules that make up the cytoskeleton can be broken down into three major groups: actin-based microfilaments associated with the cell's cortical cytoskeleton (adhesion complexes), stiff hollow tubulin-based microtubules that radiate from an organizational center (centrosome), and thick intermediate filaments, such as lamin, vimentin, and keratin, that integrate with the cell's nucleus and attachment sites (desmosomes). Formation of larger and stronger cytoskeletal structures is possible when these molecules are supplemented by other proteins, such as actin-bound myosin chains, which combine to form stress fibers around the cell periphery. Altogether, these structural elements organize throughout the cell interior to form a complex network (Ingber, 2003a).

A "Tensional Integrity" (or tensegrity) model, one that assumes compressive-bearing struts (microtubules and ECM adhesions) is resisted by the pull of surrounding tensile elements (microfilaments, intermediate filaments), has been espoused by Donald Ingber and his colleagues, and serves as the most prominent cytoskeletal structure-function model to date. This model adapts the principles used in the design of a ship mast and riggings, as linear reinforcing elements can be linked together to form tension-resistant scaffolding around a hull (cell membrane). Following this model of tensegrity, the majority of structural elements need only to have good tension resistance to maintain shape and stability, while avoiding the need for buckling or compression resistance of large rigid struts by creating a network of triangulated shorter members (Ingber, 1997; Boal, 2002).

These intracellular structures, not unlike the tissues they inhabit, have been regarded as soft materials because of their shrinking and stiffening response to temperature and their relative ease of deformation. Cytoskeletal elements can exchange energy with their surroundings, permitting their shapes to fluctuate as they bend and twist in response to transverse loading. Whatever the deformation mode, energy is required to distort the filament from its "natural" shape (Boal, 2002). According to the tensegrity model, many cytoskeletal molecules in their natural state have a certain level of prestress or isometric tension, generated by the contractile function of actin and myosin sliding, osmotic forces, and/or forming new ECM adhesions. This prestress can be visually confirmed when the plating of cells on a flexible substrate does not lead to distortion, or when the cutting of a cell leads to spontaneous retraction of its intracellular cytoskeleton (Ingber, 2003a).

The structural assembly of these elements, in addition to being designed for optimal structural stability, has also been designed for effective transport of molecules throughout the cell; therefore, their configuration must resemble that of a spider web or city plan. This two-dimensional network, common to all three main groups of molecules, can be found attached to the plasma or nuclear membrane, and exhibits many deformation modes in response to an applied force. Actin filaments and microtubules require linking proteins, such as Actin-binding Proteins (ABPs), in order to form these cross-link networks and composite structures (Boal, 2002). Permanent cross-links and a high intracellular density of filaments will add a compressive and shear resistance to cells.

Mechanochemical Transduction

Mechanochemical transduction (mechanotransduction) is the process whereby cells sense and respond to external stimuli. One widely held belief is that ECM proteins and integrins will undergo conformational changes in response to mechanical stimuli. Another belief is that intracellular perturbations of the preexisting cytoskeletal tension will initialize a response. It would seem as though the two are not mutually exclusive events, but, instead, coincide to facilitate signal transduction and to produce intracellular change.

Ligands and Cryptic Binding Sites

As an immature tissue develops and grows, the ECM must have a role in regulating cell interactions. For instance, it is believed that local changes in ECM structure and mechanics will alter the adhesion characteristics of epithelial cells, as they encounter new sections of the emergent basement membrane. In this instance, high turnover in growing sections of new ECM may lead to relatively thinner and more flexible (compliant) regions of the basement membrane (Ingber, 2003b). In response, fibronectin attached to these flexible regions will undergo large strains, leading to a possible exposure of cryptic self-association sites necessary for fibronectin polymerization (Geiger et al., 2001). It has been shown that forces as low as 3–5 pN are sufficient to unfold these cryptic subdomains in fibronectin, which can then lead to fibronectin fibril formation. In summary, once extracellular proteins are altered, integrin binding and cytoskeletal signaling will presumably occur (Huang et al., 2004).

Mechanosensitive Ion Channels

The lipid bilayer is believed to be one of the major mechanosensory (MS) components of a cell. When a membrane, such as the plasma membrane in eukaryotic cells, is deformed by a force, two changes may occur. Disturbance of the lateral force balance around a lipid bilayer may first lead to conformational changes of transmembrane proteins, with or without necessarily activating a second receptor (Janmey and Weitz, 2004). A second change may occur as membrane forces trigger opposing local curvatures that could reorganize the membrane chemically. It is believed that these changes lead to activated ion channels, which, in turn, respond with changes in their permeability (Hamill and Martinac, 2001; Kung, 2005). Whether the forces affect membrane proteins or lipids, and whether the magnitude of these forces can be correlated to channel properties is not clearly understood (Janmey and Weitz, 2004).

The original study of channel gating (regulatory open-and-close mechanisms) in eukaryotic cells was heavily focused on mechanosensory neurons, in which gated channels of *Xenopus* oocytes could be activated for a latent response with pressure-clamp techniques (McBride and Hamill, 1999). After more than 20 years of research, patch-clamp studies have illustrated not only the prevalence of these channels across many eukaryotic species but also their key influences on cell volume regulation and the possibility that tight seal formation could lead to mechanosensitivity in focused K^+ channels (Hamill and Martinac, 2001). Many of these experiments controlled membrane tension by suction pressure in a micropipette attached to a small region of the cell membrane. In these experiments, increased pressure to just below that which would cause the membrane to rupture was shown to increase pore dimension on the order of 0.5 nm in MS channels of large conductance. This "pressure relief valve" mechanism, in addition to its role in MS, can be seen as a cell's natural defense against large osmotic gradients. Similar mechanisms have been linked to calcium ion (Ca^{2+}) channel activation in the stereocilia of hair cells in the inner ear and fluctuations in intracellular ion concentration of endothelial cells (ECs) (Hamill and Martinac, 2001; Huang et al., 2004).

Although the existence of stretch-activated ion channels has been well documented for "specialized" cells (i.e. human cells associated with auditory function, visual function, etc.) and also for non-specialized cells, its mechanisms are not clear, nor are its connection with cytoskeleton-related mechanisms (Hamill and Martinac, 2001). Furthermore, a study in which a Triton buffer was used to remove the cytoplasm and apical cellular membrane showed a binding of paxillin, pp125FAK, and p130CAS to the Triton-insoluble cytoskeleton following a 10% stretch of collagen substrate (Sawada and Sheetz, 2002). These and other results indicate that, in addition to ion channels, there are other mechanisms that enable cells to sense physical forces. These mechanisms, in conjunction with transmembrane proteins discussed earlier, are part of the underlying cell cytoskeletal structure.

Altering Intracellular Mechanics

The major intracellular change that seems to occur as a result of external forces is in cytoskeletal molecular mechanics, as the internal lattice endures global change. Immediately after mechanical signals are sensed by surface integrins, the cytoskeleton will realign in the direction of the applied tensional stimulus, through deformations of the cytoskeletal lattice and nuclear scaffolds (Ingber, 1997). In keeping with the tensegrity model, forces transmitted by integrins to microfilaments in focal adhesions can be passed to microtubules at distant sites via intermediate filament connections (Ingber, 2003a).

Actin and tubulin are dynamic polymers; their fundamental protein building blocks can both polymerize and depolymerize, depending on the conditions, changing the length of the filament in the process. Rapid depolymerization releases the contents of the microtubule to the cytoplasm and permits it, or nearby microtubules, to start reconstruction elsewhere (Boal, 2002). These mechanisms allow a cell to constantly adjust its internal prestress, and thus alter the tightness with which the cytoskeletal lattice is held together, if tensegrity is indeed at play. To account for this change, one must first recognize that many of the enzymes and substrates that mediate protein synthesis, glycolysis, and signal transduction appear to be immobilized on the insoluble cytoskeleton. It is believed that if the cytoskeletal molecules and their immobilized proteins distort without breaking following focal adhesion stimulation, then those attached molecules must similarly change shape. Altered biophysical properties may result in altered local thermodynamic properties, or altered kinetic behavior, just as a spring would change its vibration frequency following distortion (Ingber, 2003b). Similarly, the altered cytoskeleton has been shown to influence protein synthesis by destabilizing cytoskeleton-associated mRNAs at the intersections of actin filaments, and through polymerization at vertices within highly triangulated microfilament networks (Bassell et al., 1994; Ingber, 1997).

The Hard-Wired Nucleus

One major principle of the tensegrity model is that structural hierarchies exist on many levels between muscles and bones of connective tissue, between cells and the ECM, connecting surface receptors to the cytoskeletal elements, and sub-structures within the cytoskeleton, including a nuclear scaffold. The tensed intermediate filaments that connect to the nucleus and its proximal network may be a route by which the signals transduced through surface-level integrin complexes are delivered (Ingber, 2003a). Nuclear scaffolds may be "hard wired" to the integrins, such that distortions of adhesion complexes result in synchronized realignment of structural elements (mainly intermediate filaments) to the nuclear envelope, via the underlying laminin network (Huang et al., 2004). To prove this hypothesis, ligand-coated beads were used to pull focal adhesions of cultured ECs at very high rates, and then the nucleoli were shown to deform and elongate in the direction of applied force. This same phenomenon was also observed in cultures with extracted membranes and intracellular components, suggesting that this signal was transduced directly through the cytoskeletal lattice, and not through a signaling cascade (Maniotis et al., 1997).

Cell Fates: Growth, Differentiation, and Apoptosis

Cell proliferation (multiplication through mitosis), differentiation (changes in phenotype and matrix production), and apoptosis (programmed cell death) are all heavily reliant on the signaling mechanisms which were previously discussed. Most importantly, experimental observations combined with the tensegrity model account for how cellular interactions with their local environment, whether from other cells or ECM, can cause these different modes to be triggered. As previously explained, damaged or reconstructed ECM exhibit degraded mechanical properties. Following injury of normal tissue and subsequent loss of cellular elements, residual ECM will remain intact and will promote organized cellular tissue regrowth, ensuring the correct cellular placement and alignment. The intracellular changes that result from local alteration in ECM permit the cells to respond to soluble growth factors and other mitogens, thereby driving changes in tissue phenotype (Ingber, 2003b).

Qualitative data on these intracellular changes serve to corroborate the mechanotransduction models. In a study by Chen et al., cell shape seemed to determine whether individual cells proliferate or undergo apoptosis, independent of the growth factor used to stabilize cell adhesions. The study first confirmed that attached capillary ECs display a flattened nucleus with an extended morphology when cultured on flat beads, as opposed to suspended cells which remain small and spherical. More importantly, the study also confirmed that, while most cells survived when spread on larger beads coated in fibronectin, cells seeded on the same beads with decreased diameter became more rounded, and matched the apoptotic pattern of non-adherent cells. The molecular trigger for this programming may be linked to the "hard-wired" nuclear mechanisms, affected by cytoskeleton rearrangement (Chen et al., 1997). The results of a more recent study would also indicate that the formation of ECM, via cooperative interactions between integrins, the cytoskeleton, and three-dimensional tissue organization, confers prevention of apoptosis. Furthermore, this study showed that laminin-induced integrin ligation directs tissue polarity and promotes resistance to apoptosis, regardless of growth status (Weaver et al., 2002).

It has been shown that transitions between growth and differentiation stimulated by these mitogens can also be influenced by the elasticity and geometry of the growth substrate, like the developing ECM. For example, a study by Mochitate et al. (1991) showed that collagen gels seeded with human fibroblasts which underwent stress relaxation (transient hypercontraction followed by dissipation) led to differences in cell morphology and biosynthetic activity, including disruption of their actin filament bundles, loss of cell surface fibronectin, and marked decrease in both cellular DNA and protein synthesis. Although mechanical stimulation appears to trigger the transition between growth and differentiation, what mechanisms are involved are not firmly established.

OVERVIEW OF BIOREACTORS

Bioreactors provide an *in vitro* environment for tissue generation and growth. Ideally they mimic the mechanochemical regulation that tissues experience *in vivo* in their native environment. The key functions of the bioreactor are to: (1) allow the seeding of uniform concentrations of cells to a scaffold; (2) control physiological conditions in the cell culture medium (e.g. temperature, pH, oxygen levels, nutrients); (3) supply sufficient metabolites; and (4) provide physiologically relevant signals in the form of mechanical loads (Altman et al., 2002; Freed et al., 2000). Since there are many types of bioreactors in current use, one must choose a tissue-appropriate design that incorporates the unique set of *in vitro* environmental and mechanical loading conditions that can produce a tissue that is as similar as possible to the native tissue. The following is a survey of some bioreactor designs in current use. For additional details on bioreactor design and specific comparisons, the reader is referred to additional sources (Barron et al., 2003; Vunjak-Novakovic et al., 2004).

Types

The simplest type of bioreactor is the static flask; a tissue construct is fixed in place in a culture medium. Gas aeration is provided by surface aeration of the culture medium. Mass transfer, therefore, occurs by molecular diffusion since there is no fluid flow at the surface of the tissue construct (Barron et al., 2003). The structures formed under static conditions tend to exhibit limited cellular ingrowth, resulting in two-dimensional tissue structures.

To better produce clinically relevant three-dimensional tissues, recent advances toward more biomimetic bioreactor designs with complex environments have been implemented. A stirred-flask bioreactor uses a magnetic stirrer to mix a dilute cell suspension around a stationary scaffold, aiding in cell distribution through the scaffold. Stirring of the culture medium produces mass transfer through turbulent convection, generates shear stresses that enhance cell and tissue growth in comparison to static incubation conditions, and improves nutrient supply through the scaffold (Barron et al., 2003; Nasseri et al., 2003; Martin et al., 2004). A wavy walled bioreactor is similar to a spinner-flask bioreactor, with the exception that the flask wall contains wavy contours that mimic baffles. This design provides a range of hydrodynamic forces, enhancing mixing of the culture medium (Bilgen et al., 2005). A rotating-wall vessel bioreactor provides a dynamic environment in which two concentric cylinders are horizontally rotated. Cells are grown on tissue constructs which are freely suspended in the annular volume, in essentially a microgravity environment, which is filled with culture medium (Nasseri et al., 2003). The tissue constructs benefit from the laminar flow, low shear stress fluid environment, and improved supply of nutrients and outflow of wastes (Barron et al., 2003; Martin et al., 2004). Direct perfusion bioreactors involve the perfusion of culture medium through tissue constructs. The perfusion can produce higher density and more uniform distribution of cells than with stirred-flask bioreactors and provide mechanical stress in the form of flow-induced shear (Martin et al., 2004). More advanced bioreactor designs include strain actuation that can apply static and dynamic loads, as found in the Flexcell line of bioreactors (Shukla et al., 2004).

Modeling

Given the need for future bioreactors to more faithfully represent the *in vivo* environment, which is very complex, the use of analytical and computational modeling will become more important. Computational fluid dynamic (CFD) software is a powerful tool to calculate flow fields, shear stresses, and mass transport within and around three-dimensional tissue constructs. CFD models have been use to study oxygen transport in a rotating-wall bioreactor, to model direct perfusion in various scaffold designs, and to evaluation the effect of pore structure and interconnectivity on tissue development (Martin et al., 2004). CFD models can help optimize bioreactor design and flow conditions (Bilgen et al., 2005). To aid in modeling efforts, additional modern tools have been employed. For example, computed tomography scanning can be used to construct computer-based models of tissue engineering scaffolds (Cioffi et al., 2005). Additional technology is showing promise for tissue engineering, including rapid prototyping, the introduction of smart materials in scaffolds, and advanced manufacturing techniques like electrospinning.

PRACTICAL EXAMPLES OF MECHANICAL DETERMINANTS

The previous reviews of mechanotransduction and general cell responses outline cellular behavior in their mechanical environment. Several examples of these phenomena are offered next, as they occur in specific tissue lineages: vasculature, bone, and cartilage. The goal is to provide a sense of the physiological loading regimes, signaling that is transduced to matrix and cells, and their application in bioreactor design.

Vasculature – Endothelial and Smooth Muscle Cell

Loading Conditions

Vasculature is made up of smooth muscle cells (SMCs), ECM (collagen and elastin fibrils), and ECs. ECs form a monolayer that covers the innermost aspects of vasculature, providing a barrier between flowing blood and the tissue wall. The SMCs and ECM provide the proper shape and size for blood flow, constrain the ECs, and provide structural integrity to withstand internal and external stresses (Davies, 1995). The three main components of vasculature, SMCs, ECM, and ECs are all subjected to stretching (strain) as a result of pulsatile blood flow. The amount of strain that is generated is directly related to blood pressure. Shear stress, due to fluid flow-generated frictional forces, is experienced principally by the ECs. These shear stress levels vary with the blood velocity profile generated during the cardiac cycle and with the shape and size characteristics of the vasculature. For example, curves in arterial walls can lead to flow separation and the development of vortices, which affects the shear stress near the vessel wall. It should be noted that pulsatile blood flow and blood elements also influence EC responses by varying the luminal concentrations of growth factors and other soluble mitogens that interact with apical surface integrin receptors (Davies, 1995). Only effects of fluid-induced shear stress on vasculature cell responses will be the focus herein.

Cell Response and Transduction Mechanisms

In response to fluid-induced shear stress, monolayers of ECs change morphology and become torpedo shaped, aligned in the fluid flow direction. It would follow that the cytoskeletal backbones of EC undergo major alterations, in which stress fibers reinforce the EC membrane (Satcher and Dewey, 1996). Many improvements in the estimation of cytoskeletal structure and strength of EC have aided in the understanding of how these and other human cells respond to external stimuli through cytoskeletal remodeling (Fung, 1993; Satcher and Dewey, 1996; Helmke and Davies, 2002). In the same way that mathematical models of whole tissue structures, based on their estimates of substructure and geometry, are used to aid in mechanical analysis of load distribution, the same tools are widely used in EC cytoskeletal analysis.

Because of cytoskeletal responses, shear stresses acting on the luminal cell membrane of ECs *in vivo* are transmitted to the basal attachment sites. It is believed that the collection of plaque proteins ABP and spectrin, used to reinforce integrins at focal complexes on the basal side of EC, work in concert with reassembly effects (Satcher and Dewey, 1996). It is unclear whether focal complex enhancement is solely driven by basal side integrin activation, or if further support is also provided by the translocation of inactive apical side integrins to the basal membrane following shear stress (Shyy and Chien, 2002). In either case, the development of focal adhesions will lead to recruitment of cytoplasmic signaling molecules and MAPK signaling pathways. Focal adhesion sites, like the cytoskeleton, align their shape parallel to the flow direction without changing their overall contact area (Helmke and Davies, 2002).

Activated luminal cell surface mechanisms (stretch-activated or potassium ion channels) have been linked to EC shear strain response. Similarly, G-protein activation due to distortions of the plasma membrane from shear has also been documented (Helmke and Davies, 2002). These and the integrin-dependent mechanisms are part of either the inside-out or outside-in signaling routes that develop from an EC's complex response to shear.

Bioreactor Design

In the engineering of cardiovascular tissue, it is believed that bioreactor design should involve laminar fluid flows that induce a uniform distribution of shear stress and laminar convective mass transfer. Rotating-wall bioreactors have been used to create engineered cardiac tissues that are structurally and functionally superior to those grown in static or mixed flasks (Barron et al., 2003; Martin et al., 2004). Other bioreactors have been used which include strain actuation, mimicking the dynamic mechanical stimuli present *in vivo*. For example, it is thought that since arteries experience axial strains through connective tissue, tubular scaffolds that represent a cardiovascular vessel should experience the same strain. In addition, circumferential strains can be provided by a pulsatile force through the tissue scaffold, mimicking pulsatile blood flow in actual arteries (McCulloch et al., 2004).

Bone – Osteocytes, Osteoblasts, and Osteoclasts

Loading Conditions

The enduring principles established by Wolff (i.e. Wolff's Law) state that the rate and degree of new bone tissue deposition is dependent on the tissue's stress levels, and that the pattern of bone architecture coincide with stress trajectories (Wolff, 1870). Because of the rigidity (overall structural stiffness) of bone tissue, its deformation resulting from gross loads is small, typically on the order of microstrain (where 10,000 microstrain is the same as a 1% change in length). One study, reporting the use of high-resolution ($\pm$0.08 microstrain) strain gauges to measure *in vivo* bone strains, showed that during the course of a day, few high-magnitude ($>$1,000 microstrain) events occur. Furthermore, daily strains which fall below 0.2% strain are the predominant contribution to the strain history (strain cycling over time) of bone, and are encountered during the body's regular posture corrections (Fritton et al., 2000). Assuming that focal adhesions of bone cells are distributed along force-bearing members of the surrounding ECM, osteocyte stretch will reflect microstrain-level deformation. However, most *in vitro* work on osteoctye response to dynamic substrates requires substrate deformations at least 1–2 orders of magnitude larger to induce changes in bone modeling (Han et al., 2004).

The contradiction between modeled results using *in vivo* parameters and *in vitro* requirements can be rationalized by noting that mechanical loads applied to bone *in vivo* can cause increased pressure gradients and interstitial fluid flow (IFF) through bone channels, as the microcollapsed pores undergo volume changes. Additionally, it is hypothesized that cytoskeletal rearrangements (intracellular actin-bundle strains) can compensate for this distortion (Han et al., 2004). It has been suggested that the bending stresses in bone cause local and opposing tension and compression forces, which generate IFF in addition to the expected microstrains (Turner and Pavalko, 1998). The fluid shear from the IFF is responsible for the larger order disturbances to osteocytes (Cowin and Weinbaum, 1998).

Cell Response and Transduction Mechanisms

The mechanosensory mechanisms of bone tissues adopt the mechanotransduction models previously discussed, but are complicated by the signaling between bone sensory cells (osteocytes) and their effector cells (osteoblasts and osteoclasts), which are ultimately responsible for bone homeostasis and adaptation to strains. These cells, not including osteoclasts, are linked through what has been called a "connected cellular network" (CCN) (Cowin et al., 1991). Normal bone remodeling involves the creation of canals (bone resorption) by osteoclasts, followed by a filling of surface sites with mineralized osteoid (fibrous organic matrix) by stem-cell-derived osteoblasts. After resorption is triggered, it is believed that receptor-based mitogen (TGF-β, etc.) signaling and production of prostaglandin and nitric oxide (NO) initializes osteoblast activity. From there, the osteoblasts will either maintain their phenotype, resting at the new bone periphery, or differentiate into osteocytes while encapsulated in the surrounding matrix.

Canaliculi, or nutrient and biochemical channels between bone cells, connect the embedded osteocytes and surrounding osteoblasts to form the CCN (Huiskes and van Rietbergen, 2005). Intracellular epigenetic mechanisms (cytoskeletal prestress, biochemistry, etc.) in the CCN allow the cells to respond to physical activity, while gap junctions are thought to act as electrical synapses to permit or block bidirectional information exchange (Cowin et al., 1991). Although damage to microstructure appears to be the source of osteoclast recruitment signaling, it is unclear whether the signals stem from breaks in the CCN pathways or from damage to the osteocyte matrix (Huiskes and van Rietbergen, 2005).

Varying mechanical factors, including stress, strain-rate, and fatigue microdamage, have been extensively investigated at the macroscopic level, both of tissue explants and *in vivo*, as to their effect on remodeling and developing bone tissue. In some cases, the effects seem to be time dependent, while other experiments seem to indicate that the amplitude of oscillatory components or peak stress/strain values has the most impact. To parallel the studies on gross loading effects, research is now focused on modeling the CCN, as a number of densely interconnected electrical processing elements, or through computational finite element modeling that are capable of organizing the multitude of mechanical inputs (You, et al., 2001; Huiskes and van Rietbergen, 2005).

As a transduction vehicle previously discussed, the distortions of bone matrix under strain may lead to cytoskeletal remodeling around the cell nucleus, via surface adhesions and distortions of the internal cytoskeletal lattice (Shyy and Chien, 1997). The mechanical stimulation of bone cells will also increase intracellular calcium levels and production of prostaglandins and NO within minutes, and has been linked to mechanosensitive ion channels (Turner and Pavalko, 1998). Furthermore, slow pulsating and/or oscillatory flow of interstitial fluids has been proven to be more effective in activating osteoctye cells, over situations of hydrostatic pressure or rapid oscillations (Huiskes and van Rietbergen, 2005).

Bioreactor Design

Several publications highlight attempts to recreate these mechanical loading regimes for engineered bone within different bioreactor environments. In one study, enhancements to static culture environments were incorporated through spinner-flask and rotating-wall bioreactors, which stimulated mesenchymal stem cells (MSCs) through bulk convective flow stimulation and centrifugal force balance, respectively. The results of dynamic culture in the spinner flask were most promising in comparison with rotating-wall vessel culture due to enhanced mixing; however, both showed marked improvement over static culture (Sikavitsas et al., 2001). Another more recent study compared static culture to spinner-flask cultures and perfused cartridge culture, one in which laminar IFF (35 μm/s) is mimicked using a gross fluid pressure differential across the sides of a cell-seeded construct. The results of this study indicated that although spinner-flask culture was the most successful at bulk generation of bone markers, the perfusion bioreactor did so in a randomly distributed manner throughout the construct's volume (Meinel et al., 2004).

To recreate load-induced IFF through indirect measures, one recent study cultured MSC on partially demineralized bone scaffolds subjected to cyclic bending loads in a custom-designed static flow bioreactor. The results of this study showed that mechanical stimulation of this nature promoted osteogenic differentiation of MSC by significantly elevating alkaline phosphatase activity and calcium deposition (known markers for bone) over static controls (Mauney et al., 2004).

Cartilage – Chondrocytes

Loading Conditions

Because of the hydrophilic aggrecan proteins, and thus the large water content (~65–85%) within cartilage tissues, the gross mechanical response to compression is somewhat like compressing a pneumatic tire, leading to stress levels varying between 0 and 20 MPa during movements. Collagen fibrils and other matrix proteins compensate for stress-bearing responsibilities under tensile and compressive loads, respectively. The response of cartilage tissue is nonlinear and time dependent (viscoelastic), meaning that deformation will increase with a constant applied stress. Like bone tissue, cartilage also behaves anisotropically and is subject to complex loads, including bending and shear. A thorough review of experimentally derived modeling considerations are outside the scope of this review, and can be found elsewhere (Mow et al., 2005).

Cell Response and Transduction Mechanisms

As in the two previous examples, chondrocytes within cartilage tissue can sense and respond to mechanical stimuli; however, chondrocytes do not rely on the stimulus/effector relationship with other cell phenotypes. Instead, the interactions of chondrocytes with their matrix seem to be critical, owing to the scarcity of chondrocytes within most cartilage tissues (~5%). Cartilage tissue is almost completely avascular and aneural, which somewhat simplifies the study of chondrocytes and their matrix in tissue engineering research. *In vivo*, static compression of the tissue to physiological strain magnitudes leads to breakdown of cartilage proteoglycan, not renewal; however cyclic loads of a higher magnitude or frequency can also be deleterious (Mow et al., 2005). Therefore, consistent and mid-level stresses appear to create favorable mechanical environments for cartilage regeneration *in vitro*.

The rapid or acute response of chondrocytes to mechanical stimulation was studied *in vitro* using a two-dimensional monolayer model, and revealed that substrate stretch, via deforming pressure gradients of 50 kPa at 0.33 Hz, induced membrane hyperpolarization within 20 min. *In vitro*, this study confirmed that tyrosine phosphorylation of both paxillin and FAK was induced within 1 min of initiation of stretch, and led to the eventual signaling cascade inducing small conductance potassium channels (Millward-Sadler and Salter, 2004).

Bioreactor Design

For tissue engineering of cartilage, *in vitro* cultivation of chondrocytes seeded on biodegradable scaffolds has been pursued using spinner-flask bioreactors. However, some research has shown that the turbulent, high-shear mixing environment in spinner flasks can lead to altered, undesirable morphology of the engineered cartilage tissue (Sucosky et al., 2004). The use of a wavy walled bioreactor can produce enhanced mixing at low shear levels, leading to increased rates of formation and size of chondrocyte aggregates in suspension cultures (Bilgen et al., 2005). Other research has shown that flow-induced shear stress can be beneficial in increasing ECM component production by chondrocytes. A bioreactor that includes direct perfusion should also exhibit enhanced convective transport of nutrients to the cells and catabolites (waste) away. The level of shear stress applied depends on the culture medium flow rate through the constructs and also the three-dimensional scaffold geometry (Cioffi et al., 2005).

CONCLUSION

Since mechanical forces play a crucial role in tissue development, function and repair *in vivo*, the design of novel bioreactors to impart complex mechanical forces to cells and tissues *in vitro* offers important options to improve functional tissue engineering. These inputs have to be considered within the context of the biomaterial scaffolds used in the bioreactors to transmit the applied forces or to handle fluid flow, to the cells used in these systems, and to the overall system needs to generate functional tissues *in vitro* for utility *in vivo*. Full restoration of a mechanical match for tissue grown *in vitro* to repair needs *in vivo* may not be required, as long as the engineered tissue satisfies temporary mechanical and related requirements until tissue regeneration and integration is achieved. It is clear that the road ahead is challenging, yet promising results and approaches as summarized here offer a glimpse into the future opportunities and therapeutic benefits.

REFERENCES

Altman, G., Lu, H., Horan, R., Calabro, T., Ryder, D., Kaplan, D., Stark, P., Martin, I., Richmond, J. and Vunjak-Novakovic, G. (2002). Advanced bioreactor with controlled application of multi-dimensional strain for tissue engineering. *J. Biomech. Eng.* **124**: 742–749.

Barron, V., Lyons, E., Stenson-Cox, C., McHugh, P.E. and Pandit, A. (2003). Bioreactors for cardiovascular cell and tissue growth: a review. *Anna. Biomed. Eng.* **31**: 1017–1030.

Bassell, G.J., Powers, C.M., Taneja, K.L. and Singer, R.H. (1994). Single mRNAs visualized by ultrastructural *in situ* hybridization are principally localized actin filament intersections in fibroblasts. *J. Cell Biol.* **126**: 863–876.

Bilgen, B., Chang-Mateu, I. and Barabino, G. (2005). Characterization of mixing in a novel wavy-walled bioreactor for tissue engineering. *Biotech. Bioeng.* **92**(7): 907–919.

Boal, D. (2002). *Mechanics of the Cell*. Cambridge, UK: Cambridge University Press.

Chen, C.S., Mrksich, M., Huang, S., Whitesides, G.M. and Ingber, D.E. (1997). Geometric control of cell life and death. *Science* **276**: 1425–1428.

Cioffi, M., Boschetti, F., Raimondi, M. and Dubini, G. (2006). Modeling evaluation of the fluid-dynamic microenvironment in tissue-engineered constructs: a micro-CT based model. Publication pending for *Biotech. Bioeng*. **93**: 500–510.

Cowin, S.C. and Weinbaum, S. (1998). Strain amplification in the bone mechanosensory system. *Am. J. Med. Sci.* **316**: 184–188.

Cowin, S.C., Moss-Salentijn, L. and Moss, M.L. (1991). Candidates for the mechanosensory system in bone. *J. Biomech. Eng.* **113**: 191–197.

Davies, P.F. (1995). Flow-mediated endothelial mechanotransduction. *Physiol. Rev.* **75**: 519–560.

Doyley, M.M., Srinivasan, S., Prendergrass, S.A., Wu, Z. and Ophir, J. (2005). Comparative evaluations of strain-based and model-based modulus elastography. *Ultrasound Med. Biol.* **31**: 787–802.

Freed, L.E. and Vunjak-Novakovic, G. (2000). Tissue Engineering Bioreactors. In: Lanza, R.P., Langer, R. and Vacanti, J. (eds.), *Principles of Tissue Engineering*. London: Academic Press, pp. 143–154.

Fritton, S.P., McLeod, K.J. and Rubin, C.T. (2000). Quantifying the strain history of bone: spatial uniformity and self-similarity of low-magnitude strains. *J. Biomech.* **33**: 317–325.

Fung, Y.C. (1993). *Biomechanics: Mechanical Properties of Living Tissues*, 2nd edn. New York: Springer-Verlag.

Geiger, B., Bershadsky, A., Pankov, R. and Yamada, K.M. (2001). Transmembrane extracellular matrix cytoskeleton cross-talk. *Nat. Rev. Mol. Cell Biol.* **2**(11): 793–805.

Giancotti, F.G. (2003). A structural view of integrin activation and signaling. *Dev. Cell* **4**: 149–151.

Giancotti, F.G. and Ruoslahti, E. (1999). Integrin signaling. *Science* **285**: 1028–1032.

Hamill, O.P. and Martinac, B. (2001). Molecular basis of mechanotransduction in living cells. *Physiol. Rev.* **81**: 685–740.

Han, Y., Cowin, S.C., Schaffler, M.B. and Weinbaum, S. (2004). Mechanotransduction and strain amplification in osteocytes cell processes. *Proc. Natl Acad. Sci. USA* **101**: 16689–16694.

Helmke, B.P. and Davies, P.F. (2002). The cytoskeleton under external fluid mechanical forces: hemodynamic forces acting on the endothelium. *Ann. Biomed. Eng.* **30**: 284–296.

Hibbeler, R.C. (2000). *Mechanics of Materials*, 4th edn. New Jersey: Prentice Hall.

Huang, H., Kamm, R.D. and Lee, R.T. (2004). Cell mechanics and mechanotransduction: pathways, probes and physiology. *Am. J. Physiol. Cell Physiol.* **287**: C1–C11.

Huiskes, R. and van Rietbergen, B. (2005). Biomechanics of Bone. In: Mow and Huiskes (eds.), *Basic Orthopaedic Biomechanics and Mechano-Biology*, 3rd edn. New York: Lippincott, Williams & Wilkins, pp. 123–179.

Ingber, D.E. (1997). Tensegrity: the architectural basis of cellular mechanotransduction. *Annu. Rev. Physiol.* **59**: 575–599.

Ingber, D.E. (2003a). Cellular tensegrity I. Cell structure and hierarchical systems biology. *J. Cell Sci.* **116**: 1157–1173.

Ingber, D.E. (2003b). Cellular tensegrity II. How structural networks influence cellular information processing networks. *J. Cell Sci.* **116**: 1397–1408.

Janmey, P.A. and Weitz, D.A. (2004). Dealing with mechanics: mechanisms of force transduction in cells. *Trends Biochem. Sci.* **29**(7): 364–370.

Juncosa, N., West, J., Galloway, M., Boivin, G. and Butler, D. (2003). *In vivo* forces used to develop design parameters for tissue engineered implants for rabbit patellar tendon repair. *J. Biomech.* **36**: 483–488.

Kaplan, D.L., Moon, R.T. and Vunjak-Novakovic, G. (2005). It takes a village to grow a tissue. *Nat. Biotechnol.* **23**: 1237–1239.

Kung, C. (2005). A possible unifying principle for mechanosensation. *Nature* **436**(7051): 647–654.

Maniotis, A.J., Chen, C.S. and Ingber, D.E. (1997). Direct evidence for mechanical connections between cell surface integrin receptors, cytoskeletal filaments, and the nucleoplasm of living cells. *Proc. Natl Acad. USA* **94**: 849–854.

Martin, I., Wendt, D. and Heberer, M. (2004). The role of bioreactors in tissue engineering. *Trends Biotechnol.* **22**(2): 80–86.

Mauney, J.R., Sjostorm, S., Blumberg, J., Horan, R., O'Leary, J.P., Vunjak-Novakovic, G., Volloch, V. and Kaplan, D.L. (2004). Mechanical stimulation promotes osteogenic differentiation of human bone marrow stromal cells on 3-D partially demineralized bone scaffolds *in vitro*. *Calcif. Tissue Int.* **74**: 458–468.

McBride Jr., D.W. and Hamill, O.P. (1999). A simplified fast pressure clamp technique for studying mechanically gated channels. *Method. Enzymol.* **294**: 482–489.

McCulloch, A., Harris, A., Sarraf, C. and Eastwood, M. (2004). New multi-cue bioreactor for tissue engineering of tubular cardiovascular samples under physiological conditions. *Tissue Eng.* **10**(3/4): 565–573.

Meinel, L., Karageorgiou, V., Fajardo, R., Snyder, B., Shinde-Patil, V., Zichner, L., Kaplan, D.L., Langer, R. and Vunjak-Novakovic, G. (2004). Bone tissue engineering using human mesenchymal stem cells: effects of scaffold material and medium flow. *Ann. Biomed. Eng.* **32**: 112–122.

Millward-Sadler and Salter, D.M. (2004). Integrin-dependent signal cascades in chondrocytes mechanotransduction. *Ann. Biomed. Eng.* **32**: 435–446.

Mitra, S.K., Hanson, D.A. and Schlaepfer, D.D. (2005). Focal adhesion kinase: in command and control of cell motility. *Nat. Rev. Mol. Cell Biol.* **6**: 56–68.

Mochitate, K., Pawekek, P. and Grinnell, F. (1991). Stress relaxation of contracted collagen gels: disruption of actin filament bundles, release of cell surface fibronectin, and down-regulation of DNA and protein synthesis. *Exp. Cell Res.* **193**: 198–207.

Mow, V.C. and Huiskes, R. (2005). *Basic Orthopaedic Biomechanics and Mechano-Biology*, 3rd edn. New York: Lippincott, Williams & Wilkins.

Mow, V.C., Gu, W.Y. and Chen, F.H. (2005). Structure and function of articular cartilage and meniscus. In: Mow and Huiskes (eds.), *Basic Orthopaedic Biomechanics and Mechano-Biology*, 3rd edn. New York: Lippincott, Williams & Wilkins, pp. 29–89.

Nasseri, B., Pomerantseva, I., Kaazempur-Mofrad, M., Sutherland, F., Perry, T., Ochoa, E., Thompson, C., Mayer, J., Oesterle, S. and Vacanti, J. (2003). Dynamic rotational seeding and cell culture system for vascular tube formation. *Tissue Eng.* **9**(2): 291–298.

Ottensmeyer, M.P. and Salisbury, J. (2001). *In vivo* data acquisition instrument for solid organ mechanical property measurement. *Lect. Not. Comp. Sci.* **2208**: 975–982.

Prendergast, P.J., Van Der Helm, F.C.T. and Duda, G.N. (2005). Analysis of muscle and joint loads. In: Mow and Huiskes (eds.), *Basic Orthopaedic Biomechanics and Mechano-Biology*, 3rd edn. New York: Lippincott, Williams & Wilkins, pp. 29–89.

Satcher, R.L. and Dewey, C.F. (1996). Theoretical estimates of mechanical properties of the endothelial cell cytoskeleton. *Biophys. J.* **71**: 109–118.

Sawada, Y. and Sheetz, M.P. (2002). Force transduction by Triton cytoskeletons. *J. Cell Biol.* **156**: 609–615.

Shukla, A., Dunn, A.R., Moses, M.A. and Van Vliet, K.J. (2004). Endothelial cells as mechanical transducers: enzymatic activity and network formation under cyclic strain. *Mol. Cell Biomech.* **1**: 279–290.

Shyy, J.Y. and Chien, S. (1997). Role of integrins in cellular responses to mechanical stress and adhesion. *Curr. Opin. Cell Biol.* **9**: 707–713.

Shyy, Y.-J. and Chien, S. (2002). Role of integrins in endothelial mechanosensing of shear stress. *Circ. Res.* **91**: 769–775.

Sikavitsas, V.I., Bancroft, G.N. and Mikos, A.G. (2001). Formation of three-dimensional cell/polymer constructs for bone tissue engineering in a spinner flask and rotating wall vessel bioreactor. *J. Biomed. Mater. Res.* **62**: 136–148.

Stephens, L.E., Sutherland, A.E., Klimanskaya, I.V., Andrieux, A., Meneses, J., Pedersen, R.A. and Damsky, C.H. (1995). Deletion of beta 1 integrins in mice results in inner cell mass failure and peri-implantation lethality. *Genes Dev.* **9**: 1883–1895.

Sucosky, P., Osorio, D., Brown, J. and Neitzel, G. (2004). Fluid mechanics of a spinner-flask bioreactor. *Biotechnol. Bioeng.* **85**(1): 34–46.

Turner, C.H. and Pavalko, F.M. (1998). Mechanotransduction and function response of the skeleton to physical stress: the mechanisms and mechanics of bone adaptation. *J. Orthop. Sci.* **3**: 346–355.

Vunjak-Novakovic, G., Altman, G., Horan, R. and Kaplan, D.L. (2004). Tissue engineering of ligaments. *Annu. Rev. Biomed. Eng.* **6**: 131–156.

Vunjak-Novakovic, G., Meinel, L., Altman, G. and Kaplan, D.L. (2005). Bioreactor cultivation of osteochondral grafts. *Orthod. Craniofac. Res.* **8**: 209–218.

Weaver, V.M., Lelievre, S., Lakins, J.N. Chrenek, M.A., Jones, J., Giancotti, F., Werb, Z. and Bissell, M.J. (2002). B4 integrin-dependent formation of polarized three-dimensional architecture confers resistance to apoptosis in normal and malignant mammary epithelium. *Cancer Cell* **2**: 205–216.

Wolff, J. (1870). Uber der innere Architektur der Knochen und ihre Bedeutung fur die Frage vom Knochenwachstum. *Arch. Pathol. Anat. Physiol. Klin. Med.* **50**: 389–453.

were established for chemotaxis, mitogenesis, and chondrogenesis, and an *in vivo* bioassay for bone formation. Although the *in vitro* assays are expedient, we monitored routinely a labor-intensive *in vivo* bioassay as it is the only valid *bona fide* bone induction assay.

A major stumbling block in the approach was that the demineralized bone matrix is insoluble and in the solid state. In view of this dissociative extractants such as 4 M guanidine HCl or 8 M urea as 1% sodium dodecyl sulfate (SDS) at pH 7.4 were used (Sampath and Reddi, 1981) to solubilize proteins. Approximately 3% of the proteins were solubilized from demineralized bone matrix, and the remaining residue was mainly insoluble type I bone collagen. The extract alone or the residue alone was incapable of new bone induction. However, addition of the extract to the residue (insoluble collagen) and then implantation in a subcutaneous site resulted in bone induction (Figure 14.4). Therefore, for optimal osteogenic activity it is essential to have a collaboration between soluble signal in the extract and the insoluble substratum of collagenous ECM (Sampath and Reddi, 1981). This bioassay was a critical advance in the ultimate purification of BMPs and led to determination of limited tryptic peptide sequences leading to the eventual cloning of BMPs (Wozney et al., 1988; Luyten et al., 1989; Ozkaynak et al., 1990). The dissociative extraction of soluble signals from the demineralized ECM of bone and its subsequent reconstitution with collagen established the cardinal principle of regenerative medicine. The key principle is that morphogenetic signals stimulate the stem cells to differentiate in the optimal scaffold microenvironment (Figure 14.5). Thus, the triumvirate of signals, stem cells, and scaffolds for regenerative medicine was conceived as a concept.

Although the basic description of bone induction was performed in rats, purification requires a larger and more abundant source of bone. A switch was made to bovine bone. Demineralized bovine bone matrix was not osteoinductive in rats and the results were variable. However, when the guanidine extracts of demineralized

Three key steps in bone morphogenesis

- Chemotaxis
- Mitosis
- Differentiation

Figure 14.3 **The three key steps in bone morphogenesis.**

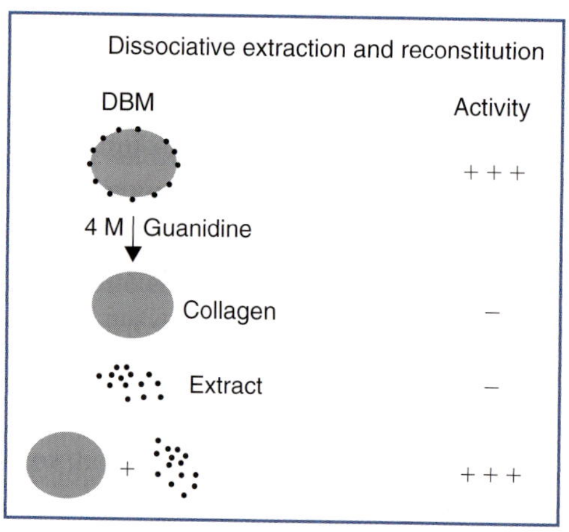

Figure 14.4 **Dissociative extraction of bone matrix by chaotropic reagents such as 4 M guanidine hydrochloride, and reconstitution of extract with collagenous matrix scaffold. The results indicate that there is a collaboration between soluble signal in the extract and the insoluble ECM of bone.**

bovine bone were fractionated on a S-200 molecular sieve column, fractions less than 50 kD were consistently osteogenic in rats when bioassayed after reconstitution with allogeneic insoluble collagen (Sampath and Reddi, 1983; Reddi, 1994). Thus, protein fractions inducing bone were not species specific and appear to be homologous in several mammals. It is likely that larger molecular mass fractions and/or the insoluble xenogeneic (bovine and human) collagens were inhibitory or immunogenic. Initial estimates revealed 1 μg of active osteogenic fraction in a kilogram of bone. Hence, over a ton of bovine bone was processed to yield optimal amounts for animo acid sequence determination. The amino acid sequences revealed homology to transforming growth factor (TGF)-β_1 (Reddi, 1994). The decisive work of Wozney et al. (1988) cloned BMP-2, BMP-2B (now called BMP-4), and BMP-3 (also called osteogenin). Ozkaynak et al. (1990) cloned osteogenic proteins 1 and 2 (OP 1 and OP 2). There are several members of this BMP family (Figure 14.6). The other members of the extended TGF-β/BMP superfamily include inhibins and activins (implicated in follicle stimulating hormone release from pituitary). Müllerian duct inhibitory substance (MIS), growth/differentiation factors (GDFs), nodal, and lefty genes implicated in establishing right/left asymmetry (Cunningham et al., 1995;

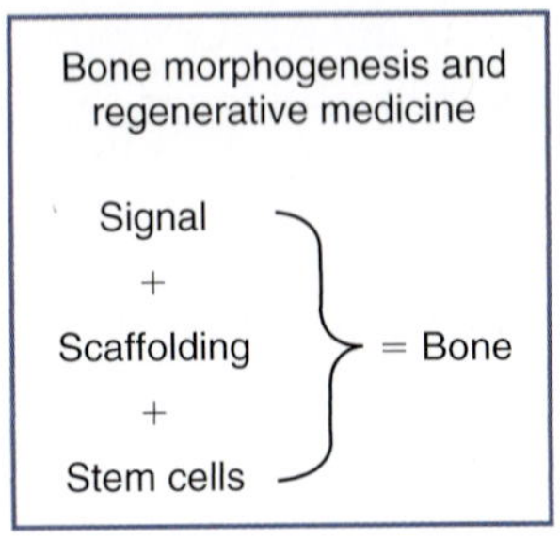

Figure 14.5 **The key principle of regenerative medicine is that signals stimulate differentiation of stem cells in the appropriate scaffold.**

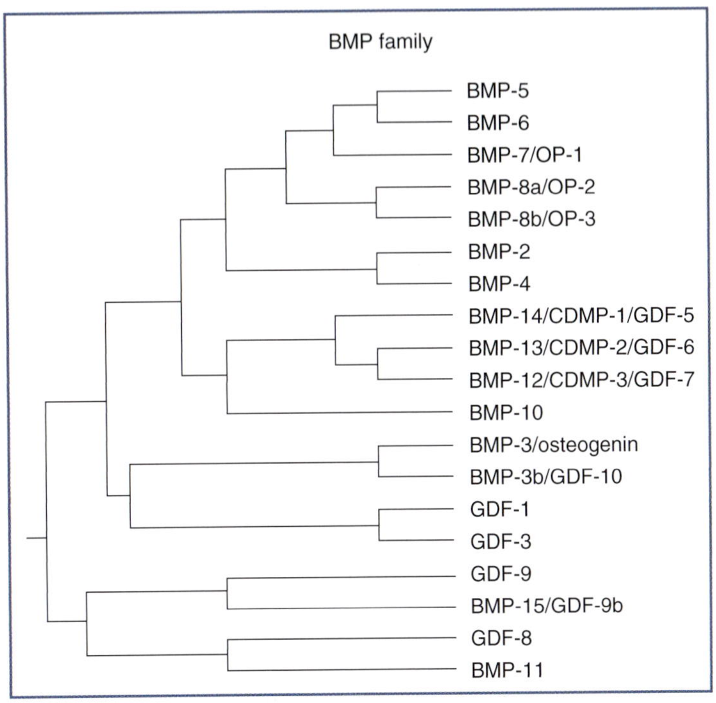

Figure 14.6 **Members of the BMP family include three main subfamilies: BMP 5, 6, and 7; BMP 2 and 4; BMP 3 and 3b; and GDF 5, 6, and 7.**

Reddi, 1997; Reddi, 1998). BMPs are also involved in embryonic induction (Melton, 1991; Lemaire and Gurdon, 1994; Lyons et al., 1995; Reddi, 1997).

BMPs are dimeric molecules and the conformation is critical for biological actions. Reduction of the single interchain disulfide bond resulted in the loss of biological activity. The mature monomer molecule consists of about 120 amino acids, with 7 canonical cysteine residues. There are three intrachain disulfides per monomer and one interchain disulfide bond in the dimer. In the critical core of the BMP monomer is the cysteine knot. The crystal structure of BMP-7 has been determined (Griffith et al., 1996).

Morphogenesis is a sequential multi-step cascade. BMPs regulate each of the key steps: chemotaxis, mitosis, and differentiation of cartilage and bone. BMPs initiate chondrogenesis in the limb (Chen et al., 1991; Duboule, 1994). The apical ectodermal ridge is the source of BMPs in the developing limb bud. The intricate dynamic, reciprocal interactions between the ectodermally derived epithelium and mesoderm-derived mesenchyme sets into motion the train of events culminating in the pattern of phalanges, radius, ulna and the humerus.

The chemotaxis of human monocytes is optimal at femtomolar concentration (Cunningham et al., 1992). The apparent affinity was 100–200 pM. The mitogenic response was optimal at 100 pM range. The initiation of differentiation was in nanomolar range in solution. However, caution should be exercised as BMPs may be sequestered by ECM components and the local concentration may be higher when BMPs are bounded on the ECM. Thus BMPs are pleiotropic regulators that act in concentration-dependent thresholds.

It is well known that ECM components play a critical role in morphogenesis. The structural macromolecules and their supramolecular assembly in the matrix do not explain their role in epithelial–mesenchymal interaction and morphogenesis. This riddle can now be explained by the binding of BMPs to heparan sulfate heparin, and type IV collagen (Paralkar et al., 1990, 1991, 1992) of the basement membranes. In fact, this might explain in part the necessity for angiogenesis prior to osteogenesis during development. In addition, the actions of activin in development of the frog, in terms of dorsal mesoderm induction, are modified to neuralization by follistatin (Hemmati-Brivanlou et al., 1994). Similarly, Chordin and Noggin from the Spemann organizer induces neuralization by binding and inactivation of BMP-4. Thus neural induction is likely to be a default pathway when BMP-4 is non-functional (Piccolo et al., 1996; Zimmerman et al., 1996). Thus, this is an emerging principle in development and morphogenesis that binding proteins can terminate a dominant morphogen's action and initiate a default pathway. Finally, the binding of a soluble morphogen to ECM converts it into an insoluble matrix bound morphogen to act locally in the solid state (Paralkar et al., 1990).

Although BMPs were isolated and cloned from bone, recent work with gene knockouts has revealed a plethora of actions beyond bone. Mice with targeted disruption of BMP-2 caused embryonic lethality. The heart development is abnormal indicating a need for BMP-2 in heart development (Zhang and Bradley, 1996). BMP-4 "knockouts" exhibit no mesoderm induction, and gastrulation is impaired (Winnier et al., 1996). Transgenic overexpression of BMP-s under the control of keratin 10 promoter leads to psoriasis. The targeted deletion of BMP-7 revealed the critical role of this molecule in kidney and eye development (Dudley et al., 1995; Luo et al., 1995; Vukicevic et al., 1996). Thus the BMPs are really true morphogens for such disparate tissues as skin, heart, kidney, and eye. In view of the emerging wider role, BMPs may be called body morphogenetic proteins (BMPs).

Recombinant human BMP-4 and BMP-7 bind to BMP receptor IA (BMPR-IA) and BMP receptor IB (BMPR-IB) (ten Dijke et al., 1994). CDMP-1 also binds to both the type I BMP receptors. There is a collaboration between type I and II BMP receptors (Nishitoh et al., 1996). The type I receptor serine/threonine kinase phosphorylates a signal-transducing protein substrate called Smad 1 or 5 (Chen et al., 1996). Smad is a term derived from fusion of Drosophila Mad gene and Caenorhabtitis elegans (nematode) Sma gene. Smads 1 and 5 signal in partnership with a common co-Smad, Smad 4 (Figure 14.7). The transcription of BMP-response genes are initiated by Smad 1/Smad 4 heterodimers. Smads are trimeric molecules as gleaned by X-ray crystallography. The phosphorylation of Smads 1 and 5 by type I BMP receptor kinase is inhibited by inhibitory

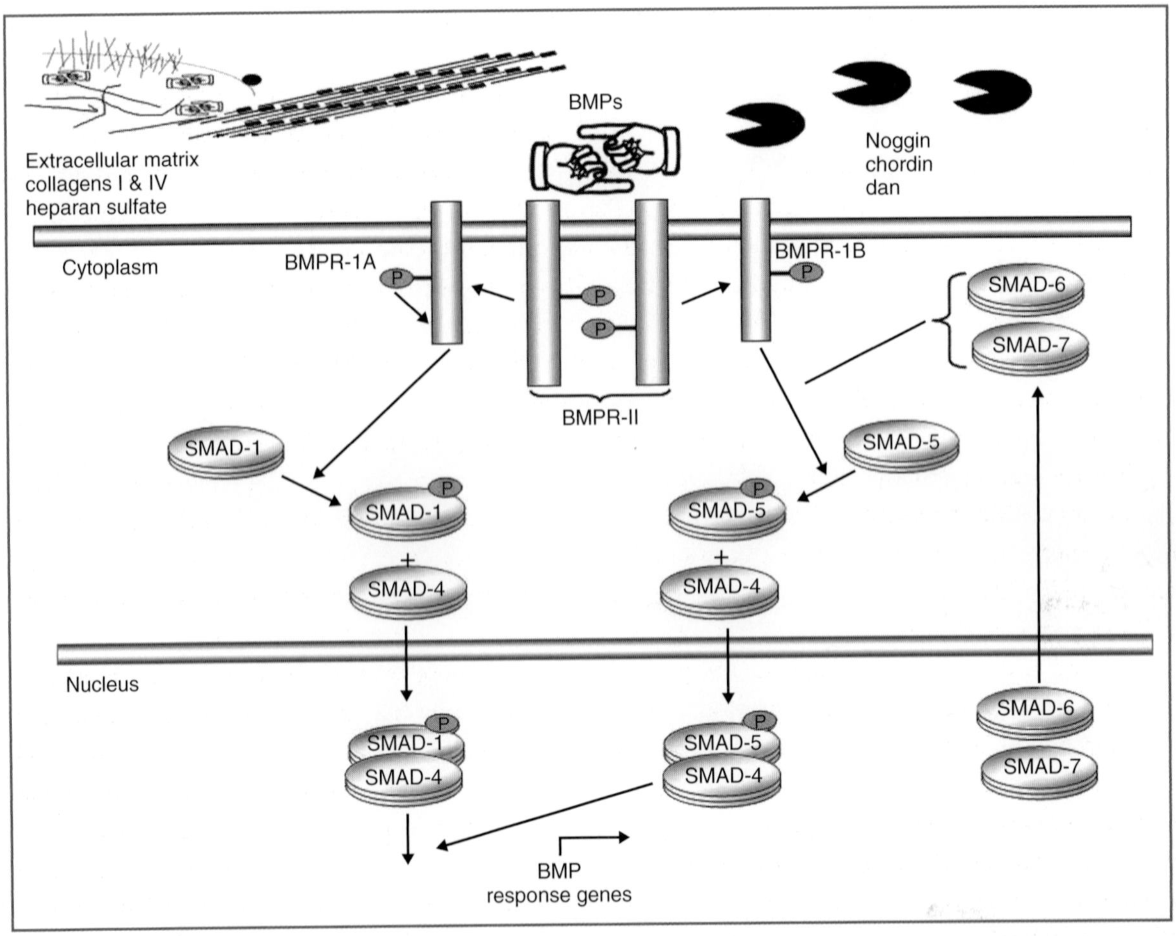

Figure 14.7 BMP receptors and signaling cascades. BMPs are dimeric ligands with cysteine knot in each monomer fold. Each monomer has two β sheets represented as two pointed fingers. In the functional dimer the fingers are oriented in opposite directions. BMPs interact with both type I and II BMP receptors. The exact stoichiometry of the receptor complex is currently being elucidated. BMPR-II phosphorylates the GS domain of BMPR-I. The collaboration between type I and II receptors forms the signal-transducing complex. BMP type I receptor kinase complex phosphorylates the trimeric signaling substrates Smad 1 or Smad 5. This phosphorylation is inhibited and modulated by inhibitory Smads 6 and 7. Phosphorylated Smad 1 or 5 interacts with Smad 4 (functional partner) and enters the nucleus to activate the transcriptional machinery for early BMP-response genes. A novel SIP may interact and modulate the binding of heteromeric Smad 1/Smad 4 complexes to the DNA.

Smads 6 and 7 (Hayashi et al., 1997). Smad interacting protein (SIP) may interact with Smad 1 and modulate BMP-response gene expression (Heldin et al., 1997; Reddi, 1997). The downstream targets of BMP signaling are likely to be homeobox genes, the cardinal genes for morphogenesis and transcription. BMPs in turn may be regulated by members of the hedgehog family of genes such as Sonic and Indian hedgehog (Johnson and Tabin, 1997).

STEM CELLS

It is well known that the embryonic mesoderm-derived mesenchymal cells are progenitors for bone, cartilage, tendons, ligaments, and muscle. However, certain stem cells in adult bone marrow, muscle, and fascia can form bone and cartilage (Figure 14.8). The identification of stem cells readily sourced from bone marrow

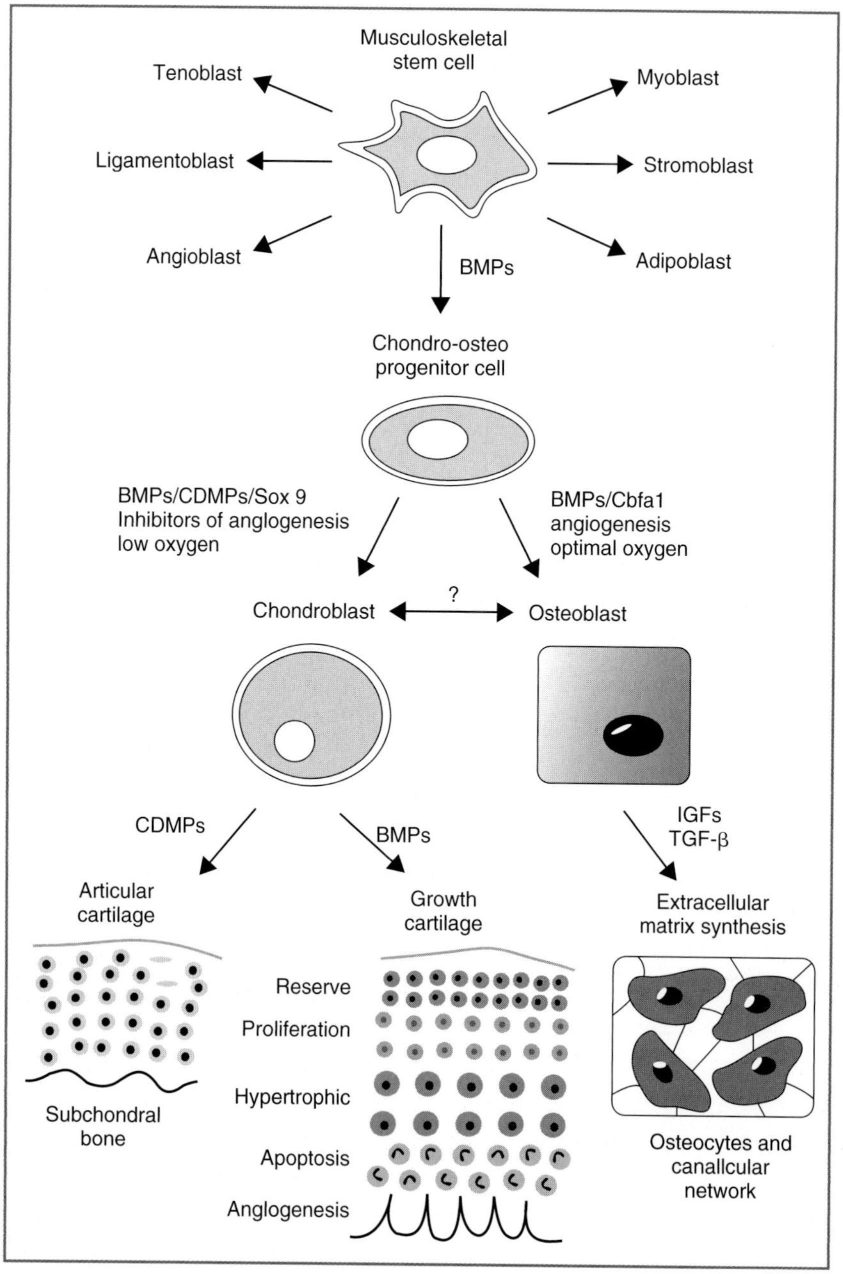

Figure 14.8 The lineages of the putative musculoskeletal stem cell. The BMPs determine the lineage into chondro/osteo progenitor cells and further specialization into articular chondrocytes growth plate chondrocytes and osteoblast lineage. BMPs are critical morphogens to direct the differentiation of cartilage and bone cells.

may lead to banks of stem cells for cell therapy and perhaps gene therapy with appropriate "homing" characteristics to bone marrow and hence to the skeleton. The pioneering work of Friedenstein et al. (1968, 1987), and Owen and Friedenstein (1988) identified bone marrow stromal stem cells. These stromal cells are distinct from the hematopoietic stem cell lineage. The bone marrow stromal stem cells consist of inducible and determined osteoprogenitors committed to osteogenesis. Determined osteogenic precursor cells have the propensity to form bone cells without any external cues or signals. On the other hand inducible osteogenic precursors require an inductive signal such as BMP or demineralized bone matrix. It is noteworthy that operational distinction between stromal stem cells and hematopoietic stem cells are getting more and more blurry!

The stromal stem cells of Friedenstein and Owen are also called mesenchymal stem cells (Caplan, 1991; Pittenger et al., 1999), with potential to form bone, cartilage, adipocytes, and myoblasts in response to cues from environment and/or intrinsic factors. Mesenechymal stem cells are present in synovium (De Bari et al., 2001), periosteum (Nakahara et al., 1991), adipose tissue (Zuk et al., 2001), and blood (Zvaifler et al., 2000). There is very recently considerable hope and anticipation that these bone marrow stromal cells may be excellent vehicles for cell and gene therapy (Prockop, 1997; Kuznetsov et al., 1997).

From a practical standpoint these stromal stem cells can be obtained by bone marrow biopsies and expanded rapidly for use in cell therapy after pre-treatment with BMPs. The potential uses in both cell and gene therapy is very promising. There are continuous improvements in the viral vectors and efficiency of gene therapy (Mulligan, 1993; Kozarsky and Wilson, 1993; Morsy et al., 1993; Bank, 1996). For example, it is possible to use BMP genes transfected in stromal stem cells to target to the bone marrow.

SCAFFOLDS OF BIOMIMETIC BIOMATERIALS

The earlier discussion of inductive signals (BMPs) responding stem cells (stromal cells) leads us to the scaffolding (the microenvironment/ECM) for optimal tissue engineering. The natural biomaterials in the composite tissue of bones and joints are collagens, proteoglycans, and glycoproteins of cell adhesion such as fibronectin and the mineral phase. The mineral phase in bone is predominantly hydroxyapatite. In native state the associated citrate, fluoride, carbonate, and trace elements constitutes the physiological hydroxyapatite. The high protein binding capacity makes hydroxyapatite a natural delivery system. Comparison of insoluble collagen, hydroxyapatite, tricalcium phosphate, glass beads, and polymethylmethacrylate as carriers revealed collagen to be an optimal delivery system for BMPs (Ma et al., 1990). It is well known that collagen is an ideal delivery system for growth factors in soft and hard tissue wound repair (McPherson, 1992).

During the course of systematic work on hydroxyapatite of two pore sizes (200 or 500 μm) in two geometrical forms (beads or disks) an unexpected observation was made. The geometry of the delivery system is critical for optimal bone induction. The disks were consistently osteoinductive with BMPs in rats; but the beads were inactive (Ripamonti et al., 1992). The chemical composition of the two hydroxyapatite configurations was identical. In certain species the hydroxyapatite alone appears to be "osteoinductive" (Ripamonti, 1996). In subhuman primates the hydroxyapatite induces bone albeit at a much slower rate. One interpretation is that osteoinductive endogenous BMPs in circulation progressively bind to implanted disk of hydroxyapatite. When an optimal threshold concentration of native BMPs is achieved the hydroxyapatite becomes osteoinductive. Strictly speaking most hydroxyapatite substrata are ideal osteoconductive materials. This example in certain species also serves to illustrate how an osteoconductive biomimetic biomaterial may progressively function as an osteoinductive substance by binding to endogenous BMPs. Thus, there is a physiological–physicochemical continuum between the hydroxyapatite alone and progressive composites with endogenous BMPs. Recognition of this experimental nuance will save unnecessary arguments amongst biomaterials scientists about the osteoinductive action of a conductive substratum such as hydroxyapatite.

Complete regeneration of baboon craniotomy defect was accomplished by recombinant human osteogenic protein (rhOP-1; human BMP-7) (Ripamonti et al., 1996). Recombinant BMP-2 was delivered by poly(-hydroxy acid) carrier for calvarial regeneration (Hollinger et al., 1996). Copolymer of polylactic and polyglycolic acid in a non-union model in rabbit ulna and the results were satisfactory (Figure 14.9) (Bostrom et al., 1996).

An important problem in the clinical application of biomimetic biomaterials with BMPs and/or other morphogens in regenerative medicine is the sterilization. Although gas (ethylene oxide) is used, one always should be concerned about reactive free radicals. Using allogeneic demineralized bone matrix with endogenous native

BMPS and tissue regeneration

- Orthopaedics
- Fractures
- Spine/fusions
- Articular cartilage repair
- Dentistry/oral surgery
- Periodontal surgery
- Craniofacial surgery
- Plastic surgery

Figure 14.9 **BMPs have wide ranging roles in regenerative medicine and surgery. The applications include but are not limited to orthopedics, plastic and reconstructive surgery, in dentistry and oral surgery. Recombinant BMP 2 has been approved by the FDA for spine fusions and non-unions of fractures.**

BMPs, as long as low temperature (4°C or less) is maintained, the samples tolerated up to 5–7 M rads of irradiation (Weintroub and Reddi, 1988; Weintroub et al., 1990). The standard dose acceptable to the Food and Drug Administration (FDA) is 2.5 M rads. This information would be useful to the biotechnology companies preparing to market recombinant BMP-based osteogenic devices. Perhaps, tissue banking industry with interest in bone grafts (Damien and Parson, 1991) could also use this critical information. The various freeze-dried and demineralized allogeneic bone may be used in the interim as satisfactory carriers for BMPs. The moral of this experiment is it is not the irradiation dose but the ambient sample temperature during irradiation is absolutely critical.

CARTILAGE-DERIVED MORPHOGENETIC PROTEINS

Morphogenesis of the cartilage is the key rate-limiting step in the dynamics of bone development. Cartilage is the initial model for the architecture of bones. Bone can form either directly from mesenchyme as in intramembranous bone formation or with an intervening cartilage stage as in endochondral bone development (Reddi, 1981). All BMPs induce, first, the cascade of chondrogenesis, and therefore they all sense are cartilage morphogenetic proteins. The hypertrophic chondrocytes in the epiphyseal growth plate mineralize and serves as a template for appositional bone morphogenesis. Cartilage morphogenesis is critical for both bone and joint morphogenesis. The two lineages of cartilage are clear-cut. The first at the ends of bone, forms articulating articular cartilage. The second is the growth plate chondrocytes which hypertrophy synthesize cartilage matrix destined to calcify prior to replacement by bone and are the "organizer" centers of longitudinal and circumferental growth of cartilage setting into motion the orderly program of endochondral bone formation. The phenotypic stability of the articular (permanent) cartilage is at the crux of the osteoarthritis problem. The "maintenance" factors for articular chondrocytes include TGF-β isoforms and the BMP isoforms (Luyten et al., 1992).

An *in vivo* chondrogenic bioassay with soluble purified proteins and insoluble collagen scored for chondrogenesis. A concurrent reverse transcription-polymerase chain reaction (RT-PCR) approach was taken with degenerate oligonucleotide primers. Two novel genes for CDMPs 1 and 2 were identified and cloned (Chang et al., 1994). CDMPs 1 and 2 are also called GDF-5 and GDF-6, respectively (Storm et al., 1994). CDMPs are related to BMPs (Figure 14.6). CDMPs are critical for cartilage and joint morphogenesis (Tsumaki et al., 1999). CDMPs stimulate proteoglycan synthesis in cartilage. GDF-7 initiates tendon and ligament morphogenesis.

REGENERATIVE MEDICINE AND SURGERY

Unlike bone with its considerable prowess for repair and even regeneration, cartilage is recalcitrant. This part may be due to relative a vascularity of hyaline cartilage, the high concentration of protease inhibitors and perhaps even cytokine inhibitors. The wound debridement phase is not optimal to prepare the cartilage wound bed for the optimal regeneration. Although cartilage has been successfully engineered to predetermined shapes (Kim et al., 1994), true repair of the tissue continues to be a real challenge in part due to hierarchical organization and geometry (Mow et al., 1992). The utility of autologous culture-expanded human chondrocytes is gaining (Brittberg et al., 1994). Also gaining increasing attention is mosaicplasty for defects in articular cartilage (Hangody et al., 2001). A continuous challenge in chondrocyte cell therapy is progressive de-differentiation and loss of characteristic cartilage phenotype. The re-differentiation and maintenance of the chondrocytes for cell therapy can be aided by BMPs, CDMPs, TGF-β isoforms, and insulin growth factors (IGFs). It is also possible to repair cartilage using muscle-derived mesenchymal stem cells (Grande et al., 1995). The potential possibility of the problems posed by cartilage proteoglycans in preventing cell immigration for repair was investigated by chondroitinase ABC and trypsin pre-treatment in partial-thickness defects (Hunzinker and Rosenberg, 1996), with and without TGF-β. Pre-treatment with chondroitinase ABC followed by TGF-β revealed a contiguous layer of cells from the synovial membrane hinting at the potential source of "repair" cells from synovium. Multiple avenues of cartilage morphogens, cell therapy with chondrocytes and stem cells from marrow and muscle and a biomaterial scaffolding may lead to an optimal tissue engineered articular cartilage.

It is inevitable during aging most humans will confront the challenges of impaired locomotion due to wear and tear in bones and joints. Therefore, the repair and possibly complete regeneration of the musculoskeletal system and other vital organs such as skin, liver, and kidney may potentially need optimal repair or a spare part for replacement. Can we create spare parts for the human body? There is much reason for optimism that tissue engineering can help patients. We are living at an extraordinary time in the biology, medicine, surgery, and computational and related technology. The confluence of advances in molecular developmental biology and attendant advances in inductive signals for morphogenesis, stem cells, and biomimetic biomaterials.

The symbiosis of biotechnology and biomaterials has set the stage for systematic advances in tissue engineering (Langer and Vacanti, 1993; Reddi, 1994; Hubbell, 1995). The recent advances in enabling platform technology include molecular imprinting (Mosbach and Ramstrom, 1996). In principle, specific recognition and catalytic sites are imprinted using templates. The applications range from biosensors, catalytic applications to antibody, and receptor recognition sites. For example, the cell binding RGD site in fibronectin (Ruoslahti and Pierschbacher, 1987) or YIGSR domain in laminin can be imprinted in complementary sites (Vukicevic et al., 1990).

The rapidly advancing frontiers in morphogenesis with BMPs, hedgehogs, homeobox genes, and a veritable cornucopia of general and specific transcription factors co-activators and repressors will lead to co-crystallization of ligand–receptor complexes, protein-DNA complexes, and other macromolecular interactions. This will lead to peptidomimetic agonists for large proteins as exemplified by erythroprotein (Livnah et al., 1996). To such advances one can add new developments in self-assembly of millimeter-scale structures floating at the interface of perfluorodecalin and water and interacting by capillary forces controlled by the pattern of wettablity (Bowden et al., 1997). The final self-assembly is due to minimization of free energy in the interface. These are truly incredible advances that will lead to man-made materials that mimic ECM in tissues. Superimpose on such chemical progress a biological platform in a bone and joint mold. Let us imagine a head of the femur and a mold is fabricated with computer-assisted design and manufacture. It faithfully reproduces the structural features and may be imprinted with morphogens, inductive signals, and cell adhesion sites. This assembly can be loaded with stem cells and BMPs and other inductive signals with a nutrient medium optimized for optimal number of cell cycles, and then predictably exit into differentiation phase to reproduce a totally new bone femoral head. In fact such a biological approach with vascularized muscle flap and BMPs yielded new bone with a defined shape and has demonstrated the proof of principle for further development and validation (Khouri et al., 1991). We indeed are entering a brave new world of prefabricated biological spare parts for the human body based on sound architectural rules of inductive signals for morphogenesis, responding stem cells with lineage control, and with growth factors immobilized on a template of biomimetic biomaterial based on ECM.

ACKNOWLEDGMENTS

This work is supported by the Lawrence Ellison Chair in Musculoskeletal Molecular Biology and the NIH grant AR4 7345-01 A2. I thank Ms. Danielle Neff for outstanding bibliographic assistance and enthusiastic help.

REFERENCES

Bank, A. (1996). Human somatic cell gene therapy. *Bioessays* **18**: 999–1007.

Bostrom, M., Lane, J.M., Tomin, E., Browne, M., Berbian, W., Turek, T., Smith, J., Wozney, J. and Schildhauer, T. (1996). Use of bone morphogenetic protein-2 in the rabbit ulnar nonunion model. *Clin. Orthop. Rel. Res.* **327**: 272–282.

Bowden, N., Terfort, A., Carbeck, J. and Whitesides, G.M. (1997). Self-assembly of mesoscale objects into ordered-two-dimensional assays. *Science* **276**: 233–235.

Brittberg, M., Lindahl, A., Nilsson, A., Ohlsson, C., Isaksson, O. and Peterson, L. (1994). Treatment of deep cartilage defects in the knee with autologous chondrocyte transplantation. *N. Engl. J. Med.* **331**: 889–895.

Caplan, A.I. (1991). Mesenchymal stem cell. *J. Orthop. Res.* **9**: 641–650.

Chang, S.C., Hoang, B., Thomas, J.T., Vukicevic, S., Luyten, F.P., Ryban, N.J.P., Kozak, C.A., Reddi, A.H. and Moos Jr., M. (1994). Cartilage-derived morphogenetic proteins. *J. Biol. Chem.* **269**: 28227–28234.

Chen, P., Carrington, J.L., Hammonds, R.G. and Reddi, A.H. (1991). Stimulation of chondrogenesis in limb bud mesodermal cells by recombinant human BMP-2B and modulation by TGF-β_1 and TGF-β_2. *Exp. Cell Res.* **195**: 509–515.

Chen, S., Rubbock, M.J. and Whitman, M. (1996). A transcriptional partner for Mad proteins in TGF-β signalling. *Nature* **383**: 691–696.

Cunningham, N.S., Jenkins, N.A., Gilbert, D.J., Copeland, N.G., Reddi, A.H. and Lee, S.-J. (1995). Growth/differentiation factor-10: a new member of the transforming growth factor-β superfamily related to bone morphogenetic protein-3. *Growth Factors* **12**: 99–109.

Cunningham, N.S., Paralkar, V. and Reddi, A.H. (1992). Osteogenin and recombinant bone morphogenetic protein-2B are chemotactic for human monocytes and stimulate transforming growth factor-β_1 mRNA expression. *Proc. Natl Acad. Sci. USA* **89**: 11740–11744.

Damien, C.J. and Parson, J.R. (1991). Bone graft and bone graft substitutes: a review of current technology and applications. *J. Appl. Biomater.* **2**: 1187–208.

De Bari, C., Dell'Accio, F., Tylzanowski, P. and Luyten, F.P. (2001). Multipotent mesenchymal stem cells from adult human synovial membrane. *Arthritis Rheum.* **44**(8): 1928–1942.

Duboule, D. (1994). How to make a limb? *Science* **266**: 575–576.

Dudley, A.T., Lyons, K.M. and Robertson, E.J. (1995). A requirement for bone morphogenetic protein-7 during development of the mammalian kidney and eye. *Gene. Dev.* **9**: 2795–2807.

Friedenstein, A.J., Petrakova, K.V., Kurolesova, A.I. and Frolora, G.P. (1968). Heterotopic transplants of bone marrow: analysis of precursor cell for osteogenic and hemopoietic tissues. *Transplantation* **6**: 230–247.

Friedenstein, A.J., Chailakhyan, R.K. and Gerasimov, U.V. (1987). Bone marrow osteogenic stem cells: *in vitro* cultivation and transplantation in diffusion chambers. *Cell Tissue Kinet.* **20**(3): 263–272.

Grande, D.A., Southerland, S.S., Manji, R., Pate, D.W., Schwartz, R.E. and Lucas, P.A. (1995). Repair of articular cartilage defects using mesenchymal stem cells. *Tissue Eng.* **1**: 345–353.

Griffith, D.L., Keck, P.C., Sampath, T.K., Rueger, D.C. and Carlson, W.D. (1996). Three-dimensional structure of recombinant human osteogenic protein-1: structural paradigm for the transforming growth factor-β superfamily. *Proc. Natl Acad. Sci. USA* **93**: 878–883.

Hangody, L., Feczko, P., Bartha, L., Bodo, G. and Kish, G. (2001). Mosaicplasty for the treatment of articular defects of the knee and ankle. *Clin. Orthop. Relat. Res.* **391**(Suppl): S328–S336 (Review).

Hayashi, H., Abdollah, S., Qui, Y., Cai, J., Xu, Y.Y., Grinnell, B.W., et al. (1997). The MAD-related protein Smad 7 associates with the TGF-β receptor and functions as an antagonist of TGF-β signalling. *Cell* **89**: 1165–1173.

Heldin, C.H., Miyazono, K. and ten Dijke, P. (1997). TGF-β signaling from cell membrane to nucleus through Smad proteins. *Nature* **390**: 465–471.

Hemmati-Brivanlou, A., Kelly, O.G. and Melton, D.A. (1994). Follistatin an antagonist of activin is expressed in the Spemann organizer and displays direct neuralizing activity. *Cell* **77**: 283–295.

Hollinger, J., Mayer, M., Buck, D., Zegzula, H., Ron, E., Smith, J., Jin, L. and Wozney, J. (1996). Poly ("-hydroxy acid") carrier for delivering recombinant human bone morphogenetic protein-2 for bone regeneration. *J. Contr. Release* **39**: 287–304.

Hubbell, J.A. (1995). Biomaterials in tissue engineering. *Biotechnology* **13**: 565–575.

Hunzinker, E.B. and Rosenberg, L.C. (1996). Repair of partial-thickness defects in articular cartilage: cell recruitment from the synovial membrane. *J. Bone Joint Surg.* **78-A**: 721–733.

Johnson, R.L. and Tabin, C.J. (1997). Molecular models for vertebrate limb development. *Cell* **90**: 979–990.

Khouri, R.K., Koudsi, B. and Reddi, A.H. (1991). Tissue transformation into bone *in vivo*. *JAMA* **266**: 1953–1955.

Kim, W.S., Vacanti, J.P., Cima, L., Mooney, D., Upton, J., Puelacher, W.C. and Vacanti, C.A. (1994). Cartilage engineered in predetermined shapes employing cell transplantation on synthetic biodegradable polymers. *Plast. Reconstr. Surg.* **94**: 233–237.

Kozarsky, K.F. and Wilson, J.M. (1993). Gene therapy: adenovirus vectors. *Curr. Opin. Genetic. Dev.* **3**: 499–503.

Kuznetsov, S.A., Friedenstein, A.J. and Robey, P.G. (1997). Factors required for bone marrow stromal fibroblast colony formation *in vitro*. *Br. J. Haematol.* **97**(3): 561–570.

Lacroix, P. (1945). Recent investigations on the growth of bone. *Nature* **156**: 576.

Langer, R. and Vacanti, J.P. (1993). Tissue engineering. *Science* **260**: 930–932.

Lemaire, P. and Gurdon, J.B. (1994). Vertebrate embryonic inductions. *Bio. Essays* **16**(9): 617–620.

Livnah, O., Stura, E.A., Johnson, D.L., Middleton, S.A., Mulcahy, L.S., Wrighton, N.D., Dower, W.J., Jolliffe, L.K. and Wilson, I.A. (1996). Functional mimicry of a protein hormone by a peptide agonist: the EPO receptor complex at 2.8 C. *Science* **273**: 464–471.

Luo, G., Hoffman, M., Bronckers, A.L.J., Sohuki, M., Bradley, A. and Karsenty, G. (1995). BMP-7 is an inducer of morphogens and is also required for eye development, and skeletal patterning. *Gene. Dev* **9**: 2808–2820.

Luyten, F., Cunningham, N.S., Ma, S., Muthukumaran, S., Hammonds, R.G., Nevins, W.B., Wood, W.I. and Reddi, A.H. (1989). Purification and partial amino acid sequence of osteogenin, a protein initiating bone differentiation. *J. Biol. Chem.* **265**: 13377–13380.

Luyten, F.P., Yu, Y.M., Yanagishita, M., Vukicevic, S., Hammonds, R.G. and Reddi, A.H. (1992). Natural bovine osteogenin and recombinant BMP-2B are equipotent in the maintenance of proteoglycans in bovine articular cartilage explant cultures. *J. Biol. Chem.* **267**: 3685–3691.

Lyons, K.M., Hogan, B.L.M. and Robertson, E.J. (1995). Colocalization of BMP-2 and BMP-7 RNA suggest that these factors cooperatively act in tissue interactions during murine development. *Mech. Dev.* **50**: 71–83.

Ma, S., Chen, G. and Reddi, A.H. (1990). Collaboration between collagenous matrix and osteogenin is required for bone induction. *Ann. NY Acad. Sci.* **580**: 524–525.

Melton, D.A. (1991). Pattern formation during animal development. *Science* **252**: 234–241.

Morsy, M.A., Mitani, K, Clemens, P. and Caskey, T. (1993). Progress toward human gene therapy. *JAMA* **270**(19): 2338–2345.

Mosbach, K. and Ramstrom, O. (1996). The emerging technique of molecular imprinting and its future impact on biotechnology. *Biotechnology* **14**: 163–170.

Mow, V.C., Ratcliffe, A. and Poole, A.R. (1992). Cartilage and diarthrodial joints as paradigms for hierarchical materials and structures. *Biomaterials* **13**: 67–97.

Mulligan, R.C. (1993). The basic science of gene therapy. *Science* **260**: 926–932.

McPherson, J.M. (1992). The utility of collagen-based vehicles in delivery of growth factors for hard and soft tissue wound repair. *Clin. Mater.* **9**: 225–234.

Nakahara, H., Goldberg, V.M. and Caplan, A.I. (1991). Culture-expanded human periosteal-derived cells exhibit osteochondral potential *in vivo*. *J. Orthop. Res.* **9**(4): 465–476.

Nishitoh, H., Ichijo, H., Kimura, M., Matsumoto, T., Makishima, F., Yamaguchi, A., Yamashita, H., Enomoto, S. and Miyazono, K. (1996). Identification of type I and type II serine/threonine kinase receptors for growth and differentiation factor-5. *J. Biol. Chem.* **271**: 21345–21352.

Owen, M.E. and Friedenstein, A.J. (1988). Stromal stem cells: marrow derived osteogenic precursors. *CIBA Foundation Symposium* **136**: 42–60.

Ozkaynak, E., Rueger, D.C., Drier, E.A., Corbett, C., Ridge, R.J., Sampath, T.K. and Opperman, H. (1990). OP-1 cDNA encodes an osteogenic protein in the TGF-β family. *EMBO J.* **9**: 2085–2093.

Paralkar, V.M., Nandedkar, A.K.N., Pointers, R.H., Kleinman, H.K. and Reddi, A.H. (1990). Interaction of osteogenin, a heparin binding bone morphogenetic protein, with type IV collagen. *J. Biol. Chem.* **265**: 17281–17284.

Paralkar, V.M., Vukicevic, S. and Reddi, A.H. (1991). Transforming growth factor-β type I binds to collagen IV of basement membrane matrix: implications for development. *Dev. Biol.* **143**: 303–308.

Paralkar, V.M., Weeks, B.S., Yu, Y.M., Kleinman, H.K. and Reddi, A.H. (1992). Recombinant human bone morphogenetic protein 2B stimulates PC12 cell differentiation: potentiation and binding to type IV collagen. *J. Cell Biol.* **119**: 1721–1728.

Piccolo, S., Sasai, Y., Lu, B. and De Robertis, E.M. (1996). Dorsoventral patterning in *Xenopus*: inhibition of ventral signals by direct binding of chordin to BMP-4. *Cell* **86**: 589–598.

Pittenger, M.F., Mackay, A.M., Beck, S.C., et al. (1999). Multilineage potential of adult human mesenchymal stem cells. *Science* **284**(5411): 143–147.

Prockop, D.J. (1997). Marrow stromal cells and stem cells for non hematopoietic tissues. *Science* **276**: 71–74.

Rath, N.C. and Reddi, A.H. (1979). Collagenous bone matrix is a local mitogen. *Nature* **278**: 855–857.

Reddi, A.H. and Huggins, C.B. (1972). Biochemical sequences in the transformation of normal fibroblasts in adolescent rat. *Proc. Natl Acad. Sci. USA* **69**: 1601–1605.

Reddi, A.H. and Anderson, W.A. (1976). Collagenous bone matrix-induced endochondral ossification and hemopoiesis. *J. Cell Biol.* **69**: 557–572.

Reddi, A.H. (1981). Cell biology and biochemistry of endochondral bone development. *Collagen Rel. Res.* **1**: 209–226.

Reddi, A.H. (1984). Extracellular matrix and development. In: Piez, K.A. and Reddi, A.H. (eds.), *Extracellular Matrix Biochemistry.* New York: Elsevier, pp. 375–412.

Reddi, A.H. (1994). Bone and cartilage differentiation. *Curr. Opinion Gen. Dev.* **4**: 937–944.

Reddi, A.H. (1997). Bone morphogenetic proteins: an unconventional approach to isolation of first mammalian morphogens. *Cytokine Growth Factor Rev.* **8**: 11–20.

Reddi, A.H. (1998). Role of morphogenetic proteins in skeletal tissue engineering and regeneration. *Nat. Biotechnol.* **16**: 247–252.

Ripamonti, U., Ma, S. and Reddi, A.H. (1992). The critical role of geometry of porus hydroxyapatite delivery system induction of bone by osteogenin, a bone morphogenetic protein. *Matrix* **12**: 202–212.

Ripamonti, U. (1996). Osteoinduction in porous hydroxyapatite implanted in heterotopic sites of different animal models. *Biomaterials* **17**: 31–35.

Ripamonti, U., Van den Heever, B., Sampath, T.K., Tucker, M.M., Rueger, D.C. and Reddi, A.H. (1996). Complete regeneration of bone in the baboon by recombinant human osteogenic protein-1 (hOP-1, bone morphogenetic protein-7). *Growth Factors* **123**: 273–289.

Ruoslahti, E. and Pierschbacher, M.D. (1987). New perspectives in cell adhesion: RGD and integrins. *Science* **238**: 491–497.

Sampath, T.K. and Reddi, A.H. (1981). Dissociative extraction and reconstitution of bone matrix components involved in local bone differentiation. *Proc. Natl Acad. Sci. USA* **78**: 7599–7603.

Sampath, T.K. and Reddi, A.H. (1983). Homology of bone inductive proteins from human, monkey, bovine, and rat extracellular matrix. *Proc. Natl Acad. Sci. USA* **80**: 6591–6595.

Senn, N. (1989). On the healing of aseptic bone cavities by implantation of antiseptic decalcified bone. *Am. J. Med. Sci.* **98**: 219–240.

Storm, E.E., Huynh, T.V., Copeland, N.G., Jenkins, N.A., Kingsley, D.M. and Lee, S.-J. (1994). Limb alterations in brachypodism mice due to mutations in a new member of TGF-β superfamily. *Nature* **368**: 639–642.

ten Dijke, P., Yamashita, H., Sampath, T.K., Reddi, A.H., Riddle, D., Heldin, C.H. and Miyazono, K. (1994). Identification of type I receptors for OP-1 and BMP-4. *J. Biol. Chem.* **269**: 16986–16988.

Tsumaki, N. Tanaka, K., Arikawa-Hirasawa, E., Nakase, T., Kimura, T., Thomas, J.T., Ochi, T., Luyten, F.P. and Yamada, Y. (1999). Role of CDMP-1 in skeletal morphogenesis: promotion of mesenchymal cell recruitment and chondrocyte differentiation. *J. Cell. Biol.* **144**: 161–173.

Urist, M.R. (1965). Bone: formation by autoinduction. *Science* **150**: 893–899.

Vukicevic, S., Luyten, F.P., Kleinman, H.K. and Reddi, A.H. (1990). Differentiation of canalicular cell processes in bone cells by basement membrane matrix components: regulation by discrete domains of laminin. *Cell* **63**: 437–445.

Vukicevic, S., Kopp, J.B., Luyten, F.P. and Sampath, K. (1996). Induction of nephrogenic mesenchyme by osteogenic protein-1 (bone morphogenetic protein-7). *Proc. Natl Acad. Sci. USA* **92**: 9021–9026.

Weintroub, S. and Reddi, A.H. (1988). Influence of irradiation on the osteoinductive potential of demineralized bone matrix. *Calcif. Tissue Int.* **42**: 255–260.

Weintroub, S., Weiss, J.F., Catravas, G.N. and Reddi, A.H. (1990). Influence of whole body irradiation and local shielding on matrix-induced endochondral bone differentiation. *Calcif. Tissue Int.* **46**: 38–45.

Weiss, R.E. and Reddi, A.H. (1980). Synthesis and localization of fibronectin during collagenous matrix mesenchymal cell interaction and differentiation of cartilage and bone *in vivo*. *Proc. Natl Acad. Sci. USA* **77**: 2074–2078.

Winnier, G., Blessing, M., Labosky, P.A. and Hogan, B.L.M. (1996). Bone morphogenetic protein-4 is required for mesoderm formation and patterning in the mouse. *Gene. Dev.* **9**: 2105–2116.

Wozney, J.M., Rosen, V., Celeste, A.J., Mitsock, L.M., Whittiers, M., Kriz, W.R., Heweick, R.M. and Wang, E.A. (1988). Novel regulators of bone formation: molecular clones and activities. *Science* **242**: 1528–1534.

Zhang, H. and Bradley, A. (1996). Mice deficient of BMP-2 are nonviable and have defects in amnion/chorion and cardiac development. *Development* **122**: 2977–2986.

Zimmerman, L.B., Jesus-Escobar, J.M. and Harland, R.M. (1996). The Spemann organizer signal Noggin binds and inactivates bone morphogenetic protein-4. *Cell* **86**: 599–606.

Zuk, P.A., Zhu, M., Mizuno, H., et al. (2001). Multilineage cells from human adipose tissue: implications for cell-based therapies. *Tissue Eng.* **7**(2): 211–228.

Zvaifler, N.J., Marinova-Mutafchieva, L., Adams, G., Edwards, C.J., Moss J., et al. (2000). Mesenchymal precursor cells in the blood of normal individuals. *Arthritis Res.* **2**(6): 477–488.

15

Physical Stress as a Factor in Tissue Growth and Remodeling

Robert E. Guldberg, Christopher S. Gemmiti, Yash Kolambkar, and Blaise Porter

INTRODUCTION

The role of physical stresses and strains in regulating tissue growth and remodeling has been of tremendous interest to investigators for well over 100 years. Although somewhat unfairly to his contemporary colleagues, Julius Wolff is often credited with the concept that tissue structure or form follows from its function (i.e. Wolff's Law). At the time, Wolff's Law was simply based on the general correspondence noted between anatomical observations of trabecular bone organization and estimations of principal stress directions due to functional loading conditions. The recognition that adaptation of tissue structure and composition is cell mediated was not made until later by other investigators. These early observations spawned the interdisciplinary field of mechanobiology, focused on identifying mechanisms by which mechanical signals are transduced into cellular activity, and emphasized the need to consider the effects of physical factors on tissue growth and remodeling as an important part of strategies for tissue regeneration.

Many different cell types from various tissues have been shown to be sensitive to mechanical stimuli in one form or another. The effects of physiological mechanical signals on cells and tissues can be beneficial, playing a central role in the maintenance of tissue structural integrity via remodeling processes. Alterations in mechanical signals can also contribute to the development of pathological conditions. For example, local shear stresses play a key role in the development and localization of atherosclerotic lesions. Likewise, the progression of osteoarthritis is due to a vicious cycle of cartilage matrix degradation and increased local stresses. In bone, the mechanical environment also has important clinical implications in the development of osteoporosis, stress fractures, total joint implant loosening, and bone loss during space flight.

Given the ability of cells to sense and respond to mechanical signals, *in vitro* and *in vivo* strategies for engineering tissues that serve a mechanical function must consider adaptational responses to physical stresses. For many tissue types, static culture conditions *in vitro* produce tissue-engineered constructs with deficient mechanical properties, typically due to reduced content and organization of structural protein constituents. Bioreactor systems that deliver tissue appropriate mechanical signals have been designed to overcome this limitation and exploit cellular adaptation responses to produce constructs that more closely resemble native tissue properties. Upon implantation, the interaction between constructs and the *in vivo* mechanical environment is a critical determinant of whether functional integration is ultimately achieved.

This chapter begins by introducing the continuum concept and the idea that structural hierarchy must be considered when studying the effects of physical stresses on cells and tissues. After defining stress and strain, an overview is provided of the role of mechanical factors in tissue growth, repair, and remodeling *in vivo*. The

Copyright © 2008 Elsevier Inc. All rights reserved

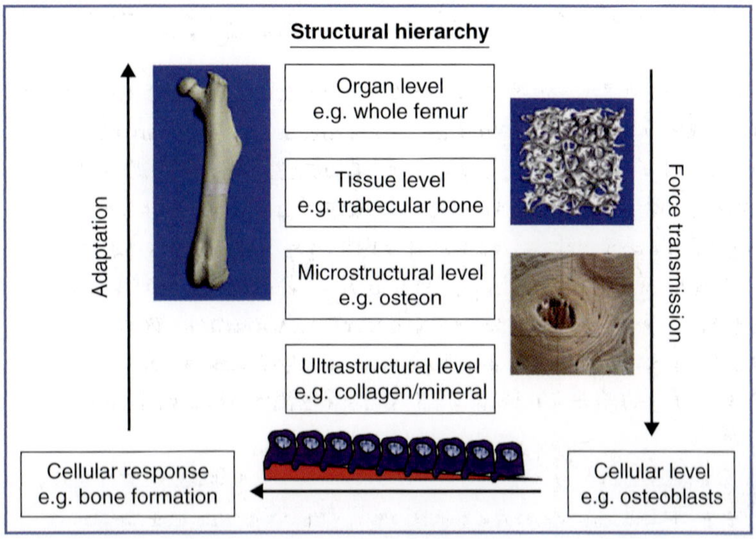

Figure 15.1 Force transmission through the structural hierarchy of bone to the cellular level resulting in cell-mediated adaptation of tissue structure and composition.

fundamental mechanisms by which cells may sense and respond to mechanical signals are then reviewed. Finally, the chapter concludes by considering the application of mechanical stimuli in bioreactor systems to produce larger and stronger tissue constructs for implantation.

STRUCTURAL HIERARCHY AND THE CONTINUUM CONCEPT

It is useful to view tissues as a structural hierarchy through which functional loads are transmitted down to the cellular level (Figure 15.1). In bone, for example, applied joint and muscle forces result in stresses and strains within the mineralized tissue that can be defined at different scale levels from the whole bone level down to sub-micron mineral crystals embedded within collagen molecules. At each hierarchical level, it is convenient to assume that everything below that level is a continuum (i.e. there is a finite mass density at every point within the material). This simplification allows material properties to be expressed at a given hierarchical level in terms of constitutive equations. As described in the next section, constitutive equations define the relationship between stresses and strains at each level. Cells sense and respond to local stresses or strains produced by forces transmitted from the macro level down through the complex structural hierarchy to the cellular level. Cell-mediated adaptational changes in tissue structure and composition subsequently alter the local stresses and strains resulting from functionally applied loads, thus providing a regulatory feedback mechanism. It is important to note that the sensitivity of the cellular response to mechanical stimuli can be altered by a variety of non-mechanical factors such as age, disease, as well as numerous biochemical factors.

STRAIN AND STRESS DEFINITIONS

Strain

Strain is a normalized measure of deformation. Consider the simple case of a thin rectangular piece of tissue being axially loaded by a force, as shown in Figure 15.2a. The axial force increases the length of the tissue, but

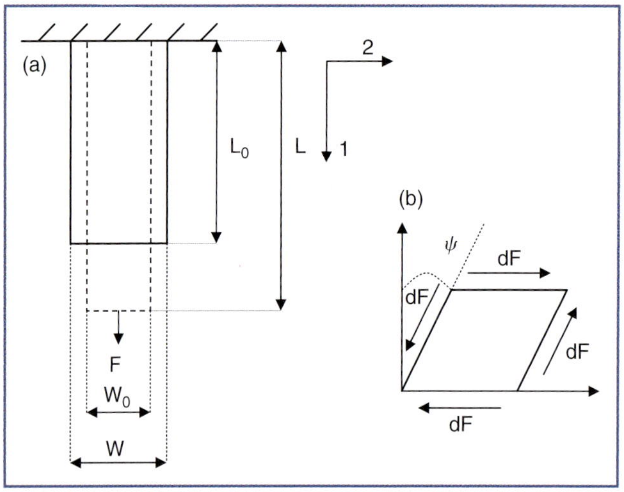

Figure 15.2 (a) Axial and transverse strains associated with uniaxial tensile loading. (b) Shear strain associated with torsional or shear loading.

at the same time decreases its width and thickness. Engineering strain is defined as the change in a dimension of the tissue normalized by its original dimension, and is given in the axial direction by:

$$\varepsilon_{11} = \frac{L - L_0}{L_0}$$

Another important deformation parameter is the Poisson's ratio ν, which is defined as the ratio of lateral strain to axial strain, and is given in this case by:

$$\nu = \frac{\varepsilon_{22}}{\varepsilon_{11}} = \frac{\dfrac{W - W_0}{W_0}}{\dfrac{L - L_0}{L_0}}$$

The Poisson ratio is a measure of the tendency for a material body to try to retain its total volume as it is deformed. When $\nu = 0.5$, the material is said to be incompressible (e.g. water), and does not undergo a volume change after deformation. The typical value of ν for tissues is between 0.2 and 0.45. Thus, a tissue subjected to tensile deformation and strain would increase in volume slightly.

In contrast to normal strains, shear deformations and strains due to shear forces dF or from pure torsional loading, for example, produce a change in shape but not volume, as shown in Figure 15.2b. Measurement of the angle of shear deformation, ψ, allows calculation of shear strain, as given by:

$$\varepsilon_{12} = \frac{\psi}{2}$$

The complex deformations created by forces acting in multiple directions necessitate the generalization of deformation to 3-D space. Deformation in 3-D can be expressed by the deformation gradient **F**. Consider the body shown in Figure 15.3a undergoing a deformation from the reference state to a deformed configuration. If one follows the particles P_1 and P_2, they move from position $\mathbf{X}_{P1}$ and $\mathbf{X}_{P2}$ to $\mathbf{x}_{P1}$ and $\mathbf{x}_{P2}$, respectively. There will also be a similar one-to-one mapping of other particles in the reference and deformed configurations. Thus the deformation of the body can be written as a function:

$$\mathbf{x} = f(\mathbf{X})$$

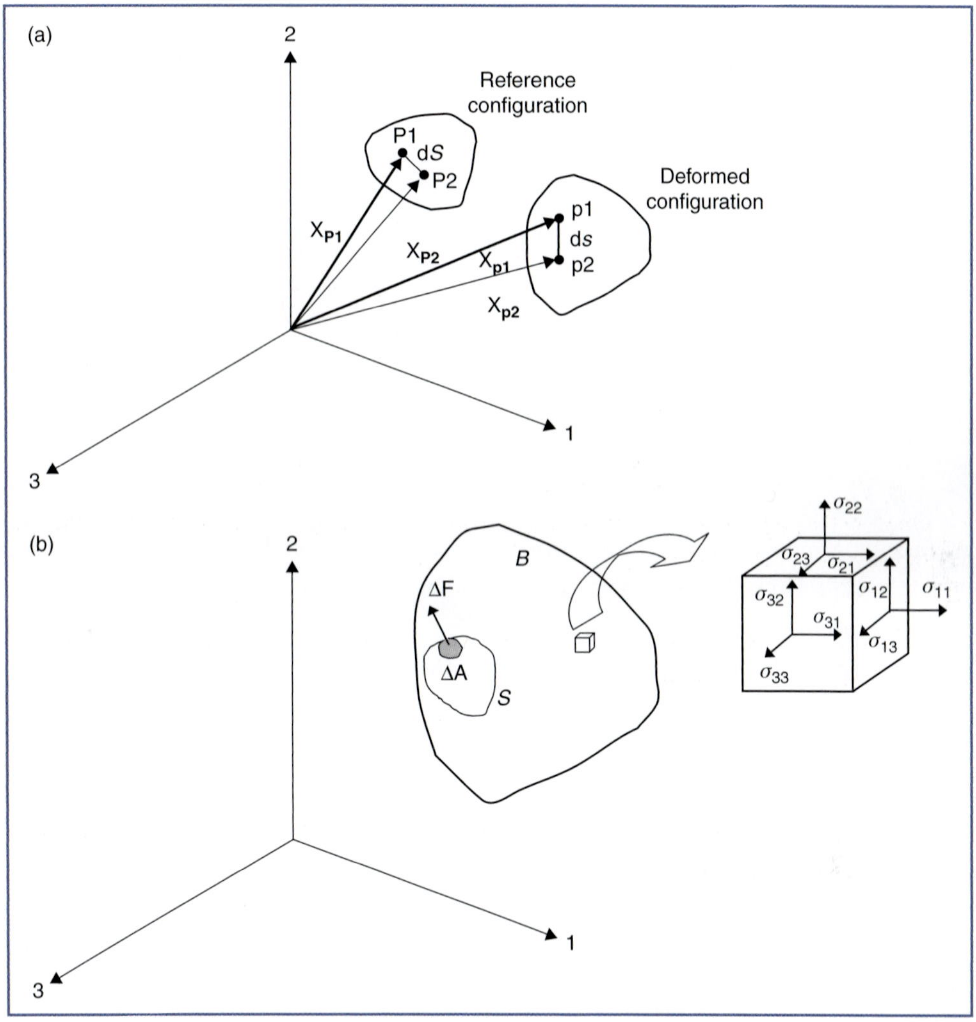

Figure 15.3 **(a) Deformation of a 3-D body from a reference configuration to a deformed configuration. (b) Stress on a surface element, and the nine stress components defining the stress state at a point.**

In scalar form, this would involve three equations:

$$x_1 = f_1(X_1, X_2, X_3)$$
$$x_2 = f_2(X_1, X_2, X_3)$$
$$x_3 = f_3(X_1, X_2, X_3)$$

where 1, 2, and 3 correspond to the three directions of the coordinate system.

The displacement vector is given by:

$$\mathbf{u} = \mathbf{x} - \mathbf{X}$$

The deformation gradient **F** is then defined as:

$$\mathbf{F} = \frac{\partial \mathbf{x}}{\partial \mathbf{X}}$$

In matrix form, the deformation gradient can be written as:

$$F = \begin{bmatrix} \frac{\partial x_1}{\partial X_1} & \frac{\partial x_1}{\partial X_2} & \frac{\partial x_1}{\partial X_3} \\ \frac{\partial x_2}{\partial X_1} & \frac{\partial x_2}{\partial X_2} & \frac{\partial x_2}{\partial X_3} \\ \frac{\partial x_3}{\partial X_1} & \frac{\partial x_3}{\partial X_2} & \frac{\partial x_3}{\partial X_3} \end{bmatrix}$$

and is related to the gradient of displacement by the following expression in which I is the unit vector:

$$F = \frac{\partial u}{\partial X} + I$$

The engineering strains as defined above are appropriate to use when the strains in the material are small (typically less than 5%). However the analysis of large deformations, as frequently observed for soft tissues under functional loading conditions, requires use of other strain measures. Consider the segment P_1P_2 of length dS that has deformed to p_1p_2 with length ds. When the deformation is large, a useful measure of deformation is the Green (i.e. Lagrangian) strain (E), which is defined as:

$$E = \frac{1}{2}\left(\frac{ds^2 - dS^2}{dS^2}\right)$$

The Green strain in the body can be expressed in terms of the gradient of displacement as:

$$E = \frac{1}{2}\left[D + D^{T} + D^{T}D\right],$$

where $D = \left[\frac{\partial u}{\partial X}\right]$ and the superscript T stands for the transpose of the matrix form of the second-order tensor.

If the deformation under consideration is small, as is typically the case for bone and most structural-engineering materials, the quadratic term in the Green strain can be neglected to give the infinitesimal (engineering) strain tensor (ε):

$$\varepsilon = \frac{1}{2}\left[D + D^{T}\right]$$

This is what gives us the familiar expression of engineering strain in a uniaxial test:

$$\varepsilon = \frac{L - L_0}{L_0}.$$

To get a feel for the relative values of these strain measures, consider the following example of uniaxial elongation of our rectangular tissue having original length of 5 cm. In one case, the tissue is stretched to a final length of 5.05 cm (small strain), whereas in the second case, it is elongated to 10 cm (large strain).

	Case I (L = 5.05 cm)	Case II (L = 10 cm)
Green strain $\left(E = \frac{1}{2}\frac{L^2 - L_0^2}{L_0^2}\right)$	0.01005	1.5
Engineering strain $\left(\varepsilon = \frac{L - L_0}{L_0}\right)$	0.01	1.0

Thus, we see that for the small deformations, the different strain definitions give approximately the same value and engineering strain is reasonably accurate. Whereas for large deformations, the strain definitions yield very different values due to neglect of the higher-order terms in the engineering strain definition.

Stress

Stress is a measure of the intensity of internal force developed in a material upon application of an external force. Consider the force $\Delta\mathbf{F}$ acting on a small surface element of area ΔA in Figure 15.3b. This element lies on the surface S, which is part of the larger body B. As ΔA tends to zero, the ratio $\frac{\Delta F}{\Delta A}$ tends to a finite limit $\frac{dF}{dA}$, which is defined as the stress on the surface element.

Consider an infinitesimal cube in the body as shown in Figure 15.3b. Due to the external force applied on the body, internal forces are applied on the surface of the cube. Each internal force can be resolved into its three components and normalized by the area to give three stress components on each face. The volume of the cube can be continuously decreased such that the cube collapses to a point. The nine stress components define the second-order stress tensor, and completely describe the stress state at this point. Using equilibrium conditions, we can show that $\sigma_{ij} = \sigma_{ji}$; thus the stress tensor has only six independent components. If a stress component acts in a direction perpendicular to the surface it acts on, it is referred to as a normal stress. On the other hand, if it is parallel to the surface, it is called a shear stress. Thus σ_{11}, σ_{22}, and σ_{33} are normal stresses, while σ_{12}, σ_{23}, and σ_{31} are shear stresses. Normal stresses tend to change the volume of the body, while shear stresses tend to modify the shape.

If the body is in the original reference configuration, ΔA represents the undeformed area and the stress is called the first Piola–Kirchoff stress tensor (**T**). In a typical experiment, the force is constantly measured, but the cross-sectional area is not. Thus the first Piola–Kirchoff stress is an easy quantity to compute as the undeformed cross-sectional area can be measured prior to loading. However, while considering force balance in the deformed body at equilibrium after external force is applied, the deformed area Δa of the surface element is required for the stress definition. The Cauchy stress is thus defined as the limit of $\sigma = \frac{\Delta F}{\Delta a}$ as Δa tends to zero. The difference between Δa and ΔA is negligible for small deformations. For large deformations, however, the stress definition again makes a significant difference.

Constitutive Equations

A constitutive equation is a mathematical model that specifies the relationship between stress and strain. Typically the model is phenomenological in nature, and is not derived from the microstructure. For example, the simplest constitutive equation is that for the linearly elastic materials, where there is a linear relationship

as vascular tissue engineering. Arteries are capable of remodeling their structure in response to changes in their mechanical environment. A chronic increase in systemic blood pressure induces an increase in vessel wall thickness and area, while reduced pressure leads to a decrease in vessel dimensions (Arner et al., 1984).

Abnormal joint loads have been shown to induce changes in composition, structure, and mechanical properties of articular cartilage. Disuse studies, for example, that use casting or other means of immobilization have demonstrated a loss of matrix constituents such as proteoglycans and a reduction in tissue thickness and mechanical properties (Akeson et al., 1987). Conversely, moderate exercise may have beneficial effects on maintaining healthy articular cartilage (Lane, 1996). However, high-impact loading or altered joint loading due to instability or injury is recognized as a significant risk factor for the development and progression of osteoarthritis (Buckwalter, 1995; Lane, 1996b). These studies suggest that there is a range of local stresses and strains that promote healthy tissue homeostasis, but loading conditions that are abnormally high or low can trigger catabolic responses and a loss of tissue function.

Several theories have been put forth to explain the relationship between mechanical stress and strain distributions and patterns of cellular differentiation and tissue formation. For musculoskeletal connective tissues, Carter et al. (1998) introduced tissue differentiation phase diagrams that shift depending on the local vascular environment. The theory asserts that bone will form directly under conditions of moderate loading and adequate local blood supply. However, high shear or tensile hydrostatic stresses will tend to stimulate fibrous tissue formation, as often seen in unstable fracture non-unions. In addition, high-compressive hydrostatic stresses as well as a poor vascular supply are predicted to shunt tissue differentiation toward a cartilage pathway. The concept of taking advantage of mechanical stimuli to promote tissue repair has been applied clinically. Whereas prolonged rest was once typically prescribed to repair injured tissues, it is now recognized that early resumption of limited physical activity can promote tissue repair and restoration of function (Buckwalter, 1995a).

Physical Stresses and Regenerative Medicine

Replacing tissues that serve a significant biomechanical function has proven exceptionally challenging (Butler et al., 2000). Musculoskeletal connective tissues such as bone, cartilage, meniscus, tendon, and ligament and cardiovascular tissues such as blood vessels and heart valves are excellent examples of tissues that are subjected to repetitive high stress conditions *in vivo*. Tissue-engineering strategies designed to replace or regenerate such tissues must provide adequate biomechanical properties and integrate with surrounding native tissues in order to restore local function.

Baseline biomechanical data for the tissue targeted for repair or replacement is essential (Butler et al., 2000). For example, the types and magnitudes of stresses and strains applied to the native tissue *in vivo* during a variety of activities must be determined. Along with measurements of native tissue mechanical properties, stress and strain history data provide design objectives for tissue-engineered constructs or regeneration strategies. Prioritization of desired mechanical properties will likely be necessary since the optimized structure–function relationships in native tissues may be difficult or impossible to duplicate. As such, a critical issue in the field is setting standards for adequate mechanical integrity (Butler et al., 2000). Such standards will certainly be tissue dependent and may even require patient-specific information such as weight or level of physical activity.

Few studies to date have attempted to directly assess the effects of *in vivo* stresses on tissue-engineered constructs following implantation. Case et al. (2003) investigated the effects of controlled intermittent compressive deformation on cellular constructs using a hydraulic bone chamber device implanted into the distal femoral metaphyses of rabbits (Figure 15.4). Constructs receiving 4 weeks of daily mechanical loading at 0.5 Hz were found to have nine-fold more new bone formation compared to contralateral control constructs

Figure 15.4 **Hydraulic bone chamber implant (top) used to apply cyclic compressive loading to tissue-engineered constructs *in vivo*. Implanted constructs receiving the mechanical stimulus (bottom right) had nine-fold more new bone formation than no load controls (bottom left).**

that did not receive loading. This study demonstrates the important role that the *in vivo* mechanical environment can play in the repair and integration of an implanted tissue-engineered construct.

MECHANOTRANSDUCTION MECHANISMS

So how are local mechanical signals transduced into cellular responses that affect tissue growth, repair, and remodeling? The process of mechanotransduction can be divided into four stages (Gooch et al., 1998), as shown in Figure 15.5. They are: (1) force transmission, (2) mechanotransduction, (3) signal propagation, and (4) cellular response. The first stage refers to the transmission of the force from the point it is applied to the cell surface. The second corresponds to the sensory action of the cells in sensing mechanical stimuli, and transducing it into a biochemical signal, which is propagated inside the cell in the third stage. Finally the cell responds to the intracellular signal by modulating gene expression, completing the mechanotransduction process.

In the first stage of mechanotransduction, applied forces are converted into local stimuli that may be detected by cells. Transmitted forces can cause direct cellular deformation by deforming the surrounding extracellular matrix (ECM). Applied forces may also result in local fluid flow and/or hydrostatic pressures. For example, compression of articular cartilage generates hydrostatic pressure that can regulate chondrocyte metabolism. Dynamic compression of cartilage induces fluid flow through the matrix and exposes cells to local shear stresses. The relative importance of these different types of local stimuli *in vivo* is not clear due

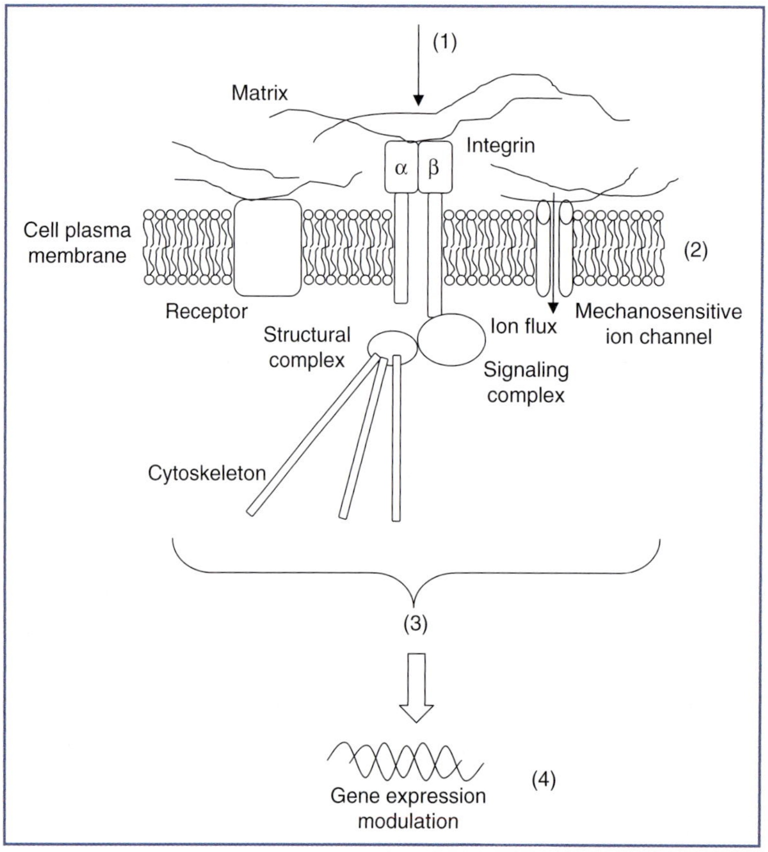

Figure 15.5 Schematic showing the four stages of mechanotransduction: (1) force transmission, (2) mechanotransduction, (3) signal propagation, and (4) cellular response. See text for details.

to the difficulty of isolating each kind of mechanical stimulus. However extensive research has been done to study the effects of various forms of mechanical stimuli on cells *in vitro*. These include tensile stretch, compression, hydrostatic pressure, and fluid-flow-induced shear stress, applied either statically or dynamically. These studies have allowed investigators to identify potential mechanotransduction mechanisms.

The next stage of mechanotransduction occurs at the plasma membrane of the cell, and it is here that the cell detects the external signal and converts it into an intracellular signal. The plasma membrane contains numerous receptors and ion channels that can serve as sensors of the mechanical stimuli. The key structures in this interaction are the mechanosensitive (also known as stretch-activated) ion channels, integrin receptors, and other plasma membrane receptors.

Mechanosensitive ion channels (Sachs, 1991; Hamill and Martinac, 2001; Martinac, 2004) are thought to be important to many cell types including chondrocytes (Wright et al., 1996; Guilak and Hung, 2005), osteoblasts (Charras and Horton, 2002), endothelial cells (Davies, 1995), and cardiac myocytes (Hu and Sachs, 1997). Experiments involving direct perturbation of the chondrocyte membrane have implicated such ion channels in the increase in concentration of cytosolic calcium ion (Guilak et al., 1999), which is a second messenger and has well-known intracellular effects (Rasmussen, 1986; Carafoli, 1987; Faber and Sah, 2003). Recently annexin V, a calcium-dependent phospholipid-binding protein, was proposed as a Ca2+channel in osteoblastic cells (Haut Donahue et al., 2004). The flux of ions through these channels also affects the membrane potential (Wright et al., 1992; Gannier et al., 1996; Zabel et al., 1996) that triggers voltage-gated ion channels (Mobasheri et al., 2002), which further change the ion concentrations inside the cell. Two models have been proposed to explain the mechanism of gating of these channels: the bilayer (Martinac et al., 1990; Hamill and Martinac, 2001) and the tethered models (Hamill and McBride, 1997; Gillespie and Walker, 2001). In the bilayer model, lipid bilayer tension alone is sufficient to activate the channels directly. The tethered model assumes that molecules in the cortical cytoskeleton and/or the extracellular domains directly interact with the channel protein to open/close the channel.

Integrins are heterodimeric transmembrane proteins that bind to ECM proteins and cluster together leading to the assembly of focal adhesions, at which the cell contacts the ECM. Focal adhesions intracellularly associate with α-actinin (Otey et al., 1993), talin (Critchley, 2004), tensin (Bockholt and Burridge, 1993), and other cytoskeletal-binding proteins as well as signaling molecules like focal adhesion kinase (FAK) (Schaller et al., 1995). Due to their associations with both structural and signaling proteins, integrins are well placed to act as transducers of physical stimuli, and have been implicated as a link between the extracellular and intracellular environments for a variety of cell types that allows transmission of inside-out and outside-in signals capable of modulating cell behavior (Wright et al., 1997; Pelham and Wang, 1999; Jalali et al., 2001; Aikawa et al., 2002; Martinez-Lemus et al., 2003). In one study, over-expression of the tumor suppressor PTEN, which inhibits outside-in integrin signaling, strongly suppressed stretch-induced activation of p38 mitogen-activated protein kinase (MAPK) in cardiac myocytes (Aikawa et al., 2002). Jalali et al. (2001) demonstrated that fluid flow over endothelial cells activates integrin-mediated adhesion in an ECM-specific manner. The shear stress-induced mechanotransduction was abolished when new integrin–ECM ligand interactions were prevented by either blocking the integrin-binding sites of ECM ligands or conjugating the integrins to immobilized antibodies. Wright et al. (1997) reported that the transduction pathways involved in the hyperpolarization response of human articular chondrocytes *in vitro* after cyclical pressure-induced strain involve $\alpha5\beta1$ integrin, which they suggest to be an important chondrocyte mechanoreceptor. Externally applied forces would cause changes in the conformations of the ECM molecules that would affect their binding to integrins, and modify the force balance within focal adhesions. It is thought that increased tension within focal adhesions can trigger increased integrin clustering and FAK phosphorylation (Sieg et al., 1999; Katsumi et al., 2004), which initiates a signal cascade resulting in altered gene expression.

In addition to integrins, the plasma membrane is host to other receptors for specific ECM proteins like collagen, aggrecan, and hyaluronic acid, which may also be able to sense extracellular forces due to their interactions with their ligands. It is also possible that G-protein-coupled receptors may act as mechanotransducers or be activated secondary to other pathways, as the consequences of G protein stimulation of phospholipase C (PLC)–inositol trisphosphate (IP_3) pathway has been observed in mechanically stimulated cells (Davies, 1995). It is very likely that the above-mentioned transducer molecules collaborate in the mechanotransduction response. In fact both integrin function and mechanosensitive ion channel activity were found to be required for chondrocyte response to cyclic pressurization (Lee et al., 2000). It has also been suggested that mechanical stimuli regulate cell behavior by a physical connection from intracellular organelles to the ECM via the cytoskeleton and the adhesion plaque (Guilak and Hung, 2005).

The third stage of mechanotransduction is signal propagation, in which the signal generated at the plasma membrane in the second stage is propagated within the cell. This is usually carried out using the same machinery that the cell uses for responding to biochemical stimuli. Signal propagation is initiated by second messengers such as Ca2+ , cAMP, and MAPK. Activated kinases subsequently phosphorylate transcription factors leading to changes in gene expression.

Cytoplasmic calcium serves as a ubiquitous signal for regulation of important cellular processes such as cell growth, differentiation, protein synthesis, and even cell death. Numerous studies have found an increase in cytosolic Ca2+ concentration due to mechanical loading in a variety of cell types (Hung et al., 1997; Edlich et al., 2001; Sharma et al., 2002; Donahue et al., 2003). This may be due to the opening of mechanosensitive

Ca2+ channels as discussed above or secondary to a mechanotransducer. The intracellular Ca2+ concentration can also be elevated by release of calcium from intracellular stores through the IP_3/diacylglycerol (DAG) pathway (Berridge, 1987). This pathway can be triggered by G-protein-coupled receptors leading to the activation of the enzyme PLC. PLC cleaves the phosphoinositide PIP_2 to generate two second messengers: DAG and IP_3. After diffusing though the cytosol, IP_3 interacts with and opens Ca2+ channels in the membrane of the endoplasmic reticulum, causing release of Ca2+ into the cytosol. One of the various cellular responses induced by a rise in cytosolic Ca2+ is recruitment of protein kinase C (PKC) to the plasma membrane, where it is activated by DAG. The activated kinase can phosphorylate various proteins, including transcription factors, leading to gene activation. Ca2+ is also known to bind to the small cytosolic protein calmodulin to form a complex that interacts with and modulates activity of other enzymes and transcription factors. Ca2+ influx is known to activate certain K+ channels thus affecting membrane potential (Wright et al., 1992; Faber and Sah, 2003), and has been shown to be necessary for integrin-dependent tyrosine phosphorylation of focal adhesion associated molecules (Alessandro et al., 1998).

The cyclic nucleotide cAMP is produced by adenylyl cyclases which are in turn activated by G-protein-coupled receptors. Protein kinase A (PKA), which consists of two catalytic subunits and two regulatory subunits, is the most well-known cAMP effector. Binding of cAMP to the regulatory subunits releases the catalytic subunits, which are then free to phosphorylate substrates (Dumaz and Marais, 2005). cAMP, along with intracellular Ca2+ , has been implicated in the regulation of gene expression in response to static compression of cartilage explants (Valhmu et al., 1998; Fitzgerald et al., 2004). Boo et al. (2002) demonstrated that shear stress stimulates phosphorylation of eNOS and thus nitric oxide (NO) production in bovine aortic endothelial cells in a PKA-dependent manner.

As discussed earlier, mechanical stimuli may be able to activate FAK and other signaling proteins via integrin receptors. In chondrocytes, these signaling proteins are known to stimulate docking proteins such as Src-homology collagen (Shc) leading to the activation of the MAPK pathway (Shakibaei et al., 1999). The MAPK family consists of an array of serine/threonine kinases (ERK1/2, p38 MAPK, etc.) that are activated by a variety of physical and biochemical stimuli. However, integrins specifically appear to be involved upstream in this mechanotransduction response, irrespective of the tissue involved. The MAPKs are known to be activated by Ras, a small G protein. Ras is a membrane anchored switch protein that is turned on by certain receptors via docking proteins (Mitin et al., 2005). After being switched on, Ras phosphorylates and consequently activates a cascade of proteins, which ultimately lead to the activation of the MAPKs. The activated MAPKs regulate several regulatory molecules in the cytoplasm and in the nucleus to initiate cellular processes such as proliferation, differentiation, and development (Seger and Krebs, 1995). Many studies have implicated MAPKs in the cellular response to fluid flow and stretch (Hung et al., 2000; You et al., 2001; Plotkin et al., 2005; Torsoni et al., 2005). For example, Hung et al. (2000) showed that fluid-induced shear stress suppression of aggrecan gene expression in culture bovine chondrocytes is mediated in part by calcium-independent MAPK regulation.

There is also evidence to show that some of the signal transduction pathways are linked. For example, Ca2+ -activated calmodulin activates the enzyme cAMP phosphodiesterase that degrades cAMP and thus terminates its effect (Kakkar et al., 1999). Also as mentioned above, IP_3 is an important mediator of cytosolic Ca2+ release from intracellular stores. It has also been shown that cAMP inhibits MAPKs in several cell types (Dumaz and Marais, 2005).

The final stage of mechanotransduction is the altered response of the cell, which may include changes in matrix synthesis/degradation, proliferation, differentiation, apoptosis, cell alignment, and migration. The effectors of the mechanotransduction pathways are the various transcription factors, which are activated by the events discussed previously. Numerous studies on vascular cells have shown activation of transcription factors like AP-1, CRE, and NF-κβ in response to cyclic strain (Kakisis et al., 2004). The activated transcription factors interact with the promoter and enhancer regions of various genes to mediate transcription. This results in an increase in expression of genes like Cox-2, VEGF, TGF-β3, and eNOS (Kakisis et al., 2004), which orchestrate the cellular responses. Lee et al. (2001) demonstrated that vascular smooth muscle cells respond to mechanical strain by increasing specific proteoglycan synthesis and aggregation. It is known that mechanical loading of osteocytes results in anabolic responses such as the expression of c-fos, insulin-like growth factor-I (IGF-I), and osteocalcin (Mikuni-Takagaki, 1999). Elevations in Ca2+ activate a Ca2+/calmodulin-dependent protein kinase that causes increased c-fos expression, which is a pro-growth transcription factor. Calcineurin, a Ca2+/calmodulin-activated phosphatase, dephosphorylates and activates the NF-AT family of transcription factors. Different NF-ATs, expressed in different cells including those of the heart, cartilage, and bone, serve as tissue-specific activators of cell growth and differentiation (Crabtree, 1999; Iqbal and Zaidi, 2005).

IN VITRO MECHANICAL CONDITIONING

The replacement of tissues which reside in a complex, dynamic mechanical environment is a daunting challenge. Articular cartilage and blood vessels, for example, must bear tremendous stress and strain over repeated loading cycles *in vivo* while maintaining normal function. To date, no engineered construct has been developed *in vitro* possessing the same biomechanical properties as its *in situ* counterpart. One approach to address this challenge is the use of physiologically inspired mechanical forces to transmit stimuli to developing constructs *in vitro*. Since these tissues normally experience a dynamic environment *in vivo*, the rationale is that the application of mechanical forces such as compression or shear stress will stimulate the cells of the engineered construct to secrete and organize the proper matrix proteins required to reproduce the native tissue mechanical function.

Delivery of controlled stresses and strains *in vitro* is achieved through mechanical devices known as bioreactors. Bioreactors have been used extensively as production vessels for engineered tissues. Many of these systems take advantage of the controlled *in vitro* environment to investigate the effects of specific biochemical or biomechanical factors on construct development. Bioreactor systems are highly diverse, but many are designed to delivery-specific mechanical signals to tissue constructs. Another common feature is they typically function to increase the mass transport of nutrients and waste through constructs via convective fluid flow. Bioreactors are also amenable to large-scale tissue production, as they are inherently scaleable and allow for increased process control, such as on-line measurement of pH or dissolved oxygen.

Perhaps the tissues of the body most subjected to mechanical forces are those of the musculoskeletal and cardiovascular origin. Consequently, orthopaedic and cardiovascular tissue-engineered constructs represent the bulk of the research in which mechanical forces have been applied to developing tissues *in vitro*. Cartilage, bone, tendon, ligament, blood vessels, heart valves, and muscle have been cultured *in vitro* under the influence of mechanical forces. The remainder of this section will discuss select examples from the orthopaedic and cardiovascular fields which use the *in vivo* environment as inspiration to mechanically condition tissue-engineered constructs *in vitro*.

Cartilage Bioreactors

Articular cartilage is the whitish, low-friction tissue which lines the ends of long bones in a diarthrodial joint. It is a highly hydrated tissue (~80% water), with type II collagen and proteoglycans constituting the majority of the solid matrix. These constituents combine to yield resilient mechanical properties which provide the shock absorption and nearly friction-free surface in joints such as the knee, shoulder, and hip. Joint-bearing surfaces regularly experience complex high-magnitude mechanical loads through activities such as media flow around or within tissue-engineered constructs have been designed to enhance nutrient and waste exchange *in vitro* (Bujia et al., 1995). In addition to enhancing mass transport, fluid flow applies shear stresses to the cells within the scaffolds. The effects of flow-mediated shear on cells have been studied in 2-D monolayer cultures. Continuous fluid flow applied to osteoblasts *in vitro* has been shown to alter bone-related gene expression and cellular phenotype (Ogata, 2000). Parallel-plate flow experiments have shown that bone cells cultured in monolayer are highly responsive to flow-mediated shear stresses. Shear stresses in the range of 5–15 dynes/cm^2 affect osteoblast proliferation as well as production of NO and prostaglandin E_2 (PGE_2), suggesting that shear stress is an important regulator of osteoblast function (McAllister et al., 2000). Pulsatile and oscillatory flow conditions applied to osteoblasts using *in vitro* parallel-plate flow chambers have also been shown to increase gene expression, intracellular calcium concentration, and the production of NO and PGE_2 in comparison to static controls (Klein-Nulend et al., 1997; Bakker et al., 2001). Furthermore, cell responsiveness has been reported to vary with fluid flow rate and frequency (Jacobs et al., 1998; Edlich et al., 2001). Proposed mechanisms for the stimulation of cells by fluid flow include increased mass transport, generation of streaming potentials, and application of shear stresses to the cell membranes (McAllister and Frangos, 1999; Bakker et al., 2001). Although these studies were performed using 2-D cell culture systems for short-term experiments, they suggest that variable flow conditions may also have differential effects in 3-D tissue culture systems.

Such tissue culture systems may be useful to engineer thicker, more uniform bone graft substitutes for implantation or as test bed models that simulate aspects of the *in vivo* environment. While many different bioreactor systems have been developed, perfusion bioreactors in particular have shown significant increases in both cell viability and mineralized matrix formation on large 3-D constructs *in vitro*. In a recent study, micro-CT has been used to quantify mineralized matrix production within perfused and statically cultured marrow progenitor cells seeded on large polymer scaffolds (6.35 mm diameter, 9 mm thick) (Porter et al., 2005). Statically cultured constructs were found to have mineralized matrix localized only to the periphery of the constructs. In contrast, perfused constructs were found to have a several fold increase in mineralized matrix production distributed throughout the constructs (Figure 15.7).

Blood Vessel Bioreactors

Following the same rationale for mechanical conditioning of orthopaedic-engineered tissues, cardiovascular tissues can also be enhanced by *in vitro* mechanical stimulation. Cardiovascular tissues reside in a dynamic environment which can be mimicked *in vitro* using bioreactors and mechanical loading systems to deliver the physiologically inspired environmental cues.

Small-diameter blood vessels (<6 mm) are of particular importance because of their potential use to alleviate complications associated with atherosclerosis. Generally, a blood vessel has three layers (intima, media, and adventitia) in a tubular shape, forming a lumen through which blood passes. The intima is comprised mostly of a confluent, tightly adherent monolayer of endothelial cells (collectively called the endothelium) which is necessary to provide a non-thrombogenic surface for the blood to flow (van Hinsbergh, 2001). The media possesses smooth muscle cells and elastin and is set between the intima and the adventitia. The adventitia contains connective tissue (i.e. collagen) with fibroblasts embedded within. The ECM produced by the smooth muscle cells – the organized, cross-linked network of collagen and elastin – gives rise to the mechanical properties (Bank et al., 1996). These layers come together to form a vital tissue which must respond to the body's complex and dynamic needs.

In vivo, the pulsatile flow of blood imparts cyclic strains and shear stresses to the vessel's constituents, which respond in a variety of ways to these mechanical signals. Endothelial cells are uniquely situated in the

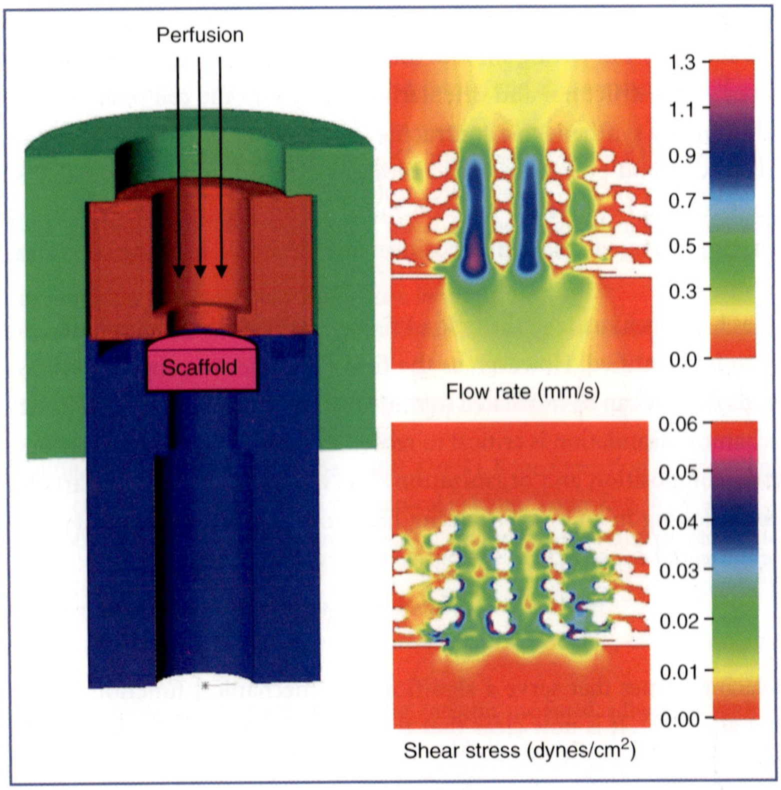

Figure 15.7 **Perfusion bioreactor system (left) for production of mineralized constructs for bone defects. Computational fluid dynamics simulation of flow rate and shear stresses within the 3-D scaffold porosity (right).**

lumen and are directly in contact with the flowing blood, which causes a shear stress to be applied to the cells. Consequently, these rapidly responding, mechanosensitive cells attain an elongated shape, aligning their long axis with the direction of flow. Sensing of the shear via cell surface receptors, ion channels, or integrins leads to secretion and/or activation of a number of signaling molecules, such as NO, endothelial nitric oxide synthase (eNOS), kinases, and transcription factors (Takahashi et al., 1997; Fisslthaler et al., 2000; Fisher et al., 2001). Perhaps most importantly, the fluid-induced shear stress confers a protective effect on the vessel by decreasing the probability of atherosclerosis (Traub and Berk, 1998). Indeed, areas of irregular blood flow (i.e. velocity, direction, and shear stress) have been implicated as sites of increased atherosclerosis (Papadaki et al., 1999). Shear stress also modulates smooth muscle cells' production of signaling molecules (such as NO) (Papadaki et al., 1998) and gene transcription levels of cell surface receptors (Papadaki et al., 1998).

Tissue-engineered vessels aim to reproduce cellular and mechanical properties of the native vessel in order to be an effective replacement. However, similar to other engineered tissues, those cultured in static conditions fall short of native tissue properties. Use of mechanical conditioning inspired by the *in vivo* environment has been shown in a variety of *in vitro* systems to modulate and improve engineered constructs. Exposing tissue-engineered vascular grafts to fluid-induced shear stress has been shown to increase endothelial cell adherence (Ott and Ballermann, 1995) and proliferation (Imberti et al., 2002) and alter tissue morphology and mechanical properties (Niklason et al., 2001). Cyclic mechanical strains cause an increase in collagen (types I and III) transcription by smooth muscle cells (Leung et al., 1976), an increase in mechanical properties (strength and stiffness), attributed to an increase in remodeling enzymes such as matrix

Gemmiti, C.V. and Guldberg R.E. (2006). Fluid flow increases type II collagen deposition and tensile mechanical properties in bioreactor-grown tissue-engineered cartilage. *Tissue Eng.* **12**(3): 469–479.

Germiller, J.A. and Goldstein, S.A. (1997). Structure and function of embryonic growth plate in the absence of functioning skeletal muscle. *J. Orthop. Res.* **15**: 362–370.

Gersbach, C.A., Byers, B.A., Pavlath, G.K., Guldberg, R.E. and Garcia, A.J. (2004). Runx2/Cbfa1-genetically engineered skeletal myoblasts mineralize collagen scaffolds *in vitro*. *Biotechnol. Bioeng.* **88**: 369–378.

Gillespie, P. and Walker, R. (2001). Molecular basis of mechanosensory transduction. *Nature* **413**: 194–202.

Gooch, K., Blunck, T., Vunjak-Novakovic, G., Langer, R., Freed, L. and Tennant, C. (eds.). 1998. *Mechanical Forces and Growth Factors Utilized in Tissue Engineering.* Elsevier Science.

Guilak, F. and Hung, C.T. (2005). *Physical Regulation of Cartilage Metabolism.* Lippincott Williams and Wilkins, Philadelphia, PA.

Guilak, F., Zell, R., Erickson, G., Grande, D., Rubin, C., McLeod, K. and Donahue, H.J. (1999). Mechanically induced calcium waves in articular chondrocytes are inhibited by gadolinium and amiloride. *J. Orthop. Res.* **17**: 421–429.

Hamill, O.P. and Martinac, B. (2001). Molecular basis of mechanotransduction in living cells. *Physiol. Rev.* **81**: 685–740.

Hamill, O.P. and McBride Jr., D.W. (1997). Induced membrane hypo/hypermechanosensitivity, a limitation of patch-clamp recording. *Annu. Rev. Physiol.* **59**: 621–631.

Haut Donahue, T., Genetos, D., Jacobs, C., Donahue, H. and Yellowley, C.E. (2004). Annexin V disruption impairs mechanically induced calcium signaling in osteoblastic cells. *Bone* **35**: 656–663.

Hu, H. and Sachs, F. (1997). Stretch-activated ion channels in the heart. *J. Mol. Cell. Cardiol.* **29**: 1511–1523.

Hung, C., Allen, F., Pollack, S., Attia, E., Hannafin, J. and Trozilli, P.A. (1997). Intracellular calcium response of ACL and MCL ligament fibroblasts to fluid-induced shear stress. *Cell. Signal.* **9**: 587–594.

Hung, C., Henshaw, D., Wang, C., Mauck, R., Raia, F., Palmer, G., Chao, P., Mow, V., Ratcliffe, A. and Valhmu, W. (2000). Mitogen-activated protein kinase signaling in bovine articular chondrocytes in response to fluid flow does not require calcium mobilization. *J. Biomech.* **33**: 73–80.

Hung, C.T., Mauck, R.L., Wang, C.C., Lima, E.G. and Ateshian, G.A. (2004). A paradigm for functional tissue engineering of articular cartilage via applied physiologic deformational loading. *Ann. Biomed. Eng.* **32**: 35–49.

Imberti, B., Seliktar, D., Nerem, R.M. and Remuzzi, A. (2002). The response of endothelial cells to fluid shear stress using a co-culture model of the arterial wall. *Endothelium* **9**: 11–23.

Iqbal, J. and Zaidi, M. (2005). Molecular regulation of mechanotransduction. *Biochem. Biophys. Res. Commun.* **328**: 751–755.

Jacobs, C.R., Yellowley, C.E., Davis, B.R., Zhou, Z., Cimbala, J.M. and Donahue, H.J. (1998). Differential effect of steady versus oscillating flow on bone cells. *J. Biomech.* **31**: 969–976.

Jalali, S., del Pozo, M, Chen, K., Miao, H., Li, Y., Schwartz, M., Shyy, J.Y. and Chien, S. (2001). Integrin-mediated mechanotransduction requires its dynamic interaction with specific extracellular matrix (ECM) ligands. *Proc. Natl. Acad. Sci. USA* **98**: 1042–1046.

Jin, M., Frank, E.H., Quinn, T.M., Hunziker, E.B. and Grodzinsky, A.J. (2001). Tissue shear deformation stimulates proteoglycan and protein biosynthesis in bovine cartilage explants. *Arch. Biochem. Biophys.* **395**: 41–48.

Kakisis, J., Liapis, C. and Sumpio, B. (2004). Effects of cyclic strain on vascular cells. *Endothelium* **11**: 17–28.

Kakkar, R., Raju, R. and Sharma, R. (1999). Calmodulin-dependent cyclic nucleotide phosphodiesterase (PDE1). *Cell. Mol. Life Sci.* **55**: 1164–1186.

Kanda, K. and Matsuda, T. (1994). Mechanical stress-induced orientation and ultrastructural change of smooth muscle cells cultured in three-dimensional collagen lattices. *Cell Transplant.* **3**: 481–492.

Katsumi, A., Orr, A.W., Tzima, E. and Schwart, M.A. (2004). Integrins in mechanotransduction. *J. Biol. Chem.* **279**: 12001–12004.

Klein-Nulend, J., Burger, E.H., Semeins, C.M., Raisz, L.G. and Pilbeam, C.C. (1997). Pulsating fluid flow stimulates prostaglandin release and inducible prostaglandin G/H synthase mRNA expression in primary mouse bone cells. *J. Bone Miner. Res.* **12**: 45–51.

Lane, N.E. (1996). Physical activity at leisure and risk of osteoarthritis. *Ann. Rheum. Dis.* **55**: 682–684.

Lee, H., Millward-Sadler, S., Wright, M., Nuki, G. and Salter, D. (2000). Integrin and mechanosensitive ion channel-dependent tyrosine phosphorylation of focal adhesion proteins and beta-catenin in human articular chondrocytes after mechanical stimulation. *J. Bone Miner. Res.* **15**: 1501–1509.

Lee, R., Yamamoto, C., Feng, Y., Potter-Perigo, S., Briggs, W., Landschulz, K., Turi, T., Thompson, J., Libby, P. and Wight, T. (2001). Mechanical strain induces specific changes in the synthesis and organization of proteoglycans by vascular smooth muscle cells. *J. Biol. Chem.* **276**: 13847–13851.

Leung, D.Y., Glagov, S. and Mathews, M.B. (1976). Cyclic stretching stimulates synthesis of matrix components by arterial smooth muscle cells *in vitro*. *Science* **191**: 475–477.

Li, K.W., Williamson, A.K., Wang, A.S. and Sah, R.L. (2001). Growth responses of cartilage to static and dynamic compression. *Clin. Orthop. Relat. Res.* **391**(Suppl 1): S34–S48.

Martin, I., Obradovic, B., Treppo, S., Grodzinsky, A.J., Langer, R., Freed, L.E. and Vunjak-Novakovic, G. (2000). Modulation of the mechanical properties of tissue engineered cartilage. *Biorheology* **37**: 141–147.

Martinac, B. (2004). Mechanosensitive ion channels: molecules of mechanotransduction. *J. Cell Sci.* **117**: 2449–2460.

Martinac, B., Adler, J. and Kung, C. (1990). Mechanosensitive ion channels of E. coli activated by amphipaths. *Nature* **348**: 261–263.

Martinez-Lemus, L., Wu, X., Wilson, E., Hill, M., Davis, G., Davis, M. and Meininger, G. (2003). Integrins as unique receptors for vascular control. *J. Vasc. Res.* **40**: 211–233.

Mauck, R.L., Soltz, M.A., Wang, C.C., Wong, D.D., Chao, P.H., Valhmu, W.B., Hung, C.T. and Ateshian, G.A. (2000). Functional tissue engineering of articular cartilage through dynamic loading of chondrocyte-seeded agarose gels. *J. Biomech. Eng.* **122**: 252–260.

McAllister, T.N. and Frangos, J.A. (1999). Steady and transient fluid shear stress stimulate NO release in osteoblasts through distinct biochemical pathways. *J. Bone Miner. Res.* **14**: 930–936.

McAllister, T.N., Du, T. and Frangos, J.A. (2000). Fluid shear stress stimulates prostaglandin and nitric oxide release in bone marrow-derived preosteoclast-like cells. *Biochem. Biophys. Res. Commun.* **270**: 643–648.

Mikuni-Takagaki, Y. (1999). Mechanical responses and signal transduction pathways in stretched osteocytes. *J. Bone Miner. Metabol.* **17**: 57–60.

Mitin, N., Rossman, K. and Der, C. (2005). Signaling interplay in Ras superfamily function. *Curr. Biol.* **15**: R563–R574.

Mobasheri, A., Carter, S., Martin-Vasallo, P. and Shakibaei, M. (2002). Integrins and stretch activated ion channels; putative components of functional cell surface mechanoreceptors in articular chondrocytes. *Cell Biol. Int.* **26**: 1–18.

Mow, V. and Ratcliffe, A. (1997). In: Mow, V. and Hayes, W. (eds.), *Basic Orthopaedic Biomechanics.* Philadelphia: Lippincott-Raven Publishers, pp. 113–177.

Niklason, L.E., Gao, J., Abbott, W.M., Hirschi, K.K., Houser, S., Marini, R. and Langer, R. (1999). Functional arteries grown *in vitro*. *Science* **284**: 489–493.

Niklason, L.E., Abbott, W., Gao, J., Klagges, B., Hirschi, K.K., Ulubayram, K., Conroy, N., Jones, R., Vasanawala, A., Sanzgiri, S. and Langer, R. (2001). Morphologic and mechanical characteristics of engineered bovine arteries. *J. Vasc. Surg.* **33**: 628–638.

Ogata, T. (2000). Fluid flow-induced tyrosine phosphorylation and participation of growth factor signaling pathway in osteoblast-like cells. *J. Cell. Biochem.* **76**: 529–538.

Otey, C., Vasquez, G., Burridge, K. and Erickson, B. (1993). Mapping of the alpha-actinin binding site within the beta 1 integrin cytoplasmic domain. *J. Biol. Chem.* **268**: 21193–21197.

Ott, M.J. and Ballermann, B.J. (1995). Shear stress-conditioned, endothelial cell-seeded vascular grafts: improved cell adherence in response to *in vitro* shear stress. *Surgery* **117**: 334–339.

Papadaki, M., Ruef, J., Nguyen, K.T., Li, F., Patterson, C., Eskin, S.G., McIntire, L.V. and Runge, M.S. (1998a). Differential regulation of protease activated receptor-1 and tissue plasminogen activator expression by shear stress in vascular smooth muscle cells. *Circ. Res.* **83**: 1027–1034.

Papadaki, M., Tilton, R.G., Eskin, S.G. and McIntire, L.V. (1998b). Nitric oxide production by cultured human aortic smooth muscle cells: stimulation by fluid flow. *Am. J. Physiol.* **274**: H616–H626.

Papadaki, M., Eskin, S.G., Ruef, J., Runge, M.S. and McIntire, L.V. (1999). Fluid shear stress as a regulator of gene expression in vascular cells: possible correlations with diabetic abnormalities. *Diabetes Res. Clin. Pract.* **45**: 89–99.

Pelham, R.J. and Wang, Y. (1999). High resolution detection of mechanical forces exerted by locomoting fibroblasts on the substrate. *Mol. Biol. Cell.* **10**: 935–945.

Piscoya, J.L., Fermor, B., Kraus, V.B., Stabler, T.V. and Guilak, F. (2005). The influence of mechanical compression on the induction of osteoarthritis-related biomarkers in articular cartilage explants. *Osteoarthr. Cartilage* **13**(2).

Plotkin, L., Mathov, I., Aguirre, J., Parfitt, A., Manolagas, S. and Bellido, T. (2005). Mechanical stimulation prevents osteocyte apoptosis: requirement of integrins, Src kinases, and ERKs. *Am. J. Physiol. Cell Physiol.* **289**: C633–C643.

Porter, B.D., Lin, A.S., Peister, A., Hutmacher, D., Guldberg, R.E. (2007). Noninvasive image analysis of 3D Construct mineralization in a perfusion bioreactor. *Biomaterials* **28**(15): 2525–2533.

Porter, B.D., Zauel, R., Hutmacher, D., Guldberg, R.E. and Fyhrie, D. (2005). Perfusion significantly increases mineral production inside 3-D PCL composite scaffolds. Presented at *Proceedings of the American Society for Mechanical Engineering Summer Bioengineering Meeting*, Vail, CO.

Rasmussen, H. (1986). The calcium messenger system. *New Engl J. Med.* **17**: 1094–1170.

Riha, G.M., Lin, P.H., Lumsden, A.B., Yao, Q. and Chen, C. (2005). Roles of hemodynamic forces in vascular cell differentiation. *Ann. Biomed. Eng.* **33**: 772–779.

Sachs, F. (1991). Mechanical transduction by membrane ion channels: a mini review. *Mol. Cell. Biochem.* **104**: 57–60.

Saini, S. and Wick, T.M. (2003). Concentric cylinder bioreactor for production of tissue engineered cartilage: effect of seeding density and hydrodynamic loading on construct development. *Biotechnol. Prog.* **19**: 510–521.

Schaller, M., Otey, C., Hildebrand, J. and Parsons, J. (1995). Focal adhesion kinase and paxillin bind to peptides mimicking beta integrin cytoplasmic domains. *J. Cell Biol.* **130**: 1181–1187.

Seger, R. and Krebs, E.G. (1995). The MAPK signaling cascade. *FASEB J.* **9**: 726–735.

Seliktar, D., Black, R.A., Vito, R.P. and Nerem, R.M. (2000). Dynamic mechanical conditioning of collagen-gel blood vessel constructs induces remodeling *in vitro*. *Ann. Biomed. Eng.* **28**: 351–362.

Seliktar, D., Nerem, R.M. and Galis, Z.S. (2001). The role of matrix metalloproteinase-2 in the remodeling of cell-seeded vascular constructs subjected to cyclic strain. *Ann. Biomed. Eng.* **29**: 923–934.

Shakibaei, M., John, T., De Souza, P., Rahmanzadeh, R. and Merker, H. (1999). Signal transduction by beta1 integrin receptors in human chondrocytes *in vitro*: collaboration with the insulin-like growth factor-I receptor. *Biochem. J.* **342**: 615–623.

Sharma, R., Yellowley, C., Civelek, M., Ainslie, K., Hodgson, L., Tarbell, J. and Donahue, H.J. (2002). Intracellular calcium changes in rat aortic smooth muscle cells in response to fluid flow. *Ann. Biomed. Eng.* **30**: 371–378.

Sieg, D., Hauck, C. and Schlaepfer, D. (1999). Required role of focal adhesion kinase (FAK) for integrin-stimulated cell migration. *J. Cell Sci.* **112**: 2677–2691.

Takahashi, M., Ishida, T., Traub, O., Corson, M.A. and Berk, B.C. (1997). Mechanotransduction in endothelial cells: temporal signaling events in response to shear stress. *J. Vasc. Res.* **34**: 212–219.

Torsoni, A., Marin, T., Velloso, L. and Franchini, K. (2005). RhoA/ROCK signaling is critical to FAK activation by cyclic stretch in cardiac myocytes. *Am. J. Physiol. Heart Circ. Physiol.* **289**: H1488–H1496.

Traub, O. and Berk, B.C. (1998). Laminar shear stress: mechanisms by which endothelial cells transduce an atheroprotective force. *Arteriosci. Throm. Vas.* **18**: 677–685.

Valhmu, W., Stazzone, E., Bachrach, N., Saed-Nejad, F., Fischer, S., Mow, V. and Ratcliffe, A. (1998). Load-controlled compression of articular cartilage induces a transient stimulation of aggrecan gene expression. *Arch. Biochem. Biophys.* **353**: 29–36.

van Hinsbergh, V.W. (2001). The endothelium: vascular control of haemostasis. *Eur. J. Obstet. Gynecol. Reprod. Biol.* **95**: 198–201.

Waldman, S.D., Spiteri, C.G., Grynpas, M.D., Pilliar, R.M. and Kandel, R.A. (2003). Long-term intermittent shear deformation improves the quality of cartilaginous tissue formed *in vitro*. *J. Orthop. Res.* **21**: 590–596.

Williams, K.A., Saini, S. and Wick, T.M. (2002). Computational fluid dynamics modeling of steady-state momentum and mass transport in a bioreactor for cartilage tissue engineering. *Biotechnol. Prog.* **18**: 951–963.

Wright, M., Stockwell, R. and Nuki, G. (1992). Response of plasma membrane to applied hydrostatic pressure in chondrocytes and fibroblasts. *Connect. Tissue Res.* **28**: 49–70.

Wright, M., Jobanputra, P., Bavington, C., Salter, D. and Nuki, G. (1996). Effects of intermittent pressure-induced strain on the electrophysiology of cultured human chondrocytes: evidence for the presence of stretch-activated membrane ion channels. *Clin. Sci.* **90**: 61–71.

Wright, M., Nishida, K., Bavington, C., Godolphin, J., Dunne, E., Walmsley, S., Jobanputra, P., Nuki, G. and Salter, D. (1997). Hyperpolarisation of cultured human chondrocytes following cyclical pressure-induced strain: evidence of a role for alpha 5 beta 1 integrin as a chondrocyte mechanoreceptor. *J. Orthop. Res.* **15**: 742–747.

You, J., Reilly, G., Zhen, X., Yellowley, C., Chen, Q., Donahue, H. and Jacobs, C. (2001). Osteopontin gene regulation by oscillatory fluid flow via intracellular calcium mobilization and activation of mitogen-activated protein kinase in MC3T3-E1 osteoblasts. *J. Biol. Chem.* **276**: 13365–13371.

Zabel, M., Koller, B., Sachs, F. and Franz, M. (1996). Stretch-induced voltage changes in the isolated beating heart: importance of the timing of stretch and implications for stretch-activated ion channels. *Cardiovasc. Res.* **32**: 120–130.

16

Engineering Cellular Microenvironments

Wendy F. Liu, Elliot E. Hui, Sangeeta N. Bhatia, and Christopher S. Chen

INTRODUCTION

Engineering cellular environments at the micrometer scale is critical for tissue engineering.

The primary strategy for engineering tissue constructs uses a combination of cells and artificial scaffolds. Obtaining an adequate source of cells is a major challenge, since many of the cell types taken from adult tissue have a limited capacity for expansion. Recent developments in stem cell biology suggest that these cells might provide a key source of cells because they have the capacity for self-renewal and differentiation into multiple lineages. While promising, these cells alone cannot form a tissue. Cells must be combined with a scaffold, which provides the initial structural support onto which the cells adhere and organize into a functioning tissue. While simple in concept, forming complex tissues such as liver, which contain many different cell types and a defined tissue architecture, is a formidable task. When cells are removed from their natural *in vivo* environment, and placed in an artificial environment they often lose their tissue-specific functions. Hepatocytes, for example, are normally rounded and do not proliferate, but when removed from the body and cultured on a plastic culture dish, they spread, dedifferentiate, and reduce their liver-specific functions (Mooney et al., 1992). Mesenchymal stem cells (MSCs), which are derived from the bone marrow, differentiate into osteoblasts or adipocytes depending on their adhesive environment (Pittenger et al., 1999; McBeath et al., 2004). Engineering a functional cellular phenotype in an artificial environment has become a major effort in tissue engineering. A greater understanding of the extracellular cues that control the behavior of cells, stem cells or others, may lead to smarter design of scaffold materials.

Biological structure and function are intricately linked at the tissue, cellular, and subcellular scales. Cells interact with soluble factors such as growth factors and cytokines, as well as insoluble factors such as extracellular matrix (ECM) proteins and other cells. The integration of soluble cues with those from both the matrix and neighboring cells plays an important role in regulating cell function. Cells are physically connected to the ECM through adhesion molecules known as integrins, which link the intracellular cytoskeleton to the ECM (Tamkun et al., 1986; Hynes, 1992). Many studies have demonstrated that binding of integrins to ECM leads to their clustering and the formation of focal adhesions, which then trigger intracellular signaling cascades and changes in numerous cellular processes (Schwartz and Ginsberg, 2002). Similarly, cells are physically connected to neighboring cells through cadherin molecules, which also serve as both mechanical linkages to the extracellular environment as well as signaling hubs to relay information to intracellular signaling pathways (Fagotto and Gumbiner, 1996; Wheelock and Johnson, 2003). Both integrin- and cadherin-mediated adhesions have been shown to modulate the ability of specific growth factor receptors to initiate intracellular signaling, induce changes in gene expression, and trigger specific cellular phenotypes. On the multicellular scale, cells within tissues are organized into functional units composed of multiple different cell types and arranged

Copyright © 2008 Elsevier Inc. All rights reserved

in a spatially defined manner. For example, the acinus, which consists of epithelial cells and fibroblasts organized in a spherical geometry, is critical for milk production in mammary glands (Nelson and Bissell, 2005). In fact, most tissues have highly defined structural components, which are indispensable for the functional integrity of the tissue. Thus, designing tissue-engineered constructs is not a simple amalgamation of cells with a scaffold, but instead requires an understanding of how cells behave in response to extracellular cues and the ability to design scaffolds with cellular scale resolution to mimic the architecture of the *in vivo* cellular environment.

Numerous recent advances in microscale fabrication technologies have enabled investigators to control the architecture of biomaterials at the cellular and multicellular scale, and the organization of cells on such materials. These tools, which have been adapted from the microfabrication industry, utilize photolithographic methods to generate microscale features on silicon wafers. Poly(dimethylsiloxane) (PDMS), a biocompatible silicone rubber, is then cast directly on the silicon wafers yielding a rubber stamp with a negative replicate of the original features (a technique termed soft lithography). PDMS stamps are then used in a variety of different applications such as microfluidic delivery of biological agents or microcontact printing of proteins. These methods allow for spatial and temporal control over the presentation of extracellular cues to cells. Furthermore, the ability to miniaturize assays using microscale technologies allows for higher throughput screening of hundreds of thousands of materials and molecules for studying cell–environmental interactions. These tools have utility not only in basic research, where they can help identify the relevant structural cues that stabilize specific cellular phenotypes, but also in applications for producing tissue constructs, where devices to manipulate cellular phenotype by extracellular cues can help to improve overall tissue function.

In the following chapter, we will examine recent efforts using microscale technologies to further advance the field of regenerative medicine. We will describe how these tools have been utilized to improve both the cellular and materials components of regenerative medicine. For engineering cells, these tools help investigators understand how the presentation of soluble cues, adhesive cues, and mechanical cues affects cellular behavior. For the biomaterials component, microfabrication can help to create spatially and structurally defined scaffolds that can be used to direct cellular function. We will also describe how these tools are being developed specifically for introducing tissue complexity in engineered cultures, such as in the examination of multiple cells or cell types, or in creating a structurally defined, three-dimensional scaffolds. While far from a complete review, this chapter will provide a glimpse into the ways in which microfabrication tools can be used to study cellular interactions and to create artificially engineered tissues for regenerative medicine.

DEFINING THE CELLULAR MICROENVIRONMENT

The ability to control the cellular microenvironment has traditionally been limited by the inability to generate spatially defined structures on the cellular scale. Here, we will describe some of the pioneering studies and recent advances in microfabrication technologies used to engineer the cellular microenvironment, including techniques to spatially control the soluble, adhesive, and mechanical environment.

Microfluidics to Spatially Control Soluble Cues

Many of the earliest studies in biology focused on understanding the role of soluble factors. Changes in media components dramatically affect simple cellular behaviors such as cell growth and proliferation. These studies were generally performed with bulk changes in the concentration of soluble factors within well-mixed media. However, it has long been known that geometric patterns and gradients of soluble factors have profound effects on cell migration and differentiation. For example, asymmetric growth factor signaling in a developing embryo determines the anterior–posterior layout of the organism. During wound healing, the release of chemokines stiffnesses of the substrates on which differentiated phenotypes were observed *in vitro* mimicked the physiologic stiffnesses of these tissues.

The stiffness of a tissue is likely inhomogeneous in nature – stiffness might vary from region to region, across micrometer length scales. Cells respond to gradients in substrate stiffness, also termed durotaxis, or the migration of cells between regions of different mechanical properties. Wang et al. (2000) demonstrated that fibroblasts migrate from regions of soft rigidity to regions of stiff rigidity, and furthermore that the processes mediating cellular responses to substrate stiffness are regulated by intracellular contractility (Pelham and Wang, 1997; Guo et al., 2006). Grey et al. (2003) extended this to the micrometer scale by adapting microengineering tools. Here, the stiffnesses of PDMS or acrylamide substrates were tuned from 1.8 to 34 kPa by varying the crosslinker density. Stiff islands were patterned among a soft substrate, and cells were observed to migrate predominantly onto the stiff regions, forming islands of cells (Figure 16.3b). These effects were observed in both endothelial cells and fibroblasts.

Currently there are only a handful of studies suggesting that cell substrate mechanics or stiffness play an important role in modulating cell behavior, but this concept is quickly gaining widespread support. A deeper understanding of how different cells respond to stiffness of their surroundings may be useful for applications in tissue engineering. Moreover, the design of materials with spatially and temporally controlled mechanical properties may be important for generating functional units of tissue.

DEFINING THE ORGANIZATION OF MULTICELLULAR CONSTRUCTS

Tissues and organs are exquisitely ordered three-dimensional structures composed of multiple cells and cell types. To a large extent, the microenvironment is defined by the local organization of cells, which secrete paracrine factors, deposit ECM, present surface ligands, and exert physical force. Therefore, fully understanding and recapitulating the microenvironment involves not only the techniques described in section "Defining the cellular microenvironment," but also additional methods to organize and study heterogeneous multicellular constructs, in both two and three dimensions.

Patterning Multicellular Constructs in Two Dimensions

Patterning of adhesive and non-adhesive regions on two-dimensional substrates has been described in section "Microengineered tools to define the adhesive environment." Once the surface is defined, uniformly seeded cells will selectively adhere to adhesive regions and form the desired pattern. Co-cultures of multiple cell types can be patterned using biochemistries specific to individual cell types. Selective chemistries are not always available, however, thus recent studies have explored more general means to pattern multiple cell types. Using microfluidics, cells can be directly delivered to desired locations on a uniform substrate (Chiu et al., 2000). Additional methods include hydrogel molding (Tang et al., 2003), layer-by-layer deposition of ionic polymers (Khademhosseini et al., 2006), and dynamically regulated surfaces (see section "Dynamically changing the adhesive environment"). Microscale control of multicellular organization has brought an unprecedented ability to study interactions between individual cells or groups of cells within a colony. Using simple but carefully planned geometries, the following examples illustrate the biological insights that can be gained using cell patterning tools.

Homotypic Interactions

Previous studies examining the role of cell–cell adhesions typically uniformly seeded cells at different densities. Cells seeded at low densities had few cell–cell contacts, while cells seeded at high densities had many cell–cell contacts. However, when seeding density is changed, other microenvironmental cues are also varied. At higher density, not only does the number of cell–cell contacts increase, but also the amount of cell spreading decreases as the cells become crowded to fill the culture dish. Furthermore, the amount of soluble paracrine signals secreted from the cells may differ across cultures with varying densities. Based on these studies, it was therefore unclear what respective roles are played by each of these factors in regulating cell function. Microfabricated tools can be used to independently vary microenvironmental factors such as the degree of cell spreading from cell–cell interactions, therefore enabling investigators to tease out the critical parameters leading to changes in cell function.

A microfabricated bowtie system was devised to decouple control of cell–cell contact and cell spreading. Bowtie-shaped regions for cell attachment were defined by patterning a non-adhesive agarose gel on a glass substrate. Each half of the bowtie allowed room for a single cell, fixing the amount of cell spreading. Pairs of cells could contact each other through the constriction at the center of the bowtie (Figure 16.4a). Cells were cultured either in pairs or as single cells occupying only half of the bowtie. Paired cells in contact demonstrated significantly higher rates of proliferation in comparison to single cells, implicating contact as an inducer of proliferation. In addition, paired cells in bowties where contact was physically blocked (Figure 16.4a) did not show greater proliferation than single cells, suggesting that paracrine signaling at close proximity was not sufficient. In fact, the authors demonstrated that specific receptors – cadherins – engaged upon cell–cell contact, and this receptor ligation induced the changes in cell function (Nelson and Chen, 2003).

Besides the biochemical signaling that occurs within a community of cells, physical forces are another important "signal" that is transmitted through cell–cell interactions. Recently, Nelson et al. (2005) utilized micropatterned cultures to bring new insight into the factors that drive tissue morphogenesis. It was observed that cell proliferation was greatest at the edges of patterned sheets of endothelial and epithelial cells. In addition, the effect was more pronounced along longer edges of a rectangular sheet of cells and was not observed on concave edges. Mechanical modeling of variously shaped cell patterns revealed distributions of tensile stress within the cell sheets that directly correlated to the observed patterns of proliferation, with higher

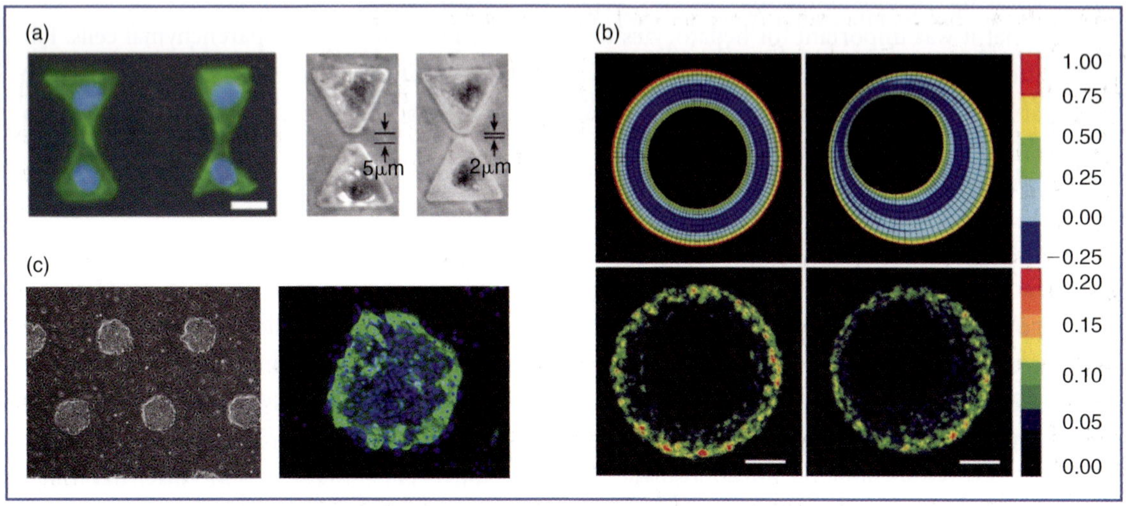

Figure 16.4 **(a) Phase contrast and fluorescence images of cell pairs plated onto bowtie structures, in contact (left) and without contact (right) (Nelson and Chen, 2002). (b) Plot of strain over FEM models of patterned cell sheets (top) corresponds to regions of rapid proliferation in cultures (bottom) (Nelson et al., 2005). (c) Phase contrast image of patterned hepatocyte islands surrounded by fibroblasts (left), and fluorescence image of albumin expression (green) localized to the periphery of a hepatocyte island (right).**

microns to centimeters. Future advances in three-dimensional micropatterning methodology promise further elucidation of tissue biology as well as continued functional improvements in engineered constructs.

DYNAMICALLY CHANGING THE MICROENVIRONMENT

While most studies have examined the cellular microenvironment in a static setting, cells are constantly experiencing dynamic changes in their natural environment. For example, during angiogenesis, or the development of new blood vessels, endothelial cells detach and migrate away from their neighboring cells, form new adhesions with the surrounding ECM, proliferate, and form tube-like structures. While numerous researchers have sought to understand the various environmental cues that affect this process, very little is known about its temporal regulation. A greater understanding of blood vessel formation could not only have broad scientific impact, but also have practical applications such as help to find new methods to vascularize engineered tissues. Studies that can modulate a temporal component are only in their infancy since methods to control the dynamics of extracellular cues are technically challenging. In the following sections, we will describe some of the recent developments in microscale technologies that have not only spatial, but also temporal control over the cellular microenvironment.

Dynamic Regulation of the Soluble Environment

Experimentally, it is difficult to dynamically regulate the soluble environment at a physiologically relevant frequency and to provide controlled, reproducible dynamic changes for systematic studies. With bulk changes in the media, the frequency is limited by the researcher's ability to change the medium. However, cells in the body experience dynamic change with frequencies that cannot be attained by manual changes in medium. For examples, chondrocytes experience changes in osmotic loading due to mechanical forces on the charged ECM; the frequency of these changes are on the order of 0.01–0.1 Hz. An advantage of microfluidic technology described earlier in section "Microfluidics to spatially control soluble cues" is that computers, pumps, and valves control the changes in media, therefore allowing much higher frequencies of loading. Chao et al. (2005) applied changes in osmotic pressures to chondrocytes using a microfluidic device with two input liquid streams of different osmotic pressures. Here, the dynamic changes in cell morphology response to osmotic loading were dependent on frequency of loading.

In addition to the ability to generate geometrically defined soluble gradients described earlier, microfluidic technology also has the ability to temporally control these gradients. Irimia et al. examined neutrophil migration response to changes in gradients of IL-8. They tracked neutrophil migration response to "step up" (increased steepness), "step down" (decreased steepness), or "flip" (reversed) changes in gradient that were achieved in less than 5 s. Neutrophils changed their velocity but not direction in response to "step up" and "step down," and changed direction in response to the "flip." These findings may provide further insight to the mechanism of neutrophil chemotaxis.

Dynamically Changing the Adhesive Environment

A variety of methods to define the adhesive microenvironment are discussed in section "Microengineered tools to define the adhesive environment," however these procedures are only applicable prior to the introduction of cells into the system. Once cells are plated, little adhesive regulation is experimentally possible short of global application of an enzymatic cleaving agent to release all cells. Only recently have groups begun to report methods to dynamically modify substrate surfaces during cell culture.

Okano et al. (1995) developed a thermally responsive material, poly(N-isopropyl acrylamide) (pNIPA), which switches from hydrophobic at 37°C to hydrophilic at 20°C. Cells can attach and spread on the surface at the higher temperature, but detach when the surface is cooled. Cheng et al. (2004) extended this strategy by fabricating a microheater array underneath a pNIPA layer to locally regulate cell adhesion.

Significantly, Yeo et al. (2003) have been able to achieve dynamic, molecular-level control of a substrate surface. Self-assembled monolayers (SAMs) were modified to present an electrically active ester, which can release and bind ligands via electrochemical redox reactions. Using this method to control the expression of the RGD peptide on a substrate, fibroblast adhesion to the surface was dynamically regulated. The RGD peptide mediates cell adhesion via integrin binding sites, thus cells were able to attach and spread on surfaces expressing this ligand. Upon application of an electrical potential, the RGD ligand was electrochemically cleaved from the SAM and released, along with the attached cells. Electrically active and non-active chemistries could be patterned together, enabling selective patterned release. In addition, following release, another RGD conjugation could be introduced to bind onto the vacated sites, rendering those regions cell adhesive once again.

Although dynamic surfaces are rapidly increasing in capability, this field is still relatively new, and to this point there has yet to be much success in applying these tools to study the biology of the cellular microenvironment. One recent example, reported by Jiang et al. (2005), employed a patterned SAM to constrain adhered cells to a teardrop shape. After applying an electric potential to electrochemically desorb the SAM, the cells were observed to migrate in the direction of the blunt end of the teardrop. Studies such as these will help to provide insight to how cells respond to dynamic changes in their local environment.

CONCLUSIONS AND FUTURE DIRECTIONS

While the field of regenerative medicine has blossomed in the past decade, there are still major obstacles that must be overcome before the dream of functional artificially engineered tissues can be achieved. Understanding how to use artificial environments to control cell function and finding suitable scaffolds to provide this control are keystones for future endeavors in tissue engineering. Microscale technologies undoubtedly will provide some of the tools necessary to achieve these goals.

Appropriately directing cell fate and function remains a critical challenge in engineering tissue constructs. Microfabricated systems as those presented here will provide an important tool in elucidating the mechanisms underlying how extracellular cues can be used to drive cell function. While much is known about how the chemical properties of these cues affect various intracellular signaling pathways, microfabricated systems have only recently revealed that physical and mechanical cues are also equally important. It is now being realized that a cell can sense the physical and mechanical parameters of its surroundings through the cytoskeleton and through numerous intertwined intracellular signaling pathways. Understanding how the spatial presentation of soluble, adhesive, and mechanical cues is integrated within cells will be a critical challenge of the near future of regenerative medicine. In addition, one must not overlook the fact that the body is composed of many different types of cells, each of which has a distinctive response that defines the phenotype of that particular cell type. A deeper understanding of how each type of cell behaves in the context of other cells and in response to multiple cues remains an enormous task that may be partially simplified by the miniaturization and screening approaches offered by microscale technologies. Moreover, the recent shift of the focus of the biomedical community toward stem cells for regenerative applications further highlights the need for microculture systems to study these rare and valuable cells. Thus, microfabricated systems may provide a critical set of tools to engineer stem cells for regenerative medicine applications.

While understanding the cellular component of tissue engineering is a major endeavor, this only constitutes one-half of the challenge, as cells are combined with different types of scaffolds to yield the desired tissue product. Microengineered tools may impact tissue engineering by increasing the physical complexity and spatial resolution of scaffold materials. As described in this chapter, microfabrication technologies can easily generate features with spatial resolution on the micrometer scale. With improvements of these tools and the advent of nanotechnologies, generation of devices with subcellular-scale resolution is on the immediate horizon. Here, we have reviewed how these tools can be used to control the geometric presentation of soluble, adhesive, and mechanical cues. In addition, microfabrication technologies can also be used to include other features such as substrate topology, and mechanically or electrically activated components that can interact with cells. It remains to be seen how and when these functionalities can be applied to scaffold engineering for regenerative medicine.

The integration of all of the above elements into a microfabricated scaffold that can support the growth and maintenance of specified cellular phenotypes, and most importantly a desirable multicellular functionality will be critical to the design of novel, serviceable tissue-engineered constructs. While currently only a budding area of study, this field offers exciting new potential to engineer devices on a level of complexity that would otherwise not be possible.

ACKNOWLEDGMENTS

The authors declare no competing financial interests. This work was supported in part by the NIH, NSF, David and Lucile Packard Foundation, and Desphande Foundation. W.F.L. acknowledges the NSF for financial support and E.E.H. acknowledges a NIH NRSA postdoctoral fellowship.

REFERENCES

Albrecht, D.R., Sah, R.L. and Bhatia, S.N. (2004). Geometric and material determinants of patterning efficiency by dielectrophoresis. *Biophys. J.* **87**: 2131–2147.

Albrecht, D.R., Tsang, V.L., Sah, R.L. and Bhatia, S.N. (2005). Photo- and electro-patterning of hydrogel-encapsulated living cell arrays. *Lab. Chip.* **5**: 111–118.

Albrecht, D.R., Underhill, G.H., Wassermann, T.B., Sah, R.L. and Bhatia, S.N. (2006). Probing the role of multicellular organization in 3D microenvironments. *Nat. Method.* **3**: 369–375.

Anderson, D.G., Levenberg, S. and Langer, R. (2004). Nanoliter-scale synthesis of arrayed biomaterials and application to human embryonic stem cells. *Nat. Biotechnol.* **22**: 863–866.

Beebe, D.J., Mensing, G.A. and Walker, G.M. (2002). Physics and applications of microfluidics in biology. *Annu. Rev. Biomed. Eng.* **4**: 261–286.

Bhatia, S.N., Balis, U.J., Yarmush, M.L. and Toner, M. (1999). Effect of cell–cell interactions in preservation of cellular phenotype: cocultivation of hepatocytes and nonparenchymal cells. *FASEB J.* **13**: 1883–1900.

Boyden, S. (1962). The chemotactic effect of mixtures of antibody and antigen on polymorphonuclear leukocytes. *J. Exp. Med.* **115**: 453–466.

Carter, S.B. (1967). Haptotactic islands: a method of confining single cells to study individual cell reactions and clone formation. *Exp. Cell Res.* **48**: 189–193.

Chao, P., Tang, Z., Angelini, E., West, A.C., Costa, K.D. and Hung, C.T. (2005). Dynamic osmotic loading of chondrocytes using a novel microfluidic device. *J. Biomech.* **38**: 1273–1281.

Chen, C.S., Mrksich, M., Huang, S., Whitesides, G.M. and Ingber, D.E. (1997). Geometric control of cell life and death. *Science.* **276**: 1425–1428.

Cheng, X., Wang, Y., Hanein, Y., Bohringer, K.F. and Ratner, B.D. (2004). Novel cell patterning using microheater-controlled thermoresponsive plasma films. *J Biomed. Mater. Res. A.* **70**: 159–168.

Chiu, D.T., Jeon, N.L., Huang, S., Kane, R.S., Wargo, C.J., Choi, I.S., Ingber, D.E. and Whitesides, G.M. (2000). Patterned deposition of cells and proteins onto surfaces by using three-dimensional microfluidic systems. *PNAS* **97**: 2408–2413.

Chung, B.G., Flanagan, L.A., Rhee, S.W., Schwartz, P.H., Lee, A.P., Monuki, E.S. and Jeon, N.L. (2005). Human neural stem cell growth and differentiation in a gradient-generating microfluidic device. *Lab. Chip.* **5**: 401–406.

Cukierman, E., Pankov, R. and Yamada, K.M. (2002). Cell interactions with three-dimensional matrices. *Curr. Opin. Cell Biol.* **14**: 633–639.

Dembo, M., Oliver, T., Ishihara, A. and Jacobson, K. (1996). Imaging the traction stresses exerted by locomoting cells with the elastic substratum method. *Biophys. J.* **70**: 2008–2022.

Deroanne, C.F., Lapiere, C.M. and Nusgens, B.V. (2001). *In vitro* tubulogenesis of endothelial cells by relaxation of the coupling extracellular matrix-cytoskeleton. *Cardiovasc. Res.* **49**: 647–658.

Engler, A.J., Griffin, M.A., Sen, S., Bonnemann, C.G., Sweeney, H.L. and Discher, D.E. (2004). Myotubes differentiate optimally on substrates with tissue-like stiffness: pathological implications for soft or stiff microenvironments. *J. Cell Biol.* **166**: 877–887.

Fagotto, F. and Gumbiner, B.M. (1996). Cell contact-dependent signaling. *Dev. Biol.* **180**: 445–454.

Falsey, J.R., Renil, M., Park, S., Li, S. and Lam, K.S. (2001). Peptide and small molecule microarray for high throughput cell adhesion and functional assays. *Bioconjug. Chem.* **12**: 346–353.

Flaim, C.J., Chien, S. and Bhatia, S.N. (2005). An extracellular matrix microarray for probing cellular differentiation. *Nat. Method.* **2**: 119–125.

Galbraith, C.G. and Sheetz, M.P. (1997). A micromachined device provides a new bend on fibroblast traction forces. *PNAS* **94**: 9114–9118.

Geiger, B. and Bershadsky, A. (2001). Assembly and mechanosensory function of focal contacts. *Curr. Opin. Cell Biol.* **13**: 584–592.

Gray, D.S., Tien, J. and Chen, C.S. (2003). Repositioning of cells by mechanotaxis on surfaces with micropatterned Young's modulus. *J. Biomed. Mater. Res. A.* **66**: 605–614.

Guo, W.-H., Frey, M.T., Burnham, N.A. and Wang, Y.-L. (2006). Substrate rigidity regulates the formation and maintenance of tissues. *Biophys. J.* **90**: 2213–2220.

Harris, A.K., Wild, P. and Stopak, D. (1980). Silicone rubber substrata: a new wrinkle in the study of cell locomotion. *Science* **208**: 177–179.

Hynes, R.O. (1992). Integrins – versatility, modulation, and signaling in cell-adhesion. *Cell* **69**: 11–25.

Ingber, D.E. and Folkman, J. (1989). Mechanochemical switching between growth and differentiation during fibroblast growth factor-stimulated angiogenesis *in vitro*: role of extracellular matrix. *J. Cell Biol.* **109**: 317–330.

Jeon, N.L., Dertinger, S.K.W., Chiu, D.T., Choi, I.S., Stroock, A.D. and Whitesides, G.M. (2000). Generation of solution and surface gradients using microfluidic systems. *Langmuir* **16**: 8311–8316.

Jiang, X., Bruzewicz, D.A., Wong, A.P., Piel, M. and Whitesides, G.M. (2005). Directing cell migration with asymmetric micropatterns. *Proc. Natl Acad. Sci. USA* **102**: 975–978.

Khademhosseini, A., Langer, R., Borenstein, J. and Vacanti, J.P. (2006). Microscale technologies for tissue engineering and biology. *Proc. Natl Acad. Sci. USA* **103**: 2480–2487.

Lee, K.Y. and Mooney, D.J. (2001). Hydrogels for tissue engineering. *Chem. Rev.* **101**: 1869–1879.

Li Jeon, N., Baskaran, H., Dertinger, S.K.W., Whitesides, G.M., Van De Water, L. and Toner, M. (2002). Neutrophil chemotaxis in linear and complex gradients of interleukin-8 formed in a microfabricated device. *Nat. Biotechnol.* **20**: 826–830.

Liu, V.A. and Bhatia, S.N. (2002). Three-dimensional photopatterning of hydrogels containing living cells. *Biomed. Microdev.* **4**: 257–266.

Lutolf, M.P. and Hubbell, J.A. (2005). Synthetic biomaterials as instructive extracellular microenvironments for morphogenesis in tissue engineering. *Nat. Biotechnol.* **23**: 47–55.

MacBeath, G. and Schreiber, S.L. (2000). Printing proteins as microarrays for high-throughput function determination. *Science* **289**: 1760–1763.

McBeath, R., Pirone, D.M., Nelson, C.M., Bhadriraju, K. and Chen, C.S. (2004). Cell shape, cytoskeletal tension, and RhoA regulate stem cell lineage commitment. *Dev. Cell* **6**: 483–495.

Mooney, D., Hansen, L., Vacanti, J., Langer, R., Farmer, S. and Ingber, D. (1992). Switching from differentiation to growth in hepatocytes: control by extracellular matrix. *J. Cell Physiol.* **151**: 497–505.

Mueller-Klieser, W. (1997). Three-dimensional cell cultures: from molecular mechanisms to clinical applications. *Am. J. Physiol.* **273**: C1109–C1123.

Nelson, C.M. and Bissell, M.J. (2005). Modeling dynamic reciprocity: engineering three-dimensional culture models of breast architecture, function, and neoplastic transformation. *Semin. Cancer Biol.* **15**: 342–352.

Nelson, C.M. and Chen, C.S. (2002). Cell–cell signaling by direct contact increases cell proliferation via a PI3K-dependent signal. *FEBS Lett.* **514**: 238–242.

Nelson, C.M. and Chen, C.S. (2003). VE-cadherin simultaneously stimulates and inhibits cell proliferation by altering cytoskeletal structure and tension. *J. Cell Sci.* **116**: 3571–3581.

Nelson, C.M., Raghavan, S., Tan, J.L. and Chen, C.S. (2003). Degradation of micropatterned surfaces by cell-dependent and -independent processes. *Langmuir* **19**: 1493–1499.

Nelson, C.M., Jean, R.P., Tan, J.L., Liu, W.F., Sniadecki, N.J., Spector, A.A. and Chen, C.S. (2005). From the cover: emergent patterns of growth controlled by multicellular form and mechanics. *PNAS* **102**: 11594–11599.

Okano, T., Yamada, N., Okuhara, M., Sakai, H. and Sakurai, Y. (1995). Mechanism of cell detachment from temperature-modulated, hydrophilic–hydrophobic polymer surfaces. *Biomaterials* **16**: 297–303.

Oliver, T., Dembo, M. and Jacobson, K. (1995). Traction forces in locomoting cells. *Cell Motil. Cytoskeleton* **31**: 225–240.

Paszek, M.J., Zahir, N., Johnson, K.R., Lakins, J.N., Rozenberg, G.I., Gefen, A., Reinhart-King, C.A., Margulies, S.S., Dembo, M., Boettiger, D., Hammer, D.A. and Weaver, V.M. (2005). Tensional homeostasis and the malignant phenotype. *Cancer Cell* **8**: 241–254.

Pelham Jr., R.J. and Wang, Y. (1997). Cell locomotion and focal adhesions are regulated by substrate flexibility. *Proc. Natl Acad. Sci. USA* **94**: 13661–13665.

Pittenger, M.F., Mackay, A.M., Beck, S.C., Jaiswal, R.K., Douglas, R., Mosca, J.D., Moorman, M.A., Simonetti, D.W., Craig, S. and Marshak, D.R. (1999). Multilineage potential of adult human mesenchymal stem cells. *Science* **284**: 143–147.

Raty, S., Walters, E.M., Davis, J., Zeringue, H., Beebe, D.J., Rodriguez-Zas, S.L. and Wheeler, M.B. (2004). Embryonic development in the mouse is enhanced via microchannel culture. *Lab. Chip.* **4**: 186–190.

Schwartz, M.A. and Ginsberg, M.H. (2002). Networks and crosstalk: integrin signalling spreads. *Nat. Cell Biol.* **4**: E65–E68.

Sia, S.K. and Whitesides, G.M. (2003). Microfluidic devices fabricated in poly(dimethylsiloxane) for biological studies. *Electrophoresis* **24**: 3563–3576.

Singhvi, R., Kumar, A., Lopez, G.P., Stephanopoulos, G.N., Wang, D.I., Whitesides, G.M. and Ingber, D.E. (1994). Engineering cell shape and function. *Science* **264**: 696–698.

Stroock, A.D., Dertinger, S.K.W., Ajdari, A., Mezic, I., Stone, H.A. and Whitesides, G.M. (2002). Chaotic mixer for microchannels. *Science* **295**: 647–651.

Takayama, S., Ostuni, E., LeDuc, P., Naruse, K., Ingber, D.E. and Whitesides, G.M. (2003). Selective chemical treatment of cellular microdomains using multiple laminar streams. *Chem. Biol.* **10**: 123–130.

Tamkun, J.W., Desimone, D.W., Fonda, D., Patel, R.S., Buck, C., Horwitz, A.F. and Hynes, R.O. (1986). Structure of integrin, a glycoprotein involved in the transmembrane linkage between fibronectin and actin. *Cell* **46**: 271–282.

Tan, J.L., Tien, J., Pirone, D.M., Gray, D.S., Bhadriraju, K. and Chen, C.S. (2003). Cells lying on a bed of microneedles: an approach to isolate mechanical force. *Proc. Natl Acad. Sci. USA* **100**: 1484–1489.

Tan, J.L., Liu, W., Nelson, C.M., Raghavan, S. and Chen, C.S. (2004). Simple approach to micropattern cells on common culture substrates by tuning substrate wettability. *Tissue Eng.* **10**: 865–872.

Tang, M.D., Golden, A.P. and Tien, J. (2003). Molding of three-dimensional microstructures of gels. *J. Am. Chem. Soc.* **125**: 12988–12989.

Taylor, A.M., Blurton-Jones, M., Rhee, S.W., Cribbs, D.H., Cotman, C.W. and Jeon, N.L. (2005). A microfluidic culture platform for CNS axonal injury, regeneration and transport. *Nat. Methods* **2**: 599–605.

Tsang, V.L. and Bhatia, S.N. (2004). Three-dimensional tissue fabrication. *Adv. Drug Deliv. Rev.* **56**: 1635–1647.

Unger, M.A., Chou, H.-P., Thorsen, T., Scherer, A. and Quake, S.R. (2000). Monolithic microfabricated valves and pumps by multilayer soft lithography. *Science* **288**: 113–116.

Wang, H.B., Dembo, M. and Wang, Y.L. (2000). Substrate flexibility regulates growth and apoptosis of normal but not transformed cells. *Am. J. Physiol. Cell Physiol.* **279**: C1345–C1350.

Weiner, O.D., Servant, G., Welch, M.D., Mitchison, T.J., Sedat, J.W. and Bourne, H.R. (1999). Spatial control of actin polymerization during neutrophil chemotaxis. *Nat. Cell Biol.* **1**: 75–81.

Wheelock, M.J. and Johnson, K.R. (2003). Cadherin-mediated cellular signaling. *Curr. Opin. Cell Biol.* **15**: 509–514.

Wilkinson, P.C. (1998). Assays of leukocyte locomotion and chemotaxis. *J. Immunol. Method.* **216**: 139–153.

Yeo, W.S., Yousaf, M.N. and Mrksich, M. (2003). Dynamic interfaces between cells and surfaces: electroactive substrates that sequentially release and attach cells. *J. Am. Chem. Soc.* **125**: 14994–14995.

Ziauddin, J. and Sabatini, D.M. (2001). Microarrays of cells expressing defined cDNAs. *Nature* **411**: 107–110.

Zicha, D., Dunn, G.A. and Brown, A.F. (1991). A new direct-viewing chemotaxis chamber. *J. Cell Sci.* **99**(Pt 4): 769–775.

Zigmond, S. (1977). Ability of polymorphonuclear leukocytes to orient in gradients of chemotactic factors. *J. Cell Biol.* **75**: 606–616.

17

Applications of Nanotechnology

Benjamin S. Harrison

INTRODUCTION

Regenerative medicine represents one of the greatest challenges in modern day science and medicine. With the goal of repairing diseased or damaged tissue to restore normal function, it has become increasingly apparent that our current understanding of biology is insufficient to reach such a lofty goal. Numerous implants, cell therapies, and engineered tissues that have been developed so far indicate that our current understanding of the superstructure and the microstructure of tissues is no longer adequate to create truly regenerative therapies. Understanding and controlling the underlying nanostructures in cells and the extracellular matrix represent key pieces in reaching the goals of regenerative medicine.

Nanotechnology is a bottom-up approach that focuses on assembling simple elements to form complex structures. At the nanometer scale, where many biological processes operate, nanotechnology can provide the tools to probe and even direct these biological processes. This means that nanotechnology can be used for repairing damaged parts, curing diseases, and even monitoring and responding to the needs of the body. Cells and the extracellular matrix possess a multitude of nanodimensionality that interplays with one another. Cells, typically microns in diameter, are composed of numerous nanosized components all working together, creating a highly organized, self-regulating machine. For example, the cell surface is composed of ion-channels that regulate the coming and going of ions such as calcium and potassium in and out of the cell. Enzyme reactions, protein dynamics, and DNA all possess some aspect of nanodimensionality. These nanodimensional components control how cells produce the extracellular matrix (ECM) including the ECM composition and architecture. The ECM that the cell interacts with also abounds with nanosize features that influence the behaviors of other cells and tissues. These nanosized features, such as fiber diameter and pores, along with the intrinsic properties of the matrix itself, control the mechanical strength, the adhesiveness of the cells to the matrix, cell proliferation, and the shape of the ECM.

Nanotechnology will provide regenerative medicine with the new multifunctional tools for imaging and monitoring the regenerative process and controlling the structure of the ECM. This is an exciting feature of nanotechnology in that it should not be thought of as a single object that has only one function. The size of nanomaterials allows multiple components to be combined and contained in a single nanocarrier unit. In addition, the small size allows nanomaterials to probe biological processes with minimal intrusion. Included in or on this nanocarrier can be therapeutic, targeting, contrast, and/or biocompatibilizing components which can be designed to meet a particular need. A description of the various components of a nanocarrier can be found in Table 17.1. Individual components, such as a therapeutic agent, can be exchanged or removed to create the desired effect without necessarily compromising the remaining components. This is significantly different compared to drug synthesis, for example, in which a single change can dramatically influence the pharmacological kinetics and potency of the drug.

Besides realizing the potential of the small size, exploration into the nano-world is revealing unique quantum phenomena that only occur on the nanoscale. These quantum effects could be exploited to provide

Copyright © 2008 Elsevier Inc. All rights reserved

Table 17.1 A typical nanocarrier of image contrast and/or therapeutic agents is composed of six components

Binder	All the different components are held together using a binder. The binder may be an inert piece of the nanocarrier, however many times it also serves another purpose. The binder may also be the image contrasting agent. For example, iron nanoparticles and QD serve as the core for the attachment of the other components. Polymers such as polyglycolic acid may serve as the binder of the therapeutic and also the biocompatiblizing agent.
Biocompatibilization	This component makes the nanocarrier compatible with the biological environment. It does this by minimizing aggregation of the nanocarrier and increasing the lifetime of the carrier by avoiding the defense mechanisms of the biological systems such as the reticuloendothelial system.
Imaging contrast	This component provides the means for imaging modalities to observe the nanocarrier. These contrasting agents may be observed using optical, magnetic, ultrasound, and scintillating methods.
Sensor	The sensor or trigger is used to alter the behavior of the nanocarrier once it has been deployed. For example, near-infrared light or electromagnetic radiation may be used to accelerate the release of a therapeutic or cause rapid localized heating as part of a therapy. Chemical sensors such as polymers that are pH or ion sensitive may also provide feedback to the nanocarrier in the delivery of its payload.
Targeting	This component provides the means of driving the nanocarrier to its desired location. There are two types of targeting: passive and active. Passive targeting incorporates only non-specific targeting agents which may be useful for determining microenvironment permeability or areas of increased angiogenesis. Active targeting uses ligands or antibodies that bind to specific receptors at the target site. Active targeting aids in obtaining higher concentrations of therapeutics and contrasting agents at the desired site. Also, multiple targeting agents can be bound to the nanocarrier, allowing lower binding affinity molecules to be used to increase binding probabilities.
Therapeutics	Bioactive agents such as drugs or DNA are typical payloads of the nanocarrier. Drugs that are incapable of penetrating cellular membranes or hydrophobic drugs which cannot be administrated systemically by themselves can be contained within the nanocarrier awaiting release in a controlled manner. Other novel properties of nanoparticles have also shown promise as hyperthermic agents.

new approaches to regenerative therapies. Such quantum effects result in high optical absorptivities coupled with large photostabilities, or unusually magnetic properties within nanomaterials. Already such nanomaterials are being explored to enhance cellular imaging (Zhang et al., 2002; Medintz et al., 2005). Besides imaging, these quantum effects will allow novel methods of drug delivery, using light, electric or magnetic fields as drug delivering triggers. While these may involve exotic materials or elements which never would be found to naturally occur in the body, the expectation of nanotechnology should only be to serve as a temporary aid to direct the regeneration process and so should be developed with a relatively short-term use in mind.

Nanotechnology's impact on regenerative medicine will be through the development of multifunctional tools to enhance the performance or capabilities of implants, cell therapies, and tissue engineering. These advances will be the result of understanding and exploiting the underlying nanodimensionality of life. Nanotechnology will play a role in the ongoing development of tools for controlling the cell and its support matrix. Since nanotechnology is at the interface of modern physical science and medicine, new and unconventional ideas will develop, capable of bringing about major revolutions in science and medicine. Therapies developed using nanotechnology will someday minimize or eliminate the side-effects of drugs through targeted delivery and will provide real-time, and even non-invasive, monitoring of the disease and tissue repair. In this chapter, we will examine the impact nanotechnology will have on regenerative medicine related to cellular therapies and biomaterial control which play an important role for implant design and tissue engineering.

NANOTECHNOLOGY AS A MULTI-FUNCTIONAL TOOL FOR CELL-BASED THERAPIES

Cell-based therapies, in particular those based on stem cells, have generated much excitement, both in the media and scientific communities, and are one of the most promising and active areas of research in regenerative medicine. One method of accelerating the pace of research is through the creation of multi-functional tools which allow the improved monitoring and modifying of cell behavior. While cancer-related research is a large part of the nanomedicine effort, there is great potential for applying nanotechnology in cell-based therapies for regenerative medicine. For example, with the enormous self-repair potential of stem cells, it is important to be able to locate, recruit, and signal these cells to begin the regeneration process.

Improving non-invasive monitoring methods is particularly desirable since current methods of evaluating cell treatments typically involve destructive or invasive techniques such as tissue biopsies. Traditional non-invasive methods such as magnetic resonance imaging (MRI) and positron emission tomography (PET), which rely heavily on contrast agents, lack the specificity or resident time to be a viable option for cell tracking. However, *in vitro* and *in vivo* visualization of nanoscale systems can be carried out using a variety of clinically relevant modalities such as fluorescence microscopy, single photon emission computed tomography (SPECT), PET, MRI, ultrasound, and radiotracing such as gamma scintigraphy. Nanoparticulate imaging probes include semiconductor quantum dots (QD), magnetic and magnetofluorescent nanoparticles, gold nanoparticles, and nanoshells among others. While there are currently few examples of nanotechnologies being applied to the understanding of important processes in tissue regeneration, relevant uses of nanoparticles for regenerative medicine such as monitoring angiogensis (Winter et al., 2003) and apoptosis are appearing (Jung et al., 2004).

QD is one type of nanomaterial that is receiving special attention. QD are inorganic nanocrystals that possess physical dimensions between 2 and 10 nm. The emission wavelength is controlled by the size of the nanocrystal and can be tuned throughout the visible spectrum to the near-infrared region (>670 nm). Early live cell experiments using fluorescent QD sparked interest in using nanoparticles for immunocytochemical and immunohistochemical assays as well as for cell tracking (Akerman et al., 2002; Tokumasu et al., 2003; Sukhanova et al., 2004). A significant advantage for QD is their increased photostability (typically 10–1,000 times more stable) compared to organic dyes. This allows QD and the cells or proteins attached to them to be tracked over longer periods of time. Tumor cells labeled with QD have been intravenously injected into mice and successfully followed using fluo-rescence microscopy (Gao et al., 2004; Voura et al., 2004). As passive imaging agents, QD can be used for image microvasular in animals since polyethylene glycols (PEG)-coated QD injected into mice have shown good tissue perfusion and appear to be biocompatible (Ballou et al., 2004). QD represent just one novel class of nanomaterials whose ability to aid in long-term imaging of cells would help develop better regenerative therapies.

Other nanoparticles are showing promise for optical cell tracking and imaging. For instance, nanosized tubes of carbon known as carbon nanotubes possess optical transitions in the near-infrared that can be utilized for tracking cells. The infrared spectrum between 900 and 1,300 nm is an important optical window for biomedical applications because of the lower optical absorption (greater penetration or depth of light) and small auto-fluorescent background. Like QD, carbon nanotubes possess good photostability and can be imaged over long periods of time using Raman scattering and fluorescence microscopy. However, unlike QD, which are typically composed of heavy metals such as cadmium, carbon nanotubes are made of carbon, an abundant element in nature. Carbon nanotubes possess large aspect ratios with nanometer diameters and lengths ranging from submicron to millimeters. These tubes can contain a single wall of carbon or multiple walls (typically 3–10) of carbon, commonly called single-wall carbon nanotubes (SWNT) or multi-wall carbon nanotubes (MWNT), respectively. SWNT dispersed in a pluronic surfactant can be readily imaged through fluorescence microscopy after being ingested by mouse peritoneal macrophage-like cells. The small size of the SWNT makes it possible for 70,000 nanotubes to be ingested where they can remain stable for weeks inside 3T3 fibroblasts and murine myoblast stems cells (Cherukuri et al., 2004; Heller et al., 2005). Having such a high concentration of carbon nanotubes within a cell without distributing the cell behavior means such probes could be used for studying cell proliferation and stem-cell differentiation, even through repeated cells. While such nanomaterials have yet to reach clinical applications, it does show the potential for non-invasive optical imaging.

Along with optical contrast agents, magnetic nanoparticles also have been used to track cells and report on cell behavior. Many nanoparticle contrasting agents are based on superparamagnetic iron oxide nanoparticles and some have already been approved as clinical MRI contrast agents. When placed into a magnetic field, magnetic nanoparticles create perturbations of the external field that significantly reduce the spin–spin relaxation time (T2) of the nearby environment generating MRI contrast. Typically, these probes consist of a magnetic iron oxide core that is surrounded by a biocompatibilizing material such as dextran. Sizes of these particles can range from one nanometer to hundreds of nanometers in diameter. When used in conjunction with HIV-Tat and polyArginine peptides, these particles are readily taken up by many cell types (Dodd et al., 2001; Zhao et al., 2002). For example, stabilizing pressure input orthosis (SPIO) labeled rat mesenchymal stem cells injected into rats could be imaged and tracked to the liver and kidneys (Bos et al., 2004). Apoptosis is commonly detected by using the binding of annexin V to externalized phosphatidylserine. This binding event is the basis of optical and radiolabels methods for detecting apoptotic cells and can be bound to iron nanoparticles for sensing using MRI. It has been demonstrated that tumor-bearing mice injected with SPIO particles bearing apoptotic sensing proteins showed a sharp decrease in the T2* weight image corresponding to the location of the tumor (Zhao et al., 2001). This demonstrated that nanomaterials can be used to create high specificity MRI contrast agents for apoptotic cells. Such results are encouraging because they show that nanomaterials can be used for not only imaging the physical location of cells, but also providing information on the biological state of cells.

While MRI has revolutionized our way of visualization *in vivo*, allowing cells to be tracked non-invasively, it is difficult to quantify the MRI signals and provide real quantification of cell numbers. The difficulty arises because MRI contrasting agents that are based on paramagnetic gadolinium and iron metals are not directly detected by the scanner but are indirectly detected by their influence on surrounding water molecules. However, the use of perfluoronated nanoparticles has recently been shown to be a new way to provide quantitative numbers to MRI since the fluorine nuclei (^{19}F) can be directly detected (Morawski et al., 2004; Ahrens et al., 2005). Since endogenous fluorine is negligible in the body, ^{19}FMRI is capable of directly detecting fluorine against a dark background similar to radiotracers and fluorescent dyes. While this has been demonstrated with dendritic cells, similar results should be obtainable using other cell types.

Nanotechnology can provide powerful new tools for non-invasive tracking of cells in engineered tissues. As was also mentioned in the outset, the real benefits of nanotechnology are the multifunctional tools that it can bring. Besides imaging enhancements, nanotechnology can produce carriers for delivery of therapeutics for aiding the regeneration process. For example, biodegradable nanoparticles can deliver drugs, growth factors, and other bioactive agents to cells and tissue (Panyam et al., 2003). Nanodelivery vehicles possess three distinct advantages over conventional drug delivery methods. First, nanoparticles, due to their small size, are able to bypass biological barriers such as cell membranes and the blood brain barrier (BBB) allowing greater concentrations of therapeutics to be delivered. Second, nanocarriers can be functionalized with active targeting agents to allow selective delivery of bioactive active agents. Third, drug delivery systems can incorporate nano-triggers for non-invasive delivery of therapeutic agents. These sensitive triggers can be activated using *in vivo* signals such as

REFERENCES

Ahrens, E.T., Flores, R., Xu, H.Y. and Morel, P.A. (2005). *In vivo* imaging platform for tracking immunotherapeutic cells. *Nat. Biotechnol.* **23**: 983–987.

Akerman, M.E., Chan, W.C.W., Laakkonen, P., Bhatia, S.N. and Ruoslahti, E. (2002). Nanocrystal targeting *in vivo*. *Proc. Natl. Acad. Sci. USA* **99**: 12617–12621.

Atala, A., Bauer, S.B., Soker, S., Yoo, J.J. and Retik, A.B. (2006). Tissue-engineered autologous bladders for patients needing cystoplasty. *Lancet* **367**: 1241–1246.

Ballou, B., Lagerholm, B.C., Ernst, L.A., Bruchez, M.P. and Waggoner, A.S. (2004). Noninvasive imaging of quantum dots in mice. *Bioconjugate Chem.* **15**: 79–86.

Bos, C., Delmas, Y., Desmouliere, A., Solanilla, A., Hauger, O., Grosset, C., Dubus, I., Ivanovic, Z., Rosenbaum, J., Charbord, P., et al. (2004). *In vivo* MR imaging of intravascularly injected magnetically labeled mesenchymal stem cells in rat kidney and liver. *Radiology* **233**: 781–789.

Bulmus, V., Woodward, M., Lin, L., Murthy, N., Stayton, P. and Hoffman, A. (2003). A new pH-responsive and glutathione-reactive, endosomal membrane-disruptive polymeric carrier for intracellular delivery of biomolecular drugs. *J. Control. Release* **93**: 105–120.

Cherukuri, P., Bachilo, S.M., Litovsky, S.H. and Weisman, R.B. (2004). Near-infrared fluorescence microscopy of single-walled carbon nanotubes in phagocytic cells. *J. Am. Chem. Soc.* **126**: 15638–15639.

Dodd, C.H., Hsu, H.C., Chu, W.J., Yang, P.G., Zhang, H.G., Mountz, J.D., Zinn, K., Forder, J., Josephson, L., Weissleder, R., et al. (2001). Normal T-cell response and *in vivo* magnetic resonance imaging of T cells loaded with HIV transactivator-peptide-derived superparamagnetic nanoparticles. *J. Immunol. Methods.* **256**: 89–105.

Du, C., Cui, F.Z., Zhu, X.D. and de Groot, K. (1999). Three-dimensional nano-HAp/collagen matrix loading with osteogenic cells in organ culture. *J. Biomed. Mater. Res.* **44**: 407–415.

Ellis-Behnke, R.G., Liang, Y.X., You, S.W., Tay, D.K.C., Zhang, S.G., So, K.F. and Schneider, G.E. (2006). Nano neuro knitting: peptide nanofiber scaffold for brain repair and axon regeneration with functional return of vision. *Proc. Natl. Acad. Sci. USA* **103**: 5054–5059.

Fujihara, K., Kotaki, M. and Ramakrishna, S. (2005). Guided bone regeneration membrane made of polycaprolactone/calcium carbonate composite nano-fibers. *Biomaterials* **26**: 4139–4147.

Gabay, T., Jakobs, E., Ben-Jacob, E. and Hanein, Y. (2005). Engineered self-organization of neural networks using carbon nanotube clusters. *Phys. Stat. Mech. Appl.* **350**: 611–621.

Gan, D.J. and Lyon, L.A. (2001). Tunable swelling kinetics in core-shell hydrogel nanoparticles. *J. Am. Chem. Soc.* **123**: 7511–7517.

Gao, X.H., Cui, Y.Y., Levenson, R.M., Chung, L.W.K. and Nie, S.M. (2004). *In vivo* cancer targeting and imaging with semiconductor quantum dots. *Nat. Biotechnol.* **22**: 969–976.

Heller, D.A., Baik, S., Eurell, T.E. and Strano, M.S. (2005). Single-walled carbon nanotube spectroscopy in live cells: towards long-term labels and optical sensors. *Adv. Mater.* **17**: 2793–+

Jordan, A., Scholz, R., Wust, P., Fahling, H. and Felix, R. (1999). Magnetic fluid hyperthermia (MFH): cancer treatment with AC magnetic field induced excitation of biocompatible superparamagnetic nanoparticles. *J. Magn. Magn. Mater.* **201**: 413–419.

Jung, H.I., Kettunen, M.I., Davletov, B. and Brindle, K.M. (2004). Detection of apoptosis using the C2A domain of synaptotagmin I. *Bioconjugate Chem.* **15**: 983–987.

Kay, S., Thapa, A., Haberstroh, K.M. and Webster, T.J. (2002). Nanostructured polymer/nanophase ceramic composites enhance osteoblast and chondrocyte adhesion. *Tissue Eng.* **8**: 753–761.

Kikuchi, M., Itoh, S., Ichinose, S., Shinomiya, K. and Tanaka, J. (2001). Self-organization mechanism in a bone-like hydroxyapatite/collagen nanocomposite synthesized *in vitro* and its biological reaction *in vivo*. *Biomaterials* **22**: 1705–1711.

Kim, S.S., Park, M.S., Jeon, O., Choi, C.Y. and Kim, B.S. (2006). Poly(lactide-co-glycolide)/hydroxyapatite composite scaffolds for bone tissue engineering. *Biomaterials* **27**: 1399–1409.

Lahann, J., Mitragotri, S., Tran, T.N., Kaido, H., Sundaram, J., Choi, I.S., Hoffer, S., Somorjai, G.A. and Langer, R. (2003). A reversibly switching surface. *Science* **299**: 371–374.

Liao, S.S., Cui, F.Z., Zhang, W. and Feng, Q.L. (2004). Hierarchically biomimetic bone scaffold materials: nano-HA/collagen/PLA composite. *J. Biomed. Mater. Res. B* **69**(B): 158–165.

Medintz, I.L., Uyeda, H.T., Goldman, E.R. and Mattoussi, H. (2005). Quantum dot bioconjugates for imaging, labelling and sensing. *Nat. Mater.* **4**: 435–446.

Morawski, A.M., Winter, P.M., Yu, X., Fuhrhop, R.W., Scott, M.J., Hockett, F., Robertson, J.D., Gaffney, P.J., Lanza, G.M. and Wickline, S.A. (2004). Quantitative "magnetic resonance immunohistochemistry" with ligand-targeted F-19 nanoparticles. *Magn. Reson. Med.* **52**: 1255–1262.

Panyam, J. and Labhasetwar, V. (2003). Biodegradable nanoparticles for drug and gene delivery to cells and tissue. *Adv. Drug Deliver. Rev.* **55**: 329–347.

Pattison, M.A., Wurster, S., Webster, T.J. and Haberstroh, K.M. (2005). Three-dimensional, nano-structured PLGA scaffolds for bladder tissue replacement applications. *Biomaterials* **26**: 2491–2500.

Price, R.L., Waid M.C., Haberstroh, K.M. and Webster, T.J. (2003). Selective bone cell adhesion on formulations containing carbon nanofibers. *Biomaterials* **24**: 1877–1887.

Ren, L. and Chow, G.M. (2003). Synthesis of nir-sensitive Au-Au2S nanocolloids for drug delivery. *Mat. Sci. Eng. C-Bio. S.* **23**: 113–116.

Silva, G.A., Czeisler, C., Niece, K.L., Beniash, E., Harrington, D.A., Kessler, J.A. and Stupp, S.I. (2004). Selective differentiation of neural progenitor cells by high-epitope density nanofibers. *Science* **303**: 1352–1355.

Sukhanova, A., Devy, M., Venteo, L., Kaplan, H., Artemyev, M., Oleinikov, V., Klinov, D., Pluot, M., Cohen, J.H.M. and Nabiev, I. (2004). Biocompatible fluorescent nanocrystals for immunolabeling of membrane proteins and cells. *Anal. Biochem.* **324**: 60–67.

Supronowicz, P.R., Ajayan, P.M., Ullmann, K.R., Arulanandam, B.P., Metzger, D.W. and Bizios, R. (2002). Novel current-conducting composite substrates for exposing osteoblasts to alternating current stimulation. *J. Biomed. Mater. Res.* **59**: 499–506.

Thapa, A., Miller, D.C., Webster, T.J. and Haberstroh, K.M. (2003). Nano-structured polymers enhance bladder smooth muscle cell function. *Biomaterials* **24**: 2915–2926.

Thapa, A., Webster, T.J. and Haberstroh, K.M. (2003). Polymers with nano-dimensional surface features enhance bladder smooth muscle cell adhesion. *J. Biomed. Mater. Res. A.* **67**(A): 1374–1383.

Tokumasu, F. and Dvorak, J. (2003). Development and application of quantum dots for immunocytochemistry of human erythrocytes. *J. Microsc-Oxford.* **211**: 256–261.

Voura, E.B., Jaiswal, J.K., Mattoussi, H. and Simon, S.M. (2004). Tracking metastatic tumor cell extravasation with quantum dot nanocrystals and fluorescence emission-scanning microscopy. *Nat. Med.* **10**: 993–998.

Winter, P.M., Caruthers, S.D., Kassner, A., Harris, T.D., Chinen, L.K., Allen, J.S., Lacy, E.K., Zhang, H.Y., Robertson, J.D., Wickline, S.A., et al. (2003). Molecular imaging of angiogenesis in nascent vx-2 rabbit tumors using a novel alpha(v)beta(3)-targeted nanoparticle and 1.5 tesla magnetic resonance imaging. *Cancer Res.* **63**: 5838–5843.

Yang, F., Murugan, R., Wang, S. and Ramakrishna, S. (2005). Electrospinning of nano/micro scale poly(L-lactic acid) aligned fibers and their potential in neural tissue engineering. *Biomaterials* **26**: 2603–2610.

Zhang, Y., Kohler, N. and Zhang, M.Q. (2002). Surface modification of superparamagnetic magnetite nanoparticles and their intracellular uptake. *Biomaterials* **23**: 1553–1561.

Zhao, M., Beauregard, D.A., Loizou, L., Davletov, B. and Brindle, K.M. (2001). Non-invasive detection of apoptosis using magnetic resonance imaging and a targeted contrast agent. *Nat. Med.* **7**: 1241–1244.

Zhao, M., Kircher, M.F., Josephson, L. and Weissleder, R. (2002). Differential conjugation of tat peptide to superparamagnetic nanoparticles and its effect on cellular uptake. *Bioconjugate Chem.* **13**: 840–844.

Zong, X.H., Bien, H., Chung, C.Y., Yin, L.H., Fang, D.F., Hsiao, B.S., Chu, B. and Entcheva, E. (2005). Electrospun fine-textured scaffolds for heart tissue constructs. *Biomaterials* **26**: 5330–5338.

Part IV

Biomaterials for Regenerative Medicine

18

Design Principles in Biomaterials and Scaffolds

Hyukjin Lee and Tae Gwan Park

INTRODUCTION

Tissue or organ transplantation is severely limited by the problems of donor shortage and immune rejection from the patients. The development of tissue engineering allows the transplantation of cells from a patient's own tissue to regenerate damaged tissue or organ without causing immune responses. For the cell transplantation, extracted cells are often required to cultivate in a large scale to attain a sufficient cell seeding density. In culturing the cells, the *in vitro* culture conditions play pivotal roles in proliferation and differentiation. Three-dimensional biomaterial scaffolds are firstly developed for the temporary substrate to grow cells in an organized fashion. Although direct injection or implantation of *in vitro* cultured cells is often performed, using cell suspension is doubtful for the successful regeneration of impaired tissues. It is also well established that the three-dimensional organization of cells often related with cellular attachments affects the fate of cellular development. As a result, biodegradable and biocompatible polymers have been widely used to fabricate three-dimensional scaffolds for tissue engineering.

In the past, biomaterial scaffolds were mainly used for temporary prosthetic devices to fill the void spaces after tissue necrosis or surgery. However, current biomaterials pursue to mimic the role of natural extracellular matrix (ECM) which can support cell adhesion, differentiation, and proliferation. ECM mimicking biomaterial scaffolds should be designed considering the following requirements. First, suitable biomaterials are selected for particular applications (Mikos and Langer, 1993; Athanasiou and Agrawal, 1996). This is analogous to the effort to build up the target-specific biological scaffolds. Second, biomaterial scaffolds require a highly open porous structure with good interconnectivity, yet possessing sufficient mechanical strength for cellular in- or outgrowth (Cima and Langer, 1991). Third, the surface of fabricated scaffolds must be able to support cellular attachment, proliferation, and differentiation (Varkey and Uludag, 2004; Peattie and Prestwich, 2006; Vasita and Katti, 2006). Fourth, drug or cytokine releasing scaffolds are ideal for modulating tissue regeneration since cytokines such as growth factors and other small molecules have fundamental roles on growing functional living tissues (Niemann, 2005; Raghunath and Seifalian, 2005; Keilhoff and Wolf, 2006). The harmony of the above considerations is essential to fulfill the requirements of excellent biological scaffolds, thereby inducing synergic effects on successful tissue repair.

This chapter focuses on recent developments on fabricating biomimetic, ECM-like porous scaffolds useful for tissue engineering. Our experiences on designing novel biomaterials and innovating scaffold fabrication techniques are highlighted here as well as other leading researchers' works. Novel fabrication methods and designing strategies are elucidated such as generating the macroporous biodegradable scaffolds, the surface modification of biodegradable scaffolds to enhance cellular attachment and biological activity, and the incorporation of bioactive molecules within the scaffold systems. A number of excellent reviews are available for synthetic biomaterials for medical applications and tissue engineering (Peppas and Langer, 1994; Ratner, 1996; Uhrich, 1999).

Copyright © 2008 Elsevier Inc. All rights reserved

SELECTION OF BIOMATERIALS

Natural biomaterials have been extensively used for tissue engineering since they have advantages over synthetic materials such as similarity with natural ECM. For example, alginate, chitosan, collagen and its derivatives, fibrin, and hyaluronic acid (HA) were investigated for the fabrication of three-dimensional scaffolds (Rosso and Barbarisi, 2005). However, the properties of natural biomaterials are difficult to adjust and the source-related immunogenicity is still unsolved. In contrast, synthetic biomaterials are man-made materials mainly composed of synthetic polymers. Although synthetic polymers often reveal poor biocompatibility, the building up process of synthetic polymer provides precise control of the properties of synthetic materials and even can give better performance over naturally occurring biomaterials. For instance, aliphatic polyesters and polyanhydrides are common synthetic polymers for tissue engineering and drug delivery system. These polymers have distinct biodegradability and biocompatibility. The combinations of hydrophilic and hydrophobic segments in the structure generate a variety of synthetic biomaterials with different mechanical properties and degradation behaviors.

BIODEGRADABLE SYNTHETIC POLYMERS

Aliphatic Polyesters

Aliphatic polyesters are Food and Drug Administration (FDA) approved synthetic biomaterials which have been widely used for biodegradable applications such as surgical sutures and bone fixing screws. Poly(α-hydroxyl esters) such as poly(L-lactic acid) (PLLA), poly(lactic-co-glycolic acid) (PLGA), and polycarprolactone can be synthesized by the ring-opening polymerization of monomers and have hydrolytically cleavable bonds along the polymer backbone. When these synthetic polymers are implanted in the body, hydrolysis of polymer backbone reduces the molecular weight of polymer and their degraded products such as lactic and glycolic acid can be metabolized in the body (Figure 18.1). In addition, based on their biocompatibility and safety record in humans, these polyesters have been used extensively in drug delivery systems and tissue engineering applications (Saltzman, 1999; Putman, 2001).

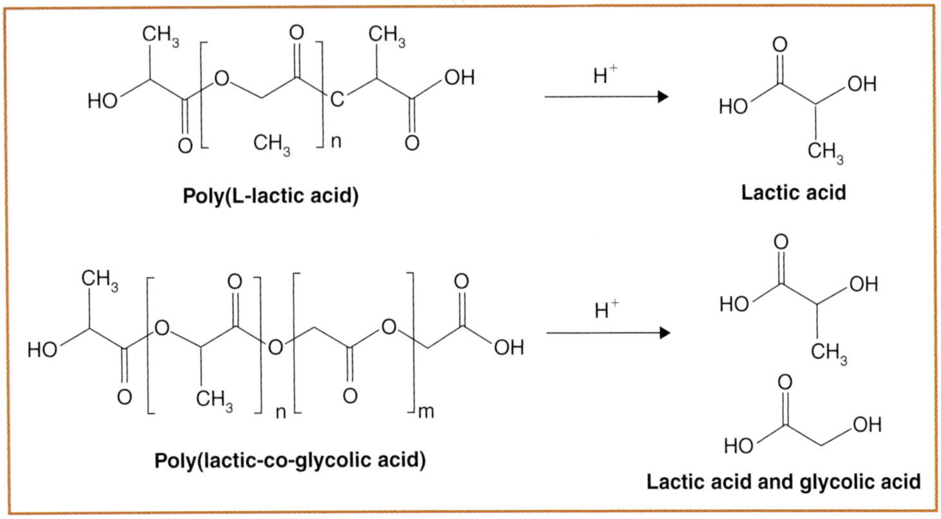

Figure 18.1 Structure of PLLA and PLGA and their degradation products; acid hydrolysis of PLLA and PLGA to give lactic and glycolic acid.

Aliphatic polyesters typically lack a chemical functionality for modification with biological molecules. As an example for the introduction of functional groups in the polymer backbone, Barrera and Langer (1993) reported the use of a novel monomer to incorporate functional amine groups into polylactic acid (PLA) polymers. Poly(lactic acid-co-lysine) was synthesized by the copolymerization of cyclic lactide and its analog containing the lysine. This novel amine containing PLA showed similar biocompatibilty while providing additional sites for further chemical modifications.

Polyanhydrides

Another class of degradable biopolymers is polyanhydrides. Unlike polyesters which predominately show a bulk-erosion process, polyanhydrides exhibit a surface-erosion process which is particularly useful for sustained drug delivery systems. Langer et al. demonstrated the use of polyanhydrides based on sebacic acid (SA) and *p*-carboxyphenoxyproane (CPP) (Leong et al., 1985). By combining hydrophilic SA and hydrophobic CPP, the rate of surface erosion can be controlled from days to years. In addition, these polyanhydrides exhibit great biocompatibility and excellent *in vivo* performance for potential biomedical applications.

DESIGN PRINCIPLES OF BIOLOGICAL SCAFFOLDS

Fabrication of Macroporous Biodegradable Scaffolds

Along with the material selection, fabrication methods are also critical for designing biological scaffolds. For tissue regeneration, highly open porous polymer scaffolds are often required for high density cell seeding, efficient nutrient and oxygen transport. There have been multiple methods to fabricate highly porous biodegradable polymer scaffolds which are listed in Table 18.1.

Briefly illustrating a few techniques, the compressed polyglycolic acid (PGA) meshes are made out of non-woven PGA fibers and these meshes have been widely used for soft tissue regeneration (Freed and Langer, 1993). Random coiling and heat treatment of PGA fibers can generate highly open porous and interconnected structures with a high surface to volume ratio. However, the mechanical strength of these meshes is insufficient for hard tissue regeneration (Mikos and Langer, 1993).

To enhance the mechanical properties of compressed PGA meshes, Mooney and Langer (1993) demonstrated that a mixed solution of PLLA and PLGA can be applied to the compressed PGA meshes. Mixture of PLLA and PLGA dissolved in organic solvent was sprayed throughout the compressed PGA meshes. As the organic solvent evaporated, dried PLLA/PLGA strengthened the cross regions of fibers and enhanced mechanical properties

Table 18.1 List of fabrication methods for preparation of highly porous biodegradable scaffolds

Fabrication methods	Materials	References
Compressed mesh of non-woven fibers	PGA, PLGA	Freed and Langer (1993), Mikos and Langer (1993), Mooney and Langer (1996)
Solvent casting/salt leaching	PLLA, PLGA	Mikos and Langer (1993), Mikos and Vacanti (1994)
CO_2 expansion	PLGA	Mooney and Langer (1996), Harris and Mooney (1998)
Emulsion freeze drying	PLGA	Whang and Nuber (1995)
Phase separation	PLLA, PLGA	LO and Leong (1996), Schugens and Teyssie (1996), Nam and Park (1999)
Three-dimensional imprinting	PLLA, PLGA	Park and Griffith (1998)

of compressed meshes. Despite the mechanical result, this method exhibited the reduction of high surface to volume ratio of compressed meshes and the difficulty of matching the degradation rate of surface coated materials and bulk materials.

In addition, the solvent casting/salt-leaching technique has been extensively exploited for fabricating scaffolds for tissue engineering (Mikos and Langer, 1993; Mikos and Vacanti, 1994). PLGA dissolved in an organic solvent with salt particles is placed in a mold to produce a polymer/salt mixture, which is immersed in water to remove salt particles to generate open pore structures. The scaffolds prepared by this method often demonstrate a dense surface layer and poor interconnectivity between macropores, which reduces cell seeding into the scaffolds *in vitro* and causes non-uniform distribution of seeded cells. Thus, poor cell viability and tissue ingrowth after *in vivo* implantation are observed.

In order to resolve the problems from salt-leaching techniques, Nam and Park (2000) utilized PLLA paste containing ammonium bicarbonate salt particles which acts as a gas-foaming agent as well as a salt-leaching porogen to fabricate highly interconnected porous biodegradable scaffolds (Figure 18.2). Sodium bicarbonate salt with acidic excipients has been widely used for effervescent gas evolving oral tablets. Since ammonium bicarbonate salt upon contact to an acidic aqueous solution such as citric acid and/or incubated at elevated temperature produces gaseous ammonia and carbon dioxide by itself, ammonium bicarbonate salt particles could be incorporated into a biodegradable gel paste prepared by dissolving high molecular weight PLLA in an organic solvent. The resultant putty paste was easy of shaping into different geometry and could be immersed in hot water solution and directly dried under vacuum oven to remove or leach out the salt particles while concurrently generated gaseous ammonia and carbon dioxide provide highly interconnected pores within a solidifying polymer scaffold. Thus, the formation of dense surface skin layer was not found on either sides of the surface of the scaffolds (Figure 18.3).

Macroporous PLGA scaffolds using gas-foaming/salt-leaching method with controlled degradation rate was also investigated (Yoon and Park, 2001). Unlike semi-crystalline PLLA, amorphous PLGA could form a gel-like paste in an organic solvent even at high concentration. PLGA was dissolved in an organic solvent such as chloroform then precipitated in a non-solvent, ethanol. Resulting precipitates exhibited a gel-like property such that the paste can be molded or hand-shaped in any desirable dimensions. In this study, instead of incubating scaffolds in hot water bath or vacuum oven, citric acid solution was used to control the porosity of scaffolds as well as mechanical property. Using citric acid, carbon dioxide and ammonia gases could be generated

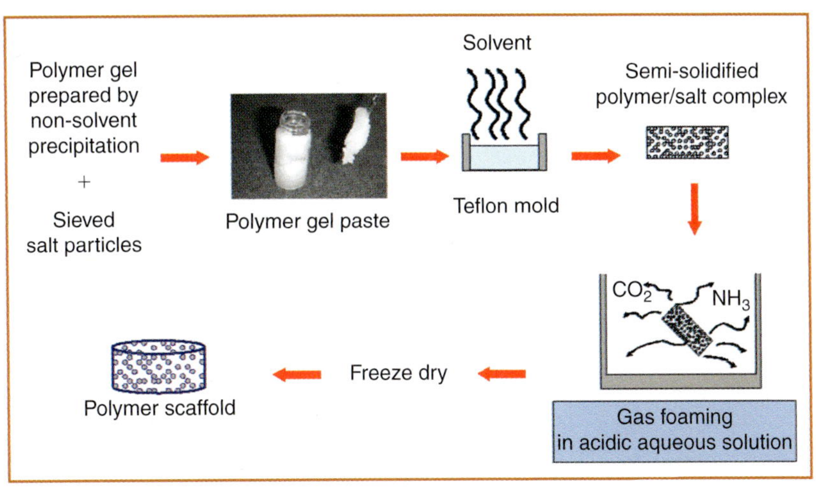

Figure 18.2 **Schematic of gas-foaming and salt-leaching process to fabricate macroporous scaffolds.**

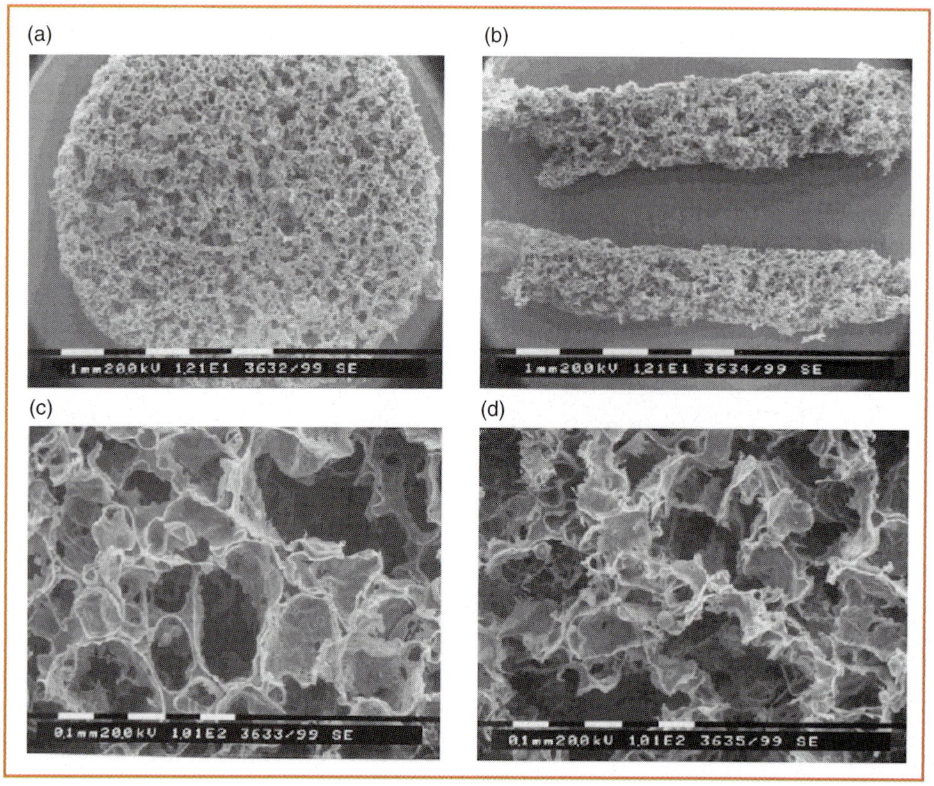

Figure 18.3 SEM images of macroporous scaffolds fabricated by gas-foaming and salt-leaching process. Uniform interconnectivity and high porosity are observed on both surface (a, c) and cross-section of scaffolds (b, d).

at room temperature and changing the concentration of citric acid in the solution enabled to control the porosity of scaffolds. The result supported that the increase in porosity of scaffolds was observed with increased citric acid concentration as high concentration citric acid are more gas generating. In addition, degradation and swelling behaviors of PLGA scaffolds with different compositions were investigated. The macroporous scaffolds with three different compositions of lactic and glycolic acid were incubated in phosphate buffered solution (pH 7.4) at 37°C. During the incubation period, significant swelling of the scaffolds was observed depending on the composition, and the change in dimension and morphology was caused by the accelerated degradation of PLGA scaffold which could generate more water adsorbing small molecular weight PLGA oligomers within the degrading scaffolds (Figure 18.4).

As an alternative to salt-leaching and gas-forming fabrication, electrospinning has received much attention for fabricating polymeric ultrafine nanofibers to build three-dimensional tissue engineering scaffolds. Nanofibrous biodegradable scaffolds would have definitive advantages for cell attachment, proliferation, and differentiation because they resemble an ECM structure. Recently, Kim and Park (2006) demonstrated ECM mimicking nanofiber mesh for tissue engineering applications. The amine terminated PLGA dissolved in a mixture of DMF/THF solvent was ejected through a nozzle by an electrostatic force, resulting in the formation of non-woven fabrics. During the electrospinning, the solvent evaporates and the charged polymer nanofibers were deposited on a grounded collector. The resultant structure was a three-dimensional, randomly oriented nanofiber network mesh with a highly nanoporous architecture (Figure 18.5). The *in vitro* cell culture revealed that the resulting nanofiber ranged from 300 to 1,000 nm provided an excellent environment for cellular attachment, proliferation, and differentiation.

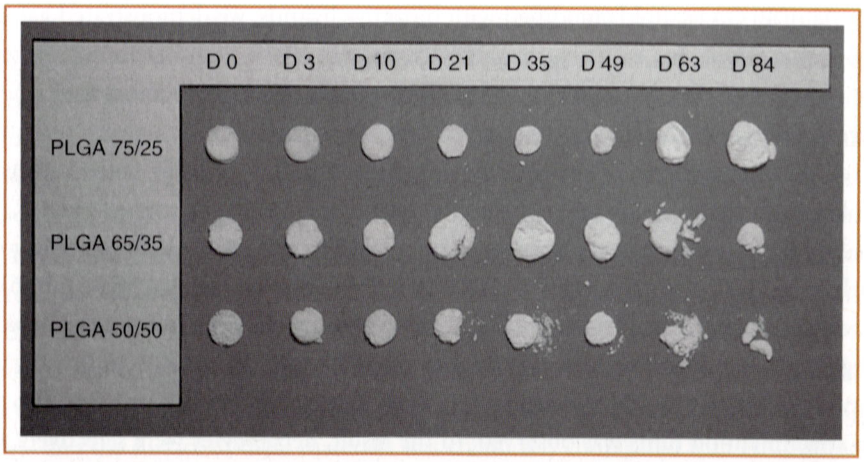

Figure 18.4 Photographs of different PLGA scaffolds after hydrolytic degradation in phosphate buffered saline (PBS) at 37°C. With increasing composition of glycolic acid, rapid degradation and swelling of scaffolds are observed.

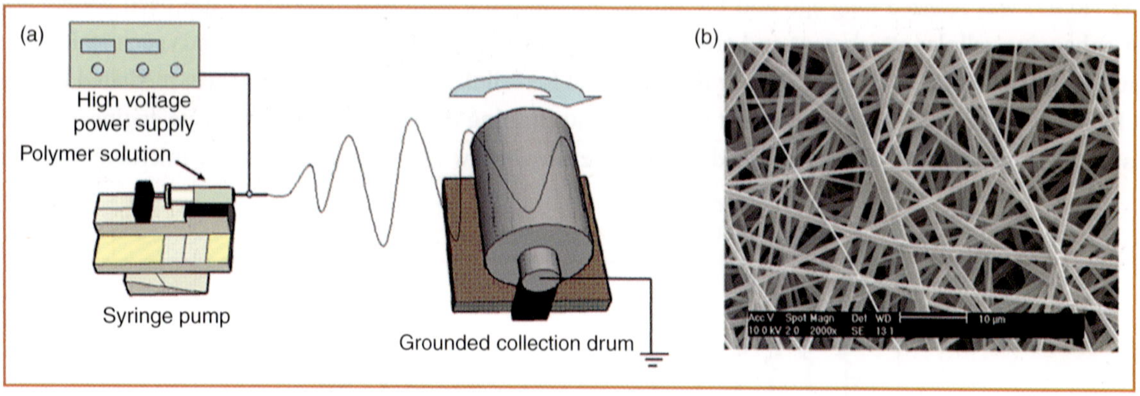

Figure 18.5 Schematic of electrospinning (a) and an SEM image of electrospun PLGA nanofiber (b).

Surface Immobilization of Bioactive Molecules on Macroporous Biodegradable Scaffolds

The surface modification of scaffolds is essential since the microenvironment of the body cannot see the bulk property of biomaterials, but the surface of biomaterials. In the past, major issues concerned with biomaterials are their biocompatibility upon the injection or implantation of materials *in vivo*. In the case of material selections, a few biomaterials are known for free of causing acute inflammation. As a result, the surfaces of fouling devices were modified with non-protein adsorbing materials such as polyethylene glycol (PEG) to stealth the implants from the body.

Since many cell adhesive peptides present abundantly in the ECM dictate cellular behaviors, the immobilization of various bioactive ligands on the surface of biomaterials was attempted for actively mimicking physiological conditions, thereby increasing cytocompatibility and biological functionality when the biomaterials are implanted in the body. A number of surface modification methods were developed such as chemical oxidation and etching, plasma and corona discharge, radiation and UV grafting, partial hydrolysis, protein adsorption, and conjugation/immobilization of bioactive ligands (Rasmussen and Whitesides, 1977; Ramsey and Binkowski, 1984; Weisz and Schnaar, 1991; Gao and Langer, 1998; Nam and Park, 1999; Otsuka and Kataoka, 2000).

Nam, Y.S. and Park, T.G. (1999b). Adhesion behaviors of hepatocytes cultured onto biodegradable polymer surface modified by alkali hydrolysis process. *J. Biomater. Sci. Polymer Ed.* **10**: 1145–1158.

Nam, Y.S. and Park, T.G. (2000). A novel fabrication method for macroporous scaffolds using gas foaming salt as porogen additive. *J. Biomed. Mater. Res.* **53**: 1–7.

Niemann, C. (2005). Controlling the stem cell niche: right time, right place, right strength. *Bio Essays* **1**: 1–5.

Otsuka, H. and Kataoka, K. (2000). Surface characterization of functional polylactide through the coating with heterobifunctional poly(ethylene glycol)/polylacide block copolymers. *Biomacromolecules* **1**: 29–48.

Park, T.G. (2002). Perfusion culture of hepatocytes within galactose-derivatized biodegradable poly(lactide-co-glycolide) scaffolds prepared by gas foaming of effervescent salts. *J. Biomed. Mater. Res.* **59**: 127–135.

Park, A. and Griffith, L.G. (1998). Integration of surface modification and 3D fabrication techniques to prepare patterned poly(L-lactide) substrates allowing regionally selective cell adhesion. *J. Biomater. Sci. Polymer Ed.* **9**: 89–110.

Peattie, R.A. and Prestwich, G.D. (2006). Dual growth factor induced angiogenesis *in vivo* using hyaluronan hydrogel implants. *Biomaterials* **9**: 1868–1875.

Peppas, N.A. and Langer, R. (1994). New challenges in biomaterials. *Science* **263**: 1715–1720.

Peter, S.J. and Mikos, A.G. (1998). Osteoblastic phenotype of rat marrow stromal cells cultured in the presence of dexamethasone, b-glycerolphosphate, and L-ascorbic acid. *J. Cell Biochem.* **71**: 55–62.

Putnam, D. (2001). Polymer-based gene delivery with low cytotoxicity by a unique balance of side-chain termini. *Proc. Natl. Acad. Sci. USA* **98**: 1200–1205.

Raghunath, J. and Seifalian, A.M. (2005). Advancing cartilage tissue engineering: the application of stem cell technology. *Current Opinion in Biotechnology* **15**: 503–509.

Ramsey, W.S. and Binkowski, N.J. (1984). Surface treatments and cell attachment. *In Vitro* **20**: 802–808.

Rasmussen, J.R. and Whitesides, G.M. (1977). Introduction, modification, and characterization of functional groups on the surface of low density polyethylene films. *J. Am. Chem. Soc.* **99**: 4736–4745.

Ratner, B.D. (1996). *Biomaterials Science*. San Diego: Academic Press, 11–35.

Reil, T.D. and Gelabert, H.A. (1999). Dexamethasone suppress vascular smooth muscle cell proliferation. *J. Surg. Res.* **85**: 109–114.

Richardson, T.P. and Mooney, D.J. (2001). Polymeric system for dual growth factor delivery. *Nat. Biotechnol.* **19**: 1029–1034.

Rosso, F. and Barbarisi, A. (2005). Smart materials as scaffolds for tissue engineering. *J. Cell Physiol.* **203**: 465–470.

Royce, S.M. and Marra, K.G. (2004). Incorporation of polymer microspheres within fibrin scaffolds for controlled delivery of FGF-1. *J. Biomater. Sci. Polymer Ed.* **15**: 1327–1336.

Saltzman, W.M. (1999). Delivering tissue regeneration. *Nat. Biotechnol.* **17**: 534–535.

Schugens, C. and Teyssie, P. (1996). Poly-lactide macroporous biodegradable implants for cell transplantation. II. Preparation of polylactide foams by liquid–liquid phase separation. *J. Biomed. Mater. Res.* **30**: 449–461.

Shakesheff, K. and Langer, R. (1998). Creating biomimetic micro-environment with synthetic polymer–peptide hybrid molecules. *J. Biomater. Sci. Polym. Ed.* **9**:507–518.

Uhrich, K.E. (1999). Polymeric systems for controlled drug release. *Chem. Rev.* **99**: 3181–3198.

Varkey, M. and Uludag, H. (2004). Growth factor delivery for bone tissue repair: an update. *Expert. Opin. Drug Deliv.* **1**: 19–36.

Vasita, R. and Katti, D.S. (2006). Growth factor delivery systems for tissue engineering: a materials perspective. *Expert. Rev. Med. Dev.* **1**: 29–47.

Weisz, O.A. and Schnaar, R.L. (1991). Hepatocyte adhesion to carbohydrate-derived surfaces II. Regulation of cytoskeletal organization and cell morphology. *J. Cell Biol.* **115**: 495–504.

Wissink, M.J.B. and Feijen, J. (2000). Improved endothelialization of vascular grafts by local release of growth factor from heparinized collagen matrices. *J. Contr. Release* **64**: 103–114.

Whang, K. and Nuber, G.A. (1995). Novel methods to fabricate bioabsorbable scaffolds. *Polymer* **36**: 837–842.

Whang, K. and Healy, K.E. (2000). A biodegradable polymer scaffold for delivery of osteotropic factors. *Biomaterials* **21**: 2535–2551.

Yoon, J.J. and Park, T.G. (2001). Degradation behaviors of biodegradable macroporous scaffolds prepared by gas foaming of effervescent salts. *J. Biomed. Mater. Res.* **55**: 401–408.

Yoon, J.J. and Park, T.G. (2002). Surface immobilization of galactose onto aliphatic biodegradable polymers for hepatocyte culture. *Biotech. Bioeng.* **78**: 1–10.

Yoon, J.J. and Park, T.G. (2003). Dexamethasone releasing biodegradable polymer scaffolds fabricated by a gas foaming/salt leaching method. *Biomaterials* **24**: 2323–2329.

Yoon, J.J. and Park, T.G. (2004). Immobilization of cell adhesive RGD peptide onto the surface of highly porous biodegradable polymer scaffolds fabricated by gas foaming/salt leaching method. *Biomaterials* **25**: 5613–5620.

Yoon, J.J. and Park, T.G. (2006). Heparin immobilized biodegradable porous scaffolds for sustained release of angiogenic growth factors. *Biomaterials* **79A**(4): 934–942.

Yoo, H.S. and Park, T.G. (2005). Hyaluronic acid modified biodegradable scaffolds for cartilage tissue engineering. *Biomaterials* **26**: 1925–1933.

19

Naturally Occurring Scaffold Materials

Stephen F. Badylak

INTRODUCTION

Most regenerative medicine approaches to the restoration and replacement of damaged or missing tissues require a scaffold material upon which cells can attach, migrate, proliferate, and/or differentiate, hopefully into a functionally and structurally appropriate tissue. A variety of scaffold materials are available including synthetic polymers, and naturally occurring polymers that are produced during the course of tissue development in both vertebrate and invertebrate species. These various materials are characterized by unique physical and mechanical properties and each material is associated with a distinctive tissue response when implanted in a mammalian host.

Synthetic scaffold materials such a poly(L)-(lactic acid) and poly(glycolic acid) have received considerable attention for tissue engineering applications and have shown promise in preclinical animal studies. Synthetic materials have predictable mechanical and physical properties and can be manufactured with great precision. However, synthetic materials tend to elicit a chronic active inflammatory response within the host tissue, which limits constructive remodeling and tissue regeneration, and promotes the deposition of fibrous connective tissue. The present chapter will not deal further with synthetic materials, but will instead focus upon naturally occurring scaffold materials.

Naturally occurring scaffold materials are defined as those that occur in nature and are produced by the cells of living organisms. These materials typically occupy an extracellular location; that is, they become part of the extracellular matrix (ECM). Individual components of the ECM such as collagen or the intact matrix itself can be harvested and prepared for use as a scaffold for a variety of regenerative medicine applications. The present chapter will describe the use of three such materials as scaffolds; specifically purified collagens, chitosan, and intact extracellular matrix. Other naturally occurring materials such as hyaluronic acid and alginates have also shown potential as useful scaffold materials, but will not be discussed herein.

COLLAGEN

The most common and abundant naturally occurring scaffold material is the structural protein collagen. Collagen is a highly conserved protein that is ubiquitous among mammalian species and accounts for approximately 30% of all body proteins (Nimni et al., 1987). Inherent common amino acid sequences and epitope structures exist within collagen molecules across species lines (Boyd et al., 1991; Garrone et al., 1993; Beier et al., 1996). These common antigens appear to account for the lack of an adverse immune response when

Copyright © 2008 Elsevier Inc. All rights reserved

xenogeneic collagen is used as an implantable scaffold material. Bovine and porcine type I collagen provide readily available sources of scaffold material for numerous clinical applications and have proven quite compatible with most human systems. Examples of collagen scaffolds include Autologen® (Collagenesis Corporation, Beverly, MA), Contigen® (C.R. Bard, Inc., Covington, GA), Zyplast® and Zyderm I® and II® (INAMED Aesthetics, formerly McGhan Medical Corporation, Fremont, CA), and the Collagen Meniscal Implant (CMI) (ReGen Biologics, Inc., Franklin Lakes, NJ).

In its native state, collagen is a natural substrate for cellular attachment, growth, and differentiation. In addition to its desirable structural properties, collagen has inherent functional properties such as the stimulation or inhibition of angiogenesis (Cornelius et al., 1998; Maeshima et al., 2000; Brennan et al., 2006), and the promotion of cellular proliferation and differentiation. For the above-mentioned reasons and others, collagen has become a favorite substrate for many tissue engineering and regenerative medicine applications.

Collagen can be extracted from tissues such as tendons and ligaments, solubilized, and then reconstituted into fine strands that can, in turn, be fashioned into a variety of shapes and sizes that mimic body structures such as heart valves, blood vessels, and skin (Berthiaume et al., 1995). The reconstituted collagen is usually stabilized by chemical cross-linking methods and must be sterilized prior to surgical use.

Allogeneic and xenogeneic collagen is generally recognized as "self" tissue when used as a biologic scaffold material regardless of its species of origin, and it is subjected to the fundamental biological processes of degradation and integration into adjacent host tissues when left in its native ultrastructure. Certain processing methods, however, can alter the mechanical and physical properties of collagen-based materials and may negatively affect the processes of host-cell attachment, proliferation, differentiation, and tissue remodeling. These methods include glutaraldehyde treatment, carbodiimide treatment, dye-mediated photooxidation, exposure to polyepoxy compounds, and glycerol treatment. Commonly used methods of terminal sterilization include gamma or electron beam irradiation, or ethylene oxide treatment. Exposure to chemical cross-linking agents can change a biocompatible collagen-based material into a form that incites a host foreign body response (Sato, 1983). Most methods of chemical cross-linking alter (i.e. usually decrease) the rate of *in vivo* degradation and change the mechanical properties (i.e. usually strengthen) of collagen.

Collagen provides considerable mechanical strength in its natural polymeric state. The necessary and required mechanical and physical properties of tissue engineered products for use in cardiovascular, orthopedic, and other body systems often depend upon the chemical manipulation of collagen-based scaffolds to achieve the desired mechanical properties. The tissue and species source of collagen and its treatment prior to use are important variables in the design of tissue engineered devices.

CHITOSAN

Chitosans are the second most abundant biopolymer in nature and represent a family of biodegradable cationic polysaccharides consisting of glucosamine and randomly distributed *N*-acetylglucosamine linked in a β(1–4) manner (Dornish et al., 2001), and have a chemical structure similar to hyaluronic acid. Chitosans are derived by the alkaline *N*-deacetylation of chitin, a component of the protective layer of shellfish. The molecular weight of chitosan ranges from 300 to over 1,000 kD, depending on the preparation procedure and the degree of deacetylation, where the degree of deacetylation is defined as the ratio of glucosamine and *N*-acetylglucosamine (Madihally and Matthew, 1999; Dornish et al., 2001). The degree of deacetylation of commercially available chitosan can vary from 50% to 90%, while degrees of deacetylation higher than 95% can be achieved using acetylation chemistry methods (Mima et al., 1983; Madihally and Matthew, 1999; Cao et al., 2005).

Chitosan, in its crystalline form, is generally insoluble in solutions with a pH of 7 and above; however, in dilute acids of pH less than 6, the free amino groups are protonated, allowing chitosan to form a viscous solution which can then be molded into various structures (e.g. blocks, tubes, beads, membranes) (Aiedeh et al., 1997; Madihally and Matthew, 1999; Dornish et al., 2001; Cao et al., 2005; Freier et al., 2005). The formation of a porous structure is generally achieved by freezing and then lyophilizing a chitosan construct, leaving pores in the space originally occupied by frozen solvent crystals. The size, distribution, and orientation of the pores can be controlled by varying the freezing method (ice crystal size, temperature gradient, and freezing rate) (Madihally and Matthew, 1999). Pore size in chitosan scaffolds is easily controlled in the range of 40–250 μm and porosities greater than 80% can be achieved (Madihally and Matthew, 1999). Porous chitosan scaffolds can also be formed by various processes which do not involve lyophilization (Chow and Khor, 2000; Ho et al., 2004; Geng et al., 2005). One such method is the freeze-gelation method, in which frozen chitosan solution is placed in an NaOH/ethanol solution at –20 C in order to adjust the pH so that gelation of the chitosan can occur at a temperature less than the freezing point of the solution, thus allowing for the formation of a chitosan scaffold while retaining a porous structure, without the necessity of lyophilization (Ho et al., 2004). In general, porous chitosan membranes possess a low elastic modulus (0.1–0.5 MPa) and tensile strength (30–60 kPa), while the extensibility can range from 30% to 110% based on pore size and orientation (Madihally and Matthew, 1999; Suh and Matthew, 2000; Di Martino et al., 2005). The mechanical properties of other configurations of chitosan (tubes, blocks, and beads) vary depending on the size, shape, and pore characteristics of the scaffold. The design and production of porous chitosan scaffolds have been extensively reviewed in Madihally and Matthew (1999).

Chitosan can be enzymatically degraded *in vitro* using chitinase, chitosanase, lysozyme, and pectinase. Some other proteolytic enzymes have also been shown to have low-level degradation effects (Tomihata and Ikada, 1997; Jolles and Muzzarelli, 1999; Khor, 2001). *In vivo*, chitosan is degraded primarily by lysozyme into oligosaccharides through the hydrolysis of acetylated residues (Tomihata and Ikada, 1997; Zhang and Neau, 2001; Huang et al., 2004). The *in vivo* degradation products of chitosan are non-toxic and non-immunogenic (Muzzarelli, 1993). The degree of deacetylation has been shown to play an important role in the rate of degradation of chitosan materials; an important consideration for tissue engineering applications. Studies have shown that, when implanted subcutaneously in a rat model, chitosan materials with a degree of deacetylation of less that 70% were readily degraded *in vivo* while those with a degree of deacetylation of greater than 70% degraded more slowly (Tomihata and Ikada, 1997; Zhang and Neau, 2001). The degree of deacetylation of chitosan materials has also been shown to be directly related to the ability of the material to support cell attachment, with a higher degree of deacetylation being more favorable for cell attachment (Mao et al., 2004; Cao et al., 2005).

Chitosan, due to its cationic nature and high charge density in solution, is able to interact with glycosaminoglycans and other negatively charged particles, including various water soluble anionic polymers (Denuziere et al., 1998; Gaserod et al., 1998; Di Martino et al., 2005; Raman et al., 2005; Chen et al., 2006; Mi et al., 2006). This property has been shown to allow the immobilization of glycosaminoglycans on the surface of chitosan (Denuziere et al., 1998; Madihally and Matthew, 1999; Raman et al., 2005; Mi et al., 2006). These glycosaminoglycans can then, via various pathways, influence cell adhesion, migration, proliferation, and differentiation (Takahashi et al., 1990). Furthermore, the *N*-acetylglucosamine moiety on chitosan is analogous to that on glycosaminoglycans and suggests that additional biological activity may be attributed to this naturally occurring scaffold.

The *in vivo* tissue response to various chitosan-based implant materials is consistent with an acute to subacute inflammatory reaction (Nishimura et al., 1984; Muzzarelli et al., 1988; Muzzarelli et al., 1989; Damour et al., 1994; Peluso et al., 1994; Muzzarelli, 1997). Chitosan oligosaccharides have been shown to modulate macrophage response through interactions via their acetylated residues (VandeVord et al., 2002). Both chitosan and chitin have been shown to be chemoattractants for neutrophils *in vitro* and *in vivo* (Leuba and

Stossel, 1986; Iida et al., 1987; Muzzarelli et al., 1990), resulting in a high concentration of neutrophils at the site of implantation during the first 7 days post implantation. However, the neutrophil population dissipates thereafter, and a chronic inflammatory response does not develop (Chen et al., 2006). In most cases, when chitosan is used *in vivo*, little or no fibrous encapsulation is observed nor does chitosan elicit the multinucleate giant cell or chronic mononuclear cell presence that is typically associated with an adverse foreign body response (Suh et al., 2000). Granulation tissue accompanied by accelerated angiogenesis in response to chitosan implantation has been reported (Chen et al., 2006). The *in vivo* response to chitosan in tissue engineering applications has been reviewed (Suh et al., 2000; Khor and Lim, 2003; Di Martino et al., 2005).

Chitosan has been used as a conduit for guided peripheral nerve regeneration (Jenq and Coggeshall, 1987; Aebischer et al., 1990; Knoops et al., 1990; Guenard et al., 1991; Kim et al., 1993; den Dunnen et al., 1995; Rodriguez et al., 1999; Wang et al., 2005) and as a scaffold for the treatment of experimentally induced skin wounds with good results (Ueno et al., 1999; Ueno, 2001a, b; Chen et al., 2002; Tanabe et al., 2002; Mizuno et al., 2003). Cartilage repair (Di Martino et al., 2005) and bone tissue engineering applications (Lee et al., 2002; Bumgardener, 2003a, b; Wang et al., 2004) of chitosan have also been investigated.

In summary, a significant body of work has been conducted with chitosan as a naturally occurring scaffold for tissue engineering applications and perhaps more is known about its chemistry, degradation, and host tissue response than any of the synthetic or naturally occurring scaffold materials.

INTACT EXTRACELLULAR MATRIX AS A SCAFFOLD MATERIAL

The use of intact ECM, derived via the decellularization of various tissues and organs, has received considerable attention in the past 15 years. The ECM consists of the naturally occurring milieu of structural and functional molecules that are secreted by the resident cells of each tissue and organ; thus, there is a unique ECM composition and ultrastructure for each tissue and organ. The ECM even varies from location to location within various tissues such as the endocrine versus exocrine loci within the pancreas, or the valvular versus mural loci within the heart. The molecular motifs for cell attachment, migration, and differentiation are tissue specific, and attempts to mimic this compositional and structural complexity by synthetic methods have achieved very limited success. Naturally occurring ECM is one of the scaffold materials that has achieved commercial success for tissue engineering applications.

ECM scaffolds derived from human dermis (Wainwright, 1995; Isch et al., 2000; Clemons et al., 2003), porcine and human urinary bladder (Duel et al., 1996; Atala, 1998; Dahms et al., 1998), porcine small intestinal submucosa (SIS) (Oelschlager et al., 2003; Wang et al., 2003; Badylak, 2004; Derwin et al., 2004; Musahl et al., 2004), porcine heart valves (Cohn et al., 1989; Hammermeister et al., 1993; Simon et al., 2003), and bovine dermis (Barber et al., 2006; Coons and Barber, 2006), among others, have all been used in human clinical applications. Methods for the decellularization of these tissues have recently been reviewed (Gilbert et al., 2006), and although complete elimination of all cellular remnants from any tissue is unlikely, the biologic response to scaffold materials composed of ECM is not characterized by immune-mediated rejection, even when the ECM is of xenogeneic origin (Allman et al., 2001; Allman et al., 2002; Palmer et al., 2002).

Few studies have examined the host immune response to ECM scaffolds. Such studies have probably been the most extensive with allogeneic and xenogeneic heart valves and SIS. These studies have shown that host immune recognition of the ECM material does indeed occur, but is of Th-2 (accommodation) type of response rather than a Th-1 (cell-mediated rejection) type of response (Allman et al., 2002). In addition, although small amounts of the galactosyl 1,3 galactose (i.e. “GAL-epitope”) can be found in ECM scaffolds of porcine origin, they are not of sufficient amount to activate a complement in human serum (McPherson et al., 2000; Raeder et al., 2002). T-lymphocyte suppression has been found in some *in vitro* studies and this phenomenon has been suggested as a contributing factor to the absence of an adverse immune response when ECM is used as a biologic scaffold material. Preclinical studies have shown that immune challenges with inactivated influenza virus, bovine serum, albumin, and other antigens cause identical responses in mice that have been implanted with SIS versus mice that have not been exposed to SIS; that is, no systemic immune suppression was found.

The *in vivo* degradation of ECM scaffolds has been most thoroughly evaluated with porcine SIS. ^{14}C-labeling studies have shown that rapid degradation occurs following *in vivo* implantation of SIS that has not been chemically cross-linked. Approximately 60% of the SIS-ECM is degraded and removed from the implantation site by 28 days and virtually 100% of the SIS-ECM has been eliminated (mostly via urinary excretion) within 60 days (Badylak et al., 1998). The rapid degradation of the ECM scaffold material is likely to be responsible for the absence of a chronic inflammatory or foreign body type of tissue response when it is used as a scaffold for tissue reconstruction.

Protein–protein cross-linking agents such as glutaraldehyde, carbodiimide, and diisocyanate convert degradable ECM scaffolds into non-degradable or slowly degradable scaffolds and, thus, elicit a chronic inflammatory or foreign body type of tissue response when implanted in mammalian hosts. Although mechanical properties (strength) can be enhanced by the use of such agents, this benefit occurs at the cost of diminished constructive remodeling in many applications (Valentin et al., in press).

ECM-based scaffolds have been extensively evaluated in preclinical animal studies for numerous applications, including lower urinary tract reconstruction (Kropp et al., 1995; Badylak et al., 1998; Kropp et al., 1998), the treatment of dermal wounds (Lindberg and Badylak, 2001), and musculoskeletal tissue reconstruction (Hodde et al., 1997; Valentin et al., in press). Human clinical studies with ECM scaffolds have also included a broad range of clinical uses (De Ugarte et al., 2004; Alpert et al., 2005; Dedecker et al., 2005; Helton et al., 2005; Jones et al., 2005a, b; Sievert et al., 2005; Smith et al., 2005; Zalavras et al., 2006).

The host response to ECM scaffolds includes angiogenesis, mononuclear cell infiltration, and the deposition of new ECM by host cells that assume residence at the site of scaffold degradation (Voytik-Harbin et al., 1997; Badylak et al., 1999; Hodde et al., 2000; Badylak et al., 2002; Badylak, 2002; Valentin et al., in press). These biologic phenomena are thought to be the result of released growth factors and cytokines during scaffold degradation and the response to biologically active degradation products of the parent molecules within ECM (Sarikaya et al., 2002; Li et al., 2004).

In summary, biologic scaffolds composed of extracellular matrix show promise for numerous surgical applications. Of the scaffolds reviewed in this chapter, both collagen and ECM biomaterials have been successfully translated into devices currently used in human patients. Optimization of the applications will depend upon a more thorough understanding of the mechanisms of action and the effect of various processing methods upon the *in vivo* remodeling outcomes.

REFERENCES

Aebischer, P., Guenard, V. and Velentini, R.F. (1990). The morphology of regenerating peripheral nerves is modulated by the surface microgeometry of polymeric guidance channels. *Brain Res.* **531**: 211–218.

Aiedeh, K., Gianasi, E., Orienti, I. and Zecchi, V. (1997). Chitosan microcapsules as controlled release systems for insulin. *J. Microencapsul.* **14**: 567–576.

Allman, A.J., McPherson, T.B., et al. (2001). Xenogeneic extracellular matrix grafts elicit a TH2-restricted immune response. *Transplantation* **71**(11): 1631–1640.

Allman, A.J., McPherson, T.B., et al. (2002). The Th2-restricted immune response to xenogeneic small intestinal submucosa does not influence systemic protective immunity to viral and bacterial pathogens. *Tissue Eng.* **8**(1): 53–62.

Alpert, S.A., Cheng, E.Y., et al. (2005). Bladder neck fistula after the complete primary repair of exstrophy: a multi-institutional experience. *J. Urol.* **174**(4 Pt 2): 1687–1689; discussion 1689–1690.

Atala, A. (1998). Tissue engineering in urologic surgery. *Urol. Clin. North Am.* **25**(1): 39–50.

Badylak, S. (2004). Xenogeneic extracellular matrix as a scaffold for tissue reconstruction. *Transpl. Immunol.* **12**(3–4): 367–377.

Badylak, S., Kokini, K., et al. (2002). Morphologic study of small intestinal submucosa as a body wall repair device. *J. Surg. Res.* **103**(2): 190–202.

Badylak, S., Liang, A., et al. (1999). Endothelial cell adherence to small intestinal submucosa: an acellular bioscaffold. *Biomaterials* **20**(23–24): 2257–2263.

Badylak, S.F. (2002). The extracellular matrix as a scaffold for tissue reconstruction. *Sem. Cell Dev. Biol.* **13**(5): 377–383.

Badylak, S.F., Kropp, B., et al. (1998). Small intestinal submucosa: a rapidly resorbed bioscaffold for augmentation cystoplasty in a dog model. *Tissue Eng.* **4**(4): 379–387.

Barber, F.A., Herbert, M.A. and Coons, D.A. (2006). Tendon augmentation grafts: biomechanical failure loads and failure patterns. *Arthroscopy* **22**(5): 534–538.

Beier, F., Eerola, I., Vuorio, E., LuValle, P., Reichenberger, E., Bertling, W., von der Mark, K. and Lammi, M.J. (1996). Variability in the upstream promoter and intron sequences of the human, mouse, and chick type X collagen genes. *Matrix Biol.* **15**(6): 415–422.

Berthiaume, F. and Yarmush, M.L. (1995). *The Biomedical Engineering Handbook*. Boca Raton, FL: CRC Press.

Boyd, C.D., Christiano, A.M., Pierce, R.A., Stolle, C.A. and Deak, S.B. (1991). Mammalian tropoelastin: multiple domains of the protein define an evolutionary divergent amino acid sequence. *Matrix* **11**(4): 235–241.

Brennan, E.P., Reing, J., Chew, D., Myers-Irvin, J.M., Young, E.J. and Badylak, S.F. (2006). Antibacterial activity within degradation products of biologic scaffolds composed of extracellular matrix. *Tissue Eng.* **12**(10): 2949–2955.

Bumgardener, J.D., Wiser, R., Elder, S.H., Jouett, R., Yang, Y. and Ong, J.L. (2003a). Contact angle, protein adsorption and osteoblast precursor cell attachment to chitosan coatings bonded to titanium. *J. Biomater. Sci. Polym. Ed.* **14**: 1401–1409.

Bumgardener, J.D., Wiser, R., Gerard, P.D., Bergin, P., Chestnutt, B., Marin, M., Ramsey, V., Elder, S.H. and Gilbert, J.A. (2003b). Chitosan: potential use as a bioactive coating for orthopaedic and craniofacial/dental implants. *J. Biomater. Sci. Polym. Ed.* **14**: 423–438.

Cao, W., Jing, D.H., Li, J.M., Gong, Y.D., Zhao, N.M. and Zhang, X.F. (2005). Effects of the degree of deacetylation on the physicochemical properties and Schwann cell affinity of chitosan films. *J. Biomater. Appl.* **20**: 157–177.

Chen, X.-G., Wang, Z., Liu, W.-S. and Park, H.-J. (2002). The effect of carboxymethyl-chitosan on proliferation and collagen secretion of normal and keloid skin fibroblasts. *Biomaterials* **23**: 4609–4614.

Chen, Y.L., Chen, H.C., Lee, H.P., Chan, H.Y. and Hu, Y.C. (2006). Rational development of GAG-augmented chitosan membranes by fractional factorial design methodology. *Biomaterials* **27**(10): 2222–2232.

Chow, K.S. and Khor, E. (2000). Novel fabrication of open-pore chitin matrixes. *Biomacromolecules* **1**: 61–67.

Clemons, J.L., Myers, D.L., et al. (2003). Vaginal paravaginal repair with an AlloDerm graft. *Am. J. Obstet. Gynecol.* **189**(6): 1612–1618; discussion 1618–1619.

Cohn, L.H., Collins, J.J., DiSesa, V.J., Couper, G.S., Peigh, P.S., Kowalker, W. and Allred, E. (1989). Fifteen-year experience with 1678 Hancock porcine bioprosthetic heart valve replacements. *Ann. Surg.* **210**(4): 435–443.

Coons, D.A. and Barber, A. (2006). Tendon graft substitutes-rotator cuff patches. *Sports Med. Arthrosc. Rev.* **14**(3): 185–190.

Cornelius, L.A., Nehring, L.C., et al. (1998). Matrix metalloproteinases generate angiostatin: effects on neovascularization. *J. Immunol.* **161**(12): 6845–6852.

Dahms, S.E., Piechota, H.J., et al. (1998). Composition and biomechanical properties of the bladder acellular matrix graft: comparative analysis in rat, pig and human. *Br. J. Urol.* **82**(3): 411–419.

Damour, O., Gueugniaud, P.Y., Berthin-Maghit, M., Rousselle, P., Berthod, F., Sahuc, F. and Collombel, C. (1994). A dermal substrate made of collagen-GAG-chitosan for deep burn coverage: first clinical uses. *Clin. Mater.* **15**(4): 273–276.

De Ugarte, D.A., Choi, E., et al. (2004). Mucosal regeneration of a duodenal defect using small intestine submucosa. *Am. Surg.* **70**(1): 49–51.

Dedecker, F., Grynberg, M., et al. (2005). Small intestinal submucosa (SIS): prospects in urogenital surgery. *Prog. Urol.* **15**(3): 405–410.

den Dunnen, W.F.A., van der Lei, B., Robinson, P.H., Holwerda, A., Pennings, A.J. and Schakenraad, J.M. (1995). Biological performance of a degradable poly(lactic acid-e-caprolactone) nerve guide: influence of tube dimensions. *J. Biomed. Mater. Res.* **29**: 757–766.

Denuziere, A., Ferrier, D., Damour, O. and Domard, A. (1998). Chitosan–chondroitin sulfate and chitosan–hyaluronate polyelectrolyte complexes: biological properties. *Biomaterials* **19**: 1275–1285.

Derwin, K., Androjna, C., et al. (2004). Porcine small intestine submucosa as a flexor tendon graft. *Clin. Orthop.* **423**: 245–252.

Di Martino, A., Sittingerc, M. and Risbud, M.V. (2005). A versatile biopolymer for orthopaedic tissue-engineering. *Biomaterials* **26**: 5983–5990.

Dornish, M., Kaplan, D. and Skaugrud, O. (2001). Standards and guidelines for biopolymers in tissue-engineered medical products. *Ann. NYAcad. Sci.* **944**: 388–397.

Duel, B.P., Hendren, W.H., et al. (1996). Reconstructive options in genitourinary rhabdomyosarcoma. *J. Urol.* **156**(5): 1798–1804.

Freier, T., Montenegro, R., Shan Koh, K. and Shoichet, M.S. (2005). Chitin-based tubes for tissue engineering in the nervous system. *Biomaterials* **26**(22): 4624–4632.

Garrone, R., Exposito, J.Y., Franc, J.M., Franc, S., Humbert-David, N., Qin, L. and Tillet, E. (1993). Phylogenesis of the extracellular matrix. *C. R. Seances Soc. Biol. Fil.* **187**(2): 114–123.

Gaserod, O., Smidsrod, O. and Skjak-Braek, G. (1998). Microcapsules of alginate–chitosan: a quantitative study of the interaction between alginate and chitosan. *Biomaterials* **19**(20): 1815–1825.

Geng, L., Feng, W., Hutmatcher, D.W., Wong., Loh, H. and Fuh, J.Y.H. (2005). Direct writing of chitosan scaffolds using a robotic system. *Rapid Prototyping J.* **11**: 90–97.

Gilbert, T.W., Sellaro T.L. and Badylak, S.F. (2006). Decellularization of tissues and organs. *Biomaterials* **27**(19): 3675–3683.

Guenard, V., Velentini, R.F. and Aebischer, P. (1991). Influence of surface texture of polymeric sheets on peripheral nerve regeneration in a two-compartment guidance system. *Biomaterials* **12**: 259–263.

Hammermeister, K.E., Sethi, G.K., Henderson, W.G., Oprian, C., Kim, T. and Rahimtoola, S. (1993). A comparison of outcomes in men 11 years after heart-valve replacement with a mechanical valve or bioprosthesis. *N. Engl. J. Med.* **328**: 1289–1296.

Helton, W.S., Fisichella, P.M., et al. (2005). Short-term outcomes with small intestinal submucosa for ventral abdominal hernia. *Arch. Surg.* **140**(6): 549–560; discussion 560–562.

Ho, M.H., Kuo, P.Y., Hsieh, H.J., Hsien, T.Y., Hou, L.T., Lai, J.Y. and Wang, D.M. (2004). Preparation of porous scaffolds by using freeze-extraction and freeze-gelation methods. *Biomaterials* **25**(1): 129–138.

Hodde, J.,Record, R., et al. (2000). Assessment of angiogenesis in an *in vitro* three-dimensional assay. *Cardiac Vasc. Regeneration* **4**: 1–10.

Hodde, J.P., Badylak, S.F., et al. (1997). The effect of range of motion on remodeling of small intestinal submucosa (SIS) when used as an Achilles' tendon repair material in the rabbit. *Tissue Eng.* **3**: 27–37.

Huang, M., Khor, E. and Lim, L.Y. (2004). Uptake and cytotoxicity of chitosan molecules and nanoparticles: effects of molecular weight and degree of deacetylation. *Pharm. Res.* **21**: 344–353.

Iida, J., Une, T., Ishihara, C., Nishimura, K., Tokura, S., Mizukoshi, N. and Azuma, I. (1987). Stimulation of non-specific host resistance against Sendai virus and *Escherichia coli* infections by chitin derivatives in mice. *Vaccine* **5**(4): 270–274.

Isch, J.A., Engum, S.A., Ruble. C.A., Davis, M.M. and Grosfeld, J.L. (2000). Patch esophagoplasty using AlloDerm as a tissue scaffold. *47th Annual International Congress of the British Association of Paediatric Surgeons*, Sorrento, Italy.

Jenq, C.B. and Coggeshall, R.E. (1987). Permeable tubes increase the length of the gap that regenerating axons can span. *Brain Res.* **4**: 239–242.

Jolles, P. and Muzzarelli, R.A. (1999). *Chitin and Chitinase*. Boston: Birkhauser Verlag.

Jones, J.S., Rackley, R.R., et al. (2005a). Porcine small intestinal submucosa as a percutaneous mid-urethral sling: 2 year results. *BJU Int.* **96**(1): 103–106.

Jones, J.S., Vasavada, S.P., et al. (2005b). Sling may hasten return of continence after radical prostatectomy. *J. Urol.* **65**(6): 1163–1167.

Khor, E. (2001). *Chitin: Fulfilling a Biomaterials Promise*. Oxford, UK: Elsevier Science.

Khor, E. and Lim, L.Y. (2003). Implantable applications of chitin and chitosan. *Biomaterials* **24**: 2339–2349.

Kim, D.H., Connolly, S.E., Zhao, S., Beuerman, R.W., Voorhies, R.M. and Kline, D.G. (1993). Comparison of macropore, semipermeable, and nonpermeable collagen conduits in nerve repair. *J. Reconstr. Microsurg.* **9**: 415–420.

Knoops, B., Hurtado, H. and van den Bosch de Aguilar, P. (1990). Rat sciatic nerve regeneration within an acrylic semipermeable tube and comparison with a silicone impermeable material. *J. Neuropathol. Exp. Neurol.* **49**: 438–448.

Kropp, B.P., Eppley, B.L., et al. (1995). Experimental assessment of small intestinal submucosa as a bladder wall substitute. *Urology* **46**(3): 396–400.

Kropp, B.P., Ludlow, J.K., et al. (1998). Rabbit urethral regeneration using small intestinal submucosa onlay grafts. *Urology* **52**(1): 138–142.

Lee, J.Y., Nam, S.H., Im, S.Y., Park, Y.J., Lee, Y.M., Seol, Y.J., Chung, C.P. and Lee, S.J. (2002). Enhanced bone formation by controlled growth factor delivery from chitosan-based biomaterials. *J. Contr. Release* **78**: 187–197.

Leuba, J.L. and Stossel, P. (1986). Chitosan and other polyamines: antifungal activity and interaction with biological membranes. In: Muzzarelli, C.J.R.A. and Gooday, G.W. (eds.), *Chitin in Nature and Technology*. New York: Plenum Publishing Corp.

Li, F., Li, W., et al. (2004). Low-molecular-weight peptides derived from extracellular matrix as chemoattractants for primary endothelial cells. *Endothelium* **11**: 199–206.

Lindberg, K. and Badylak, S.F. (2001). Porcine small intestinal submucosa (SIS): a bioscaffold supporting *in vitro* primary human epidermal cell differentiation and synthesis of basement membrane proteins. *Burns* **27**(3): 254–266.

Madihally, S.V. and Matthew, H.W.T. (1999). Porous chitosan scaffolds for tissue engineering. *Biomaterials* **20**: 1133–1142.

Maeshima, Y., Colorado, P.C., et al. (2000). Distinct antitumor properties of a type IV collagen domain derived from basement membrane. *J. Biol. Chem.* **275**(28): 21340–21348.

Mao, J.S., Cui, Y.L., Wang, X.H., Sun, Y., Yin, Y.J., Zhao, H.M. and De Yao, K. (2004). A preliminary study on chitosan and gelatin polyelectrolyte complex cytocompatibility by cell cycle and apoptosis analysis. *Biomaterials* **25**: 3973–3981.

McPherson, T.B., Liang, H., et al. (2000). Galalpha(1,3)Gal epitope in porcine small intestinal submucosa. *Tissue Eng.* **6**(3): 233–239.

Mi, F.L., Shyu, S.S., Peng, C.K., Wu, Y.B., Sung, H.W., Wang, P.S. and Huang, C.C. (2006). Fabrication of chondroitin sulfate–chitosan composite artificial extracellular matrix for stabilization of fibroblast growth factor. *J. Biomed. Mater. Res.* **76**(1): 1–15.

Mima, S., Miya, M., Iwamoto, R. and Yoshikawa, S. (1983). Highly deacetylated chitosan and its properties. *J. Appl. Polym. Sci.* **28**: 1909–1917.

Mizuno, K., Yamamura, K., Yano, K., Osada, T., Saeki, S., Takimoto, N., Sakura, T. and Nimura, Y. (2003). Effect of chitosan film containing basic fibroblast growth factor on wound healing in genetically diabetic mice. *J. Biomed. Mater. Res.* **64**: 177–181.

Musahl, V., Abramowitch, S.D., et al. (2004). The use of porcine small intestinal submucosa to enhance the healing of the medial collateral ligament: a functional tissue engineering study in rabbits. *J. Orthop. Res.* **22**(1): 214–220.

Muzzarelli, R., Baldassarre, V., Conti, F., Ferrara, P., Biagini, G., Gazzanelli, G. and Vasi, V. (1988). Biological activity of chitosan: ultrastructural study. *Biomaterials* **9**(3): 247–252.

Muzzarelli, R., Biagini, G., Pugnaloni, A., Filippini, O., Baldassarre, V., Castaldini, C. and Rizzoli, C. (1989). Reconstruction of parodontal tissue with chitosan. *Biomaterials* **10**(9): 598–603.

Muzzarelli, R., Tarsi, R., Filippini, O., Giovanetti, E., Biagini, G. and Varaldo, P.E. (1990). Antimicrobial properties of *N*-carboxybutyl chitosan. *Antimicrob. Agents Ch.*: 2019–2023.

Muzzarelli, R.A. (1993). Biochemical significance of exogenous chitins, chitosan in animals, patients. *Carbohydrate Polymers* **20**: 7–16.

Muzzarelli, R.A. (1997). Human enzymatic activities related to the therapeutic administration of chitin derivatives. *Cell. Mol. Life Sci.* **53**(2): 131–140.

Nimni, M.E., Cheung, D., Strates, B., Kodama, M. and Sheikh, K. (1987). Chemically modified collagen: a natural biomaterial for tissue replacement. *J. Biomed. Mater. Res.* **21**: 741–771.

Nishimura, K., Nishimaura, S., Nishi, N., Saiki, I., Tokura, S. and Azuma, I. (1984). Immunological activity of chitin and its derivatives. *Vaccine* **2**(1): 93–99.

Oelschlager, B.K., Barreca, M., et al. (2003). The use of small intestine submucosa in the repair of paraesophageal hernias: initial observations of a new technique. *Am. J. Surg.* **186**: 4–8.

Palmer, E.M., Beilfuss, B.A., et al. (2002). Human helper T cell activation and differentiation is suppressed by porcine small intestinal submucosa. *Tissue Eng.* **8**(5): 893–900.

Peluso, G., Petillo, O., Ranieri, M., Santin, M., Ambrosio, L., Calabro, D., Avallone, B. and Balsamo, G. (1994). Chitosan-mediated stimulation of macrophage function. *Biomaterials* **15**(15): 1215–1220.

Raeder, R.H., Badylak, S.F., et al. (2002). Natural anti-galactose alpha1,3 galactose antibodies delay, but do not prevent the acceptance of extracellular matrix xenografts. *Transpl. Immunol.* **10**(1): 15–24.

Raman, R., Saisekharan, V. and Saisekharan, R. (2005). Structural insights into biological roles of protein–glycosaminoglycan interactions. *Chem. Biol.* **12**: 267–277.

Rodriguez, F.J., Gomez, N., Perego, G. and Navarro, X. (1999). Highly permeable polylactide–caprolactone nerve guides enhance peripheral nerve regeneration through long gaps. *Biomaterials* **20**: 1489–1500.

Sarikaya, A., Record, R., et al. (2002). Antimicrobial activity associated with extracellular matrices. *Tissue Eng.* **8**(1): 63–71.

Sato, K. (1983). Radiation sterilization of medical products. *Radioisotopes* **32**: 431.

Sievert, K.D., Nagele, U., et al. (2005). Off-shelf commercially available acellular collagen matrix SIS® by Cook for urethral reconstruction. *Eur. Urol. Suppl.* **4**(3): 242 (Abstract 958).

Simon, P., Kasimir, M.T., Seebacher, G., Weigel, G., Ullrich, R., Salzer-Muhar, U., Rieder, E. and Wolner E. (2003). Early failure of the tissue engineered porcine heart valve SYNERGRAFTTM in pediatric patients. *Eur. J. Cardiothorac. Surg.* **23**: 1002–1006.

Smith, A.M., Walsh, R.M., et al. (2005). Novel bile duct repair for bleeding biliary anastomotic varices: case report and literature review. *J. Gastroint. Surg.* **9**(6): 832–836.

Suh, J.K.F. and Matthew, H.W.T. (2000). Application of chitosan-based polysaccharide biomaterials in cartilage tissue engineering: a review. *Biomaterials* **21**: 2589–2598.

Takahashi, T., Takayama, K., Machida, Y. and Nagai, T. (1990). Characteristics of polyion complexes of chitosan with sodium alginate and sodium polyacrylate. *Int. J. Pharm.* **61**: 35–41.

Tanabe, T., Okitsu, N., Tachibana, A. and Yamauchi, K. (2002). Preparation and characterization of keratin–chitosan composite film. *Biomaterials* **23**: 817–825.

Tomihata, K. and Ikada, Y. (1997). *In vitro* and *in vivo* degradation of films of chitin and its deacetylated derivatives. *Biomaterials* **18**: 567–575.

Ueno, H., Murakami, M., Okumura, M., Kadosawa, T., Uede, T. and Fujinaga, T. (2001a). Chitosan accelerates the production of osteopontin from polymorphonuclear leukocytes. *Biomaterials* **22**: 1667–1673.

Ueno, H., Nakamura, F., Murakami, M., Okumura, M., Kadosawa, T. and Fujinaga, T. (2001b). Evaluation effects of chitosan for the extracellular matrix production by fibroblasts and the growth factors production by macrophages. *Biomaterials* **22**: 2125–2130.

Ueno, H., Yamada, H., Tanaka, I., Kaba, N., Matsuura, M., Okumura, M. Kadosawa, T. and Fujinaga, T. (1999). Accelerating effects of chitosan for healing at early phase of experimental open wound in dogs. *Biomaterials* **20**: 1407–1414.

Valentin, J.E., Badylak, J.S., McCabe, G.P. and Badylak, S.F. Extracellular matrix bioscaffolds for orthopaedic applications: a comparative histologic study. *J.Bone Joint Surg. Am. Vol.* (In press).

VandeVord, P.J., Matthew, H.W., DeSilva, S.P., Mayton, L., Wu, B. and Wooley, P.H. (2002). Evaluation of the biocompatibility of a chitosan scaffold in mice. *J. Biomed. Mater. Res.* **59**: 585–590.

Voytik-Harbin, S.L., Brightman, A.O., et al. (1997). Identification of extractable growth factors from small intestinal submucosa. *J. Cell. Biochem.* **67**(4): 478–491.

Wainwright, D.J. (1995). Use of an acellular allograft dermal matrix (AlloDerm) in the management of full-thickness burns. *Burns* **21**(4): 243–248.

Wang, J., de Boer, J. and de Groot, K. (2004). Preparation and characterization of electrodepo sited calcium phosphate/chitosan coating on Ti6Al4V plates. *J. Dent. Res.* **83**: 296–301.

Wang, X.D., Hu, W., Cao, Y., Yao, J., Wu, J. and Gu, X.S. (2005). Dog sciatic nerve regeneration across a 30-mm defect bridged by a chitosan/PGA artificial nerve graft. *Brain* **128**: 1897–1910.

Wang, Z.Q., Watanabe, Y., et al. (2003). Experimental assessment of small intestinal submucosa as a small bowel graft in a rat model. *J. Pediatr. Surg.* **38**(11): 1596–1601.

Zalavras, C., Gardocki, R., et al. (2006). Reconstruction of large rotator cuff tendon defects with porcine small intestinal submucosa in an animal model. *J. Shoulder Elbow Surgery* **15**: 225–231.

Zhang, H. and Neau, S.H. (2001). *In vitro* degradation of chitosan by a commercial enzyme preparation: effect of molecular weight and degree of deacetylation. *Biomaterials* **22**: 1653–1658.

20

Synthetic Polymers

M.C. Hacker and A.G. Mikos

INTRODUCTION

Regenerative medicine is an emerging, interdisciplinary approach to repairing or replacing damaged or diseased tissues and organs. In order to re-establish tissue and organ function impaired by disease, trauma, or congenital abnormalities, regenerative medicine employs cellular therapies, tissue engineering strategies, and artificial or biohybrid organ devices. Typically, these techniques rely on combinations of cells, genes, morphogens, or other biological building blocks with bioengineered materials and technologies to address tissue or organ insufficiency.

Materials used in these approaches range from metals and ceramics, to natural and synthetic polymers, as well as micro- and nanocomposites thereof. When used in a three-dimensional context, these materials are processed into micro- and/or nanoporous cell carriers, typically addressed as scaffolds, of various structures and properties, a topic that is discussed elsewhere in this book. This chapter focuses exclusively on synthetic polymers used in regenerative medicine. Some synthetic derivatives of natural materials are briefly discussed where appropriate. Accompanying the various facets of regenerative medicine, a plethora of synthetic polymers with different compositions and physicochemical properties have already been developed and investigated; however, research is still ongoing. Synthetic materials play a key role in many applications of regenerative medicine, including implants, tissue engineering scaffolds, and orthopedic fixation devices. In a broader sense, sutures, drug delivery systems, non-viral gene delivery vectors, and sensors made from synthetic polymers are further examples.

This chapter provides a structural overview of these synthetic polymers and discusses their physicochemical characteristics, structure property relationships, applications, and limitations. Synthetic polymers that are hydrolytically labile and erode (biodegradable polymers) as well as those that are bioinert and remain unchanged after implantation (non-degradable polymers) are considered. It is the authors' intention to provide a thorough overview over the synthetic material classes available. Some polymer classes are briefly mentioned and their chemical structures are provided, other more relevant materials are discussed in more detail. For most polymer classes and properties, reviews are referenced to present guidance to further reading.

Biomaterial history in general can be best organized into four eras: prehistory, the era of the surgeon hero (first generation biomaterials), designed biomaterials and engineered devices (second generation biomaterials), and the contemporary era leading into the new millennium (third generation biomaterials) (Hench and Polak, 2002; Ratner, 2004). As far back as 600 AD, the use of dental implants made from materials like seashells or iron was reported. Also, there is evidence that sutures have been used for as long as 32,000 years to close large wounds. The word "biomaterials," however, was first introduced within the last 50 years. Almost at the same time, aided by rapid advancements in industrial polymer development and synthesis, the exploration of synthetic polymers for biomedical applications began. The development of plastic contact lenses, utilizing primarily poly(methyl methacrylate) (PMMA), started around 1936, and the first data on implantation of nylon as a suture was reported in 1941. This development was accompanied by studies on the biocompatibility of the new materials. From the beginning, differences in foreign body reaction to materials like nylon and Teflon®,

Copyright © 2008 Elsevier Inc. All rights reserved

which exhibited a very mild response, became apparent. Additives such as plasticizers, unpolymerized reactants, and degradation products were discussed as possible causes leading to awareness of polymer quality for biomedical applications and biocompatibility testing.

At the end of World War II, a wide variety of durable high-performance metal, ceramic, and especially polymeric materials was available inspiring surgeons to break new grounds in replacing diseased or damaged body parts. Materials including silicones, polyurethanes (PUs), Teflon®, nylon, methacrylates, titanium, and stainless steel were available "off-the-shelf" for surgeons to apply to medical problems (Ratner, 2004). Primarily medical and dental practitioners, driven by the vision to replace lost organ or tissue functionality, made use of minimal government regulatory activity and negligible human subject protections to develop and improvise replacements, bridges, conduits, and even organ systems based on such materials. Those pioneering approaches laid the foundation for novel procedures and engineered biomaterials. Such early implants made from materials available "off-the-shelf" in part proved to be either pathogenic or toxic. With a developing understanding of the immune system and foreign body reaction, a first generation of materials was developed during the 1960s and 1970s by engineers and scientists for use inside the human body. The primary goal of early biomaterial development was to achieve a suitable combination of physical properties to match those of the replaced tissue with a minimal toxic response in the host (Hench, 1980). Following this paradigm, more than fifty implanted devices made from forty different materials were in clinical use in 1980. In the early 1980s, research began to shift from materials that exclusively exhibited a bioinert tissue response to materials that actively interacted with their environment. Another advance in this second generation was the development of biodegradable materials that exhibited controllable chemical breakdown into non-toxic degradation products, which were either metabolized or directly eliminated. Biodegradable synthetic polymers were designed to resolve the interface problem, since the foreign material is ultimately replaced by regenerating tissues and eventually the regeneration site is histologically indistinguishable from the host tissue. Resorbable polymers were routinely used clinically as sutures by 1984. Other applications in fracture fixation aids or drug delivery devices emerged quickly. Despite considerable clinical success of bioinert, bioactive, and resorbable implants, there is still a high long-term prostheses failure rate and need for revision surgery (Ratner, 2004).

Improvements of first and second generation biomaterials have been limited for one main reason: unlike living tissue, artificial biomaterials cannot respond to changing physiological loads or biochemical stimuli. This limits the lifetime of artificial body parts. To overcome these limitations, a third generation of biomaterials is being developed that involves molecular tailoring of resorbable polymers for specific cellular responses. By immobilizing specific biomolecules, such as signaling molecules or cell-specific adhesion peptides or proteins, onto a material it is possible to mimic the extracellular matrix (ECM) environment and provide a cell-adhesive surface (Hench and Polak, 2002; Drotleff et al., 2004; Lutolf and Hubbell, 2005). Biomimetic surfaces are promising tools to control cell adhesion, implant integration, cell differentiation, and tissue development. Synthetic polymer matrices can also be tailored to deliver drug, signaling molecules, and genetic code and thus provide versatile technologies for regenerative medicine (Saltzman and Olbricht, 2002; Segura and Shea, 2002; Tabata, 2003). Constantly expanding knowledge of the basic biology of stem cell differentiation and the corresponding signaling pathways as well as tissue development provide the basis for molecular design of scaffolds. In tissue engineering attempts, which aim at regenerating lost or defective tissue by transplanting *in vitro* engineered tissue constructs based on a patient's own cells, one no longer attempts to closely match scaffold mechanical properties to those of the replaced tissue. It is rather considered important that the transplanted construct is engineered to be steadily remodeled *in vivo* to resemble the histological and mechanical properties of the surrounding tissue (Nerem, 2006). Due to this paradigm shift, mechanically labile hydrogels, especially injectable systems that can be used to directly encapsulate cells, have gained great importance as basis for biomimetic cell carriers. Hydrogels are characterized by a high water content that allows encapsulated cells to survive and enables

introduced to orthopedic surgery in the mid-1950s (Saha and Pal, 1984). PMMA tissue biocompatibility became further apparent when Plexiglas fragments were accidentally implanted in the eyes and other body tissues of World War II fighter pilots during aircraft crashes.

PMMA can be *in situ* polymerized and crosslinked from a slurry containing PMMA and MMA monomers and is so used as a common bone grafting material mainly in the fixation of orthopedic prosthetic materials for hips, knees, and shoulders (Kenny and Buggy, 2003). PMMA-based bone cements can be mixed with inorganic ceramics or bioactive glass to modulate curing kinetics and enforce mechanical properties. Antibiotics can be loaded within the cement to reduce the risk of prosthesis-related infection. Significant drawbacks of self-curing PMMA cements include that they are not degraded, that their high curing temperatures and toxic monomers can cause necrosis of the surrounding tissue, and that the cements show limited interactions with the surrounding bone (Hendriks et al., 2004). Therefore, development of alternative injectable bone cements is directed toward biodegradable materials with improved curing properties and osteoconductive interfaces (Yaszemski et al., 1996; Hendriks et al., 2004).

Due to its excellent bio- and hemocompatibility and ease of manipulation, PMMA is used in many medical devices, including blood pumps and dialyzers. Its optical properties make it a candidate material for implantable ocular lenses and hard contact lenses (Lloyd et al., 2001). PMMA also offers physical and coloring properties that are beneficial for denture fabrication (Hendriks et al., 2004).

Poly(2-hydroxyethyl methacrylate)

PHEMA (Figure 20.1g) was the first hydrogel successfully employed for biological use (Wichterle and Lim, 1960). PHEMA has become the major component of most soft contact lenses and is also part of intraocular lenses (Lloyd et al., 2001). Due to their free hydroxyl groups, PHEMA gels contain relatively high amounts of water, facilitating the diffusion of solutes and oxygen. PHEMA has excellent biocompatibility which initiated the development of a plethora of HEMA-containing copolymers. Hydrogels fabricated from PHEMA and copolymers have been intensively characterized for controlled drug delivery applications (Mack et al., 1987; Lu and Anseth, 1999) and employed for biomedical uses. PHEMA gels, which have limited mechanical properties, have been used in attempts to reconstruct female breasts, nasal cartilages, and as artificial corneas as well as wound dressings (Young et al., 1998).

In a subcutaneous rabbit model, porous PHEMA sponges promoted significant cellular ingrowth and neovascularization in combination with good cytocompatibility (Chirila et al., 1993). Recently, a mineralization technique has been demonstrated that exposes carboxylate groups on crosslinked PHEMA hydrogel scaffolds, promoting calcification (Song et al., 2003).

*Poly(*N*-isopropylacrylamide)*

Poly(*N*-isopropylacrylamide) (PNiPAAm) (Figure 20.1h) has gained great significance for injectable applications in drug and cell delivery using minimally invasive techniques due to its unique physicochemical properties (Hoffman, 2002). PNiPAAm undergoes (lower critical) phase separation resulting in the formation of an opaque hydrogel in response to a temperature above 32°C, the material's lower critical solution temperature (LCST). This thermoresponsive behavior is the result of strong hydrogen bonds between the polymer and water molecules and the specific molecular orientations of these bonds due to the molecular structure of the polymer. The formation of hydrogen bonds between the polymer and the solvent lowers the free energy of the solution. Due to the hydrophobic *N*-isopropyl residues in PNiPAAm, the hydrogen bonds between water and the amide functionality require specific molecular orientations, which lead to negative entropy changes and positive contributions to the free energy. Since the enthalpic contribution to the free energy is temperature dependent, the formation of strong but specifically oriented hydrogen bonds is no longer thermodynamically favored above a

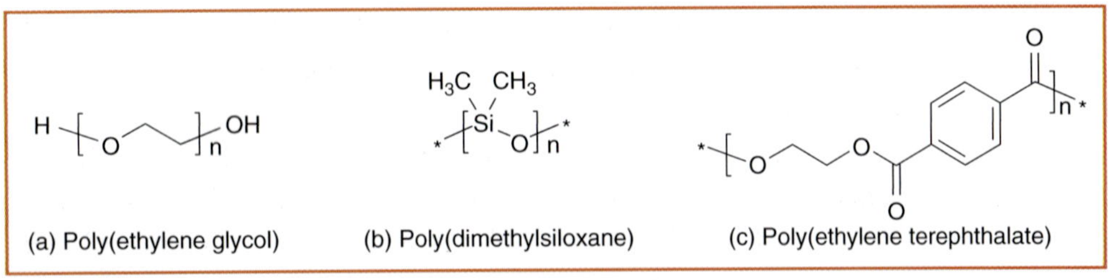

Figure 20.2 **Chemical structures of non-degradable synthetic polymers (II).**

certain temperature. Consequently, PNiPAAm dissolves in water below the LCST. At and above the LCST, the polymer chains partially desolvate and undergo a coil-to-globule transition resulting in colloidal aggregation that may lead to gel formation or polymer precipitation (Schild and Tirrell, 1990; Schild, 1992). Hydrogels formed by linear PNiPAAm at 32°C are instable and collapse substantially as the temperature is increased above the LCST. The synthesis of crosslinked networks and copolymers, typically with hydrophilic building blocks, has resulted in materials that demonstrate reversible thermogelation and form hydrogels without significant syneresis at body temperature. Different PNiPAAm-containing copolymers for cell delivery have been synthesized with acrylic acid, poly(ethylene glycol) (PEG), hyaluronic acid, and gelatin (Stile et al., 1999; Ohya et al., 2001; Hoffman, 2002; Morikawa and Matsuda, 2002). Detailed information is available for the *in vitro* and *in vivo* use of gelatin–PNiPAAm conjugates for the regeneration of articular cartilage (Ibusuki et al., 2003a, b).

Polyethers

PEG (Figure 20.2a), often also called poly(ethylene oxide) (PEO), is a non-degradable polyether of the monomer ethylene glycol. Technically, PEG and PEO should not be used as synonyms, since PEO is synthesized from the monomer ethylene oxide and typically terminated by only one hydroxyl group and an initiator fragment. Commonly, PEG is often used to refer to the polymer with molecular weight less than 50,000 Da while PEO is used for higher molecular weights. PEG is water soluble and solutions of its high molecular weight form can be categorized as a hydrogel. PEG hydrogels for biomedical applications are typically comprised of polymer chains that are crosslinked. These crosslinked networks frequently contain chemical bonds between the PEG chains and the crosslinkable moieties, which are prone to aqueous hydrolysis and are therefore characterized as biodegradable system. The molecular weight of the PEG chains crosslinked in such hydrogels is below a threshold molecular weight to allow for complete resorption by renal elimination of the individual chains. Consequently, these systems are discussed with biodegradable polymers in section "Biodegradable crosslinked polymer networks."

Favorable characteristics of PEG and PEO are their high hydrophilicity, bioinertness, and outstanding biocompatibility, which make them candidate biomaterials. PEG and PEO are frequently used as hydrophilic polymeric building blocks in copolymers with more hydrophobic degradable or non-degradable polymers for drug delivery (Jeong et al., 1997), gene delivery, tissue engineering scaffolds, medical devices, and implants. PEG has also been immobilized on polymeric biomaterial surfaces to make them resistant to protein absorption and cell adhesion. These effects are attributed to highly hydrated PEG chains on the polymer surfaces that exhibit steric repulsion based on an osmotic or entropic mechanism. Attempts to benefit from this phenomenon include the design of long-circulating nanoparticles or liposomes (Gref et al., 1997, 2000; Photos et al., 2003; Vonarbourg et al., 2006) and PEGylated enzymes or proteins with prolonged functional residence time *in vivo* compared to unmodified biomolecules (Roberts et al., 2002; Harris and Chess, 2003).

A variety of PEG-containing block copolymers for injectable drug delivery have been developed over the last decades. The most prominent class are triblock copolymers composed of two hydrophilic PEO blocks and one hydrophobic poly(propylene oxide) (PPO), also known as Pluronics® or poloxamers. These materials are designed to show similar phase transition behavior as the thermogelling PNiPAAm-containing materials (section "Poly(*N*-isopropylacrylamide)"). Poloxamers have been intensively investigated for the delivery of drugs and proteins (Jeong et al., 2002). Since poloxamers are non-degradable, biodegradable structural analogs have been synthesized.

Polysiloxanes

Polysiloxanes, or silicones, are a general category of polymers consisting of a silicon–oxygen backbone with organic groups, typically methyl groups, attached to the silicon atoms (Colas and Curtis, 2004). Certain organic side groups can be used to link two or more chains together. By varying the —Si—O— chain length, side groups, and crosslinking extent, silicone with properties ranging from liquids to hard plastics can be synthesized. Silicone synthesis typically involves the hydrolysis of chlorosilanes into linear or cyclic siloxane oligomers, which are then polymerized into polysiloxanes by polycondensation or polymerization, respectively. The most common polysiloxane is linear poly(dimethylsiloxane) (PDMS) (Figure 20.2b).

Polysiloxanes, which are characterized by unique material properties combining biocompatibility and biodurability, have found widespread application in health care (Curtis and Colas, 2004). The material's high biodurability is a result of other material properties such as hydrophobicity, low surface tension, and chemical and thermal stability. Silicone surfaces have been found to inhibit blood from clogging for many hours and have been therefore used for the fabrication of silicone coated needles, syringes, and other blood-collecting instruments. Silicone materials have also been employed as heart valves and as components in kidney dialysis, blood-oxygenator, and heart-bypass machines due to their hemocompatibility. Silicone elastomers have found application in numerous catheters, shunts, drains, and tubular implants, such as artificial urethra. Significant orthopedic applications of silicone are hand and foot joint implants. The most prominent application of silicones is their extensive use as cosmetic implants in esthetic and reconstructive plastic surgery. Prosthetic silicone implants are available for the breast, scrotum, chin, nose, cheek, calf, and buttocks. Different silicone materials, including slightly crosslinked silicone gels, are combined to achieve a natural feel. Controversy aroused regarding the safety of popular silicone gel-filled breast implants in early 1990s. These discussions initially involved increased risk for breast cancer, then progressed to autoimmune connective tissue disease, and continued to evolve to the frequency of local or surgical complications such as rupture, infection, or capsular contracture. To date, no epidemiology study has indicated that the rate of breast cancer has significantly increased in women with silicone breast implants (Silverman et al., 1996). Similarly, studies on autoimmune or connective tissue disease agreed on a lack of causal association between breast implants and these diseases (Sanchez-Guerrero et al., 1995; Lewin and Miller, 1997). A safety concern that has been controversially discussed recently involves the amount of platinum (part of catalysts used during silicone synthesis) that is released from silicone implants and accumulated in the host organism (Arepalli et al., 2002; Brook, 2006). Other mentioned complications, especially implant rupture, are persisting problems; in 1992, the FDA restricted the use of silicone gel-filled implants. Since that time, the implants may be used only under certain controlled conditions. The pre-market approval, an application for marketing a device, has only been approved for two saline-filled breast implants and no silicone gel-filled implants by the FDA as of 2004 (US FDA, 2004).

Polysiloxane gels, combining the high oxygen permeability of silicone and the comfort and clinical performance of conventional, polyacrylate hydrogels, enabled the fabrication of soft, gas permeable contact lenses for extended wear. In contrast to conventional hydrogels, silicone gels make the lens surface highly hydrophobic and less "wettable," which frequently results in discomfort and dryness during lens wear. Surface modifications of the silicones or the addition of conventional hydrogels are suitable strategies to compensate for the hydrophobicity.

Overall, polysiloxanes have displayed expanded medical application since the 1960s and today are one of the most thoroughly tested and important biomaterials.

Other Non-degradable Polymers

Poly(ethylene terephthalate)

Poly(ethylene terephthalate) (PET) (Figure 20.2c), a linear polyester synthesized by polycondensation of terephthalic acid and ethylene glycol, is typically processed into fiber meshes. These meshes are applied as vascular grafts (Xue and Greisler, 2003) or used to reinforce prostheses.

Hydrolytically Stable Polyurethanes

PUs are a heterogeneous class of polymers that consist of organic units joined by urethane links (Figure 20.3). Generally, PUs can be synthesized from a bischloroformate and a diamine or by reacting a diisocyanate with a dihydroxy component. PUs used in biomedical applications typically have a segmented structure that results in useful physicochemical properties (Boretos and Pierce, 1967). Such segmented PUs or PU copolymers are elastomers composed of alternating polydispersed "soft" and "hard" segments. These two segments are thermodynamically incompatible and phase-segregate, resulting in discrete, crystalline domains of the associated "hard" segments surrounded by a continuous, amorphous phase of "soft" segments. The segregated domains

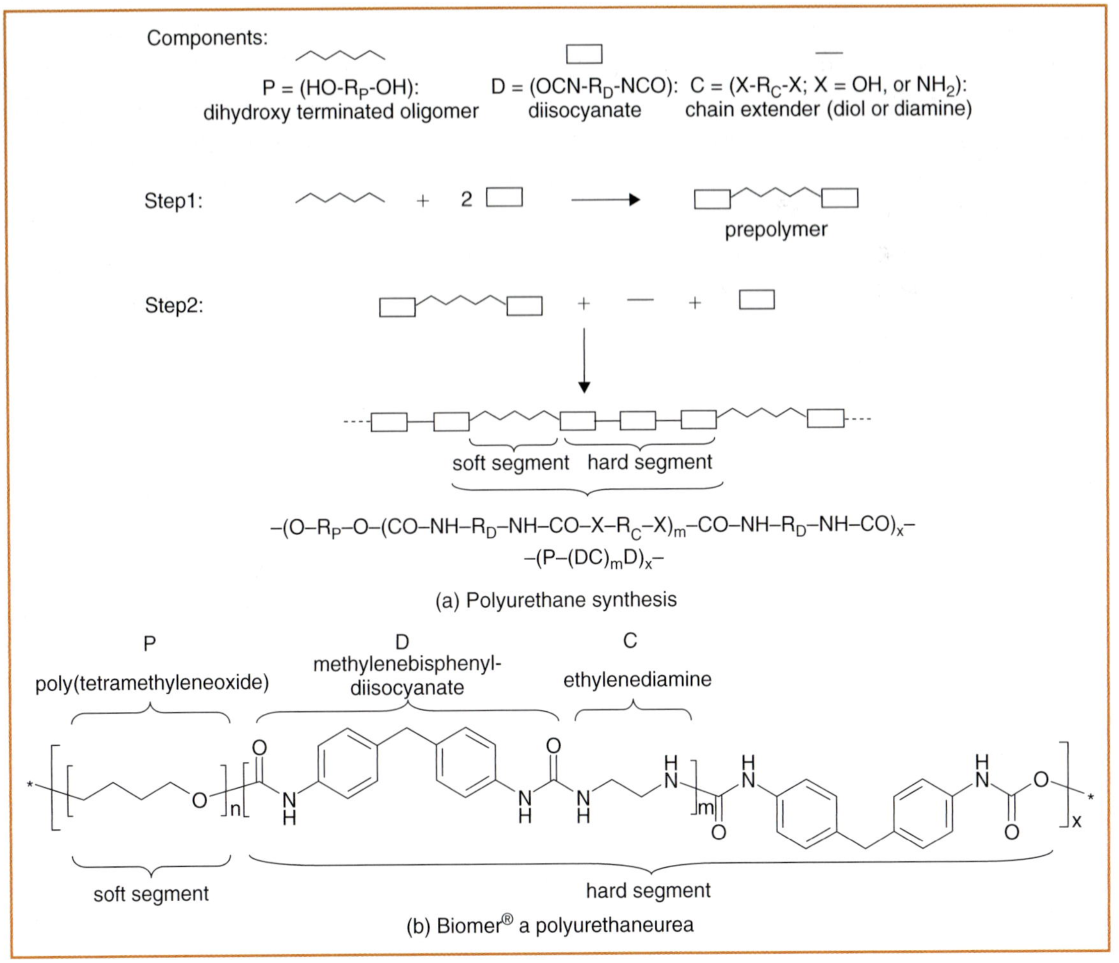

Figure 20.3 General synthesis scheme (a) and an example structure (b) for polyurethanes.

are stabilized by interchain hydrogen bonds and are responsible for the materials' mechanical properties (Gunatillake et al., 2003). Segmented PUs are synthesized in a two-step process that provides control over polymer architecture (Figure 20.3a). The first step involves the synthesis of an isocyanate-terminated prepolymer from a diisocyanate (D in Figure 20.3) and a hydroxyl group terminated polyether or polyester (P in Figure 20.3). The prepolymer and excess diisocyanate is then reacted with a hydroxy or amine group terminated chain extender (C in Figure 20.3) to generate the final PU (Figure 20.3a).

A chain extender terminated with hydroxy groups yields segmented PUs, while a diamine extender yields polyurethaneurea (Figure 20.3b). The "hard" segment of the PU copolymer is comprised of the diisocyanate and the chain extender, while the "soft" segment contains the polymeric segment introduced during the first step. The extent of phase separation is dependent on molecular weights, chemistry, and relative percentages of the building blocks (Fromstein and Woodhouse, 2006).

After almost 50 years of use in biomedical applications, PUs remain one of the most popular group of biomaterials for the fabrication of medical devices. Their popularity results from a wide range of versatility with regard to tailoring their physicochemical and mechanical properties, blood and tissue compatibility, and degradative properties by altering block copolymer composition. PUs are traditionally applied as synthetic polymers in numerous medical devices, such as breast implants, catheters, vascular, and aortic grafts, pacemaker leads, artificial heart valves, and artificial hearts. For such applications, traditional PUs, such as Biomer® (P: polytetramethylene oxide; D: methylene bisphenylenediisocyanate; C: ethylenediamine) (Figure 20.3b), were materials of first choice. However, the assumption of polyetherurethane non-degradability had to be revised following well-documented failures of pacemaker leads and breast implant coatings containing PUs in the late 1980s. Although PUs can be designed to be stable against hydrolysis, these materials have been shown to degrade in the biological environment by mechanisms including oxidation and enzyme and cell-mediated degradation (Howard, 2002; Santerre et al., 2005; Fromstein and Woodhouse, 2006). Oxidation of PUs can be initiated by peroxides, free radicals, and enzymes. Metal-catalyzed oxidation was found to be most frequently associated with pacemaker lead failure. Another important oxidation driven problem with long-term PU implants is environmental stress cracking. It has also been found that PU surfaces become coated with a protein layer that enhances the adhesion of macrophages. The macrophages, activated by proteins of the complement family, release oxidative factors that accelerate degradation of the polymer (Stokes et al., 1995).

Chemical design criteria for biostable PUs have been identified. To increase the degree of interchain hydrogen bonding, on which biostability depends in part, low molecular weight oligomeric diols (P) are preferred as building blocks. To avoid oligomer hydrolysis, oligoethers are favored over oligoesters. Aromatic diisocyanates (D) have been found to yield more biostable PUs than aliphatic diisocyanates. The use of a diamine chain extender (C) instead of a dihydroxy-terminated one typically results in stronger polyurethaneurea, but polymer fabrication is often hampered due to solubility problems. Using soft segment building blocks with high crystallinity, such as polycaprolactone, or employing silicone-based oligomers are also assumed to improve polymer biostability (Fromstein and Woodhouse, 2006).

Biomedical PUs were found to perform well in a variety of *in vivo* applications and to generally have better blood and tissue compatibilities in comparison to numerous other synthetic polymers. The efficient removal of impurities from the polymer synthesis, such as catalyst residues and low molecular weight oligomers, has been found to critically determine PU biocompatibility (Gogolewski, 1989).

PUs can be surface modified to reduce the risk of thrombosis or improve the interactions with cells and tissues. Different strategies, including adsorption, covalent grafting, or the use of self-assembled monolayers, have been applied to distribute proteins, such as fibronectin, or adhesion peptides, which contain the integrin-binding peptide motif RGD, across the PU surface (Lin et al., 1994; Fromstein and Woodhouse, 2006).

BIODEGRADABLE SYNTHETIC POLYMERS FOR REGENERATIVE MEDICINE

Biodegradable synthetic polymers offer a number of advantages over non-degradable materials for applications in regenerative medicine. Like all synthetic polymers, they can be synthesized at reproducible quality and purity and fabricated into various shapes with desired bulk and surface properties. Specific advantages include the ability to tailor mechanical properties and degradation kinetics to suit various applications. Clinical applications for biodegradable synthetic polymers are manifold and traditionally include resorbable sutures, drug delivery systems, and orthopedic fixation devices such as pins, rods, and screws (Behravesh et al., 1999). More recently, synthetic biodegradables were widely explored as artificial matrices for tissue engineering applications (Seal et al., 2001; Nguyen and West, 2002; Salgado et al., 2004). For such applications, the mechanical properties of the scaffolds, which are determined by the constitutive polymer, should functionally mimic the properties of the tissue to be regenerated. Ultimately, the polymeric support is designed to degrade while transplanted or invading cells proliferate, lay down ECM, and form coherent tissue that, in the ideal case, is functionally, histologically, and mechanically indistinguishable from the surrounding tissue. To engineer scaffolds suitable for different applications, a wide variety of biodegradable polymers is required ranging from pliable, elastic materials for soft tissue regeneration to stiff materials that can be used in load-bearing tissues such as bone. In addition to the mechanical properties, the degradation kinetics of polymer and ultimately scaffold also have to be tailored to suit various applications.

The major classes of synthetic, biodegradable polymers are briefly reviewed and their potential in regenerative medicine is discussed below.

Polyesters

Polyesters have been attractive for biomedical applications because of their ease of degradation by primarily non-enzymatic hydrolysis of ester linkages along the backbone. Additionally, degradation products can be resorbed through the metabolic pathways in most cases, and there is the potential to tailor the structure to alter degradation rates (Gunatillake and Adhikari, 2003).

A vast majority of biodegradable polymers studied belong to the polyester family (Middleton and Tipton, 2000). Polyester fibers, which also became popular with the textile industry, were used as resorbable sutures (Freed et al., 1994). Promising observations regarding biocompatibility of the materials lead to applications in drug delivery, orthopedic implants, and most recently tissue engineering scaffolds, particularly for orthopedic applications (Heller, 1984; Amecke et al., 1992; Hubbell, 1995; Behravesh et al., 1999; Webb et al., 2004).

Poly(α-hydroxy acids)

The family of polyesters can be subdivided according to the structure of the monomers. In poly(α-hydroxy acids) each monomer carries two functionalities, a carboxylic acid and a hydroxyl group, located at the carbon atom next to the carboxylic acid (α-position), that form ester bonds. Poly(α-hydroxy acids) are linear thermoplastic elastomers that are typically synthesized by ring-opening polymerization of cyclic dimers of the building blocks. Poly(lactic acid) (PLA) (Figure 20.4a), poly(glycolic acid) (PGA) (Figure 20.4b), and a range of their copolymers (poly(lactic-*co*-glycolic acid), PLGA) (Figure 20.4c) are prominent representatives of not only biodegradable polyesters but of biodegradables in general. Poly(α-hydroxy acids) have a long history of use as synthetic biodegradable materials in a number of clinical applications. Initially, resorbable sutures were made from these materials (Cutright et al., 1971). Later, poly(α-hydroxy acids) were the basis for controlled release systems for drugs and proteins (Juni and Nakano, 1987; Brannon-Peppas, 1995; Jain, 2000) and orthopedic fixation devices. Langer and coworkers have pioneered the development of these polymers in the form of porous scaffolds for tissue engineering (Langer and Vacanti, 1993).

Polylactones

The most prominent and thoroughly investigated polylactone is poly(ε-caprolactone) (PCL) (Figure 20.4e), an aliphatic, semicrystalline polyester with an interestingly low glass transition temperature (–60°C) and melting temperature (59–64°C) (Middleton and Tipton, 2000). PCL is considered biocompatible (Matsuda et al., 2003). PCL is prepared by the ring-opening polymerization of the cyclic monomer ε-caprolactone, and is compatible with a range of other polymers. Catalysts, such as stannous octoate, are used to catalyze the polymerization and low molecular weight alcohols can be used as initiator and to control the molecular weight of the polymer. ε-caprolactone can be copolymerized with numerous other monomers. Copolymers with PLA and PEG are probably the most noteworthy and have been investigated extensively (Pitt C.G. et al., 1979; Pitt G.G. et al., 1981; Cerrai et al., 1994; Petrova et al., 1998). PCL degrades at a much slower rate than PLA and is therefore most suitable for the development of long-term, implantable drug delivery systems. Aforementioned copolymers of caprolactone with dilactide were synthesized to accelerate degradation rates (Middleton and Tipton, 2000). Tubular, highly permeable poly($_L$-lactide-*co*-ε-caprolactone) guides were found to be suitable for regeneration and functional reinnervation of large gaps in injured nerves (Rodriguez et al., 1999). While this study focuses on tissue regeneration, the application of PCL in drug-delivery devices is still far more common (Sinha et al., 2004). With increasing popularity of electrospinning, a laboratory-scale technique that allows for the fabrication of non-woven meshes composed of nano- and/or micro-fibers (Pham et al., 2006), PCL might find its way into cell-based therapies since slowly degrading polymers are preferred for this technique to ensure sufficient stability of the fibers (Yoshimoto et al., 2003).

Poly(*p*-dioxanone) (Figure 20.4f), another polylactone, and its copolymers with lactide, glycolide, and/or trimethylene carbonate are synthesized by catalyzed ring-opening polymerization and have been used in a number of clinical applications ranging from suture materials to bone fixation devices (Wang et al., 1998; Yang et al., 2002).

Polyorthoesters

Polyorthoesters (POEs) (Figure 20.4g) have been developed by the Alza Corporation and SRI International in the 1970 in search of a new biodegradable polymer for drug delivery applications (Heller et al., 2002). Since then, polymer synthesis has been improved over four generations. POEs are synthesized by condensation or addition reactions typically involving dialcohols and monomeric orthoester or diketene acetals, respectively. The use of triethylene glycol as the diol component produced predominantly hydrophilic polymers, whereas hydrophobic materials could be obtained by using 1,10-decanediol. Orthoester is a functional group containing three alkoxy groups attached to one carbon atom. In POEs two of the three alkoxy groups are typically part of a cyclic acetal (Figure 20.4g).

POEs were synthesized that degrade by surface erosion, which is characterized by a constant decrease of bulk mass while polymer molecular weight within the polymer bulk is preserved (Burkersroda et al., 2002). It is known that materials built from functional groups with short hydrolysis half lives and low water diffusivity tend to be surface eroding. Polymers that exhibit surface erosion can be used to fabricate drug delivery systems that, at a high aspect to volume ratio (e.g., as for wafers), release loaded drugs at a constant rate.

The addition of lactide segments to the POE structure resulted in self-catalyzed erosion and allowed for tunable degradation times ranging from weeks to months (Ng et al., 1997). POEs provide the material platform for a variety of drug delivery applications including the treatment of postsurgical pain, osteoarthritis, and ophthalmic diseases as well as the delivery of proteins, and DNA. Block copolymers of POE and PEG have been prepared, and their use as drug delivery matrices or as colloidal structures for tumor targeting are being explored (Heller et al., 2002).

Initial biocompatibility studies revealed that POEs provoked little inflammation and were largely absorbed by 4 weeks. In contrast, $P_{D,L}$LA degraded slower and provoked a chronic inflammation with multinuclear giant cells, macrophages with engulfed material, and proliferating fibroblasts within the same model. Ossicles with bone marrow had formed in the implants of PEO in combination with demineralized bone. In PLA/demineralized bone implants the bone formation was inhibited (Andriano et al., 1999; Solheim et al., 2000).

Polycarbonates

Polycarbonates have become interesting biomaterials due to their excellent mechanical strength and good processability. Since pure polycarbonates degrade extremely slowly under physiological conditions, polyiminocarbonates (Kohn and Langer, 1986) and tyrosine-based polycarbonates (Pulapura and Kohn, 1992) (Figure 20.4h) have been engineered to yield biodegradable polymers of good mechanical strength (Engelberg and Kohn, 1991) for use in drug delivery and orthopedic applications. Degradation of most polycarbonates is controlled by the hydrolysis of the carbonate group which yields two alcohols and carbon dioxide thus alleviating the problem of acid bursting seen in polyesters (Gunatillake and Adhikari, 2003). Structural variation of the pendant side groups allows for the preparation of polymers with different mechanical properties, degradation rates, as well as cellular response. Polycarbonates that contain a pendant ethyl ester group have been shown to be osteoconductive and to possess mechanical properties sufficient for load-bearing bone fixation. Long-term (48 week) *in vivo* degradation kinetics and host bone response to tyrosine-derived polycarbonates were investigated using a canine bone chamber model (Choueka et al., 1996). Histological sections revealed intimate contact between bone and the tested polycarbonates. It was concluded that, from a degradation–biocompatibility perspective, the tyrosine-derived polycarbonates appear to be comparable, if not superior, to PLA in this model.

Amino Acid-Derived Polymers, Poly(amino acids), and Peptides

Amino acids are an interesting building block for polymers due to the biocompatibility of the degradation products and the degradability of the amide or ester bonds by which amino acids are typically polymerized or integrated in copolymers. Early studies on pure poly(amino acids)s revealed significant concerns with the materials immunogenicity and mechanical properties (Bourke and Kohn, 2003). To improve those unfavorable properties, amino acids have been used as monomeric building blocks in polymers that have a backbone structure different from natural peptides. Based on polymer structure and chemistry, four major groups have been used to classify such "non-peptide amino acid-based polymers." As for the above described tyrosine-derived polycarbonates, L-tyrosine is the predominantly employed amino acid for the formation of tyrosine-derived polyarylates and polyesters. These polymers exhibit excellent engineering properties, and polymer systems can be designed whose members show exceptional strength (polycarbonates), flexibility and elastomeric behavior (polyarylates), or water-solubility and self-assembly properties (copolymers with PEG). Poly(DTE carbonate) (DTE: desaminotyrosyl-tyrosine ethyl ester) (Figure 20.4h, R: —CH_2CH_3) exhibits a high degree of tissue compatibility and is currently being evaluated for possible clinical uses by the US Federal Drug Administration (Bourke and Kohn, 2003).

Solid-phase peptide synthesis, pioneered by Merrifield, and genetic engineering allow for the automated and highly efficient synthesis of peptides of a predefined sequence. In contrast to synthetic poly(amino acid)s, which are traditionally composed of a single amino acid and were found to be highly immunogenic in most cases, synthetic peptides have become an important polymer class for biomedical applications. Specifically peptides and peptide-amphiphiles that undergo self-assembly-driven *in situ* gelation in response to temperature, pH, or chemical stimuli are of interest as these materials can be minimally invasively implanted starting from aqueous solutions (Stupp et al., 1997; Meyer and Chilkoti, 1999; Hartgerink et al., 2001).

Genetically engineered elastin-like polypeptides, which are composed of a pentapeptide repeat and undergo inverse temperature phase transition, have been used to encapsulate chondrocytes. The cell culture studies showed that cartilaginous tissue formation, characterized by the biosynthesis of sulfated glycosaminoglycans (GAGs) and collagen, was supported (Betre et al., 2002).

Self-assembled peptide-amphiphiles, which form hydrogels composed of nanofibers resembling the native ECM components, have been demonstrated to be cytocompatible in cell encapsulation studies (Beniash et al., 2005). Recently, peptide-amphiphile nanofibers were shown to promote *in vitro* proliferation and osteogenic differentiation of marrow stromal cells (MSCs) (Hosseinkhani et al., 2006).

Polyurethanes

As outlined in section "Hydrolytically stable polyurethanes," PUs represent a major class of synthetic elastomers that have excellent mechanical properties and good biocompatibility. PUs have been evaluated for a variety of medical devices and implants, particularly for long-term implants.

Knowledge gained about the mechanisms of PU biodegradation in response to implant failures throughout the 1990s has been translated to form a new class of bioresorbable materials (Santerre et al., 2005). Recent research has utilized the flexible chemistry and diverse mechanical properties of PUs to design degradable polymers for a variety of regenerative applications. Segmented PUs with varied molecular structure have been synthesized to control rates of hydrolysis (Skarja and Woodhouse, 2001; Santerre et al., 2005). To obtain biodegradable, segmented PUs significant changes were required to the structural components historically used for their synthesis. Traditional aromatic diisocyanates (D in Figure 20.3) can yield toxic or carcinogenic degradation products when part of a degradable PU; therefore, linear diisocyanates, such as lysine-diisocyanate that yields the non-toxic degradation product lysine, are preferred. The soft segment, typically comprised of an oligomeric diol (P in Figure 20.3), is typically the block of the PU used to modify the degradation rate. Biodegradable PUs have been synthesized with a variety of soft segments including PEO, degradable polyesters such as PLA, PGA, or PCL, and combinations thereof. Other strategies focus on the copolymers' hard segments. PUs were synthesized that contain enzyme sensitive linkages introduced with the chain extender (C in Figure 20.3). For example, the use of a phenylalanine diester chain extender yielded a PU that showed susceptibility to enzyme-mediated degradation upon exposure to chymotrypsin and trypsin.

Saad et al. investigated cell and tissue interactions with a series of degradable polyesterurethanes. *In vivo* investigations showed that all test polymers exhibited favorable tissue compatibility and degraded significantly during the course of 1 year (Saad et al., 1997). Polyurethaneurea matrices were shown to allow vascularization and tissue infiltration *in vivo* (Ganta et al., 2003). The flexible chemistry and diverse mechanical properties of PU materials allowed researches to design degradable polymers for the regeneration of tissues as varied as neurons, vasculature, smooth muscle, cartilage, and bone (Xue and Greisler, 2003; Zhang et al., 2003; Santerre et al., 2005).

Block Copolymers of Polyesters or Polyamides with PEG

Amphiphilic block copolymers of biodegradable polymers with PEG have become popular materials for injectable drug delivery applications (Jeong et al., 2002). Inspired by the thermoresponsive behavior observed for non-degradable A–B–A type triblock copolymers composed of hydrophilic PEO (block A) and hydrophobic PPO (block B), polymer development focused on synthesizing biodegradable analogs of these poloxamers (or Pluronics®) that were water soluble at ambient temperature and formed stable hydrogels at body temperature. Biodegradable block copolymers were synthesized by substituting the hydrophobic PPO block with a biodegradable polymer block, such as PLA or PCL (Jeong et al., 1997; Lee et al., 2001; Ruel-Gariepy and Leroux, 2004).

Biodegradable, physically crosslinkable block copolymers of inverse structure, that is, B–A–B triblock copolymers with two biodegradable hydrophobic polymer blocks (block B) and a hydrophilic PEO block, have also been investigated as protein delivery systems (Kissel et al., 2002).

Polyanhydrides

Drug delivery technologies rely on engineered polymers that degrade in a well controllable and adjustable fashion (Langer, 1990). Increasing understanding of erosion mechanisms led to a demand for synthetic polymers that contain a hydrolytically labile backbone while limiting water diffusion within the polymer bulk significantly to confine erosion to the polymer–water interface. Such surface eroding polymers allow for the fabrication of drug delivery devices that erode at constant velocity at any time during erosion, thereby, releasing incorporated drugs at constant rates (Gopferich and Tessmar, 2002). Polyanhydrides were engineered following this paradigm by selecting the anhydride linkage, one of the least hydrolytically stable chemical bonds available, to connect the building hydrophobic monomers.

Polyanhydrides (Figure 20.4i) have been synthesized by various techniques, including melt condensation, ring-opening polymerization, interfacial condensation, dehydrochlorination, and dehydrative coupling agents (Kumar et al., 2002). Solution polymerization traditionally yielded low molecular weight polymers. Different dicarboxylic acid monomers have been polymerized to yield polyanhydrides with various physicochemical properties. Examples are linear, aromatic, fatty acid-based dicarboxylic acid monomers, and fatty acid terminated polyanhydrides. Polyanhydrides made from linear sebacic acid (SA) and aromatic 1,3-bis(*p*-carboxyphenoxy) propane (CPP) (Figure 20.4i) have been engineered to deliver carmustine (1,3-bis(2-chloroethyl)-1-nitrosourea (BCNU)), an anticancer drug, to sites in the brain following primary resection of a malignant glioma (Westphal et al., 2003). Poly(SA–CPP) hydrolyzes into non-toxic degradation products and the local chemotherapy with BCNU wafers was shown to be well tolerated and to offer a survival benefit to patients with newly diagnosed malignant glioma.

The chemical composition of a polyanhydride can be used to custom-design its degradation properties. While polyanhydrides from linear monomers, such as poly(SA), degrade within a few days, polymerized aromatic dicarboxylic acids, such as poly(1,6-bis(*p*-carboxyphenoxy) hexane), degrade much more slowly (up to a year) (Temenoff and Mikos, 2000). The structural versatility of polyanhydrides in combination with their unique degradation and erosion properties make them precious materials for numerous medical, biomedical, and pharmaceutical applications in which degradable polymers that allow for a perfect erosion control are needed (Gopferich and Tessmar, 2002). With regard to tissue engineering applications, polyanhydrides have also been interesting polymers due to their degradative properties and their good biocompatibility (Katti et al., 2002). The use of polyanhydrides in load-bearing orthopedic applications, however, is restricted due to limited mechanical properties. Poly(anhydrides-co-imides) which were developed in order to combine the good mechanical properties of polyimides with the degradative properties of polyanhydrides were shown to meet compressive strengths comparable to human bone (Uhrich et al., 1995) and displayed good osteocompatibility (Ibim et al., 1998).

Photopolymerizable polyanhydrides have been synthesized with the objective to combine high strength, controlled degradation, and minimal invasive techniques for orthopedic applications and were shown to be osteocompatible (Anseth et al., 1999). Depending on the chemical composition, these materials reached compressive and tensile strengths similar to those of cancelleous bone (Muggli et al., 1999).

Polyphosphazenes

Polyphosphazenes (Figure 20.4j), which are polymers containing a high molecular weight backbone of alternating phosphorus and nitrogen atoms with two organic side groups attached to each phosphorus atom, is a relatively new heterogenic class of biomaterials. Because different synthetic pathways allow for a tremendous variety of substituents, phosphazene polymers exhibit a very diverse spectrum of chemical and physical properties. This spectrum makes them suitable for many biomedical applications ranging from templates for nerve regeneration, cardiovascular, and dental uses to implantable and controlled release devices (Langone et al., 1995; Schacht et al., 1996; Andrianov and Payne, 1998).

The best studied and most important route to polyphosphazenes, whose synthesis is generally more involved than that for most petrochemical biomaterials but offers unique flexibility, is a macromolecular substitution route. A reactive polymeric intermediate, poly(dichlorophosphazene), is typically synthesized by a thermal ring-opening cationic polymerization of hexachlorocyclotriphosphazene in bulk at 250°C that yields a polydisperse high molecular weight product. The intermediate is reacted with low molecular weight organic nucleophiles resulting in stable, substituted polyphosphazenes, which in this case are also addressed as poly(organo)phosphazenes. Depending on the substituent chemistry, the polyphosphazene is more or less susceptible to hydrolysis. Biodegradable hydrophobic polyphosphazenes have been synthesized using imidazolyl, ethylamino, oligopeptides, amino acid esters, and depsipeptide groups (dimers composed of an amino acid and a glycolic or lactic ester) as hydrolysis sensitive side groups. Hydrolytic degradation products include free side group units, phosphate, and ammonia due to backbone degradation (Andrianov and Payne, 1998). Hydrogel-forming, hydrophilic polyphosphazenes can be synthesized through the introduction of small, hydrophilic side groups, such as glucosyl, glyceryl, or methylamino side groups. Ionic side groups yield polymers that form hydrogels upon ionic complexation with multivalent ions (Allcock and Kwon, 1989). Hydrophilic, water-soluble polyphosphazenes with amphiphilic side groups, such as poly(bis(methoxyethoxyethoxy)phosphazene) (Figure 20.4j, R,R′: —$OCH_2CH_2OCH_2CH_2OCH_3$), display a LCST (section "Poly(*N*-isopropylacrylamide)") and are responsive to changes in temperature and ionic strength (Lee, 1999). Both hydrophilic and hydrophobic polyphosphazenes have demonstrated their potential as biocompatible materials for controlled protein delivery. Ionic polyphosphazenes have been explored as vaccine delivery systems and poly(di(carboxylatophenoxy)phosphazene) has demonstrated a remarkable adjuvant activity on the immunogenicity of inactivated influenza virions and commercial trivalent influenza vaccine in the soluble state (Andrianov and Payne, 1998).

Porous scaffolds from biodegradable polyphosphazenes have been shown to be good substrates for osteoblast-like cell attachment and growth with regard to skeletal tissue regeneration (Laurencin et al., 1996). Tubular polyphosphazene nerve guides were investigated in a rat sciatic nerve defect. After 45 days, a regenerated nerve fiber bundle was found bridging the nerve stumps in all cases (Langone et al., 1995).

Biodegradable Crosslinked Polymer Networks

The chemical crosslinking of individual, linear polymer chains results in networks of increased stability. This concept has been extensively explored for applications in regenerative medicine and most likely represents the concept of choice for modern biomaterial research, especially if polymer crosslinking can be conducted inside a tissue defect (Temenoff and Mikos, 2000). The crosslinking of hydrophobic polymers or monomers results in tough polymer networks that can be used for orthopedic fixation. PMMA (Figure 20.1f), the main component in injectable bone cements, is the most prominent example. Due to their hydrophobicity, the precursors are typically injected as a moldable liquid or paste free of additional solvents. *In situ* crosslinking can be initiated thermally or photo-chemically by UV-rich light. Both ways of initiation are also applicable to hydrophilic injectable systems that form highly swollen gels (hydrogels) as a result of precursor crosslinking. In contrast to hydrophobic networks that scarcely swell in the presence of water, injectable hydrogels are characterized by a high water content and diffusivity, which allow for the direct encapsulation of cells and sufficient transport of oxygen, nutrients, and waste. Hydrophobic networks, however, often require the addition of a leachable porogen, such as salt particles, to facilitate cell migration and tissue ingrowth. Generally, injectable polymer systems have considerable advantages over pre-fabricated implants or tissue engineering scaffolds, which include the ability to fill irregularly shaped defects with minimal surgical intervention (Peter et al., 1998a).

A number of demanding requirements have to be fulfilled by synthetic materials for applications in regenerative medicine. Not only do the physicochemical properties have to be adjusted to the application site, but also the polymer and any adjuvant component required to formulate an *in situ* crosslinkable system have to be biocompatible. Ideally, the resulting network should also have the ability to support cell growth and proliferation early in the tissue regeneration process (Temenoff and Mikos, 2000).

The crosslinkable synthetic polymers that will be discussed in the following sections are reactive polyesters. The main chemical functionality involved in the chemical crosslinking mechanisms is the polarized, electron-poor double bond, such as in vinylsulfones and in esters of acrylic acid, methacrylic acid, and fumaric acid. Other chemically or thermally crosslinkable macromonomer functional groups are styryl, coumarin, and phenylazide and will not be discussed here (Hou et al., 2004).

Crosslinked Polyesters

Fumarate-based polymers: The development of fumarate-based polyesters for biomedical applications started around 20 years ago. Fumaric acid is a naturally occurring metabolite, which is found in the tri-carboxylate cycle (Krebs cycle), and is comprised of a reactive double bound available for chemically crosslinking reactions. These characteristics make fumaric acid a candidate building block for crosslinkable polymers. The first and most comprehensively investigated fumarate-based copolymer is the biodegradable copolyester poly(propylene fumarate) (PPF) (Figure 20.5a). PPF was first polymerized from fumaric acid and propylene oxide (Domb et al., 1990). Mikos and coworkers optimized the synthesis of PPF and broadly investigated tissue

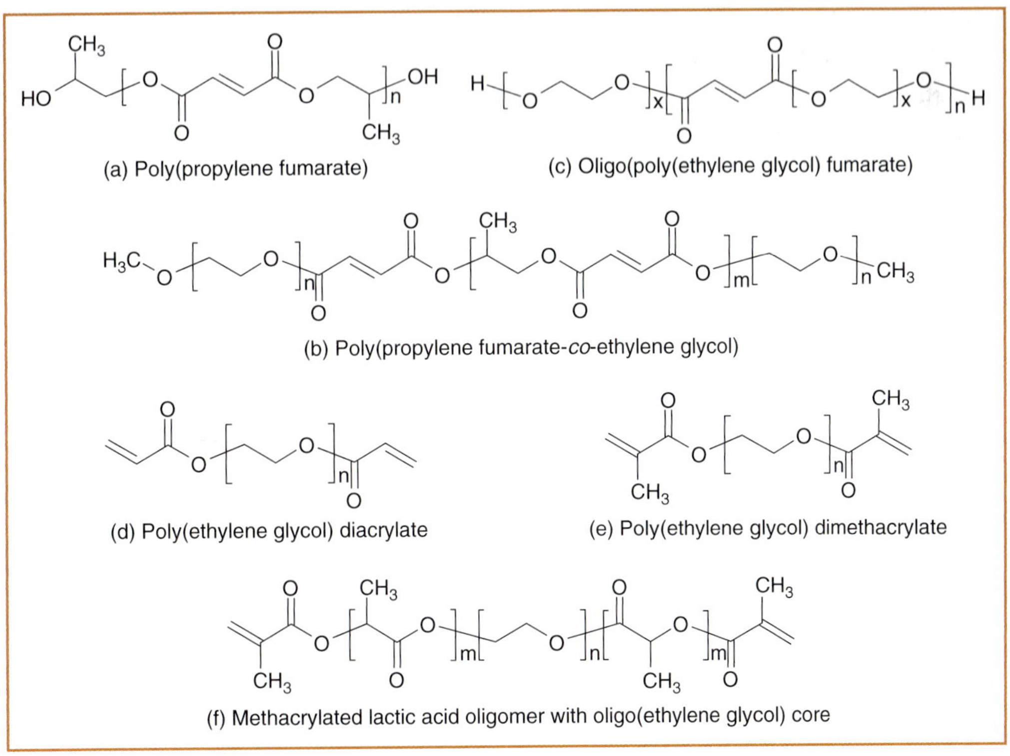

Figure 20.5 Chemical structures of synthetic polymers for the fabrication of crosslinked biodegradable networks.

compatibility and applications of PPF both *in vitro* and *in vivo*. Synthesis progressed to copolymerizing fumaryl chloride with 1,2-propanediol (propylene glycol) (Peter et al., 1999b) and now involves the transesterification of diethylfumarate with propylene glycol and subsequent polycondensation of the diester intermediate bis(2-hydroxypropyl) fumarate (PF) (Shung et al., 2003). A variety of methods to synthesize PPF have been explored, and each results in different polymer molecular weights and properties (Peter et al., 1997a). PPF has been developed as an alternative to PMMA bone cements. PPF can be injected as a viscous liquid and thermally crosslinked *in vivo* eliminating the need for direct exposure of the defect site to light. Typically, PPF is crosslinked with either MMA or *N*-vinyl pyrrolidone (NVP) monomers and benzoyl peroxide as a radical initiator (Gresser et al., 1995; Frazier et al., 1997). Depending on the ratio of initiator, monomer, and PPF, the curing time can be controlled between 1 and 121 min. Compared to PMMA, which is not resorbable and suffers from the fact that its high curing temperatures (94°C) can cause necrosis of the surrounding tissue, the curing temperature of PPF has been shown to never exceed 48°C (Peter et al., 1997b, 1999a). PPF can also be photo-crosslinked along the electron-poor double bonds along the backbone. Typical formulations include NVP, diethylfumarate, or PF-diacrylate (DA) as co-monomers together with a photoinitiator, such as bis(2,4,6-trimethylbenzoyl) phenylphosphine oxide (Fisher et al., 2001, 2002a; He et al., 2001). The mechanical properties of PPF, which are dependent on composition, synthesis condition, and crosslinking density, are already promising. However, these materials are probably not sufficient for load-bearing applications, especially when used as macro-porous scaffolds (Peter et al., 1998a; Fisher et al., 2002a; Timmer et al., 2003). One strategy to further strengthen PPF scaffolds includes the incorporation of nanoparticulate fillers. Reinforced PPF composites have been synthesized using aluminum oxide-based ceramic nanoparticles and modified single walled carbon nanotubes. For just 0.05 wt% loading with the latter, a 74% increase was recorded for the compressive modulus and a 69% increase for the flexural modulus as compared to plain PPF/PF–DA (Shi et al., 2005). The chemical integrations of alumoxane nanoparticles in crosslinked PPF/PF–DA networks resulted in a significantly increased flexural modulus (Horch et al., 2004). Micro-particulate ceramic materials, such as β-tricalcium phosphate (β-TCP), have also been employed as inorganic filler to improve mechanical properties of composite scaffolds and to improve the material's osteoconductivity (Peter et al., 2000). The composite scaffolds exhibit increased compressive strengths in the range of 2–30 MPa, and β-TCP reinforcement delayed scaffold disintegration significantly *in vivo* (Peter et al., 1998b). This subcutaneous rat implantation study also revealed a mild initial inflammatory response and formation of a fibrous capsule around the implant at 12 weeks. A deleterious long-term inflammatory response was not observed. Rabbit *in vivo* studies also revealed biocompatibility of photo-crosslinked PPF scaffolds in both soft and hard tissues (Fisher et al., 2002b).

PPF hydrolytically degrades along the ester bond in its backbone. Degradation time was found to be dependent on polymer structure as well as other components, such as fillers. *In vitro* studies identified the time needed to reach 20% original mass ranging from around 84 (PPF/β-TCP composite) to over 200 days (PPF/$CaSO_4$ composite) (Temenoff and Mikos, 2000).

In order to broaden the application spectrum for *in situ* crosslinkable PPF, block copolymers with hydrophilic PEG of different compositions were synthesized. Poly(propylene fumarate-*co*-ethylene glycol) (P(PF-*co*-EG)) (Figure 20.5b) was synthesized from PPF and PEG in a transesterification reaction catalyzed by antimony trioxide; propylene glycol was removed by condensation (Suggs et al., 1997). Behravesh et al. have modified the synthesis to yield well-defined ABA-type triblock copolymers from two moles monomethoxy-PEG and one mole PPF (Behravesh et al., 2002a). Generally, P(PF-*co*-EG) copolymers are hydrophilic polymers with specific properties including crystallinity and mechanical characteristics being dependent on the molecular weights of the individual blocks and the copolymer. As a result, platelet attachment to P(PF-*co*-EG) hydrogels was significantly reduced as compared to the PPF homopolymer making these copolymers candidate materials when direct biomaterial–blood contact is inevitable, such as for vascular grafts (Suggs et al., 1999b).

Most P(PF-*co*-EG) copolymers are soluble in water making them candidate materials for injectable applications. ABA-type copolymers were found to show thermoreversible properties, comparable to other PEG-containing triblock copolymers discussed above. The thermogelling properties of P(PF-*co*-EG) were dependent on the PEG molecular weight and salt concentration and the physical gelation temperature could be adjusted to values below body temperature (Behravesh et al., 2002a). In addition, the hydrophobic PPF block is highly unsaturated and available for additional chemical crosslinking, which could result in stiff crosslinked networks suitable for the fabrication of prefabricated cell carriers. *In vitro* degradation studies of macro-porous, crosslinked P(PF-*co*-EG) scaffolds revealed considerable mass loss and swelling over 12 weeks. In these studies the degradation rate was mainly dependent on content of the PEG–DA crosslinker and almost unaffected by construct porosity. Overall, the results indicated a bulk degradation mechanism of the macro-porous constructs (Behravesh et al., 2002b). In a subcutaneous rat model, P(PF-*co*-EG) hydrogels demonstrated good initial biocompatibility, showing an acute inflammatory response characterized by infiltration of neutrophils, followed by development and maturation of a fibrous capsule, characteristic of biomaterial implants (Suggs et al., 1999a). Overall, the reported *in vitro* cytotoxicity and *in vivo* biocompatibility assays suggest that P(PF-*co*-EG) hydrogels have potential for use as injectable biomaterials. Fisher et al. have demonstrated the suitability of thermoresponsive P(PF-*co*-EG) hydrogels for chondrocyte delivery toward the regeneration of articular cartilage defects (Fisher et al., 2004).

Similar to previously discussed, stealthy, PEG-containing biodegradables, PEG-content and hydrophilicity of crosslinked P(PF-*co*-EG) hydrogels are critical factors affecting cell adhesion (Tanahashi and Mikos, 2002). Low-adhesive hydrogels allow for a controlled surface or bulk modification with adhesion molecules to specifically enhance cell adhesion. P(PF-*co*-EG) hydrogels have been modified by covalent integration of agmatine (Tanahashi and Mikos, 2003) and the adhesion peptide GRGDS (Behravesh et al., 2003). Significantly increased numbers of smooth muscle cells and MSCs were found adhered as compared to the unmodified networks.

An exclusively hydrophilic fumarate-based macromer is oligo(poly(ethylene glycol) fumarate) (OPF) (Figure 20.5c). OPF macromers have been synthesized from PEG and fumaryl chloride by a simple condensation reaction in the presence of triethylamine. OPF crosslinking, with or without the addition of crosslinker such as PEG–DA, can be initiated photo-chemically (Jo et al., 2001) or thermally (Temenoff et al., 2002). In contrast to chemically crosslinked PPF and P(PF-*co*-EG), which both form rigid scarcely swelling polymer networks, crosslinked OPF gels exhibit typical properties of hydrogels, which were dependent on the molecular weight of PEG and reactant ratio (Jo et al., 2001). Crosslinked OPF hydrogels degrade hydrolytically along the ester bonds between fumaric acid and PEG resulting in increased polymer swelling and decreased dry weight. The weight loss of OPF hydrogels was dependent on their crosslinking density (Shin et al., 2003c). Studies investigating the mechanical properties revealed that crosslinked OPF hydrogels made from low molecular weight PEG (1,000 Da), swelled less, were stiffer, and elongated less before fracture when compared to hydrogels comprised of longer PEG chains. OPF hydrogels can also be combined in layers to form biphasic gels, with each phase having different material properties (Temenoff et al., 2002). *In vitro* investigation of the cytotoxicity of each component of OPF hydrogel formulations and the resulting crosslinked network were conducted employing MSCs. After 24 h, the MSCs maintained more than 75% viability except for OPF concentrations higher than 25% (w/v). A high molecular weight (3,400 Da) PEG–DA crosslinker demonstrated significantly higher viability compared to lower molecular weight (575 Da) PEG–DA. Leachable products from crosslinked OPF hydrogels were found to have minimal adverse effects on MSC viability (Shin et al., 2003a). The *in vivo* bone and soft tissue compatibility of OPF hydrogels was demonstrated using a rabbit model (Shin et al., 2003c). Based on these promising biocompatibility data, OPF-based hydrogels were investigated as injectable drug, DNA, and cell delivery devices. Crosslinked OPF hydrogels which encapsulated gelatin microparticles were developed as a means of simultaneously delivering two chondrogenic proteins, insulin-like growth factor-1 (IGF-1) and transforming growth factor-β1 (TGF-β1) (Holland et al., 2005b). Similar systems were implanted into osteochondral defects in the rabbit model. No evidence of prolonged inflammation was observed, and hyaline cartilage was found filling the chondral region of the defect at 14 weeks. The subchondral region was filled with bony tissue and completely integrated with the surrounding bone. The newly formed surface tissue stained positive for Safranin O and displayed promising chondrocyte organization (Holland et al., 2005a). Kasper et al. developed and characterized composites of OPF and cationized gelatin microspheres that release plasmid DNA in a sustained, controlled manner *in vivo* (Kasper et al., 2005). In order to control cell adhesion to the hydrophilic hydrogels, RGD adhesion peptide modified OPF hydrogels have been developed (Shin et al., 2002). OPF hydrogels have also been shown useful as injectable cell delivery vehicles for bone regeneration. MSCs were directly combined with the OPF hydrogel precursors and encapsulated during thermal crosslinking. In the presence of osteogenic supplements, MSC differentiation in these hydrogels was apparent by day 21. At day 28, mineralized matrix could be seen throughout the hydrogels (Temenoff et al., 2004a). Hydrogel properties have been identified to affect osteogenic differentiation within these systems (Temenoff et al., 2004b). Recent studies focused on the combination of cell and growth factor delivery using injectable OPF formulations (Park et al., 2005).

Polymers-containing acrylate, methacrylate, or vinylsulfone functionalities: Precursors for crosslinked biodegradable polyester networks that bear vinylsulfone, acrylate, or methacrylate functionalities include PEG–DA (Figure 20.5d), PEG–dimethacrylate (Figure 20.5e), PEG vinylsulfones, diacrylated PLA–PEG–PLA block copolymers, acrylic modified polyvinyl alcohol (PVA), methacrylate-modified dextran, and acrylated chitosan (Hoffman, 2002; Nguyen and West, 2002; Hou et al., 2004). Since the last two are synthetic derivatives of natural macromolecules, they are not discussed further. Besides such hydrophilic, natural macromolecules, which are considered candidate building blocks based on their inherent biocompatibility, PEG is the most prominent synthetic component of crosslinked polymer networks due to its biocompatibility and inertness. As described above, PEG is hydrophilic and does not promote cell adhesion. To improve cell adhesion to crosslinked PEG hydrogels, adhesion peptides containing the tripeptide motif RGD have been incorporated (Hern and Hubbell, 1998; Burdick and Anseth, 2002; Gonzalez et al., 2004). Recent research on engineered hydrogels has been focused on mimicking the invasive characteristics of native ECMs by including substrates for matrix metalloproteinases (MMP) in addition to integrin-binding sites. PEG hydrogels crosslinked in part by MMP sensitive linkers were made degradable and invasive for cells via cell-secreted MMPs (Lutolf et al., 2003a). Critical-sized defects in rat crania were completely infiltrated by cells and were remodeled into bony tissue within 5 weeks when above-mentioned gels were loaded with recombinant human bone morphogenetic protein-2 and implanted in the defect site. As in natural ECMs, that sequester a variety of cellular growth factors and act as a local depot for them, invading cells were presented with a mitogen that, in this case, specifically promoted bone regeneration (Lutolf et al., 2003b). The PEG-based hydrogels used in these studies were fabricated by a conjugate addition reaction between vinylsufone-functionalized branched PEG and thiol-bearing peptides under almost physiological conditions.

In order to enhance the initial mechanical stability and biodegradability of crosslinked PEG-based hydrogels, oligomeric biodegradable lipophilic blocks, such as oligo(lactic acid) (Burdick et al., 2001) (Figure 20.5f) and oligo(ε-caprolactone) (Davis et al., 2003), were included in the crosslinkable polymeric precursors. In a critical size cranial defect model, porous crosslinked poly(ethylene glycol(2)-lactic acid(10)) scaffolds in combination with osteoinductive growth factors have shown potential as an *in situ* forming synthetic bone graft material (Burdick et al., 2003).

Photopolymerized (meth)acrylated biodegradable hydrogels have been used in a wide range of biomedical applications. As described above, limited interactions with proteins are characteristic for hydrophilic surfaces. Consequently, applications such as the use of crosslinked hydrogels as barriers following tissue injury in order to improve wound healing and as cell encapsulation materials to immunoisolate transplanted cells capitalize of this property (Cruise et al., 1999; Nguyen and West, 2002). Islets of Langerhans encapsulated in PEG–DA hydrogels and transplanted in order to develop a bioartificial endocrine pancreas are a prominent example for the later application. The hydrogels are permeable for nutrients, oxygen, and metabolic products allowing for the entrapped islets to survive and to secrete insulin that is released by diffusion. Hydrophilic tissue barriers from crosslinked polyesters, such as poly(ethylene glycol-*co*-lactic acid) DA, have been used to prevent thrombosis and re-stenosis following vascular injury and postoperative adhesion formation following many abdominal and pelvic surgical procedures.

Crosslinked hydrophilic polyesters are also promising depots for local drug delivery because of their compatibility with hydrophilic, macromolecular drugs, such as proteins or oligonucleotides. The materials' good tissue and hemocompatibility even allows for intravascular applications (An and Hubbell, 2000). Drug release from crosslinked hydrogels generally can be well controlled by adjusting swelling, crosslink density, and polymer degradation (Peppas et al., 1999, 2000; Davis and Anseth, 2002).

Photopolymerized (meth)acrylated polymer networks have also been widely explored for injectable tissue engineering (Hoffman, 2002; Varghese and Elisseeff, 2006). Elisseeff and coworkers employed PEG–DA scaffolds for cartilage engineering by encapsulating chondrocytes, MSCs, and embryonic stem cells. In these studies, the crosslinked PEG-based hydrogels served as an efficient scaffold for anchorage-independent cells and promoted tissue formation. Photogelation, which offers good spatial and temporal control of hydrogel curing, has been used to control the spatial organization of different cell types within a three-dimensional system for osteochondral defect regeneration by sequentially polymerizing multiple cell/hydrogel layers. In an attempt to promote hydrogel–tissue integration, a tissue-initiated polymerization technique has been developed that utilizes *in situ* generated tyrosyl radicals to initiate photogelation of an injectable macromer solution (Varghese and Elisseeff, 2006).

Traditionally, photopolymerization occurs by directly exposing materials to UV or visible light in accessible cavities or during invasive surgery. For PEG–dimethacrylate hydrogels, it has been shown that light, which penetrates tissue including skin, can cause a photopolymerization indirectly (transdermal photopolymerization). *In vivo* studies revealed that gels can be polymerized in 3 min with no harm to imbedded chondrocytes and subsequent cartilaginous tissue formation as indicated by increasing GAG and collagen contents (Elisseeff et al., 1999). In deep crevices, as they may be found in larger orthopedic defects, problems are expected to arise from limited light penetration and inconsistent photopolymerization. For those applications, thermally induced crosslinking techniques appear to be advantageous (Temenoff and Mikos, 2000).

APPLICATIONS OF SYNTHETIC POLYMERS

Synthetic polymers play a vital role in biomedical applications, including nano-, micro-, and macroscopic drug and gene delivery devices (Brannon-Peppas, 1995; Hubbell, 1998; Uhrich et al., 1999; Panyam and Labhasetwar, 2003), orthopedic fixation devices (Bostman and Pihlajamaki, 2000), cosmetic, and prosthetic implants (Behravesh et al., 1999), and as artificial matrices for tissue engineering applications (Seal et al., 2001). The interested reader may be directed to the referenced reviews that provide in-depth insight in current trends and technologies. Researches have sought to develop and clinically explore third generation biomaterials (Hench and Polak, 2002) that are designed to control protein adsorption, cell adhesion, and differentiation, implant integration, foreign body reaction, and to develop biomimetic synthetic materials (Shin et al., 2003b; Drotleff et al., 2004; Lutolf and Hubbell, 2005).

CONCLUSION/SUMMARY

Synthetic biomaterials have progressed from testing "off-the-shelf" plastics not developed for biomedical purposes, to a field of synergistic research by engineers, scientists, and physicians dedicated to tailoring material properties for specific applications. Most recent trends shift the focus toward biology in order to first understand and then mimic physiological interactions and signaling.

Hydrogels, especially injectable systems, enjoy increasing attention due to the comfort of their application, their structural similarity to native ECM, and their good compatibility for direct cell encapsulation due to high water contents. It is no longer believed in tissue engineering that the biomaterial itself has to provide mechanical properties comparable to the diseased tissue; the polymer rather has to promote defect site remodeling and tissue regeneration *in vivo* in a way that the regenerated tissue is histologically and functionally indistinguishable from the surrounding tissue. Hydrogels might be superior to hydrophobic polymers in that regard, as they can degrade faster resolving the problem of non-functional fibrous tissue formation on the polymer–tissue interface. Also, hydrogel breakdown can be synchronized with cell proliferation and migration by using enzymatically cleavable crosslinker.

Besides providing tailored degradative properties, synthetic materials for regenerative medicine should allow for minimally invasive application techniques, integrate well with the surrounding tissue, and promote cell adhesion, migration, and finally differentiation. The development and thorough characterization of injectable biodegradables provides the foundation for injectable tissue regeneration. *In situ* gelation or polymerization concepts will still have to be developed and optimized with regard to cytocompatibility and stability of the resulting construct. The implementation of biomimetic design strategies will allow to control and custom-design cell–biomaterial interactions in order to guide tissue formation from transplanted cells. Strategies based on gene delivery or gene-activating biomaterials also have great potential in regenerative medicine but the long-term safety of such therapies remains to be proven.

Overall, the advances that have been made in the field of biomaterial synthesis and design of physicochemical properties during the last 50 years in conjunction with the rapidly increasing knowledge in adult and stem cell biology concerning adhesion, migration, differentiation, and signaling will reveal design concepts for improved injectable, biomimetic polymer-based formulations for tissue engineering applications.

REFERENCES

Agrawal, C.M., Athanasiou, K.A. and Heckman, J.D. (2000). Biodegradable PLA–PGA polymers for tissue engineering in orthopaedics. *Mater. Sci. Forum* 1997; **250**: 115–128.

Allcock, H.R. and Kwon, S. (1989). Ionically cross-linkable polyphosphazene: poly[bis(carboxylatophenoxy)phosphazene] and its hydrogels and membranes. *Macromolecules* **22**: 75–79.

Amecke, B., Bendix, D. and Entenmann, G. (1992). Resorbable polyesters: composition, properties, applications. *Clin. Mater.* **10**: 47–50.

An, Y. and Hubbell, J.A. (2000). Intraarterial protein delivery via intimally-adherent bilayer hydrogels. *J. Control. Release* **64**: 205–215.

Anderson, J.M. and Shive, M.S. (1997). Biodegradation and biocompatibility of PLA and PLGA microspheres. *Adv. Drug Deliver. Rev.* **28**: 5–24.

Andriano, K.P., Tabata, Y., Ikada, Y. and Heller, J. (1999). *In vitro* and *in vivo* comparison of bulk and surface hydrolysis in absorbable polymer scaffolds for tissue engineering. *J. Biomed. Mater. Res.* **48**: 602–612.

Andrianov, A.K. and Payne, L.G. (1998). Protein release from polyphosphazene matrices. *Adv. Drug Deliver. Rev.* **31**: 185–196.

Anseth, K.S., Shastri, V.R. and Langer, R. (1999). Photopolymerizable degradable polyanhydrides with osteocompatibility. *Nat. Biotechnol.* **17**: 156–159.

Arepalli, S.R., Bezabeh, S. and Brown, S.L. (2002). Allergic reaction to platinum in silicone breast implants. *J. Long Term. Eff. Med. Implants* **12**: 299–306.

Athanasiou, K.A., Niederauer, G.G. and Agrawal, C.M. (1996). Sterilization, toxicity, biocompatibility and clinical applications of polylactic acid/polyglycolic acid copolymers. *Biomaterials* **17**: 93–102.

Bazile, D., Prud'homme, C., Bassoullet, M.T., Marlard, M., Spenlehauer, G. and Veillard, M. (1995). Stealth Me.PEG–PLA nanoparticles avoid uptake by the mononuclear phagocytes system. *J. Pharm. Sci.* **84**: 493–498.

Behravesh, E., Yasko, A.W., Engel, P.S. and Mikos, A.G. (1999). Synthetic biodegradable polymers for orthopaedic applications. *Clin. Orthop. Relat. Res.* **367**(Suppl.): S118–S129.

Behravesh, E., Shung, A.K., Jo, S. and Mikos, A.G. (2002a). Synthesis and characterization of triblock copolymers of methoxy poly(ethylene glycol) and poly(propylene fumarate). *Biomacromolecules* **3**: 153–158.

Behravesh, E., Timmer, M.D., Lemoine, J.J., Liebschner, M.A.K. and Mikos, A.G. (2002b). Evaluation of the *in vitro* degradation of macroporous hydrogels using gravimetry, confined compression testing, and microcomputed tomography. *Biomacromolecules* **3**: 1263–1270.

Behravesh, E., Zygourakis, K. and Mikos, A.G. (2003). Adhesion and migration of marrow-derived osteoblasts on injectable *in situ* crosslinkable poly(propylene fumarate-co-ethylene glycol)-based hydrogels with a covalently linked RGDS peptide. *J. Biomed. Mater. Res. A* **65**: 260–270.

Beniash, E., Hartgerink, J.D., Storrie, H., Stendahl, J.C. and Stupp, S.I. (2005). Self-assembling peptide amphiphile nanofiber matrices for cell entrapment. *Acta Biomater.* **1**: 387–397.

Betre, H., Setton, L.A., Meyer, D.E. and Chilkoti, A. (2002). Characterization of a genetically engineered elastin-like polypeptide for cartilaginous tissue repair. *Biomacromolecules* **3**: 910–916.

Bobyn, J.D., Mortimer, E.S., Glassman, A.H., Engh, C.A., Miller, J.E. and Brooks, C.E. (1992). Producing and avoiding stress shielding: laboratory and clinical observations of noncemented total hip arthroplasty. *Clin. Orthop. Relat. Res.* **274**: 79–96.

Boretos, J.W. and Pierce, W.S. (1967). Segmented polyurethane: a new elastomer for biomedical applications. *Science* **158**: 1481–1482.

Bostman, O. and Pihlajamaki, H. (2000). Clinical biocompatibility of biodegradable orthopaedic implants for internal fixation: a review. *Biomaterials* **21**: 2615–2621.

Bouillot, P., Petit, A. and Dellacherie, E. (1998). Protein encapsulation in biodegradable amphiphilic microspheres. I. Polymer synthesis and characterization and microsphere elaboration. *J. Appl. Polym. Sci.* **68**: 1695–1702.

Bourke, S.L. and Kohn, J. (2003). Polymers derived from the amino acid tyrosine: polycarbonates, polyarylates and copolymers with poly(ethylene glycol). *Adv. Drug Deliver. Rev.* **55**: 447–466.

Brannon-Peppas, L. (1995). Recent advances on the use of biodegradable microparticles and nanoparticles in controlled drug delivery. *Int. J. Pharm.* **116**: 1–9.

Brook, M.A. (2006). Platinum in silicone breast implants. *Biomaterials* **27**: 3274–3286.

Brunner, A., Mader, K. and Gopferich, A. (1999). pH and osmotic pressure inside biodegradable microspheres during erosion. *Pharm. Res.* **16**: 847–853.

Burdick, J.A. and Anseth, K.S. (2002). Photoencapsulation of osteoblasts in injectable RGD-modified PEG hydrogels for bone tissue engineering. *Biomaterials* **23**: 4315–4323.

Burdick, J.A., Philpott, L.M. and Anseth, K.S. (2001). Synthesis and characterization of tetrafunctional lactic acid oligomers: a potential *in situ* forming degradable orthopaedic biomaterial. *J. Polym. Sci. A* **39**: 683–692.

Burdick, J.A., Frankel, D., Dernell, W.S. and Anseth, K.S. (2003). An initial investigation of photocurable three-dimensional lactic acid based scaffolds in a critical-sized cranial defect. *Biomaterials* **24**: 1613–1620.

Burkersroda, F.v., Schedl, L. and Gopferich, A. (2002). Why degradable polymers undergo surface erosion or bulk erosion. *Biomaterials* **23**: 4221–4231.

Byrne, M.E., Park, K. and Peppas, N.A. (2002). Molecular imprinting within hydrogels. *Adv. Drug Deliver. Rev.* **54**: 149–161.

Cannizzaro, S.M., Padera, R.F., Langer, R., Rogers, R.A., Black, F.E., Davies, M.C., Tendler, S.J. and Shakesheff, K.M. (1998). A novel biotinylated degradable polymer for cell-interactive applications. *Biotechnol. Bioeng.* **58**: 529–520.

Cerrai, P., Guerra, G.D., Lelli, L., Tricoli, M., Sbarbati Del Guerra, R., Cascone, M.G. and Giusti, P. (1994). Poly(ester-ether-ester) block copolymers as biomaterials. *J. Mater. Sci. Mater. Med.* **5**: 33–39.

Chandrasekaran, S.K., Capozza, R. and Wong, P.S.L. (1978). Therapeutic systems and controlled drug delivery. *J. Membrane Sci.* **3**: 271–286.

Chirila, T.V., Constable, I.J., Crawford, G.J., Vijayasekaran, S., Thompson, D.E., Chen, Y.C., Fletcher, W.A. and Griffin, B.J. (1993). Poly(2-hydroxyethyl methacrylate) sponges as implant materials: *in vivo* and *in vitro* evaluation of cellular invasion. *Biomaterials* **14**: 26–38.

Choueka, J., Charvet, J.L., Koval, K.J., Alexander, H., James, K.S., Hooper, K.A. and Kohn, J. (1996). Canine bone response to tyrosine-derived polycarbonates and poly(L-lactic acid). *J. Biomed. Mater. Res.* **31**: 35–41.

Colas, A. and Curtis, J. (2004). Silicone biomaterials: history and chemistry. In: Ratner, B.D., Hoffman, A.S., Schoen, F.J. and Lemons, J.E. (eds.), *Biomaterials Science. An Introduction to Materials in Medicine*, 2nd edn. San Diego, CA: Academic Press, pp. 80–86.

Cruise, G.M., Hegre, O.D., Lamberti, F.V., Hager, S.R., Hill, R., Scharp, D.S. and Hubbell, J.A. (1999). *In vitro* and *in vivo* performance of porcine islets encapsulated in interfacially photopolymerized poly(ethylene glycol) diacrylate membranes. *Cell Transplant.* **8**: 293–306.

Curtis, J. and Colas, A. (2004). Medical applications of silicones. In: Ratner, B.D., Hoffman, A.S., Schoen, F.J. and Lemons, J.E. (eds.), *Biomaterials Science. An Introduction to Materials in Medicine*, 2nd edn. San Diego, CA: Academic Press, pp. 697–707.

Cutright, D.E., Beasley, I.I.I. and Perez, B. (1971). Histologic comparison of polylactic and polyglycolic acid sutures. *Oral Surg. Oral Med. Oral Pathol.* **32**: 165–173.

Davis, K.A. and Anseth, K.S. (2002). Controlled release from crosslinked degradable networks. *Crit. Rev. Ther. Drug Carrier Syst.* **19**: 385–423.

Davis, K.A., Burdick, J.A. and Anseth, K.S. (2003). Photoinitiated crosslinked degradable copolymer networks for tissue engineering applications. *Biomaterials* **24**: 2485–2495.

Domb, A.J., Laurencin, C.T., Israeli, O., Gerhart, T.N. and Langer, R. (1990). Formation of propylene fumarate oligomers for use in bioerodible bone cement composites. *J. Polym. Sci. A* **28**: 973–985.

Drotleff, S., Lungwitz, U., Breunig, M., Dennis, A., Blunk, T., Tessmar, J. and Gopferich, A. (2004). Biomimetic polymers in pharmaceutical and biomedical sciences. *Eur. J. Pharm. Biopharm.* **58**: 385–407.

Elisseeff, J., Anseth, K., Sims, D., McIntosh, W., Randolph, M. and Langer, R. (1999). Transdermal photopolymerization for minimally invasive implantation. *Proc. Natl. Acad. Sci. USA* **96**: 3104–3107.

Engelberg, I. and Kohn, J. (1991). Physico-mechanical properties of degradable polymers used in medical applications: a comparative study. *Biomaterials* **12**: 292–304.

Fisher, J.P., Holland, T.A., Dean, D., Engel, P.S., and Mikos, A.G. (2001). Synthesis and properties of photocross-linked poly(propylene fumarate) scaffolds. *J. Biomater. Sci. Polym. Ed.* **12**: 673–687.

Fisher, J.P., Dean, D. and Mikos, A.G. (2002a). Photocrosslinking characteristics and mechanical properties of diethyl fumarate/poly(propylene fumarate) biomaterials. *Biomaterials* **23**: 4333–4343.

Fisher, J.P., Vehof, J.W.M., Dean, D., van der Waerden, J.P., Holland, T.A., Mikos, A.G. and Jansen, J.A. (2002b). Soft and hard tissue response to photocrosslinked poly(propylene fumarate) scaffolds in a rabbit model. *J. Biomed. Mater. Res.* **59**: 547–556.

Fisher, J.P., Jo, S., Mikos, A.G. and Reddi, A.H. (2004). Thermoreversible hydrogel scaffolds for articular cartilage engineering. *J. Biomed. Mater. Res. A* **71**: 268–274.

Frazier, D.D., Lathi, V.K., Gerhart, T.N. and Hayes, W.C. (1997). *Ex vivo* degradation of a poly(propylene glycol-fumarate) biodegradable particulate composite bone cement. *J. Biomed. Mater. Res.* **35**: 383–389.

Freed, L.E., Vunjak, N.G., Biron, R.J., Eagles, D.B., Lesnoy, D.C., Barlow, S.K. and Langer, R. (1994). Biodegradable polymer scaffolds for tissue engineering. *Biotechnology (NY)* **12**: 689–693.

Freed, L.E., Langer, R., Martin, I., Pellis, N.R. and Vunjak-Novakovic, G. (1997). Tissue engineering of cartilage in space. *Proc. Natl. Acad. Sci. USA* **94**: 13885–13890.

Fromstein, J.D. and Woodhouse, K.A. (2006). Polyurethane biomaterials. In: Wnek, G.E. and Bowlin, G.L. (eds.), *Encyclopedia of Biomaterials and Biomedical Engineering.* New York: Marcel Dekker.

Ganta, S.R., Piesco, N.P., Long, P., Gassner, R., Motta, L.F., Papworth, G.D., Stolz, D.B., Watkins, S.C. and Agarwal, S. (2003). Vascularization and tissue infiltration of a biodegradable polyurethane matrix. *J. Biomed. Mater. Res. A* **64**: 242–248.

Goepferich, A. (1996). Polymer degradation and erosion. Mechanisms and applications. *Eur. J. Pharm. Biopharm.* **42**: 1–11.

Goepferich, A. and Tessmar, J. (2002). Polyanhydride degradation and erosion. *Adv. Drug Deliver. Rev.* **54**: 911–931.

Gogolewski, S. (1989). Selected topics in biomedical polyurethanes. A review. *Colloid Polym. Sci.* **267**: 757–785.

Gonzalez, A.L., Gobin, A.S., West, J.L., McIntire, L.V. and Smith, C.W. (2004). Integrin interactions with immobilized peptides in polyethylene glycol diacrylate hydrogels. *Tissue Eng.* **10**: 1775–1786.

Gref, R., Minamitake, Y., Peracchia, M.T., Trubetskoy, V., Torchilin, V. and Langer, R. (1994). Biodegradable long-circulating polymeric nanospheres. *Science* **263**: 1600–1603.

Gref, R., Minamitake, Y., Peracchia, M.T., Domb, A., Trubetskoy, V., Torchilin, V. and Langer, R. (1997). Poly(ethylene glycol)-coated nanospheres: potential carriers for intravenous drug administration. *Pharm. Biotechnol.* **10**: 167–198.

Gref, R., Luck, M., Quellec, P., Marchand, M., Dellacherie, E., Harnisch, S., Blunk, T. and Muller, R.H. (2000). "Stealth" corona-core nanoparticles surface modified by polyethylene glycol (PEG): influences of the corona (PEG chain length and surface density) and of the core composition on phagocytic uptake and plasma protein adsorption. *Colloids Surf. B* **18**: 301–313.

Gresser, J.D., Hsu, S.H., Nagaoka, H., Lyons, C.M., Nieratko, D.P., Wise, D.L., Barabino, G.A. and Trantolo, D.J. (1995). Analysis of a vinyl pyrrolidone/poly(propylene fumarate) resorbable bone cement. *J. Biomed. Mater. Res.* **29**: 1241–1247.

Gunatillake, P.A. and Adhikari, R. (2003). Biodegradable synthetic polymers for tissue engineering. *Eur. Cell. Mater.* **5**: 1–16.

Gunatillake, P.A., Martin, D.J., Meijs, G.F., McCarthy, S.J. and Adhikari, R. (2003). Designing biostable polyurethane elastomers for biomedical implants. *Aust. J. Chem.* **56**: 545–557.

Hacker, M., Tessmar, J., Neubauer, M., Blaimer, A., Blunk, T., Gopferich, A. and Schulz, M.B. (2003). Towards biomimetic scaffolds: anhydrous scaffold fabrication from biodegradable amine-reactive diblock copolymers. *Biomaterials* **24**: 4459–4473.

Harris, J.M. and Chess, R.B. (2003). Effect of pegylation on pharmaceuticals. *Nat. Rev. Drug Discov.* **2**: 214–221.

Hartgerink, J.D., Beniash, E. and Stupp, S.I. (2001). Self-assembly and mineralization of peptide-amphiphile nanofibers. *Science* **294**: 1684–1688.

Hasegawa, M., Sudo, A., Shikinami, Y. and Uchida, A. (1999). Biological performance of a three-dimensional fabric as artificial cartilage in the repair of large osteochondral defects in rabbit. *Biomaterials* **20**: 1969–1975.

He, S., Timmer, M.D., Yaszemski, M.J., Yasko, A.W., Engel, P.S. and Mikos, A.G. (2001). Synthesis of biodegradable poly(propylene fumarate) networks with poly(propylene fumarate)-diacrylate macromers as crosslinking agents and characterization of their degradation products. *Polymer* **42**: 1251–1260.

Heller, J. (1984). Biodegradable polymers in controlled drug delivery. *Crit. Rev. Ther. Drug Carrier Syst.* **1**: 39–90.

Heller, J., Barr, J., Ng, S.Y., Abdellauoi, K.S. and Gurny, R. (2002). Poly(ortho esters): synthesis, characterization, properties and uses. *Adv. Drug Deliver. Rev.* **54**: 1015–1039.

Hench, L.L. (1980). Biomaterials. *Science* **208**: 826–831.

Hench, L.L. and Polak, J.M. (2002). Third-generation biomedical materials. *Science* **295**: 1014–1017.

Hendriks, J.G.E., van Horn, J.R., van der Mei, H.C. and Busscher, H.J. (2004). Backgrounds of antibiotic-loaded bone cement and prosthesis-related infection. *Biomaterials* **25**: 545–556.

Hern, D.L. and Hubbell, J.A. (1998). Incorporation of adhesion peptides into nonadhesive hydrogels useful for tissue resurfacing. *J. Biomed. Mater. Res.* **39**: 266–276.

Hoffman, A.S. (2002). Hydrogels for biomedical applications. *Adv. Drug Deliver. Rev.* **54**: 3–12.

Holland, T.A., Bodde, E.W.H., Baggett, L.S., Tabata, Y., Mikos, A.G. and Jansen, J.A. (2005a). Osteochondral repair in the rabbit model utilizing bilayered, degradable oligo(poly(ethylene glycol) fumarate) hydrogel scaffolds. *J. Biomed. Mater. Res. A* **75**: 156–167.

Holland, T.A., Tabata, Y. and Mikos, A.G. (2005b). Dual growth factor delivery from degradable oligo(poly(ethylene glycol) fumarate) hydrogel scaffolds for cartilage tissue engineering. *J. Control. Release* **101**: 111–125.

Horch, R.A., Shahid, N., Mistry, A.S., Timmer, M.D., Mikos, A.G. and Barron, A.R. (2004). Nanoreinforcement of poly(propylene fumarate)-based networks with surface modified alumoxane nanoparticles for bone tissue engineering. *Biomacromolecules* **5**: 1990–1998.

21

Surface Modification of Biomaterials

Andrés J. García

INTRODUCTION

Biomaterial Interfaces in Regenerative Medicine

Biomaterials, either synthetic (e.g. polymers, metals, ceramics) or natural (e.g. proteins, polysaccharides), play central roles in tissue engineering and regenerative medicine applications by providing (i) three-dimensional scaffolds to support cellular activities; (ii) matrices for delivery of therapeutic agents (e.g. drugs, proteins, DNA, siRNA); and (iii) functional device components (e.g. mechanical supports, sensing/stimulating elements, non-thrombogenic surfaces, diffusional barriers). The bulk properties of the biomaterial are critical determinants of the biological performance of the material (Ratner et al., 2004). For example, the mechanical properties of a vascular substitute, including elastic modulus, ultimate tensile stress, and compliance, dictate the ability of this tissue construct to support the applied mechanical loads associated with blood flow. On the other hand, the biological response to a biomaterial is governed by the material surface properties, primarily surface chemistry and structure. Protein adsorption/activation and cell adhesion, events that regulate host responses to materials, occur at the biomaterial–tissue interface, and the physicochemical properties of the material surface modulate these biological events (Anderson, 2001). For instance, the chemical properties of the surface of a vascular substitute control blood compatibility (i.e. protein adsorption, platelet adhesion, thrombogenicity, patency). Hence, modification of biomaterial surfaces represents a promising route to engineer biofunctionality at the material–tissue interface in order to modulate biological responses without altering material bulk properties.

Overview of Surface Modification Strategies

Numerous surface modification approaches have been developed for all classes of materials to modulate biological responses and improve device performance. Applications include reduction of protein adsorption and thrombogenicity, control of cell adhesion, growth and differentiation, modulation of fibrous encapsulation and osseointegration, improved wear and/or corrosion resistance, and potentiation of electrical conductivity (Ratner et al., 2004). Surface modifications fall into two general categories: (i) physicochemical modifications involving alterations to the atoms, compounds, or molecules on the surface; and (ii) surface coatings consisting of a different material from the underlying support. Physicochemical modifications include chemical reactions (e.g. oxidation, reduction, silanization, acetylation), etching, and mechanical roughening/polishing and patterning (Figure 21.1). Overcoating alterations comprise grafting (including tethering of biomolecules), non-covalent and covalent coatings, and thin film deposition (Figure 21.2).

While the specific requirements of the surface modification approach vary with application, several characteristics are generally desirable. Thin surface modifications are preferred for most applications since thicker

Copyright © 2008 Elsevier Inc. All rights reserved

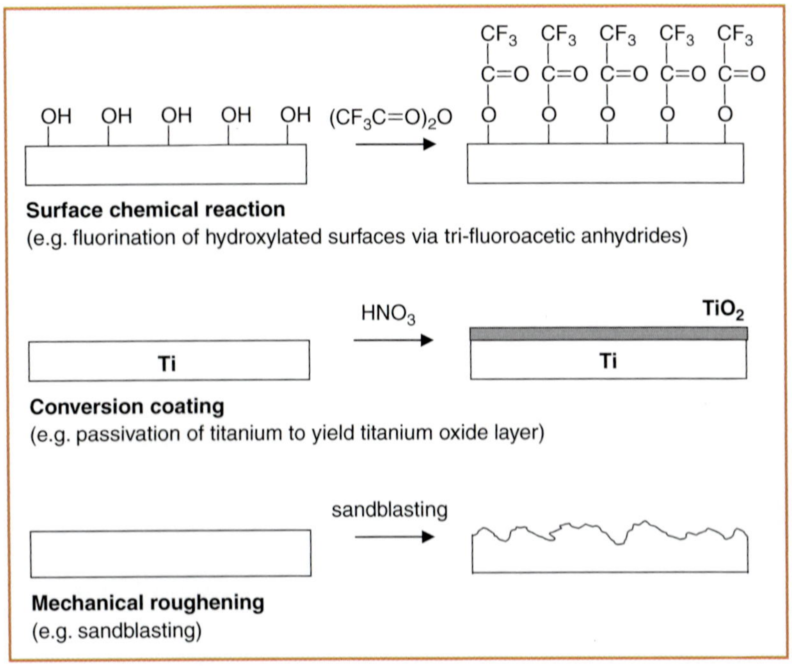

Figure 21.1 **Schematic representations of common physicochemical surface modifications of biomaterials.**

coatings often negatively influence the mechanical and functional properties of the material. Ideally, the surface modification should be confined to the outermost molecular layer (~10–15 Å), but in practice, thicker layers (10–100 nm) are used to ensure uniformity, durability, and functionality. Stability of the modified surface is a critical requirement for adequate biological performance. Surface stability not only refers to mechanical durability (i.e. resistance to cracking, delamination, debonding) but also chemical stability, especially in aggressive, chemically active environments such as biological milieu. Several types of surface rearrangements, such as translation of surface atoms or molecules in response to environmental factors and mobility of bulk molecules to the surface and vice versa, readily occur in polymers and ceramics following exposure to biological fluids.

Given the uniquely reactive nature and mobility/rearrangement of surfaces, as well as the tendency of surfaces to readily contaminate, rigorous analyses of surface treatments are essential to surface modification strategies. Surface analyses technologies generally focus on characterizing topography, chemistry/composition, and surface energy (Woodruff and Delchar, 1994) (Table 21.1). Important considerations for these surface analysis technologies include operational principles (impact of high-energy particles/X-rays under ultrahigh vacuum, adsorption or emission spectroscopies), depth of analysis, sensitivity, and resolution. For most applications, several analysis techniques must be used to obtain a complete description of the surface.

PHYSICOCHEMICAL SURFACE MODIFICATIONS

Physicochemical modifications involve alterations to the atoms, compounds, or molecules on the material surface (Figure 21.1).

Chemical Modifications

Countless chemical reactions, including UV/laser irradiation and etching reactions to clean, alter or cross-link surface groups, have been developed to modify biomaterial surfaces (Ratner and Hoffman, 2004). Non-specific reactions yield a distribution of chemically distinct groups at the surface, and the resulting surface is complex and difficult to characterize due to the presence of different chemical species in various concentrations. Nevertheless,

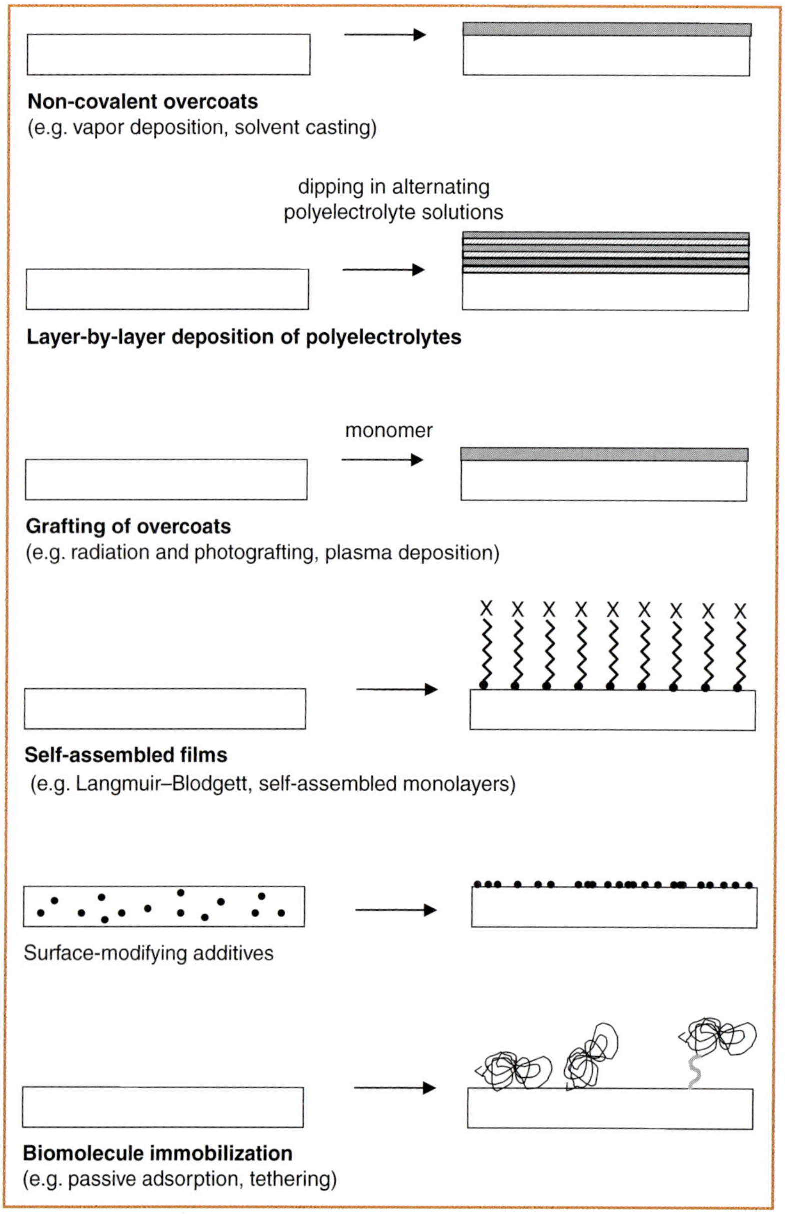

Figure 21.2 **Schematic representations of common overcoating technologies for surface modification.**

non-specific chemical reactions are widely used in biomaterials processing. Examples of non-specific reactions include radio-frequency glow discharge (RFGD) in different plasmas (e.g. oxygen, nitrogen, argon), corona discharge in air, oxidation of metals, and acid–base treatments of polymers. In contrast, specific chemical reactions target particular chemical moieties on the surface to convert them into another functional group with few side (unwanted) reactions. Acetylation, fluorination of hydroxylated surfaces via tri-fluoroacetic anhydrides, silanization of hydroxylated surfaces, and incorporation of glycidyl groups into polysiloxanes are examples of specific chemical reactions. In addition, various chemical methods exist to tether biomacromolecules onto available anchoring groups on surfaces, as described in section "Biological Modification of Surfaces."

Reaction of metal surfaces to produce an oxide-rich layer that conveys corrosion resistance, passivation, and improved wear and adhesive properties (also referred to as conversion coatings) are common surface modifications in metallic biomaterials. For example, nitric acid treatment of titanium and titanium alloys to generate titanium oxide layers is regularly performed on titanium-based medical devices, and the excellent

Table 21.1 Common surface analysis techniques

	Principle	Operation	Spatial resolution	Information depth	Sensitivity	Texture	Chemical composition information			
							Elements	Compounds	Isotopes	Additional
Contact angle	Liquid wetting of surfaces	Air Liquid	NA	3–20 Å	NA	Indirect				Surface energy
AFM	Records interatomic forces between tip and sample.	Air Aqueous	Atomic	NA	Single atom	Yes	No	No	No	
SEM	Secondary electron emission caused by electron bombardment is imaged	Vacuum	40 Å	5–10 Å	High	Yes	No	No	No	Crystal-linity
EDXA	X-ray emission caused by electron bombardment	Vacuum	40 Å	1 μm	10^{-7} g/cm^2	No	$Z > 5$	No	No	
AES	Auger electron emission caused by electron bombardment	Vacuum	100 Å	15–50 Å	10^{-10} g/cm^2 0.1 atom%	No	$Z > 3$	Chemical shift	No	
XPS	X-rays cause emission of photoelectrons with characteristic energies	Vacuum	10 μm	10–150 Å	10^{-10} g/cm^2 0.1 atom%	No	$Z > 3$	Chemical shift (excellent)	No	
SIMS	Ion bombardment causes secondary ion emission	Vacuum	3–10 μm	10 Å	10^{-13} g/cm^2	No	All	Yes	Yes	
FTIR	Molecular vibrations resulting from adsorption of IR radiation	Air Aqueous (ATR)	10 μm	<1 μm	1 mol%	No	Indirect	Vibration frequency	No	Monolayer orientation

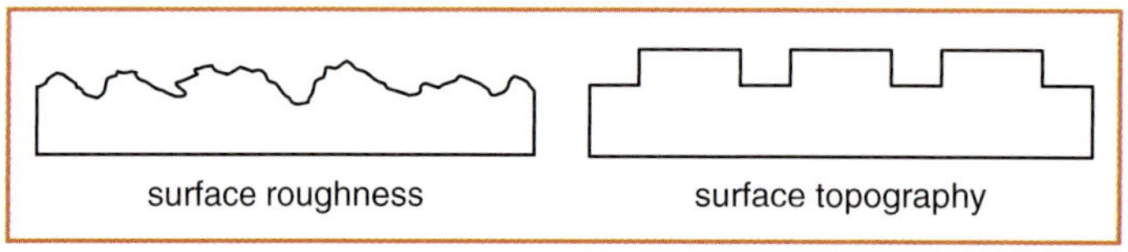

Figure 21.3 Surface roughness and topography.

biocompatibility properties of titanium are attributed to this oxide layer (Albrektsson et al., 1983). Implantation of ions into surfaces via a beam of accelerated ions has been applied to modify the surface properties of mostly metals and ceramics. For example, ion beam implantation of nitrogen into titanium and boron and carbon into stainless steel improves wear resistance and fatigue life, respectively (Sioshansi, 1987). In addition, recent evidence suggests that ion beam implantation of silicone and silver can also enhance the blood compatibility and infection resistance of silicone rubber catheters (Bambauer et al., 2004).

Topographical Modifications

The size and shape of topographical features on a surface influence cellular and host responses to the material. For example, surface macro- and micro-texture alters cell adhesion, spreading, and alignment (Curtis and Wilkinson, 1998; Flemming et al., 1999) and can regulate cell phenotypic activities, including neurite extension and osteoblastic differentiation (Boyan et al., 1996; Jansen and von Recum, 2004). Moreover, surface topography can have significant *in vivo* effects. For instance, implant porosity modulates bone and soft tissue ingrowth (Yamamoto et al., 2000; Pilliar, 2005), and surface texture alters epithelial downgrowth responses to percutaneous devices and inflammatory reactions and fibrous encapsulation to materials implanted subcutaneously (Chehroudi et al., 1989; Brauker et al., 1995; Chehroudi and Brunette, 2002). While specific surface texture parameters that elicit particular biological responses have been identified in several cases, the mechanisms generating these behaviors remain poorly understood.

Methods for generating surface texture can be grouped into approaches for engineering either roughness or topography (Figure 21.3). Surface roughness indicates a random or complex pattern of features of varying amplitude and spacing, typically on a scale smaller than a cell (10–20 μm). On the other hand, surface topography refers to patterns of well-defined, controlled features on the surface. Surface roughness has been traditionally modified via sandblasting, plasma spraying, and mechanical polishing, and it is the non-specific nature of these processes that renders surfaces with random or complex topographies. Ion beam and electric arc (for conductive materials) texturing approaches have also been applied to modulate surface roughness. For generating controlled topographies, micro- and nano-machining techniques have been exploited using silicon, glass, and polymers as substrate materials (Flemming et al., 1999). Photolithography combined with reactive plasma and ion etching has been extensively applied to generate surfaces with well-defined topographies. This technique allows the preparation of machined silicon and polymeric substrates and silicon templates which can then be used as molds to transfer features to polymers via solvent casting or injection molding. Similarly, LIGA (German for "Lithographie, Galvanoformung, Abformung"), electron beam, and laser machining have been used to manufacture defined topographical features on various materials. More recently, hot embossing imprint lithography has been applied for low cost and rapid fabrication of micro- and nano-scale features on biomedically relevant polymers (Charest et al., 2004).

OVERCOATING TECHNOLOGIES

Coating strategies rely on the deposition of a surface layer consisting of a different composition from the underlying base material (Figure 21.2). These surface modification approaches include non-covalent and covalent coatings (Ratner and Hoffman, 2004).

Non-covalent Coatings

Major advantages of non-covalent coatings include simple application and the ability to coat a variety of different base materials. Examples of common non-covalent coating methods are solvent casting, and vapor deposition of metals, parylene, and carbons. In the Langmuir–Blodgett deposition method, one or more highly ordered layers of surfactant molecules (e.g. phospholipids, amphiphiles) are placed at the surface of the base material via assembly at the air–water interface and compression of the surfactant molecules. Langmuir–Blodgett films exhibit high order and uniformity and provide flexibility in incorporating a wide range of chemistries. The stability of these films can be improved by cross-linking or internally polymerizing the surfactant molecules following film formation. Another surface modification strategy that takes advantage of intermolecular interactions is the deposition of multilayer polyelectrolytes (e.g. poly(styrenesulfonate)/poly(allylamine), hyaluronic acid/chitosan). In this simple layer-by-layer method, a charged surface is sequentially dipped into alternating aqueous solutions of polyelectrolytes of opposite charge in order to deposit multilayers of a polyelectrolyte complex. Another elegant strategy for surface modification is the use of surface-modifying additives. These molecules are blended in the bulk material during fabrication but will spontaneously rise to and concentrate at the surface due to the driving force to minimize interfacial energy.

Covalent Coatings

Covalent coating methodologies rely on direct tethering of overcoats onto the base material to improve film stability and adherence. Radiation grafting, both with ionizing radiation and high-energy electron beams, and photografting have been extensively pursued to modify polymer substrates in order to introduce chemically reactable groups into inert hydrophobic polymers and polymerize overcoats onto the base support (Ratner and Hoffman, 2004). In principle, the radiation breaks chemical bonds in the base material into free radicals and other reactive species, which are then exposed to a monomer. The monomer reacts with the reactive species at the surface and propagates as a free radical chain reaction into a surface grafted polymer. These strategies allow for generation of a wide range of surface chemistries, and unique graft co-polymers can be synthesized by combining different monomers. Plasma deposition (also referred to as glow discharge deposition) via radio frequency or microwave has also been extensively applied to biomaterial surface modification (Hoffman, 1988). In particular, RFGD plasma deposition has received considerable attention because it can generate continuous (relatively free of pin holes and voids) conformal coatings that can be applied to many different types of supports (metals, ceramics, polymers) with complex geometries. In addition, these films exhibit good adherence to the substrate and can be engineered to present different functionalities, although the resulting chemistry is complex and ill-defined. In contrast to these relatively low-energy/low-temperature plasmas, high-energy/high-temperature plasmas have also been used to apply inorganic surface modifications onto inorganic substrates. For example, calcium phosphate ceramic particles, such as hydroxyapatite, have been deposited via flame spraying onto titanium and cobalt chrome orthopedic implants to improve osseointegration (Gruner, 2005).

Coatings consisting of self-assembled monolayers (SAMs) have gained significant attention as robust surface modification agents (Ulman, 1991; Mrksich and Whitesides, 1995). These films spontaneously assemble, form highly ordered, well-defined surfaces with excellent chemical stability, and provide a wide range of available surface functionalities. The basic structure of molecules that form SAMs is an anchoring "head" group, organic chain backbone, and functional "tail" group. Common SAM systems are alkanethiols on coinage metals (gold, silver), *n*-alkyl silanes on hydroxylated supports (glass, silica), and phosphoric acid or phosphate groups on titanium or tantalum surfaces. Assembly of these organic chains into highly ordered structures is driven by the strong adsorption of the anchoring "head" group of the monolayer constituent to the surface and van der Waals interactions of the backbone chains. The order and stability of the SAMs are strongly influenced by the length of the backbone chain, and in the case of alkanethiols, molecules with backbones between

9 and 24 methylene groups assemble well on gold. Importantly, the terminal functional group is presented at the surface–solution interface and controls the physicochemical properties of the SAM.

BIOLOGICAL MODIFICATION OF SURFACES

Biomolecules (e.g. cell receptor ligands, enzymes, antibodies, pharmacological agents, lipids, nucleic acids) have been immobilized onto and within biomaterial supports for numerous therapeutic, diagnostic, and bioprocess applications. Table 21.2 lists several examples of biological modifications to surfaces for biomedical and biotechnological applications. The rationale for these hybrid materials integrating synthetic and biological components is to convey biofunctionality and hence engineer materials that elicit desired biological responses or have attributes associated with biosystems. One of the earliest examples of this strategy is the immobilization of heparin onto polymer surfaces to improve blood compatibility. More recently, drug eluting stents (stents coated with a polymeric layer loaded with anti-hyperplasia drugs) have been developed to reduce restenosis and improve patency. Another example of a widely used biological modification strategy is the immobilization of adhesive ligands, either adsorbed proteins (e.g. fibronectin, laminin) or tethered synthetic oligopeptides (e.g. RGD), on synthetic and natural supports to promote cell adhesion and function in various tissue engineering and regenerative medicine applications (Lutolf and Hubbell, 2005).

Three major methods are used to immobilize biomolecules onto biomaterial surfaces: physical adsorption, physical entrapment, and covalent immobilization (Figure 21.4) (Hoffman and Hubbell, 2004). Passive physisorption of biomacromolecules (i.e. proteins, polysaccharides, nucleic acids) is a simple yet efficient method to render surfaces biologically active. Everyday applications include coating of synthetic materials with extracellular matrix proteins, such as fibronectin and collagen, to improve cell adhesion. Protein adsorption is a complex, dynamic energy-driven process involving hydrophobic interactions, electrostatic interactions, hydrogen bonding, and van der Waals forces. Protein parameters such as primary structure, size, and structural stability as well as surface properties including surface energy, and chemistry influence the biological activity of the adsorbed biomacromolecules. It is important to point out these biologically modified surfaces can undergo further modifications, such as displacement of adsorbed proteins and cell-mediated deposition and remodeling of matrix components, in the biological milieu. As an approach to improve the stability of these modified surfaces, the biological molecules can be cross-linked following adsorption. Finally, the use of high-affinity interactions, for example avidin–biotin and antibody–antigen, represents a special case of these physical immobilization methods that is particularly important in diagnostics and bioprocessing.

Table 21.2 Biomedical and biotechnological applications of immobilized biomolecules

Biomolecule	Applications
Heparin	Blood-compatible surfaces; growth factor immobilization
Fibronectin, collagen RGD peptides	Cell adhesion and function in biosensors, arrays, devices and tissue-engineered constructs
Antibodies	Biosensors; bioseparations; anti-cancer treatments
DNA plasmids anti-sense oligonucleotides siRNA	Gene therapy for a multitude of diseases; DNA probes
Growth factor proteins and peptides	Anti-cancer treatments; treatments for auto-immune and inflammatory conditions; enhanced wound repair
Enzymes	Biosensors; bioreactors; anti-cancer treatments; anti-thrombotic surfaces
Drugs and antibiotics	Anti-thrombotic agents; anti-cancer treatments; anti-hyperplasia treatments; anti-infection/inflammation treatments
Polysaccharides	Non-fouling supports for biosensors and bioseparations

Physical entrapment methods rely on diffusive barriers or matrix systems to control the transport or availability of the biomolecule. For example, entrapment of enzymes within sol-gels with nano-scale porosites and drug or protein therapeutics within encapsulation matrices provides technologies for enhanced stability, separation or recovery of the biological agent, and regulated delivery kinetics. The encapsulation systems can be engineered to permanently isolate the biomolecule or degrade in non-specific (e.g. hydrolysis) or specific (e.g. enzymatic degradation) fashions for controlled release kinetics.

An extensive and diverse group of strategies has been developed to covalently immobilize or tether biomolecules to soluble or solid supports (Figure 21.4) (Weetall, 1976; Hoffman and Hubbell, 2004). Soluble polymers functionalized with biomolecules can then be polymerized into a network or grafted onto a solid support. These strategies rely on coupling reactions between groups in the biomolecule (—NH_2,—COOH,—SH) and the biomaterial support, and often involve cross-linkers or coupling agents such as CNBr, carbodiimides, and *N*-hydroxysulfosuccinimide. In many instances, the biomolecule is covalently immobilized via a spacer arm

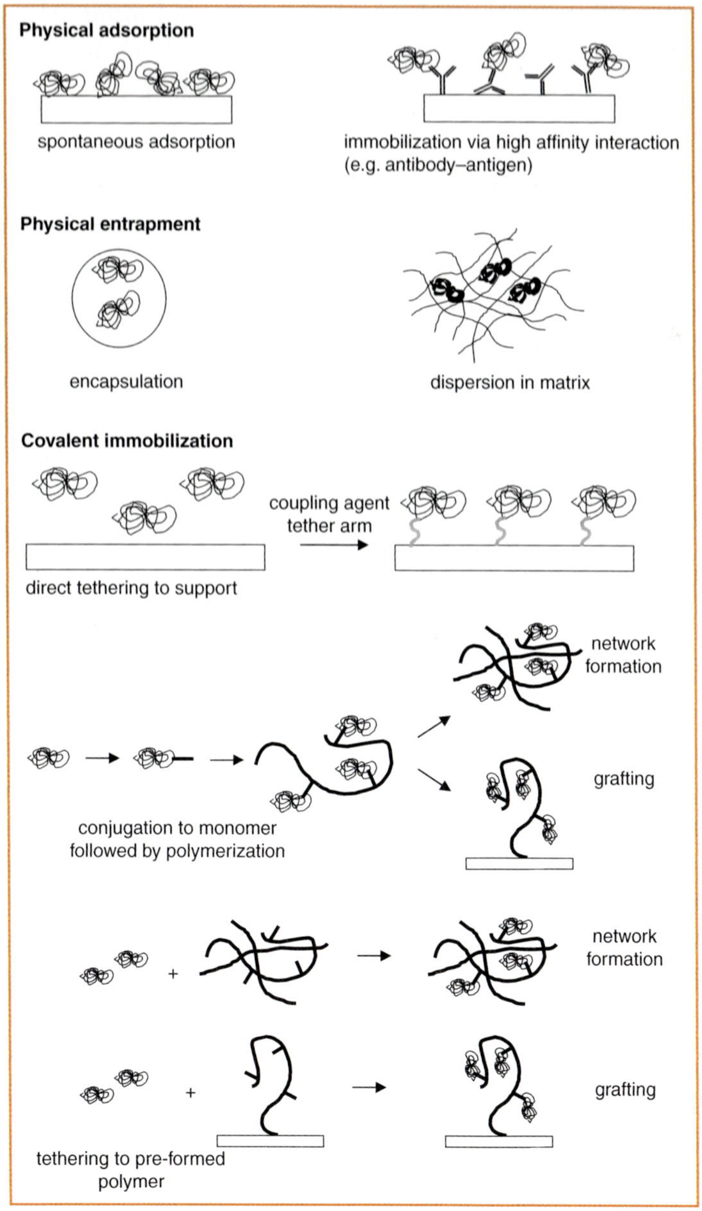

Figure 21.4 **Schematic diagram of methods for immobilizing biomolecules onto and within biomaterials.**

Table 22.2 Important chemical mediators of inflammation derived from plasma, cells, or injured tissue

Mediators	Examples
Vasoactive agents	Histamine, serotonin, adenosine, endothelial derived relaxing factor (EDRF), prostacyclin, endothelin, thromboxane a_2
Plasma proteases	
Kinin system	Bradykinin, kallikrein
Complement system	C3a, C5a, C3b, C5b–C9
Coagulation/fibrinolytic system	Fibrin degradation products, activated Hageman factor (FXIIA), tissue plasminogen activator (tPA)
Leukotrienes	Leukotriene B_4 (LTB_4), hydroxyeicosatetranoic acid (HETE)
Lysosomal proteases	Collagenase, elastase
Oxygen-derived free radicals	H_2O_2, superoxide anion, nitric oxide
Platelet activating factors	Cell membrane lipids
Cytokines	Interleukin-1 (IL-1), TNF
Growth factors	PDGF, fibroblast growth factor (FGF), transforming growth factor (TGF-α or TGF-β), epithelial growth factor (EGF)

points must be noted in order to understand the inflammatory response and how it relates to biomaterials. First, although chemical mediators are classified on a structural or functional basis, different mediator systems interact and provide a system of checks and balances regarding their respective activities and functions. Second, chemical mediators are quickly inactivated or destroyed, suggesting that their action is predominantly local (i.e. at the implant site). Third, generally acid, lyosomal proteases and oxygen-derived free radicals produce the most significant damage or injury. These chemical mediators are also important in the degradation of biomaterials.

The predominant cell type present in the inflammatory response varies with the age of the injury. In general, neutrophils, commonly called polymorphonuclear leukocytes or polys, predominate during the first several days following injury and then are replaced by monocytes as the predominant cell type. Three factors account for this change in cell type: (i) Neutrophils are short-lived and disintegrate and disappear after 24–48 h; neutrophil emigration is of short duration because chemotactic factors for neutrophil migration are activated early in the inflammatory response. (ii) Following emigration from the vasculature, monocytes differentiate into macrophages, and these cells are very long-lived (up to months). (iii) Monocyte emigration may continue for days to weeks, depending on the injury and implanted biomaterial, and chemotactic factors for monocytes are activated over longer periods of time.

Provisional Matrix Formation

Injury to vascularized tissue in the implantation procedure leads to immediate development of the provisional matrix at the implant site. This provisional matrix consists of fibrin, produced by activation of the coagulative and thrombosis systems, and inflammatory products released by the complement system, activated platelets, inflammatory cells, and endothelial cells (Clark et al., 1982; Tang et al., 1993; Tang, 1998). These events occur early, within minutes to hours following implantation of a medical device. Components within or released from the provisional matrix, that is, fibrin network (thrombosis or clot), initiate the resolution, reorganization, and repair processes such as inflammatory cell and fibroblast recruitment. Platelets, activated during the fibrin network formation, release platelet factor 4, platelet-derived growth factor (PDGF), and transforming growth factor β (TGF-β), which contribute to fibroblast recruitment (Wahl et al., 1989; Riches, 1998). Monocytes and lymphocytes, upon activation, generate additional chemotactic factors including LTB_4, PDGF, and TGF-β to recruit fibroblasts.

The provisional matrix is composed of adhesive molecules such as fibronectin and thrombospondin bound to fibrin as well as platelet granule components released during platelet aggregation. Platelet granule components include thrombospondin, released from the platelet α-granule, and cytokines including TGF-α, TGF-β, PDGF, platelet factor 4, and platelet-derived endothelial cell growth factor. The provisional matrix is stabilized by the cross-linking of fibrin by factor XIIIa.

The provisional matrix appears to provide both structural and biochemical components to the process of wound healing. The complex three-dimensional structure of the fibrin network with attached adhesive proteins provides a substrate for cell adhesion and migration. The presence of mitogens, chemoattractants, cytokines, and growth factors within the provisional matrix provide for a rich milieu of activating and inhibiting substances for various cellular proliferative and synthetic processes.

The provisional matrix may be viewed as a naturally derived, biodegradable, sustained release system in which mitogens, chemoattractants, cytokines, and growth factors are released to control subsequent wound healing processes (Dvorak et al., 1987; Ignotz et al., 1987; Muller et al., 1987; Wahl et al., 1987; Madri et al., 1988; Sporn and Roberts, 1988; Broadley et al., 1989). In spite of the rapid increase in our knowledge of the provisional matrix and its capabilities, our knowledge of the control of the formation of the provisional matrix and its effect on subsequent wound healing events is poor.

Temporal Sequence of Inflammation and Wound Healing

Inflammation is generally defined as the reaction of vascularized living tissue to local injury. Inflammation serves to contain, neutralize, dilute, or wall off the injurious agent or process. In addition, it sets into motion a series of events that may heal and reconstitute the implant site through replacement of the injured tissue by regeneration of native parenchymal cells, formation of fibroblastic scar tissue, or a combination of these two processes (Cotran et al., 1999; Gallin and Synderman, 1999).

The sequence of events following implantation of a biomaterial is illustrated in Figure 22.1. The size, shape, and chemical and physical properties of the biomaterial and the physical dimensions and properties of the prosthesis or device may be responsible for variations in the intensity and time duration of the inflammatory and wound healing processes. Thus, intensity and/or time duration of inflammatory reaction may characterize the biocompatibility of a biomaterial, or device.

Classically, the biocompatibility of an implanted material has been described in terms of the morphological appearance of the inflammatory reaction to the material; however, the inflammatory response is a series of complex reactions involving various types of cells, the densities, activities, and functions of which are controlled by various endogenous and autocoid mediators. The simplistic view of the acute inflammatory response progressing to the chronic inflammatory response may be misleading with respect to biocompatibility studies and the inflammatory response to implants. *In vivo* studies using the cage implant system show that monocytes and macrophages are present in highest concentrations when neutrophils are also at their highest concentrations, that is, the acute inflammatory response (Marchant et al., 1983; Spilizewski et al., 1985). Neutrophils have short lifetimes – hours to days – and disappear from the exudates more rapidly than do macrophages, which have lifetimes of days to weeks to months. Eventually macrophages become the predominant cell type in the exudates, resulting in a chronic inflammatory response. Monocytes rapidly differentiate into macrophages, the cells principally responsible for normal wound healing in the foreign body reaction. Classically, the development of granulation tissue has been considered to be part of chronic inflammation, but because of unique tissue–material interactions, it is preferable to differentiate the foreign body reaction – with its varying degree of granulation tissue development, including macrophages, fibroblasts, and capillary formation – from chronic inflammation.

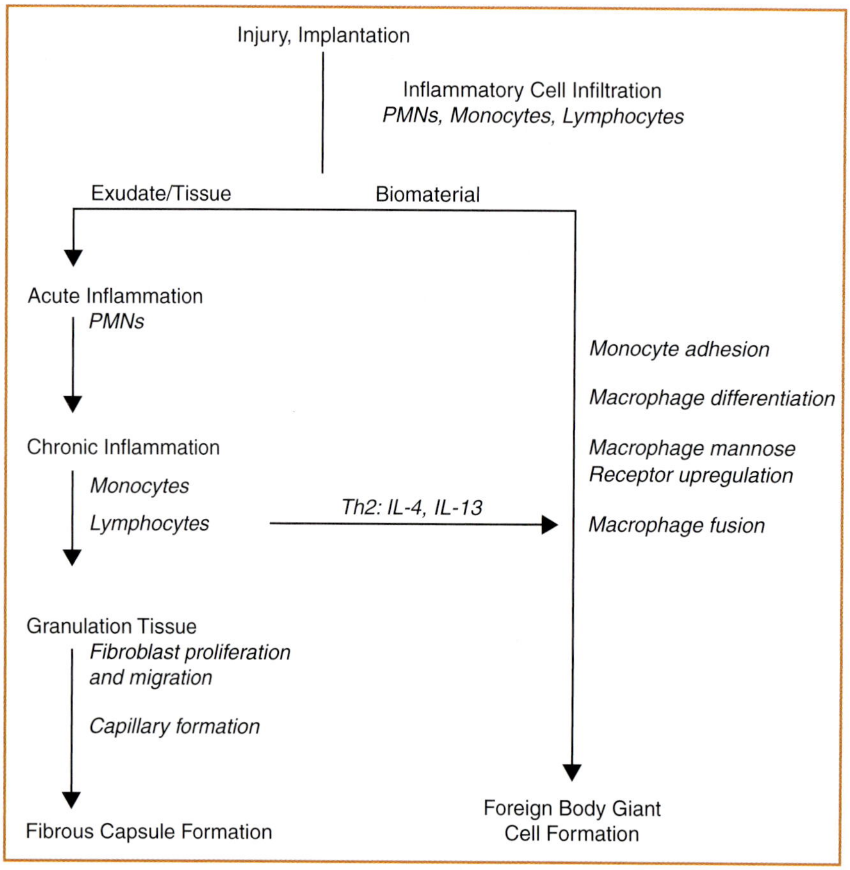

Figure 22.1 **Sequence of events involved in inflammatory and wound healing responses leading to FBGC formation. This shows the importance of Th2 lymphocytes in the transient chronic inflammatory phase with the production of IL-4 and IL-3 that can induce monocyte/macrophage fusion to form FBGCs.**

Acute Inflammation

Acute inflammation is of relatively short duration, lasting from minutes to days, depending on the extent of injury. The main characteristics of acute inflammation are the exudation of fluid and plasma proteins (edema) and the emigration of leukocytes (predominantly neutrophils). Neutrophils and other motile white cells emigrate or move from the blood vessels to the perivascular tissues and the injury (implant) site (Henson et al., 1987; Malech et al., 1987; Ganz, 1988).

The accumulation of leukocytes, in particular neutrophils and monocytes, is the most important feature of the inflammatory reaction. Leukocytes accumulate through a series of processes including margination, adhesion, emigration, phagocytosis, and extracellular release of leukocyte products (Jutila, 1990). Increased leukocytic adhesion in inflammation involves specific interactions between complementary "adhesion molecules" present on the leukocyte and endothelial surfaces (Cotran and Pober, 1990; Pober and Cotran, 1990). The surface expression of these adhesion molecules is modulated by inflammatory agents; mechanisms of interaction include stimulation of leukocyte adhesion molecules (C5a, LTB_4), stimulation of endothelial adhesion molecules (IL-1), or both effects tumor necrosis factor-α (TNF-α). Integrins comprise a family of transmembrane glycoproteins that modulate cell–matrix and cell–cell relationships by acting as receptors to extracellular protein ligands and also as direct adhesion molecules (Hynes, 1992). An important group of integrins (adhesion molecules) on leukocytes include the CD11/CD18 family of adhesion molecules.

Inflammatory mediators (i.e. cytokines) stimulate a rapid increase in these adhesion molecules on the leukocyte surface as well as increased leukocyte adhesion to endothelium. Leukocyte–endothelial cell interactions are also controlled by endothelial–leukocyte adhesion molecules (ELAMs, E-selectins) or intracellular adhesion molecules (ICAM-1, ICAM-2, and vascular cell adhesion molecules (VCAMs)) on endothelial cells (Butcher, 1991).

Inflammatory cell emigration is controlled in part by chemotaxis, which is the unidirectional migration of cells along a chemical gradient. A wide variety of exogenous and endogenous substances have been identified as chemotactic agents (Henson, 1971, 1980; Weisman et al., 1980; Henson et al., 1987; Malech and Gallin, 1987; Ganz, 1988; Weiss, 1989; Cotran and Pober, 1990; Jutila, 1990; Paty et al., 1990; Pober and Cotran, 1990; Butcher, 1991; Hynes, 1992). Important to the emigration or movement of leukocytes is the presence of specific receptors for chemotactic agents on the cell membranes of leukocytes. These and other receptors may also play a role in the activation of leukocytes. Following localization of leukocytes at the injury (implant) site, phagocytosis and the release of enzymes occur following activation of neutrophils and macrophages. The major role of the neutrophils in acute inflammation is to phagocytose microorganisms and foreign materials. Phagocytosis is seen as a three-step process in which the injurious agent undergoes recognition and neutrophil attachment, engulfment, and killing or degradation. With regard to biomaterials, engulfment and degradation may or may not occur depending on the properties of the biomaterial.

Although biomaterials are not generally phagocytosed by neutrophils or macrophages because of the size disparity (i.e. the surface of the biomaterial is greater than the size of the cell), certain events in phagocytosis may occur. The process of recognition and attachment is expedited when the injurious agent is coated by naturally occurring serum factors called opsonins. The two major opsonins are IgG and the complement-activated fragment, C3b. Both of these plasma-derived proteins are known to adsorb to biomaterials, and neutrophils and macrophages have corresponding cell membrane receptors for these opsonization proteins. These receptors may also play a role in the activation of the attached neutrophil or macrophage. Because of the size disparity between the biomaterial surface and the attached cell, "frustrated phagocytosis" may occur (Henson, 1971, 1980). This process does not involve engulfment of the biomaterial but does cause the extracellular release of leukocyte products in an attempt to degrade the biomaterial. Neutrophils adherent to complement-coated and immunoglobulin-coated non-phagocytosable surfaces may release enzymes by direct extrusion or exocytosis from the cell (Henson, 1971, 1980). The amount of enzyme released during this process depends on the size of the polymer particle, with larger particles inducing greater amounts of enzyme release. This suggests that the specific mode of cell activation in the inflammatory response in tissue is dependent upon the size of the implant and that a material in a phagocytosable form (e.g. powder or particulate) may provoke a degree of inflammatory response different from that of the same material in a non-phagocytosable form (e.g. film).

Tissue-engineered constructs containing biomaterial scaffolds alone, or with cells and/or chemokines, growth factors, or other biological components are thus subjected to an aggressive microenvironment that may quickly compromise the intended function of the construct (Babensee et al., 1998).

Chronic Inflammation

Chronic inflammation is less uniform histologically than is acute inflammation. In general, chronic inflammation is characterized by the presence of monocytes and lymphocytes with the early proliferation of blood vessels and connective tissue (Williams et al., 1983; Johnston, 1988; Cotran et al., 1999; Gallin and Synderman, 1999). It must be noted that many factors modify the course and histological appearance of chronic inflammation.

Persistent inflammatory stimuli lead to chronic inflammation. Although the chemical and physical properties of the biomaterial may lead to chronic inflammation, motion in the implant site by the biomaterial may also produce chronic inflammation. The chronic inflammatory response to biomaterials is confined to the implant site. Inflammation with the presence of mononuclear cells, including lymphocytes and plasma cells, is given the designation chronic inflammation, whereas the foreign body reaction with granulation tissue development is considered the normal wound healing response to implanted biomaterials (i.e. the normal foreign body reaction). Chronic inflammation with biocompatible materials is usually of very short duration (i.e. a few days).

Lymphocytes and plasma cells are involved principally in immune reactions and are key mediators of antibody production and delayed hypersensitivity responses. Their roles in non-immunological injuries and inflammation are largely unknown. Little is known regarding immune responses and cell-mediated immunity to synthetic biomaterials. The role of macrophages must be considered in the possible development of immune responses to synthetic biomaterials. Macrophages process and present the antigen to immunocompetent cells and thus are key mediators in the development of immune reactions.

The macrophage is probably the most important cell in chronic inflammation because of the great number of biologically active products its produces (Johnston, 1988). Important classes of products produced and secreted by macrophages include neutral proteases, chemotactic factors, arachidonic acid metabolites, reactive oxygen metabolites, complement components, coagulation factors, growth-promoting factors, and cytokines.

Growth factors such as PDGF, FGF, TGF-β, TGF-α/EGF, and IL-1 or TNF are important to the growth of fibroblasts and blood vessels and the regeneration of epithelial cells. Growth factors, released by activated cells, stimulated production of a wide variety of cells; initiate cell migration, differentiation, and tissue remodeling; and may be involved in various stages of wound healing (Mustoe et al., 1987; Wahl et al., 1989; Fong et al., 1990; Sporn and Roberts, 1990; Golden et al., 1991; Kovacs, 1991). It is clear that there is a lack of information regarding interaction and synergy among various cytokines and growth factors and their abilities to exhibit chemotactic, mitogenic, and angiogenic properties.

Granulation Tissue

Within 1 day following implantation of a biomaterial (i.e. injury), the healing response is initiated by the action of monocytes and macrophages, followed by proliferation of fibroblasts and vascular endothelial cells at the implant site, leading to the formation of granulation tissue, the hallmark of healing inflammation. Granulation tissue derives its name from the pink, soft granular appearance on the surface of healing wounds, and its characteristic histological features include the proliferation of new small blood vessels and fibroblasts. Depending on the extent of injury, granulation tissue may be seen as early as 3–5 days following implantation of a biomaterial.

The new small blood vessels are formed by budding or sprouting of pre-existing vessels in a process known as neovascularization or angiogenesis (Ziats et al., 1985; Thompson et al., 1988; Maciag, 1990). This process involves proliferation, maturation, and organization of endothelial cells into capillary tubes. Fibroblasts also proliferate in developing granulation tissue and are active in synthesizing collagen and proteoglycans. In the early stages of granulation tissue development, proteoglycans predominate; later, however, collagen – especially type I collagen – predominates and forms the fibrous capsule. Some fibroblasts in developing granulation tissue may have features of smooth muscle cells. These cells are called myofibroblasts and are considered to be responsible for the wound contraction seen during the development of granulation tissue.

Macrophage Interactions

The inflammatory and immune systems overlap considerably through the activity and phenotypic expression of macrophages that are derived from blood-borne monocytes. Monocytes and macrophages belong to the mononuclear phagocytic system (MPS), Table 22.3. Cells in the MPS may be considered as resident macrophages in the respective tissues that take on specialized functions that are dependent on their tissue environment. From this perspective, the host defense system may be seen as blood-borne or circulating inflammatory and immune cells as well as mononuclear phagocytic cells that reside in specific tissues with specialized functions. In the inflammatory and immune responses, the macrophage plays a pivotal role in both the induction and effector phases of these responses.

Two factors that play a role in monocyte/macrophage adhesion and activation and foreign body giant cell (FBGC) formation are the surface chemistry of the substrate onto which the cells adhere and the protein adsorption that occurs before cell adhesion. These two factors have been hypothesized to play significant roles in the inflammatory and wound healing responses to biomaterials and medical devices *in vivo*.

Macrophage interactions with biomaterials are initiated when blood-borne monocytes in the early, transient responses migrate to the implant site and adhere to the blood protein adsorbed biomaterial through monocyte–integrin interactions. Following adhesion, adherent monocytes differentiate into macrophages that may then fuse to form FBGCs. Figure 22.2 demonstrates the progression from circulating blood monocyte to tissue macrophage to FBGC development that is most commonly observed. Because of the progression of monocytes to macrophages to FBGCs (Figure 22.2), the following discussion of macrophage interactions also includes perspectives on how macrophages are formed (i.e. monocyte adhesion) and what happens to macrophages on biomaterial surfaces (i.e. FBGC formation) (McNally et al., 1994; McNally et al., 1995).

Table 22.3 The mononuclear phagocytic system

Tissues	Cells
Implant sites	Inflammatory macrophages, FBGCs
Liver	Kupffer cells
Lung	Alveolar macrophages
Connective tissue	Histiocytes
Bone marrow	Macrophages
Spleen and lymph nodes	Fixed and free macrophages
Serous cavities	Pleural and peritoneal macrophages
Nervous system	Microglial cells
Bone	Osteoclasts
Skin	Langerhans' cells, dendritic cells
Lymphoid tissue	Dendritic cells

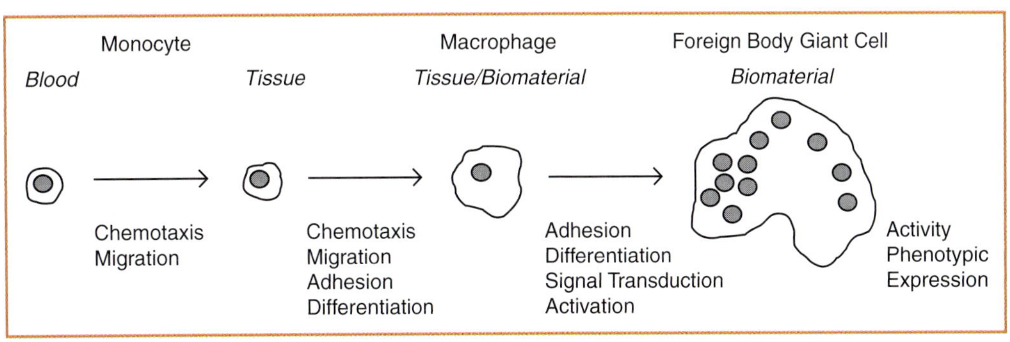

Figure 22.2 ***In vivo*** **transition from blood-borne monocyte to biomaterial adherent monocyte/macrophage to FBGC at the tissue/biomaterial interface. Little is known regarding the indicated biological responses that are considered to play important roles in the transition to FBGC development.**

Material surface property-dependent blood protein adsorption occurs immediately upon surgical implantation of a biomaterial and it is the protein-modified biomaterial that inflammatory cells subsequently encounter. Monocytes express receptors for various blood components, but they recognize naturally occurring foreign surfaces by receptors for opsonins such as fragments of complement component C3. Complement activation by biomaterials has been well documented. Exposure to blood during biomaterial implantation may permit extensive opsonization with the labile fragment C3b and the rapid conversion of C3b to its hemolytically inactive but nevertheless opsonic and more stable form, C3bi. C3b is bound by the CD35 receptor, but C3bi is recognized by distinct receptors, CD11b/CD18 and CD11c/CD18 on monocytes (McNally et al., 1994). Fibrinogen, a major plasma protein that adsorbs to biomaterials, is another ligand for these receptors that together with CD11a/CD18 constitutes a subfamily of integrins that is restricted to leukocytes (McNally et al., 1994, 1995). Studies with monoclonal antibodies to their common β_2 subunit (CD 18) and distinct α chains have implicated CD11b/CD18 and CD11c/CD18 in monocyte/macrophage responses. Other potential adhesion-mediating proteins that adsorb to biomaterials include IgG, which may interact with monocytes via various receptors and fibronectin, for which monocytes also express multiple types of receptors (Jenney and Anderson, 2000; McNally and Anderson, 2002).

FBGC Formation and Interactions

The foreign body reaction is composed of FBGCs and the components of granulation tissue, which consist of macrophages, fibroblasts, and capillaries in varying amounts, depending upon the form and topography of the implanted material. Relatively flat and smooth surfaces, such as those found on breast prostheses, have a foreign body reaction that is composed of a layer of macrophages one to two cells in thickness. Relatively rough surfaces, such as those found on the outer surfaces of expanded poly(tetrafluroethylene) (ePTFE) vascular prostheses or poly(methyl methacrylate) (PMMA) bone cement, have a foreign body reaction composed of several layers of macrophages and FBGCs at the surface. Fabric materials generally have a surface response composed of macrophages and FBGCs with varying degrees of granulation tissue subjacent to the surface response.

As previously discussed, the form and topography of the surface of the biomaterial determines the composition of the foreign body reaction. With biocompatible materials, the composition of the foreign body reaction in the implant site may be controlled by the surface properties of the biomaterial, the form of the implant, and the relationship between the surface area of the biomaterial and the volume of the implant. For example, high surface-to-volume implants such as fabrics or porous materials will have higher ratios of macrophages and FBGCs in the implant site than will smooth-surface implants, which will have fibrosis as a significant component of the implant site.

The foreign body reaction consisting mainly of macrophages and/or FBGCs may persist at the tissue–implant interface for the lifetime of the implant (Chambers and Spector, 1982; Rae, 1986; Anderson, 1988, 1993, 2000; Greisler, 1988). Generally, fibrosis (i.e. fibrous encapsulation) surrounds the biomaterial or implant with its interfacial foreign body reaction, isolating the implant and foreign body reaction from the local tissue environment. Early in the inflammatory and wound healing response, the macrophages are activated upon adherence to the material surface. Although it is generally considered that the chemical and physical properties of the biomaterial are responsible for macrophage activation, the nature of the subsequent events regarding the activity of macrophages at the surface is not clear. Tissue macrophages, derived from circulating blood monocytes, may coalesce to form multinucleated FBGCs. FBGCs containing large numbers of nuclei are typically present on the surface of biomaterials. Although these FBGCs may persist for the lifetime of the implant, it is not known if they remain activated, releasing their lysosomal constituents, or become quiescent. FBGCs have been implicated in the biodegradation of polymeric medical devices (Zhao et al., 1990, 1991; Wiggins et al., 2001).

Figure 22.1 demonstrates the sequence of events involved in inflammation and wound healing when medical devices are implanted. In general, the neutrophil (PMN) predominant acute inflammatory response and the lymphocyte/monocyte predominant chronic inflammatory response resolve quickly (i.e. within 2 weeks) depending on the type and location of implant. Studies utilizing IL-4 demonstrate the role for Th2 helper lymphocytes in the development of the foreign body reaction at the tissue/material interface. Th2 helper lymphocytes have been described as "anti-inflammatory" based on their cytokine profile of which IL-4 is a significant component. Th2 helper lymphocytes also produce IL-13 that has a similar effect to IL-4 on FBGC formation. In this regard, it is noteworthy that anti-IL-4 antibody does not inhibit IL-13 induced FBGC formation nor does anti-IL-13 antibody inhibit IL-4 induced FBGC formation. In IL-4 and IL-13 FBGC culture systems, the macrophage mannose receptor (MMR) has been identified as critical to the fusion of macrophages in the formation of FBGC (McNally et al., 1996; DeFife et al., 1997). FBGC formation can be prevented by competitive inhibitors of MMR activity (i.e. α-mannan) or inhibitors of glycoprotein processing that restrict MMR surface expression.

FIBROSIS AND FIBROUS ENCAPSULATION

The end-stage healing response to biomaterials is generally fibrosis or fibrous encapsulation. However, tissue-engineered devices may be exceptions to this general statement (e.g. porous materials inoculated with parenchymal cells or porous materials implanted into bone).

Repair of implant sites involves two distinct processes: regeneration, which is the replacement of injury tissue by parenchymal cells of the same type, or replacement by connective tissue that constitutes the fibrous capsule. These processes are generally controlled by either (i) the proliferative capacity of the cells in the tissue receiving the implant and the extent of injury as it relates to the destruction or (ii) persistence of the tissue framework of the implant site. The regenerative capacity of cells permits classification into three groups: labile, stable (or expanding), and permanent (or static) cells. Labile cells continue to proliferate throughout life, stable cells retain this capacity but do not normally replicate, and permanent cells cannot reproduce themselves after birth. Perfect repair with restitution of normal structure theoretically occurs only in tissue consisting of stable and labile cells, whereas all injuries to tissues composed of permanent cells may give rise to fibrosis and fibrous capsule formation with very little restitution of the normal tissue or organ structure. Tissues composed of permanent cells (e.g. nerve cells, skeletal muscle cells, and cardiac muscle cells) most commonly undergo an organization of the inflammatory exudates, leading to fibrosis. Tissues composed of stable cells (e.g. parenchymal cells of the liver, kidney, and pancreas), mesenchymal cells (e.g. fibroblasts, smooth muscle cells, osteoblasts, and chondroblasts), and vascular endothelial and labile cells (e.g. epithelial cells and lymphoid and hematopoietic cells) may also follow this pathway to fibrosis or may undergo resolution of the inflammatory exudates, leading to restitution of the normal tissue structure. The condition of the underlying framework or supporting stroma of the parenchymal cells following an injury plays an important role in the restoration of normal tissue structure. Retention of the framework may lead to restitution of the normal tissue structure, whereas destruction of the framework most commonly leads to fibrosis. It is important to consider the species-dependent nature of the regenerative capacity of cells. For example, cells from the same organ or tissue but from different species may exhibit different regenerative capacities and/or connective tissue repair.

The extent of provisional matrix formation is an important factor as it is related to wound healing by first or second intention. First intention (primary union) wound healing occurs when there is minimal to no space between the tissue and device whereas second intention (secondary union) wound healing occurs when a large space, providing for extensive provisional matrix formation, is present. Obviously, inappropriate or inadequate

Table 22.4 Common components in the inflammatory (innate) and immune (adaptive) responses

Components
Complement cascade components
Immunoglobulins
Cellular components
Macrophages
NK (natural killer) cells
Dendritic cells
Cells with dual phagocytic and antigen presenting capabilities

preparation of the implant site leading to extensive provisional matrix formation may predispose the implant to failure through mechanisms related to fibrous capsule formation.

The inflammatory (innate) and immune (adaptive) responses have common components. It is possible to have inflammatory responses only with no adaptive immune response. In this situation, both humoral and cellular components that are shared by both types of responses may only participate in the inflammatory response. Table 22.4 indicates the common components to the inflammatory (innate) and immune (adaptive) responses. Macrophages and dendritic cells are known as professional antigen-presenting cells responsible for the initiation of the adaptive immune response.

IMMUNOTOXICITY (ACQUIRED IMMUNITY)

The acquired or adaptive immune system acts to protect the host from foreign agents or materials and is usually initiated through specific recognition mechanisms and the ability of humoral and cellular components to recognize the foreign agent or material as being "non-self" (Coligan et al., 1992; Burleson et al., 1995; Smialowicz and Holsapple, 1996; Janeway and Travers, 1997; Rose et al., 1997). Generally, the adaptive immune system may be considered as having two components: humoral or cellular. Humoral components include antibodies, complement components, cytokines, chemokines, growth factors, and other soluble mediators. These components are synthesized by cells of the immune response and, in turn, function to regulate the activity of these same cells and provide for communication between different cells in the cellular component of the adaptive immune response. Cells of the immune system arise from stem cells in the bone marrow (B lymphocytes) or the thymus (T lymphocytes) and differ from each other in morphology, function, and the expression of cell-surface antigens. They share the common features of maintaining cell-surface receptors that assist in the recognition and/or elimination of foreign materials. Regarding tissue-engineered devices, the adaptive immune response may recognize the biological components, modifications of the biological components, or degradation products of the biological components, commonly known as antigens, and initiate immune response through humoral or cellular mechanisms.

Components of the humoral immune system play important roles in the inflammatory responses to foreign materials. Antibodies and complement components C3b and C3bi adhere to foreign materials, act as opsonins and facilitate phagocytosis of the foreign materials by neutrophils and macrophages that have cell-surface receptors for C3b. Complement component C5a is a chemotactic agent for neutrophils, monocytes, and other inflammatory cells and facilitate the immigration of these cells to the implant site. The complement system is composed of classic and alternative pathways that eventuate in a common pathway to produce the membrane attack complex (MAC), which is capable of lysing microbial agents. The complement system (i.e.complement cascade) is closely controlled by protein inhibitors in the host cell membrane that may prevent damage to host cells. This inhibitory mechanism may not function when non-host cells are used in tissue-engineered devices.

T (thymus-derived) lymphocytes are significant cells in the cell-mediated adaptive immune response and their cell-adhesion molecules play a significant role in lymphocyte migration, activation, and effector function. The specific interaction of cell membrane adhesion molecules, sometimes also called ligands or antigens, with antigen-presenting cells (APCs) produce specific types of lymphocytes with specific functions. Table 22.5 indicates cell types and function in the adaptive immune response. Obviously, the functions of these cells are more numerous than that indicated in Table 22.5 but the major function of these cells is provided to indicate similarities and differences in the interaction and responsiveness of these cells. Effector T-cells (Table 22.6) are produced when their antigen-specific receptors and either the CD4 or the CD8 co-receptors bind to peptide-MHC (major histocompatibility complex) complexes. A second, co-stimulatory signal is also required and this is provided by the interaction of the CD28 receptor on the T-cell and the B7.1 and B7.2 glycoproteins of the immunoglobulin superfamily present on APCs. B lymphocytes bind soluble antigens through their cell-surface immunoglobulin and thus can function as professional APCs by internalizing the soluble antigens and presenting peptide fragments of these antigens as MHC: peptide complexes. Once activated, T-cells can

Table 22.5 Cell types and function in the adaptive immune system

Cell type	Motor function
Macrophages (APC)	Process and present antigen to immunocompetent T-cells Phagocytosis Activated by cytokines (i.e. IFN-γ) from other immune cells
T-cells	Interact with APCs and are activated through two required cell membrane interactions Facilitate target cell apoptosis Participate in transplant rejection (type IV hypersensitivity)
B-cells	Form plasma cells that secrete immunoglobulins (IgG, IgA, and IgE) Participate in antigen–antibody complex mediated tissue damage (type III hypersensitivity)
Dendritic cells (APC)	Process and present antigen to immunocompetent T-cells Utilize Fc receptors for IgG to trap antigen–antibody complexes
NK cells (non-T, non-B lymphocytes)	Innate ability to lyse tumor, virus infected, and other cells without previous sensitization Mediates T- and B-cell function by secretion of IFN-γ

Table 22.6 Effector T lymphocytes in adaptive immunity

Thl helper cells	CD4+ Pro-inflammatory Activation of macrophages Produces IL-2, interferon-γ (IFN-γ), IL-3, TNF-α, GM-CSF, macrophage chemotactic factor (MCF), migration inhibitor factor (MIF) Induce IgG2a
Th2 helper cells	CD4+ Anti-inflammatory Activation of B-cells to make antibodies Produces IL-4, IL-5, IL-6, IL-10, IL-3, GM-CSF, and IL-13 Induce IgG1
Cytotoxic T-cells (CTL)	CD8+ Induce apoptosis of target cells Produce IFN-γ, TNF-β, and TNF-α Release cytotoxic proteins

synthesize the T-cell growth factor IL-2 and its receptor. Thus, activated T-cells secrete and respond to IL-2 to promote T-cell growth in an autocrine fashion.

Cytokines are the messenger molecules of the immune system. Most cytokines have a wide spectrum of effects, reacting with many different cell types, and some are produced by several different cell types. Table 22.7 presents common categories of cytokines and lists some of their general properties. It should be noted that while cytokines can be subdivided into functional groups, many cytokines such as IL-1, TNF-α, and IFN-γ are pleotropic in their effects and regulate, mediate, and activate numerous responses by various cells.

Immunotoxicity is any adverse effect on the function or structure of the immune system or other systems as a result of an immune system dysfunction (Langone, 1998). Adverse or immunotoxic effects occur when humoral or cellular immunity needed by the host to defend itself against infections or neoplastic disease (immunosuppression) or unnecessary tissue damage (chronic inflammation, hypersensitivity, or autoimmunity) is compromised. Potential immunological effects and responses that may be associated with one or more of these effects are presented in Table 22.8. Hypersensitivity responses are classified on the basis of the immunological mechanism that mediates the response. There are four types: type I (anaphylactic), type II (cytotoxic), type III (immune complex), and type IV (cell-mediated delayed hypersensitivity). Hypersensitivity is considered to be increased reactivity to an antigen to which a human or animal has been previously exposed, with an adverse rather than a protective effect. Hypersensitivity is a synonym for allergy. Type I (anaphylactic) reactions and type IV (cell-mediated delayed hypersensitivity) reactions are the most common. Types II and III reactions are relatively rare and are less likely to occur with medical devices and biomaterials, however, with tissue-engineered

Table 22.7 Selected cytokines and their effects

Cytokine	Effect
IL-1, TNF-α, INF-γ, IL-6	Mediate natural immunity
IL-1, TNF-α, IL-6	Initiate non-specific inflammatory responses
IL-2, IL-4, IL-5, IL-12, IL-15 and TGF-β	Regulate lymphocyte growth, activation, and differentiation
IL-2 and IL-4	Promote lymphocyte growth and differentiation
IL-10 and TGF-β	Down-regulate immune responses
IL-1, INF-γ, TNF-α, and MIF	Activate inflammatory cells
IL-8	Produced by activated macrophages and endothelial cells Chemoattractant for neutrophils
MCP-1, MIP-α, and RANTES	Chemoattractant for monocytes and lymphocytes
GM-CSF and G-CSF	Stimulate hematopoiesis
IL-4 and IL-13	Promote macrophage fusion and foreign body giant cell formation

Table 22.8 Potential immunological effects and responses

Effects	Responses
Hypersensitivity	Histopathological changes
Type I – anaphylactic	Humoral responses
Type II – cytotoxic	Host resistance
Type III – immune complex	Clinical symptoms
Type IV – cell-mediated (delayed)	Cellular responses
Chronic inflammation	T-cells
Immunosuppression	NK cells
Immunostimulation	Macrophages
Autoimmunity	Granulocytes

devices containing potential antigens (i.e. proteins), extracellular matrix (ECM) components, and/or cells, types II and III reactions must be considered in biological response evaluations.

Type I (anaphylactic) hypersensitivity reactions are mediated by IgE antibodies which are cytotropic and affect the immediate release of basoactive amines and other mediators from basophils and mast cells followed by recruitment of other inflammatory cells. Type IV cell-mediated (delayed) hypersensitivity responses involve sensitized T lymphocytes that release cytokines and other mediators that lead to cellular and tissue injury. Type IV hypersensitivity (cell-mediated) reactions are initiated by specifically sensitized T lymphocytes. This reaction includes the classic delayed-type hypersensitivity reaction initiated by CD4+ T-cells and direct cell cytotoxicity mediated by CD8+ T-cells. The less common type II (cytotoxic) hypersensitivity involves the formation and binding IgG and/or IgM to antigens on target cell surfaces that facilitate phagocytosis of the target cell or lysis of the target cell by activated complement components. Type II hypersensitivity (cytotoxic) is mediated by antibodies directed toward antigens present on the surface of cells or other tissue components. Three different antibody-dependent mechanisms may be involved in this type of reaction: complement-dependent reactions, antibody-dependent, cell-mediated cytotoxicity, or antibody-mediated cellular dysfunction. Type III immune complex hypersensitivity is present when circulating antigen–antibody complexes activate complement whose components are chemotactic for neutrophils that release enzymes and other toxic moieties and mediators leading to cellular and tissue injury.

Immunological reactions that occur with organ transplant rejection also offer insight into potential immune responses to tissue-engineered devices. Mechanisms involved in organ transplant rejection include T-cell-mediated reactions by direct and indirect pathways and antibody-mediated reactions. Immune responses may be avoided or diminished by using autologous or isogeneic cells in cell/polymer scaffold constructs. The use of allogeneic or xenogenic cells incorporated into the device requires prevention of immune rejection by immune suppression of the host, induction of tolerance in the host, or immunomodulation of the tissue-engineered construct. The development of tissue-engineered constructs by immunoisolation using polymer membranes and the use of non-host cells have been compromised by immune responses. In this concept, a polymer membrane is used to encapsulate non-host cells or tissues thus separating them from the host immune system. However, antigens shed by encapsulated cells were released from the device and initiated immune responses (Brauker, 1992; Brauker et al., 1995; Babensee et al., 1998).

Although exceptionally minimal and superficial in its presentation, the previously discussed humoral and cell-mediated immune responses demonstrate the possibility that any known tissue-engineered construct may undergo immunological tissue injury. To date, our understanding of immune mechanisms and their interactions with tissue-engineered constructs is markedly limited. One of the obvious problems is that preliminary studies are generally carried out with non-human tissues and immune reactions result when tissue-engineered constructs from one species are used in testing the device in another species. Ideally, tissue-engineered constructs would be prepared from cells and tissues of a given species and subsequently tested in that species. While this approach does not guarantee that immune responses will not be present, the probability of immune responses in this type of situation is markedly decreased.

The following examples provide perspective to these issues. They further demonstrate the detailed and in-depth approach that must be taken to appropriately and adequately evaluate tissue-engineered constructs or devices and their potential adverse responses.

The inflammatory response considered to be immunotoxic is persistent chronic inflammation. With biomaterials, controlled release systems and tissue-engineered devices, potential antigens capable of stimulating the immune response may be present and these agents may facilitate a chronic inflammatory response that is of extended duration (weeks, months). Regarding immunotoxicity, it is this persistent chronic inflammation that is of concern as immune granuloma formation and other serious immunological reactions such as autoimmune disease may occur. Thus, in biological response evaluation, it is important to discriminate between the short-lived chronic inflammation that is a component of the normal inflammatory and healing responses versus long-term, persistent chronic inflammation that may indicate an adverse immunological response.

Immunosuppression may occur when antibody and T-cell responses (adaptive immune response) are inhibited. Potentially significant consequences of this type of response are frequent and serious infections resulting from reduced host defense.

Immunostimulation may occur when unintended or inappropriate antigen-specific or non-specific activation of the immune system is present. From a biomaterial and controlled release system perspective, antibody and/or cellular immune responses to a foreign protein may lead to unintended immunogenicity. Enhancement of the immune response to an antigen by a biomaterial with which it is mixed *ex vivo* or *in situ* may lead to adjuvancy, which is a form of immunostimulation. This effect must be considered when biodegradable controlled release systems are designed and developed for use as vaccines.

Autoimmunity is the immune response to the body's own constituents, which are considered in this response to be autoantigens. An autoimmune response, indicated by the presence of autoantibodies or T lymphocytes that are reactive with host tissue or cellular antigens may, but not necessarily, result in autoimmune disease with chronic, debilitating and sometimes life-threatening tissue and organ injury.

Representative tests for the evaluation of immune responses are given in Table 22.9. Table 22.9 is not all-inclusive and other tests may be applicable. The examples presented in Table 22.9 are only representative of the large number of tests that are currently available (Coligan et al., 1992; Burleson et al., 1995; Smialowicz and Holsapple, 1996; Rose et al., 1997). Table 22.9 is informative but incomplete as in the future direct and indirect markers of immune response may be validated and their predictive value documented thus providing new tests for immunotoxicity. Direct measures of immune system activity by functional assays are the most important types of test for immunotoxicity. Functional assays are generally more important than tests for soluble mediators, which are more important than phenotyping. Signs of illness may be important in *in vivo* experiments but symptoms may also have a significant role in studies of immune function in clinical trials and postmarket studies.

As with any type of test for biological response evaluation, immunotoxicity tests should be valid and have been shown to provide accurate, reproducible results that are indicative of the effect being studied and are useful in a statistical analysis. This implies that appropriate control groups are also included in the study design.

Table 22.9 Representative tests for the evaluation of immune responses

Functional assays	**Soluble mediators**
Skin testing	Antibodies
Immunoassays (e.g. ELISA)	Complement
Lymphocyte proliferation	Immune complexes
Plaque-forming cells	Cytokine patterns (T-cell subsets)
Local lymph node assay	Cytokines (IL-1, IL-1ra, TNF-α, IL-6, TGF-β, IL-4, IL-13)
Mixed lymphocyte reaction	Chemokines
Tumor cytotoxicity	Basoactive amines
Antigen presentation	**Signs of illness**
Phagocytosis	Allergy
Degranulation	Skin rash
Resistance to bacteria, viruses, and tumors	Urticaria
Phenotyping	Edema
Cell-surface markers	Lymphadenopathy
MHC markers	

Immunogenicity involving a specific immune response to a biomaterial is an important consideration as it may lead to serious adverse effects. For example, a foreign, non-human, protein may induce IgE antibodies that cause an anaphylactic (type I) hypersensitivity reaction. An example of this type of response is latex protein found in latex gloves. Low molecular weight compounds such as chemical accelerators used in the manufacture of latex gloves may also induce a T-cell-mediated (type IV) reaction resulting in contact dermatitis. Tests for type I (e.g. antigen-specific IgE) and type IV (e.g. guinea pig) maximization tests, hypersensitivity should be considered for materials with the potential to cause these allergic reactions. In addition to hypersensitivity reactions, a device may elicit autoimmune responses (i.e. antibodies or T-cells) that react with the body's own constituents. An autoimmune response may lead to the pathological consequences of an autoimmune disease. For example, a foreign protein may induce IgG or IgM antibodies that cross-react with a human protein and cause tissue damage by activating the complement system. In a similar fashion, a biomaterial or controlled release system which has a gel or oil constituent may act as an adjuvant leading to the induction of an autoimmune response. Even if an autoimmune response (autoantibodies and/or autoreactive T lymphocytes) is suggested in preclinical testing, it is difficult to obtain convincing evidence that a biomaterial or controlled release system causes autoimmune disease in animals. Therefore, routine testing for induction of autoimmune disease in animal models is not recommended.

Babensee and co-workers have tested the hypothesis that the biomaterial component of a medical device, by promoting an inflammatory response can recruit APCs (e.g. macrophages and dendritic cells) and induce their activation, thus acting as an adjuvant in the immune response to foreign antigens originating from the histological component of the device (Babensee et al., 2002; Matzell and Babensee, 2004). Utilizing polystyrene and polylactic-glycolic acid microparticles and polylactic-glycolic scaffolds together with their model antigen, ovalbumin, in a mouse model for 18 weeks, Babensee et al. demonstrated that a persistent humoral immune response that was Th2 helper T-cell dependent, as determined by the IgG1, was present. These findings indicated that activation of CD4+ T-cells and the proliferation and isotype switching of B-cells had occurred. A Th1 immune response characterized by the presence of IgG2a was not identified. Moreover, the humoral immune responses for all three types of microparticles were similar indicating that the production of antigen-specific antibodies was not material chemistry-dependent in this model. Babensee suggests that the presence of the biomaterial functions as an adjuvant for initiation and promotion of the immune response and augments the phagocytosis of the antigen with expression of MHC class II and co-stimulatory molecules on APCs with the presentation of antigen to CD4+ T-cells.

Babensee and co-workers have identified differential levels of dendritic cell maturation on different biomaterials used in combination products (Babensee and Paranjpe, 2005; Bennewitz and Babensee, 2005). The effect of biomaterials on dendritic cell maturation, and the associated adjuvant effect, is a novel biocompatibility selection and design criteria for biomaterials to be used in combination products in which immune consequences are potential complications or outcomes.

Badylak and colleagues have carried out extensive studies on the utilization of xenogeneic ECM as a scaffold for tissue reconstruction (Allman et al., 2002; Badylak, 2004). Use of the small intestinal submucosa (SIS) ECM in animals has indicated a restricted Th2-type immune response. The presence of natural antibodies to the terminal galactose-α1,3-galactose (α-gal) epitope is considered to be a major barrier to xenotransplantation in humans. Cell membranes of all animals except those of the humans express this epitope and naturally occurring antibodies mediate hyperacute or delayed rejection of transplanted organs through complement fixation or antibody dependence cell-mediated cytotoxicity. While ECM derived from porcine tissues, SIS, contain small amounts of the gal epitope, it appears that the quantity or distribution of this epitope and/or the subtype of immunoglobulin response to the epitope is such that complement activation does not occur (McPherson et al., 2000). In addition, the resorbable characteristics of this non-chemically cross-linked ECM

Gallin, J.I. and Synderman, R. (eds.) (1999). *Inflammation: Basic Principles and Clinical Correlates*, New York: Raven Press.

Ganz, T. (1988). Neutrophil receptors. In: Lehrer, R.I. moderator, *Neutrophils and Host Defense. Ann. Intern. Med.* **109**: 127–142.

Golden, M.A., Au, Y.P., Kirkman, T.R., Wilcox, J.N., Raines, E.W., Ross, R. and Clowes, A.W. (1991). Platelet-derived growth factor activity and RNA expression in healing vascular grafts in baboons. *J. Clin. Invest.* **87**: 406–414.

Greenwald, R.A. and Diamond, H.S. (eds.) (1988). *Handbook of Animal Models for the Rheumatic Diseases*, Vol. I. Boca Raton, FL: CRC Press.

Greisler, H. (1988). Macrophage-biomaterial interactions with bioresorbable vascular prostheses. *Trans. Am. Soc. Artif. Intern. Organs.* **34**: 1051–1057.

Henson, P.M. (1971). The immunologic release of constituents from neutrophil leukocytes. II. Mechanisms of release during phagocytosis, and adherence to nonphagocytosable surfaces. *J. Immunol.* **197**: 1547–1557.

Henson, P.M. (1980). Mechanisms of exocytosis in phagocytic inflammatory cells. *Am. J. Pathol.* **101**: 494–511.

Henson, P.M. and Johnston Jr., R.B. (1987). Tissue injury in inflammation: oxidants, proteinases, and cationic proteins. *J. Clin. Invest.* **79**: 669–674.

Hynes, R.O. (1992). Integrins: versatility, modulation, and signaling in cell adhesion. *Cell* **69**: 11–25.

Ignotz, R., Endo, T. and Massague, J. (1987). Regulation of fibronectin and type I collagen mRNA levels by transforming growth factor-beta. *J. Biol. Chem.* **262**(14): 6443–6446.

Janeway Jr., C.A. and Travers, P. (1997). *Immunobiology: The Immune System in Health and Disease*, 3rd edn. Current Biology. New York: Garland.

Jenney, C.R. and Anderson, J.M. (2000). Adsorbed IgG: a potent adhesive susbstrate for human macrophages. *J. Biomed. Mater. Res.* **50**: 281–290.

Johnston Jr., R.B. (1988). Monocytes and macrophages. *N. Engl. J. Med.* **318**: 747–752.

Jutila, M.A. (1990). Leukocyte traffic to sites of inflammation. *APMIS* **100**: 191–201.

Kovacs, E.J. (1991). Fibrogenic cytokines: the role of immune mediators in the development of scar tissue. *Immunol. Today*. **12**: 17–23.

Langone, J.J. (1998). *Immunotoxicity Testing Guidance*, Draft Document. Office of Science and Technology, Center for Devices and Radiological Health, Food and Drug Administration.

Maciag, T. (1990). Molecular and cellular mechanisms of angiogenesis. In: DeVita, V T., Hellman, S. and Rosenberg, S. (eds.), *Important Advances in Oncology*. Lippincott, Philadelphia. pp. 85–103.

Madri, J.A., Pratt, B.M. and Tucker, A.M. (1988). Phenotypic modulation of endothelial cells by transforming growth factor-beta depends upon the composition and organization of the extracellular matrix. *J. Cell Biol.* **106**(4): 1375–1384.

Malech, H.L. and Gallin, J.I. (1987). Current concepts: immunology. neutrophils in human diseases. *N. Engl. J. Med.* **317**: 687–694.

Marchant, R., Hiltner, A., Hamlin, C., Rabinovitch, A., Slobodkin, R. and Anderson, J.M. (1983). *In vivo* biocompatibility studies: I. The cage implant system and a biodegradable hydrogel. *J. Biomed. Mater. Res.* **17**: 301–325.

Matzell, M.M. and Babensee, J.E. (2004). Humoral immune responses to model antigen co-delivered with biomaterials used in tissue engineering. *Biomaterials* **25**: 295–304.

McNally, A.K. and Anderson, J.M. (1994). Complement C3 participation in monocyte adhesion to different surfaces. *Proc. Natl Acad. Sci. USA*. **91**: 10119–10123.

McNally, A.K. and Anderson, J.M. (1995). Interleukin-4 induces foreign body giant cells from human monocytes/macrophages. Differential lymphokine regulation of macrophage fusion leads to morphological variants of multinucleated giant cells. *Am. J. Pathol.* **147**: 1487–1499.

McNally, A.K., DeFife, K.M. and Anderson, J.M. (1996). Interleukin-4-induced macrophage fusion is prevented by inhibitors of mannose receptor activity. *Am. J. Pathol.* **149**: 975–985.

McNally, A.K. and Anderson, J.M. (2002). Beta1 and beta2 integrins mediate adhesion during macrophage fusion and multinucleated foreign body giant cell formation. *Am. J. Pathol.* **160**: 621–630.

McPherson, T.B., Liang, H., Record, R.D. and Badylak, S.F. (2000). Galα (1,3)Gal epitope in porcine small intestinal submucosa. *Tissue Eng.* **6**(3): 233–239.

Muller, G., Behrens, J., Nussbaumer, U., Böhlen, P. and Birchmeier, W. (1987). Inhibitor action of transforming growth factor beta on endothelial cells. *Proc. Natl Acad. Sci. USA*. **84**(16): 5600–5604.

Mustoe, T.A., Pierce, G.F., Thomason, A., Gramats, P., Sporn, M.B. and Deuel, T.F. (1987). Accelerated healing of incisional wounds in rats induced by transforming growth factor. *Science* **237**: 1333–1336.

Paty, P.B., Graeff, R.W., Mathes, S.J. and Hunt, T.K. (1990). Superoxide production by wound neutrophils: evidence for increased activity of the NADPH oxidase. *Arch. Surg.* **125**: 65–69.

Pober, J.S. and Cotran, R.S. (1990). The role of endothelial cells in inflammation. *Transplantation* **50**: 537–544.

Rae, T. (1986). The macrophage response to implant materials. *Crit. Rev. Biocompatibility* **2**: 97–126.

Riches, D.W.F. (1988). Macrophage involvement in wound repair, remodeling, and fibrosis. In: Clark, R.A.P. and Henson, P M. (eds.), *The Molecular and Cellular Biology of Wound Repair.* New York: Plenum Press, pp. 213–239.

Rose, N.R. (1997). Immunologic diagnosis of autoimmune disease. In: Lefell, M.S., Donnenberg, A.D. and Rose, N.R. (eds.), *Handbook of Human Immunology.* Boca Raton, FL: CRC Press, pp. 111–123.

Rose, N.R., de Mecario, E.C., Folds, J.D., Lane, H.C. and Nakamura, R.M. (eds.) (1997). *Manual of Clinical Laboratory Immunology*, Washington, D.C.: ASM Press.

Salthouse, T.N. (1976). Cellular enzyme activity at the polymer–tissue interface: a review. *J. Biomed. Mater. Res.* **10**: 197–229.

Smialowicz, R.J. and Holsapple, M.P. (eds.) (1996). *Experimental Immunotoxicology*, Boca Raton, FL: CRC Press.

Spilizewski, K.L., Marchant, R.E., Hamlin, C.R., Anderson, M., Tice, T.R., Dappert, T.O. and Meyers, W.E. (1985). The effect of hydrocortisone acetate loaded poly(DL-lactide) films on the inflammatory response. *J. Contr. Rel.* **2**: 197–203.

Sporn, M.B. and Roberts, A.B. (1988). Peptide growth factors are multifunctional. *Nature* **332**(6161): 217–219.

Sporn, M.B. and Roberts, A.B. (eds.) (1990). *Peptide Growth Factors and Their Receptors I*, New York: Springer

Stewart, T.A., Hollinghead, P.G., Pitts, S.L., Chang, R., Martin, L.E. and Oakley, H. (1989). Transgenic mice as a model to test the immunogenicity of proteins altered by site-specific mutagenesis. *Mol. Biol. Med.* **6**: 275–281.

Stewart, T.A. (1993). Models of human endocrine disorders in transgenic rodents. *Trends Endocrinol. Metab.* **4**: 136–141.

Tang, L. and Eaton, J.W. (1993). Fibrin(ogen) mediates acute inflammatory responses to biomaterials. *J. Exp. Med.* **178**: 2147–2156.

Tang, L. (1998). Mechanism of pro-inflammatory fibrinogen-biomaterial interactions. *J. Biomater. Sci. Polymer Ed.* **9**: 1257–1266.

Thompson, J.A., Anderson, K.D., DiPetro, J.M., Zweibel, J.A., Zmaetta, M., Anderson, W.F. and Maciag, T. (1988). Site-directed neovessel formation *in vivo*. *Science* **241**: 1349–1352.

Wahl, S.M., Hunt, D.A., Wakefield, L.M., Roberts, A.B. and Sporn, M.B. (1987). Transforming growth factor-beta (TGF-β) induces monocyte chemotaxis and growth factor production. *Proc. Natl Acad. Sci. USA*. **84**(16): 5788–5792.

Wahl, S.M., Wong, H. and McCartney-Francis, N. (1989). Role of growth factors in inflammation and repair. *J. Cell Biochem*. **40**(2) 193–199.

Weisman, G., Smolen, J.E. and Krochak, H.M. (1980). Release of inflammatory mediators from stimulated neutrophils. *N. Engl. J. Med.* **303**: 27–34.

Weiss, S.J. (1989). Tissue destruction by neutrophils. *N. Engl. J. Med.* **320**: 365–376.

Wiggins, M.J., Wilkoff, B., Anderson, J.M. and Hiltner, A. (2001). Biodegradation of polyether polyurethane inner insulation in bipolar pacemaker leads. *J. Biomed. Mater. Res. (Appl. Biomater.)* **58**: 302–307.

Williams, G.T. and Williams, W.J. (1983). Granulomatous inflammation – a review. *J. Clin. Pathol.* **36**: 723–733.

Zhao, Q., Agger, M.P., Fitzpatrick, M., Anderson, J.M., Hiltner, A., Stokes, K. and Urbanski, P. (1990). Cellular interactions with biomaterials: *in vivo* cracking of pre-stressed pellethane 2363–80A. *J. Biomed. Mater. Res.* **24**: 621–637.

Zhao, Q., Topham, N., Anderson, J.M., Hiltner, A., Lodoen, G. and Payet, C.R. (1991). Foreign-body giant cells and polyurethane biostability: *in vivo* correlation of cell adhesion and surface cracking. *J. Biomed. Mater. Res.* **25**: 177–183.

Ziats, N.P., Miller, K.M. and Anderson, J.M. (1985). *In vitro* and *in vivo* interactions of cells with biomaterials. *Biomaterials* **9**: 5–13.

Part V

Therapeutic Applications: Cell Therapy

23

Islet Cell Transplantation

Juliet A. Emamaullee and A.M. James Shapiro

INTRODUCTION

Background

Diabetes is a disease that results from impaired glucose metabolism. Approximately 90% of diabetes is caused by a defect in insulin production and/or utilization (Type 2 diabetes mellitus; "T2DM"), while the more severe form, Type 1 diabetes mellitus ("T1DM"), is caused by a complete loss of the insulin-producing β-cells within the islets of Langerhans of the pancreas. Diabetes currently affects more than 200 million patients worldwide and is projected to afflict at least 5% of the global adult population by the year 2025 (King et al., 1998). As the incidence of diabetes increases, the cost of treating these patients has skyrocketed, consuming between 7% and 13% of health-care expenditure in developed countries (WHO, 2002). Since the discovery of insulin in 1921, diabetes has become a treatable condition, and the life expectancy of patients with diabetes has been greatly improved. However, even with diligent blood glucose monitoring and insulin administration, the metabolic abnormalities associated with diabetes can lead to many chronic secondary complications, including nephropathy, retinopathy, peripheral neuropathy, coronary ischemia, stroke, amputation, erectile dysfunction, and gastroparesis (National Diabetes Data Group (US) et al., 1995). In the US, patients with diabetes represent 8% of those who are legally blind, 30% of all patients on dialysis due to end-stage renal disease, and 20% of all patients receiving kidney transplants (National Diabetes Data Group (US) et al., 1995). The Diabetes Control and Complications Trial (DCCT) was conducted to determine if intensive blood glucose regulation by frequent insulin injection or pump could prevent these long-term complications in patients with diabetes (DCCT Research Group, 1990, 1993; Keen, 1994). Results from the DCCT and subsequent Epidemiology of Diabetes Interventions and Complications (EDIC) study have clearly demonstrated that this approach improved but did not normalize glycosylated hemoglobin levels (HbA1C) and significantly protected against cardiovascular disease, nephropathy, neuropathy, and retinopathy (DCCT Research Group, 1990; Keen, 1994; Nathan et al., 2003, 2005). However, the consequence of improved glycemic control was a threefold increased risk of serious hypoglycemic reactions leading to recurrent seizures and coma (Keen, 1994; DCCT Research Group, 1995). Recent improvements in the size and sensitivity of insulin pumps have increased their utility, but the creation of implantable devices has been more challenging. Also, while insulin pump therapy can improve HbA1C levels compared to multiple daily injections of insulin, pumps may malfunction and thus still necessitate frequent blood glucose monitoring by the user (Owen, 2006).

While advances in the formulation, half-life, and administration of insulin have markedly improved the quality of life- and long-term survival of patients with diabetes, it has long been recognized that the restoration of an adequate islet mass would provide the maximum benefit to diabetic patients, leading to a true physiological correction of the diabetic state. In the early 1960s, great advances were made in the field of renal transplantation due to improved immunosuppressive therapies (azathioprine and corticosteroids), which prompted the first attempts in whole pancreas transplantation (Merrill et al., 1963; Murray et al., 1963).

 Copyright © 2008 Elsevier Inc. All rights reserved

First introduced by Kelly and Lillehei in 1966, early attempts were associated with high mortality rates and poor graft survival, with <3% graft function at 1 year post-transplant (Kelly et al., 1967). The risk profile and long-term outcomes in whole pancreas transplantation have been greatly improved by recent improvements in surgical technique, including portal venous and enteric endocrine drainage, and steroid-free maintenance immunosuppression (Newell et al., 1996; Kendall et al., 1997). To date more than 25,000 pancreas transplants have been performed worldwide for end-stage renal disease (simultaneous kidney pancreas or pancreas after kidney transplantation) or less frequently for severe hypoglycemic unawareness (pancreas transplant alone). Data collected in the International Pancreas Transplant Registry (IPTR) have shown that only 50% of patients who have undergone pancreas-alone transplantation remain insulin independent at 5 years, despite recent improvements in surgical technique and immunosuppression (Larsen, 2004; Gruessner and Sutherland, 2005). Also, <30% of the approximately 6,000 cadaveric pancreata donated each year are transplanted due to strict donor criteria and requirements for short cold ischemic time (Larsen, 2004, 2005b). The surgical risks and requirement for lifelong immunosuppression have reserved pancreas-alone transplantation only for those diabetic patients with the most severe and life-threatening disease, despite strong evidence that the procedure can prolong life, reverse established nephropathy, and improve quality of life. Since the major surgical complications in whole pancreas transplantation are related to the exocrine function of the pancreas, which is not necessary to restore euglycemia in diabetic patients, it has long been recognized that β-cell replacement could be achieved with implantation of isolated pancreatic islets. Since this approach involves transplantation of a cellular graft that would be implanted using minimally invasive techniques, it would avoid the risks associated with major surgery, resulting in a more widely available treatment for patients with diabetes.

History of Islet Transplantation

The concept of islet transplantation actually preceded the discovery of insulin in 1921 by nearly 30 years (Figure 23.1). In 1893, physicians in Bristol attempted to treat a young boy suffering from diabetic ketoacidosis by transplanting fragments of a freshly slaughtered sheep's pancreas (Williams, 1894). While the graft ultimately failed in the absence of immunosuppression, the patient's health did temporarily improve, which suggested that cells within the pancreas could restore euglycemia. After the discovery of insulin, it was thought that exogenous insulin replacement would be an effective treatment for patients with T1DM, and therefore islet transplantation was not actively pursued. However, as insulin therapy transformed T1DM from an acute health crisis to a chronic disease, it became apparent that insulin injections could not prevent the onset of debilitating and life-threatening secondary complications. As the first series of whole pancreas transplants in the late 1960s were associated with poor morbidity and mortality, isolated islet transplantation gained a renewed interest (Sutherland et al., 2001). The first successful islet isolations and subsequent transplantation into chemically induced diabetic rodents were pioneered by Dr. Paul Lacy at Washington University in St. Louis, which immediately sparked interest in the implementation of clinical trials (Lacy and Kostianovsky, 1967; Ballinger and Lacy, 1972; Kemp et al., 1973; Reckard et al., 1973). While euglycemia was routinely obtained in animal models of islet transplantation, clinical islet transplantation struggled to find success for most of the 1970s and 1980s. During this time, unpurified islets were infused into the portal vein, leading to many serious complications including portal vein thrombosis, portal hypertension, and disseminated intravascular coagulation (Walsh et al., 1982). While working in Lacy's group, Dr. Camillo Ricordi developed the "automated method" for high-yield islet isolation in 1989 (Ricordi et al., 1989). This represented a major turning point in the field and led to the report that Lacy's group had achieved short-lived insulin independence in a patient with T1DM who had received an islet graft following a previous kidney transplant (Scharp et al., 1990). The following year, the group led by Ricordi at the University of Pittsburgh reported the first series of clinical islet allografts that demonstrated improved insulin-independence rates of 50% at 1 year, in

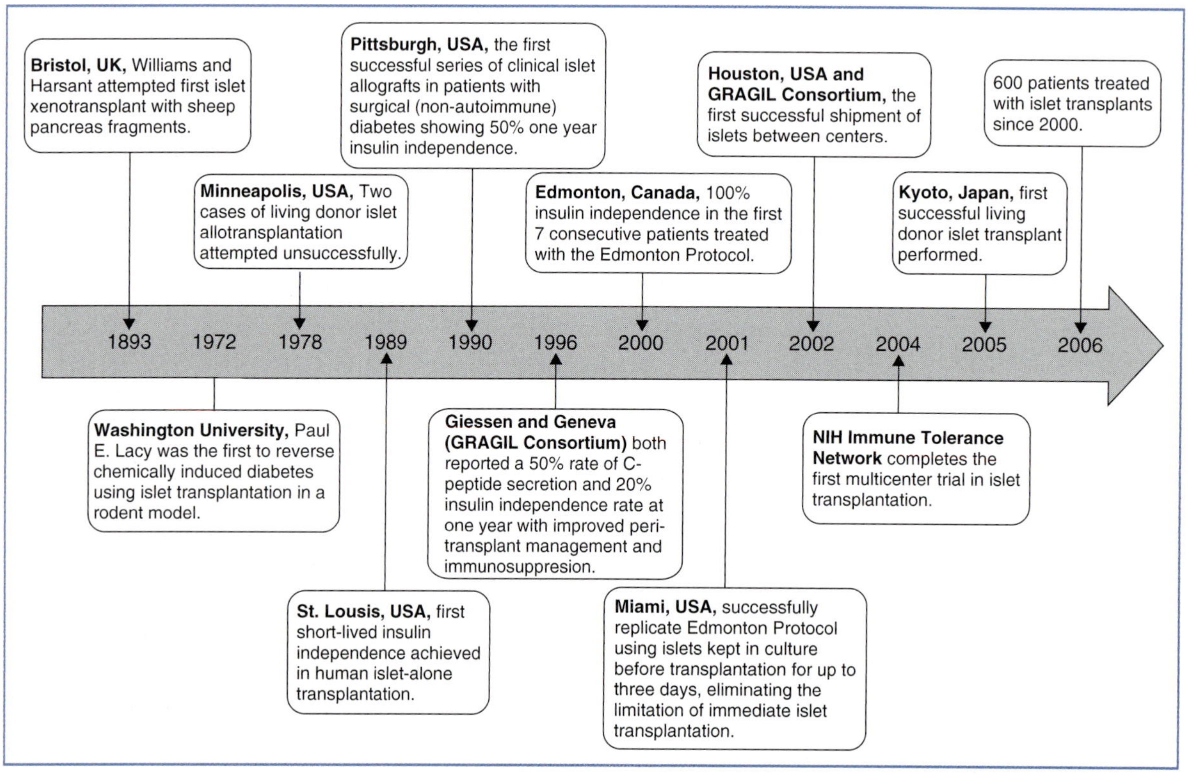

Figure 23.1 Timeline of notable advances in the history of islet transplantation.

subjects who underwent cluster islet–liver transplants for abdominal malignancies in the setting of surgical-induced (non-autoimmune) diabetes (Ricordi et al., 1989; Tzakis et al., 1990). Although this represented a major advance in the field of islet transplantation, these results could not be reproduced in patients with T1DM, the key patient population in need of β-cell replacement (Ricordi et al., 1992). In the late 1990s, the European GRAGIL consortium reported the first modestly successful insulin-independence rates of 20% at 1 year in patients with T1DM, which could be attributed to improved peritransplant management and immunosuppressive drug regimens (Benhamou et al., 2001). Since the results from Pittsburgh and the GRAGIL consortium were obtained in patients who had previously received a kidney transplant, there was no additional risk in terms of immunosuppression to the patients after receiving an islet graft (Ricordi et al., 1992; Benhamou et al., 2001). An international registry held in Giessen, Germany, has maintained a comprehensive record of previous clinical attempts at islet transplantation globally, and of the total world experience of over 450 attempts at clinical islet transplantation prior to 2000, <8% of subjects achieved insulin independence (Brendel, 2001). After three decades of research, the 1 year insulin-independence rates in clinical islet transplantation were still too low to justify the risks associated with portal infusion and lifelong immunosuppression in the majority of patients with T1DM (Secchi et al., 1991; Gross et al., 1998; Hering, 1999; Benhamou et al., 2001; Brendel, 2001).

The Edmonton Protocol

Shapiro and colleagues at the University of Alberta developed a new protocol in 1999 that was designed for patients with "brittle diabetes" who experienced extreme difficulty in managing their blood glucose levels ("glucose lability") and/or severe hypoglycemic unawareness (Shapiro et al., 2000). The so-called "Edmonton Protocol" was unique compared to previous attempts in clinical islet transplantation in its high-targeted islet mass, with a mean of approximately 13,000 islet equivalents (IE)/kg recipient body weight, often derived from two (or occasionally more) fresh islet preparations, and in its immunosuppression strategy, with emphasized avoidance of corticosteroids and use of potent immunosuppression with combined sirolimus, tacrolimus, and anti-CD25 antibody to protect against rejection and recurrent autoimmunity (Shapiro et al., 2000). This approach lead to dramatic improvements in islet allograft survival, with all of the first seven patients achieved sustained independence from insulin (Shapiro et al., 2000). More than 85 consecutive patients have received islet transplants at the University of Alberta since 1999, and the 1 year insulin-independence rate remains steady at approximately 80% after completed transplants (<13,000 IE/kg). The results obtained at the University of Alberta have been replicated at other centers as part of an international multicenter trial through the Immune Tolerance Network, but each center's success has varied greatly depending on its previous experience and skill in islet isolation and immunosuppressive management (Shapiro et al., 2003). The Miami group has demonstrated that islets can be cultured for up to 3 days pre-transplant or shipped and transplanted at a remote facility (Houston) with similar success as freshly isolated islets when transplanted using Edmonton-like immunosuppression (Goss et al., 2002, 2004). The GRAGIL Network (a Swiss-French consortium) has also demonstrated the benefits of centralized islet processing facilities which can service a broader network of centers throughout Europe (Benhamou et al., 2001; Kempf et al., 2005).

Based upon the success of the Edmonton group, islet transplantation has been funded in Alberta, Canada, as accepted clinical standard of care since 2001. Progress in this area has been slower in the United States, but large registration trials are currently moving forward to secure a Biological License and therefore reimbursement, which will make a significant difference to the availability of islets for transplantation in that country. The recent success of clinical islet transplantation has encouraged many centers around the world to implement a program, and since 2000 more than 550 patients have been transplanted using recent variants of the Edmonton Protocol in almost 50 centers worldwide (International Islet Transplant Registry, 2005a).

Despite this success, the current requirement for lifelong immunosuppression in islet-alone transplantation has restricted its availability to patients with T1DM and severe hypoglycemia or glycemic lability. The benefit of islet transplantation in patients with T2DM has not been determined, since many of these patients are overweight and/or insulin resistant and thus would require a large islet mass to meet their metabolic demands. Most patients require two or occasionally three islet implant procedures in order to achieve insulin independence, although insulin independence following single donor infusion has been reported in a cohort of patients at the University of Minnesota (Hering et al., 2004, 2005). While C-peptide secretion (>0.5 ng/ml) has been maintained in 88% of islet graft recipients beyond 3 years in Edmonton, emerging data on the long-term insulin-independence rates have shown that only 50% of recipients remain off insulin at 3 years, with <10% off insulin at 5 years post-transplant (Ryan et al., 2005). Although the exact cause of the discrepancy between insulin independence and maintenance of C-peptide status is not fully understood, it is likely that there are multiple events which hinder graft function and survival over time. While rejection (acute or chronic) and recurrent autoimmunity may be responsible for graft loss, it is probable that other, non-immune-mediated damage occurs, such as chronic toxicity from sirolimus/tacrolimus and failure of islet regeneration or transdifferentiation due to the anti-proliferative effects of sirolimus. Perhaps the most important component of decaying graft function over time is the concept of islet "burn-out" from constant metabolic stimulation, since only a marginal mass of islets actually engraft in most subjects. In clinical islet transplantation thus far, the risks of malignancy, post-transplant lymphoma and life-threatening sepsis have been minimal, but fears of these complications limit a broader application in patients with less severe forms of diabetes including children. Moreover, a number of immunosuppression-related side effects have been encountered, including dyslipidemia, mouth ulceration, peripheral edema, fatigue, ovarian cysts, and menstrual irregularities in female subjects, which can be dose or drug limiting in some patients (Ryan et al., 2002).

Thus, while dramatic improvements in outcomes following islet transplantation have been observed, extensive refinements in clinical protocols are needed both to improve safety and to enhance success with single donor islet infusions.

CLINICAL ISLET TRANSPLANTATION

Patient Assessment and Selection

Clinical islet transplantation is associated with a number of risks, including procedural complications such as bleeding or portal vein thrombosis, or those associated with lifelong immunosuppression (i.e. infection or malignancy). For these reasons, patients selected as islet recipients must have severe, life-threatening diabetic complications that justify the risks of transplantation. Two T1DM patient populations have been identified as suitable candidates for islet transplantation: those individuals that experience frequent, severe and recurrent hypoglycemic unawareness, or those patients with highly unstable blood glucose control despite an optimized insulin regimen (glycemic lability).

When patients are evaluated for islet transplantation, their metabolic status and diabetes-related secondary complications should be carefully characterized so that those patients who would receive the greatest benefit despite the requirement for lifelong immunosuppression are selected. First and foremost, islet transplantation is reserved for patients with C-peptide negative (<0.3ng/ml) T1DM. Recipients with elevated body mass index (BMI) (>30kg/m^2) or those >90kg are generally excluded, as their metabolic demand may not be met by the transplanted islet mass. As mentioned previously, the current indications for islet-alone transplantation include severe hypoglycemic unawareness and/or glycemic lability. To assess these symptoms, Ryan et al. developed an objective scoring system to measure the severity of both hypoglycemia (the HYPO score), and the lability index (LI), which is based upon the changes in blood glucose over time (Ryan et al., 2004b). Current selection criteria for islet-alone transplantation include a HYPO score >1047 (90th percentile), LI >433mmol/L^2/h/week (90th percentile), or a composite with the HYPO score >423 (75th percentile), and LI >329 (75th percentile) (Ryan et al., 2005). Since patients with poor diabetes compliance or an inadequate baseline insulin regimen are likely to benefit from improved design of their insulin dosing regimens, patients selected for transplant should have a plasma HbA1C <10%. In an effort to reduce the risk of serious procedural and immunosuppressive drug-related complications, the patient's cardiac and renal function should be carefully assessed. Selected recipients should have adequate cardiac function including blood pressure <160/100mmHg, no evidence of myocardial infarction in the 6 months prior to assessment, no angiographic evidence of non-correctable coronary artery disease, and left ventricular ejection fraction (LVEF) >30% as measured by echocardiogram. To eliminate patients who are better candidates for simultaneous kidney–pancreas transplantation or those who may experience adverse renal function as a result of tacrolimus or sirolimus therapy, selected recipients should have no evidence of macroscopic proteinuria (<300mg/24h) and a calculated glomerular filtration rate (GFR) >80 (>70 in females) ml/min/1.73m^2. Proliferative retinopathy should be stabilized prior to transplantation, as acute correction of glycemic control may lead to accelerated retinopathy. Finally, to reduce the risk of antibody-mediated graft rejection, potential recipients should be screened for panel reactive antibody assays (PRA) and determined to be <20%.

Islet Transplantation Procedure

Although several locations have been tested as potential implantation sites for islet grafts, the high level of graft function and ease of delivery associated with infusion into the portal circulation of the liver have led to this being the transplantation site of choice in clinical protocols (Kemp et al., 1973). There are two accepted approaches for

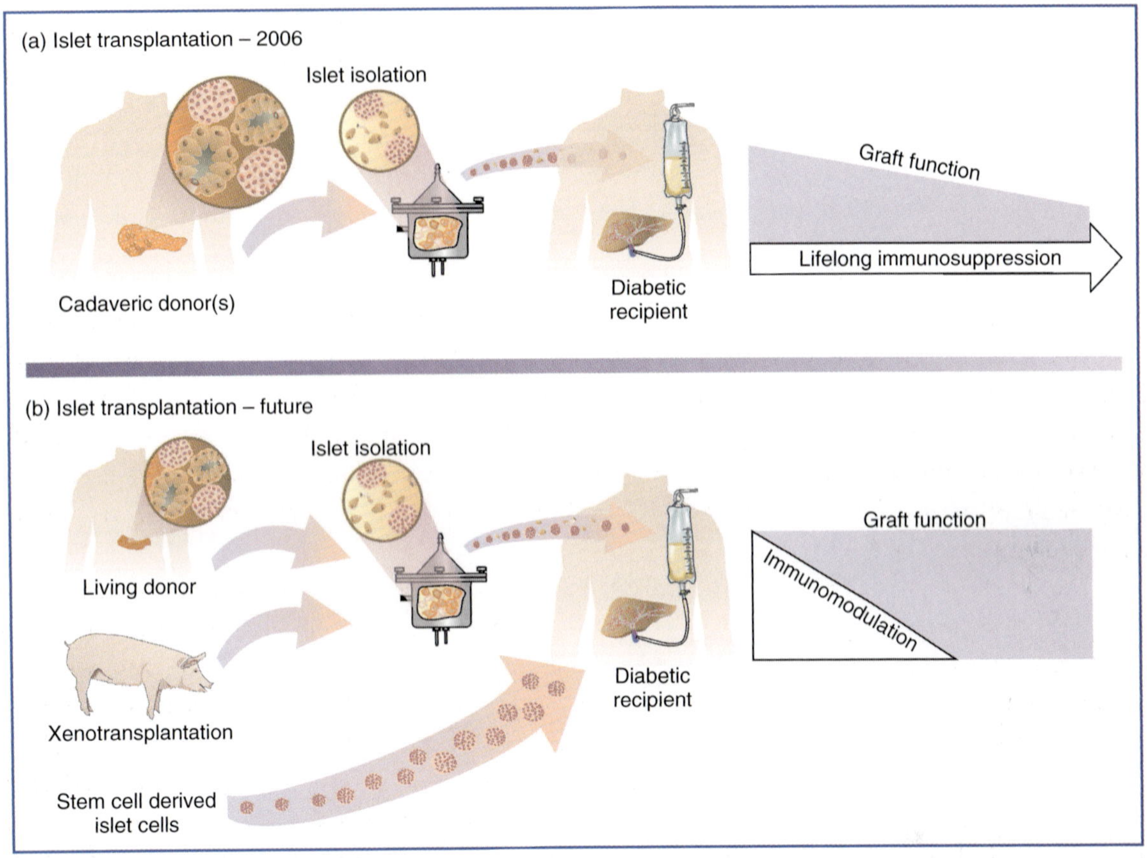

Figure 23.2 The islet transplant procedure-present and future. Islet transplantation, in its current form (a), has provided insulin independence in most diabetic patients at one year post-transplant, but this procedure is currently limited by the availability of suitable cadaveric donors and the requirement for lifelong immunosuppression. In the future (b), islet transplantation could be made available to a broader range of diabetic patients through the usage of alternative tissue sources, such as living donors, xenogeneic donors, or stem-cell derived β-cells. Also, as novel immunomodulatory therapies are identified, tolerance induction strategies can be developed that will prolong graft function and allow for the reduction or complete withdrawal of immunosuppressive drug therapy.

implanting purified islets into the liver by way of the portal vein. While surgical laparotomy and cannulation of the portal vein was most often used in the early islet transplant programs, current protocols routinely employ the percutaneous transhepatic approach to implant donor islets in cadaveric islet transplantation (Figure 23.2a) (Ryan et al., 2005). Compared to surgical laparotomy, this procedure is minimally invasive and thus can be performed using local anesthesia, combined with opiate analgesia and hypnotics given as pre-medication. Access to the portal vein is achieved by percutaneous transhepatic approach using a combination of ultrasound and fluoroscopy to guide the radiologist. A branch of the right portal vein is cannulated, and a catheter is positioned proximal to the confluence of the portal vein, which is confirmed with a portal venogram (Owen et al., 2003). The risk of portal vein thrombosis is reduced by inclusion of unfractionated heparin (70 units/kg) in the islet preparation. Islets are then infused, aseptically, into the main portal vein under gravity, with regular monitoring of portal venous pressure (by an indirect pressure transducer) before, during, and after the infusion. An ultrasound examination should be performed at 1 day and 1 week post-transplant to rule out intraperitoneal hemorrhage and to confirm that the portal vein is patent and has normal flow.

If a patient must be anti-coagulated prior to transplantation or if a hemangioma is present on the right side of the liver that may be at risk for puncture and bleeding if the percutaneous approach were to be used, surgical laparotomy and cannulation of a mesenteric venous tributary of the portal system should be considered. In this situation, complete surgical control is in place to prevent uncontrolled bleeding. Another advantage includes the potential for use of a dual lumen catheter for cannulation of a mesenteric vein (i.e. dual lumen 9Fr Broviac line), which allows for continuous monitoring of portal pressure during islet infusion. Still, this surgical approach should only be considered when the percutaneous transhepatic approach cannot be utilized, as it does present several major disadvantages, including the requirement for a surgical incision, formation of adhesions, and the risk of wound infection and wound herniation, which may be exacerbated when the drug sirolimus is used post-transplant, as this drug interferes with wound healing.

RISKS TO THE RECIPIENT

Surgical Complications

There are two potentially serious procedural complications in islet transplantation: bleeding from the catheter tract created by the percutaneous transhepatic approach, and portal vein thrombosis, particularly when large volumes of tissue are infused. Adverse bleeding events were noted early in the development of the Edmonton program, but these have been completely avoided in the past 40 consecutive procedures with the routine use of effective methods to seal and ablated the transhepatic portal catheter tract on egress when the catheter is withdrawn. The combination of coils and tissue fibrin glue (Tisseel®) was used previously, but more recently has been replaced by Avitene® paste (1 g Avitene powder mixed with 3 ml of radiological contrast media and 3 ml of saline – approximately 0.5–1.0 ml of this paste is injected into the liver tract) (Villiger et al., 2005). The use of purified islet allograft preparations has not resulted in main portal vein thrombosis in the Edmonton program, but thrombosis of a right or left branch, or peripheral segmental vein has been encountered in approximately 5% of patients. Other rarely observed procedural side effects have included fine needle gallbladder puncture, arteriovenous fistulae (which may require selective embolisation) or steatosis in the hepatic parenchyma, which generally does not present any clinical complications or require intervention (Bhargava et al., 2004).

Immunosuppressive Therapy and Complications

Islet transplantation for T1DM represents a unique challenge in immunosuppression, as both alloimmunity and islet-specific autoimmunity must be effectively controlled to preserve graft function. An additional important consideration is that many of the immunosuppressive agents used in solid organ transplantation since the 1960s, particularly corticosteroids, are known to be toxic to islets. In the current version of the Edmonton Protocol, the induction agent daclizumab (anti-CD25 (IL-2R) antibody) is administered intravenously immediately prior to transplantation and again at 2 weeks post-transplant (1 mg/kg). Maintenance immunosuppression is achieved using sirolimus with a low dose of tacrolimus, as sirolimus appears to be associated with less nephrotoxicity and diabetogenicity than calcineurin inhibitors (i.e. cyclosporine and tacrolimus). A loading dose of sirolimus (0.2 mg/kg) is given prior to transplant, followed by 0.15 mg/kg, which is then adjusted subsequently to achieve trough levels between 10–12 ng/ml for the first 3 months and 7–10 ng/ml thereafter. Tacrolimus is adjusted to maintain trough levels between 3 and 6 ng/ml. This regimen, described initially at the University of Alberta, has been successfully replicated at other centers as part of a multicenter ITN trial (Shapiro et al., 2003, 2005b).

In addition to the Edmonton Protocol immunosuppression described above, alternative regimens have been reported. The Minnesota Group, led by Dr. Bernhard Hering, has utilized anti-thymocyte globulin and etanercept (anti-tumor necrosis factor-α (TNFα) antibody) induction with a combination of sirolimus and mycophenolate mofetil $\pm$ low-dose tacrolimus for maintenance, or hOKT3γ1(Ala–Ala) (humanized anti-CD3 antibody) and sirolimus induction with sirolimus and reduced-dose tacrolimus for maintenance (Hering et al., 2004, 2005). In some instances, alternative immunosuppressive agents have been used because of drug intolerance or other side effects. Islet patients often possess mild preexisting renal impairment as a result of longstanding diabetes, and this renal dysfunction may be exacerbated with calcineurin inhibitor therapy, even at the low doses involved in the Edmonton Protocol. The drug sirolimus may also have nephrotoxic side effects, which may be compounded when used in combination with a calcineurin inhibitor drug (Kaplan et al., 2004; Senior et al., 2005). For these reasons, renal status must be monitored diligently in all patients following islet transplantation. In addition to its recognized nephrotoxicity, tacrolimus is associated with gastrointestinal side effects which may lead to episodic diarrhea. Neurotoxicity may be seen with tacrolimus but is often avoided in low-dose regimens (Gruessner et al., 1996). Sirolimus is associated with neutropenia and mouth ulceration, but these side effects can be reduced with lower target trough levels and tablet formulations. In the context of islet transplantation, sirolimus has been linked to a number of side effects including dyslipidemia, small bowel ulceration, peripheral edema, and the development of ovarian cysts or menstrual cycle irregularities in female recipients (Molinari et al., 2005; Ryan et al., 2005).

While chronically immunosuppressed patients are at risk for developing all types of malignancy, squamous epithelial cancers most commonly occur and are most readily treatable. The lifetime risk of lymphoma is estimated to be 1–2% in transplant recipients, but this risk is likely to be reduced in islet recipients, as these patients are generally not treated with glucocorticoids or OKT3.

FUTURE CHALLENGES

Overcoming Tissue Shortage

In its current form, islet transplantation is reserved for patients with the most severe forms of diabetes, which in reality constitute a small fraction of all patients with T1DM. Even with the relatively small patient population selected for islet transplantation, the waitlist time for patients in Edmonton, which has access to organs from a large geographic region, ranges from 6 months to 2 years depending on blood group. As islet transplantation becomes more suitable for a broader range of diabetic patients and as the incidence of diabetes increases, there will be an even more severe shortage of islet tissue for transplantation. Presently, clinical islet programs rely on the scarce supply of pancreas organs derived exclusively from heart-beating, brain-dead cadavers. Compared to organs procured for whole pancreas transplantation, which must fall within very strict donor criteria, organs obtained for islet transplantation tend to be more "marginal" and come from older, less stable donors. Furthermore, the pancreas is particularly susceptible to toxicity from the circulating products of severe brain injury, hemodynamic instability, and inotropic support in a brain-dead organ donor. The quality of the pancreas is further degraded by cold ischemic injury during transportation, which inevitably results in islet damage and loss. Contreras et al. demonstrated a marked reduction in islet recovery and in islet viability in experimental islet transplantation using tissue derived following brain death compared to healthy rodent donors, highlighting this issue, and recently his group has confirmed these findings using human islets (Contreras et al., 2003). Similarly, Lakey et al. demonstrated a strong relationship between islet recovery and donor stability (Lakey et al., 1996). Once the pancreas is in the isolation laboratory, the extensive processing and purification steps during processing result in further islet destruction and loss, often resulting in at best 60% recovery of the estimated 10^7 IE/pancreas (Tsujimura et al., 2004). As a result, nearly all islet recipients require islets derived from two cadaveric donors. Thus, a rapidly growing area of islet transplant research involves the development of improved cadaveric or alternative islet tissue sources for transplantation.

Living Donor Islet Transplantation

One approach to alleviating islet tissue demand would be to make use of living donors for islet transplantation. Living donor programs in kidney, liver, and lung transplantation have moved forward successfully at most leading transplant centers worldwide, in an attempt to meet the growing demand for donor organs and to improve clinical outcomes. Given the rapid, global acceptance of cadaveric islet transplantation over the past 5 years, it is likely that living donor islet transplantation will soon be offered to patients listed in cadaveric islet transplant programs. Despite remarkable progress in clinical islet transplantation since 1999, islet supply and functional viability remain to be significant challenges when islets are derived from cadaveric organ donors, even at the most experienced centers (Contreras et al., 2003). In the living donor setting, the distal half pancreas could be procured under "ideal" circumstances, without exposure of the pancreas to hemodynamic instability or inotropic drugs, and the pancreas would be processed immediately without prolonged cold ischemia. Thus, the potency of islets derived from a living donor source is assumed to be far superior to cadaveric tissue. Living donor islet transplantation represents a unique opportunity to overcome donor organ shortage and procure the islet tissue under perfect conditions, with closer human leukocyte antigen (HLA) matching between donor and recipient. Furthermore, the living donor islet transplant setting will provide a unique opportunity to develop protocols for pre-transplant recipient conditioning for donor-specific tolerance induction.

While cadaveric islet transplantation has been an active area of clinical research involving more than 1,000 patients in the past 30 years, only three cases of living-donor islet allo-transplantation have been reported (Sutherland et al., 1980; Matsumoto et al., 2005). The first two clinical attempts at living donor islet allo-transplantation were carried out in 1978 by Sutherland and colleagues at the University of Minnesota (Sutherland et al., 1980). While neither recipient achieved sustained islet function, these pioneering efforts were truly remarkable given the early stage of clinical islet transplant development at the time. The immunosuppression available was primitive by current standards (azathioprine and high-dose steroids), and the islets were isolated using suboptimal conditions, prior to the development of the Ricordi chamber and the sophisticated purification schemes currently used in clinical islet transplantation. The dramatic improvement in clinical outcomes obtained in cadaveric islet transplantation since 2000 has renewed interest in the development of living donor islet transplantation. The first living donor islet transplantation case attempted since the introduction of the Edmonton Protocol was carried out at the University of Kyoto in early 2005, as a collaboration between the Japanese and Edmonton programs (Matsumoto et al., 2005). The recipient, a 27-year-old female, developed C-peptide negative, unstable diabetes following chronic pancreatitis as a child. Her 56-year-old mother was approved to be the donor, and islets were purified from the distal pancreas (47% as measured pre-operatively by computed tomography (CT) volumetry) obtained during an open laparotomy. There were no surgical complications in either donor or recipient. The unpurified islet mass (408,114 IE (8,200 IE/kg) in a volume of 9.5 ml after tissue digestion) was transplanted into the portal vein using the percutaneous approach under full systemic heparinisation. Edmonton Protocol-style immunosuppression was started pre-transplant using sirolimus and low-dose tacrolimus (started 7 days pre-transplant), anti-IL2R antibody (given 4 days pre-transplant and on the day of transplant) and anti-TNFα blockade induction (infliximab; given 1 day pre-transplant). Insulin therapy in the recipient was discontinued at 22 days post-transplant, and this patient continues to be insulin independent with excellent glycemic control and a normal HbA1C more than 1 year post-transplant. The donor has presented no evidence of glucose intolerance and has maintained normal HbA1C values since the procedure.

While no definitive conclusions can be drawn from this single successful case of living donor islet allo-transplantation, results from living donor islet auto-transplantation suggest that the insulin independence may be achieved routinely with significantly less IE/kg recipient body weight than has been required for cadaveric allografts thus far. It is widely accepted that over 70% of patients will remain insulin free following islet auto-transplantation if an islet mass exceeding 300,000 IE ($\geq$2,500 IE/kg) is transplanted, compared to the 13,000 IE/kg that is often required to achieve insulin independence with cadaveric islet preparations (Gruessner et al., 2004). It must be noted, however, that robust long-term follow-up of patients receiving islet autografts has not been reported to date. Despite the potential risks for a living donor in terms of surgically induced diabetes and surgical complications, the demand for islet tissue and relative ease of implementation of living donor protocols into established islet transplant programs is likely to move this approach forward rapidly.

Xenotransplantation

Living donor islet transplantation may circumvent the wait for suitable donor tissue in some diabetic patients, but the risks to the donor and the possibility of insufficient islet yield to obtain insulin dependence remain to be significant concerns. Identification of a renewable xenogeneic source of islets would avoid the requirement for human islet donors altogether and could provide enough tissue to transplant diabetic patients as often as required. Pigs are particularly attractive as a xenogeneic islet donor since they are widely available, produce insulin that is functional in humans, and could be selected for certain donor characteristics. Of all types of experimental xenotransplantation, islet transplantation is probably the closest to clinical application. Over the past decade, a number of small clinical trials in islet transplantation using porcine islets have been reported, but few have resulted in reduced insulin requirements and no patients have achieved prolonged insulin independence (Groth et al., 1994; Elliott et al., 2000; Valdes-Gonzalez et al., 2005). Despite these set-backs, islet xenotransplantation using porcine tissue has remained an active area of research, and progress has been made over the past several years in experimental islet xenotransplantation using pre-clinical non-human primate models (Cardona et al., 2006; Hering et al., 2006; Rood et al., 2006). The generation of α1,3-galactosyltransferase-deficient pigs has provided a source of islet tissue lacking the major xenoantigens causing hyperacute rejection in pig-to-human xenotransplantation (Phelps et al., 2003). Still, it remains to be determined whether the transmission of endogenous retroviruses or other zoonotic infections from pig to human can be completely avoided in xenotransplantation, even with the establishment of highly monitored "clean" pig colonies (Fishman and Patience, 2004).

While significant advances have been made in the area of islet xenotransplantation, it is unclear whether enough data has been generated to justify the move toward large scale clinical trials. However, there are verbal reports that clinical trials are ongoing in centers in China and Russia (Rood and Cooper, 2006).

Stem-Cell Transplantation

Unlike solid organ transplantation, which requires a complex vascularized tissue structure to restore function in a recipient, islet transplantation could be achieved through the development of a renewable source of stem-cell derived β-cells. Substantial research efforts have been made in identifying suitable islet precursor cells that could be differentiated into an unlimited source of insulin-producing β-cells, but difficulties in producing physiologically regulated insulin secretion and control of proliferation have delayed progress in this area (reviewed in Bonner-Weir and Weir, 2005; Otonkoski et al., 2005). Some exciting data has been reported using genetically modified human fetal hepatocytes, but data in large animal models is lacking (Zalzman et al., 2003, 2005). The challenge of reproducing the highly differentiated neuroendocrine β-cell phenotype is significant, and more investigation in this area is required before stem-cell derived islets will see clinical application. Even as progress is made in this area, political and ethical issues may prevent the timely application of this technology in human subjects.

Improving Engraftment Post-transplant

In clinical islet transplantation, islets derived from multiple donors are often required to achieve insulin independence, which suggests that a significant portion of the transplanted islets must fail to engraft and become functional. It has been estimated that up to 70% of the transplanted β-cell mass may be destroyed in the early post-transplant period (Davalli et al., 1995; Biarnes et al., 2002; Ryan et al., 2005). Since this profound loss has been observed in both immunodeficient and syngeneic islet transplantation models, islet survival is likely regulated by non-immune-mediated stimuli. Following isolation, the islet microvasculature is completely disrupted, and upon implantation into the portal circulation, hypoxia persists while the islets revascularize, which can take up to 2 weeks (Dionne et al., 1993; Carlsson et al., 2001, 2002; Giuliani et al., 2005). During this engraftment period, the islets are continuously exposed to immunosuppressive drugs including tacrolimus and sirolimus, which are known to adversely impact β-cell survival and function (Hyder et al., 2005). These negative effects are likely compounded by the proximity of the transplanted islets and high concentrations of these drugs in the hepatoportal circulation, further degrading β-cell mass over time (Desai et al., 2003; Shapiro et al., 2005a).

Another process which may influence islet engraftment and survival in the early post-transplant period has been termed the "instant blood-mediated inflammatory reaction" (IBMIR). Islets have been shown to naturally express tissue factor, a protein which acts as a receptor and cofactor for Factor VII, an important mediator of the coagulation cascade (Moberg et al., 2002). Isolated human islets release tissue factor along with glucagon and insulin, which ultimately leads to platelet activation and binding at the surface of the islets. This causes the formation of a fibrin capsule around the islet and disruption of the islet morphology (Bennet et al., 1999; Moberg et al., 2002; Ozmen et al., 2002). Most of this process has been characterized using an *in vitro* tubing loop model, so the true impact of this process in the clinical setting has yet to be fully characterized. However, examination of serum in patients undergoing islet transplantation has shown that a statistically significant increase in the serum concentration of thrombin/anti-thrombin complexes is present almost immediately following portal infusion, with peak levels occurring at 15 min, even when there was no clinical evidence of portal hypertension or intraportal thrombosis (Moberg et al., 2002). Given that platelet activation is one of the primary contributing factors in the generation of an inflammatory response, IBMIR is probably one of the important early processes in islet transplantation that elicits an immune response (Rabinovitch and Suarez-Pinzon, 1998; Moberg et al., 2002).

Many studies targeted at enhancing islet survival during the early post-transplant period have been published, and a variety of different strategies have been tested. Some groups have aimed to enhance revascularization with vascular endothelial growth factor (VEGF), but these studies have not yet demonstrated that this approach significantly improves islet graft survival (Narang et al., 2004). Anti-coagulation strategies using injection of activated protein C or inhibition of thrombin have been studied as a means to inhibit IBMIR, but these interventions have shown only a modest benefit in a limited series of *in vivo* studies in animal models (Ozmen et al., 2002; Contreras et al., 2004). Clinical studies designed to prevent IBMIR are currently under investigation and should provide more insight into this area.

Since the processes described above involve both extracellular (i.e. IBMIR) and intracellular (i.e. hypoxia) stimuli leading to β-cell death, another approach to preserve β-cell mass in the early post-transplant period has been to directly inhibit the apoptotic triggers which ultimately lead to loss of islet mass post-transplant. A variety of strategies have been explored in the experimental setting, and while promising data has been generated *in vitro*, demonstration of *in vivo* benefit to islet graft survival has been more elusive (Dupraz et al., 1999, 2000; Cottet et al., 2001, 2002; Cattan et al., 2003; Klein et al., 2004). Many studies have described inhibition of a variety of apoptosis-associated proteins, including cFLIP (cellular FLICE-inhibitory protein; prevents caspase-8 activation), A20 (inhibits NF-κB activation), Bcl-2, and Bcl-XL (mitochondria-associated anti-apoptotic proteins) (Dupraz et al., 1999, 2000; Grey et al., 1999, 2003; Cottet et al., 2001, 2002; Klein et al., 2004). A20 has shown promise, as its overexpression reduced the islet mass required in syngeneic islet transplantation in mice (Grey et al., 1999, 2003). Recently investigations using XIAP (X-linked inhibitor of apoptosis protein), which inhibits the downstream effector caspases that function in the final common pathway of apoptosis, have demonstrated promise in both human and rodent models of engraftment and in promoting murine islet allograft survival (Emamaullee et al., 2005a, b; Plesner et al., 2005). However, this area of research is currently limited by its requirement for genetic manipulation of islet tissue pre-transplant, which has proven to be quite variable and difficult to achieve in human islets. Also, these genetic alterations are most often regulated with viral vectors, which represent a highly controversial reagent for clinical use, especially in immunosuppressed transplant recipients. As pharmacological compounds which can reproduce or stimulate the expression of anti-apoptotic mediators are identified, this area of research will likely have a positive impact in clinical islet transplantation.

Improved Immunomodulation: Toward Donor-Specific Tolerance

One unique component of islet transplantation in patients with T1DM is the possibility of recurrent autoimmunity, which may elevate the demand for immunosuppression. Indeed, it has been well established using a rodent model of T1DM, the non-obese diabetic (NOD) mouse, that control of recurrent autoimmune reactivity to β-cells is one of the most difficult obstacles to overcome in islet transplantation (reviewed in Rossini et al., 2001; Pearson et al., 2003). Although it has been quite challenging to study recurrent autoimmunity in clinical patients, some evidence exists to suggest that levels of autoantibodies to glutamic acid decarboxylase (GAD) and IA-2 increase following islet transplantation, although the direct impact of this phenomenon on graft survival is not yet clear (Jaeger et al., 2000; Bosi et al., 2001). If recurrent autoimmunity does alter immunosuppressive drug functional thresholds, this presents yet another problem in the context of islet transplantation, as many of the drugs are directly β-cell toxic. In fact, up to 15% of non-diabetic patients who receive solid organ grafts can develop post-transplant diabetes as a result of calcineurin inhibitor therapy (i.e. tacrolimus) or steroids (i.e. prednisone) (Jindal et al., 1997; Djamali et al., 2003). Most patients that are candidates for islet transplantation have had disregulated diabetes for many years, and as such their renal status may be somewhat impaired (Shapiro et al., 2000). This leads to an increased susceptibility to the deleterious renal side effects of these immunosuppressive drugs, and thus limits the extent to which the dose can be increased to preserve graft function (Ryan et al., 2004a). It is therefore likely that immunosuppressive drugs either contribute directly to β-cell loss over time via toxicity, or indirectly by incomplete protection against recurrent autoimmunity and/or alloreactivity.

Direct control of recurrent autoimmunity may enhance long-term graft function in islet transplantation. Attempts have been made to control autoimmunity at the time of diabetes onset, using various immunosuppressive agents such as azathioprine, prednisone, cyclosporin A, or anti-thymocytic globulin, but no significant benefit was observed (Elliott et al., 1981; Eisenbarth et al., 1985; Silverstein et al., 1988; Bougneres et al., 1990). Recent clinical studies using a modified anti-CD3 (hOKT3γ1(Ala–Ala) in patients with new onset T1DM have demonstrated that this treatment significantly improved C-peptide responses in these patients, which persisted for up to 2 years following treatment (Herold et al., 2005). Incorporation of this induction agent into clinical islet transplant protocols has suggested that it may enhance insulin-independence rates following single donor infusion, which may be related to its ability to curtail β-cell autoimmunity in these patients (Hering et al., 2004). Continued development of therapies targeted at regulation of autoimmunity will allow further refinement of immunosuppression protocols for islet transplantation in the future.

In all types of transplantation, the ultimate goal is to develop therapeutic protocols that involve a brief period of treatment only during the initial post-transplant period, followed by the complete withdrawal of all immunosuppressive drugs. This phenomenon has been termed "operational tolerance," since it may involve a passive ignorance of the graft or a more active T-cell tolerance to the graft antigens. In experimental transplantation, the difference in these two types of response is quite important and can be measured using retransplantation of donor type or third party tissue, with tolerance resulting in acceptance of the donor-type graft and rejection of the third party graft. In the clinical setting, however, the distinction may not be so critical, as both ignorance and tolerance would allow for reduction or withdrawal of immunosuppressive therapies. The most widely studied pathway to tolerance induction involves the inhibition of T-cell costimulation following T-cell receptor ligation. During an immune response, a T-cell must receive "signal 2" through interactions between its surface molecule CD28 and CD80 or CD86 on the antigen presenting cell to become fully activated. In order to disrupt this interaction, the extracellular portion of CTLA-4, which has a higher affinity for CD80/CD86 than CD28, has been artificially fused with human Fcγ to produce the soluble molecule CTLA4-Ig, designed for therapeutic purposes. CTLA4-Ig has been recognized for its potent immunoregulatory activity in murine models of T1DM, where treatment of young NOD mice dramatically reduced the incidence of T1DM (Lenschow et al., 1995). Our laboratory and others have demonstrated that CTLA4-Ig treatment in allogeneic islet transplantation can prolong graft survival but does not induce tolerance (Kirk et al., 1997; Levisetti et al., 1997; Benhamou, 2002; Casey et al., 2002). A new high-affinity version of CTLA4-Ig called belatacept or LEA29Y has been developed for clinical use and has shown considerable promise in promoting allograft survival in non-human primates and in clinical renal transplantation (Adams et al., 2002, 2005; Vincenti et al., 2005). These studies have generated considerable excitement for this approach and have prompted initiation of clinical trials using belatacept in clinical islet transplantation.

A second costimulatory pathway that has been examined in transplantation involves the interaction between CD40 on antigen presenting cells and CD40L (CD154) on T-cells, leading to T-cell activation. This interaction also promotes B-cell differentiation and the activation of antigen presenting cells including macrophages and dendritic cells. Blockade of this pathway using anti-CD154 therapies demonstrated considerable promise in promoting tolerance induction in primate models early on, but further testing of the potent anti-CD154 blocking antibody (Hu5C8) has been halted due to unexpected thromboembolic complications in clinical trials (Kenyon et al., 1999; Kirk et al., 1999, 1997; Kawai et al., 2000). Recent development of therapeutic antibodies targeting the CD40 molecule appears to avoid this negative side effect and should prove to be important in future clinical tolerance induction protocols in islet transplantation (Adams et al., 2005).

SUMMARY AND CONCLUSIONS

β-cell replacement through islet transplantation presents the best opportunity to treat T1DM and prevent the long-term serious complications associated with this disease. The concept of islet transplantation is not new, but investigators struggled to find success in achieving insulin independence until the introduction of the Edmonton Protocol in 2000. This has provided hope for many patients with diabetes, but islet transplantation, in its current form, is reserved only for those patients with the most severe disease. While up to 80% of recipients may attain and maintain insulin independence at 1 year post-transplant, insulin independence has not been sustainable over time, with the most recent Edmonton data suggesting that nearly 90% of recipients will have resumed insulin therapy at 5 years post-transplant, albeit with a much lower insulin requirement than before receiving an islet graft. Also, most patients continue to exhibit partial islet function with C-peptide secretion in sufficient amounts to avoid both glycemic lability and hypoglycemic unawareness, which greatly improves the quality of life for many patients.

However, the current requirement for islets derived from two or more cadaveric donors severely limits the current availability of this procedure. There are multiple opportunities for intervention throughout the entire process, from pancreas procurement, shipment, and islet processing, through to strategies for enhanced islet survival after implantation. Priority areas for clinical trials currently include expansion of living donor protocols, interventions to impede the IBMIR process, and the use of non-diabetogenic and more "islet-friendly" immunosuppressive and tolerance induction strategies to effectively control both auto- and alloimmunity. Strategies targeted at preserving β-cell mass throughout the process will have a substantial and immediate impact on islet transplantation by reducing the amount of islet tissue necessary to reverse diabetes. Once some of these obstacles are overcome, islet transplantation will become available to a broader population of patients with T1DM, especially those early in the progression of their disease who will benefit most as the development of serious chronic secondary complications could be avoided.

REFERENCES

Adams, A.B., et al. (2002). Calcineurin inhibitor-free CD28 blockade-based protocol protects allogeneic islets in non-human primates. *Diabetes* **51**: 265–270.

Adams, A.B., et al. (2005). Development of a chimeric anti-CD40 monoclonal antibody that synergizes with LEA29Y to prolong islet allograft survival. *J. Immunol.* **174**: 542–550.

Ballinger, W.F. and Lacy, P.E. (1972). Transplantation of intact pancreatic islets in rats. *Surgery* **72**: 175–186.

Benhamou, P.Y. (2002). Immunomodulation with CTLA4-Ig in islet transplantation. *Transplantation* **73**: S40–S42.

Benhamou, P.Y., et al. (2001). Human islet transplantation network for the treatment of type I diabetes: first data from the Swiss-French GRAGIL consortium (1999–2000). Groupe de Recherche Rhin Rhjne Alpes Geneve pour la transplantation d'Ilots de Langerhans. *Diabetologia* **44**: 859–864.

Bennet, W., et al. (1999). Incompatibility between human blood and isolated islets of langerhans: a finding with implications for clinical intraportal islet transplantation? *Diabetes* **48**: 1907–1914.

Bhargava, R., et al. (2004). Prevalence of hepatic steatosis after islet transplantation and its relation to graft function. *Diabetes* **53**: 1311–1317.

Biarnes, M., et al. (2002). Beta-cell death and mass in syngeneically transplanted islets exposed to short- and long-term hyperglycemia. *Diabetes* **51**: 66–72.

Bonner-Weir, S. and Weir, G.C. (2005). New sources of pancreatic beta-cells. *Nat. Biotechnol.* **23**: 857–861.

Bosi, E., et al. (2001). Autoantibody response to islet transplantation in type 1 diabetes. *Diabetes* **50**: 2464–2471.

Bougneres, P.F., et al. (1990). Limited duration of remission of insulin dependency in children with recent overt type I diabetes treated with low-dose cyclosporin. *Diabetes* **39**: 1264–1272.

Brendel, M.H.B., Schulz, A. and Bretzel, R. (2001). International Islet Transplant Registry Report. University of Giessen, Germany. 1.

Cardona, K., et al. (2006). Long-term survival of neonatal porcine islets in nonhuman primates by targeting costimulation pathways. *Nat. Med.* **12**: 304–306.

Carlsson, P.O., et al. (2001). Markedly decreased oxygen tension in transplanted rat pancreatic islets irrespective of the implantation site. *Diabetes* **50**: 489–495.

Carlsson, P.O., et al. (2002). Low revascularization of experimentally transplanted human pancreatic islets. *J. Clin. Endocrinol. Metab.* **87**: 5418–5423.

Casey, J.J., et al. (2002). Portal venous pressure changes after sequential clinical islet transplantation. *Transplantation* **74**: 913–915.

Cattan, P., et al. (2003). Destruction of conditional insulinoma cell lines in NOD mice: a role for autoimmunity. *Diabetologia* **46**: 504–510.

Contreras, J.L., et al. (2003). Brain death significantly reduces isolated pancreatic islet yields and functionality *in vitro* and *in vivo* after transplantation in rats. *Diabetes* **52**: 2935–2942.

Contreras, J.L., et al. (2004). Activated protein C preserves functional islet mass after intraportal transplantation: a novel link between endothelial cell activation, thrombosis, inflammation, and islet cell death. *Diabetes* **53**: 2804–2814.

Cottet, S., et al. (2001). SOCS-1 protein prevents Janus Kinase/STAT-dependent inhibition of beta cell insulin gene transcription and secretion in response to interferon-gamma. *J. Biol. Chem.* **276**: 25862–25870.

Cottet, S., et al. (2002). cFLIP protein prevents tumor necrosis factor-alpha-mediated induction of caspase-8-dependent apoptosis in insulin-secreting betaTc-Tet cells. *Diabetes* **51**: 1805–1814.

Davalli, A.M., et al. (1995). A selective decrease in the beta cell mass of human islets transplanted into diabetic nude mice. *Transplantation* **59**: 817–820.

DCCT Research Group (1990) *Diabetes Care* **13**: 427–433.

DCCT Research Group (1993). The effect of intensive treatment of diabetes on the development and progression of long-term complications in insulin-dependent diabetes mellitus. *N. Engl. J. Med.* **329**: 977–986.

This explosion of trials lies in the simplicity, straightforward nature, and timing of cell therapy. First, cell therapy offers an opportunity for repair of injury, rather than simply an augmentation of the remaining uninjured heart, often to its detriment. In other words, cell therapy provides hope of a permanent solution to a previously unsolvable problem.

Next, the treatment makes sense to clinicians and patients. When cells die after AMI leading to HF (Abbate et al., 2006), it seems logical to prevent them from dying, as is a goal with current bone marrow mononuclear cells (BMNCs) trials early post-AMI, or replace them with new, undamaged cells, a current goal of virtually all HF trials. Because cell therapy primarily involves autologous cells, patients endorse it – since ethical concerns (such as with embryonic stem cell research or live allogeneic donors) are not an issue – and often seek out participation in clinical trials. In addition, as is often the case with anything cutting edge in medicine, it is exciting and prestigious to participate both for investigators and patients.

In addition, cell therapy came just as the hope of angiogenic therapies (vascular endothelial growth factor, fibroblast growth factor, etc.) failed to deliver the new blood vessel growth they so enticingly promised (Henry et al., 2003; Simons, 2005). When therapeutic angiogenesis did not become the mainstream one-shot-fix-all treatment, there was an urgent need for a new frontier: for something that would fulfill the promise and become the "clinical product." Cell-based therapies became "what's next" in cardiovascular medicine. As it did so, the anticipated 5.3 billion dollar market potential in CVD has led scientists, companies, investors, clinicians, and patients into a plethora of first-in-man studies. As a result, within 10 years, we have moved from conception of a field to completed Phase I clinical studies, completed, and ongoing Phase II studies and additional "next generation" bench and pre-clinical studies.

But has cell therapy arrived? The good news is the field is moving rapidly, and the possibility of cell-based therapies joining the clinical armamentarium to repair myocardium has begun to be supported by several Phase II patient studies. The bad news is the field is sometimes moving too fast, without a critical evaluation of the science behind the data or with trials that are outdated even by the time they begin to enroll patients. So it is not at all surprising that we have conflicting clinical outcomes: both negative and positive results with the same cells, in the same patients, and in similar pathophysiological contexts. We do not **yet** understand all the relationships among cells, engraftment, delivery, mechanism of action, and outcomes. We have **yet** to ask all the right questions. The field is just emerging, and with it our knowledge changes fast. Within the next few years, we should witness the completion of both surgical and percutaneous catheter-based Phase II studies with bone marrow, blood, muscle and possibly even fat-derived progenitor cells. We will likely see the initiation of at least one clinical trial utilizing resident cardiac progenitor cells. It is very likely that as this process further unfolds, the applicability and the maximum benefits of the cells types will segregate with a specific disease state; and as we gain more molecular/mechanistic insights, we will be enabled to choose the right cell for the right stage of the CVD continuum. The likely major impact will be provided by Phase II (and even Phase III) data regarding the use of BMNCs in the acute and subacute setting of AMI and the use of SKMBs or mesenchymal cells (MSCs) in HF patients. So has cell therapy arrived? The short answer is that it is **arriving**.

In summary, this is a dynamic time in a new and exciting field: treatment strategies are being developed and modified almost every week; new cell types are being reported, novel delivery strategies are emerging, and slowly we are dissecting the mechanistic components of cellular cardiomyoplasty.

CELL-BASED REPAIR AS A MEANS OF REGENERATION IN CVD

The goal of regeneration as a therapeutic process is to repopulate diseased area of the tissue with exogenous (by direct application) or endogenous (by stimulation of production and homing) cells that restore function of the organ and halt the disease process that resulted in tissue damage. In AMI and HF, cardiac regeneration – restoration of myocardial function and cessation of LV remodelling – is the ultimate goal. Functional vasculature has to develop and electrical conductance has to occur in underperfused areas and/or fibrous scar to provide optimal contractile force. However, full regeneration is currently not yet achievable, although it may be possible in the future through tissue engineering methods, such as application of cell-based patches to critically injured areas (Liu et al., 2004). Futurism apart, what **can** be claimed as a realistic goal of cell therapy in the next 5 years is **restoring at least some degree of mechanical function and perfusion to the injured, dilated, and decompensated myocardium** – even in the absence of full regeneration.

Endogenous tissue repair is inherent in the human body. We realize now that virtually every organ in the body, including heart, is capable of ongoing maintenance throughout much of our lifespan, and only with ongoing disease, aging, or when overwhelmed by a catastrophic event does this process fail. Yet, repair in the heart and vasculature appears to fail more often than in other organs. As we understand more about the elements of this endogenous cardiac process – what initiates it, what controls it, what allows it to work, and what leads to failure – we will refine our targets for repair. Manipulating those targets first pre-clinically and then clinically will become increasingly important as the field advances. Meanwhile, what can we learn from non-cardiac repair that could apply to heart?

Successful endogenous cell-based repair of many non-cardiac tissues routinely occurs after injury. For example, in bone, endogenous repair occurs **if** three simple premises are fulfilled: reduction of inflammation, fixation of new cells, and perfusion of the tissue (Ott and Taylor, 2006; Taylor et al., 2006). Although bone injuries are not usually ischemic in origin, the cascade of wound healing is somewhat similar to the process in myocardium: inflammation leads to clearing of necrotic (bone) tissue and formation of a fibrous scar. However, unlike in the heart where the scar stays (Kwong et al., 2006), in bone the scar is replaced by regenerating osteoclasts and osteoblasts (Maddi et al., 2006; Wutzl et al., 2006), and in a relatively short period the bone is completely healed (Karladani et al., 2001). So why does endogenous cardiac repair fail? In bone repair, **mechanical stress has to be minimized.** Patients' fractured extremities are routinely immobilized to allow healing. When an injured bone is insufficiently immobilized, an unstable scar forms (and this process frequently takes longer than formation of a stable scar), which could be compared to the process of LV remodeling and progression to HF. **Next, perfusion has to be maintained.** When perfusion is compromised, tissue necrosis leads to sequestration, chronic inflammation, and pain – a process not unlike chronic post-AMI angina (Hansen-Algenstaedt et al., 2006). Furthermore, in the absence of adequate perfusion, a mature bone does not arise (Hansen-Algenstaedt et al., 2005, 2006), similar to the need for functional angiogenesis to support new muscle formation in the myocardium (Simons, 2005). Finally, **cells have to home to the site of the lesion.** Bone regeneration is ultimately performed by endogenous progenitor cells (osteoblasts and osteoclasts) that migrate into the lesion (Wright et al., 2005). In the injured myocardium, this process breaks down. Rather than intensive colonization with the desired cardiac progenitor cells, the lesion is colonized primarily by fibroblasts, resulting in collagen deposition and scar formation, instead of nascent myocardium (Lu et al., 2004). This comparison with bone makes endogenous cell-based repair in the heart seem difficult, as the need for immobilization cannot be achieved, we do not understand all the components of homing of cells to the lesion, and adequate perfusion is a product of endothelial health, function, and regulation, which is often missing in CVD.

VASCULAR INTEGRITY IS A BALANCING ACT IN WHICH INFLAMMATION IS KEY: INJURY VERSUS REPAIR

What we do understand, however, is that atherosclerosis represents the failure of **attempts** of endogenous vascular repair in response to repeated injuries to the vessel wall (Goldschmidt-Clermont et al., 2005). After

Ross' seminal work introducing the "response to injury" hypothesis whereby the **primum movens** in atherogenesis was endothelial denudation (Ross, 1993), it became apparent that endothelial cells directly regulate vascular function, transport of solutes, and antithrombotic properties of the blood–tissue interface (Hansson, 2005; Feletou and Vanhoutte, 2006; Schmieder, 2006). More recently, the revised "response to injury" hypothesis focuses on endothelial dysfunction rather than denudation as a trigger for the inflammatory responses and progression of atherosclerosis (Sun et al., 2000; Hansson, 2005). Once regarded simply as a static barrier between tissue and blood, vascular endothelium is now known to play a key regulatory and integrating role in the initiation and progression of atherosclerosis (Feletou and Vanhoutte, 2006; Schmieder, 2006). We now understand that vasodilatory, antiplatelet, and antithrombotic processes are primarily regulated at the endothelial level, and the loss of normal endothelial function may be the most important driver of the balance in favor of inflammation and thrombosis (Feletou and Vanhoutte, 2006; Schmieder, 2006), which, in turn, contributes to the transition between stable and unstable angina (Kostner et al., 2006). Indeed, atherosclerotic plaques have been shown to be more thrombogenic when they express higher levels of tissue factor, a key endothelial regulator in the extrinsic clotting cascade (Marmur et al., 1993; Libby and Theroux, 2005). Therefore, in CVD, the repair of endothelium with exogenously applied cells must be one of the main steps to achieve restoration of perfusion and slowing and possibly halting both LV remodeling and atherosclerotic lesion progression.

We believe that ongoing endogenous repair is a process that reflects the balance between protective and detrimental factors (Goldschmidt-Clermont et al., 2005), and that by providing reparative cells it is possible to tip this balance to allow for repair of the injured tissues (Figure 24.1). Vascular repair is most likely a stepwise process. First, the tissue undergoes injury (e.g. ischemic insult), as shown in Figure 24.1, accompanied by inflammatory response (release of cytokines and chemokines). The inflammatory milieu and injury-specific mediators recruit "detrimental" BMNCs (e.g. $CD45^+$, $CD11^+$, $CD3^+$) and exacerbate injury (Yu et al., 2002; Abbate et al., 2004). At the same time, mobilization of "reparative" BMNCs (e.g. $AC133^+$, $CD34^+$, $CD31^+$, KDR^+) occurs in an endogenous attempt to promote repair (Fadini et al., 2006; Haider, 2006). If sufficient amounts of reparative cells are recruited (or supplied) the balance tips toward repair and inflammations is halted. Three factors play a crucial role: availability of the "reparative" BMNCs, the capability of these cells to home to the site of injury, and their functional capacity to initiate and propagate repair. Obviously, any of these factors could be the weakest link, and that alone would make repair inefficient allowing injury to prevail. The balance between the injury and repair may also reside in the quantitative, functional, and mechanistic/paracrine relationships of **both** "protective" and "detrimental" BMNCs at the site of injury (Figure 24.2). Therefore, from the balance point of view, we believe that these relationships are different in mild, moderate, and severe disease (Figure 24.2). Recent studies have shown that the availability of the bone marrow progenitors is reduced in number in severe CVD (Werner et al., 2005; Kunz et al., 2006). Similar findings as well as a profound reduction of functional capacity of these cells have been observed in patients with HF (Valgimigli et al., 2004). The decline in number and function is most likely not **the** exclusive feature of the advanced stages of atherosclerotic process because similar results have been reported much earlier in the disease process, when endothelial damage is only provoked by hypertension or type 2 diabetes (Hill et al., 2003; Fadini et al., 2005; van Zonneveld, 2006). Inverse associations with aging itself have also been shown (Thorin-Trescases et al., 2005; Shaffer et al., 2006). Our preliminary data in ApoE $^{-/-}$ mice as well as the recently published WISE study (Bairey Merz et al., 2006) indicate that males and females display different speeds of progression of disease that correlate with progenitor cell profiles. If these data hold, gender differences in repair will need to be extensively studied as the results may have major implications not only for the field of cell therapy but for the entire cardiovascular medicine. Those insights could not only provide guidance for developing next generation clinical studies, but could greatly advance our understanding of pathophysiological changes in men and women.

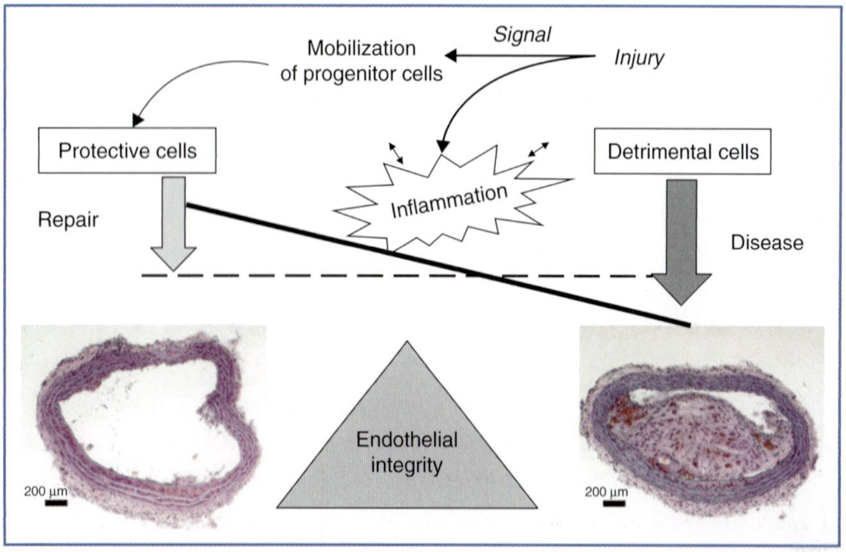

Figure 24.1 Schematic representation of the main factors influencing the balance between positive and negative factors (cells, cytokines, chemokines) influencing progression of disease and endogenous repair.

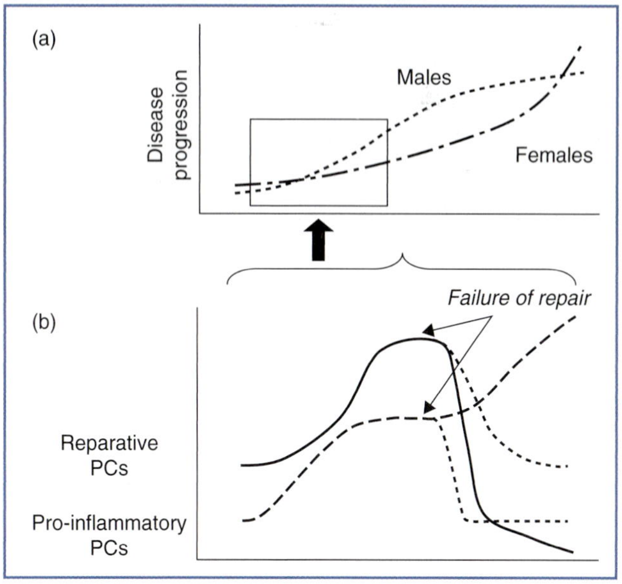

Figure 24.2 Schematic representation of relationships between CVD progression (a) and pro-inflammatory (solid line) and reparative progenitor cells (dashed line) number/function (b). As ongoing injury occurs (panel b, left side), the number of pro-inflammatory cells increases (reflecting inflammation), and the reparative progenitor cells compensates in attempt to repair. However, with failure of repair (shown by arrow), reparatory cells falls, and pro-inflammatory cells increase leading to disease progression (panel b, right side). Dotted lines indicate the proposed dynamic of both reparatory and pro-inflammatory cells when repair is successful. Data from animal experiments in our laboratory and data from the Women's Ischemia Syndrome Evaluation (WISE) study (Bairey Merz et al., 2006) were used to reflect the time course of CVD in (a). Relationships of pro-inflammatory and reparative cells were portrayed based on recent publications and the preliminary data of our laboratory (b). *Abbreviations.* PC: progenitor cells. See text for a detailed discussion.

Overall, there are several important aspects in cardiac repair. First, the failure of repair occurs earlier in the disease process than previously thought (i.e. before a clinical event) suggesting that owing to age, atherosclerosis, and risk factors, the availability of the progenitor cells and their function decline (Figure 24.2) (Hill et al., 2003). Secondly, repair is most likely evident in the mild–moderate/moderate phase of the disease, where it ultimately fails and injury progresses (Goldschmidt-Clermont et al., 2005). Unfortunately (for detection of abnormalities) many patients in this stage are asymptomatic (Blumenthal et al., 2006). Finally, mobilization of these cells does occur following the acute injury of AMI (Lev et al., 2005; Li et al., 2005; Massa et al., 2005), however, the response is not sufficient for two main reasons: (a) the availability of the cells and their function are greatly reduced (Thum et al., 2006) and (b) a recent study showed that revascularization abolish endogenous mobilization (Muller-Ehmsen et al., 2005).

We also now understand that the reduction of inflammation is one of the key steps in successful vascular repair. Recent research has demonstrated that inflammation plays a key role in CVD (Shin et al., 2002; Hansson, 2005; Libby and Theroux, 2005). Indeed, endothelial-dependent relaxation is impaired early in atherogenesis (Feletou and Vanhoutte, 2006). In addition, local administration of proinflammatory cytokines impairs endothelium-dependent dilatation in humans (Bhagat and Vallance, 1997; Shin et al., 2002). Furthermore, clinical measures of coronary endothelial dysfunction are associated with myocardial ischemia in the absence of flow-limiting lesions (Suwaidi et al., 2000) and predict cardiovascular events, including stroke (Targonski et al., 2003). In the early atherosclerotic lesion, immune cells predominate, and their effector molecules significantly contribute to the progression of the plaque (Paoletti et al., 2004; Goldschmidt-Clermont et al., 2005; Libby and Theroux, 2005). Inflammatory response has been shown to be a large part of acute coronary syndromes (ACS)/AMI (Ikeda, 2003; Paoletti et al., 2004; Libby and Theroux, 2005). The cascade of events leading to clinical atherosclerosis begins with lipid deposition and oxidation, and initiation of an arterial lesion site within the endothelial lining (Libby and Theroux, 2005). These acute processes trigger platelet aggregation and monocyte/macrophage infiltration (Ferns and Avades, 2000; Ikeda, 2003). Smooth muscle cell (SMC) apoptosis is observed within the first few days following the initial assault (Sata et al., 2000). During this time, there is upregulation of pro-inflammatory that recruit various cells including monocytes, macrophages, neutrophils, leukocytes, and smooth muscle precursors to the injury site (Simon et al., 2000; Fujiyama et al., 2003; Ikeda, 2003; Schober et al., 2003; Weber et al., 2004). Studies have demonstrated that by blocking these cytokine signals, a significant reduction in lesion formation is achieved, suggesting that inflammation and/or the cells these signals recruit are a cause of plaque progression (Rogers et al., 1998; Simon et al., 2000; Schober et al., 2003; Weber et al., 2004). A reduction of these signals is associated with re-endothelialization of the artery and the "end to the injury" process (Ferns and Avades, 2000; Schwartz et al., 2004). Extracellular lipids and foam cells form the center of the atherosclerotic plaque, which is surrounded by SMCs and collagen matrix (Paoletti et al., 2004; Libby and Theroux, 2005). However, it is the T-cells, macrophages, and mast cells present in the shoulder region of the plaque ensure atheroma growth (Frostegard et al., 1999). Many of the immune cells show the signs of activation and release pro-inflammatory cytokines (Shin et al., 2002). Recent analysis of soluble inflammatory mediators has been driven by the expectation that they may be used as indicators of presence and/or progression of CVD (Baldus et al., 2003; Labarrere and Zaloga, 2004; Paoletti et al., 2004). ACS occurs when activation of the atherosclerotic plaque leads to accelerated thrombogenesis and blocks the blood flow to the myocardium downstream from the lesion (Libby and Theroux, 2005). Plaque activation is a complex inflammatory process, in which metalloproteinases along with a number of cytokines directly attack collagen and other components of the tissue matrix, which surrounds the lipid core of the plaque (van der Wal et al., 1994). One of the indicators of inflammation is C-reactive protein, which has been found to correlate with (and reflect) the presence of vascular inflammation and endothelial activation (Labarrere and Zaloga, 2004). CRP is elevated in patients at a high risk and/or experiencing ACS (Tsimikas et al., 2006). Increased levels of CRP indicate upregulation of many inflammatory cytokines (IL-6, IL1β, TNFα) and adhesion molecules (ICAM-1, VCAM-1, E-selectin), facilitating the inflammatory process, and promoting cell recruitment and attachment, ultimately advancing CVD (Shin et al., 2002; Paoletti et al., 2004; Jain and Ridker, 2005; Tsimikas et al., 2006). Other inflammatory markers such as MPO, SDF-1, and MCP-1 have also been associated with CVD (Shin et al., 2002; Baldus et al., 2003; Schober et al., 2003).

However, reduction of inflammation in clinical settings is problematic. There is no single drug therapy which addresses all components of inflammation or results in the "end to the injury" response. Statins have been shown to be beneficial in reduction of hs-CRP, MCP-1 and some inflammatory cytokines (Jain and Ridker, 2005; Chello et al., 2006). However, whether the effect of statins on inflammatory biomarkers is a function of reduction of low-density lipoprotein (LDL) cholesterol or represents a separate pleiotropic effect (including possible mobilization of progenitor cells), represents a matter of considerable debate (Davidson, 2005). Other therapeutics, such as aspirin and ACE inhibitors have also been shown to elicit a beneficial effect on some components of inflammatory process, but their administration does not abolish the entire ravaging cytokine and soluble marker cascade (Peeters et al., 1998; Lauten et al., 2003; Tsikouris et al., 2004). Non-steroidal agents, cyclooxygenase-2 inhibitors, and glucocorticosteroids are potent anti-inflammatory drugs, but their administration alters the course of ACS and combat restenosis yielded more adverse effects than clinical benefits (Lee et al., 1999; Mukherjee et al., 2001; Niederberger et al., 2001; Tamai et al., 2002; Niederberger et al., 2004; Khan and Mehta, 2005; Krotz et al., 2005; Nicolae et al., 2005; Williams et al., 2005).

We have recently shown that repeated intravenous injection of BMNCs from young ApoE$^{-/-}$ mice prevented further progression of atherosclerotic plaque lesions in old ApoE null mice (Rauscher et al., 2003). Injected cells differentiated into endothelial cells and engrafted in atherosclerotic lesions of recipient animals. Comparison of bone marrow progenitor cell (BMC) profiles showed a specific depletion of intermediate vascular progenitor cells (CD31^{+}/CD45^{+}), without parallel changes in more primitive stem cells (sca-1^{+}, c-kit^{+}, or CD34^{+}) or mature vascular cells (VEGFR-2^{+}), most likely accounted for the age-related loss of bone-marrow-derived vascular repair capacity. In addition, the treatment with BMNCs reduced IL-6, and the effect was maintained up until day 15 post-injection, and then the levels would increase again (Rauscher et al., 2003). Significant regression of plaque that occurred in concert with the reduction of IL-6 demonstrated that exogenous BMNCs might mediate regression of atherosclerotic plaque via anti-inflammatory action/positive interference with immune responses that are associated with the atherosclerotic process. However, when BMNCs were given in sex-matched and mismatched fashion only males that received female marrow showed the highest degree of plaque regression (Nelson, unpublished data). Further dissection of the cytokines involved in BMNC-mediated atheroprotection is in its final stages in our laboratory. Preliminary data suggests that plaque regression may be mediated by a distinct group of cytokines, rather than by IL-6 alone (Zenovich, unpublished data). Because atherosclerosis is an inflammatory disease (Paoletti et al., 2004; Hansson, 2005; Libby and Theroux, 2005), the anti-inflammatory properties of BMNC administration and resulting plaque regression offers hope that the treatment with exogenous progenitor cells truly interferes into the pathogenesis of the disease. In addition, recent data on the regression of atherosclerosis with rosuvastatin have demonstrated that this drug is capable of reduction of plaque volume when measured by intravascular ultrasound technique (Nissen et al., 2006). Hopefully one day in the future, a cocktail of a statin drug and bone marrow progenitors will be a therapeutic option in interventional cardiology.

While we only have a conceptual understanding of repair, and many questions are yet unanswered, mediation of this process by progenitor cells makes sense. After all, if cells that originate from the bone marrow did not have the ability to fight inflammation and/or pathogens and participate in healing, the organism would not survive for too long. When the number and/or function of these cells decreases because of a systemic disease (such as type 2 diabetes) (Fadini et al., 2005), aging or toxic substances (such as smoking) (Raupach et al.,

2006), healthy and functioning endothelium fails and adverse consequences manifest clinically (as atherosclerosis, thrombosis and its micro- and macrovascular sequelae) (Feletou and Vanhoutte, 2006).

A recent data analysis modeled the potential health effects of bone-marrow-derived progenitor cell therapy using the follow-up data (1950–1996) of the Framingham Study. To model CVD mortality, progenitor cell therapy was applied at age 30, with the effect assumed to be a 10-year delay in atherosclerosis progression. This study suggests that progenitor cell therapy might increase life expectancy in the population by as much as the complete elimination of cancer (in females, an additional 3.67 versus 3.37 years; in males, an additional 5.94 versus 2.86 years, respectively) (Kravchenko et al., 2005). These exciting findings fuel enthusiasm for cell therapy to halt and ultimately prevent CVD.

In summary, in contrast to other tissues in the human body that successfully accomplish the process of repair, cardiac repair becomes inefficient prior to AMI or HF. This failure to endogenously repair/regenerate the vasculature stems from the reduced availability of the circulating bone marrow progenitors and existent inflammatory milieu. Exogenous BMNCs show attractive qualities of plaque regression and reduction of inflammation. Interventions to tip the balance of injury and repair to promote repair may need to be applied earlier in the disease process to ensure efficacy. Further research is needed to optimize that target, as well as to define specific inflammatory mechanisms that play a major role in vascular repair.

REPARATIVE POTENTIAL OF CELL THERAPY

Autologous SKMBs

The idea of using skeletal muscle to repair the heart evolved well before cell therapy emerged as a possible treatment option. In 1987, after being preconditioned by chronic pacing, the latissimus dorsi muscle was surgically wrapped around the failing heart to provide contractile support to the left ventricle – a procedure named "dynamic cardiomyoplasty" (Chachques et al., 1987). Thereafter, cellular cardiomyoplasty was introduced, when cells derived from the C_2C_{12} SKMB transformed cell line were successfully transplanted into **normal** mouse hearts (Koh et al., 1993).

SKMBs, derived from muscle "satellite cells," were first described in 1961 as cells that regenerate damaged skeletal muscle (Mauro, 1961). SKMBs expand and form neofibers after muscle injury. It is not surprising that these cells were the first candidates for cardiac repair. In 1994, Magovern et al. reported the first successful transplantation of SKMBs into an **injured** heart (Zibaitis et al., 1994). The critical finding that transplanted SKMBs survived and formed striated muscle grafts within the damaged cardiac tissue was followed by several independent experimental studies investigating the engraftment potential of these cells.

When we showed in 1998, for the first time, that the successful engraftment of SKMBs into injured myocardium improved LV function and attenuated remodeling (Taylor et al., 1998), it was deemed novel by many and unrealistic by some. The mechanisms of how these successfully engrafted SKMBs improved function was unclear. It appeared that the muscle cells could improve contractility of the scarred heart without strict transdifferentiating into cardiomyocytes. SKMBs appeared to yield two populations of cells in injured heart: myogenin-positive transplanted skeletal muscle-like cells in the center of the scar and a second population of myogenin-negative more primitive cardiac muscle-like cells "recruited" around the scar periphery (Atkins et al., 1999a). The transplanted SKMBs appeared to adapt to the surrounding myocardium by forming slow twitch myofibers that were electrically isolated from host cardiomyocytes and yet to improve LV performance (Murry et al., 1996; Atkins et al., 1999b).

These results, stating an improved cardiac function without full integration of transplanted cells into the host myocardium and with recruitment of endogenous cells, raised questions about potential mechanisms that provide functional benefit. Numerous mechanisms have been suggested – from changes in LV wall stress and/or geometry to active mechanically induced contraction of the injected cells – but the exact mechanism(s) underlying the beneficial effect are still the matter of considerable debate (Ott and Taylor, 2006). It is likely that the improvement of function is a result of both a direct effect of the transplanted SKMBs on LV performance, and an indirect "paracrine" effect on endogenous cell recruitment and on LV remodeling (van den Bos and Taylor, 2003). As the mechanisms underlying the positive effects of SKMB transplantation are still not fully understood, there is some discordance of thought as to whether autologous myoblasts improve contraction or just prevent further deterioration of the injured heart – despite much preclinical data showing a direct positive effect of the cells above that seen with other or sham treatments (van den Bos et al., 2004). However, from a clinical point of view, whether the cells do one or the other, may not be the right question to ask, as even attenuation of remodeling is a therapeutic avenue much needed in the HF armamentarium, since the current drug therapies are only capable of a very moderate effect on LV remodeling, only approximately 30%–40% (Reiffel, 2005). Despite these and other remaining questions, there are approximately 15 years of preclinical data available have shown that autologous SKMBs transplantation can augment both diastolic and systolic myocardial performance in a number of animal models after both acute and chronic injury. These preclinical data opened the field of CVD to this new therapeutic approach.

The advantages of autologous SKMBs for treating patients with CVD/HF extend beyond the evidence of benefit in preclinical studies. By using self-derived SKMBs, it is possible to overcome the major limitations associated with allogeneic cell-based treatments: a critical shortage of donor tissue and the clinical complexities of immunosuppression. By using adult-derived cells, it is possible to avoid the ethical dilemma associated with embryonic stem cells. By using primary cells rather than immortalized or totipotent stem cells, the likelihood of tumor formation after SKMB transplantation is decreased (Tremblay et al., 1991). By using muscle-forming, myogenic cells, the regeneration of contractile muscle in an infarcted cardiac region is more likely. By using relatively ischemia-resistant SKMBs rather than cardiocytes, a higher level of engraftment and survival is likely to occur in infarcted regions – where transplanted cardiocytes perish (Reffelmann et al., 2003). Based on these advantages and the suggestions that regeneration of functional muscle in infarct is possible after autologous SKMBs in pre-clinical models (Taylor et al., 1998; Hutcheson et al., 2000; Fuchs, S. et al., 2001; Ohno et al., 2003; Thompson et al., 2003; Agbulut et al., 2004; Hiasa et al., 2004; Horackova et al., 2004; Ott et al., 2004; Ott et al., 2006) (Table 24.1), clinical studies in both Europe and the United States took off (Field and Reinlib, 2000).

Although the initial clinical data (Herreros et al., 2003; Menasche et al., 2003; Smits et al., 2003; Chachques et al., 2004b; Ince et al., 2004; Siminiak et al., 2004; Dib et al., 2005; Siminiak et al., 2005; Gavira et al., 2006) (Table 24.2) appear encouraging, myoblast transplantation is not without potential limitations. The first limitation is associated with any autologous cell that has to be expanded in the laboratory (i.e. myoblasts, mesenchymal stem cells, endothelial progenitor cells (EPCs)) that the use of autologous cells necessitates sufficient time between injury and injection to allow cell expansion *in vitro*. In normal healthy donors and in many HF patients, this ranges from several days to several weeks depending on the cell type, which does not seem problematic if the treatment can wait. In the case of AMI, where an early treatment may be beneficial, either alternative cells can be employed, or if myoblasts are truly superior, allogeneic cells may ultimately offer a solution after the acuity has subsided. Of note, the 2–3-week period after AMI is one of the flourishing inflammation (Paoletti et al., 2004; Libby and Theroux, 2005), which in preclinical studies has been associated with increased loss of cells after transplantation (Suzuki et al., 2004). Thus, a delayed treatment with autologous SKMBs could still be beneficial.

A second potential limitation to any cardiac cell therapy is the (**in**)ability of the transplanted cells to electrically integrate with native tissue (Abraham et al., 2005). Myoblasts are the most well studied cell type, yet it is not clear if, and/or how, do they integrate into surrounding myocardium. Nor is it understood what impact various degrees of integration may have on either LV function or continuous normal sinus rhythm. Menasche

Table 24.1 Selected Clinically-Relevant Pre-Clinical Models Used for Cell Therapy

Cell Type(s)	Species	Model	Outcomes	Investigator, Country
SKMBs	Rabbit	Cryoinjury	LV systolic function and diastolic relaxation improved in 7/12 rabbits, correlated with engraftment of cells.	Taylor et al. (1998) (US)
SKMBs	Pig (targeted cell placement with Da Vinci robotic system)	LAD or LCX embolization	Cells successfully transplanted into apical, anterior and lateral target segments; LVEF, wall thickening, regional wall motion, LV end-diastolic volume improved.	Ott et al. (2006) (US)
SKMBs vs. FBs	Rabbit	Cryoinjury	Diastolic performance improved with FBs and SKMBs. FBs reduced but SKMBs increased systolic function.	(Hutcheson et al., 2000) (US)
SKMBs vs. BMNCs	Rabbit	Cryoinjury	Similar improvement of LV systolic function and equal degree of engraftment with SKMBs and BMNCs. Subset of BMNCs differentiated towards a myogenic phenotype.	Thompson et al. (2003) (US)
SKMBs vs. CFS vs.CMs	Guinea pig	LAD Ligation	CMs and CFs concentrated in infarction territory; CMs formed gap junctions with native cells, CFs did not; SKMBs proliferated and partially differentiated into cardiac phenotype by 2–3 weeks post-procedure. Gap junctions present.	Horackova et al. (2004) (Canada)
Human SKMBs vs. Human AC133$^+$	Rat	LAD Ligation	SKMBs increased LV EF by 15±5%, AC133$^+$ by 7±3% (controls reduced by 8±4%). Engraftment larger with SKMBs.	Agbulut et al. (2004) (France)
SKMBs vs. BMNCs	Rat	LAD Ligation	Combination of cell types prevented LV remodeling at 8 weeks post-procedure.	Ott et al. (2004) (Austria)
SKMBs vs. VSMCs	Hamster	Cardiomyopathy (Δ-sarcoglycan-deficiency)	Attenuation of LV remodeling was greater with SKMBs vs. VSMCs.	Ohno et al. (2003) (Canada)
BMNCs	Mouse	LAD Ligation	Reduction of myocardial infarction size.	Hiasa et al. (2004) (Japan)
BMNCs	Pig	Ameroid Occluder, Placement on LCX	BMNCs secreted angiogenic growth factors, augmented perfusion and LV function.	Fuchs et al. (2001) (US)

Abbreviations. AC133$^+$: immature endothelial progenitor cells; CF: cardiac fibroblast; CM: cardiomyocyte; EF: ejection fraction; FB: fibroblast; LAD: left anterior descending artery; LCX: left coronary circumflex artery; LV: left ventricle; VSMC: vascular smooth muscle cell.

and colleagues did not observe any electrical integration of myoblasts pre-clinically (Scorsin et al., 2000). In support of that, Suzuki et al. (2001) reported that in the absence of connexin-43 overexpression, SKMBs did not couple very well with surrounding myocardium. Yet the cells appear to synchronously contract in with surrounding tissue and contribute to overall cardiac performance. Early clinical data (Table 24.2) suggest that

Table 24.2 Cell Therapy Trials with SKMBs

Year	Investigator, Country	Patients No.	Diagnosis	Average SKMB Dose ($\times 10^6$)	Delivery Route	Outcomes
2003	Menasche et al. (2003) (France)	10	Post-AMI HF	871	Transepicardial without CABG	NYHA class improved to 1.6 ± 0.1 from 2.7 ± 0.2; LVEF increased to 32 ± 1% from 24 ± 1%.
2003	Herreros et al. (2003) (Spain)	11	Ischemia + prior AMI	221	Transepicardial with CABG	LVEF increased to 53.5 ± 5% from 35.5 ± 2.3%, regional wall motion (by E) and viability (glucose update by PET) improved.
2003	Smits et al. (2003) (Netherlands)	5	Post-AMI (anterior) HF	196	Transendocardial guided by electromechanical mapping of LV	LVEF increased to 41 ± 9% from 36 ± 11%, regional contractility (by MRI) significantly increased.
2004	Chachques et al. (2004b) (France)	20	Post-AMI	300	Transepicardial without CABG	LVEF normalized to 52.0 ± 4.7% (baseline = 28 ± 3%), wall motion score improved, glucose update (by PET) increased.
2004	Siminiak et al. (2004) (Poland)	10	>3-month post-AMI	0.4-50 (range)	Transepicardial with CABG	Mean LVEF improved to 42% from 35.2%
2004	Ince et al. (2004) (Germany)	6 (+6 controls)	Ischemic HF	210	Transendocardial	LVEF rose to 32.2 ± 10.2% from 24.3 ± 6.7% (in controls decreased to 21.0 ± 4.0%), walking distance and NYHA class significantly improved.
2005	Dib et al. (2005) (US)	30	Post-AMI HF	2.2-300	Transepicardial with CABG or LVAD	LVEF increased to 36% at year 2 post-procedure vs. 35% at year 1 vs. 28% at baseline.
2005	(Siminiak et al., 2005) (Poland)	9	Post-AMI HF	17-106 (range)	Transcoronary	LVEF improved 3–8% in 6 of 9 patients; NYHA class improved in all patients.
2006	Gavira et al. (2006) (Spain)	12 (+14 historical controls)	Post-AMI	221	Transepicardial with CABG	LVEF rose to 55.1 ± 8.2% from 35.5 ± 2.3% at baseline (controls 38. 6 ± 11.0%); wall motion score (by E) improved; myocardial viability (by PET) increased.

Studies with ≥5 patients enrolled.

Abbreviations. E: echocardiography; ICM: ischemic cardiomyopathy; LVED: left ventricular end-diastolic diameter; MRI: magnetic resonance imaging; NYHA class: New York Heart Association functional class; PET: positron emission tomography.

Note. (Range) denotes the minimum and the maximum amount of cells given in a trial. Dose-escalation or variation was used in studies where range is provided. For methodological and other details, please refer to original publications.

inflammation. To date, EPCs have not been shown to induce or play a role in neomyogenesis within the injured myocardium, but several paracrine properties have recently been attributed to these cells (Kinnaird et al., 2004).

Human EPCs are typically thought to primarily express CD133 (AC133), CD34, and VEGF-R2. The quantity of the EPCs circulating in humans decreases with age, and mirrors a rapid increase in CVD-related deaths (Hill et al., 2003; Werner et al., 2005). It has been suggested that this correlation is due to the EPCs contribution to maintaining vascular integrity (Hill et al., 2003). Recent data have shown that the number of circulating EPCs and their ability to migrate is decreased in patients at a high risk for clinical CVD, including AMI (Hill et al., 2003). The reduction in the number and/or functional capacity of EPCs may be a critical factor in the development of major cardiovascular events (Werner et al., 2005). Our group recently published data showing that a reduction in $CD31^{+}CD45^{-}$ vascular progenitor cells, thought to be related to EPCs, is associated with aging and disease state in the mouse $ApoE^{-/-}$ model of atherosclerosis (Rauscher et al., 2003). We showed that delivery of functionally viable cells could prevent the progression of atherosclerosis. Completed studies using EPCs and other cell types to treat CVD in humans are shown in Table 24.3.

Umbilical cord blood cells: A relatively new source for progenitor cells is umbilical cord blood, which contains most of the bone-marrow-derived cell types. Cord blood cells are easily obtained, albeit not in large volumes, have the potential to develop into multiple lineages, do not pose a myriad of the ethical questions and are less immunogenic than their bone marrow counterparts. As a result, a larger proportion of the population could receive cells from appropriately matched donors. Further, if cord blood cells are isolated and stored at birth, these cells could provide an autologous source of stem cells to treat myocardial damage later in life. Current studies in animal models show that unfractionated cord blood cells injected directly into the infarcted myocardium improve LV ejection fraction, anteroseptal wall thickening, and dP/dt (max), while decreasing infarct size (Henning et al., 2004). In addition, intravenous injection of cord blood cells in mice following ligation-induced injury resulted in an approximately 20% higher capillary density in the border zones of the infarction – a finding not observed in untreated animals (Ma et al., 2005). Recent data have suggested that human cord blood-derived $CD34^{+}$ cells may be capable both of preventing injury progression in nude rats and of partially reversing systolic and diastolic dysfunction in the failing heart, if administered shortly after AMI (Leor et al., 2006). No evidence **yet** suggests that cord blood cells injected into the infarcted myocardium are able to produce mature cardiomyocytes. Overall, however, it appears that cord blood cells may appear to be an interesting cell of choice to be used in further studies of treatment of myocardial injury.

Cardiac-Derived Stem Cells

Within the past several years, cardiac-derived stem cells (CSCs) have been identified and are now considered a potential option for cardiac repair. Although the evidence for cardiac repair with these cells is limited, their potential to mature into cardiomyocytes makes them a promising candidate (Laugwitz et al., 2005). These cells have primarily been isolated from neonatal heart (Laugwitz et al., 2005) and, to a very limited extent, from adult myocardium (Anversa and Nadal-Ginard, 2002; Oh et al., 2003). The results of the pre-clinical use are intriguing and suggest that the future of cardiac repair may involve endogenous mobilization or recruitment of these cells – if they can be found in reasonable numbers in the adult myocardium, or can demonstrate adequate transdifferentiation when transplanted into an infarction milieu.

CSCs can be isolated from neonatal rat hearts using LIM-homeodomain transcription factor islet-1 (Laugwitz et al., 2005). It is possible to expand these cells *in vitro* when coupled with a cardiac mesenchymal layer. Further, when these cells are co-cultured with neonatal cardiomyocytes, they are able to electrically integrate with myocardial cells *in vitro* by forming gap junctions (Laugwitz et al., 2005). CSCs isolated from adult hearts, including those from acutely infarcted, failing, and even the hearts destined to be replaced by transplantation, have been

Table 24.3 Cell Therapy Trials with BMNCs, MSCs, EPCs, and CPCs

Year	Investigator (country)	Patient's number (type of cell)	Diagnosis	Average SKMB dose ($\times 10^6$)	Delivery route	LVEF and other outcomes (maximum follow-up time, month/year)
2001	Hamano et al. (2001) (Japan)	5 (BMNCs)	Advanced CAD	300–2,200 (range)	Transepi during CABG	Perfusion improved in three out of five patients (by S) (1 year)
2002	Strauer et al. (2002) (Germany)	10 (BMNCs) (+10 controls)	AMI, 5–9 days post	28	IC (after standard Tx)	Infarcted region decreased to 12 ± 7% from 30 ± 13% (by V); LV contractility and EDV improved, perfusion increased (by DE, RV, RC) (3 months)
2002	*TOPCARE-AMI*: Assmus et al. (2002); Britten et al. (2003) (Germany)	10 (CPCs) + 9 (BMNCs) (+ 11 controls)	AMI (reperfused)	CPCs: 13; BMNCs: 238	IC, 4-day post-AMI	LVEF improved to 60% from 51.6%; ESV reduced; wall motion in the infarction zone improved (all by V and DE); similar LV functional data by MRI; myocardial viability increased (by PET); CPCs and BMNCs behaved similarly. Migratory capacity predicted LV remodeling in multivariate analysis (4 months)
2003	Fuchs, S. et al. (2003) (US)	10 (BMNCs)	Refractory Angina	78	Transendo EMM-guided	Perfusion improved (by SPECT); CCS angina score decreased to 2.0 ± 0.9 from 3.1 ± 0.3 (3 months)
2003	Tse et al. (2003) (China)	8 (BMNCs)	Advanced CAD	40 ml BMNC (0.6–8.9% $CD34^+$)	Transendo EMM-guided	Regional wall motion, thickening improved, hypoperfused areas lessened (by MRI); angina reduced to 16.4 from 26.5 episodes per week (3 months)
2003	Stamm et al. (2003) (Germany)	6 (AC 133^+)	AMI	1.02–1.57 (purified)	Transepi with CABG	LVEF, EDV, EDD improved (by E); perfusion (area at-risk) improved in five out of six patients (3 months)
2003	Perin et al. (2003) (study conducted in Brazil)	14 (+7 controls)	Severe Ischemic HF	25.5	Transendo EMM guided	Mean LVEF increased to 35.5% from 30% (by E); perfusion improved (by SPECT); NYHA class decreased to 1.1 ± 0.4 from 2.2 ± 0.9; CCS angina reduced to 1.3 ± 0.6 from 2.6 ± 0.8 class. (2 and 4 months)

(Continued)

Table 24.3 (Continued)

Year	Investigator (country)	Patient's number (type of cell)	Diagnosis	Average SKMB dose ($\times 10^6$)	Delivery route	LVEF and other outcomes (maximum follow-up time, month/year)
2004	*TOPCARE-AMI*: Schachinger et al. (2004); Schachinger et al. (2006a) (Germany)	30 (CPCs) + 29 (BMNCs)	AMI	CPCs: 13; BMNCs: 238	IC, 4.9 days after AMI	At 1 year, one patient in each cell group died due to cardiogenic shock, no other MACE or malignant arrhythmias; LV functional data similar to prior report for additional patients. MRI at 1 year showed maintenance of LVEF and reducion of infarct size, no reactive LV hypertrophy. Coronary flow normalized in infarct-related arteries (1 year)
2004	Chen et al. (2004) (China)	34 (MSCs) (+35 controls)	AMI	8,000–10,000 (range)	IC, 8 ± 4 h after AMI	LVEF and regional wall motion increased (by V); per fusion defects decreased (by PET); real-time electromechanical mapping of LV showed improvements in mechanical capabilities, electrical properties and functional indices (6 months)
2004	*BOOST*: Wollert et al. (2004); Meyer et al. (2006); Schaefer et al. (2006) (Germany)	30 (BMNCs) (+30 controls)	STEMI	2,460	IC 4.8 ± 1.3-days post-PCI	LVEF increased by 6.7% mostly due to improved regional wall motion in the peripheral area (by MRI) LVEDV and infarct size decreased. Diastolic function improved (by E). LV functional benefits did not persist at 1 year (1 year)
2004	Fernandez-Aviles et al. (2004) (Spain)	20 (BMNCs)	Extensive AMI (reperfused)	78	IC, 13.5 ± 5.5-day post-AMI	LVEF improved by mean of 6%, contractile reserve increased; ESV decreased in (by MRI, DE) (6 months)
2004	Perin et al. (2004) (study conducted in Brazil)	11 (BMNCs) (+9 controls)	End-stage ICM	15 injections, 0.2 ml/each (50 ml aspirated)	Transendo EMM guided	LVEF increased at 2 m, did not change at 6 and 12 m, perfusion improved (by SPECT), NYHA class decreased to a mean of 1.4 from 2.2 and CCS fell to 1.2 from 2.6 class. Exercise capacity improved (by treadmill) (6 and 12 months)
2004	Kuethe et al. (2004) (Germany)	5 (BMNCs)	AMI (reperfused, stented)	39	IC, 6.3 ± 0.4 days post-PCI	LVEF and regional wall motion did not change (by E). Coronary flow (by IC Doppler) and contractility indices (by DE) remained similar at follow-up (3 and 12 months)

2004	Silva et al. (2004) (US)	5 (BMNCs)	Pre-transplantation HF	15 injections, 0.2 ml/each (50 ml aspirated)	Transendo EMM guided	Exercise capacity (by treadmill oxygen consumption) improved in four out of five patients, disqualifying them from listing for transplantation (2 and 6 months)
2005	Bartunek et al. (2005) (Belgium)	19 ($AC133^+$) (+16 controls)	AMI	12.6	IC, 11.6 ± 1.4 day post-AMI	LVEF increased to 52.1% from 45% similar to controls (by E), perfusion improved (by SPECT); seven patients in cell therapy group developed restenosis (versus four in control group), two patients in cell therapy group had de novo lesions (4 months)
2005	Dohmann et al. (2005) (Brasil)	14 (BMNCs) (+7 controls)	Severe CAD + HF	25.5	Transendo EMM guided	Perfusion increased (by S), NYHA class, functional capacity, and function improved. (2 and 6 months)
2005	*IACT*: Strauer et al. (2005) (Germany)	18 (BMNCs)	Post-AMI (5m–8.5y)	15–22 (range) each infusion, 4–6 total	IC	LVEF increased by 15% (by V), infarct size fell by 30% (by SPECT), myocardial viability of infarcted zone increased by 15% (by PET) (3 months)
2005	Blatt et al. (2005) (Israel)	6 (BMNCs)	ICM	50 ml of aspirated BMNCs	IC, after induction of ischemia by balloon inflation for 3 min	LVEF improved from mean of 25% to 28% (by E); wall motion (by DE) increased but only in segments with baseline hibernation. NYHA class fell to mean of 2.3 from 3.5; one patient developed post-procedure hypotension and troponin increase (4 months)
2006	*ASTAMI*: Lunde et al. (2006) (Norway)	47 (BMNCs) (+50 controls)	AMI treated with PCI	54–130 (range)	IC	No differences in LVEF (by MRI), perfusion (by SPECT), trend toward infarct size reduction (by MRI) (6 months)
2006	*REPAIR-AMI*: Schachinger et al. (2006b) (Germany)	101 (BMNCs) (+103 controls)	STEMI (reperfused)	236	IC, 3–6 days after AMI	LVEF increased by a mean of 5.5% (by V), patients with LVEF <49% benefited most. At 1 year, BMNC-treated patients exhibited reduction in a combined primary end-point (death, AMI recurrence, revascularization) (4 and 12 months)

(*Continued*)

Table 24.3 (Continued)

Year	Investigator (country)	Patient's number (type of cell)	Diagnosis	Average SKMB dose ($\times 10^6$)	Delivery route	LVEF and other outcomes (maximum follow-up time, month/year)
2006	*TOPCARE-CHD*: Assmus et al. (2006) (Germany)	34 (CPCs) + 35 (BMNCs) (+23 controls)	Prior AMI (>3 m)	CPCs: 22; BMNCs: 205	IC, with cross-over to the other cell type	LVEF increased significantly in patients that crossed over to BMNCs, absolute increase = 2.9% (by MRI). No changes in LVEF with CPCs. NYHA class improved in BMNC group – reduction of 2.0 ± 0.7 from 2.2 ± 0.6, no improvement with CPCs (3 months).
2006	Fuchs et al. (2006) (Israel)	27 (BMNCs)	Refractory Angina + Ischemia	28	TransendoEMM guided	CCS angina class improved to 2.0 ± 0.9 from 3.2 ± 0.5; exercise duration increased to 489 ± 142 s from 418 ± 136 s; ischemia lessened (by SPECT). At 1 year, five patients had revascularization procedures, functional and symptomatic improvements were maintained in other patients. (3 and 12 months)
2006	Tse et al. (2006) (China)	12 (BMNCs)	Advanced CAD	12–16	TransendoEMM guided	No significant changes in LVEF at 3 or 6 months (baseline LVEF 60%). At long-term follow-up, two patients died, one patient received CABG. (3, 6 and 44 ± 10 months)

Studies with ⩾ 5 patients enrolled.

Abbreviations. CAD: coronary artery disease; CCS: Canadian Cardiovascular Society; DE: dobutamine echocardiography; EDD: end-diastolic (LV) dimension; EDV: end-diastolic (LV) volume; EMM: electromechanical mapping of LV; ; IC: intracoronary; ICM: ischemic cardiomyopathy MACE: major adverse cardiac events; NYHA class: New York Heart Association functional class; PET: positron emission tomography; RC: right heart catheterization; RV: radionuclide ventriculography; S: scintigraphy; SPECT: single photon emission tomography; STEMI: acute ST-elevation myocardial infarction; Transendo: transendocardial; Transepi: transepicardial; Tx: therapy/treatment; V: ventriculography;

Note: (range) denotes the minimum and the maximum amount of cells given in a trial. Dose-escalation/variation was used in studies where range is provided. Controls comprise historical and active randomized participants and those patients who received placebo. For methodological and other details, please refer to original publications.

identified by their expression of c-kit, MRD-1, and Sca-1 and by their lack of expression of hematopoietic lineage markers (Urbanek et al., 2005). These cells have shown the ability to differentiate down myocyte, smooth muscle, and endothelial cell pathways, but their ability to form mature cells of these types (or cardiomyocytes) is not yet unknown. Endogenous Sca-1^+ CSCs may differentiate into functional cardiomyocytes (Oh et al., 2003), but such potential within an infarct scar has not been elucidated. To date, methods for harvest, expansion, and *in vitro* growth of these precursors are very limited. Therefore, because of these factors, it is difficult to judge the clinical potential of these cells. Nonetheless, CSCs biology is interesting enough to make future developments be anticipated with interest and hope. For example, CSCs expanded from endomyocardial biopsies and predifferentiated *in vitro* could become very strong candidates for cardiac repair.

CLINICAL STUDIES

Clinical trials with SKMBs (Table 24.2), BMNCs, MSCs, EPCs and CPCs (Hamano et al., 2001; Assmus et al., 2002; Strauer et al., 2002; Britten et al., 2003; Fuchs et al., 2003; Perin et al., 2003; Stamm et al., 2003; Tse et al., 2003; Chen et al., 2004; Fernandez-Aviles et al., 2004; Kuethe et al., 2004; Perin et al., 2004; Schachinger et al., 2004; Silva et al., 2004; Wollert et al., 2004; Strauer et al., 2005; Bartunek et al., 2005; Blatt et al., 2005; Dohmann et al., 2005; Assmus et al., 2006; Fuchs et al., 2006; Lunde et al., 2006; Meyer et al., 2006; Schachinger et al., 2006a, b; Schaefer et al., 2006; Tse et al., 2006) (Table 24.3) published to date have been summarized. Ongoing trials are listed on the Internet (http://www.clinicaltrials.gov; http://www.thescientist.com). To highlight important points, we chose to comment on several published trials.

SKMBs: The first clinical trial using cell therapy to treat CVD was initiated by Menasche (2003) in 2000. In this trial, an average of 871×10^6 cells (at least 85% were identified as SKMBs by a positive staining for CD56) was injected into non-revascularizable scarred portion of LV as an adjunct to CABG. Over several years following transplantation, significant improvements in LV ejection fraction (EF) and regional wall thickening were observed. Unfortunately, there was no control group. Nonetheless, the data are encouraging. However, 4 out of 10 patients experienced ventricular tachycardia requiring ICD implantation. Fortunately, none of the patients experienced intractable/fatal ventricular tachycardia/fibrillation. The data suggests that concurrent administration of amiodarone can minimize untoward electrical events without compromising efficacy of the cells. The data of Menasche et al. (2003) provided the impetus to begin a new trial. The Myoblast Autologous Graft in Ischemic Cardiomyopathy trial is a Phase II randomized clinical trial to examine the efficacy and safety of CABG + SKMBs versus CABG alone in approximately 300 patients in North America and Europe. The trial was halted in 2006 in part because the design of the trial was no longer considered state of the art (as the number of CABG cases declined), and as a result recruitment was below projected targets. Although no increase in mortality was reported, the published results of this large study are greatly anticipated.

In a separate US trial (Dib et al., 2005), SKMBs were injected concurrently with CABG ($n = 12$) or LV assist device as a bridge to transplantation ($n = 6$), myocardial perfusion improved and left ventricular ejection fraction (LVEF) increased. Upon examination of the explanted hearts (for indicated cardiac transplantation), four of the five specimens showed areas of engrafted myoblasts within the infarcted regions. In another clinical trial (Smits et al., 2003), $196+/-10^5 \times 10^6$ SKMBs were injected directly into the infarcted area (via a NOGA-guided catheter system) as sole therapy in HF patients. These patients showed improved regional wall motion and a trend toward increased LV EF over 3–6 months. Taken together, these data suggest that SKMBs can be delivered in HF patients and survive within the infarcted myocardium to achieve improved LV function.

Early reports of electrical instability in patients after receipt of autologous SKMBs have led to doubts and overt clinical skepticism about the safety of these cells as a treatment option. However, conflicting data exists, and therefore, several considerations should be made. First, patients who received SKMBs in the earliest clinical

Massa, M., Rosti, V., Ferrario, M., Campanelli, R., Ramajoli, I., Rosso, R., De Ferrari, G. M., Ferlini, M., Goffredo, L., Bertoletti, A., Klersy, C., Pecci, A., Moratti, R. and Tavazzi, L. (2005). Increased circulating hematopoietic and endothelial progenitor cells in the early phase of acute myocardial infarction. *Blood* **105**: 199–206.

Mauro, A. (1961). Satellite cell of skeletal muscle fibers. *J. Biophys. Biochem. Cytol.* **9**: 493–495.

Menasche, P., Hagege, A.A., Scorsin, M., Pouzet, B., Desnos, M., Duboc, D., Schwartz, K., Vilquin, J.T. and Marolleau, J.P. (2001). Myoblast transplantation for heart failure. *Lancet* **357**: 279–280.

Menasche, P., Hagege, A.A., Vilquin, J.T., Desnos, M., Abergel, E., Pouzet, B., Bel, A., Sarateanu, S., Scorsin, M., Schwartz, K., Bruneval, P., Benbunan, M., Marolleau, J.P. and Duboc, D. (2003). Autologous skeletal myoblast transplantation for severe postinfarction left ventricular dysfunction. *J. Am. Coll. Cardiol.* **41**: 1078–1083.

Meyer, G., Wollert, K., Lotz, J., Steffens, J., Lippolt, P., Fichtner, S., Hecker, H., Schaefer, A., Arseniev, L., Hertenstein, B., Ganser, A. and Drexler, H. (2006). Intracoronary bone marrow cell transfer after myocardial infarction: eighteen months' follow-up data from the randomized, controlled BOOST (BOne marrOw transfer to enhance ST-elevation infarct regeneration) trial. *Circulation* **113**: 1287–1294.

Miller, L.W. and Missov, E.D. (2001). Epidemiology of heart failure. *Cardiol. Clin.* **19**: 547–555.

Moss, A.J., Zareba, W., Hall, W.J., Klein, H., Wilber, D.J., Cannom, D.S., Daubert, J.P., Higgins, S.L., Brown, M.W. and Andrews, M.L. (2002). Prophylactic implantation of a defibrillator in patients with myocardial infarction and reduced ejection fraction. *N. Engl. J. Med.* **346**: 877–883.

Mukherjee, D., Nissen, S.E. and Topol, E.J. (2001). Risk of cardiovascular events associated with selective COX-2 inhibitors. *JAMA* **286**: 954–959.

Muller-Ehmsen, J., Scheid, C., Grundmann, F., Hirsch, I., Turan, G., Tossios, P., Mehlhorn, U. and Schwinger, R. (2005). The mobilization of CD34 positive mononuclear cells after myocardial infarction is abolished by revascularization of the culprit vessel. *Int. J. Cardiol.* **103**: 7–11.

Murry, C.E., Wiseman, R.W., Schwartz, S.M. and Hauschka, S.D. (1996). Skeletal myoblast transplantation for repair of myocardial necrosis. *J. Clin. Invest.* **98**: 2512–2523.

Murry, C.E., Soonpaa, M.H., Reinecke, H., Nakajima, H., Nakajima, H.O., Rubart, M., Pasumarthi, K.B., Virag, J.I., Bartelmez, S.H., Poppa, V., Bradford, G., Dowell, J.D., Williams, D.A. and Field, L.J. (2004). Hematopoietic stem cells do not transdifferentiate into cardiac myocytes in myocardial infarcts. *Nature* **428**: 664–668.

Nicolae, M., Tircol, M. and Alexandru, D. (2005). Inhibitory effect of acetylsalicylic acid on matrix metalloproteinase – 2 activity in human endothelial cells exposed to high glucose. *J. Cell Mol. Med.* **9**: 953–960.

Niederberger, E., Tegeder, I., Vetter, G., Schmidtko, A., Schmidt, H., Euchenhofer, C., Brautigam, L., Grosch, S. and Geisslinger, G. (2001). Celecoxib loses its anti-inflammatory efficacy at high doses through activation of NF-kappaB. *FASEB J.* **15**: 1622–1624.

Niederberger, E., Manderscheid, C., Grosch, S., Schmidt, H., Ehnert, C. and Geisslinger, G. (2004). Effects of the selective COX-2 inhibitors celecoxib and rofecoxib on human vascular cells. *Biochem. Pharmacol.* **68**: 341–350.

Nissen, S., Nicholls, S., Sipahi, I., Libby, P., Raichlen, J., Ballantyne, C., Davignon, J., Erbel, R., Fruchart, J., Tardif, J., Schoenhagen, P., Crowe, T., Cain, V., Wolski, K., Goormastic, M. and Tuzcu, E.A.I. (2006). Effect of very high-intensity statin therapy on regression of coronary atherosclerosis: the ASTEROID trial. *JAMA* **295**: 1556–1565.

Oh, H., Bradfute, S.B., Gallardo, T.D., Nakamura, T., Gaussin, V., Mishina, Y., Pocius, J., Michael, L.H., Behringer, R.R., Garry, D.J., Entman, M.L. and Schneider, M.D. (2003). Cardiac progenitor cells from adult myocardium: homing, differentiation, and fusion after infarction. *Proc. Natl Acad. Sci. USA* **100**: 12313–12318.

Ohno, N., Fedak, P., Weisel, R., Mickle, D., Fujii, T. and Li, R. (2003). Transplantation of cryopreserved muscle cells in dilated cardiomyopathy: effects on left ventricular geometry and function. *J. Thorac. Cardiovasc. Surg.* **126**: 1537–1548.

Ott, H.C. and Taylor, D.A. (2006). From cardiac repair to cardiac regeneration – ready to translate? *Expert Opin. Biol. Ther.* **6**: 867–878.

Ott, H.C., Bonaros, N., Marksteiner, R., Wolf, D., Margreiter, E., Schachner, T., Laufer, G. and Hering, S. (2004). Combined transplantation of skeletal myoblasts and bone marrow stem cells for myocardial repair in rats. *Eur. J. Cardiothor. Surg.* **25**: 627–634.

Ott, H.C., Brechtken, J., Swingen, C., Feldberg, T., Matthiesen, T., Nelson, W.D. and Taylor, D. (2006). Robotic minimally invasive cell transplantation for heart failure. *J. Thorac. Cardiovasc. Surg.* **132**: 170–173.

Paoletti, R., Gotto Jr., A.M. and Hajjar, D.P. (2004). Inflammation in atherosclerosis and implications for therapy. *Circulation* **109**: III20–III26.

Pearson, T.A., Blair, S.N., Daniels, S.R., Eckel, R.H., Fair, J.M., Fortmann, S.P., Franklin, B.A., Goldstein, L.B., Greenland, P., Grundy, S.M., Hong, Y., Miller, N.H., Lauer, R.M., Ockene, I.S., Sacco, R.L., Sallis Jr., J.F., Smith Jr., S.C., Stone, N.J. and Taubert, K.A. (2002). AHA guidelines for primary prevention of cardiovascular disease and stroke: 2002 Update: consensus panel guide to comprehensive risk reduction for adult patients without coronary or other atherosclerotic vascular diseases. American Heart Association Science Advisory and Coordinating Committee. *Circulation* **106**: 388–391.

Peeters, A., Netea, M., Kullberg, B., Thien, T. and van der Meer, J. (1998). The effect of renin-angiotensin system inhibitors on pro- and anti-inflammatory cytokine production. *Immunology* **94**: 376–379.

Perin, E.C., ohmann, H.F., Borojevic, R., Silva, S.A., Sousa, A.L., Mesquita, C.T., Rossi, M.I., Carvalho, A.C., Dutra, H.S., Dohmann, H.J., Silva, G.V., Belem, L., Vivacqua, R., Rangel, F.O., Esporcatte, R., Geng, Y.J., Vaughn, W.K., Assad, J.A., Mesquita, E.T. and Willerson, J.T. (2003). Transendocardial, autologous bone marrow cell transplantation for severe, chronic ischemic heart failure. *Circulation* **107**: 2294–2302.

Perin, E.C., Dohmann, H.F., Borojevic, R., Silva, S.A., Sousa, A.L., Silva, G.V., Mesquita, C.T., Belem, L., Vaughn, W.K., Rangel, F.O., Assad, J.A., Carvalho, A.C., Branco, R.V., Rossi, M.I., Dohmann, H.J. and Willerson, J.T. (2004). Improved exercise capacity and ischemia 6 and 12 months after transendocardial injection of autologous bone marrow mononuclear cells for ischemic cardiomyopathy. *Circulation* **110** II213–II218.

Raupach, T., Schafer, K., Konstantinides, S. and Andreas, S. (2006). Smoke as an acute threat for the cardiovascular system: a change in paradigm. *Eur. Heart J.* **27**: 386–392.

Rauscher, F.M., Goldschmidt-Clermont, P.J., Davis, B.H., Wang, T., Gregg, D., Ramaswami, P., Pippen, A.M., Annex, B.H., Dong, C. and Taylor, D.A. (2003). Aging, progenitor cell exhaustion, and atherosclerosis. *Circulation* **108**: 457–463.

Reffelmann, T., Leor, J., Muller-Ehmsen, J., Kedes, L. and Kloner, R.A. (2003). Cardiomyocyte transplantation into the failing heart – new therapeutic approach for heart failure? *Heart Fail. Rev.* **8**: 201–211.

Reiffel, J. (2005). Practical algorithms for pharmacologic management of the post myocardial infarction patient. *Clin. Cardiol.* **28**: I28–I37.

Rogers, C., Edelman, E.R. and Simon, D.I. (1998). A mAb to the beta2-leukocyte integrin Mac-1 (CD11b/CD18) reduces intimal thickening after angioplasty or stent implantation in rabbits. *Proc. Natl Acad. Sci. USA* **95**: 10134–10139.

Ross, R. (1993). The pathogenesis of atherosclerosis: a perspective for the 1990s. *Nature* **362**: 801–809.

Sablayrolles, S. and Le Grand, B. (2006). Drug evaluation: dronedarone, a novel non-iodinated anti-arrhythmic agent. *Curr. Opin. Invest. Drug.* **7**: 842–849.

Sata, M., Maejima, Y., Adachi, F., Fukino, K., Saiura, A., Sugiura, S., Aoyagi, T., Imai, Y., Kurihara, H., Kimura, K., Omata, M., Makuuchi, M., Hirata, Y. and Nagai, R. (2000). A mouse model of vascular injury that induces rapid onset of medial cell apoptosis followed by reproducible neointimal hyperplasia. *J. Mol. Cell Cardiol.* **32**: 2097–2104.

Saulnier, N., Di Campli, C., Zocco, M., Di Gioacchino, G., Novi, M. and Gasbarrini, A. (2005). From stem cell to solid organ. Bone marrow, peripheral blood or umbilical cord blood as favorable source? *Eur. Rev. Med. Pharmacol. Sci.* **9**: 315–324.

Schachinger, V., Assmus, B., Britten, M.B., Honold, J., Lehmann, R., Teupe, C., Abolmaali, N.D., Vogl, T.J., Hofmann, W.K., Martin, H., Dimmeler, S. and Zeiher, A.M. (2004). Transplantation of progenitor cells and regeneration enhancement in acute myocardial infarction: final one-year results of the TOPCARE-AMI Trial. *J. Am. Coll. Cardiol.* **44**: 1690–1699.

Schachinger, V., Assmus, B., Honold, J., Lehmann, R., Hofmann, W., Martin, H., Dimmeler, S. and Zeiher, A. (2006a). Normalization of coronary blood flow in the infarct-related artery after intracoronary progenitor cell therapy: intracoronary Doppler substudy of the TOPCARE-AMI trial. *Clin. Res. Cardiol.* **95**: 13–22.

Schachinger, V., Erbs, S., Elsasser, A., Haberbosch, W., Hambrecht, R., Holschermann, H., Yu, J., Corti, R., Mathey, D., Hamm, C., Suselbeck, T., Assmu, s.B., Tonn, T., Dimmeler, S., Zeiher, A. and REPAIR-AMI Investigators. (2006b). Intracoronary bone marrow-derived progenitor cells in acute myocardial infarction. *N. Engl. J. Med.* **355**: 1210–1221.

Schaefer, A., Meyer, G., Fuchs, M., Klein, G., Kaplan, M., Wollert, K. and Drexler, H. (2006). Impact of intracoronary bone marrow cell transfer on diastolic function in patients after acute myocardial infarction: results from the BOOST trial. *Eur. Heart J.* **27**: 929–935.

Schmieder, R.E. (2006). Endothelial dysfunction: how can one intervene at the beginning of the cardiovascular continuum? *J. Hypertens Suppl.* **24**: S31–S35.

Schober, A., Knarren, S., Lietz, M., Lin, E.A. and Weber, C. (2003). Crucial role of stromal cell-derived factor-1alpha in neointima formation after vascular injury in apolipoprotein E-deficient mice. *Circulation* **108**: 2491–2497.

Schuster, M.D., Kocher, A.A., Seki, T., Martens, T.P., Xiang, G., Homma, S. and Itescu, S. (2004). Myocardial neovascularization by bone marrow angioblasts results in cardiomyocyte regeneration. *Am. J. Physiol. Heart Circ. Physiol.* **287**: H525–H532.

Schwartz, R.S., Chronos, N.A. and Virmani, R. (2004). Preclinical restenosis models and drug-eluting stents still important, still much to learn. *J. Am. Coll. Cardiol.* **44**: 1373–1385.

Scorsin, M., Hagege, A., Vilquin, J.T., Fiszman, M., Marotte, F., Samuel, J.L., Rappaport, L., Schwartz, K. and Menasche, P. (2000). Comparison of the effects of fetal cardiomyocyte and skeletal myoblast transplantation on postinfarction left ventricular function. *J. Thorac. Cardiovasc. Surg.* **119**: 1169–1175.

Shaffer, R.G., Greene, S., Arshi, A., Supple, G., Bantly, A., Moore, J.S. and Mohler III, E.R. (2006). Flow cytometric measurement of circulating endothelial cells: the effect of age and peripheral arterial disease on baseline levels of mature and progenitor populations. *Cytometr. B Clin. Cytom.* **70**: 56–62.

Shin, W., Szuba, A. and Rockson, S. (2002). The role of chemokines in human cardiovascular pathology: enhanced biological insights. *Atherosclerosis* **160**: 91–102.

Shintani, S., Murohara, T., Ikeda, H., Ueno, T., Honma, T., Katoh, A., Sasaki, K., Shimada, T., Oike, Y. and Imaizumi, T. (2001). Mobilization of endothelial progenitor cells in patients with acute myocardial infarction. *Circulation* **103**: 2776–2779.

Silva, G., Perin, E., Dohmann, H., Borojevic, R., Silva, S., Sousa, A., Assad, J., Vaughn, W., Mesquita, C., Belem, L., Carvalho, A., Dohmann, H., Barroso do Amaral, E., Coutinho, J., Branco, R., Oliveira, E. and Willerson, J. (2004). Catheter-based transendocardial delivery of autologous bone-marrow-derived mononuclear cells in patients listed for heart transplantation. *Tex. Heart Inst. J.* **31**: 214–219.

Siminiak, T., Kalawski, R., Fiszer, D., Jerzykowska, O., Rzezniczak, J., Rozwadowska, N. and Kurpisz, M. (2004). Autologous skeletal myoblast transplantation for the treatment of postinfarction myocardial injury: phase I clinical study with 12 months of follow-up. *Am. Heart J.* **148**: 531–537.

Siminiak, T., Fiszer, D., Jerzykowska, O., Grygielska, B., Rozwadowska, N., Kalmucki, P. and Kurpisz, M. (2005). Percutaneous trans-coronary-venous transplantation of autologous skeletal myoblasts in the treatment of post-infarction myocardial contractility impairment: the POZNAN trial. *Eur. Heart J.* **26:,** 1188–1195.

Simon, D.I., Dhen, Z., Seifert, P., Edelman, E.R., Ballantyne, C.M. and Rogers, C. (2000). Decreased neointimal formation in Mac-1(–/–) mice reveals a role for inflammation in vascular repair after angioplasty. *J. Clin. Invest.* **105**: 293–300.

Simons, M. (2005). Angiogenesis: where do we stand now? *Circulation* **111**: 1556–1566.

Smith Jr., S.C., Allen, J., Blair, S.N., Bonow, R.O., Brass, L.M., Fonarow, G.C., Grundy, S.M., Hiratzka, L., Jones, D., Krumholz, H.M., Mosca, L., Pasternak, R.C., Pearson, T., Pfeffer, M.A., Taubert, K.A., Aha/Acc, National Heart, L. and Blood, I. (2006). AHA/ACC guidelines for secondary prevention for patients with coronary and other atherosclerotic vascular disease: 2006 update: endorsed by the National Heart, Lung, and Blood Institute.(erratum appears in *Circulation* 2006 Jun 6, **113**(22): e847). *Circulation* **113**: 2363–2372.

Smits, P.C., van Geuns, R.J., Poldermans, D., Bountioukos, M., Onderwater, E.E., Lee, C.H., Maat, A.P. and Serruys, P.W. (2003). Catheter-based intramyocardial injection of autologous skeletal myoblasts as a primary treatment of ischemic heart failure: clinical experience with six-month follow-up. *J. Am. Coll. Cardiol.* **42**: 2063–2069.

Stamm, C., Westphal, B., Kleine, H.D., Petzsch, M., Kittner, C., Klinge, H., Schumichen, C., Nienaber, C.A., Freund, M. and Steinhoff, G. (2003). Autologous bone-marrow stem-cell transplantation for myocardial regeneration. *Lancet* **361**:-45–46.

Strauer, B., Brehm, M., Zeus, T., Bartsch, T., Schannwell, C., Antke, C., Sorg, R., Kogler, G., Wernet, P., Muller, H. and Kostering, M. (2005). Regeneration of human infarcted heart muscle by intracoronary autologous bone marrow cell transplantation in chronic coronary artery disease: the IACT Study. *J. Am. Coll. Cardiol.* **46**: 1651–1658.

Strauer, B.E., Brehm, M., Zeus, T., Kostering, M., Hernandez, A., Sorg, R.V., Kogler, G. and Wernet, P. (2002). Repair of infarcted myocardium by autologous intracoronary mononuclear bone marrow cell transplantation in humans. *Circulation* **106**: 1913–1918.

Sun, P., Dwyer, K.M., Merz, C.N., Sun, W., Johnson, C.A., Shircore, A.M. and Dwyer, J.H. (2000). Blood pressure, LDL cholesterol, and intima-media thickness: a test of the "response to injury" hypothesis of atherosclerosis. *Arterioscl. Thromb. Vasc. Biol.* **20**: 2005–2010.

Suwaidi, J.A., Hamasaki, S., Higano, S.T., Nishimura, R.A., Holmes Jr., D.R. and Lerman, A. (2000). Long-term follow-up of patients with mild coronary artery disease and endothelial dysfunction. *Circulation* **101**: 948–954.

Suzuki, K., Brand, N.J., Allen, S., Khan, M.A., Farrell, A.O., Murtuza, B., Oakley, R.E. and Yacoub, M.H. (2001). Overexpression of connexin 43 in skeletal myoblasts: relevance to cell transplantation to the heart. *J. Thorac. Cardiovasc. Surg.* **122**: 759–766.

Suzuki, K., Murtuza, B., Beauchamp, J.R., Smolenski, R.T., Varela-Carver, A., Fukushima, S., Coppen, S.R., Partridge, T.A. and Yacoub, M.H. (2004). Dynamics and mediators of acute graft attrition after myoblast transplantation to the heart. *FASEB* **18**: 1153–1155.

Tamai, H., Katoh, K., Yamaguchi, T., Hayakawa, H., Kanmatsuse, K., Haze, K., Aizawa, T., Nakanishi, S., Suzuki, S., Suzuki, T., Takase, S., Nishikawa, H. and Katoh, O. (2002). The impact of tranilast on restenosis after coronary angioplasty: the second tranilast restenosis following angioplasty trial (TREAT-2). *Am. Heart J.* **143**: 506–513.

Targonski, P.V., Bonetti, P.O., Pumper, G.M., Higano, S.T., Holmes Jr., D.R. and Lerman, A. (2003). Coronary endothelial dysfunction is associated with an increased risk of cerebrovascular events. *Circulation* **107**:-2805–2809.

Tateishi-Yuyama, E., Matsubara, H., Murohara, T., Ikeda, U., Shintani, S., Masaki, H., Amano, K., Kishimoto, Y., Yoshimoto, K., Akashi, H., Shimada, K., Iwasaka, T. and Imaizumi, T. (2002). Therapeutic angiogenesis for patients with limb ischaemia by autologous transplantation of bone-marrow cells: a pilot study and a randomised controlled trial. *Lancet* **360**: 427–435.

Taylor, D.A., Atkins, B.Z., Hungspreugs, P., Jones, T.R., Reedy, M.C., Hutcheson, K.A., Glower, D.D. and Kraus, W.E. (1998). Regenerating functional myocardium: improved performance after skeletal myoblast transplantation. *Nat. Med.* **4**: 929–933.

Taylor, D.A., Ott, H.C. and Serruys P.W. (2006). Cell transplantation for cardiovascular repair. In: *Textbook of Interventional Cardiovascular Pharmacology*, Vol. . Taylor & Francis Books LTD. pp. 419–437.

Thattassery, E. and Gheorghiade, M. (2004). Beta blocker therapy after acute myocardial infarction in patients with heart failure and systolic dysfunction. *Heart Fail. Rev.* **9**: 107–113.

Thom, T., Haase, N., Rosamond, W., Howard, V.J., Rumsfeld, J., Manolio, T., Zheng, Z.J., Flegal, K., O'Donnell, C., Kittner, S., Lloyd-Jones, D., Goff Jr., D.C., Hong, Y., Adams, R., Friday, G., Furie, K., Gorelick, P., Kissela, B., Marler, J., Meigs, J., Roger, V., Sidney, S., Sorlie, P., Steinberger, J., Wasserthiel-Smoller, S., Wilson, M., Wolf, P., American Heart Association Statistics, C. and Stroke Statistics, S. (2006). Heart disease and stroke statistics – 2006 update: a report from the American Heart Association Statistics Committee and Stroke Statistics Subcommittee. *Circulation* **113**: e85–e151.

Thompson, R.B., Emani, S.M., Davis, B.H., van den Bos, E.J., Morimoto, Y., Craig, D., Glower, D. and Taylor, D.A. (2003). Comparison of intracardiac cell transplantation: autologous skeletal myoblasts versus bone marrow cells. *Circulation* **108**(Suppl 1): II264–II271.

Thorin-Trescases, N., Voghel, G., Gendron, M., Krummen, S., Farhat, N., Drouin, A., Perrault, L. and Thorin, E. (2005). Pathological aging of the vascular endothelium: are endothelial progenitor cells the sentinels of the cardiovascular system? *Can. J. Cardiol.* **21**: 1019–1024.

Thum, T., Fraccarollo, D., Galuppo, P., Tsikas, D., Frantz, S., Ertl, G. and Bauersachs, J. (2006). Bone marrow molecular alterations after myocardial infarction: impact on endothelial progenitor cells. *Cardiovasc. Res.* **70**: 50–60.

Till, J., Price, G., Lan, S. and McCulloch, E. (1978). Factors affecting the differentiation of blood cells. *Symp. Soc. Dev. Biol.* **35**: 133–147.

Torp-Pedersen, C., Poole-Wilson, P.A., Swedberg, K., Cleland, J.G., Di Lenarda, A., Hanrath, P., Komajda, M., Lutiger, B., Metra, M., Remme, W.J., Scherhag, A. and Skene, A. (2005). Effects of metoprolol and carvedilol on cause-specific mortality and morbidity in patients with chronic heart failure – COMET. *Am. Heart J.* **149**: 370–376.

Tremblay, J.P., Roy, B. and Goulet, M. (1991). Human myoblast transplantation: a simple assay for tumorigenicity. *Neuromuscul. Disord.* **1**: 341–343.

Tse, H.F., Kwong, Y.L., Chan, J.K., Lo, G., Ho, C.L. and Lau, C.P. (2003). Angiogenesis in ischaemic myocardium by intramyocardial autologous bone marrow mononuclear cell implantation. *Lancet* **361**: 47–49.

Tse, H.F., Thambar, S., Kwong, Y.L., Rowlings, P., Bellamy, G., McCrohon, J., Bastian, B., Chan, J.K., Lo, G., Ho, C. L., and Lau, C.P. (2006). Safety of catheter-based intramyocardial autologous bone marrow cells implantation for therapeutic angiogenesis. *Am. J. Cardiol.* **98**: 60–62.

Tsikouris, J.P., Suarez, J.A., Simoni, J.S., Ziska, M. and Meyrrose, G.E. (2004). Exploring the effects of ACE inhibitor tissue penetration on vascular inflammation following acute myocardial infarction. *Coron. Artery Dis.* **15**: 211–217.

Tsimikas, S., Willerson, J.T. and Ridker, P.M. (2006). C-reactive protein and other emerging blood biomarkers to optimize risk stratification of vulnerable patients. *J. Am. Coll. Cardiol.* **47**: C19–C31.

Urbanek, K., Torella, D., Sheikh, F., De Angelis, A., Nurzynska, D., Silvestri, F., Beltrami, C.A., Bussani, R., Beltrami, A.P., Quaini, F., Bolli, R., Leri, A., Kajstura, J. and Anversa, P. (2005). Myocardial regeneration by activation of multipotent cardiac stem cells in ischemic heart failure. *Proc. Natl Acad. Sci. USA.* **102**: 8692–8697.

Valgimigli, M., Rigolin, G.M., Fucili, A., Porta, M.D., Soukhomovskaia, O., Malagutti, P., Bugli, A.M., Bragotti, L.Z., Francolini, G., Mauro, E., Castoldi, G. and Ferrari, R. (2004). CD34+ and endothelial progenitor cells in patients with various degrees of congestive heart failure. *Circulation* **110**: 1209–1212.

van den Bos, E.J. and Taylor, D.A. (2003). Cardiac transplantation of skeletal myoblasts for heart failure. *Minerva Cardioangiol.* **51**: 227–243.

van den Bos, E.J., Davis, B.H. and Taylor, D.A. (2004). Transplantation of skeletal myoblasts for cardiac repair. *J. Heart Lung Transplant.* **23**: 1217–1227.

van der Wal, A.C., Becker, A.E., van der Loos, C.M., Tigges, A.J. and Das, P.K. (1994). Fibrous and lipid-rich atherosclerotic plaques are part of interchangeable morphologies related to inflammation: a concept. *Coron. Artery Dis.* **5**: 463–469.

van Zonneveld, A.R.T. (2006). Endothelial progenitor cells: biology and therapeutic potential in hypertension. *Curr. Opin. Nephrol. Hypertens.* **15**: 167–172.

Verfaillie, C.M. (2005). Multipotent adult progenitor cells: an update. *Novartis Found. Symp.* **265**: 55–61.

Vulliet, P.R., Greeley, M., Halloran, S.M., MacDonald, K.A. and Kittleson, M.D. (2004). Intra-coronary arterial injection of mesenchymal stromal cells and microinfarction in dogs. *Lancet* **363**: 783–784.

Weber, C., Schober, A. and Zernecke, A. (2004). Chemokines. Key regulators of mononuclear cell recruitment in atherosclerotic vascular disease. *Arterioscler. Thromb. Vasc. Biol.* **24**: 1891–1896.

Werner, N., Kosiol, S., Schiegl, T., Ahlers, P., Walenta, K., Link, A., Bohm, M. and Nickenig, G. (2005). Circulating endothelial progenitor cells and cardiovascular outcomes. *N. Engl. J. Med.* **353**: 999–1007.

Wilke, N., Zenovich, A., Jerosch-Herold, M. and Henry, T. (2001). Cardiac magnetic resonance imaging for the assessment of myocardial angiogenesis. *Curr. Interv. Cardiol. Rep.* **3**: 205–212.

Williams, P.C., Coffey, M.J., Coles, B., Sanchez, S., Morrow, J.D., Cockcroft, J.R., Lewis, M.J. and O'Donnell, V.B. (2005). *In vivo* aspirin supplementation inhibits nitric oxide consumption by human platelets. *Blood* **106**: 2737–2743.

Wollert, K.C., Meyer, G.P., Lotz, J., Ringes-Lichtenberg, S., Lippolt, P., Breidenbach, C., Fichtner, S., Korte, T., Hornig, B., Messinger, D., Arseniev, L., Hertenstein, B., Ganser, A. and Drexler, H. (2004). Intracoronary autologous bone-marrow cell transfer after myocardial infarction: the BOOST randomised controlled clinical trial. *Lancet* **364**: 141–148.

Wright, L., Maloney, W., Yu, X., Kindle, L., Collin-Osdoby, P. and Osdoby P. (2005). Stromal cell-derived factor-1 binding to its chemokine receptor CXCR4 on precursor cells promotes the chemotactic recruitment, development and survival of human osteoclasts. *Bone.* **36**: 840–853.

Wutzl, A., Brozek, W., Lernbass, I., Rauner, M., Hofbauer, G., Schopper, C., Watzinger, F., Peterlik, M. and Pietschmann, P. (2006). Bone morphogenetic proteins 5 and 6 stimulate osteoclast generation. *J. Biomed. Mater. Res.* **77**:-75–83.

Wyatt, S.B., Winters, K.P. and Dubbert, P.M. (2006). Overweight and obesity: prevalence, consequences, and causes of a growing public health problem. *Am. J. Med. Sci.* **331**: 166–174.

Yan, L.L., Liu, K., Daviglus, M.L., Colangelo, L.A., Kiefe, C.I., Sidney, S., Matthews, K.A. and Greenland, P. (2006). Education, 15-year risk factor progression, and coronary artery calcium in young adulthood and early middle age: the coronary artery risk development in young adults study. *JAMA* **295**: 1793–1800.

Yancy, C. and Fonarow, G. (2004). Quality of care and outcomes in acute decompensated heart failure: The ADHERE Registry. *Curr. Heart Fail. Rep.* **1**: 121–128.

Yeh, E.T., Zhang, S., Wu, H.D., Korbling, M., Willerson, J.T. and Estrov, Z. (2003). Transdifferentiation of human peripheral blood CD34+-enriched cell population into cardiomyocytes, endothelial cells, and smooth muscle cells *in vivo*. *Circulation* **108**: 2070–2073.

Yu, J., Rudic, R.D. and Sessa, W.C. (2002). Nitric oxide-releasing aspirin decreases vascular injury by reducing inflammation and promoting apoptosis. *Lab. Invest.* **82**: 825–832.

Zhai, Q., Qiu, L., Li, Q., Meng, H., Han, J., Herzig, R. and Han, Z. (2004). Short-term *ex vivo* expansion sustains the homing-related properties of umbilical cord blood hematopoietic stem and progenitor cells. *Haematologica* **89**: 265–273.

Zhang, M., Methot, D., Poppa, V., Fujio, Y., Walsh, K. and Murry, C.E. (2001). Cardiomyocyte grafting for cardiac repair: graft cell death and anti-death strategies. *J. Mol. Cell. Cardiol.* **33**: 907–921.

Zibaitis, A., Greentree, D., Ma, F., Marelli, D., Duong, M. and Chiu, R.C. (1994). Myocardial regeneration with satellite cell implantation. *Transplant. Proc.* **26**: 3294.

25

Cell Therapies for Bone Regeneration

Rehan N. Khanzada, Chantal E. Holy, F. Jerry Volenec, and Scott P. Bruder

INTRODUCTION

Historical Overview

Bone damage, either due to pathology or trauma, is a very common occurrence that requires costly medical and/or surgical intervention, and is associated with significant morbidity (Cancedda et al., 2004). Of all fractures that occur in the United States each year, about 15% require some type of bone grafting to improve the healing process. To date, graft materials include autograft (bone taken from one part of the patient's body and replaced in another site that requires bone healing), allograft (bone taken from a donor) or synthetic materials.

The earliest evidence of an orthotopic autograft dates back to the Bronze Age. A circular disk of bone was removed from a human's calvarium to relieve intracranial pressure and placed elsewhere as an autograft. Written accounts from Egypt, China, and India dating back many centuries describe similar autograft-based experimentation. One Indian text from 700 CE describes a procedure for nasal reconstruction that is very similar to modern methods. While autograft is currently considered the gold standard for bone regeneration due to its success rate, it requires secondary bone harvesting procedures that can cause high morbidity (Gupta et al., 2001) and is the only available in small supply.

The first use of allografts and xenografts (bone from a donor of different species) dates back to over 300 years ago, when Job van Meekeren historically performed the first bone graft procedure using canine xenograft in 1668. The need for bone grafting became critical during World War II, as the US Navy established bone banks to treat fractures sustained in war. However, despite significant progress in allograft preparation and cleansing technologies, allografts still carry the risk of disease transmission.

Synthetic grafts of all types have therefore been developed. While these grafts are available in high volumes and do not carry risks of disease transmission, their effectiveness *in vivo* does not consistently meet that of autograft. Research on synthetic grafts for bone regeneration has thus evolved into state-of-the-art science, especially after the discovery of mesenchymal stem cells (MSCs) capable of forming bone (Friedenstein et al., 1968) and bone morphogenetic proteins (BMPs) (Urist, 1965).

The Clinical Need for Therapeutic Solutions to Bone Regeneration

One of the reasons for so many graft choices is the vast quantity of bone graft required: an estimated 1.5 million bone graft operations were performed in the United States in 2004 to enhance the healing of spinal fusions, internal fixation of fractures, maxillofacial reconstruction, long-bone repair, and lost bone due to trauma or ablative surgery. However, selecting the right graft for the right patient is one of the key challenges

Copyright © 2008 Elsevier Inc. All rights reserved

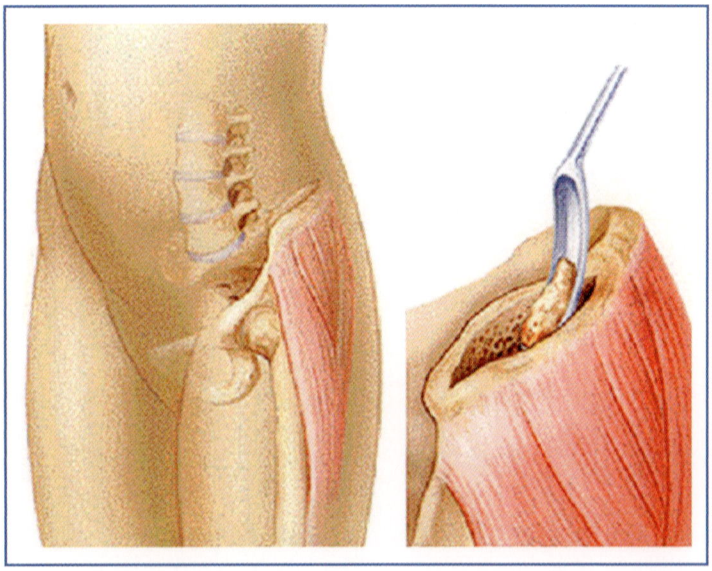

Figure 25.1 **Schematic representation of iliac crest autograft harvesting. The iliac crest represents the largest source of autologous bone, and requires a secondary surgery that can lead to morbidity.**

for orthopedic and spine surgeons. In fact, the choice of a bone graft is based on four main factors: the size of the defect, the location of the defect, the biology of the defect site, and whether structural support is required (Gamradt and Lieberman, 2003).

Autograft harvested from the iliac crest is most often used in treating these conditions, as it is histocompatible, and does not transport any diseases. A schematic representation of iliac crest autograft harvesting is shown in Figure 25.1. Hydroxyapatite and collagen within the native bone serve as osteoconductive frameworks, while stromal cells within the bone marrow, and to some extent, along the microcavities of the bone, contain osteogenic cells that lead to reproducible bone formation when placed in a surgical site. In addition, growth factors within the bone and adjacent hematoma provide osteoinductive factors (Sutherland and Bostrom, 2005). There are some drawbacks and potential complications associated with autograft harvested from the iliac crest. Although severe complications from iliac crest bone harvesting are rare, the incidence of donor site pain reported in the literature ranges from 25% to 49%, with 19% to 27% of patients experiencing chronic site pain 2 years postoperatively (Younger and Chapman, 1989; Fernyhough et al., 1992). To better understand the causes of this morbidity, Gupta et al. reviewed literature reports spanning 34 years and including 1,020 patients. The authors found no correlation between the patients' pain ratings and any of the following parameters: incision site, surgical approach, harvesting technique, or demographics including patient age or gender.

In addition to the issue of unpredictable morbidity, limited harvest supply of autograft is sometimes problematic for patients undergoing procedures that require large graft volumes. Autograft also has poor handling characteristics, as it is typically morcelized during harvesting and does not have any structural integrity (Figure 25.2).

Cadaveric allograft is sometimes used but there are continued concerns about graft resorption, inadequate revascularization, and possible transmission of blood-borne diseases. Allografts may be demineralized to expose native growth factors, which increase the grafts' *in vivo* efficacy (Zhang et al., 1997). Structural allografts, on the other hand, are frozen or freeze-dried, which destroys cells within the allografts, thereby reducing potential complications from immune responses but also destroying the grafts osteogenic activity (Goldberg and Stevenson, 1987).

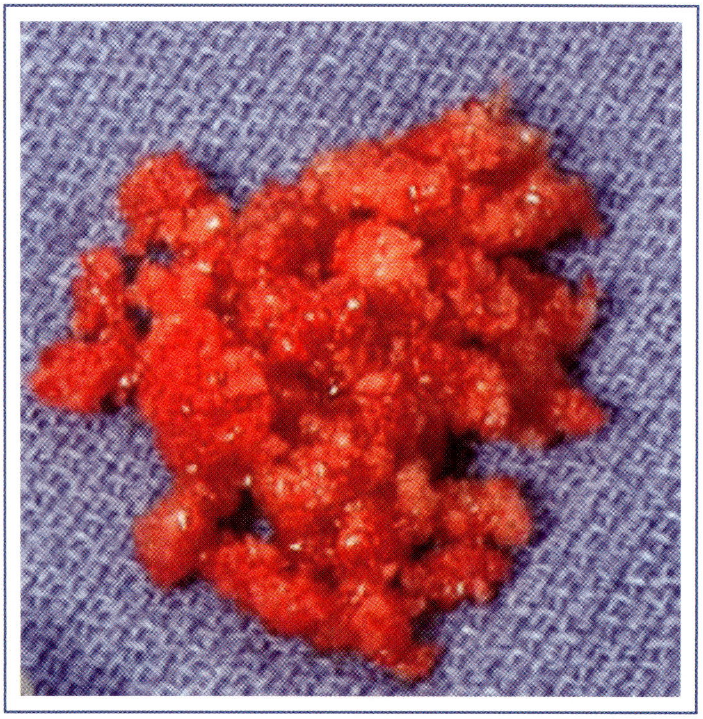

Figure 25.2 **Photograph of human iliac crest autograft, morcelized, and ready for re-implantation. Autograft from the iliac crest is the "gold standard" for bone regeneration.**

To address shortcomings of both autograft and allograft, completely synthetic options are being developed with an eye toward creating synthetics that would mimic autograft, and thus a strong understanding of the biological processes required for bone formation has become critical.

Biological Ingredients

Bone repair and regeneration is a complex process consisting of a tightly regulated cascade of cellular interactions. As part of the acute inflammatory response, bioactive molecules are released, which promote the influx of MSCs to the fracture site. These MSCs adhere to osteoconductive scaffolding within the fracture site and, in response to local growth factors, proliferate and differentiate into osteoblasts capable of secreting osteoid, which is subsequently mineralized to form new bone (Whang and Lieberman, 2003). Thus, the ideal bone graft for bone repair and regeneration requires three key ingredients: (1) surface areas allowing cell attachment (i.e. osteoconductive scaffold), (2) cells capable of forming bone (i.e. osteogenic cells), and (3) biological stimulants.

While synthetic methods can be used to develop large amounts of osteoconductive surfaces, finding enough osteogenic cells to populate grafts may be seen as a limiting factor for success. The search for rich and easily accessible sources of osteogenic cells is therefore spurring significant interest.

Delivery of Osteogenic Cells

Early work by Burwell (1964) demonstrated that the main repository of cells capable of forming bone within iliac crest bone grafts was the bone marrow. Owen's studies (1985) using *in vitro* cell growth confirmed that isolated cells from bone marrow had osteogenic and adipogenic potential (Figure 25.3). The term osteogenic was thus coined to define a cell capable of forming bone or capable of differentiating into a bone-forming cell; osteogenicity further referred to materials containing osteogenic cells. Following the report of bone

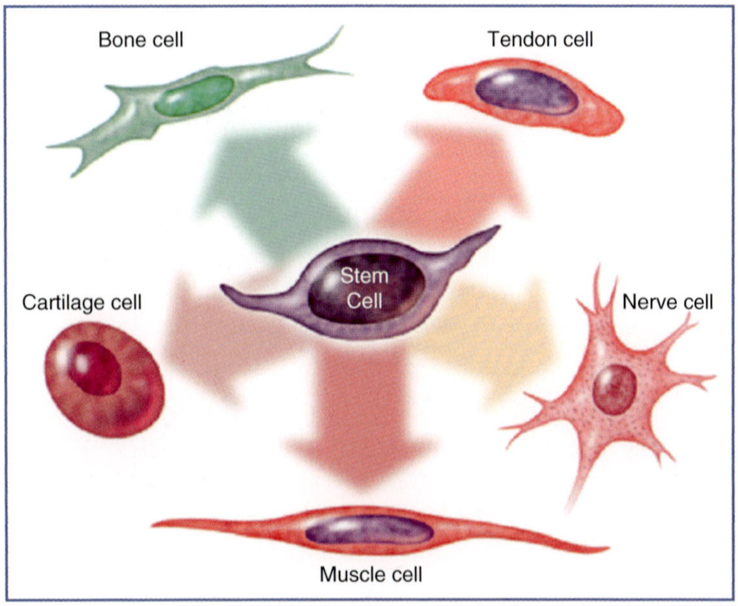

Figure 25.3 **Schematic representation of mesenchymal stem cell development pathways. As discovered in the 1980s by Owen et al. mesenchymal stem cells can develop along multiple pathways.**

marrow's osteogenicity, orthopedic surgeons since the 1980s routinely used fresh bone marrow for repairing large bone defects.

As bone marrow was analyzed for osteogenic cell content, Muschler and Midura (2002) and others demonstrated that less than 1% cells within the marrow had osteogenic potential. In addition, Muschler et al. (2001) demonstrated that the number of osteogenic cells was variable from one patient to another, especially as a function of age and gender. Therefore, new methodologies that would take advantage of bone marrow's osteogenicity and alleviate issues of cell count variability were also investigated, as described below.

As bone marrow has been defined as an easily accessible source of osteogenic cells, two additional ingredients for bone regeneration are thus required: an optimized carrier and biological stimulants.

Carriers and Growth Factors

Osteoconductive graft materials refer to scaffolds that provide the appropriate framework for bone growth and osteoblast attachment. These scaffolds provide appropriate three-dimensional shape and structure to restrict cell movement in an implant site. For successful bone healing, these scaffolds need to have direct contact with viable bone and support bony ingrowth and vascularization without excessive inflammatory response. Examples of osteoconductive scaffolds include naturally occurring materials such as mineralized cancellous chips and fibrin clots, and synthetics such as tricalcium phosphates, hydroxyapatites, collagen sponges, and various polymers. The most appropriate scaffold for a given clinical application depends on the pathological condition being treated, its anatomic location, and the biomechanical stresses and loads that apply to that specific site.

Osteopromotive graft materials have the ability to provide stimulatory signals at various stages and enhance the bone repair and regeneration process. These materials do not have the capacity to induce new bone growth by themselves and work best with osteoinductive and osteogenic graft materials in orthotopic applications. An example is platelet-rich plasma (PRP), which is prepared by collecting and concentrating platelets from a patient's whole blood immediately before surgery. These platelets contain a rich source of various growth factors that play an important role in bone repair and regeneration (Kevy and Jacobsen, 2004).

Osteoinductive graft materials have the capacity to induce bone growth in ectopic sites. These materials function as biological stimulants (i.e. growth factors), which activate MSCs toward chemotaxis, proliferation, and differentiation into osteoblasts that leads to new bone growth (Urist, 1965). BMPs are the key osteoinductive proteins identified to date. These are either available at native levels in allogeneic demineralized bone matrix (DBM) or as recombinant human proteins. Osteoinductivity of DBM varies between donors and as a function of bone processing (Zhang et al., 1997); therefore, ongoing osteoinductivity testing of DBM is required to ensure high potency of these graft materials. The BMPs are members of the TGF-β superfamily of growth factors known to play a critical role in initiating endochondral bone formation. They are low molecular weight proteins that induce a quick biological response at the implantation site. The US Food and Drug Administration has recently approved recombinant human BMP-2 and BMP-7 for specific and limited clinical indications.

Since these are potent osteogenic agents, they require an optimized delivery system in order to provide appropriate biological response. Currently, supraphysiologic doses of these recombinant BMPs are required for induction of bone formation (Yoon and Boden, 2002). Current delivery methods for these growth factors are either collagen-based sponges alone or calcium–phosphate granules, used as bulking agents, surrounded with collagen sponges (Barnes et al., 2005). While these combination matrices may be more effective than the collagen alone, these delivery systems still require significant optimization as current recommended clinical doses are excessively high, very costly, and have unknown long-term effects (Gamradt and Lieberman, 2004). Despite such high doses, BMPs may not produce sufficient osteogenic response where there is poor bone quality, scar tissue, large defect size, or inadequate vascularization (Cook et al., 1995). While more research is required to optimize the use of BMPs, the availability of these growth factors represents progress for bone grafting, as grafts containing all three ingredients (matrix, cells, and biological stimulants) could thus be envisioned. Practical and effective approaches for the preparation of such grafts have therefore been investigated, as described below, starting with optimized methods for obtaining osteogenic cells.

CURRENT SOURCES OF OSTEOGENIC CELLS AND CELL ISOLATION TECHNIQUES

Source of Osteogenic Cells

Osteogenic cells, as mentioned above, can be defined as cells that are, or will differentiate into, osteoblasts capable of forming bone. There are two major sources for osteogenic cells: (1) tissues containing MSCs (e.g. bone marrow) and (2) differentiated bone tissue. MSCs, the main source of osteogenic cells, can be further defined as cells that retain the capacity to differentiate along osteogenic, adipogenic, fibroblastic, and chondrogenic lines (Lennon et al., 1996; Bruder and Caplan, 2000). The most accessible source for MSCs is the bone marrow. More recently, differentiated connective tissues have also been shown to contain osteogenic cells that could form bone in specific culture conditions (Zuk et al., 2002).

A secondary source of osteogenic cells that was recently described consisted of mature bone fragments, obtained as debris from reaming procedures (Wenisch et al., 2005). These samples did not contain any bone marrow, and thus the osteogenicity of the bone fragments was described as specifically due to bone-lining cells. This source of osteogenic cells has not been widely investigated, and thus will not be further described in this manuscript.

Autologous Bone Marrow

As described above, bone marrow was shown in the mid-1960s to contain both hematopoietic stem cells and MSCs (Friedenstein et al., 1968). Recently, a third type of precursor cell was identified within bone marrow: the "side population" (SP); these cells are defined by their ability to regenerate the hematopoietic compartment as well as to differentiate into osteoblasts through a mesenchymal intermediate (Olmsted-Davis et al., 2002). These findings suggest a population of cells, precursor to MSCs, and elucidates yet another step in the early development of osteoblasts.

Unlike the bony site of an autograft harvest, bone marrow is self-renewing and can be obtained from the iliac crest in a non-invasive procedure with a simple needle aspiration. This usually does not cause morbidity (Connolly, 1995) and is the most inexpensive method for bone repair and regeneration. To increase the effectiveness of bone marrow, Muschler et al. (1997) investigated bone marrow aspiration techniques, and described a methodology to maximize the osteogenic cell content within a given bone marrow aspirate. This technique involved aspirating no more than 2 ml of bone marrow from a given site. The aspiration needle was then moved either further into the bone marrow cavity or at a different location, in which an additional 2 ml could then be aspirated. Aspirating more than 2 ml per site resulted in dilution of the bone marrow with peripheral blood, and thus diluting the cellularity of the final aspirate.

While this technique was shown to ensure the highest possible cellularity within the bone marrow aspirate, researchers are also looking at other methods to completely move away from potential dilution issues by developing osteogenic cell banks that would provide the same number and efficacy of cells without requiring aspiration of bone marrow from patients prior to bone grafting.

Allogeneic Bone Marrow

Allogeneic stem cell sources hold great promises as universal cell banks that may be developed for bone and other tissue repair. It was hypothesized early on that allogeneic MSCs might be applicable for bone repair and regeneration if one could successfully mute immunoreactive groups on the MSCs. However, *in vivo* preclinical studies seemed to indicate that, surprisingly, allogeneic MSC implantation failed to provoke an immune response. In one instance, analysis of circulating antibody levels against MSCs 9-week postimplantation in a canine cranial site supported the hypothesis that neither autologous nor allogeneic MSCs induced a systemic response by the host. Authors concluded that autologous and allogeneic MSCs had the capacity to regenerate bone within craniofacial defects (De Kok et al., 2003). More recently, undifferentiated human MSCs were shown not to express immunologically relevant cell surface markers. They also seemed to inhibit the proliferation of allogeneic T-cells *in vitro.* Evidence seemed to indicate that these cells did not elicit an immune response after allogeneic or xenogenic transplantation. Thus, MSC could be described as immunoprivileged or immunomodulating cells (Niemeyer et al., 2004). These findings confirmed that allogeneic stem cells may indeed become a possible therapeutic tool and as such are currently being developed for bone and soft tissue repair. However, due to the potentially arduous regulatory path required for allogeneic stem cells to meet approval by federal agencies, other autologous sources of osteogenic cells are also being investigated.

Novel Tissue Sources for Osteogenic Stem Cells: Muscle, Fat and Other Connective Tissues

Cell derived from connective tissues such as muscle and fat were shown to "behave" similarly to bone-marrow-derived MSCs. These cells had the ability to differentiate into bone under appropriate biological cues (Betz et al., 2005). Both muscle and fat cells were found to contain MSCs that were readily expanded in culture and underwent osteogenic differentiation. These MSCs were obtained conveniently from muscle biopsy or liposuction, procedures involving less morbidity than traditional bone graft harvesting. Interestingly, these cells were shown to differentiate not only into bone, but also fat, cartilage, and muscle tissues, with growth factors specific to each culture condition (Zuk et al., 2002). Muscle-derived stem cells were also retrovirally transduced to express osteogenic factors BMP-2 and BMP-4, and were capable of differentiating into bone and accelerating repair of skull defects in mice (Huard and Peng, 2004).

Fat was found to be the most convenient tissue other than bone marrow for harvesting MSCs, as it is easily biopsied, cultured, expanded, and transduced. Moreover, adipose tissue was shown to contain proliferative properties that did not decline with age. In addition, this tissue was described as a richer, and more effective, source of osteoprogenitor cells than bone marrow, when genetically modified to express BMP-2 (Dragoo, 2003). More recently, Cowan et al. used these cells to investigate the *in vivo* osteogenic capability of adipose-derived stromal (ADAS) cells to heal critical-sized mouse calvarial defects. These ADAS cells were harvested from the subcutaneous anterior abdominal wall yielding ~800 mg of fat tissue, and were shown to be multipotent, and available in large numbers. *In vitro*, they were observed to attach and proliferate rapidly. Authors also reported a yield of cells that, by itself, was much higher than that of typical bone marrow; however, it is worthwhile to note that mouse bone-marrow-derived cells are technically difficult to isolate and manipulate, and that this fact is not usually observed with tissues from other species. *In vivo*, ADAS cells seeded onto apatite coated PLGA scaffolds regenerated bone in critical sized calvarial defects (Cowan et al., 2004).

With osteogenic cells available and clearly identifiable *in vitro*, researchers also investigated the potential to genetically modify these cells to secrete the biological stimulants required for bone formation. That way, two of the three ingredients for bone formation could be provided within a given cell population. This gene therapy approach is briefly described below and in other chapters.

Gene Therapy

Gene therapy deals with the transfer of genetic material into cells, which in turn, secrete specific proteins in selected sites. In this model, growth factor(s) are synthesized *in situ* as a result of gene transfer and would be presented to the surrounding tissue in a natural, cell-based manner (Nussenbaum and Krebsbach, 2004). Local MSCs would then undergo osteogenic differentiation, or form another appropriate tissue (Lou, 2004).

Gene therapy involves three fundamental elements: a sequence of DNA encoding a protein of interest, a vector that facilitates the entry of genetic material into cells, and target cells into which the gene is inserted. Two different types of therapeutic conditions can be envisioned for gene therapy: (1) conditions that require continuous, sustained delivery of specific proteins and (2) conditions that require a transient bone inducing agent (e.g. trauma cases and spinal fusions).

A clinical example of a case requiring continuous delivery of proteins includes osteogenesis imperfecta. In this case, a patient would be implanted with MSCs genetically modified with a retrovirus containing the gene for normal type I collagen. These cells would re-establish themselves in the bone marrow and thus provide mesenchymal progenitors with the appropriate collagen Type I building capabilities (Pereira et al., 1995). Retroviruses are currently the key vectors for continuous gene expression but, as they depend on cell replication for transcription, they can only be used in highly proliferative cells (Tibor, 2003).

For one-time bone repair applications (e.g. bone fractures, spinal fusions), gene therapy must be limited to short-term gene expression, and thus, non-viral vectors are currently under investigation. These are typically easier to generate, more stable than viruses, and less immunogenic (Gamradt and Lieberman, 2004). These vectors are however far less effective than viral vectors to transduce cells. Others options to transduce cells also include adenoviruses, whose limitation include the potential to provoke immune responses (Musgrave et al., 2002).

In addition to short- and long-term protein expression, two types of gene therapy approaches are currently under investigations: a so-called *in vivo* approach, as well as an *ex vivo* approach. *In vivo* gene therapy involves the direct transfer of genes into patients, while the *ex vivo* gene therapy involves transducing cells *in vitro* and then implanting those cells at a specific site. Both *in vivo* and *ex vivo* gene therapies for bone formation have been successfully demonstrated in several different animal models, including rat femurs, mice skulls, and anterior spinal fusion in pigs. But while successful in preclinical trials for one-time bone repair applications, the use of gene therapy for these particular conditions may be excessive, since biologically simpler and more economical approaches can effectively treat bone conditions like fractures and fusions. Gene therapy, therefore, may be better suited for pathologies that, like osteogenesis imperfecta, require continuous protein expression and for which no satisfactory treatment option exists.

Cell Isolation Techniques

Use of Autologous Bone Marrow on Optimized Matrices

As far back as the late 1960s, bone marrow was used as a research tool to isolate MSCs that would form bone *in vitro*. Maniatopoulos et al. (1988) first described a methodology to isolate osteoprogenitor cells from rat bone marrow. This technique involved explanting the entire femur of the rats and flushing the bone marrow in an osteogenic cell media. After 7–14 days, alkaline–phosphatase positive colonies would become visible, indicating potential differentiation of osteogenic cells.

Cell isolation methodologies were then developed to culture human bone-marrow-derived cells. Jaiswal et al. (1998) first established a reproducible system for the *in vitro* osteogenic differentiation of human marrow-derived MSCs. Muschler et al. (1997) utilized similar cell isolation techniques to quantify osteogenic precursors in bone marrow aspirates of patients.

Both rodent and human cultures were found to be strongly sensitive to media as well as surface conditions. The term "osteoconductive" was found to be particularly important *in vitro* (as well as *in vivo*), since cell growth on specific polymers was found to be inhibited, while that on, for example, collagen or poly-L-lysine coated surfaces, it was found to be optimal (Liu et al., 1999; Karp et al., 2003). These studies highlighted the importance of osteoconduction and optimized carriers for cell proliferation and differentiation.

Point-of-care Osteogenic Cell Enrichment

As discussed previously, age, disease, and other factors can reduce bone marrow cellularity prompting the idea of cellular enrichment methodologies that could be used at point of care. Three major approaches were described in the literature to increase cell numbers within graft materials: (1) enzymatic tissue digestion, (2) bone marrow centrifugation, and (3) selective cell retention.

Enzymatic Tissue Digestion

Enzymatic tissue digestion so far has strictly been used as an *in vitro* method to release osteoblasts from bone tissue. It can be hypothesized that enzymatic tissue digestion could be used to release cells from bone fragments obtained during surgery, for re-implantation in defect sites.

In brief, enzymatic tissue digestion as described in the literature involves mincing bone tissues (typically from rodent femurs or calvaria) and washing those minced fragments with series of collagenase/trypsin enzyme solutions. These solutions degrade connective tissues between cells and release osteoblasts (Thomas et al., 2004). This technique was shown to offer high-level cellular yields containing precursor, differentiating, and osteoblastic cells in enzyme-enriched cell preparations (Vinay et al., 1981). Released osteoblasts and committed osteoprogenitor cells exhibited properties of bone that included characteristic morphology, synthesis of bone-related proteins, and calcification after 3–4 weeks in culture (Webster et al., 2000). In another example, rat calvarial cells formed mineralized nodules within 2 weeks *in vivo* (Irie et al., 1998). Similarly, osteoblasts obtained from canine diaphyseal bones developed bone-like tissue *in vitro* (Boyan et al., 1999).

While successful in yielding osteoprogenitors, this method was also shown to have significant shortfalls: the collagenase had the potential to harm osteoblasts by removing proteins from their membrane, thereby affecting their ability for attachment. When compared to other cell isolation techniques, enzymatic tissue digestion therefore produced the lowest amount of functional osteoblasts.

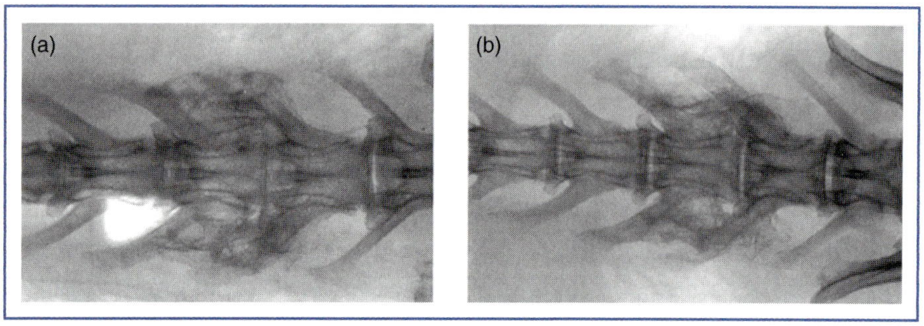

Figure 25.5 Postmortem radiographs of rabbit spine segments implanted with: (a) autograft; (b) bone marrow on a carrier (courtesy: Tay et al., 1998). Animals were sacrificed 12-week postsurgery for further analyses of the fusion sites. Autograft and bone marrow on carrier (Healos®) were equivalent in this study.

regenerate bone when combined with autologous bone marrow (Figure 25.5 – Tay et al., 1998). This last study, however, was the subject of a controversy that highlights the need for a clear understanding of bone marrow cellularity in animal models. Tay et al. used a rabbit posterolateral fusion model to evaluate the carrier with bone marrow and showed that, when combined with heparinized or non-heparinized bone marrow, bone marrow constructs yielded fusion rates comparable to autograft. As with all animal models, several challenges needed to be addressed to generate clinically relevant data; in particular, rabbits were known as suboptimal species for bone marrow research, their iliac crest having poor cellularity. Tay et al. therefore created a secondary surgical site to harvest bone marrow from the rabbits' long bone. The cellularity of the marrow aspirate obtained by Tay averaged 238 million cells/ml. Unlike humans, this bony site was shown to provide the most cellular marrow.

In a recent publication using the same carrier, Kraiwattanapong et al. (2005) went back to the rabbit model to compare the efficacy of bone marrow aspirate to human recombinant BMP-2 (rhBMP-2). Two groups were compared: the first group was implanted with the carrier and autologous bone marrow from the iliac crest of the rabbits, while the second group was implanted with BMP-2 on a collagen–ceramic material. The average cellularity of the marrow aspirate was 30 million cells/ml. The authors used radiographs and manual palpation to report no fusion in the bone marrow group and 100% fusion in the rhBMP-2 animals.

In a second publication involving rhBMP-2 and bone marrow, Minamide et al. (2005) evaluated four different graft materials in groups of seven animals each: (1) autologous bone; (2) collagen–ceramic material with rhBMP-2; (3) bone-marrow-derived, culture-expanded cells at a concentration of 1 million cells/ml on hydroxyapatite; and (4) bone-marrow-derived, culture-expanded cells at a concentration of 100 million cells/ml on hydroxyapatite. Manual palpation and radiography indicated 4/7 fusions in the autograft group, 7/7 fusions in the rhBMP-2 group, 0/7 fusions in the 1 million cells/ml group and 5/7 fusions in the 100 million cells/ml group. Minamide et al. (2005) thus concluded that, if expanded, bone marrow cells were capable of forming bone similar to autograft.

Interestingly, while conclusions of these 3 studies seem contradictory at first, a closer look in the use of bone marrow and the cellularity of the marrow aspirates indicated that in fact, these papers demonstrated the same message: bone marrow requires osteogenic cells to form bone, and the use of suboptimal marrow, depleted in cells, did not result in bone formation: Both Kraiwattanapong and Minamide used bone marrow from the rabbits iliac crest that had very low cell counts. When Minamide increased the cellularity of the grafts to 100 million cells/ml by culture expansion, fusion rates were comparable to autograft. Using the rabbit's long bone, Tay obtained fresh bone marrow with 238 million cells/ml and observed a fusion rate similar to autograft without the need to culture-expanded cells.

Interestingly as well, these publications may seem to indicate that the bone healing performance of the carrier with the highly cellular bone marrow, as seen by Tay et al. was comparable to that seen with the rhBMP-2, as reported by Kraiwattanapong. However, in absence of a true head-to-head comparison between rhBMP-2 and optimized bone marrow grafts, the relative efficacy of those two types of graft cannot be effectively demonstrated.

Others however have attempted to understand how bone marrow graft would perform compared to BMPs. The earliest study evaluating bone marrow versus BMP-2 dates back to 1981 (Takagi and Urist, 1982). In this early rat segmental defect study, grafts containing rhBMP alone or bone marrow alone did not perform as well as grafts containing both rhBMP and bone marrow.

More recently, Den Boer et al. (2003) investigated the healing potential of ceramic grafts containing either bone marrow or OP-1 (rhBMP-7), in a 3-cm segmental bone defect in sheep tibia. Five treatment groups were included: no implant, autograft, hydroxyapatite alone, hydroxyapatite loaded with rhOP-1, and hydroxyapatite loaded with autologous bone marrow. At 12 weeks, torsional strength and stiffness of the healing tibiae were about two to three times higher for autograft and hydroxyapatite plus rhOP-1 or bone marrow compared to hydroxyapatite alone and empty defects. The mean values of both combination groups were comparable to those of autograft. Healing of bone defects, treated with porous hydroxyapatite was enhanced by the addition of rhOP-1 or autologous bone marrow. The results of these composite biosynthetic grafts were equivalent to those of autograft.

Cell-Enriched Grafts

Connolly et al. (1989) first suggested concentrating bone-marrow-derived cells on bone grafts for performance enhancement. A 4× cell concentrate of rabbit bone marrow using the centrifugation method significantly improved the bone-forming rate *in vivo* in a rabbit intraperitoneal chamber model.

The selective cell retention technology developed by Muschler was facilitated by a novel, single-use, disposable device that could be used at the point of care (Figure 25.6). Prototypes of this device were tested in many preclinical studies, as described below. Muschler et al. utilized the selective cell retention technology to create bone grafts for posterolateral fusion in dogs. In a preliminary study, the authors tested the hypothesis that the biologic milieu of a bone marrow clot significantly would improve the efficacy of such a graft. An established posterior spinal fusion model was used to compare cell-enriched cancellous bone alone, cancellous bone plus a bone marrow clot, and a cell-enriched cancellous bone plus bone marrow clot. Results from union score, quantitative computed tomography, and mechanical testing all demonstrated that the bone matrix plus enriched bone marrow clot was superior to all other groups. These data also confirm that cell enrichment significantly improved graft performance.

In a subsequent study, the actual cell concentrate was compared directly to whole bone marrow. Groups included (1) matrix alone (demineralized cortical bone powder), (2) matrix plus marrow, and (3) matrix with enriched marrow cells. Enriched matrix grafts delivered a mean of 2.3 times more cells and approximately 5.6 times more progenitors than matrix mixed with bone marrow. Again, union scores and fusion volumes both confirmed that selective cell retention improved healing outcomes (Muschler et al., 2005).

Using the same selective cell retention methodology, Brodke et al. used a canine critical-sized segmental defect to evaluate the healing efficacy of cell-enriched grafts versus autograft (Brodke et al., pending). Canine demineralized bone matrix (cDBM) and cancellous chips were enriched in osteoprogenitors and placed in 21-mm long osteoperiosteal femoral defects for 16 weeks, at which point the animals were sacrificed and the femurs removed and analyzed (Figure 25.7). The results showed equivalency between both the cell-enriched grafts and autograft. Both resulted in 100% bridging bone across the defect spans. Histology sections also demonstrated bone formation across the defects in all autograft and cell-enriched cases (Figure 25.8).

strategy would alleviate the need for a primary cell biopsy from a patient followed by a cell growth phase, since off-the-shelf cell concentrates would be available at all times.

In vivo studies thus evaluated the use of allogeneic and xenogeneic cells for bone formation. Discoveries that MSCs did not express immunologically relevant cell surface markers further heightened enthusiasm for this strategy. In a recent study, Arinzeh et al. investigated the effectiveness of allogeneic MSCs to heal a critical-sized bone defect in the femoral diaphysis in dogs without the use of immunosuppressive therapy. Similar to that used by Bruder et al., a critical-sized segmental bone defect of 21 mm in length was created in the mid-portion of the femoral diaphysis of 12 adult dogs. Each defect was treated with allogeneic MSCs loaded onto a hollow ceramic cylinder such that a complete mismatch between donor stem cells and recipient dogs was achieved. For defects treated with allogeneic mesenchymal stem cell implants, no adverse host response could be detected at any time point. Histologically, no lymphocytic infiltration occurred and no antibodies against allogeneic cells were seen. In addition, at 16 weeks, new bone had formed throughout the cell-enriched implants. These results demonstrated that allogeneic MSCs loaded on ceramic implants did not generate an immune response and were effective for bone repair (Arinzeh et al., 2003).

The positive results obtained in preclinical settings with either plain bone marrow or cell-enriched marrow prompted surgeons to use bone marrow in clinical applications. This has been facilitated by the fact that no regulatory or indeed, risk/complication concerns deterred surgeons from this procedure. As a result, while no long-term, multi-center, prospective, randomized, blinded clinical studies have been completed on the efficacy of bone marrow versus autograft, multiple clinical reports have described the use of bone marrow and bone marrow cell-enriched grafts to improve bone healing.

Clinical Studies

Autologous Bone Marrow with Optimized Matrices

In one of the first published studies, Salama and Weissman (1978) published a preliminary reported on 28 patients undergoing long-bone repair under conditions covering a wide range of indications. Bone marrow was implanted and in all cases provided very satisfactory results. Connolly et al. injected bone marrow directly into bone grafting sites, thereby alleviating the need for open surgery. Comparing healing patterns in 100 patients, the study reported an 80% healing rate following marrow grafting (Connolly, 1998). In a clinical study for collagen–calcium phosphate graft material (Collagraft®), Chapman et al. (1997) described the efficacy of bone marrow with the carrier in the treatment of long-bone fractures. No significant difference between the autograft and the bone marrow carrier groups was observed. In a more challenging clinical application, Garg et al. (1993) reported the use of percutaneous autogenous bone marrow grafting in 20 cases of non-united fracture. After 5 months, 17 cases progressed to healing. Delayed union and non-union cases were also treated with bone marrow by Sim et al. (1993), who described 11 cases that healed within a median time of 10 weeks following injection of bone marrow.

Bone marrow with allograft was used in a variety of pediatric cases, including cysts, fibromas, long-bone non-unions, and tibia lengthening procedures (Wientroub et al., 1989). In this study, all cases showed good new bone formation with no adverse reaction. In a subsequent pediatric tibial non-ossifying fibromas case, Tiedeman et al. (1991) also reported successful healing after injection of demineralized bone powder with autologous bone marrow.

Grafting bone marrow was also proven effective in medically compromised patients, for example, cancer patients, with delayed union or non-unions. Healey et al. (1990) reported bone marrow injections in eight patients with primary sarcomas. Bone formation was observed in seven patients after marrow injection, while complete healing was observed in five patients. Healey et al. concluded that these encouraging results warranted further clinical studies and that his findings suggested a useful technique for the treatment of delayed unions and non-unions in difficult clinical circumstances.

In clinical spinal fusion applications, bone marrow combined with either DBM or osteoconductive substrates also resulted in improved bone fusion rates. Results similar to autograft were observed in a retrospective posterior spinal fusion study examining 88 consecutive patients and comparing: (1) autologous iliac crest bone graft, (2) freeze-dried corticocancellous bone without marrow, and (3) DBM plus autologous bone marrow. Success rates of 88% and 89% were observed in the autologous iliac crest bone graft and the DBM plus autologous bone marrow groups, respectively. The highest failure rate (28%) was obtained in the freeze-dried corticocancellous bone without marrow group. The authors concluded that augmentation of demineralized bone with bone marrow resulted fusion rates similar to those of iliac crest bone graft (Price et al., 2003).

Similar findings have been reported with the use of synthetic grafting materials. Bone marrow aspirate combined with a mineralized collagen matrix (HEALOS®) produced similar fusion rates to those observed with autograft in a posterior spine fusion study (Kitchel et al., pending).

Cell-Enriched Grafts

The concept of using cell-enriched grafts was initially based on the assumption that a minimum number of osteoprogenitor cells was required to successfully form bone, and that in some severely compromised patients, this cell number may not be reached using whole bone marrow. This hypothesis was investigated by Hernigou et al. (2005), who evaluated 60 non-union patients implanted with concentrated autologous bone marrow. In this study, concentration was achieved by centrifugation, and a 4.2-fold cell concentration ratio was typically achieved (from 612 ± 134 progenitors/cm^3 before concentration to an average of 2,579 ± 1,121 progenitors/cm^3 after concentration). Union was obtained in 88% cases (53 patients), and the bone marrow that had been injected into the non-unions of those patients contained an average of 54,962 ± 17,431 progenitors, or more than 1,500 progenitors/cm^3. In contrast, the total number of osteoprogenitors (19,324 ± 6,843 or less than 700 progenitors/cm^3) injected into the non-union sites of the seven patients who did not heal was significantly lower ($p < 0.01$) than that of patients who healed. Therefore, in this study, a minimum of 1,000 progenitors/cm^3 and 30,000 progenitors in total seemed to be required to achieve healing. This study represented the first clinical attempt to quantify the required cell numbers for successful union.

Muschler's selective cell retention technique, based on a 3- to 4-fold increase in osteoprogenitors, was also evaluated *in vivo*. A pilot clinical trial conducted at the Cleveland Clinic reviewed spinal fusion outcomes of 21 patients that received DBM enriched using Muschler's technique; 20 out of 21 patients showed radiographic evidence of fusion at 12 months (Lieberman, 2004).

This study was followed by a prospective, multi-center, randomized study; 51 patients across 5 centers were included in the study. All underwent one or two-level posterolateral fusions. Grafts were prepared using iliac crest bone marrow aspirate. Selective cell retention of the marrow was prepared on DBM. After 12 months, VAS scores were decreased favorably by an average of 55% for back pain, 58.5% for the right leg pain and 65.7% for the left leg pain. Fusion rates were 84.2% (Wang et al., 2005).

While encouraging, there is still a need for additional data to fully demonstrate the potential of cell-enriched grafts, and their role in bone graft surgery.

CONCLUSION: FUTURE DEVELOPMENTS AND CHALLENGES

The osteogenic potential of bone marrow and its role and efficacy in long-bone repair and spinal fusion procedures has been demonstrated in a large body of preclinical and clinical studies over the past 50 years. Bone marrow, combined with an osteoconductive substrate, was shown to produce fusion rates similar to those reported with the use of iliac crest bone graft.

As our understanding of the complex phenomena of bone formation increases, there will be an increasing number of potent grafts and cell therapies available to help in bone-related surgical procedures.

Short- and medium-term research may include new osteopromotive and osteoinductive growth factors for bone repair and bone fusion, as well as improved delivery systems for existing growth factors. In the long term, focus may be shifted to injectable formulations that can form bone *in situ* and would altogether alleviate the need for invasive surgeries.

Another example of ongoing research includes so-called "cell-painting" technologies, which involve introduction of specific proteins on the outer surface of selected cells to guide and target them to a defect site (Dennis et al., 2004; Caplan, 2005). These powerful technologies, as well as advances in gene therapy, may address severe pathologies for which no other satisfactory cure currently exists.

REFERENCES

Arinzeh, T.L., Peter, S.J., Archambault, M.P., van den Bos, C., Gordon, S., Kraus, K., Smith, A. and Kadiyala, S. (2003). Allogeneic mesenchymal stem cells regenerate bone in a critical-sized canine segmental defect. *J. Bone Joint Surg.* **85A**: 1927–1935.

Barnes, B., Boden, S.D., Louis-Ugbo, J., Tomak, P.R., Park, J.S., Park, M.S. and Minamide, A. (2005). Lower dose of rhBMP-2 achieves spine fusion when combined with an osteoconductive bulking agent in non-human primates. *Spine* **30**(10): 1127–1133.

Betz, O., Vrahas, M., Baltzer, A., Lieberman, J.R., Robbins, P.D. and Evans, C.H. (2005). Gene transfer approaches to enhancing bone healing. In: Lieberman, J.R. and Friedlaender, G.E. (eds.), *Bone Regeneration and Repair*. Totowa, NJ: Humana Press, pp. 158–162.

Boyan, B.D., Caplan, A.I., Heckman, J.D., Lennon, D.P., Ehler, W. and Schwartz, Z. (1999). Osteochondral progenitor cells in acute and chronic canine nonunions. *J. Orthop. Res.* **17**: 246–255.

Brodke, D., Pedrozo, H.A., Kapur, T.A., Attawia, M., Kraus, K.H., Holy, C.E., Kadiyala, S. and Bruder, S.P. (2006). Bone grafts prepared with selective cell retention technology heal canine segmental defects as effectively as autograft. *J. Orthop. Res.* **24**(5): 857–866.

Bruder, S.P. and Caplan, A.I. (2000). Bone regeneration through cellular engineering. In: Lanza, R.P., Langer, R. and Vacanti, J. (eds.), *Principles of Tissue Engineering*. San Diego, CA: Academic Press, pp. 683–693.

Bruder, S.P., Jaiswal, N. and Haynesworth, S.F. (1997). Growth kinetics, self-renewal, and the osteogenic potential of purified human mesenchymal stem cells during extensive subcultivation and following cryopreservation. *J. Cell. Biochem.* **64**: 278–294.

Bruder, S.P., Jaiswal, N., Ricalton, N.S., Mosca, J.D., Kraus, K.H. and Kadiyala, S. (1998). Mesenchymal stem cells in osteobiology and applied bone regeneration. *Clinical Orthopaedics & Related Research* **355**(Suppl): S247–S256.

Burwell, R.G. (1964). Studies in the transplantation of bone VII. The fresh composite homograft–autograft of cancellous bone: an analysis of factors leading to osteogenesis in marrow transplants and in marrow-containing bone grafts. *J. Bone Joint Surg. Br.* **46**(1): 110–140.

Cancedda, R., Quarto, R., Bianchi, G., Mastrogiacomo, M. and Muraglia, A. (2004). Engineered cells in scaffolds heal bone. In: Sandell, L.J. and Grodzinsky, A.J. (eds.), *Tissue Engineering in Musculoskeletal Clinical Practice*. Rosemont, IL: AAOS, p. 115.

Caplan, A.I. (2005). Mesenchymal stem cells: cell-based reconstructive therapy in orthopedics. *Tissue Eng.* **11**(7–8): 1198–1211.

Chapman, M.W., Bucholz, R. and Cornell, C. (1997). Treatment of acute fractures with a collagen–calcium phosphate graft material. *J.Bone Joint Surg.* **79A**(4): 495–502.

Connolly, J.F. (1995). Injectable bone marrow preparations to stimulate osteogenic repair. *Clin. Orthop.* **313**: 8–18.

Connolly, J.F., Guse, R., Lippiello, L. and Dehne, R. (1989). Development of an osteogenic bone-marrow preparation. *J. Bone Joint Surg. Am.* **71**: 684–691.

Cannolly, J.F. (1998). Clinical use of marrow osteoprogenitor cells to stimulate osteogenesis. *Clinical Orthopaedics & Related Research*. **355**(Suppl.): S257–S266.

Cook, S.D., Wolfe, M.W., Salkeld, S.L. and Rueger, D.C. (1995). Effect of recombinant human osteogenic protein-1 on healing of segmental defects in non-human primates. *J. Bone Joint Surg. Am.* **77**: 734–750.

Cowan, C.M., Shi, Y.Y., Aalami, O.O., Chou, Y.F., Carina, M., Thomas, R., Quarto, N., Contag, C.H., Wu, B. and Longaker, M.T. (2004). Adipose-derived adult stromal cells heal critical-size mouse calvarial defects. *Nat. Biotechnol.* **22**(5): 560–567.

De Kok, I.J., Peter, S.J., Archambault, M., Van den Bos, C., Kadiyala, S., Aukhil, I. and Cooper, L.F. (2003). Investigation of allogeneic mesenchymal stem cell-based alveolar bone formation: preliminary findings. *Clin. Oral Implant. Res.* **14**(4): 481–489.

Den Boer, F.C., Wippermann, B.W., Blokhuis, T.J., Patka, P., Bakker, F.C. and Haarman, H.J. (2003). Healing of segmental bone defects with granular porous hydroxyapatite augmented with recombinant human osteogenic protein-1 or autologous bone marrow. *J. Orthop. Res.* **21**: 521–528.

Dennis, J.E., Cohen, N., Caplan, A.I. and Goldberg, V.M. (2004). Targeted delivery of progenitor cells for cartilage repair. *J. Orthop. Res.* **22**: 735.

Dragoo, J.L. (2003). Bone induction by BMP-2 transduced stem cells derived from human fat. *J. Orthop. Res.* **21**(4): 622–629.

Fernyhough, J.C., Schimandle, J.J., Weigel, M.C., Edwards, C.C. and Levine, A.M. (1992). Chronic donor site pain complicating bone graft harvesting from the posterior iliac crest for spinal fusion. *Spine* **17**(12): 1474–1480.

Fialkov, J.A., Holy, C.H., Shoichet, M.S. and Davies, J.E. (2003). *In vivo* bone engineering in a rabbit femur. *J. Cranifac. Surg.* **14**(3): 324–332.

Friedenstein, A.J., Petrakova, K.V., Kurolesova, A.I. and Frolova, G.P. (1968) Heterotopic transplants of bone marrow. Analysis of precursor cells for osteogenic and hematopoietic tissues. *Transplantation* **6**(2): 230–247.

Gamradt, S.C. and Lieberman, J.R. (2003). Bone graft for revision hip arthroplasty. *Clin. Orthop. Relat. Res.* **417**: 183–194.

Gamradt, S.C. and Lieberman, J.R. (2004). Genetic modification of stem cells to enhance bone repair. *Ann. Biomed. Eng.* **32**: 136–147.

Garg, N.K., Gaur, S. and Sharma, S. (1993). Percutaneous autologenous bone marrow grafting in 20 cases of ununited fracture. *Acta. Orthop. Scand.* **64**(6): 671–672

Goldberg, V.M. and Stevenson, S. (1987). Natural history of autografts and allografts. *Clin. Orthop.* **225**: 7–16.

Gupta, A.R., Shah, N.R., Patel, T.C. and Grauer, J.N. (2001). Perioperative and long-term complications of iliac crest bone graft harvesting for spinal surgery: a quantitative review of the literature. *Int. Med. J.* **8**(3): 163–166.

Gupta, M.C., Theerajunyaporn, T., Schmid, M.B., Holy, C.E., Kadiyala, S. and Bruder, S.P. (2004). Use of mesenchymal stem cells enriched grafts in an ovine posterolateral lumbar spine model. IMAST.

Healey, J.H., Zimmerman, P.A., Jessop, A.B., McDonnel, M. and Lane, J.M. (1990). Percutaneous bone marrow grafting of delayed union and non-union in cancer patients. *Clin. Orthop. Relat. Res.* **256**: 280–285.

Hernigou, Ph., Poignard, A., Beaujean, F. and Rouard, H. (2005). Percutaneous autologous *bone-marrow* grafting for nonunions: influence of the number and concentration of progenitor cells. *J. Bone Joint Surg. Am.* **87**: 1430–1437

Huard, J. and Peng, H. (2004). Induction of bone formation by stem cells. In: Sandell, L.J. and Sandell, A.J. (eds.), *Tissue Engineering in Musculoskeletal Clinical Practice*. Rosemont, IL: AAOS, p. 131.

Irie, K., Zalzal, S., Ozawa, H., McKee, M. and Nanci, A. (1998). Morphological and immunocytochemical characterization of primary osteogenic cell cultures derived from fetal rat cranial tissue. *Anat. Rec.* **252**(4): 554–567.

Karp, J.M., Shoichet, M.S. and Davies, J.E. (2003). Bone formation on two-dimensional poly(DL-lactide-co-glycolide) (PLGA) films and three-dimensional PLGA tissue engineering scaffolds *in vitro*. *J. Biomed. Mater. Res. A* **64**(2): 388–396.

Kevy, S.V. and Jacobson, M.S. (2004) Comparison of methods for point of care preparation of autologous platelet gel. *J. Extra-Corp. Technol.* **36**(1): 28–35.

Kitchel, S.H. (2006). A preliminary comparative study of radiographic results using mineralized collagen and bone marrow aspirate vs. autologous bone in the same patients undergoing posterior lumbar interbody fusion with instrumented posterolateral lumbar fusion. *Spine J.: Official Journal of the North American Spine Society* **6**(4): 405–411.

Kraiwattanapong, C., Boden, S.D., Louis-Ugbo, J., Attallah, E., Barnes, B. and Hutton, W.C. (2005). Comparison of Healos/bone marrow to INFUSE(rhBMP-2/ACS) with a collagen-ceramic sponge bulking agent as graft substitutes for lumbar spine fusion. *Spine* **30**(9): 1001–1007.

Lennon, D.P., Haynesworth, S.E., Bruder, S.P., Jaiswal, N. and Caplan, A.I. (1996). Human and animal mesenchymal progenitor cells from bone marrow: identification of serum for optimal selection and proliferation. *In Vitro Cell. Dev. Biol. Anim.* **32**: 602–611.

Lieberman, I. (2004). Local cell delivery strategies. Bone Summit, Cleveland, OH.

Liu, L.S., Thompson, A.Y., Heidaran, M.A., Poser, J.W. and Spiro, R.C. (1999). An osteoconductive collagen/hyaluronate matrix for bone regeneration. *Biomaterials* **20**(12): 1097–1108.

Lou, J. (2004). Bone engineering with mesenchymal stem cells and gene therapy. In: Sandell, L.J. and Grodzinsky, A.J. (eds.), *Tissue Engineering in Musculoskeletal Clinical Practice*. Rosemont, IL: AAOS, p. 123.

Maniatopoulos, C., Sodek, J. and Melcher, A.H. (1988). Bone formation *in vitro* by stromal cells obtained from bone marrow of young adult rats. *Cell Tissue Res.* **254**: 317–330.

Minamide, A., Yoshida, M., Kawakami, M., Yamasaki, S., Kojima, H., Hashizume, H. and Boden, S.D. (2005). The use of cultured bone marrow cells in Type I collagen gel and porous hydroxyapatite for posterolateral lumbar spine fusion. *Spine* **30**(10): 1134–1138.

Muschler, G.F. and Midura, R.J. (2002) Connective tissue progenitors: practical concepts for clinical applications. *Clin. Orthop.* **395**: 66–80.

Muschler, G.F., Boehm, C. and Easley, K. (1997) Aspiration to obtain osteoblast progenitor cells from human bone marrow: the influence of aspiration volume. *J. Bone Joint Surg. Am.* **79**(11): 1699–1709.

Muschler, G.F., Nitto, H., Boehm, C.A. and Easley, K.A. (2001). Age-and gender-related changes in the cellularity of human bone marrow and the prevalence of osteoblastic progenitors. *J. Orthop. Res.* **19**(1): 117–125.

Muschler, G.F., Nitto, H., Matsukura, Y., Boehm, C., Valdevit, A., Kambic, H., Davros, W. Powell, K. and Easley, K. (2005). Selective retention of bone marrow-derived cells to enhance spinal fusion. *Clin. Orthop. Rel. Res.* **432**: 242–251.

Musgrave, D.S., Fu, F.H. and Huard, J. (2002). Gene therapy and tissue engineering in orthopedic surgery. *J. Am. Acad. Orthop. Surg.* **10**: 6–15.

Niemeyer, P., Seckinger, A., Simank, H.G., Kasten, P., Sudkamp, N. and Krause, U. (2004). Allogenic transplantation of human mesenchymal stem cells for tissue engineering purposes: an *in vitro* study. *Orthopade* **33**(12): 1346–1353.

Nussenbaum, B. and Krebsbach, P.H. (2004). Practical matters in the application of tissue engineered products for skeletal regeneration in the head and neck region. In: Sandell, L.J. and Grodzinsky, J. (eds.), *Tissue Engineering in Musculoskeletal Clinical Practice*. Rosemont, IL: AAOS, p. 154.

Ohgushi, H., Goldberg, A.I. and Caplan, A.I. (1989). Repair of bone defects with marrow cells and porous ceramic. Experiments in rats. *Acta. Orthop. Scand.* **60**: 334–339.

Olmsted-Davis, E.A., Gugala, Z., Gannon, F.H., Yotnda, P., McAlhany, R.E., Lindsey, R.W. and Davis A.R. (2002). Use of a chimeric adenovirus vector enhances BMP2 production and bone formation. *Human Gene Ther.* **13**(11): 1337–1347.

Owen, M. (1985). Lineage of osteogenic cells and their relationship to the stromal system. In: Peck, W.A. (ed.), *Bone and Mineral*. Amsterdam: Elsevier, pp. 1–25.

Pereira, R.F., Halford, K.W., O'Hara, M.D., Leeper, D.B., Sokolov, B.P., Pollard, M.D., Bagasva, O. and Prockop, D.J. (1995). Cultured adherent cells from marrow can serve as long-lasting precursor cells for bone, cartilage, and lung in irradiated mice. *Proc. Natl Acad. Sci. USA* **92**(11): 4857–4861.

Price, C.T., Connolly, J.F., Carantzas, A.C. and Ilyas, I. (2003). Comparison of bone grafts for posterior spinal fusion in adolescent idiopathic scoliosis. *Spine* **28**(8): 793–798.

Salama, R. and Weissman, S.L. (1978). The clinical use of combined xenografts of bone and autologous red marrow. *J.Bone Joint Surg.* **60B**(1): 111–115.

Sim, R., Liang, T.S. and Tay, B.K. (1993). Autologous marrow injection in the treatment of delayed and non-union in long bones. *Singapore Med. J.* **34**: 412–417.

Sutherland, D. and Bostrom, M. (2005). Grafts and bone graft substitutes. In: Lieberman, J.R. and Friedlaender, G.E. (eds.), *Bone Regeneration and Repair*. Totowa, NJ: Humana Press, pp. 133–136.

Takagi, K. and Urist, M.R. (1982). The role of bone marrow in BMP-induced repair of femoral massive diaphyseal defects. *Clin. Orthop. Rel. Res.* **171**: 224–231.

Tay, B.K., Le, A.X., Heilman, M., Lotz, J. and Bradford, D.S. (1998). Use of a collagen–hydroxyapatite matrix in spinal fusion. *Spine* **23**(21): 2276–2281.

Thomas, C.B., Kellam, J.F. and Burg, K. (2004). Comparative study of bone cell culture methods for tissue engineering applications. *J. ASTM Int.* **1**: 1–17.

Tibor, T.G. (2003). Short overview of potential gene therapy approaches in orthopedic spine surgery. *Spine* **28**(3): 207–208.

Tiedeman, J.J., Connolly, J.F., Strates, B.S. and Lippiello, L. (1991). Treatment of nonunion by percutaneous injection of bone marrow and demineralized bone matrix. An experimental study in dogs. *Clin. Orthop.* **268**: 294–302.

Tiedeman, J.J., Huurman, W.W., Connolly, J.F. and Strates, B.S. (1991). Healing of a large nonossifying fibroma after grafting with bone matrix and marrow. *Clin. Orthop*. **265**: 302–305.

Urist, M.R. (1965). Bone: formation by autoinduction. *Science* **150**: 893–899.

Vinay, P., Gougoux, A. and Lemieux, G. (1981). Isolation of a pure suspension of rat proximal tubules. *Am. J. Physiol.* **241**: F403–F411.

Wang, J.C., Youssef, J.A., Lieberman, I.H., Brodke, D.S., Lauryssen, C., Haynesworth, S.E. and Muschler, G.F. (2005). A prospective, multi-center study of selective osteoprogenitor cell retention for enhancement of lumbar spinal fusion. IMAST, Banff, Canada.

Webster, T.J., Ergun, C., Doremus, R.H., Siegel, R.W. and Bizios, R. (2000). Enhanced functions of osteoblasts on nano-phase ceramics. *Biomaterials*, **21**(17): 1803–1810.

Wenisch, S., Trinkaus, K., Hild, A., Hose, D., Herde K., Heiss, C., Kilian, O., Alt, V. and Schettler, R. (2005). Human reaming debris: a source of multipotent stem cells. *Bone*, **36**(1): 74–83.

Whang, P.G. and Lieberman, J.R. (2003). Clinical issues in the development of cellular systems for use as bone graft substitutes. In: Laurencin C.T. (ed.), *Bone Graft Substitutes*. West Conshohocken, PA: ASTM International, pp. 142–155.

Wientroub, S., Goodwin, D., Khermosh, O. and Salama, R. (1989). The clinical use of autologous marrow to improve osteogenic potential of bone grafts in pediatric orthopedics. *J. Pediatr. Orthop.* **9**(2):186–190.

Yoon, S.T. and Boden, S.D. (2002). Osteoinductive molecules in orthopaedics: basic science and preclinical studies. *Clin. Orthop.* **395**: 33–43.

Younger, E.M. and Chapman, M. (1989). Morbidity at bone graft donor sites. *J. Orthop. Trauma* **3**: 192–195.

Zhang, M., Powers, R.M. and Wolfinbarger, L. (1997). Effect(s) of the demineralization process on the osteoinductivity of demineralized bone matrix. *J. Periodontol.* **68**: 1085–1092.

Zuk, P.A., Zhu, M., Ashjian, P., De Ugarte, D.A., Huang, J.I., Mizuno, H., Alfonso, Z.C., Fraser, J.K., Benhaim, P. and Hedrick, M.H. (2002). Human adipose tissue is a source of multipotent stem cells. *Mol. Biol. Cell* **13**: 4279–4295.

26

Cell-Based Therapies for Musculoskeletal Repair

Wan-Ju Li, Kiran Gollapudi, David P. Patterson, George T.-J. Huang, and Rocky S. Tuan

INTRODUCTION

Common musculoskeletal disorders include osteoarthritis (OA), rheumatoid arthritis (RA), intervertebral disk (IVD) degeneration, anterior cruciate ligament (ACL) injuries, and muscular dystrophy. These diseased conditions can result from trauma, work-related injuries, immunological malfunction, aging, or genetics. Arthritis is highly prevalent among adults. As the population ages, arthritis is expected to affect an estimated 67 million adults in the United States by 2030, according to the report by Centers for Disease Control and Prevention (CDC, 2006). Approximately 13–16 million people are diagnosed with OA per year in the US or Europe, and joints are replaced due to OA at the rate of one every 1–2 min. A total of ~300,000 and ~500,000 joint replacements are performed per year in Europe and the US, respectively. Although OA is not normally life threatening, it is progressive, disabling, and can greatly impact an individual's quality of life.

Current approaches to musculoskeletal care emphasize prevention, medical treatment, and surgical intervention as a final resort. Progress has been made in the past decades with biological therapeutic approaches to reverse or slow disease progression by specifically targeting molecules involved in the disease process. Despite advances in these approaches, once tissue damage reaches a certain stage, self-repair does not take place. Particularly with cartilage, the repair mechanism is almost non-existent. Currently, there is no available remedy to repair the eroded cartilage in the arthritic joints and the degenerated IVD. While joint replacements provide significant functional restoration and symptomatic improvement, they are compromised by the possibility of prosthetic failure and associated complications (e.g. peri-implant osteolysis).

The emerging disciplines of cell-based therapy and tissue engineering have suggested the prospect of regenerative medicine as a promising approach to the treatment of damaged and diseased musculoskeletal tissues. This chapter will outline mesenchymal cell biology in the context of musculoskeletal tissues, including osteogenec, chondrogenec, myogenec, tenogenec, and ligament cell types. Particular emphasis will be made on the definition of mesenchymal stem cells (MSCs), their niches, isolation and *in vitro* characterization, lineage differentiation and regulation, and immunomodulatory properties. Cell-based applications using these cells to produce specific tissues will be reviewed, covering clinical disorders, gene therapy approaches, *in vivo* studies, and current applications, as well as potential pitfalls and future improvement.

BIOLOGY OF CELLS IN MUSCULOSKELETAL TISSUES

Osteoblasts and Osteocytes

Bone provides structural support for the body, facilitates movement, protects internal organs, and acts as a mineral reserve. Derived from embryonic mesoderm, bone contains a matrix consisting of hydroxyapatite

Copyright © 2008 Elsevier Inc. All rights reserved

mineral and macromolecular extracellular matrix (ECM) components. Osteoblasts (OB) are the bone forming cells found on all surfaces of bone. They produce osteoid consisting of collagen type I, fibronectin, proteoglycans, and other specialized proteins, which mineralizes by the deposition of hydroxyapatite crystals [$Ca_{10}(OH)_2(PO_4)_6$]. Osteocytes are former OB that have become encased by the bone matrix and are the most abundant cells in mature bone. Their presence highlights the fact that bone is a living, dynamic tissue. Osteocytes cease to produce osteoid but communicate with other cells (osteocytes and OB) through small pores called canaliculi, and are believed to play a role in mechanotransduction activities of bone with feedback to the remodeling process.

Osteoprogenitors, located in the bone marrow and periosteum, can be induced to become OB via growth factors, such as the transforming growth factor-β (TGF- β) superfamily and in particular the bone morphogenetic proteins (BMPs). BMPs regulate chemotaxis, mitosis and differentiation, and are critical in initiating fracture healing. The transcription factors Cbfa1/Runx2 and Osterix are essential for OB differentiation in both intramembranous and endochondral ossification. Runx2 is involved in activating gene expression of collagen type I and other bone proteins, such as osteopontin and osteocalcin. Osterix is believed to be downstream of Runx2 as it is not expressed in Runx2 null mice. Other transcription factors shown to be involved in OB proliferation and differentiation include Msx and Dlx proteins. However, mice containing null alleles for these genes do produce bone, unlike Runx2 and Osterix null mice (Eames et al., 2003).

Chondrocytes

There are three main types of cartilage: hyaline cartilage, elastic cartilage, and fibrocartilage. The external ear is an example of elastic cartilage, and the meniscus and annulus fibrosus of the IVD, which will be discussed later, are examples of fibrocartilaginous tissues. Articular cartilage is an example of hyaline cartilage, the most common type. The major function of articular cartilage is to provide a smooth surface for reduced friction and to support large loads during movement. Articular cartilage consists of a fluid phase composed of water and electrolytes as well as a solid phase consisting of ECM and chondrocytes. Chondrocytes, making up less than 10% by volume of articular cartilage, are the only cell type in the tissue and are responsible for the maintenance of the ECM. Collagen type II is the predominant collagen in articular cartilage and is responsible for its tensile strength; however, other minor collagens, such as collagen types V, VI, IX, and X, are also present. The most abundant component of articular cartilage is water (60–85% by volume), which is held in place by the highly charged proteoglycans, and allows for its compressive behavior. Thus, collagens and proteoglycans provide cartilage with its mechanical properties, and their content is a key component in assessing the functional quality of engineered articular cartilage.

During developmental chondrogenesis, mesenchymal cells are recruited and migrate to areas of chondrogenesis, and subsequent mesenchymal–epithelial cell and cell–cell matrix interactions promote cellular condensation. Mesenchymal cells then differentiate into chondrocytes, heralding the deposition of ECM proteins and the activation of transcription factors such as Sox-9. Sox-9 is required for the expression of cartilage-specific collagen type II in normal skeletal development (Bi et al., 2001). L-Sox-5 and Sox-6, which are other members of the Sox family, cooperate with Sox-9 to turn on the collagen type II gene and are also essential for cartilage formation (Lefebvre and Smits, 2005).

Muscle Cells

Muscle is a specialized contractile tissue, derived from embryonic mesoderm and allows for force exertion and locomotion. There are three types of muscle: skeletal (voluntary) muscle found in the musculoskeletal system, smooth muscle (involuntary) found within walls of organs and structures, and cardiac muscle, a specialized tissue found only in the heart. Skeletal muscle cell differentiation from embryonic mesoderm is driven by transcription factors MyoD, Myf5, myogenin, Mrf4, and MEF2 in a highly coordinated fashion (Sartorelli and Caretti, 2005). Skeletal muscle is composed of bundles of dense muscle fibers that are highly oriented and able to generate longitudinal contraction. These bundles (muscle cells) are multinucleated and derived from myoblasts (muscle cell precursors). Satellite cells are specialized myoblast populations capable of muscle regeneration. They can be obtained and cultured from muscle biopsy. Upon initiation by local growth factors, these normally quiescent and undifferentiated cells become mitotic, differentiate, and eventually self-assemble into muscle fibers themselves. M-Cadherin, Pax7, and neural cell adhesion molecule (NCAM) are known satellite cell markers that can be used to localize and follow satellite cells *in vivo*. Pax7 null mice lacked satellite cells but retained a unique population of interstitial stem cells in muscle that express the stem cell markers, CD 34, and Sca-I (Tamaki et al., 2002). These stem cells, tracked by expression of green fluorescent protein (GFP), were found to originate in the bone marrow (Dreyfus et al., 2004).

Tendon and Ligament Cells

Tendons and ligaments are dense fibrous structures that connect muscle to bone and bone to bone, respectively. Both are composed mostly of collagen type I, produced by specialized elongated fibroblasts (known as tenocytes in tendons) that lie between the collagen fibers. These tissues are very hypocellular compared to other connective tissues, presenting problems for injury repair. Both arise from mesodermal compartments distinct from those that give rise to myogeneic cells. Not many cellular markers have been identified, but the transcription factor Scleraxis appears promising as a specific and early marker for tendons and ligaments (Tozer and Duprez, 2005).

EMBRYONIC AND ADULT STEM CELLS

Embryonic Stem Cells

Embryonic stem (ES) cells are pluripotent cells, derived experimentally from the inner cell mass of the embryonic blastocyst. Human ES (hES) cells are typically obtained from 4- to 5-day-old blastocysts of embryos after *in vitro* fertilization. These cells are potentially immortal *in vitro* without loss of differentiation potential, and when reimplanted into a host embryo, they give rise to pluripotent daughter cells that differentiate into all tissue types. The use of hES cells for research and clinical applications is complicated by controversies surrounding the legal and ethical status of human embryos and is currently restricted by regulations on federal funding. Despite these challenges, both mouse and hES cells have been examined for applications in musculoskeletal regeneration, albeit limited.

Human ES cell differentiation and organization can be influenced by a supportive three-dimensional (3D) environment such as poly(lactic-co-glycolic acid) (PLGA) polymer scaffolds and directed by growth factors such as retinoic acid, TGF-β, activin-A, or insulin-like growth factor (IGF). These growth factors induce differentiation into 3D structures with characteristics of developing neural tissues, cartilage, or liver, respectively (Levenberg et al., 2003). hES cells have been injected into the joint space of immunocompromised rats to promote cartilage repair (Wakitani et al., 2004). Osteogeneic potential of hES cells in the presence of chemical stimuli *in vitro* has also been demonstrated (Karp et al., 2006). In a mouse model, chondrogene differentiation was observed using mES cells via embryoid bodies (EBs) modulated by members of the TGF-β family (TGF-β1, BMP-2 and -4) (Kramer et al., 2000). mES cells differentiate into chondrocytes, which progressively develop into hypertrophic and calcifying cells. At a terminal differentiation stage, cells expressing an OB-like phenotype appear either by transdifferentiation from hypertrophic chondrocytes or directly from OB precursor cells. Under the influence of ascorbic acid, β-glycerophosphate, and 1,25-dihydroxy vitamin-D_3, mES cells are induced to differentiate *in vitro* into OB that produce a mineralized matrix. In 3D scaffold systems such as hydrogels, both hES and mES cells were shown to have chondrogene capacity (Elisseeff et al., 2006).

Mesenchymal Stem Cells

While ES cells hold promise as a cell source for tissue regeneration, ethical issues and the possibility of teratoma formation need to be resolved before they can be used in clinical applications. Notably, our body has other stem cell populations that possess the capabilities of self-renewal and multidifferentiation to repair damaged tissues or maintain tissue homeostasis when the repair is needed. With our improved knowledge of stem cells combined with advances in culture techniques, it is possible that we can harness their potential for the treatment of degenerative musculoskeletal diseases. Unlike pluripotent ES cells which have the ability to form tissues from all three germ layers, adult stem cell populations are believed to be limited in their differentiation capacity. In general, stem cells derived from a particular tissue are programmed to differentiate into various progenies that belong to the same developmental germ layer origin. Although recent studies suggest that adult stem cells can differentiate across germ line boundaries, it remains debatable if differentiation plasticity between the cells of different germ layers actually exists or is simply an artifact resulting from contamination of heterogeneous cell populations or cell fusion. More evidence is needed to assess the possibility of differentiation across the three germ layers. However, plasticity within a germ layer is more strongly supported by evidence from several research groups. The ethically acceptable nature of adult stem cells, combined with proven differentiation abilities, as compared to ES cells and tissue-committed cells, makes them an attractive option for use in cell-based therapy.

For cell-based musculoskeletal tissue regeneration, adult MSCs are an attractive candidate progenitor cell type since they may be isolated from various adult tissues, and can differentiate into different mesenchymal lineage cells, such as bone, cartilage, fat, muscle, ligament, tendon, and stroma (reviewed by Tuan et al., 2003) (Figure 26.1). Reflecting their origin and cell functions, these cells have also been named or described as

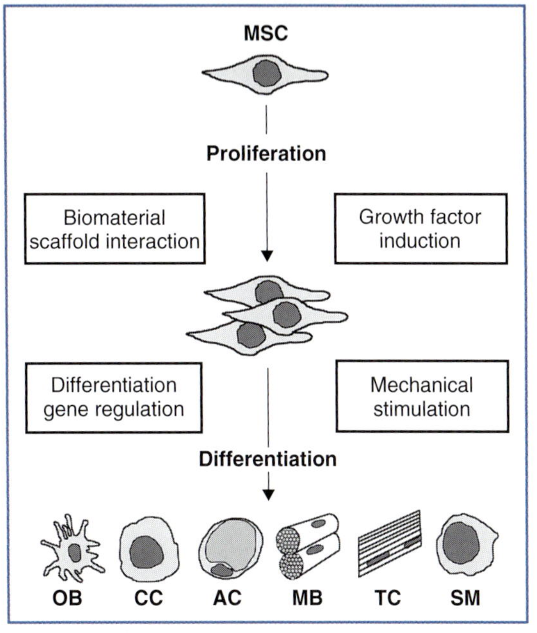

Figure 26.1 Multidifferentiation potential of MSCs and factors regulating their biological activities *in vitro*. CC, chondrocyte; AC, adipocyte; MB, myoblast; TC, tenocyte; SM, stroma.

marrow stromal cells, or mesenchymal stromal cells, which are all abbreviated as MSCs. MSCs were first discovered by a German pathologist in the 1860s and he described the cell morphology as "fibroblast-like." More than a century later, in 1976, Friedenstein further identified MSCs as colony-forming unit-fibroblasts, which were able to commit to osteogene differentiation. The multidifferentiation potential of MSCs was demonstrated *in vivo* as early as in 1980, in which MSCs were induced to become bone and cartilage. Caplan (1991) and Pittenger et al. (1999) showed that MSCs underwent osteogene, chondrogene, and adipogene differentiation in response to different biochemical signals. Numerous subsequent studies have produced a significant body of evidence demonstrating the phenotypic and functional characteristics of MSCs. According to the statistic data retrieved from Medline, research publications containing the key word "MSC" have increased 15-fold in the past 5 years, highlighting the increasing interest in MSC studies. While many studies are focused on characterization and differentiation potential of MSCs, few are concentrated on the molecular regulation of MSCs. Future MSC research should focus on the intrinsic and extrinsic mechanisms mediating the molecular switch between undifferentiated and differentiated MSCs.

MSC Isolation

MSCs can be isolated from a variety of mesenchymal tissues such as bone marrow, fat, trabecular bone, cartilage, muscle, peripheral blood, and umbilical cord blood (reviewed by Tuan et al., 2003). Dependent on the species and tissue types, different isolation protocols and culture methods have been developed. Among these MSC sources, bone marrow is the best studied tissue. The isolation process for bone marrow-derived MSCs includes several steps aimed at reducing contamination by other cell types. Erythrocytes can be removed by density gradient centrifugation using Percoll or Ficoll after bone marrow aspirate is obtained from the iliac crest, tibia, or femur. Fluorescence-activated cell sorting (FACS) and magnetic-activated cell sorting (MACS) are sometimes used to select a more defined cell population based on cell surface markers. With their ability to adhere to tissue culture plastic, MSCs are further discriminated from non-adherent hematopoietic cells after several medium changes. The low frequency of 1 MSC per 10,000–100,000 bone marrow cells *in vitro* indicates that the MSC is a rare cell population (Pittenger et al., 1999). Previous studies have shown that MSC yield is affected by age and health of a donor. The trend is that MSC yield decreases with donor age. Patients with degenerative diseases, such as osteoporosis and OA, tend to have lower MSC yield. Unfortunately, it is this group of people who would benefit most from MSC-based treatment. Therefore, an alternative could be the use of allogeneic MSCs. Although immune reaction is a concern associated with using allogeneic cells, the finding that MSCs have low immunogeneic potential as well as immunosuppressive properties suggests that this concern may not be significant. Immunoregulation of MSCs will be discussed in detail later.

Another MSC source gaining recent attention is adipose tissue, since fat is abundant, easy to access, and considered surgical waste during a cosmetic surgery operation. The procedure of marrow harvesting is painful, and over-harvesting bone marrow may be a risk to the patient's health. In contrast, liposuction is less painful and considered a relatively safe procedure. The method to isolate MSCs is similar to a typical primary cell isolation, in which MSCs are released from collagenase-digested fat and purified after being plated in plastic culture to remove unattached hematopoietic cells. A study comparing MSC proliferation between bone marrow and fat tissue suggests that fat tissue is a more effective source than bone marrow for MSC yield (Lee et al., 2004).

As mentioned above, bone marrow-derived MSC number, proliferation, and differentiation potential may decrease with the donor age. A possible solution is to isolate MSCs from tissue of a younger donor. Fetal tissues, such as umbilical cord blood (Mareschi et al., 2001), cord vein (Romanov et al., 2003), placenta (Fukuchi et al., 2004), and amniotic fluid (In't Anker et al., 2003), have been processed to isolate MSCs. A comprehensive study by Kern et al. (2006) in which they compared the morphology and functions of MSCs isolated from bone marrow, adipose tissue, and umbilical cord blood showed that MSCs isolated from these three tissue sources are morphologically and immunophenotypically similar. However, umbilical cord blood-derived MSCs form the fewest colonies and show the highest proliferation capacity, whereas adipose tissue-derived MSCs form the greatest number of colonies and bone marrow-derived MSCs show the lowest proliferation capacity. The findings suggest both adipose tissue and umbilical cord blood can be used as alternatives to bone marrow for MSC isolation.

Recently, various tissue-specific adult stem cells have been successfully isolated and characterized from non-mesenchymal tissues. These cells from ectoderm (epidermis) or endoderm origin (pancreas) morphologically and functionally behave similarly to bone marrow-derived MSCs. Although the lack of specific markers makes verification of their identity as MSCs difficult, and their phenotype could be a result of *in vitro* culture conditions, the possibility of MSCs residing in different tissues throughout the body raises a number of interesting questions. For example, how and where do these cells reside in the different tissues? Is there a common tissue pool, such as bone marrow, that houses the MSCs which are able to integrate to different tissues in response to specific biological needs or activities?

In Vitro MSC Behavior

After being plated in culture, MSCs adhere to the substratum and start to divide, forming colonies. The typical growth pattern is that MSCs experience a few days of a lag phase before undergoing a log phase of growth and then reach a stationary phase. Generally, these cells can undergo 50 population doublings in 10 weeks without losing their multidifferentiation capability. Colter et al. (2000) reported that the initial cell seeding density affects the expansion capacity and doubling time of mouse MSCs. When plated at low initial plating density (1.5–3 cells/cm^2), MSCs have 2,000-fold expansion in 10 days, whereas when plated at high density (12 cells/cm^2), the cell number only increases 60-fold in the same period of time. In addition, the doubling time increases from 12 to 24 h with high cell seeding density. MSCs are a heterogeneous population, demonstrated by the finding that there are two MSC morphologies in a colony; small spindle-shaped and large flat fibroblast-like cells. Interestingly, a third type of extremely small, rapidly self-renewing (RS) cells was recently identified in MSC colonies. During the three phases of growth, RS cells give rise to large flat cells in the log phase. It is believed that the large flat MSCs are more mature cells, replicating slowly, whereas small RS cells proliferate rapidly.

MSC Identification

Despite a significant number of studies, to date, there is no specific marker available for the identification of MSCs. There have been some cell surface markers available for MSC identification, but none are exclusive to MSCs. Because MSCs can not be positively identified, their biological activities *in vivo* cannot be verified as with hematopoietic stem cells (HSCs). Our knowledge of MSCs is thus primarily from *in vitro* experimental results. The *in vitro* results of MSC identification are likely dependent on culture environment and purity of cell population, which could explain the differences between the findings reported in the literature. Nevertheless, the use of multiple markers, such as cell surface cluster of differentiation (CD) markers, ECM proteins, cell adhesion molecules, integrins, and cytokines as well as genetic or proteomic fingerprinting can help one identify MSCs.

CD cell surface molecules are the most commonly used markers to identify MSCs. Both positive and negative CD markers have been used to identify MSCs and exclude endothelial cells, HSCs, and hematopoietic lineage cells. Positive MSC markers include Stro-1, SH2 (CD105), SH3 (CD73), SH4 (CD73), CD 29, CD 44, CD 54, CD 90, CD 105, CD 133, CD 166, and p75LNGFR, whereas negative markers are CD 11, CD 14, CD 19, CD 31, CD 34, CD 45, CD 79, and HLA-DR (Deans and Moseley, 2000). To shorten the list, the International Society for Cellular Therapy (ISCT) has provided minimum criteria for defining MSCs

(Dominici et al., 2006). Acceptable MSCs meet the minimum requirements of CD 73, CD 90, and CD 105 positive and CD 14, CD 34, CD 45, and HLA-DR negative expression.

MSC Niche

The term MSC niche refers to both physical and chemical environment that MSCs reside in, which includes other cell types, ECM molecules directly or indirectly contacting with MSCs, and soluble factors regulating MSC activities. The cellular and non-cellular components interact with each other in this highly complex 3D environment, responsible for the maintenance of MSC stemness properties as well as the regulation of symmetric and asymmetric cell division. The concept of the stem cell niche was first introduced in the 1970s (Schofield, 1978) and has been elucidated by *in vitro* co-culture experiments. Ball et al. (2004) demonstrated that, by co-culturing MSCs and endothelial cells, MSCs were induced to a phenotype similar to smooth muscle cells, whereas MSCs became myofibroblast-like cells when co-cultured with dermal fibroblasts. Their study suggests that the interaction between MSCs and their neighboring cells may regulate the fate of MSCs and that the type of progeny of MSCs may be determined by the interacting cells in the niche.

The bone marrow microenvironment is a principal MSC niche in the body. It is not only a complex 3D structure, but also allows for many interactions between cellular and non-cellular components, such as HSCs, MSCs, stroma, hematopoietic cells, mesenchymal origin cells, ECM components, growth factors, and cytokines. Tellingly, the success rate of HSC engraftment improves when co-transplanting with MSCs *in vivo*. The *ex vivo* expansion of HSCs increases dramatically when MSCs are co-cultured with HSCs, suggesting that HSCs and MSCs maintain a close biological interaction in the naïve marrow niche. The signals needed for MSC proliferation and differentiation come from soluble factors as well as both cell–cell and cell–matrix interactions. OB have been known to play an important role in the regulation of hematopoiesis but their role in MSC osteogenesis still remains inconclusively defined. A previously published report shows that OB have synergistic interactions on MSC proliferation and alkaline phosphatase activity but not calcium deposition (Kim et al., 2003). In comparison, our preliminary results show that MSC co-culturing with OB enhances their osteogene differentiation (unpublished observation).

MSC–ECM interactions, both physical and chemical, are likely to be critical in the regulation of MSC physiology in the niche. Matsubara et al. (2004) cultured and maintained MSCs on basement membrane-like ECM and observed profound effects on MSC proliferation and differentiation. With the support of basement membrane-like ECM, MSCs better maintain their multidifferentiation potential after many cell divisions, which suggests that the interactions between MSCs and basement membrane-like ECM may recapitulate some of the MSC–ECM interactions in bone marrow. Cell–ECM interactions are also the center of the study in biomaterial-based cell therapy and tissue engineering. Artificial ECM, a biomaterial scaffold used to replace damaged or malfunctioning ECM, is designed to fully function as native ECM and interact with MSCs for successful tissue regeneration. The interactions of MSC and a biomaterial scaffold will be discussed later.

MSC Regulation

Soluble factors such as growth factors and cytokines play a significant role in physiological regulation of MSCs in the niche. These biochemical signals guide MSCs either to stay as undifferentiated cells or to differentiate into tissue-specific progenitor cells by activating specific signal pathways. A recent important finding is the involvement of the Wnt signaling pathway. Gregory et al. (2003) demonstrate that MSCs enter the cell cycle and inhibit osteogeneic differentiation after Dickkopf-1 (Dkk-1) deactivates the Wnt pathway. Boland et al. (2004) further identify that Wnt 3a working through the canonical pathway promotes MSC proliferation but discourages osteogenesis, whereas Wnt 5a via the non-canonical pathway promotes osteogenesis. Chondrogenesis of MSCs also involves the activation of the Wnt signaling pathway (Tuli et al., 2003). During chondrogene differentiation of MSCs, TGF-β1 activates the mitogen activated protein (MAP) kinase pathway which is demonstrated to be involved in cross-talk signaling with the Wnt signaling pathway. The regulation of the Wnt signaling pathway likely induces chondrogenesis by enhancing cell–cell interaction through N-cadherin expression.

The most important and valuable characteristic of MSCs is their multipotential differentiation capacity. Previous studies have shown that MSC differentiation can be induced and regulated by soluble signal factors, both protein- and non-protein-based molecules. TGF-βs and BMPs induce MSCs to undergo chondrogenesis in a serum-free medium. Each member of the TGF-β family has a different level of induction efficiency. TGF-β3 has a higher efficiency in inducing chondrogenesis compared to TGF-β1 but they both contribute similarly to chondrogenesis in long-term culture. To induce osteogenesis, β-glycero-2-phosphate, ascorbic acid, dexamethasone, and 1,25-dihydroxy vitamin-D_3 are required to enhance alkaline phosphatase activity and matrix mineralization. For adipogenesis, MSCs are treated with isobutyl-1-methylxanthine and insulin, resulting in adipocytes with the presence of lipid droplets in the cytoplasm (Pittenger et al., 1999).

In addition to biochemical factors, physical factors, such as mechanical loading as well as matrix geometry and elasticity, have also been found to play a role in MSC biology. Cells generally receive and transduce physical cues from the surrounding environment to the cell nucleus through cytoskeletal changes or signaling pathways. McBeath et al. (2004) demonstrated that cell shape regulates commitment of MSCs by using a micropatterned substrate to control cell shape and size of cultured MSCs. They found that cell shape is a key regulator in MSC differentiation with the shape-dependent control of lineage commitment mediated by ROCK-mediated cytoskeletal tension. MSCs forced to spread and flatten in the large substrate pattern differentiate into OB, whereas those forced to unspread and become round in the small pattern differentiate into adipocytes. Recently, Engler et al. (2006) demonstrated the effect of matrix stiffness on differentiation of MSCs, mediated by non-muscle myosin II. MSCs on a soft matrix with stiffness similar to brain stiffness differentiate into neurons, on a matrix with intermediate stiffness close to muscle stiffness become myoblasts, and on a stiff matrix comparable to bone commit to OB. In addition, they also show that typical soluble factors for lineage commitment do not alter the differentiation lineage previously activated by matrix stiffness. This study suggests matrix stiffness appears to be important in MSC lineage commitment.

MSC Immunoregulation

Immunoregulation by and of MSCs can be viewed from two perspectives: (1) immunosuppressive effects of allogeneic MSCs, and (2) inflammatory cytokine effect on MSC activity and differentiation. Due to the interest in using allogeneic or xenogeneic MSCs to compensate for the paucity and time constraints associated with expanding autologous MSCs, there has been considerable progress in the understanding of the MSC immunoregulatory effect. While xenogeneic MSCs are rejected by the host after transplantation, allogeneic MSCs are well tolerated by the recipient hosts. Many *in vivo* studies have confirmed the immunosuppressive effects of MSCs (Chen XI et al., 2006). The potential mechanisms underlying this immunosuppression can be explained by downregulation of T, dendritic, natural killer (NK), and B cells. This immunosuppressive characteristic suggests that MSCs can be potentially used *in vivo* for enhancing the engraftment of other tissues (e.g. HSCs), or for prophylactic prevention and even possibly as a treatment of graft-versus-host-disease or autoimmune diseases such as RA (Jorgensen et al., 2003).

Limited information is available on the effects of pro-inflammatory cytokines on MSCs. Liu and Hwang (2005) demonstrated that human cord blood-derived MSCs secrete cytokines and growth factors. Most importantly, continuous supplementation of IL-1β in the cord blood-derived-MSC culture facilitates adipogenec maturation of cord blood-MSCs. A preliminary study using porcine MSCs showed that interferon-α-2b may act to differentiate MSCs into OB (Abukawa et al., 2006). Our recent study suggests that MSCs are relatively resilient to pro-inflammatory cytokines in terms of apoptotic response (Okafor et al., 2006). In the context of autologous implantation for chondrogenesis, a study has shown that inflammatory reactions against scaffold materials and serum components lead to the production of cytokines such as IL-1α that may inhibit cartilage tissue formation (Rotter et al., 2005). These few studies suggest the importance of understanding the effect of tissue inflammation, either caused by diseases or in response to scaffold materials after implantation, on the differentiation and cell behavior of MSCs.

CELL-BASED THERAPIES AND TISSUE ENGINEERING

Tissue Engineering

Three general strategies have been adopted for the engineering of new tissues: (1) delivery of cells, (2) local or systemic delivery of tissue inducing agents, and (3) delivery of biomaterial scaffolds containing both cells and inductive agents. The use of a biomaterial scaffold-based carrier facilitates the delivery of therapeutic cells or agents to the target site and subsequent growth and regeneration of new tissue. A general strategy is that target cells (differentiated/undifferentiated) expanded *in vitro* are cultured in 3D, biomaterial scaffolds (natural/synthetic) under conditions that favor the desired phenotype. After the appropriate culture period, the cell-seeded composite exposed to biological and physical stimuli develops into a natural tissue-like cellular construct (Figure 26.2). The biomaterial scaffolds are believed to play a critical role in the tissue engineering process, by providing a 3D structure for cellular functions such as attachment, migration, proliferation, and differentiation. The ultimate success of this process is determined by the biological and functional similarity of the engineered tissue to native tissue.

Despite the promising prospects of tissue engineering, regenerating tissues that serve a predominantly biomechanical function, such as bone, articular cartilage, and tendon, presents significant challenges. The

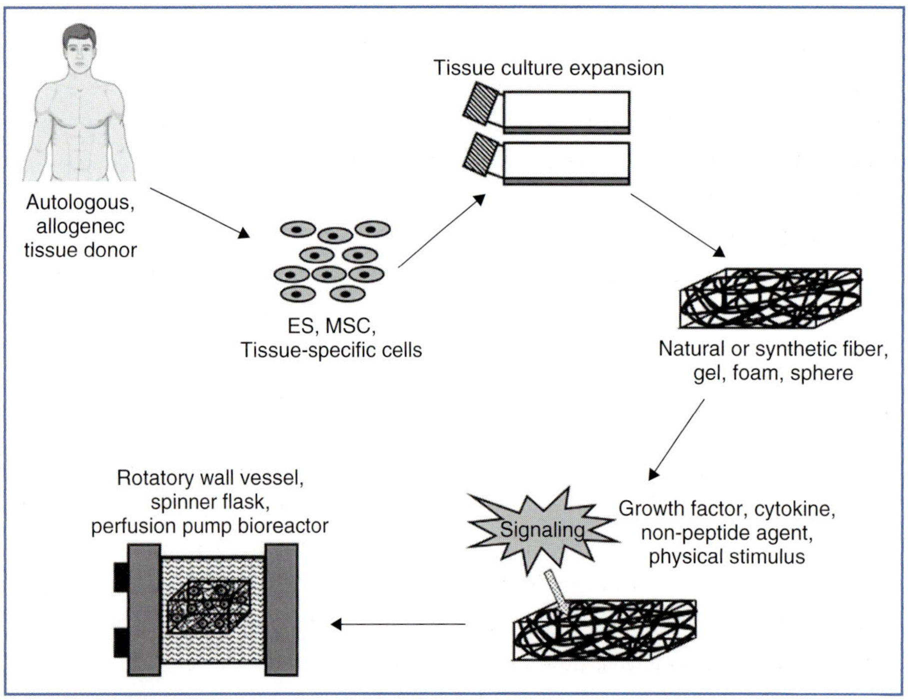

Figure 26.2 General strategy of cell-based therapy and tissue engineering.

concept of "functional tissue engineering" has emerged, specifically prescribing the production of a tissue that meets functional and in our case especially, mechanical requirements of the repair tissue. This requires a clear delineation of the structure and functions of living tissues, design of biomaterial scaffolds, and mechanical optimization of regenerative tissues. It is necessary to incorporate each of these steps to regenerate a mechanically sound tissue for safer and more efficacious repairs and replacements for the patient.

Gene Therapy

Gene therapy involves the gene-based modification of cells for the correction of defective genes or the regulation of gene expression. It has a great potential in cell therapy-based treatments due to the fact that a number of musculoskeletal diseases are caused by genetically malfunctioning cells; thus successful tissue engineering strategies must include the repair of the "defective" cells in any engineered construct.

The self-renewal capacity and multipotentiality of MSCs suggest their suitability for cell-based and gene therapy applications in regenerative medicine. For gene therapy, viral transduction has a high efficiency of delivering genes into MSCs. Adeno-associated viral-mediated gene transfer has been tested to be effective in delivering genes into MSCs and to repair bone disorders such as osteogenesis imperfecta (Chamberlain et al., 2004). Lentiviral vectors have also been shown to be effective in delivering genes into MSCs (Lu et al., 2005). Non-viral methods such as transfection using Nucleofection™ has been demonstrated to be promising in delivering functional genes into MSCs (Haleem-Smith et al., 2005).

We will now discuss the application of these cell-based strategies towards regenerating musculoskeletal tissues. A brief list of representative studies in the field of musculoskeletal regeneration is provided in Table 26.1.

Bone

A wide variety of patients with significant bone defects that necessitated amputation in the past now benefit from various orthopedic strategies. Congenital defects of bone, growth plate fractures and defects, fractures resulting in malunion or non-union, the genetic disorder osteogenesis imperfecta (brittle bone disease), and bone loss from tumor resection (primary bone tumors or tumors metastatic to bone) are just a handful of musculoskeletal problems that could be addressed by regenerative medicine. Currently, using orthopedic prosthetics is a severe but highly functional option. Distraction osteogenesis, a surgical procedure for bone reconstruction and lengthening, was developed in the 1950s by Ilizarov and is still used today. Bone autografting is a therapeutic option developed in the 19th century and considered to be the current gold standard, but has limitations, particularly in the size of the defect to be grafted. Autografts contain the patient's own OB and osteocytes, but require a second surgical site for the bone harvesting, most often the iliac crest of the pelvis. This increases patient morbidity such as post-operative pain and risk of infection. Allograft bone from bone banks or cadavers avoids the pitfalls of autograft bone but does not possess the strength or cellularity of autograft bone. As the number of surgeries requiring bone grafting continues to rise, the development of functional tissue-engineered bone grafts becomes increasingly significant.

Four critical factors to successful bone tissue engineering are osteoconduction, osteoproduction, osteoinduction, and mechanical stimulation. Osteoconduction refers to the integration of the scaffold or graft material into the site and its eventual remodeling and replacement. Osteoproduction is the production of bone material by cells, and osteoinduction is the use of growth factors that draw additional osteogene cells to the site. For both *in vivo* and *ex vivo* bone tissue, mechanical stimulation appears to be a critical factor in the development of biologically and mechanically optimal bone tissue. Tissue engineering of bone must take into account the tremendous mechanical strength and elasticity of bone . For load-bearing long bones such as the femur, mechanical stability of the construct is crucial, whereas for finer tissues such as fingers or

Table 26.1 Representative reports of *in vivo* cell-based musculoskeletal repair

Tissue type	Model	Cells	Strategy and observation	Reference
Bone	Rat	MSCs	Cells transduced with BMP-2 improved healing of a critical defect	Lieberman et al. (1999)
	Human	MSCs	Porous ceramic seeded with MSCs repaired large bone defects	Quarto et al. (2001)
	Human	MSCs	Direct grafting of cells to non-union defects achieved union	Hernigou et al. (2005)
	Human	MSCs	Cells with platelet-rich plasma were polymerized and used successfully for alveolar graft osteoplasty	Hibi et al. (2006)
Growth plate	Rabbit	MSCs	Direct loading of cells into growth plate defects reduced growth arrest in the tibia	Chen et al. (2003)
Cartilage	Human	Chondrocytes	Autologous cells injected into deep cartilage defects produced good to excellent results in 14 of 16 patients	Brittberg et al. (1994)
	Pig	Chondrocytes	Cells seeded in gelatin microbeads mixed with type I collagen gel repaired full-thickness cartilage defects	Chiang et al. (2005)
	Human	MSCs	Collagen gel containing cells implanted into cartilage defects improved arthroscopic and histologic scores	Wakitani et al. (2002)
Muscle	Rat	MSCs	Myogenic-induced cells were directly injected to damaged muscle, contributing to muscle repair	Dezawa et al. (2005)
	Dog	Mesangioblasts	Vascular delivery of wild-type mesangioblasts led to significant clinical amelioration of muscular dystrophy	Sampaolesi et al. (2006)
Tendon	Rabbit	MSCs	Collagen gels seeded with MSCs implanted into patellar tendon defects showed improved biomechanics	Awad et al. (2003)
Meniscus	Rat	MSCs	MSCs embedded in fibrin glue contributed to the healing of meniscal defects	Izuta et al. (2005)
Intervertebral disk	Human	Chondrocytes	Autologous cells delivered to the nucleus pulposus improved pain and disability scores	Meisel et al. (2006)

craniofacial applications, plasticity takes an increased significance. Tissue-engineered bone used for clinical applications should meet both biological and mechanical requirements.

Various scaffold strategies have been used for MSC-based bone tissue engineering. Matching the strength of bone is a leading concern, and many strategies have been employed. Fully or partially demineralized bone matrix (DBM) from processed allograft bone contains collagen, growth factors, and other proteins and has been seeded with MSCs to create promising engineered constructs. DBM shares many structural and functional similarities to autologous bone and, as expected, supported osteogene function of MSCs (Mauney et al., 2004). Coral, composed mostly of calcium carbonate and with a similar structure to bone, has been seeded with periosteum as a therapeutic strategy (Vacanti et al., 2001). Porous ceramics, such as those composed of tricalcium phosphate and hydroxyapatite, have been used in conjunction with MSCs to produce bone replacement tissues successfully in patients who failed traditional therapies (Quarto et al., 2001). Optimization of the scaffold strategy will require understanding the mechanism of its action. DBM is capable of withstanding shear forces and does not impair elasticity in the implant, and partially mimics the autologous environment in bone, although allogeneic antigens and pathogens may not have been fully removed. Ceramics provide good osteoconductivity and good integration into the defect site by bonding to tissues without rejection or inflammatory reactions, but unfortunately lack tensile strength, limiting applications involving torsion, shear stress, or bending. Natural coral has been investigated for decades as a bone graft substitute, and is biocompatible, osteoconductive and biodegradable. Improved *ex vivo* construct manufacturing requires combining biomaterial strategies with bioreactors that can produce shear and compressive forces to provide a dynamic culture system. Dynamic culture of cell-seeded scaffolds, for example, using spinner flasks, has been shown to result in more even cell distribution and a 121% increase in cell density (Mauney et al., 2004). Direct cell therapy has also been tested for musculoskeletal applications. Percutaneous autologous bone-marrow grafting, the re-introduction of aspirated bone marrow directly to the site of a non-union in the tibia, has been described in human patients with good results (Hernigou et al., 2005). Growth plate (physis) injuries in children can result in shortening or angular deformity with the formation of bony bridges across the growth plate between the epiphysis and metaphysis. Direct implantation of MSCs into growth plate defects resulted in a significant reduction of growth arrest in rabbit tibia (Chen et al., 2003).

Gene therapy also holds promise for bone tissue engineering. A number of strategies have been tested for bone repair. Proof of concept was established with improved healing of a critical defect in a rat femur with delivery of rat MSCs transduced with the gene for BMP-2 to the site of the defect (Lieberman et al., 1999). In mice, it was demonstrated that systemically injected MSCs transduced with IGF-1 established themselves in bone marrow. The MSCs demonstrated chemotactic ability by responding to the local fracture environment and locating preferentially to the fracture site, where they also accelerated the healing process (Shen et al., 2002). Gene therapy with an MSC-based vehicle is also being harnessed for the treatment of a genetic disease. Engineered adeno-associated viral vectors were successfully used to disrupt the expression of mutated collagen type I gene in MSCs derived from individuals with osteogenesis imperfecta. Subcutaneous implantation of transduced MSCs into immunodeficient mice produced improved bone matrix (Chamberlain et al., 2004). If host MSCs could be augmented or replaced, future OB could then produce osteoid of higher quality. A gene therapy approach has also been developed with muscle-derived mesenchymal progenitor cells. These cells act as vehicles producing osteoinductive proteins and have been demonstrated to heal critically sized bone defects (Young et al., 2002). Mouse primary myoblasts over-expressing Runx2 via aretroviral system were implanted in conjunction with collagen scaffolds in the hind legs of mice and resulted in trabecular bone growth (Gersbach et al., 2006).

Future improvement of bone tissue engineering depends critically on understanding the biological signals necessary for bone induction and optimizing the pharmacokinetics of their delivery. Optimized vascularization is essential, as cell labeling experiments show a considerable loss of OB in the first week following transplantation in porous cancellous bone matrices, presumably due to suboptimal initial vascularization (Kneser et al., 2006). Scaffolds that incorporate growth factors such as vascular endothelial growth factor (VEGF) as well as endothelial cells have been shown to increase vascular formation in constructs *in vivo*, but integration with the host vascular system remains a challenge (Rouwkema et al., 2006).

Articular Cartilage

The demand for engineered articular cartilage arises from the prevalence of traumatic injuries and degenerative diseases of articular cartilage such as OA. Other than total joint arthroplasty, which is associated with surgical risks as well as a finite life span, current therapeutic modalities for OA patients include pharmacologic intervention, lavage, shaving, laser abrasion, drilling or microfracture of subchondral bone to stimulate healing, autologous periosteal and perichondrial grafting, autogeneic or allogeneic osteochondral transplantation, and autologous chondrocyte implantation. These procedures, although variably effective, often cannot repair cartilage to a disease-free state. With an increasing elderly population and the predicted rise in the incidence of OA, novel cell-based therapies are a promising avenue to meet the therapeutic needs of these patients.

Ultimate success of a tissue-engineered cartilage construct requires the presence of cells which can produce the proper cartilaginous ECM. In principle, fully differentiated chondrocytes are the ideal cell candidate, since they are programmed for the cartilage phenotype. Indeed, Brittberg et al. (1994) showed that full thickness chondral defects could be repaired using autologous chondrocytes derived from a minor load-bearing area and injected under a periosteal flap. This procedure, commonly referred to as autologous chondrocyte implantation/transplantation, is marketed as Carticel™ by Genzyme Biosurgery and is the only Food and Drug Administration (FDA) approved cell-based therapy for cartilage repair. This therapy has had promising results; however, the cost effectiveness as well as the superiority of this procedure over other available procedures is debatable. Furthermore, the use of chondrocytes is hampered by their limited availability, the potential donor site morbidity, and the propensity for chondrocytes to dedifferentiate when in monolayer culture. Alternatively, stem cells, including ES cells and MSCs, may be induced to differentiate into the chondrogene lineage. EB-derived ES cells cultured as pellets with TGF-β3 showed increased gene expression of cartilage-specific ECM markers as well as a significant increase in collagen and proteoglycan production after 14 days compared to untreated controls. These changes toward a chondrogenec phenotype were accompanied by a downregulation of hematopoietic and neural markers. MSCs derived from various sources present another stem cell source for cartilage repair. Wakitani et al. (2002) showed that in OA patients undergoing high tibial osteotomy, autologous bone marrow MSCs seeded in a collagen gel had histological and arthroscopic improvement compared to controls; however, there was no statistically significant clinical improvement.

A variety of both natural and synthetic scaffolds have been tested to serve as carriers for the aforementioned cells (reviewed by Kuo et al., 2006). The advantage of using a carrier scaffold is demonstrated by a study by Chiang et al. (2005) in which chondrocytes seeded in gelatin microbeads mixed with a collagen type I gel improved repair in a porcine cartilage defect model and maintained a chondrocyte phenotype better than cells delivered alone. Examples of scaffolds that have been studied for cartilage tissue engineering include fibrous scaffolds of biodegradable polymers such as poly-glycolic acid (PGA), poly-lactic acid (PLA), and their copolymer PLGA, as well as fibrin, agarose, collagen, alginate, gelatin, poly-ethylene glycol (PEG), and hyaluronan-based gels. Combinations of these materials have also been used. However, the currently available scaffolds do not fully meet the necessary requirements and can have variable effects on cell behavior and function. Novel techniques are being developed to engineer scaffolds with biomimetic properties to optimize biomaterial framework for cartilage regeneration. For example, bovine chondrocytes cultured in a PEG-based hydrogel cross-linked with a matrix metalloproteinase (MMP) sensitive peptide, which allows for remodeling by the seeded cells, had increased expression of collagen type II and aggrecan, compared to hydrogels lacking the MMP sensitive peptide (Park et al., 2004). In addition, electrospun nanofibers, structurally similar to natural ECM, have recently been developed as a novel scaffold for tissue engineering. MSCs seeded into a poly(ε-caprolactone) (PCL) nanofibrous scaffold and treated with TGF-β1 were able to differentiate into a chondrocytic phenotype (Li et al., 2005b). More recently it has been shown that changing the fiber diameter in these scaffolds can influence chondrocyte proliferation and ECM matrix production, demonstrating that even slight dimensional changes in scaffolds of the same material can affect seeded cells (Li et al., 2006).

Supplementation of growth medium with specific signaling molecules and growth factors is commonly used to promote cell proliferation, differentiation, and ECM production. These molecules include members of the TGF-β superfamily (TGF-βs), BMPs, and growth differentiation factors (GDFs), IGFs, platelet-derived growth factors (PDGFs), and Wnts. These factors activate intracellular signaling pathways that are presumably similar to those involved in developmental morphogenesis (reviewed by Chen FH et al., 2006). Isoforms of TGF-β as well as BMPs are some of the most potent positive modulators used in tissue engineering. Although TGF-β1 has been widely used as an anabolic factor for both chondrocytes and MSCs, TGF-β2 and TGF-β3 have been shown to be superior inducers of chondrogenesis in MSCs (Barry et al., 2001). Among the BMPs, BMP-2 has been shown to be more effective than other BMPs (-12 and -13, and -4 and -6) for chondrocyte and MSC-based engineered constructs, respectively (Sekiya et al., 2005). Similar to scaffold design, the current trend in growth factor supplementation is to use a combination approach; however, not all combinations are favorable. For example, FGF-2 and IGF-1 showed no synergism, and actually decreased protein synthetic rate by canine articular chondrocytes seeded in a collagen type II–glycosaminoglycan scaffold (Veilleux and Spector, 2005). The challenge of applying soluble factors for cartilage regeneration is that the effects of these factors depend not only on their optimal combination but also on their amount, the timing of administration, and the target cell type.

Since cartilage is a tissue whose function and form are related to physical stimuli, recent attempts to optimize the functional properties of engineered cartilage have focused on using bioreactors to incorporate mechanical loading environments *in vitro* to mimic *in vivo* environments. Mechanical loading environments tested include dynamic deformation, hydrostatic pressure, fluid flow, and shear stress. Beneficial effects of mechanical loading on cartilage constructs have been widely reported. For example, dynamic deformational loading of cell-seeded agarose disks improved the mechanical properties, as well as the sulfated glycosaminoglycan (sGAG) and hydroxyproline contents, compared to controls (Mauck et al., 2000). Recent evidence also supports a positive role for mechanical loading on MSCs. Human MSCs grown in pellet culture with TGF-β3 showed increased expression of cartilage markers in the presence of cyclic hydrostatic pressure. These positive effects were dependent on both the dose and length of time cells were exposed to loads (Miyanishi et al., 2006). This study underscores the need for further studies to determine the optimal type, timing, and amount of mechanical loading, and to establish the appropriate cells, scaffolds, and soluble factors with which it should be combined in order to regenerate truly functional cartilage.

Gene therapy is gaining recognition as a tool for cartilage regeneration. Lapine articular chondrocytes transfected with plasmid vectors expressing IGF-1 were combined with alginate and implanted into a rabbit osteochondral defect model (Madry et al., 2005). IGF-1 expressing implants improved articular cartilage and subchondral bone repair compared to controls. Although gene therapy strategies are promising, long-term studies are needed to evaluate repaired cartilage. It is likely that in the future, optimal cartilage regeneration will result from the combination of genetically modified cells with tissue engineering.

Meniscus

The menisci are two semilunar fibrocartilaginous structures located between the tibia and femur in the knee joint. The meniscus functions as a shock absorber, joint stabilizer, and joint lubricator. Furthermore, it has an outer vascular area which possesses the ability to repair itself and an inner avascular zone where the majority of tears occur. The cells in meniscus have both fibroblast- and chondrocyte-like properties and are termed fibrochondrocytes. Also, the cell phenotype as well as the ECM components vary between the different zones. In addition to having a different cell type, fibrocartilage differs from hyaline cartilage in that the cells predominantly secrete collagen type I and it has a relatively lower sGAG content giving it different mechanical properties. Despite these differences, many attempts at meniscal regeneration have incorporated some of the same scaffolds and growth factors used in articular cartilage repair and have been tested in a variety of animal models. Recent evidence has pointed to the potential for the use of MSCs in meniscal regeneration. GFP-labeled rat MSCs were seeded in fibrin glue and used to treat meniscal defects (Izuta et al., 2005). GFP-positive cells were detected up to 8 weeks post implantation and promoted meniscal repair. Successful cell-based meniscal repair would be of substantial therapeutic value, but further optimization of regenerative conditions and factors is still needed. Another challenge for meniscus tissue engineering is to regenerate a meniscus exhibiting significant anisotropic mechanical properties reflective of a highly oriented underlying ECM. We recently applied electrospinning technology to fabricate a biomaterial scaffold mimicking meniscal ECM fiber alignment, which directs fibrochondrocyte orientation and has controllable, anisotropic properties. These findings suggest the potential application of aligned-nanofiber-based scaffolds for meniscal tissue engineering (Li et al., 2007).

Osteochondral Tissue

Defects of both articular cartilage and the underlying subchondral bone are often associated with pain and joint instability, risk factors for the development of OA. Using a tissue-engineered osteochondral graft is a promising alternative to the current use of autologous osteochondral grafts which are limited by tissue availability, lack of appropriate geometric configuration, and donor site morbidity. Engineering an osteochondral graft has been challenging due to the technical difficulties involved in generating a single unit consisting of two tissues with different properties, which naturally require different conditions for development and optimal functionality. General approaches to osteochondral tissue engineering have involved choosing a scaffold for the bone layer alone, two separate scaffolds for the bone and cartilage layer, or a single scaffold for both layers and combining these scaffolds with one or two cell sources. These cell sources may have either chondrogene, osteogene, or bipotential differentiation capacity. The disadvantage to using two separate scaffolds or cells is that there may be impaired integration between the osteo and chondral components. Various approaches have been employed to circumvent this problem. For instance, a pellet of trabecular bone-derived mesenchymal progenitor cells previously induced to undergo chondrogenesis was press-coated onto a PLA scaffold (Tuli et al., 2004). Progenitor cells from the same patient which were induced to undergo osteogenesis were then seeded onto the other end of the scaffold, followed by culturing under conditions that supported both osteogenesis and chondrogenesis, resulting in an interface resembling the native osteochondral junction. Ideally, a single cell source would be able to differentiate into both lineages on a single scaffold. Recent evidence that electrospun PCL nanofibrous scaffolds allow MSCs to undergo multilineage differentiation suggests that this scaffold is a promising candidate for tissue engineering osteochondral grafts (Li et al., 2005a).

Skeletal Muscle

While muscle-based disorders are not as prevalent as OA or bone defects, there is a serious clinical need for muscle tissue engineering. These needs include muscle atrophy and muscular dystrophy, as well as muscle loss from trauma or surgical resection. Several disease states, such as cancer, infectious disease, heart failure and AIDS can produce a body wasting syndrome known as cachexia, in which muscle atrophy is severe. Muscular dystrophy is a group of mostly inherited neuromuscular disorders that cause muscle wasting, and can lead to death in patients with severe mutations in the dystrophin gene. For the treatment of traumatic muscle loss, free tissue transfer is an option, but autologous muscle transfer causes not only donor site morbidity, but can produce loss of function at the donor site.

Requirements of functional engineering of muscle would be the recapitulation of functional motion and integration with host connective tissues. On the cellular level, parallel alignment of fibers and integration of functional neuromuscular junctions are also important to achieve. As with other musculoskeletal tissues, mechanical stimulation is essential during myogenesis, and influences metabolic activity and gene expression, as well as fiber alignment. Also, myoblasts, similar to many types of cells, become difficult to differentiate as they are culture-expanded. Myoblast cell lines, once immortalized, cannot approximate myogenesis as well as primary myoblasts. Therefore, understanding the biology of precursor cell types, such as satellite cells and other muscle-derived stem cells, is crucial. While satellite cells act as the local regenerative cells in muscle, their limited expansion potential *in vitro* limits their current usefulness relative to MSCs. The therapeutic use of MSCs in muscle disorders was demonstrated when GFP-labeled human bone marrow-derived MSCs were induced to undergo myogenesis and then transplanted by local injection into muscles of immunosuppressed rats. Histological section 4 weeks later showed mature muscle characteristics in most GFP-positive myofibers. In addition, some of the cells appeared to become Pax7 expressing satellite cells, which could respond to local damage and contribute to muscle repair (Dezawa et al., 2005). For genetic disorders such as muscular dystrophy, there have been many attempts to reintroduce a wild type dystrophin gene into muscle via cell-based *ex vivo* gene therapy. Mesangioblasts, a type of vessel-associated stem cell, from wild type canine were delivered intra-arterially in a muscular dystrophy canine model with significant recovery of dystrophin gene expression, muscle morphology, and muscle function. Autologous genetically corrected mesangioblasts were not as effective (Sampaolesi et al., 2006).

As with all tissue engineering, uncovering the optimal growth factors and stimulation environment to effectively produce muscle tissue is essential. As would be expected for the tissue responsible for motor functions, mechanical stimulation is crucial in the development of *in vitro* functional muscle tissue. Mechanical forces applied *in vitro* yield significant differences in morphological and functional appearance of muscle tissues. Mean myofiber diameter and elasticity both improve, as well as the ratio of muscle fiber to ECM (Bach et al., 2004). Electrical stimulation mimicking nerve stimulation during myogenesis and regeneration of injured skeletal muscle also drives differentiation. Chronic electrical stimulation can change gene expression of muscle-specific genes, with increased VEGF expression and blood flow also shown after stimulation. Due to the unique contractile requirement of skeletal muscle, optimal scaffold strategies may differ from those of bone or cartilage production. PGA meshes, collagens, and alginates have been used, as in other tissues. Acellular muscle has also been exploited as a potential scaffold, and 3 weeks after seeding with myoblasts, isometric contractile force testing revealed longitudinal contractile forces upon stimulation (Borschel et al., 2004). Additionally, co-culture of fibroblasts with myoblasts has been explored as a way to produce a primary matrix.

Tendon and Ligament

Injuries to ligaments and tendons heal by forming tissues of inferior quality, due to hypocellularity of the tissue as well as scar formation that is weaker than normal tissue. Current strategies to replace tendons and ligaments consist of tissue autografts and allografts. Both have the traditional problems associated with graft technology, but additionally exhibit slow or poor functional integration to the surgical site, requiring at minimum several months for recovery. In addition, tendons and ligaments are prime examples of how in the musculoskeletal system, mechanical and structural properties are crucial to the function of the tissue. There are significant challenges in developing tissue engineering strategies, due to the fact that using autologous tenocytes can cause a tendon defect, limiting options for cell-based therapy.

MSCs have been used to produce engineered tendon and ligament tissues. Collagen gels seeded with autologous MSCs were implanted into full thickness, full length, central defects in patellar tendons of rabbits.

This approach significantly improved the biomechanical properties of tendon repair tissues as compared to natural repair controls (Awad et al., 2003). In one approach to overcoming poor functional integration of graft tissues, MSCs were used at the tendon–bone junction in tendon grafting for ACL repair in the rabbit knee. This resulted in a zone of fibrocartilage closely resembling that of the normal ACL (Lim et al., 2004). Dermal fibroblasts have also been used to engineer tendon. Autologous dermal fibroblasts, seeded on PGA fibers, were implanted in a porcine model and shared similar tensile strength with constructs developed with autologous tenocytes (Liu et al., 2006). Gene therapy using MSCs as a vehicle was demonstrated with transduction of BMP-12 into a mouse MSC cell line, leading to tendon and cartilage-like tissue formation after injection in the thigh muscle of nude mice (Lou et al., 1999).

Tensile strength and stretch loading are believed to be vital in producing the proper alignment of ligament and tendon tissues in *ex vivo* engineering. It is known that collagen is arranged along the axis of the loading force, implying its requirement for structurally competent tissues (Liu et al., 2006). Tensional and torsional stimulation of MSC-based constructs, mimicking cues the cells receive *in vivo*, enhanced ECM fiber deposition. Our recent findings reveal that both mechanical and biological stimuli act to regulate differentiation and matrix synthesis and remodeling during MSC tenogenesis (Kuo and Tuan, submitted for publication). Designing bioreactors that replicate the most essential of these conditions is critical to allow for *ex vivo* production of ready-to-implant tendons and ligaments with increased load to failure over endogenous tissues.

Intervertebral Disk

Degenerative disk disease is a leading cause of back pain and disability. The IVD has a complex structure comprised of a proteoglycan-rich nucleus pulposus and an outer annulus fibrosus. Disk degeneration is characterized by proteoglycan loss in the nucleus pulposus with concurrent degradation of the annulus fibrosus. As cell-based regenerative efforts for the IVD are still in its infancy, appropriate cell, scaffold, and growth factor combinations are under investigation and have been tested in various animal models. In Europe, based on results from a canine model and pilot studies in humans, a prospective, randomized trial for comparing autologous disk chondrocyte transplantation (ADCT) plus discectomy to discectomy alone is being performed. Nucleus pulposus cells derived from therapeutic discectomy are cultured and delivered 12 weeks after discectomy. An interim analysis after 2 years shows promising results with sustained pain relief and improvement of disability scores in the ADCT patients (Meisel et al., 2006). Autologous disk cells are, however, in limited supply. MSCs have been recently shown to differentiate into cells similar to IVD cells, and present an alternative cell source for regenerative purposes (Steck et al., 2005). In addition, gene therapy strategies can be used to repair IVD, as demonstrated by a study in which adenovirus expressing Sox-9 was used to infect human IVD cells *in vivo*, and was also injected directly into rabbit IVD following a stab injury (Paul et al., 2003). *In vitro*, infected disk cells exhibited increased production of collagen type II, and *in vivo*, IVDs injected with these cells showed decreased scarring compared to those that were untreated or injected with mock adenovirus.

Craniofacial Tissues

The temporomandibular joints (TMJ) in the craniofacial system, connecting the mandible to the cranium, play a vital role in coordinating our eating and speech activities. Osteoarthritic TMJ is correlated with not only aging but also dental function. Craniofacial defects can result from trauma, neoplasm, and most commonly from infection. Additionally, teeth penetrating through the mucosal epithelial barrier are exposed to the moist harsh environment of the mouth and subject to infection via dental caries and periodontal pathogens. Loss of teeth resulting from severe periodontal diseases or root canal infection is a common cause of losing the jaw bone. Recent clinical success of the dental implant renders the need for augmenting lost mandible before implant even more important, because the success of the dental implant is dependent upon sufficient quality bone.

Cell-based therapy to regenerate periodontal defects using autologous MSCs has been recently demonstrated in a dog model (Hasegawa et al., 2006), in which 4 weeks after MSC transplantation, the periodontal defects were almost regenerated with periodontal tissue. Cementoblasts, OB, and osteocytes are found in the regenerated periodontal tissue, suggesting that transplanted MSCs could survive and differentiate into periodontal tissue cells and repair the tissue. Similar accomplishments have also been observed in humans except the assessment is only performed clinically with radiographs and periodontal probing (Yamada et al., 2006). More extensive mandibular repair using cell therapy has been recently demonstrated in both animal models and humans. Mandibular distraction osteogenesis is enhanced by direct delivery of MSCs into the defect in rats (Qi et al., 2006). Autologous MSCs from patient's iliac crest plus platelet-rich plasma (PRP) was applied to reconstruct an alveolar cleft defect or to augment alveolar bone with success (Hibi et al., 2006). In addition, the mixture of MSCs and PRP facilitates osteointegration in dental implants, and may replace the use of autologous particulate cancellous bone and marrow for the same purpose (Yamada et al., 2004).

Ex vivo gene therapy for *de novo* jaw bone regeneration has been demonstrated in a swine model using adenovirus BMP-2-mediated gene transfer to expanded autologous MSCs. Functional bone capable of sustaining axial compressive loads is formed in the maxillary bone defect filled with the BMP-2-transduced cells cast in collagen gel (Chang et al., 2003). Jaw bone regeneration utilizing recombinant protein therapy is considered more straightforward and simpler. The recombinant human BMP-2/collagen sponge implant converts undifferentiated MSCs into OB and promotes an intense local neovascular response. The type of bone synthesis depends upon the mesenchymal substrate and the local cellular environment. Using this simple technique, bone defects can be resynthesized with good outcomes and a significant reduction in donor site morbidity. A drawback of recombinant protein therapy is that a single dose of exogenous protein is not as robust for osteoinduction when compared to the results from preclinical animal studies (Nussenbaum and Krebsbach, 2006).

A proof of concept to regenerate cartilage of TMJ has been demonstrated using rat MSCs to engineer human-shaped mandibular condyle in immunocompromised mice (Alhadlaq and Mao, 2003). Challenges for TMJ tissue engineering include promotion of ECM synthesis and tissue maturation by stem cell-derived chondrogene and osteogene cells that have been encapsulated in biocompatible and bioactive scaffolds. Enhancement of the mechanical properties of a tissue-engineered mandibular condyle for ultimate *in situ* implantation into the human TMJ is another challenge for tissue engineers. Tooth regeneration with cells from tooth buds, and other dental tissue regeneration including dentin and periodontal ligament using stem cells from the respective tissues, has also been proposed and tested in animal models (Ohazama et al., 2004).

CONCLUSIONS AND PROSPECTS

It appears promising that regenerative medicine via cell-based therapy and tissue engineering will become a widespread therapeutic modality. The essential procedures consist of utilizing and manipulating *ex vivo* expanded multipotent stem cells, and delivering them into hosts under a designed condition or package to grow new musculoskeletal tissues (Figure 26.3). However, there are many challenges before reaching this goal. Specifically, we need to understand (1) the native environment of stem cells, that is, the niche and stem cell markers, to allow stem cell isolation and culture expansion in a more specific and predictable manner; (2) the molecular regulation at various stages of the stem cell differentiation program such that a better control of cell

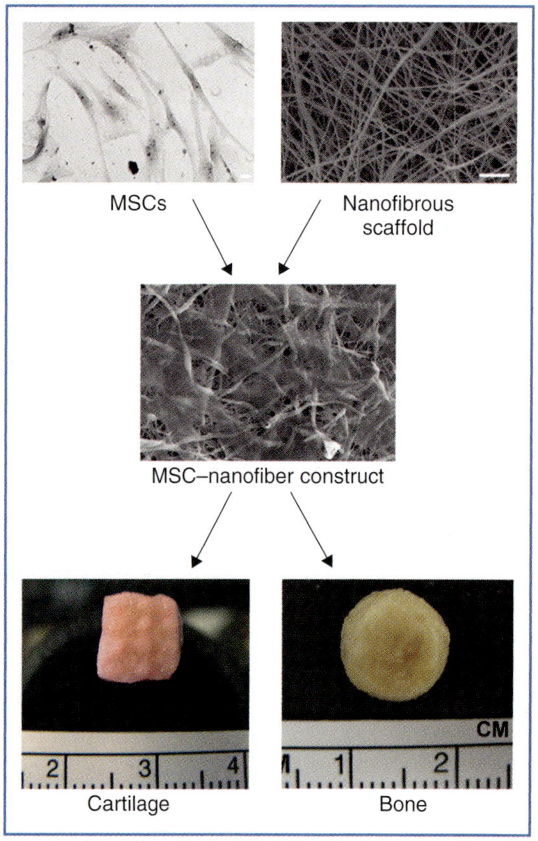

Figure 26.3 Production of tissue-engineered cartilage and bone based on MSC-seeded nanofibrous scaffolds. MSCs were cultured in chondrogene and osteogene media for 3 and 6 weeks, respectively. Bar = 25 mm.

lineage commitment can be achieved; (3) which delivery method or scaffold system is optimal for regeneration of a specific tissue; and (4) the interactions of cells with their carrier or scaffold system and the newly produced ECM, so that optimal tissue regeneration can be accomplished. There is also a need for testing the long-term compatibility of allogeneic MSCs in the host before cell banks can be established to provide an adequate cell source for cell-based applications and tissue engineering. There has been significant experience accumulated in the regeneration of bone, whereas other musculoskeletal tissue regeneration, such as cartilage, ligament, and tendon, is still at its infancy. Another aspect of tissue engineering that requires further development is the design of bioreactors that can replicate conditions to allow *ex vivo* production of ready-to-implant musculoskeletal tissues, which is critical to achieve functional tissue engineering.

ACKNOWLEDGMENTS

Supported by NIH NIAMS Intramural Research Program (AR Z0141131) and Howard Hughes Medical Institute-NIH Research Scholar Program.

REFERENCES

Abukawa, H., Kaban, L.B., Williams, W.B., Terada, S., Vacanti, J.P. and Troulis, M.J. (2006). Effect of interferon-alpha-2b on porcine mesenchymal stem cells. *J. Oral Maxillofac. Surg.* **64**: 1214–1220.

Alhadlaq, A. and Mao, J.J. (2003). Tissue-engineered neogenesis of human-shaped mandibular condyle from rat mesenchymal stemcells. *J. Dent. Res.* **82**: 951–956.

Awad, H.A., Boivin, G.P., Dressler, M.R., Smith, F.N., Young, R.G. and Butler, D.L. (2003). Repair of patellar tendon injuries using a cell–collagen composite. *J. Orthop. Res.* **21**: 420–431.

Bach, A.D., Beier, J.P., Stern-Staeter, J. and Horch, R.E. (2004). Skeletal muscle tissue engineering. *J. Cell. Mol. Med.* **8**: 413–422.

Ball, S.G., Shuttleworth, A.C. and Kielty, C.M. (2004). Direct cell contact influences bone marrow mesenchymal stem cell fate. *Int. J. Biochem. Cell Biol.* **36**: 714–727.

Barry, F., Boynton, R.E., Liu, B. and Murphy, J.M. (2001). Chondrogenic differentiation of mesenchymal stem cells from bone marrow: differentiation-dependent gene expression of matrix components. *Exp. Cell Res.* **268**: 189–200.

Bi, W., Huang, W., Whitworth, D.J., Deng, J.M., Zhang, Z., Behringer, R.R. and de Crombrugghe, B. (2001). Haploinsufficiency of Sox9 results in defective cartilage primordia and premature skeletal mineralization. *Proc. Natl. Acad. Sci. USA* **98**: 6698–6703.

Boland, G.M., Perkins, G., Hall, D.J. and Tuan, R.S. (2004). Wnt 3a promotes proliferation and suppresses osteogenic differentiation of adult human mesenchymal stem cells. *J. Cell. Biochem.* **93**: 1210–1230.

Borschel, G.H., Dennis, R.G. and Kuzon Jr., W.M. (2004). Contractile skeletal muscle tissue-engineered on a cellular scaffold. *Plast. Reconstr. Surg.* **113**: 595–602; discussion 603–604.

Brittberg, M., Lindahl, A., Nilsson, A., Ohlsson, C., Isaksson, O. and Peterson, L. (1994). Treatment of deep cartilage defects in the knee with autologous chondrocyte transplantation. *N. Engl. J. Med.* **331**: 889–895.

Caplan, A.I. (1991). Mesenchymal stem cells. *J. Orthop. Res.* **9**: 641–650.

CDC (2006). Prevalence of doctor-diagnosed arthritis and arthritis-attributable activity limitation – United States, 2003–2005. *MMWR Morb. Mortal. Wkly Rep.* **55**: 1089–1092.

Chamberlain, J.R., Schwarze, U., Wang, P.R., Hirata, R.K., Hankenson, K.D., Pace, J.M., Underwood, R.A., Song, K.M., Sussman, M., Byers, P.H., et al. (2004). Gene targeting in stem cells from individuals with osteogenesis imperfecta. *Science* **303**: 1198–1201.

Chang, S.C., Chuang, H.L., Chen, Y.R., Chen, J.K., Chung, H.Y., Lu, Y.L., Lin, H.Y., Tai, C.L. and Lou, J. (2003). *Ex vivo* gene therapy in autologous bone marrow stromal stem cells for tissue-engineered maxillofacial bone regeneration. *Gene Ther.* **10**: 2013–2019.

Chen, F., Hui, J.H., Chan, W.K. and Lee, E.H. (2003). Cultured mesenchymal stem cell transfers in the treatment of partial growth arrest. *J. Pediatr. Orthop.* **23**: 425–429.

Chen, F.H., Rousche, K.T. and Tuan, R.S. (2006). Technology insight: adult stem cells in cartilage regeneration and tissue engineering. *Nat. Clin. Pract. Rheumatol.* **2**: 373–382.

Chen, X.I., Armstrong, M.A. and Li, G. (2006). Mesenchymal stem cells in immunoregulation. *Immunol. Cell Biol.* **84**: 413–421.

Chiang, H., Kuo, T.F., Tsai, C.C., Lin, M.C., She, B.R., Huang, Y.Y., Lee, H.S., Shieh, C.S., Chen, M.H., Ramshaw, J.A., et al. (2005). Repair of porcine articular cartilage defect with autologous chondrocyte transplantation. *J. Orthop. Res.* **23**: 584–593.

Colter, D.C., Class, R., DiGirolamo, C.M. and Prockop, D.J. (2000). Rapid expansion of recycling stem cells in cultures of plastic-adherent cells from human bone marrow. *Proc. Natl Acad. Sci. USA* **97**: 3213–3218.

Deans, R.J. and Moseley, A.B. (2000). Mesenchymal stem cells: biology and potential clinical uses. *Exp. Hematol.* **28**: 875–884.

Dezawa, M., Ishikawa, H., Itokazu, Y., Yoshihara, T., Hoshino, M., Takeda, S., Ide, C. and Nabeshima, Y. (2005). Bone marrow stromal cells generate muscle cells and repair muscle degeneration. *Science* **309**: 314–317.

Dominici, M., Le Blanc, K., Mueller, I., Slaper-Cortenbach, I., Marini, F., Krause, D., Deans, R., Keating, A., Prockop, D. and Horwitz, E. (2006). Minimal criteria for defining multipotent mesenchymal stromal cells. The International Society for Cellular Therapy position statement. *Cytotherapy* **8**: 315–317.

Dreyfus, P.A., Chretien, F., Chazaud, B., Kirova, Y., Caramelle, P., Garcia, L., Butler-Browne, G. and Gherardi, R.K. (2004). Adult bone marrow-derived stem cells in muscle connective tissue and satellite cell niches. *Am. J. Pathol.* **164**: 773–779.

Eames, B.F., de la Fuente, L. and Helms, J.A. (2003). Molecular ontogeny of the skeleton. *Birth Defects Res. C Embryo Today* **69**: 93–101.

Elisseeff, J., Ferran, A., Hwang, S., Varghese, S. and Zhang, Z. (2006). The role of biomaterials in stem cell differentiation: applications in the musculoskeletal system. *Stem Cells Dev.* **15**: 295–303.

Engler, A.J., Sen, S., Sweeney, H.L. and Discher, D.E. (2006). Matrix elasticity directs stem cell lineage specification. *Cell* **126**: 677–689.

Fukuchi, Y., Nakajima, H., Sugiyama, D., Hirose, I., Kitamura, T. and Tsuji, K. (2004). Human placenta-derived cells have mesenchymal stem/progenitor cell potential. *Stem Cells* **22**: 649–658.

Gersbach, C.A., Le Doux, J.M., Guldberg, R.E. and Garcia, A.J. (2006). Inducible regulation of Runx2-stimulated osteogenesis. *Gene Ther.* **13**: 873–882.

Gregory, C.A., Singh, H., Perry, A.S. and Prockop, D.J. (2003). The Wnt signaling inhibitor dickkopf-1 is required for reentry into the cell cycle of human adult stem cells from bone marrow. *J. Biol. Chem.* **278**: 28067–28078.

Haleem-Smith, H., Derfoul, A., Okafor, C., Tuli, R., Olsen, D., Hall, D.J. and Tuan, R.S. (2005). Optimization of high-efficiency transfection of adult human mesenchymal stem cells *in vitro*. *Mol. Biotechnol.* **30**: 9–20.

Hasegawa, N., Kawaguchi, H., Hirachi, A., Takeda, K., Mizuno, N., Nishimura, M., Koike, C., Tsuji, K., Iba, H., Kato, Y., et al. (2006). Behavior of transplanted bone marrow-derived mesenchymal stem cells in periodontal defects. *J. Periodontol.* **77**: 1003–1007.

Hernigou, P., Poignard, A., Beaujean, F. and Rouard, H. (2005). Percutaneous autologous bone-marrow grafting for nonunions. Influence of the number and concentration of progenitor cells. *J. Bone Joint Surg. Am.* **87**: 1430–1437.

Hibi, H., Yamada, Y., Ueda, M. and Endo, Y. (2006). Alveolar cleft osteoplasty using tissue-engineered osteogenic material. *Int. J. Oral Maxillofac. Surg.* **35**: 551–555.

In 't Anker, P.S., Scherjon, S.A., Kleijburg-van der Keur, C., Noort, W.A., Claas, F.H., Willemze, R., Fibbe, W.E. and Kanhai, H.H. (2003). Amniotic fluid as a novel source of mesenchymal stem cells for therapeutic transplantation. *Blood* **102**: 1548–1549.

Izuta, Y., Ochi, M., Adachi, N., Deie, M., Yamasaki, T. and Shinomiya, R. (2005). Meniscal repair using bone marrow-derived mesenchymal stem cells: experimental study using green fluorescent protein transgenic rats. *Knee* **12**: 217–223.

Jorgensen, C., Djouad, F., Fritz, V., Apparailly, F., Plence, P. and Noel, D. (2003). Mesenchymal stem cells and rheumatoid arthritis. *Joint Bone Spine* **70**: 483–485.

Karp, J.M., Ferreira, L.S., Khademhosseini, A., Kwon, A.H., Yeh, J. and Langer, R.S. (2006). Cultivation of human embryonic stem cells without the embryoid body step enhances osteogenesis *in vitro*. *Stem Cells* **24**: 835–843.

Kern, S., Eichler, H., Stoeve, J., Kluter, H. and Bieback, K. (2006). Comparative analysis of mesenchymal stem cells from bone marrow, umbilical cord blood, or adipose tissue. *Stem Cells* **24**: 1294–1301.

Kim, H., Lee, J.H. and Suh, H. (2003). Interaction of mesenchymal stem cells and osteoblasts for *in vitro* osteogenesis. *Yonsei Med. J.* **44**: 187–197.

Kneser, U., Stangenberg, L., Ohnolz, J., Buettner, O., Stern-Straeter, J., Mobest, D., Horch, R.E., Stark, G.B. and Schaefer, D.J. (2006). Evaluation of processed bovine cancellous bone matrix seeded with syngenic osteoblasts in a critical size calvarial defect rat model. *J. Cell. Mol. Med.* **10**: 695–707.

Kramer, J., Hegert, C., Guan, K., Wobus, A.M., Muller, P.K. and Rohwedel, J. (2000). Embryonic stem cell-derived chondrogenic differentiation *in vitro*: activation by BMP-2 and BMP-4. *Mech. Dev.* **92**: 193–205.

Kuo, C.K., Li, W.J., Mauck, R.L. and Tuan, R.S. (2006). Cartilage tissue engineering: its potential and uses. *Curr. Opin. Rheumatol.* **18**: 64–73.

Lee, R.H., Kim, B., Choi, I., Kim, H., Choi, H.S., Suh, K., Bae, Y.C. and Jung, J.S. (2004). Characterization and expression analysis of mesenchymal stem cells from human bone marrow and adipose tissue. *Cell. Physiol. Biochem.* **14**: 311–324.

Lefebvre, V. and Smits, P. (2005). Transcriptional control of chondrocyte fate and differentiation. *Birth Defects Res. C Embryo Today* **75**: 200–212.

Levenberg, S., Huang, N.F., Lavik, E., Rogers, A.B., Itskovitz-Eldor, J. and Langer, R. (2003). Differentiation of human embryonic stem cells on three-dimensional polymer scaffolds. *Proc. Natl Acad. Sci. USA* **100**: 12741–12746.

Li, W.J., Tuli, R., Huang, X., Laquerriere, P. and Tuan, R.S. (2005a). Multilineage differentiation of human mesenchymal stem cells in a three-dimensional nanofibrous scaffold. *Biomaterials* **26**: 5158–5166.

Li, W.J., Tuli, R., Okafor, C., Derfoul, A., Danielson, K.G., Hall, D.J. and Tuan, R.S. (2005b). A three-dimensional nanofibrous scaffold for cartilage tissue engineering using human mesenchymal stem cells. *Biomaterials* **26**: 599–609.

Li, W.J., Jiang, Y.J. and Tuan, R.S. (2006). Chondrocyte phenotype in engineered fibrous matrix is regulated by fiber size. *Tissue Eng.* **12**: 1775–1785.

Li, W.J., Mauck, R.L., Cooper, J.A., Yuan, X. and Tuan, R.S. (2007). Engineering controllable anisotropy in electrospun biodegradable nanofibrous scaffolds for musculoskeletal tissue engineering. *J. Biomech.* **40**: 1686–1693.

Lieberman, J.R., Daluiski, A., Stevenson, S., Wu, L., McAllister, P., Lee, Y.P., Kabo, J.M., Finerman, G.A., Berk, A.J. and Witte, O.N. (1999). The effect of regional gene therapy with bone morphogenetic protein-2-producing bone-marrow cells on the repair of segmental femoral defects in rats. *J. Bone Joint Surg. Am.* **81**: 905–917.

Lim, J.K., Hui, J., Li, L., Thambyah, A., Goh, J. and Lee, E.H. (2004). Enhancement of tendon graft osteointegration using mesenchymal stem cells in a rabbit model of anterior cruciate ligament reconstruction. *Arthroscopy* **20**: 899–910.

Liu, C.-H. and Hwang, S.-M. (2005). Cytokine interactions in mesenchymal stem cells from cord blood. *Cytokine* **32**: 270–279.

Liu, W., Chen, B., Deng, D., Xu, F., Cui, L. and Cao, Y. (2006). Repair of tendon defect with dermal fibroblast engineered tendon in a porcine model. *Tissue Eng.* **12**: 775–788.

Lou, J., Tu, Y., Ludwig, F.J., Zhang, J. and Manske, P.R. (1999). Effect of bone morphogenetic protein-12 gene transfer on mesenchymal progenitor cells. *Clin. Orthop. Relat. Res.* 333–339.

Lu, F.-Z., Fujino, M., Kitazawa, Y., Uyama, T., Hara, Y., Funeshima, N., Jiang, J.-Y., Umezawa, A. and Li, X.-K. (2005). Characterization and gene transfer in mesenchymal stem cells derived from human umbilical-cord blood. *J. Lab. Clin. Med.* **146**: 271–278.

Madry, H., Kaul, G., Cucchiarini, M., Stein, U., Zurakowski, D., Remberger, K., Menger, M.D., Kohn, D. and Trippel, S.B. (2005). Enhanced repair of articular cartilage defects *in vivo* by transplanted chondrocytes overexpressing insulin-like growth factor I (IGF-I). *Gene Ther.* **12**: 1171–1179.

Mareschi, K., Biasin, E., Piacibello, W., Aglietta, M., Madon, E. and Fagioli, F. (2001). Isolation of human mesenchymal stem cells: bone marrow versus umbilical cord blood. *Haematologica* **86**: 1099–1100.

Matsubara, T., Tsutsumi, S., Pan, H., Hiraoka, H., Oda, R., Nishimura, M., Kawaguchi, H., Nakamura, K. and Kato, Y. (2004). A new technique to expand human mesenchymal stem cells using basement membrane extracellular matrix. *Biochem. Biophys. Res. Commun.* **313**: 503–508.

Mauck, R.L., Soltz, M.A., Wang, C.C., Wong, D.D., Chao, P.H., Valhmu, W.B., Hung, C.T. and Ateshian, G.A. (2000). Functional tissue engineering of articular cartilage through dynamic loading of chondrocyte-seeded agarose gels. *J. Biomech. Eng.* **122**: 252–260.

Mauney, J.R., Blumberg, J., Pirun, M., Volloch, V., Vunjak-Novakovic, G. and Kaplan, D.L. (2004). Osteogenic differentiation of human bone marrow stromal cells on partially demineralized bone scaffolds *in vitro*. *Tissue Eng.* **10**: 81–92.

McBeath, R., Pirone, D.M., Nelson, C.M., Bhadriraju, K. and Chen, C.S. (2004). Cell shape, cytoskeletal tension, and RhoA regulate stem cell lineage commitment. *Dev. Cell* **6**: 483–495.

Meisel, H.J., Ganey, T., Hutton, W.C., Libera, J., Minkus, Y. and Alasevic, O. (2006). Clinical experience in cell-based therapeutics: intervention and outcome. *Eur. Spine J.* **15**(Suppl 15): 397–405.

Miyanishi, K., Trindade, M.C., Lindsey, D.P., Beaupre, G.S., Carter, D.R., Goodman, S.B., Schurman, D.J. and Smith, R.L. (2006). Dose- and time-dependent effects of cyclic hydrostatic pressure on transforming growth factor-beta3-induced chondrogenesis by adult human mesenchymal stem cells *in vitro*. *Tissue Eng.* **12**: 2253–2262.

Nussenbaum, B. and Krebsbach, P.H. (2006). The role of gene therapy for craniofacial and dental tissue engineering. *Adv. Drug Deliv. Rev.* **58**: 577–591.

Ohazama, A., Modino, S.A., Miletich, I. and Sharpe, P.T. (2004). Stem-cell-based tissue engineering of murine teeth. *J. Dent. Res.* **83**: 518–522.

Okafor, C.C., Haleem-Smith, H., Laqueriere, P., Manner, P.A. and Tuan, R.S. (2006). Particulate endocytosis mediates biological responses of human mesenchymal stem cells to titanium wear debris. *J. Orthop. Res.* **24**: 461–473.

Park, Y., Lutolf, M.P., Hubbell, J.A., Hunziker, E.B. and Wong, M. (2004). Bovine primary chondrocyte culture in synthetic matrix metalloproteinase-sensitive poly(ethylene glycol)-based hydrogels as a scaffold for cartilage repair. *Tissue Eng.* **10**: 515–522.

Paul, R., Haydon, R.C., Cheng, H., Ishikawa, A., Nenadovich, N., Jiang, W., Zhou, L., Breyer, B., Feng, T., Gupta, P., et al. (2003). Potential use of Sox9 gene therapy for intervertebral degenerative disc disease. *Spine* **28**: 755–763.

Pittenger, M.F., Mackay, A.M., Beck, S.C., Jaiswal, R.K., Douglas, R., Mosca, J.D., Moorman, M.A., Simonetti, D.W., Craig, S. and Marshak, D.R. (1999). Multilineage potential of adult human mesenchymal stem cells. *Science* **284**: 143–147.

Qi, M., Hu, J., Zou, S., Zhou, H. and Han, L. (2006). Mandibular distraction osteogenesis enhanced by bone marrow mesenchymal stem cells in rats. *J. Craniomaxillofac. Surg.* **34**: 283–289.

Quarto, R., Mastrogiacomo, M., Cancedda, R., Kutepov, S.M., Mukhachev, V., Lavroukov, A., Kon, E. and Marcacci, M. (2001). Repair of large bone defects with the use of autologous bone marrow stromal cells. *N. Engl. J. Med.* **344**: 385–386.

Romanov, Y.A., Svintsitskaya, V.A. and Smirnov, V.N. (2003). Searching for alternative sources of postnatal human mesenchymal stem cells: candidate MSC-like cells from umbilical cord. *Stem Cells* **21**: 105–110.

Rotter, N., Ung, F., Roy, A.K., Vacanti, M., Eavey, R.D., Vacanti, C.A. and Bonassar, L.J. (2005). Role for interleukin 1alpha in the inhibition of chondrogenesis in autologous implants using polyglycolic acid–polylactic acid scaffolds. *Tissue Eng.* **11**: 192–200.

Rouwkema, J., de Boer, J. and Van Blitterswijk, C.A. (2006). Endothelial cells assemble into a 3-dimensional prevascular network in a bone tissue engineering construct. *Tissue Eng.* **12**: 2685–2693.

Sampaolesi, M., Blot, S., D'Antona, G., Granger, N., Tonlorenzi, R., Innocenzi, A., Mognol, P., Thibaud, J.L., Galvez, B.G., Barthelemy, I., et al. (2006). Mesangioblast stem cells ameliorate muscle function in dystrophic dogs. *Nature* **444**: 574–579.

Sartorelli, V. and Caretti, G. (2005). Mechanisms underlying the transcriptional regulation of skeletal myogenesis. *Curr. Opin. Genet. Dev.* **15**: 528–535.

Schofield, R. (1978). The relationship between the spleen colony-forming cell and the haemopoietic stem cell. *Blood Cells* **4**: 7–25.

Sekiya, I., Larson, B.L., Vuoristo, J.T., Reger, R.L. and Prockop, D.J. (2005). Comparison of effect of BMP-2, -4, and -6 on *in vitro* cartilage formation of human adult stem cells from bone marrow stroma. *Cell Tissue Res.* **320**: 269–276.

Shen, F.H., Visger, J.M., Balian, G., Hurwitz, S.R. and Diduch, D.R. (2002). Systemically administered mesenchymal stromal cells transduced with insulin-like growth factor-I localize to a fracture site and potentiate healing. *J. Orthop. Trauma* **16**: 651–659.

Steck, E., Bertram, H., Abel, R., Chen, B., Winter, A. and Richter, W. (2005). Induction of intervertebral disc-like cells from adult mesenchymal stem cells. *Stem Cells* **23**: 403–411.

Tamaki, T., Akatsuka, A., Ando, K., Nakamura, Y., Matsuzawa, H., Hotta, T., Roy, R.R. and Edgerton, V.R. (2002). Identification of myogenic-endothelial progenitor cells in the interstitial spaces of skeletal muscle. *J. Cell Biol.* **157**: 571–577.

Tozer, S. and Duprez, D. (2005). Tendon and ligament: development, repair and disease. *Birth Defects Res. C Embryo Today* **75**: 226–236.

Tuan, R.S., Boland, G. and Tuli, R. (2003). Adult mesenchymal stem cells and cell-based tissue engineering. *Arthr. Res. Ther.* **5**: 32–45.

Tuli, R., Tuli, S., Nandi, S., Huang, X., Manner, P.A., Hozack, W.J., Danielson, K.G., Hall, D.J. and Tuan, R.S. (2003). Transforming growth factor-beta-mediated chondrogenesis of human mesenchymal progenitor cells involves N-cadherin and mitogen-activated protein kinase and Wnt signaling cross-talk. *J. Biol. Chem.* **278**: 41227–41236.

Tuli, R., Nandi, S., Li, W.J., Tuli, S., Huang, X., Manner, P.A., Laquerriere, P., Noth, U., Hall, D.J. and Tuan, R.S. (2004). Human mesenchymal progenitor cell-based tissue engineering of a single-unit osteochondral construct. *Tissue Eng.* **10**: 1169–1179.

Vacanti, C.A., Bonassar, L.J., Vacanti, M.P. and Shufflebarger, J. (2001). Replacement of an avulsed phalanx with tissue-engineered bone. *N. Engl. J. Med.* **344**: 1511–1514.

Veilleux, N. and Spector, M. (2005). Effects of FGF-2 and IGF-1 on adult canine articular chondrocytes in type II collagen–glycosaminoglycan scaffolds *in vitro*. *Osteoarthr. Cartilage* **13**: 278–286.

Wakitani, S., Imoto, K., Yamamoto, T., Saito, M., Murata, N. and Yoneda, M. (2002). Human autologous culture expanded bone marrow mesenchymal cell transplantation for repair of cartilage defects in osteoarthritic knees. *Osteoarthr. Cartilage* **10**: 199–206.

Wakitani, S., Aoki, H., Harada, Y., Sonobe, M., Morita, Y., Mu, Y., Tomita, N., Nakamura, Y., Takeda, S., Watanabe, T.K., et al. (2004). Embryonic stem cells form articular cartilage, not teratomas, in osteochondral defects of rat joints. *Cell Transplant.* **13**: 331–336.

Yamada, Y., Ueda, M., Naiki, T. and Nagasaka, T. (2004). Tissue-engineered injectable bone regeneration for osteointegrated dental implants. *Clin. Oral Implants Res.* **15**: 589–597.

Yamada, Y., Ueda, M., Hibi, H. and Baba, S. (2006). A novel approach to periodontal tissue regeneration with mesenchymal stem cells and platelet-rich plasma using tissue engineering technology: a clinical case report. *Int. J. Periodont. Rest. Dent.* **26**: 363–369.

Young, B.H., Peng, H. and Huard, J. (2002). Muscle-based gene therapy and tissue engineering to improve bone healing. *Clin. Orthop. Relat. Res.* S243–S251.

27

Hepatocyte Transplantation

Stephen C. Strom and Ewa C.S. Ellis

INTRODUCTION

The concept of regenerative medicine implies that the clinician works with the innate healing and regenerative process of the body to affect an improvement in a patient's health. Perhaps more than with any other organ, the liver offers the greatest opportunity for regenerative medicine. This is because, unlike most other tissues, the liver has the capacity to regenerate following massive chemical or physical insult and tissue loss (Michalopoulos and DeFrances, 1997). Our very existence may well rely on the ability to regenerate liver mass. The liver is an incredibly complex organ which performs quite diverse biological functions, from glycogen storage and catabolism to maintain blood sugar levels, to the production and secretion of critical plasma proteins including albumin, clotting factors, and protease inhibitors. In addition the liver is the major site in the body for the metabolism and excretion of hormones, metabolic waste products such as ammonia as well as exogenous compounds such as toxins, drugs, and a variety of other compounds to which we are exposed through odiet and environment. These processes are so critical to survival that the loss of any of these functions has serious and often lethal consequences for the individual.

Until recently, the only option for treating chronic liver disease or metabolic defects in liver function has been whole organ transplantation. Recently, hepatocyte transplantation has been performed. Although still an experimental therapy, there are some potential advantages for a cell therapy approach to treat liver disease. Some of the advantages and problems with the current treatments for liver disease are listed in Table 27.1.

Despite the unquestioned success of this technique orthotopic liver transplantation (OLT) requires major surgery and has a significantly long recovery period. The financial costs associated with OLT and lifelong immunosuppression is considerable. There is a high incidence of complications from the surgical procedure and the

Table 27.1 Current treatments for liver disease

Orthotopic liver transplantation
Major and expensive surgery
Extensive recovery period
High incidence of complications
Expensive maintenance therapy
Shortage of donor organs
Timing is critical
Hepatocyte transplantation
Less invasive and less costly procedure
Complications, fewer, and less severe
Timing of procedure is easier
Alternative cell sources
Patient retains native liver
Graft loss is not necessarily lethal
Option remains for whole organ transplant

Copyright © 2008 Elsevier Inc. All rights reserved

concomitant immunosuppression which is required following the organ transplant. Complications can range from simple infections to renal failure, hyperlipidemia, and an increased incidence of skin and other types of cancers following long-term immunosuppression. As with all other organs, the number of liver donors does not nearly equal the number of patients on the waiting list. Patients may wait 2 or more years for a liver transplant, and there is a death rate of greater than 10% per year of patients on the waiting list. Timing is critical for whole organ transplant. An ABO-compatible liver donor must be available when a patient requires the transplant. Some of the limitations associated with whole organ transplants are addressed with hepatocyte transplants (Table 27.1). Hepatocyte transplants do not require major surgical procedures as they are performed by infusion of cells into the blood supply to an organ such as the liver or spleen. Thus, hepatocyte transplants are less invasive and less costly procedures. Because major surgery is not required there are fewer complications associated with the procedure. Since cell infusions are minor procedures, there is essentially no recovery period needed. If patients were healthy prior to the procedures, such as a stable metabolic disease patient, they would likely feel no adverse effects from the procedure other than from the placement of a catheter. Hepatocytes can be banked and cryopreserved, so theoretically, cells could be available anytime for a patient transplant. The timing of a hepatocyte transplant depends on the status of the patient rather than on the availability of a suitable organ. Currently, the source of hepatocytes for hepatocyte transplants is mainly discard organs not suitable for whole organ transplant (Nakazawa et al., 2002). Currently, there are not enough hepatocytes to transplant all recipients who would likely benefit from the procedure. However, some inventive new ideas have been proposed, such as to use segment IV, which can be made available from a split-liver procedure (Mitry et al., 2004) to make more hepatocytes available for transplants. Alternative sources of hepatocytes could also be available in the future. Although many options are discussed, the most prominent sources are xenotransplants from pigs or other species, immortalized hepatocytes and most recently stem cell-derived hepatocytes (Strom and Fisher, 2003). Future developments in these areas may make the number of cells available for hepatocyte transplants virtually unlimited.

A significant benefit of hepatocyte transplantation is that the patients retain their native liver. In cases of cell transplants for metabolic disease, the patient's native liver still performs all of the liver functions with the exception of the function which initiates the disease. Patients with ornithine transcarbamylase deficiency (OTC) have mutation in an enzyme involved in the urea cycle which prevents the metabolism and elimination of ammonia. Although the native liver is not proficient in ammonia metabolism, it is still capable of performing other liver functions including the secretion of clotting factors, albumin, drug metabolism, and all other metabolic and synthetic processes. A cell transplant need only support the ammonia metabolism for the patient, and will not be required to provide complete liver support. Because all liver functions are not dependent on donor cells, loss of the cell graft or failure of the cells to function properly will not necessarily be life threatening, especially for a stable metabolic disease patient. Finally, a whole organ transplant always remains as an option for the cell transplant patient. Even if the cell transplant fails to function or is rejected, nothing done as part of the cell transplant procedure would likely interfere with a subsequent whole organ transplant. Fisher et al. (1998) reported that prior hepatocyte transplantation did not sensitize the cell transplant recipient to either the donor cells or to an eventual liver graft. Thus, despite sometimes transplanting hepatocytes directly into an immunological response organ, the spleen, no immunological reactions are initiated which are deleterious to the cell transplant or an eventual whole organ transplant.

There are potential disadvantages of hepatocyte transplants as well. First, there are no reports of long-term complete corrections of metabolic liver disease in patients following cell transplantation alone. Because it is a new field, much additional experimentation will be required to determine the full efficacy of cell therapy of liver disease and the length of time the cell graft will function. Also, like whole organ transplants, it is believed that cell transplant recipients will require the administration of immunosuppressive drugs. It is likely that lower doses of the drugs will be needed to prevent rejection of cell transplants than are required for whole organ transplants.

Because of this, fewer and less severe side effects from immunosuppressive drugs would be expected, but definitive studies are lacking.

BACKGROUND STUDIES

Choice of Sites for Hepatocyte Transplantation

Hepatocyte transplants have been conducted for over 20 years. A number of good reviews are available for details of the experiments and the original references which may be omitted in this review (Strom et al., 1999, 2006; Malhi and Gupta, 2001; Ohashi et al., 2001; Fox, 2002; Fox and Roy-Chowdhury, 2004). The large numbers of preclinical studies conducted on hepatocyte transplants firmly establish that the transplants are safe and effective. The most common sites for the transplantation of hepatocytes are the spleen and the liver; however, transplants to the peritoneal cavity, stomach, or omentum have been reported. Long-term survival of the cells is readily measured following transplants into the spleen or liver. The majority of cells transplanted into the peritoneal cavity intellectual property (IP) are rapidly lost. Following IP transplants, only those cells which nidate near blood vessels and can attract sufficient nutrition survive long term. Despite the ease of the procedure, IP transplants of hepatocytes have only limited efficacy. Transplants of hepatocytes to the spleen or the liver have been shown to function for the lifetime of the recipient (Mito et al., 1979; Gupta et al., 1991; Ponder et al., 1991; Holzman et al., 1993). Studies by Mito and coworkers clearly show long-term survival of hepatocytes and that over time the spleen of an animal can be "hepatized" to where 80% of the mass of the organ can replaced with hepatocytes (Mito et al., 1978, 1979; Kusano et al., 1981, 1992; Kusano and Mito, 1982).

The concept of establishing ectopic liver function in the spleen is similar in theory to the bioartificial liver (BAL). In BAL, the hepatocytes are seeded into and maintained in some form of an extracorporal device. The patient's blood or plasma is pumped to the device where it interacts with the hepatocytes across membrane barriers and is then returned to the patient by a second series of pumps. There are reports that BAL can provide short-term synthetic and metabolic support (Gerlach et al., 2003; Demetriou et al., 2004). The ease of transplant of hepatocytes and the abundance of the patient's own natural basement membrane components coupled with the naturally high blood flow make the spleen a useful site for the establishment of short- or long-term ectopic liver function. It is likely that hepatocyte transplants will be easier, cheaper, more efficient, and will provide the same, or better, level of support as extracorporal devices.

For transplants into the liver, the preferred route for administration of cells is via the portal vein. Cells are infused into the blood supply which feeds the liver and the hepatocytes are distributed to the different lobes in proportion to the blood flow they receive from the portal vein. Portal vein injections are difficult in small animals, so an alternative method is used in these studies. Hepatocytes are injected directly into the splenic pulp. The proportion of the cells which remain in the spleen is determined by the extent to which the outflow through splenic veins is impeded. In the studies of Mito et al. (1979), where the spleen was "hepatized" the authors briefly occlude the splenic outflow which helps retain the cells in the spleen. Alternatively, when the spleen is used as a method to affect a portal vein injection, the splenic veins are left open. It was reported that up to 52% of the cells injected into the spleen traverse to the liver via the splenic and portal veins within a few minutes (Gupta et al., 1991; Ponder et al., 1991).

Integration of Hepatocytes Following Transplantation

Integration of hepatocytes into recipient liver is a complex process which requires the interaction of donor and native hepatocytes to form an integrated tissue. The process may be considered in four steps (Table 27.2) (Gupta et al., 1995, 1999b, 2000; Koenig et al., 2005). Although they are presented as separate, there is considerable overlap of the steps in both time and space. Some of the most spectacular photographs of the entire

Table 27.2 Integration of donor hepatocytes into native liver following transplantation

Filling vascular spaces with donor cells
Disruption of the sinusoidal endothelium
Donor cell integration in host parenchyma
Remodeling of liver via modulation of extracellular matrix

process are provided by Koenig et al. (2005). Following infusion into the portal vein hepatocytes must traverse the endothelium to escape the vascular system. Although the liver has fenestrated endothelium, under normal conditions the pores which are in the range of 150 nm are far too small to provide a simple transit of parenchymal hepatocytes which range in size of 20–50 μm. Infusions of hepatocytes quickly fill the portal veins and embolize secondary and tertiary portal radicals (Gupta et al., 1999a). Portal pressures increase as flow is restricted by hepatocyte plugs in the portal veins. Venograms which were normal prior to cell transplantation become markedly attenuated and show greater filling of vessels proximal to the portal vein including the mesenteric and splenic vein. If the number of hepatocytes transplanted is in the range of 5% of the total number of hepatocytes in the native liver, the portal hypertension is transient and resolves within minutes to hours.

A proportion of transplanted cells begin to fill sinusoidal spaces and the space of Disse as the endothelium in the region of the transplanted cells begins to degenerate. It is likely that both physical and humoral (growth factors, cytokines) factors are involved in this process. Microscopic analysis of tissue sections reveal that endothelium is breached in many places and donor hepatocytes leave the portal veins in regions where endothelium is incomplete and broken. Reports suggest that most of the hepatocytes which eventually integrate into recipient liver will have traversed the endothelial barrier by 24 h post transplant. Cells which remain in the portal vessels are eventually removed by macrophages between 16 and 24 h post transplant. Other reports suggest that cells may continue to integrate into parenchyma for 2–3 days following transplantation (Shani-Peretz et al., 2005). Transient hypoxia in the region of the occluded vessels leads to changes in both the endothelium as well as both recipient and donor hepatocytes. Endothelium and donor and native hepatocytes all express vascular endothelial growth factor (VEGF) in the areas of hepatocyte integration (Gupta et al., 1999b; Shani-Peretz et al., 2005) a factor known to be induced by hypoxia. It is interesting that VEGF was previously known as vascular permeability factor (VPF). Expression and secretion of VEGF/VPF a potent angiogenesis factor is thought to contribute to the reformation of new sinusoids and restoration of the endothelial barrier following cell transplantation.

Passage through the endothelial barrier allows donor hepatocytes to become integrated into recipient parenchyma. Full integration of donor hepatocytes and restoration of full hepatic function is difficult to ascertain. However, careful studies of the expression of antigens and activities localized to specific membrane fractions clearly demonstrate that donor hepatocytes fully integrate into the hepatic plate of native liver and for hybrid structures between native and donor cells within 3–5 days following transplantation. The antibody to CD26 recognizes the dipeptidylpeptidase IV (DPPIV) antigen which is localized to the basolateral membrane of hepatocytes. Antibodies to connexin 32 can be used to visualize gap junctions between adjacent hepatocytes. Likewise, canicular ATPase activity can be used to identify bile cannicular regions between adjacent hepatocytes. The proper localization of these different antigens and activities requires that the hepatocyte be fully integrated into the hepatic plate and polarized. By 3–7 days post transplant, hybrid structures could be visualized in recipient liver containing both donor (DPPIV) hepatocytes and recipient ATPase activity (Gupta et al., 1995) or donor DPPIV co-localized with connexin 32 (Koenig et al., 2005). Both studies clearly demonstrate proper integration of donor hepatocytes as well as the reestablishment of intracellular communication (connexin 32) between donor and recipient hepatocytes. Hybrid structures between donor and recipient hepatocytes were shown to be functional as shown by the transport and excretion of a fluorescent conjugated bile acid (Gupta et al., 1995). Hepatic transport of indocyanine and sulfobromothalein into the bile following hepatocyte transplantation was also reported by

Table 27.3 Opportunities for hepatocyte transplantation

"Bridge" for patients to whole organ transplantation
Cell support for acute liver failure
"Cell therapy" for metabolic disease

D-galactosamine (Sutherland et al., 1977; Sommer et al., 1979; Makowka et al., 1981; Baumgartner et al., 1983); 90% hepatectomy (Cuervas-Mons et al., 1984; Demetriou et al., 1988; Mito et al., 1993; Kobayashi et al., 2000), or ischemic liver injury (Takeshita et al., 1993; Arkadopoulos et al., 1998a).

There are now reports of reversal of acute liver failure in four patients following hepatocyte transplantation (Fisher et al., 2000; Soriano, 2002; Fisher and Strom, 2006; Ott et al., 2006). The causes of acute liver failure ranged from hepatitis B-induced liver failure to acetaminophen intoxication, to liver toxicity following eating poisonous mushrooms to liver failure of unknown etiology in a pediatric patient. In each case patients presented with classic symptoms of acute liver failure, and most were immediately listed for OLT. The number of cells transplanted varied between different procedures but ranged from approximately 1 to 5 billion total viable cells. In all cases cells were transplanted into the portal vein to get a direct transplant into the liver. In general, patients were given fresh frozen plasma prior to placement of the catheter to prevent bleeding. The results presented by Fisher et al. (2000) are typical of the response to hepatocyte transplantation. There is usually a rapid fall in ammonia levels following the transplant. Circulating levels of clotting factors stabilize following the transplant and then slowly increase over the next 2 weeks. Fisher et al. reports that Factor VII levels were 1% of normal prior to transplant and increased to 25% by 7 days and 64% of normal by week 2 post cell transplant. The recovery of the clotting factors is usually rapid enough that following the cell transplant, no additional fresh frozen plasma is required.

Patients are generally discharged within 2–4 weeks and are judged to experience a complete recovery. The cell transplant recipients ranged in age from 3 to 64 years in age, indicating that even older patients have sufficient regenerative capacity to be supported by hepatocyte transplantation.

As is observed with donor tissue allografts, hepatocyte allografts produce and secrete human leukocyte antigen-I (sHLA-I) immediately upon implantation. If there is a mismatch between the donor and recipient the donor specific sHLA-I can be detected in the circulation and quantified by enzyme-linked immunosorbent assay (ELISA). Donor specific HLA class I alleles can be identified and quantified by polymerase chain reaction (PCR) analysis of tissue samples taken at biopsy. When it is determined that the preponderance of cells in the patients liver are native, the patients can slowly be removed from immunosuppressive therapy as was described by Fisher et al. (2000). In the cases described to date, the patients recovered completely from liver failure following hepatocyte transplantation without serious adverse consequences and without whole organ transplant and lifelong immunosuppression. Although the numbers of patients are small, the treatment of acute liver failure by hepatocyte transplant has some significant advantages which make further investigation of this novel therapy appropriate (Table 27.3).

Hepatocyte Transplantation for Metabolic Liver Disease

A common indication for whole organ transplantation in pediatric patients is metabolic liver disease. In these cases, there is usually a genetic defect in an enzyme or protein which is produced in the liver which inactivates a critical liver function. Although all other liver functions are generally normal, the liver is removed and replaced with a liver which can perform the missing function. Because there is usually only one genetic defect associated with each metabolic liver disease, a gene therapy approach to correct the defect would seem appropriate. Unfortunately, gene therapy has met with considerable problems which have prevented successful use of this experimental technique. Hepatocyte transplantation has been used in attempts to correct the metabolic defects associated with several types of metabolic liver disease (Table 27.4).

Table 27.4 Clinical transplants for metabolic liver disease

Familial hypercholesterolemia (3)
Crigler–Najjar(5)
Ornithine transcarbamylase deficiency (4)
Arginosuccinate lyase deficiency (1)
Factor VII deficiency (2)
Glycogen storage disease (2)
Infantile Refsum disease (1)
Progressive familial intrahepatic cholestasis (2)
Alpha-1 antitrypsin deficiency (2)

In an approach similar to gene therapy, with hepatocyte transplants one tries to seed the patient's liver with cells which are proficient in the enzyme or function missing in the native liver. The goal is to repopulate the liver of the transplant recipient with sufficient numbers of hepatocytes to provide the missing liver function by donor cells.

Large numbers of hepatocytes cannot be infused into the portal system because of the problems with embolism of the portal veins and portal hypertension. We have used as a general rule to infuse approximately 2×10^8 cells/kg body weight of the recipient (Fox et al., 1998; Horslen et al., 2003). Infusions of these cell numbers has not resulted in any long-term complications. There is always a transient increase in portal pressures which resolves within hours (Strom et al., 1997a; Fox et al., 1998; Bohnen et al., 2000; Soriano, 2002; Horslen et al., 2003; Sokal et al., 2003; Horslen and Fox, 2004). While quite experimental, this number was arrived at by an extrapolation from preclinical studies with non-human primates. Grossman et al. (1992) reported that the infusion of between $1–2 \times 10^8$ cells/kg into baboons who had previously received a left or right lobectomy was accomplished without serious complications and with only transient increases in portal pressures. Because only a few percent of liver mass can be transplanted at any one time, single hepatocyte transplants cannot be expected to replace a large percentage of liver with donor cells. For this reason, the metabolic diseases which are candidates for cell transplants are those in which the restoration of 10% less of total liver function or activity is likely to correct the disease. The liver has highly redundant functions. Thus, it is recognized that 10% of a normal amount of gene product or enzyme activity would likely correct the symptoms of most metabolic liver diseases. Exceptions exist, like hypercholesterolemia, where more than 50% replacement of liver with donor cells would likely be needed to correct circulating low density lipoprotein levels. However, for most metabolic liver disease and all of those listed in Table 27.4, it is believed that the replacement of the liver with 10% donor hepatocytes would either be completely corrective or at least ameliorate most of the symptoms of the disease.

In general, hepatocyte transplants work best when the donor cells have a selective growth advantage. There are a number of animal models of liver disease where the native hepatocytes show an increased death rate as compared to normal liver (Sandgren et al., 1991; Rhim et al., 1994; Overturf et al., 1996; De Vree et al., 2000). In these situations, when cells without the defect are transplanted into the diseased liver, the donor cells have a strong and selective growth advantage as compared to the native hepatocytes. Over time the liver may become nearly completely replaced with donor cells. In certain human diseases there might be sufficient selective pressure to strongly favor the replacement of large parts of the liver with donor cells. Such diseases include tyrosinemia Type 1, Wilson's disease (Irani et al., 2001), progressive familial intrahepatic cholestasis (PFIC) (De Vree et al., 2000), alpha-1 antitrypsin deficiency (A1AT) (Rudnick and Perlmutter, 2005). In these diseases, integration of only a small proportion of liver mass by hepatocyte transplantation would likely be necessary because the donor cells would be expected to continue to proliferate in the host liver, and over time replace the diseased cells. Although there are clear examples of this in studies of transplants of laboratory animals, there are no studies with human patients showing comparable results.

Most metabolic diseases such as Crigler–Najjar (CN), OTC deficiency, and all of those diseases listed in Table 27.4 would not be expected to show such selective growth pressure for donor cells. For diseases such as these, multiple transplants over time will be required to populate the liver with 10% donor cells (Rozga et al., 1995).

A large number of studies with different animal models have shown the efficacy of hepatocyte transplantation to correct metabolic liver disease (reviewed in Malhi and Gupta, 2001 and Strom et al., 2006). Metabolic defects in bilirubin metabolism (Matas et al., 1976; Groth et al., 1977; Vroemen et al., 1986; Demetriou et al., 1988; Moscioni et al., 1989; Holzman et al., 1993; Hamaguchi et al., 1994), albumin secretion (Mito et al., 1979; Kusano and Mito, 1982; Demetriou et al., 1993; Rozga et al., 1995; Moscioni et al., 1996; Oren et al., 1999), ascorbic acid production (Onodera et al., 1995; Nakazawa et al., 1996), tyrosinemia Type 1 (Overturf et al., 1996), copper excretion (Yoshida et al., 1996; Irani et al., 2001; Allen et al., 2004), PFIC (De Vree et al., 2000) as well as other defects in biliary transport similar to Dubin–Johnson syndrome in humans (Hamaguchi et al., 1994) have been shown to be amenable to correction by hepatocyte transplantation. These encouraging results suggested that similar defects in human patients could be corrected by hepatocyte transplantation. The diseases listed in Table 27.4 have been the focus of human trials of hepatocyte transplants. The numbers in parenthesis are the number of patients who have received transplants.

Hepatocyte transplants were previously shown to result in a rapid correction of ammonia levels (Strom et al., 1997b, 1999; Bilir et al., 2000; Soriano, 2002). For this reason, urea cycle defects which result in life-threatening hyperammonemia were the first metabolic disease target for hepatocyte transplants (Strom et al., 1997b; Bohnen et al., 2000). In the initial study, 1 billion viable cells were transplanted into the portal vein of a 5-year-old recipient. Portal pressures increased from 11 cm of water prior to cell transplant to 19 cm immediately following the cell infusion, but recovered rapidly. The patient's ammonia levels normalized without medical intervention within 48 h of cell infusion and his glutamine levels returned to normal. Although OTC activity was undetectable prior to cell transplant, measurable OTC activity was detected in a biopsy performed at 28 days. In these studies 10% of the cells were labeled with indium^{111} prior to infusion into the patient to monitor distribution of the cells. Quantitative analysis of the scientigraphic images showed an average distribution ratio of liver:spleen of 9.5:1. Measurements made prior to cell infusion indicated that free indium was released from hepatocytes at a rate of 10% per hour, and free indium is rapidly cleared from circulation by reticuloendothelial systems such as the spleen. Thus, most of the tracer in the spleen following cell infusion were thought to be free indium, not hepatocytes. Pulmonary radiotracer uptake was consistent with background counts, indicating the absence of portosystemic shunting despite the modest increase in portal pressures observed at the time of transplant. This first transplant for metabolic liver disease indicated that hepatocyte transplantation into the portal vein could be conducted safely in patients with no significant liver pathology with only a moderate and reversible increase in portal pressures. From the rapid normalization of ammonia levels following hepatocyte transplant, it was concluded that cell transplantation can partially correct the hyperammonemia associated with the disease. Subsequent studies have verified that partial corrections of ammonia levels are possible by cell transplants alone (Horslen et al., 2003; Dhawan et al., 2004; Stephenne et al., 2005). While complete corrections of OTC deficiency have not been accomplished these studies indicate that cell transplants provide much needed metabolic control of ammonia levels. Even in the absence of complete correction, liver cell transplantation should be considered as a bridge to whole organ transplantation for OTC patients to prevent the neurological problems associated with uncontrolled hyperammonemia (Bohnen et al., 2000; Stephenne et al., 2005).

A number of groups have attempted to correct CN syndrome, Type 1 with hepatocyte transplants. The first case was in many ways typical of the results obtained by other groups and will be discussed in greater detail (Fox et al., 1998). This disease is caused by a defect in the enzyme which is responsible for the conjugation and eventual excretion of bilirubin. The absence of the enzyme results in severe hyperbilirubinemia which can lead to central nervous system (CNS) toxicity including kernicterus. Following the transplantation of approximately 7.5 billion cells into the liver of a 10-year-old female, there was a slow and continuous decrease in circulating bilirubin levels over the first 30–40 days, and bilirubin conjugates were readily detected in the bile. Overall, there was approximately a 60–65% decrease in bilirubin levels as compared to pretransplant levels. Because the bilirubin conjugates could only be produced by the donor cells, their detection in the bile demonstrates the robust biochemical function of the transplanted cells and established that donor hepatocytes integrated into the hepatic parenchyma and quickly established connections with the recipient's biliary tree.

Several important finding were gained from this transplant. First, large numbers of hepatocytes could be safely transplanted into the portal vein without complication. Although the total numbers of hepatocytes in liver are difficult to assess, a transplant of 7.5 billion cells represents and estimated 3.5–7.5% of the liver mass, which was transplanted without complication over approximately a 15-h period. Second, the apparent engraftment and function of hepatocytes in the clinical trials seems to exceed that found in previous animal studies. The transplantation of 3.5–7.5% of liver mass resulted in the restoration of approximately 5% of a normal amount of bilirubin conjugation capacity in the liver. Third, a long-term correction in bilirubin levels was observed. This patient was followed for more than 1.5 years. Fourth, single transplants of hepatocytes are effective in creating partial corrections of the disease, but given the limitation of transplanting 2×10^8 cells/kg body weight, one cannot transplant sufficient numbers of hepatocytes to achieve a complete correction of metabolic liver disease with one transplant. It is estimated that complete corrections would require 2–4 transplants if each were as successful and efficient as the first. Finally, this was the first unequivocal demonstration of the long-term success of hepatocyte transplantation. Although patients were bridged to transplant and clinical parameters such as ammonia levels rapidly changed following transplantation, many of the previous patients underwent subsequent OLT the long-term metabolic function of the transplanted cells was difficult to assess. These studies firmly established that hepatocyte transplants were an effective means to correct metabolic liver disease. The results of hepatocyte transplants of other patients with CN largely confirm those seen with the first patient (Dhawan et al., 2004; Ambrosino et al., 2005).

Muraca et al. (2002) reported partial correction of glycogen storage disease, Type 1 following hepatocyte transplantation. Improvement was documented by the patient's ability to maintain blood glucose between meals as well as sustained and higher glucose levels with meals. Sokal et al. (2003) employed hepatocyte transplants to achieve a partial correction of infantile Refsum disease an autosomal recessive inborn error in peroxisome metabolism of very-long chain fatty acid metabolism, bile acid, and pipecolic acid. The authors reported improvement in fatty acids metabolism, a reduction in circulating pipecolic acid and bile salt levels. An overall improvement in the health of the patient was evident by the report of significant increase in muscle strength and weight gain. Dhawan et al. (2004) reported that hepatocyte transplantation partially corrected a severe deficiency in the production and secretion of coagulation Factor VII. Following cell transplant, the Factor VII requirement decreased nearly 80% of that administered prior to HTx. Most recently, Stephenne et al. (2006) reported the complete correction of a 3.5-year-old female patient with neonatal onset arginosuccinate lyase (ASL) deficiency. Like OTC deficiency, ASL patients are at risk of brain damage from hyperammonemia. The patient received three-sequential hepatocyte transplants over a 5-month period. Both freshly isolated and previously cryopreserved hepatocytes were used. At 1 year post transplant the patient displayed 3% of normal ASL activity in hepatic biopsy samples. Engraftment of donor cells could be demonstrated by fluorescence *in situ* hybridization for Y chromosome. These results confirm that hepatocyte transplantation can achieve sustained engraftment of donor cells and sustained metabolic and clinical control.

HEPATOCYTE TRANSPLANTATION NOVEL USES, CHALLENGES, AND FUTURE DIRECTIONS

Hepatocyte Transplants for Non-organ Transplant Candidates

Most of the patients who have received a hepatocyte transplant were already listed for a whole organ transplant. The need for liver support is not limited to this group. There are large numbers of patients for whom OLT is not an option. Patients in this group could include alcoholic cirrhotic patients who have not met the required abstinence period, acute liver failure patients resulting from suicide attempts and cancer patients. Early case reports suggested that hepatocyte transplants into the spleen could be useful to restore liver function to end-stage cirrhotic patients (Strom et al., 1999). Although both of the patients in the reported study eventually died of concomitant renal failure which was left untreated, the patients were sufficiently improved following the cell transplants that they were able to be discharged from the hospital. Fox and coworkers created an animal model to study the efficacy of hepatocyte transplants to support liver function in cirrhosis in a more controlled setting. Their studies clearly demonstrated that hepatocyte transplants significantly improve liver function and survival of rats experiencing chronic liver failure following repeated injections of carbon tetrachloride (Ahmad et al., 2002; Cai et al., 2002; Nagata et al., 2003a). With millions of patients currently infected with hepatitis viruses there is clearly a need for additional means to support liver function in these patients. Not withstanding the difficulties of such clinical studies in cirrhotic patients, cell transplantation should be thoroughly evaluated as possible support therapy.

The single most important factor preventing the use of hepatocyte transplants in additional medical centers is the limited availability of hepatocytes. The normal source of cells for hepatocyte transplants are livers with greater than 50% steatosis, vascular plaques, or other factors which render the tissue unsuitable for whole organ transplantation (Strom et al., 1997a, b; Fox et al., 1998; Bilir et al., 2000; Fisher et al., 2000; Muraca et al., 2002; Nakazawa et al., 2002; Soriano, 2002; Horslen et al., 2003; Mitry et al., 2003; Strom and Fisher, 2003; Ott et al., 2004). Hepatocyte transplants will not be able to progress past the small proof of concept studies in humans until sufficient numbers of hepatocytes become available (Strom and Fisher, 2003). Xenotransplants (Nagata et al., 2003a), immortalized human hepatocytes (Kobayashi et al., 2000; Cai et al., 2002; Wege et al., 2003a, b) and stem cell-derived hepatocytes (Avital et al., 2002; Miki et al., 2002, 2005; Davila et al., 2004; Ruhnke et al., 2005) and fetal hepatocytes have been proposed as alternative sources of cells for clinical transplants. To date, no alternative cell source has been found which meets all of the requirements for safety and efficacy. Because of the increased interest in stem cell-derived hepatocytes and scientific investigations into their production, it is likely that they will be a significant source of cells for future hepatocyte transplants.

Better utilization of existing liver tissue could increase the numbers of hepatocytes available immediately. In the United States there are no regulations requiring that donor organs be allocated to transplantation research centers for hepatocyte isolation, and relatively few organs go to centers where hepatocyte transplant is a possibility. Most of the organs not used for whole organ transplant are provided to commercial firms where hepatocytes are isolated for resale or for in-house metabolism and toxicology studies. While most uses of donor liver tissue have merit, simple allocation procedures could be instituted to route the organs to transplant centers for initial review and selection of the most suitable cases for cell isolation. Split-liver procedures have made it possible to use caudate lobe and segment IV for hepatocyte isolation. Depending on the surgical procedure, these portions of liver tissue may remain untransplanted and have been shown to be useful for hepatocyte isolation (Mitry et al., 2004). Although, currently quite hypothetical, in the future most or all livers which are currently transplanted could be split. A portion such as the left lateral segment or the entire left lobe could be made available for cell isolation while the remaining liver tissue is utilized as a tissue graft. Because hepatocyte transplantation is not currently the standard of care, such proposals are not currently feasible. However, if the efficacy of hepatocyte transplants were firmly established, the risk and the extra time needed for the split procedure would be outweighed by the benefit of the cell transplants. Cell transplants rather than OLT could free-up the organs which are now used for acute liver failure and metabolic disease patients.

SUMMARY

Hepatocyte transplantation studies conducted in animal models of liver failure and liver-based metabolic disease have proven safe and effective means to provide short- or long-term synthetic and metabolic support of liver function. For certain organ transplant candidates such as those with metabolic liver disease, cell transplantation alone could provide relief of the clinical symptoms. Cell transplant studies in patients with acute or chronic liver failure or genetic defects in liver function clearly demonstrate the efficacy of hepatocyte transplantation to treat liver disease. In virtually all cases a clinical improvement in the condition of the patient could be documented. No serious complications of hepatocyte transplant have been reported. Although all of the initial reports concerning hepatocyte transplants are encouraging, it must be realized that there are still no reports of long-term and complete corrections of any metabolic disease in patients. The recent report of a complete correction of a patient with a urea cycle defect is most encouraging; however, the length of time that human hepatocytes will function following transplantation has not been determined. Studies in animal models of liver disease have documented that donor hepatocytes transplanted into the spleen or the liver function for the lifetime of the recipient and participate in normal regenerative events. Although it is likely that human hepatocyte transplantation will result in lifelong and normal function of donor cells, this needs to be clearly demonstrated in a clinical study.

Future work will have to be conducted to establish optimal transplant and immunosuppression protocols to minimize complications and maximize engraftment and function. A major problem for clinical hepatocyte transplant is the inability to track donor cells following transplantation. Except for the short-term tracking of hepatocytes pre-labeled with radioactive substances such as indium^{111} (Bohnen et al., 2000), there are no reports of quantitative and facile methods to detect donor cells. Relatively non-invasive methods will be needed to optimize transplant and immunosuppressive protocols as well as for day-to-day monitoring of the cell graft. None of the problems cited here seem insurmountable. There are now reports of successful hepatocyte transplants from laboratories in many different countries. The cooperative spirit which has developed between the investigators at the different transplant centers should benefit the research field and especially the future recipients of hepatocyte transplants.

REFERENCES

Ahmad, T.A., Eguchi, S., Yanaga, K., Miyamoto, S., Kamohara, Y., Fujioka, H., Furui, J. and Kanematsu, T. (2002). Role of intrasplenic hepatocyte transplantation in improving survival and liver regeneration after hepatic resection in cirrhotic rats. *Cell Transplant.* **11**: 399–402.

Allen, K.J., Cheah, D.M., Wright, P.F., Gazeas, S., Pettigrew-Buck, N.E., Deal, Y.H., Mercer, J.F. and Williamson, R. (2004). Liver cell transplantation leads to repopulation and functional correction in a mouse model of Wilson's disease. *J. Gastroenterol. Hepatol.* **19**: 1283–1290.

Ambrosino, G., Varotto, S., Strom, S.C., Guariso, G., Franchin, E., Miotto, D., Caenazzo, L., Basso, S., Carraro, P., Valente, M.L., D'Amico, D., Zancan, L. and D'Antiga, L. (2005). Isolated hepatocyte transplantation for Crigler-Najjar syndrome type 1. *Cell Transplant.* **14**: 151–157.

Aoki, T., Jin, Z., Nishino, N., Kato, H., Shimizu, Y., Niiya, T., Murai, N., Enami, Y., Mitamura, K., Koizumi, T., Yasuda, D., Izumida, Y., Avital, I., Umehara, Y., Demetriou, A.A., Rozga, J. and Kusano, M. (2005). Intrasplenic transplantation of encapsulated hepatocytes decreases mortality and improves liver functions in fulminant hepatic failure from 90% partial hepatectomy in rats. *Transplantation* **79**: 783–790.

Arkadopoulos, N., Chen, S.C., Khalili, T.M., Detry, O., Hewitt, W.R., Lilja, H., Kamachi, H., Petrovic, L., Mullon, C.J., Demetriou, A.A. and Rozga, J. (1998a). Transplantation of hepatocytes for prevention of intracranial hypertension in pigs with ischemic liver failure. *Cell Transplant.* **7**: 357–363.

Arkadopoulos, N., Lilja, H., Suh, K.S., Demetriou, A.A. and Rozga, J. (1998b). Intrasplenic transplantation of allogeneic hepatocytes prolongs survival in anhepatic rats. *Hepatology* **28**: 1365–1370.

Avital, I., Feraresso, C., Aoki, T., Hui, T., Rozga, J., Demetriou, A. and Muraca, M. (2002). Bone marrow-derived liver stem cell and mature hepatocyte engraftment in livers undergoing rejection. *Surgery* **132**: 384–390.

Baumgartner, D., LaPlante-O'Neill, P.M., Sutherland, D.E. and Najarian, J.S. (1983). Effects of intrasplenic injection of hepatocytes, hepatocyte fragments and hepatocyte culture supernatants on D-galactosamine-induced liver failure in rats. *Eur. Surg. Res.* **15**: 129–135.

Bilir, B.M., Guinette, D., Karrer, F., Kumpe, D.A., Krysl, J., Stephens, J., McGavran, L., Ostrowska, A. and Durham, J. (2000). Hepatocyte transplantation in acute liver failure. *Liver Transplant.* **6**: 32–40.

Bohnen, N.I., Charron, M., Reyes, J., Rubinstein, W., Strom, S.C., Swanson, D. and Towbin, R. (2000). Use of indium-111-labeled hepatocytes to determine the biodistribution of transplanted hepatocytes through portal vein infusion. *Clin. Nucl. Med.* **25**: 447–450.

Cai, J., Ito, M., Nagata, H., Westerman, K.A., Lafleur, D., Chowdhury, J.R., Leboulch, P. and Fox, I.J. (2002). Treatment of liver failure in rats with end-stage cirrhosis by transplantation of immortalized hepatocytes. *Hepatology* **36**: 386–394.

Cuervas-Mons, V., Cienfuegos, J.A., Maganto, P., Golitsin, A., Eroles, G., Castillo-Olivares, J. and Segovia de Arana, J.M. (1984). Time-related efficacy of liver cell isografts in fulminant hepatic failure. *Transplantation* **38**: 23–25.

Davila, J.C., Cezar, G.G., Thiede, M., Strom, S., Miki, T. and Trosko, J. (2004). Use and application of stem cells in toxicology. *Toxicol. Sci.* **79**: 214–223.

De Vree, J.M., Ottenhoff, R., Bosma, P.J., Smith, A.J., Aten, J. and Oude Elferink, R.P. (2000). Correction of liver disease by hepatocyte transplantation in a mouse model of progressive familial intrahepatic cholestasis. *Gastroenterology* **119**: 1720–1730.

Demetriou, A.A., Reisner, A., Sanchez, J., Levenson, S.M., Moscioni, A.D. and Chowdhury, J.R. (1988). Transplantation of microcarrier-attached hepatocytes into 90% partially hepatectomized rats. *Hepatology* **8**: 1006–1009.

Demetriou, A.A., Holzman, M., Moscioni, A.D. and Rozga, J. (1993). Hepatic cell transplantation. *Adv. Vet. Sci. Comp. Med.* **37**: 313–332.

Demetriou, A.A., Brown Jr., R.S., Busuttil, R.W., Fair, J., McGuire, B.M., Rosenthal, P., Am Esch II, J.S., Lerut, J., Nyberg, S.L., Salizzoni, M., Fagan, E. A., de Hemptinne, B., Broelsch, C.E., Muraca, M., Salmeron, J.M., Rabkin, J.M., Metselaar, H.J., Pratt, D., De La Mata, M., McChesney, L.P., Everson, G.T., Lavin, P.T., Stevens, A.C., Pitkin, Z. and Solomon, B.A. (2004). Prospective, randomized, multicenter, controlled trial of a bioartificial liver in treating acute liver failure. *Ann. Surg.* **239**: 660–667; discussion 667–670.

Dhawan, A., Mitry, R.R., Hughes, R.D., Lehec, S., Terry, C., Bansal, S., Arya, R., Wade, J.J., Verma, A., Heaton, N.D., Rela, M. and Mieli-Vergani, G. (2004). Hepatocyte transplantation for inherited factor VII deficiency. *Transplantation* **78**: 1812–1814.

Fisher, R.A. (2004). Adult human hepatocyte transplantation. *7th International Congress of Cell Transplantation Society, Boston* 304.

Fisher, R.A. and Strom, S.C. (2006). Human hepatocyte transplantation: worldwide results. *Transplantation* **82**: 441–449.

Fisher, R.A., Kimball, P.M., Thompson, M., Saggi, B., Wolfe, L. and Posner, M. (1998). An immunologic study of liver failure in patients bridged with hepatocyte infusions. *Transplantation* **65**: S45.

Fisher, R.A., Bu, D., Thompson, M., Tisnado, J., Prasad, U., Sterling, R., Posner, M. and Strom, S. (2000). Defining hepatocellular chimerism in a liver failure patient bridged with hepatocyte infusion. *Transplantation* **69**: 303–307.

Fox, I.J. (2002). Transplantation into and inside the liver. *Hepatology* **36**: 249–251.

Fox, I.J. and Roy-Chowdhury, J. (2004). Hepatocyte transplantation. *J. Hepatol.* **40**: 878–886.

Fox, I.J., Chowdhury, J.R., Kaufman, S.S., Goertzen, T.C., Chowdhury, N.R., Warkentin, P.I., Dorko, K., Sauter, B.V. and Strom, S.C. (1998). Treatment of the Crigler-Najjar syndrome type I with hepatocyte transplantation. *N. Engl. J. Med.* **338**: 1422–1426.

Gerlach, J.C., Mutig, K., Sauer, I.M., Schrade, P., Efimova, E., Mieder, T., Naumann, G., Grunwald, A., Pless, G., Mas, A., Bachmann, S., Neuhaus, P. and Zeilinger, K. (2003). Use of primary human liver cells originating from discarded grafts in a bioreactor for liver support therapy and the prospects of culturing adult liver stem cells in bioreactors: a morphologic study. *Transplantation* **76**: 781–786.

Grossman, M., Raper, S.E. and Wilson, J.M. (1992). Transplantation of genetically modified autologous hepatocytes into nonhuman primates: feasibility and short-term toxicity. *Hum. Gene. Ther.* **3**: 501–510.

Groth, C.G., Arborgh, B., Bjorken, C., Sundberg, B. and Lundgren, G. (1977). Correction of hyperbilirubinemia in the glucuronyltransferase-deficient rat by intraportal hepatocyte transplantation. *Transplant. Proc.* **9**: 313–316.

Gupta, S., Aragona, E., Vemuru, R.P., Bhargava, K.K., Burk, R.D. and Chowdhury, J.R. (1991). Permanent engraftment and function of hepatocytes delivered to the liver: implications for gene therapy and liver repopulation. *Hepatology* **14**: 144–149.

Gupta, S., Yerneni, P.R., Vemuru, R.P., Lee, C.D., Yellin, E.L. and Bhargava, K.K. (1993). Studies on the safety of intrasplenic hepatocyte transplantation: relevance to ex vivo gene therapy and liver repopulation in acute hepatic failure. *Hum. Gene. Ther.* **4**: 249–257.

Gupta, S., Rajvanshi, P. and Lee, C. D. (1995). Integration of transplanted hepatocytes into host liver plates demonstrated with dipeptidyl peptidase IV-deficient rats. *Proc. Natl. Acad. Sci. USA* **92**: 5860–5864.

Gupta, S., Rajvanshi, P., Aragona, E., Lee, C.D., Yerneni, P.R. and Burk, R.D. (1999a). Transplanted hepatocytes proliferate differently after CCl_4 treatment and hepatocyte growth factor infusion. *Am. J. Physiol.* **276**: G629–G638.

Gupta, S., Rajvanshi, P., Sokhi, R., Slehria, S., Yam, A., Kerr, A. and Novikoff, P.M. (1999b). Entry and integration of transplanted hepatocytes in rat liver plates occur by disruption of hepatic sinusoidal endothelium. *Hepatology* **29:** 509–519.

Gupta, S., Rajvanshi, P., Malhi, H., Slehria, S., Sokhi, R.P., Vasa, S.R., Dabeva, M., Shafritz, D.A. and Kerr. (2000). Cell transplantation causes loss of gap junctions and activates GGT expression permanently in host liver. *Am. J. Physiol. Gastrointest. Liver Physiol.* **279**: G815–G826.

Habibullah, C.M., Syed, I.H., Qamar, A. and Taher-Uz, Z. (1994). Human fetal hepatocyte transplantation in patients with fulminant hepatic failure. *Transplantation* **58**: 951–952.

Hamaguchi, H., Yamaguchi, Y., Goto, M., Misumi, M., Hisama, N., Miyanari, N., Mori, K. and Ogawa, M. (1994). Hepatic biliary transport after hepatocyte transplantation in Eizai hyperbilirubinemic rats. *Hepatology* **20**: 220–224.

Holzman, M.D., Rozga, J., Neuzil, D.F., Griffin, D., Moscioni, A.D. and Demetriou, A.A. (1993). Selective intraportal hepatocyte transplantation in analbuminemic and Gunn rats. *Transplantation* **55**: 1213–1219.

Horslen, S.P. and Fox, I.J. (2004). Hepatocyte transplantation. *Transplantation* **77**: 1481–1486.

Horslen, S.P., McCowan, T.C., Goertzen, T.C., Warkentin, P.I., Cai, H.B., Strom, S.C. and Fox, I.J. (2003). Isolated hepatocyte transplantation in an infant with a severe urea cycle disorder. *Pediatrics* **111**: 1262–1267.

Irani, A.N., Malhi, H., Slehria, S., Gorla, G.R., Volenberg, I., Schilsky, M.L. and Gupta, S. (2001). Correction of liver disease following transplantation of normal rat hepatocytes into Long-Evans Cinnamon rats modeling Wilson's disease. *Mol. Ther.* **3**: 302–309.

Kobayashi, N., Fujiwara, T., Westerman, K.A., Inoue, Y., Sakaguchi, M., Noguchi, H., Miyazaki, M., Cai, J., Tanaka, N., Fox, I.J. and Leboulch, P. (2000). Prevention of acute liver failure in rats with reversibly immortalized human hepatocytes. *Science* **287**: 1258–1262.

Koenig, S., Stoesser, C., Krause, P., Becker, H. and Markus, P.M. (2005). Liver repopulation after hepatocellular transplantation: integration and interaction of transplanted hepatocytes in the host. *Cell Transplant.* **14**: 31–40.

Kusano, M. and Mito, M. (1982). Observations on the fine structure of long-survived isolated hepatocytes inoculated into rat spleen. *Gastroenterology* **82**: 616–628.

Kusano, M., Ebata, H., Onishi, T., Saito, T. and Mito, M. (1981). Transplantation of cryopreserved isolated hepatocytes into the rat spleen. *Transplant. Proc.* **13**: 848–854.

Kusano, M., Sawa, M., Jiang, B., Kino, S., Itoh, K., Sakata, H., Katoh, K. and Mito, M. (1992). Proliferation and differentiation of fetal liver cells transplanted into rat spleen. *Transplant. Proc.* **24**: 2960–2961.

Makowka, L., Rotstein, L.E., Falk, R.E., Falk, J.A., Zuk, R., Langer, B., Blendis, L.M. and Phillips, M.J. (1981). Studies into the mechanism of reversal of experimental acute hepatic failure by hepatocyte transplantation. 1. *Can. J. Surg.* **24**: 39–44.

Malhi, H. and Gupta, S. (2001). Hepatocyte transplantation: new horizons and challenges. *J Hepatobiliary Pancreat Surg* **8**: 40–50.

Matas, A.J., Sutherland, D.E., Steffes, M.W., Mauer, S.M., Sowe, A., Simmons, R.L. and Najarian, J.S. (1976). Hepatocellular transplantation for metabolic deficiencies: decrease of plasms bilirubin in Gunn rats. *Science* **192**: 892–894.

Michalopoulos, G.K. and DeFrances, M.C. (1997). Liver regeneration. *Science* **276**: 60–66.

Miki, T., Cai, H., Lehmann, T. and Strom, S. (2002). Production of hepatocytes from human amniotic stem cells. *Hepatology* **36**: 171A.

Miki, T., Lehmann, T., Cai, H., Stolz, D.B. and Strom, S.C. (2005). Stem cell characteristics of amniotic epithelial cells. *Stem Cells* **23**: 1549–1559.

Mito, M., Kusano, M., Onishi, T., Saito, T. and Ebata, H. (1978). Hepatocellular transplantation – morphological study on hepatocytes transplanted into rat spleen. *Gastroenterol Jpn.* **13**: 480–490.

Mito, M., Ebata, H., Kusano, M., Onishi, T., Saito, T. and Sakamoto, S. (1979). Morphology and function of isolated hepatocytes transplanted into rat spleen. *Transplantation* **28**: 499–505.

Mito, M., Kusano, M. and Sawa, M. (1993). Hepatocyte transplantation for hepatic failure. *Transplant Rev.* **7**: 35.

Mitry, R.R., Hughes, R.D., Aw, M.M., Terry, C., Mieli-Vergani, G., Girlanda, R., Muiesan, P., Rela, M., Heaton, N.D. and Dhawan, A. (2003). Human hepatocyte isolation and relationship of cell viability to early graft function. *Cell Transplant.* **12**: 69–74.

Mitry, R.R., Dhawan, A., Hughes, R.D., Bansal, S., Lehec, S., Terry, C., Heaton, N.D., Karani, J.B., Mieli-Vergani, G. and Rela, M. (2004). One liver, three recipients: segment IV from split-liver procedures as a source of hepatocytes for cell transplantation. *Transplantation* **77**: 1614–1616.

Moscioni, A.D., Roy-Chowdhury, J., Barbour, R., Brown, L.L., Roy-Chowdhury, N., Competiello, L.S., Lahiri, P. and Demetriou, A.A. (1989). Human liver cell transplantation. Prolonged function in athymic-Gunn and athymic-analbuminemic hybrid rats. *Gastroenterology* **96**: 1546–1551.

Moscioni, A.D., Rozga, J., Chen, S., Naim, A., Scott, H.S. and Demetriou, A.A. (1996). Long-term correction of albumin levels in the Nagase analbuminemic rat: repopulation of the liver by transplanted normal hepatocytes under a regeneration response. *Cell Transplant.* **5**: 499–503.

Muraca, M., Gerunda, G., Neri, D., Vilei, M.T., Granato, A., Feltracco, P., Meroni, M., Giron, G. and Burlina, A.B. (2002). Hepatocyte transplantation as a treatment for glycogen storage disease type 1a. *Lancet* **359**: 317–318.

Nagata, H., Ito, M., Cai, J., Edge, A.S., Platt, J.L. and Fox, I.J. (2003a). Treatment of cirrhosis and liver failure in rats by hepatocyte xenotransplantation. *Gastroenterology* **124**: 422–431.

Nagata, H., Ito, M., Shirota, C., Edge, A., McCowan, T.C. and Fox, I.J. (2003b). Route of hepatocyte delivery affects hepatocyte engraftment in the spleen. *Transplantation* **76**: 732–734.

Nakazawa, F., Onodera, K., Kato, K., Sawa, M., Kino, Y., Imai, M., Kasai, S., Mito, M., Matsushita, T. and Funatsu, K. (1996). Multilocational hepatocyte transplantation for treatment of congenital ascorbic acid deficiency rats. *Cell Transplant.* **5**: S23–S25.

Nakazawa, F., Cai, H., Miki, T., Dorko, K., Abdelmeguid, A., Walldorf, J., Lehmann, T. and Strom, S. (2002). Human hepatocyte isolation from cadaver donor liver. *In Proceedings of Falk Symposium, Hepatocyte Transplantation*, Vol. 126, Kouwer Academic Publishers, Lancaster, UK, pp. 147–158.

Oertel, M., Menthena, A., Dabeva, M.D. and Shafritz, D.A. (2006). Cell competition leads to a high level of normal liver reconstitution by transplanted fetal liver stem/progenitor cells. *Gastroenterology* **130**: 507–520; quiz 590.

Ohashi, K., Park, F. and Kay, M.A. (2001). Hepatocyte transplantation: clinical and experimental application. *J. Mol. Med.* **79**: 617–630.

Onodera, K., Kasai, S., Kato, K., Nakazawa, F. and Mito, M. (1995). Long-term effect of intrasplenic hepatocyte transplantation in congenitally ascorbic acid biosynthetic enzyme-deficient rats. *Cell Transplant.* **4**(Suppl 1): S41–S43.

Oren, R., Dabeva, M.D., Petkov, P.M., Hurston, E., Laconi, E. and Shafritz, D.A. (1999). Restoration of serum albumin levels in nagase analbuminemic rats by hepatocyte transplantation. *Hepatology* **29**: 75–81.

Ott, M.C., Barthold, M., Alexandrova, K., Griesel, C., Shchneider, A., Attaran, M., Arsenieva, M., Penkov, B., Net, M., Peralta, V., Bredehorn, T., Manyalich, M., Kafert-Kasting, S., Manns, M. P., Dimitrova, V., Nachkov, Y. and LArseniev, L. (2004). Isolation of human hepatocytes from donor organs under cgmp conditions and clinical application in patients with liver disease. *7th International Congress of Cell Transplantation Society, Boston* 142.

Ott, M., Schneider, A., Attaran, M. and Manns, M.P. (2006). Transplantation of hepatocytes in liver failure. *Dtsch. Med. Wochenschr.* **131**: 888–891.

Overturf, K., Al-Dhalimy, M., Tanguay, R., Brantly, M., Ou, C.N., Finegold, M. and Grompe, M. (1996). Hepatocytes corrected by gene therapy are selected *in vivo* in a murine model of hereditary tyrosinaemia type I. *Nat. Genet.* **12**: 266–273.

Ponder, K.P., Gupta, S., Leland, F., Darlington, G., Finegold, M., DeMayo, J., Ledley, F.D., Chowdhury, J.R. and Woo, S.L. (1991). Mouse hepatocytes migrate to liver parenchyma and function indefinitely after intrasplenic transplantation. *Proc. Natl Acad. Sci. USA* **88**: 1217–1221.

Rhim, J.A., Sandgren, E.P., Degen, J.L., Palmiter, R.D. and Brinster, R.L. (1994). Replacement of diseased mouse liver by hepatic cell transplantation. *Science* **263**: 1149–1152.

Rozga, J., Holzman, M., Moscioni, A.D., Fujioka, H., Morsiani, E. and Demetriou, A.A. (1995). Repeated intraportal hepatocyte transplantation in analbuminemic rats. *Cell Transplant.* **4**: 237–243.

Rudnick, D.A. and Perlmutter, D.H. (2005). Alpha-1-antitrypsin deficiency: a new paradigm for hepatocellular carcinoma in genetic liver disease. *Hepatology* **42**: 514–521.

Ruhnke, M., Nussler, A.K., Ungefroren, H., Hengstler, J.G., Kremer, B., Hoeckh, W., Gottwald, T., Heeckt, P. and Fandrich, F. (2005). Human monocyte-derived neohepatocytes: a promising alternative to primary human hepatocytes for autologous cell therapy. *Transplantation* **79**: 1097–1103.

Sandgren, E.P., Palmiter, R.D., Heckel, J.L., Daugherty, C.C., Brinster, R.L. and Degen, J.L. (1991). Complete hepatic regeneration after somatic deletion of an albumin-plasminogen activator transgene. *Cell* **66**: 245–256.

Shani-Peretz, H., Tsiperson, V., Shoshani, G., Veitzman, E., Neufeld, G. and Baruch, Y. (2005). HVEGF165 increases survival of transplanted hepatocytes within portal radicles: suggested mechanism for early cell engraftment. *Cell Transplant.* **14**: 49–57.

Sokal, E.M., Smets, F., Bourgois, A., Van Maldergem, L., Buts, J.P., Reding, R., Bernard Otte, J., Evrard, V., Latinne, D., Vincent, M.F., Moser, A. and Soriano, H.E. (2003). Hepatocyte transplantation in a 4-year-old girl with peroxisomal biogenesis disease: technique, safety, and metabolic follow-up. *Transplantation* **76**: 735–738.

Sommer, B.G., Sutherland, D.E., Matas, A.J., Simmons, R.L. and Najarian, J.S. (1979). Hepatocellular transplantation for treatment of D-galactosamine-induced acute liver failure in rats. *Transplant. Proc.* **11**: 578–584.

Soriano, H.E. (2002). Liver cell transplantation: human applications in adults and children. *In Proceedings of Falk Symposium, Hepatocyte Transplantation*, Vol. 126, pp. 99–115. Kouwer Academic Publishers, Lancaster, UK.

Soriano, H.E., Kang, D.C., Finegold, M.J., Hicks, M.J., Wang, N.D., Harrison, W. and Darlington, G.J. (1998). Lack of C/EBP alpha gene expression results in increased DNA synthesis and an increased frequency of immortalization of freshly isolated mice [correction of rat] hepatocytes. *Hepatology* **27**: 392–401.

Stephenne, X., Najimi, M., Smets, F., Reding, R., de Ville de Goyet, J. and Sokal, E.M. (2005). Cryopreserved liver cell transplantation controls ornithine transcarbamylase deficient patient while awaiting liver transplantation. *Am. J. Transplant.* **5**: 2058–2061.

Stephenne, X., Najimi, M., Sibille, C., Nassogne, M.C., Smets, F. and Sokal, E.M. (2006). Sustained engraftment and tissue enzyme activity after liver cell transplantation for argininosuccinate lyase deficiency. *Gastroenterology* **130**: 1317–1323.

Strom, S. and Fisher, R. (2003). Hepatocyte transplantation: new possibilities for therapy. *Gastroenterology* **124**: 568–571.

Strom, S.C., Fisher, R.A., Rubinstein, W.S., Barranger, J.A., Towbin, R.B., Charron, M., Mieles, L., Pisarov, L.A., Dorko, K., Thompson, M.T. and Reyes, J. (1997a). Transplantation of human hepatocytes. *Transplant. Proc.* **29**: 2103–2106.

Strom, S.C., Fisher, R.A., Thompson, M.T., Sanyal, A.J., Cole, P.E., Ham, J.M. and Posner, M.P. (1997b). Hepatocyte transplantation as a bridge to orthotopic liver transplantation in terminal liver failure. *Transplantation* **63**: 559–569.

Strom, S.C., Chowdhury, J.R. and Fox, I.J. (1999). Hepatocyte transplantation for the treatment of human disease. *Semin. Liver Dis.* **19**: 39–48.

Strom, S., Bruzzone, P., Cai, H., Ellis, E., Lehmann, T., Mitamura, K. and Miki, T. (2006). Hepatocyte Transplantation: Clinical Experience and Potential for Future Use. *Cell Transplantation.* **15:** S105–S110.

Sutherland, D.E., Numata, M., Matas, A.J., Simmons, R.L. and Najarian, J.S. (1977). Hepatocellular transplantation in acute liver failure. *Surgery* **82**: 124–132.

Takeshita, K., Ishibashi, H., Suzuki, M. and Kodama, M. (1993). Hepatocellular transplantation for metabolic support in experimental acute ischemic liver failure in rats. *Cell Transplant.* **2**: 319–324.

Vroemen, J.P., Buurman, W.A., Heirwegh, K.P., van der Linden, C.J. and Kootstra, G. (1986). Hepatocyte transplantation for enzyme deficiency disease in congenic rats. *Transplantation* **42**: 130–135.

Wege, H., Chui, M.S., Le, H.T., Strom, S. and Zern, M.A. (2003a). *In vitro* expansion of human hepatocytes is restricted by telomere-dependent replicative aging. *Cell Transplant.* **12**: 897–906.

Wege, H., Le, H.T., Chui, M.S., Liu, L., Wu, J., Giri, R., Malhi, H., Sappal, B.S., Kumaran, V., Gupta, S. and Zern, M.A. (2003b). Telomerase reconstitution immortalizes human fetal hepatocytes without disrupting their differentiation potential. *Gastroenterology* **124**: 432–444.

Yoshida, Y., Tokusashi, Y., Lee, G.H. and Ogawa, K. (1996). Intrahepatic transplantation of normal hepatocytes prevents Wilson's disease in Long-Evans cinnamon rats. *Gastroenterology* **111**: 1654–1660.

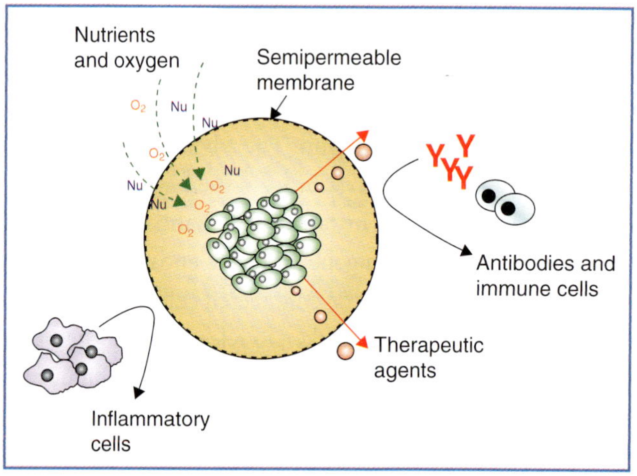

Figure 28.1 Schematic diagram of cell encapsulation. Nutrients and oxygen diffuse across the membrane, whereas inflammatory cells, antibodies, and immune cells are excluded.

and the continuous and controlled delivery of drugs. This technique consists of enclosing the biologically active material within a polymeric matrix surrounded by a semipermeable membrane that is designed to circumvent immune rejection. The capsule membrane allows the bi-directional diffusion of nutrients, oxygen, and waste and the secretion of the therapeutic product. It has the advantage of preventing immune cells and antibodies, which might destroy the enclosed cells, from entering the capsule (Figure 28.1).

The capsules deliver large molecular weight proteins like insulin through routes other than conventional injection. In this case, insulin is being manufactured not in the drug company's facilities, but in the transplanted cells, and insulin is delivered directly to the patient in response to glucose levels in the blood.

The rising concept of cell-based therapeutics requires advances in cell encapsulation technology, and there have been successful efforts in applying this technology for the treatment of human diseases including renal failure, neurological disorders, cancers, and liver diseases (Aebischer et al., 1991; Emerich et al., 1994; Hasse et al., 1997; Liu and Chang, 2002; Brodie and Humes, 2005).

Parameters for Cell Encapsulation

Since Chang proposed the idea of using ultrathin polymer membrane microcapsules of the immunoprotection of transplanted cells in 1964 (Chang et al., 1964), a great number of techniques of cell encapsulation have been developed. Such increased interest in this field started when Lim and Sun published their results on the pancreatic islet encapsulation in the way of mild electrostatic cross-linking of sodium alginate and its complexation by poly(L-lysine), which is now the most commonly used cell encapsulation technique (Lim et al., 1980). The study showed that microencapsulated islets implanted in rats corrected the diabetic state for several weeks by producing insulin.

Following this technique, several polymeric encapsulation systems have been developed and are currently being tested in clinical trials. For example, Novocell, Inc. has developed a photopolymerizable poly(ethylene glycol) polymer to encapsulate individual cells or cell clusters. Although much effort has been focused on identifying alternative systems to alginate/poly(L-lysine) chemistry, none have overcome all of the disadvantages of the poly(L-lysine). Therefore, although the polycation is required for xenotransplantation, and this renders the solution problematic, the allografts are likely to revert to the uncoated alginate beads. There are a variety of cell encapsulation methods using polymeric materials for treatment of disease and relevant methods and these are listed in Table 28.1.

Table 28.1 **Cells and materials used for cell encapsulation**

Cells	Functions	Materials
Human and animal primary cells		
Pancreatic islets	Diabetes	Alginate–poly(L-lysine)–alginate (Sun et al., 1996) Alginate–aminopropylsilicate–alginate (Sakai et al., 2001) Alginate–poly(L-ornithine) (Calafiore et al., 1999) Alginate–cellulose sulfate-poly(methylene-co-guanidine) (Wang et al., 1997) Agarose–poly(styrene sulfonic acid) (Iwata et al., 1994) Poly(*N*-isopropylacrylamide-co-acrylic acid) (Gappa et al., 2001) Aginate–poly(L-lysine)–poly(ethyleneimine)–protamine–heparin (Tatarkiewicz et al., 1994) Alginate–chitosan–polyethylene glycol (Chandy et al., 1999)
Hepatocytes	Liver transplantation	Alginate–chitosan (Haque et al., 2005) Hydroxyethyl methacrylate–methyl methacrylate (Sefton et al., 1997)
Kidney cells	Erythropoietin	Alginate–poly(L-lysine)–alginate (Koo et al., 1993)
Parathyroid cells	Parathyroid hormone	Alginate (Hasse et al., 1997)
Chromaffin cells	Catecholamines	Alginate–poly(L-lysine)–alginate (Aebischer et al., 1991; Kim et al., 2004)
Chondrocytes	Chondrocyte transplantation	Alginate (Grunder et al., 2004)
Hybridomas	Antibody production	Alginate–agarose (Orive et al., 2003)
Stem cells		
BMSC	Improve hepatocyte survival	Alginate–poly(L-lysine)–alginate (Liu and Chang, 2002, 2005)
Embryonic cells	Epilepsy	Polyethersulfone hollow fiber (Lindvall et al., 2004)
Mesenchymal stem cells	Tissue repair	Collagen–agarose (Batorsky et al., 2005)
Genetically engineered cells		
BHK cells	hNGF	Poly(acrylonitrile–vinyl chloride) (PAN–PVC) (Winn et al., 1994)
BHK cells	VEGF	Polysulfone hollow fiber (Yano et al., 2005)
BHK cells	Human ciliary neurotrophic factor (hCNTF)	Polyethersulfone (Bachoud-Levi et al., 2000)
Myoblasts	Mouse growth hormone (GH)	Alginate–poly(L-lysine)–alginate (Al-Hendy et al., 1996)
Mouse C_2C_{12} myoblasts	Adenosine	Polyethersulfone (Guttinger et al., 2005)
SK2 hybridoma cells	Anti-human interleukin-6 (hIL-6) monoclonal antibodies	Alginate–poly(L-lysine)–alginate (Okada et al., 1997)
Mouse Ltk fibroblast	Human growth hormone (hGH)	Alginate–poly(L-lysine)–alginate (Basic et al., 1996)
Xenogeneic tumor cells (Neuro2A)	Beta-endorphin	Polyethersulfone hollow fiber (Saitoh et al., 1995)
iNOS-expressing cells	Inducible nitric oxide synthase gene (iNOS) for tumor suppression	Alginate–poly(L-lysine) (Xu et al., 2002)

Membrane permeability is a function of both transport and thermodynamic properties, which are dependent upon the molecular characteristics of both the membrane and solute population. Thus materials for cell encapsulation should be selected or designed for each specific therapeutic device, as one may engineer several different membranes with required membrane properties for a desired application. The use of different membranes allows for variations in permeability, mass transfer, mechanical stability, buffering capability, biocompatibility, and other characteristics. A balance, however, has to be maintained among the physical properties of capsule membranes so as to support the entrapped cells' survival.

The mass transport properties of a membrane are critical since the influx rate of molecules, essential for cell survival, and the outflow rate of metabolic waste ultimately determine the viability of entrapped cells. Ordinarily the desired capsule permeability is determined by the molecular weight cutoff (MWCO) and is application-dependent. The MWCO is the maximum molecular weight of a molecule that is allowed passage through the pores of the capsule membrane. For transplantation, the MWCO must be high enough to allow passage of nutrients but low enough to reject antibodies and other immune molecules (Uludag et al., 2000).

Table 28.2 summarizes the MWCO of membranes and related therapeutic molecules. Recent efforts at defining the membrane permeability of biologically relevant proteins, rather than the use of arbitrary markers of varying molecular size, will likely have greater predictive capacity with respect to *in vivo* performance.

Cell encapsulation technology plays important roles in not only by providing immune protection by isolating encapsulated cells from host tissue but by maintaining the phenotype of cells by providing a proper 3D environment and subsequently enhancing the production of therapeutic biologics from cells. For example, when autologous chondrocytes were expanded in capsule *in vitro*, these cells did not dedifferentiate and maintained their phenotype by high expression of type I collagen and a decrease in type II collagen expression (Grunder et al., 2004).

A co-encapsulation approach is widely used to increase the duration of viability and function of cells. For example, co-encapsulated hepatocytes with BMSCs resulted in increased viability of the hepatocytes *in vitro* and *in vivo*, and also significantly prolonged the lowering of high systemic bilirubin levels in congenital Gunn rats with defects in the liver enzyme uridine diphosphate glucuronosyltransferase (UDPGT) (Liu and Chang, 2002).

Challenges in Cell Capsule Technology

In spite of a great promise of cell encapsulation concepts, there have been continuous challenges in cell capsule-based therapeutic delivery. The major challenge is long-term cell survival or prolonged cell viability in capsules. However, the encapsulated cells have a limitation due to the supply of nutrients and oxygen. Nutrients typically include low molecular weight solutes such as glucose, macromolecules such as albumin, and transferrin for iron uptake. Growth factors may also be required. Although the transport limitations for macromolecules have not yet been quantified, it is likely that oxygen supply limitations are the most serious.

A class of microporous membranes that induce neovascularization membrane is in direct contact with the bloodstream at an arterial pO_2 of ~100 mmHg. By contrast, extravascular devices implanted intraperitoneally or in subcutaneous tissue are exposed to the average pO_2 of the microvasculature (~40 mmHg). Implantation in soft tissue is further disadvantaged if a foreign-body response occurs, in which an avascular layer typically ~100 μm thick is produced adjacent to the membrane. This fibrotic tissue increases the distance between blood vessels and the implant, and the fibroblasts in the avascular layer consume oxygen.

Researchers dealing with the limitations in oxygen transport attempted using cross-linked hemoglobin (Hb-C), and inclusion of materials that induce neovascularization in the vicinity of the implant use (Chae et al., 2004). Prolonged glucose normalization of streptozotocin-induced diabetic mice was observed by transplantation of rat islets co-encapsulated with cross-linked hemoglobin, while the mice that received the conventional control islet microcapsule (without Hb-C) transplant showed graft failure in 4 weeks, exhibited by hyperglycemia, weight loss, and deteriorated glucose tolerance.

Table 28.2 MWCOs of semipermeable cell capsule membranes (Prakash et al., 2005)

MWCOs of semipermeable membranes	Molecules (molecular weight)
	Leukocytes IgM (950 kd) Urease (482.7 kd) C19 (410 kd) Fibrinogen (339 kd) Phenylalanine NH3 lyase (320 kd) Catalase (247 kd) C4 (210 kd)
Hollow fiber (>200 kd)	C5 (195 kd) IgE (190 kd) Human leukocytes antigen (180–210 kd) C3 (185 kd) IgA (170–720 kd) C2 (170 kd) C8 (163 kd) IgD (160 kd) IgG (150 kd) Tyrosinase (128 kd) C6 (110 kd) C7 (100 kd) Transferrin (81 kd) C9 (79 kd)
Alginate–Poly(L-lysine)–Alginate (60–70 kd)	Albumin (66.248 kd) Hemoglobin (64 kd) FactorX (55 kd) Tumor necrosis factor (TNF) (51 kd) Platelet-derived growth factor-C (46/30 kd) Superoxide dismutase (31.187 kd)
Cellulose nitrate or polyamide (30 kd)	Fibroblast growth factor-7 (28 kd) Vascular endothelial growth factor (21/42 kd) Bone morphogenic proteins-4 (18/33 kd) Interleukin-beta (17 kd) Fibroblast growth factor-2 (17 kd) Insulin-like growth factor-1 (17 kd) Fibroblast growth factor-1 (15.5 kd) Platelet-derived growth factor-B (14/33 kd) Platelet-derived growth factor-A (14 kd) Nerve growth factor (13 kd) C3a (9000 d) Epidermal growth factor (6 kd) Insulin (5.7 kd) Beta-endorphin (3.4 kd)
Lipid-complexed polymer (100–200 d)	Glucose (180 d) Tyrosine (163 d) Phenylalanine (147 d) Glutamine (128 d) Aspargine (114 d) Creatinine (113 d)
Lipid vesicles (lipophilic)	Urea (60 d) Carbon dioxide (44 d) Ammonia (17 d) Oxygen (16 d)

Device geometry also critically affects the local pO_2 to which cells are exposed. A spherical geometry is known to be most advantageous because of the high surface area to volume ratio. Thus, an islet (~150 μm in diameter) microencapsulated in an alginate bead (600–800 μm in diameter) was shown to be less susceptible to oxygen mass transfer than a tubular or planar diffusion chamber (Colton, 1995).

CELL-BASED PROTEIN FACTORY

Cells can be manufactured to produce a therapeutic protein as a protein factory *in vivo* for controlled release. One of these approaches include manipulating cells to deliver growth factor, that would allow the stable incorporation of growth factors within a cell in-growth matrix in a manner such that local enzymatic activity associated with tissue regeneration could trigger growth factor release.

A research group (Sakiyama-Elbert et al., 2001) investigated this approach in the context of peripheral nerve regeneration by designing modified beta-nerve growth factor (NGF) fusion proteins and testing their ability to promote neurite extension. They selected fibrin as the cell in-growth matrix, and the transglutaminase activity of factor XIIIa to covalently incorporate NGF fusion proteins within fibrin matrices as shown in Figure 28.3. Novel NGF fusion proteins, which contained an exogenous factor XIIIa substrate to allow incorporation into fibrin matrices, were expressed recombinantly. An intervening plasmin substrate domain

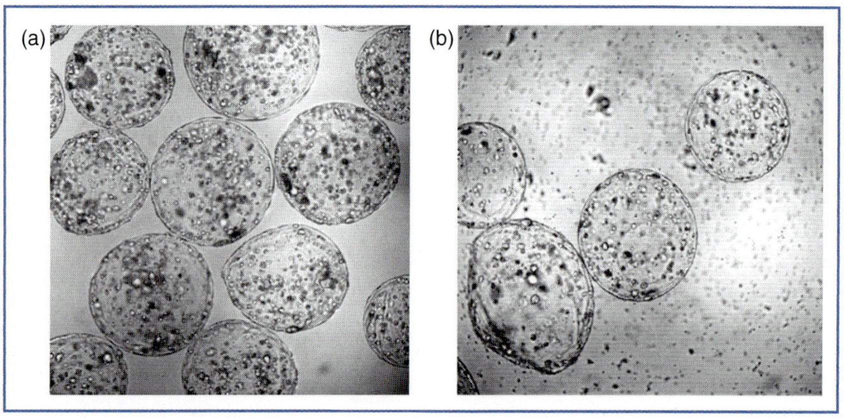

Figure 28.2 Confocal images of microcapsulated chromaffin cells. (a) Before implantation of microencapsulated chromaffin cells. (b) After retrieval of the microcapsules from the subarachnoid space 30 days after implantation. Images were captured with a confocal laser scanning microscope (×100 magnification) (Kim et al., 2004).

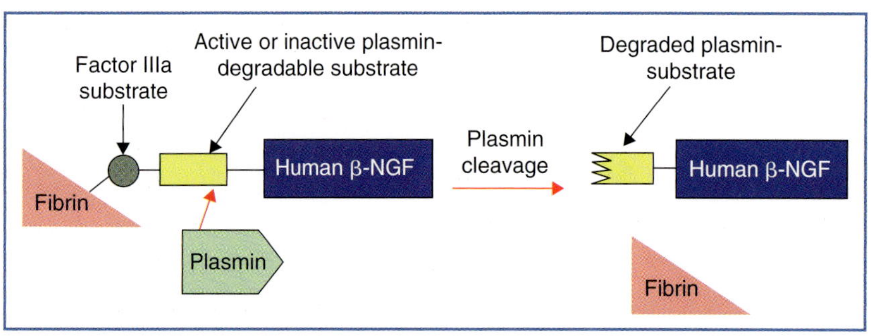

Figure 28.3 Cell triggered growth factor delivery. β-NGF fusion proteins with exogenous domains for growth factor immobilization via the transglutaminase factor XIIIa and cell-triggered release via the proteolytic activity of plasmin (Sakiyama-Elbert et al., 2001).

was placed between the factor XIIIa substrate and the NGF domain to allow cell-mediated growth factor release in response to plasmin, which is generated by invading cells.

The results showed that by placing an enzymatically degradable linker between the cross-linking substrate and the growth factor domain in the fusion protein, growth factors can be delivered in an active form in response to cell-regulated processes. It further suggested that the release of immobilized growth factors in a manner that can be temporally and spatially regulated by cell-associated enzymatic processes may be important in the context of wound healing. Thus, delivery systems that allow drug release to be regulated by the progress of wound healing through a cell in-growth matrix could prove to be more effective in promoting successful tissue regeneration.

As another example, implantable protein factory (ImPACT) products have been created by Cell Based Delivery (CBD), Inc., using muscle cells. CBD's ImPACT™ products deliver predictable, therapeutic levels of proteins, such as vascular endothelial growth factor (VEGF) and fibroblast growth factor (FGF) to stimulate rapid, sustained angiogenesis. It was reported that the growth factors were shown to be functional for an extended period in the body, and the use of these products for various cardiovascular diseases, hormonal growth deficiency, musculoskeletal disease, and solid tumors are in preclinical trials with animals.

Rugged and compact, ImPACT™ tissue implants measure only 20 mm long and 1–2 mm wide. Using standard catheter or minimally invasive techniques, interventional cardiologists and cardiothoracic surgeons can position ImPACT™ at selected sites with fibrin glues or stents. ImPACT products have survived in animals for up to 6 months with no sign of tissue loss, indicating that long-lasting drug delivery can be achieved with just a single procedure. Currently available protein-delivery systems, such as biodegradable microspheres, provide several weeks of continuous dosing. Moreover, because muscle cells live for years, autologous ImPACT products could ultimately last a year or longer. For example, products placed below the skin could deliver proteins systemically for the treatment of chronic diseases such as hemophilia or anemia.

For production of therapeutic substances for suppressing cancer, cell-based delivery system is currently in use. hES secreting cells were engineered using BHK-21 for cancer therapy by a research group (Joki et al., 2001). It was found that cell-based delivery of endostatin, an inhibitor of angiogenesis, more effectively reverses the growth of blood vessels feeding a tumor rather than do discrete injections of the same protein.

Therefore rather than expressing a therapeutic protein in cultured cells, then purifying it into the patient, it would make it easier for patients by implanting cells directly once or a few times a year rather than taking a pill or injection daily. Protein delivery based on cells is promising and attracts many scientists as an alternative therapeutic administration.

DRUG-LOADED TUMOR CELL SYSTEM

Cell-based drug delivery system does not necessarily need living cells only. Dead cells can be used for controlled delivery of therapeutic molecules. A drug-loaded tumor cell (DLTC) system has been developed for lung metastasis-targeting drug delivery. Doxorubicin was loaded into B16-F10 murine melanoma, and the loading process led to the death of all the carrier cells. The diameter of DLTC was approximately 15 μm (Shao et al., 2001). The amount and rate of doxorubicin being released from the DLTC mainly depended on the drug loading and carrier cell concentration. Over 6 month storage in phosphate buffered saline (PBS) at 4°C, the decrease in intracellular drug concentration and the carrier cell numbers were less than 25% and 5%, respectively. After a bolus injection of 30 μg doxorubicin in either DLTC form or free solution into the mice tail veins, drug deposit in the lung from DLTC was about 4-fold of that achieved by free drug solution.

In spite of potential problems associated with using dead cells as drug carriers, the finding from extensive research strongly suggested the DLTC system possessed a lung-targeting activity that may be partially attributed to its specific surface characteristics.

Table 28.3 Clinical applications of cell-based drug delivery

Year	Cell types	Formulation	Target diseases	References
1994	Pancreas islets	Alginate high in guluronic acid	Diabetes	Soon-Shiong et al.
1994	Encapsulated xenogeneic cells	Alginate–poly(L-lysine)–alginate	Amyotrophic lateral sclerosis	Aebischer et al.
1996	BHK cells	Alginate–poly(L-lysine)–alginate	Amyotrophic lateral sclerosis	Aebischer et al.
1999	Xenogeneic cells	Polymer-based hollow fibers	Amyotrophic lateral sclerosis and chronic cancer pain	Abicht and Lochmuller
2000	BHK cells	Hollow fiber	Amyotrophic lateral sclerosis	Zurn et al.
2001	Allogeneic CYP2B1-expressing cells	Cellulose sulfate	Pancreatic tumor	Lohr et al.
2001	Parathyroid cells	Polytetrafluroethylene (PTFE) membrane	Chronic hypoparathyroidism	Tibell et al.
2004	BHK cells	Thermoplastic polyethersulfone	Huntington's disease	Bloch et al.

Table 28.4 Selected companies working on cell-based therapies

Company	Technology	Major disease focus
Cell-based drug delivery		
BioHybrid Technologies (Shrewsbury, MA)	Encapsulation system for allografts	Encapsulated pancreatic islet cell allografts; therapeutic protein delivery
Islet Sheet Medical (San Francisco, CA)	Encapsulated pancreatic islet cells	Retrievable bioartificial pancreas for diabetes
Ixion Biotechnology (Alachua, FL)	Pancreatic islet-producing human stem cells	Unencapsulated islet cell allografts for diabetes
Neurotech (Evry, France)	Encapsulation system for allografts	Therapeutic protein delivery to eye and brain
Novocell (Irvine, CA)	Individually polymer-coated pancreatic islet cells	Encapsulated islet cell allografts for diabetes
Oxford BioMedica (Oxford, UK)	MacroGen (macrophages as gene delivery systems)	Prodrug activating enzyme (CYP2B6) for treating cancers
Layton Biosciences (Sunnyvale, CA)	Human neuronal stem cells	Central nervous system (CNS) disorders (stroke, tumors, Parkinson's disease, Alzheimer's disease)
Cell Based Delivery (Providence, RI)	Implantable protein-expressing muscle tissue allografts	Therapeutic protein delivery for chronic diseases
Sertoli Technologies (Cranston, RI)	Sertoli cells to protect implanted allografts	Pancreatic islet cell allografts, therapeutic protein delivery, and cell-based gene therapy
Cell-based immunotherapy		
Aastrom Bioscience (Ann Arbor, MI)	Autologous cell processing system, bone marrow, and cord blood stem cells	Dendritic cell-based cancer vaccine, solid tissue, and blood regeneration with stem cells
CellExSys (Seattle, WA)	*Ex vivo* production of cytotoxic T lymphocytes	Cell-based treatments for hepatitis B and C, cancer
Geron (Menlo Park, CA)	Human ES cells; dendritic cell vaccines	Cell-based treatments for cancer, diabetes, osteoarthritis

Table 28.4 (Continued)

Company	Technology	Major disease focus
Xcyte Therapies (Seattle, WA)	*Ex vivo* expansion of T-cells	Autologous cytotoxic T-cell generation to treat cancer, infectious diseases
Immuno-Designed Molecules (Paris, France)	Immunotherapeutics, and dendritic cells used as vaccines; cell processor technologies	Cell drugs, to enhance immunity to treat cancer
Cell retrieval and expansion		
Gamida-Cell (Jerusalem, Israel)	System for expanding stem cell populations *ex vivo*	Hematopoietic stem cells-derived umbilical cord blood for use in high dose chemotherapy
Nexell Therapeutics (Irvine, CA)	System for isolating hematopoietic stem cells	Stem cell therapy for chronic granulomatous disease and other hereditary blood disorders; cancer vaccines
TEI Biosciences (Boston, MA)	Signaling molecules to induce stem cell differentiation	Tissue engineering using derived cell types
Progenitor Cell Therapy (Saddle Brook, NJ)	Cell therapy manufacturing services	Good manufacture practice (GMP) factory and distribution system to grow and deliver autologous therapies nationwide (USA)
Cytomatrix (Woburn, MA)	A 3D matrix, an artificial thymus, for the growth and maturation of T-cells	Cell culture devices for bench research and bioreactors for commercial production of cells
Select Therapeutics (Woburn, MA)	A 3D matrix for growing cells *ex vivo*	Expansion of hematopoietic stem cells for bone marrow transplants, cytotoxic T-cells to treat cancer

SUMMARY

Delivery of biological products from living cells in response to biological need is a physiologically attractive approach. There have been successful business and clinical attempts for producing therapeutic proteins from various types of cells, as is summarized in Tables 28.3 and 28.4. Cell-based delivery system might allow a lower total drug dose to be incorporated within the delivery system, and spatial regulation of release could permit a greater percentage of the drug to be released at the time and place of greatest cellular activity. The significance of cell-based drug delivery is using biological feedback control in drug release, which could overcome the limit of a polymer-based drug delivery system. However, issues on long-term viability, risk of immune development, related safety, and retrieval of the unwanted cells should be addressed to further explore their possible clinical applications.

Many experimental applications of drug delivery systems are easing their way into the clinic, and the hope that cells may be used for therapeutics seems increasingly likely to be realized.

REFERENCES

Abicht, A. and Lochmuller, H. (1999). Technology evaluation: CRIB (CNTF delivery) CytoTherapeutics Inc. *Curr. Opin. Mol. Ther.* **1**: 645–650.

Aebischer, P., Tresco, P.A., Sangen, J. and Winn, S.R. (1991). Transplantation of microencapsulated bovine chromaffin cells reduces lesion-induced rotational asymmetry in rats. *Brain Res.* **560**: 43–49.

Aebischer, P., Buchser, E., Joseph, J.M., Favre, J., de Tribolet, N., Lysaght, M., Rudnick, S. and Goddard, M. (1994). Transplantation in humans of encapsulated xenogeneic cells without immunosuppression. A preliminary report. *Transplantation* **58**: 1275–1277.

Aebischer, P., Schluep, M., Deglon, N., Joseph, J.M., Hirt, L., Heyd, B., Goddard, M., Hammang, J.P., Zurn, A.D., Kato, A.C., Regli, F. and Baetge, E.E. (1996). Intrathecal delivery of CNTF using encapsulated genetically modified xenogeneic cells in amyotrophic lateral sclerosis patients. *Nat. Med.* **2**: 696–699.

Al-Hendy, A., Hortelano, G., Tannenbaum, G.S. and Chang, P.L. (1996). Growth retardation – an unexpected outcome from growth hormone gene therapy in normal mice with microencapsulated myoblasts. *Hum. Gene Ther.* **7**: 61–70.

Bachoud-Levi, A.C., Deglon, N., Nguyen, J.P., Bloch, J., Bourdet, C., Winkel, L., Remy, P., Goddard, M., Lefaucheur, J.P., Brugieres, P., Baudic, S., Cesaro, P., Peschanski, M. and Aebischer, P. (2000). Neuroprotective gene therapy for Huntington's disease using a polymer encapsulated BHK cell line engineered to secrete human CNTF. *Hum. Gene Ther.* **11**: 1723–1729.

Basic, D., Vacek, I. and Sun, A.M. (1996). Microencapsulation and transplantation of genetically engineered cells: a new approach to somatic gene therapy. *Arif. Cells Blood Substit. Immobil. Biotechnol.* **24**: 219–255.

Batorsky, A., Liao, J., Lund, A.W., Plopper, G.E. and Stegemann, J.P. (2005). Encapsulation of adult human mesenchymal stem cells within collagen–agarose microenvironments. *Biotech. Bioeng.* **92**: 492–500.

Bloch, J., Bachoud-Levi, A.C., Deglon, N., Lefaucheur, J.P., Winkel, L., Palfi, S., Nguyen, J.P., Bourdet, C., Gaura, V., Remy, P., Brugieres, P., Boisse, M.F., Baudic, S., Cesaro, P., Hantraye, P., Aebischer, P. and Peschanski, M. (2004). Neuroprotective gene therapy for Huntington's disease, using polymer-encapsulated cells engineered to secrete human ciliary neurotrophic factor: results of a phase I study. *Hum. Gene Ther.* **15**: 968–975.

Brodie, J.C. and Humes, H.D. (2005). Stem cell approaches for the treatment of renal failure. *Pharmacol. Rev.* **57**(3): 299–313.

Chae, S.Y., Kim, Y.Y., Kim, S.W., and Bae, Y.H. (2004). Prolonged glucose normalization of streptozotocin-induced diabetic mice by transplantation of rat islets coencapsulated with crosslinked hemoglobin. *Transplantation.* **78**: 392–397.

Chandy, T., Mooradian, D.L. and Rao, G.H. (1999). Evaluation of modified alginate–chitosan–polyethylene glycol microcapsules for cell encapsulation. *Artif. Organs* **23**: 894–903.

Chang, T.M.S. (1964). Semipermeable microcapsules. *Science.* **146**: 524–525.

Chen Calafiore, R., Basta, G., Luca, G., Boselli, C., Bufalari, A., Cassarani, M.P., Giustozzi, G.M. and Brunetti, P. (1999). Transplantation of pancreatic islets contained in minimal volume microcapsules in diabetic high mammalians. *Ann. NY Acad. Sci.* **875**: 219–232.

Colton, C.K. (1995). Implantable biohybrid artificial organs. *Cell Transplant.* **4**: 415–436.

Chen, R.R. and Mooney, D.J. (2003). Polymeric growth factor delivery strategies for tissue engineering. *Pharm. Res.* **20**: 1103–1112.

Emerich, D.F., Winn, S.R., Harper, J., Hammang, J.P., Baetge, E.E. and Kordower, J.H. (1994). Implants of polymer-encapsulated human NGF-secreting cells in the nonhuman primate: rescue and sprouting of degenerating cholinergic basal forebrain neurons. *J. Comp. Neurol.* **349**: 148–164.

Gappa, H., Baudys, M., Koh, J.J., Kim, S.W. and Bae, Y.H. (2001). The effect of zinc-crystallized glucagon-like peptide-1 on insulin secretion of macroencapsulated pancreatic islets. *Tissue Eng.* **7**: 35–44.Grunder, T., Gaissmaier, C., Fritz, J., Stoop, R., Hortschansky, P., Mollenhauer, J. and Aicher, W.K. (2004). Bone morphogenetic protein-2 enhances the expression of type II collagen and aggrecan in chondrocytes embedded in alginate beads. *Osteoarthritis Cartilage* **12**: 559–567.

Guttinger, M., Padrun, V., Pralong, W.F. and Boison, D. (2005). Seizure suppression and lack of adenosine A1 receptor desensitization after focal long-term delivery of adenosine by encapsulated myoblasts. *Exp. Neurol.* **193**: 53–54.

Haque, T., Chen, H., Ouyang, W., Martoni, C., Lawuyi, B., Urbanska, A.M. and Prakash, S. (2005). *In vitro* study of alginate–chitosan microcapsules: an alternative to liver cell transplants for the treatment of liver failure. *Biotechnol. Lett.* **27**: 317–322.

Hasse, C., Klock, G., Schlosser, A., Zimmermann, U. and Rothmund, M. (1997). Parathyroid allotransplantation without immunosuppression. *Lancet* **350**: 1296–1297.

Huber, A., Padrun, V., Deglon, N., Aebischer, P., Mohler, H., and Boison, D. (2001). Grafts or adenosine-releasing cells suppress seizures in kindling epilepsy. *Proc. Natl. Acad. Sci. U. S. A.* **98**: 7611–7616

Iwata, H., Takai, T., Kobayashi, K., Oka, T., Tsuji, T. and Ito, F. (1994). Strategy for developing microbeads applicable to islet xenotransplantation into a spontaneous diabetic NOD mouse. *J. Biomed. Mater. Res.* **28**: 1201–1207.

Joki, T., Machluf, M., Atala, A., Zhu, J., Seyfried, N., Dunn, I., Abe, T., Carroll, R. and Black, P. (2001). Continuous release of endostatin from microencapsulated engineered cells for tumor therapy. *Nat. Biotechnol.* **19**: 35–39.

Kim, Y.M., Jeon, Y.H., Jin, G.C., Lim, J.O. and Baek, W.Y. (2004). Immunoisolated chromaffin cells implanted into the subarachnoid space of rats reduce cold allodynia in a model of neuropathic pain: a novel application of microencapsulation technology. *Artif. Organs* **28**: 1059–1066.

Koo, J. and Chang, T.M. (1993). Secretion erythropoietin from microencapsulated rat kidney cells: preliminary results. *Int. J. Artif. Organs* **16**: 557–560.

Langer, R. (1990). New methods of drug delivery. *Science* **249**: 1527–1533.

Lee, M. and Kim, S.W. (2005). Polyethylene glycol-conjugated copolymers for plasmid DNA delivery. *Pharm. Res.* **22**: 1–10.

Lim, F. and Sun, A.M. (1980). Microencapsulated islets as bioartificial endocrine pancreas. *Science.* **210**: 908–910.

Liu, Z.C. and Chang, T.M. (2002). Increased viability of transplanted hepatocytes when hepatocytes are co-encapsulated with bone marrow stem cells using a novel method. *Artif. Cells Blood Substit. Immobil. Biotechnol.* **30**: 99–112.

Liu, Z.C. and Chang, T.M. (2005). Transplantation of bioencapsulated bone marrow stem cells improves hepatic regeneration and survival of 90% hepatectomized rats: a preliminary report. *Artif. Cells Blood Substit. Immobil. Biotechnol.* **33**: 405–410.

Lohr, M., Hoffmeyer, A., Kroger, J., Freund, M., Hain, J., Holle, A., Karle, P., Knofel, W.T., Liebe, S., Muller, P., Nizze, H., Renner, M., Saller, R.M., Wagner, T., Hauenstein, K., Gunzburg, W.H. and Salmons, B. (2001). Microencapsulated cell-mediated treatment of inoperable pancreatic carcinoma. *Lancet* **357**: 1591–1592.

Okada, N., Miyamoto, H., Yoshioka, T., Katsume, A., Saito, H., Yorozu, K., Ueda, O., Itoh, N., Mizuguchi, H., Nakagawa, S., Ohsugi, Y. and Mayumi, T. (1997). Cytomedical therapy for IgG1 plasmacytosis in human interleukin-6 transgenic mice using hybridoma cells microencapsulated in alginate-poly(L)lysine–alginate membrane.*Biochim. Biophys. Acta.* **1360**: 53–63.

Orive, G., Hernandez, R.M., Gascon, A.R., Calafiore, R., Chang, T.M., De Vos, P., Hortelano, G., Hunkeler, D., Lacik, I., Shapiro, A.M. and Pedraz, J.L. (2003). Cell encapsulation: promise and progress. *Nat. Med.* **9**: 104–107.

Prakash, S. and Jones, M.L. (2005). Artificial cell therapy: new strategies for the therapeutic delivery of live bacteria. *J Biomed Biotech.* **1**: 44–56.

Saitoh, Y., Taki, T., Arita, N., Ohnishi, T. and Hayakawa, T. (1995). Cell therapy with encapsulated xenogenic tumor cells secreting beta-endorphin for treatment of peripheral pain. *Cell Transplant.* **1**: S13–S17.

Sakai, S., Ono, T., Ijima, H. and Kawakami, K. (2001). Synthesis and transport characterization of alginate/aminopropyl-silicate/alginate microcapsule: application to bioartificial pancreas. *Biomaterials* **22**: 2827–2834.

Sakiyama-Elbert, S.E., Panitch, A. and Hubbell, J.A. (2001). Development of growth factor fusion proteins for cell triggered drug delivery. *FASEB J.* **15**: 1300–1302.

Sefton, M.V., Hwang, J.R. and Babensee, J.E. (1997). Selected aspects of the microencapsulation of mammalian cells in HEMA–MMA. *Ann. NY Acad. Sci.* **831**: 260–270.

Shao, J., DeHaven, J., Lamm, D., Weissman, D.N., Runyan, K., Malanga, C.J., Rojanasakul, Y. and Ma, J.K. (2001). A cell-based drug delivery system for lung targeting: I. Preparation and pharmacokinetics. *Drug Deliv.* **8**: 61–69.

Shao, J., DeHaven, J., Lamm, D., Weissman, D.N., Malanga, C.J., Rojanasakul, Y. and Ma, J.K. (2001). A cell-based drug delivery system for lung targeting: II. Therapeutic activities on B16-F10 melanoma in mouse lungs. *Drug Deliv.* **8**: 71–76.

Soon-Shiong, P., Heintz, R.E., Merideth, N., Yao, Q.X., Yao, Z., Zheng, T., Murphy, M., Moloney, M.K., Schmehl, M. and Harris, M. (1994). Insulin independence in a type 1 diabetic patient after encapsulated islet transplantation. *Lancet* **343**: 950–951.

Stayton, P.S., El-Sayed, M.E., Murthy, N., Bulmus, V., Lackey, C., Cheung, C. and Hoffman, A.S. (2005). "Smart" delivery systems for biomolecular therapeutics. *Orthod. Craniofac. Res.* **8**: 219–225.

Sun, Y., Ma, X., Zhou, D., Vacek, I. and Sun, A.M. (1996). Normalization of diabetes in spontaneously diabetic cynomologus monkeys by xenografts of microencapsulated porcine islets without immunosuppression. *J. Clin. Invest.* **98**: 1417–1422.

Tatarkiewicz, K., Sitarek, E., Fiedor, P., Sabat, M. and Orlowski, T. (1994). *In vitro* and *in vivo* evaluation of protamine–heparin membrane for microencapsulation of rat Langerhans islets. *Artif. Organs* **18**: 736–739.

Tibell, A., Rafael, E., Wennberg, L., Nordenstrom, J., Bergstrom, M., Geller, R.L., Loudovaris, T., Johnson, R.C., Brauker, J.H., Neuenfeldt, S. and Wernerson, A. (2001). Survival of macroencapsulated allogeneic parathyroid tissue one year after transplantation in nonimmunosuppressed humans. *Cell Transplant.* **10**: 591–599.

Uludag, H., De Vos, P., and Tresco, P.A. (2000). Technology of mammalian cell encapsulation. *Adv Drug Deliv Rev.* **42**: 29–64.

Wang, T., Lacik, I., Brissova, M., Anikumar, A.V., Prokop, A., Hunkeler, D., Green, R., Shahrokhi, K. and Powers, A.C. (1997). An encapsulation system for the immunoisolation of pancreatic islets. *Nat. Biotechnol.* **15**: 362–385.

Winn, S.R., Hammang, J.P., Emerich, D.F., Lee, A., Palmiter, R.D. and Baetge, E.E. (1994). Polymer-encapsulated cells genetically modified to secrete human nerve growth factor promote the survival of axotomized septal cholinergic neurons. *Proc. Natl. Acad. Sci. USA* **91**: 2324–2328.

Xu, W., Liu, L. and Charles, I.G. (2002). Microencapsulated iNOS-expressing cells cause tumor suppression in mice. *FASEB J.* **16**: 213–215.

Yano, A., Shingo, T., Takeuchi, A., Yasuhara, T., Kobayashi, K., Takahashi, K., Muraoka, K., Matsui, T., Miyoshi, Y., Hamada, H. and Date, I. (2005). Encapsulated vascular endothelial growth factor-secreting cell grafts have neuroprotective and angiogenic effects on focal cerebral ischemia. *J. Neurosurg.* **103**: 104–114.

Zurn, A.D., Henry, H., Schluep, M., Aubert, V., Winkel, L., Eilers, B., Bachmann, C. and Aebischer, P. (2000). Evaluation of an intrathecal immune response in amyotrophic lateral sclerosis patients implanted with encapsulated genetically engineered xenogeneic cells. *Cell Transplant.* **9**: 471–484.

Part VI

Therapeutic Applications: Tissue Therapy

29

Engineering of Large Diameter Vessels

Saami K. Yazdani and George J. Christ

INTRODUCTION

Vascular disease affects millions of people worldwide, occurs at all levels of the vascular tree, and represents a major cause of morbidity and mortality. When the extent of vascular disease is severe and requires vessel bypass or replacement, the available supply of healthy native collateral vessels is frequently inadequate. The only currently available clinical alternative is the use of synthetic vascular grafts. While synthetic grafts have been reasonably successful for larger diameter vessels (i.e. ⩾6 mm), they have faired poorly in smaller caliber vessels. Tissue-engineered blood vessels (TEBV) have been forwarded as a viable clinical alternative for both indications, with a majority of the clinical success limited to large caliber vessels, while the preclinical work to date has focused primarily on smaller caliber vessels. However, the virtually epidemic increase in end-stage renal disease (ESRD) has highlighted the deficiencies of synthetic grafts, even when used for relatively large caliber vessels. In this scenario, a huge demand for improved dialysis vascular access is anticipated (Figure 29.1). Certainly this clinical indication requires larger caliber TEBV (>6 mm). The goal of this report, therefore, is to briefly review the status of TEBV research, and moreover, to describe the challenges and opportunities associated with creating large caliber TEBV, such as those that might be used for improved dialysis vascular access. In so doing, we will pay special attention to the importance of the vascular smooth muscle cell (SMC) to TEBV. To date, relatively little attention has been paid to the importance of the medial SMC layer to both vessel function and accelerated vessel maturation (both *in vitro* and *in vivo*). Both of these beneficial properties of smooth muscle have important implications for the further development and clinical translation of vascular tissue engineering. As such, the creation of TEBV for dialysis vascular access provides an extraordinary opportunity to further examine the role of the SMC in TEBV. To this end, we will address how the presence of the SMC can help meet the physiological characteristics/demands of the bioengineered vessels that would be required for such clinical success, and finally, outline one currently envisioned strategy for achieving this end.

PREVALENCE AND IMPACT OF VASCULAR DISEASE

Vascular diseases are the second leading cause of morbidity and mortality in the United States. (www.americanheart.org). Abnormal vascular function contributes to coronary artery disease, stroke, peripheral arterial disease, renal insufficiency, and diabetic neuropathy. In 2003 alone, nearly 500,000 coronary artery bypass graft surgeries were performed and over 100,000 lower extremity bypass procedures are performed (www.americanheart.org, Birkmeyer et al., 2002). Important risk factors for vascular disease include older age, hypertension, hyperlipidemia, smoking, diabetes, and chronic renal insufficiency (Collins et al., 2003a). Population trends are unfavorable with respect to vascular disease, as the US population is ageing, diabetes is reaching epidemic proportions, and chronic renal disease, especially ESRD, is now epidemic (McClellan, 1994;

Copyright © 2008 Elsevier Inc. All rights reserved

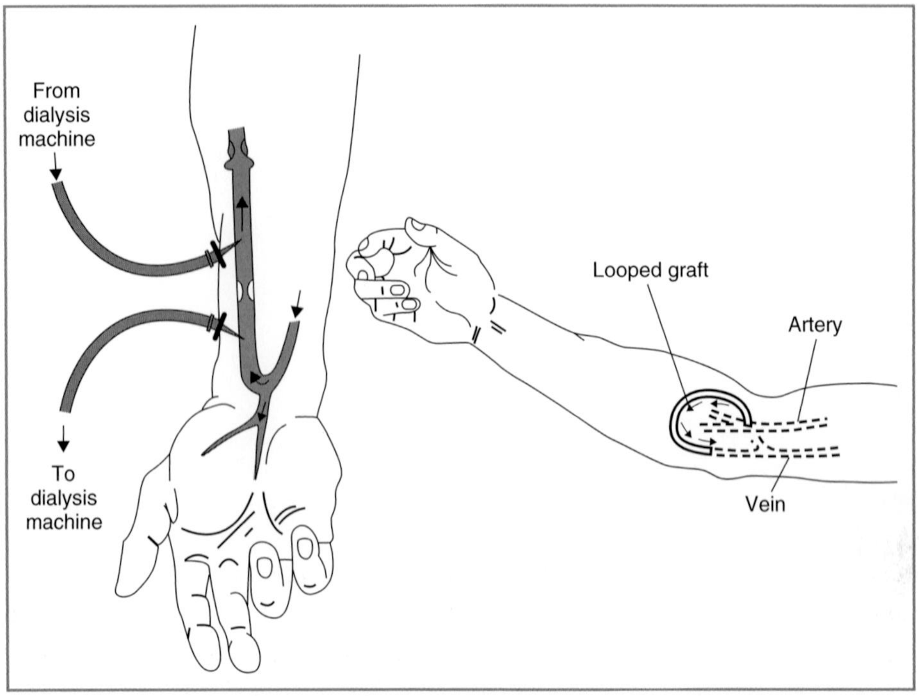

Figure 29.1 Schematic illustration of dialysis vascular access. The common two scenarios of dialysis vascular access are demonstrated here (adopted from the National Kidney and Urological Disease Information Clearing house, http://kidney.niddk.nih.gov/).

Gilbertson et al., 2005). With respect to ESRD, there is a significant unmet medical need for autologous dialysis vascular access graft. Such grafts are clearly of relatively large diameter when mature (>6mm) and thus, represent an important target for the relatively large diameter TEBVs that are the subject of this report.

THE NEED FOR IMPROVED DIALYSIS VASCULAR ACCESS

In the United States, 297,928 individuals received chronic dialysis therapy in December 2002 (Rafii and Lyden, 2003). This number is projected to increase to 712,290 by 2015, more than a doubling of the dialysis population in just over one decade (Gilbertson et al., 2005). Presently only a mature native radial artery to cephalic vein fistula achieves the ideal access route of blood circulation for hemodialysis. A close alternative is another site of native arteriovenous fistula (AVF) within the upper extremity, for example, an upper arm brachial artery to cephalic or basilic vein fistulas. Regardless, only 33% of hemodialysis patients in the United States achieve dialysis via a native AVF while the majority requires a prosthetic polytetrafluoroethylene (PTFE) artery to vein bypass grafts (AVBG, 41%) or chronic indwelling central venous catheters (McClellan, 1994; National Kidney Foundation, 2000, 2001; Hsu et al., 2004). A detailed discussion of the limitations of PTFE is well beyond the scope of this report. Suffice it to say, that stenosis is the most common problem, and moreover, the presence of the PTFE creates a foreign body response (Kohler and Kirkman, 1999; Huber et al., 2003, 2004). In addition, endothelialization occurs only within the first 1–2 cm at anastomoses, and furthermore, prosthetic materials are prone to infection. In fact, chronic cannulation with needles inserted through the skin and left in place for hours during dialysis predisposes to frequent graft infection (National Kidney Foundation, 2002; Basaran et al., 2003; Huber et al., 2004; Neville et al., 2004). Failure of the lumen surface to heal in PTFE grafts may also predispose to hematogenous seeding. Finally, as PTFE does not regenerate, the graft wall deteriorates over time from chronic puncture, predisposing to pseudoaneurysm formation, skin breakdown, cannulation site bleeding, and graft infection.

Table 29.1 Properties of the IDEAL vascular graft for hemodialysis access

Anti-thrombogenic
Anti-inflammatory
Resistant to injury and intimal proliferation (i.e. stenosis)
Resists degradation of the scaffold
Maintains structural integrity and adverse remodeling under a wide range of pressure conditions
Pharmacological/physiological and mechanical similarities to native vessels
Durability
Rapid replacement (i.e. increased maturation rate)
Easy accessibility
Durability and structural integrity in the face of repeated punction over a prolonged period of time (years)
Suitable geometry (diameter and length) to achieve high-volume blood flow

For all of the aforementioned reasons, creating an autologous blood vessel of the appropriate geometry for AVBG directly addresses many of the limitations of the PTFE grafts currently used for dialysis access. Certainly, a cellularized vessel wall with luminal endothelial coverage is likely to be more resistant to thrombosis and infection. Furthermore, the cellularized wall of a mature bioengineered vessel should allow healing at puncture sites to prevent vessel wall deterioration and provide resistance to infection superior to PTFE. Moreover, the engineered blood vessel is likely to have a compliance profile better matched to the outflow vein than PTFE, which in turn should reduce the extent of outflow venous stenosis (Kohler and Kirkman, 1999). All of these properties are prerequisites for the next generation of dialysis vascular access, and are summarized in Table 29.1.

VASCULAR PHYSIOLOGY RELEVANT TO TEBVs

Blood is carried from the heart to the capillaries by the arteries, and then returned via the venous circulation. The magnitude of the cardiovascular problems described above has certainly served to focus most TEBV research on the arterial side of the vascular tree, which will also remain the subject of this report. In that regard, the arterial vascular tree can be subdivided into three general types of arteries based both on their location in the vascular tree, as well as the functions they serve. As blood is moved away from the heart, it moves from large elastic arteries that have a strictly conduit function (e.g. aorta) to more medium-sized muscular arteries that have a distributive function, and eventually to small muscular arteries and arterioles, which provide the majority of the resistive function. The lumen to wall ratio decreases as one moves down the vascular tree away from the heart, and similarly, so does the ratio of the elastic component versus the smooth muscle component (Boulpaep, 2003). Regardless of the considerable differences in function, the vessel wall in all three types of arteries possess three distinct layers (tunics) which are the intima, media, and adventitia (Figure 29.2). The innermost layer encountered traversing the vessel wall from the luminal side is the tunica intima, which is in direct contact with moving blood. The intima is covered by the endothelium, which in turn, resides on a thick basement membrane referred to as the internal elastic lamina. The endothelium provides the anti-thrombogenic surface that ensures continuous laminar blood flow. The middle layer in the vessel wall is the tunica media. The media is composed of SMCs embedded in a matrix of collagen, elastin, and proteoglycans, the ratio and composition of which varies along the vascular tree (see below). The media resides between the internal elastic lamina and the tunica externa (i.e. adventitia). The adventitia represents the outermost portion of the vessel wall, and is primarily comprised of loose connective tissue, fibroblasts, and small nerve fibers. Of note, nerve fibers rarely penetrate the adventitial–medial SMC border.

The physiological characteristics of each vessel depend on their location in the vascular tree. Of note, there is no native vessel that mimics the physiological characteristics of the proposed dialysis vascular access

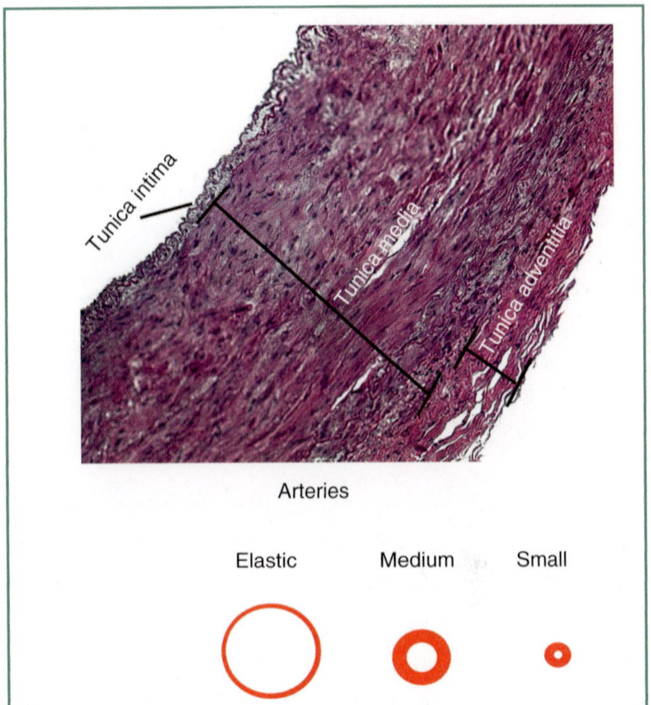

	Elastic	Medium	Small
Internal diameter	8–12 mm	2–5 mm	0.1–1 mm
Wall thickness	1–2 mm	0.5–1 mm	0.1–0.25 mm
Elastic fibers	++++	++	+
Smooth muscle cells	++	++++	+++
Collagen fibers	++++	++	++

Figure 29.2 Structure and composition of the arterial wall. Representative H&E staining illustrating the major components of the vessel wall (Boulpaep, 2003).

graft (i.e. AVF). While arteriovenous anastomoses are quite common in the circulation (e.g. for rapid shunting of blood in the skin for heat exchange), categorizing the behavior of the AVF as proposed herein is somewhat unique. In fact, arteriovenous anastomoses naturally occur between small muscular arteries and venules to bypass the capillary network and provide rapid shunting of blood. The proposed bioengineered AVF described herein would be a much larger vessel (>6 mm), and therefore, has some unique characteristics. Thus, the ideal AVF must possess some hybrid characteristics, for example, the compliance of large elastic arteries and perhaps the tone of large- to medium-sized muscular arteries. The main goal of these bioengineered vessels is to maintain a non-thrombogenic and non-proliferative surface, while retaining the ability to adapt and remodel to external stimuli, and moreover, be able to heal in response to repetitive puncture wounds (i.e. 3×/week). Clearly, to incorporate all of these features will require the presence of both SMCs and endothelial cells (ECs). A brief review of the phenotypic and functional characteristics of these two vascular wall cell types most directly pertinent to TEBV is provided below.

ECs

There are many excellent reviews on ECs and the reader is referred to a few of these for more details (Cines et al., 1998; Michiels, 2003; Aird, 2006). ECs line the entire vascular tree and provide a functional barrier between blood and the vascular wall cells and tissue parenchyma. Perhaps more importantly, they serve as a biologically active lining of the blood vessels and play a critical role in the control of vascular tone. Regulation of vascular tone is accomplished via a variety of endothelium-derived vasoactive substances. Some important endothelium-derived vasorelaxants include nitric oxide (NO), prostacyclin (PGI_2), and endothelium-derived hyperpolarizing factor (EDHF). The endothelium also provides an important source of constrictor substances such as endothelin-1, superoxide anions/radicals, angiotensin II, thromboxane A_2, and endoperoxides. These are synthesized and released in response to a wide variety of environmental and mechanical stimuli. In addition to regulation of vascular tone, the endothelium is also responsible for the maintenance of vessel wall permeability (i.e. regulating the flow of nutrients, biological molecules), as well as the balance between coagulation and fibrinolysis, the composition of the subendothelial matrix, the adhesion and extravasation of leukocytes, and mediation of inflammatory processes in the vascular wall. Prevention of thrombotic events is accomplished by maintaining a healthy monolayer of ECs that retain the ability to secret anti-thrombotic agents such as NO, PGI_2, tissue plasminogen activator (tPA), and thrombomodulin. All of these EC functions are controlled via membrane bound proteins, junctional proteins, and a variety of cell surface receptors, and are critical to circulatory homeostasis, and thus, normal organ function.

Smooth Muscle Cells

Vascular myocytes are interposed between the variable autonomic innervation on one side of the vessel (adventitial or abluminal side), and the endothelium on the other. This anatomical arrangement has important mechanistic implications for coordinated vessel function, as the size of the medial SMC layer varies from a single cell in the terminal arteriole to numerous relatively concentric layers of muscle such as those that encircle the large elastic and muscular arteries. Nonetheless, the role of myocytes in most vessels is similar, that is, to maintain vessel tone at some partial level of contractility, with the ability to become further constricted, or relaxed, as the physiological necessities of the vessels dictate. More importantly, contraction and relaxation of individual myocytes in the vessel wall must be coordinated both across the width of the muscle layer (i.e. perpendicular axis to the vessel wall), as well as along the length (i.e. longitudinal axis) of the blood vessel. The exact mechanism(s) that endow the vascular myocyte with the ability to accomplish this task differs throughout the vascular tree, and the details of such are well beyond the scope of this report. Those mechanisms pertinent to the conduit-type bioengineered vessels that are the subject of this report are described briefly below.

It is hard to overestimate the importance of the vascular SMC to circulatory homeostasis and function. In this regard, vascular SMCs make at least two major contributions to TEBV function: (1) contractility/tone and (2) accelerated tissue maturation/formation. Both of these properties are illustrated in Figure 29.3 and are discussed in more detail below.

The "tone" or contractility provided by the presence of SMCs in the vessel wall ensures that the TEBV will not be passively dilated in the presence of increased systemic pressure. In fact, a direct contribution of SMC tone to vascular diameter and/or compliance has been demonstrated *in vitro* (Figure 29.3) and *in vivo* in both human vessels and animal models at all levels of the vascular tree (Barra et al., 1993; Bank et al., 1995; Kuecherer et al., 2000; Safar et al., 2000; Moosmang et al., 2003; Jarajapu and Knot, 2005). Examples include modulation of pulse pressure and compliance in large elastic conduit vessels such as the aorta, as well as autoregulation of blood flow in specialized circulations (i.e. cerebral arterioles). Control of medial SMC tone is modulated by intravascular pressure and filling (myogenic response in muscular arteries and arterioles), circulating neurotransmitters and hormones (neurogenic response), as well as factors released from surrounding tissues (metabolic response). There are a variety of neurotransmitters known to regulate vasoconstriction (e.g. neuropeptide Y (NPY), norepinephrine (NE), and ATP (i.e. purinergic signaling)) as well vasorelaxation (e.g. vasoactive intestinal polypeptide (VIP), calcitonin gene related peptide (CGRP), and NO (Christ and Barr, 2000; Christ and Wingard, 2005; del Valle-Rodriguez et al., 2006)). Furthermore, as noted above, ECs also release both relaxing and contracting factors (see section above). As described in

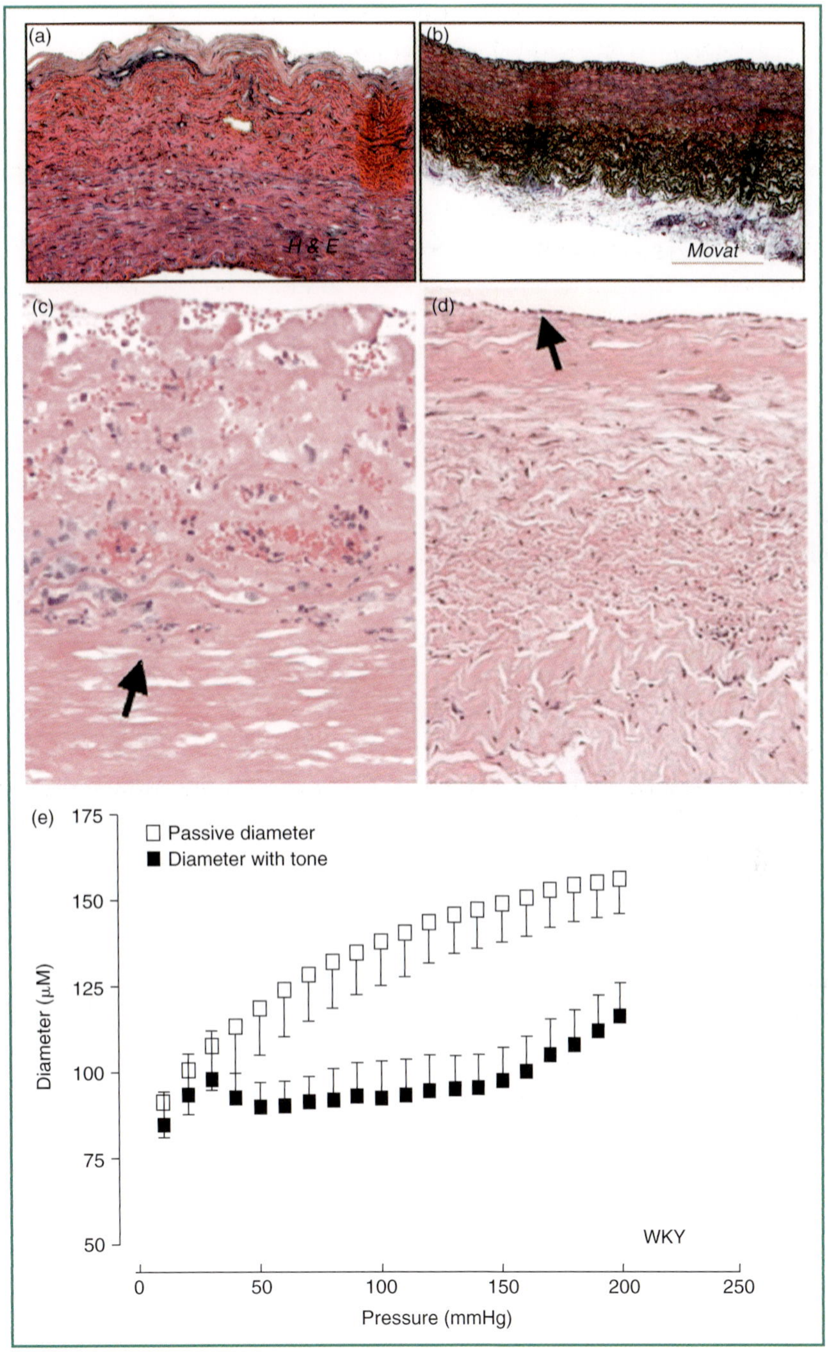

Figure 29.3 **Vascular SMC infiltration and function. (a) The EC and SMC cell seeded engineered grafts 2 weeks after implantation in sheep contained uniform cellularity throughout the vascular walls. (b) Abundant elastin fibers were observed in the entire arterial wall with a prominent distribution in the serosa and luminal surface. (c) In the EC-only graft 15 days post-implantation the vessel showed a poorly organized thrombotic deposit (arrow). (d) In the EC-only seeded graft 130 days post-implantation, the vessel architecture looked relatively normal. The vessel lumen in the lower panels is in the center (direction arrows point). (e) In the presence of SMC contraction, the heightened contractile response of the muscle cell resist the passive dilation due to the increase of intramural pressure, resulting in constant diameter within the 50–150 mmHg range. However, calcium depletion ablates SMC contraction and leads to passive vessel dilation over the same pressure range. This phenomenon clearly documents the importance of vascular muscle tone to vascular function. (The authors would like to thank Dr. Yagna P.R. Jarajapu for providing Figure 29.3e).**

detail elsewhere, all of these processes can be further integrated via intercellular communication through gap junctions (Christ et al., 1996, 1999; Brink et al., 2000; Wang et al., 2001; Lagaud et al., 2002a, b; Haefliger et al., 2004; Haddock and Hill, 2005). In fact, gap junction (Cx40, Cx43, and Cx45) appear to play a role in the control of vascular tone in a variety of ways. First, they can help coordinate locally restricted signals arising across the vessel wall (i.e. integrating neural and endothelial signals that originate on opposite sides of the vessel). Second, they can help orchestrate responses along the length of the blood vessel (up- and down-stream vasodilation or constriction). Third, they can provide a safety factor ensuring syncytial SMC responses, even when not all cells in the vessel wall can respond to any given stimulus (i.e. cellular heterogeneity).

However, in addition to tone/compliance as discussed above, the presence of SMCs in TEBV also appears to accelerate vascular tissue maturation/formation. This point is also illustrated in Figure 29.3, where the anatomy/histology of the blood vessel appears almost "normal" only 2 weeks after implantation; as opposed to the relatively immature looking vessel observed at the same time point in an endothelial-only seeded implant. In this scenario, the presence of the SMC layer may confer a third advantage of special significance to the TEBV for dialysis vascular access. That is, the repeated puncturing of the vessel wall (i.e. typically 3×/week) may require the presence of the additional cell type for tissue/wound healing. Finally, it would seem that the presence of the SMC and the commensurate cell-to-cell interactions with the endothelium would be required to confer the full range of phenotype(s) and function(s) characteristic of the native vessel wall.

TISSUE-ENGINEERED VASCULAR GRAFTS: A BRIEF REVIEW OF THE LITERATURE

From the aforementioned discussion it is clear that tissue engineering of vascular conduits directly addresses a critical need for improved treatment options for vascular disease. There have been a significant number of review publications on the topic of TEBV (Ratcliffe, 2000; Tiwari et al., 2001; Rabkin and Schoen, 2002; Teebken and Haverich, 2002; Sales et al., 2005; Vara et al., 2005; Isenberg et al., 2006) and many of the primary studies are summarized in Table 29.2. Provided below are brief summaries of some of the seminal research findings.

Small diameter TEBV (≤4 mm): Clinical studies have indicated that EC-seeded synthetic grafts have high patency rates in human coronary artery bypass grafts and in lower extremity artery bypass grafts (Deutsch et al., 1999; Laube et al., 2000). In the last two decades many attempts have been made to engineer endothelial-lined, patent 4–6 mm arterial substitutes. Weinberg and Bell (1986) were the first to engineer blood vessel substitutes by seeding ECs, SMCs, and fibroblasts on preformed collagen gels. However, mechanical and burst strengths were poor, precluding *in vivo* implantation. A similar approach was taken by L'Heureux et al. (1998), who used SMCs, fibroblasts, and EC to engineer a polymer-free blood vessel that had better mechanical properties and performed reasonably well *in vivo* (three out of six implanted grafts remained patent after 7 days). In addition, Niklason et al. (1999) described seeding SMCs and ECs on biodegradable polymers made of polyglycolic acid (PGA) and the implanted grafts remained patent up to 24 days. Most recently, L'Heureux et al. (2006) have implanted autologous TEBV extracted from fibroblasts for up to 8 months in rats, canine, and primate models. These grafts demonstrated tissue integration, suitable mechanical properties, and formation of vasa vasorum. Badylak et al. (1989) introduced the concept of a native collagen-rich matrix (small intestinal submucosa) as a vascular graft, and Huynh et al. (1999) showed that these grafts, in the absence of cells, were fully endothelialized within 3 months and impregnated with SMC, improving long-term patency. Kaushal et al. (2001) showed similar results by maintaining vascular graft patency for greater than 4 months by seeding decellularized porcine arterial segments with ECs from circulating progenitor cells. In fact, the explanted grafts exhibited contractile activity and NO-mediated vascular relaxation similar to the native carotid artery (Kaushal et al., 2001). These early studies demonstrated the capabilities of collagen matrices to mature and remodel via cell infiltration of SMC in the vessel wall and EC coverage of the lumen, leading to development of vasomotor tone and responsiveness.

Large diameter TEBV (*≥6 mm*): There is significantly less information available concerning the investigation and development of large diameter TEBV. However, as noted above, with increased longevity worldwide, and the rapidly expanding number of patients with diabetes and renal disease that will require dialysis, a need to create a functional, patent, autologous large diameter graft that can remodel and regenerate is clearly emerging. Shin'oka et al. (2005) have developed a tissue-engineered graft from a PGA/PLLA or poly (L-lactide) scaffold seeded with autologous bone marrow cells on the luminal surface to treat pediatric patients with congenital heart defects. The performance of these grafts was first evaluated in animal models (Watanabe et al., 2001). The results of the animal studies revealed that seeded TEBV remain patent for up to 6 months with no sign of stenosis or dilation. Moreover, when retrieved, the endothelium of the vessel stained positive for functional endothelial-specific surface marker (Factor VIII). In a groundbreaking clinical study, the peripheral pulmonary arteries of 23 patients were replaced with large diameter autologous seeded biodegradable scaffolds (PGA/PLLA + autologous bone marrow cells). Long-term follow-up of these seminal clinical studies (32 months) have shown no complications such as thrombosis, stenosis, or obstruction associated with the implants. Importantly, these results demonstrate the potential of TEBV to remodel, grow, and remain patent in a growing patient.

Opitz et al. (2004a) investigated the development of a tissue-engineered graft for aortic replacement. The challenges of a bioengineered aorta clearly present a significant departure from the TEBV investigations that have been conducted elsewhere in the vascular system. The scaffold for these studies was constructed from poly-4-hydroxybutyrate (P-4-HB, Tepha Inc., Cambridge, MA) and endothelialized and impregnated with SMC within a bioreactor system. Dynamic preconditioning of the scaffold for 2 weeks resulted in a TEBV with a rupture force of approximately 80% of the native ovine aorta, the target replacement arterial segment. *In vivo* experiments of the TEBV demonstrated that the implanted grafts remained patent up to 3 months, followed by significant dilation and thrombus formation of the graft likely due to insufficient elastic fiber synthesis.

The development of large diameter vessels within our group has blossomed from the knowledge gained from past experiences in developing small diameter TEBVs (Kaushal et al., 2001; Amiel et al., 2006; Stitzel et al., 2006). Despite the obvious differences in the clinical application, the approach in developing both large and small TEBV share many common features. One strategy for so doing is outlined below.

TISSUE-ENGINEERED VASCULAR GRAFTS: A BRIEF REVIEW OF THE PROCESS

An approach to the construction of relatively large diameter tissue-engineered vessels is illustrated in Figure 29.4, and reflects the general approach taken by several groups for the development of both large and small diameter TEBV. This over-simplified conceptual framework does not depict the numerous complexities associated with this process. In fact, each step in the TEBV process, from selection of the scaffold, to cell source (i.e. isolation of progenitor cells, etc.), cell seeding conditions and bioreactor TEBV preconditioning protocols, to selection of the appropriate animal model for implantation needs to be thoroughly evaluated. Each of these steps has a critical impact on the TEBV process that will likely vary with each TEBV for each indication. Certainly, with respect to the best "recipe" for TEBV, the devil is in the details. Below we provide some basic concepts, features, and requirements for each step in the process.

Scaffolds

Various synthetic and naturally derived biomaterials have been used in constructing vascular grafts but none have proven entirely satisfactory. The goal is always the same, that is, to develop a reproducible, biocompatible scaffold similar to that characteristic of native vasculature. With respect to the synthetic constructs, polymers

Table 29.2 Summary of TEBV studies

Authors	Year	Graft type	Diameter (mm)	Cell(s)	Methods	
					Static	Dynamic
Weinberg et al.	1986	Collagen	6	Bovine aortic SMC, EC, fibroblasts	1 week rotation @ 1RPM	n/a
Badylak et al.	1989	Small intestine submucosa	10	None	n/a	n/a
L'Heureux et al.	1998	VSMC and fibroblast sheet	3	SMC (human umbilical cord)	3 months of total maturation (SMC, fibroblast, EC)	n/a
Campbell et al.	1999	Myofibroblast tube	3,5	Mesothellum, myofibroblast (autologous)	n/a	n/a
Huynh et al.	1999	Intestinal collagen layer (small intestine submucosa)	4	None	n/a	n/a
Shum-Tim et al.	1999	PGA–PHA	7	Ovine carotid SMC, EC, fibroblasts	1 week of incubation (mixed cell population)	n/a
Niklason et al.	1999	PGA	3.1	EC, SMC (bovine aorta, porcine carotid artery (in vivo))	30 min of static seeding	8 weeks of pulsatile conditions
Teebken et al.	2000	Acellular porcine artery	n/a	EC, myofibroblasts (both human saphenous vein)	60 min of incubation	4 day pulsatile condition
Hoerstrup et al.	2001	PGA/P-4-HB	5	EC, myofibroblast (ovine carotid artery)	4 days of static seeding	Up to 28 days of pulsatile conditions
Niklason et al.	2001	PGA	3.1	EC, SMC (bovine aorta)	Rotation for 30 min	Up to 8 weeks of pulsatile conditions
Kaushal et al.	2001	Acellular porcine artery	4	EPC (ovine peripheral blood)	Rotation for 6 h	Steady flow
Teebken et al.	2001	Acellular porcine artery	4	EC (porcine external jugular vein)	60 min of incubation	n/a
Watanabe et al.	2001	PGA-CL/LA	10	Canine femoral vein SMC, fibroblasts	1 week of incubation (mixed cell population)	n/a
Mckee et al.	2002	PGA	3.1	EC (HUVEC), SMC (Human aortic)	16 h of static seeding	7 weeks of pulsatile condition
Berglund et al.	2003	Hybrid collagen	3	EC (human coronary EC)	60 min of incubation	n/a
Nasseri et al.	2003	PGA/P-4-HB	5,12	Myofibroblast (ovine carotid artery)	Rotation @ 5 RPM 5–10 days	n/a
Yu et al.	2003	PTFE	4	EC, SMC (both Rabbit Jugular vein)	Rotation at 1 RPM for 2 h	n/a
Shirota et al.	2003	Polyurethane	1.5	EPC (human peripheral	Rotation @ 120 degrees each	n/a

In vivo model	Outcome	In vitro Outcome
n/a	n/a	EC and SMC were seeded with success
Canine (infrarenal aorta)	100% patency for up to 52 week ($n = 9$)	n/a
Canine (femoral artery)	Three out of six grafts remained patent at 7 days	EC and SMC were organized successfully to mimic the structure of native artery
Rat (aorta), rabbit (carotid)	Rats: 67% patency at 4 months ($n = 30$), Rabbit: 70% patency at 4 months ($n = 20$)	n/a
Rabbit (carotid artery)	100% patency at 28 days ($n = 9$), 53 days ($n = 4$), and 90 days ($n = 4$)	n/a
Ovine (aortic replacement)	100% patency at 10 days ($n = 1$), 84 days ($n = 3$), 150 days ($n = 3$)	n/a
Porcine (saphenous artery)	100% patency at 4 weeks for preconditioned graft ($n = 1$), non-preconditioned vessels occluded at 3 weeks ($n = 2$)	Endothelium layer was achieved, SMC impregnation of the scaffold was achieved
n/a	n/a	EC were seeded with success
n/a	n/a	Endothelium layer was achieved
n/a	n/a	EC and SMC were seeded with success
Sheep (carotid artery)	100% patency at 15 days and 130 days after implantation ($n = 7$)	Endothelium layer was achieved
Sheep (carotid artery)	54% patency at 1 week ($n = 8$) and 71% patency at 4 months ($n = 8$) for seeded graft	n/a
Canine (Inferior vena cava)	100% patency at 3 ($n = 1$), 4 ($n = 1$), 5 ($n = 1$), 6 months ($n = 1$)	n/a
n/a	n/a	EC and SMC were seeded with success
n/a	n/a	EC were seeded with success dynamic rotation seeding can culture myofibroblasts onto tubular polymer scaffold
n/a	n/a	
Rabbit (aorta shunt)	Retention rate of EC at 1 h is 65% and 1 day (51%), EC/SMC at 1 h (98%), and 1 day (90%)	EC were seeded with success
n/a	n/a	EPCs were seeded with success

(*Continued*)

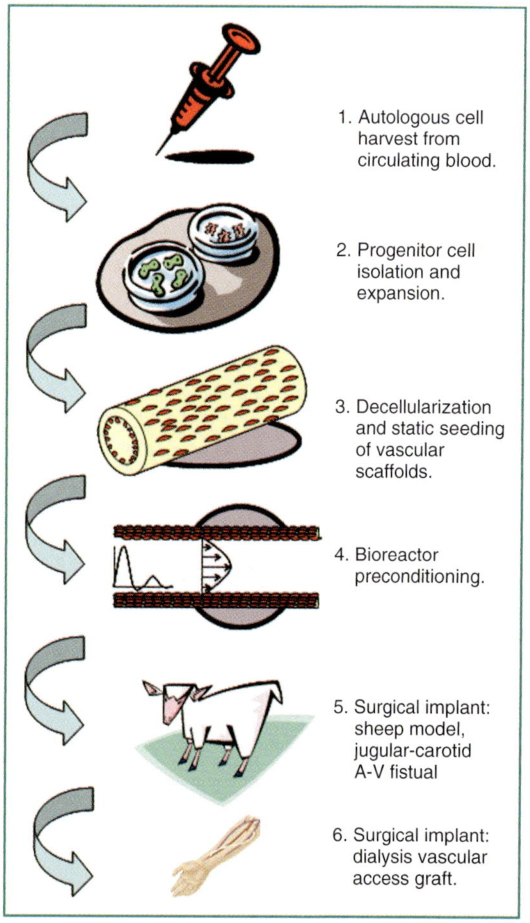

Figure 29.4 **Schematic depiction of the TEBV process.**

and electrospun scaffolds are both very attractive options due to the control one has over composition, architecture, and the reproducibility of the manufacturing process. The current generations of polymers are mostly biodegradable and include polylactic acid (PLA), PGA, polyhydroxyalkanoate (PHA), and polydioxanone (PDS). These polymers can be used singly, or in combination to optimize the desired mechanical performance and biocompatibility of the graft. Similar to polymers, electrospinning techniques can take advantage of a variety of materials to create scaffolds. Electrospinning involves creation of an electromagnetic field with a high-voltage source. Exposure to high voltage causes polymers in volatile solvents to elongate and splay into small fibers and be drawn/sprayed onto a grounded surface (i.e. a mandrel) where they can be spun into tubular structures. By controlling the characteristics of individual fiber formation during the electrospinning process, as well as the rotational speed of the mandrel (see Stitzel et al., 2006) structural characteristics such as porosity and geometry can be precisely controlled. Thus, from a commercial perspective, synthetic scaffolds are very attractive for the clinical translation of TEBV.

However, from a biological perspective, decellularized vessels (i.e. natural scaffolds), possess a biochemical composition, ultrastructural architecture, and biomechanics similar to native vessels. Not surprisingly, decellularized collagen-based vascular scaffolds derived from porcine blood vessels have been successfully used for TEBV *in vivo* (Kaushal et al., 2001). Similar approaches have been used in a variety of clinical applications for developing tissue-engineered vascular patches (Cho et al., 2005), heart valves (Lichtenberg et al., 2006), and bladders (Gabouev et al., 2003). To summarize, while synthetic scaffolds will undoubtedly

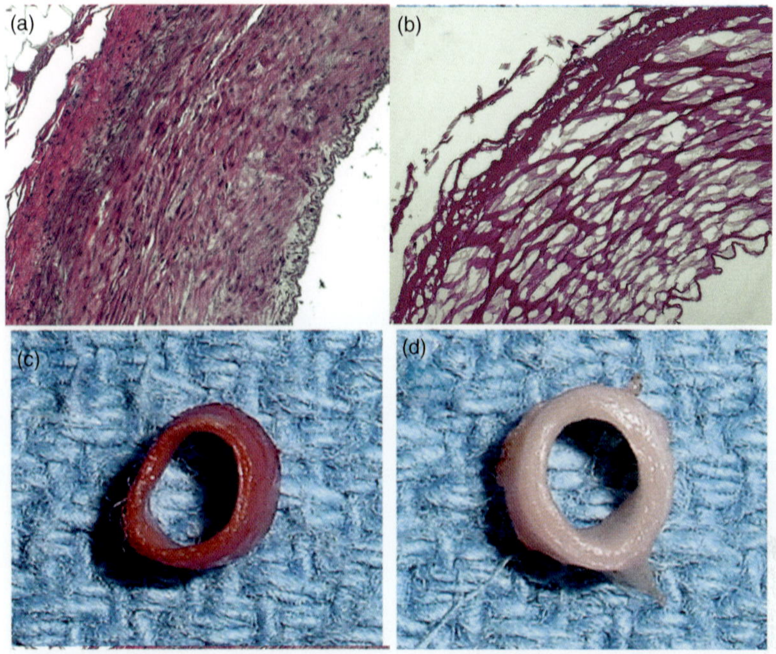

	Percent
Collagen type I	60.20%
Collagen type II	5.30%
Collagen type III	14.80%
Elastin	19.70%

Figure 29.5 Natural scaffolds derived from porcine arterial segments. (a) H&E of native porcine carotid artery. (b) H&E of decellularized porcine carotid artery. (c) Segment of native porcine carotid artery. (d) Segment of retrieved TEBV following in vivo implantation. The collagen and elastin composition of the decellularized porcine carotid artery are provided in the table below.

provide an important source of "off the shelf" scaffold material for clinical TEBV, the natural scaffold still provide the ultimate "gold" standard with respect to the biological requirements and characteristics of native vessels required to guide the development of the TEBV *in vivo*. The TEBV strategy outlined below utilizes the decellularized scaffold.

Step 1: Removal of cells from mature arteries produces a collagen-based scaffold that is amenable for seeding and growth of vascular cells. Prior work has established a working protocol for preparation of scaffolds from animal arteries using a multi-step decellularization process. Details of the procedure can be found in previous literature that shows the overall concept (Kaushal et al., 2001; Amiel et al., 2006). As shown in Figure 29.5, decellularized scaffolds preserve their extracellular matrix architecture, including internal and external elastin layers and several layers of collagen. Moreover, the decellularization process removed all cellular components, maintaining only collagen and elastin components. The quantity and distribution of collagen and elastin in a vascular scaffold is vital information in consideration for scaffold material in developing TEBV.

Mechanical characteristics of vascular grafts play a significant influence in long-term patency of the implant. In fact, compliance mismatch is thought to be one of the most important factors predisposing prosthetic vascular grafts to intimal hyperplasia, thrombosis, and occlusion. If the TEBV is stiff then flow disturbances and tissue

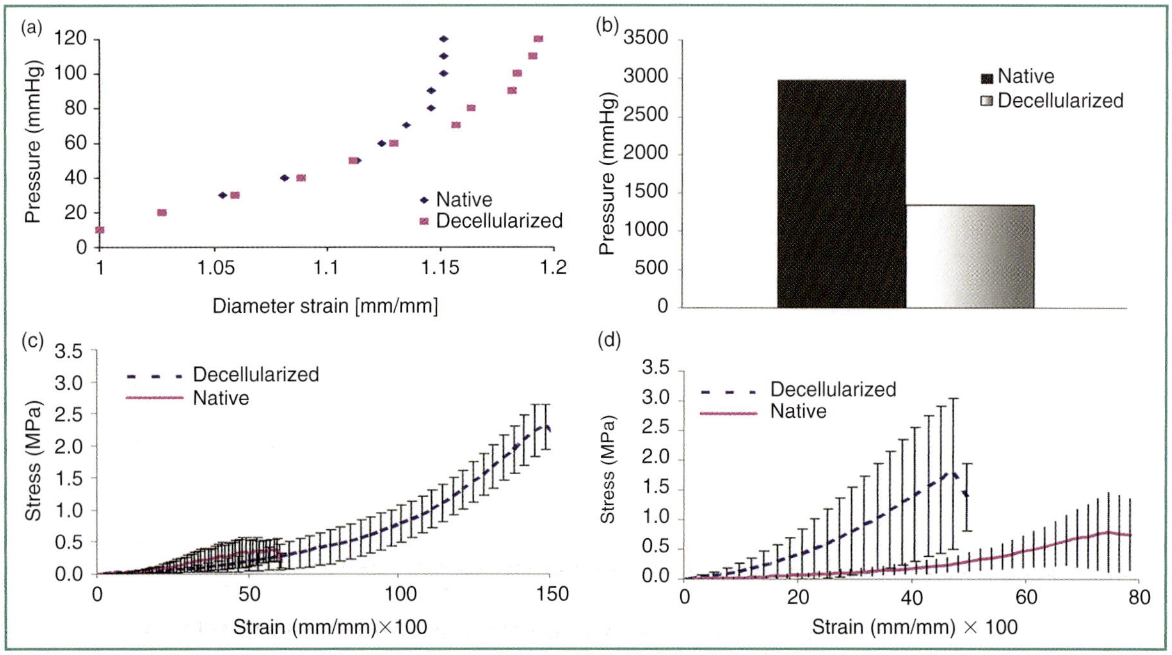

Figure 29.6 Mechanical behavior of native and decellularized grafts. (a) Pressure versus diameter measurements of the decellularized vessel compared to native. (b) Burst pressure measurements of native and decellularized vessels. (c) Stress versus strain measurements in the axial direction. (d) Stress versus strain measurements in the circumferential direction.

vibration may predispose to hyperplasia. Conversely, a TEBV that is too compliant may result in the formation of an aneurysm. As such, we have rigorously analyzed the biomechanical characteristics of the decellularized scaffolds. To measure compliance, decellularized vascular scaffolds were immersed in a water bath, cannulated at one end, and pressurized, while recording the diameter change using a digital camera. Figure 29.6 summarized the data and demonstrates that the decellularized scaffolds are similar to that of the native artery. Moreover, burst strength testing and stress–strain measurements, demonstrate that the decellularization process does not disturb the mechanical integrity to the extent that failure might occur *in vivo* (Figure 29.6).

Cell Source

Step 2: There are numerous potential cell sources available for cellularizing the synthetic or naturally derived scaffolds. The strategy that we are currently pursuing is to isolate progenitor cells from circulating blood and expand them to obtain the EC and SMC that are required for TEBV, as outlined in Figure 29.7. The overall concept is to utilize cell-selective markers to isolate and expand the progenitor cells prior to differentiation and further proliferation for seeding purposes. This process is well characterized with respect to differentiation of ECs from endothelial progenitor cells, but further research is required for obtaining similar procedures for derivation of SMCs from circulating muscle progenitor cells. The latter work is ongoing in our group.

Cell Seeding and Preconditioning

Steps 3 and 4: The final steps in creating TEBV involve the development of a bioreactor system for cell seeding and preconditioning; that is to expose TEBV to *in vivo* conditions they will face upon implantation. Seeding TEBV consists of depositing cells (EC and/or SMC) onto a three-dimensional scaffold to achieve a confluent monolayer of EC at the inner surface and/or SMC on the outside. A variety of approaches have been attempted

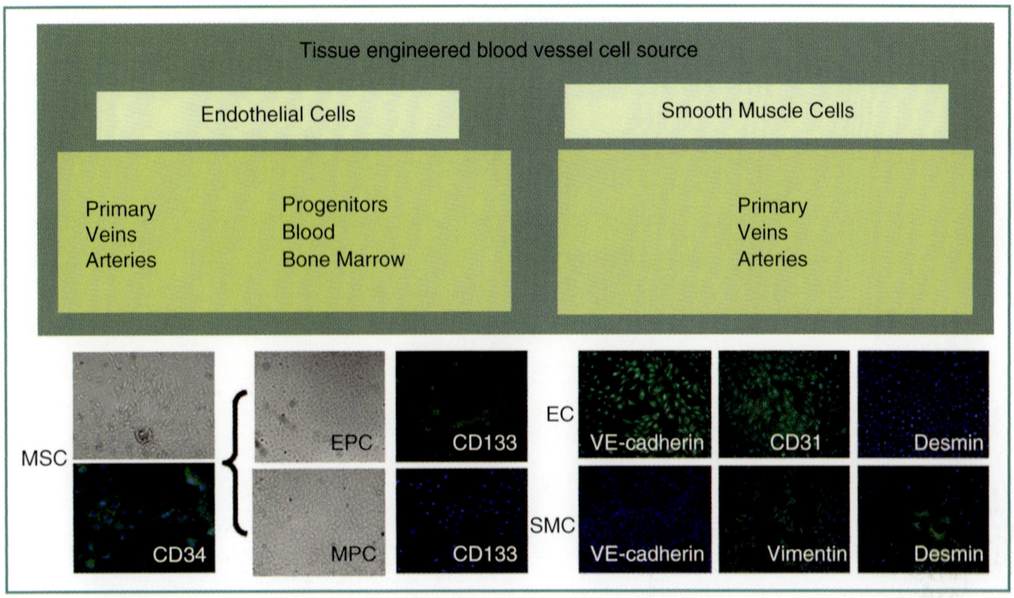

Figure 29.7 Identification of progenitor-derived EC and SMCs. As illustrated, mesenchymal cells (MS) are collected from sheep blood and separated into endothelial progenitor (EP) and muscle progenitor (MP) cell populations. The cells are then subcultured into differentiated SMC and EC types.

in seeding both the endothelium and SMCs, and recent published studies have demonstrated highly evolved bioreactor systems to produce and monitor the mechanical forces required for cell seeding and/or preconditioning (Thompson et al., 2002; Barron et al., 2003; Mironov et al., 2003; McCulloch et al., 2004; Narita et al., 2004; Williams and Wick, 2004; Portner et al., 2005; Soletti et al., 2006).

The theory behind the use of bioreactors for TEBV derives from studies demonstrating that mechanical stress accelerated cell and tissue growth and phenotypic differentiation (Braddon et al., 2002; Nerem, 2003; Jeong et al., 2005; Kurpinski et al., 2006). In this regard, a properly designed bioreactor system provides physiologically relevant stress in a three-dimensional tissue, accelerating tissue maturation, and development functional properties. While we are unaware of any published studies documenting that bioreactor preconditioning *per se* is capable of producing a relatively mature and fully functional vessel *in vitro*, this certainly seems an area worthy of further investigation. It corresponds to intuition that implantation of a more mature functional TEBV would accelerate tissue formation and maturation *in vivo*; thereby providing for quicker restoration of function, and presumably, promoting more widespread clinical applications. Regardless of the precise operational concept, a bioreactor system for development of TEBV should be capable of the following functions:

- Permitting static and/or dynamic seeding.
- Providing and monitoring physiological flow rate and pressure.
- Capable of dynamic data display and recording (archival).
- Providing physiological axial and circumferential stress.
- Providing an external bath.
- Maintaining desired concentration of gases and nutrients in the culture medium.
- Maintaining temperature and sterility.
- Be easily portable and accessible for transportation and use in surgical procedure.

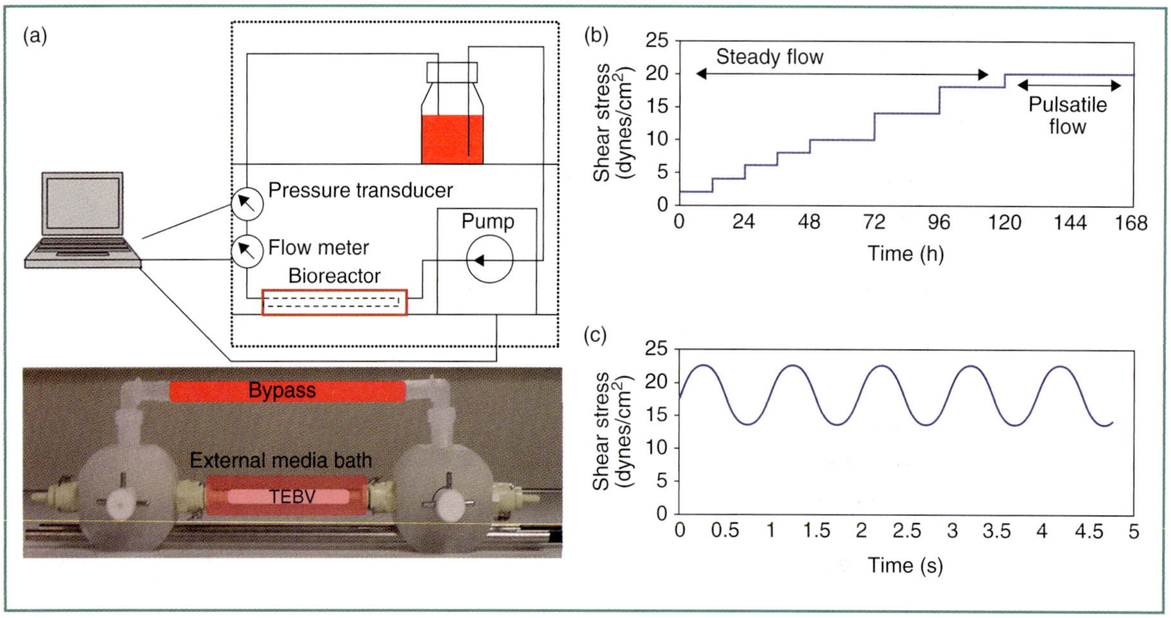

Figure 29.8 **Bioreactor system. (a) Image of the bioreactor flow system. The bioreactor provides an external media bath, optical access, a bypass system, control over flow and pressure conditions, and the ability to maintain sterility. (b) Summary of the 7 day preconditioning protocol of the TEBV. (c) Pulsatile shear conditions during the final 48 h of preconditioning.**

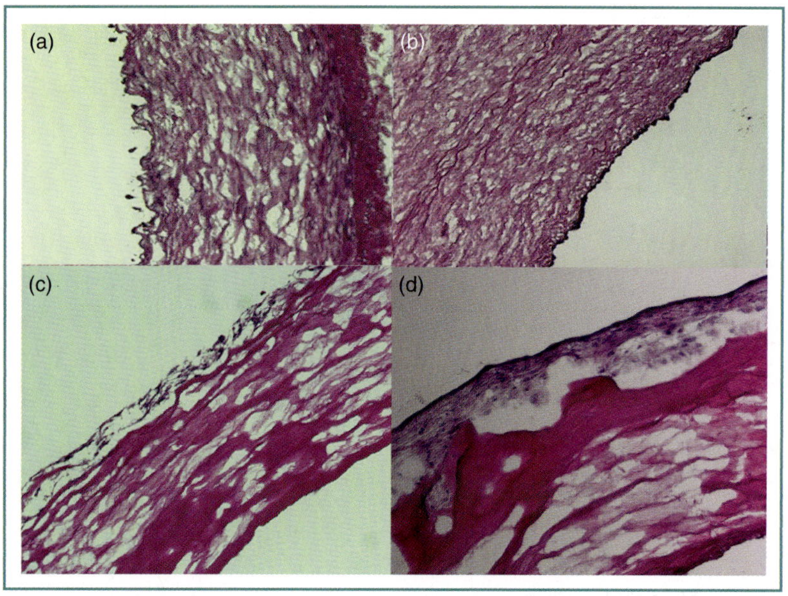

Figure 29.9 **Cell seeding of decellularized scaffolds. (a) H&E staining of the decellularized vessel after static EC seeding. (b) H&E staining illustrating the presence of a confluent monolayer of EC within the lumen of the decellularized vessel after 7 days in the bioreactor. (c) Static seeding of vascular SMCs after 48 h. (d) One week bioreactor preconditioned decellularized scaffold seeded with vascular SMCs.**

Obviously, the optimal preconditioning protocol(s) required to seed and mature TEBV are still being developed. However, Figure 29.8 shows the general features of a bioreactor system, while Figure 29.9 shows some preliminary results with both EC and SMC seeding on decellularized scaffolds. We are currently investigating the impact of various bioreactor protocols on the efficiency of cell seeding and the phenotypic differentiation of ECs and SMCs. Major parameters of interest include rotational speed of scaffold during seeding, optimal cell seeding density and time course of cell seeding protocol, and duration of bioreactor preconditioning period (i.e. days or weeks). Clearly further development and refinement of the bioreactor system is required, but unequivocally, such development holds intrinsic scientific value, and moreover, will likely be required to ensure the widespread clinical application of TEBV.

REFERENCES

National Kidney Foundation (2000). NKF-DOQI clinical practice guidelines for hemodialysis adequacy. *Am. J. Kidney Dis.* **37**: S7–S64.

National Kidney Foundation (2001). NKF-DOQI clinical practice guidelines for vascular access, 2000. *Am. J. Kidney Dis.* **37**: S137–S181.

National Kidney Foundation (2002). Annual report, end stage renal disease clinical performance measures project. *Am. J. Kidney Dis.* **42**: S1–S96.

Aird, W.C. (2006). Endothelial cell heterogeneity and atherosclerosis. *Curr. Atheroscler. Rep.* **8**: 69–75.

Amiel, G.E., Komua, M., Shapira, O., Yoo, J.J., Yazdani, S., Berry, J., Kaushal, S., Bischoff, J., Atala, A. and Soker, S. (2006). Engineering of blood vessels from acellular collagen matrices coated with human endothelial cells. *Tissue Eng.* **12**: 1–11.

Badylak, S.F., Lantz, G.C., Coffey, A. and Geddes, L.A. (1989). Small intestinal submucosa as a large diameter vascular graft in the dog. *J. Surg. Res.* **47**: 74–80.

Baguneid, M., Murray, D., Salacinski, H.J., Fuller, B., Hamilton, G., Walker, M. and Seifalian, A.M. (2004). Shear-stress preconditioning and tissue-engineering-based paradigms for generating arterial substitutes. *Biotechnol. Appl. Biochem.* **39**: 151–157.

Bank, A.J., Wilson, R.F., Kubo, S.H., Holte, J.E., Dresing, T.J. and Wang, H. (1995). Direct effects of smooth muscle relaxation and contraction on *in vivo* human brachial artery elastic properties. *Circ. Res.* **77**: 1008–1116.

Barra, J.G., Armentano, R.L., Levenson, J., Fischer, E.I., Pichel, R.H. and Simon, A. (1993). Assessment of smooth muscle contribution to descending thoracic aortic elastic mechanics in conscious dogs. *Circ. Res.* **73**: 1040–1050.

Barron, V., Lyons, E., Stenson-Cox, C., McHugh, P.E. and Pandit, A. (2003). Bioreactor for cardiovascular cell and tissue growth: a review. *Ann. Biomed. Eng.* **31**: 1017–1030.

Basaran, O., Karakayali, H., Emiroglu, R., Belli, S. and Haberal, M. (2003). Complications and long-term follow-up of 4416 vascular access procedures. *Transplant. Proc.* **35**: 2578–2579.

Bergland, J.D., Mohseni, M.M., Nerem, R.M. and Sambanis, A. (2003). A biological hybrid model for collagen-based tissue engineered vascular constructs. *Biomaterials* **24**: 1241–1254.

Birkmeyer, J.D., Siewers, A.E., Finlayson, V.A., Stukel, T.A., Lucas, F.L., Batista, I., Welch, G. and Wennberg, D.E. (2002). Hospital volume and surgical mortality in the United States. *N. Engl. J. Med.* **346**: 1128–1137.

Borschel, G.H., Huang, Y.C., Calve, S., Arruda, E.M., Lynch, J.B., Down, D.E., Kuzon, W.M., Dennis, R.G. and Brown, D.L. (2005). Tissue engineering of recellularized small-diameter vascular grafts. *Tissue Eng.* **11**: 778–786.

Boulpaep, E.L. (2003). Arteries and veins. In: Boulpaep, E.L. and Boron, W.F. (eds.), *Medical Physiology*. NY: Saunders, pp. 447–462.

Braddon, L.G., Karoyli, D., Harrison, D.G. and Nerem, R.M. (2002). Maintenance of a functional endothelial cell monolayer on a fibroblast/polymer substrate under physiologically relevant shear stress conditions. *Tissue Eng.* **8**: 695–708.

Brink, P.R., Ricotta, J. and Christ, G.J. (2000). Biophysical characteristics of gap junctions in vascular wall cells: implications for vascular biology and disease. *Braz. J. Med. Biol. Res.* **33**: 415–422.

Campbell, J.H., Efendy, J.L. and Campbell, G.R. (1999). Novel vascular graft grown within recipient's own peritoneal cavity. *Circ. Res.* **85**: 1173–1178.

Cho, S.W., Park, H.J., Ryu, J.H., Kim, S.H., Kim, Y.H., Choi, C.Y., Lee, M.J., Kim, J.S., Jang, I.S., Kim, D.I. and Kim, B.S. (2005). Vascular patches tissue-engineered with autologous bone marrow-derived cells and decellularized tissue matrices. *Biomaterials* **26**: 1915–1924.

Christ, G. and Wingard, C. (2005). Calcium sensitization as a pharmacological target in vascular smooth muscle regulation. *Curr. Opin. Invest. Drug* **6**: 920–933.

Christ, G.J. and Barr, L. (2000). Neural control of smooth muscle. In: Barr, L. and Christ, G.J. (eds.), *A View of Smooth Muscle*. JAI Press, Greenwich, CT pp. 345–395.

Christ, G.J., Spray, D.C., el-Sabban, M., Moore, L.K. and Brink, P.R. (1996). Gap junctions in vascular tissues. Evaluating the role of intracellular communication in the modulation of vasomotor tone. *Circ. Res.* **79**: 631–646.

Christ, G.J., Wang, H.Z., Venkateswarlu, K., Zhao, W. and Day, N.S. (1999). Ion channels and gap junctions: their role in erectile physiology, dysfunction, and future therapy. *Mol. Urol.* **3**: 61–73.

Cines, D.B., Pollk, E.S., Buck, C.A., Loscalzo, J., Zimmerman G.A., McEver, R.P., Pober, J.S., Wick, T.M., Konkle, B.A., Schwartz, B.S., et al. (1998). Endothelial cells in physiology and in the pathophysiology of vascular disorders. *Blood* **91**: 3527–3561.

Collins, A.J., Kasiske, B., Herzog, C., Chen, S.C., Everson, S., Constantini, E., Grimm, R., McBean, M., Xue, J. and Chavers, B. (2003a). Excerpts from the United States Renal Data System 2003 Annual Data Report: atlas of end-stage renal disease in the United States. *Am. J. Kidney Dis.* **42**: A5–A7.

Collins, A.J., Li, S., Gilbertson, D.T., Liu, J., Chen, S.C. and Herzog, C.A. (2003b). Chronic kidney disease and cardiovascular disease in the medicare population. *Kidney Int. Suppl.* **87**: S24–S31.

del Valle-Rodriguez, A., Calderon, E., Ruiz, M., Ordonez, A., Lopez-Barneo, J. and Urena, J. (2006). Metabotropic Ca^{2+} channel-induced Ca^{2+} release and ATP-dependent facilitation of arterial myocyte contraction. *PCNS* **103**: 4316–4321.

Deutsch, M., Meinhart, J., Fischlein, T., Preiss, P. and Zilla, P. (1999). Clinical autologous *in vitro* endothelialization of infrainguinal ePTFE grafts in 100 patients: a 9-year experience. *Surgery* **126**: 847–855.

Gabouev, A.I., Schulthesiss, D., Mertsching, H., Koppe, M., Schlote, N., Wefer, J., Jonas, U. and Stief, C.G. (2003). *In vitro* construction of urinary bladder wall using porcine primary cells reseeded on acellularized bladder matrix and small intestinal submucosa. *Int. J. Artif. Organs* **26**: 935–942.

Gilbertson, D.T., Liu, J., Xue, J.L., Louis, T.A., Solid, C.A., Ebben, J.P. and Collins, A.J. (2005). Projecting the number of patients with end-stage renal disease in the United States to the year 2015. *J. Am. Soc. Nephrol.* **16**: 3736–3741.

Haddock, R.E. and Hill, C.E. (2005). Rhythmicity in arterial smooth muscle. *J. Physiol.* **566**: 645–656.

Haefliger, J.A., Nicod, P. and Meda, P. (2004). Contribution of connexins to the function of the vascular wall. *Cardiovasc. Res.* **62**: 345–356.

Hibino, N., Shin'oka, T., Matsumura, G., Ikada, Y. and Kurosawa, H. (2005). The tissue-engineered vascular graft using bone marrow without culture. *J. Thorac. Cardiovasc. Surg.* **129**: 1064–1070.

Hoerstrup, S.P., Zund, G., Sodian, R., Schnell, A.M., Grunenfelder, J. and Turina, M.I. (2001). Tissue engineering of small caliber vascular grafts. *Eur. J. Cardiothorac. Surg.* **20**: 164–169.

Hoerstrup, S.P., Cummings, I.G., Lachat, M., Schoen, F.J., Jenni, R., Leschka, S., Neuenschwander, S., Schmidt, D., Mol, A., Gunter, C., Gossi, M., Genoni, M. and Zund, G. (2006). Functional growth in tissue-engineered living, vascular grafts: follow-up at 100 weeks in a large animal model. *Circulation* **4**: I159–I166.

Hsu, C.Y., Vittinghoff, E., Lin, F. and Shlipak, M.G. (2004). The incidence of end-stage renal disease is increasing faster that the prevalence of chronic renal insufficiency. *Ann. Intern. Med.* **141**: 95–101.

Huber, T.S., Carter, J.W., Carter, R.L. and Seeger, J.M. (2003). Patency of autogenous and polytetrafluoroethylene upper extremity arteriovenous hemodialysis accesses: a systematic review. *J. Vasc. Surg.* **38**: 1005–1011.

Huber, T.S., Hirneise, C.M., Lee, W.A., Flynn, T.C. and Seeger, J.M. (2004). Outcome after autogenous brachial–axillary translocated superficial femoropopliteal vein hemodialysis access. *J. Vasc. Surg.* **40**: 311–318.

Huynh, T., Abraham, G., Murray, J., Brockbank, K., Hagen, P.O. and Sullivan, S. (1999). Remodeling of an acellular graft into a physiologically responsive neovessel. *Nat. Biotechnol.* **17**: 1083–1086.

Isenberg, B.C., Williams, C. and Tranquillo, R.T. (2006). Small-diameter artificial arteries engineered *in vitro*. *Circ. Res.* **98**: 25–35.

Jarajapu, Y. and Knot, H. (2005). Relative contribution of rho kinase and protein kinase C to myogenic tone in rat cerebral arteries in hypertension. *Am. J. Physiol. Heart Circ. Physiol.* **289**: 1917–1922.

Jeong, S.I., Kwon, J.H., Lim, J.I., Cho, S.W., Jung, Y., Kim, S.H., Kim, Y.H., Lee, Y.M., Kim, B.S., Choi, C.Y. and Kim, S.J. (2005). Mechano-active tissue engineering of vascular smooth muscle using pulsatile perfusion bioreactors and elastic PLCL scaffolds. *Biomaterials* **26**: 1405–1411.

Kaushal, S., Amiel, G.E., Guleserian, K.J., Shapira, O.M., Perry, T., Sutherland, F.W., Rabkin, E., Moran, A.M., Schoen, F.J., Atala, A., Soker, S., Bischoff, J., Mayer Jr., J.E. and Kohler, T.R. (2001). Functional small-diameter neovessels created using endothelial progenitor cells expanded *ex vivo*. *Nat. Med.* **7**: 1035–1040.

Kohler, T.R. and Kirkman, T.R. (1999). Dialysis access failure: a sheep model of rapid stenosis. *J. Vasc. Surg.* **30**: 744–751.

Kuecherer, H.F., Just, A. and Kirchheim, H. (2000). Evaluation of aortic compliance in humans. *Am. J. Physiol. Heart Circ. Physiol.* **278**: 1411–1413.

Kurpinski, K., Park, J., Thakar, R.G. and Li, S. (2006). Regulation of vascular smooth muscle cells and mesenchymal stem cells by mechanical strain. *Mol. Cell Biomech.* **3**: 21–34.

Laflamme, K., Roberge, C.J., Labonte, J., Pouliot, S., D'Orleans-Juste, P., Auger, F.A. and Germain, L. (2005). Tissue-engineered human vascular media with a functional endothelin system. *Circulation* **111**: 459–464.

Laflamme, K., Roberge, C.J., Pouliot, S., D'Orleans-Juste, P., Auger, F.A. and Germain, L. (2006). Tissue-engineered human vascular media produced *in vitro* by the self-assembly approach present functional properties similar to those of their native blood vessels. *Tissue Eng.* **12**: 2275–2281.

Lagaud, G., Davies, K.P., Venkateswarlu, K. and Christ, G.J. (2002a). The physiology, pathophysiology and therapeutic potential of gap junctions in smooth muscle. *Curr. Drug Targets* **3**: 427–440.

Lagaud, G., Karicheti, V., Knot, H.J., Christ, G.J. and Laher, I. (2002b). Inhibitors of gap junctions attenuate myogenic tone in cerebral arteries. *Am. J. Physiol. Heart Circ. Physiol.* **283**: H2177–H2186.

Laube, H.R., Duwe, J., Rutsch, W. and Konertz, W. (2000). Clinical experience with autologous endothelial cell-seeded polytetrafluoroethylene coronary artery bypass grafts. *J. Thorac. Cardiovasc. Surg.* **120**: 134–141.

Leyh, R.G., Wilhelmi, M., Rebe, P., Ciboutari, S., Haverich, A. and Mertsching, H. (2006). Tissue engineering of viable pulmonary arteries for surgical correction of congenital heart defects. *Ann. Thorac. Surg.* **81**: 1466–1470.

L'Heureux, N., Paquet, S., Labbe, R., Germain, L. and Auger, F.A. (1998). A completely biological tissue-engineered human blood vessel. *FASEB J.* **12**: 47–56.

L'Heureux, N., Dusserre, N., Konig, G., Victor, B., Keire, P., Wight, T.N., Chronos, N.A., Kyles, A.E., Gregory, C.R., Hoyt, G., Robbins, R.C. and McAllister, T.N. (2006). Human tissue-engineered blood vessels for adult arterial revascularization. *Nat. Med.* **12**: 361–365.

Lichtenberg, A., Tudorache, I., Cebotari, S., Suprunov, M., Tudorache, G., Goerler, H., Park, J.K., Hilfiker-Kleiner, D., Ringes-Lichtenberg, S., Karck, M., et al. (2006). Preclinical trials of tissue-engineered heart valves re-endothelialized under simulated physiological conditions. *Circulation* **114**: I559–I565.

McClellan, W.M. (1994). Epidemic end-stage renal disease in the United States. *Artif. Organs* **18**: 413–415.

McCulloch, A.D., Harris, A.B., Sarraf, C.E. and Eastwood, M. (2004). New multi-cue bioreactor for tissue engineering of tubular cardiovascular samples under physiological conditions. *Tissue Eng.* **10**: 565–573.

McFetridge, P.S., Bodamyali, T., Horrocks, M. and Chaudhuri, J.B. (2004). Endothelial and smooth muscle cell seeding onto processed *ex vivo* arterial scaffolds using 3D vascular bioreactors. *ASAIO J.* **50**: 591–600.

McKee, J.A., Banik, S.S., Boyer, M.J., Hamad, N.M., Lawson, J.H., Niklason, L.E. and Counter, C.M. (2003). Human arteries engineered *in vitro*. *EMBO Rep.* **4**: 633–638.

Michiels, C. (2003). Endothelial cell functions. *J. Cell Physiol.* **196**: 430–443.

Mironov, V., Kasyanov, V., McAllister, K., Oliver, S., Sistino, J. and Markwald, R. (2003). Perfusion bioreactor for vascular tissue engineering with capacities for longitudinal stretch. *J. Craniofac. Surg.* **14**: 340–347.

Moosmang, S., Schulla, V., Welling, A., Feil, R., Feil, S., Wegener, J.W., Hofmann, F., Klugbauer, N. (2003). Dominant role of smooth muscle L-type calcium channel $Ca_v1.2$ for blood pressure regulation. *EMBO J.* **22**: 6027–6034.

Naito, Y., Imai, Y., Shin'oka, T., Kashiwagi, J., Aoki, M., Watanabe, M., Matsumura, G., Kosaka, Y., Konuma, T., Hibino, N., Murata, A., Miyake, T. and Kurosawa, H. (2003). Successful clinical application of tissue-engineered graft for extracardiac Fontan operation. *J. Thorac. Cardiovasc. Surg.* **125**: 419–420.

Narita, Y., Hata, K.I., Kagami, H., Usui, A., Ueda, M. and Ueda, Y. (2004). Novel pulse duplicating bioreactor system for tissue-engineered vascular construct. *Tissue Eng.* **10**: 1224–1233.

Nasseri, B.A., Pomerantseval, L., Kaazempur-Mofrad, M.R., Sutherland, F.W., Perry, T., Ochoa, E., Thompson, C.A., Mayer Jr., J.E., Oesterle, S.N. and Vacanti, J.P. (2003). Dynamic rotational seeding and cell culture system for vascular tube formation. *Tissue Eng.* **9**: 291–299.

Nerem, R.M. (2003). Role of mechanics in vascular tissue engineering. *Biorheology* **40**: 281–287.

Neville, R.F., Abularrage, C.J., White, P.W. and Sidawy, A.N. (2004). Venous hypertension associated with arteriovenous hemodialysis access. *Semin. Vasc. Surg.* **17**: 50–56.

Niklason, L.E., Gao, J., Abbott, W.M., Hirschi, K.K., Houser, S., Marini, R. and Langer, R. (1999). *Science* **285**: 489–493.

Niklason, L.E., Abbott, W., Gao, J., Klagges, B., Ulubayram, K., Conroy, N., Jones, R., Vasanawala, A., Sanzgiri, S. and Langer, R. (2001). Morphological and mechanical characteristics of engineered bovine arteries. *J. Vasc. Surg.* **33**: 628–638.

Opitz, F., Schenke-Layland, K., Richter, W., Martin, D.P., Degenkolbe, I., Wahlers, T. and Stock, U.A. (2004a). Tissue engineering of ovine aortic blood vessel substitutes using applied shear stress and enzymatically derived vascular smooth muscle cells. *Ann. Biomed. Eng.* **32**: 212–222.

Opitz, F., Schenke-Layland, K., Cohnert, T.U., Halbhuber, K.J., Martin, D.P. and Stock, U.A. (2004b). Tissue engineering of aortic tissue: dire consequence of suboptimal elastic fiber synthesis *in vivo*. *Cardiovasc. Res.* **63**: 719–730.

Poh, M., Boyer, M., Solan, A., Dahl, S.L., Pedrotty, D., Banik, S.S., McKee, J.A., Klinger, R.Y., Counter, C.M. and Niklason, L.E. (2005). Blood vessels engineered from human cells. *Lancet* **364**: 2122–2124.

Portner, R., Nagel-Heyer, S., Goepfert, C., Adamietz, P. and Meenen, N.M. (2005). Bioreactor design for tissue engineering. *J. Biosci. Bioeng.* **100**: 235–245.

Rabkin, E. and Schoen, F.J. (2002). Cardiovascular tissue engineering. *Cardiovasc. Pathol.* **11**: 305–317.

Rafii, S. and Lyden, D. (2003). Therapeutic stem and progenitor cell transplantation for organ vascularization and regeneration. *Nat. Med.* **9**: 702–712.

Ratcliffe, A. (2000). Tissue engineering of vascular grafts. *Matrix Biol.* **19**: 353–357.

Safar, M.E., Blacher, J., Mourad J.J. and London, G.M. (2000). Stiffness of carotid artery wall material and blood pressure in humans, application to antihypertensive therapy and stroke prevention. *Stroke* **31**: 782–790.

Sales, K.M., Salacinski, H.J., Alobaid, N., Mikhail, M., Balakrishnan, V. and Seifalian, A.M. (2005). Advancing vascular tissue engineering: the role of stem cell technology. *Trends Biotechnol.* **23**: 461–467.

Shin'oka, T., Matsumura, G., Hibino, N., Naito, Y., Watanabe, M., Konuma, T., Sakamoto, T., Nagatsu, M. and Kurosawa, H. (2005). Midterm clinical result of tissue-engineered vascular autografts seeded with autologous bone marrow cells. *J. Thorac. Cardiovasc. Surg.* **129**: 1330–1338.

Shirota, T., He, H., Yasui, H. and Matsuda, T. (2003). Human endothelial progenitor cell-seeded hybrid graft: proliferative and antithrombogenic potentials *in vitro* and fabrication processing. *Tissue Eng.* **9**: 127–136.

Shum-Tim, D., Stock, U., Hrkach, J., Shin'oka, T., Lien, J., Moses, J., Stamp, A., Taylor, G., Moran, A.M., Landis, W., Langer, R., Vacanti, J.P. and Mayer Jr., J.E. (1999). Tissue engineering of autologous aorta using a new biodegradable polymer. *Ann. Thorac. Surg.* **68**: 2298–2305.

Soletti, L., Niepnice, A., Guan, J., Stankus, J.J., Wanger, W.R. and Vorp, D.A. (2006). A seeding device for tissue engineered tubular structures. *Biomaterials* **27**: 4863–4870.

Stitzel, J., Liu, J., Lee, S.J., Komua, M., Berry, J., Soker, S., Lim, G., Van Dyke, M., Czerw, R., Yoo, J.J. and Atala, A. (2006). Controlled fabrication of a biological vascular substitute. *Biomaterials* **27**: 1008–1094.

Teebken, O.E. and Haverich, A. (2002). Tissue engineering of small diameter vascular grafts. *Eur. J. Endovasc. Surg.* **23**: 475–485.

Teebken, O.E., Bader, A., Steinhoff, G. and Haverich, A. (2000). Tissue engineering of vascular grafts: human cell seeding of decellularized porcine matrix. *Eur. J. Vasc. Endovasc. Surg.* **19**: 381–386.

Teebken, O.E., Pichlmaier, A.M. and Haverich, A. (2001). Cell seeded decellularized allogeneic matrix grafts and biodegradable polydioxanone-prostheses compared with arterial autografts in a porcine model. *Eur. J. Vasc. Endovasc. Surg.* **22**: 139–145.

Thompson, C.A., Colon-Hernandez, P., Pomerantseva, I., MacNeil, B.D., Nasseri, B., Vacanti, J.P. and Oesterle, S.N. (2002). A novel pulsatile, laminar flow bioreactor for the development of tissue-engineered vascular structures. *Tissue Eng.* **8**: 1083–1088.

Tiwari, A., Salacinski, H.J., Hamilton, G. and Seifalian, A.M. (2001). Tissue engineering of vascular bypass grafts: role of endothelial cell extraction. *Eur. J. Vasc. Endovasc. Surg.* **21**: 193–201.

Vara, D.S., Salacinski, H.J., Kanna, R.Y., Bordenave, L., Hamilton, G. and Seifalian, A.M. (2005). Cardiovascular tissue engineering: state of the art. *Pathol. Biol.* **53**: 599–612.

Wang, H.Z., Day, N., Valcic, M., Hsieh, K., Serels, S., Brink, P.R. and Christ, G.J. (2001). Intracellular communication in cultured human vascular smooth muscle cells. *Am. J. Physiol. Cell Physiol.* **281**: C75–C88.

Watanabe, M., Shin'oka, T., Tohyama, S., Hibino, N., Konuma, T., Matsumura, G., Kosaka, Y., Ishida, T., Imai, Y., Yamakawa, M., Ikada, Y. and Morita, S. (2001). Tissue engineered vascular autograft: inferior vena cava replacement in a dog model. *Tissue Eng.* **7**: 429–439.

Weinberg, C.B. and Bell, E. (1986). A blood vessel model constructed from collagen and cultured vascular cells. *Science* **231**: 397–400.

Williams, C. and Wick, T.M. (2004). Perfusion bioreactor for small diameter tissue-engineered arteries. *Tissue Eng.* **10**: 930–941.

Williams, C. and Wick, T.M. (2005). Endothelial cell-smooth muscle cell co-culture in a perfusion bioreactor system. *Ann. Biomed. Eng.* **33**: 920–928.

Xu, J., Ge, H., Zhou, X., Yang, D., Guo, T., He, J., Li, Q. and Hao, Z. (2005). Tissue-engineered vessel strengthens quickly under physiological deformation: application of a new perfusion bioreactor with machine vision. *J. Vasc. Res.* **42**: 503–508.

Yang, J., Motlagh, D., Webb, A.R. and Ameer, G.A. (2005). Novel biphasic elastomeric scaffold for small-diameter blood vessel tissue engineering. *Tissue Eng.* **11**: 1876–1886.

Yu, H., Dai, W., Yang, Z., Kirkman, P., Weaver, F.A., Eton, D. and Rowe, V.L. (2003). Smooth muscle cells improve endothelial cell retention on polytetrafluoroethylene graft *in vivo*. *J. Vasc. Surg.* **38**: 557–563.

30

Cardiac Tissue

Milica Radisic and Michael V. Sefton

INTRODUCTION: FROM TISSUES TO ORGANS: KEY GOALS AND ISSUES

Nearly 8 million people in the United States have suffered from myocardial infarction, with 800,000 new cases occurring each year (American Heart Association, 2004). Myocardial infarction results in the substantial death of cardiomyocytes in the infarct zone followed by pathological remodeling of the heart. The remodeling process involves cardiac dilation, wall thinning and severe deterioration of contractile function leading to congestive heart failure in more than 500,000 patients in the United States each year (American Heart Association, 2002). Conventional therapies are limited by the inability of myocardium to regenerate after injury (Soonpaa and Field, 1998) and the shortage of organs available for transplantation. This chapter will focus on describing cell and tissue-based therapies that have been considered as novel treatment options (Reinlib and Field, 2000).

Regardless of the approach to regenerative medicine or the scope of the application (a vascular graft, a pediatric valve or an entire heart) there are three overlapping therapeutic goals – the three R's:

- Make tissue and organ **replacement** safer, more effective and more widely available.
- **Repair** tissues and organs without having to replace them.
- Enable tissues and organs to **regenerate** so that repair and regeneration become one and the same.

Furthermore, the problems of reaching these goals can be summarized (Table 30.1) in three categories (here largely in the context of tissue engineering) (Sefton, 2002; Sefton et al., 2005):

- **Cell number:** What is the source of cells to be used and how will large numbers be generated? How will they be supplied with nutrients and oxygen (and have wastes removed) within a device of reasonable volume?
- **Cell function:** How will the scaffold, extracellular matrix, and diffusible factors interact to generate the desired cell phenotype? How will the engineered tissue/organ function integrate with the host to ensure a functional outcome?
- **Cell durability:** What will happen over the long term as remodeling and/or the host immune/inflammatory system responds to the new tissue?

In order to replace, repair, or regenerate cardiovascular tissue, these central issues of regenerative medicine will need to be addressed. Some of these issues (Table 30.1) reflect the fundamental nature of how an organ is different from a tissue: the large size and 3-dimensional (3D) structure and the presence of multiple cell types that work in unison. Beyond these largely scientific challenges, there are the no less critical, practical questions of manufacturing, sterilization, storage and distribution, and the regulatory and public policy issues that will need to be addressed before such therapies can be made available to the patients who are expected to benefit. Furthermore we will also need new imaging or other non-invasive strategies to monitor the success (or not) of these therapies (i.e. to enable the translation into clinical practice).

Copyright © 2008 Elsevier Inc. All rights reserved

Table 30.1 Critical issues associated with tissue engineering a heart (with permission from Sabiston and Spencer, Sefton et al., 2005)

	Objective	Critical issues
Cell number	• ~300 g of cells (3×10^{11} cells) • ~200 mL O_2/h[a]	• Cell source/purity • Vascularization
Function	• Cellular phenotype (multiple cell types) • Co-ordinated muscle contraction • Pump blood • Connect to circulation	• Microenvironment (soluble and insoluble factors) • Pacemaker and electrical conduction • Valves and conduits • Biomechanical elasticity and strength • Non-thrombogenicity
Durability	• Fatigue resistance • Hypoxia and disease tolerance • Host tolerance	• Biocompatibility • Remodeling • Innate/adaptive immune response

- Manufacturing and quality control
- Ethical, legal and social issues
- Imaging and non-invasive diagnostics
- Regulatory and public policy issues

[a] Based on moderate activity, (Burton, 1972)

CELL AND GENE THERAPY

Cell Therapy

Treatment options for heart failure and myocardial infraction (MI) are limited by the inability of adult cardiomyocytes to proliferate and regenerate injured myocardium. Cell injection, has thus emerged as an alternative treatment option. In animal models, injection of fetal or neonatal cardiomyocytes improved left ventricular (LV) function and ventricle thickness, thus attenuating pathological ventricular remodeling (Reinecke et al., 1999; Muller-Ehmsen et al., 2002a, b). Differentiated cardiomyocytes are indeed an ideal cell source for injection or tissue engineering, since they contain a developed contractile apparatus and can integrate through gap junctions and intercalated disks with the host cardiomyocytes. However, large numbers of clinically relevant autologous cardiomyocytes are unavailable.

In searching for an appropriate cell source (Table 30.2), regeneration of infarcted myocardium has been attempted in animal models by transplantation of skeletal myoblasts (Dorfman et al., 1998), as well as cardiomyocytes derived from embryonic stem (ES) cells (Klug et al., 1996) and bone-marrow-derived mesenchymal stem cells (MSCs) (Toma et al., 2002b). For a review of cell therapy approaches see Laflamme and Murry (2005).

The obvious advantage of skeletal myoblasts is that they can be harvested from the patient and expanded *in vitro*. However, mature skeletal myoblasts do not express gap junctional proteins, thus they are incapable of functionally integrating with the host myocardium. This was the most likely reason for the occurrence of arrhythmias in four out of ten patients in a Phase 1 clinical trial of autologous skeletal myoblast transplantation (Menasche et al., 2003). For further information on myoblast clinical trails see Laflamme and Murry (2005).

Hematopoietic stem (HS) cells from bone marrow were tested in their ability to contribute to the regeneration of infarcted myocardium. The general consensus on the effect of injection or the mechanism of action has not been reached yet. Anversa and colleagues (Orlic et al., 2001) reported that HS cells injected into

Table 30.2 **Cell sources for cardiac tissue engineering, and some of their advantages and disadvantages (with permission, Sefton et al., 2005)**

Cell sources	Advantages	Disadvantages
Adult cardiac cells	Target cell source	Little proliferative or developmental potential, limited resource
Fetal cardiac cells	Some proliferative potential, appropriate developmental potential; demonstrated efficacy	Limited resource; ethical considerations
Endothelial progenitor cells	Some proliferative potential; may elicit *in vivo* healing through indirect mechanisms	Appropriate developmental potential yet to be demonstrated; may not be appropriate for larger tissue replacement or *in vitro* tissue engineering
Adult bone-marrow-derived cells	Significant *in vitro* proliferative potential; some demonstration of efficacy	Appropriate developmental potential to be demonstrated; safety tolerance after *in vitro* culture to be determined
ES cells	Significant *in vitro* proliferative potential; demonstration of efficacy; appropriate developmental potential; sustainable resource	*In vitro* culture may introduce genetic changes; safety tolerance after *in vitro* culture and differentiation to be determined

the peri-infarct zone in mice with acute MI gave rise to cardiomyocytes regenerating ~68% of the infarct. These results could not be reproduced by other groups (Balsam et al., 2004; Murry et al., 2004). Instead the studies suggest that HS cells differentiate into blood cells (Murry et al., 2004; Nygren et al., 2004), and occasionally fuse with host cardiomyocytes. The discrepancy may lie in the different techniques used.

Bone marrow MSCs have also been considered as a cell source for myocardial repair. When injected directly into the hearts of mice (Toma et al., 2002a) and pigs (Shake et al., 2002) post-infarction, the cells attenuated pathological ventricle remodeling and expressed cardiac markers. Contribution of cell fusion to these events remains to be determined. Bone marrow mononuclear cells (BMNCs) (consisting of both HS and MSC) were evaluated in clinical trials (for a review see Dimmeler et al., 2005). In general, the initial clinical studies indicate that bone marrow transplantation is safe and contributed to the increase in ejection fraction (Chen et al., 2004; Wollert et al., 2004) although the mechanism of the effect is unclear. The main advantage of bone marrow as a cell source is that it can be harvested from the patient; however, the frequency of stem cells is generally low (<0.1%).

Recent emerging work suggests that the heart may contain resident progenitor cells. This is an exciting possibility, as resident progenitors may be an ideal source of autologous cardiomyocytes. However, it appears that there is more than one heart cell subpopulation that fits the description of a cardiac progenitor. C-kit+ cells isolated from adult rat hearts and expanded under limited dilution gave rise to cardiomyocytes, smooth muscle, and endothelial cells (ECs) when injected into ischemic myocardium (Beltrami et al., 2003). Oh et al. (2003) reported Sca-1 as a marker of resident cardiac progenitors, and expression of cardiac markers upon treatment with 5-azacytidine. LIM homeodomain islet 1 transcription factor (isl1+) was also identified as a marker of resident cardiac progenitor cells (Laugwitz et al., 2005). The isl1+ cells from mouse hearts were propagated in culture and they differentiated into functional cardiac myocytes when in contact with terminally differentiated cardiomyocytes. It remains to be determined if the progenitors, regardless of their marker, can be isolated from adult human biopsies and if sufficient numbers of cardiomyocytes ($>10^8$ cells/patient) can be generated *in vitro*.

ES cells have enormous proliferative potential, and in combination with nuclear transfer can generate autologous cells. However, the main technical concern in utilization of ES cells is that the presence of a single undifferentiated cell *in vivo* can potentially yield teratomas (Laflamme and Murry, 2005). Highly pure populations of cardiomyocytes (~99.6%) can be generated using a neomyocin-resistant transgene driven by a cardiac marker promoter (Klug et al., 1996; Zandstra et al., 2003). Upon injection into hearts, the ES cell-derived cardiomyocytes formed stable intracardiac grafts (Klug et al., 1996) and improved contractile function (Etzion et al., 2001). Electromechanical integration of the cardiomyocytes derived from human ES cells with the host myocardium was also reported (Kehat et al., 2001).

Besides focusing on restoration of contractile function through injection of myogenic cells, regeneration of infarcted myocardium has also been attempted through injection of EC progenitors (Kocher et al., 2001). The regeneration is based on the improvements in infarct neovasculature that lead to improved perfusion and ultimately improved LV function.

In most cases described above, the cells were suspended in an appropriate liquid (saline or culture medium) followed by intramyocardial or coronary injection. The main challenges associated with this procedure are poor survival of the injected cells (Muller-Ehmsen et al., 2002b) and washout from the injection site (Reffelmann and Kloner, 2003). According to some estimates, 90% of the cell delivered through a needle leak out of the injection site (Muller-Ehmsen et al., 2002a, b). In addition, significant number of cells (~90%) die within days after injection (Zhang et al., 2001; Muller-Ehmsen et al., 2002b). Thus developing improved delivery and localization methods (e.g. hydrogels) and effective anti-death strategies (e.g. heat shock treatment) could significantly improve effectiveness of cell injection procedures.

Gene therapy

Gene therapy approaches are based on either delivering exogenous genes capable of expressing therapeutic proteins or on the blockade of genes involved in pathological process. The genes can be delivered using non-viral vectors (such as naked plasmids, liposome formulation, and synthetic peptides) or recombinant viruses. Replication defective recombinant viruses are significantly more effective in gene transfer to myocardium compared to the non-viral vectors that are limited by high degradation rate and low genomic integration (Melo et al., 2004a). However, viruses sometimes lead to immune reaction, and there is a small risk that they may become proliferative.

In an early work aimed at converting the non-contractile scar tissue into tissue capable of contraction, Murry et al. (1996) used adenovirus to transfer MyoD, a myogenic determination gene, into granulation tissue of rat myocardium post-infraction. *In vitro*, gene transfer converted fibroblasts into skeletal muscle cells. Similar results (i.e. expression of MyoD, myogenin, and embryonic isoform of myosin heavy chain) were observed *in vivo* after transfection with high doses of virus (10^{10}pfu). Restoration of contractile function has also been attempted by normalization of β-adregenic receptor signaling. In rabbits, intracoronary delivery of β2-adregenic receptor gene led to improvements in LV and hemodynamic function (Maurice et al., 1999). Using similar approach, the β-adregenic receptor signaling was rescued in ventricular myocytes from patients with heart failure. Calcium signaling was another target for gene therapy aimed at restoration of contractile function (review in Hajjar et al., 2000). Intracoronary delivery of SERCA2a genes in a rat model of heart failure improved long-term survival, restored systolic and diastolic function, and improved Ca2+ ATP-ase activity (del Monte et al., 2001). Antisense inhibition of phospholamban was shown to improve contractility of cardiomyocytes from end-stage heart failure patients (del Monte et al., 2002).

Gene therapies for acute MI were limited by the available delivery techniques. In general, the time it takes for transcription and translation is too long for a successful intervention in acute MI (Melo et al., 2004b). However, individuals at risk may benefit from preventive strategies that protect from ischemia/reperfusion injury. In that respect, overexpression of antioxidant enzyme systems (HO-1), heat shock proteins and survival genes (Bcl-2 Akt) was demonstrated to be beneficial in small animal models (Melo et al., 2004b).

Most recently, a novel gene therapy approach was reported for treatment of acute MI and chronic ischemia. Intramyocardial injection of naked DNA encoding human sonic hedgehog preserved LV function, enhanced neovascularization, and reduced fibrosis and cardiac apoptosis. Sonic hedgehog is a morphogen and a crucial regulator of organ development during embryogenesis, thus transient reconstruction of embryonic signaling had beneficial effect on tissue repair and neovascularization (Kusano et al., 2005).

Gene therapy was also utilized to treat ischeamia in patients with coronary artery disease who were not eligible for standard treatment options such as percutaneous angioblasty or surgical vascularization. A number of pre-clinical and clinical trails focused on overexpression of vascular endothelial growth factor (VEGF), fibroblast growth factor (FGF) and hepatocyte growth factor in an attempt to improve collateral blood vessel formation (Melo et al., 2004b). Although functional improvements were reported in large animals, phase II and III clinical trials failed to conclusively prove efficacy and the long-term therapeutic effect (Yla-Herttuala et al., 2004; Markkanen et al., 2005). Although the safety record was excellent in all of the trails, the following reasons were considered as possible causes for the disappointing results in efficacy: a wrong dose, a less-than-optimal route of administration, an inefficient delivery system, an insufficient duration of the treatment, selection of an appropriate animal model in pre-clinical trials as well as selection of an appropriate patient group.

While all of the above-mentioned limitations are technical in nature, targeting a single gene, as most commonly used in gene therapy, may have conceptual limitations as well. Most pathological processes are complex and involve expression or down-regulation of multiple genes. In many instances this genetic complexity is not well understood and thus it is difficult to predict a prior what the ultimate effect of overexpression or blockade of a single gene will be. In this respect combination of gene and cell therapy may be a preferred approach in the treatment of heart diseases.

One of the major limitations of cell therapy approaches is low cell survival. Thus, transfecting the injected cells with agents that enhance angiogenesis or cell survival may benefit the cell injection procedure. Once in the appropriate location, the cells may contribute to the contractile function and adjust appropriately to the complex physiological stimuli of the local milieu. Li and colleagues demonstrated that injection of $VEGF_{165}$ transfected cardiomyocytes into cryoinjured rat myocardium sustained VEGF expression and increased capillary density in the border zone as well as regional blood flow within the scar (Sakai et al., 2001). Most other studies focused on the injection of cardiomyocytes expressing growth factors (for review see Fazel et al., 2005) consistently reported that a combination of cell and gene therapy results in improved angiogenesis and functional properties in comparison to cell therapy alone.

SCAFFOLD-BASED APPROACHES

While small infarcts may be treated with cell therapy, larger areas of damaged tissue will require excision and replacement with a cardiac patch. The time post-infarction is critical in the success of any regeneration strategy. Upon myocardial infarction, a vigorous inflammatory response is elicited and dead cells are removed by marrow-derived macrophages. Over the subsequent weeks to months, fibroblasts and ECs proliferate forming granulation tissue and ultimately dense collagenous scar. Formation of scar tissue severely reduces contractile function of the myocardium and leads to ventricle wall thinning and dilatation, remodeling, and ultimately heart failure. The best regeneration strategy thus depends on the time post-infarction, that is, new and old infarcts most likely cannot be treated using the same approach.

Cell injection strategies will work best if applied shortly after MI. Application of cells and growth factors within hours and days after MI has a potential of directing the wound repair process so that the minimum amount of scar tissue is formed, the contractile function is maintained in the border zone, and pathological remodeling is attenuated. Tissue engineering strategies will work in the acute phase as well, but may be more necessary after scar has formed. In that case, larger areas of heart must be replaced or augmented and this is potentially where a scaffold-based approach may be most useful.

Cell-free cardiac patches

Patients with large transmural akinetic scars often benefit from the Dor procedure (endoventricular circular patch plasty) (Dor et al., 1989; Di Donato et al., 1997). In some cases, however, the success of this procedure is temporary, thus motivating the need for viable tissue patches. In this procedure the scar tissue is excised and the ventricle is closed using a circular Dacron (polyethylene terephthalate) patch lined with endocardium.

Another strategy to address pathological remodeling and prevent heart failure is a CorCap cardiac support device. CorCap is an implant-grade polyethylene terephthalate mesh that is wrapped around the heart ventricle to prevent further dilatation and support contractile function. In clinical trials, it was demonstrated that it results in improved quality of life, as well as improved heart size and shape (Starling and Jessup, 2004).

Cell-based cardiac patches

Self-assembly: In cardiac tissue engineering approaches, most studies suggest that some type of scaffold, an inductive 3D matrix, is necessary to support assembly of cardiac tissue *in vitro*. An important scaffold-free approach includes stacking of confluent monolayers of cardiomyocytes (Shimizu et al., 2002). Although cardiac patches obtained in this way generate high active force, engineering patches more than 2–3 cell layer thick remains a problem. Recently, 24-mm long and 100-μm thick contractile cardiac organoids were fabricated by self-organization (Baar et al., 2005). Cardiomyocytes were cultivated on a poly(dimethylsiloxane) (PDMS) surface coated with laminin. As laminin degraded, the confluent monolayer detached from the periphery of the substrate moving towards the center and wrapping around a string placed in the center of the plate until a cylindrical contractile organoid was formed.

The scaffold approaches can be divided into: (i) hydrogel approaches where cells are either encapsulated and cultivated *in vitro* or injected directly into MI without pre-culture and (ii) porous and fibrous 3D scaffold approaches where scaffolds are seeded with cells and in most cases cultivated *in vitro* prior to the utilization as cardiac patches.

Hydrogels

The most important example of hydrogel-based cardiac tissue engineering includes the work of Eschenhagen and colleagues. Cardiomyocytes were cast in growth factor supplemented collagen gels and cultivated in the presence of cyclic mechanical stretch (Eschenhagen et al., 1997; Fink et al., 2000; Zimmermann et al., 2000; Zimmermann et al., 2002a, b). The main advantage of the hydrogel approach is the higher active force generated by such cardiac tissues, compared to the force generated by tissues on porous or fibrous 3D scaffolds. In addition, collagen and laminin are the main components of the myocardial extracellular matrix, thus they are supportive of cardiomyocyte attachment and elongation. However, the main remaining challenge is tailoring the shape and dimensions of such tissues. One interesting approach to address this issue is the use of extruded collagen type I tubes (Yost et al., 2004).

A technique that can potentially combine the advantages of the hydrogel approach with ease in tailoring tissue shape and size is inkjet printing. Cardiac constructs based on feline cardiomyocytes were created by printing cell solution onto alginate and using calcium as a cross-linking agent. This approach may be particularly useful for co-culture (Tao et al., 2004) as it enables precise control over cell location in the tissue construct.

Without pre-culture, hydrogels were utilized to provide structural stability and deliver cells for regeneration of infarcted myocardium. Various cell types were injected into myocardium using a biomaterial that crosslinks upon injection such as Matrigel (Balsam et al., 2004), fibrin glue (Christman et al., 2004a, b; Ryu et al., 2005), or self-assembled peptide hydrogels (Davis et al., 2005). In general, the studies report prevention of ventricle dilatation and improvement of fractional shortening as well as improved angiogenesis.

Kofidis et al. (Balsam et al., 2004) reported that injection of Matrigel or Matrigel and ES cells into infarcted rat hearts resulted in structural stabilization, prevented ventricular wall thinning and improved fractional shortening. Chirsman et al. (2004a, b) demonstrated that injection of skeletal myoblasts into myocardial infarcts using fibrin glue increased cell localization within the infarct after 5 weeks, reduced infarct size and increased vascularization of the scar without causing a significant inflammatory response or foreign body reaction. Similarly, Ryu et al. (2005) found that injection of BMNCs into cryoinjured rat myocardium using fibrin matrix increased the amount of viable tissue and microvessel formation and reduced the amount of fibrous tissue in comparison to the injection of BMNC in culture medium or culture medium alone. Recently, it was demonstrated that a synthetic material, self-assembling peptide hydrogel, can also be utilized for cell injection into the myocardium (Davis et al., 2005). Upon injection, the peptide formed a nanofibrous structure that promoted recruitment of endogenous cells expressing endothelial markers and supported survival of injected cardiomyocytes.

Porous scaffolds

Three-dimensional cardiac tissue constructs were successfully cultivated in dishes using a variety of scaffolds amongst which collagen sponges were the most common. In the pioneering approach of Li et al. (1999), fetal rat ventricular cardiac myocytes were expanded after isolation, inoculated into collagen sponges and cultivated in static dishes for up to 4 weeks. The cells proliferated with time in culture and expressed multiple sarcomeres. Adult human ventricular cells were used in a similar system, although they exhibited no proliferation (Li et al., 2000).

Fetal cardiac cells were also cultivated on porous alginate scaffolds in static 96-well plates. After 4 days in culture the cells formed spontaneously beating aggregates in the scaffold pores (Leor et al., 2000). Cell seeding densities of the order of 10^8 cells/cm^3 were achieved in the alginate scaffolds using centrifugal forces during seeding (Dar et al., 2002). Neonatal rat cardiomyocytes formed spontaneously contracting constructs when inoculated in collagen sponges (Tissue Fleece) within 36 h after seeding (Kofidis et al., 2003) and maintained their activity for up to 12 weeks. The contractile force increased upon addition of Ca^{2+} and epinephrine.

Fibrous scaffolds

In a classical tissue engineering approach, fibrous polyglycolic acid (PGA) (Figure 30.1a) scaffolds were combined with neonatal rat cardiomyocytes and cultivated in spinner flasks and rotating vessels (Carrier et al., 1999). The scaffold was 97% porous and consisted of non-woven PGA fibers 14 μm in diameter. This material has advantages from a clinical stand point since it is FDA approved and found in biodegradable sutures.

Neonatal rat or embryonic chick ventricular myocytes were seeded onto (PGA) scaffolds by placing a dilute cell suspension in the spinner flasks and mixing for 3 days (50 rpm) (Carrier et al., 1999). Mixing in the spinner flasks (0, 50, or 90rpm) had a significant effect on the construct metabolism and cellularity. Constructs cultivated in well mixed flasks had significantly higher cellularity index and metabolic activity compared to the constructs cultivated in the static flasks. After 1 week of culture, constructs seeded with neonatal heart cells contained a peripheral tissue-like region (50–70 mm thick) in which cells stained positive for tropomyosin and organized in multiple layers in a 3-D configuration (Bursac et al., 1999) (Figure 30.1a and b). Electrophysiological studies conducted using a linear array of extracellular electrodes showed that the peripheral layer of the constructs exhibited relatively homogeneous electrical properties and sustained macroscopically continuous impulse propagation on a centimeter size scale (Bursac et al., 1999). Constructs based on the cardiomyocytes enriched by preplating exhibited lower excitation threshold (ET), higher

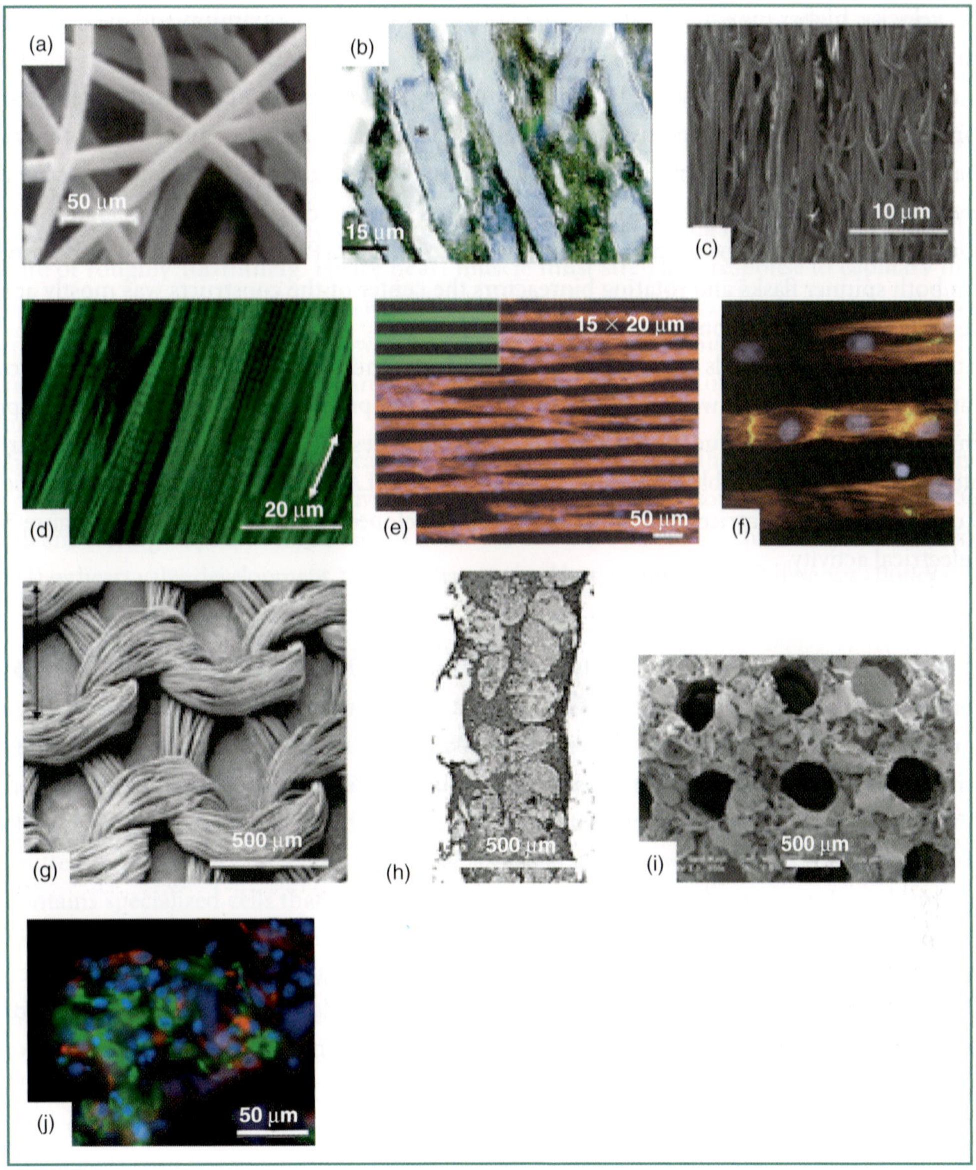

Figure 30.1 Representative scaffolds used in cardiac tissue engineering. (a) Scanning electron micrograph of a non-woven fibrous PGA scaffold used in a classical approach by Freed and colleagues. (b) Immunohistochemical staining for tropomyosin in constructs based on surface-hydrolyzed PGA seeded with neonatal rat cardiomyocytes and cultivated in rotating vessels for 1 week (with permission from Papadaki et al., 2001). (c) Scanning electron micrograph of a fibrous PLA scaffold obtained by electrospinning followed by uniaxial stretching. (d) Neonatal rat cardiomyocytes cultured on oriented PLA scaffolds exhibited well-developed contractile apparatus (actin – green) (with permission from Bui et al., 2005). (e) Thin polylactide-co-glycolic acid (PLGA) films patterned with laminin using microcontact printing (inset: 15 μm laminin lanes spaced 20 μm apart) and seeded with neonatal rat cardiomyocytes (actin filaments – red, nuclei – blue). (f) Immunohistochemical staining illustrates elements of intercalated disks (N-cadherin-yellow, actin filaments-red) (with permission from McDevitt et al., 2002). (g) Scanning electron micrograph of the knitted Hylonect fabric; arrow indicates the direction of cyclic stretch applied during culture. (h) Cross-section of a construct sampled 2 h after cell seeding, showing the multifilament yarn and immunohistochemical staining for cardiac troponin I. Neonatal rat cardiomyocytes were inoculated into the scaffold using fibrin (with permission from Boublik et al., 2005). (i) Parallel channel array bored in the PGS scaffolds using CO_2 laser/scanning engraving system. (j) Neonatal rat heart cells seeded onto channeled PGS scaffolds using Matrigel™ and cultivated in perfusion with 5.4 vol% perfluorocarbon emulsion supplemented culture medium (vimentin stained fibroblasts – red, troponin I stained cardiomyocytes – green, nuclei – blue) (with permission from Radisic et al., 2006).

clinical success (Meinhart et al., 2001) has renewed enthusiasm for seeding grafts with ECs. In some protocols, many of the pre-seeded cells are lost on implantation due to insufficient adhesion (Williams, 1995) and thus the protection from thrombosis provided by the cells is limited due to the incomplete cell coverage. The potential to exploit the presence of circulating EC progenitors has only begun to be explored (Rafii, 2000).

It is also worth noting the effects of the endothelium on the neighboring tissue and the corresponding effects on EC phenotype. With vascular smooth muscle cell (VSMC), this bidirectional cross-talk is thought to be a critical regulator of vascular homeostasis (Korff et al., 2001): secretion and expression of molecules such as nitric oxide (Palmer et al., 1988), prostacyclin (Moncada, 1982), and endothelin (Mawji and Marsden, 2003) act on VSMC to regulate vessel tone. Meanwhile, VSMC inhibits EC endothelin 1 (ET-1) production to increase EC NO and eNOS expression (Di Luozzo et al., 2000). Many other relevant systems (e.g. matrix metalloprotease (MMP) secretion and matrix remodeling) are also affected by the interactions between EC and other cell types.

Vascularization

The intrinsic nature of large cell-based constructs and the corresponding difficulty of supplying cells deep within the construct with nutrients is yet another problem. Diffusion is fine for 100 μm or so and low cell densities can extend this limit, but at the cost of making constructs too large to be useful. Thin or essentially 2D (e.g. a tube) constructs are feasible without an internal blood/nutrient supply. However it is hard to combine cells at tissue densities ($>10^8$ cells/cm^3) into large tissues without some sort of prevascularization or its alternative. Thus, a capillary network (and a lymphatic network) needs to be "engineered" as part of the creation of a larger structure.

In a cell-free approach, vascularization and improvement of LV function following MI were achieved by sustained release of basic FGF (bFGF) incorporated into gelatin microspheres (Sakakibara et al., 2003), aFGF from ethylene vinyl acetate copolymer (Sellke and Simons, 1999) and bFGF from heparin-alginate beads (Harada et al., 1994). Mooney and colleagues have incorporated an EC mitogen (VEGF) into three-dimensional porous poly(lactide-co-glycolide) (PLG) scaffolds during fabrication (Sheridan et al., 2000) to promote scaffold vascularization. Sustained delivery of bioactive VEGF translated into a significant increase in blood vessel ingrowth in mice and the vessels appeared to integrate with the host vasculature. We are using microencapsulated $VEGF_{165}$ secreting cells (prepared by transfection of L929 cells) as a means of exploring this strategy, at least for microcapsules (Vallbacka et al., 2001). Of course VEGF is but one angiogenic factor (Ahrendt et al., 1998) and issues associated with the functional maturity of the vessels and the need for multiple factors may limit this strategy. In a third approach, Vacanti et al., micromachined a hierarchical branched network mimicking the vascular system in 2D. Silicon and Pyrex surfaces were etched with branching channels ranging from 500–10 μm in diameter (Kaihara et al., 2000) that were then seeded with rat hepatocytes and microvascular ECs.

Most recently, prevascularized skeletal muscle was created (Levenberg et al., 2005) by co-culturing skeletal muscle cells with ES-cell-derived EC and fibroblasts. It appeared that up to 40% of the engineered blood vessels "connected" to the host vasculature upon implantation, at least in this small animal model. Finally we note that there are initial attempts at adapting endothelial seeding approaches in a modular approach to create scalable and vascularized tissue constructs (Figure 30.2b). ECs were seeded onto sub-mm sized collagen gel cylindrical modules that contained a second cell (e.g. HepG2 or smooth muscle cells or in the future perhaps cardiomyocytes). These modules were packed into a larger tube, thereby creating interconnected channels lined with ECs. These channels permitted the perfusion of whole blood, creating a means of producing uniform, scaleable tissue constructs with an internal vascular supply (McCuigan and Sefton, 2006).

Host Response and Biocompatibility

Questions related to the immune and inflammatory response to tissue constructs are starting to draw attention. The host response to a tissue engineered construct is manifested by the innate and adaptive immune

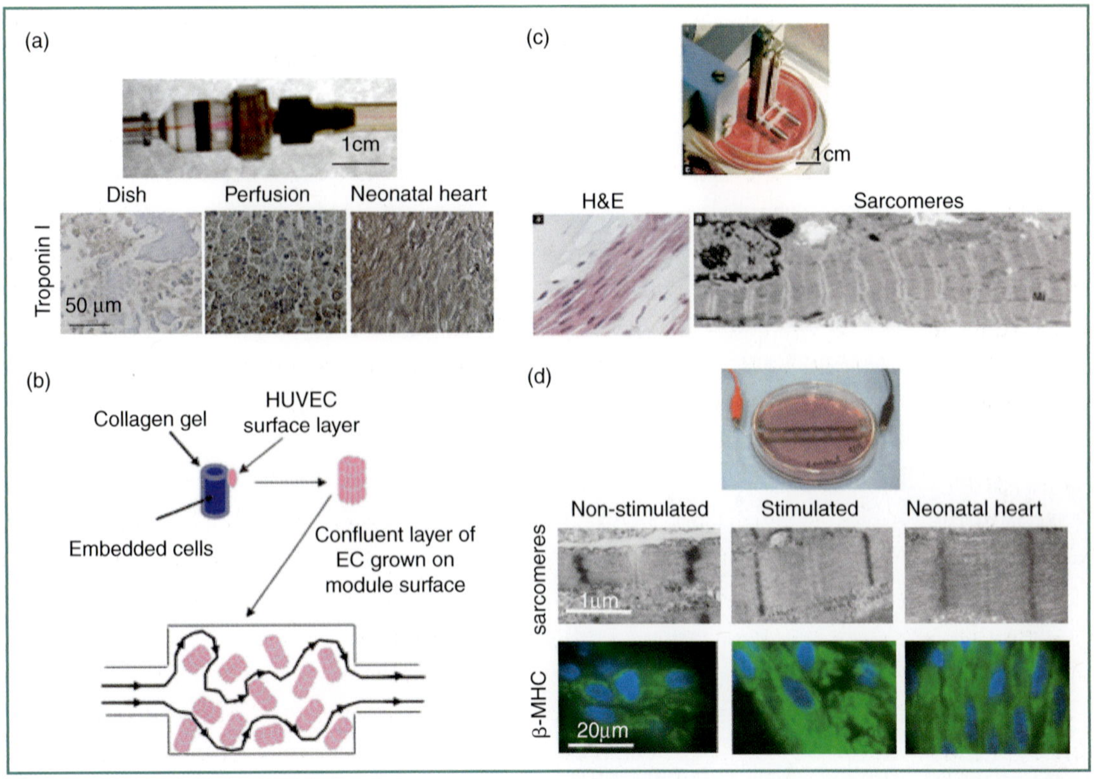

Figure 30.2 Cardiac tissue engineering culture systems focus on achieving adequate oxygen supply for highly metabolically active cells (a, b) and providing appropriate physical cues that lead to differentiated phenotype (c, d). (a) Direct culture medium perfusion of constructs based on neonatal rat cardiomyocytes inoculated into collagen sponges using Matrigel. Medium perfusion resulted in uniform cell distribution and maintenance of cell viability. Immunohistochemical staining illustrated cross-sectional distribution of cells expressing cardiac Troponin I (with permission from Radisic et al., 2004). (b) Modular tissue engineering approach using sub-mm sized EC seeded collagen modules assembled into a larger tube or construct (with permission from McGuigan and Sefton, 2006). (c) Zimmermann and Eschenhagen designed a bioreactor that provides cyclic mechanical stretch to engineered heart tissue based on neonatal rat cardiomyocytes and collagen gel. Mechanical stimulation yielded elongated cardiomyocytes with remarkably well-developed contractile apparatus (with permission from Zimmermann et al., 2002). (d) Cardiac-like electrical field stimulation was applied to collagen sponges inoculated with suspension of neonatal rat cardiomyocytes in Matrigel, resulting in differentiated phenotype and improved tissue assembly (with permission from Radisic et al., 2004).

systems, involving both plasma (e.g. complement) and cellular components (e.g. macrophages, T cells, etc.), that are directed against engineered cells and grafts or the materials used in tissue constructs. This potent immune response is most often mediated by major histocompatability complex mismatches between donor and host tissue in allogeneic transplantations. This response can also be manifested in situations where autologous cells or tissues are engineered to express therapeutic but foreign factors or if these autologous cells are placed in tissue constructs that themselves negatively impact immune consequences (Mikos et al., 1998).

Immunosuppressants have enabled the successful transplantation of kidneys, hearts, and other organs. With the advent of tissue engineering, new configurations of tissues and organs (often with an added biomaterial component) are being developed and our understanding of the immune and inflammatory response to these new therapies is being shown to be inadequate. Some xenogeneic cell transplants (mice to rat) survive in situations of cardiac repair despite the species differences (Saito et al., 2002) although this may be specific to the animal model or to cardiac repair. The longevity of a transplant is also dependent on the ability of somatic cells to withstand and respond to the stresses of implantation, rejection, and other injuries (Halloran and Melk, 2001). The classic "foreign body reaction" to biomaterials is well known, but the details of the molecular signals (complement regulatory proteins, MMPs) that accompany this phenomenon (in the context of biomaterials) are only beginning to be defined.

A variety of approaches have been undertaken or are in development to generate or to improve upon graft acceptance (Rossini et al., 1999). These approaches include methods to block the innate immune response such as by use of drugs or transferred genes to block NFkB signaling pathways, for example. Other methods to block the innate response include the use of antibodies to IL-1 or TNF or the use of anti-adhesion and anti-elastase antibodies. Perhaps nuclear transfer and therapeutic cloning strategies (McLaren, 2000) may be necessary assuming the various ethical issues can be resolved. We must better understand the mechanism of the host response itself so that we can design better biomaterials, select or engineer more suitable cells or devise better strategies for controlling both innate and adaptive immune responses and enable a functional integration of the new tissue with the host.

BIOREACTORS AND CONDITIONING

Major efforts in the development of bioreactors for tissue engineering of myocardium focus on (a) providing sufficient oxygen supply for the highly metabolically active cardiomyocytes and (b) providing appropriate physical stimuli necessary to reproduce complex structure at various length scales (subcellular to tissue).

The most common culture vessels utilized for tissue engineering of the myocardium include static or mixed dishes, static or mixed flasks, and rotating vessels. These bioreactors offer three distinct flow conditions (static, turbulent, and laminar) and therefore differ significantly in the rate of oxygen supply to the surface of the tissue construct. Oxygen transport is a key factor for myocardial tissue engineering due to the high cell density, very limited cell proliferation, and low tolerance of cardiac myocytes for hypoxia. In all configurations oxygen is supplied only by diffusion from the surface to the interior of the tissue construct, yielding $\sim 100\,\mu m$ thick surface layer of compact tissue capable of electrical signal propagation and an acellular interior (Radisic et al., 2005).

Oxygen Supply

In an attempt to enhance mass transport within cultured constructs, a perfusion bioreactor that provides interstitial medium flow through the cultured construct at velocities similar to those found in native myocardium ($\sim$400–500 $\mu m/s$) was developed (Radisic et al., 2004b). In such a system oxygen and nutrients were supplied to the construct interior by both diffusion and convection (Figure 30.2a). Interstitial flow of culture medium through the central 5 mm diameter $\times$ 1.5 mm thick region resulted in physiologic density of viable and differentiated, aerobically metabolizing cells. In response to electrical stimulation, perfused constructs contracted synchronously had lower ET and recovered their baseline function levels of ET and MCR following treatment with a gap junctional blocker; dish-grown constructs exhibited arrhythmic contractile patterns and failed to recover their baseline MCR levels. These studies suggested that the immediate establishment and maintenance of interstitial medium flow markedly enhanced the control of oxygen supply to the cells and thereby enabled engineering of compact constructs. However, most cells in perfused constructs were round and mononucleated, indicating that some of the regulatory signals, either molecular or physical, were not present in the culture environment.

In another approach, a separate compartment for medium flow was created by perfusing channeled scaffolds in a configuration resembling the capillary network *in vivo*. Neonatal rat heart cells were inoculated into the pores of an elastic, highly porous scaffold (PGS) with a parallel channel array and perfused with a synthetic oxygen carrier (Oxygent™ in culture medium, PFC emulsion) (Radisic et al., 2006). Constructs cultivated with PFC emulsion had significantly higher DNA content, significantly lower ET and higher relative presence of cardiac markers troponin I and connexin-43 (Western blot) compared to the culture medium alone. Cells were present throughout the construct volume. In this configuration, the presence of PFC emulsion further enhanced the oxygen supply to the cells by improving both axial (convective term) and radial (effective diffusivity) transport properties (Radisic et al., 2005).

Differentiation

Mechanical Stimulation

One significant approach to cardiac tissue engineering, established by Eschenhagen and colleagues (Eschenhagen et al., 1997; Fink et al., 2000; Zimmermann et al., 2000, 2002b) involves the cultivation of neonatal rat heart cells in collagen gel or Matrigel, in the presence of growth factors. The cultured tissues are subjected to sustained mechanical strain. Under these conditions, cardiomyocytes and non-myocytes form 3D cardiac organoids, consisting of a well-organized and highly differentiated cardiac muscle syncytium, that exhibit contractile and electrophysiological properties of working myocardium. First implantation experiments in healthy rats showed survival, strong vascularization, and signs of terminal differentiation of cardiac tissue grafts (Zimmermann et al., 2002a).

In the state of the art approach by Eschenhagen and colleagues neonatal rat cardiac cells were suspended in the collagen/Matrigel mix and cast into circular molds (Zimmermann et al., 2002b). After 7 days of static culture, the strips of cardiac tissue were placed around two rods of a custom made mechanical stretcher and subjected to either unidirectional or cyclic stretch (Figure 30.2c). Histology and immunohistochemistry revealed the formation of intensively interconnected, longitudinally oriented cardiac muscle bundles with morphological features resembling adult rather than immature native tissue. Primitive capillary structures were also detected. Cardiomyocytes exhibited well-developed ultrastructural features: sarcomeres arranged in myofibrils, with well-developed Z, I, A, H, and M bands, specialized cell–cell junctions, T-tubules as well as well-developed basement membrane. Contractile properties were similar to those measured for native tissue, with a high ratio of twitch to resting tension and strong β-adrenegenic response. Action potentials characteristic of rat ventricular myocytes were recorded.

Electrical Stimulation

In native heart, mechanical stretch is induced by electrical signals. Contraction of the cardiac muscle is driven by the waves of electrical excitation (generated by pacing cells) that spread rapidly along the membranes of adjoining cardiac myocytes and trigger release of calcium, which in turn stimulates contraction of the myofibrils. Electromechanical coupling of the myocytes is crucial for their synchronous response to electrical pacing signals, resulting in contractile function and pumping of blood (Severs, 2000).

In a recent study, (Radisic et al., 2004a) cardiac constructs prepared by seeding collagen sponges with neonatal rat ventricular cells were electrically stimulated using suprathreshold square biphasic pulses (2 ms duration, 1 Hz, 5 V). The stimulation was initiated after 1–5 days of scaffold seeding (3-day period was optimal) and applied for up to 8 days. Over only 8 days *in vitro*, electrical field stimulation induced cell alignment and coupling, increased the amplitude of synchronous construct contractions by a factor of 7 and resulted in a remarkable level of ultrastructural organization. Development of conductive and contractile properties of cardiac constructs was concurrent, with strong dependence on the initiation and duration of electrical stimulation. Aligned myofibers expressing cardiac markers were present in stimulated samples and neonatal heart (Figure 30.2d). Stimulated samples had sarcomeres with clearly visible M, Z lines, H, I and A bands. In most cells, Z lines were aligned, and the intercalated disks were positioned between two Z lines. Mitochondria (between myofibrils) and abundant glycogen were detected. In contrast, non-stimulated constructs had poorly developed cardiac-specific organelles and poor organization of ultrastructural features.

Hence the *in vitro* application of a single, but key *in vivo* factor, progressively enhanced the functional tissue assembly and improved the properties of engineered myocardium at the cellular, ultrastructural, and tissue levels.

IMPLANTATION OF CARDIAC PATCHES

While significant progress has been made in constructing *in vitro* cultivation systems and biomaterial scaffolds, very few studies have focused on implantation of cell-based cardiac patches onto viable or injured myocardium (Figure 30.3). In a pioneering study, Li et al. (1999) implanted a construct based on neonatal rat cardiomyocytes and collagen sponges onto the surface of the cryoinjured myocardium of Lewis rats (Figure 30.3). The grafts were implanted 3 weeks post-infraction. After 5 weeks *in vivo*, the cells survived, supported by the blood vessel ingrowth and integrated with the surrounding tissue. However, the graft did not improve LV function.

Attenuation of pathological remodeling (i.e. prevention of ventricle dilatation and maintenance of contractile function) was observed in a study by Leor et al. (2000), where cardiac constructs based on neonatal rat cardiomyocytes and porous alginate scaffolds were implanted onto myocardium of Sprague-Dawley rats that underwent permanent main coronary artery occlusion (Figure 30.3). The grafts were implanted 7 days after MI. After 9 weeks of implantation, the grafts demonstrated integration with host myocardium at the anchorage sites as well as inflammatory infiltrates and presence of fibrous collagen.

Zimmerman et al. (2002a) placed cardiac tissue rings cultivated in the presence of mechanical stimulation onto uninjured hearts of Fisher 344 rats for 14 days (Figure 30.3). They noticed that although both cells and collagen were isolated from Fisher rats, immunosuppression was required for maintenance of heart tissue upon implantation. In the absence of immunosuppression, even in the syngeneic approach, cardiac

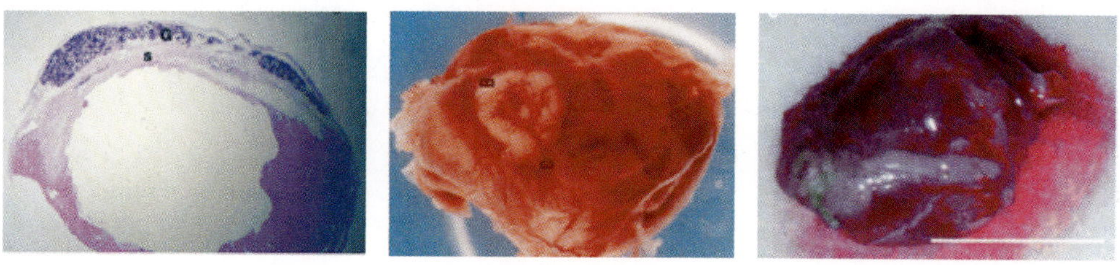

Cryoinjured myocardium of Lewis rats	Coronary artery occlusion in Sprague-Dawley rat myocardium	Uninjured heart of Fisher 344 rats
Constructs implanted after 3 weeks	Constructs implanted after 1 week	
Fetal (Lewis) rat cardiomyocytes in collagen sponge Cultivated under static condition for 7 days	Fetal (Sprague-Dawley) rat cardyomyocytes in porous aglinate scaffolds Cultivated under static conditions for 4 days	Neonatal (Fisher 344) rat cardiomyocytes in collagen type I gel Cultivation with mechanical stimulation for 12 days
After 5 weeks *in vivo* no significant difference compared to the controls	After 9 weeks *in vivo* attenuation of LV function and maintenance of contractile function in comparison to controls without construct	After 14 weeks *in vivo* the implant vascularized and improved the level of maturation. Immunosupression was required
Li et al. (1999)	Leor et al. (2000)	Zimmermann et al. (2002)

Figure 30.3 Representative studies investigating the effect of implantation of the cardiomyocyte-based constructs on the function of injured or viable hearts.

constructs completely degraded after only 2 weeks *in vivo*. It is unknown what exactly caused the response; it is possible that it was the remainder of serum or chick extract. Regardless, the finding has significant implications in the potential implantation of cardiac patches in clinical settings.

Limitations related to the source of autologous cardiomyocytes motivated the studies that utilized non-myocyte-based patches for MI repair. Smooth muscle cells seeded with poly (ε-caprolactone-co-L-lactide) sponge reinforced with poly-L-actide fabric were used in a modified endoventricular circular patch plasty procedure (Dor procedure). Cell seeded grafts resulted in improved LV function (as assessed by echocardiography) compared to cell-free controls (Matsubayashi et al., 2003). A patch made of dermal fibroblasts seeded onto knitted Vicryl mesh (Dermagraft) was used in an attempt to increase angiogenesis upon MI. When placed over the infracted regions on the hearts of severe combined immunodeficient (SCID) mice, the grafts improved microvessel density within the damaged myocardium (Kellar et al., 2001).

These studies demonstrated feasibility of cardiac patch implantation, but further studies are necessary to estimate the effect of culture conditions and scaffold type on the *in vivo* outcome. Although, significant progress has been made in the area of biomaterials and bioreactors, it is currently unknown which cultivation conditions and what biomaterial will best preserve contractile function and prevent pathological remodeling upon implantation. Thus studies that investigate this in a systematic fashion and correlate *in vitro* parameters (e.g. force of contraction) to *in vivo* outcomes (e.g. fractional shortening) are required.

SUMMARY

Overall, the field of cardiac tissue engineering is very much in its infancy. Although the results to date are exceedingly encouraging, much remains to be done in order to develop clinically relevant approaches, let alone move towards a whole heart. Not surprisingly a NIH task force (National Institutes of Health, 1999) has emphasized the development of heart components such as a cardiac patch or a valve before "graduating" to whole heart engineering.

However, significant progress has been made ever since the LIFE initiative embarked on the creation of the artificial heart in 1999. Functional viable cardiac patches have been engineered based on neonatal rat cardiomyocytes and more recently based on ES-cell-derived cardiomyocytes. Various biomaterials have been tested for this purpose and *in vitro* culture systems have been developed that enhance cardiac construct differentiation (mechanical and electrical stimulation) as well as improve cardiomyocyte survival at high density (medium perfusion). Exciting new findings on resident progenitor cells have also emerged. While the completely artificial heart will remain a dream, the near future will bring a clinically relevant autologous cardiac patch.

REFERENCES

Ahrendt, G., Chickering, D.E. and Ranieri, J.P. (1998). Angiogenic growth factors: a review for tissue engineering. *Tissue Eng.* **4**: 117–130.

Akins, R.E. (2000). Prospects for use of cell implantation, gene therapy and tissue engineering in the treatment of myocardial disease and congenital heart defects. In: Sames, K. (ed.), *Medizinische Regeneration und Tissue Engineering*, EcoMed, Landsberg, Germany.

Alperin, C., Zandstra, P.W. and Woodhouse, K.A. (2005). Polyurethane films seeded with embryonic stem cell-derived cardiomyocytes for use in cardiac tissue engineering applications. *Biomaterials* **26**: 7377–7786.

American Heart Association (2002). Congestive heart failure.

American Heart Association (2004). Heart Disease and Stroke Statistics-2004 (Update).

Augustin-Voss, H.G., Johnson, R.C. and Pauli, B.U. (1991). Modulation of endothelial cell surface glycoconjugate expression by organ-derived biomatrices. *Exp. Cell Res.* **192**: 346–351.

Baar, K., Birla, R., Boluyt, M.O., Borschel, G.H., Arruda, E.M. and Dennis, R.G. (2005). Self-organization of rat cardiac cells into contractile 3-D cardiac tissue. *FASEB J.* **19**: 275–277.

Balsam, L.B., Wagers, A.J., Christensen, J.L., Kofidis, T., Weissman, I.L. and Robbins, R.C. (2004). Haematopoietic stem cells adopt mature haematopoietic fates in ischaemic myocardium. *Nature* **428**: 668–673.

Beltrami, A.P., Barlucchi, L., Torella, D., Baker, M., Limana, F., Chimenti, S., Kasahara, H., Rota, M., Musso, E., Urbanek, K., Leri, A., Kajstura, J., Nadal-Ginard, B. and Anversa, P. (2003). Adult cardiac stem cells are multipotent and support myocardial regeneration. *Cell* **114**: 763–776.

Boublik, J., Park, H., Radisic, M., Tognana, E., Chen, F., Pei, M., Vunjak-Novakovic, G. and Freed, L.E. (2005). Mechanical properties and remodeling of hybrid cardiac constructs made from heart cells, fibrin, and biodegradable, elastomeric knitted fabric. *Tissue Eng.* **11**: 1122–1132.

Bursac, N., Papadaki, M., Cohen, R.J., Schoen, F.J., Eisenberg, S.R., Carrier, R., Vunjak-Novakovic, G. and Freed, L.E. (1999). Cardiac muscle tissue engineering: toward an *in vitro* model for electrophysiological studies. *Am. J. Physiol.: Heart Circ. Physiol.* **277**: H433–H444.

Burton, A.C. (1972). Biophysical Basis of the Circulation. St. Louis, MO: Yearbook Medical Publishers.

Butler, D.L., Goldstein, S.A. and Guilak, F. (2000). Functional tissue engineering: the role of biomechanics. *J. Biomech. Eng.* **122**: 570–575.

Carrier, R.L., Papadaki, M., Rupnick, M., Schoen, F.J., Bursac, N., Langer, R., Freed, L.E. and Vunjak-Novakovic, G. (1999). Cardiac tissue engineering: cell seeding, cultivation parameters and tissue construct characterization. *Biotechnol. Bioeng.* **64**: 580–589.

Cenni, E., Ciapetti, G., Cavedagna, D., Di Leo, A. and Pizzoferrato, A. (1993). Production of prostacyclin and fibrinolysis modulators by endothelial cells cultured in the presence of polyethylene terephthalate. *J. Biomed. Mater. Res.* **27**: 1161–1164.

Cenni, E., Granchi, D., Ciapetti, G., Savarino, L., Corradini, A. and Di Leo, A. (2000). Cytokine expression *in vitro* by cultured human endothelial cells in contact with polyethylene terephthalate coated with pyrolytic carbon and collagen. *J. Biomed. Mater. Res.* **50**: 483–489.

Chen, S.L., Fang, W.W., Ye, F., Liu, Y.H., Qian, J., Shan, S.J., Zhang, J.J., Chunhua, R.Z., Liao, L.M., Lin, S. and Sun, J.P. (2004). Effect on left ventricular function of intracoronary transplantation of autologous bone marrow mesenchymal stem cell in patients with acute myocardial infarction. *Am. J. Cardiol.* **94**: 92–95.

Christman, K.L., Fok, H.H., Sievers, R.E., Fang, Q. and Lee, R.J. (2004a). Fibrin glue alone and skeletal myoblasts in a fibrin scaffold preserve cardiac function after myocardial infarction. *Tissue Eng.* **10**: 403–409.

Christman, K.L., Vardanian, A.J., Fang, Q., Sievers, R.E., Fok, H.H. and Lee, R.J. (2004b). Injectable fibrin scaffold improves cell transplant survival, reduces infarct expansion, and induces neovasculature formation in ischemic myocardium. *J. Am. Coll. Cardiol.* **44**: 654–660.

Dar, A., Shachar, M., Leor, J. and Cohen, S. (2002). Cardiac tissue engineering Optimization of cardiac cell seeding and distribution in 3D porous alginate scaffolds. *Biotechnol. Bioeng.* **80**: 305–312.

Davis, M.E., Motion, J.P., Narmoneva, D.A., Takahashi, T., Hakuno, D., Kamm, R.D., Zhang, S. and Lee, R.T. (2005). Injectable self-assembling peptide nanofibers create intramyocardial microenvironments for endothelial cells. *Circulation* **111**: 442–450.

del Monte, F., Williams, E., Lebeche, D., Schmidt, U., Rosenzweig, A., Gwathmey, J.K., Lewandowski, E.D. and Hajjar, R.J. (2001). Improvement in survival and cardiac metabolism after gene transfer of sarcoplasmic reticulum Ca(2+)-ATPase in a rat model of heart failure. *Circulation* **104**: 1424–1429.

del Monte, F., Harding, S.E., Dec, G.W., Gwathmey, J.K. and Hajjar, R.J. (2002). Targeting phospholamban by gene transfer in human heart failure. *Circulation* **105**: 904–907.

Di Donato, M., Sabatier, M., Dor, V., Toso, A., Maioli, M. and Fantini, F. (1997). Akinetic versus dyskinetic postinfarction scar: relation to surgical outcome in patients undergoing endoventricular circular patch plasty repair. *J. Am. Coll. Cardiol.* **29**: 1569–1575.

Di Luozzo, G., Bhargava, J. and Powell, R.J. (2000). Vascular smooth muscle cell effect on endothelial cell endothelin-1 production. *J. Vasc. Surg.* **31**: 781–789.

Dimmeler, S., Zeiher, A.M. and Schneider, M.D. (2005). Unchain my heart: the scientific foundations of cardiac repair. *J. Clin. Invest.* **115**: 572–583.

Dor, V., Saab, M., Coste, P., Kornaszewska, M. and Montiglio, F. (1989). Left ventricular aneurysm: a new surgical approach. *Thorac. Cardiovasc. Surg.* **37**: 11–9.

Dorfman, J., Duong, M., Zibaitis, A., Pelletier, M.P., Shum-Tim, D., Li, C. and Chiu, R.C. (1998). Myocardial tissue engineering with autologous myoblast implantation. *J. Thoracic. Cardiovasc. Surg.* **116**: 744–751.

Eschenhagen, T., Fink, C., Remmers, U., Scholz, H., Wattchow, J., Woil, J., Zimmermann, W., Dohmen, H.H., Schafer, H., Bishopric, N., Wakatsuki, T. and Elson, E. (1997). Three-dimensional reconstitution of embryonic cardiomyocytes in a collagen matrix: a new heart model system. *FASEB J.* **11**: 683–694.

Eskin, S.G., Ives, C.L., McIntire, L.V. and Navarro, L.T. (1984). Response of cultured endothelial cells to steady flow. *Microvasc. Res.* **28**: 87–94.

Esmon, C.T. (2000). Regulation of blood coagulation. *Biochim. Biophys. Acta* **1477**: 349–360.

Etzion, S., Battler, A., Barbash, I.M., Cagnano, E., Zarin, P., Granot, Y., Kedes, L.H., Kloner, R.A. and Leor, J. (2001). Influence of embryonic cardiomyocyte transplantation on the progression of heart failure in a rat model of extensive myocardial infarction. *J. Mol. Cell. Cardiol.* **33**: 1321–30.

Fazel, S., Tang, G.H., Angoulvant, D., Cimini, M., Weisel, R.D., Li, R.K., Yau, T.M. (2005). Current status of cellular therapy for ischemic heart disease. *Ann. Thorac. Surg.* **79**: S2238–S2247.

Fink, C., Ergun, S., Kralisch, D., Remmers, U., Weil, J. and Eschenhagen, T. (2000). Chronic stretch of engineered heart tissue induces hypertrophy and functional improvement. *FASEB J.* **14**: 669–679.

Fung, Y.C. (1993). Biomechanics: Mechanical Properties of Living Tissues. 2nd edition, p433.

Hajjar, R.J., del Monte, F., Matsui, T. and Rosenzweig, A. (2000). Prospects for gene therapy for heart failure. *Circ. Res.* **86**: 616–621.

Halloran, P.F. and Melk, A. (2001). Renal senescence, cellular senescence, and their relevance to nephrology and transplantation. *Adv. Nephrol. Necker Hosp.* **31**: 273–283.

Harada, K., Grossman, W., Friedman, M., Edelman, E.R., Prasad, P.V., Keighley, C.S., Manning, W.J., Sellke, F.W. and Simons, M. (1994). Basic fibroblast growth factor improves myocardial function in chronically ischemic porcine hearts. *J. Clin. Invest.* **94**: 623–630.

Ives, C.L., Eskin, S.G. and McIntire, L.V. (1986). Mechanical effects on endothelial cell morphology: *in vitro* assessment. *In Vitro Cell Dev. Biol.* **22**: 500–507.

Kaihara, S., Borenstein, J., Koka, R., Lalan, S., Ochoa, E.R., Ravens, M., Pien, H., Cunningham, B. and Vacanti, J.P. (2000). Silicon micromachining to tissue engineer branched vascular channels for liver fabrication. *Tissue Eng.* **6**: 105–117.

Kehat, I., Kenyagin-Karsenti, D., Snir, M., Segev, H., Amit, M., Gepstein, A., Livne, E., Binah, O., Itskovitz-Eldor, J. and Gepstein, L. (2001). Human embryonic stem cells can differentiate into myocytes with structural and functional properties of cardiomyocytes. *J. Clin. Invest.* **108**: 407-14.

Kellar, R.S., Landeen, L.K., Shepherd, B.R., Naughton, G.K., Ratcliffe, A. and Williams, S.K. (2001). Scaffold-based three-dimensional human fibroblast culture provides a structural matrix that supports angiogenesis in infarcted heart tissue. *Circulation* **104**: 2063–2068.

Klug, M.G., Soonpaa, M.H., Koh, G.Y. and Field, L.J. (1996). Genetically selected cardiomyocytes from differentiating embryonic stem cells form stable intracardiac grafts. *J. Clin. Invest.* **98**: 216–224.

Kocher, A.A., Schuster, M.D., Szabolcs, M.J., Takuma, S., Burkhoff, D., Wang, J., Homma, S., Edwards, N.M. and Itescu, S. (2001). Neovascularization of ischemic myocardium by human bone-marrow-derived angioblasts prevents cardiomyocyte apoptosis, reduces remodeling and improves cardiac function. *Nat. Med.* **7**: 430–436.

Kofidis, T., Akhyari, P., Wachsmann, B., Mueller-Stahl, K., Boublik, J., Ruhparwar, A., Mertsching, H., Balsam, L., Robbins, R. and Haverich, A. (2003). Clinically established hemostatic scaffold (tissue fleece) as biomatrix in tissue- and organ-engineering research. *Tissue Eng.* **9**: 517–523.

Korff, T., Kimmina, S., Martiny-Baron, G. and Augustin, H.G. (2001). Blood vessel maturation in a 3-dimensional spheroidal coculture model: direct contact with smooth muscle cells regulates endothelial cell quiescence and abrogates VEGF responsiveness. *FASEB J.* **15**: 447–457.

Kusano, K.F., Pola, R., Murayama, T., Curry, C., Kawamoto, A., Iwakura, A., Shintani, S., Ii, M., Asai, J., Tkebuchava, T., Thorne, T., Takenaka, H., Aikawa, R., Goukassian, D., von Samson, P., Hamada, H., Yoon, Y.S., Silver, M., Eaton, E., Ma, H., Heyd, L., Kearney, M., Munger, W., Porter, J.A., Kishore, R. and Losordo, D.W. (2005). Sonic hedgehog myocardial gene therapy: tissue repair through transient reconstitution of embryonic signaling. *Nat. Med.* **11**: 1197–1204.

Laflamme, M.A. and Murry, C.E. (2005). Regenerating the heart. *Nat. Biotechnol.* **23**: 845–856.

Laugwitz, K.L., Moretti, A., Lam, J., Gruber, P., Chen, Y., Woodard, S., Lin, L.Z., Cai, C.L., Lu, M.M., Reth, M., Platoshyn, O., Yuan, J.X., Evans, S. and Chien, K.R. (2005). Postnatal isl1+ cardioblasts enter fully differentiated cardiomyocyte lineages. *Nature* **433**: 647–653.

Leor, J., Aboulafia-Etzion, S., Dar, A., Shapiro, L., Barbash, I.M., Battler, A., Granot, Y. and Cohen, S. (2000). Bioengineered cardiac grafts: a new approach to repair the infarcted myocardium? *Circulation* **102**: III56–III61.

Levenberg, S., Rouwkema, J., Macdonald, M., Garfein, E.S., Kohane, D.S., Darland, D.C., Marini, R., van Blitterswijk, C.A., Mulligan, R.C., D'Amore, P.A. and Langer, R. (2005). Engineering vascularized skeletal muscle tissue. *Nat. Biotechnol.* **23**: 879–884.

Li, J.M., Menconi, M.J., Wheeler, H.B., Rohrer, M.J., Klassen, V.A., Ansell, J.E. and Appel, M.C. (1992). Precoating expanded polytetrafluoroethylene grafts alters production of endothelial cell-derived thrombomodulators. *J. Vasc. Surg.* **15**: 1010–1017.

Li, R.-K., Jia, Z.Q., Weisel, R.D., Mickle, D.A.G., Choi, A. and Yau, T.M. (1999). Survival and function of bioengineered cardiac grafts. *Circulation* **100**: II63–II69.

Li, R.K., Yau, T.M., Weisel, R.D., Mickle, D.A., Sakai, T., Choi, A. and Jia, Z.Q. (2000). Construction of a bioengineered cardiac graft. *J. Thorac. Cardiovasc. Surg.* **119**: 368–375.

Lipton, B.H., Bensch, K.G. and Karasek, M.A. (1991). Microvessel endothelial cell transdifferentiation: phenotypic characterization. *Differentiation* **46**: 117–133.

Lu, A. and Sipehia, R. (2001). Antithrombotic and fibrinolytic system of human endothelial cells seeded on PTFE: the effects of surface modification of PTFE by ammonia plasma treatment and ECM protein coatings. *Biomaterials* **22**: 1439–1446.

Marcum, J.A., McKenney, J.B. and Rosenberg, R.D. (1984). Acceleration of thrombin–antithrombin complex formation in rat hindquarters via heparin-like molecules bound to the endothelium. *J. Clin. Invest.* **74**: 341–350.

Markkanen, J.E., Rissanen, T.T., Kivela, A. and Yla-Herttuala, S. (2005). Growth factor-induced therapeutic angiogenesis and arteriogenesis in the heart–gene therapy. *Cardiovasc. Res.* **65**: 656–664.

Matsubayashi, K., Fedak, P.W., Mickle, D.A., Weisel, R.D., Ozawa, T. and Li, R.K. (2003). Improved left ventricular aneurysm repair with bioengineered vascular smooth muscle grafts. *Circulation* **108**(Suppl 1): II219–II225.

Maurice, J.P., Hata, J.A., Shah, A.S., White, D.C., McDonald, P.H., Dolber, P.C., Wilson, K.H., Lefkowitz, R.J., Glower, D.D. and Koch, W.J. (1999). Enhancement of cardiac function after adenoviral-mediated *in vivo* intracoronary beta2-adrenergic receptor gene delivery. *J. Clin. Invest.* **104**: 21–29.

Mawji, I.A. and Marsden, P.A. (2003). Perturbations in paracrine control of the circulation: role of the endothelial-derived vasomediators, endothelin-1 and nitric oxide. *Microsc. Res. Tech.* **60**: 46–58.

McCormick, S.M., Eskin, S.G., McIntire, L.V., Teng, C.L., Lu, C.M., Russell, C.G. and Chittur, K.K. (2001). DNA microarray reveals changes in gene expression of shear stressed human umbilical vein endothelial cells. *Proc. Natl. Acad. Sci. USA* **98**: 8955–8960.

McCuigan, A.P. and Sefton, M.V. (2006). Vascularized organoid engineering by modular assembly enables blood perfusion. *Proc. Natl Acad. Sci. USA* **103**: 11461–11466.

McDevitt, T.C., Angello, J.C., Whitney, M.L., Reinecke, H., Hauschka, S.D., Murry, C.E. and Stayton, P.S. (2002). *In vitro* generation of differentiated cardiac myofibers on micropatterned laminin surfaces. *J. Biomed. Mater. Res.* **60**: 472–479.

McDevitt, T.C., Woodhouse, K.A., Hauschka, S.D., Murry, C.E. and Stayton, P.S. (2003). Spatially organized layers of cardiomyocytes on biodegradable polyurethane films for myocardial repair. *J. Biomed. Mater. Res. A* **66**: 586–595.

McLaren, A. (2000). Cloning: pathways to a pluripotent future. *Science* **288**: 1775–1780.

Meinhart, J.G., Deutsch, M., Fischlein, T., Howanietz, N., Froschl, A. and Zilla, P. (2001). Clinical autologous *in vitro* endothelialization of 153 infrainguinal ePTFE grafts. *Ann. Thorac. Surg.* **71**: S327–S331.

Melo, L.G., Pachori, A.S., Kong, D., Gnecchi, M., Wang, K., Pratt, R.E., Dzau and V.J. (2004a). Gene and cell-based therapies for heart disease. *FASEB J.* **18**: 648–663.

Melo, L.G., Pachori, A.S., Kong, D., Gnecchi, M., Wang, K., Pratt, R.E. and Dzau, V.J. (2004b). Molecular and cell-based therapies for protection, rescue, and repair of ischemic myocardium: reasons for cautious optimism. *Circulation* **109**: 2386–2393.

Menasche, P., Hagege, A.A., Vilquin, J.T., Desnos, M., Abergel, E., Pouzet, B., Bel, A., Sarateanu, S., Scorsin, M., Schwartz, K., Bruneval, P., Benbunan, M., Marolleau, J.P. and Duboc, D. (2003). Autologous skeletal myoblast transplantation for severe postinfarction left ventricular dysfunction. *J. Am. Coll. Cardiol.* **41**: 1078–1083.

Mikos, A.G., McIntire, L.V., Anderson, J.M. and Babensee, J.E. (1998). Host response to tissue engineered devices. *Adv. Drug Deliv. Rev.* **33**: 111–139.

Moncada, S. (1982). Eighth Gaddum memorial lecture. University of London Institute of Education, December 1980. Biological importance of prostacyclin. *Br. J. Pharmacol.* **76**: 3–31.

Muller-Ehmsen, J., Peterson, K.L., Kedes, L., Whittaker, P., Dow, J.S., Long, T.I., Laird, P.W. and Kloner, R.A. (2002a). Rebuilding a damaged heart: long-term survival of transplanted neonatal rat cardiomyocytes after myocardial infarction and effect on cardiac function. *Circulation* **105**: 1720–1726.

Muller-Ehmsen, J., Whittaker, P., Kloner, R.A., Dow, J.S., Sakoda, T., Long, T.I., Laird, P.W. and Kedes, L. (2002b). Survival and development of neonatal rat cardiomyocytes transplanted into adult myocardium. *J. Mol. Cell. Cardiol.* **34**: 107–116.

Murry, C.E., Kay, M.A., Bartosek, T., Hauschka, S.D. and Schwartz, S.M. (1996). Muscle differentiation during repair of myocardial necrosis in rats via gene transfer with MyoD. *J. Clin. Invest.* **98**: 2209–2217.

Murry, C.E., Soonpaa, M.H., Reinecke, H., Nakajima, H., Nakajima, H.O., Rubart, M., Pasumarthi, K.B., Virag, J.I., Bartelmez, S.H., Poppa, V., Bradford, G., Dowell, J.D., Williams, D.A. and Field, L.J. (2004). Haematopoietic stem cells do not transdifferentiate into cardiac myocytes in myocardial infarcts. *Nature* **428**: 664–668.

National Institutes of Health (NIH) (1999). Working Group on Tissuegenesis and organogenesis for Heart, Lung and Blood Applications.

Nerem, R.M. (2003). Role of mechanics in vascular tissue engineering. *Biorheology* **40**: 281–287.

Nygren, J.M., Jovinge, S., Breitbach, M., Sawen, P., Roll, W., Hescheler, J., Taneera, J., Fleischmann, B.K. and Jacobsen, S.E. (2004). Bone marrow-derived hematopoietic cells generate cardiomyocytes at a low frequency through cell fusion, but not transdifferentiation. *Nat. Med.* **10**: 494–501.

Oh, H., Bradfute, S.B., Gallardo, T.D., Nakamura, T., Gaussin, V., Mishina, Y., Pocius, J., Michael, L.H., Behringer, R.R., Garry, D.J., Entman, M.L. and Schneider, M.D. (2003). Cardiac progenitor cells from adult myocardium: homing, differentiation, and fusion after infarction. *Proc. Natl Acad. Sci. USA* **100**: 12313–12318.

Orlic, D., Kajstura, J., Chimenti, S., Jakoniuk, I., Anderson, S.M., Li, B., Pickel, J., McKay, R., Nadal-Ginard, B., Bodine, D.M., Leri, A. and Anversa, P. (2001). Bone marrow cells regenerate infarcted myocardium. *Nature* **410**: 701–705.

Palmer, R.M., Ashton, D.S. and Moncada, S. (1988). Vascular endothelial cells synthesize nitric oxide from L-arginine. *Nature* **333**: 664–666.

Papadaki, M., Bursac, N., Langer, R., Merok, J., Vunjak-Novakovic, G. and Freed, L.E. (2001). Tissue engineering of functional cardiac muscle: molecular, structural and electrophysiological studies. *Am. J. Physiol. Heart Circ. Physiol.* **280**: H168–H178.

Radisic, M., Park, H., Shing, H., Consi, T., Schoen, F.J., Langer, R., Freed, L.E. and Vunjak-Novakovic, G. (2004a). Functional assembly of engineered myocardium by electrical stimulation of cardiac myocytes cultured on scaffolds. *Proc. Natl Acad. Sci. USA* **101**: 18129–18134.

Radisic, M., Yang, L., Boublik, J., Cohen, R.J., Langer, R., Freed, L.E. and Vunjak-Novakovic, G. (2004b). Medium perfusion enables engineering of compact and contractile cardiac tissue. *Am. J. Physiol. Heart Circ. Physiol.* **286**: H507–H516.

Radisic, M., Malda, J., Epping, E., Geng, W., Langer, R. and Vunjak-Novakovic, G. (2005). Oxygen gradients correlate with cell density and cell viability in engineered cardiac tissue. *Biotechnol. Bioeng.* **93**: 332–343.

Radisic, M., Park, H., Chen, F., Salazar-Lazzaro, J.E., Wang, Y., Dennis, R.G., Langer, R., Freed, L.E. and Vunjak-Novakovic, G. (2006). Biomimetic approach to cardiac tissue engineering: oxygen carriers and channeled scaffolds. *Tissue Eng.* **12**: 2077–2091.

Rafii, S. (2000). Circulating endothelial precursors: mystery, reality, and promise. *J. Clin. Invest.* **105**: 17–19.

Reffelmann, T. and Kloner, R.A. (2003). Cellular cardiomyoplasty – cardiomyocytes, skeletal myoblasts, or stem cells for regenerating myocardium and treatment of heart failure? *Cardiovasc. Res.* **58**: 358–368.

Reinecke, H., Zhang, M., Bartosek, T. and Murry, C.E. (1999). Survival, integration, and differentiation of cardiomyocyte grafts: a study in normal and injured rat hearts. *Circulation* **100**: 193–202.

Reinlib, L. and Field, L. (2000). Cell transplantation as future therapy for cardiovascular disease?: a workshop of the National Heart, Lung, and Blood Institute. *Circulation* **101**: e182–e187.

Risau, W. (1995). Differentiation of endothelium. *FASEB J.* **9**: 926–933.

31

Intracorporeal Kidney Support

James J. Yoo, Akira Joraku, and Anthony Atala

INTRODUCTION

End stage renal failure is a devastating disease which involves multiple organs in affected individuals. Although currently available treatment modalities, including dialysis and transplantation, can prolong survival for many patients, problems such as donor shortage, complications, and graft failure remain a continued concern (Chazan et al., 1991; Feldman et al., 1996; Amiel et al., 2000). Numerous investigative efforts have been commenced in order to improve, restore, or replace renal function. Cell-based approaches for kidney tissue regeneration have been proposed recently as an alternative method. In this chapter, we describe various cell-based approaches to achieve functional intracorporeal kidney support.

The kidney is considered as one of the more challenging organs to reconstruct in the genitourinary system, due to its complex structure and function. Normal renal function includes synthesis of 1,25-vitamin D3, erythropoietin, glutathione, and free radical scavenging enzymes. The kidney also participates in the catabolism of low molecular weight proteins and in the production and regulation of cytokines (Frank et al., 1993; Stadnyk, 1994). Because these functions can not be replaced with dialysis therapy, long-term consequences, such as anemia and malnutrition, are prevalent in these patients.

Limitation of current therapies for renal functional augmentation has led investigators to pursue alternative therapeutic modalities. The concept of cell transplantation using tissue engineering techniques has been proposed as a method to improve, restore, or replace renal function (Atala, 1997, 1999; Amiel and Atala, 1999; Humes et al., 1999). The emergence of tissue engineering and regenerative medicine strategies has presented alternative possibilities for the management of pathologic renal conditions (Figure 31.1). Augmentation of either isolated or total renal function with kidney cell expansion may be a feasible solution. We have followed an approach which involves the development of intracorporeal support systems for renal functional replacement.

BASIC PRINCIPLES OF KIDNEY TISSUE REGENERATION

Components required to achieve partial or total kidney function are renal cells, three-dimensional scaffolds, and an *in vivo* environment. The challenge associated with renal cell culture is due to the unique structural and cellular heterogeneity present within the kidney. The system of nephrons and collecting ducts is composed of multiple functionally and phenotypically distinct segments. For this reason, appropriate conditions need to be provided for the long-term survival, differentiation, and growth of the cells. Extensive research has been performed in order to determine optimal growth conditions for renal cell enrichment (Milici et al., 1985; Carley et al., 1988; Horikoshi et al., 1988; Humes and Cieslinski, 1992; Schena, 1998). Isolation of particular cell types that produce specific factors, such as erythropoietin, may be a feasible approach for selective cell therapies. However, total renal function would not be achieved if specific cell types were separately isolated. To reconstitute kidney tissue that would deliver renal function, cells composing the functional

 Copyright © 2008 Elsevier Inc. All rights reserved

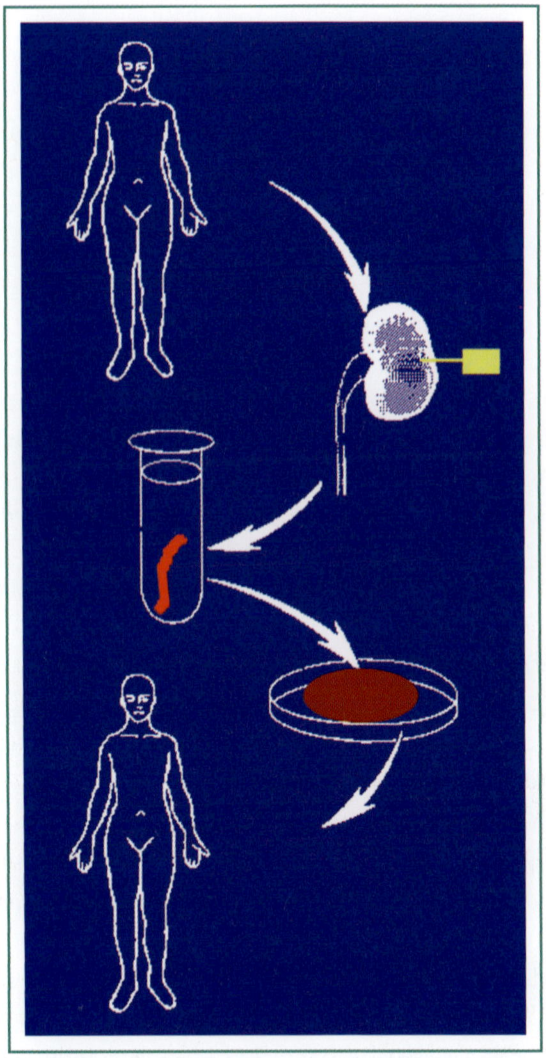

Figure 31.1 **A strategy for engineering of renal tissue. A patient with end stage renal failure undergoes a percutaneous biopsy. Renal cells are grown, expanded in culture, and seeded onto three-dimensional support system, which is then implanted into the same patient.**

nephron units may be preferable. Based on the literature and our experience, we were able to obtain optimal growth conditions for a stable cell culture system for kidney tissue reconstitution. Renal cells grown in culture are able to maintain their cellular characteristics (Lanza et al., 2002). When primary renal cells are placed in a collagen-based three-dimensional culture system, they are able to reconstitute into renal structures.

Recent efforts in the area of kidney tissue regeneration were focused toward the development of a reliable cell source. Multipotent or progenitor cells have been proposed as a promising source due to their potential to differentiate into several cell lineages. Bone marrow-derived human mesenchymal stem cells have been shown to exhibit plasticity and differentiation potential into several different cell types (Prockop, 1997). These cells have been shown to participate in the kidney development when they are placed in a rat embryonic niche that allows for continued exposure to repertoire of nephrogenic signals (Yokoo et al., 2005). Another potential cell source is circulating stem cells, which are also known to participate at the site of kidney regeneration such as tubular and glomerular epithelial cells, podocytes, mesangial cells, and interstitial cells after renal injury (Ito et al., 2001; Poulsom et al., 2001; Gupta et al., 2002; Iwano et al., 2002; Kale et al., 2003; Lin et al., 2003; Rookmaaker et al., 2003). Although bone marrow cells were found to contribute to regeneration of damaged glomerular endothelial cells, the major cell source of kidney regeneration was found to originate from intrarenal cells in an ischemic renal injury model (Ikarashi et al., 2005; Lin et al., 2005).

Although isolated renal cells are able to retain their phenotypic and functional characteristics in culture, transplantation of these cells *in vivo* may not result in structural remodeling. In addition, cell or tissue components may not be implanted in large volumes due to the limited diffusion (Folkman and Hochberg, 1973). Thus, a cell-support matrix is necessary to allow diffusion of nutrients across the entire implant. A variety of synthetic and naturally derived materials has been examined in order to determine the ideal support structures for the regeneration of urologic tissue (Tachibana et al., 1985; Atala et al., 1995, 2006; Oberpenning et al., 1999; El-Kassaby et al., 2003). Biodegradable synthetic materials, such as poly-lactic and glycolic acid polymers, have been used to provide structural support for cells. Synthetic materials can be easily fabricated and configured in a controlled manner. Naturally derived materials, such as collagen, laminin, and fibronectin, are biocompatible and provide a similar extracellular matrix environment as normal tissue. For this reason, collagen-based scaffolds have been preferred in many tissue applications (Hubbell et al., 1991; Wald et al., 1993; Freed et al., 1994; Mooney et al., 1996).

CREATION OF RENAL STRUCTURES *IN VIVO*

The kidney is responsible not only for urine excretion but also for several other important metabolic functions. Our initial study involved in investigating the feasibility of achieving renal cell growth, expansion, and *in vivo* reconstitution using tissue engineering techniques (Atala et al., 1995). New Zealand white rabbits underwent nephrectomy and renal artery perfusion with a non-oxide solution which promoted iron particle entrapment in the glomeruli. Homogenization of the renal cortex and fractionation in 83 and 210 μm sieves with subsequent magnetic extraction yielded three separate purified suspensions of distal tubules, glomeruli, and proximal tubules. The cells were plated separately *in vitro* and seeded onto biodegradable polyglycolic acid scaffolds. Polymer scaffolds were implanted subcutaneously into host athymic mice. This included implants of proximal tubular cells, glomeruli, distal tubular cells, and a mixture of all three cell types. Polymers alone served as controls. Animals were sacrificed at 1 week, 2 weeks, and 1 month after implantation and the retrieved scaffolds were analyzed. An acute inflammatory phase and a chronic foreign body reaction were seen, accompanied by vascular ingrowth by 7 days after implantation. Histologic examination demonstrated progressive formation and organization of the nephron segments within the polymer fibers with time. Renal cell proliferation in the cell–polymer scaffolds was detected by *in vivo* labeling of replicating cells with the thymidine analog bromodeoxyuridine (BrdU). BrdU incorporation into renal cell DNA was identified immunocytochemically with monoclonal anti-BrdU antibodies. These results demonstrated that renal specific cells can be successfully harvested, survive in culture, and attach to artificial biodegradable polymers. The renal cell–polymer scaffolds can be implanted into host animals where the cells replicate and organize into nephron segments, as the polymer, which acts as a cell delivery vehicle, undergoes biodegradation.

The initial experiments demonstrated that implanted cell–polymer scaffolds gave rise to renal tubular structures. However, it was unclear whether the tubular structures reconstituted *de novo* from dispersed renal elements, or they merely represented fragments of donor tubules, which survived intact. Further investigation was conducted in order to examine the tubular reconstitution process (Fung et al., 1996). Mouse renal cells were harvested, grown, and expanded in culture. Subsequently, single isolated cells were seeded on biodegradable polymers and implanted into syngeneic hosts. Renal epithelial cells were observed to reconstitute into tubular structures *in vivo*. Sequential analyses of the retrieved implants over time demonstrated that renal epithelial cells first organized into a cord-like structure with a solid center. Subsequent canalization into a hollow tube could be seen by 2 weeks. Histologic examination with nephron segment specific lactins showed successful reconstitution of proximal tubules, distal tubules, loop of Henle, collecting tubules, and collecting ducts. These results showed that single suspended cells are capable of reconstituting into tubular structures, with homogeneous cell types within each tubule.

REGENERATION OF FUNCTIONAL RENAL TISSUE *IN VIVO*

The kidneys are critical to body homeostasis because of their excretory, regulatory, and endocrinologic functions. The excretory function is initiated by filtration of blood at the glomerulus, and the regulatory function is provided by the tubular segments. Although our prior studies demonstrated that renal cells seeded on biodegradable polymer scaffolds are able to reconstitute into renal structures *in vivo*, complete renal function could not be achieved due to the type and structural configuration of polymers used. In our subsequent study we explored the feasibility of creating a functional artificial renal unit, wherein urine production could be achieved (Yoo et al., 1996). Mouse renal cells were harvested and expanded in culture. The cells were seeded onto a tubular device constructed from polycarbonate (4 μm pore size), connected at one end with a silastic catheter which terminated into a reservoir. The device was implanted in the subcutaneous space of athymic mice. Animals were sacrificed at 1, 2, 3, 4, and 8 weeks after implantation and the retrieved specimens were examined histologically and immunocytochemically. Fluid was collected from inside the implant, and uric acid and creatinine levels were determined.

Histologic examination of the implanted device revealed extensive vascularization, formation of glomeruli, and highly organized tubule-like structures. Immunocytochemical staining with anti-osteopontin antibody, which is secreted by proximal and distal tubular cells and the cells of the thin ascending loop of Henle, stained the tubular sections. Immunohistochemical staining for alkaline phosphatase stained proximal tubule-like structures. Uniform staining for fibronectin in the extracellular matrix of newly formed tubes was observed. The fluid collected from the reservoir was yellow and contained 66 mg/dl uric acid (as compared to 2 mg/dl in plasma) suggesting that these tubules are capable of unidirectional secretion and concentration of uric acid. The creatinine assay performed on the collected fluid showed an 8.2-fold increase in concentration, as compared to serum. These results demonstrated that single cells form multicellular structures and become organized into functional renal units that are able to excrete high levels of solutes through a urine-like fluid.

To determine whether renal tissue could be formed using an alternative cell source, nuclear transplantation was performed to generate histocompatible tissues. The feasibility of engineering syngeneic renal tissues *in vivo* using cloned cells was investigated (Lanza et al., 2002). In this study nuclear material from bovine dermal fibroblasts were transferred into unfertilized enucleated donor bovine eggs. Renal cells from the cloned embryos were harvested, expanded *in vitro*, and seeded onto three-dimensional renal devices (Figure 31.2a). The devices were implanted into the back of the same steer from which the cells were cloned, and were retrieved 12 weeks later. Functioning renal units were created from cells cloned from bovine fibroblasts. Urine production and viability were demonstrated after transplantation back into the nuclear donor animal despite expressing a different mtDNA haplotype (Figure 31.2b). Chemical analysis suggested unidirectional secretion and concentration of urea nitrogen and creatinine. The devices revealed formation of organized glomeruli and tubular structures (Figure 31.2c). Immunohistochemical and reverse transcription-polymerase chain reaction (RT-PCR) analysis confirmed the expression of renal mRNA and proteins, whereas delayed-type hypersensitivity testing and *in vitro* proliferative assays showed that there was no rejection response to the cloned cells. This study indicates that the cloned renal cells are able to form and organize into functional tissue structures, which are genetically same as the host. Generating immune-compatible cells using therapeutic cloning techniques is feasible and may be useful for the engineering of renal tissues for autologous applications.

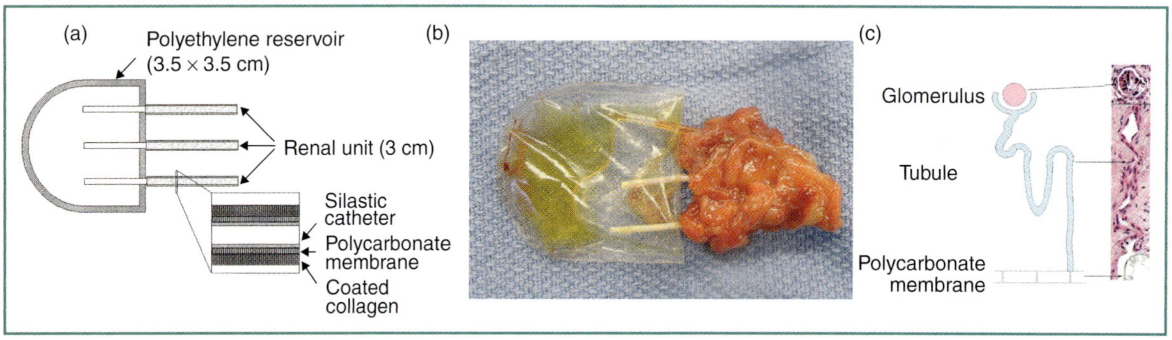

Figure 31.2 Formation of functional renal tissue *in vivo*. (a) Renal device. (b) Tissue-engineered renal unit shows the accumulation of urine-like fluid. (c) There was a clear unidirectional continuity between the mature glomeruli, their tubules, and the polycarbonate membrane.

In our previous study, we showed that renal cells seeded on synthetic renal devices with a collecting system are able to form functional renal structures with urine-like fluid excretion. However, a naturally derived tissue matrix with existing three-dimensional kidney architecture would be preferable, in that it would allow for transplantation of a large number of cells for the creation of greater renal tissue volumes. We developed an acellular collagen-based kidney matrix, which is identical to the native renal architecture. In a subsequent study we investigated whether the collagen-based matrices could accommodate large volumes of renal cells which could proliferate and form kidney structures *in vivo* (Amiel et al., 2000).

Acellular collagen matrices, derived from porcine kidneys, were obtained through a multiple step decellularization process. Serial evaluation of the matrix for cellular remnants was performed using histochemistry, scanning electron microscopy (SEM), and RT-PCR. Mouse renal cells were harvested, grown, and seeded on 80 collagen matrices at a concentration of 30×10^6 cells/ml. Forty cell–matrix constructs grown *in vitro* were analyzed 3 days, 1, 2, 4, and 6 weeks after seeding. The remaining 40 cell–matrices were implanted in the subcutaneous space of 20 athymic mice. The animals were sacrificed 3 days, 1, 2, 4, 8, and 24 weeks after implantation for analyses. Gross, SEM, histochemical, immunocytochemical, and biochemical analyses were performed.

SEM and histologic examination confirmed the acellularity of the processed matrix. RT-PCR performed on the kidney matrices demonstrated the absence of any RNA residues. Renal cells seeded on the matrix adhered to the inner surface and proliferated to confluency 7 days after seeding, as demonstrated by SEM. Histochemical and immunocytochemical analyses performed using H & E, periodic acid Schiff, alkaline phosphatase, anti-osteopontin and anti-CD-31 identified stromal, endothelial, and tubular epithelial cell phenotypes within the matrix. Renal tubular and glomeruli-like structures were observed 8 weeks after implantation. MTT proliferation and radioactive thymidine incorporation assays performed 6 weeks after cell seeding demonstrated a cell population increase of 116% and 92%, respectively, as compared to the 2-week time points. This study demonstrates that renal cells are able to adhere and proliferate on the collagen-based kidney matrices. The renal cells reconstitute renal tubular and glomeruli-like structures. The collagen-based kidney matrix system seeded with renal cells may be useful in the future for augmenting renal function.

Our prior studies demonstrated that culture expanded primary renal cells seeded on artificial renal devices with a collecting system are able to form functional renal structures with urine-like fluid excretion. However, creation of renal structures without the use of an artificial device system would be preferable. In addition, implantation procedures are invasive and may result in unnecessary complications. In a subsequent study we investigated the feasibility of creating three-dimensional renal structures for *in situ* implantation within the native kidney tissue. Primary renal cells from 4-week-old mice were grown and expanded in

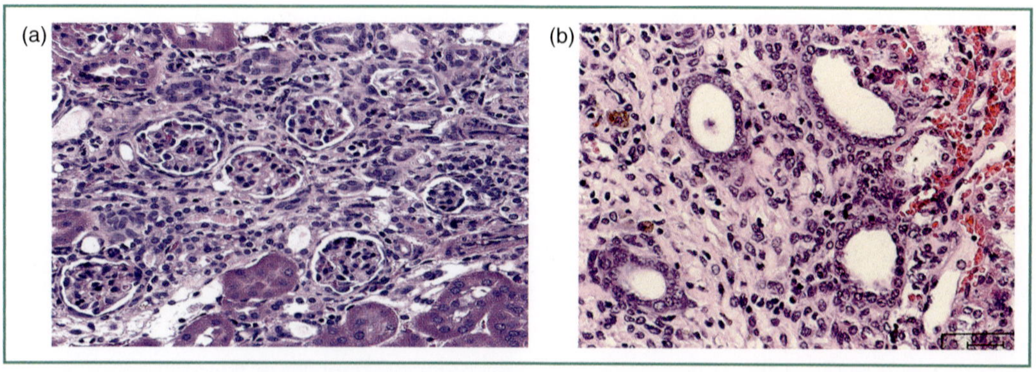

Figure 31.3 **Reconstitution of kidney structures. The implanted renal cells self-assembled into (a) glomerular and (b) tubular structures within the kidney tissue.**

culture. Culture expanded renal cells were labeled with fluorescent markers and injected into mouse kidneys in a collagen gel for *in vivo* formation of renal tissues. Collagen injection without cells and sham operated animals served as controls. *In vitro* reconstituted renal structures and *in vivo* implanted cells were retrieved and analyzed.

The implanted renal cells formed tubular and glomerular structures within the kidney tissue, as confirmed by the fluorescent markers. There was no evidence of renal tissue formation in the control and the sham operated groups. These results demonstrate that single renal cells are able to reconstitute into organized kidney structures when placed in a collagen-based scaffolding system. The implanted renal cells are able to self-assemble into tubular and glomerular structures within the kidney tissue (Figure 31.3). These findings suggest that this system may be the preferred approach to engineer functional kidney tissues for the treatment of end stage renal disease.

SUMMARY

Renal transplantation remains as the gold standard of treatment for end stage renal failure. The increasing demand and shortage of donor organs have ignited tremendous efforts to seek for alternative treatment modalities. Cell-based approaches have been proposed recently as an alternative method for kidney tissue regeneration. In this chapter, various tissue engineering and regenerative medicine approaches were presented in an effort to achieve functional intracorporeal kidney support. Research progress in the regeneration of kidney tissues has been somewhat successful toward augmenting tissue function. However, clinical application of this technology is still distant.

Although it has been demonstrated that renal cells are able to reconstitute into functional kidney tissues *in vivo*, numerous challenges need to be worked out for clinical translation. Some of these include the generation of a large tissue mass that would augment systemic renal function, integration of engineered renal tissue into the host with adequate vascularization and excretory systems, and development of a reliable renal failure model system for testing cell-based technologies. Current work in our institute is aimed at addressing these challenges.

REFERENCES

Amiel, G.E. and Atala, A. (1999). Current and future modalities for functional renal replacement. *Urol. Clin. North Am.* **26**: 235–246, xi.

Amiel, G.E., Yoo, J.J. and Atala, A. (2000). Renal therapy using tissue-engineered constructs and gene delivery. *World J. Urol.* **18**: 71–79.

Atala, A. (1997). *Tissue Engineering in the Genitourinary System*, Chap. 8. Boston: Birkhauser Press.

Atala, A. (1999). Future perspectives in reconstructive surgery using tissue engineering. *Urol. Clin. North Am.* **26**: 157–165, ix–x.

Atala, A., Schlussel, R.N. and Retik, A.B. (1995). Renal cell growth *in vivo* after attachment to biodegradable polymer scaffolds. *J. Urol.* **153**: 4.

Atala, A., Bauer, S.B., Soker, S., Yoo, J.J. and Retik, A.B. (2006). Tissue-engineered autologous bladders for patients needing cystoplasty. *Lancet* **367**: 1241–1246.

Carley, W.W., Milici, A.J. and Madri, J.A. (1988). Extracellular matrix specificity for the differentiation of capillary endothelial cells. *Exp. Cell Res.* **178**: 426–434.

Chazan, J.A., Libbey, N.P., London, M.R., Pono, L. and Abuelo, J.G. (1991). The clinical spectrum of renal osteodystrophy in 57 chronic hemodialysis patients: a correlation between biochemical parameters and bone pathology findings. *Clin. Nephrol.* **35**: 78–85.

El-Kassaby, A.W., Retik, A.B., Yoo, J.J. and Atala, A. (2003). Urethral stricture repair with an off-the-shelf collagen matrix. *J. Urol.* **169**: 170–173; discussion 173.

Feldman, H.I., Kobrin, S. and Wasserstein, A. (1996). Hemodialysis vascular access morbidity. *J. Am. Soc. Nephrol.* **7**: 523–535.

Folkman, J. and Hochberg, M. (1973). Self-regulation of growth in three dimensions. *J. Exp. Med.* **138**: 745–753.

Frank, J., Engler-Blum, G., Rodemann, H.P. and Muller, G.A. (1993). Human renal tubular cells as a cytokine source: PDGF-B, GM-CSF and IL-6 mRNA expression *in vitro*. *Exp. Nephrol.* **1**: 26–35.

Freed, L.E., Vunjak-Novakovic, G., Biron, R.J., Eagles, D.B., Lesnoy, D.C., Barlow, S.K. and Langer, R. (1994). Biodegradable polymer scaffolds for tissue engineering. *Biotechnology (NY)* **12**: 689–693.

Fung, L.C.T., Elenius, K., Freeman, M., Donovan, M.J. and Atala, A. (1996). Reconstitution of poor EGFr-poor renal epithelial cells into tubular structures on biodegradable polymer scaffold. *Pediatrics* **98**(Suppl): S631.

Gupta, S., Verfaillie, C., Chmielewski, D., Kim, Y. and Rosenberg, M.E. (2002). A role for extrarenal cells in the regeneration following acute renal failure. *Kidney Int.* **62**: 1285–1290.

Horikoshi, S., Koide, H. and Shirai, T. (1988). Monoclonal antibodies against laminin A chain and B chain in the human and mouse kidneys. *Lab. Invest.* **58**: 532–538.

Hubbell, J.A., Massia, S.P., Desai, N.P. and Drumheller, P.D. (1991). Endothelial cell-selective materials for tissue engineering in the vascular graft via a new receptor. *Biotechnology (NY)* **9**: 568–572.

Humes, H.D. and Cieslinski, D.A. (1992). Interaction between growth factors and retinoic acid in the induction of kidney tubulogenesis in tissue culture. *Exp. Cell Res.* **201**: 8–15.

Humes, H.D., Buffington, D.A., MacKay, S.M., Funke, A.J. and Weitzel, W.F. (1999). Replacement of renal function in uremic animals with a tissue-engineered kidney. *Nat. Biotechnol.* **17**: 451–455.

Ikarashi, K., Li, B., Suwa, M., Kawamura, K., Morioka, T., Yao, J., Khan, F., Uchiyama, M. and Oite, T. (2005). Bone marrow cells contribute to regeneration of damaged glomerular endothelial cells. *Kidney Int.* **67**: 1925–1933.

Ito, T., Suzuki, A., Imai, E., Okabe, M. and Hori, M. (2001). Bone marrow is a reservoir of repopulating mesangial cells during glomerular remodeling. *J. Am. Soc. Nephrol.* **12**: 2625–2635.

Iwano, M., Plieth, D., Danoff, T.M., Xue, C., Okada, H. and Neilson, E.G. (2002). Evidence that fibroblasts derive from epithelium during tissue fibrosis. *J. Clin. Invest.* **110**: 341–350.

Kale, S., Karihaloo, A., Clark, P.R., Kashgarian, M., Krause, D.S. and Cantley, L.G. (2003). Bone marrow stem cells contribute to repair of the ischemically injured renal tubule. *J. Clin. Invest.* **112**: 42–49.

Lanza, R.P., Chung, H.Y., Yoo, J.J., Wettstein, P.J., Blackwell, C., Borson, N., Hofmeister, E., Schuch, G., Soker, S., Moraes, C.T., et al. (2002). Generation of histocompatible tissues using nuclear transplantation. *Nat. Biotechnol.* **20**: 689–696.

Lin, F., Cordes, K., Li, L., Hood, L., Couser, W.G., Shankland, S.J. and Igarashi, P. (2003). Hematopoietic stem cells contribute to the regeneration of renal tubules after renal ischemia-reperfusion injury in mice. *J. Am. Soc. Nephrol.* **14**: 1188–1199.

Lin, F., Moran, A. and Igarashi, P. (2005). Intrarenal cells, not bone marrow-derived cells, are the major source for regeneration in postischemic kidney. *J. Clin. Invest.* **115**: 1756–1764.

Milici, A.J., Furie, M.B. and Carley, W.W. (1985). The formation of fenestrations and channels by capillary endothelium *in vitro*. *Proc. Natl. Acad. Sci. USA* **82**: 6181–6185.

Mooney, D.J., Mazzoni, C.L., Breuer, C., McNamara, K., Hern, D., Vacanti, J.P. and Langer, R. (1996). Stabilized polyglycolic acid fibre-based tubes for tissue engineering. *Biomaterials* **17**: 115–124.

Oberpenning, F., Meng, J., Yoo, J.J. and Atala, A. (1999). *De novo* reconstitution of a functional mammalian urinary bladder by tissue engineering. *Nat. Biotechnol.* **17**: 149–155.

Poulsom, R., Forbes, S.J., Hodivala-Dilke, K., Ryan, E., Wyles, S., Navaratnarasah, S., Jeffery, R., Hunt, T., Alison, M., Cook, T., et al. (2001). Bone marrow contributes to renal parenchymal turnover and regeneration. *J. Pathol.* **195**: 229–235.

Prockop, D.J. (1997). Marrow stromal cells as stem cells for nonhematopoietic tissues. *Science* **276**: 71–74.

Rookmaaker, M.B., Smits, A.M., Tolboom, H., Van't Wout, K., Martens, A.C., Goldschmeding, R., Joles, J.A., Van Zonneveld, A.J., Grone, H.J., Rabelink, T.J. and Verhaar, M.C. (2003). Bone-marrow-derived cells contribute to glomerular endothelial repair in experimental glomerulonephritis. *Am. J. Pathol.* **163**: 553–562.

Schena, F.P. (1998). Role of growth factors in acute renal failure. *Kidney Int. Suppl.* **66**: S11–S15.

Stadnyk, A.W. (1994). Cytokine production by epithelial cells. *Faseb J.* **8**: 1041–1047.

Tachibana, M., Nagamatsu, G.R. and Addonizio, J.C. (1985). Ureteral replacement using collagen sponge tube grafts. *J. Urol.* **133**: 866–869.

Wald, H.L., Sarakinos, G., Lyman, M.D., Mikos, A.G., Vacanti, J.P. and Langer, R. (1993). Cell seeding in porous transplantation devices. *Biomaterials* **14**: 270–278.

Yokoo, T., Ohashi, T., Shen, J.S., Sakurai, K., Miyazaki, Y., Utsunomiya, Y., Takahashi, M., Terada, Y., Eto, Y., Kawamura, T., et al. (2005). Human mesenchymal stem cells in rodent whole-embryo culture are reprogrammed to contribute to kidney tissues. *Proc. Natl. Acad. Sci. USA* **102**: 3296–3300.

Yoo, J.J., Ashkar, S. and Atala, A. (1996). Creation of functional kidney structures with excretion of kidney-like fluid *in vivo*. *Pediatrics* **98S**: 605.

32

Genitourinary System

Anthony Atala

INTRODUCTION

From the time the fetus develops, the genitourinary system may be exposed to a variety of possible injuries. Individuals may suffer from infection, congenital disorders, trauma, cancer, inflammation, iatrogenic injuries, or other conditions that may lead to genitourinary organ damage or loss and necessitate eventual reconstruction. Whenever there is a lack of native urologic tissue, reconstruction may be performed with native non-urologic tissues (such as skin, gastrointestinal segments, or mucosa from multiple body sites), homologous tissues (cadaver fascia, cadaver, or donor kidney), heterologous tissues (bovine collagen), or artificial materials (silicone, polyurethane, Teflon). The tissues used for reconstruction may lead to complications because of their inherently different functional parameters. In most cases, the replacement of lost or deficient tissues with functionally equivalent tissues would improve the outcome for these patients. This goal may be attainable with the use of tissue-engineering techniques.

RECONSTITUTION STRATEGIES

The goal of tissue engineering is to develop biologic substitutes that can restore and maintain normal function. Tissue engineering may involve matrices alone, wherein the body's natural ability to regenerate is used to orient or direct new tissue growth, or it may use matrices with cells. When cells are used for tissue engineering, donor tissue (heterologous, allogeneic, or autologous) is dissociated into individual cells, which are implanted directly into the host or expanded in culture, attached to a support matrix, and reimplanted after expansion. Ideally, this approach allows lost tissue function to be restored or replaced *in toto* and with limited complications (Atala et al., 1992, 1993a, b; Atala et al., 1994; Atala, 1996, 1997, 1998, 1999; Cilento et al., 1994; Yoo, 1997; Fauza et al., 1998a, b; Machluf, 1998a; Yoo et al., 1998; Amiel, 1999; Kershen, 1999; Oberpenning et al., 1999; Park et al., 1999).

THE ROLE OF BIOMATERIALS

Biomaterials in genitourinary tissue-engineering function as an artificial extracellular matrix (ECM) and elicit biologic and mechanical functions of native ECM found in body tissues. Biomaterials facilitate the localization and delivery of cells and/or bioactive factors (such as cell adhesion peptides and growth factors) to desired sites in the body; define a three-dimensional space for the formation of new tissues with appropriate structure; and guide the development of new tissues with appropriate function (Kim, 1998). While direct injection of cell suspensions without biomaterial matrices has been used (Ponder et al., 1991; Brittberg et al., 1994). it is difficult to control the localization of transplanted cells.

The ideal biomaterial should be biocompatible, promote cellular interaction and tissue development, and possess proper mechanical and physical properties. Generally, three classes of biomaterials have been used for engineering of genitourinary tissues: naturally derived materials, such as collagen and alginate; acellular

 Copyright © 2008 Elsevier Inc. All rights reserved

tissue matrices, such as bladder submucosa and small-intestinal submucosa (SIS) and synthetic polymers, such as polyglycolic acid (PGA), polylactic acid (PLA), and poly(lactic-co-glycolic acid) (PLGA). While naturally derived materials and acellular tissue matrices have the potential advantage of biologic recognition, synthetic polymers can be produced reproducibly on a large scale with controlled properties of strength, degradation rate, and microstructure.

VASCULARIZATION

A restriction of tissue engineering is that cells cannot be implanted in volumes exceeding 3 mm^3 because of the limitations of nutrition and gas exchange (Folkman, 1973). To achieve the goals of engineering large complex tissues, and possibly internal organs, vascularization of the regenerating cells is essential. Three approaches have been used for vascularization of bioengineered tissue: (1) incorporation of angiogeneic factors in the bioengineered tissue; (2) seeding embryonal carcinoma (EC) with other cell types in the bioengineered tissue; and (3) prevascularization of the matrix prior to cell seeding. Angiogeneic growth factors may be incorporated into the bioengineered tissue prior to implantation, in order to attract host capillaries and to enhance neovascularization of the implanted tissue. Many obstacles must be overcome before large entire tissue-engineered solid organs are produced. Recent developments in angiogenesis research may provide important knowledge and essential materials to accomplish this goal.

PROGRESS IN TISSUE ENGINEERING OF UROLOGIC STRUCTURES

Tissue-engineering techniques are currently being investigated for the replacement of lost or deficient genitourinary structures, including urethra, bladder, male and female genital tissues, ureter, and renal structures.

Urethra

Various strategies have been proposed to regenerate urethral tissue. Woven meshes of PGA, without cells, have been used to reconstruct urethras in dogs (Bazeed et al., 1983; Olsen et al., 1992). PGA has been used as a cell transplantation vehicle to engineer tubular urothelium *in vivo* (Atala et al., 1992). SIS without cells was used as an onlay patch graft for urethroplasty in rabbits (Kropp et al., 1998), and a homologous free graft of acellular urethral matrix was also used in a rabbit model (Sievert et al., 2000).

Bladder-derived acellular collagen matrix has proven to be a suitable graft for repairing urethral defects in rabbits. The created neourethras demonstrated a normal urothelial luminal lining and organized muscle bundles (Chen, 1999). Results were confirmed clinically in a series of patients with a history of failed hypospadias reconstruction whose urethral defects were repaired with human bladder acellular collagen matrices (Figure 32.1) (Atala, 1999). An advantage of this material over non-genital tissue grafts for urethroplasty is that it is "off the shelf," eliminating the need for additional surgical procedures for graft harvesting and decreasing operative time and potential morbidity from the harvest procedure.

The above techniques, using non-seeded acellular matrices, were successfully applied experimentally and clinically for onlay urethral repairs. However, when tubularized repairs were attempted experimentally, adequate urethral tissue regeneration was not achieved, and complications, such as graft contracture and stricture formation ensued (le Roux, 2005). Seeded tubularized collagen matrices have performed better than their non-seeded counterparts in animal studies. In a rabbit model, entire urethral segments were resected, and urethroplasties were performed with tubularized collagen matrices, either seeded or non-seeded. The tubularized collagen matrices seeded with autologous cells formed new tissue which was histologically similar to native urethra. Those without cells lead to poor tissue development, fibrosis, and stricture formation. These findings were recently confirmed clinically (DeFilippo et al., 2002).

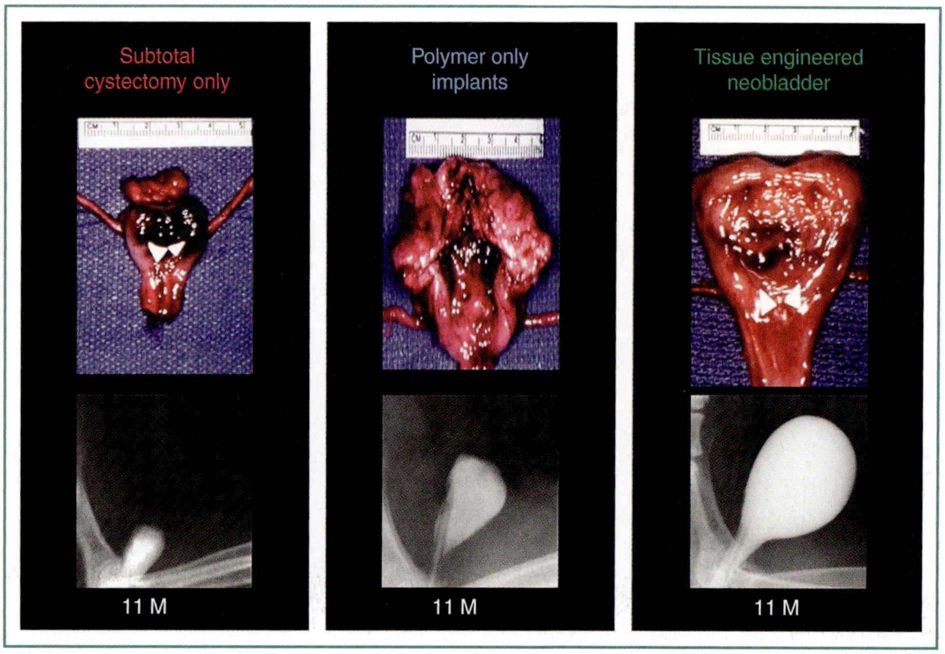

Figure 32.2 **Comparison of tissue-engineered neo-bladders. Gross specimens and cystograms at 11 months of the cystectomy-only, non-seeded controls, and cell-seeded tissue-engineered bladder replacements. The cystectomy-only bladder had a capacity of 22% of the preoperative value and a decrease in bladder compliance to 10% of the preoperative value. The non-seeded controls showed significant scarring with a capacity of 46% of the preoperative value and a decrease in bladder compliance to 42% of the preoperative value. An average bladder capacity of 95% of the original precystectomy volume was achieved in the cell-seeded tissue-engineered bladder replacements and the compliance showed almost no difference from preoperative values that were measured when the native bladder was present (106%).**

Corporal Smooth Muscle

Because one of the major components of the phallus is corporal smooth muscle, the creation of autologous functional and structural corporal tissue *de novo* would be beneficial. To examine functional parameters of engineered corpora, acellular corporal collagen matrices were obtained from donor rabbit penis, and autologous corpus cavernosal smooth muscle, and endothelial cells were harvested, expanded, and seeded on the matrices. The entire rabbit corpora was removed and replaced with engineered scaffolds. The experimental corporal bodies demonstrated intact structural integrity by cavernosography and showed similar pressure by cavernosometry when compared with normal controls. The control rabbits (without cells) failed to show normal erectile function throughout the study. Mating activity in the animals with the engineered corpora appeared normal by 1 month after implantation. The presence of sperm was confirmed during mating and was present in all the rabbits with the engineered corpora. The female rabbits mated with the animals implanted with engineered corpora and also conceived and delivered healthy pups. Animals implanted with the matrix alone were unable to demonstrate normal mating activity and failed to ejaculate into the vagina (Chen, 2005).

Engineered Penile Prostheses

Although silicone is an accepted biomaterial for penile prostheses, biocompatibility remains a concern (Thomalla et al., 1987; Nukui et al., 1997). The use of a natural prosthesis composed of autologous cells may be advantageous.

A recent study using an autologous system investigated the feasibility of applying the engineered cartilage rods *in situ* (Yoo et al., 1999). Autologous chondrocytes harvested from rabbit ear were grown and expanded in culture. The cells were seeded onto biodegradable poly-L-lactic acid-coated PGA polymer rods and implanted into the corporal spaces of rabbits. Examination at retrieval 1 month later showed the presence of well-formed, milky-white cartilage structures within the corpora. All polymers were fully degraded by 2 months. There was no evidence of erosion or infection in any of the implantation sites.

Subsequent studies assessed the long-term functionality of the cartilage penile rods *in vivo* (Yoo et al., 1999). To date, the animals have done well and can copulate and impregnate their female partners without problems.

Female Genital Tissues

Congenital malformations of the uterus may have profound implications clinically. Patients with cloacal exstrophy or intersex disorders may not have sufficient uterine tissue for future reproduction.

We investigated the possibility of engineering functional uterine tissue using autologous cells (Wang et al., 2003). Autologous rabbit uterine smooth muscle and epithelial cells were harvested, then grown, and expanded in culture. These cells were seeded onto preconfigured uterine-shaped biodegradable polymer scaffolds, which were then used for subtotal uterine tissue replacement in the corresponding autologous animals. Upon retrieval 6 months after implantation, histological, immunocytochemical, and Western blot analyses confirmed the presence of normal uterine tissue components. Biomechanical analyses and organ bath studies showed that the functional characteristics of these tissues were similar to those of normal uterine tissue. Breeding studies using these engineered uteri are currently being performed.

Several pathologic conditions, including congenital malformations and malignancy, can adversely affect normal vaginal development or anatomy. Vaginal reconstruction has traditionally been challenging due to the paucity of available native tissue.

Vaginal epithelial and smooth muscle cells of female rabbits were harvested, grown, and expanded in culture. The cells were seeded onto biodegradable polymer scaffolds, which were then implanted into nude mice for up to 6 weeks. Immunocytochemical, histological, and Western blot analyses confirmed the presence of vaginal tissue phenotypes. Electrical field stimulation studies in the tissue-engineered constructs showed similar functional properties to those of normal vaginal tissue. When these constructs were used for autologous total vaginal replacement, patent vaginal structures were noted in the tissue-engineered specimens, while the non-seeded structures were noted to be stenotic (De Filippo, 2003).

Ureter

Ureteral non-seeded matrices have been used as a scaffold for the ingrowth of ureteral tissue in rats. On implantation, the acellular matrices promoted the regeneration of the ureteral wall components (Dahms et al., 1997). Ureteral replacement with polytetrafluoroethylene (Teflon) grafts was also attempted in dogs, but with poor functional results (Baltaci, 1998). In a more recent study, non-seeded ureteral collagen acellular matrices were tabularized, but attempts to use them to replace 3 cm segments of canine ureters were unsuccessful (Osman, 2004).

Cell-seeded biodegradable polymer scaffolds have been used with more success to reconstruct ureteral tissues. In one study, urothelial and smooth muscle cells isolated from bladders and expanded *in vitro* were seeded onto PGA scaffolds with tubular configurations and implanted subcutaneously into athymic mice. After implantation, the urothelial cells proliferated to form a multilayered luminal lining of tubular structures, while the smooth muscle cells organized into multilayered structures surrounding the urothelial cells. Abundant angiogenesis was evident. Polymer scaffold degradation resulted in the eventual formation of natural urothelial tissues. This approach has also been used to replace ureters in dogs (Yoo, 1995).

Renal Structures

Due to its complex structure and function, the kidney is possibly the most challenging organ in the genitourinary system to reconstruct using tissue-engineering techniques. However concepts for a bioartificial kidney are emerging. Some investigators are pursuing the replacement of isolated kidney function parameters with the use of extracorporeal units, while others are working toward the replacement of total renal function by tissue-engineered bioartificial structures.

Ex vivo Renal Units

Although dialysis is currently the most prevalent form of renal replacement therapy, the relatively high rates of morbidity and mortality have spurred investigators to seek alternative solutions involving *ex vivo* systems.

To assess the viability and physiologic functionality of a cell-seeded device to replace the filtration, transport, metabolic, and endocrinologic functions of the kidney in acutely uremic dogs, researchers introduced a synthetic hemofiltration device combined with a renal tubular cell therapy device (containing porcine renal tubules in an extracorporeal perfusion circuit). Levels of potassium and blood urea nitrogen (BUN) were controlled during treatment with the device. The fractional reabsorption of sodium and water was possible, and active transport of potassium, bicarbonate, and glucose and a gradual ability to excrete ammonia was observed. These results demonstrated the technologic feasibility of an extracorporeal assist device that is reinforced by the use of proximal tubular cells (Humes et al., 1999).

Using similar techniques, a tissue-engineered bioartificial kidney – consisting of a conventional hemofiltration cartridge in series with a renal tubule assist device containing human renal proximal tubule cells – was used in patients with acute renal failure in the intensive care unit. Initial clinical experience with the bioartificial kidney and the renal tubule assist device suggests that such therapy may provide a dynamic and individualized treatment program as assessed by acute physiologic and biochemical indices (Humes et al., 2003).

In vivo Renal Structures

Another method of improving renal function involves augmenting renal tissue with kidney cell expansion *in vitro* and subsequent autologous transplantation. The feasibility of achieving renal cell growth, expansion, and *in vivo* reconstitution with tissue-engineering techniques has been explored.

Most recently, an attempt was made to harness the reconstitution of renal epithelial cells to generate functional nephron units. Renal cells harvested and expanded in culture were seeded onto a tubular device constructed from a polycarbonate membrane connected at one end to a Silastic catheter terminating into a reservoir. The device was implanted into athymic mice. Histologic examination of the implanted devices over time revealed extensive vascularization, with formation of glomeruli and highly organized tubule-like structures. Immunocytochemical staining confirmed the renal phenotype. Yellow fluid consistent with the makeup of dilute urine in its creatinine and uric acid concentrations was retrieved from inside the implant (Yoo, 1996). Further studies using nuclear transfer techniques have been performed showing the formation of renal structures in cows (Figure 32.3) (Lanza et al., 2002). Challenges facing this technology include the expansion to larger, three-dimensional structures.

ADDITIONAL APPLICATIONS

Tissue engineering and cell therapy hold promise for a number of additional genitourinary applications.

Fetal Tissue Engineering

Improved prenatal diagnostic techniques have led to the use of intervention before birth to reverse potentially life-threatening processes. Several strategies may be pursued to facilitate the future prenatal management of

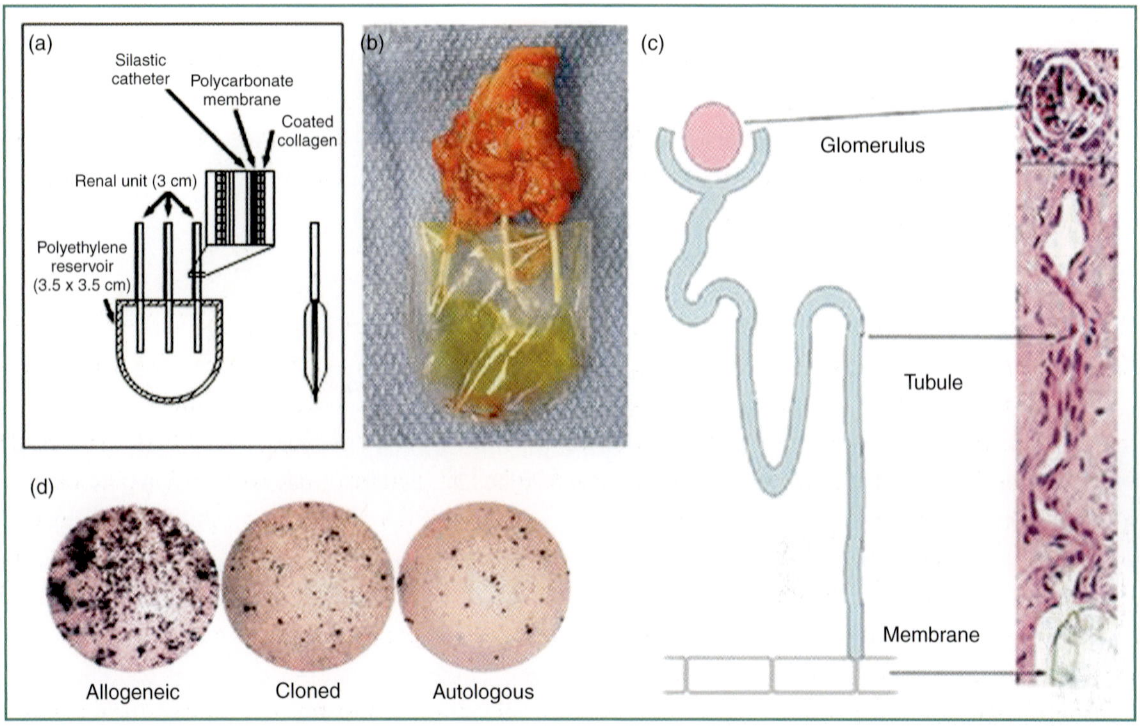

Figure 32.3 Creation of kidney tissue from therapeutic cloning and tissue-engineering strategies. (a) Illustration of the tissue-engineered renal unit. (b) Renal unit seeded with cloned cells, 3 months after implantation, showing the accumulation of urine-like fluid. (c) There was a clear unidirectional continuity between the mature glomeruli, their tubules, and the polycarbonate membrane. (d) Elispot analyses of the frequencies of T cells that secrete IFN-gamma after primary and secondary stimulation with allogeneic renal cells, cloned renal cells, or nuclear donor fibroblasts.

urologic disease. Having a ready supply of urologic-associated tissue for surgical reconstruction at birth may be advantageous.

Theoretically, once the diagnosis of the pathologic condition is confirmed prenatally, a small tissue biopsy could be obtained under US guidance. These biopsy materials could then be processed and the various cell types expanded *in vitro*. Using tissue-engineering techniques, reconstituted structures *in vitro* could then be readily available at the time of birth for reconstruction (Fauza et al., 1998a).

Injectable Therapies

Both urinary incontinence and vesicoureteral reflux (VUR) are common conditions affecting the genitourinary system that can be treated with injectable bulking agents. The ideal substance for endoscopic treatment of VUR and incontinence should be injectable, non-antigenic, non-migratory, volume stable, and safe for human use.

Animal studies have shown that chondrocytes can be easily harvested and combined with alginate *in vitro*; the suspension can be easily injected cystoscopically; and the elastic cartilage tissue formed is able to correct VUR without any evidence of obstruction (Atala, 1994). The first human application of cell-based tissue-engineering technology for urologic applications occurred with the injection of chondrocytes for the correction of VUR in children and for urinary incontinence in adults (Figure 32.4) (Diamond, 1999; Bent et al., 2001).

Using cell therapy techniques, the use of autologous smooth muscle cells was explored for both urinary incontinence and VUR applications (Cilento, 1995). The potential use of injectable, cultured myoblasts for

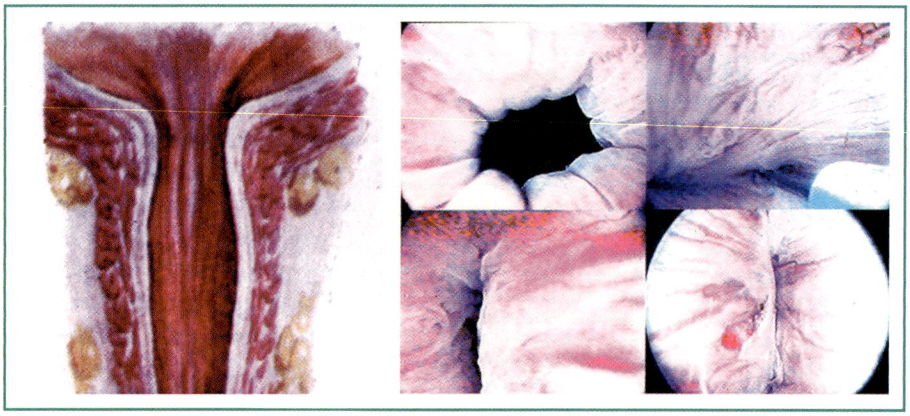

Figure 32.4 Tissue-engineered bulking agent. Chondrocytes are harvested and combined with alginate *in vitro*, and the suspension is injected cystoscopically as a bulking agent to treat urinary incontinence.

the treatment of stress urinary incontinence (SUI) has also been investigated (Yokoyama et al., 1999; Chancellor et al., 2000).

Use of injectable muscle precursor cells has also been studied for treatment of urinary incontinence due to irreversible urethral sphincter injury or maldevelopment (Yiou, 2003). A clinical trial involving the use muscle-derived stem cells (MDSC) to treat SUI has also been performed with good results. Biopsies of skeletal muscle were obtained, and autologous myoblasts and fibroblasts were cultured. Under US guidance, myoblasts were injected into the rhabdosphincter, and fibroblasts mixed with collagen were injected into the submucosa. One year following injection, the thickness and function of the rhabdosphincter had significantly increased, and all patients were continent (Strasser et al., 2004). These are the first demonstrations of the replacement of both sphincter muscle tissue and its innervation by the injection of muscle precursor cells.

In addition, injectable muscle-based gene therapy and tissue engineering were combined to improve detrusor function in a bladder injury model, and may have potential as a novel treatment option for urinary incontinence (Huard et al., 2002).

Testicular Hormone Replacement

Patients with testicular dysfunction require androgen replacement for somatic development. Conventional treatment consists of periodic intramuscular injections of chemically modified testosterone or, more recently, skin patch applications. However, long-term non-pulsatile testosterone therapy is not optimal and can cause multiple problems, including erythropoiesis and bone density changes.

A system was designed wherein Leydig cells were microencapsulated for controlled testosterone replacement. Purified Leydig cells were isolated and encapsulated in an alginate-poly-L-lysine solution. The encapsulated Leydig cells were injected into castrated animals, and serum testosterone was measured serially; the animals were able to maintain testosterone levels in the long term (Machluf et al., 1998b). These studies suggest that microencapsulated Leydig cells may be able to replace or supplement testosterone in situations where anorchia or testicular failure is present.

CONCLUSION

Tissue-engineering efforts are currently being undertaken for every type of tissue and organ within the urinary system. Most of the effort expended to engineer genitourinary tissues has occurred within the last decade. Tissue-engineering techniques require a cell culture facility designed for human application. Recent progress suggests that engineered urologic tissues and cell therapy may have clinical applicability. Personnel who have mastered the techniques of cell harvest, culture, and expansion as well as polymer design are essential for the successful application of this technology. Before these engineering techniques can be applied to humans, further studies need to be performed in many of the tissues described.

ACKNOWLEDGMENTS

The author would like to express his appreciation for editorial assistance provided by Diane Q. Mann, M.S.

Dr. Atala is the W.H. Boyce Professor and Chair, Department of Urology, and Director, Wake Forest Institute for Regenerative Medicine, Wake Forest University School of Medicine, Winston-Salem, NC.

REFERENCES

Amiel, G.E. and Atala, A. (1999). Current and future modalities for functional renal replacement. *Urol. Clin. N. Am.* **26**: 235–246.

Atala, A. (1994). Use of non-autologous substances in VUR and incontinence treatment. *Dial. Pediatr. Urol.* **17**: 11–12.

Atala, A. (1995). Commentary on the replacement of urologic associated mucosa. *J. Urol.* **156**: 338–339.

Atala, A. (1997). Tissue engineering in the genitourinary system. In: Atala, A., Mooney, D. (eds.), *Tissue Engineering.* Boston: Birkhauser Press, p. 149.

Atala, A. (1998). Autologous cell transplantation for urologic reconstruction. *J. Urol.* **159**: 2–3.

Atala, A. (1999). Future perspectives in reconstructive surgery using tissue engineering. *Urol. Clin. N. Am.* **26**: 157–165.

Atala, A., Cima, L.G., Kim, W., Paige, K.T., Vacanti, J.P., Retik, A.B. and Vacanti, C.A. (1993a). Injectable alginate seeded with chondrocytes as a potential treatment for vesicoureteral reflux. *J. Urol.* **150**: 745–747.

Atala, A., Freeman, M.R., Vacanti, J.P., Shepard, J. and Retik, A.B. (1993b) Implantation *in vivo* and retrieval of artificial structures consisting of rabbit and human urothelium and human bladder muscle. *J. Urol.* **150**: 608–612.

Atala, A., Guzman, L. and Retik, A. (1999). A novel inert collagen matrix for hypospadias repair. *J. Urol.* **162**: 1148–1151.

Atala, A., Kim, W., Paige, K.T., Vacanti, C.A. and Retik, A.B. (1994). Endoscopic treatment of vesicoureteral reflux with chondrocyte-alginate suspension. *J. Urol.* **152**: 641–643.

Atala, A., Vacanti, J.P., Peters, C.A., Mandell J., Retick, J.B. and Freeman, M.R. (1992). Formation of urothelial structures *in vivo* from dissociated cells attached to biodegradable polymer scaffolds *in vitro*. *J. Urol.* **148**: 658–662.

Badylak, S.F., Lantz, G.C., Coffey, A. and Geddes, L.A. (1989). Small intestinal submucosa as a large diameter vascular graft in the dog. *J. Surg. Res.* **47**: 74–80.

Baltaci, S., Ozer, G., Ozer, E., Soygur, T., Besalti, O. and Anafarta, K. (1998). Failure of ureteral replacement with Gore-Tex tube grafts. *Urology* **51**: 400–403.

Bazeed, M.A., Thüroff, J.W., Schmidt, R.A. and Tanagho, E.A. (1983). New treatment for urethral strictures. *Urology* **21**: 53–57.

Bent, A., Tutrone, R., McLennan, M., Lloyd, L.K., Kennelly, M.J. and Badlani, G. (2001). Treatment of intrinsic sphincter deficiency using autologous ear chondrocytes as a bulking agent. *Neurourol. Urodynam.* **20**: 157–165.

Blandy, J.P. (1961). Neal pouch with transitional epithelium and anal sphincter as a continent urinary reservoir. *J. Urol.* **86**: 749–767.

Blandy, J.P. (1964). The feasibility of preparing an ideal substitute for the urinary bladder. *Ann. R. Coll. Surg.* **35**: 287–311.

Brittberg, M., Lindahl, A., Nilsson, A., Ohlsson, C., Isaksson, O. and Peterson, L. (1994). Treatment of deep cartilage defects in the knee with autologous chondrocyte transplantation. *N. Engl. J. Med.* **331**: 889–895.

Chancellor, M.B., Yokoyama, T., Tirney, S., Mattes, C.E., Ozawa, H., Yoshimura, N., de Groat, W.C. and Huard, J. (2000). Preliminary results of myoblast injection into the urethra and bladder wall: a possible method for the treatment of stress urinary incontinence and impaired detrusor contractility. *Neurourol. Urodynam.* **19**: 279–287.

Chen, F., Yoo, J.J. and Atala, A. (1999). Acellular collagen matrix as a possible off the shelf biomaterial for urethral repair. *Urology* **54**: 407–410.

Chen, K.L., Yoo, J.J. and Atala, A. (2005). Total penile corpora cavernosa replacement using tissue engineering techniques. *Regenerate* (abstract).

Cheng, E., Rento, R., Grayhack, T.J., Oyasu, R. and McVary, K.T. (1994). Reversed seromuscular flaps in the urinary tract in dogs. *J. Urol.* **152**: 2252–2257.

Cilento, B.G. and Atala, A. (1995). Treatment of reflux and incontinence with autologous chondrocytes and bladder muscle cells. *Dial. Pediatr. Urol.* **18**: 11–15.

Cilento, B.G., Freeman, M.R., Schneck, F.X., Retik, A.B. and Atala, A. (1994). Phenotypic and cytogenetic characterization of human bladder urothelia expanded *in vitro*. *J. Urol.* **152**: 655–670.

Dahms, S.E., Piechota, H.J., Nunes, L., Dahiya, R., Lue, T.F. and Tanagho, E.A. (1997). Free ureteral replacement in rats: Regeneration of ureteral wall components in the acellular matrix graft. *Urology* **50**: 818–825.

DeFilippo, R.E., Pohl, H.G., Yoo, J.J. and Atala, A. (2002). Total penile urethra replacement with autologous cell-seeded collagen matrices. **167**: 152–153.

DeFilippo, R.E., Yoo, J.J. and Atala, A. (2003). Engineering of vaginal tissue *in vivo*. *Tissue Eng.* **9**: 301–306.

Dewan, P.A. (1998). Autoaugmentation demucosalized enterocystoplasty. *World J. Urol.* **16**: 255–261.

Diamond, D.A. and Caldamone, A.A. (1999). Endoscopic correction of vesicoureteral reflux in children using autologous chondrocytes: preliminary results. *J. Urol.* **162**: 1185–1188.

Fauza, D.O., Fishman, S., Mehegan, K. and Atala, A. (1998a). Videofetoscopically assisted fetal tissue engineering: bladder augmentation. *J. Pediatr. Surg.* **33**: 7–12.

Fauza, D.O., Fishman, S, Mehegan, K. and Atala, A. (1998b). Videofetoscopically assisted fetal tissue engineering: skin replacement. *J. Pediatr. Surg.* **33**: 357–361.

Folkman, J. and Hochberg, M. (1973). Self-regulation of growth in three dimensions. *J. Exp. Med.* **138**: 745–753.

Harada, N., Yano, H., Ohkawa, T., Misse, T., Kurita, T. and Nagahara, A. (1965). New surgical treatment of bladder tumors: mucosal denudation of the bladder. *Br. J. Urol.* **37**: 545–547.

Huard, J., Yokoyama, T., Pruchnic, R., Ou, A., Li, Y., Lee, J.Y., Somogyi, G.T., de Groat, W.C. and Chancellor, M.B. (2002). Muscle-derived cell-mediated *ex vivo* gene therapy for urological dysfunction. *Gene Ther.* **9**: 1617–1626.

Humes, H.D., Buffington, D.A., MacKay, S.M., Funke, A.J. and Weitzel, W.F. (1999). Replacement of renal function in uremic animals with a tissue engineered kidney. *Nat. Biotech.* **17**: 451–455.

Humes, H.D., Weitzel, W.F., Bartlett, R.H., Swaniker, F.C. and Paganini, E.P. (2003). Renal cell therapy is associated with dynamic and individualized responses in patients with acute renal failure. *Blood Purif.* **21**: 64–71.

Kaefer, M., Hendren, H., Bauer, S., Goldenblatt, P., Peters, C.A., Atala, A. and Retik, A.B. (1998). Reservoir calculi: a comparison of reservoirs constructed from stomach and other enteric segments. *J .Urol.* **160**: 2187–2190.

Kaefer, M., Tobin, M., Hendren, H., Bauer, S.B., Peters, C.A., Atala, A., Colodny, A.H., Mandell, J. and Retik, A.B. (1997). Continent urinary diversion: the Children's Hospital experience. *J. Urol.* **157**: 1394–1399.

Kershen, R.T. and Atala, A. (1999). Advances in injectable therapies for the treatment of incontinence and vesicoureteral reflux. *Urol. Clin. N. Am.* **26**: 81–94.

Kim, B.S. and Mooney, D.J. (1998). Development of biocompatible synthetic extracellular matrices for tissue engineering. *Trends Biotechnol.* **16**: 224–230.

Kropp, B.P., Cheng, E.Y., Lin, H.K. and Zhang, Y. (2004). Reliable and reproducible bladder regeneration using unseeded distal small intestinal submucosa. *J. Urol.* **172**:1710–1713.

Kropp, B.P., Ludlow, J.K., Spicer, D., Rippy, M.K., Badylak, S.F., Adams, M.C., Keating, M.A., Rink, R.C., Birhle, R. and Thor, K.B. (1998). Rabbit urethral regeneration using small intestinal submucosa onlay grafts. *Urology* **52**: 138–142.

Kropp, B.P., Rippy, M.K., Badylak, S.F., Adams, M.C., Keating, M.A., Rink, R.C. and Thor, K.B. (1996). Small intestinal submucosa: urodynamic and histopathologic evaluation in long term canine bladder augmentations. *J. Urol.* **155**: 2098–2104.

Lailas, N.G., Cilento, B. and Atala, A. (1996). Progressive ureteral dilation for subsequent ureterocystoplasty. *J. Urol.* **156**: 1151– 1153.

Landman, J., Olweny, E., Sundaram, C.P., Andreoni, C., Collyer, W.C., Rehman, J., Jerde, T.J., Lin, H.K., Lee, D.I., Nunlist, E.H., et al. (2004). Laparoscopic mid sagittal hemicystectomy and bladder reconstruction with small intestinal submucosa and reimplantation of ureter into small intestinal submucosa: 1-year followup. *J. Urol.* **171**: 2450–2455.

Lanza, R.P., Chung, H.Y., Yoo, J.J., Wettstein, P.J., Blackwell, C., Borson, N., Hofmeister, E., Schuch, G., Soker, S., Moraes, C.T., et al. (2002). Generation of histocompatible tissues using nuclear transplantation. *Nat. Biotechnol.* **20**: 689–696.

le Roux, P.J. (2005). Endoscopic urethroplasty with unseeded small intestinal submucosa collagen matrix grafts: a pilot study. *J. Urol.* **173**: 140–143.

Machluf, M. and Atala, A. (1998a). Emerging concepts for tissue and organ transplantation. *Graft* **1**: 31–37.

Machluf, M., Boorjian, S., Caffaratti, J., Kershen, R., Orsola, A. and Atala, A. (1998b). Microencapsulation of Leydig cells: a new system for the therapeutic delivery of testosterone. *Pediatrics* **102**(Suppl.): 32.

McDougal, W.S. (1992). Metabolic complications of urinary intestinal diversion. *J. Urol.* **147**: 1199–1208.

Nukui, F., Okamoto, S., Nagata, M., Kurukawa, J. and Fukui, J. (1997). Complications and reimplantation of penile implants. *Int. J. Urol.* **4**: 52–54.

Oberpenning, F.O., Meng, J., Yoo, J. and Atala, A. (1999). *De novo* reconstitution of a functional urinary bladder by tissue engineering. *Nat. Biotechnol.* **17**: 149–155.

Oesch, I. (1988). Neourothelium in bladder augmentation: an experimental study in rats. *Eur. Urol.* **14**: 328–329.

Olsen, L., Bowald, S., Busch, C., Carlsten, J. and Eriksson, I. (1992). Urethral reconstruction with a new synthetic absorbable device. *Scan. J. Urol. Nephrol.* **26**: 323–326.

Osman, Y., Shokeir, A., Gabr, M., El-Tabey, N., Mohsen, T. and El-Baz, M. (2004). Canine ureteral replacement with long acellular matrix tube: is it clinically applicable? *J. Urol.* **172**: 1151–1154.

Park, H.J., Kershen, R., Yoo, J. and Atala, A. (1999). Reconstitution of human corporal smooth muscle and endothelial cells *in vivo*. *J. Urol.* **162**: 1106–1109.

Piechota, H.J., Dahms, S.E., Nunes, L.S., Dahiya, R., Lue, T.F. and Tanagho, E.A. (1998). *In vitro* functional properties of the rat bladder regenerated by the bladder acellular matrix graft. *J. Urol.* **159**: 1717–1724.

Ponder, K.P., Gupta, S., Leland, F., Darlington, G., Finegold, M., DeMayo, J., Ledley, F.D., Chowdbury, J.R. and Woo, S.L. (1991). Mouse hepatocytes migrate to liver parenchyma and function indefinitely after intrasplenic transplantation. *Proc. Natl. Acad. Sci. USA* **88**: 1217–1221.

Portis, A.J., Elbahnasy, A.M., Shalhav, A.L., Brewer, A., Humphrey, P., McDougall, E.M. and Clayman, R.V. (2000). Laparoscopic augmentation cystoplasty with different biodegradable grafts in an animal model. *J. Urol.* **164**: 1405–1411.

Probst, M., Dahiya, R., Carrier, S. and Tanagho, E.A. (1997). Reproduction of functional smooth muscle tissue and partial bladder replacement. *Br. J. Urol.* **79**: 505–515.

Salle, J., Fraga, C., Lucib, A., Lampertz, M., Jobim, G., Jobim, G. and Putten, A. (1990). Seromuscular enterocystoplasty in dogs. *J. Urol.* **144**: 454–456.

Satar, N., Yoo, J. and Atala, A. (1999). Progressive bladder dilation for subsequent augmentation cystoplasty. *J. Urol.* **162**: 829–831.

Sievert, K.D., Bakircioglu, M.E., Nunes, L., Tu, R., Dahiya, R. and Tanagho, E.A. (2000). Homologous acellular matrix graft for urethral reconstruction in the rabbit: histological and functional evaluation. *J. Urol.* **163**: 1958–1965.

Strasser, H., Berjukow, S., Markstiner, R., Margreiter, E., Hering, S., Bartsch, G. and Hering, S. (2004). Stem cell therapy for urinary stress incontinence. *Exp. Gerontol.* **39**: 1259–1265.

Sutherland, R.S., Baskin, L.S., Hayward, S.W. and Cunha, G.R. (1996). Regeneration of bladder urothelium, smooth muscle, blood vessels and nerves into an acellular tissue matrix. *J. Urol.* **156**: 571–577.

Thomalla, J.V., Thompson, S.T., Rowland, R.G. and Mulcahy, J.J. (1987). Infectious complications of penile prosthetic implants. *J. Urol.* **138**: 65–67.

Wang, T., Koh, C.J. and Yoo, J.J. (2003). Creation of an engineered uterus for surgical reconstruction *Proceedings of the American Academy of Pediatrics Section on Urology*, New Orleans, LA.

Yiou, R., Yoo, J.Y. and Atala, A. (2003). Restoration of functional motor units in a rat model of sphincter injury by muscle precursor cell autografts. *Transplantation* **76**: 1053–1060.

Yokoyama, T., Chancellor, M.B. and Watanabe, T. (1999). Primary myoblasts injection into the urethra and bladder as a potential treatment of stress urinary incontinence and impaired detrusor contractility: long term survival without significant cytotoxicity. *J. Urol.* **161**: 307.

Yoo, J.J., Ashkar, S. and Atala, A. (1996). Creation of functional kidney structures with excretion of urine-like fluid *in vivo*. *Pediatrics* **98**(Suppl.): 605.

Yoo, J.J. and Atala, A. (1997). A novel gene delivery system using urothelial tissue engineered neo-organs. *J. Urol.* **158**: 1066–1070.

Yoo, J.J., Meng, J., Oberpenning, F. and Atala, A. (1998). Bladder augmentation using allogenic bladder submucosa seeded with cells. *Urology* **51**: 221–225.

Yoo, J., Park, H.J., Lee, I. and Atala, A. (1999). Autologous engineered cartilage rods for penile reconstruction. *J. Urol.* **162**: 1119–1121.

Yoo, J.J., Satar, N., Retik, A.B. and Atala, A. (1995). Ureteral replacement using biodegradable polymer scaffolds seeded with urothelial and smooth muscle cells. *J. Urol.* **153**(Suppl.): 375A.

33

Tissue Engineering of the Reproductive System

Stefano Giuliani, Laura Perin, Sargis Sedrakyan, and Roger De Filippo

INTRODUCTION

Human organ replacement is limited by a donor shortage, and problems of tissue compatibility, and rejection. Tissue engineering combines the principles and methods of the life sciences with those of engineering to elucidate a fundamental understanding of structure–function relationships of normal and diseased tissues, in order to facilitate the development of materials and methods to repair damaged or diseased tissues, and to create entire tissue replacements.

A large number of materials, including naturally derived and synthetic polymers have been utilized to facilitate prostheses for genitourinary system.

Porous, absorbable matrices made of natural or synthetic polymers are currently being investigated as scaffolds for genitourinary tissue transplantation. These biodegradable polymers include poly(glycolic acid) (PGA), polylactide (PLA), poly(glycolide-co-lactide) (PGLA), poly(caprolactone) (PCL), poly (glycolide-co-ε-caprolactone), collagen, alginate, hyaluronate, and laminin. These scaffolds require proper biocompatibility, degradability, mechanical stability, high surface area/volume ratio, and proper pore size. A highly porous scaffold is desirable to allow large number of cell seeding or migration throughout the material and the pore size affects both tissue ingrowth and the internal surface area available for cell attachment (Mikos et al., 1994).

The success of using cell transplantation strategies for genitourinary reconstruction depends on our ability to use donor tissue efficiently and to provide the right conditions for long-term survival, differentiation, and growth (Patrick et al.,1998).

Tissue engineering follows the principles of cell transplantation, materials science, and engineering toward the development of biological substitutes, which would restore and maintain normal function. Tissue engineering may involve matrices alone, wherein the body's natural ability to regenerate is used to orientate or direct new tissue growth, or the use of matrices with cells. When cells are used for tissue engineering, donor tissue is dissociated into individual cells, which are either implanted directly into the host, or expanded in culture, attached to a support matrix, and reimplanted after expansion. The implanted tissue can be heterologous, allogeneic, or autologous. Ideally, this approach might allow lost tissue function to be restored or replaced in total and with limited complications (Atala, 1997). The use of autologous cells would avoid rejection, wherein a biopsy of tissue is obtained from the host, the cells are dissociated and expanded *in vitro*, reattached to a matrix, and implanted into the same host.

Copyright © 2008 Elsevier Inc. All rights reserved

MALE

Urethra

Congenital or acquired disorders of the urethra remain a challenge in the urology field. Various urethral conditions such as strictures, traumatic defects, congenital defects, and cancer often require additional tissue for reconstruction. Under circumstances in which there is a lack of urethral mucosa for adequate reconstruction, tissue from other sources has been used, such as genital and extragenital skin flaps or grafts (Xu et al., 2002). Complications such as hair growth, graft shrinkage, stricture, stone formation, and diverticuli have been associated with skin grafts (Hendran and Reda, 1986; Ozcan and Kahveci, 1987).

Anatomy

In men the urethra is divided into three parts, named after the location: the prostatic urethra crosses through the prostate gland; the membranous urethra is a small (1 or 2 cm) portion passing through the external urethral sphincter, this is the narrowest part of the urethra; the spongy (or penile) urethra runs along the length of the penis on its ventral (underneath) surface. It is about 15–16 cm in length, and travels through the corpus spongiosum (Figure 33.1).

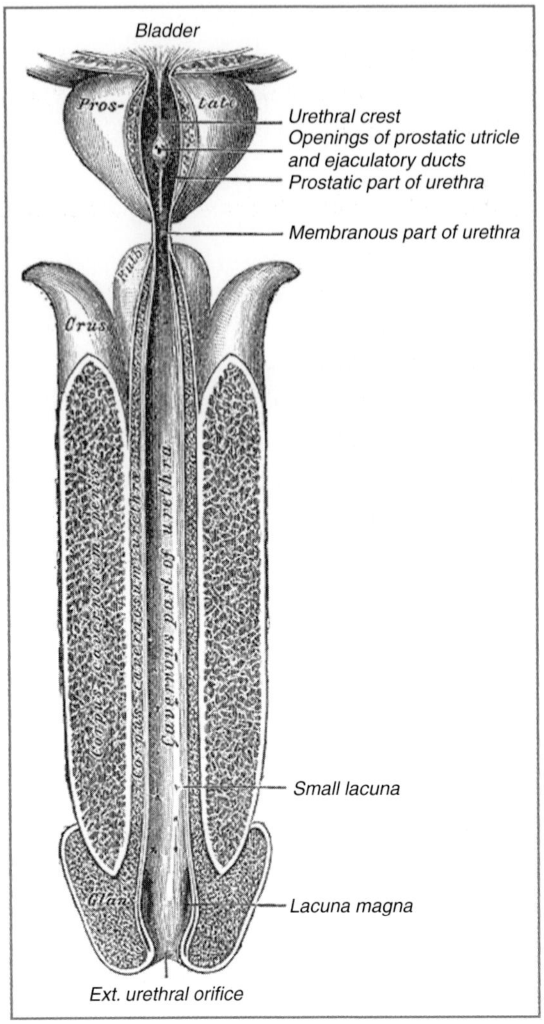

Figure 33.1 Anatomy of Urethra.

The epithelium of the urethra starts off as transitional cells as it exits the bladder. Further along the urethra there are stratified columnar cells, then stratified squamous cells near the external meatus (exit hole). There are small mucus secreting urethral glands that help to protect the epithelium from the corrosive urine.

Cell Growth

One of the initial limitations of applying cell-based tissue engineering techniques to urologic organs had been the previously encountered inherent difficulty of growing genitourinary-associated cells in large quantities. Normal urothelial cells could be grown in the laboratory setting, but with limited expansion. Several protocols were developed over the last two decades, which improved urothelial growth and expansion (Cilento et al., 1994; Liebert et al., 1997; Scriven et al., 1997; Puthenveettil et al., 1999). A system of urothelial cell harvest was developed which does not use any enzymes or serum and has a large expansion potential. Using these methods of cell culture, it is possible to expand an urothelial strain from a single specimen which initially covers a surface area of 1 cm^2 to one covering a surface area of 4202 m^2 within 8 weeks. These studies indicated that it should be possible to collect autologous urothelial cells from human patients, expand them in culture, and return them to the human donor in sufficient quantities for reconstructive purposes.

Normal human bladder epithelial and muscle cells can be efficiently harvested from surgical material, extensively expanded in culture, and their differentiation characteristics, growth requirements, and other biological properties can be studied (see Protocols I.A and B) (Liebert et al., 1991; Tobin et al., 1994). The cells were seeded onto non-woven meshes of PGA (Cilento et al., 1995). Partial urethrectomies were performed in rabbits and a segment of the polymer mesh of the appropriate diameter was interposed to form the neourethra in each animal. There was no evidence of voiding difficulties or any other complications. Retrograde urethrograms showed no evidence of stricture formation. Histologic examination of the neourethras demonstrated complete reepithelialization of the polymer mesh implanted sites by day 14, and reepithelialization continued for the entire duration of the study. Polymer fiber degradation was evident 14 days after implantation.

Protocol I.A: Urothelial Cell Culture

1. Materials and medium
   (a) Tissue source: bladder tissue.
   (b) Medium: keratinocyte and serum-free medium (Gibco/BRL), bovine pituitary extract (25 mg/500 ml medium), and recombinant epidermal growth factor (2.5 g/500 ml medium).
2. Tissue harvest
   (a) Obtain bladder specimen.
   (b) Gently rinse the specimen several times with medium in culture plates.
   (c) Mechanically scrape urothelial surface gently with a No. 10 scalpel blade. Be sure to use gentle short strokes and avoid cutting into the specimen. Urothelial cell clumps should be visible as tiny opaque material dispersing into the medium.
   (d) Aspirate urothelial cell-medium suspension and plate the cells in a 24-well cell culture plate with approximately 0.5 ml of the suspension in each well. Add an additional 0.5 ml to make a final volume of 1 ml. Incubate cells at 37°C with 5% CO_2.
   (e) On the following day, aspirate the medium from the wells and replace with fresh medium.
   (f) Centrifuge the cells in the aspirate medium at 1000 rpm for 4 min.
   (g) Remove the supernatant and resuspend the cells in 3–4.5 ml of fresh medium. Replate the cells in new wells.
3. Maintenance of urothelial cells
   (a) Replace the medium with fresh warm (37°C) medium every 3 days, depending on the cell density.
   (b) Tripsinize cells when they are 80–90% confluent.

4. Subculture of corporal smooth muscle cells
   (a) Remove medium and add 1 ml of phosphate-buffered saline–ethylenediaminetetraacetic acid (PBS–EDTA) (0.5 M) to each well or 10 ml to each 10-cm culture plate. Observe the cells under a phase contrast microscope.
   (b) When cell–cell junctions are separated for the majority of the cells, remove PBS–EDTA and add 300 μl of trypsin–EDTA to each well or 5 ml to each 10-cm culture plate.
   (c) Periodically agitated the plates. When 80–90% of the cells are detached, add 30 μl of soybean trypsin inhibitor (Gibco/BRL, 294 mg of inibitor) to 20 ml of PBS to each well, or 700 μl to each 10-cm plate. Add 0.5 ml of medium to each well or 3 ml to each 10-cm plate.
   (d) Aspirate and centrifuge the cell suspension at 1000 rpm for 4 min, and remove the supernatant.
   (e) Resuspend cells and count the number of viable cells by means of trypan blue exclusion.
   (f) Aliquot the desired number of cells on the plate and place the cells in the incubator.

Protocol I.B: Bladder Smooth Muscle Cell Culture

1. Materials and medium
   (a) Tissue source: bladder tissue.
   (b) Medium: Dulbecco's Modified Eagle's Medium (DMEM), 10% fetal bovine serum (FBS), and antibiotic (penicillin (100 U/ml)-streptomycin (100 l/ml), and amphotericin B (0.25 l/ml)).
2. Tissue harvest
   (a) Obtain fresh bladder tissue specimen.
   (b) Use sharp tenotomy scissors to cut muscle tissue into small fragments (2–3 mm).
   (c) Space muscle fragments evenly onto a cell culture plate (100 mm).
   (d) Allow muscle fragments to dry and adhere to the plate (5–10 min).
   (e) Add 15 ml of DMEM and incubate for 5 days.
   (f) Change medium on the sixth day and remove non-adherent tissue fragments.
   (g) When small islands of cell colony are formed, remove the tissue fragments and the change medium.
   (h) When sufficient cells are grown, trypsinize, count, and plate them onto new plates.
3. Maintenance of bladder smooth muscle cells
   (a) Feed cells every 3 days, depending on the cell density.
   (b) Trypsinize cells when they are 80–90% confluent.
4. Subculture of bladder smooth muscle cells
   (a) Remove medium and add 10 ml of PBS–EDTA (0.5 M) for 4 min. Confirm the separation of cell junction under a phase microscope.
   (b) Remove PBS–EDTA and add 5 ml of trypsin–EDTA.
   (c) Add 5 ml of medium when 80–90% of the cells lift under microscope.
   (d) Aspirate the cell suspension into a 15-ml test tube.
   (e) Aliquot the desired number of cells on the plate and make the volume of medium to a total of 10 ml.
   (f) Place the cells in the incubator.

A variety of synthetic grafts composed of silicone, Teflon, or polyvinyl have been proposed for urethral reconstruction and erosion, dislodgment, fistula, stenosis, extravasation, and calcification have been associated with synthetic grafts (Guzman, 1999; Vozzi et al., 2002). In the past years, in case of limited urethral mucosa for adequate reconstruction, tissues from other sources have been used, such as genital and extragenital skin flaps or grafts, mucosal grafts from the bladder or buccal regions, tunica vaginalis, and peritoneal grafts (Humby, 1941; Ehrlich et al., 1989; Dessanti et al., 1992). Various biomaterials without cells have been used experimentally (in animal models) for the regeneration of urethral tissue, including PGA, and acellular collagen-based matrices from small intestine and bladder (Bazeed et al., 1983; Atala et al., 1992; Olsen et al., 1992; Kropp et al., 1998; Chen et al., 1999; Sievert et al., 2000). Some of these biomaterials, like acellular collagen matrices derived from bladder submucosa, have also been seeded with autologous cells for urethral reconstruction. Atala's laboratory has been able to replace tubularized urethral segments with cell-seeded collagen matrices. Acellular collagen matrices derived from bladder submucosa have been used experimentally and clinically. In animal studies, segments of the urethra were resected and replaced with acellular matrix grafts in an onlay fashion. The animals were able to void through the neourethras (Chen et al., 1999). These results were confirmed clinically in a series of patients with hypospadias and urethral stricture disease (Atala, 1999; El kassaby et al., 2003). Cadaveric bladders were microdissected and the submucosal layers isolated. The submucosa was washed and decellularized. The matrix was used for urethral repair in patients with stricture disease (n = 33; 28 adults, 5 children) and hypospadias (n = 7 children). The matrices were trimmed to size and the neourethras were created by anastomosing the matrix in an onlay fashion to the urethral plate. The size of the neourethras ranged from 2 to 16 cm. Voiding histories, physical examination, retrograde urethrography, uroflowmetry, and cystoscopies were performed serially, pre- and postoperatively, with up to a 7-year follow-up. After a 4- to 7-year follow-up, 34 of the 40 patients had a successful outcome. Six patients with a urethral stricture had a recurrence, and one patient with hypospadias developed a fistula. The mean maximum urine flow rate significantly increased postoperatively. Cystoscopic studies showed adequate caliber conduits. Histologic examination of the biopsies showed the typical urethral epithelium. The use of an off the shelf matrix appears to be beneficial for patients with abnormal urethral conditions, and obviates the need for obtaining autologous grafts, thus decreasing operative time and eliminating donor site morbidity.

Unfortunately, the above techniques are not applicable for tubularized urethral repairs. The collagen matrices are able to replace urethral segments when used in an onlay fashion. However, if a tubularized repair is needed, the collagen matrices need to be seeded with autologous cells (De Fillipo et al., 2002a, b). Autologous bladder epithelial and smooth muscle cells from male rabbits were grown and seeded on to preconfigured tubular matrices. The entire anterior urethra was resected and urethroplasties were performed with tubularized collagen matrices seeded with cells in nine animals, and without cells in six animals. Serial urethrograms showed a wide urethral caliber without strictures in the animals implanted with the cell-seeded matrices, and collapsed urethral segments with strictures within the unseeded scaffolds. Gross examination of the urethral implants seeded with cells showed normal appearing tissue without any evidence of fibrosis. Histologically, a transitional cell layer surrounded by muscle cell fiber bundles with increasing cellular organization over time were observed on the cell-seeded constructs. The epithelial and muscle phenotypes were confirmed with pAE1/AE3 and smooth muscle specific a-actin antibodies. A transitional cell layer with scant unorganized muscle fiber bundles and large areas of fibrosis were present at the anastomotic sites on the unseeded constructs. Therefore, tubularized collagen matrices seeded with autologous cells can be used successfully for total penile urethra replacement; whereas, tubularized collagen matrices without cells lead to poor tissue development and stricture formation. The cell-seeded collagen matrices form new tissue, which is histologically similar to native urethra. This technology may be applicable to patients requiring tubularized urethral repair.

A variety of synthetic grafts composed of silicone, Teflon, and Dacron have been proposed for urethral reconstruction and then left behind because of their complications. These materials have been associated with erosion, dislodgment, fistula, stenosis, extravasation, or calcification (Hakky, 1976, 1977; Anwar et al., 1984).

Biodegradable substitutes like a polyglactin fiber mesh tube coated with poly(hydroxybutyric acid) and hyaluronan benzyl ester has been used experimentally. Complete regeneration of the urethral epithelium and the adjacent connective tissue was achieved as a consequence of the fact that the scaffolds guided urothelial and connective tissue regeneration (Olsen et al., 1992; Italiano et al., 1997). Although several innovative tissues have been proposed as possible free grafts for urethral repair, it is evident that all have specific advantages and disadvantages.

Free grafts of tubularized peritoneum were used as urethral tissue substitutes experimentally in rabbit. Organized multilayered graft epithelialization occurred; however, fistula formed in two of the animals (Shaul et al., 1996). Later, porcine small intestine submucosa (SIS) was used for urethral repair in a rabbit model to determine whether this material can evoke urethral regeneration. The SIS onlay grafts were shown to promote regeneration of the normal rabbit epithelium supported by a vascularized collagen and smooth muscle backing (Kropp et al., 1998).

More recently, Nuininga et al. partially resected a 0.5–1 cm segment of the native urethra in 24 rabbits and a novel molecularly defined collagen-based biocompatible and biodegradable matrix graft was sewn into place and compared with SIS. They did not notice any differences between the two biomatrics and the major advantage is that the new biometrics proposed can be modulated in different ways such as variation in the porous matrix structures, incorporation of growth factors and binding of glycosaminoglycans (Nuininga et al., 2003).

A naturally derived acellular collagen-based tissue substitute was developed from donor porcine bladder (see Protocol II). The acellular collagen matrix had been initially developed in our laboratory as a biomaterial for bladder augmentation. The results from this study demonstrated that the acellular matrix was biocompatible and was able, upon *in vivo* implantation, to form bladder tissue similar to the native bladder (Yoo et al., 1998).

Protocol II: Acellular Collagen Matrix Preparation

1. Obtain donor bladder tissue.
2. Isolate the submucosa from the muscular and serosal layers means of microdissection techniques.
3. Treat tissue with distilled water in a magnetic stirring flask set at moderate speed for 24–48 h at 4°C.
4. Remove distilled water and treat with Triton X-100 (0.5%) and ammonium hydroxide (0.05%) in fresh distilled water for 72 h in a stirring flask at 4°C.
5. Wash with distilled water in a stirring flask for 24–48 h at 4°C. After this washing step, take a small piece of tissue for histological analysis to confirm any cellular remnants. Tissue matrix is usually decellularized at this time.
6. After confirmation of decellularization, wash with distilled water in a stirring flask for 24–48 h at 4°C. Tissue retaining cellular components should undergo an additional cycle of treatment. Repeat Steps 4 and 5, and perform another histological analysis.
7. After the washing cycle with distilled water, rinse with 1 × PBS overnight.
8. Freeze-dry the tissue sample overnight.
9. Pack the samples and sterilize in ethylene oxide.
10. Store until used. When ready to use equilibrate the tissue in 1 × PBS or normal saline.

Penis

The indication for extended phalloplastic procedures results from the severe congenital malformation, penile tissue loss from malignancies, trauma or other diseases, and gender dysphoria. Owing to the shortage of autologous penile tissue, multiple staged surgeries using non-genital tissues and silicone prostheses have been the mainstay in phallic reconstruction. However, graft failure and prothesis-related complications remain a problem. Creation of penile structures composed of autologous tissue would be preferable treatment approach for these patients. Replacement of penile tissue with alternative materials has been a challenge due to the unique anatomical architecture of the corporal bodies.

One of the major limitations of penile tissue reconstruction is the availability of sufficient autologous tissue. Non-genital tissue sources have been used over the years; however, complications such as infection, graft failure,

3. Treat tissue with distilled water in a magnetic stirring flask (moderate speed) for 24–48 h at 4°C.
4. Remove distilled water and treat with TritonX-100 (0.5%) and ammonium hydroxide (0.05%) in fresh distilled water for 72 h in a stirring flask at 4°C.
5. Wash with distilled water in a stirring flask for 24–48 h at 4°C. After this washing step, take a small piece of tissue for histology to confirm any cellular remnants.
6. A small tissue mass is usually decellularized at this time. After confirmation of decellularization, repeat Step 5. Dense tissue may require another cycle of treatment. Repeat Steps 4 and 5, and perform another histological analysis.
7. After washing the tissue with distilled water, rinse with 1 × PBS overnight.
8. Freeze-dry the tissue samples overnight.
9. Pack and sterilize in ethylene oxide.
10. Store until used. When ready to use, equilibrate the tissue in culture medium overnight prior to cell seeding.

Protocol I.C: Human Endothelial Cell Culture from Foreskin

1. Materials and media
   (a) Medim A (for primary culture and first passage after UEA-I bead selection): 38.5 ml of endothelial basal medium 131 (Clonetics Corp., cat. No. CC 3121), 10 ml of 20% FBS, 0.5 ml (2 mM) L-glutamine, 0.5 ml of PFS (antibiotic–antimycotic) (Gibco, cat. No. 600-5240AG), 0.5 ml (0.5 mM) dibutyryl cyclic adenosine-3′,5′-cyclic monophosphate (AMP) (Sigma, cat. No. D-0627), and 50 μl (1 μg/ml) hydrocortisone (Sigma, cat.No. H-0888).
   (b) Medium B (for passage 2 and all following passages): endothelial basal medium 131, 1 × GPS, 10% FBS, and 2 μg/ml basic fibroblast growth factor (25 μg/ml stock solution) (Scios Nova).
   (c) Gelatin coating (1% Difco Bacto Gelatin in PBS): dissolve gelatin in PBS; autoclave to sterilize, and filter to remove particles.
2. Processing foreskin
   (a) Prepare foreskin collecting medium: Four hundred and fifty milliliter of DMEM, 25 ml of PBS (5%), 20 ml of antibiotic–antimycotic (400 U/ml penicillin, 400 μg/ml streptomycin, 1 μg/ml fungizone), 5 ml of L-glutamine (2 mM), and 1 ml of gentamicin sulfate (100 μg/ml).
   (b) Place the collecting medium with the foreskin in a culture plate (100 mm) in a tissue culture hood.
   (c) Rinse two or three times with the collecting medium.
   (d) Add 30 ml of collecting medium to a new 50-ml Falcon tube. Add an additional 2 ml of antibiotic–antimycotic.
   (e) Separate the skin and subcutaneous tissue with a sterile scalpel blade and transfer the segments into the collecting medium in a 50-ml tube.
   (f) Agitate the segments in the collecting medium at room temperature for at least 4–5 h to kill bacteria and spores that reside on the skin.
3. Isolation of endothelial cells
   (a) Prepare digestion solution: 7.5 ml of 1:250 trypsin, 2.7 ml of 0.5 M EDTA, pH 8.0, and 40 ml of Hanks' Balanced Salt Solution (HBSS).
   (b) Prepare 10 × HBSS without Ca^{2+} and Mg^{2+}: 40 g of NaCl, 2 g of KCl, 240 mg of Na_2HPO_4, 300 mg of KH_2PO_4, 1750 mg of $NaHCO_3$, 5 g of glucose, and 100 mg of phenol red.
   (c) Prepare wash solution (HBBS with 1 × Ca^{2+} and Mg^{2+}): 50 ml of 10 × HBSS, 92.7 mg of $CaCl_2$ $2H_2O$ (1.26 mM final), 100 mg of $MgSO_4$ $7H_2O$ (0.8 mM final), 25 ml of FBS (5% final), and 5 ml of PSF (antibiotic–antimycotic).
   (d) Coat a petri dish (100 mm) for each one or two foreskins with 8.0 ml of 1% gelatin–PBS. Remove excessive gelatin before plating.

(e) Autoclave a Teflon homogenizer (2.5 cm diameter) and gauze.
(f) Remove the collecting medium from the foreskin segments.
(g) Transfer the tissue segments into a sterile culture plate (100 mm).
(h) Cut the foreskin segments into 4-mm^2 fragments with a sterile scalpel blade.
(i) Transfer the tissue fragments to a sterile 50 ml Falcon tube and add 6.0 ml of digestion solution for 1–2 foreskins. Agitate vigorously at 37°C for 10 min.
(j) Allow the skin fragments to sediment by gravitational force and aspirate the digestion medium. Wash once with 20 ml of wash solution, swirl vigorously, and remove the wash solution.
(k) Add 10 ml of fresh wash solution and squeeze the fragments with the homogenizer.
(l) Filter through 8–10 layers of sterile gauze into a 50 ml Falcon tube (mesh filter).
(m) Repeat Steps k and l, and collect the expelled cells into the same Falcon tube.
(n) Centrifugate cells at 1000 rpm for 10 min at room temperature.
(o) Aspirate the supernatant and plate the cells with 10 ml of endothelial cell basal medium (EBM) 131 (culture medium A) in a gelatin-coated culture dish (100 mm). Place the cells in an incubator overnight with 5% CO_2.
(p) Wash the cells vigorously three or four times with PBS. Feed the cells with 10 ml of culture medium A.
(q) Change the medium every 2 days. The primary culture will be subconfluent after 7–8 days. They will be ready for the UEA-I isolation procedure at this point.

4. UEA-I selection of endothelial cells
   (a) Coating of Dynabeads with UEA: mix together 250 μl of Dynabeads (4×10^8 beads/ml) (Dynal, cat. No. 140.03), M-450, tosylactivated, 50 μg of unconjugated UEA-I (Vector, cat. No. L-1060), and 225 μl of 0.5 M boric acid, pH 9.5. The bead/lectin ratio should be 2.0×10^6 beads per microgram of lectin. The volume ratio of Dynabeads to boric acid with lectin should be 1:1.
   (b) Reconstitute the UEA-I with 1 ml of sterile PBS-0.1 mM $CaCl_2$ to 2 mg/ml and store at 4°C (UEA-I is quite stable); 50 μg = 25 μl.
   (c) Mix Dynabeads, lectin, and boric acid in a sterile 2.0-ml screw-cap tube and agitate on a rotor at room temperature overnight.
   (d) Pipette the bead–lectin mixture (in 10 ml of HBSS) into a 15-ml Falcon tube. Wash with 10 ml of HBSS (plus Ca^{2+}/Mg^{2+}, 1% BSA) on the rotator for 15 min at room temperature.
   (e) Place the tube in a magnetic particle concentrator (MPC) (MPC-1, Dynal, cat. No. 12001) and wait 1 min for the beads to be collected onto the magnet. Aspirate the supernatant with a Pasteur pipette. Take the tube out of the MCP, rinse three times at room temperature for 15 min, and once overnight at 4°C.
   (f) Resuspend the beads in 250 μl of HBSS (plus Ca^{2+}/Mg^{2+}, 5% FBS, $1 \times$ PBS) and store at 4°C in a sterile 2.0-ml screw-cap tube. The beads will be stable for several months.

5. Purification of endothelial cells from primary cultures
   (a) Trypsinize subconfluent cell cultures (7–8 days) with $1 \times$ trypsin–EDTA.
   (b) Centrifuge the trypsinize cells at 208 g (1000 rpm) for 10 min.
   (c) Resuspend the cell pellet from one 100-mm petri dish in 190 μl of HBSS buffer. Pipette up and down several times with a 200 μl pipetman to break up the cell clusters. Transfer the cell suspension into a sterile 2-ml screw-cap tube and add 5 μl UEA-I-coated Dynabeads.
   (d) Incubate cells and the beads for 3–5 min. Hold the tube in your hand and roll it between your palms gently to keep the beads in suspension. Endothelial cells and beads will form visible tiny clusters.
   (e) Transfer the cell–bead mixture to a 15-ml Falcon tube. Add 5 ml of HBSS buffer and pipette the cells several times up and down with the buffer. Place the Falcon tube into the MPC and collect the beads onto the magnet for about 1 min. Aspirate the wash solution with a Pasteur pipette while the tube is in the MCP. Take the tube out of the MCP. Repeat this wash four times with 5 ml HBSS wash buffer.

hosts. Full details of this work have not survived, and is difficult to evaluate its outcome. Berhold was the first to report on a successful testicular transplant, since he used autografts and avoided the risk of rejection. When he replaced the testes of capons in their own body cavity, he found that the growth of comb. and plumage, and courting behavior, all of which are androgen dependent, were maintained (Berhold, 1849). A century later, interest in testicular transplant increased as a result of the misapprehension that somatic aging is caused by withdrawal of sex hormones. Lydston had published a series of testicular transplantation experiments performed in his patients (Lydston, 1916). He and others following him believed that transplanted of midlife and later. sex glands produced a hormone that was a "cell stimulant, nutrient and regenerator" capable of prolonging life and restoring waning sexual functions, arteriosclerosis and other infirmities

Voronoff in 1923 was the first to use chimpanzee and baboon organs for treating patients (Brinster and Zimmermann, 1994). This approach was taken by other surgeons, but none of them used microsurgery to join blood vessels of the graft to the host's circulation, resulting in ischemic necrosis preceded by organ rejection. Later, successful testicular transplantation could be achieved when ischemia time was reduced to less than an hour by using vascular anastomosis in dogs (Attaran et al., 1966; Lee et al., 1971; Gittes et al., 1972). The first convincing human testicular transplant was published by Silber (1978), who grafted a patient with a testis from the patient's genetically identical twin brother (Silber, 1978). However, with time the stringent requirements for success have precluded a surge in demand for this operation. Moreover, carefully conducted grafting trials failed to confirm former claim; the new synthetic sex steroids were shown not to affect the life span of experimental animals (Parkes, 1966). Nevertheless, testicular transplantation may still be regarded as having clinical potential for example, who in carriers of genetic disease, who can receive normal germ cells from donors.

Transplantation of Testicular Tissue

The problems arising from the size of the testis and its fibrous capsule led some transplanters to use sliced or minced organs. Kearns (1941), who reimplanted testicular tissue subcutaneously in a victim of accidental castration, reported the most plausible case (Kearns, 1941). According to this report, testosterone was being produced by the autograft, but without the normal architecture of the seminiferous epithelium, it hard to understand how germ cell transfer could have restored spermatogenesis. Furthermore, injecting spermatogonial stem cells from donor testes into atrophic tubules is daunting the testes must produce millions of spermatozoa per day to be fertile. Therefore, efforts to develop tissue grafting for the purpose of improving testosterone levels in hypogonadal men are more likely to succeed that are attempts at restoring fertility. The former goal appears to be simple, requiring the transfer of interstitial cells (Leydig cells), which are readily isolated from the donor testes by means of collagenase. Interstitial cells grafted in castrated rodents resulted in partial restoration of body weight, and testosterone levels above those controls (Fox et al., 1973; Boyle et al., 1975; Tai et al., 1992). A number of vehicles and several implantation sites for interstitial cells have been tried, but none fully replaced testicular androgen production.

Testosterone Delivery Systems

The main goal of androgen replacement therapy is to maintain physiological levels of serum testosterone and also its metabolites, dihydrotestosterone, and estradiol. Hypogonadal states secondary to hypothalamic–pituitary disorders, gonadal abnormalities, and defects in androgen action or secretion may benefit from androgen replacement.

Androgen replacement modalities include oral administration of testosterone tablets, or capsules (Franchimont et al., 1978; Snyder and Lawrence, 1980; Sokol et al., 1982; Canteril, 1984; Fujioka et al., 1986; Stuenkel et al., 1991; Ferrini and Barret-Conner, 1998; Wilson et al., 1998; McClella and Goa, 1998; Bennet, 1998). When taken orally, testosterone preparations are largely rendered metabolically inactive during the "first pass" through the liver. This metabolic inactivation requires large oral doses of testosterone (>200 ng/day) to reach normal serum levels. These large doses of testosterone may be toxic to the liver and may lead to hepatitis, hepatoma, or hepatocarcinoma (Snyder and Lawrence, 1980; Gooren, 1994; Bagatelle and Bremner, 1996).

Parenteral depot preparation include testosterone enanthante (delatestryl) and testosterone cypionate (depot testosterone cypionate). These preparations are based in 17B-hydroxyl esters, which are given intramuscularly (IM), with slow-release, oil-based injection vehicles every 10–21 days. Testosterone levels with these preparations rise to supernormal levels for 1 or 2 days, after which they gradually fall within the normal range for 10–12 days, reaching baseline at approximately 21 days. This fluctuation in testosterone levels may produce significant swings in mood, libido, and sexual function (Sokol et al., 1982; Bhasin and Bremner, 1997).

Transdermal testosterone therapy includes both scrotal and non-scrotal patches. Testoderm and androderm are multilayered skin patches that deliver measured doses of testosterone across the scrotal skin acting thanks to the 5α-reductase activity present within this site. When used in non-scrotal skin, the patch has to be applied twice daily reducing frequency of administration. However, despite these advantages, the transdermal systems have been associated with adverse effects, such as transient erythema, pruritis, induration, burning, rash, and skin necrosis (Hogan and Maibach, 1990; McClellan and Goa, 1998; Bennet, 1998).

Long-term exogenous testosterone therapy has been associated with several complications, such as fluid and nitrogen retention, erythropoiesis, hypertension, and bone-density changes. In addition, fluctuating serum testosterone levels may occur, and frequent treatments may be required. Due to these problems, alternate treatment modalities, involving more physiological and longer-acting systems for androgen delivery, have been pursued.

Cell Encapsulation for Testosterone Therapy

Cell transplantation has long been proposed as a treatment for several diseases involving hormone or protein deficiencies. Cell rejection by the host immune system, however, has limited the use of this strategy. Encapsulation of living cells in a protective, biocompatible, and semipermeable polymeric membrane has been proven to be an effective method of immunoprotection of the desired cells, regardless of the type of recipient (allograft, xenograft) (Chang, 1993). A majority of the implantation work using microencapsulated cells as delivery vehicles employs two polymers: sodium alginate and poly(L-lysine) (PLL) (Lim and Sun, 1980). Alginate microcapsule have been used for various applications (Chang, 1998; Joki et al., 2001) particularly for the encapsulation of the pancreatic islet cells/or insulin delivery (Lim and Sun, 1980; Wang et al., 1997) and recombinant cells have served for the delivery of therapeutic gene products (Tai and Sun, 1993).

The Leydig cells of the testes are the major source of testosterone in men (95%). Implantation of heterologous Leydig cells has been proposed as a method for chronic testosterone replacement. However, these approaches were limited by tissue and cell failure to produce long-term testosterone and dissemination of the implanted cells. Therefore, encapsulation of Leydig cells might be useful for testosterone replacement therapy. Such a system might be able to stimulate the normal diurnal pattern of testosterone release by the testes, therapy avoiding side effects such as those associated with chemically modified testosterone administration. Leydig cell transplantation may be also beneficial not only for testosterone replacement but also for the secretion of other associated hormones and growth factors such as melanocytes, β-andorpilin, prostaglandins, insulin-like growth factor 1 (IGF-1), and interleukins (Verhoeven, 1992).

Studies in our laboratory have been focused on the encapsulation and implantation of isolated Leydig cells for long-term testosterone delivery. Leydig cells were isolated from male Sprague–Dawley rats, 56–70 days old, by means of collagenase and Percoll gradient separation. The isolated Leydig cells were encapsulated within microspheres composed of calcium-alginate, coated with the positively charged polyelectrolyte PLL, and recoated with alginate. Based on the molecular weight of testosterone (300 Da), PLL having a molecular weight of 21 kDa and 1.2% sodium alginate with a high glucuronic acid content (>65%) were chosen. PLL with a molecular weight ranging between 16 and 22 kDa produced a semipermeable membrane with a molecular weight cutoff of 70 kDa (i.e. preventing the diffusion of cells and metabolites). Methyl(this) tetrazole assays performed daily on the microencapsulated cells showed that the cells remained viable during the experiments. Testosterone secretion from cultured encapsulated Leydig cells in response to human chorionic gonadotropin (hCG) was the highest 24 h after hCG stimulation (0.06 IU/ml). There was no significant difference in testosterone secretion when cells were cultured at either 32 or 37°C. The encapsulated and non-encapsulated Leydig cells were found to be resistant to temperature changes. This finding broadens the list of possible *in vivo* sites for Leydig cell transplantation. The intraperitoneal cavity was the first site chosen for cell implantation because of its generous vascular and nutritional capabilities. *In vivo* studies performed in castrated rats showed that the total testosterone levels measured in the serum of castrated rats that were injected intraperitonally with 5×10^6 encapsulated Leydig cells were between 0.23 and 0.51 ng/ml for more than 3 weeks. These animals did not receive any exogenous hCG stimulation. Similar testosterone levels (0.24–0.48 ng/ml) were obtained when encapsulated Leydig cells were injected subcutaneously. However, testosterone was detected for a longer time period in the subcutaneous group (43 days) than in the intraperitoneal group (35 days). These testosterone levels were lower than the ones detected in the control rats (1.7 ng/ml), which were not castrated. However, only 5×10^6 microencapsulated cells were implanted in each animal, representing only 10% of the normal adult rat Leydig cell population (Machluf et al., 2003).

Methods for Encapsulation

Microencapsulation is currently the optimal immuno-isolation technique. Different approaches and polymers are being used for encapsulating cells and tissue for therapeutic applications. The technique of microencapsulation used by our laboratory utilizes two polymers: highly purified calcium-alginate (Pronova, Norway) and low molecular weight (23.6 kDa, Sigma) PLL. This procedure is described as follows.

Protocol I: Cell Encapsulation

1. Isolated cells are suspended in sodium alginate (1.2%) (60% glucuronic acid content) in 0.9% saline for 5 min.
2. The cell–alginate suspension is extruded through a 22-gauge airjet-needle into a calcium chloride–4-(2-hydroxyethyl)-1-piperazineethanesulfonic acid ($CaCl_2$–HEPES) solution (1.5%).
3. The beads are stirred for 20 min in the $CaCl_2$–HEPES solution.
4. Gelled droplets are transferred to ecno-colums (Bio-Rad) and decanted.
5. The columns are filled with 15 ml of PLL solution in 0.9% saline, sealed, and rotated gently for 12 min.
6. The PLL solution is decanted from the columns and washed three times with HEPES solution.
7. A 0.125% alginate solution is added, and the mixture is rotated for 10 min. Then the alginate solution is decanted and the supernatant is washed three times with HEPES prior to culturing.

FEMALE

Vagina

A variety of pathological and congenital disorders affect the vagina and require extensive surgical intervention (Machluf et al., 2003). The choice of operation and outcome depend critically on correct identification of the underlying disorder. Three basic categories of anomalies must be distinguished, namely, vaginal agenesis and its variants, ambiguous genitalia, and imperforate anus and urogenital sinus variants. Vaginal reconstruction is an uncommon and a challenging procedure that varies considerably by specialty, with plastic surgeons and gynecologists generally recommending skin graft/dilation procedures and pediatric urologists recommending bowel vaginoplasty (Rajimwale et al., 2004). Various procedures have been used in the past for vaginal reconstruction and different tissue sources have been employed for reconstructive surgery. Traditionally, the reconstructions have been performed with non-urologic tissues or synthetic prostheses. The non-urologic tissues include gastrointestinal segments (Leong and Ong, 1972; Hendren and Atala, 1994), skin (Draper and Stark, 1956), peritoneum (Hutschenreiter et al., 1978), fascia (Neuhof, 1917), omentum (Goldstein et al., 1933), pericardium (Kambic et al., 1992), and dura (Kelami, 1971). The majority of surgical options require the use of non-genital tissues for vaginal replacement too. However, the use of non-vaginal tissue for surgical reconstruction is not ideal in terms of normal vaginal function (Machluf et al., 2003). Tissue engineering may offer a solution for challenging cases when shortage of local tissue exists. While tissue engineering has been applied to many tissue–organ reconstructions, there is a paucity of information regarding the engineering of female reproductive and genital tissues.

This chapter summarizes the known and recently developed tissue engineering applications for total vaginal reconstruction.

Anatomy

The vagina (Kelami, 1971) is a muscular, highly expandable, tubular cavity that connects the vulva at the outside to the cervix of the uterus on the inside. The vagina consists of an internal mucous lining and a muscular coat separated by a layer of erectile tissue. It does not have any glands and is kept moist by the lubrication provided by the cervical and uterine glands.

The mucous membrane (tunica mucosa) is continuous above with that lining the uterus. Its inner surface presents two longitudinal ridges, one on its anterior and one on its posterior wall. These ridges are called the columns of the vagina and from them numerous transverse ridges or rugae extend outward on either side. These rugæ are divided by furrows of variable depth, giving to the mucous membrane the appearance of being studded over with conical projections or papillae; they are most numerous near the orifice of the vagina, especially before parturition. The epithelium covering the mucous membrane is of the stratified squamous variety. The submucous tissue is very loose, and contains numerous large veins, which by their anastomoses form a plexus, together with smooth muscular fibers derived from the muscular coat; it is regarded by Gussenbauer as an erectile tissue. It contains a number of mucous crypts, but no true glands.

The muscular coat (tunica muscularis) consists of two layers: an external longitudinal, which is by far the stronger, and an internal circular layer. The longitudinal fibers are continuous with the superficial muscular fibers of the uterus. The strongest fasciculi are those attached to the rectovesical fascia on either side. The two layers are not distinctly separable from each other, but are connected by oblique decussating fasciculi, which pass from the one layer to the other. In addition to this, the vagina at its lower end is surrounded by a band of striped muscular fibers, the Bulbocavernosus. External to the muscular coat is a layer of connective tissue, containing a large plexus of blood vessels.

The erectile tissue consists of a layer of loose connective tissue, situated between the mucous membrane and the muscular coat; imbedded in it is a plexus of large veins, and numerous bundles of unstriped muscular fibers, derived from the circular muscular layer. The arrangement of the veins is similar to that found in other erectile tissues.

Vaginal Tissue Engineering

Clinically related studies have already demonstrated encouraging results with regard to the applicability of tissue engineering in genitourinary reconstruction (Atala, 1999). Atala et al. have also demonstrated that *in vitro* expansion of vaginal epithelial and smooth muscle cells followed by seeding them onto synthetic matrices and placing them *in vivo* can reconstitute to *de novo* tissues (Mikos, et al., 1994). In this study expanded cells of muscle and epithelial cells seeded onto PGA scaffolds at a concentration of 10×10^6 and 20×10^6 cells/cm^3 were co-cultured at 37°C with 5% CO_2 for 24 to 48 h and implanted subcutaneously into athymic mice. The cells were able to survive and replicate *in vivo* for prolonged periods and could self-organize toward seemingly normal structural orientation. By the sixthth week of implantation the constructs have shown to organize into a distinguishable layer of both the vaginal epithelial and smooth muscle cell types. Penetrating native vasculature was also observed. Further analysis of the tissue engineered vaginal constructs has shown to produce contractile forces similar to those seen with native vaginal tissue when simulated with a series of electrical impulses.

Protocol: Methods of Cell Culture

1. Materials and medium
   - (a) Tissue source: vaginal tissue from New Zealand White rabbits
   - (b) Medium: (i) Smooth muscle cells
     DMEM supplemented with 10% FBS
     (ii) Epithelial cells – keratinocytes
     Keratinocyte serum-free medium (K-SFM) supplemented with bovine pituitary extract and epidermal growth factor.
2. Tissue harvest and cell culture
   - (a) Obtain vaginal tissue.
   - (b) Wash the specimen several times with PBS containing EDTA.
3. Smooth muscle
   - (a) Mechanically microdissect the muscle from the seromuscular layer with sterile instruments under loop magnification.
   - (b) Individually place small portions of the dissected samples onto culture dishes, and allow them to dry and adhere to the surface.
   - (c) Incubate the pieces with medium at 37°C in air and 5% CO_2 undisturbed until a sufficient colony of progenitor cells grows from the tissue islets.
   - (d) Remove the tissue explants by gentle suction when sufficient amount of cells are established.
4. Epithelial cells
   - (a) Digest the vaginal specimen with collagenase type IV by immersing them into the enzymatic solution and shake vigorously for 30 min at 37°C.
   - (b) Centrifuge the cell-fluid suspension at low revolutions for 5 min.
   - (c) Resuspend the supernatant in K-SFM and distribute onto culture dishes.

C. Cell expansion
   - (a) Remove the culture medium and wash the cells with PBS–EDTA.
   - (b) Incubate the cells with a 0.05% trypsin–EDTA solution, (0.5 g trypsin and 0.2 g EDTA per 1.0 L of stock solution) and monitor under the microscope until cell separation is observed.
   - (c) With a pipette gently transfer the cell–trypsin solution in to a 50-ml Falcon tube with serum containing medium to inactivate the trypsin.
   - (d) Centrifuge the cells at 1500-rpm for 5 min.
   - (e) Resuspend the cell pellet into a predetermined volume of fresh medium and partition equally among several more culture dishes for expansion.

D. Cell maintenance
   - (a) Replace the medium with fresh warm (37°C) medium every 24–48 h.

Uterus

Tissue engineering is a relatively new and rapidly expanding field of biological research. It is also a clinically applicable discipline that aims to provide a repository of alternative tissue sources when reconstructive surgery is necessary (Skalak and Fox, 1998). Congenital malformations of the uterus may have profound implications clinically. Patients with cloacal exstrophy or intersex conditions may not have sufficient uterine tissue for future reproduction. With developing aspects of tissue engineering it may be possible to solve this kind of problems in the future. (Figure 33.2)

Anatomy

The uterus (Gray, 1918) is a hollow, thick-walled, muscular organ situated deeply in the pelvic cavity between the bladder and rectum. Into its upper part the uterine tubes open, one on either side, while below, its cavity communicates with that of the vagina.

The uterus measures about 7.5 cm in length, 5 cm in breadth, at its upper part, and nearly 2.5 cm in thickness; it weighs from 30 to 40 g. It is divisible into two portions. On the surface, about midway between the apex and base, is a slight constriction, known as the isthmus, and corresponding to this in the interior is a narrowing of the uterine cavity, the internal orifice of the uterus. The portion above the isthmus is termed the body, and that below, the cervix. The part of the body, which lies above a plane passing through the points of entrance of the uterine tubes, is known as the fundus.

The cavity of the uterus is small in comparison with the size of the organ. It is a mere slit, flattened anteroposteriorly. It is triangular in shape, the base being formed by the internal surface of the fundus between the orifices of the uterine tubes, the apex by the internal orifice of the uterus through which the cavity of the body communicates with the canal of the cervix.

The canal of the cervix (canalis cervicis uteri) is somewhat fusiform, flattened from before backward, and broader at the middle than at either extremity. It communicates above through the internal orifice with the cavity of the body, and below through the external orifice with the vaginal cavity. The wall of the canal presents an anterior and a posterior longitudinal ridge, from each of which proceed a number of small oblique columns, the palmate folds, giving the appearance of branches from the stem of a tree; to this arrangement the name

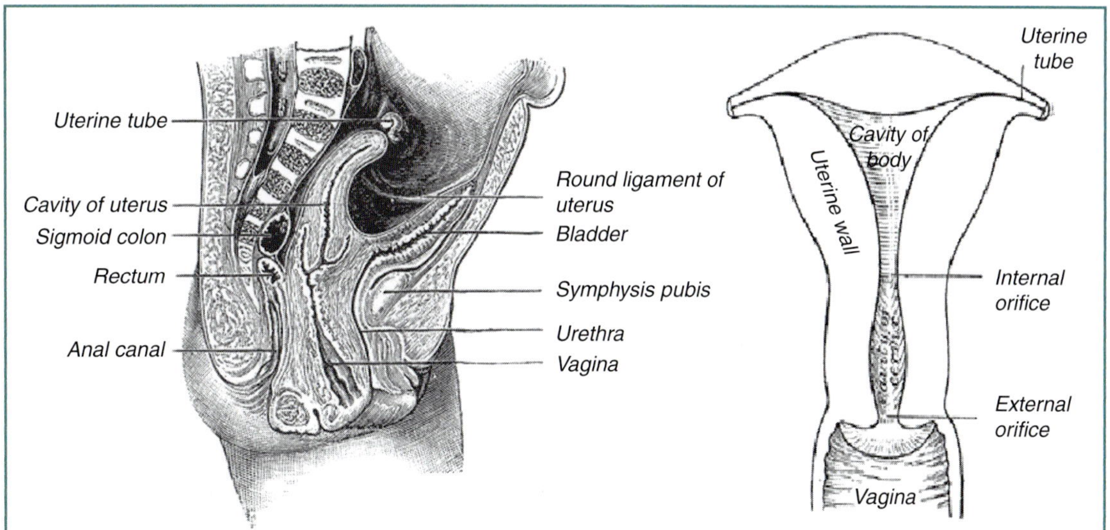

Figure 33.2 Uterus.

arbor vitae uterina is applied. The folds on the two walls are not exactly opposed, but fit between one another so as to close the cervical canal.

The uterus is composed of three coats: an external or serous, a middle or muscular, and an internal or mucous. The serous coat (*tunica serosa*) is derived from the peritoneum; it invests the fundus and the whole of the intestinal surface of the uterus; but covers the vesical surface only as far as the junction of the body and cervix. The muscular coat (*tunica muscularis*) forms the chief bulk of the substance of the uterus. It consists of bundles of unstriped muscular fibers, disposed in layers, intermixed with areolar tissue, bloodvessels, lymphatic vessels, and nerves. The layers are three in number: external, middle, and internal. The external and middle layers constitute the muscular coat proper, while the inner layer is a greatly hypertrophied muscularis mucosæ.

The arteries of the uterus are the uterine, from the hypogastric; and the ovarian, from the abdominal aorta. They are remarkable for their tortuous course in the substance of the organ, and for their frequent anastomoses. The veins are of large size, and correspond with the arteries. They end in the uterine plexuses. In the impregnated uterus the arteries carry the blood to, and the veins convey it away from, the intervillous space of the placenta.

Uterine Tissue Reconstruction

The first report of tissue engineering of human uterine smooth muscle cells was reported in 2003 (Atala, 2004). In this study, primary cell lines were initiated from human myomerium obtained at the time of term cesarean delivery. After several passages the cells were seeded onto a polyglactin-910 (Vicryl) mesh and maintained in culture. This system provides a three-dimensional myocyte culture where cells are attached to each other instead of to a culture dish. The resulting "tissue" contains cells in an environment that approximates whole tissue, but grown under controlled conditions. Similar experiments have been reported for urinary bladder (Vozzi, et al., 2002) and vascular smooth muscle cells (Dessanti, et al., 1992).

In addition to this, double-mesh experiments were performed to build thicker sections of tissue. The mechanical strength of the bridging myocytes was determined by hanging the two-mesh complexes in the muscle bath, with one mesh fixed and the other attached to the force transducer. The meshes were subjected to increasing force until separation. The constructs were able to maintain a maximum force of 5 g/cm^2. The bridging myocytes were also tested for contractile activity by hanging a two-mesh complex in the muscle bath and applying a $2 \times g$ of force. Addition of oxytocin (100 nM) to the bath produced small, irregular contractions, which remained stable for 25 min. Addition of 140 mM KCl to a final concentration of about 50 mM resulted in loss of contractile behavior. Although no repetitive pattern reminiscent of human labor was observed, these observations represent the first example of a group of cultured human uterine myocytes exhibiting coordinated contraction.

Protocol: Uterine Cell Culture

1. Materials and medium
   (a) Tissue source: human myometrium.
   (b) Medium: Dulbecco's modified essential medium supplemented with 10% FBS
2. Tissue harvest
   (a) Obtain human myocytes from the upper margin of the uterine incision
   (b) Mince the collected tissue
   (c) Perform double digestion at 37°C for 45 min. each
      (i) Prepare and perform the first digestion containing collagenase-dispase (1.5 mg/ml), trypsin inhibitor (1 mg/ml), and bovine serum albumin (2 mg/ml) in calcium-free Hanks' solution.
      (ii) Prepare and perform the second digestion containing collagenase (1 mg/ml), trypsin inhibitor (0.3 mg/ml), and bovine serum albumin (2 mg/ml) in the same Hanks' solution.

(d) Centrifuge the cell-digestion solution mix at low revolutions for 5 min wash with PBS, and resuspend in culture medium
(e) Culture the cells onto culture flasks in an atmosphere of 95% O_2 and 5% CO_2 at 37°C.

3. Cell expansion
   (a) Follow the protocol for vaginal cell culture expansion.
4. Cell maintenance
   (a) Replace the medium with fresh warm (37°C) medium every 2–3 days.

In the subsequent study the possibility of engineering functional uterine tissue using autologous cells was investigated (Kim et al., 1999). Autologous rabbit uterine smooth muscle and epithelial cells were harvested, then grown and expanded in culture. These cells were seeded onto preconfigured, uterine-shaped, biodegradable polymer scaffolds, which were then used for subtotal uterine tissue replacement in the corresponding autologous animals. Upon retrieval 6 months after implantation, histologic, immunocytochemical, and Western blot analyses confirmed the presence of normal uterine tissue components. Biomechanical analyses and organ bath studies showed that the functional characteristics of these tissues were similar to those of normal uterine tissue. Breeding studies using these engineered uteri are currently being performed.

Ovary

Anatomy

The ovaries (Gray, 1918) are homologous with the testes in the male. They are two nodular bodies, situated one on either side of the uterus in relation to the lateral wall of the pelvis, and attached to the back of the broad ligament of the uterus. The surface of the ovary is covered by a layer of columnar cells, which constitutes the germinal epithelium of Waldeyer. This epithelium gives to the ovary a dull gray color as compared with the shining smoothness of the peritoneum; and the transition between the squamous epithelium of the peritoneum and the columnar cells, which cover the ovary, is usually marked by a line around the anterior border of the ovary. The ovary consists of a number of vesicular ovarian follicles imbedded in the meshes of a stroma or frame-work.

The development and maturation of the follicles and ova continue uninterruptedly from puberty to the end of the fruitful period of woman's life, while their formation commences before birth. Before puberty the ovaries are small and the follicles contained in them are disposed in a comparatively thick layer in the cortical substance; here they present the appearance of a large number of minute closed vesicles, constituting the early condition of the follicles; many, however, never attain full development, but shrink and disappear. At puberty the ovaries enlarge and become more vascular, the follicles are developed in greater abundance, and their ova are capable of fecundation (Figure 33.3).

The follicles, after attaining a certain stage of development, gradually approach the surface of the ovary and burst; the ovum and fluid contents of the follicle are liberated on the exterior of the ovary, and carried into the uterine tube by currents set up by the movements of the cilia covering the mucous membrane of the fimbriæ.

After the discharge of the ovum the lining of the follicle is thrown into folds, and vascular processes grow inward from the surrounding tissue. In this way the space is filled up and the corpus luteum formed.

The arteries of the ovaries, each anastomoses freely in the mesosalpinx, which traverse the mesovarium and enter the hilum of the ovary.

In Vitro Culture of Ovarian Follicles

The fundamental role of the ovary is to produce oocytes capable of fertilization and subsequent development into viable offspring (Wang et al., 2003). Number of pathological conditions such as polycystic ovarian syndrome (PCOS), premature ovarian failure, or definitive sterility (postoncotherapy) may affect ovarian function

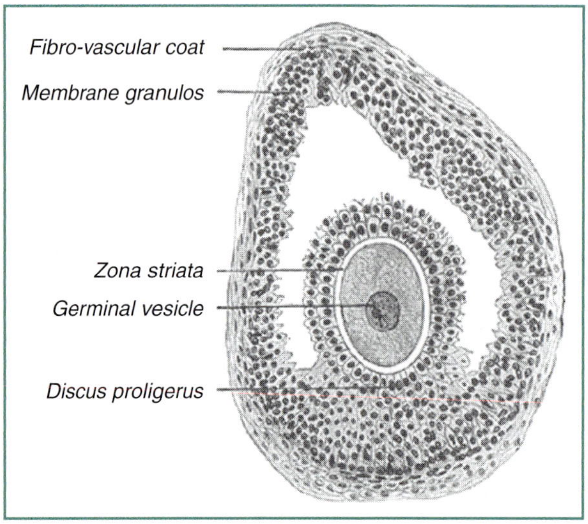

Figure 33.3 The Ovary.

and severely compromise their reproductive potential. Recently, new *in vitro* culture methods involving tissue engineered matrices have been developed to study the maturation of ovarian follicles (Pangas et al., 2003). Unlike the two-dimensional culture systems supporting the production of immature mouse follicles or granulose cell-oocyte complexes where the granulose cells attach to the culture substrate, and migrate away from the oocyte (Spears et al., 1994; Cortvrindt et al., 1996; Rowley et al., 1999; Smitz and Cortvrind, 2002; O'brien et al., 2003; Kreeger et al., 2006). This research study has developed a three-dimensional culture system for mouse granulose-oocyte complexes, which maintains cell–cell connections and provides an environment that supports follicle development (Wang et al., 2003).

Protocol: Follicle Isolation And Culture

1. Materials and medium
   (a) Tissue source: C57BL/6 × CBA F_1mouse
   (b) Medium: αMEM supplemented with 3 mg/ml BSA, 5 μg/ml insulin, 5 μg/ml transferring, and 5 μg/ml selenium.
2. Tissue harvest and culture
   (a) Obtain two-layered (100–130 μm) and multilayered secondary follicles (150–180 μm) using insulin gauge needles in L-15 media, while maintaining them at 37°C and pH 7.
   (b) Encapsulate the follicles into alginate or alginate-ECM matrices
      (i) Suspend droplets (~2–3 μL) of alginate or alginate-ECM solution on a polypropylene mesh (0.1 mm opening)
      (ii) Pipette a single follicle into each droplet in a minimal amount of media.
   (c) After all the droplets are filled, immerse the mesh in sterile 50 mM $CaCl_2$ for 2 min.
   (d) Rinse the mesh in L-15 media.
   (e) Plate individual beads in 96 well plates in 100 μl of culture media.
   (f) Culture the follicles at 37°C in 5% CO_2 for 8 days.
   (g) Change half of the media volume every 2 days.

For the preservation of fertility for women or young girls, cryopreservation of ovaries has been proposed; however, there is a critical limitation in obtaining a sufficient supply of meiotically competent oocytes (Cortvrindt et al., 1996). By merging principles from tissue engineering with those from follicle biology, this research team has developed synthetic matrices that promote follicle maturation to produce meiotically competent oocytes, which may provide mechanisms to preserve fertility. It was concluded that, this alginate culture system serves as a tool for fundamental studies that correlate the composition of the cellular microenvironment to the properties of the developing tissue, which may ultimately provide design principles for scaffold-based approaches to tissue engineering.

In the subsequent study, similarly, *in vitro* cultures of immature ovarian follicles were used to examine the factors that regulate the follicle development (Wang et al., 2003). In this system, individual granulose cell-oocyte complexes were incorporated into a three-dimensional culture system based on an alginate hydrogel. Briefly, ovaries from 12 day old mice were dissected and dissociated in μ-MEM containing 0.3% bovine serum albumin, 0.1% type I collagenase, and 0.02% deoxyribonuclease I at 37°C and 5% O_2, 5% CO_2, 90% N_2 for approximately 1 hour. GOCs were collected and washed 3 times in α-MEM, manually counted and encapsulated into alginate. The growth medium consisted of α-MEM supplemented with 0.3% BSA, bovine pancreatic insulin (5 μg/ml), human transferring (5 μg/ml), sodium selenite (ITS, 5 ng/ml), penicillin (5 U/ml), and streptomycin (5 μg/ml). Alginate beads were then cultured, and fed every other day by replacement of one-third the volume of growth medium.

After 10 days the morphology of GOCs was assessed by TEM. They showed no signs of degeneration, had cortical granules around the periphery, and contained an intact zona pellucida. No evidence of cellular apoptosis was detected in intact GOCs containing an oocyte. At this time point the GOCs were also analyzed for their ability to undergo *in vitro* maturation. On average, 40% of oocytes retrieved from *in vitro* growth in alginate beads underwent germinal vesicle breakdown and proceeded to meiosis II.

In these studies, alginate exhibits minimal cellular interactions with mammalian cells, and thus likely provides only mechanical support. However, cell adhesion ligands can be incorporated onto the alginate backbone (Huet et al., 2001), which are known to influence granulose cell morphology, differentiation, and signaling (Huet et al., 2001). In conclusion, this three-dimensional culture system allows immature GOCs to be maintained in culture.

REFERENCES

Atala, A. (2004). Tissue engineering and regenerative medicine: concepts for clinical application. *Rejuvenation Res.* **7**(1): 15–31.

Anwar, H., Dave, B. and Seebode, J.J. (1984). Replacement of partially resected canine urethra by polytetrafluoroethylene. *Urology* **24**: 583.

Atala, A. (1997). Tissue engineering in the genitourinary system. In: Atala, A. and Mooney, D. (eds.), *Tissue Engineering.* p. 149.

Atala, A. (1999). Engineering tissues and organs. *Curr. Opin. Urol.* **9**(6): 517–526.

Atala, A. (1999). Future perspectives in reconstructive surgery using tissue engineering. *Urol. Clin. N. Am.* **26**: 157.

Atala, A. (1999). Tissue engineering applications for erectile dysfunction. *Int. J. Impot. Res.* **11**(Suppl 1): S41.

Atala, A., Cima, L.G., Kim, W., et al. (1993). Injectable alginate seeded with chondrocytes as a potential treatment for vesicoureteral reflux. *J. Urol.* **150**: 745.

Atala, A., Kim, W., Paige, K.T., et al. (1994). Endoscopic treatment of vesicoureteral reflux with a chondrocyte-alginate suspension. *J. Urol.* **152**: 641.

Atala, A., Vacanti, J.P., Peters, C.A., et al. (1992).: Formation of urothelial structures *in vivo* from dissociated cells attached to biodegradable polymer scaffolds *in vitro*. *J. Urol.* **148**: 658.

Attaran, S.E., Hodges, C.V., Crary Jr., L.S., et al. (1966). Homotransplants of the testis. *J. Urol.* **95**: 387.

Bagatelle, C. and Bremner, W. (1996). Drug therapy: androgen in men, use and abuses. *N. Eng. J. Med.* **334**: 707.

Bazeed, M.A., Thuroff, J.W., Schmidt, R.A., et al. (1983). New treatment for urethral strictures. *Urology* **21**: 53.

Bennett, N.J. (1998). A burn-like lesion caused by a testosterone transdermal system. *Burns* **24**: 478.

Berhold, A. (1849). Transplantation der hoden. *Arch. Anat. Physiol. Wiss. Med.* **16**: 42.

Bhasin, S. and Bremner, W.J. (1997). Clinical review 85: Emerging issues in androgen replacement therapy. *J. Clin. Endocrinol. Metab.* **82**: 3.

Boyle, P.F., Fox, M. and Slater, D. (1975). Transplantation of interstitial cells of the testis: effect of implant site, graft mass and ischaemia. *Br. J. Urol.* **47**: 891.

Bretan Jr., P.N. (1989). History of the prosthetic treatment of impotence. *Urol. Clin. N. Am.* **16**: 1.

Brinster, R.L. and Zimmermann, J.W. (1994). Spermatogenesis following male germ-cell transplantation. *Proc. Natl. Acad. Sci. USA* **91**: 11298.

Canteril, J., al, e. (1984). Which testosterone therapy? *Clin. Endocrinol.* **21**: 97.

Chang, T.M. (1993). Bioencapsulation in biotechnology. *Biomater. Artif. Cell Immobil. Biotechnol.* **21**: 291.

Chang, T.M. (1998). Pharmaceutical and therapeutic applications of artificial cells including microencapsulation. *Eur. J. Pharm. Biopharm.* **45**: 3.

Chang, T.S. and Hwang, W.Y. (1984). Forearm flap in one-stage reconstruction of the penis. *Plast. Reconstr. Surg.* **74**: 251.

Chen, F., Yoo, J.J. and Atala, A. (1999). Acellular collagen matrix as a possible off the shelf biomaterial for urethral repair. *Urology* **54**: 407.

Cilento, B.G., Freeman, M.R., Schneck, F.X., et al. (1994). Phenotypic and cytogenetic characterization of human bladder urothelia expanded *in vitro*. *J. Urol.* **152**: 665.

Cilento, B., Retik, A. and Atala, A. (1995). Uretheral reconstruction using a polymer mesh. *J. Urol.* **153**: 371A.

Cortvrindt, R., Smitz, J. and Van Steirteghem, A.C. (1996). *In-vitro* maturation , fertilization and embryo development of immature oocytes from early preantral follicles from prepuberal mice in a simplified culture system. *Hum. Reprod.* **11**(12): 2656–2666.

De Filippo, R.E., Yoo, J.J. and Atala, A. (2002a). Urethral replacement using cell seeded tubularized collagen matrices. *J. Urol.* **168**: 1789.

De Filippo, R., Pohl, H.G., Yoo, J., et al. (2002b). Total penile urethral replacement with autologous cell-seeded collagen matrices. *J. Urol.* **168**: 1789 (abstract).

De Filippo, R.E., Yoo, J.J. and Atala, A. (2003). Engineering of Vaginal Tissue *in Vivo*. *Tissue Eng.* **9**: 301–306.

Dessanti, A., Rigamonti, W., Merulla, V., et al. (1992). Autologous buccal mucosa graft for hypospadias repair: an initial report. *J. Urol.* **147**: 1081.

Draper, J.W. and Stark, R.B. (1956) End results in the replacement of mucous membrane of theurinary bladder with thick-split grafts of skin. *Surgery* **39**(3): 434–440.

Ehrlich, R., Reda, E. and Kyle, M. (1989). Complications of bladder mucosal graft. *J. Urol.* **142**: 626.

El kassaby, A., Retik, A., Yoo, J., et al. (2003). Urethral stricture repair with an "off the shelf" collagen matrix. *J. Urol.* **169**: 170.

Falke, G., Yoo, J.J., Kwon, T.G., et al. (2003). Formation of corporal tissue architecture *in vivo* using human cavernosal muscle and endothelial cells seeded on collagen matrices. *Tissue Eng.* **9**: 871.

Ferrini, R.L. and Barrett-Connor, E. (1998). Sex hormones and age: a cross-sectional study of testosterone and estradiol and their bioavailable fractions in community-dwelling men. *Am. J. Epidemiol.* **147**: 750.

Fox, M., Boyle, P.F. and Hammonds, J.C. (1973). Transplantation of interstitial cells of the testis. *Br. J. Urol.* **45**: 696.

Franchimont, P., Kicovic, P.M., Mattei, A., et al. (1978). Effects of oral testosterone undecanoate in hypogonadal male patients. *Clin. Endocrinol.* (*Oxf.*) **9**: 313.

Frumpkin, A. (1944). Reconstruction of male genitalia. *Am. Rev. Sov. Med.* **2**: 14.

Fujioka, M., Shinohara, Y., Baba, S., et al. (1986). Pharmacokinetic properties of testosterone propionate in normal men. *J. Clin. Endocrinol. Metab.* **63**: 1361.

Gilbert, D.A., Williams, M.W., Horton, C.E., et al. (1988). Phallic reinnervation via the pudendal nerve. *J. Urol.* **140**: 295.

Gittes, R.F., Altwein, J.E., Yen, S.S., et al. (1972). Testicular transplantation in the rat: long-term gonadotropin and testosterone radioimmunoassays. *Surgery* **72**: 187.

Goldstein, M.B., Dearden, L.C. and Gualtieri, V. (1967). Regeneration of subtotally cystectomized bladder patched with omentum: an experimental study in rabbits. *J. Urol.* **97**(4): 664–668.

Goodwin, W.E., Scott, W.W. (1952). Phalloplasty. *J. Urol.* **68**: 903.

Gooren, L.J. (1994). A ten-year safety study of the oral androgen testosterone undecanoate. *J. Androl.* **15**: 212.

Gray, H. (1918). *Anatomy of the Human Body.* Philadelphia: Lea & Febiger,

Griffen, J. and Willson, J. (1998). Disorders of the testes and the male reproductive tract. In: Willson J.D., et al (eds.), *Williams Textbook of Endocrinology.* , Philadelphia, PA: Saunders, p. 819.:,

Guzman, L. (1999). Neourethra with rectum, posterior sagittal approach. In: Ehrlich, R.M.and Alter, G.J. (eds.), *Reconstructive and plastic surgery of the external genitalia: adult and pediatric.* Philadelphia, PA: Saunders, p. 101.

Hakky, S.I. (1976). Urethral replacement by dacron mesh. *Lancet* **2**: 1192.

Hakky, S.I. (1977). The use of fine double siliconised dacron in urethral replacement. *Br. J. Urol.* **49**: 167.

Hendren, W.H. and Reda, E.F. (1986). Bladder mucosa graft for construction of male urethra. *J. Pediatr. Surg.* **21**: 189.

Hendren, W.H. and Atala, A. (1994). Use of bowel for vaginal reconstruction. *J. Urol.* **152**(2 Pt 2): 752–755; discussion 756–757.

Hogan, D.J. and Maibach, H.I. (1990). Adverse dermatologic reactions to transdermal drug delivery systems. *J. Am. Acad. Dermatol.* **22**: 811.

Horton, C.E. and Dean, J.A. (1990). Reconstruction of traumatically acquired defects of the phallus. *World J. Surg.* **14**: 757.

Huet, C., Pisselet, C., Mandon-Pepin, B., Monget, P. and Monniaux, D. (2001). Extracellular matrix regulates ovine granulose cell survival, proliferation and steroidogenesis: Relationships between cell shape and function. *J. Endocrinal.* **169**: 347.

Humby, G. (1941). A one-stage operation for hypospadia. *Br. J. Surg.* **29**: 84.

Hutschenreiter, G., Rumpelt, H.J., Klippel, K.F. and Hohenfellner, R. (1978). The free peritoneal transplant as substitute for the urinary bladder wall. *Invest. Urol.* **15**(5): 375.

Italiano, G., Abatangelo Jr., G., Calabro, A., et al. (1997). Reconstructive surgery of the urethra: a pilot study in the rabbit on the use of hyaluronan benzyl ester (Hyaff-11) biodegradable grafts. *Urol. Res.* **25**: 137.

Jackson, C.J., Garbett, P.K., Nissen, B., et al. (1990). Binding of human endothelium to Ulex europaeus I-coated Dynabeads: application to the isolation of microvascular endothelium. *J. Cell Sci.* **96**(Pt 2): 257.

Joki, T., Machluf, M., Atala, A., et al. (2001). Continuous release of endostatin from microencapsulated engineered cells for tumor therapy. *Nat. Biotechnol.* **19**: 35.

Kambic, H., Kay, R., Chen, J.F., Matsushita, M., Harasaki, H. and Zilber, S. (1992). Biodegradable pericardial implants for bladder augmentation: a 2.5-year study in dogs. *J. Urol.* **148**(2 Pt 2): 539–543.

Kardar, A. and Pettersson, B.A. (1995). Penile gangrene: a complication of penile prosthesis. *Scand. J. Urol. Nephrol.* **29**: 355.

Kearns, W. (1941). Successful autoplastic graft following accidental castration. *Ann. Surg.* **114**: 886.

Kelami, A. (1971). Lyophilized human dura as a bladder wall substitute: experimental and clinical results. *J. Urol.* **105**(4): 518–522.

Kim, B.S., Nikolovski, J., Bonadio, J., Smiley, E. and Mooney, D.J. (1999). Engineered smooth muscle tissues: Regulating cell phenotye with the scaffold. *Exp. Cell Res.* **251**: 318.

Kraling, B.M. and Bischoff, J. (1998). A simplified method for growth of human microvascular endothelial cells results in decreased senescence and continued responsiveness to cytokines and growth factors. *In Vitro Cell Dev. Biol. Anim.* **34**: 308.

Kreeger, P.K., Deck, J.W., Woodruff, T.K. and Shea, L.D. (2006). The *in vitro* regulation of ovarian follicle development using alginate-extracellular matrix gels. *Biomaterials.* **27**(5): 714–723 (Epub August 1, 2005).

Kropp, B.P., Ludlow, J.K., Spicer, D., et al. (1998). Rabbit urethral regeneration using small intestinal submucosa onlay grafts. *Urology* **52**: 138.

Lee, S., Tung, K.S. and Orloff, M.J. (1971). Testicular transplantation in the rat. *Transplant Proc.* **3**: 586.

Leong, C.H. and Ong, G.B. (1972). Gastrocystoplasty in dogs. *Aust. NZ. J. Surg.* **41**(3): 272–279.

Liebert, M., Hubbel, A., Chung, M., et al. (1997). Expression of mal is associated with urothelial differentiation *in vitro*: identification by differential display reverse-transcriptase polymerase chain reaction. *Differentiation.* **61**: 177.

Liebert, M., Wedemeyer, G., Abruzzo, L.V., et al. (1991). Stimulated urothelial cells produce cytokines and express an activated cell surface antigenic phenotype. *Semin. Urol.* **9**: 124.

Lim, F. and Sun, A.M. (1980). Microencapsulated islets as bioartificial endocrine pancreas. *Science* **210**: 908.

Lydston, G. (1916). Sex gland implantation. Additional cases and conclusions to date. *JAMA* **66**: 1540.

Machluf, M., Orsola, A., Boorjian, S., et al. (2003). Microencapsulation of Leydig cells: a system for testosterone supplementation. *Endocrinology* **144**: 4975.

McClellan, K.J. and Goa, K.L. (1998). Transdermal testosterone. *Drugs* **55**: 253.

Mikos, A.G., Lyman, M.D., Freed, L.E., et al. (1994). Wetting of poly(L-lactic acid) and poly(DL-lactic-co-glycolic acid) foams for tissue culture. *Biomaterials* **15**: 55.

Nuininga, J.E., van Moerkerk, H., Hanssen, A., et al. (2003).: Rabbit urethra replacement with a defined biomatrix or small intestinal submucosa. *Eur. Urol.* **44**: 266.

Nukui, F., Okamoto, S., Nagata, M., et al. (1997). Complications and reimplantation of penile implants. *Int. J. Urol.* **4**: 52.

O'brien, M.J., Pendola, J.K. and Eppig, J.J. (2003). A revised protocol for *in vitro* development of mouse oocytes from primordial follicles dramatically improves their developmental competence. *Biol. Reporod.* **68**(5): 1682–1686.

Olsen, L., Bowald, S., Busch, C., et al. (1992). Urethral reconstruction with a new synthetic absorbable device. An experimental study. *Scand. J. Urol. Nephrol.* **26**: 323.

Ozcan, M. and Kahveci, R. (1987). One-Stage repair of distal and midpenile hypospadia by modified Hodgson III technique. *Eur. J. Plast. Surg.* **10**: 159.

Pangas, S.A., Saudye, H., Shea, L.D., Woodruff, T.K. (2003). Novel approach for the three-dimensional culture of granulosa cell-oocyte complexes. *Tissue Eng.* **9**(5): 1013–1021.

Park, H.J., Yoo, J.J., Kershen, R.T., et al. (1999). Reconstitution of human corporal smooth muscle and endothelial cells *in vivo*. *J. Urol.* **162**: 1106.

Parkes, A.S. (1966). The rise of reproductive endocrinology. *J. Endocrinol.* **34**: xx. 1926–1940.

Patrick, C., Mikos, A. and Mcintire, L. (1998). *Frontiers in Tissue Engineering: Tissue Engineering in Urology*. New York: Elsevier Science Inc. p. 649.

Puckett, C.L. and Montie, J.E. (1978). Construction of male genitalia in the transsexual, using a tubed groin flap for the penis and a hydraulic inflation device. *Plast. Reconstr. Surg.* **61**: 523.

Puthenveettil, J.A., Burger, M.S. and Reznikoff, C.A. (1999). Replicative senescence in human uroepithelial cells. *Adv. Exp. Med. Biol.* **462**: 83.

Rajimwale, A., Furness III, P.D., Brant, W.O. and Koyle, M.A. (2004). Vaginal construction using sigmoid colon in children and young adults. *BJU Int.* **94**: 115–119.

Rowley, J.A., Madlambayan, G. and Mooney, D.J. (1999). Alginate hydrogels as synthetic extracellular matrixmaterials. *Biomaterials* **20**: 45.

Scriven, S.D., Booth, C., Thomas, D.F., et al. (1997). Reconstitution of human urothelium from monolayer cultures. *J. Urol.* **158**: 1147.

Sharaby, J.S., Benet, A.E. and Melman, A. (1995). Penile revascularization. *Urol. Clin. N. Am.* **22**: 821.

Shaul, D.B., Xie, H.W., Diaz, J.F., et al. (1996). Use of tubularized peritoneal free grafts as urethral substitutes in the rabbit. *J. Pediatr. Surg.* **31**: 225.

Sievert, K.D., Bakircioglu, M.E., Nunes, L., et al. (2000). Homologous acellular matrix graft for urethral reconstruction in the rabbit: histological and functional evaluation. *J. Urol.* **163**: 1958.

Silber, S.J. (1978). Transplantation of a human testis for anorchia. *Fertil. Steril.* **30**: 181.

Skalak, R. and Fox, C. (1988). *Tissue Engineering*, New York: Alan R. Liss,

Small, M.P. (1976). Small-Carrion penile prosthesis. A new implant for management of impotence. Mayo *Clin. Proc.* **51**: 336.

Smitz, J.E. and Cortvrind, R.G. (2002). The earliest stages of folliculogenesis *in vitro*. *Reproduction* **123**(2): 185–202.

Snyder, P.J. and Lawrence, D.A. (1980). Treatment of male hypogonadism with testosterone enanthate. *J. Clin. Endocrinol. Metab.* **51**: 1335.

Sokol, R.Z., Palacios, A., Campfield, L.A., et al. (1982). Comparison of the kinetics of injectable testosterone in eugonadal and hypogonadal men. *Fertil. Steril.* **37**: 425.

Spears, N., Boldand, N.I., Murray, A.A. and Gosden, R.G. (1994). Mouse oocytes derived from *in vitro* grown primary ovarian follicles are fertile. *Hum. Reprod.* **9**(3): 527–532.

Stuenkel, C.A., Dudley, R.E. and Yen, S.S. (1991). Sublingual administration of testosterone-hydroxypropyl-beta-cyclodextrin inclusion complex simulates episodic androgen release in hypogonadal men. *J. Clin. Endocrinol. Metab.* **72**: 1054.

Tai, I.T. and Sun, A.M. (1993). Microencapsulation of recombinant cells: a new delivery system for gene therapy. *Faseb. J.* **7**: 1061.

Tai, J., Tze, W.J., Yip, S., et al. (1992). Transplantation studies of Leydig cells in castrated inbred rats. *Transplant. Proc.* **24**: 2917.

Tobin, M., Freeman, M. and Atala, A. (1994). Maturation response of normal human urothelial cells in culture is dependent on extracellular matrix and serum additives. *Surg. Forum* **45**: 786.

Verhoeven, G. (1992). Local control system within the testis. In: Tindall, B. (ed.), *Bailliere's Clinical Endocrinology and Metablolism*. p. 313.

Vozzi, G., Flaim, C., F, B., et al. (2002). Microfabricated PLGA scaffolds: a comparative study for application to tissue engineering. *Mat. Sci. Eng.* **20**: 43.

Wang, T., Koh, C.J., Yoo, J.J., et al. (2003). Creation of an engineered uterus for surgical reconstruction. Presented at: The American Academy of Pediatrics Section on Urology; 4-7, New Orleans, La

Wang, T., Lacik, I., Brissova, M., et al. (1997). An encapsulation system for the immunoisolation of pancreatic islets. *Nat. Biotechnol.* **15**: 358.

Wilson, D.E., Meikle, A.W., Boike, S.C., et al. (1998). Bioequivalence assessment of a single 5 mg/day testosterone transdermal system versus two 2.5 mg/day systems in hypogonadal men. *J. Clin. Pharmacol.* **38**: 54.

Xu, Y., Qiao, Y., Sa, Y., et al. (2002). An experimental study of colonic mucosal graft for urethral reconstruction. *Chin. Med. J. (Engl)* **115**: 1163.

Yoo, J.J., Lee, I. and Atala, A. (1998). Cartilage rods as a potential material for penile reconstruction. *J. Urol.* **160**: 1164.

Yoo, J.J., Meng, J., Oberpenning, F., et al. (1998).: Bladder augmentation using allogenic bladder submucosa seeded with cells. *Urology* **51**: 221.

Yoo, J.J., Park, H.J. and Atala, A. (2000). Tissue-engineering applications for phallic reconstruction. *World J. Urol.* **18**: 62.

Yoo, J.J., Park, H.J., Lee, I., et al. (1999). Autologous engineered cartilage rods for penile reconstruction. *J. Urol.* **162**: 1119.

Yoo, J.J. and Atala, A. (2000). Tissue engineering applications in the genitourinary tract system. *Yonsei Med.* **41**: 789.

Young, R.C., Schumann, R. and Zhang, P. (2003). Three-dimensional culture of human uterine smooth muscle myocytes on a resorbable scaffolding. *Tissue Eng.* **9**(3): 451–459.

nuclear transfer techniques to generate stem cells that can be used to engineer tissues, such as skeletal muscle, cardiac muscle, and kidney tissues (Lanza et al., 2002). While the use of embryonic stem cells is ethically controversial, adult stem cells are being proposed as a new source. Although their potential and limitations are not fully evaluated, stem cells from fat, skin, muscle, placenta, umbilical cord, amniotic, and bone marrow have been used for various cell and tissue therapies (Bartsch et al., 2005; Deasy et al., 2005; Moise, 2005; Tholpady et al., 2006).

A variety of synthetic and naturally derived materials have been examined in order to determine the ideal support systems for the regeneration of tissues (Kim and Mooney, 1998; Isogai et al., 1999; Amiel et al., 2001; Vacanti et al., 2001; Pariente et al., 2002; El-Kassaby et al., 2003; Falke et al., 2003; Derwin et al., 2004; Kim et al., 2004; Atala et al., 2006; Murray et al., 2006). Biodegradable synthetic materials, such as poly-lactic and glycolic acid polymers have been used to provide structural support for cells. Synthetic materials can be easily fabricated and configured in a reproducible manner. Naturally derived materials, such as collagen, laminin, and fibronectin are biocompatible and provide a similar extracellular matrix (ECM) environment as normal tissue.

Biomaterials used for regenerative medicine are designed to replicate the biologic and mechanical function of native tissue structures and their ECM. They provide three-dimensional architecture for the cells to reconstitute into new tissues with appropriate structure and function, and allow for the delivery of cells and appropriate bioactive factors (e.g. cell-adhesion peptides, growth factors) to desired sites in the body (Kim and Mooney, 1998). As the majority of mammalian cell types are anchorage-dependent, biomaterials provide a cell-adhesion substrate that can deliver cells to specific regions of the body with high loading efficiency. Biomaterials also provide mechanical support against *in vivo* forces such that the predefined three-dimensional structure is maintained during tissue development. Furthermore, bioactive signals such as cell-adhesion peptides and growth factors can be loaded along with cells to help regulate cellular function.

The ideal biomaterial for digit reconstruction should be biodegradable and bioabsorbable without eliciting inflammatory responses that interfere with tissue formation. Incompatible materials are destined for an inflammatory or foreign-body response that eventually leads to rejection and/or necrosis. Since biomaterials provide temporary mechanical support while the cells undergo spatial tissue organization, a properly chosen biomaterial should allow the engineered tissue to maintain adequate mechanical integrity to support itself in early development.

Both the synthetic polymers (e.g. polyglycolic acid (PGA)) and naturally derived materials (e.g. bladder submucosa and small intestinal submucosa) have been used as biomaterials for digit reconstruction. These materials have been shown to be biocompatible and suitable for tissue engineering applications when appropriately configured (Pariente et al., 2002). Synthetic biomaterials have been used successfully for phalangeal reconstruction, which include PGA, PCL (poly ε-caprolactone) and PLLA (poly-L-lactate) (Rosenberg, 1971; Isogai et al., 2004). Non-woven PGA has porosity of greater than 95%, thus is able to accommodate a large quantity of cells and offers adequate environment for bone and cartilage formation. Further, PGA can easily be configured to structures similar to native tissue. Naturally derived materials, such as hydroxyapatite and small intestine submucosa (SIS) have been used to engineer phalangeal tissues (Vacanti et al., 2001; Derwin et al., 2004).

Engineering of a viable tissue is the prime goal of cell-based technology, thus, obtaining adequate vascular supply is critical to cell viability and the development of tissues. Although the body has the ability to form new vessels over time, it is critical to provide nutrients and oxygen to the cell-constructs within the limited space, as in distal phalanges. In addition, it is a general conception that cell or tissue components may not be implanted in large volumes due to the limited diffusion (Folkman and Hochberg, 1973). Numerous efforts have been made to overcome this limitation and attempts to enhance vascularization within the host tissue have been pursued using several approaches. These include the use of angiogeneic factors, such as use of vascular endothelial growth factors (VEGF) and endothelial cells (EC), and cell-support matrices that permit enhanced easy diffusion of nutrients across the entire implant (Nomi et al., 2002; De Coppi et al., 2005; Kaigler et al., 2006). In one study, VEGF secreting myoblasts were implanted with vascular EC to enhance neovascularization of engineered tissues. In this study skeletal myoblasts from were cultured and transfected with an adenovirus encoding VEGF and combined with human vascular EC (De Coppi et al., 2005). The cell mixtures were injected subcutaneously in athymic mice, and the engineered tissues were retrieved up to 8 weeks after implantation. The transfected cells formed vascularized muscle tissue mass with evidence of adequate neovascularization by histology and immunohistochemical probing. The engineered muscle tissue, composed of non-transfected cells had a significantly smaller tissue mass with decreased muscle volume and less neovascularization. These results indicate that a combination of VEGF and EC may be useful for inducing neovascularization and volume preservation in the engineered tissues.

In addition to the biological approach, incorporation of bioactive factors that enhance vascularization into the support matrices has been tried (Schuch et al., 2002; Kim et al., 2004; Ennett et al., 2006; Kaigler et al., 2006). VEGF was incorporated into poly(lactide-co-glycolide) (PLG) scaffolds or pre-encapsulated in PLG microspheres for therapeutic angiogenesis. In this study VEGF was positioned predominantly adjacent to scaffold pores when incorporated directly and was released rapidly. Pre-encapsulation led to the VEGF being more deeply embedded and resulted in a delayed release. *In vivo*, the released VEGF generated local protein concentrations for the 21 days of the experiment, with negligible release into the systemic circulation, and significantly enhanced local angiogenesis. These data indicate that VEGF can be administered in a sustained and localized fashion *in vivo* (Ennett et al., 2006). This angiogenic factor delivery system was applied to bone regeneration application and was demonstrated that VEGF scaffolds have the ability to enhance neovascularization and bone regeneration in irradiated osseous defects, which outlines a novel approach for engineering tissues in hypovascular environments (Kaigler et al., 2006).

CREATION OF COMPOSITE TISSUE STRUCTURES

The goal of engineering of the digit is to achieve esthetic and functional tissues that would allow for adequate dexterity for daily activities. To permit this capability, composite tissues with all cellular and tissue components are necessary. Toward this goal, initial efforts were focused on engineering of individual tissues, such as the skin, bone, cartilage, tendon, muscle, fat, and nerve (Amiel et al., 2001; Lanza et al., 2002; Kim et al., 2002, 2004; Yiou et al., 2003; Derwin et al., 2004; Lee et al., 2006; Wood, 2006). In a case report, a tissue engineered distal phalanx was used to replace a distal bone in a patient who had a partial avulsion of the thumb. The procedure resulted in the functional restoration of a stable and biomechanically sound thumb of normal length, without the pain and complications that are usually associated with harvesting a bone graft (Vacanti et al., 2001).

Although the formation of individual tissues has been successfully demonstrated experimentally and clinically, engineering of a more complex tissue, consisting of multiple cell types, remains a challenge. The first attempt to engineer a phalanx was demonstrated through the use of cartilage, bone, and tendon on a single construct (Isogai et al., 1999). In an *ex situ* model, bovine chondrocytes and tenocytes were seeded on biodegradable polymer scaffolds that were configured to the shape of human phalanx bones and wrapped with the bovine periosteum to serve as a joint. Subsequently, these constructs were implanted into the subcutaneous space of athymic mice and followed for up to 60 weeks. The retrieved phalangeal constructs showed the formation of bone, cartilage and tendon tissues, which were confirmed using polymerase chain reaction (PCR), histology, and immunohistochemistry. More importantly, the implanted phalanges maintained the shape of human phalanges (Landis et al., 2005). This experimentation demonstrated that phalangeal tissues can be engineered *ex situ* using a composite cell system and provide the possibility of engineering morphologically adequate phalanx. However, considerations for engineering of functional phalangeal units have not been suggested.

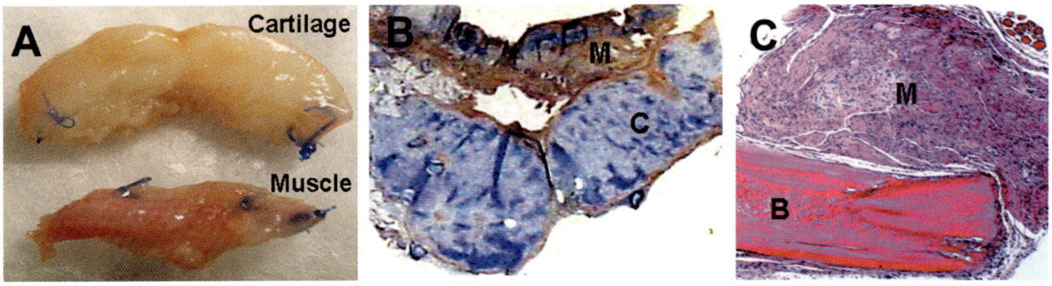

Figure 34.1 Engineered composite tissue *in vivo*. (a) Gross examination of the explanted muscle and cartilage tissue. (b) Histology of cartilage and muscle tissue composite (H&E). (c) Formation of bone (B) and organized muscle bundles (M) was observed in the bone/muscle composite tissue.

In an effort to engineer functional digits for reconstruction, we have designed scaffolds that would permit the delivery of different cell types to target sites and would allow for joint movement. The PGA polymer was chosen to generate the rigid portion of the digit, made from cartilage or bone, and acellular collagen matrix derived from bladder submucosa to allow movement by muscle. The PGA polymer was composed of two tubular pieces connected with non-absorbable sutures, which served as a hinge between the two bony scaffolds. Multiple thin strips of acellular collagen matrix were attached to the distal ends of the bony scaffolds. This configuration was designed to allow for contraction, relaxation, and movement of the composite engineered digit-like structure. In the initial experiments, the PGA polymer scaffolds were seeded with bovine chondrocytes/osteoblasts, and bovine skeletal muscle cells were seeded onto the collagen matrix. The cell-seeded composites were incubated *in vitro* for 2 weeks. Histologically, abundant muscle cells were present throughout the collagen matrix and the PGA polymers seeded with chondrocytes showed the presence of whitish ECM. Western blot analysis confirmed the tissue identity and showed expression of muscle specific genes such as actin, desmin, and tropomyosin in the muscle component and expression of collagen-1 in the chondrocyte-seeded PGA.

Subsequently, the feasibility of engineering skeletal muscle and cartilage/bone composite structure was studied *in vivo*. Composite scaffolds consisting of synthetic PGA polymers ($1.0 \times 0.5 \times 0.5$ cm) and naturally derived collagen matrices obtained from the bladder submucosa ($1.4 \times 0.4 \times 0.3$ cm) were constructed. Bovine chondrocytes or osteoblasts were seeded onto the PGA polymer matrices and skeletal muscle cells onto the collagen matrices. The scaffolds containing both cell types were analyzed *in vitro* for cell viability and tissue formation. The engineered digits were implanted subcutaneously in athymic mice ($n = 36$) and followed for up to 6 months. The cells seeded on the composite digit constructs readily attached to their designated region of the scaffold and remained viable. Grossly, the implanted scaffolds formed muscle and cartilage or bone tissues adjacent to each other (Figure 34.1). Each tissue type was confirmed histo- and immunohistochemically using cell specific antibodies. Biomechanical studies showed that the cartilage tissue was elastic and could withstand high degrees of pressure, which demonstrate the ability to preserve its structural integrity. Physiologic organ bath studies of the retrieved muscle tissues showed adequate contractility in response to electric field stimulation. These findings show that different tissue types can be engineered simultaneously using a composite scaffold system. The tissues retained their respective phenotypic and functional characteristics independent of the other. This study demonstrates that the engineering of functional digit tissues may be feasible.

To determine the feasibility of replacing a missing digit segment, *in situ* implantation of digit segments were performed in a rabbit model. Autologous muscle cells and chondrocytes were grown, expanded, and seeded on the composite scaffolds, consisting of interconnecting bony segments, muscle, and tendon. The digit

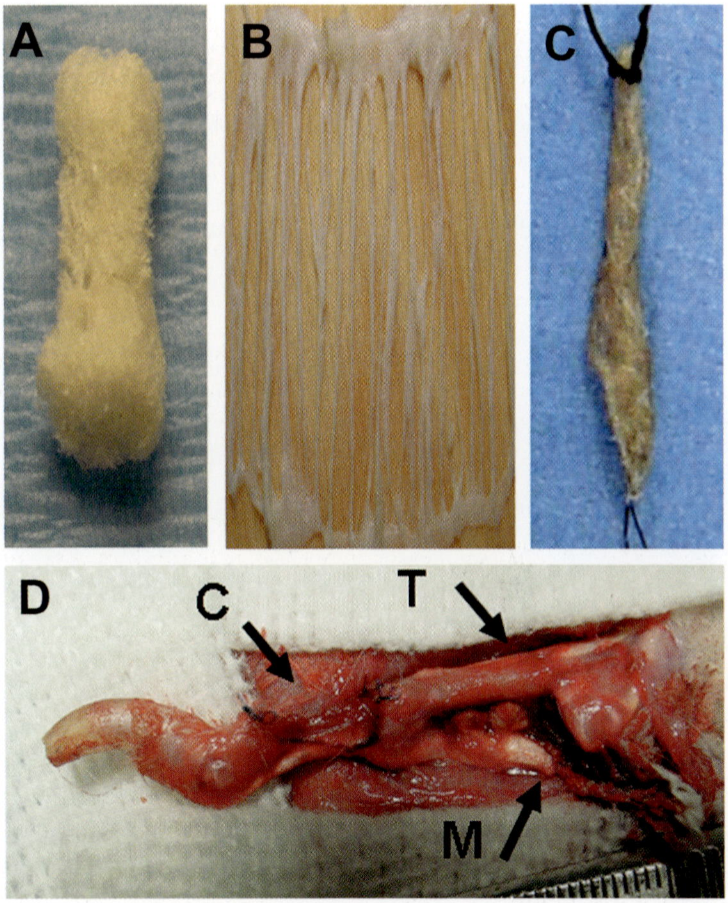

Figure 34.2 Engineering of phalangeal tissues *in vivo*. (a) Bone scaffold, (b) muscle scaffolds composed of multiple thin collagen fiber strips, (c) tendon scaffold, (d) engineered phalangeal tissue at retrieval shows the formation of cartilage, muscle and tendon in a rabbit model.

segments with an interconnected joint were excised and the engineered digit segments of the same length and caliber were replaced in rabbits. The forelimb with the engineered digit was placed in a cast for approximately 4 weeks in order to protect the wound and enhance tissue maturation *in situ*. The engineered digit segments were able to form cartilage, muscle, and tendon at retrieval. Scaffolds without cells failed to form tissue structures (Figure 34.2).

SUMMARY AND FUTURE PERSPECTIVES IN CLINICAL TRANSLATION

Efforts in engineering of the digit are focused toward the eventual clinical application. Engineering of the phalangeal tissues presents a unique challenge in tissue reconstruction due to the complexity of the functions required. Although individual or partial replacement of phalanges has been demonstrated experimentally and clinically, achieving a fully functional engineered digit requires continued investigations. One of the unsolved tasks that would accelerate the progress of this research is finding methods to integrate all tissue components, including the vascular and neural network, into a compact compartment of the distal limb. In addition, development of an intelligent composite bioscaffold system that would allow for the enhanced formation of individual tissue types in a controlled manner is critical. Recent progress made towards the engineering of phalanx suggests that achieving functional phalangeal tissues may have an expanded role in clinical medicine.

REFERENCES

Amiel, G.E., Yoo, J.J., Kim, B.S. and Atala, A. (2001). Tissue engineered stents created from chondrocytes. *J. Urol.* **165**: 2091–2095.

Atala, A., Bauer, S.B., Soker, S., Yoo, J.J. and Retik, A.B. (2006). Tissue-engineered autologous bladders for patients needing cystoplasty. *Lancet* **367**: 1241–1246.

Bartsch, G., Yoo, J.J., De Coppi, P., Siddiqui, M.M., Schuch, G., Pohl, H.G., Fuhr, J., Perin, L., Soker, S. and Atala, A. (2005). Propagation, expansion, and multilineage differentiation of human somatic stem cells from dermal progenitors. *Stem Cell. Dev.* **14**: 337–348.

Brivanlou, A.H., Gage, F.H., Jaenisch, R., Jessell, T., Melton, D. and Rossant, J. (2003). Stem cells. Setting standards for human embryonic stem cells. *Science* **300**: 913–916.

Campbell, C.J., Ishida, H., Takahashi, H. and Kelly, F. (1963). The transplantation of articular cartilage. An experimental study in dogs. *J. Bone Joint Surg.* **45**: 1579–1592.

Campbell, C.J., Ishida, H., Takahashi, H. and Kelly, F. (1972). Homotransplantation of a half or whole joint. *Clin. Orthop. Relat. Res.* **87**: 146–155.

De Coppi, P., Delo, D., Farrugia, L., Udompanyanan, K., Yoo, J.J., Nomi, M., Atala, A. and Soker, S. (2005). Angiogenic gene-modified muscle cells for enhancement of tissue formation. *Tissue Eng.* **11**: 1034–1044.

De Filippo, R.E., Yoo, J.J. and Atala, A. (2003). Engineering of vaginal tissue *in vivo*. *Tissue Eng.* **9**: 301–306.

Deasy, B.M., Gharaibeh, B.M., Pollett, J.B., Jones, M.M., Lucas, M.A., Kanda, Y. and Huard, J. (2005). Long-term self-renewal of postnatal muscle-derived stem cells. *Mol. Biol. Cell* **16**: 3323–3333.

Derwin, K., Androjna, C., Spencer, E., Safran, O., Bauer, T.W., Hunt, T., Caplan, A. and Iannotti, J. (2004). Porcine small intestine submucosa as a flexor tendon graft. *Clin. Orthop. Relat. Res.* 245–252.

Dubernard, J.M., Owen, E., Herzberg, G., Lanzetta, M., Martin, X., Kapila, H., Dawahra, M. and Hakim, N.S. (1999). Human hand allograft: report on first 6 months. *Lancet* **353**: 1315–1320.

El-Kassaby, A.W., Retik, A.B., Yoo, J.J. and Atala, A. (2003). Urethral stricture repair with an off-the-shelf collagen matrix. *J. Urol.* **169**: 170–173; discussion 173.

Ennett, A.B., Kaigler, D. and Mooney, D.J. (2006). Temporally regulated delivery of VEGF *in vitro* and *in vivo*. *J. Biomed. Mater. Res. A*.

Falke, G., Yoo, J.J., Kwon, T.G., Moreland, R. and Atala, A. (2003). Formation of corporal tissue architecture *in vivo* using human cavernosal muscle and endothelial cells seeded on collagen matrices. *Tissue Eng.* **9**: 871–879.

Folkman, J. and Hochberg, M. (1973). Self-regulation of growth in three dimensions. *J. Exp. Med.* **138**: 745–753.

Isogai, N., Asamura, S., Higashi, T., Ikada, Y., Morita, S., Hillyer, J., Jacquet, R. and Landis, W.J. (2004). Tissue engineering of an auricular cartilage model utilizing cultured chondrocyte-poly(L-lactide-epsilon-caprolactone) scaffolds. *Tissue Eng.* **10**: 673–687.

Isogai, N., Landis, W., Kim, T.H., Gerstenfeld, L.C., Upton, J. and Vacanti, J.P. (1999). Formation of phalanges and small joints by tissue-engineering. *J. Bone Joint Surg.* **81**: 306–316.

Jovanovic, S., Wertheimer, B., Zelic, Z. and Getos, Z. (1999). Wartime amputations. *Mil. Med.* **164**: 44–47.

Kaigler, D., Wang, Z., Horger, K., Mooney, D.J. and Krebsbach, P.H. (2006). VEGF scaffolds enhance angiogenesis and bone regeneration in irradiated osseous defects. *J. Bone Miner. Res.* **21**: 735–744.

Kim, B.S. and Mooney, D.J. (1998). Development of biocompatible synthetic extracellular matrices for tissue engineering. *Trends Biotechnol.* **16**: 224–230.

Kim, B.S., Yoo, J.J. and Atala, A. (2002). Engineering of human cartilage rods: potential application for penile prostheses. *J. Urol.* **168**: 1794–1797.

Kim, B.S., Yoo, J.J. and Atala, A. (2004). Peripheral nerve regeneration using acellular nerve grafts. *J. Biomed. Mater. Res. A* **68**: 201–209.

Kwon, T.G., Yoo, J.J. and Atala, A. (2002). Autologous penile corpora cavernosa replacement using tissue engineering techniques. *J. Urol.* **168**: 1754–1758.

Landis, W.J., Jacquet, R., Hillyer, J., Lowder, E., Yanke, A., Siperko, L., Asamura, S., Kusuhara, H., Enjo, M., Chubinskaya, S., et al. (2005). Design and assessment of a tissue-engineered model of human phalanges and a small joint. *Orthod. Craniofac. Res.* **8**: 303–312.

Lanza, R.P., Chung, H.Y., Yoo, J.J., Wettstein, P.J., Blackwell, C., Borson, N., Hofmeister, E., Schuch, G., Soker, S., Moraes, C.T., et al. (2002). Generation of histocompatible tissues using nuclear transplantation. *Nat. Biotechnol.* **20**: 689–696.

Lee, S.J., Lim, G.J., Lee, J.W., Atala, A. and Yoo, J.J. (2006). *In vitro* evaluation of a poly(lactide-co-glycolide)-collagen composite scaffold for bone regeneration. *Biomaterials* **27**: 3466–3472.

Moise Jr., K.J. (2005). Umbilical cord stem cells. *Obstet. Gynecol.* **106**: 1393–1407.

Murray, M.M., Forsythe, B., Chen, F., Lee, S.J., Yoo, J.J., Atala, A. and Steinert, A. (2006). The effect of thrombin on ACL fibroblast interactions with collagen hydrogels. *J. Orthop. Res.* **24**: 508–515.

Nomi, M., Atala, A., Coppi, P.D. and Soker, S. (2002). Principals of neovascularization for tissue engineering. *Mol. Aspect. Med.* **23**: 463–483.

Oberpenning, F., Meng, J., Yoo, J.J. and Atala, A. (1999). *De novo* reconstitution of a functional mammalian urinary bladder by tissue engineering. *Nat. Biotechnol.* **17**: 149–155.

Pariente, J.L., Kim, B.S. and Atala, A. (2002). *In vitro* biocompatibility evaluation of naturally derived and synthetic biomaterials using normal human bladder smooth muscle cells. *J. Urol.* **167**: 1867–1871.

Patradul, A., Ngarmukos, C. and Parkpian, V. (1998). Distal digital replantations and revascularizations. 237 digits in 192 patients. *J. Hand Surg.* (Edinburgh, Lothian) **23**: 578–582.

Pereira, B.P., Kour, A.K., Leow, E.L. and Pho, R.W. (1996). Benefits and use of digital prostheses. *J. Hand Surg.* **21**: 222–228.

Rosenberg, L. (1971). Chemical basis for the histological use of safranin O in the study of articular cartilage. *J. Bone Joint Surg.* **53**: 69–82.

Sandzen Jr., S.C. (1985). Classification and functional management of congenital central defect of the hand. *Hand Clin.* **1**: 483–498.

Schuch, G., Machluf, M., Bartsch Jr., G., Nomi, M., Richard, H., Atala, A. and Soker, S. (2002). *In vivo* administration of vascular endothelial growth factor (VEGF) and its antagonist, soluble neuropilin-1, predicts a role of VEGF in the progression of acute myeloid leukemia *in vivo*. *Blood* **100**: 4622–4628.

Tholpady, S.S., Llull, R., Ogle, R.C., Rubin, J.P., Futrell, J.W. and Katz, A.J. (2006). A dipose tissue: stem cells and beyond. *Clin. Plast. Surg.* **33**: 55–62, vi.

Vacanti, C.A., Bonassar, L.J., Vacanti, M.P. and Shufflebarger, J. (2001). Replacement of an avulsed phalanx with tissue-engineered bone. *N. Engl. J. Med.* **344**: 1511–1514.

Wood, F.M., Kolybaba, M. L., Allen, P. (2006). The use of cultured epithelial autograft in the treatment of major burn wounds: Eleven years of clinical experience. *Burns* **32**: 538–544.

Yiou, R., Yoo, J.J. and Atala, A. (2003). Restoration of functional motor units in a rat model of sphincter injury by muscle precursor cell autografts. *Transplantation* **76**: 1053—1060.

Yoo, J.J., Park, H.J., Lee, I. and Atala, A. (1999). Autologous engineered cartilage rods for penile reconstruction. *J. Urol.* **162**: 1119–1121.

35

Functional Tissue Engineering of Ligament and Tendon Injuries

Savio L.-Y. Woo, Alejandro J. Almarza, Sinan Karaoglu, and Steven D. Abramowitch

INTRODUCTION

Tendons and ligaments are soft connective tissues composed of closely packed, parallel collagen fiber bundles which connect bone to muscle and bone to bone, respectively. These unique tissues serve essential roles in the musculoskeletal system by transferring tensile loads to guide motion and stabilize diarthrodial joints. Injuries to tendons, such as the patellar tendon (PT) of the knee, or ligaments, such as the collateral and cruciate ligaments of the knee, upset the balance between mobility and stability of this joint. These injuries are often manifested in abnormal knee kinematics and damage to other tissues in and around the joint such as meniscus and articular cartilage, which may lead to morbidity, pain, and osteoarthritis. With the high incidence of ligament and tendon injuries in sports and work related activities, improvements on healing and repair of these tissues are of great interest (Beaty 1999).

Interestingly, there is a dramatic variability in the propensity for healing of ligaments within the same knee joint, namely the medial collateral ligament (MCL) and anterior cruciate ligament (ACL). Clinical and laboratory studies have shown that injuries to the MCL generally heal sufficiently well such that non-surgical management has become the treatment of choice (Frank et al., 1983; Indelicato, 1983; Jokl et al., 1984; Woo et al., 1987b; Kannus, 1988; Scheffler et al., 2001). While most structural properties of the femur–MCL-tibia complex (FMTC) are restored within weeks, the mechanical properties of the healed MCL (i.e. the stress–strain curve) remain much different from those of the normal MCL, as are the altered histomorphological appearance (e.g. uniform distribution of small collagen fibrils) and biochemical composition (e.g. elevated type III and V collagens) (Adachi and Hayashi, 1986; Birk et al., 1990; Weiss et al., 1991; Frank et al., 1992; Hart et al., 1992; Marchant et al., 1996; Frank et al., 1997; Hart et al., 2000; Nakamura et al., 2000; Niyibizi et al., 2000; Birk, 2001).

For the ACL, it is well known that a midsubstance tear would not heal and the success of non-surgical management is limited. Thus, surgical reconstruction of the ACL using autografts harvested from the PT or hamstring tendons is recommended. Issues affecting patient outcome from the use of bone–PT–bone (BPTP) autografts include a persistent palpable defect in the tendon, anterior knee pain, arthrofibrosis, changes to the remaining PT, and PT adhesion to adjacent tissues (i.e. the fat pad) (Coupens et al., 1992; Svensson et al., 2005). The problems associated with hamstrings tendon autografts include slower healing due to a development of a soft tissue to bone interphase, less long-term stability of the knee (Freedman et al., 2003), significant hamstring muscle weakness (Marder 1991; Aune et al., 2001), as well as the increased prevalence of bone tunnel enlargement after reconstruction (Nebelung et al., 1998; Clatworthy et al., 1999; Jansson et al., 1999;

Copyright © 2008 Elsevier Inc. All rights reserved

Feller and Webster, 2003). Hence, functional tissue engineering (FTE) efforts are aiming to improve the sub-optimal properties of the healing MCL, as well as the issues related to ACL graft harvest and healing following reconstruction. With the knowledge gained, it is hoped that the same principles could be applied to aid the repair of other ligaments and tendons (Huang et al., 1993; Badylak et al., 1995; Hildebrand and Frank, 1998; Woo et al., 1999; Nakamura et al., 2000; Spindler et al., 2002; Shimomura et al., 2003).

Thus, FTE offers many attractive approaches to enhance ligament and tendon healing. The goal is not only to restore the normal histomorphological appearance, biochemistry, and mechanical properties of the healing ligament, but most importantly restore its normal joint function. In this chapter, we will review the properties of normal and healing ligaments and tendons, and discuss the current FTE methods, which include the use of growth factors, gene delivery, stem cell therapy, and the use of scaffolding as well as external mechanical stimuli, aimed at enhancing tendon and ligament healing. To conclude, new technologies and research avenues that have the potential to enhance treatment strategies for ligament and tendon injuries are suggested.

NORMAL LIGAMENTS AND TENDONS

Biology

Ligaments and tendons consist of collagen, proteoglycans, elastin, glycolipids, water (65% to 70% of the total weight), and cells. Both tissues are hypocellular with less than 5% of the total volume occupied by cells (Woo et al., 2000; Lo et al., 2002; Hildebrand et al., 2004) and are relatively hypovascular (Manske, 1988; Bray et al., 1996; Lo et al., 2002). The cells in these tissues, fibroblasts and tenocytes, are arranged in rows along the fibers of the extracellular matrix (ECM). Both cell types produce both the fibrillar and the non-fibrillar components of the ECM and may also reabsorb collagen fibers (Birk and Trelstad, 1984; Maffulli, 1999).

Roughly 70% to 80% of the dry weight of normal tendon or ligament is composed of type I collagen. Histologically, collagen displays a crimp pattern, which refers to a regular, wavy pattern of the matrix when viewed in unloaded conditions (Amiel et al., 1984; Woo et al., 2000; Thornton et al., 2002; Hildebrand et al., 2004). Both in ligaments and tendons, there is a bimodal distribution of collagen fibril diameters. One group of fibrils measures between 40 and 75 nm in diameter, the other is between 100 and 150 nm (Dyer and Enna, 1976; Eyden and Tzaphlidou, 2001; Goh et al., 2003). It has been proposed that a bimodal diameter distribution endows tendons and ligaments with better functional properties. The incorporation of a high fraction of small diameter fibrils would ensure a better interfibrillar binding by virtue of their higher surface/volume ratio, whereas the strength requirements would be satisfied by the inclusion of large diameter fibrils. A bimodal distribution would also improve fibril packing, the smaller fibrils wedging themselves in the spaces left among the larger ones (Ottani et al., 2001).

There are many other collagen types, including III, V, X, XI, and XII, which exist only in minor amounts in ligaments and tendons. The significance of some of these minor collagen types has recently been elucidated. For example, type V collagen is believed to exist in association with type I collagen and serves as a regulator of collagen fibril diameter (Linsenmayer et al., 1993; Birk and Mayne, 1997), whereas type III collagen is needed for wound healing (Liu et al., 1995). Our research has further identified that type XII collagen provides lubrication between collagen fibers (Niyibizi et al., 1995). Lastly, collagen types IX, X, and XI have been identified to exist with type II collagen at the fibrocartilaginous zone of the ligament–bone and tendon–bone interface (Niyibizi et al., 1996; Sagarriga Visconti et al., 1996; Fukuta et al., 1998). It is hypothesized that these collagens exist in this zone to minimize the stress concentrations when loads are transmitted from soft tissue to bone (Cooper and Misol, 1970; Matyas et al., 1995).

The ground substance constituents of tendons or ligaments make up only a small percentage of the total dry tissue weight but are nevertheless quite significant because of their ability to imbibe water. The water and proteoglycans provide lubrication and spacing which are crucial to the gliding function of fibers in the tissue matrix. Elastin, which is present in ligaments and tendons in a few percent by weight, allows the tissue to return to its prestretched length following physiological loading, but the detailed significance has yet to be elucidated. Collectively, these constituents serve to maintain fiber orientation and separation for optimal load distributions.

Although ligaments and tendons are morphologically similar to each other, there are some biochemical differences. When compared with tendons, ligaments are more metabolically active. They have more cellular nuclei, a higher DNA content, and greater amounts of reducible cross-links between collagen fibers (Amiel et al., 1984). Ligaments are composed primarily of water (65–70% of wet weight), cells and collagen (70–80% of dry weight). The most abundant collagen is type I collagen (90% dry weight). Type III collagen (8% dry weight) and type V collagen (12% dry weight) are other major components (Linsenmayer et al., 1993; Birk and Mayne, 1997). Collagen types II, IX, X, XI, and XII have also been found to be present (Niyibizi et al., 1996; Sagarriga Visconti et al., 1996; Fukuta et al., 1998; Woo et al., 2006). Tendons, on the other hand, generally contain less water (55% of wet weight), and slightly more type I collagen (85% of dry weight), along with much smaller amounts of other collagens, such as collagens type III, V, XII, and XIV (Goh et al., 2003).

Biomechanics

The major function of ligaments and tendons include maintaining the proper anatomic alignment of the skeleton and guiding joint motions. They accomplish this by transmitting forces along their longitudinal axis; hence their biomechanical properties are measured in uniaxial tension. They demonstrate non-linear behavior, which is governed by the recruitment of collagen. This allows ligaments to maintain normal joint laxity in response to low loads, and also stiffen dramatically in response to high loads, preventing excessive joint displacements. Ligaments and tendons also exhibit time- and history-dependent viscoelastic behavior which could be attributed to the complex interactions of tissue constituents such as collagen, proteoglycans, water, and ground substance (Fung et al., 1972; Woo et al., 1981b). Viscoelastic properties of ligaments and tendons are important and clinically relevant. For instance, during regular activities such as walking and jogging, tissues have the ability to soften over time. This phenomenon reduces the susceptibility to damage related to fatigue (Frank and Jackson, 1997).

However, following injury, ligaments and tendons generally fail to recover their normal mechanical and viscoelastic behaviors. Thus, abnormal joint kinematics result which can directly lead to excessive forces in surrounding tissues (e.g. articular cartilage). This can lead to further injury to other structures either through traumatic mechanisms or degeneration (i.e. osteoarthritis). As the ultimate goal of FTE is to restore the function of ligaments and tendons, and thereby the function of the injured joint, it is necessary to understand their normal mechanical behavior and contributions to joint function. Testing methodologies for this purpose include (1) functional testing, which involves determining the contribution of the ligament or tendon to joint kinematics (i.e. the *in situ* forces in response to external loading conditions), and (2) tensile testing, which provides an assessment of the structural properties of the bone–ligament–bone complex and mechanical properties of the tissue substance.

Tensile Testing

Tendons are generally long and can be tested in their isolated state using sinusoidal-shape or frozen grips to limit slippage. Isolated ligaments, on the other hand, are short in length, making it difficult to clamp them independently. Hence, a tensile test is generally conducted on the entire bone–ligament–bone complex (e.g. FMTC) with tissue insertion sites left anatomically intact. With cross-sectional area (CSA) measurements and the utilization of tissue markers to measure tissue strain, the structural properties of the bone–ligament–bone complex as well as mechanical properties of the ligament substance can be measured from a load to failure test (Woo et al., 1983; Lee and Woo, 1988).

Structural properties (Figure 35.1) of the bone–ligament–bone complex (i.e. a load-elongation curve) are generally described by four parameters including stiffness (slope of the linear portion of the load-elongation curve), ultimate load (maximum load at which the complex fails), ultimate elongation (elongation corresponding to the maximum load), and energy absorbed at failure (area under the curve to the maximum load). These data reflect behavior of the entire bone–ligament–bone complex which includes tissue size, orientation of collagen fibers to applied loads as well as the contribution of the bony insertions (Woo et al., 1983).

Measuring the mechanical properties (Figure 35.2) of the ligament substance (i.e. a stress–strain curve), on the other hand, requires knowledge of the CSA of the ligament, commonly measured using a laser micrometer system (Lee and Woo, 1988), and tissue strain, commonly measured using video techniques to track two or more reflective markers placed on the tissue midsubstance (Scheffler et al., 2001). Stress in the tissue is obtained by dividing load by the CSA and strain is obtained by computing change in marker distance during the test relative to their original distance. Parameters describing the mechanical properties of the ligaments and tendons (Figure 35.2) include tangent modulus (slope of the linear portion of the stress–strain curve), tensile strength (stress at failure), ultimate strain (strain corresponding with the tensile strength), and strain energy density (area under the stress–strain curve until failure). These data represent the quality of the tissue, irrespective of tissue size.

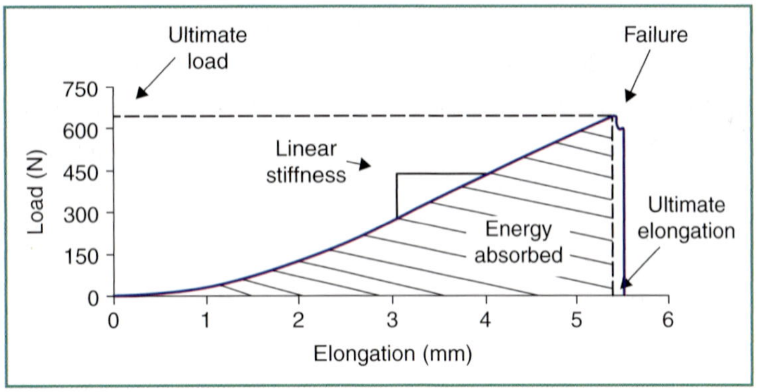

Figure 35.1 **A typical load-elongation curve representing the structural properties of the femur–anterior–medial bundle-tibia complex of the human anterior cruciate ligament.**

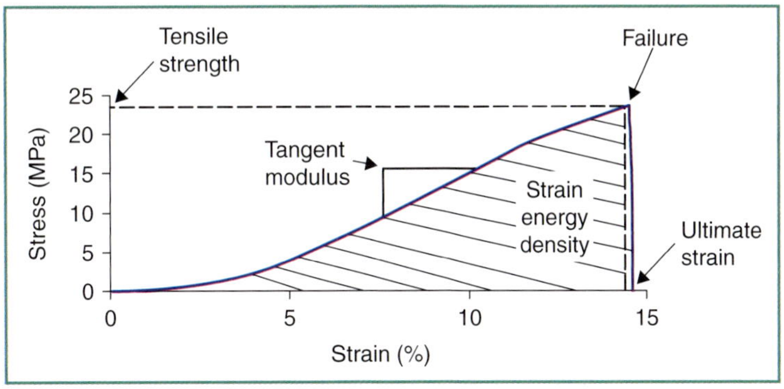

Figure 35.2 **A typical stress–strain curve representing the mechanical properties of the anterior–medial bundle of the human anterior cruciate ligament.**

following cellular events. Triggered by cytokines released within the clot, polymononuclear leukocytes and lymphocytes appear within several hours. These cells respond to autocrine and paracrine signals to expand the inflammatory response and recruit other types of cells to the wound (Frank et al., 1994).

The reparative phase takes place over the first couple of weeks to months following the injury. During this phase fibroblasts recruited to the injury site start forming healing tissue. Growth factors, including TGF-β and platelet-derived growth factor (PDGF) isoforms, are likely to be involved in modulating the healing environment in favor of effectively repairing the damaged ligament substance (Murphy et al., 1994). Meanwhile, increased neovascularization brings in circulating cells and nutrients to further enhance the healing process. The blood clot quickly turns into newly formed healing tissue which is composed of an aggregation of cells surrounded by a matrix whose histomorphological appearance and biochemical composition is different from that of the uninjured ligament. It is notably characterized by a homogenous distribution of smaller diameter collagen fibrils which is in stark contrast to the bimodal distribution of the normal ligament (Frank et al., 1992, 1997; Hart et al., 1992, 2000; Nakamura et al., 2000). Biochemically, it contains increased amount of proteoglycans, a higher ratio of type V to type I collagen, and a decrease in the number of mature collagen cross-links.

The proliferative phase gives way to the remodeling phase. It occurs from months to years after the injury and is characterized by decreasing cellularity, decreasing levels of collagen type III, and realigning of the matrix to respond better to the forces applied to the tissues. On the other hand, the diameter of collagen fibrils have been found to remain small and levels of collagen type V have been found to remain elevated for years after injury (Adachi and Hayashi, 1986; Birk et al., 1990; Frank et al., 1992, 1997; Hart et al., 1992, 2000; Marchant et al., 1996; Nakamura et al., 2000; Niyibizi et al., 2000; Birk, 2001). Interestingly, the type V collagen has been shown to play a central role in the regulation of the lateral growth of collagen fibrils. The elevated type V collagen could be involved in the lack of large collagen fibrils which in turn are associated with the inferior mechanical properties of healing tissue (Parry et al., 1978; Doillon et al., 1985).

MCL OF THE KNEE

The healing process of the MCL follows this general wound healing pathway and has been well studied. Thus, it serves as a good model for the histological, biochemical, and biomechanical events. The process of ligament healing is also greatly impacted by the selection of treatment (Clayton et al., 1968; Tipton et al., 1970; O'Donoghue et al., 1971; Woo et al., 1987b, 1990). Laboratory and clinical studies have shown mobilization is superior to immobilization (Woo et al., 1987a; Inoue et al., 1990; Weiss et al., 1991). Interestingly, non-operative repairs have an equivalent healing outcome to surgical repairs. A severe "mop-end" injury model in the rabbit, developed in our research center, that causes a midsubstance tear and damage at the insertion sites (Weiss et al., 1991), along with non-operative treatment with mobilization was used to compare to surgical repair with mobilization. After 52 weeks of healing, there were no significant differences in varus–valgus rotation of the knee, *in situ* force of the MCL, or tensile properties between repaired and non-repaired MCL (Weiss et al., 1991). Based on these and other studies clinical management has shifted from surgical repair with immobilization to non-operative management with early controlled range-of-motion exercises as soon as pain subsides (Reider et al., 1994; Indelicato, 1995).

While the MCL heals with non-operative treatment and the stiffness of the healing FMTC begins to approach normal levels by 52 weeks after injury, the CSA of the healed tissue increases with time, measuring as much as 2.5 times its normal size (Inoue et al., 1990). Thus, the mechanical properties of the healing MCL midsubstance remain consistently inferior to those of the normal ligament and do not change with time. Thus, the healing process involves making a larger quantity of lesser quality ligamentous tissue. Moreover, the rate of healing between the ligament midsubstance and the insertion sites is asynchronous with the insertion sites demonstrating a lower stiffness and strength resulting from injury as well as a lack of stress during the healing process.

There has been evidence that indicates that activity level influences the rate of healing (Abramowitch et al., 2003a). The goat model is a more clinically relevant model to study ligament healing than the rabbit due to its more robust activity level. In addition, its large size and the previously published success of ACL reconstructions using this animal (Ng et al., 1996), make it attractive to study more complex multiple-ligament injuries. Generally, these models display similar trends. However, it was noted that the stiffness and ultimate load of the healing goat FMTC are closer to control values at earlier time periods when compared to data from the rabbit model, suggesting that activity level may influence the healing response.

ACL Reconstruction

With the ultimate goal of ACL reconstruction being to restore knee function, the success of these procedures is dependent on a number of surgical, biomechanical, and biological factors. The most popular choice being autografts from the PT (i.e. bone-central third of the patellar tendon-bone (BPTB)), or hamstring tendons (i.e. semitendinosous plus gracilis tendons). Allografts, including the Achilles tendon, BPTB, and hamstring tendons, have seen limited use except in revision surgery or for multiple ligamentous injuries. BPTB grafts are generally considered the "gold standard" for ACL reconstruction because it facilitates better fixation and bone-to-bone healing inside the bone tunnels (Jones, 1970; Lambert, 1983; Noyes et al., 1984; Kurosaka et al., 1987; Aglietti et al., 1992; Cooper et al., 1993). However, the major drawback is that the open defect of the donor site remains visible and is not completely healed for months (Coupens et al., 1992; Rubinstein et al., 1994; Cerullo et al., 1995; Nixon et al., 1995). This contributes to a higher incidence of complications including donor site morbidity, patella baja, arthrofibrosis, adhesion to the fat pad, and patellofemoral pain (Paulos et al., 1987; Tibone and Antich, 1988; Sachs et al., 1989; Shelbourne et al., 1991; Breitfuss et al., 1996; Kartus et al., 1999).

Efforts have been made to examine the healing PT after harvest, using animal models, after removal of the central third. Studies have found a deterioration of PT structural properties with a concomitant increase in the CSA of the PT tissue (Kamps et al., 1994; Linder et al., 1994; Beynnon et al., 1995; Awad et al., 2003; Tohyama et al., 2003). Specifically in the rabbit model, the ultimate load of the entire BPTB complex decreased by 38% (Beynnon et al., 1995), while a CSA increase of 83–108% was observed at 12 weeks post-harvesting (Awad et al., 2003; Tohyama et al., 2003). For the central healing tissue, its tangent modulus and ultimate tensile strength were only measured to be 15% and 18% of controls after 26 weeks, respectively (Awad et al., 2003). The mechanical properties of the remaining PT tissues also deteriorated compared to sham controls after 24 weeks (Tohyama et al., 2003).

Following implantation, the autograft becomes inflamed and necrotic leading to a decrease in graft stiffness and strength (Tohyama and Yasuda, 2000). The graft undergoes revascularization and repopulation with fibroblasts followed by a remodeling period with restructure of collagen fibers and proteoglycans (Arnoczky et al., 1982). Further, bone-to-bone healing and tendon-to-bone healing within the femoral and tibial tunnels revealed that there was complete incorporation of the bone block by 6 weeks, but incomplete incorporation at the tendon–bone interface (Papageorgiou et al., 2001). For the latter, the failure mode of the femur–graft–tibia complex (FGTC) consistently occurred as a pull-out from the tibial tunnel. Over time, however, experimental animal studies show that the FGTC gradually shows improvement (Ballock et al., 1989; Butler et al., 1989; Gerich et al., 1996), but its structural properties failed to be restored to levels of the intact femur–ACL–tibia complex (FATC) even after 12 months (Clancy et al., 1981; Ballock et al., 1989; Butler et al., 1989). It is agreed that accelerating graft incorporation and healing may lead to an earlier return to sports and normal activities, and therefore has become a goal of tissue engineering efforts, as discussed later in this chapter.

In addition to the graft selection, other important surgical decisions include tunnel placement, graft tension, and fixation. There has been a substantial amount of research which is focused on the impact of these variables at time zero and following various periods of healing both in human and animal models (Yagi et al., 2002; Abramowitch et al., 2003b; Loh et al., 2003). All these parameters can lead to various degrees of graft tunnel motion which may impact graft integration and graft healing. Ultimately, these factors impact the post-operative rehabilitation and the return to normal activities and sports (Tsuda et al., 2002). Thus, it is important to take into account the large changes in properties of the graft after implantation.

Combined Ligamentous Injuries

Combined ACL/MCL injuries occur frequently and the best methods for treatment are still controversial. Some surgeons elect to surgically reconstruct the ACL without addressing the MCL, while others advocate reconstruction of the ACL with repair of the MCL. Regardless of the treatment modality, clinical and basic science studies continue to show that the outcome of this injury is worse than for an isolated MCL injury.

Our research center has elucidated the effects of ACL deficiency on the healing of the injured MCL using canine, rabbit, and goat model (Woo et al., 1990; Ohno et al., 1995; Yamaji et al., 1996; Abramowitch et al., 2003c), thus ACL reconstruction has been suggested. Further, repairing the MCL in combination with ACL reconstruction resulted in reduced valgus laxity and improved the structural properties of the FMTC in early stages, but the long-term effect became minimal (Ohno et al., 1995; Yamaji et al., 1996). These data further suggest that only reconstruction of the ACL is necessary for successful healing of the MCL after a combined ACL/MCL injury.

Recently, a larger animal model (i.e. the goat knee) was used in our research center to examine the function of the knee and quality of the healing MCL after a combined ACL/MCL injury treated with ACL reconstruction (Abramowitch et al., 2003c). These results confirmed that valgus rotation was twice that for an isolated MCL injury. Moreover, the structural properties of the FMTC and tangent modulus of the MCL substance are all substantially lower than that for the isolated MCL injury (Scheffler et al., 2001; Abramowitch et al., 2003c). These results demonstrate a clear need to enhance ligament healing after such a severe knee injury, requiring improved treatment strategies.

APPLICATION OF FTE

FTE emphasizes the importance of biomechanical considerations in the design and development of cell and matrix-based implants for soft and hard tissue repair. Musculoskeletal tissues, especially ligaments and tendons, are accustomed to being mechanically challenged, therefore tissue engineered constructs used to replace these tissues after injury or disease must meet these requirements. By combining the fields of molecular biology, biochemistry, and biomechanics, novel therapeutic approaches (e.g. growth factors, gene transfer/gene therapy, cell therapy, and biological scaffolds) offer the possibilities for improvement of the treatment of ligament and tendon injuries. The following will be a brief review of the current available approaches to enhance ligament and tendon healing.

Growth Factors

Growth factors can induce wide ranging effects on cell function including migration, proliferation, and protein synthesis. The application of exogenous growth factors is based on the premise that they can promote ligament regeneration that will lead to a biologically and biomechanically superior healed ligament substance. Many studies, *in vitro* and *in vivo*, have tried to define the role of growth factors in ligament and tendon healing and to determine appropriate strategies for the use of growth factors for these structures (Steenfos, 1994; Duffy et al., 1995; Panossian et al., 1997; Sciore et al., 1998).

In Vitro Studies

Cell culture or tissue explant methodologies involving the addition of exogenous growth factors have been the major study designs. Measured responses include cell proliferation, synthesis of ECM proteins such as collagen, proteoglycans, tissue remodeling enzymes, and cell migration or chemotaxis. In our research center, the effects of eight different growth factors on the MCL and ACL fibroblast culture were determined for proliferation and ECM production (Ohno et al., 1995; Deie et al., 1997; Marui et al., 1997; Scherping et al., 1997).

In terms of cell proliferation, PDGF-BB, epidermal growth factor (EGF), basic fibroblast growth factor (bFGF) have been found to have a significant effect on cell proliferation and caused greater proliferation in MCL fibroblasts versus ACL fibroblasts (Scherping et al., 1997). We found that the proliferation of MCL and ACL fibroblasts from skeletally immature rabbits increased by 7.6 times in response to EGF and 5.6 times in response to bFGF (Ohno et al., 1995). The same study in skeletally mature rabbits showed that insulin-like growth factor IGF and bFGF also had significant effects on fibroblast proliferation in both cell types, but the difference was less pronounced (Scherping et al., 1997). The biological role of TGF-β1 in ligament healing has also been recently addressed. Studies in our research center on fibroblast proliferation in skeletally immature rabbits demonstrated that TGF-β1 stimulated proliferation of MCL fibroblasts 1.3–1.4 times greater than in ACL fibroblasts (Ohno et al., 1995). Subsequent studies on skeletally mature rabbits showed little effect of TGF-β1 on cell proliferation for either fibroblast type. Comparison of these results suggests that age has a significant effect on the ability of growth factors to stimulate fibroblast proliferation (Scherping et al., 1997). The effect of TGF-β1 on canine ACL fibroblast proliferation was shown to be dose-dependent because smaller doses acted synergistically with PDGF whereas higher concentrations inhibited the stimulatory effect of PDGF (Desrosiers et al., 1995). These findings show the complex interactions of growth factors to enhance proliferation of fibroblasts.

In terms of *in vitro* protein synthesis in MCL and ACL fibroblasts, collagen synthesis increased 160% over controls in both MCL and ACL fibroblasts treated with TGF-β1, and the majority of this increase was for type I collagen (Marui et al., 1997). The relative increase in protein production was similar for both cell types, but the absolute increase in protein synthesis was twice as much for MCL fibroblasts as for ACL fibroblasts. These data suggest that TGF-β1 may improve ligament healing by increasing matrix synthesis during the proliferative and remodeling phases (Marui et al., 1997). Similar results have also been found by other investigators (Desrosiers et al., 1995). These studies illustrate the ability of TGF-β1 to increase the production of ECM by fibroblasts *in vitro*.

In vitro models, however, are limited in the extent that they cannot reproduce the complex interplay of signals affected by growth factors in the intricate process of ligament or tendon healing. Differences in the effects of different growth factors on cell proliferation and matrix synthesis suggest that wound healing depends on a highly integrated biochemical network of cell signaling events with intrinsic stimulatory and inhibitory feedback loops. Thus, in addition to providing a better physiological model, *in vivo* studies are critical to defining the interaction of biology and biomechanics and the degree to which healed ligament or tendon substance restores the structural and mechanical properties of the native tissue.

In Vivo Studies

In vitro studies showed that EGF and PDGF-BB have the greatest effect on ligament fibroblast proliferation, whereas TGF-β1 superiorly promotes ECM production. These growth factors were then applied *in vivo* at different dosages, in isolation and in combination, for an MCL injury in the rabbit model. It was found that a higher dose of PDGF-BB improved the structural properties of the FMTC compared to a lower dose of PDGF-BB, demonstrating that the effects of PDGF-BB were dose-dependent (Woo et al., 1998). However, the mechanical properties of the ligament substance remained unchanged from untreated controls demonstrating that the improved structural properties resulted from a larger quantity of tissue instead of tissue with improved quality. In contrast, the combination of EGF or PDGF-BB plus TGF-β1 did not lead to additional improvements in MCL healing compared with PDGF-BB alone.

Other investigators have also found that higher doses of PDGF demonstrated a plateau effect in improving the structural properties of the healed ligament. Furthermore, administration of PDGF for more than 24 h after the injury markedly decreased the efficacy of growth factors in improving MCL healing (Batten et al., 1996). Additionally, administration of TGF-β2 to the healing rabbit MCL improved the mechanical stiffness, but not the load at failure, of the ligament scar (Spindler et al., 2003). These data show that *in vivo* application of growth factors is indeed more complex than *in vitro* studies.

One possible approach to improve the *in vivo* application might be to combine growth factors with gene transfer technology. Adenoviral bone morphogenetic protein-2 (AdBMP-2) delivered to the bone–tendon interface using a gene transfer technique has been shown to improve the integration of semitendinosous tendon grafts in rabbits (Martinek et al., 2002). The stiffness (29.0 ± 7.1 N/mm versus 16.7 ± 8.3 N/mm) and the ultimate load (108.8 ± 50.8 N versus 45.0 ± 18.0) were also significantly increased in the specimens with AdBMP-2 compared to untreated controls. Hence, the enhancement of tendon to bone healing is a promising approach to accelerate the patient's return to activity.

Based on these studies, an optimal therapy of introducing growth factors to injury sites is still an open question. While promising, timing of application, mode of delivery, and dosage, remain as major hurdles that need to be crossed before success can be achieved *in vivo*.

Gene Therapy

Gene therapy is a potential approach to improve ligament and tendon healing. Foreign nucleic acids, gene transfer, can be introduced into cells to alter protein synthesis or induce the expression of therapeutic proteins. Modern gene therapy relies on mammalian viruses and cationic liposomes as delivery vectors, and both have been developed to deliver genes into host tissue via the direct and indirect methods. Direct gene transfer involves *in vivo* injection of the delivery vector into the host tissue. Indirect gene transfer involves *in vitro* transduction of host cells with the desired gene, and subsequent replantation of these cultured cells *in vivo*. Studies have shown that PT fibroblasts can be transduced with the LacZ marker gene both directly using an adenovirus liposomal vector, and directly using a retrovirus, with the expression of the transferred genes persisting for 6 weeks following the application (Gerich et al., 1996). PT fibroblasts staining positively for β-galactosidase were subsequently found to migrate and incorporate into the tendon tissue following injection. In our research center, we sought to determine if genes could be transduced into MCL and ACL fibroblasts and if ligament injury affected gene transfer and expression (Hildebrand et al., 1999). When both the direct and indirect methods were employed using adenovirus and BAG retrovirus, respectively, it was found that both techniques resulted in expression of the LacZ marker gene by fibroblasts from intact as well as injured ligaments. Gene expression lasted longer (6 weeks) with the direct method as compared to the indirect technique (3 weeks). Fibroblasts from injured ligaments showed transduction both in the wound site and in the ligament substance as well. There was no difference in the duration of gene expression by fibroblasts from intact and injured ligaments suggesting that injury does not affect gene transfer or expression (Hildebrand et al., 1999).

Newer techniques for gene transfer have recently been reported. Gene transfer using liposomal vectors may reduce the adverse immune responses seen when using viral vectors. Antisense gene therapy involving blocking the transcription or translation of specific genes which may be excessively expressed within healing tissue has been proposed. By the binding of antisense oligodeoxynucleotides (ODN) to target DNA investigators have performed direct transfer of an HVJ–liposome complex containing a labeled ODN for the protein decorin, which has been shown to inhibit type I collagen fibril formation in prior *in vitro* studies (Nakamura et al., 2000).

Histological analysis was performed 24 h after direct injection into rabbit MCL specimens which were 2 weeks post-injury. Quantitative analysis revealed a transduction efficiency of 62% at 1 day and 23% at 7 days. Significant suppression of decorin mRNA expression was seen at both 2 days (42.7%) and 2 weeks (60.3%) (Nakamura et al., 2000).

In our research center, we have evaluated the efficacy of utilizing ODNs to regulate the overproduction of collagens III and V (Shimomura et al., 2003; Woo et al., 2004). Normal human patellar tendon fibroblasts (HPTFs) were transfected with antisense collagen III or V ODNs by mixing with lipofectamine. The uptake of the ODNs was detected as early as 1 h and as late as 3 days after delivery. The relative expression of collagen V mRNA was reduced to 67.8 $\pm$ 5.1% of missense levels. Also, preliminary reverse transcription-polymerase chain reaction (RT-PCR) results showed that the inhibitory effects of the collagen III antisense ODNs were most dominant at 1 day as the type III collagen mRNA level was 38.9 $\pm$ 19.6% of missense controls. At 3 and 7 days, differences could not be observed. These results suggested that antisense gene therapy can indeed be a potential FTE approach to enhance the quality of ligaments and tendons.

Despite these promising results, several obstacles currently impede the practical implementation of gene transfer as a biological intervention in ligament healing. The immune reaction against these antigens decreases the expression of the introduced gene (Tripathy et al., 1996). In addition, retroviral infection of fibroblasts often leads to shut-off of the promoter region, which adversely affects expression of the incorporated gene (Krall et al., 1994). Thus, delivering the ODNs to the appropriate target and reproducibility of the results remains a great challenge.

Newer strategies in the evolving field of gene transfer include the search for more effective and less immunogenic vectors, modification of promoters to ensure gene expression after incorporation, and temporary and self-limiting gene expression regulation tailored to the changing environment of the healing ligament. As the complex steps involved in gene expression and regulation are further elucidated, the potential therapeutic efficacy of gene transfer is likely to enjoy practical application.

Cell Therapy

Cell therapy is another potential strategy to enhance ligament and tendon healing. Studies have focused on the application of mesenchymal stem cells (MSCs), bone marrow-derived cells (BMDCs) and synovial tissue-derived fibroblasts into the healing site (Young et al., 1998; Watanabe et al., 2002). BMDCs have been shown to play an important role in wound healing (Badiavas et al., 2003; Galiano et al., 2004; Mathews et al., 2004) and can be obtained in high numbers with relative ease (Awad et al., 2003; Juncosa-Melvin et al., 2005). In one study, autologous marrow-derived progenitor cells were seeded on a collagen gel, and subsequently contracted onto a pretensioned suture (Young et al., 1998). The resulting tissue prosthesis was then implanted into the rabbit Achilles tendon gap defect. Significantly greater structural and mechanical properties were seen after the implantation. The treated tissues had a significantly larger CSA, and their collagen fibers appeared to be better aligned than those in the matched controls.

Recently, new methods geared toward PT healing have tried to fill the central third PT defect with collagen gels filled with BMDCs, in which BMDCs were expanded *in vitro*. These MSC–collagen composites were implanted into full thickness, full length, central defects created in the PTs of rabbits. The healing PTs treated with MSC–collagen gels were one-fourth of the maximum stress of the normal central portion of the PT. The modulus and maximum stress of the repair tissues grafted with MSC–collagen gels increased at significantly faster rates than did natural repairs over time. Thus, overall improved mechanical properties were seen when compared to non-treated defects (Awad et al., 1999, 2003; Butler, 2005; Juncosa-Melvin et al., 2005). This particular cell therapy is attractive because the use of autogenous cells would minimize the immune response at the injury site.

Alternatively, fibroblasts, myoblasts, and bone marrow cells have also been transplanted into injured ligaments following the induction of marker genes or stimulation by growth factors *in vitro*. These results show great potential *in vivo* (Day et al., 1997; Caplan et al., 1998; Hildebrand et al., 1999; Watanabe et al., 2002). However, issues remain on cell therapy. MSCs from bone marrow are relatively few in number and their numbers decrease further after transplantation. Thus, it is essential to develop *in vitro* techniques to expand MSCs without altering their differentiation potential. Further, scaffolds may help to prevent the loss of these cells and might further enhance the effect already provided by this strategy.

Scaffolding

An ideal scaffold should provide a suitable mechanical and biological environment for cells to migrate into, guide the formation of the newly synthesized ECM, and then slowly degrade such that the new matrix begins to bear the mechanical loads. Both synthetic and naturally occurring biological scaffolds are commonly utilized in FTE of ligaments and tendons (Aragona et al., 1983; Dunn et al., 1992; Badylak et al., 1999; Bourke et al., 2004). The major advantage of using synthetic polymers as scaffolds is their ease of fabrication and reproducibility. A structure can be created that mimics the structure of a ligament or tendon, and appropriate proteins (e.g. growth factors) can be incorporated into the scaffold during manufacturing. However, performance of current synthetic grafts *in vivo* have been disappointing (Bellincampi et al., 1998; Guidoin et al., 2000); thus, current work has focused on the use of synthetic scaffolds seeded with fibroblasts (Cao et al., 1994; Lin et al., 1999), or alternatively, a number of naturally occurring biological scaffolds, including bovine pericardium (Integra Life Science), human dermal collagen (Alloderm), and porcine small intestinal submucosa (SIS) have been used. The SIS has shown most promising results in enhancing healing of both ligaments and tendons (Badylak et al., 1995, 1999; Dejardin et al., 1999, 2001; Musahl et al., 2004; Liang et al., 2006), as it possesses a structural hierarchy that is naturally arranged, and it is mostly comprised of collagen type I. Further, 40% of the SIS degrades within 1 month *in vivo* (Record et al., 2001) and its byproducts have been shown to be chemoattractants for cells (including BMDCs) (Badylak et al., 2001; Li et al., 2004; Zantop et al., 2005). Moreover, it contains many bioactive agents (growth factors, fibronectin, and so on) (Voytik-Harbin et al., 1997; McPherson et al., 2000; Hodde et al., 2002) and causes a limited inflammatory reaction (Allman et al., 2002).

Based on these positive findings, our research center conducted multidisciplinary studies to determine the effect of SIS treatment on MCL in the short- and long-term (12 weeks and 26 weeks). The mechanistic hypothesis is that SIS would act as a guidance of neo-ligament tissue formation, limiting the cross-sectional growth of the healing tissue, thereby increasing the mechanical demand. The tissue would respond by decreasing the production of collagen type V which in turn would lead to an increase in collagen fibril diameters and improved collagen orientation in the healing ligament. Finally, these changes would result in an improvement in the mechanical properties of SIS-treated MCLs (Figure 35.5).

It was observed that the histomorphological appearance and biochemical composition of the healing ligament were indeed closer to the normal ligament when compared to the non-treated ligament. Biochemical analysis revealed that collagen type V/I ratio decreased with SIS treatment, while TEM (transmission electron microscopy) showed a heterogeneous distribution of large collagen fibril diameters (Figure 35.6), and ultimately mechanical properties of the healing MCL were improved. Most importantly, the effects of SIS treatment persisted even up to 26 weeks (Liang et al., 2006). In addition, CSA in the SIS treated group decreased by 28% and tangent modulus increased by 33% compared to non-treated group, also stress at failure was 49% higher than non-treated. These findings demonstrate that a layer of SIS can act as a guide for neoligament formation by inducing better organization and limiting cross-sectional growth of the healing ligament. This, in turn, requires the mechanical properties of the SIS treated ligament to improve through the formation of larger collagen fibrils resulting from a lower collagen type V/I ratio.

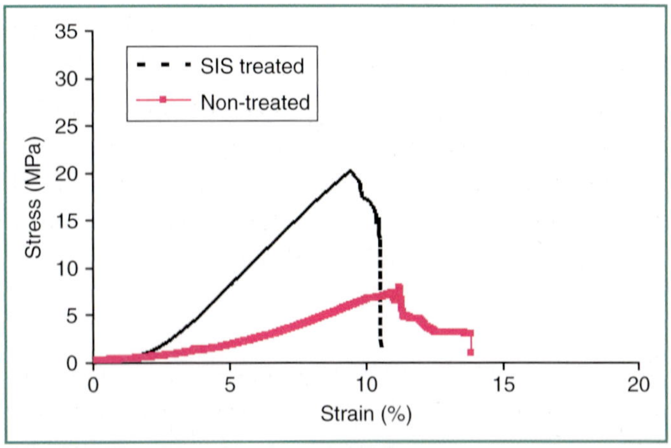

Figure 35.5 Typical stress–strain curves for SIS-treated and non-treated groups at 12 weeks post-injury (permission granted by Interscience-Wiley).

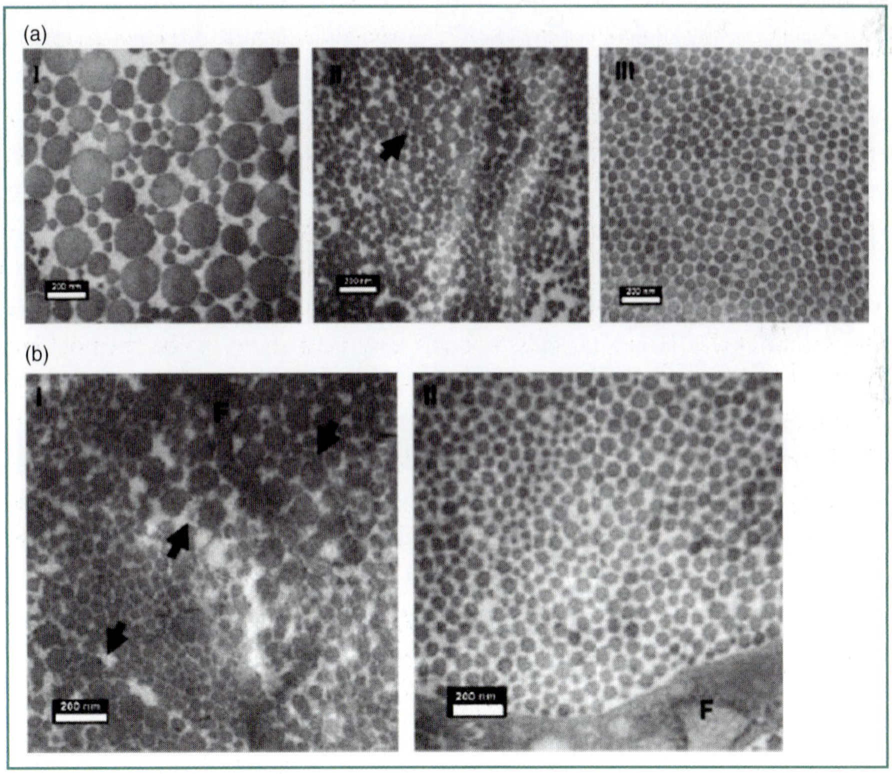

Figure 35.6 Transmission electron micrographs (×70,000) of collagen fibrils in (a) sham operated MCL (I), SIS-treatedMCL (II) and non-treatedMCL(III) at 26 weeks post-injury. The arrow indicates the appearance of large fibrils between cells in the SIS-treated MCL. (b) The TEM appearance of both large and small fibrils (heterogeneity) in the pericellular area in the SIS-treated MCL (I) and non-treated MCL (II). The arrow indicates the large fibrils surrounding a cell process. F indicates fibroblast (permission granted by Interscience-Wiley).

We have extended the use of SIS on PT healing after the central third was harvested for ACL reconstruction (Figure 35.7). Since SIS has a preferred collagen alignment (Sacks and Gloeckner, 1999), it has the potential of contact guidance and of promoting cells to produce a newly more aligned deposited matrix (Brunette, 1986; Clark et al., 1990; Chen et al., 1998; Walboomers et al., 1999); and as a result, a concomitant set of improved

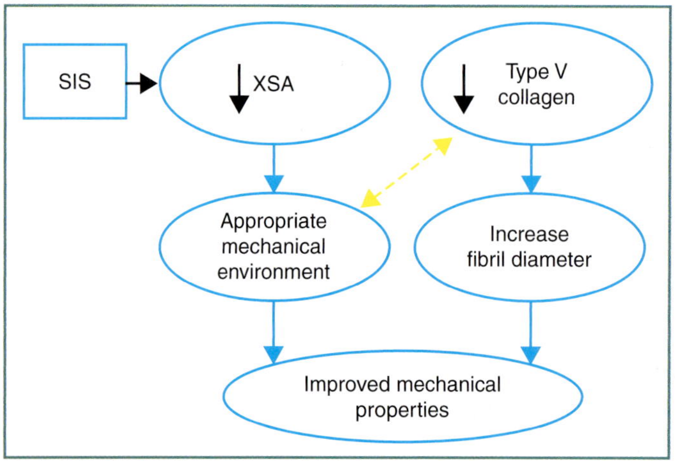

Figure 35.7 **Schematic of possible mechanism of action of SIS on MCL healing response.**

mechanical and viscoelastic properties. Additionally, the chemoattractant degradation products (Li et al., 2004) and bioactive agents of SIS could enhance the rate of healing. Finally, the SIS scaffold could also form a barrier between the healing PT and the underlying fat pad, limiting adhesion formation to permit motion between them to take place. The maintenance of stress and motion would help the homeostasis of the remaining PT (Woo et al., 1981c, 1982). Together, these effects will limit problems associated with poor healing, such as excessive hypertrophy of the remaining PT, and limiting the deterioration of its mechanical properties.

In a preliminary study, the effects of SIS on healing of a central third defect (3 mm width) of the rabbit PT was performed. By 12 weeks, the SIS treated group contained a large number of spindle shaped cells with more organized collagen matrix, while the non-treated group had a sparse distribution of cells with only patches of collagen. After the healing PT tissue was dissected, the CSA was 61% greater in the SIS-treated group compared to those in the non-treated group ($5.0 \pm 2.0\,mm^2$ versus $3.1 \pm 1.2\,mm^2$, respectively). SIS-treatment also showed higher stiffness (33.9 ± 14.0 versus 24.3 ± 14.9 N/mm, or 38%) and ultimate load (67.7 ± 25.8 versus 43.8 ± 27.4 N, or 58%) compared to non-treatment. This study demonstrated that SIS-treatment shows the potential to increase the quantity of healing PT tissue and structural properties of the healing central BPTB complex after a surgically created central third PT defect. Thus, the results of morphology, histology, and structural properties are very encouraging for further investigation of this application.

Recent studies have also shown the feasibility of enhancing ACL graft integration following reconstruction using a tri-phasic scaffold (Spalazzi et al., 2006). Fibroblast and osteoblasts were seeded on a section of the scaffold that mimicked the native environment of the ligament insertion and bone, respectively. It was found that the specific cell types populated and thrived in their respective phase and also each migrated into the middle phase, while each cell type expressed the appropriate type of genes for its particular matrix. Collectively these approaches demonstrate that scaffolds have many potential applications for improving the treatment for injured ligaments and tendons.

Mechanical Factors

Progress in FTE also has included the elucidation of the importance of mechanical stimuli on cells and on tissue development and remodeling (Huang et al., 1993; Banes et al., 1995; Eastwood et al., 1998; Hsieh et al., 2000; Altman et al., 2002; Wang et al., 2003b). Cyclic stretching of cells from ligaments and tendons *in vitro* has been shown to cause increases in collagen synthesis (Desrosiers et al., 1995; Hsieh et al., 2000) and changes in intra-cellular processes (i.e. different regulation of metabolic and inflammatory genes and calcium signaling) (Banes et al., 1995, 1999; Archambault et al., 2002; Ralphs et al., 2002; Wang et al., 2003a; Yang et al., 2004).

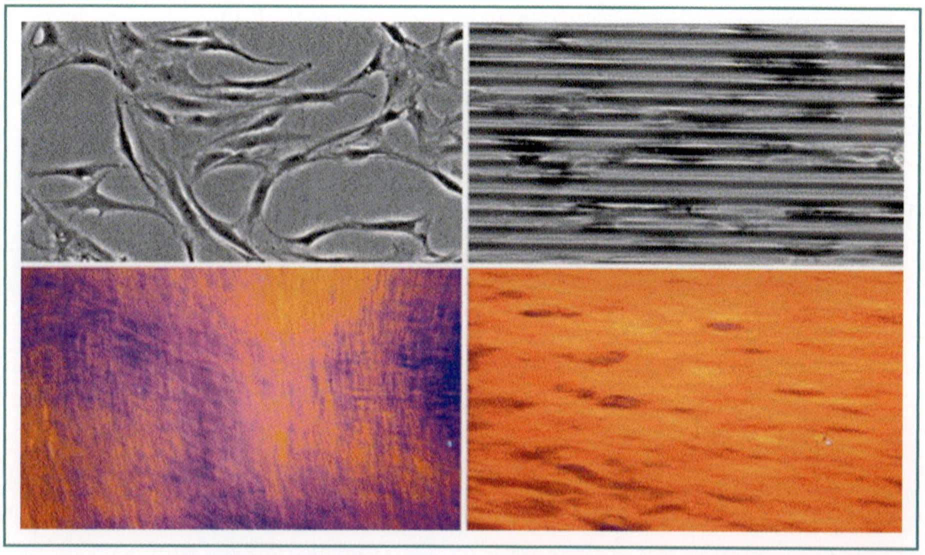

Figure 35.8 **Randomly aligned cells cultured on a smooth dish (upper left). Aligned cellsculture on dish etched with microgroove (upper right). Randomly aligned matrix produced by cells cultured on a smooth dish (lower left). Aligned matrix produced by cells culture on dish etched with microgrooves (lower right) (permission granted by Wang et al., 2003a).**

Interestingly, fibroblasts alignment can be a result of the external mechanical environment. When fibroblasts are grown in microgrooved silicone surfaces instead of smooth culture surfaces, they become elongated and aligned within the microgrooves through contact guidance. Most importantly, the ECM that the cells produce is also aligned along the microgroove direction (Figure 35.8) (Wang et al., 2003a).

Cells also align along the direction of the maximum principal strain in collagen gels (Eastwood et al., 1998). Similarly, *in vitro* studies have shown that multidimensional mechanical strains applied to BMDCs embedded in a collagen gel upregulated the gene expression of collagen types I and III and tenascin-C, which are typically expressed in fibroblasts (Altman et al., 2002).

Recently, our research center developed a uni-axial stretching system to study the effects of the mechanical stimuli on cells seeded on the bio-scaffold (SIS). In order to first understand the effect of elongating the SIS on the alignment of cells, the scaffold was stretched for 24 h to 15% of its original length and was then seeded with cells. After 5 days, collagen fibers of the scaffold were more aligned compared to non-stretched controls. In addition, the cells seeded on the scaffold demonstrated a preferred alignment along the direction of stretch. Preliminary results have also been obtained for continuous cyclic stretching (15% at 1 Hz for 4 h/day for 5 days) of SIS. Again, it was observed that the collagen fiber organization of the SIS improved to a more aligned state when the SIS was both seeded and stretched, and the cells were also aligned along the stretching direction. Gene expression analysis is underway to determine the differences in matrix protein expression between the cyclic group and the constant elongated group. Since tendon and ligament fibroblasts are aligned with collagen fibrils *in vivo,* it is hoped that mechanical stimuli can align cells within the scaffold and produce a better organized collagen matrix that may further enhance the healing response when implanted *in vivo*.

SUMMARY AND FUTURE DIRECTIONS

In this review, the biomechanical and biological problems facing healing and repair of ligament and tendon injuries were discussed. There have already been tremendous improvements to clinical treatment paradigms based on studies that have established a fundamental understanding of healing following ligament or tendon injury and the benefits of controlled mobilization. Nevertheless, many issues still remain.

For ligaments and tendons that display healing potential after injury, the major challenges are recovering their normal ultrastructural appearance, biochemical composition, and mechanical properties. Specifically, increasing the fibril diameters of healing tissues by limiting the production of type V collagen and decorin as well as improving the alignment of healing tissue by guiding the organization of newly produced matrix are important steps to be taken. By manipulating the healing response at the molecular and cellular level and guiding tissue formation, the following FTE approaches may offer the potential to restore the properties of healing tissue to normal levels.

We are particularly interested in bioscaffolds, such as the porcine SIS. When applied to a healing ligament or tendon *in vivo*, it serves as a substrate that provides contact guidance for cells to form more aligned collagen fibers with a concomitant improvement in mechanical and viscoelastic properties when compared to non-treated controls. Further, the chemoattractant degradation products and bioactive agents of SIS could enhance the rate of healing (Li et al., 2004), allowing better maintenance of stress and motion dependent homeostasis. More excitingly, the SIS can be modified *in vitro* by seeding BMDCs on the scaffold and applying cyclic stretching in order to increase its alignment. Hence, when applied *in vivo*, the tissue engineered scaffold could serve to accelerate the initiation of the healing process by improving the production and orientation of collagen that ultimately will help to make a better neo-ligament or tendon.

On the other hand, for ligaments and tendons that do not heal following injury and require surgical reconstruction using replacement grafts (e.g. ACL reconstruction), the major challenge is to promote a remodeling response such that the graft maintains sufficient stiffness and strength to provide functional stability of the joint. Most importantly, enhancing the rate of integration of tendon–bone interfaces during early graft incorporation that may permit an earlier and more aggressive post-operative rehabilitation (Chen et al., 2002). These complex issues may require a combination of approaches including gene and cell therapies as well as biological scaffolds. Indeed, grafts treated with AdBMP-2 has shown some potential (Martinek et al., 2002) in both canine and rabbit models. Additionally, other biological tissues such as periosteum have also been used to enhance the interface between tendon and bone with some success (Chen et al., 2002). All these results suggest an exciting potential for clinical application.

Indeed, FTE has generated many exciting developments. Further the development of stem cell-based therapies presents both opportunities and challenges. To translate the knowledge gained about a particular gene, protein, or cell to a clinical application will require that expertise from many disciplines to work in a seamless fashion. One of the roles of biomedical engineers within this framework would be to help link interactions of the functions of molecules to cells, cells to tissues, tissues to organs, and organs to body. When biologists, biomedical engineers, clinicians, as well as experts from other disciplines, work together this would result in better therapies that lead to the injured ligaments and tendons to heal with properties closer to those of normal ligaments and tendons. Efforts of such a team-based approach on the new developments of FTE will bring a bright future to the outcome of healing of ligaments and tendon injuries.

ACKNOWLEDGMENT

The financial support provided by the National Institute of Health Grants AR41820 and AR39683.

REFERENCES

Abramowitch, S.D., Papageorgiou, C.D., Debski, R.E., Clineff, T.D. and Woo, S.L. (2003a). A biomechanical and histological evaluation of the structure and function of the healing medial collateral ligament in a goat model. *Knee Surg. Sports Traumatol. Arthrosc.* **11**: 155–162.

Abramowitch, S.D., Papageorgiou, C.D., Withrow, J.D., Gilbert, T.W. and Woo, S.L. (2003b). The effect of initial graft tension on the biomechanical properties of a healing ACL replacement graft: a study in goats. *J. Orthop. Res.* **21**: 708–715.

Abramowitch, S.D., Yagi, M., Tsuda,E. and Woo, S.L. (2003c). The healing medial collateral ligament following a combined anterior cruciate and medial collateral ligament injury – a biomechanical study in a goat model. *J. Orthop. Res.* **21**: 1124–1130.

Adachi, E. and Hayashi, T. (1986). *In vitro* formation of hybrid fibrils of type V collagen and type I collagen. Limited growth of type I collagen into thick fibrils by type V collagen. *Connect. Tissue Res.* **14**: 257–266.

Aglietti, P., Buzzi, R., D'Andria, S. and Zaccherotti, G. (1992). Arthroscopic anterior cruciate ligament reconstruction with patellar tendon. *Arthroscopy* **8**: 510–516.

Allman, A.J., McPherson, T.B., Merrill, L.C., Badylak, S.F. and Metzger, D.W. (2002). The Th 2-restricted immune response to xenogeneic small intestinal submucosa does not influence systemic protective immunity to viral and bacterial pathogens. *Tissue Eng.* **8**: 53–62.

Altman, G.H., Horan, R.L., Martin, I., Farhadi, J., Stark, P.R., Volloch, V., Richmond, J.C., Vunjak-Novakovic, G. and Kaplan, D.L. (2002). Cell differentiation by mechanical stress. *FASEB J.* **16**: 270–272.

Amiel, D., Frank, C., Harwood, F., Fronek, J. and Akeson, W. (1984). Tendons and ligaments: a morphological and biochemical comparison. *J. Orthop. Res.* **1**: 257–265.

Aragona, J., Parsons, J.R., Alexander, H. and Weiss, A.B. (1983). Medial collateral ligament replacement with a partially absorbable tissue scaffold. *Am. J. Sports Med.* **11**: 228–233.

Archambault, J., Tsuzaki, M., Herzog, W. and Banes, A.J. (2002). Stretch and interleukin-1beta induce matrix metalloproteinases in rabbit tendon cells *in vitro*. *J. Orthop. Res.* **20**: 36–39.

Arnoczky, S.P., Tarvin, G.B. and Marshall, J.L. (1982). Anterior cruciate ligament replacement using patellar tendon. An evaluation of graft revascularization in the dog. *J. Bone Joint Surg. Am.* **64**: 217–224.

Aune, A.K., Holm, I., Risberg, M.A., Jensen, H.K. and Steen, H. (2001). Four-strand hamstring tendon autograft compared with patellar tendon–bone autograft for anterior cruciate ligament reconstruction. A randomized study with two-year follow-up. *Am. J. Sports Med.* **29**: 722–728.

Awad, H.A., Butler, D.L., Boivin, G.P., Smith, F.N., Malaviya, P., Huibregtse, B. and Caplan, A.I. (1999). Autologous mesenchymal stem cell-mediated repair of tendon. *Tissue Eng.* **5**: 267–277.

Awad, H.A., Boivin, G.P., Dressler, M.R., Smith, F.N., Young, R.G. and Butler, D.L. (2003). Repair of patellar tendon injuries using a cell–collagen composite. *J. Orthop. Res.* **21**: 420–431.

Badiavas, E.V., Abedi, M., Butmarc, J., Falanga, V. and Quesenberry, P. (2003). Participation of bone marrow derived cells in cutaneous wound healing. *J. Cell Physiol.* **196**: 245–250.

Badylak, S., Arnoczky, S., Plouhar, P., Haut, R., Mendenhall, V., Clarke, R. and Horvath, C. (1999). Naturally occurring extracellular matrix as a scaffold for musculoskeletal repair. *Clin. Orthop. Relat. Res.* S333–S343.

Badylak, S.F., Tullius, R., Kokini, K., Shelbourne, K.D., Klootwyk, T., Voytik, S.L., Kraine, M.R. and Simmons, C. (1995). The use of xenogeneic small intestinal submucosa as a biomaterial for Achilles tendon repair in a dog model. *J. Biomed. Mater. Res.* **29**: 977–985.

Badylak, S.F., Park, K., Peppas, N., McCabe, G. and Yoder, M. (2001). Marrow-derived cells populate scaffolds composed of xenogeneic extracellular matrix. *Exp. Hematol.* **29**: 1310–1318.

Ballock, R.T., Woo, S.L., Lyon, R.M., Hollis, J.M. and Akeson, W.H. (1989). Use of patellar tendon autograft for anterior cruciate ligament reconstruction in the rabbit: a long-term histologic and biomechanical study. *J. Orthop. Res.* **7**: 474–485.

Banes, A.J., Tsuzaki, M., Hu, P., Brigman, B., Brown, T., Almekinders, L., Lawrence, W.T. and Fischer, T. (1995). PDGF-BB, IGF-I and mechanical load stimulate DNA synthesis in avian tendon fibroblasts *in vitro*. *J. Biomech.* **28**: 1505–1513.

Banes, A.J., Weinhold, P., Yang, X., Tsuzaki, M., Bynum, D., Bottlang, M. and Brown, T. (1999). Gap junctions regulate responses of tendon cells *ex vivo* to mechanical loading. *Clin. Orthop. Relat. Res.* S356–S370.

Batten, M.L., Hansen, J.C. and Dahners, L.E. (1996). Influence of dosage and timing of application of platelet-derived growth factor on early healing of the rat medial collateral ligament. *J. Orthop. Res.* **14**: 736–741.

Beaty, J. (1999). Knee and leg: soft tissue trauma. OKU orthopaedic knowledge update. (Arendt, E.A.): xix, 442. American Academy of Orthopaedic Surgeons. Rosemont, IL.

Bellincampi, L.D., Closkey, R.F., Prasad, R., Zawadsky, J.P. and Dunn, M.G. (1998). Viability of fibroblast-seeded ligament analogs after autogenous implantation. *J. Orthop. Res.* **16**: 414–420.

Musahl, V., Abramowitch, S.D., Gilbert, T.W., Tsuda, E., Wang, J.H., Badylak, S.F. and Woo, S.L. (2004). The use of porcine small intestinal submucosa to enhance the healing of the medial collateral ligament – a functional tissue engineering study in rabbits. *J. Orthop. Res.* **22**: 214–220.

Nakamura, N., Hart, D.A., Boorman, R.S., Kaneda, Y., Shrive, N.G., Marchuk, L.L., Shino, K., Ochi, T. and Frank, C.B. (2000). Decorin antisense gene therapy improves functional healing of early rabbit ligament scar with enhanced collagen fibrillogenesis *in vivo. J. Orthop. Res.* **18**: 517–523.

Nebelung, W., Becker, R., Merkel, M. and Ropke, M. (1998). Bone tunnel enlargement after anterior cruciate ligament reconstruction with semitendinosus tendon using endobutton fixation on the femoral side. *Arthroscopy* **14**: 810–815.

Ng, G.Y., Oakes, B.W., Deacon, O.W., McLean, I.D. and Eyre, D.R. (1996). Long-term study of the biochemistry and biomechanics of anterior cruciate ligament-patellar tendon autografts in goats. *J. Orthop. Res.* **14**: 851–856.

Nixon, R.G., SeGall, G.K., Sax, S.L., Cain, T.E. and Tullos, H.S. (1995). Reconstitution of the patellar tendon donor site after graft harvest. *Clin. Orthop. Relat. Res.* **226**: 162–171.

Niyibizi, C., Visconti, C.S., Kavalkovich, K. and Woo, S.L. (1995). Collagens in an adult bovine medial collateral ligament: immunofluorescence localization by confocal microscopy reveals that type XIV collagen predominates at the ligament–bone junction. *Matrix Biol.* **14**: 743–751.

Niyibizi, C., Sagarrigo Visconti, C., Gibson, G. and Kavalkovich, K. (1996). Identification and immunolocalization of type X collagen at the ligament–bone interface. *Biochem. Biophys. Res. Commun.* **222**: 584–589.

Niyibizi, C., Kavalkovich, K., Yamaji, T. and Woo, S.L. (2000). Type V collagen is increased during rabbit medial collateral ligament healing. *Knee Surg. Sports Traumatol. Arthrosc.* **8**: 281–285.

Noyes, F.R., Butler, D.L., Grood, E.S., Zernicke, R.F. and Hefzy, M.S. (1984). Biomechanical analysis of human ligament grafts used in knee-ligament repairs and reconstructions. *J. Bone Joint Surg. Am.* **66**: 344–352.

O'Donoghue, D.H., Frank, G.R., Jeter, G.L., Johnson, W., Zeiders, J.W. and Kenyon, R. (1971). Repair and reconstruction of the anterior cruciate ligament in dogs. Factors influencing long-term results. *J. Bone Joint Surg. Am.* **53**: 710–718.

Ohno, K., Pomaybo, A.S., Schmidt, C.C., Levine, R.E., Ohland, K.J. and Woo, S.L. (1995). Healing of the medial collateral ligament after a combined medial collateral and anterior cruciate ligament injury and reconstruction of the anterior cruciate ligament: comparison of repair and nonrepair of medial collateral ligament tears in rabbits. *J. Orthop. Res.* **13**: 442–449.

Ottani, V., Raspanti, M. and Ruggeri, A. (2001). Collagen structure and functional implications. *Micron* **32**: 251–260.

Panossian, V., Liu, S.H., Lane, J.M. and Finerman, G.A. (1997). Fibroblast growth factor and epidermal growth factor receptors in ligament healing. *Clin. Orthop. Relat. Res.* 173–180.

Papageorgiou, C.D., Ma, C.B., Abramowitch, S.D., Clineff, T.D. and Woo, S.L. (2001). A multidisciplinary study of the healing of an intraarticular anterior cruciate ligament graft in a goat model. *Am. J. Sports Med.* **29**: 620–626.

Parry, D., Barnes, G. and Craig, A. (1978). A comparison of the size distribution of collagen fibrils in connective tissues as a function of age and a possible relation between fibril size distribution and mechanical properties. *Proc. R. Soc. Lond.* (Biol). 305–321.

Paulos, L.E., Rosenberg, T.D., Drawbert, J., Manning, J. and Abbott, P. (1987). Infrapatellar contracture syndrome. An unrecognized cause of knee stiffness with patella entrapment and patella infera. *Am. J. Sports Med.* **15**: 331–341.

Ralphs, J.R., Waggett, A.D. and Benjamin, M. (2002). Actin stress fibres and cell–cell adhesion molecules in tendons: organisation *in vivo* and response to mechanical loading of tendon cells *in vitro. Matrix Biol.* **21**: 67–74.

Record, R.D., Hillegonds, D., Simmons, C., Tullius, R., Rickey, F.A., Elmore, D. and Badylak, S.F. (2001). *In vivo* degradation of 14C-labeled small intestinal submucosa (SIS) when used for urinary bladder repair. *Biomaterials* **22**: 2653–2659.

Reider, B., Sathy, M.R., Talkington, J., Blyznak, N. and Kollias, S. (1994). Treatment of isolated medial collateral ligament injuries in athletes with early functional rehabilitation. A five-year follow-up study. *Am. J .Sports Med.* **22**: 470–477.

Rubinstein Jr., R.A., Shelbourne, K.D., VanMeter, C.D., McCarroll, J.C. and Rettig, A.C. (1994). Isolated autogenous bone–patellar tendon–bone graft site morbidity. *Am. J. Sports Med.* **22**: 324–327.

Rudy, T.W., Livesay, G.A., Woo, S.L. and Fu, F.H. (1996). A combined robotic/universal force sensor approach to determine *in situ* forces of knee ligaments. *J. Biomech.* **29**: 1357–1360.

Sachs, R.A., Daniel, D.M., Stone, M.L. and Garfein, R.F. (1989). Patellofemoral problems after anterior cruciate ligament reconstruction. *Am. J.Sports Med.* **17**: 760–765.

Sacks, M.S. and Gloeckner, D.C. (1999). Quantification of the fiber architecture and biaxial mechanical behavior of porcine intestinal submucosa. *J. Biomed. Mater. Res.* **46**: 1–10.

Sagarriga Visconti, C., Kavalkovich, K., Wu, J. and Niyibizi, C. (1996). Biochemical analysis of collagens at the ligament–bone interface reveals presence of cartilage-specific collagens. *Arch. Biochem. Biophys.* **328**: 135–142.

Scheffler, S.U., Clineff, T.D., Papageorgiou, C.D., Debski, R.E., Benjamin, C. and Woo, S.L. (2001). Structure and function of the healing medial collateral ligament in a goat model. *Ann. Biomed. Eng.* **29**: 173–180.

Scherping Jr., S.C., Schmidt, C.C., Georgescu, H.I., Kwoh, C.K., Evans, C.H. and Woo, S.L. (1997). Effect of growth factors on the proliferation of ligament fibroblasts from skeletally mature rabbits. *Connect. Tissue Res.* **36**: 1–8.

Sciore, P., Boykiw, R. and Hart, D.A. (1998). Semiquantitative reverse transcription-polymerase chain reaction analysis of mRNA for growth factors and growth factor receptors from normal and healing rabbit medial collateral ligament tissue. *J. Orthop. Res.* **16**: 429–437.

Shelbourne, K.D., Wilckens, J.H., Mollabashy, A. and DeCarlo, M. (1991). Arthrofibrosis in acute anterior cruciate ligament reconstruction. The effect of timing of reconstruction and rehabilitation. *Am. J. Sports Med.* **19**: 332–336.

Shimomura, T., Jia, F., Niyibizi, C. and Woo, S.L. (2003). Antisense oligonucleotides reduce synthesis of procollagen alpha1 (V) chain in human patellar tendon fibroblasts: potential application in healing ligaments and tendons. *Connect. Tissue Res.* **44**: 167–172.

Spalazzi, J.P., Doty, S.B., Levine, W.N. and Lu, H.H. (2006). *Co-culture of Human Fibroblasts and Osteoblasts on a Tri-Phasic Scaffold for Interface Tissue Engineering.* Orthopaedic Research Society: Chicago.

Spindler, K.P., Dawson, J.M., Stahlman, G.C., Davidson, J.M. and Nanney, L.B. (2002). Collagen expression and biomechanical response to human recombinant transforming growth factor beta (rhTGF-beta2) in the healing rabbit MCL. *J. Orthop. Res.* **20**: 318–324.

Spindler, K.P., Murray, M.M., Detwiler, K.B., Tarter, J.T., Dawson, J.M., Nanney, L.B. and Davidson, J.M. (2003). The biomechanical response to doses of TGF-beta 2 in the healing rabbit medial collateral ligament. *J. Orthop. Res.* **21**: 245–249.

Steenfos, H.H. (1994). Growth factors and wound healing. *Scand. J. Plast. Reconstr. Surg. Hand Surg.* **28**: 95–105.

Svensson, M., Kartus Jdogiannakis, N. and Karlsson, J. (2005). A long-term serial histological evaluation of the patellar tendon in humans after harvesting its central third. *Knee Surg. Sports Traumatol. Arthrosc.* **13**: 398–404.

Thornton, G.M., Boorman, R.S., Shrive, N.G. and Frank, C.B. (2002). Medial collateral ligament autografts have increased creep response for at least two years and early immobilization makes this worse. *J. Orthop. Res.* **20**: 346–352.

Tibone, J.E. and Antich, T.J. (1988). A biomechanical analysis of anterior cruciate ligament reconstruction with the patellar tendon. A two year followup. *Am. J. Sports Med.* **16**: 332–335.

Tipton, C.M., James, S.L., Mergner, W. and Tcheng, T.K. (1970). Influence of exercise on strength of medial collateral knee ligaments of dogs. *Am. J. Physiol.* **218**: 894–902.

Tohyama, H. and Yasuda, K. (2000). The effects of stress enhancement on the extracellular matrix and fibroblasts in the patellar tendon. *J. Biomech.* **33**: 559–565.

Tohyama, H., Yasuda, K., Kitamura, Y., Yamamoto, E. and Hayashi, K. (2003). The changes in mechanical properties of regenerated and residual tissues in the patellar tendon after removal of its central portion. *Clin. Biomech. (Bristol, Avon)* **18**: 765–772.

Tripathy, S.K., Black, H.B., Goldwasser, E. and Leiden, J.M. (1996). Immune responses to transgene-encoded proteins limit the stability of gene expression after injection of replication-defective adenovirus vectors. *Nat. Med.* **2**: 545–550.

Tsuda, E., Fukuda, Y., Loh, J.C., Debski, R.E., Fu, F.H. and Woo, S.L. (2002). The effect of soft-tissue graft fixation in anterior cruciate ligament reconstruction on graft-tunnel motion under anterior tibial loading. *Arthroscopy* **18**: 960–967.

Voytik-Harbin, S.L., Brightman, A.O., Kraine, M.R., Waisner, B. and Badylak, S.F. (1997). Identification of extractable growth factors from small intestinal submucosa. *J. Cell Biochem.* **67**: 478–491.

Walboomers, X.F., Croes, H.J., Ginsel, L.A. and Jansen, J.A. (1999). Contact guidance of rat fibroblasts on various implant materials. *J. Biomed. Mater. Res.* **47**: 204–212.

Wang, J.H., Jia, F., Gilbert, T.W. and Woo, S.L. (2003a). Cell orientation determines the alignment of cell-produced collagenous matrix. *J. Biomech.* **36**: 97–102.

Wang, J.H., Jia, F., Yang, G., Yang, S., Campbell, B.H., Stone, D. and Woo, S.L. (2003b). Cyclic mechanical stretching of human tendon fibroblasts increases the production of prostaglandin E2 and levels of cyclooxygenase expression: a novel *in vitro* model study. *Connect. Tissue Res.* **44**: 128–133.

Watanabe, N., Woo, S.L., Papageorgiou, C., Celechovsky, C. and Takai, S. (2002). Fate of donor bone marrow cells in medial collateral ligament after simulated autologous transplantation. *Microsc. Res. Tech.* **58**: 39–44.

Weiss, J.A., Woo, S.L., Ohland, K.J., Horibe, S. and Newton, P.O. (1991). Evaluation of a new injury model to study medial collateral ligament healing: primary repair versus nonoperative treatment. *J. Orthop. Res.* **9**: 516–528.

Woo, S., An, K.-N., Frank, C., Livesay, G., Ma, C. and Zeminski, J. (2000). Orthopaedic basic science : biology and biomechanics of the musculoskeletal system. In Einhorn, T.A., Simon, S.R. and American Academy of Orthopaedic Surgeons. (eds.), Anatomy, Biology and Biomechanics of Tendon and Ligaments. Rosemont, Ill: American Academy of Orthopaedic Surgeons, pp. 581–616.

Woo, S.L., Gelberman, R.H., Cobb, N.G., Amiel, D., Lothringer, K. and Akeson, W.H. (1981a). The importance of controlled passive mobilization on flexor tendon healing. A biomechanical study. *Acta Orthop. Scand.* **52**: 615–622.

Woo, S.L., Gomez, M.A. and Akeson, W.H. (1981b). The time and history-dependent viscoelastic properties of the canine medical collateral ligament. *J. Biomech. Eng.* **103**: 293–298.

Woo, S.L., Gomez, M.A., Amiel, D., Ritter, M.A., Gelberman, R.H. and Akeson, W.H. (1981c). The effects of exercise on the biomechanical and biochemical properties of swine digital flexor tendons. *J. Biomech. Eng.* **103**: 51–56.

Woo, S.L., Gomez, M.A., Woo, Y.K. and Akeson, W.H. (1982). Mechanical properties of tendons and ligaments. II. The relationships of immobilization and exercise on tissue remodeling. *Biorheology* **19**: 397–408.

Woo, S.L., Gomez, M.A., Seguchi, Y., Endo, C.M. and Akeson, W.H. (1983). Measurement of mechanical properties of ligament substance from a bone–ligament–bone preparation. *J. Orthop. Res.* **1**: 22–29.

Woo, S.L., Orlando, C.A., Camp, J.F. and Akeson, W.H. (1986a). Effects of postmortem storage by freezing on ligament tensile behavior. *J. Biomech.* **19**: 399–404.

Woo, S.L., Orlando, C.A., Gomez, M.A., Frank, C.B. and Akeson, W.H. (1986b). Tensile properties of the medial collateral ligament as a function of age. *J. Orthop. Res.* **4**: 133–141.

Woo, S.L., Gomez, M.A., Sites, T.J., Newton, P.O., Orlando, C.A. and Akeson, W.H. (1987a). The biomechanical and morphological changes in the medial collateral ligament of the rabbit after immobilization and remobilization. *J. Bone Joint Surg. Am.* **69**: 1200–1211.

Woo, S.L., Inoue, M., McGurk-Burleson, E. and Gomez, M.A. (1987b). Treatment of the medial collateral ligament injury. II: Structure and function of canine knees in response to differing treatment regimens. *Am. J. Sports Med.* **15**: 22–29.

Woo, S.L., Young, E.P., Ohland, K.J., Marcin, J.P., Horibe, S. and Lin, H.C. (1990). The effects of transection of the anterior cruciate ligament on healing of the medial collateral ligament. A biomechanical study of the knee in dogs. *J. Bone Joint Surg. Am. Vol.* **72**: 382–392.

Woo, S.L., Hollis, J.M., Adams, D.J., Lyon, R.M. and Takai, S. (1991). Tensile properties of the human femur-anterior cruciate ligament-tibia complex. The effects of specimen age and orientation. *Am. J. Sports Med.* **19**: 217–225.

Woo, S.L., Smith, D.W., Hildebrand, K.A., Zeminski, J.A. and Johnson, L.A. (1998). Engineering the healing of the rabbit medial collateral ligament. *Med. Biol. Eng. Comput.* **36**: 359–364.

Woo, S.L., Hildebrand, K., Watanabe, N., Fenwick, J.A., Papageorgiou, C.D. and Wang, J. H. (1999). Tissue engineering of ligament and tendon healing. *Clin. Orthop. Relat. Res.* S312–S323.

Woo, S.L., Kanamori, A., Zeminski, J., Yagi, M., Papageorgiou, C. and Fu, F.H. (2002). The effectiveness of reconstruction of the anterior cruciate ligament with hamstrings and patellar tendon . A cadaveric study comparing anterior tibial and rotational loads. *J. Bone Joint Surg. Am.* **84A**: 907–914.

Woo, S.L., Jia, F., Zou, L. and Gabriel, M.T. (2004). Functional tissue engineering for ligament healing: potential of antisense gene therapy. *Ann. Biomed. Eng.* **32**: 342–351.

Woo, S.L., Abramowitch, S.D., Kilger, R. and Liang, R. (2006). Biomechanics of knee ligaments: injury, healing, and repair. *J. Biomech.* **39**: 1–20.

Yagi, M., Wong, E.K., Kanamori, A., Debski, R.E., Fu, F.H. and Woo, S.L. (2002). Biomechanical analysis of an anatomic anterior cruciate ligament reconstruction. *Am. J. Sports Med.* **30**: 660–666.

Yamaji, T., Levine, R.E., Woo, S.L., Niyibizi, C., Kavalkovich, K.W. and Weaver-Green, C.M. (1996). Medial collateral ligament healing one year after a concurrent medial collateral ligament and anterior cruciate ligament injury: an interdisciplinary study in rabbits. *J. Orthop. Res.* **14**: 223–227.

Yang, G., Crawford, R.C. and Wang, J.H. (2004). Proliferation and collagen production of human patellar tendon fibroblasts in response to cyclic uniaxial stretching in serum-free conditions. *J. Biomech.* **37**: 1543–1550.

Young, R.G., Butler, D.L., Weber, W., Caplan, A.I., Gordon, S.L. and Fink, D.J. (1998). Use of mesenchymal stem cells in a collagen matrix for Achilles tendon repair. *J. Orthop. Res.* **16**: 406–413.

Zantop, T., Gilbert, T.W., Yoder, M. and Badylak, S. (2005). Extracellular matrix scaffolds attract bone marrow- derived cells in a mouse model of Achilles tendon reconstruction. *J. Orthop. Res.* **24**: 1299–1309.

36

Tissue Therapy: Implications of Regenerative Medicine for Skeletal Muscle

Shen Wei and Johnny Huard

INTRODUCTION

Skeletal muscle is the largest tissue mass in the human body, constituting 40%–45% of total body weight (Garrett and Best, 1994). Up to 55% of all injuries sustained in sports are muscle-related injuries (Kaariainen et al., 2000). Therefore, muscle injury is of major concern in traumatology and sports medicine. The treatment of such injuries can pose challenging problems, particularly because skeletal muscle injuries heal slowly and often result in incomplete functional recovery.

After skeletal muscle injury, the traumatized muscle undergoes sequential and overlapping phases of healing, including degeneration, inflammation, and regeneration. In cases involving severe injuries, such as those resulting from burns or laceration, the healing muscle also may undergo fibrosis. Each of these phases is unique in skeletal muscle healing, and all are potential targets of efforts to treat skeletal muscle injury. The optimization of each phase of muscle healing could facilitate full functional recovery of injured skeletal muscle.

In this chapter, we discuss the natural healing process of injured skeletal muscle and the tissue-level therapeutic approaches targeted toward each phase of the healing process. The utilization of tissue engineering and gene and cell therapy approaches is discussed in other chapters of this book.

THE NATURAL HEALING PROCESS OF SKELETAL MUSCLE

Skeletal muscle is a composite structure mainly consisting of muscle cells and myofibers, nerves, blood vessels, and extracellular connective tissue matrix. The skeletal muscle fiber (or **myofiber**) is a syncytium derived from the fusion of myoblasts; individual myofibers are surrounded by a connective tissue layer called endomysium. Fascicles or bundles of myofibers are grouped together and surrounded by perimysium. The epimysium is the outermost connective tissue layer surrounding the skeletal muscle (Garrett and Best, 1994). After muscle injury, satellite cells, the myogenic precursor cells located between the basal lamina and plasma membrane (Hurme and Kalimo, 1992), are activated to proliferate and differentiate into myoblasts and play a key role in muscle regeneration by fusing with damaged myofibers or forming new myofibers. The appearance of centralized nuclei within myofibers is an indication of muscle regeneration. Upon myofiber maturation, the nuclei move into the subsarcolemmal layer (Mauro, 1961; Campion, 1984) (Figure 36.1).

Skeletal muscle injuries are common results of sporting events and athletic endeavors (Kaariainen et al., 2000). However, they also can result from daily life activities or military combat. Skeletal muscle injuries range

Copyright © 2008 Elsevier Inc. All rights reserved

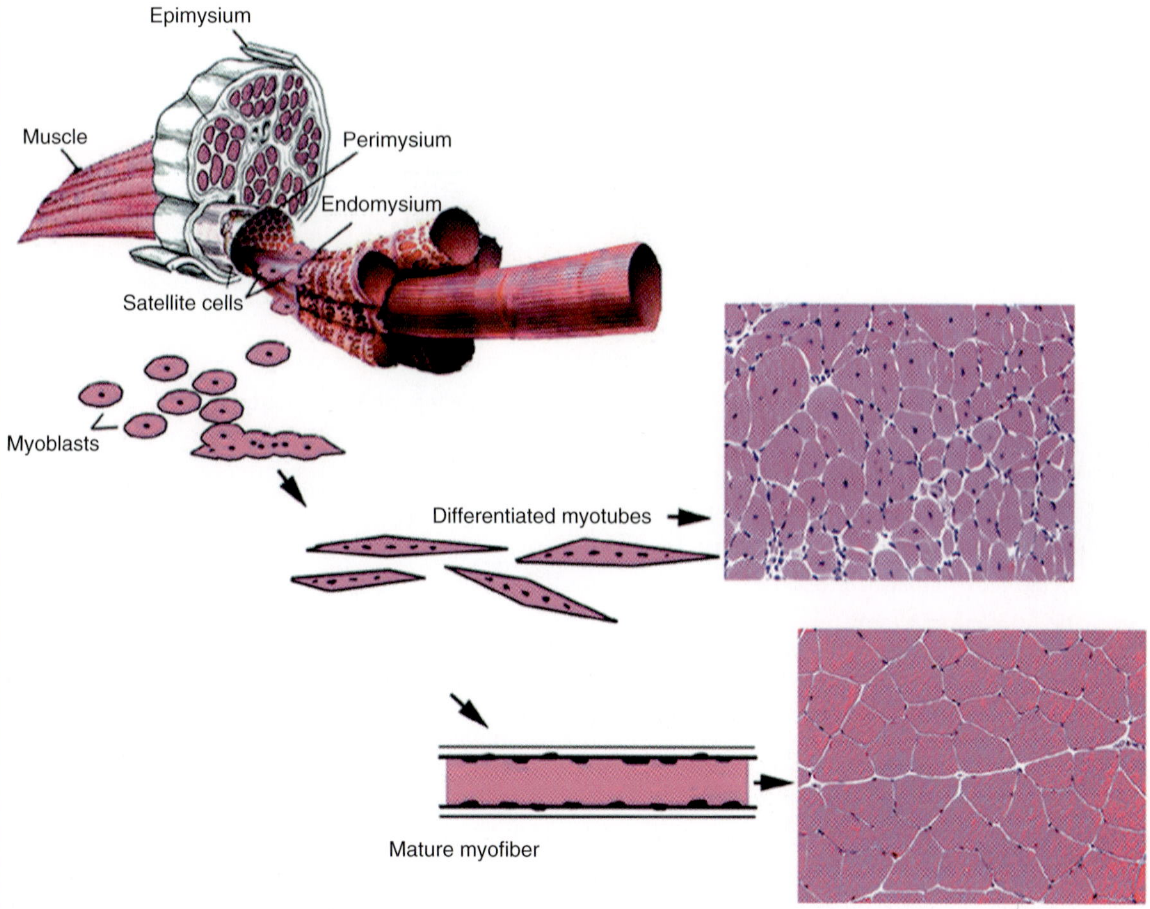

Figure 36.1 **Schematic of the structure of skeletal muscle. In injured skeletal muscle, satellite cells are released and are activated to become myoblasts. They fuse with damaged myofibers or with each other to regenerate muscle tissue. Regenerating myofibers contain centrally located nuclei that migrate to the periphery of the myofiber upon myofiber maturation. Reprinted with permission from *The Journal of Bone and Joint Surgery*, Inc.**

in severity from common and relatively minor muscle strains and contusions to more unusual and particularly devastating muscle lacerations and burns. In addition, muscle injuries also can result indirectly from ischemia and neurologic dysfunction (Campion, 1984).

Injury to skeletal muscle induces a healing response comprising various stages. The traumatized muscle sequentially undergoes degeneration, inflammation, regeneration, and, in cases of severe trauma, fibrotic tissue formation. Generally, the degeneration and inflammation phases start within hours after injury, the regeneration phase begins 3–5 days after injury, and fibrotic tissue first appears approximately 2 weeks after injury (Li et al., 2001; Huard et al., 2002) (Figure 36.2). However, the timing of these events can vary depending on the type of injury and the severity of damage. Furthermore, these healing stages are not distinct; they overlap with each other and interrelate mechanistically.

Degeneration

Trauma to muscle tissue destroys its normal structure. Proteases initiate the autodigestion of disrupted and necrotic myofibers (Ebisui et al., 1995; Mbebi et al., 1999). Cytokines up-regulate vascular permeability and

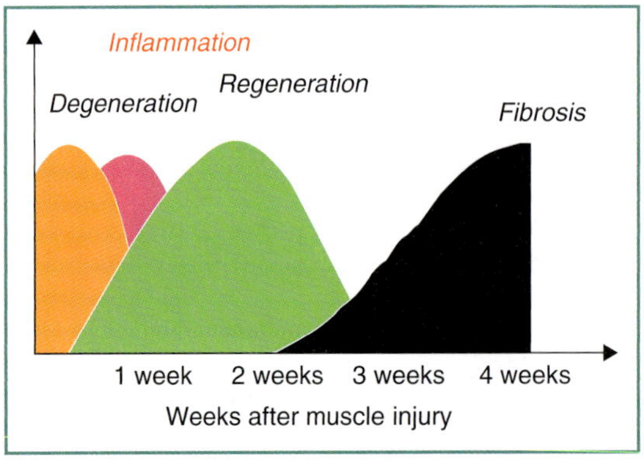

Figure 36.2 The temporal sequence of four interrelated healing phases: degeneration, inflammation, regeneration, and fibrosis.

blood flow, which in turn lead is to the ensuing inflammatory response. Toxic free radical species develop and impair the already damaged muscle tissue and the healthy tissue nearby (Clanton et al., 1999). Researchers have made little effort to improve muscle healing by limiting muscle degeneration. The use of anti-oxidant or free radical scavengers has failed to elicit promising results (Childs et al., 2001; Beaton et al., 2002). The technical difficulty impeding such approaches is the fact that degeneration begins immediately after injury occurs; therefore, efforts to interrupt this phase are nearly impossible, and preventive techniques may have to be used.

Inflammation

Inflammation begins shortly after injury and overlaps with both degeneration and regeneration. Within hours after injury, neutrophils begin to infiltrate and migrate to the injury site. Different populations of macrophages shortly follow the neutrophils. Some macrophage populations are mainly involved in phagocytosis of tissue debris and may evoke more tissue damage (Lapointe et al., 2002), whereas other populations are primarily responsible for the production of early growth factors and cytokines (McLennan, 1993; St Pierre and Tidball, 1994). In addition to macrophages, myogenic cells can release growth factors and cytokines (Huard et al., 2002) that influence the subsequent regeneration and fibrosis phases. Limiting inflammation may reduce symptoms like pain and edema induced by cytokines and prostaglandins. However, the reduced production of growth factors and cytokines and prostaglandins may lead to delayed regeneration and overgrowth of fibrotic tissue.

Regeneration

Although satellite cell activation can occur as soon as 24 h after injury (Rantanen et al., 1995), the regeneration phase, evidenced by the appearance of centrally nucleated regenerating myofibers, usually starts 3–5 days after injury (depending on the injury type and severity). Regeneration is carried out mainly by satellite cells. In uninjured muscle, these cells reside quiescently in their niche between the basal lamina and the sarcolemma (Hurme and Kalimo, 1992). In response to injury-induced disruption of normal muscle structure, quiescent satellite cells are activated and begin to proliferate and differentiate in response to various growth factors, including insulin-like growth factor 1 (IGF-1), hepatocyte growth factor (HGF), basic fibroblast growth factor (bFGF), and transforming growth factor beta 1 (TGF-β1). Finally, the satellite cells fuse with existing myofibers in the injury area or fuse with each other to form regenerating myofibers (Li et al., 2001; Huard et al., 2002). Recent reports indicate that prostaglandins play key roles in satellite cell fusion, a point we will discuss in the next section.

Fibrosis

Not all growth factors released at the injury site promote muscle regeneration. Some also stimulate cells to differentiate toward the fibroblastic lineage and form fibrotic tissue. As the fibrotic tissue becomes increasingly dense, it both hinders the regeneration of myofibers and prevents new axons from reaching myofibers and forming neuromuscular junctions (Kaariainen et al., 2000). Dense fibrotic tissue also causes decreased muscle contractility and range of movement (Shanmugasundaram, 1980). Most importantly, the presence of such tissue makes the repaired muscle more susceptible to re-injury. Research results have implicated TGF-β1 in pathogenic fibrosis within many tissues (Gaedeke et al., 2001; Ihn, 2002). After muscle injury, TGF-β1 stimulates muscle-derived stem cells to differentiate into fibroblast-like cells (Li and Huard, 2002; Li et al., 2004) and produce collagen type I, the major component of fibrous tissue (Ghosh, 2002). Our previous studies of lacerated skeletal muscle have shown up-regulation of TGF-β1 that persisted from 3 days after injury until 14 days after injury (Li et al., 2004; Shen et al., 2005).

Each of these healing phases plays a unique role in skeletal muscle repair. By interfering with any of them, we can alter the progression of natural healing in skeletal muscle. All these healing phases are interrelated and should be regarded as integral components of the skeletal muscle healing process. For example, inhibiting inflammation may delay the regeneration phase and therefore slow the whole healing process (Shen et al., 2005). The promotion of regeneration cannot alone elicit complete recovery of severely injured muscle because such muscle also undergoes fibrosis (Sato et al., 2003). Efforts to improve skeletal muscle healing should incorporate different approaches aimed at each phase and must account for the interaction between the different phases of healing.

THE ROLE OF INFLAMMATION IN SKELETAL MUSCLE HEALING

Inflammation phase, symptomized by "Rubor," "Tumor," "Calor," and "Dolor," is an important phase of natural healing in many injured tissues. During the inflammation phase, the injured tissue and coagulation cascade release various cytokines that increase the permeability of blood vessels and the chemotaxis of inflammatory cells. First neutrophils and then macrophages rapidly invade the injury site. The infiltration of inflammatory cells persists for several days after injury. These cells play pivotal roles in the inflammatory phase and also influence the subsequent phases. Inflammatory cells participate in the phagocytosis of necrotic tissue debris, which is necessary for the growth of regenerating tissue. The cytokines and growth factors released by these cells can initiate regeneration and fibrosis or cause further damage to the muscle (Tidball, 2005).

Inflammatory Cells

Studies have revealed a complex picture in which inflammatory cells promote both injury and repair through the combined actions of free radicals, growth factors, and cytokines. Neutrophils invade the muscle injury site as early as 1 h after injury and remain for up to 5 days (Fielding et al., 1993). Studies have shown that the invading neutrophils are phagocytic and can help to remove cellular debris (Lowe et al., 1995), a necessary step in skeletal muscle repair. However, most research on this topic has shown that neutrophils exacerbate muscle injuries by releasing reactive oxygen intermediates, including peroxides, hypochlorite, and superoxide (Jolly et al., 1986; Korthuis et al., 1988; Smith et al., 1989; Fielding et al., 1993; Tidball, 2002). It is not yet clear whether neutrophil-mediated damage is a necessary step in muscle repair.

In contrast, macrophages mediate the regenerative process in several important ways. Macrophages secrete growth factors and cytokines that act in a paracrine fashion to stimulate myoblast-related regenerative events (Cantini et al., 1994; Massimino et al., 1997; Lescaudron et al., 1999), but no study has definitively shown which factors macrophages release *in vivo* to promote muscle healing. Our research has shown that the administration of NS-398, a COX-2 inhibitor, reduces the infiltration of macrophages, especially at the early time point of 24 h after injury (Shen et al., 2005). However, by reducing the number of infiltrating macrophages, the administration of a COX-2-specific inhibitor may impede the proliferation of satellite cells and reduce the secretion of some growth factors and cytokines necessary for regeneration. We believe that this inhibition of the inflammatory response was at least partially responsible for the delayed myofiber regeneration observed during the healing process in this study. A recent report also has shown that direct contact with macrophages can rescue myogenic precursor cells from apoptosis after muscle injury. The rescued cells can act synergistically with macrophages to amplify chemotaxis and enhance muscle growth (Chazaud et al., 2003). Furthermore, macrophages may fuse with myofibers directly to promote regeneration (Camargo et al., 2003).

Cyclooxygenase Pathway and Prostaglandins

Because the symptoms of inflammation are uncomfortable, attempts are always made to shorten or eliminate this phase. Non-steroidal anti-inflammatory drugs (NSAIDs) reduce prostaglandin synthesis by selectively or non-selectively inhibiting the cyclooxygenase (COX) enzymes. COX enzyme has three known isoforms: COX-1, COX-2, and COX-3. COX-1 is constitutively expressed and controls homeostasis in many tissues. COX-2 is an inducible isoform of COX that is expressed mostly in pathologic scenarios (McCormack, 1998). COX-3 is involved with the control of fever via the central nervous system (Warner and Mitchell, 2002).

Studies have reported favorable effects of using NSAIDs to reduce muscle weakness and loss of function (Almekinders, 1999; Trappe et al., 2001). However, some research suggests that NSAIDs only have short-term beneficial effects on muscle healing and that NSAIDs treatment beyond the first week after injury has either no effect or a detrimental effect on the recovery of muscle strength (Almekinders and Gilbert, 1986; Obremsky et al., 1994; Mishra et al., 1995). Due to the different animal models, NSAIDs, and dosages and administration routes used by different research groups, outcomes of studies focused on NSAID treatment vary making it extremely difficult to compare the results. COX-2-selective inhibitors (such as celecoxib and valdecoxib) are new variants of NSAIDs. They have similar analgesic and anti-inflammatory properties and reduced side-effect profiles when compared with non-selective NSAIDs. By avoiding COX-1 inhibition, clinicians can essentially avoid interfering with homeostatic functions and side effects like platelet inhibition and gastric mucosal injury. Studies have shown that COX-2-selective NSAIDs have detrimental effects on muscle healing (Bondesen et al., 2004; Mendias et al., 2004; Shen et al., 2005). It has been reported that repetitive use of skeletal muscle up-regulates the COX-2 enzyme and prostaglandins. NSAIDs, whether selective or non-selective, can suppress the up-regulated synthesis of prostaglandins (Vandenburgh et al., 1995; Trappe et al., 2001). Thus, it is likely that the COX-2 enzyme and the prostaglandins play a pivotal role in the healing process of skeletal muscle. NSAIDs appear to delay muscle regeneration by blocking the COX-2 enzyme and the production of prostaglandins, thereby impeding the long-term functional recovery of injured muscle. The early functional improvement observed after NSAID administration is likely due to the reduced inflammatory symptoms.

Recent studies have clearly demonstrated the involvement of COX-2 and prostaglandins in muscle repair and regeneration (Bondesen et al., 2004; Shen et al., 2005). Prostaglandins are potent mediators of inflammatory response and are involved in multiple aspects of muscle regeneration after injury (Prisk and Huard, 2003). The prostaglandins are synthesized from arachidonic acid released from membrane phospholipids

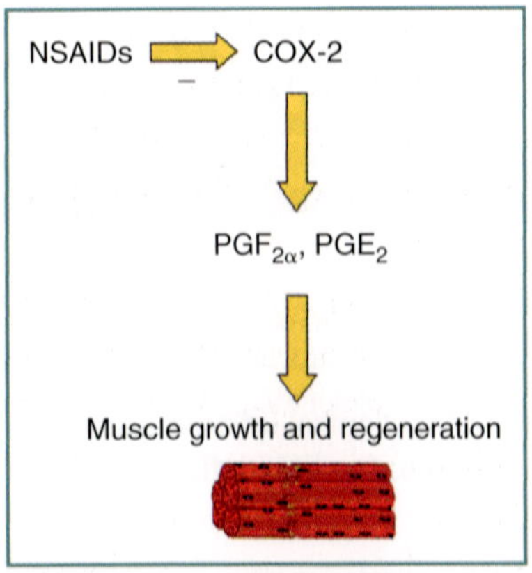

Figure 36.3 **The COX pathway and its final products, PGE_2 and $PGF_{2\alpha}$, are very important for skeletal muscle regeneration. The use of NSAIDs can delay skeletal muscle healing after injury by blocking the production of PGE_2 and $PGF_{2\alpha}$.**

by phospholipase. COX enzymes are key to converting arachidonic acid into the prostaglandin H2 (PGH_2). From PGH_2, specific synthases are responsible for the further conversion into various forms of prostaglandins, including PGE_2 and $PGF_{2\alpha}$ (Hochberg, 1989). Generally, PGE_2 and $PGF_{2\alpha}$ mediate fever, pain, and smooth muscle contraction. In injured skeletal muscle, PGE_2 and $PGF_{2\alpha}$ play their specific receptor-mediated roles in inflammation, regeneration, and nociception (Hatae et al., 2002; Prisk and Huard, 2003; Sakamoto et al., 2004). Furthermore, they help to regulate muscle protein synthesis and degradation (Palmer, 1990) (Figure 36.3). $PGF_{2\alpha}$, in particular, can promote the growth of skeletal muscle by stimulating the secondary fusion between single muscle cells and nascent myotubes (Horsley and Pavlath, 2003, 2004). PGE_2 appears to play multiple roles in the muscle inflammation phase, including chemotaxis of inflammatory cells, increasing vascular permeability and vasodilation, and induction of nitric oxide synthase and pro-inflammatory cytokine expression (Stenson et al., 1986; Tetsuka et al., 1994; Sakamoto et al., 2004). In addition, PGE_2 is a potent down-regulator of TGF-β1-stimulated fibroblast proliferation and collagen synthesis (Frungieri et al., 2002; Scheuren et al., 2002). The production of PGE_2 also may be important for inhibiting fibrosis, and the increased fibrosis observed after NSAID treatment may be due to the inhibited expression of PGE_2 (Shen et al., 2005).

Inflammation and Muscular Dystrophy

Duchenne muscular dystrophy (DMD) is caused by mutations in the dystrophin gene and leads to loss of the dystrophin-glycoprotein complex (Hoffman et al., 1987). DMD results in membrane instability of myofibers, and persons with DMD develop progressive muscle weakness beginning at birth. Previous work has revealed possible involvement of uncontrolled chronic inflammation in DMD (Porter et al., 2002; Monici et al., 2003; Grounds and Torrisi, 2004), which is one of the main causes of its secondary effects. Acting on this concept, researchers studying DMD have developed therapies targeted at inflammatory components. Recent studies have shown that treatment with corticosteroids significantly slows the progression of DMD. The use of corticosteroids may prolong ambulation and upper limb and pulmonary function (Balaban et al., 2005). Although the administration of corticosteroids is accompanied by negative side effects, intermittent prednisone treatment generally is recognized as an effective way to preserve motor functions in ambulant persons with DMD (Beenakker et al., 2005).

Gosselin, L.E., Williams, J.E., Deering, M., Brazeau, D., Koury, S. and Martinez, D.A. (2004). Localization and early time course of TGF-beta 1 mRNA expression in dystrophic muscle. *Muscle Nerve* **30**(5): 645–653.

Gregorevic, P., Lynch, G.S. and Williams, D.A. (2000). Hyperbaric oxygen improves contractile function of regenerating rat skeletal muscle after myotoxic injury. *J. Appl. Physiol.* **89**(4): 1477–1482.

Grounds, M.D. and Torrisi, J. (2004). Anti-TNFalpha (Remicade) therapy protects dystrophic skeletal muscle from necrosis. *FASEB J.* **18**(6): 676–682.

Haapaniemi, T., Sirsjo, A., Nylander, G. and Larsson, J. (1995). Hyperbaric oxygen treatment attenuates glutathione depletion and improves metabolic restitution in postischemic skeletal muscle. *Free Radical Res.* **23**(2): 91–101.

Harrison, B.C., Robinson, D., Davison, B.J., Foley, B., Seda, E. and Byrnes, W.C. (2001). Treatment of exercise-induced muscle injury via hyperbaric oxygen therapy. *Med. Sci. Sports Exerc.* **33**(1): 36–42.

Hatae, N., Sugimoto, Y. and Ichikawa, A. (2002). Prostaglandin receptors: Advances in the study of EP3 receptor signaling. *J. Biochem. Tokyo* **131**(6): 781–784.

Hochberg, M.C. (1989). NSAIDS: Mechanisms and pathways of action. *Hosp. Pract.* (Off Ed) **24**(3): 185–190, 195, 198.

Hoffman, E.P., Brown Jr., R.H., and Kunkel, L.M. (1987). Dystrophin: the protein product of the Duchenne muscular dystrophy locus. *Cell* **51**(6): 919–928.

Horsley, V. and Pavlath, G.K. (2003). Prostaglandin F2(alpha) stimulates growth of skeletal muscle cells via an NFATC2-dependent pathway. *J. Cell Biol.* **161**(1): 111–118.

Horsley, V. and Pavlath, G.K. (2004). Forming a multinucleated cell: Molecules that regulate myoblast fusion. *Cells Tissues Organs* **176**(1–3): 67–78.

Huard, J., Li, Y. and Fu, F.H. (2002). Muscle injuries and repair: current trends in research. *J. Bone Joint Surg. Am.* **84-A**(5): 822–832.

Hurme, T. and Kalimo, H. (1992). Activation of myogenic precursor cells after muscle injury. *Med. Sci. Sports Exerc.* **24**(2): 197–205.

Ihn, H. (2002). Pathogenesis of fibrosis: Role of TGF-beta and CTGF. *Curr. Opin. Rheumatol.* **14**(6): 681–685.

Jolly, S.R., Kane, W.J., Hook, B.G., Abrams, G.D., Kunkel, S.L. and Lucchesi, B.R. (1986). Reduction of myocardial infarct size by neutrophil depletion: Effect of duration of occlusion. *Am. Heart J.* **112**(4): 682–690.

Jones, J.I. and Clemmons, D.R. (1995). Insulin-like growth factors and their binding proteins: biological actions. *Endocr. Rev.* **16**(1): 3–34.

Kaariainen, M., Jarvinen, T., Jarvinen, M., Rantanen, J. and Kalimo, H. (2000). Relation between myofibers and connective tissue during muscle injury repair. *Scand. J. Med. Sci. Sports* **10**(6): 332–337.

Karnes, J.L. and Burton, H.W. (2002). Continuous therapeutic ultrasound accelerates repair of contraction-induced skeletal muscle damage in rats. *Arch. Phys. Med. Rehabil.* **83**(1): 1–4.

Kasemkijwattana, C., Menetrey, J., Somogyl, G., Moreland, M.S., Fu, F.H., Buranapanitkit, B., Watkins, S.C. and Huard, J. (1998). Development of approaches to improve the healing following muscle contusion. *Cell Transplant.* **7**(6): 585–598.

Korthuis, R.J., Grisham, M.B. and Granger, D.N. (1988). Leukocyte depletion attenuates vascular injury in postischemic skeletal muscle. *Am. J. Physiol.* **254**(5 Part 2): H823–H827.

Lapointe, B.M., Frenette, J. and Cote, C.H. (2002). Lengthening contraction-induced inflammation is linked to secondary damage but devoid of neutrophil invasion. *J. Appl. Physiol.* **92**(5): 1995–2004.

Lescaudron, L., Peltekian, E., Fontaine-Perus, J., Paulin, D., Zampieri, M., Garcia, L. and Parrish, E. (1999). Blood borne macrophages are essential for the triggering of muscle regeneration following muscle transplant. *Neuromuscul. Disord.* **9**(2): 72–80.

Li, Y. and Huard, J. (2002). Differentiation of muscle-derived cells into myofibroblasts in injured skeletal muscle. *Am. J. Pathol.* **161**(3): 895–907.

Li, Y., Cummins, J. and Huard, J. (2001). Muscle injury and repair. *Curr. Opin. Orthop.* **12**: 409–415.

Li, Y., Foster, W., Deasy, B.M., Chan, Y., Prisk, V., Tang, Y., Cummins, J. and Huard, J. (2004). Transforming growth factor-beta1 induces the differentiation of myogenic cells into fibrotic cells in injured skeletal muscle: a key event in muscle fibrogenesis. *Am. J. Pathol.* **164**(3): 1007–1019.

Lowe, D.A., Warren, G.L., Ingalls, C.P., Boorstein, D.B. and Armstrong, R.B. (1995). Muscle function and protein metabolism after initiation of eccentric contraction-induced injury. *J. Appl. Physiol.* **79**(4): 1260–1270.

Markert, C.D., Merrick, M.A., Kirby, T.E. and Devor, S.T. (2005). Nonthermal ultrasound and exercise in skeletal muscle regeneration. *Arch. Phys. Med. Rehabil.* **86**(7): 1304–1310.

Massimino, M.L., Rapizzi, E., Cantini, M., Libera, L.D., Mazzoleni, F., Arslan, P. and Carraro, U. (1997). ED2+ macrophages increase selectively myoblast proliferation in muscle cultures. *Biochem. Biophys. Res. Comm.* **235**(3): 754–759.

Mauro, A. (1961). Satellite cell of skeletal muscle fibers. *J. Biophys. Biochem. Cytol.* **9**: 493–495.

Mbebi, C., Hantai, D., Jandrot-Perrus, M., Doyennette, M.A. and Verdiere-Sahuque, M. (1999). Protease nexin I expression is up-regulated in human skeletal muscle by injury-related factors. *J. Cell. Physiol.* **179**(3): 305–314.

McCormack, K. (1998). Roles of COX-1 and COX-2. *J. Rheumatol.* **25**(11): 2279–2281.

McCroskery, S., Thomas, M., Platt, L., Hennebry, A., Nishimura, T., McLeay, L., Sharma, M. and Kambadur, R. (2005). Improved muscle healing through enhanced regeneration and reduced fibrosis in myostatin-null mice. *J. Cell Sci.* **118**(Part 15): 3531–3541.

McLennan, I.S. (1993). Resident macrophages (ED2- and ED3-positive) do not phagocytose degenerating rat skeletal muscle fibres. *Cell Tissue Res.* **272**(1): 193–196.

McPherron, A.C., Lawler, A.M. and Lee, S.J. (1997). Regulation of skeletal muscle mass in mice by a new TGF-beta superfamily member. *Nature* **387**(6628): 83–90.

Mendias, C.L., Tatsumi, R. and Allen, R.E. (2004). Role of cyclooxygenase-1 and -2 in satellite cell proliferation, differentiation, and fusion. *Muscle Nerve* **30**(4): 497–500.

Menetrey, J., Kasemkijwattana, C., Day, C.S., Bosch, P., Vogt, M., Fu, F.H., Moreland, M.S. and Huard, J. (2000). Growth factors improve muscle healing *in vivo*. *J. Bone Joint Surg. Br.* **82**(1): 131–137.

Mishra, D.K., Friden, J., Schmitz, M.C. and Lieber, R.L. (1995). Anti-inflammatory medication after muscle injury. A treatment resulting in short-term improvement but subsequent loss of muscle function. *J. Bone Joint Surg. Am.* **77**(10): 1510–1519.

Monici, M.C., Aguennouz, M., Mazzeo, A., Messina, C. and Vita, G. (2003). Activation of nuclear factor-kappaB in inflammatory myopathies and Duchenne muscular dystrophy. *Neurology* **60**(6): 993–997.

Negishi, S., Li, Y., Usas, A., Fu, F.H. and Huard, J. (2005). The effect of relaxin treatment on skeletal muscle injuries. *Am. J. Sports Med.*

Obremsky, W.T., Seaber, A.V., Ribbeck, B.M. and Garrett Jr., W.E. (1994). Biomechanical and histologic assessment of a controlled muscle strain injury treated with piroxicam. *Am. J. Sports Med.* **22**(4): 558–561.

Palmer, R.M. (1990). Prostaglandins and the control of muscle protein synthesis and degradation. *Prostag. Leukotr. Ess. Fatty Acids* **39**(2): 95–104.

Porter, J.D., Khanna, S., Kaminski, H.J., Rao, J.S., Merriam, A.P., Richmonds, C.R., Leahy, P., Li, J., Guo, W. and Andrade, F.H. (2002). A chronic inflammatory response dominates the skeletal muscle molecular signature in dystrophin-deficient mdx mice. *Hum. Mol. Genet.* **11**(3): 263–272.

Prisk, V. and Huard, J. (2003). Muscle injuries and repair: the role of prostaglandins and inflammation. *Histol. Histopathol.* **18**(4): 1243–1256.

Rantanen, J., Hurme, T., Lukka, R., Heino, J. and Kalimo, H. (1995). Satellite cell proliferation and the expression of myogenin and desmin in regenerating skeletal muscle: evidence for two different populations of satellite cells. *Lab. Invest.* **72**(3): 341–347.

Rantanen, J., Thorsson, O., Wollmer, P., Hurme, T. and Kalimo, H. (1999). Effects of therapeutic ultrasound on the regeneration of skeletal myofibers after experimental muscle injury. *Am. J. Sports Med.* **27**(1): 54–59.

Roberts, P. and McGeachie, J.K. (1992). The effects of pre- and posttransplantation exercise on satellite cell activation and the regeneration of skeletal muscle transplants: a morphometric and autoradiographic study in mice. *J. Anat.* **180** (Part 1): 67–74.

Sakamoto, A., Matsumura, J., Mii, S., Gotoh, Y. and Ogawa, R. (2004). A prostaglandin E2 receptor subtype EP4 agonist attenuates cardiovascular depression in endotoxin shock by inhibiting inflammatory cytokines and nitric oxide production. *Shock* **22**(1): 76–81.

Sato, K., Li, Y., Foster, W., Fukushima, K., Badlani, N., Adachi, N., Usas, A., Fu, F.H. and Huard, J. (2003). Improvement of muscle healing through enhancement of muscle regeneration and prevention of fibrosis. *Muscle Nerve* **28**(3): 365–372.

Scheuren, N., Jacobs, M., Ertl, G. and Schorb, W. (2002). Cyclooxygenase-2 in myocardium stimulation by angiotensin-II in cultured cardiac fibroblasts and role at acute myocardial infarction. *J. Mol. Cell. Cardiol.* **34**(1): 29–37.

Shanmugasundaram, T.K. (1980). Post-injection fibrosis of skeletal muscle: A clinical problem. A personal series of 169 cases. *Int. Orthop.* **4**(1): 31–37.

Shen, W., Li, Y., Tang, Y., Cummins, J. and Huard, J. (2005). NS-398, a cyclooxygenase-2-specific inhibitor, delays skeletal muscle healing by decreasing regeneration and promoting fibrosis. *Am. J. Pathol.* **167**(4): 1105–1117.

Sirsjo, A., Lehr, H.A., Nolte, D., Haapaniemi, T., Lewis, D.H., Nylander, G. and Messmer, K. (1993). Hyperbaric oxygen treatment enhances the recovery of blood flow and functional capillary density in postischemic striated muscle. *Circ. Shock* **40**(1): 9–13.

Skyhar, M.J., Hargens, A.R., Strauss, M.B., Gershuni, D.H., Hart, G.B. and Akeson, W.H. (1986). Hyperbaric oxygen reduces edema and necrosis of skeletal muscle in compartment syndromes associated with hemorrhagic hypotension. *J. Bone Joint Surg. Am.* **68**(8): 1218–1224.

Smith, H.K., Maxwell, L., Rodgers, C.D., McKee, N.H. and Plyley, M.J. (2001). Exercise-enhanced satellite cell proliferation and new myonuclear accretion in rat skeletal muscle. *J. Appl. Physiol.* **90**(4): 1407–1414.

Smith, J.K., Grisham, M.B., Granger, D.N. and Korthuis, R.J. (1989). Free radical defense mechanisms and neutrophil infiltration in postischemic skeletal muscle. *Am. J. Physiol.* **256**(3 Part 2): H789–H793.

St Pierre, B.A. and Tidball, J.G. (1994). Differential response of macrophage subpopulations to soleus muscle reloading after rat hindlimb suspension. *J. Appl. Physiol.* **77**(1): 290–297.

Stenson, W.F., Chang, K. and Williamson, J.R. (1986). Tissue differences in vascular permeability induced by leukotriene B4 and prostaglandin E2 in the rat. *Prostaglandins* **32**(1): 5–17.

Strauss, M.B., Hargens, A.R., Gershuni, D.H., Greenberg, D.A., Crenshaw, A.G., Hart, G.B. and Akeson, W.H. (1983). Reduction of skeletal muscle necrosis using intermittent hyperbaric oxygen in a model compartment syndrome. *J. Bone Joint Surg. Am.* **65**(5): 656–662.

Tetsuka, T., Daphna-Iken, D., Srivastava, S.K., Baier, L.D., DuMaine, J. and Morrison, A.R. (1994). Cross-talk between cyclooxygenase and nitric oxide pathways: Prostaglandin E2 negatively modulates induction of nitric oxide synthase by interleukin 1. *Proc. Natl. Acad. Sci. USA* **91**(25): 12168–12172.

Tidball, J.G. (2002). Interactions between muscle and the immune system during modified musculoskeletal loading. *Clin. Orthop. Relat. Res.* **403**(Suppl): S100–S109.

Tidball, J.G. (2005). Inflammatory processes in muscle injury and repair. *Am. J. Physiol. Regul. Integr. Comp. Physiol.* **288**(2): R345–R353.

Trappe, T.A., Fluckey, J.D., White, F., Lambert, C.P. and Evans, W.J. (2001). Skeletal muscle PGF(2)(alpha) and PGE(2) in response to eccentric resistance exercise: influence of ibuprofen acetaminophen. *J. Clin. Endocrinol. Metabol.* **86**(10): 5067–5070.

Ulloa, L., Doody, J. and Massague, J. (1999). Inhibition of transforming growth factor-beta/SMAD signalling by the interferon-gamma/STAT pathway. *Nature* **397**(6721): 710–713.

Unemori, E.N. and Amento, E.P. (1990). Relaxin modulates synthesis and secretion of procollagenase and collagen by human dermal fibroblasts. *J. Biol. Chem.* **265**(18): 10681–10685.

Unemori, E.N., Pickford, L.B., Salles, A.L., Piercy, C.E., Grove, B.H., Erikson, M.E. and Amento, E.P. (1996). Relaxin induces an extracellular matrix-degrading phenotype in human lung fibroblasts *in vitro* and inhibits lung fibrosis in a murine model *in vivo*. *J. Clin. Investig.* **98**(12): 2739–2745.

Vandenburgh, H.H., Shansky, J., Solerssi, R. and Chromiak, J. (1995). Mechanical stimulation of skeletal muscle increases prostaglandin F2 alpha production, cyclooxygenase activity, and cell growth by a pertussis toxin sensitive mechanism. *J. Cell Physiol.* **163**(2): 285–294.

Wanek, L.J. and Snow, M.H. (2000). Activity-induced fiber regeneration in rat soleus muscle. *Anat. Rec.* **258**(2): 176–185.

Warner, T.D. and Mitchell, J.A. (2002). Cyclooxygenase-3 (COX-3): Filling in the gaps toward a COX continuum? *Proc. Natl. Acad. Sci. USA* **99**(21): 13371–13373.

Webster, A.L., Syrotuik, D.G., Bell, G.J., Jones, R.L. and Hanstock, C.C. (2002). Effects of hyperbaric oxygen on recovery from exercise-induced muscle damage in humans. *Clin. J. Sport Med.* **12**(3): 139–150.

Wilkin, L.D., Merrick, M.A., Kirby, T.E. and Devor, S.T. (2004). Influence of therapeutic ultrasound on skeletal muscle regeneration following blunt contusion. *Int. J. Sports Med.* **25**(1): 73–77.

Williams, E.J., Benyon, R.C., Trim, N., Hadwin, R., Grove, B.H., Arthur, M.J., Unemori, E.N. and Iredale, J.P. (2001). Relaxin inhibits effective collagen deposition by cultured hepatic stellate cells and decreases rat liver fibrosis *in vivo. Gut* **49**(4): 577–583.

Yamaguchi, Y., Mann, D.M. and Ruoslahti, E. (1990). Negative regulation of transforming growth factor-beta by the proteoglycan decorin. *Nature* **346**(6281): 281–284.

Yamazaki, M., Minota, S., Sakurai, H., Miyazono, K., Yamada, A., Kanazawa, I. and Kawai, M. (1994). Expression of transforming growth factor-beta 1 and its relation to endomysial fibrosis in progressive muscular dystrophy. *Am. J. Pathol.* **144**(2): 221–226.

Zdanowicz, M.M., Moyse, J., Wingertzahn, M.A., O'Connor, M., Teichberg, S. and Slonim, A.E. (1995). Effect of insulin-like growth factor I in murine muscular dystrophy. *Endocrinology* **136**(11): 4880–4886.

37

Peripheral Nerve Regeneration

Mahesh C. Dodla and Ravi V. Bellamkonda

INTRODUCTION

Problems and Challenges of Peripheral Nerve Injuries

Injuries to the peripheral nervous system (PNS) occur frequently, and are a major source of disabilities. PNS injuries impair the ability to move muscles, to feel normal sensations, and result in painful neuropathies.

PNS injuries are classified as being traumatic, non-traumatic, or surgical in nature. Traumatic nerve injuries result from collisions, motor vehicle accidents, gunshot wounds, fractures, lacerations, or other forms of penetrating trauma. In 2002, more than 250,000 patients suffered traumatic peripheral nerve injuries in the United States (AxoGen Inc., 2006). Of these patients, only 15% could be treated due to difficulties in treatment. Even among the patients who received treatment, more than 50% did not show any measurable signs of recovery or suffered from drastically reduced muscle strength. Most of the non-traumatic peripheral nerve injuries are attributed to nerve compression and adhesion. In 2002, more than 400,000 repair procedures were done in the United States to correct carpal tunnel syndrome, a non-traumatic nerve injury. Although treatments for non-traumatic nerve injuries have higher efficacy than traumatic nerve injuries, patients still suffer from pain, loss of muscle strength, and dexterity for several weeks to months. Surgical injuries result from procedures, such as prostatectomy, to remove prostate tumors. Prostatectomy procedures most often require sacrificing one or both of the cavernosal nerves, adversely affecting erectile function and bladder control. In 2002, more than 260,000 patients in the United States suffered major injuries to cavernosal nerves due to prostatectomy procedures (AxoGen Inc., 2006). In order to repair these nerve injuries, several techniques have been used.

Historical Background

The simplest technique for nerve repair, in the case of nerve transection injuries, is coaptation of the two ends of the nerve using sutures or fibrin glue. However, in many cases, there is loss of nerve segment due to injury, or there might be a time lag between injury and surgical repair during which the nerve ends might retract, resulting in a nerve gap. In such cases, end-to-end nerve suturing cannot be done without creating tension in the nerve segment, resulting in a poor regeneration outcome (Terzis et al., 1975). To overcome this problem, the two nerve ends are approximated using grafts, such as nerve autografts/allografts, muscle grafts, vein grafts, muscle–vein grafts, and synthetic nerve guidance conduits/channels (NGCs).

During the 19th and early 20th centuries, various materials were used to promote nerve repair, such as bone (Gluck, 1880), metal tubes (Payr, 1900), blood vessels (Weiss and Taylor, 1946), and fat sheaths (Kirk and Lewis, 1915). The use of autologous nerve grafts was also first reported during this time (Albert, 1885). However, due to improper surgical techniques, anatomical repair rarely led to an appreciable return of

Copyright © 2008 Elsevier Inc. All rights reserved

function. During World War II, surges in the number of nerve injuries led to advances in microsurgical techniques and instrumentation. Further refinements in microsurgical techniques, revolutionized by Millesi (Millesi et al., 1972), and drug therapies have had beneficial effects. Significant advances in surgical techniques have been achieved, and now biological factors rather than surgical techniques are limiting improvements in nerve regeneration. Further advances may come from greater understanding of the molecular mechanisms of nerve regeneration, advances in nerve cell culture, development of new biomaterials and improved genetic techniques.

Current Strategies for Regeneration

At present, nerve autografts are considered the "gold standard" for bridging nerve gaps in the PNS (Lundborg, 1988). Autologous tissue grafts possess several advantages. They are likely to be more biocompatible than artificial materials, are less toxic and provide a support structure to promote cell adhesion and migration (Evans, 2001). However, there are several disadvantages with this technique. Obtaining natural graft leads to loss of function, as well as potential neuroma formation at donor site, needs multiple surgeries, and multiple small grafts are needed for a long nerve gap. There can be a size mismatch between the donor nerve graft and injured nerve. In addition, complete functional recovery is seldom obtained with autografts (Kline et al., 1998). Therefore, there is interest in developing techniques to not only enhance the performance of autografts, but also to synthesize alternatives with better functionality than autografts. Analytically, natural or artificial grafts used to bridge nerve gaps can be thought to have four central components germane to regeneration: (1) scaffold/substrate, (2) growth factors, (3) extracellular matrix (ECM) molecules and (4) cells. A graft may have any combination of the four components. In this analytical framework, the presence of these components and their spatio-temporal distribution determines the efficacy of the graft. In this chapter, based on the distribution of these four components within the graft, the grafts are classified as isotropic or anisotropic. In isotropic grafts, the components are distributed uniformly within the graft, with no directional cues. In anisotropic grafts one or more of these components are distributed anisotropically, usually along the direction of regeneration, to direct the axonal growth toward the distal target (Figure 37.1).

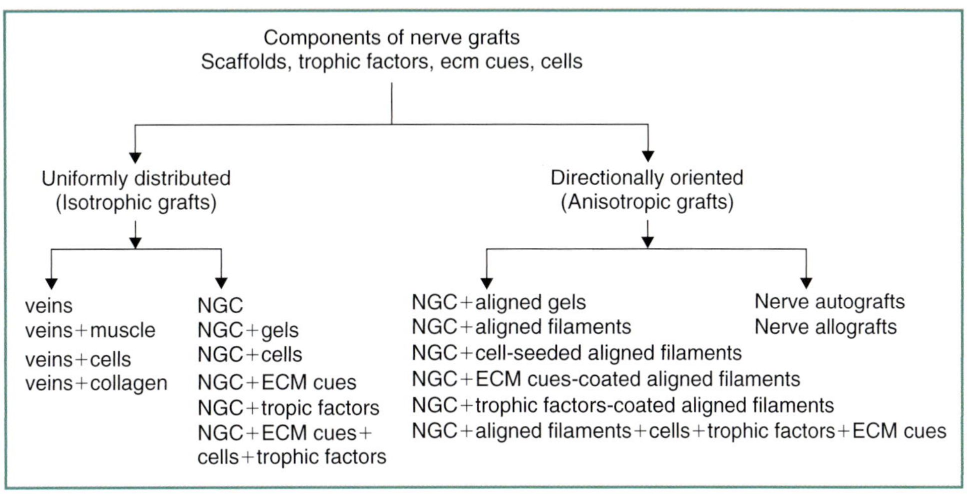

Figure 37.1 Classification of nerve grafts as isotropic or anisotropic. Four basic components make up a nerve graft. All nerve grafts consist of either one or more of these components. In isotropic grafts, the components are distributed uniformly. In anisotropic grafts, the components are aligned longitudinally or in an increasing gradient within the graft.

ISOTROPIC NERVE GRAFTS FOR REGENERATION

This section on isotropic grafts has been divided into four sub-sections, to discuss the four components influencing peripheral nerve regeneration.

Natural and Synthetic Scaffolds for Nerve Repair

A scaffold can consist of two components. The first is a tubular structure serving as a "guidance channel" and the second consists of scaffold elements that are inside the tubular structure. In general, scaffolds for nerve repair should support axonal proliferation, have low antigenicity, support vascularization, be porous for oxygen diffusion, and avoid long-term compression. The scaffold can be made from natural or synthetic materials.

Natural Materials as Scaffolds

Isotropic natural materials used as scaffolds include veins, skeletal muscle fibers, and collagen. Although these materials support nerve regeneration, they do not provide any direction to the axons. Autologous vein grafts have been shown to provide a good environment for axonal regeneration in short nerve gaps (Wang et al., 1993; Ferrari et al., 1999). However, use of vein grafts for long nerve gaps has been less successful because of the collapse of veins due to their thin walls, and constriction due to the surrounding scar tissue (Chiu and Strauch, 1990). In order to prevent vein grafts from collapsing and to improve their performance, intraluminal space fillers such as autologous Schwann cells (SCs), collagen, and muscle fibers have been used. Collagen-filled vein grafts were found to promote better axonal growth than empty vein grafts for a 15 mm nerve gap in rabbits (Choi et al., 2005b). Similarly, SC-seeded venous grafts supported axonal growth and performed better than unseeded grafts in repairing 40 mm nerve gaps (Zhang et al., 2002) and 60 mm nerve gaps in rabbits (Strauch et al., 2001). The principal drawback of this approach is that it requires the availability of a relevant amount of live autologous SCs (up to 8 million cells/ml) that are difficult to obtain. Muscle–vein combined grafts, in which the muscle fibers are inserted in veins, were used in 10 mm long nerve defects in rats and found to promote axonal regeneration comparable to that of syngenic nerve grafts (Geuna et al., 2004). Although the muscle–vein grafts were able to promote nerve regeneration in 55 mm long nerve defects in rabbits, they were not comparable to nerve autografts (Geuna et al., 2004). Autologous muscle–vein combined grafts have been used clinically in humans to bridge nerve gaps ranging from 5 to 60 mm. The results were scored as "poor", "satisfactory", "good", and "very good", based on the recovery of sensory and motor functions. Of the 21 lesions repaired (in 20 patients), 10 were lesions of the sensory nerves and 11 were mixed nerve lesions. All lesions in the sensory nerves, except one greater than 30 mm, showed "good" to "very good" recovery. All lesions in the mixed nerves showed "satisfactory" to "good" recovery of motor and sensory functions (Battiston et al., 2000).

Although autogenous/natural materials have shown encouraging results when used for nerve repair, they still have certain drawbacks. Autogenous grafts, require a second surgery, and result in the loss of function at donor site, and neuropathic pain at donor site. Allografts have problems related to preservation and immuno-rejection. In order to avoid these problems, grafts made of artificial/synthetic materials have been used.

Synthetic Scaffolds for Nerve Repair

Among the artificial materials, synthetic tubular NGCs have shown the most promising results so far (Figure 37.2). Some of the commonly used synthetic scaffolds are given in Table 37.1. The use of NGCs reduces tension at the suture line, protects the regenerating axons from the infiltrating scar tissue, and directs

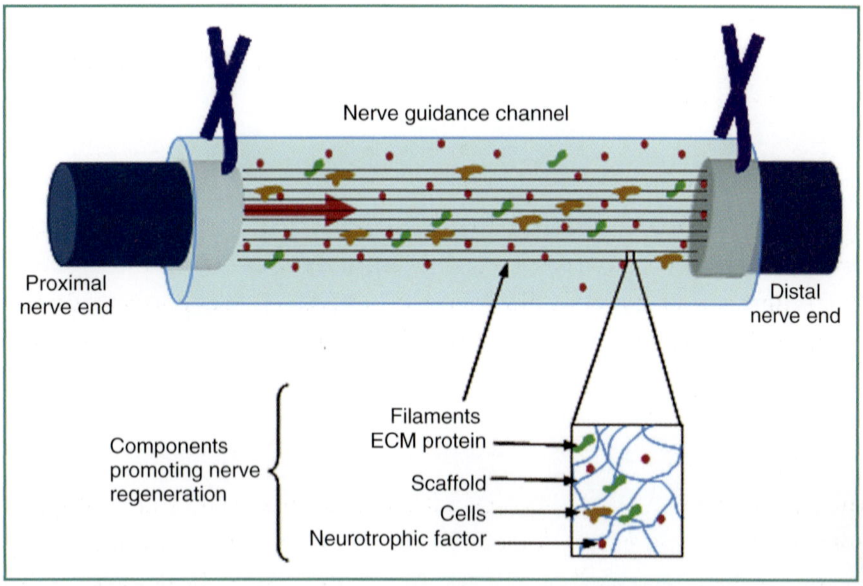

Figure 37.2 A schematic of a synthetic NGC. The NGC, sutured to the nerve ends, is filled with hydrogel, filaments, cells, neurotrophic factors, and ECM proteins. For an isotropic graft, there would be no filaments and the other components would be distributed uniformly. For an anisotropic graft, there may be filaments, and the other components would be aligned longitudinally or in increasing concentration from proximal to distal nerve end.

Table 37.1 Classification of nerve grafts

Classifications	Examples
Isotropic grafts	
Have uniform distribution of one or more of the four components	
A. Scaffolds	
1. Natural materials	Veins (Wang et al., 1993 and Ferrari et al., 1999); muscle fibers (Geuna et al., 2004)
2. Synthetic materials	
a. NGCs	PLA (Cai et al., 2005), PLLA, PGA; AN/PVC (Uzman and Villegas, 1983)
b. Gels	Agarose (Yu and Bellamkonda, 2003); alginate (Suzuki et al., 1999)
B. Neurotrophic factors	NGF (Levi-Montalcini, 1987; Thoenen et al., 1987), BDNF (Sendtner et al., 1992), IGF (Glazner et al., 1993); FGF (Gospodarowicz et al., 1987)
C. ECM proteins	Laminin (Yu and Bellamkonda, 2003), fibronectin (Chen et al., 2000); collagen (Choi et al., 2005b)
D. Support cells	SCs (Guenard et al., 1992), fibroblasts (Nakahara, et al., 1996); stem cells (Ansselin et al., 1997; Choi et al., 2005a)
Anisotropic grafts	
Have directional distribution of one or more of the four components	
A. Scaffolds	
1. Aligned filaments	Collagen (Yoshii et al., 2003; Matsumoto et al., 2000); PLLA (Ngo et al., 2003)

(Continued)

Table 37.1 (Continued)

Classifications	Examples
2. Magnetically aligned gels	Fibrin (Dubey et al., 2001); collagen (Ceballos et al., 1999; Dubey et al., 1999)
B. Neurotrophic factors	NGF (Cao and Shoichet, 2003; Kapur and Shoichet, 2004), BDNF (Cao and Shoichet, 2003); CNTF; FGF
C. ECM proteins	Laminin (Kam et al., 2001; Saneinejad and Shoichet, 1998); fibronectin; collagen
D. Support cells	SCs (Hadlock, et al., 2000; Rutkowski et al., 2004); fibroblasts; stem cells
Autologous nerve grafts (Gospodarowicz et al., 1987; Nichols et al., 2004) Have all the four components: scaffolds, neurotrophic factors, ECM proteins, and cells	
Nerve allografts (Evans et al., 1999; Midha et al., 2001) Acellular grafts, but structurally similar to autologous nerve graft	

the sprouting axons toward their distal targets. The luminal space of NGCs can be filled with growth promoting matrix, growth factors, and/or appropriate cells. In some cases of nerve repair, NGCs have been used to intentionally leave a small gap between the injured nerve ends, to allow accumulation of cytokines, growth factors, and cells (Dahlin and Lundborg, 2001). The NGCs can be used as an excellent experimental tool, to precisely control the distance between the nerve stumps, test the fluid and tissue entering the channel, and vary the properties of the channel. Although NGCs prevent regenerating nerve fibers from wandering, they do not direct axonal growth microscopically. Hence, for the purposes of this chapter, NGCs have been considered as isotropic scaffolds.

Nerve regeneration in silicone NGCs has been studied in detail (Williams et al., 1983). Within a few hours of implantation the tube fills with serum exuded by the cut blood vessels in the nerve ends. This fluid contains neurotrophic factors, as well as several cytokines and inflammatory cells, such as macrophages. The macrophages help remove the myelin and axonal debris formed due to injury. The fluid also contains the clot-forming protein fibrin. Within days, the fibrin coalesces and forms a longitudinally oriented fibrin cable bridging the two nerve ends. Without the formation of the fibrin cable, axonal regeneration cannot occur, thus making the fibrin cable formation a critical step. The fibrin cable is then invaded by cells migrating from the proximal and distal nerve stumps, including fibroblasts, macrophages, SCs, and endothelial cells (which form capillaries and larger vessels). Axons from the proximal end grow into the fibrin matrix and are engulfed in the cytoplasm of SCs. Some of these axons then reach the distal nerve end and get myelinated. In inert silicone tubes of 10 mm or shorter, these processes occur spontaneously. However, it is generally accepted that impermeable, inert NGCs such as silicone do not support regeneration across defects larger than 10 mm without the presence of exogenous growth factors. The regeneration process can be improved by changing the properties of the tube (permeability, porosity, texture, and electric charge characteristics), and the addition of matrices, neurotrophic factors, ECM molecules, and cells (Valentini and Aebischer, 1997).

Table 37.2 Classification of NGCs based on porosity and degradability

Porosity	Degradability	Examples
Impermeable	Non-degradable	Silicone (Lundborg et al., 1982)
Semipermeable	Non-degradable	PS (Yu and Bellamkonda, 2003), PAN/PVC (Uzman and Villegas, 1983; Aebischer et al.,1989)
Resorbable	Degradable	PLA (Cai et al., 2005), PGA

Based on the porosity and/or degradability of the material used, NGCs can be classified as impermeable, semipermeable and resorbable (Table 37.2). The silicone tube is an example of an impermeable NGC since it does not permit movement of molecules across the tube walls. Porosity affects the movement of soluble factors, oxygen, and waste products, into and out of the NGCs, which is vital for nerve regeneration. Examples of semipermeable tubes are polysulphone (PS) and polyacrylonitrile/polyvinylchloride (PAN/PVC). Nerves regenerated in semipermeable tubes featured more myelinated axons and less connective tissue (Uzman and Villegas, 1983; Aebischer et al., 1989). PAN/PVC channels with a molecular weight cutoff of 50 kD support regeneration even in the absence of a distal nerve stump (Aebischer et al., 1989). Examples of bioresorbable tubes are polylactic acid (PLA), polyglycolic acid (PGA), poly(L-lactide-co-glycolide) (PLGA), poly(lactide-co-caprolactone) (PLC), and poly(3-hydroxybutyrate) (PHB). The use of bioresorbable tubes negates the need for a second surgery to remove the implant and prevents long-term compression of the nerve. However, it is critical that the degradation of the tube not allow fibroblasts to invade the lumen space before regeneration occurs, as this may prevent axons from regenerating.

Inclusion of Hydrogels As Scaffolds

NGCs can be filled with gels to support axonal elongation. Here we briefly describe some of the isotropic gels used for nerve regeneration.

Agarose Gels

Agarose is a polysaccharide derived from red agar and is widely used in gel electrophoresis and gel chromatography. SeaPrep® agarose hydrogel has been shown to support neurite extension from a variety of neurons in a non-immunogenic manner (Bellamkonda et al., 1995; Labrador et al., 1998; Dillon et al., 2000). Agarose gels also allow molecules to be covalently linked to the gels through functional groups on their polysaccharide chains. For example, laminin protein or fragments of laminin can be covalently coupled to SeaPrep® agarose gels to enhance their ability to support neurite extension (Yu et al., 1999). Although agarose gels support neurite growth on their own, coupling of molecules, such as laminin, significantly enhances the gels' ability to promote neurite extension.

Collagen Gels

Collagen gels and filaments have been used to promote PNS regeneration (scaffolds with collagen filaments will be discussed later in the anisotropic scaffolds section). Collagen gel can be used to fill the intra-luminal space of a vein graft to prevent it from collapsing and improve its nerve repair efficiency. In collagen-filled vein grafts, the number and diameter of myelinated axons was significantly increased compared to vein grafts without collagen gel (Choi et al., 2005b). Nerve repair with silicone tubes can be significantly improved by filling them with collagen gel. Collagen tubes filled with collagen gel have promoted more rapid nerve sprouting, and better morphology, than saline-filled collagen tubes (Satou et al., 1986). In some cases, collagen gels have hindered regeneration (Valentini et al., 1987). This negative effect, presumably due to gel remnants blocking diffusion and axonal elongation, might be overcome by reducing the density of the collagen gel (Labrador et al., 1998).

Hyaluronic acid, an ECM component, is associated with decreased scarring and improved fibrin matrix formation. It is hypothesized that during the fibrin matrix phase of regeneration, hyaluronic acid organizes the ECM into a hydrated open lattice, thereby facilitating migration of the regenerated axons (Seckel et al., 1995). Hyaluronan-based tubular conduits, used for peripheral nerve regeneration, resulted in more myelinated axons and higher nerve conduction velocities than silicone tubes filled with saline (Wang et al., 1998), with little cytotoxicity (Jansen et al., 2004) upon degradation.

Other gels used to promote nerve regeneration *in vivo* include Matrigel, alginate gels, fibrin gels, and heparin sulfate gels (Madison et al., 1988; Suzuki et al., 1999; Dubey et al., 2001).

Neurotrophic Factors for Nerve Regeneration

Neurotrophic factors are produced in the target organs and by SCs in response to injury. These neurotrophic factors help maintain the target organ-nerve synapse. A nerve injury usually results in disruption of communication between the target organs and the neuronal cell body, and leads to Wallerian degeneration (breakdown of myelin sheath and axons). Due to cytokines released during Wallerian degeneration, SCs are activated and produce neurotrophins such as nerve growth factor (NGF) and brain-derived neurotrophic factor (BDNF). Although many other trophic factors, including insulin-like growth factor (IGF), fibroblast growth factor (FGF), and ciliary neurotrophic factor (CNTF), have been shown to be involved in the promotion of nerve regeneration (Gospodarowicz et al., 1987; Glazner et al., 1993), it is believed that they are released from SCs following mechanical damage to the cells.

NGF is produced in the target organs of sensory and sympathetic nerves in the PNS, and has been shown to stimulate and promote the survival of sensory ganglia and nerves, including spinal sensory nerves and sciatic nerves (Levi-Montalcini, 1987; Thoenen et al., 1987). BDNF is expressed in very low levels in intact adult peripheral nerves, but is upregulated following injury. BDNF is effective in promoting the survival and outgrowth of not only sensory and sympathetic nerves, but also motor nerves (Sendtner et al., 1992).

Neurotrophic factors are likely an important part of future clinical therapies for peripheral nerve injuries/diseases. Diseases in which the functions of SCs are severely suppressed (multiple sclerosis, for example) or when acellular grafts (containing no viable SCs) are used, application of neurotrophic factors could be highly effective in facilitating nerve regeneration. Various studies have utilized the functions of NGF to promote nerve regeneration. Hubbell and Sakiyama-Elbert have developed a fibrin matrix that immobilizes heparin molecules by electrostatic interactions, which in turn immobilizes heparin-binding growth factors. The fibrin matrix, when implanted *in vivo*, releases the bound growth factor due to fibrin degradation. This system was used to deliver NGF (Lee et al., 2003) for peripheral nerve regeneration *in vivo*, and basic fibroblast growth factor (bFGF) (Sakiyama-Elbert and Hubbell, 2000) for neurite extension from chick dorsal root ganglia (DRG) *in vitro*. Fibrin–heparin–NGF matrix was observed to promote nerve regeneration comparable to syngenic nerve grafts over a 13 mm nerve gap in rats. Fibrin matrix that released bFGF enhanced neurite extension from DRGs by 100% compared to unmodified fibrin matrix.

ECM Molecules for Nerve Regeneration

Insoluble ECM molecules, such as laminin, fibronectin, and certain forms of collagen, promote axonal extension and therefore, are excellent candidates for incorporation into the lumen of NGCs. Agarose gels crosslinked with laminin showed enhanced neurite extension from chick DRG *in vitro* (Yu et al., 1999). Agarose gels crosslinked with laminin and soluble NGF showed nerve regeneration comparable to autografts over a 10 mm gap in rats (Yu and Bellamkonda, 2003). However, axonal extension in the laminin gels depends on gel density. High concentrations of laminin hinder regeneration (Labrador et al., 1998). Matrigel, a gel containing collagen type IV, laminin, and glycosaminoglycans, supports some degree of regeneration over a long nerve gap in adult rats, when introduced into the lumen of NGCs (Madison et al., 1988). Similarly, a gel mixture containing laminin, collagen, and fibronectin significantly improved nerve regeneration compared to saline-filled silicone channels (Chen et al., 2000).

Seeding Neuronal Support Cells for Nerve Regeneration

In the PNS, SCs are support cells that wrap around the axons. SCs form a multilamellar sheath of myelin, a phospholipid-containing substance around axons that serves as an insulator and increases nerve conduction velocity. An individual SC may ensheath several unmyelinated axons, but only one myelinated axon within its cytoplasm.

In NGCs used for nerve regeneration, formation of fibrin cable, migration of SCs and longitudinal arrangement of SCs (known as bands of Büngner) are necessary processes for axonal regeneration. For nerve gaps less than the critical length (10 mm) these processes occur spontaneously, leading to axonal regeneration. However, for nerve gaps greater than 10 mm, spontaneous nerve regeneration does not occur, due to lack of formation of a fibrin cable and the bands of Büngner (Lundborg, 1988). SCs of uninjured nerves are quiescent. Following nerve axotomy, the SCs become "reactive" and produce a number of neurotrophic factors, including NGF, BDNF, and CNTF (Thanos et al., 1998). They also synthesize and secrete ECM molecules, such as laminin, which is known to modulate neurite outgrowth and express a variety of other cell adhesion molecules. All these components have been known to play roles in supporting neuronal survival and axonal regeneration.

Using SCs in NGCs bypasses the fibrin cable formation step, accelerates the formation of bands of Büngner, and introduces a persistent source of neurotrophic factors, leading to more efficient nerve repair. This could decrease the time required by the axons to reconnect to the target organ, as well as increase the distance over which regeneration occurs. SCs isolated from the peripheral nerve of a patient, and expanded *in vitro*, could be used to treat the patient's nerve injuries. Addition of SCs has been shown to significantly improve the performance of various scaffolds, such as empty NGCs, collagen gels, venous nerve grafts, and muscle grafts, as compared to control scaffolds without SCs (Ansselin et al., 1997; Strauch et al., 2001; Keilhoff et al., 2005). The ability of SC-seeded NGCs to promote regeneration was found to be dependent on the SCs seeding density, and the immunocompatibility between donors and host (Guenard et al., 1992). For syngenic SCs, it was observed that increasing the seeding density improves the nerve regeneration outcome. Heterologous SCs elicited a strong immune reaction, impeding the nerve regeneration. The performance of SC-seeded NGCs was further improved by designing longitudinally aligned channels in the tube to resemble acellular nerve grafts. This will be discussed in more detail in the section on anisotropic scaffolds.

As an alternative to SCs, other cells could be used, as they are genetically modified, to produce desired levels of neurotrophic factors, or to express specific ECM molecules. Fibroblasts, genetically modified to produce NGF, BDNF, neurotrophin 3 (NT-3), and bFGF, showed promising results in central nervous system (CNS) regeneration (Nakahara et al., 1996). Olfactory ensheathing cells have been shown to promote regeneration of cut nerves in the adult rat spinal cord (Li et al., 2003). Although these are examples of CNS regeneration, genetically modified cells can also be used for PNS regeneration. The addition of bone marrow stromal cells to the NGCs has shown improved regeneration over empty NGCs (Choi et al., 2005a). Similarly, pluripotent stem cells derived from hair follicles have shown improvements in rats (Amoh et al., 2005). However, the

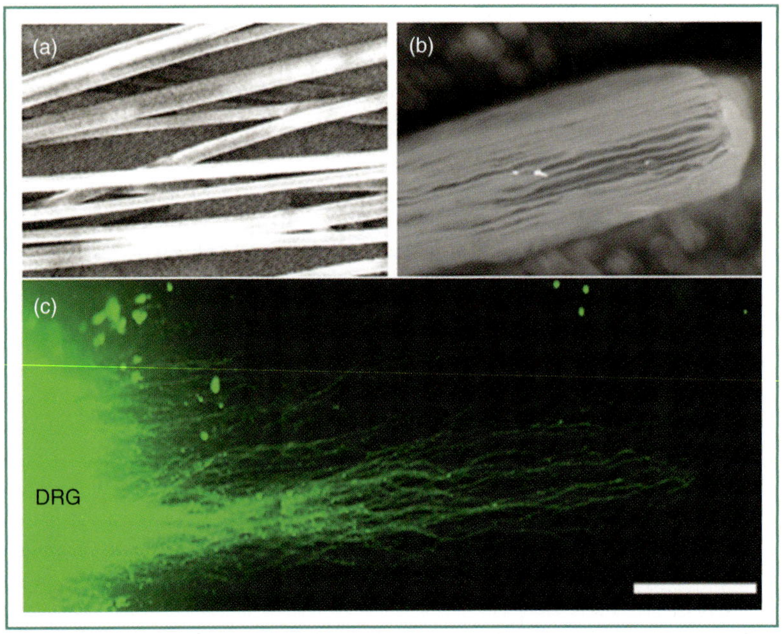

Figure 37.3 **Nanofilaments for contact guidance mediated growth. (a) SEM image of PAN-MA nanofilaments, with diameter of 400–700 nm. (b) Three-dimensional nanofilaments-based scaffold, along with agarose gel, embedded in a polysulfone NGC can be used to direct neurite growth *in vitro*, or axonal growth *in vivo*. (c) Chick DRG extending neurites along horizontally oriented PAN-MA filaments, *in vitro* (Scale bar = 1 mm). (Images courtesy: Dr. Young-tae Kim, Georgia Institute of Technology, Atlanta, GA.)**

difficulties of isolating and culturing these cells from the patient prior to surgery could limit this approach for some surgical procedures.

ANISOTROPIC NERVE GRAFTS

The four essential elements of nerve grafts, scaffolds, neurotrophic factors, ECM molecules, and cells, can be presented in an aligned fashion so as to orient the regenerating axons toward their distal targets. In this section, studies involving nerve grafts that provide directional guidance are discussed.

Aligned Anisotropic Scaffolds

We hypothesize that the superior performance of autologous nerve grafts is due to its cellular components and its longitudinally aligned structure. The longitudinally aligned structure of the degenerating nerves in the autografts provides contact guidance and direction to the regenerating nerves. In an attempt to mimic autografts, longitudinally patterned or oriented gels and filaments to guide and accelerate the regenerating axons have been designed (Figure 37.3). It has been seen that a poly (acrylonitrile-co-methylacrylate) (PAN-MA) nanofilament-based scaffold by itself can facilitate regeneration across a 17 mm nerve gap in rats (Kim et al.). Many other combinations of materials have been used, such as collagen filaments embedded in collagen tubes (Yoshii et al., 2003), laminin-coated collagen fibers in collagen tubes (Matsumoto et al., 2000), laminin–fibronectin double-coated collagen fibers in collagen tubes (Tong et al., 1994), poly(L-lactide) (PLLA) filaments in silicone tubes and PLA tubes (Ngo et al., 2003; Cai et al., 2005), and PGA fibers in chitosan tubes (Wang et al., 2005). All have been found to significantly improve regeneration compared to saline-filled tubes. In addition to synthetic filaments, magnetically aligned fibrin and collagen type I gels have also been used to provide directional guidance to neurites *in vitro* (Dubey et al., 1999, 2001) and axons *in vivo* (Ceballos et al., 1999).

Neurotrophic Factors

Neurotrophic factors have been delivered *in vivo* mostly in an isotropic manner. However, *in vitro* studies have suggested that gradients of neurotrophic factors can direct growth cones toward the source of neurotrophic factor (Gundersen and Barrett, 1979). Insoluble and soluble gradients of NGF, NT-3, and BDNF have been shown to direct the growth of neurites from PC12 cells toward increasing concentrations of neurotrophic factors (Cao and Shoichet, 2003; Kapur and Shoichet, 2004). Therefore, anisotropic scaffolds having gradients of neurotrophic factors, along with other components, might be important tools for PNS regeneration.

ECM Molecules

Gels containing ECM molecules, such as laminin, collagen, fibronectin, and glycosaminoglycans, have been widely used to make isotropic scaffolds for nerve regeneration (discussed earlier in "Current Strategies for Regeneration"). ECM molecules promote axonal growth by the mechanism of differential adhesion, wherein axons preferentially grow on substrates of ECM molecules due to the presence of specific cell surface receptors. *In vitro*, experiments with spatial patterns of whole ECM molecules (Kam et al., 2001) or their peptide derivates (Saneinejad and Soichet, 1998) have been used to direct the growth of neurites, as well as enhance neurite extension. *In vivo*, ECM protein-coated fibers have been used to enhance nerve regeneration, where the fibers provide the contact guidance for regenerating axons and the ECM protein provides the adhesive substrate (Tong et al., 1994; Matsumoto et al., 2000). Experiments performed *in vitro* have demonstrated that gradients of ECM proteins could orient and enhance neurite outgrowth toward increasing concentrations of ECM molecules (Dertinger et al., 2002; Adams et al., 2005). However, this technique is yet to be utilized to enhance nerve regeneration *in vivo*, due to the difficulties in making gradients of proteins in three-dimensional scaffolds.

Cell-Seeded Longitudinally Aligned NGCs

Neuronal growth supporting cells can be incorporated with longitudinally aligned filaments and gels in NGCs to enhance nerve regeneration. Since support cells synthesize ECM proteins and neurotrophic factors, aligned cells often result in directionally aligned ECM. Biodegradable conduits of a copolymer of lactic and PLGA with longitudinally aligned channels have been used for nerve regeneration (Hadlock et al., 2000). The channels, with the lumen coated with laminin and seeded with SCs, showed regeneration comparable to nerve autografts over a 7 mm nerve gap in rats. PLA tubes, with a micropatterned inner lumen and seeded with SCs, showed better nerve regeneration compared to unpatterned tubes with SCs (Rutkowski et al., 2004). The disadvantages of cell-seeded NGCs include the need for prolonged isolation and cell culture to prepare cells for implantation, high cell yield, and high cellular morbidity.

NATURAL NERVE GRAFTS

A common source of nerve grafts is the sural nerve, which is easy to obtain, has the appropriate diameter for most grafting needs, and is relatively dispensable. Other graft sources include the anterior branch of the medial ante-brachial cutaneous nerve, the lateral femoral cutaneous nerve, and the superficial radial sensory nerve (Sunderland, 1991). However, a motor nerve has a preference for a motor pathway (i.e. motor nerve graft) and shows inferior regeneration if a sensory nerve graft, such as sural nerve, is used. Similarly, a mixed nerve shows superior regeneration with either a mixed nerve or a motor nerve graft as compared to a sensory nerve graft (Nichols et al., 2004). Therefore, clinical outcomes might be improved by using alternatives to sensory nerve grafts in the reconstruction of a mixed nerve. However, there are relatively few expendable motor/mixed nerves in the human body that could be used as graft materials. Therefore, a more feasible alternative would be to use nerve allografts or biosynthetic graft materials. Cadavers are a source of graft materials and avoid the complications of harvesting autografts. However, cadaveric nerve allografts require maintenance and can be used only with immunosuppressive therapy. The withdrawal of the immunosuppressant leads to profound loss of axons in the allografts. The axonal loss is most profound in mixed nerve allografts as compared to motor nerve allografts, followed by sensory nerve allografts (Midha et al., 2001).

Allografts, cold-preserved and/or freeze-thawed to prevent immuno-rejection by the host body, perform better than fresh allografts in terms of axon density, fiber diameter, and nerve conduction velocity (Evans et al., 1999). Using natural materials (nerve grafts) for regeneration is ideal. However, it has been shown that if autografts or allografts are preserved for too long, their ability to support regeneration is compromised (Gulati, 1996). Also, the pre-treated allografts do not perform as well as autografts (Evans et al., 1999).

Although nerve autografts are used as a "gold standard," the lack of functional recovery even with autografts remains an important clinical problem. Techniques utilized to improve the performance of the nerve autografts include treatments to either remove inhibitory molecules like chondroitin sulfate proteoglycans (CSPGs), or provide factors for axonal growth, such as neurotrophins 4/5 (NT-4/5) or BDNF. CSPG molecules have a core protein structure with glycosaminoglycan (GAG) side chains composed of chondroitin sulfate. Due to their large size and negative charge, the GAGs of CSPGs are thought to hinder neurite access to growth-promoting matrix molecules and also to repel the axons, thereby inhibiting their growth (Properzi et al., 2003). It has been shown that CSPGs are upregulated almost seven-fold in the distal segment of peripheral nerve gaps following transection nerve injury (Zuo et al., 1998). The upregulated CSPGs contribute significantly to inhibition of neurite sprouting and, consequently, growth into the distal nerve segment. Treatment of the injury site with chondroitinase ABC, which digests away the inhibitory CSPGs, increases the neurite ingrowth into the distal nerve segment several fold, as compared to untreated controls without any chondroitinase ABC treatment (Zuo et al., 2002). Treatment with chondroitinase ABC, however, did not improve neurite ingrowth in a crush injury model, suggesting that CSPGs are not upregulated in a crush injury.

Syngenic nerve grafts treated with chondroitinase ABC, heparinase I, heparinase III, or keratanase enzymes have shown significantly improved axonal ingrowth from the proximal nerve end into the nerve graft, as compared to untreated controls (Groves et al., 2005). Autografts treated with a combination of all these four enzymes showed the most significant neurite growth into the graft. However, the combination treatment was not significantly different from the arithmetic sum of the individual treatments. This suggests that molecules such as heparan sulfate proteoglycan and keratan sulfate proteoglycan also contribute to inhibition of neurite growth apart from CSPGs, and the pathways/mechanisms of inhibition for each of these molecules might be independent of each other. In another study, an exogenous supply of BDNF and NT-4/5 was found to increase the number of axons regenerating into the nerve graft as compared to untreated nerve grafts (English et al., 2005). These techniques, used in clinical applications, could lead to better results with autografts/allografts.

ANIMAL MODELS

Traditionally, nerve regeneration studies have involved the use of various animal models, such as mouse, rat, swine, canine, sheep, and non-human primates. Rat or mouse models are used initially to determine the efficacy of the various treatments. If the results are encouraging, they are followed by experiments with larger animal models. For PNS regeneration studies, the most commonly studied nerve models are the sciatic nerve and its branches, the tibial and the peroneal nerves. Other models include the cavernosal nerve and the facial nerve. The most common nerve injury model is the single-anastomosis model, where the injury and repair are done on one sciatic nerve and the contralateral sciatic nerve is used as a control. This model is useful when the nerve gap is less than 20 mm. The second version is the cross (double) anastomosis model, where both contralateral sciatic nerves are transected; the proximal end of the right sciatic nerve is then sutured to one end of an implanted tube, and the distal end of the left sciatic nerve is inserted into the other end of the tube (Lundborg et al., 1982). This model allows the study of gaps in excess of 25 mm. Although very convenient, the rat and mouse models suffer from the serious drawback that they present only short nerve gaps for regeneration studies. In order for a regeneration technique to be successfully applied in clinical trials, the nerve gap model has to be more than 40 mm in length. To create a long nerve gap model, rabbits (Geuna et al., 2004), cats (Suzuki et al., 1999), dogs (Matsumoto et al., 2000), sheep (Lawson and Glasby, 1998), and non-human primates (Ahmed et al., 1999) have been used. The large animal models are an important intermediary step before clinical application of experimental therapeutic approaches.

SUMMARY

In spite of significant advances in research to the development of synthetic NGCs, nerve autografts are still considered the first-choice strategy for nerve repair, especially in the case of long nerve gaps. However, even the performance of autografts has been unsatisfactory. Using autografts generally results in a good recovery of sensory functions, but negligible return of motor functions. Hence, there will be continued interest in ideas to further enhance the performance of autografts by various treatments, such as chondroitinase ABC, NT-4/5, and BDNF. However, shortage of autografts and allografts is a hindrance to their usage. This shortage can be overcome only by developing synthetic alternatives to autografts. Modulating the spatio-temporal distribution of the four components of grafts germane to regeneration can potentially improve the potential outcomes with these grafts. Ongoing rapid advances in cell biology, cell culture techniques, genetic engineering, and biomaterials research are likely to provide new tools to improve regeneration using NGCs, and the day an engineered construct performs as well as autografts may be near.

REFERENCES

Adams, D.N., Kao, E.Y., Hypolite, C.L., Distefano, M.D., Hu, W.S. and Letourneau, P.C. (2005). Growth cones turn and migrate up an immobilized gradient of the laminin IKVAV peptide. *J. Neurobiol.* **62**(1): 134–147.

Aebischer, P., Guenard, V. and Brace, S. (1989). Peripheral nerve regeneration through blind-ended semipermeable guidance channels: effect of molecular weight cutoff. *J. Neurosci.* **9**: 3590–3595.

Ahmed, Z., Brown, R.A., Ljungberg, C., Wiberg, M. and Terenghi, G. (1999). Nerve growth factor enhances peripheral nerve regeneration in non-human primates. *Scand. J. Plast. Reconstr. Surg. Hand Surg.* **33**(4): 393–401.

Albert, E. (1885). Einige Operationen an Nerven. *Wien Med.* **26**: 1285.

Amoh, Y., Li, L., Campillo, R., Kawahara, K., Katsuoka, K., Penman, S. and Hoffman, R.M. (2005). Implanted hair follicle stem cells form Schwann cells that support repair of severed peripheral nerves. *Proc. Natl. Acad. Sci. USA* **102**(49): 17734–17738.

Ansselin, A.D., Fink, T. and Davey, D.F. (1997). Peripheral nerve regeneration through nerve guides seeded with adult Schwann cells. *Neuropathol. Appl. Neurobiol.* **23**(5): 387–398.

AxoGen Inc. (2006). Retrieved January 2, 2006, from www.axogeninc.com.

Battiston, B., Tos, P., Cushway, T.R. and Geuna, S. (2000). Nerve repair by means of vein filled with muscle grafts I. Clinical results. *Microsurgery* **20**: 32–36.

Bellamkonda, R.V., Ranieri, J.P., Bouche, N. and Aebischer, P. (1995). Hydrogel-based three-dimensional matrix for neural cells. *J. Biomed. Mater. Res.* **29**: 663–671.

Cai, J., Peng, X., Nelson, K.D., Eberhart, R. and Smith, G.M. (2005). Permeable guidance channels containing microfilament scaffolds enhance axon growth and maturation. *J. Biomed. Mater. Res. A.* **75**(2): 374–386.

Cao, X. and Shoichet, M.S. (2003). Investigating the synergistic effect of combined neurotrophic factor concentration gradients to guide axonal growth. *Neuroscience* **122**(2): 381–389.

Ceballos, D., Navarro, X., Dubey, N., Wendelschafer-Crabb, G., Kennedy, W.R. and Tranquillo, R.T. (1999). Magnetically aligned collagen gel filling a collagen nerve guide improves peripheral nerve regeneration. *Exp. Neurol.* **158**(2): 290–300.

Chen, Y.S., Hsieh, C.L., Tsai, C.C., Chen, T.H., Cheng, W.C., Hu, C.L. and Yao, C.H. (2000). Peripheral nerve regeneration using silicone rubber chambers filled with collagen, laminin and fibronectin. *Biomaterials* **21**(15): 1541–1547.

Chiu, D.T. and Strauch, B. (1990). A prospective clinical evaluation of autogenous vein grafts used as a nerve conduit for distal sensory nerve defects of 3 cm or less. *Plast. Reconstr. Surg.* **86**(5): 928–934.

Choi, B.H., Zhu, S.J., Kim, B.Y., Huh, J.Y., Lee, S.H. and Jung, J.H. (2005a). Transplantation of cultured bone marrow stromal cells to improve peripheral nerve regeneration. *Int. J. Oral Maxillofac. Surg.* **34**(5): 537–542.

Choi, B.H., Zhu, S.J., Kim, S.H., Kim, B.Y., Huh, J.H., Lee, S.H. and Jung, J.H. (2005b). Nerve repair using a vein graft filled with collagen gel. *J. Reconstr. Microsurg.* **21**(4): 267–272.

Dahlin, L.B. and Lundborg, G. (2001). Use of tubes in peripheral nerve repair. *Neurosurg. Clin. N. Am.* **12**(2): 341–352.

Dertinger, S.K., Jiang, X., Li, Z., Murthy, V.N. and Whitesides, G.M. (2002). Gradients of substrate-bound laminin orient axonal specification of neurons. *Proc. Natl Acad. Sci. USA* **99**(20): 12542–12547.

Dillon, G.P., Yu, X. and Bellamkonda, R.V. (2000). The polarity and magnitude of ambient charge influences three-dimensional neurite extension from DRGs. *J. Biomed. Mater. Res.* **51**(3): 510–519.

Dubey, N., Letourneau, P.C. and Tranquillo, R.T. (1999). Guided neurite elongation and Schwann cell invasion into magnetically aligned collagen in simulated peripheral nerve regeneration. *Exp. Neurol.* **158**(2): 338–350.

Dubey, N., Letourneau, P.C. and Tranquillo, R.T. (2001). Neuronal contact guidance in magnetically aligned fibrin gels: effect of variation in gel mechano-structural properties. *Biomaterials* **22**(10): 1065–1075.

English, A.W., Meador, W. and Carrasco, D.I. (2005). Neurotrophin-4/5 is required for the early growth of regenerating axons in peripheral nerves. *Eur. J. Neurosci.* **21**(10): 2624–2634.

Evans, G.R. (2001). Peripheral nerve injury: a review and approach to tissue engineered constructs. *Anat. Rec.* **263**(4): 396–404.

Evans, P.J., MacKinnon, S.E., Midha, R., Wade, J.A., Hunter, D.A., Nakao, Y. and Hare, G.M. (1999). Regeneration across cold preserved peripheral nerve allografts. *Microsurgery* **19**(3): 115–127.

Ferrari, F., De Castro Rodrigues, A., Malvezzi, C.K., Dal Pai Silava, M. and Padvoni, C.R. (1999). Inside-out vs. standard vein graft to repair a sensory nerve in rats. *Anat. Rec.* **256**: 227–232.

Geuna, S., Tos, P., Battiston, B. and Giacobini-Robecchi, M.G. (2004). Bridging peripheral nerve defects with muscle–vein combined guides. *Neurol Res.* **26**(2): 139–144.

Glazner, G.W., Lupien, S., Miller, J.A. and Ishii, D.N. (1993). Insulin-like growth factor II increases the rate of sciatic nerve regeneration in rats. *Neuroscience* **54**(3): 791–797.

Gluck, T. (1880). Ueber Neuroplastik auf dem Wege der Transplantation. *Arch. Klin. Chir.* **25**: 606–616.

Gospodarowicz, D., Ferrara, N., Schweigerer, L. and Neufeld, G. (1987). Structural characterization and biological functions of fibroblast growth factor. *Endocr. Rev.* **8**(2): 95–114.

Groves, M.L., McKeon, R., Werner, E., Nagarsheth, M., Meador, W. and English, A.W. (2005). Axon regeneration in peripheral nerves is enhanced by proteoglycan degradation. *Exp. Neurol.* **195**: 278–292.

Guenard, V., Kleitman, N., Morrissey, T.K., Bunge, R.P. and Aebischer, P. (1992). Syngeneic Schwann cells derived from adult nerves seeded in semipermeable guidance channels enhance peripheral nerve regeneration. *J. Neurosci.* **12**(9): 3310–3320.

Gulati, A.K. (1996). Peripheral nerve regeneration through short- and long-term degenerated nerve transplants. *Brain Res.* **742**(1–2): 265–270.

Gundersen, R.W. and Barrett, J.N. (1979). Neuronal chemotaxis: chick dorsal-root axons turn toward high concentrations of nerve growth factor. *Science* **206**(4422): 1079–1080.

Hadlock, T., Sundback, C., Hunter, D., Cheney, M. and Vacanti, J.P. (2000). A polymer foam conduit seeded with Schwann cells promotes guided peripheral nerve regeneration. *Tissue Eng.* **6**(2): 119–127.

Jansen, K., van der Werff, J.F., van Wachem, P.B., Nicolai, J.P., de Leij, L.F. and van Luyn, M.J. (2004). A hyaluronan-based nerve guide: *in vitro* cytotoxicity, subcutaneous tissue reactions, and degradation in the rat. *Biomaterials* **25**(3): 483–489.

Kam, L., Shain, W., Turner, J.N. and Bizios, R. (2001). Axonal outgrowth of hippocampal neurons on micro-scale networks of polylysine-conjugated laminin. *Biomaterials* **22**(10): 1049–1054.

Kapur, T.A. and Shoichet, M.S. (2004). Immobilized concentration gradients of nerve growth factor guide neurite outgrowth. *J Biomed. Mater. Res. A* **68**(2): 235–243.

Keilhoff, G., Pratsch, F., Wolf, G. and Fansa, H. (2005). Bridging extra large defects of peripheral nerves: possibilities and limitations of alternative biological grafts from a cellular muscle and Schwann cells. *Tissue Eng.* **11**(7–8): 1004–1114.

Kim, Y., Haftel, V.K., Kumar, S. and Bellamkonda, R.V. Oriented nanoscaffolds match the performance of autografts in facilitating regeneration across long peripheral nerve gaps. *Nat. Biotechnol.* (in review).

Kirk, E.G. and Lewis, D. (1915). Fascial tubulization in the repair of nerve defects. *J. Am. Med. Assoc.* **65**: 486–492.

Kline, D.G., Kim, D., Midha, R., Harsh, C. and Tiel, R. (1998). Management and results of sciatic nerve injuries: a 24-year experience. *J. Neurosurgery* **89**: 13–23.

Labrador, R.O., Buti, M. and Navarro, X. (1998). Influence of collagen and laminin gels concentration on nerve regeneration after resection and tube repair. *Exp. Neurol.* **149**(1): 243–252.

Lawson, G.M. and Glasby, M.A. (1998). Peripheral nerve reconstruction using freeze-thawed muscle grafts: a comparison with group fascicular nerve grafts in a large animal model. *J. Roy. Coll. Surg. Edin.* **43**(5): 295–302.

Lee, A.C., Yu, V.M., Lowe III, J.B., Brenner, M.J., Hunter, D.A., Mackinnon, S.E. and Sakiyama-Elbert, S.E. (2003). Controlled release of nerve growth factor enhances sciatic nerve regeneration. *Exp. Neurol.* **184**(1): 295–303.

Levi-Montalcini, R. (1987). The nerve growth factor 35 years later. *Science* **237**(4819): 1154–1162.

Li, Y., Decherchi, P. and Raisman, G. (2003). Transplantation of olfactory ensheathing cells into spinal cord lesions restores breathing and climbing. *J. Neurosci.* **23**(3): 727–731.

Lundborg, G. (1988). *Nerve Injury and Repair.* New York: Longman Group UK Ltd.

Lundborg, G., Dahlin, L.B., Danielsen, N., Gelberman, R.H., Longo, F.M., Powell, H.C. and Varon, S. (1982). Nerve regeneration in silicone chambers: influence of gap length and of distal stump components. *Exp. Neurol.* **76**(2): 361–375.

Madison, R.D., Da Silva, C.F. and Dikkes, P. (1988). Entubulation repair with protein additives increases the maximum nerve gap distance successfully bridged with tubular prostheses. *Brain Res.* **447**(2): 325–334.

Matsumoto, K., Ohnishi, K., Kiyotani, T., Sekine, T., Ueda, H., Nakamura, T., Endo, K. and Shimizu, Y. (2000). Peripheral nerve regeneration across an 80-mm gap bridged by a polyglycolic acid (PGA)-collagen tube filled with laminin-coated collagen fibers: a histological and electrophysiological evaluation of regenerated nerves. *Brain Res.* **868**(2): 315–328.

Midha, R., Nag, S., Munro, C.A. and Ang, L C. (2001). Differential response of sensory and motor axons in nerve allografts after withdrawal of immunosuppressive therapy. *J. Neurosurg.* **94**(1): 102–110.

Millesi, H., Meissl, G. and Berger, A. (1972). The interfascicular nerve-grafting of the median and ulnar nerves. *J. Bone Joint Surg. Am.* **54**: 7727–7750.

Nakahara, Y., Gage, F.H. and Tuszynski, M.H. (1996). Grafts of fibroblasts genetically modified to secrete NGF, BDNF, NT-3, or basic FGF elicit differential responses in the adult spinal cord. *Cell Transplant.* **5**(2): 191–204.

Ngo, T.T., Waggoner, P.J., Romero, A.A., Nelson, K.D., Eberhart, R.C. and Smith, G.M. (2003). Poly(L-lactide) microfilaments enhance peripheral nerve regeneration across extended nerve lesions. *J. Neurosci. Res.* **72**(2): 227–238.

Nichols, C.M., Brenner, M.J., Fox, I.K., Tung, T.H., Hunter, D.A., Rickman, S.R. and Mackinnon, S.E. (2004). Effects of motor versus sensory nerve grafts on peripheral nerve regeneration. *Exp. Neurol.* **190**(2): 347–355.

Payr, E. (1900). Beitrage zur Technik der Blutgefass und Nervennaht nebst Mittheilungen uber die Vervendung eines resorbibaren Metalles in der Chirurgie. *Arch. Klin. Chir.* **62**: 67.

Properzi, F., Asher, R.A. and Fawcett, J.W. (2003). Chondroitin sulphate proteoglycans in the central nervous system: changes and synthesis after injury. *Biochem. Soc. Trans.* **31**(2): 335–336.

Rutkowski, G.E., Miller, C.A., Jeftinija, S. and Mallapragada, S.K. (2004). Synergistic effects of micropatterned biodegradable conduits and Schwann cells on sciatic nerve regeneration. *J. Neural. Eng.* **1**(3): 151–157.

Sakiyama-Elbert, S.E. and Hubbell, J.A. (2000). Development of fibrin derivatives for controlled release of heparin-binding growth factors. *J. Control. Release* **65**(3): 389–402.

Saneinejad, S. and Shoichet, M.S. (1998). Patterned glass surfaces direct cell adhesion and process outgrowth of primary neurons of the central nervous system. *J. Biomed. Mater. Res.* **42**(1): 13–19.

Satou, T., Nishida, S., Hiruma, S., Tanji, K., Takahashi, M., Fujita, S., Mizuhara, Y., Akai, F. and Hashimoto, S. (1986). A morphological study on the effects of collagen gel matrix on regeneration of severed rat sciatic nerve in silicone tubes. *Acta. Pathol. Jpn.* **36**(2): 199–208.

Seckel, B.R., Jones, D., Hekimian, K.J., Wang, K.K., Chakalis, D.P. and Costas, P.D. (1995). Hyaluronic acid through a new injectable nerve guide delivery system enhances peripheral nerve regeneration in the rat. *J. Neurosci. Res.* **40**(3): 318–324.

Sendtner, M., Holtmann, B., Kolbeck, R., Thoenen, H. and Barde, Y.A. (1992). Brain-derived neurotrophic factor prevents the death of motoneurons in newborn rats after nerve section. *Nature* **360**(6406): 757–759.

Strauch, B., Rodriguez, D.M., Diaz, J., Yu, H.L., Kaplan, G. and Weinstein, D.E. (2001). Autologous Schwann cells drive regeneration through a 6-cm autogenous venous nerve conduit. *J. Reconstr. Microsurg.* **17**: 589–595.

Sunderland, S. (1991). *Nerve Injuries and Their Repair: A Critical Appraisal.* New York: Churchill Livingstone.

Suzuki, Y., Tanihara, M., Ohnishi, K., Suzuki, K., Endo, K. and Nishimura, Y. (1999). Cat peripheral nerve regeneration across 50 mm gap repaired with a novel nerve guide composed of freeze-dried alginate gel. *Neurosci. Lett.* **259**(2): 75–78.

Terzis, J., Faibisoff, B. and Williams, B. (1975). The nerve gap: suture under tension vs. graft. *Plast. Reconstr. Surg.* **56**(2): 166–170.

Thanos, P.K., Okajima, S. and Terzis, J.K. (1998). Ultrastructure and cellular biology of nerve regeneration. *J. Reconstr. Microsurg.* **14**: 423–436.

Thoenen, H., Barde, Y.A., Davies, A.M. and Johnson, J.E. (1987). Neurotrophic factors and neuronal death. *Ciba Found. Symp.* **126**: 82–95.

Tong, X.J., Hirai, K., Shimada, H., Mizutani, Y., Izumi, T., Toda, N. and Yu, P. (1994). Sciatic nerve regeneration navigated by laminin–fibronectin double coated biodegradable collagen grafts in rats. *Brain Res.* **663**(1): 155–162.

Uzman, B.G. and Villegas, G.M. (1983). Mouse sciatic nerve regeneration through semi-permeable tubes: a quantitative model. *J. Neurosci.* **9**: 325–338.

Valentini, R.F. and Aebischer, P. (1997). Strategies for the engineering of peripheral nervous tissue regeneration. In: Lanza, R.P., Langer, R., Chick, W.L. (eds.), *Principles of Tissue Engineering*. Austin, TX: R.G. Landes Company, pp. 671–684.

Valentini, R.F., Aebischer, P., Winn, S.R. and Galletti, P.M. (1987). Collagen- and laminin-containing gels impede peripheral nerve regeneration through semipermeable nerve guidance channels. *Exp. Neurol.* **98**(2): 350–356.

Wang, K.K., Costas, P.D., Bryan, D.J., Jones, D.S. and Seckel, B.R. (1993). Inside-out vein graft promotes improved nerve regeneration in rats. *J. Reconstr. Microsurg.* **14**: 608–618.

Wang, K.K., Nemeth, I.R., Seckel, B.R., Chakalis-Haley, D.P., Swann, D.A., Kuo, J.W., Bryan, D.J. and Cetrulo Jr., C.L. (1998). Hyaluronic acid enhances peripheral nerve regeneration *in vivo*. *Microsurgery* **18**(4): 270–275.

Wang, X., Hu, W., Cao, Y., Yao, J., Wu, J. and Gu, X. (2005). Dog sciatic nerve regeneration across a 30-mm defect bridged by a chitosan/PGA artificial nerve graft. *Brain* **128**(Pt 8): 1897–1910.

Weiss, P. and Taylor, A.C. (1946). Guides for nerve regeneration across nerve gaps. *J. Neurosurg.* **3**: 275–282.

Williams, L.R., Longo, F.M., Powell, H.C., Lundborg, G. and Varon, S. (1983). Spatial–temporal progress of peripheral nerve regeneration within a silicone chamber: parameters for a bioassay. *J. Comp. Neurol.* **218**(4): 460–470.

Yoshii, S., Oka, M., Shima, M., Taniguchi, A. and Akagi, M. (2003). Bridging a 30-mm nerve defect using collagen filaments. *J. Biomed. Mater. Res. A* **67**(2): 467–474.

Yu, X. and Bellamkonda, R.V. (2003). Tissue-engineered scaffolds are effective alternatives to autografts for bridging peripheral nerve gaps. *Tissue Eng.* **9**(3): 421–430.

Yu, X., Dillon, G.P. and Bellamkonda, R.B. (1999). A laminin and nerve growth factor-laden three-dimensional scaffold for enhanced neurite extension. *Tissue Eng.* **5**(4): 291–304.

Zhang, F., Blain, B., Beck, J., Zhang, J., Chen, Z., Chen, Z.W. and Leineaweaver, W.C. (2002). Autogenous venous graft with one-stage prepared Schwann cells as a conduit for repair of long segmental nerve defects. *J. Reconstr. Microsurg.* **18**(4): 295–300.

Zuo, J., Hernandez, Y.J. and Muir, D. (1998). Chondroitin sulfate proteoglycan with neurite-inhibiting activity is up-regulated following peripheral nerve injury. *J. Neurobiol.* **34**(1): 41–54.

Zuo, J., Neubauer, D., Graham, J., Krekoski, C.A., Ferguson, T.A. and Muir, D. (2002). Regeneration of axons after nerve transection repair is enhanced by degradation of chondroitin sulfate proteoglycan. *Exp. Neurol.* **176**(1): 221–228.

38

Innovative Regenerative Medicine Approaches to Skin Cell-Based Therapy for Patients with Burn Injuries

Jörg C. Gerlach, Steven E. Wolf, Christa Johnen, and Bernd Hartmann

INTRODUCTION

The clinical need for improved therapy of burn victims is evident; mortality is still high (Caldwell et al., 1996; Raff et al., 1996; Germann et al., 1997). Regenerative medicine, which has achieved initial success in the utilization of human cells for tissue repair and regeneration, represents a great potential for the treatment of patients with severe burns. Technologies are developing rapidly, with the ultimate goal to deliver advanced skin cell-based therapies as safely and efficiently as possible. Developments in the area of technical devices, cell biological methods, and methods in medical practice have to be considered. Research in this field includes surgical techniques, skin cell procurement, cell culture, stem cell biology, cell application, wound healing support, and wound dressings. This requires an interdisciplinary cooperation of cell biologists, surgeons, and bioengineers.

Our chapter attempts to give an overview of the application of skin cells after burn injuries and focuses on utilizing autologous skin cells from a healthy area of the patient's skin. Our topics are current commercial products, problems in the field, skin cell isolation and culture, skin progenitor cells, cell spray transplantation methods, planning clinical studies, case reports, and regulatory issues.

CONVENTIONAL THERAPY AND CURRENT COMMERCIAL PRODUCTS

Wound healing in victims suffering from full-thickness III° burns is a challenge. The therapy of choice, surgical skin auto grafting, is limited by the availability of healthy skin cell area and donor site. The standard treatment is the split-skin transplantation method, whereby the use of autologous split-skin (Stark and Kaiser, 1994) is superior to allogenic (Stark and Kaiser, 1994) split-skin application, as rejection eventually occurs. If autologous split skin is not sufficiently available, however, allogeneic skin is indicated to support the wound until an autograft is available.

During the past two decades, various commercial products for wound treatment have been applied in clinical studies, whereby one can differentiate between cell-free products and products applied with cell cultures.

Integra Life Sciences developed Integra™, which consists of a collagen layer in a matrix with a silicon overlay is a widely accepted product. Indications are described as deep II° and III° burns. Application is performed

Copyright © 2008 Elsevier Inc. All rights reserved

after excising necrotic tissue to temporarily cover the wound while the dermal part engrafts. Tissue engineering aspects were published (Michaeli et al., 1990; Stern et al., 1990; Grzesiak et al., 1997) and initial clinical studies described (Stern et al., 1990; Kinner et al., 1992; Cameron, 1997; Helvig, 1997; King et al., 1997; Lorenz et al., 1997; Besner et al., 1998; Clayton et al., 1998; Pandya et al., 1998; Senior, 1999). This product is useful clinically for temporary coverage of excised wounds when autograft is not available. Furthermore, it has the potential to act as a dermal equivalent that could presumably decrease scar contracture and improve functional and cosmetic outcomes after grafting.

Mylan Lab introduced Biobrane™, which is a nylon film with a silicone layer that is coated with a porcine xenogeneic type I collagen. Indications include temporary coverage of superficial burns. Clinical studies were described for the treatment of burns (Bradley et al., 1995; Erdmann et al., 1996; Jones, 1998; Ou et al., 1998; Still et al., 1998), wound coverage of transplanted areas (Kirwan,1995; Wang et al., 1996; Vander et al., 1997; Levy and Salomon, 1998), and applications in plastic surgery (Sakai et al., 1996). This product is used for the temporary coverage of superficial wounds or donor sites, but cannot be construed as a dermal equivalent. If used in full-thickness wounds, the wound must be eventually covered with autograft.

LifeCell Corporation developed Alloderm™, which is a cell-free human allogeneic dermal substitute with basement membrane preservation. Indications include dermal substitution for skin defects. Clinical studies are described for deep burns (Wainwright, 1995; Lattari et al., 1997) and plastic tissue replacement (Kridel et al., 1998; Tobin and Karas, 1998). It has the potential benefit of replacing the dermis, and thus inhibiting wound contracture after full wound closure. However, this product is only a dermal equivalent without epidermis; therefore, it cannot be used as an artificial skin. It must be used as a component with an epidermal graft.

Initial commercial products that utilize allogeneic cell cultures for the generation of the products were clinically introduced several years ago. *Advanced Tissue Sciences* developed Trans Cyte™/Dermagraft-TC™. *Smith & Nephew* introduced this into clinical studies. Trans Cyte was developed for acute burns. Dermagraft was developed for chronic skin wounds, including diabetic ulcers. These dressings have a "dermal" layer of allogenic fibroblasts on polyglycol acid and polyglactin tissue. Clinical studies were published for the treatment of diabetic ulcers (Gentzkow et al., 1996; Edmonds et al., 1997; Naughton et al., 1997; Grey et al., 1998; Mansbridge et al., 1998), and burns (Hansbrough et al., 1992, 1997; Economou et al., 1995; Hansbrough, 1997; Herndon, 1997; Purdue et al., 1997; Spielvogel, 1997). The reimbursement situation was described (Parente, 1997; Purdue, 1997), and tissue engineering aspects were discussed (Rennekam et al., 1996; Advanced Tissue Sciences and Smith & Nephew, 1997; Sacks et al., 1997; Jiang et al., 1998). These products, like Biobrane listed above, can only be used for temporary coverage, but the inclusion of live fibroblasts is thought to accelerate wound healing by the production of growth factors.

Organogenesis developed Apligraft™, which was introduced into clinical studies by *Novartis*. It is characterized as a double-layered composite graft of bovine collagen with allogenic fibroblasts and allogenic keratinocytes. It was developed for venous and arterial diabetic ulcer treatment. Tissue engineering aspects (Eaglstein and Falanga, 1997, 1998; Trent and Kirsner, 1998) and studies for the treatment of venous ulcers were described (Alvarez et al., 1998; Fahey, 1998).

Although these products have been shown to improve clinical results in specific clinical situations, a generally accepted wound treatment product for severe burns is not available. Some have used Integra successfully even in massive burns, but this product has probably not been tested side by side with autograft for utility in wound closure or for the prevention of long-term contractures. The same can be said for AlloDerm in terms of utility of the product. However, the purported benefits of accelerated wound closure and decreased scarring have not been verified in clinical trials, which then continue to promote further research efforts toward innovative therapies in the field of skin cell-based therapies in regenerative medicine (Leigh and Watt, 1994; Boyce, 2001).

Table 38.1 **Skin cell isolation methods**

Keywords	Author	Year
Dissocation dermis/	Stenn et al.	1989
epidermis dispase	Green	1991
Collagenase-based isolation	Sugihara et al.	2001
	Dunnwald et al.	2001
Trypsin-based isolation	Sugihara et al.	2001
	Johnen et al.	2006

Table 38.2 **Methods for skin cell *in vitro* culture**

Keywords	Author	Year
Standard keratinocyte culture methods	Rheinwald and Green	1975a, b; 1977
	Green	1991
	Boyce and Ham	1983
	Andreassi	1992
Keratinocyte culture as sheets with serum	O'Connor et al.	1981
	Gallico et al.	1984
	Tenchini et al.	1992
	De Luca and Cancedda	1992
Keratinocytes as monolayer serum free culture	Boyce and Ham	1983
	Tenchini et al.	1992
	Johnen et al	2006
Keratinocytes as suspension	Sebök et al.	1990
	Kaiser et al.	1994
	Harris et al.	1998
	Horch et al.	1998
	Bannasch et al.	2000
Fibroblast culture	Hoeller et al.	2001
	Johnen et al.	2006
Human derived fetal skin cells	Hohlfield et al.	2005

Methods for the isolation and proliferation of autologous keratinocytes were established around 30 years ago; the isolation of fibroblasts in parallel followed. A literature overview is given in Tables 38.1 and 38.2. Studies on culture media, critical for the expansion of autologous skin cells, are summarized in Table 38.3. Wound covering procedures through the application of keratinocytes, multi-layered epidermal carrier, or fibrin-glue immobilized transplants are described.

Genzyme Biosurgery introduced Epicel™, which is an autologous keratinocyte expansion service under industrial Good Manufacturing Practice (GMP) conditions. Cultured epidermal autografts are delivered in the form of sheets for patients with severe burns. There have been many reports of its use (Williamson et al., 1995; Munster,1996; Carsin et al., 2000) mostly in those with burns over 60% of the total body surface area (TBSA). One study compared clinical outcomes of patients with greater than 90% TBSA burns treated with either cultured epithelial autografts or standard skin grafting methods, and found no differences in mortality but increased costs with Epicel mostly related to increased hospital stay and increased numbers of reconstructive procedures. However, long-term cosmetic outcomes were superior in the Epicel group (Barret et al., 2000) giving the promise that such methods may actually give the best long-term outcomes if they could be refined.

Table 38.3 Culture media for *in vitro* culture of skin cells

Keywords	Author	Year
Standard culture medium	Rheinwald and Green	1975, 1977
	Boyce and Ham	1983
Culture in conditioned medium	Paini et al.	1997
	Huang et al.	2006
Medium with pituitary extract	Wang et al.	1995
	González-Castro et al.	1997
Medium with growth factors	Finch et al.	1989
	Hammar et al.	1990
	Girolomoni et al.	1993
	Tavakkol et al.	1999
	Gibbs et al.	2000
	Marchese et al.	2001
	Swope et al.	2001

PROBLEMS IN THE FIELD

The speed of wound healing in extensively burned skin represents one of the major factors contributing to the patient's survival. Patients with larger areas of burns often die during the critical time period of 4 weeks after trauma due to delays in wound closure.

To cover wounds, semi-synthetic biomaterials, for example bovine collagen engineered with aminopolysaccharides, such as chondroitin sulfate and chitosan have been used, either alone or in combination with a skin graft (Bell et al., 1981a, b; Burke et al., 1981; Bell et al., 1983; Murphy et al., 1990; Yannas and Burke, 1990; Yannas et al., 1990). At the same time, many groups focus on using the above-mentioned *in vitro* proliferation of skin keratinocytes to provide an adequate cell number for autologous cell transplantation. Cultured keratinocyte grafting was established as a treatment option for severe burn injuries and in the clinical management of chronic venous ulcers (Srivastava et al., 1990; Phillips and Pachas, 1994; Sabolinski et al., 1996). The cells are taken from the patient's healthy skin areas and expanded as 3–5 cell layers to cell sheets in Petri dishes, then transferred as confluent and stratified sheets. The functional and cosmetic results, however, require further improvement and innovative methods are of interest. Disadvantages of keratinocyte sheets include long *in vitro* expansion times of 2–4 weeks due to high passages to reach the required cell number, and the differentiation of basal keratinocytes with increasing culture time. Confluent growing cells in sheets in comparison to single cell cultures are divisionally less active, start proliferation later after transfer, and show a reduced migration activity.

These available cell-based methods are not yet advanced enough as a routine treatment to help patients in more severe cases of burn disease. Problems include serous fluid blisters between the transplanted tissue sheets and the wound, and the subsequent loss of cell areas, non-optimal cell nutrition in the early phase after cell transfer, water and electrolyte derangements in the wound, toxin accumulation, pH derangements, and infections. In fact, it is thought that the eventual amount of skin that remains after grafting is approximately 25% (Barret et al., 2000). Consequently, the biomatrix for the transplanted cells in the wound is non-physiological, and the cell sheet grafting and cell proliferation are not optimal (Figure 38.1). Furthermore, the transplanted keratinocytes only produce a very thin neo-skin with low protective quality. Therefore, further improvements are of interest.

A clinical problem of wound healing is scar tissue formation. In connective tissue, fibroblast remodeling and contraction of granulation tissue of wounds lead to scar tissue formation (Figure 38.2). This imperfect

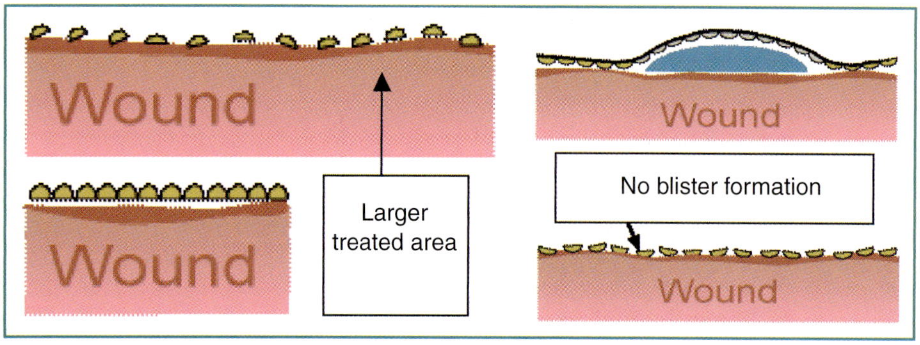

Figure 38.1 **Using sprayed cells result in the need of fewer cells while in a larger treatment surface can be enabled for therapy. Blister formation is avoided. Reducing the cell number speeds up application time, reduces *in vitro* differentiation and therefore better preserves basal keratinocyte progenitor cells.**

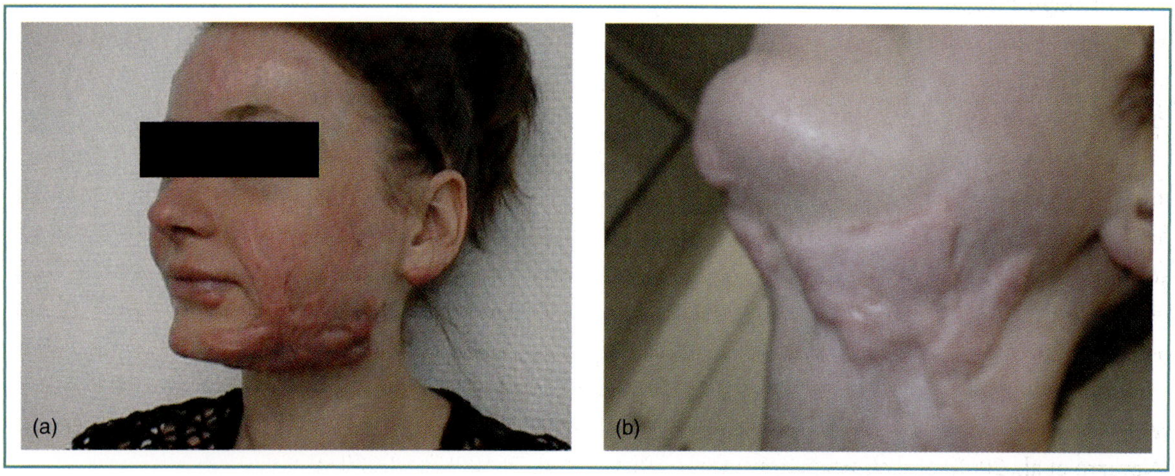

Figure 38.2 **Clinical example for large, deep facial burns. (a) Hypertrophic facial scaring after deep IIb° dermal burn. (b) Hypertrophic scaring on the neck after deep IIb° dermal burn.**

repair is characterized by disorganized collagen, which does not completely recapitulate the natural structure and function of the tissue (Yamaguchi and Yoshikawa, 2001; Hollander et al., 2003). Excess collagen deposition and altered contraction and remodeling lead to aberrant scarring outcomes, such as hypertrophic scars and keloids, in the dermis. The degree of scarring has been linked to early stages of wound healing in which inflammatory mediators play a crucial role in guiding fibroblast activity in the wound bed. In addition, the new scar tissue contracts for 1–2 years after wound closure, producing functional abnormalities when the scar extends over joints or facial tissues. Most burn scar reconstruction occurs for these reasons, making the inhibition of scar contracture an even bigger clinical issue.

CELL APPLICATION INTO THE WOUND BY SPRAYING

Distributing single cells over a skin wound area via an aerosol is of interest to maximize the transfer of viable cells onto the wound surface. Currently, cells are laid down in a sheet layer; however, it is conceivable that delivering cells by spray has significant advantages, including distributing the same number of cells over a larger space, and thus decreasing the necessary time to gain enough cells for wound closure. Table 38.4 summarizes literature describing methods of spraying cells into a skin wound.

Table 38.4 Methods of spraying cells into a skin wound

Keywords	Author	Year
With fibrin-glue	Kaiser et al.	1994
	Bannasch et al.	2000
	Currie et al.	2003
Without fibrin-glue	Navarro et al.	2000
	Wood	2001
	Wood	2003
	Currie et al.	2003

The simplest way to distribute liquid medium into a surrounding gaseous phase without generating a jet is through dripping down from a pipe. Additional force and utilization of a nozzle generate sputtering into a discontinuous stream. When scattering, a liquid stream executes wave-like oscillations that cause its breakdown into numerous droplets in the spray. When nebulizing into a spray, the fluid distribution occurs through the nozzle/nebulizer in a stream at high speed into regular individual drops of comparable size.

Scattering, spraying, or atomization can occur by generating a liquid stream and applying pressure through a pipe and the subsequent breakdown of the stream in or behind a nozzle/nebulizer. Here, the diffused fluid immerges into a latent gaseous phase. Alternatively, a dual substance spray head with a fluid flow and a gas flow can be applied, in which the fluid in the nozzle is induced into the spray head at a 90° angle to the speed direction of the air stream. Depending on the kind of air compression used, different types of flow occur and the drop formation occurs inside or directly outside the nozzle.

Disadvantages of keratinocyte-sheet transplantation can be addressed by using a sprayed suspension of subconfluently cultivated keratinocytes. Reduced hypertrophic scaring was observed in pre-clinical studies (Navarro et al., 2000) and clinical applications (Dietch et al., 1983; Wood, 2003).

Spraying cells has several advantages over sheet application. In comparison to cell sheet transplantation, blister formation can be avoided by using sprayed cells. The use of a cell spray method for cell distribution enables a larger wound treatment area (Figure 38.3). Thus, using sprayed cells results in the need for fewer cells while a larger treatment surface can be enabled for therapy. Reducing the cell number during cell expansion speeds up the application time.

When applying primary cell suspensions that are isolated from autologous skin and immediately applied onto the patient, natural pigmentation of the skin area is possible because primary isolated skin cells contain all cell types occurring in the dermis (Navarro et al., 2000; Stoner and Wood, 2000). Also, it is conceivable that the proper phenotype of keratinocytes is more available because of decreased passaging of cells which will favor the non-differentiated proliferative type over the differentiating phenotype that would presumably have greater tendency to adhere, and thus provide the basis for better and more secure keratinocyte expansion.

PATIENT GROUPS AND SELECTION CRITERIA FOR CLINICAL STUDIES IN THE FIELD OF SKIN CELL-BASED THERAPY DEVELOPMENT IN BURN DISEASE PATIENTS

Planning clinical studies for the development of regenerative therapy methods for severe burns is problematic. The question arises of how to determine selection criteria for burn patients when testing innovative skin cell therapies, including an ethical approach. Based on the frequently occurring various degrees of burns on different areas of the body, it is possible to apply different therapeutic methods on different parts of the body independently from one another. Clinically, we were interested in testing the application of keratinocytes via

510(k) submission: Must demonstrate that the device to be marketed is as safe and effective as an existing, legally marketed (predicate) device, for example in the FDA's terminology, that the *new* device is "substantially equivalent" to the predicate device.

GMP Requirement: Production of a device must meet the FDA's good manufacturing requirements as set forth in the Quality System Regulations (21 CFR 820). Using materials for production coming from an FDA registered manufacturer may allow reference to a corresponding master file, if one exists and permission is obtained from the Master File Holder.

Clinical trials: An investigational device exemption (IDE) allows a device to be used in a clinical study for the purpose of collecting safety and effectiveness data required to support a PMA application or a Pre-market Notification (510(k)) submission to the FDA. Clinical studies are most often conducted to support a PMA. Only a small percentage of 510(k)'s require clinical data to support the application.

Humanitarian use device (HUD) medical device: A HUD can be developed and marketed by a company in a more streamlined process than a typical new medical device, if it is intended to benefit patients in the treatment of a disease affecting fewer than 4,000 patients per year in the United States. It is required that the patient population is not being addressed by an approved device or therapy.

This avenue can be followed if the manufacturer is not making a profit on the HDE; he can only recover R&D cost, manufacturing, and handling. The HUD must demonstrate safety and probable benefit (not efficacy standard for normal products).

Compassionate use of a medical device: A compassionate use allows clinicians to use an "unapproved" device on a patient or patients with a life-threatening disease that does not qualify for an existing clinical trial and has no comparable or satisfactory alternative therapy available. Usually the therapy is in a clinical trial and has demonstrated acceptable performance in its intended patient population. For example, a drug that has completed Phase II or Phase III IND testing may qualify, but a drug in Phase I will not qualify because it has not demonstrated safety or any efficacy yet.

The device, therefore, must be entered into a clinical trial via an Investigational Device Exemption (IDE) for a device or an IND for a drug. The Committee for Human Research and the FDA will have to approve the application for compassionate use.

In well-defined cases of intra-operative cell harvest without cell alteration and immediate application to the same patient no FDA-specific regulations apply (see below); consulting the FDA is required. Autologous cell transplantation may be considered as unregulated in cases where cells are taken from the patient and transplanted with minimal cell manipulation at the point of care, for example in the same operation by the responsible surgeon.

Most commercial applications of skin cells or tissues, however, may be considered as biologics.

The FDA's Center for Biologics Evaluation and Research regulates biological products under Sections 351 and 361 of the Public Health Service (PHS) Act and under specific sections of the Federal Food, Drug and Cosmetic Act (FD&C Act). Under these authorities, CBER is responsible for ensuring:

- the safety of the US blood supply and the products derived from it;
- the production and approval of safe and effective childhood vaccines, including any future AIDS vaccines;
- the oversight of human tissue for transplantation;
- an adequate and safe supply of allergenic materials and anti-toxins;
- the safety and efficacy of biological therapeutics.

Licensing of Biologics

The PHS Act requires individuals or companies who manufacture biologics for introduction into interstate commerce to hold a license for the products. These licenses are issued by CBER. Licensing of biologic products under the PHS Act is very similar to the new drug approval process for human drugs. Following initial laboratory and animal testing, a biological product is studied in clinical trials in humans under an investigational new drug application (IND). If the data generated by the studies demonstrate that the product is safe and effective for its intended use, the data are submitted to CBER as part of a biologics license application (BLA) for review and approval for marketing.

Regulation of HCT/Ps

Human cells, tissue, and cellular and tissue-based products (HCT/Ps) that are intended for implantation, transplantation, infusion, or transfer into a human recipient are regulated by CBER under 21 CFR Parts 1270 and 1271. Examples of such tissues are bone, skin, corneas, ligaments, tendons, dura mater, heart valves, hematopoietic stem/progenitor cells derived from peripheral and cord blood, oocytes, and semen. Note that vascularized organs, bone marrow, blood, and non-human cells are not considered HCT/Ps, because they are regulated under other laws/agencies.

If the HCT/Ps are for homologous use, are only minimally manipulated, and meet the additional criteria listed below in 21 CFR 1271.10, marketing may proceed so long as the requirements of the 21 CFR Parts 1270 and 1271 are met.

The FDA has defined "minimal manipulation" (21 CFR 1271.3(f)) as follows:

1. For structural tissue – processing that does not alter the original relevant characteristics of the tissue relating to the tissue's utility for reconstruction, repair, or replacement.
2. For cells or non-structural tissues – processing that does not alter the relevant biological characteristics of the cells or tissues.

The FDA has defined "homologous use" (21 CFR 1271.3(c)) as follows:

1. The repair, reconstruction, replacement, or supplementation of a recipient's cells or tissues with an HCT/P that performs the same basic function or functions in the recipient as in the donor.

21 CFR Parts 1270 (Human Tissue Intended for Transplantation) and 1271 (HCT/Ps):

The FDA requires tissue establishments to meet the requirements of 21 CFR 1270 and 1271, which include:

- screen and test donors;
- prepare and follow written procedures for the prevention of the spread of communicable disease;
- maintain records;
- register and list their HCT/Ps with FDA;
- follow current Good Tissue Practices (GTP) for HCT/Ps.

The following criteria in 21 CFR 1271.10 are referenced in determining whether an HCT/P is subject to Part 351 or Part 361 of the PHS Act, as described in the following.

21 CFR 1271.10(a):

1. The HCT/P is minimally manipulated (see examples in Table 38.7).
2. The HCT/P is intended for homologous use only, as reflected by the labeling, advertising, or other indications of the manufacturer's objective intent (see examples in Table 38.8).
3. The manufacture of the HCT/P does not involve the combination of the cells or tissues with another article, except for water, crystalloids, or a sterilizing, preserving, or storage agent, provided that the addition of water, crystalloids, or the sterilizing, preserving, or storage agent does not raise new clinical safety concerns with respect to the HCT/P (see examples in Table 38.9).

Table 38.7 HCT/Pss

"351" HCT/P	"361" HCT/P
An HCT/P regulated under Section 351 of the PHS Act is defined as a product that does not meet one or more of the criteria in 21 CFR 1271.10 (listed above). • "351" HCT/Ps will be subject to premarket review and approval, as appropriate based on the product: – Biologics License Application (BLA), if a biological product. Clinical study – Investigational New Drug (IND) process. – New Drug Application (NDA), if a new drug. Clinical study – Investigational New Drug (IND) Process. – Premarket Approval (PMA) Application, or Premarket Notification (510(k)), if a new device. Clinical study – Investigational Device Exemption (IDE) process. • Pre-license approval inspection will be conducted by FDA. • Both investigational and licensed "351" HCT/Ps must comply with 21 CFR 1271 (Subparts A, B, C, and D (although if in compliance with cGMPs in 21 CFR 210/211, compliance with GTPs is likely). • Investigational "351" HCT/Ps also must comply with: – 21 CFR 312 – IND Application – 21 CFR 210/211 – cGMP (as defined) – 21 CFR 50 – Protection of Human Subjects (Informed Consent) – 21 CFR 56 – Institutional Review Boards • Licensed "351" HCT/Ps also must comply with: – 21 CFR 201 – Labeling – 21 CFR 202 – Advertising – 21 CFR 210/211 – cGMP – 21 CFR 600 – Biological Products (includes reporting of adverse experiences and biological deviations) – 21 CFR 601 – Licensing (BLA) – 21 CFR 610 – General Biologics Standards.	An HCT/P is regulated solely under Section 361 of the PHS Act and 21 CFR 1271 if it meets all of the criteria noted above in 21 CFR 1271.10. • "361" HCT/Ps are not subject to premarket review. No application to FDA is required. • FDA inspections will be conducted to confirm compliance. Compliance with 21 CFR Part 1271 is required. This regulation includes the following subparts: A – General Provisions and Definitions B – Registration and Listing C – Donor Eligibility D – Current GTP (as defined) E – Additional Requirements (Reporting and Labeling) F – Inspection and Enforcement

Table 38.8 Defining manipulations

HCT/P category	Minimal manipulation	More than minimal manipulation
Structural tissue	1. Fascia or dermis processed into particulate form. 2. Allogeneic dehydrated and decellularized amniotic membrane intended for wound covering. 3. Cutting, grinding, shaping of bone.	1. Allogeneic, decellularized human arteries, veins, heart valves, or valve conduits intended to replace dysfunctional cardiovascular tissue.

(Continued)

Table 38.8 (Continued)

HCT/P category	Minimal manipulation	More than minimal manipulation
Cell	1. CD34+ selection of allogeneic peripheral blood hematopoietic stem or progenitor cells. 2. Density gradient separation to remove a particular type of cell from a mixture of cells.	1. Autologous cultured (expanded *ex vivo*) epithelial cells isolated from skin biopsies and intended to cover burns. 2. Gene therapy
HCT/P category	**Homologous use**	**Non-homologous use**
Structural tissue	1. Demineralized bone matrix used as a bone void filler during orthopedic surgery. 2. Bone recovered from a limb, used as a bone dowel for spinal surgery.	1. Allogeneic veins or arteries intended for use as arterio-venous access (A-V shunts) for hemodialysis. 2. Cartilage tissue used in the bladder for treatment of reflux.
Cell	1. Allogeneic placental/umbilical hematopoietic stem/progenitor cells used for hematopoietic reconstitution. 2. Pancreatic islet cells used for treatment of type 1 diabetes.	1. Autologous bone marrow hematopoietic stem/progenitor cells used for myocardial repair. 2. Nasal mucosal cells used to regenerate nerve tissue.

Table 38.9 Defining combinations

"Not combined with"	"Combined with"
1. Lyophilized pericardium and a vial of saline packaged together in a kit, for reconstitution by the physician. 2. Any HCT/P to which a sterilizing (e.g. antibiotic), preserving (e.g. Optisol), or storage (e.g. DMSO) agent is added.	1. Demineralized bone matrix combined with another article to create a paste or putty used to fill bone defects. 2. Tendon allograft combined with a suture for use in ligament reconstruction.

4. Either:
   (i) the HCT/P does not have a systemic effect and is not dependent upon the metabolic activity of living cells for its primary function or
   (ii) the HCT/P has a systemic effect or is dependent upon the metabolic activity of living cells for its primary function, and:
      (*a*) is for autologous use;
      (*b*) is for allogeneic use in a first-degree or second-degree blood relative; or
      (*c*) is for reproductive use.

"351" HCT/P or "361" HCT/P

HCT/Ps are described as either "351" HCT/P or "361" HCT/P, in reference to the PHS Act sections to which these products must comply, as outlined in Table 38.7.

Table 38.8 provides examples to assist in defining and distinguishing "minimal manipulation" with "more than minimal manipulation," and "homologous" with "non-homologous" for the two categories of HCT/P tissues (structural and cells/non-structural).

ACKNOWLEDGMENT

The information, support, and co-editing of this text by Patsy J. Trisler, J.D., RAC, Regulatory Consultant – Medical Devices (patsy.trisler@verizon.net) is greatly appreciated.

REFERENCES

Advanced Tissue Sciences and Smith & Nephew present Dermagraft data (1997). *Ostomy Wound Manage* **43**(4): 77–78, 80.

Alvarez, O.M., Fahey, C.B., Auletta, M.J. and Fernandez-Obregon, A. (1998). A novel treatment for venous leg ulcers. *J. Foot Ankle Surg.* **37**(4): 319–324.

Andreassi, L. (1992). History of keratinocyte cultivation. *Burns* **18**: 52–54.

Bannasch, H., Horch, R.E., Tanczocs, E. and Stark, G.B. (2000). Treatment of chronic wounds with cultured autologous keratinocytes as suspension in fibrin glue. *Zentralbl. Chir.* **125**(Suppl 1): 79–81.

Barret, J.P., Wolf, S.E., Desai, M.H. and Herndon, D.N. (2000). Cost-efficacy of cultured epidermal autografts in massive pediatric burns. *Ann. Surg.* **231**(6): 869–876.

Bell, E., Ehrlich, H.P., Buttle, D.J. and Nakatsuji, T. (1981a). Living tissue formed *in vitro* and accepted as skin equivalent tissue of full-thickness. *Science* **211**: 1052–1054.

Bell, E., Ehrlich, H.P., Sher, S., Merrill, C., Sarber, R., Hall, B., Nakatsuji, T., Church, D. and Buttle, D.J. (1981b). Development and use of a living skin equivalent. *Past. Reconstr. Surg.* **67**: 386–392.

Bell, E., Sher, S., Hall, B., Merrill, C., Rosen, S., Chamson, A., Asselineau, D., Dubertret, L., Coulomb, B., Lapiere, C., Nusgens, B. and Neveux, Y. (1983). The reconstruction of living skin. *J. Investig. Dermatol.* **81**: S2–S10.

Besner, G.E., et al. (1998). Integra artificial skin as a useful adjunct in the treatment of purpura fulminans. *J. Burn Care Rehabil.* **19**(4): 324–329.

Bickenbach, J.R. and Chism, E. (1998). Selection and extended growth of murine epidermal stem cells in culture. *Exp. Cell Res.* **244**: 184–195.

Boyce, S.T. (2001). Design principles for composition and performance of cultured skin substitutes. *Burns* **27**: 523–533.

Boyce, S.T. and Ham, R.G. (1983). Calcium-regulated differentiation of normal human epidermal keratinocytes in chemically defined clonal culture and serum-free serial culture. *J. Investig. Dermatol.* **81**: S33–S40.

Bradley, T., Brown, R.E., Kucan, J.O., Smoot III, E.C. and Hussmann, J. (1995). Toxic epidermal necrolysis: a review and report of the successful use of Biobrane for early wound coverage. *Ann. Plast. Surg.* **35**(2): 124–132.

Brouard, M. and Barrandon, Y. (2003). Controlling skin morphogenesis: hope and despair. *Curr. Opin. Biotechnol.* **14**: 520–525.

Burke, J.F., Yannas, I.V., Quinby Jr., W.C., Bondoc, C.C. and Jung, W.K. (1981). Successful use of a physiologically acceptable artificial skin in treatment of extensive burn injured. *Ann. Surg.* **194**: 413–428.

Caldwell Jr., F.T., Wallace, B.H. and Cone, J.B. (1996). Sequential excision and grafting of the burn injuries of 1507 patients treated between 1967 and 1986: end results and the determinants of death. *J. Burn Care Rehabil.* **17**(2): 137–146.

Cameron, S. (1997). Changes in burn patient care. *Br. J. Theatre Nurs.* **7**(5): 5–7 (review).

Carsin, H., Ainaud, P., Le Bever, H., Rives, J., Lakhel, A., Stephanazzi, J., Lambert, F. and Perrot, J. (2000). Cultured epithelial autografts in extensive burn coverage of severely traumatized patients: a five year single–center experience with 30 patients. *Burns* **26**(4): 379–387.

Chen, J.D., Kim, J.P., Zhang, K., Sarret, Y., Wynn, K.C., Kramer, R.H. and Woodley, D.T. (1993). Epidermal growth factor (EGF) promotes human keratinocyte locomotion on collagen by increasing the alpha2 integrin subunit. *Exp. Cell Res.* **209**: 216–223.

Christiano, A.M. (2004). Epithelial stem cells: stepping out of their niche. *Cell* **118**: 530–532.

Clark, R.A.F. (ed.) (1996). *The Molecular and Cellular Biology of Wound Repair*, 2nd edn. New York: Plenum Press.

Claudinot, S., Nicolas, M., Oshima, H., Rochat, A. and Barrandon, Y. (2005). Long-term renewal of hair follicles from clonogenic multipotent stem cells. *PNAS* **11**: 14677–14682.

Clayton, M.C., et al. (1998). Perioperative and postoperative dressing techniques for integra artificial skin: views from two medical centers. *J. Burn Care Rehabil.* **19**(4): 358–363.

Currie, L.J., Martin, R., Sharpe, J.R. and James, S.E. (2003). A comparison of keratinocyte cell sprays with and without fibrin glue. *Burns* **29**: 677–685.

De Luca, M. and Cancedda, R. (1992). Culture of human epithelium. *Burns* **18**: S5–S10.

Dietch, E.A., Wheelan, T.M. and Rose, M.P. (1983). Hypertrophic burn scars: analysis of variables. *J. Trauma* **23**(10): 895–898.

Dostal, G.H. and Gamelli, R.L. (1993). Fetal wound healing. *Surg. Gynecol. Obstet.* **176**: 299–306.

Dunnwald, M., Tomanek-Chalkley, A., Alexandrunas, D., Fishbaugh, J., Bickenbach, J.R. (2001). Isolating a pure population of epidermal stem cells for use in tissue engineering. *Exp. Dermatol.* **10**: 45–54.

Eaglstein, W.H. and Falanga, V. (1997). Tissue engineering and the development of Apligraf, a human skin equivalent. *Clin. Ther.* **19**(5): 894–905.

Eaglstein, W.H. and Falanga, V. (1998). Tissue engineering and the development of Apligraf, a human skin equivalent. *Cutis* **62**(Suppl 1): 1–8.

Economou, T.P., et al. (1995). An experimental study to determine the effects of Dermagraft on skin graft viability in the presence of bacterial wound contamination. *J. Burn Care Rehabil.* **16**(1): 27–30.

Edmonds, M.E., et al. (1997). "Dermagraft": a new treatment for diabetic foot ulcers. *Diabet. Med.* **14**(12): 1010–1011.

Erdmann, D., Hussmann, J. and Kucan, J.O. (1996). Treatment of a severe alkali burn. *Burns* **22**(2): 141–146.

Fahey, C. (1998). Experience with a new human skin equivalent for healing venous leg ulcers. *J. Vasc. Nurs.* **16**(1): 11–15.

Finch, P.W., Rubin, J.S., Miki, T, Ron, D. and Aaronson, S.A. (1989). Human KGF is FGF-related with properties of a paracrine effector of epithelial cell growth. *Science* **245**: 752–755.

Fuchs, E. and Raghavan. S. (2002). Getting under the skin of epidermal morphogenesis. *Nature* **31**: 199–209.

Gallico III, G.G., O'Connor, N.E., Compton, C.C., Kehinde, O. and Green, H. (1984). Permanent coverage of large burn wounds with autologous cultured human epithelium. *N. Engl. J. Med.* **311**: 448–451.

Gentzkow, G.D., et al. (1996). Use of Dermagraft, a cultured human dermis, to treat diabetic foot ulcers. *Diabetes Care* **19**(4): 350–354.

Germann, G., Barthold, U., Lefering, R., Raff, T. and Hartmann, B.(1997). The impact of risk factors and pre-existing conditions on the mortality of burn patients and the precision of predictive admission-scoring systems. *Burns* **23**(3): 195–203.

Gibbs, S., Pinto, A.N.S., Murli, S., Huber, M., Hohl, D. and Ponec, M. (2000). Epidermal growth factor and keratinocytes growth factor differentially regulate epidermal migration, growth, and differentiation. *Wound Repair. Regen.* **8**: 192–203.

Girolomoni, G. and Phillips, J.J. (1993). Prolactin stimulates proliferation of cultured human keratinocytes. *J. Invest. Dermatol.* **101**: 275–279.

González-Castro, U., Castells-Rodellas, A. and Krueger, F.G. (1997). Identification of a novel keratinocytes mitogen derived from bovine pituitary glands. *Arch. Dermatol. Res.* **289**: 309–316.

Green, H. (1991). Cultured cells for treatment of disease. *Sci. Am.* November: 64–70.

Grey, J.E., et al. (1998). The use of cultured dermis in the treatment of diabetic foot ulcers. *J. Wound Care* **7**(7): 324–325.

Grzesiak, J.J., et al. (1997). Enhancement of cell interactions with collagen/glycosaminoglycan matrices by RGD derivatization. *Biomaterials* **18**(24): 1625–1632.

Häkkinen, L., Koivisto, L. and Larjava, H. (2002). An improved method for culture of epidermal keratinocytes from newborn mouse skin. *Meth. Cell Sci.* **23**: 189–196.

Hammar, H., Acevedo, F. and Naito, S. (1990). Transferrin and epidermal growth. *Acta Derm. Venereol.* **70**(1): 11–17.

Hansbrough, J. (1997). Dermagraft-TC for partial-thickness burns: a clinical evaluation. *J. Burn Care Rehabil.* **18**(1 Pt 2): S25–S28.

Hansbrough, J.F., et al. (1992). Clinical trials of a living dermal tissue replacement placed beneath meshed, split-thickness skin grafts on excised burn wounds. *J. Burn Care Rehabil.* **13**(5): 519–529.

Hansbrough, J.F., et al. (1997). Clinical trials of a biosynthetic temporary skin replacement, Dermagraft-Transitional Covering, compared with cryopreserved human cadaver skin for temporary coverage of excised burn wounds. *J. Burn Care Rehabil.* **18**(1 Pt 1): 43–51.

Harris, P.A., Leigh, I.M. and Navsaria, H.A. (1998). Pre-confluent keratinocyte grafting: the future for cultured skin replacement? *Burns* **24**: 591–593.

Hartmann, B., Ekkemkamp, A., Johnen, C., Gerlach, J.C., Belfekroun, C. and Küntscher, M.V. (2007). Sprayed cultured epithelial autografts for deep dermal burns of the face and neck. *Ann Plast Surg.* January **58**(1): 70–73.

Hinz, B., Alt, W., Johnen, C., Herzog, V. and Kaiser, H.W. (1999). Quantifying lamella dynamics of cultered cells by SACED, a new computer-assisted motion analysis. *Exp. Cell Res.* **251**: 234–243.

Helvig, E.I. (1997). Dermal replacement: an update. *Semin. Perioper. Nurs.* **6**(4): 233–235 (review).

Herndon, D.N. (1997). Multiple piece use of Dermagraft-TC. *J. Burn Care Rehabil.* **18**(1 Pt 2): S14–S15.

Hoeller, D., Huppertz, B., Roos, T.C., Gutiérrez, P.P., Merk, H.F., Frank, J. and Jugert, F.K. (2001). An improved and rapid method to construct skin equivalents from human hair follicles and fibroblasts. *Exp. Dermatol.* **10**: 264–274.

Hohlfield, J., Hirt-Burri, N., Gerber, S., Scaletta, C., Hohlfield, P. and Applegate, L.A. (2005). Tissue engineered fetal skin constructs for paediatric burns. *Lancet* **366**: 840–842.

Hollander, D.A., Erli. H.J., Theisen, A., et al. (2003). Standardized qualitative evaluation of scar tissue properties in an animal wound healing model. *Wound Repair Regen.* **11**(2): 150–157.

Horch, R.E., Bannasch, H., Kopp, J., Andree, C. and Stark, G.B. (1998). Single-cell suspension of cultured human keratinocytes in fibrin-glue reconstitute the epidermis. *Cell Transplant.* **7**: 309–317.

Huang, Y.C., Wang, T.W., Sun, J.S. and Lin, F.H. (2006). Investigation of mitomycin-C treated fibroblasts in 3-D collagen gel and conditioned medium for keratinocytes proliferation. *Artifi. Organs* **30**(3): 150–159.

Jahoda, C.A.B. (2003). Cell movement in the hair follicle dermis-more than a two-way street? *J. Investig. Dermatol.* **121**:

Jahoda, C.A.B. and Reynolds, A.J. (2001). Hair follicle dermal sheath cells: unsung participants in wound healing. *Lancet* **358**: 1445–1448.

Jahoda, C.A.B., Whitehouse, J., Reynolds, A.J. and Hole, N. (2003). Hair follicle dermal cells differentiate into adipogenic and osteogenic lineages. *Exp. Dermatol.* **12**: 849–859.

Janes, S.M., Lowell, S. and Hutter, C. (2002). Epidermal stem cells. *J. Pathol.* **197**: 479–491.

Jiang, W.G., et al. (1998). Enhancement of wound tissue expansion and angiogenesis by matrix-embedded fibroblast (Dermagraft), a role of hepatocyte growth factor/scatter factor. *Int. J. Mol. Med.* **2**(2): 203–210.

Johnen, C., Steffen, I., Beichelt, D., Bräutlgam, K., Witascheck, T., Toman, N., Moser, V., Ottomann, C., Hartmann, B. and Gerlach, J.C. (2007). Biodegradable and digestable transfer-membrane-based culture of human fibroblasts and keratinocytes. (Accepted) Burns.

Johnen, C., Hartmann, B., Steffen, I., Bräutigam, K., Witascheck, T., Toman, N., Küntscher, M.V. and Gerlach, J.C. (2006). Skin cell isolation and expansion for cell transplantation is limited in patients using tobacco, alcohol, or are exhibiting diabetes mellitus. *Burns* **32**(2): 194–200.

Jones, L.M. (1998). The Biobrane stent. *J. Burn Care Rehabil.* **19**(4): 352–353.

Kaiser, H.W., Stark, G.B., Kopp, J., Balcerkiewicz, A., Spilker, G. and Kreysel, H.W. (1994). Culture autologous keratinocytes in fibrin glue suspension. Exclusively and combined with STS-allograft (preliminary clinical and histological report of a new technique). *Burns* **20**: 23–29.

King, WW., et al. (1997). Evaluation of artificial skin (Integra) in a rodent model. *Burns* March 23 (Suppl 1): S30–S32.

Kinner, M.A., et al. (1992). Skin transplantation. *Crit. Care Nurs. Clin. N. Am.* **4**(2): 173–178 (review).

Kirwan, L. (1995). Management of difficult wounds with Biobrane. *Conn. Med.* **59**(9): 523–529.

Klima, J., Smetana Jr., K., Motlik J., et al. (2005). Comparative phenotypic characterization of keratinocytes originating from hair follicles. *J. Mol. Histol.* **36**: 89–96.

Kobayashi, K., Rochat, A. and Barrandon, Y. (1993). Segregation of keratinocyte colony-forming cells in the bulge of the rat vibrissa. *Proc. Natl Acad. Sci. USA* **90**: 7391–7395.

Kridel, R.W., Foda, H. and Lunde, K.C. (1998). Septal perforation repair with acellular human dermal allograft. *Arch. Otolaryngol. Head Neck Surg.* **124**(1): 73–78.

Krugluger, W., Rohrbacher, W., Laciak, K., et al. (2005). Reorganization of hair follicles in human skin organ culture induced by cultured human follicle-derived cells. *Exp. Dermatol.* **14**: 580–585.

Kurata, S., Itami, S., Terashi, H. and Takayasu, S. (1994). Successful transplantation of cultured human outer root sheath cells as epithelium. *Ann. Plast. Surg.* **33**: 290–294.

Lattari, V., Jones, L.M., Varcelotti, J.R., Latenser, B.A., Sherman, H.F. and Barrette, R.R. (1997). The use of a permanent dermal allograft in full-thickness burns of the hand and foot: a report of three cases. *J. Burn Care Rehabil.* **18**(2): 147–155.

Leigh, I.M. and Watt, F.M. (1994). The culture of human epidermal keratinocytes. *The Keratinocyte Handbook*. Cambridge, MA: University Press Cambridge, pp. 43–51.

Lenoir, M.C., Bernard, B.A., Pautrat, G., Darmon, M. and Shroot, B. (1998). Outer root sheath cells of human hair follicle are able to regenerate a fully differentiated epidermis *in vitro*. *Dev. Biol.* **130**: 610–620.

Levy, L., Broad, S., Diekmann, D., Evans, R.D. and Watt, F.M. (2000). Beta1 integrins regulate keratinocyte adhesion and differentiation by distinct mechanisms. *Mol. Biol. Cell* **11**: 453–466.

Levy, P.M. and Salomon. D. (1998). Use of Biobrane after laser resurfacing. *Dermatol. Surg.* **24**(7): 729–734.

Li, A., Simmons, P.J. and Kaur, P. (1998). Identification and isolation of candidate human keratinocyte stem cells based on cell surface phenotype. *Proc. Natl Acad. Sci. USA* **95**: 3902–3907.

Limat, A. and Hunziker, T. (2002). Use of epidermal equivalents generated from follicular outer root sheath cells *in vitro* and for autologous grafting of chronic wounds. *Cells Tissues Organs* **172**: 79–85.

Limat, A. and Noser, F.K. (1986). Serial cultivation of single keratinocytes from the outer root sheath of human scalp hair follicles. *J. Investig. Dermatol.* **87**: 485–488.

Limat, A., Breitkreutz, D., Hunziker, T., et al. (1991). Restoration of the epidermal phenotype by follicular outer root sheath cells in recombinant culture with dermal fibroblasts. *Exp. Cell Res.* **194**: 218–227.

Limat, A., French, L.E, Blal, L., et al. (2003). Organotypic cultures of autologous hair follicle keratinocytes for the treatment of recurrent leg ulcers. *J. Am. Acad. Dermatol.* **48**: 207–214.

Lorenz, H.P., Lin, R.Y., Longaker, M.T., Whitby, D.J. and Adzick, N.S. (1995). The fetal fibroblast: the effector cell for scarless fetal skin repair. *Plast. Reconstr. Surg.* **96**: 1251–1261.

Lorenz, C., et al. (1997). Early wound closure and early reconstruction. Experience with a dermal substitute in a child with 60 per cent surface area burn. *Burns* **23**(6): 505–508.

Lyle, S., Christofidou-Solomidou, M., Liu, Y., et al. (1998). The C8/144B monoclonal antibody recognizes cytokeratin 15 and defines the location of human hair follicle stem cells. *J. Cell Sci.* **111**(Pt 21): 3179–3188.

Mansbridge, J., et al. (1998). Three-dimensional fibroblast culture implant for the treatment of diabetic foot ulcers: metabolic activity and therapeutic range. *Tissue Eng.* **4**(4): 403–414.

Marchese, C., Felici, A., Visco, V., Lucania, G., Igarashi, M., Picaro, M., Frati, L. and Torrisi, M.R. (2001). Fibroblast growth factor 10 induces proliferation and differentiation of human primari cultured keratinocytes. *J. Investig. Dermatol.* **116**(4): 623–628.

McElwee, K.J., Kissling, S., Wenzel, E., Huth, A. and Hoffmann, R. (2003). Cultured peribulbar dermal sheath cells can induce hair follicle development and contribute to the dermal sheath and dermal papilla. *J. Investig. Dermatol.* **121**: 1267–1275.

Michaeli, D., et al. (1990). Immunologic study of artificial skin used in the treatment of thermal injuries. *J. Burn Care Rehabil.* **11**(1): 21–26.

Michel, M., Torok, N., Godbout, M.J., et al. (1996). Keratin 19 as a biochemical marker of skin stem cells *in vivo* and *in vitro*: keratin 19 expressing cells are differentially localized in function of anatomic sites, and their number varies with donor age and culture stage. *J. Cell Sci.* **109**(Pt 5): 1017–1028.

Morasso, M.I. and Tomic-Canic, M. (2005). Epidermal stem cells: the cradle of epidermal determination, differentiation and wound healing. *Biol. Cell* **97**: 173–183.

Munster, A.M. (1996). Cultured skin for massive burn prospective, controlled trial. *Ann. Surg.* **224**(3): 372–375; discussion 375–377.

Murphy, G.F., Orgill, D.P. and Yannas, I.V. (1990). Partial dermal regeneration is induced by biodegradable collagen–glycosaminoglycan grafts. *Lab. Invest.* **63**: 305–313.

Naughton, G., Mansbridge, J. and Gentzkow, G. (1997). A metabolically active human dermal replacement for the treatment of diabetic foot ulcers. *Artif. Organs* **21**(11): 1203–1210.

Navarro, F.A., Stoner, M.L., Park, C.S., et al. (2000). Sprayed keratinocyte suspensions accelerate epidermal coverage in a porcine microwound model. *J. Burn Care Rehabil.* **21**(6): 513–518.

Navsaria, H.A., Ojeh, N.O., Moiemen, N., Griffiths, M.A. and Frame, J.D. (2004). Reepithelialization of a full-thickness burn from stem cells of hair follicles micrografted into a tissue-engineered dermal template (Integra). *Plast. Reconstr. Surg.* **113**: 978–981.

Nemeth, A.J. (ed.) (1993). Dermatologic clinics: wound healing. Philadelphia, PA: W.B. Saunders.

O'Connor, N., Mulliken, J., Banks-Schlegel, S., Kehinde, O. and Green, H. (1981). Grafting of burns with cultured epithelium prepared from autologous epidermal cells. *Lancet* **317**: 75–78.

Ohyama, M., Terunuma, A., Tock, C.L., Radonovich, M.F., Pise-Masison, C.A., Hopping, S.B., Brady, J.N., Udey, M.C. and Vogel, J.C. (2006). Characterization and isolation of stem cell-enriched human hair follicle bulge cells. *J. Clin. Investig.* **116**: 249–260.

Oshima, H., Rochat, A., Kedzia, C., Kobayashi, K. and Barrandon, Y. (2001). Morphogenesis and renewal of hair follicles from adult multipotent stem cells. *Cell* **104**: 233–245.

Ou, L.F., Lee, S.Y., Chen, Y.C., Yang, R.S. and Tang, Y.W. (1998). Use of Biobrane in pediatric scald burns – experience in 106 children. *Burns* **24**(1): 49–53.

Paini, C., Savant, F., Malcovati, M. and Tenchini, M.L. (1997). Induction of keratinocyte proliferation by a short treatment with keratinocyte-conditioned medium. *Cell Biol. Int.* **21**: 477–482.

Pandya, A.N., et al. (1998). The use of cultured autologous keratinocytes with integra in the resurfacing of acute burns. *Plast. Reconstr. Surg.* **102**(3): 825–828; discussion 829–830.

Parente, S.T. (1997). Estimating the economic cost offsets of using Dermagraft-TC as an alternative to cadaver allograft in the treatment of graftable burns. *J. Burn Care Rehabil.* **18**(1 Pt 2): S18–S24.

Pellegrini, G., Dellambra, E., Golisano O., et al. (2001). p63 identifies keratinocyte stem cells. *Proc. Natl Acad. Sci. USA* **98**: 3156–3161.

Phillips, T.J and Pachas, W. (1994). Clinical trial of culture autologous keratinocytes grafts in the treatment of long-standing pressure ulcers. *Wounds* **6**: 113–119.

Purdue, G.F. (1997). Dermagraft-TC pivotal efficacy and safety study. *J. Burn Care Rehabil.* **18**(1 Pt 2): S13–S14.

Purdue, G.F., et al. (1997). A multicenter clinical trial of a biosynthetic skin replacement, Dermagraft-TC, compared with cryopreserved human cadaver skin for temporary coverage of excised burn wounds. *J. Burn Care Rehabil.* **18**(1 Pt 1): 52–57.

Raff, T., Germann, G. and Barthold, U. (1996). Factors influencing the early prediction of outcome from burns. *Acta Chir. Plast.* **38**(4): 122–127.

Randall, V.A., Hibberts, N.A. and Hamada, K. (1996). A comparison of the culture and growth of dermal papilla cells from hair follicles from non-balding and balding (androgenetic alopecia) scalp. *Br. J. Dermatol.* **134**: 437–444.

Rennekampff, H.O., et al. (1996). Integrin and matrix molecule expression in cultured skin replacements. *J. Burn Care Rehabil.* **17**(3): 213–221.

Reynolds, A.J. and Jahoda, C.A. (1994). Hair follicle reconstruction *in vitro*. *J. Dermatol. Sci.* 7(Suppl): S84–S97.

Rheinwald, J.G. and Green, H. (1975a). Formation of a keratinizing epithelium in culture by a cloned cell line derived from a teratoma. *Cell* **6**: 317–330.

Rheinwald, J.G. and Green, H. (1975b). Serial cultivation of strains of human epidermal keratinocytes. The formation of keratinizing colonies from single cells. *Cell* **6**: 331–343.

Rheinwald, J.G. and Green, H. (1977). Epidermal growth factor and the multiplication of cultured human epidermal keratinocytes. *Nature* **265**: 421–424.

Richardson, G.D., Arnott, E.C., Whitehause, C.J., Lawrence, C.M., Reynolds Hole, N. and Jahoda, C.A.B. (2005). Plasticity of rodent and human hair follicle dermal cells: implication for cell therapy and tissue engineering. *J. Investig. Dermatol. Symp. Proc.* **10**: 180–183.

Rizvi, A.Z. and Wong, M.H. (2005). Epithelial stem cells and their niche: there's no place like home. *Stem Cells* **23**: 150–165.

Sabolinski, M.L., Alvarez, O., Auletta, M., Mulder, G. and Parenteau, N.L. (1996). Cultered skin as a smart material for healing wounds: experience in venous ulcers. *Biomaterials* **17**: 311–320.

Schlabe, J., Johnen, C., Schwartländer, R., Moser, V., Hartmann, B., Gerlach, J.C. and Küntscher, M.V. (2007): Isolation and Culture of Different cell-types from the human scalp tissue for development of skin substitute in spray technique. (Accepted) Burns.

Sacks, M.S., et al. (1997). Collagen fiber architecture of a cultured dermal tissue. *J. Biomech. Eng.* **119**(1): 124–127.

Sakai, T., Hayashi, N., Kimoto, T., Kitagawa, M., Noguchi, M., Sano, A. and Ishii, Y. (1996). Life-threatening esophageal fistula: treatment with expandable metallic stents covered by biosynthetic skin. *J. Vasc. Interv-Radiol.* **7**(4): 569–572.

Sebök, B., Törok, Z. and Schneider, I. (1990). Chirurgische entfernung von Tätowierungen, Verbesserung der Ergebnisse mittels Zellsuspensions-Wundbehandlung. *Akt. Dermatol.* **16**: 295–297.

Senior, K.A. (1999). Positive approach to burn care. *Lancet* **353**: 1248.

Spielvogel, R.L. (1997). A histological study of Dermagraft-TC in patients' burn wounds. *J. Burn Care Rehabil.* **18**(1 Pt 2): S16–S18.

Srivastava, S., Gorham, S.D., French, D.A., Shivas, A.A. and Courtney, J.M. (1990). *In vivo* evaluation and comparison of collagen, acetylated collagen/glycosaminoglycan composite films and sponge as candidate biomaterials. *Biomaterials* **11**: 155–161.

Stark, G.B. and Kaiser, H.W. (1994). Cologne burn centre experience with glycerol-preserved allogenic skin: Part II: Combination with autologous cultured keratinocytes. *Burns* **20**: 34–38.

Stenn, K.S., Link, R., Moellmann, G., Madri, J. and Kuklinski, E. (1989). Dispase, a neutral protease from Bacillus Polymyxa, is a powerful fibronectinase and type IV collagenase. *J. Invest. Dermatol.* **93**: 287–290.

Stern, R., et al. (1990). Histologic study of artificial skin used in the treatment of full-thickness thermal injury. *J. Burn Care Rehabil.* **11**(1): 7–13.

Still, J., Craft-Coffman, B., Law, E., Colon-Santini, J. and Grant, J. (1998). Burns of children caused by electric stoves. *J. Burn Care Rehabil.* **19**(4): 364–365.

Stoner, M.L. and Wood, F.M. (2000). The treatment of hypopigmented lesions with cultured epithelial autograft. *J. Burn Care Rehabil.* **21**(1): 41–46.

Sugihara, H., Toda, S., Yonemitsu, N. and Watanabe, K. (2001). Effects of fat cells on keratinocytes and fibroblasts in a reconstructed rat model using collagen gel matrix. *Br. J. Dermatol.* **144**: 244–253.

Swope, V.B., Supp, A.P., Greenhalgh, D.G., Warden, G.D. and Boyce, S.T. (2001). Expression of insulin-like growth factor I by cultured skin substitutes does not replace the physiologic requirement for insulin *in vitro*. *J. Investig. Dermatol.* **116**(5): 650–657.

Tani, H., Morris, R.J. and Kaur, P. (2000). Enrichment for murine keratinocyte stem cells based on cell surface phenotype. *Proc. Natl Acad. Sci. USA* **97**: 10960–10965.

Tausche, A.K., Skaria, M., Bohlen, L., et al. (2003). An autologous epidermal equivalent tissue-engineered from follicular outer root sheath keratinocytes is as effective as split-thickness skin autograft in recalcitrant vascular leg ulcers. *Wound Repair Regen.* **11**: 248–252.

Tavakkol, A., Varani, J., Elder, J.T. and Zouboulis, C.C. (1999). Maintenance of human skin in organ culture: role for insulin-like growth factor receptor. *Arch. Dermatol. Res.* **291**: 643–651.

Tenchini, M.L., Ranzati, C. and Malcovati, M. (1992). Culture techniques for human keratinocytes. *Burns* **18**: 11–15.

Trempus, C.S., Morris, R.J., Bortner, C.D., et al. (2003). Enrichment for living murine keratinocytes from the hair follicle bulge with the cell surface marker CD34. *J. Investig. Dermatol.* **120**: 501–511.

Trent, J.F. and Kirsner, R.S. (1998). Tissue engineered skin: Apligraf, a bi-layered living skin equivalent. *Int. J. Clin. Pract.* **52**(6): 408–413.

Tobin, H.A. and Karas, N.D. (1998). Lip augmentation using an alloderm graft. *J. Oral Maxillofac. Surg.* **56**(6): 722–727.

Toma, J.G., Akhavan, M., Fernandes, K.J., et al. (2001). Isolation of multipotent adult stem cells from the dermis of mammalian skin. *Nat. Cell Biol.* **3**: 778–784.

Tumbar, T., Guasch, G., Greco, V., Blanpain, C., Lowry, W.E., Rendl, M. and Fuchs, E. (2004). Defining the epithelial stem cell niche in skin. *Science* **303**: 359–363.

VanderKam, V.M., Achauer, B.M. and Finnie, G. (1997). Use of a semipermeable dressing (Biobrane) following laser resurfacing of the face. *Plast. Surg. Nurs.* **17**(3): 177–179.

Wainwright, D.J. (1995). Use of an acellular allograft dermal matrix (AlloDerm) in the management of full-thickness burns. *Burns* **21**(4): 243–248.

Wang, H.J., Chen. T.M., Cheng, L.F., Cheng, T.Y. and Tung, Y.M. (1993). Human keratinocyte culture using porcine pituitary extract in serum-free medium. *Burns* **21**: 503–506.

Wang, H.J., Wan, H.L., Yang, T.S., Wang, D.S., Chen, T.M. and Chang, D.M. (1996). Acceleration of skin graft healing by growth factors. *Burns* **22**(1): 10–14.

Watt, F.M. (2002). Role of integrins in regulating epidermal adhesion, growth and differentiation. *Embo J.* **21**: 3919–3926.

Watt, F.M. and Hogan, B.L. (2000). Out of Eden: stem cells and their niches. *Science* **287**: 1427–1430.

Webb, A., Li, A. and Kaur, P. (2004). Location and phenotype of human adult keratinocyte stem cells of the skin. *Differentiation* **72**: 387–395.

Williamson, J.S., Snelling, C.F., Clugston, P., Macdonald, I.B. and Germann, E. (1995). Cultured epithelial autograft: five years of clinical experience with twenty-eight patients. *J. Trauma* **39**(2): 309–319.

Wood, F.M. (2001). The first seven years of the west Australian skin culture laboratory. *Cultured Human Keratinocytes and Tissue Engineered Skin Substitutes*. pp. 275–283.

Wood, F.M. (2003). Clinical potential of autologous epithelial suspension. *Wounds* **15**(1): 16–22.

Wood, F.M. and Allen, P. (2003). The use of cultured epidermal autograft in the treatment of major burn injuries. *J. Burn Care Rehabil.* **13**(1): 154–157.

Wood, F.M., Kolybaba, M.L. and Allen, P. (2006). The use of cultured epithelial autograft in the treatment of major burn wounds: eleven years of clinical experience. *Burns* **32**(5): 538–544 (Epub June 14).

Wu, J.J., Liu, R.Q., Lu, Y.G., et al. (2005). Enzyme digestion to isolate and culture human scalp dermal papilla cells: a more efficient method. *Arch. Dermatol. Res.* **297**: 60–67.

Yamaguchi, Y. and Yoshikawa, K. (2001). Cutaneous wound healing: an update. *J. Dermatol.* **28**(10): 521–534.

Yannas, I.V. and Burke, J.F. (1990). Design of an artificial skin. I. Basic design principles. *J. Biomed. Mater. Res.* **14**: 65–69.

Yannas, I.V., Burke, J.F., Gordon, P.L., Huang, C. and Rubenstein, R.H. (1990). Design of an artificial skin. II. Control of chemical composition. *J. Biomed. Mater. Res.* **14**: 107–132.

Part VII

Regulations and Ethics

39

Ethical Considerations

Louis M. Guenin

As natural phenomena are to the scientist, so are arguments to the philosopher. The philosopher Richard Hare once said, "I like to give arguments for my position. They come in handy when people don't agree with me." Consider then the inclination of many among us—perhaps by virtue of being busy, or for other reasons—to pronounce a verdict, when a moral controversy comes along, by consulting aphorisms or slogans. By following this inclination, earnest people may unwittingly betray the moral views that they aspire to uphold. For they may fail to take account of the depth and subtlety of their respective moral views. This chapter presents arguments offered in support of using human embryos in research and therapy, sorting them into arguments that work and arguments that do not. I first review six arguments that I place in the latter category. Each of these purports to justify research that I happen to support. But inasmuch as a good case is not made better by overstatement, and no case is made by an unsound argument, I am going to disavow those six arguments. I urge other supporters of donated embryo use to disavow them as well, because, as Bernard Williams once said, openness to criticism is the homage that candor pays to truth. To support my own view, I shall go on to sketch arguments that, so I shall suggest, are sound. I shall then say a bit more about cloning in particular and shall close by remarking on the risk of abuses.

ARGUMENTS THAT DO NOT WORK

Imminent Death as Justification for Killing

Consider the argument that the imminent death of an embryo—for instance, a surplus embryo in a fertility clinic—justifies its consumption in research. A more extreme proposal would have us define a concept of embryo death according to which embryos not destined for intrauterine transfer are dead. This concept of death seems to defy common sense. The embryos that I contemplate as research subjects are alive. About this concept of death I doubt that it is necessary to say more.

Let us consider the plain argument that imminent death of an admittedly living embryo justifies killing it in research. In refutation of that argument, consider the following Wild West example. One day the notorious varmint Hatfield is riding about on his horse. Feeling tired, Hatfield elects to dismount beside the railroad tracks. He sits down and eventually dozes off, stretching out between the tracks. Sometime later, as Hatfield lies sound asleep, a train approaches at high speed. Whereupon there happens to ride onto the scene Hatfield's archenemy, McCoy. Spotting Hatfield, McCoy gallops to the tracks, dismounts, and—just in the nick of time before the train arrives—yanks Hatfield from the tracks. McCoy then immediately pulls out his rifle, trains it on Hatfield, and kills Hatfield. In this case, even though Hatfield would have died under the wheels of the train a moment earlier, we hold McCoy guilty of wrongful killing. In general, the imminent death of a victim does not justify its killing. Embryo research will not be justifiable solely on the ground of imminent embryo death.

 Copyright © 2008 Elsevier Inc. All rights reserved

Nonindividuation Argument Against Zygotic Personhood

The nonindividuation argument against zygotic personhood runs as follows. Prior to formation of an embryo's primitive streak at day 14 of development, it can happen that the embryo splits into monozygotic twins. If twinning occurs, the twins can fuse. Hence it has been suggested that in respect of any embryo, one cannot say until day 14 whether there exists one individual or more. If one cannot say how many persons exist, it is untenable to say that any person exists. Another version of the nonindividuation argument begins from the premise that in twinning, an embryo vanishes and leaves no earthly remains. How could an individual person die leaving no earthly remains? If no corpse ever exists, there could not earlier have existed a person.

Thus does the nonindividuation argument characterize zygotic personhood as metaphysically impossible. The argument's biological sophistication has led many scientists to regard it as a clinching argument for embryo use. Despite that sophistication, the nonindividuation argument is susceptible to the following refutation owed to David Oderberg (Oderberg, 1997). Suppose at time $t0$ a somatic cell x.By t_1, x has divided, and there exist x's daughter cells d_1 and d_2. The process by which this has occurred, namely mitosis, is routine. Notice that as we look back at the history of x up to t_1, we do not have any doubt that x was an individual cell. Plainly x had the capability of dividing, and in fact, x did divide, but it is not incoherent to say, and we unhesitatingly do say that for so long as x existed, x was an individual. As for the apparent puzzle of dying without leaving earthly remains, again a reflection on mitosis sheds light. Necrosis is not the only means by which a life form may cease to exist. Dividing is another means. It happens that after division, there is no corpse. So it is not metaphysically incoherent to say that an embryo capable of dividing is an individual. Or that an embryo that does divide was an individual before it divided.

To rehabilitate the nonindividuation argument, a proponent might contend that indivisibility is somehow intrinsic to the individuality of a person as it may not be to the individuality of a cell, so that a being that is divisible cannot be or correspond to a person (Guenin, 2006). To buttress this claim, the proponent might offer the example that an adult individual cannot divide into surviving individuals. Or the proponent might contend that even if adult individual x could be split into surviving individuals—say, by a brain split and transplant operation in which, as imagined by philosophers of mind, x's brain is split and each half is transferred into a new body so that each successor retains memories and otherwise achieves psychological continuity with x—it would not be the case that x is the same individual as either of its successors. One reply to this, owed to the whimsy of Peter King, is that it is possible to survive with only half a brain, though in such case one is restricted to a career in politics. But we may leave aside what adults can do.

We may reply to the proponent of the nonindividuation argument with two thoughts. First, what is feasible for an adult ought not constrain our thinking, because we know of the remarkable ability of an early embryo to split into surviving individuals. Second, the individuality of a being does not depend on its being the same individual, if it happens to split, as any of its successors. When an embryo has split, we may simply say that it was one individual until it split, that the individual ceased to exist when it split, and that two individuals have succeeded it. Thus may we render not only the possibility but the actuality of twinning consistent with individuality of the original embryo and with individuality of its twin successors. Given that the case of an embryo that does split resolves in this way, no impediment arises to individuality of an embryo that has not split.

According to this analysis, the nonindividuation argument fails to establish that an embryo cannot be a person. When monozygotic twinning occurs, it may be said that two persons succeed one person that ceases. But even if we defeat the contention that an embryo cannot be a person, there remains the question whether, for purposes of the duty not to kill, we should treat every embryo as a person. That a being "is a person" is not an empirical observation or an *a priori* truth. Calling a being a person is a shorthand reply to the moral question, "How should we treat it for this purpose?" The shorthand signifies our conclusion that we should classify the being among those to whom we think all of us should accord a particular treatment. Whereupon we may fairly be asked what argument supports that conclusion. Failure of the nonindividuation argument leaves the door open to introduce, or oppose, arguments that we are obliged to treat all embryos as persons for purposes of the duty not to kill.

The Utilitarian Defense of Embryo Use

Utilitarianism commands us to maximize aggregate preference satisfaction for the universe of affected sentient beings. A familiar argument is that if, in regenerative medicine, we sacrifice a relatively small number of embryos in order to help millions, perhaps billions, of suffering people, we can achieve higher aggregate preference satisfaction than we would achieve were we to classify every embryo as a person for purposes of the duty not to kill.

John Stuart Mill, who with Bentham brought utilitarianism to prominence, learned calculus at the age of 5 years, but Mill did not envision the mathematical defect of his moral theory. The infirmity came to light through the work of economists in the 20th century. Consider that the number of affected sentient beings in respect of many policy issues, embryo use among them, is enormous. Collecting utility data from so many people would be a monumental task. A more fundamental failing is that there obtains no method of measuring utility. Utilitarianism presupposes a utility function for each member of the set of affected beings. A utility function is nowadays understood as a real-valued order homomorphism representing a transitive and connected binary relation defined by an individual on a set of alternatives. This understanding no doubt explains why many scientists have felt comfortable with utilitarianism and with a utilitarian defense of embryo research. Utilitarianism seems empirical, quantitative, precise. For physician scientists, utilitarianism evokes some of the thinking to which outcome studies, comparing benefits and costs, have accustomed them.

But if, for two or more individuals, a utilitarian seeks to sum the utilities of a given alternative, there obtains no common unit of measure. There does not exist a standard measure even for a single person. While it is easy to define, as a representation of an individual's positioning of alternatives, a real-valued order homomorphism, any of infinitely many other functions formed by affine transformations of that order homomorphism will also represent the positioning.

Though the problem of interpersonal incommensurability of utilities remains unsolved, on many occasions a utilitarian's audience either will be unaware of the problem, or willing to overlook it. For example, if a hospital were to propose construction of a new facility, and if that project would require demolition of the homes of 100 people, displacing those residents in exchange for reimbursement of their homes' fair market value and their moving expenses, a utilitarian might argue that future gains in utility produced by the project for perhaps hundreds of thousands of patients in future decades will exceed the disutility of the 100 who must presently relocate. Listeners will follow an argument such as this without worrying much about whether the utility calculation has been performed. For it may seem in such a case that for any plausible conversion ratio of units of measure, the comparison of utility between alternatives will be lopsided.

But when an advocated alternative is the killing of a life form that some people sincerely believe to be a person, not much tolerance will be found for an argument whose proponents cannot produce the calculation on which the argument purportedly rests. The root of the difficulty is the same as in the blinkered attempt to order, on the basis of supposed measures of quality, incommensurable college football teams.

Appeal to "Fact-Based Reasoning" Alone

The next argument proceeds by asserting that public advocacy concerning embryo use should appeal to "fact-based reasoning" alone, and that it therefore follows that objections resting on any other ground must give way to the progress of biomedical science. This argument serves as a euphemism for saying that appeals to religion and various "insular" moral views do not have a place in public debate about science. In reply, we must observe that the view that scientific work grounded in fact-based reasoning should go forward without obstruction by moral views is itself a moral view. When an objection lies on grounds of wrongful killing, it is neither appropriate nor feasible to oust religion or moral views from the conversation. Of course we all agree that we should rely on facts as opposed to errors, but given facts, a normative discussion awaits.

A more sophisticated cousin of the foregoing argument, the call to "public reason" by John Rawls, would have us employ in public discourse only premises that can be supported without appeal to any particular comprehensive moral or religious view (Rawls 1993). That, in a moment, is what I shall try to do in stating my own view.

Noncomplicity Defenses

It has been suggested that even if the consumption of embryos in research is wrong, a government could support derivative research by eschewing complicity in the destruction of embryos. As it has been put by proponents of this move, one could distinguish between embryonic stem cell derivation and embryonic stem cell use. The National Institutes of Health (NIH) at one time adopted this view (65 Fed. Reg. 51976—51981 [2000]). It announced that it would fund projects classified as embryonic stem cell use. The notion seemed to be that this would avoid complicity in wrongful embryo-destructive derivation of stem cells.

In another scheme for purportedly conducting embryo research without complicity in embryo killing, fertility clinicians would perform the immunosurgery by which embryonic stem cells are derived from embryos. This scheme is probably impractical, as fertility clinicians do not do that kind of work. They do not customarily derive cell lines from embryos; they customarily nurture and transfer embryos so as to achieve pregnancies. Even if the scheme were practical, it shares with the derivation-use distinction the problem that when a chain of supply runs from someone who sacrifices an embryo to someone who experiments with the sacrificed embryo's derivatives, we seem compelled to say from a moral point of view that the source and the recipient ride in the same boat. It is untenable to say that the experimenter is not complicit in the work of the supplier (Guenin, 2005b).

Another noncomplicity strategy, recently played out in the United States, might be called "government surprise." It would begin as a government announces that it will not fund research that effects or is consequent on destruction of embryos. Then, after this policy has become widely known, the government suddenly announces—in the United States, we saw this happen on August 9, 2001—that in the future it will disperse public funds for studies using derivatives of embryos already then sacrificed. The most cogent philosophical defense of this gambit that occurs to me might be to say that the government had not, prior to the second announcement, induced any destruction of embryos. It would say this on the supposition that theretofore, the government had given everyone to believe that it would not support such research. (The supposition may not be true with respect to the recent history of US government policy. Between the first and second announcements came the just mentioned NIH announcement that it would fund research on embryonic derivatives. What I am here calling "the first announcement" was a prohibition enacted by Congress as a rider to an appropriations bill [Pub. L. No. 104–99, Title I, §128, 110 Stat. 26, 34 (1996)]. But apart from this historical contingency, the government surprise scheme succumbs to the same objection that lodges against the two other schemes that I have just mentioned. The government surprise scheme would place those who participate in funded embryonic stem cell studies in the same boat with those who participate in nonfunded embryo-destructive stem cell derivation (Guenin 2004b, 2005b).

Denying That Clones Are Embryos

The last on this roster of arguments that do not work consists in the claim that somatic cell nuclear transfer performed in research is not cloning and does not produce clones or embryos, this because a suggested new semantic regimen would withhold the term "cloning" from any instance of somatic cell nuclear transfer in research, would instead call that process "nuclear transplantation," and would withhold "clone" and "embryo" from the products of that process.

For this context, I have elsewhere tried to sort out the relevant entities and events in a manner informed by biological and moral considerations (Guenin, 2003). My analysis leads me to reject the proposed semantic regimen as to both process and products. The proposed terminology risks the appearance of trying to smuggle in a morally significant event—initiation of embryogenesis—by not mentioning it. In trying to withhold "cloning" and "clone" from processes and products of research, a proponent of the proposed terminology would contradict the ordinary and morally significant understanding of cloning as a genetic event, an event completed shortly after oocyte activation, regardless whether transfer to a uterus ensues. The goal that has motivated the proposed terminology, the goal of offering the public a sharp distinction between research and producing children, does not require legerdemain to attain. Rather we may implement a simple distinction, a distinction between "procreative cloning" and "nonprocreative cloning." The distinction turns on a single observable event, intrauterine transfer.

As for "embryo," any being that is of a kind capable of developing into a neonate upon transfer to a uterus is an object of moral concern. We implicitly acknowledge that concern when we classify every prefetal developing organism as an embryo. (In both scientific writing and popular speech, we have abandoned the textbook definition according to which "embryo" applies only to a conceptus older than 2 weeks.) Recognition of a being as an embryo does not end our moral investigation. We may conclude—for reasons that I shall shortly present, I believe that we should conclude—that we are not obliged to treat every embryo in the same way. But in respect of a product of somatic cell nuclear transfer, recognizing its inclusion within a discussant's universe of moral concern is the place from which to begin our discussion with one who does not agree with us (Guenin, 2004a).

ARGUMENTS THAT WORK

Argument from Nonenablement

To advance the proposition that embryonic stem cell research is virtuous if not obligatory, I have presented an argument that I call "the argument from nonenablement" (Guenin, 2001 Guenin [forthcoming]). I refer to an embryo that will never enter a uterus as an "unenabled" embryo. In the first instance, I have in mind a situation, which often arises with fertility patients, in which the one person in the world who, together with her partner, is empowered to decide about intrauterine transfer of an embryo formed from her oocyte decides that neither does she wish to have that embryo transferred into her nor does she wish to give the embryo to anyone else for intrauterine transfer.

There is no moral view of which I know that asserts a duty of intrauterine embryo transfer into oneself. That is to say that there is no moral view that asserts that a woman lies under a duty to undergo a transfer into her of an embryo that lies outside her. About such a procedure, we respect her autonomy. Imagine, therefore, that a woman declines intrauterine transfer, and in fact, forbids it. She, with her partner, donates to medicine either an embryo created during her fertility treatment, or an embryo that will be created by a scientist from their donated cells. Let us assume that this decision is final and that the embryo has left parental control. Such a donation to medicine I call an "epidosembryo." I take this name from the Greek *epidosis* for a citizen's great beneficence to the common weal.

A distinction obtains between the developmental potential of an embryo that lies in a petri dish and will remain there, and an embryo that lies in a uterus, however it got there. In consequence of parental instructions that an epidosembryo shall be used in research or therapy and shall not be transferred to a uterus, there does not obtain any morally significant chance that from such an embryo, an infant will develop. To put the matter in language that I owe to Hare, no possible person corresponds to an epidosembryo (Hare, 1993). We also know, and this is purely empirical knowledge, that an embryo is not sentient. And that for lack of a cortex, an embryo cannot form preferences. Nor can an embryo adopt ends. Therefore nothing that we might do to an epidosembryo can cause it discomfort or frustrate it. Under these circumstances, and when we consider the duty of mutual aid asserted within each of the leading moral views, I claim that it is permissible to use some embryos, namely epidosembryos, in medicine.

As the preceding discussion makes clear, I rest the permissibility and virtuousness of epidosembryo use on the autonomous decisions of couples from whose cells such embryos originate. The moral analysis flows entirely from what it is that they decide. If it is permissible for those donations to be made, then it is permissible for recipient scientists to use epidosembryos as instructed.

Suppose that someone interjects that precisely because epidosembryos cannot form preferences, it is our obligation to act according to their advantage. I reply that we cannot promote any advantage of epidosembryos. Were we to refrain from using epidosembryos in research, we could not gain anything for them. Entry into the only environment by which they could attain birth has been forbidden by the only persons in the world empowered to decide on entry into that environment.

The argument from nonenablement is a consensus argument. It commands assent across a wide range of moral views. The argument from nonenablement differs from an appeal to imminent death as a means of justifying a killing. Nonenablement precludes a conceptus from attaining any of the attributes—autonomy, ability to feel pain, preferences, and, according to the traditional Thomist—Aristotelian teaching of Christianity, the attainment of a soul—whose infringement makes killing wrong. Nonenablement entails that there does not even correspond to an epidosembryo a possible person. It is not that death is imminent, but that development is bounded.

Replies to Objections

Let me now reply to some objections. The first objection asserts that the sacrifice of an embryo violates the second form of Kant's categorical imperative, the precept that we should treat "humanity. . .always at the same time as an end, never merely as a means" (Kant 4: 429). In this formulation, by "humanity" Kant understands rational nature. For this reason, the objection misapprehends Kant's view, which applies only to rational beings. The second form of the categorical imperative does not apply to mentally incompetent adults, infants, or embryos. That is not to say that Kant would take a cavalier view of the helpless. Kant would analyze our moral obligations toward these non-rational human beings as he would analyze moral questions in general. ("Human being," by the way, is not a decisive moral classification; the phrase obviously applies to any being of the species *Homo sapiens*, including any somatic cell.) Kant would ask whether we could without contradiction of the will adopt as a universal law whatever maxim we propose about how we shall treat such a being. We do not contradict our will by adopting as a universal law that we shall use epidosembryos, at no cost in potential lives, to provide to those who suffer the aid that we would wish were we in their shoes.

A second objection is the simple declaration that a zygote is a person. The most influential version of that objection appears in the official teaching issued within the latter part of the 20th century by the Sacred Congregation for the Doctrine of the Faith of the Roman Catholic Church. Its reasoning begins from the premise that all artificial methods of reproduction, including *in vitro* fertilization, are illicit. In adopting that stance, the magisterium anticipated a situation now before us—there exist surplus embryos whose creators have effectively consigned them, as the magisterium puts it, to an "absurd fate." Such embryos will either perish as waste or be frozen indefinitely, never entering a womb. So when the magisterium condemns embryo destruction, it speaks consistently. It condemns destruction of embryos in research just as it condemns artificial reproductive practices that inexorably lead to destruction of embryos as waste. Others who approve of *in vitro* fertilization as presently practiced but oppose use of embryos in research fall into inconsistency. They implicitly approve the destruction of surplus embryos as waste while condemning the use of surplus embryos to help others.

Although the official Catholic view cannot be accused of inconsistency in the foregoing respect, on what ground does that view rest? Why condemn the destruction of an unenabled embryo in research? As one studies the magisterial instructions and looks for arguments, one finds a single argument. The argument is not scriptural. The ancients did not understand fertilization or embryogenesis and were not thinking of embryos in petri dishes. The argument purports to be biological. The argument is that because fertilization creates a new genome, fertilization creates a person (Sacred Congregation for the Doctrine of the Faith, 1987).

This is an argument whose premise one must admit—the biology is correct, fertilization does produce a new genome—but whose conclusion does not follow. The argument presupposes a radical version of genetic reductionism. To say that a person is a genome is a view that even a materialist would not venture. It is a view contradicting the belief, held within the religious tradition from within which the argument is offered, that a person is a *corpore et anima unus*, a union of body and soul. I imagine that one reason that many people have not heard this argument is that it cannot be maintained consistently with the rest of Christian teaching. I suspect that eventually the argument will fade.

Another interpretation of Catholicism might lay stress on the notion, also advanced by the magisterium, that because we do not know whether an embryo is a person, we should exercise caution and act as if it were a person. But suppose that we could have a conversation with God. We would report that in 1998, we discovered how to culture embryonic stem cells. We explain that we have plans to relieve human suffering by the use of embryos that will never enter a womb. Is it plausible that He would tell us that He regards embryos that will never enter a womb as persons in the sense that He includes them in the universe of beings that He wishes us not to consume? I do not know of a tenable argument according to which an all-merciful and omniscient God would assert that preference. For He would know that unenabled embryos will never develop into infants. He would know that our efforts to aid actual lives would exact no cost in potential lives. Under these circumstances, there inexorably come to bear Christian social teachings, including the duty to love thy neighbor as thyself and "the law of charity" (Sacred Congregation for the Doctrine of the Faith, 1974).

Cloning as a Special Case

You may have heard some people say that the justification of embryonic stem cell research lies in the circumstance that the embryo donors initially intended procreation when they created the embryos now regarded as surplus. The argument from nonenablement does not rely on any assumption of initial procreative intent. Hence the argument from nonenablement justifies not only embryonic stem cell research, but also nonprocreative cloning.

An objection peculiar to cloning might be this. An oocyte is created for a purpose, namely to issue in offspring, and it is wrong to divert an oocyte to any other purpose. This objection rests on an Aristotelian teleology that, since Darwin, does not exert much grip on our thought. We have learned from the history of medicine how mistaken we humans have often been in inferring purposes of various cells and structures of the human body. Our forbears would have said that bones are what hold us up; today we think of the marrow as a blood factory, and think it appropriate to transfer marrow from one patient to another. We have learned the remarkable adaptability of tissues and cells. We now actively engage in directing tissues and cells to serve chosen purposes in aid of sick patients, calling this treatment "conventional" drug therapy. It is unpersuasive to say that an oocyte can or should serve only one purpose.

40

To Make is to Know: The Ethical Issues in Human Tissue Engineering

Laurie Zoloth

CASE STUDY

The small black box holds a perfectly shaped ear. The scientist at the front of the room explains how it was made. There is a scaffold of nanoparticles that support fibroblast cells that grew over the form, and the ear now looks and feels actual: it can be transplanted to tissue. For burn patients, this represents an enormous chance and change. It is the prototype of a new genre of medicine, one that uses powerful technologies and methods of bioengineering and cellular biology to transform the matter of the world.

The ear in the box is not a freak example of a new technique. It is, in fact, one of a number of new devices that utilize the convergent technologies of several different fields of science and engineering to create tissue that can mimic the structure and function of the natural world. Other examples include the creation of skin grafts, corneas, bone, cartilage, and in some pilot studies, bladders. Based on the new technologies of genomics, informatics, nanoscale engineering, molecular biology, and stem cell research, tissue engineering can be said to alter the concept of medicine itself. Instead of treating ailing tissues or organs with drugs intended to repair their structure or function, tissue engineering aims at replacing the diseased or injured parts of aging tissues of the body with entirely new ones, made from component parts of materials in the world, both naturally occurring and synthetic.

Such an advance heralds a remarkable ability to heal, a long-awaited solution to several intractable problems, and a serious alternative to cadaveric or living donor whole organ transplants, which have long been an ethically challenged sector of medicine.[1] Yet such a remarkable construction of the human body asks a great deal of any social world into which it is introduced, for it is the body that is the place of the self, the location of the acts of the sacred, and the sensory arbiter of the real. In fact, tissue engineering queries two of the very aspects of our humanity that we consider distinctive: our integral embodiment and our finitude. If we are indeed a collection of replaceable and adaptable parts, some reason, what is it that separates us from any other engineered machine? If we can engineer, for example, a synthetic and improved lymphatic system, might we improve our chances to adapt to and overcome infections disease? What other capacities for healing or alteration of our bodies might be prudent? How do we ensure that such changes are indeed ethical?

It is this query which has greeted the new biotechnologies of the body, one based, this chapter will argue, in social reactions largely shaped by culture both ancient and contemporary. We will then ask: what are the

[1] *Transplantation Ethics*, Caplan, et al.; also, Fox, Renee, "Leaving the field".

Copyright © 2008 Elsevier Inc. All rights reserved

ethical challenges to the field of tissue engineering? Does tissue engineering raise new ethical issues, or is it a description of one of the modalities enabled by the convergence of other technologies that have been understood to be individually ethically freighted?

In this chapter, we will use an established ethical framework that was suggested in 1999 by a committee of the American Association for the Advancement of Science on inheritable genetic germline modification, and was used far more widely in the field of bioethics to assess new technologies. We will then review the responses given to new technologies in the past from a variety of sources in bioethics, philosophy, and theology, and finally, we will reflect on how the legal and regulatory structure for tissue engineering has an impact on our reflections on ethical norms.[2]

Research Evaluation

Are there reasons in principle why performing the basic research should be impermissible?
What contextual factors should be taken into account and do any of these prevent development and use of the research?
What purposes, techniques, or applications would be permissible and under what circumstances?
What procedures, structures, involving what policies, should be used to decide on appropriate techniques and uses?
Adapted from the AAAS Working Group on Human Inheritable Genetic Modifications 1998–2000.[3]

ARE THERE REASONS IN PRINCIPLE WHY PERFORMING THE BASIC RESEARCH SHOULD BE IMPERMISSIBLE?

Principled reasons for objections to basic research are extremely difficult to conceive in research that is, by its very nature, intended to be translational and clinical. Yet, ethical objections to the manipulation, replacement, and engineering of human tissue can be seen as part of a long continuum of dissent about medical technology that began to assume full voice in the 1970s when successful genetic manipulation of bacterial genomes became possible.[4]

All new technology raises new challenges – in particular, technology that refashions the embodied self, becomes a part of the "self" and the identity of the subject, and seems to raise the deepest anxieties. Even tissue engineering, an emerging field with clear targets, clinical successes, and patient needs will raise familiar concerns.

First among these is the argument that humans possess an essential nature and live within an essential natural order that cannot be altered without harm. For C.S. Lewis, this is expressed as a concern that the very acts of rational science: dissection, analysis, and quantification are a violation of the sacred integrity that lies behind all of nature:

"Now I take it that when we understand a thing analytically, and then dominate and use it for our own convenience, we reduce it to the level of 'nature', we suspend our judgments of value about it, ignore its final cause (if any) and treat it in terms of quantity. This repression of elements in what would otherwise be our total reaction to it is sometimes very noticeable and even painful: something has to be overcome before we can cut up a dead man or a live animal in a dissecting room."[5]

---

[2] Chapman, et al., *Designing Our Descendants.*
[3] Ibid.
[4] Walters, LeRoy, *Enhancing Human Traits.*
[5] Lewis, C.S "The abolition of man," *On Moral Medicine*, edited by Lammers and Vehey, p. 274.

For Lewis, the understanding of the body as replaceable is disturbing: "The real objection," he says, "is that if man chooses to treat himself as raw material, raw material he will be, not raw material to be manipulated by himself as he fondly imagined, but by mere appetite."[6] (Lewis imagines that new transformative technology will be manipulated by "controllers" who will eventually transform man into mere matter.)

Callahan echoes his concern, both in the sense that limits need to be placed on what is decent to do to nature, and in the sense that such action is part of a larger danger – that power in the hands of medicine to heal is really power in the hands of the elite, or the state to manipulate and control. He argues:

"The word No perfectly sums up what I mean by a limit – a boundary point beyond which one should no go … There at least two reasons why a science of technological limits is needed. First, limits need to be set to the boundless hopes and expectations, constantly escalating, which technology has engendered. Advanced technology has promised transcendence of the human condition. That is a false promise, incapable of fulfillment … Second … limits (are) necessary in order that the social pathologies resulting from technologies can be controlled … while it can and does care, save, and free, it can also become the vehicle for the introduction of new repressions in society."[7]

These objections, made over 30 years ago, are still made (despite, one may note with some irony, that 30 years of medicine have indeed seen rapid and successful advances, without them being used by the state for repression and without any fundamental change in the capacities for intellectual and spiritual self-possession). Nevertheless, powerful arguments of opposition to the manipulation and replacement of tissues and organs continue, with some worried that perfection itself is sought when healing is the goal. Such critics, many from the disability community, raise principled objections to the use of tissue engineering if the goal is to alter the disability. Activists in the Deaf community, for example, defend their disability as a culture and a language exchange, not as a loss of function. Others are concerned that our societies' focus on "fixing things" will allow a devaluation of the persons that currently bear the broken bodies and parts. Adrienne Asche suggests that there is a "troubling side to every cure, that those of us who are uncured are seen as less valuable, perhaps even expendible." For Gerry McKinney, a community needs to embrace brokenness, and to "deny that the worth of one's life is determined by how closely one conforms to societal standards of bodily perfection." McKinney is also concerned that if medicine is successful, it will create a social and economic system that "virtually demands that we be independent of the need to care for others. …"[8]

DUTY AND HEALING: NATURAL MAKERS IN A BROKEN WORLD

While the opposition to medical technology has indeed been persistent, it has not been unchallenged. For many, the response lies in the nature of brokenness and the human duty to respond to the need of the suffering other.[9,10] The principle at stake in the assessment of tissue engineering as an ethical act is not how its use might potentially violate an abstract community in the future, but the actual problem of what one must do as a moral being when one's neighbor is in need? In this important sense, the duty to heal cannot be overridden by a "sense" of discomfort (as Lewis notes in the quote above).

It is the nature, goal, and meaning of science to address the human condition in all its yearning and capacity for defeat and failure. In this sense, there is no principle objection to the science of tissue engineering, and in fact, there may well be a strong moral imperative to develop the technology.

---

[6] Lewis, *ibid.*

[7] Callahan, Daniel, "Science, limits and prohibitions," *The Hastings Center Report*, Vol. 3, No. 5, November 1973, pp. 5–7.

[8] McKinney, Gerald P., "Bioethics, the body, and the legacy of Bacon," *To Relieve the Human Condition: Bioethics, Technology and the Body*, Albany, NY: State University of New York Press.

[9] Freedman, Benjamin, *The Duty of Healing*, Routledge, New York, 1998.

[10] Zoloth, Laurie, "Difficult freedoms," *God and the Human Embryo*.

Noted Joseph Fletcher, in speaking of an earlier generation of medical technology, and answering the critics of science:

"The belief that God is at work directly or indirectly in all natural phenomena is a form of animism or simple pantheism. If we took it really seriously, all science, including medicine would die away because we would be afraid to 'dissect God' or tamper with His activity ..."[11] "Every widening and deepening of our knowledge of reality and of our control of its forces are the ingredients of both freedom and responsibility."[12]

McKinney, Childress, and Lewis use an argument that is rooted in Christian moral theology: that since human persons are fallen creatures in a fallen world, we cannot really be counted upon to know the right and the good. God, who is transcendent from this fallenness, has set us in this place, not essentially to alter it toward our own transonic, but to find its meaning and purpose. Yet, other faith traditions differ. For Jewish and Islamic theorists, the world is morally neutral. Humans may – and will – fail in their aspirations, but can be trusted to have the capacity for moral behavior and moral yearnings. Finding meaning in suffering is not the core task – the task is to alleviate suffering, which is understood as chaotic, meaningless, and agonistic. Hence, many of the core objects in principle are rooted in religious constructions and understandings.

TO MAKE IS TO KNOW: NOTES ON AN OLD PROBLEM ABOUT KNOWLEDGE

The classic debates of the 1970s are not the only set of problems engendered in the history of ethical responses to the technological gesture at the heart of tissue engineering – at stake as well is the special kind of knowledge that such a making implies. For Aristotle, and the Hellenists, useful knowledge, "practical wisdom," was *phronesis. Phronesis* implied actually doing an act, making, in order to know. The act of making, not the act of perception or contemplation alone, what-how wisdom, and indeed, rationality and power, were achieved. Hence, making new tissue is a somewhat different moral gesture than curing the body by altering it with drugs that essentially allow the body to heal itself.

Secondly, the use of technology within the body of the patient is a different matter than the use of technology to essentially enhance the body of the practitioner. For all earlier technology, the thing that was changed or enhanced was the sense perceptions of the doctor. Stethoscopes and otoscopes allow the sounds of the body to be more audible. X-rays, CT scans, and MRIs allow the inner vistas of the body to be revealed, EEG and EKG, allow the electrical currents that animate the central and peripheral nervous system to be charted in quantifiable units. Microscopes allow the invasive bacteria to be seen at the microscopic and, increasingly, molecular level. These earlier technologies extended the reach of what Bacon increasingly trusted, and that the Greeks did not – the perception and observation of the phenomena of the world and the perception of the outcome of its deliberate perturbation:

"Bacon's method presupposes a double empirical and rational starting-point. True knowledge is acquired if we proceed from lower certainty to higher liberty and from lower liberty to higher certainty. The rule of certainty and liberty in Bacon converges ... For Bacon, making is knowing and knowing is making (cf. Bacon IV [1901], 109–110). Following the maxim "command nature ... by obeying her" (Sessions, 1999, 136; cf. Gaukroger, 2001, 139 ff.), the exclusion of superstition, imposture, error, and confusion are obligatory. Bacon introduces variations into "the maker's knowledge tradition" when the discovery of the forms of a given nature provide him with the task of developing his method for acquiring factual and proven knowledge.'[13]

Thus, the world is known by understanding the parts of the world and from that, theorizing (knowing) by induction to principles or axioms or laws of nature, physics, and chemistry. In contemporary science,

---

[11] Fletcher, Joseph, "Technological devices in medical care," *Who Shall Lie*, edited by Kenneth Vaux, 1970, Fortress Press.

[12] *Ibid.*

[13] Stanford Encyclopedia of Philosophy, http://plato.stanford.edu/entries/francis-bacon/#5.

knowing is done largely by "unmaking" – by the deconstruction of the component parts in a way scientists of Bacon's era were unable to imagine. Many of these "unmaking" techniques, such as the splicing of alternative DNA, or the manipulation of cellular structures, allow a sense of inherent interchangeability, as if the real and the person were merely a set of Lego parts, awaiting clever recombination.

WHAT IS A THING: THE PERILS OF DECONSTRUCTION?

Making actual tissue in mimesis of the real tissue of the actual body extends the Baconian act in radical ways. Here, the experimental perturbation is the unmaking of tissue and the remaking of tissue, only in a more controllable form. This cannot help but excite concern about the nearly infinite possibilities for technological shaping of the self. Heidegger asks: "what is a thing?" and in so reflecting, understands a thing as an object separate from the self. But what of a made thing, an object that becomes the self?

The technology of the alteration of the patient is distinctive. Devices for altering the functioning of the body that become a part of the body, and are actually a tissue of the body, are a step beyond the idea of a device held within the body. Why this is important in any ethical assessment of the technology is that the patient's consent and participation is needed for the final act of the technology to be completed. Such an event only happens in a specific context, for technologies, patients, and practioners operate in a social, religious, and economic contexts. Here we turn to the second ethical consideration.

WHAT CONTEXTUAL FACTORS SURROUNDING THE TECHNOLOGY HAVE TO BE TAKEN INTO ACCOUNT AND DO ANY OF THESE PREVENT THE DEVELOPMENT AND USE OF THE TECHNOLOGY?

Tissue engineering is a complex procedure still in the experimental stages. Yet, to be an ethical technology, it must be directed toward accessibility, just distribution, and efficacy. Hence, the troubling context of widespread healthcare disparity is a problem not only for this advanced technology, but for all newly emerging technologies. Emergence into an unjust world asks certain moral questions of new technological advances. First among these is the query about burdensomeness versus benefit in a context in which the vast majority of the world's people suffer from easily treatable infectious diseases, tuberculosis, malaria, AIDS, and infant diarrhea. How can tissue engineering be justly promoted in the face of other, pressing needs?

This objection can be typically met by noting that it would be deeply inappropriate to withhold medical knowledge until the world is entirely perfected, and that applications will not only be increasingly available to the poor (as in vaccines, once rare, now increasingly available) but also that the very process of research will typically uncover new and useful ways to understand disease.

The goal of tissue engineering is the widespread use of the technique. Unlike solid organ transplants, which would always require significant resources far outside the capacities of developing world clinics, the use of tissue replacement, stem cell therapies, and other transportable therapies are designed for widespread use. The possibility to create a method for allografts, that uses the patient's own cells, and the possibility for allologous cells to allow for an "off the shelf" source of tissue may allow for the basis for access – but only if research priorities are discussed in advance of the design process – a process, as we will describe later, that will need careful support and monitoring. The question of how to achieve this and how to enable a more just use of each technology has not yet been solved.

The second contextual factor for tissue engineering is that all human tissue is marked by its genomic identity. It is the very nature of cells, that thing which allows them to copy and reproduced, to carry identifying markers linked to some person somewhere. In the past, such use has raised serious objections. Such tissue can be traced and known, which may have implications for the person who is the source of the tissue, raising significant new issues in genetic privacy for the donor.

Further, whose is the tissue that is derived from the cells of a particular body? Who should have the rights to, and a fair share of, the profits derived from its use? In the seminal case in the field, Moore versus the Regents of the University of California (51 Cal 3d 120 271 Cal Reporter, 146, P2d 479, 1990), the issue of ownership was addressed. In this case, Mr. Moore had his T-cell lymphocytes taken from his spleen, during the course of his treatment for hairy-cell leukemia, cells that proved effective in deriving resistant cell cultures. Patented after manipulation to make a new "product", the cells were indeed profitable. Mr. Moore's complaint was that he was not informed of, much less, a part of the scientific enterprise and the lucrative payout for his cells. The case was decided in favor of the research laboratories, but in the insuring decades, alert patients with unique cell types or unusual cancers sought for research are selling their materials as personal possessions to the laboratory that wishes to procure them. Ownership is limited, however, by the constraints of the common law of the US and the EU which limits the ability to claim tissue as property. The goals of such restraints were put into place to prohibit the buying and selling of human tissue and organs for fear that, given the desperation of the poor, selling the bodies of the poor would become permissible and lead to their exploitation. Thus, the entire process that allows for the derivation of tissue sources needs to be noted. The current context for tissue donation is a mixed system. Organs, tissues such as blood, corneas, and marrow are donated or exchanged without compensation. Gametes, however, are another matter entirely. Because the use of human sperm and eggs emerged in the context of fertility treatment and because this treatment was largely conducted in stand-alone, private clinics that functioned without public oversight or regulation, the marketplace standards prevailed. What originally began as a compassionate exchange between family members of gametes when an infertile couple could not conceive, quickly changed into a robust marketplace in human gametes. As of this writing, the international standards prohibit the use of marketplace incentives for gametes, or embryos.[14]

The final context for the debate about the ethics of tissue engineering in general is the special case of human stem cells to make tissues. Because some applications of tissue engineering use stem cells as a part of the method of treatment (Egan, 2006), the debate about the ethics of the use of human stem cells is directly adjacent to this technology. For many, the origins of tissue matter a great deal. For some Christians, many Roman Catholics, and some Hindu sects, the destruction of the human embryo, even at the blastocyst stage, is tantamount to killing. For these faith traditions, the derivation of stem cells from embryos is always impermissible. For many other faith traditions, such as Judaism, Islam, Jainism, Buddhism, Confucianism, and Daoism, the use of these cells is permissible within certain constraints, as we will see below. For all faith traditions, however, the manipulation of adult somatic cells, in their precursor form, is completely sanctioned. Precursor cells are not as flexible as pluripotent cells, and it is that very pluripotency and immortality that are important in tissue engineering. These factors raise concern. Yet the contextual factors alone do not entirely prohibit the use of this technology, for justice in distribution, the possibility of the loss of genetic privacy, and the controversy over stem cell research when pluripotent embryonic cells are used affects many aspects of the new techniques in medicine. Hence, we turn to the third major issue.

WHAT PURPOSES, TECHNIQUES, OR APPLICATIONS WOULD BE PERMISSIBLE AND UNDER WHAT CIRCUMSTANCES?

Many of the salient, justifying arguments for the use of tissue engineering hinge on the telos, or goal, of the treatment: if the goal is to cure or treat human disease, then the benefits will outweigh the burdens of the work – controversy, cost, and difficulty. Clearly, then, tissue engineering ought not to be used in a trivial or wasteful fashion. Human tissue is understood by many as deserving a special sort of "respect."[15]

[14] International Society for Stem Cell Research, Draft Guidelines, 2006, www. Isscr.org/ task force report.

[15] Geron Ethics Advisory Board, "The ethics of human stem cell research," *The Hastings Center Report*, January 1999; see also Nelson, Larry, "The elusive nature of respect," *The Hastings Center Report*, 2002.

This proviso may not be so simple, for a core problem in genetic engineering has been the use of the technique for "enhancement" of human characteristics or traits. The initial ethical discussions about therapeutic uses of medicine versus cosmetic ones imagined ethical bright lines that would define the boundary between the use of such technology to restore "species normal functions"[16] for each tissue and for the person as a whole. Yet, medical practice has long gone beyond these lines, using surgery, for example, for cosmetic purpose. Will it be possible to restrict tissue replacement to burn victims, spinal cord injury victims, and diabetics? How can such a distinction be made?

Some tissue replacement therapies, such as the use of skin grafts for full face transplant, may also raise questions about the nature of identity. Indeed, the notion of a full face transplant alerted us to the depth of resistance to identity altering tissue replacements. (Could persons use any face? What if persons in need of facial transplants wished to change ethnicity? Should faces "match" and why?) Like many other aspects of this technology, this tension about identity was not new, only heightened. For example, the first years of organ transplant raised the same issues for recipients of hearts – a key aspect of identity in many cultures. If the face is the key determinant of the self in modernity, and even more so, if the brain is such, then how are we to understand the use of tissue engineering to transform identity?

Hence, linked inexorably to this technology are larger considerations of the use of tissue engineering for neuroscience – both for therapy and for enhancement. The applications of tissue transplant in Parkinson's are important – yet will there be concern about this use of the neurons of a stranger in the brain of the self? Of all the possible uses of tissue engineering, the ones that may alter consciousness and memory are the most troubling (what capacities or memories could neurons store?). Here, the need for restrictions on applications may be the clearest, yet it is not clear who ought to decide and who ought to ensure that the restrictions on unethical applications are maintained. By what criteria will such limits be set?

New research possibilities also offer applications to engineer gametes for use and storage. Engineered follicles may now be saved, frozen, matured, and used in animal models to create the possibility of human fertility after cancer chemotherapy, or other environmental risk.[17]

With this, as with all such technology, there will have to be careful attention to how the market may drive technology toward specific research goals rather than others, or that research goals will be framed only by the values of profit and efficacy, and not ones of more general interest, compassion, healing, and solidarity. The powerful applications, and the potential for widespread use itself creates the possibility for serious conflicts of interest, as serious market forces may be the core drivers of technology, especially in an aging population with increasing needs for all manner of new tissues and organs. This turns us toward the consideration of our final set of issues.

WHAT PROCEDURES, STRUCTURES, INVOLVING WHAT POLICIES, SHOULD BE USED TO DECIDE ON APPROPRIATE TECHNIQUES AND USES?

Much of the first reviews of the ethical issues in tissue engineering have in fact focused on the issues of policy – safety, patents, and gating. Products and drugs are typically controlled via four levels of restraints. The first is elaborate pre-market gating, first involving animal models, then typically done for pharmaceuticals in a decade-long series of tests, phased to test the drug on an increasing, but controllable number of human subjects. Such subjects must be gender balanced, must have full informed consent, and must be able to leave the trial at any time (which may be difficult for implanted tissues).

The next gating is the system of intellectual property. Patents and licensing control the use of the products, even the replication of the experiments. The next gating is that of financial backing, To do the enormous

[16] Used first by Norman Daniels in the debates about health care justice and the reasonable amount of medical care a person would be entitled to. See Daniels, Norman.

[17] Woodruff, et al. "Mouse follicles matured after tissue," 2006.

clinical trials, to do pre-market investigation, and of course, to actually make and sell the product requires a productive apparatus, which must be assembled and supported. Finally, each drug or device must be approved for use by the insurers.[18] As David Smith notes, tissue engineering faces a gamut of issues and a "new order of magnitude in interactions and science patents." Additionally, notes Smith, the "things" engineered are hybrids of two jurisdictions – that of drugs and that of devices.

Are genetically engineered insulin cells a drug like insulin, a device like a stent, or a biologic? Unlike stents, which are entirely synthetic, tissue engineering uses actual human cells – only manipulated in *de novo* ways.

Standards will need to be set for safety, efficacy, and fair use – standard for clinical use, standards for clinical trial, and standards for tissue stability and purity will be needed for the research and application to be safe.

Getting informed consent in this case will present significant challenges. Patients in need of organs, for example, are particularly desperate and their consent may be deeply affected by their utter lack of options. Of the medical system, 8% is already devoted to organ transplantation and the lack of organs is an overwhelming problem for nearly half of the patients hoping for transplants.[19] Yet, the first year of the use of engineered tissue will be experimental and will need to be conducted under the strongest possible set of National Institutes of Health (NIH) guidelines. How the first trials of engineered tissue are conducted will set the tone and the future for all subsequent use.

The question of policy and the regulation of policy are manifested in many of the first documents that evaluate the ethical and legal implications of tissue engineering. While, as Smith notes, the US faces a complex regulatory system. The European Union has regulated such research products as medical products, and these will fall under the regulatory gaze of the European Medical Evaluation Agency (EMEA). In both, the synthetic nature of tissue engineering, the very *de novo* quality of the work and the uneasy greeting that met genetically modified food has created serious political opposition. Policies need to be crafted with transparency and full public participation for such research needs not only public funding but public understanding of the complex theory and practice of tissue engineering – what promises it can hold, and what cautions need to be applied prior to use. Policy makers will need to attend to calls for justice in distribution, as was noted above, and will need to set in place structures for regulation.

How can new technologies best be regulated? I would contend that a full array of regulatory structures can be employed. First among these are local committees, IRBs and local review boards. The National Academies have played a large role in policy writing for both recombinant DNA and for stem cell research, and, in both instances, called for special, national ongoing oversight on such research. It would be prudent to reflect on the need for such a process for tissue engineering, for established structures largely address issues involving the use of donated tissue, not engineered tissue. Structures that protect human subjects also need strong enforcement, as noted above, both for the donors and the recipients of tissues.

But regulation, government oversight, and market forces can only go so far in shaping just research goals and commitments. The goal of ethics is to develop moral agents who are aware of a constancy of duty toward subjects and to humanity, who not only follow rules correctly, but who, given the chance and grace to work at the frontiers of science, act with courage and decency in their research.

CONCLUSION

Tissue engineering suggests that an old dream – the replacement of human body parts – may be realized. While any sober and reflective scientist understand the long way to success of this idea, the science described in this volume clearly suggests that our society is on the road to the enactment of the possibility.

---

[18] Smith, David, "Legal and regulatory issues in tissue engineering." *National Science Foundation Report on Tissue Engineering*, 2004.

[19] Lysaught, M.J. and O' Leagh, J.A., The growth of tissue engineering. *Tissue Eng.* **7**(15): 485–493.

information package submitted prior to the meeting, is provided in "Guidance for Industry: Formal Meetings with Sponsors and Applicants for PDUFA Products" (US FDA, 2002b). Early stage device meetings are addressed in "Early Collaboration Meetings under the FDA Modernization Act (FDAMA), Final Guidance for Industry and CDRH Staff" (US FDA, 2001e).

FDA also interacts with organizations representing a group of interested parties (e.g. International Society for Cellular Therapy, American Association for Blood Banks, and Pharmaceutical Research and Manufacturers of America), which provides an opportunity to discuss topics of interest to FDA and the organization. These interactions can be very valuable for FDA and stakeholders as they are a way to better understand general issues of concern as opposed to product-specific discussions with individual firms. In addition to such interactions and meetings with individual sponsors, FDA also has various advisory committees that review available data and information, and make recommendations related to a variety of issues, many of which are pertinent to the field of regenerative medicine. Advisory committees will be discussed further in Section "Advisory Committee Meetings."

REGULATIONS AND GUIDANCE OF SPECIAL INTEREST FOR REGENERATIVE MEDICINE

The topics discussed thus far have been of general applicability for medical product regulation: marketing pathways, clinical trial regulation, meetings, guidance development, and related topics. This section will review a few topics of particular interest to the scientific community engaged in development of regenerative medicine products: FDA regulations on human tissue products, product characterization for cellular products, FDA policy and guidance on xenotransplantation, and gene therapy.

Regenerative medicine products often face some unique product development challenges because of their complexity. Some of these products contain metabolically active cells and tissue, making the manufacturing, characterization, and study of these products a challenge because even a small variation in manufacturing may impact product safety and effectiveness. Scaffolds themselves may be difficult to characterize; for example, some of the materials used during development have complex three-dimensional structures. For cell–scaffold combination products the challenges are multiplied; as such a product is not defined by the components alone. For example, when a combination product is chemically or physically combined, product assembly is an important step in product manufacture and there may be other processing steps or further cell–scaffold interactions which will in turn further define the characteristics of the final product. Packaging and shelf life are critical concerns. The preclinical development plan needs careful attention. A particular question for cell–scaffold combinations is to determine which tests need to be conducted on individual components prior to assembly and which are most relevant after product assembly.

For many innovative products, such as cell–scaffold combinations and other regenerative medicine products, the final product, and instructions for use, can be expected to undergo iterative modifications over time. Consequently, refinement of the product and review of product modification will be an ongoing process for the sponsor and for FDA, respectively. It will be critical for the sponsor to have a good understanding of its product and the key scientific or clinical issues that could affect safety and effectiveness of the product, including establishment of manufacturing controls. Demonstrating comparability of a biological product after a manufacturing change is a question that FDA routinely asks as the manufacturer modifies their processing methods; for this reason product characterization is extremely important and should be considered early during product development.

Regulation of Human Cells and Tissues Intended for Transplantation

An understanding of the regulations applicable to cells and tissues is important for developers of regenerative medicine products since human cells or tissues comprise the whole, or a key component, of many products.

In 1997, noting the fragmented approach to regulation of human cell and tissue-based products the FDA issued the "Proposed Approach to the Regulation of Cellular and Tissue-Based Products" (US FDA, 1997b). This document proposed a tiered risk-based approach to regulation of these products. According to the proposed approach, products posing less risk would be subject to the rules designed to minimize communicable disease risks, and additional regulatory requirements would be imposed on those products posing additional risk. The proposed approach to regulation of human tissues was implemented in three parts, collectively referred to as the tissue rules: Registration and Listing, Donor Eligibility, and Good Tissue Practices (GTP). These complete set of rules went into effect on May 25, 2005. The tissue rules focus on control of infectious disease in products containing human cells or tissue. Thus the tissue rules apply to all human cellular and tissue-based products. It is important for sponsors of regenerative medicine products to be aware of these rules, as well as the specific requirements for biologics or devices that may apply depending on the particular regulatory pathway applicable to their products.

The tissue rules are published as regulations at 21 CFR Parts 1270 and 1271. With some exceptions that are noted in the tissue rules, human cells or tissue intended for implantation, transplantation, infusion, or transfer into a human recipient are regulated as a human cell, tissue, and cellular and tissue-based product (HCT/P). Examples of HCT/Ps are: musculoskeletal tissue, skin, ocular tissue, human heart valves, dura mater, reproductive tissue, and hematopoietic stem/progenitor cells. Tissues specifically excluded are: vascularized organs, minimally manipulated bone marrow, blood products, xenografts, secreted or extracted products such as human milk and collagen, ancillary products, and *in vitro* diagnostic products.

The tissue rules require tissue establishments:

- To register and list their HCT/Ps with FDA (21 CFR 1271 Subparts A and B).
- To evaluate donors through screening and testing, to reduce the transmission of infectious diseases through tissue transplantation (21 CFR 1270 and 1271 Subpart C).
- To follow GTP to prevent the spread of communicable disease (21 CFR 1271 Subpart D).

Additional requirements regarding reporting, labeling, inspections, importation, and enforcement are described in 21 CFR 1271 Subparts E and F. The rules also define the circumstances under which a product would be subject to the tissue rules only; and when additional oversight such as the need for a BLA, PMA, or other marketing application would be required (21 CFR 1271.20). Products that meet the following conditions are regulated by FDA solely under the tissue rules: the tissue is not more than minimally manipulated, is intended for homologous use, is not combined with a drug or device (with certain exceptions), and does not have a systemic effect and is not dependent upon the metabolic activity of living cells for its primary function (except for autologous use or allogeneic use in a first or second degree blood relative, or reproductive use). If all four of these conditions are not met, a marketing application is required. Additional information and documents regarding these rules, as well as electronic forms for registration and listing, can be found on the FDA website (US FDA, 2006a).

FDA has issued "Draft Guidance for Industry: Eligibility Determination for Donors of Human Cells, Tissues, and Cellular and Tissue-Based Products (HCT/Ps)" which provides guidance for donor testing and screening that is recommended in making donor eligibility determination for donors of HCT/Ps (US FDA, 2004b). FDA has also issued a draft guidance entitled "Guidance for Industry, Preventive Measures to Reduce the Possible Risk of Transmission of Creutzfeldt–Jakob Disease (CJD) and Variant Creutzfeldt–Jakob Disease (vCJD) by Human Cells, Tissues, and Cellular and Tissue-Based Products (HCT/Ps)" (US FDA, 2002a). FDA plans to issue a one final guidance document on eligibility determination to reduce the possibility of infectious disease transmission for HCT/Ps that will incorporate both draft guidance topics.

Human Cellular Therapies

Many products in development for tissue repair or replacement are comprised of cells or cells combined with a scaffold. The cell or tissue source and manufacturing process may vary greatly for different products.

Despite the diversity in products, there are regulatory considerations that apply to all cellular preparations being developed as investigational regenerative medicine products intended for early phase clinical studies. Among these considerations are three that will be discussed briefly: control of the source material, demonstrated control of the manufacturing process, and characterizations of the cellular product that results from the manufacturing process.

The cell source will vary for different products and may be autologous or allogeneic, undifferentiated stem/progenitor cells, or terminally differentiated cells. Assuring the safety of source cellular materials used during manufacture of an investigational regenerative medicine product begins by determining the eligibility of the donors selected to provide the source material through screening and testing. This screening and testing is part of the tissue rules described earlier in this chapter. Although autologous products are not required to comply with the donor screening and testing requirements in the tissue rules, for autologous tissue that is either positive for specific pathogens, or that has not been screened or tested, it is recommended that manufacturers document if tissue culture methods could propagate or spread viruses or other adventitious agents to persons other than the recipient (US FDA, 2003a). Donor eligibility determination is required for all allogeneic donors of cells and tissues.

The cell source may raise substantial concerns in addition to the possibility of infectious disease transmission. In addition to screening and testing donors for communicable disease agents, according to the document entitled "ICH Guidance on Quality of Biotechnological/Biological Products: Derivation and Characterization of Cell Substrates Used for Production of Biotechnological/Biological Products," FDA has suggested that Sponsors consider the importance of evaluating donor medical history information and the relevance of conducting specified molecular genetic testing as part of an overall comprehensive assessment program to establish the suitability of a specific cellular preparation for use in the manufacture of a regenerative medicine somatic cellular product (US FDA, 1998c). The rationale and feasibility for collecting additional information about molecular genetic testing was discussed in a public meeting of the FDA Biological Response Modifiers Advisory Committee (now known as Cellular, Tissue and Gene Therapies Advisory Committee (CTGTAC)) convened July 13–14, 2000 on the topic of "Human Stem Cells as Cellular Replacement Therapies for Neurological Disorders" (US FDA, 2006c). A description of the physiological source of the cellular material, including tissue of origin and phenotype such as hematopoietic, neuronal, fetal, or embryonic conveys important information about the cells and their critical attributes.

Control of the manufacturing process helps provide assurance of the consistent, reproducible production of the cellular component. Often, manufacturing will involve a multi-step process that must be performed using aseptic techniques to prevent introduction of microbial contamination (US FDA, 2004d). Many types of reagents may be used to manufacture the cellular component of a product including those that promote cellular replication, induce differentiation, and those used to select targeted cell populations, specifically, serum, culture medium, peptides, cytokines, and monoclonal antibodies. It is essential that reagents be properly qualified (US FDA, 1993, 1997a, 2003a, 2004a). Demonstration of manufacturing control is evidenced by strict adherence to standard operating procedures and quality control assessment of manufacturing intermediates as well as testing of the final cellular preparation.

Due to inherent biological complexity it is unlikely that a unique biomarker or other single analytical test will be sufficient to permit full characterization of a cellular product. Accordingly, as recommended in the "Guidance for Industry: Guidance for Human Somatic Cell Therapy and Gene Therapy," FDA asks Sponsors to provide documentation that their testing paradigm developed for the final cell product encompasses a multiparametric approach that may involve biological, biochemical/biophysical, and/or functional characterization (US FDA, 1998a, 2003a, 2004a). Tests developed to demonstrate identity of the cell product (physical and chemical characteristics, identify the product as being what is designated on the label), purity (freedom from contaminants including residual reagents and unintended cell populations), and potency/biological activity (the specific ability of the cells, as indicated by appropriate laboratory tests, to effect a given result) should be conceived to determine the degree to which the characteristics of the manufactured cell preparation conform to desired and specified criteria (US FDA, 1998a, 2003a, 2004a). This process can be challenging for a number of reasons. For example, the mechanism of action associated with a cell product may be incompletely understood and thus constrains the ability to develop a specific potency assay. Direct assessment of potency for a cellular preparation may not be possible due to a lack of appropriate *in vitro* or *in vivo* assay systems. On February 9–10, 2006, the FDA CTGTAC discussed this challenging topic and obtained input on alternative approaches for performing potency assessments of cellular therapy products (US FDA, 2006c).

In summary, assuring the safety of cell products that in and of themselves constitute a regenerative medicine product or that constitute a component of a product requires demonstrated control over each facet of the manufacturing process. This assurance begins with acquisition of the source material and is carried forward through manufacturing and characterization of the final cellular preparation using specified analytical tests based in large measure on the intrinsic biological properties of the cell product.

Xenotransplantation

The success of allogeneic organ transplantation has increased the demand for human cells, tissues, and organs. Scientific advances in the areas of immunology and molecular biology coupled with the growing worldwide shortage of transplantable organs have lead to increased interest in xenotransplantation. In addition to the potential use of xenotransplantation to address the shortage of human organs for transplantation, there are increasing efforts to utilize xenotransplantation in the treatment of other chronic diseases as well as end organ failure. Along with the promise of xenotransplantation are a number of challenges, including the potential risk of transmission of infectious agents from source animals to patients, and the spread of any zoonotic disease to the general public. Agents pathogenic for humans may not be pathogenic or detectable in the source animal host. In addition, the potential exists for recombination or reassortment of source animal infectious agents, such as viruses, with non-pathogenic or endogenous human infectious agents, to form new pathogenic entities. These considerations demonstrate the need to proceed with caution in this area.

The United States PHS Agencies including FDA, National Institutes of Health (NIH), Centers for Disease Control and Prevention (CDC), and Health Resources and Services Administration (HRSA) have worked together to address the risk of infectious disease transmission, publishing the "PHS Guideline on Infectious Disease Issues in Xenotransplantation" (US PHS, 2001). This Guideline discusses xenotransplantation protocols, animal source, clinical issues, and public health issues. Following publication of the PHS Guideline, FDA published a Guidance document entitled "Guidance for Industry: Source Animal, Product, Preclinical and Clinical Issues Concerning the Use of Xenotransplantation Products in Humans" to build on the concepts in the PHS Guideline, and provide more specific advice regarding xenotransplantation product development and production, and xenotransplantation clinical trials (US FDA, 2003b).

Xenotransplantation is defined in the PHS Guideline and the FDA Guidance as any procedure that involves the transplantation, implantation, or infusion into a human recipient of either live cells, tissues, or organs from a non-human animal source or human body fluids, cells, tissues, or organs that have had *ex vivo* contact with live non-human animal cells, tissues, or organs (US PHS, 2001 and US FDA, 2003b). Xenotransplantation products are defined as live cells, tissues, or organs used in xenotransplantation.

Examples of xenotransplantation products provided in the FDA Guidance are:

- Transplantation of xenogeneic hearts, kidneys, or pancreatic tissue to treat organ failure, implantation of neural cells to ameliorate neurological degenerative diseases.

- Administration of human cells previously cultured *ex vivo* with live non-human animal antigen-presenting or feeder cells.
- Extracorporeal perfusion of a patient's blood or blood component through an intact animal organ or isolated cells contained in a device to treat liver failure.

FDA encourages any potential sponsor of a xenotransplantation product to familiarize themselves with available documents that can be found on the FDA website (US FDA, 2006b).

Gene Therapy

FDA regulates human gene therapy products as biological products. The field of gene therapy holds great promise for treating a wide array of illnesses; from genetically inherited diseases such as cystic fibrosis or hemophilia, to heart disease, wound healing, AIDS, graft versus host disease, and cancer. In addition, the use of gene therapy in the area of tissue repair and tissue engineering is also being investigated.

There are a number of safety issues associated with gene therapy, some of which are unique to this area. Safety issues specific to gene therapy trials include generation of replication competent virus, vector as well as transgene associated immunity, toxicity associated with transgene expression, and inadvertent germline transmission of vector. Two examples of gene therapy-specific risks are instructive: (1) high doses of adenovirus vector particles have been shown to induce toxicity under certain circumstances, and resulted in the death to a study subject in 1999; (2) genomic integration of retroviral vectors has been shown to result in genotoxicity, such that three children developed leukemia, and one died, as a direct result of altered gene expression after vector integration. Detailed recommendations from FDA regarding what type of information to submit in an early phase study of gene therapy products are available in the FDA "Draft Guidance for FDA Review Staff and Sponsors: Content and Review of CMC Information for Human Gene Therapy INDs Applications" (US FDA, 2004a). This draft guidance covers product manufacturing and characterization information (including components and procedures), product testing (including microbiological testing, identity, purity, potency, and other testing), final release testing criteria, and product stability, giving specifics in these areas that are pertinent to gene therapy. Suggested preclinical testing includes tests designed to describe localization, and persistence of gene expression. For vectors intended for direct *in vivo* administration, demonstration of the extent of dissemination and gonadal distribution is suggested.

Gene therapies may differ from conventional drugs in that vector and transgene expression may persist for the lifetime of the subject. In these cases, there is a risk of delayed adverse effects. Indeed, the previously mentioned leukemias in a clinical study of gene therapy for the treatment of X-linked severe combined immunodeficient (SCID) did not occur until approximately 3 years after exposure to the retroviral vector. These events highlight the need for assessment of long-term risk in research subjects. FDA has discussed these issues, noting that the assessment of risk is based on the persistence of vector sequences, integration into the host genome, and transgene-specific effects. FDA has recently published the "Draft Guidance for Industry: Gene Therapy Clinical Trials – Observing Participants for Delayed Adverse Events" which addresses the duration and types of observations to be performed based on the patient population and the risks presented by the gene therapy product, and hopes to finalize such guidance in the near future (US FDA, 2005a).

Although regulatory responsibility for gene therapy trials rests with FDA, NIH serves an important complementary role. In addition to funding a number of gene therapy research studies, NIH also provides an important forum for open public deliberation on the scientific, ethical, and legal issues raised by recombinant DNA technologies and its basic and clinical research applications through the Recombinant DNA Advisory Committee (RAC), an expert advisory committee to the NIH Director (US NIH, 2005). Clinical studies discussed in this forum include studies funded by NIH, as well as industry funded studies conducted at clinical sites receiving NIH funding for DNA recombinant research.

The Preclinical Development Plan

For device and drug clinical trials, the goal of preclinical development studies is to establish a scientific rationale for the clinical investigation and to demonstrate an acceptable safety profile. Traditional pharmacology/toxicology safety studies are important to identify potential toxicity in target organs and tissues, and to obtain information on effective safe starting doses in humans as well as establishing a safety profile for dose escalation and/or clinical monitoring. For cellular therapy, gene therapy, and cell–scaffold combination products there are frequently additional product-specific safety questions that might need to be addressed prior to initiation of a clinical trial. For example, what is their potential to undergo unanticipated undesired changes in their characteristics, such as malignant transformation? For cell–scaffold constructs, are there safety issues associated with the implantation procedure or potential construct failure? Animal models have limitations that are confounded by anatomical as well as physiological differences between the animals and man. Many of these products, because of their novelty, do not have an established paradigm for preclinical evaluation and Sponsors are therefore encouraged to discuss their development plan with FDA early in the development process.

In several specific clinical applications FDA has either had public discussion at an Advisory Committee meeting or has published Guidance documents, and in either case it is valuable for Sponsors to be aware of these discussions or publications. For example, in March 2005 the CTGTAC discussed the manufacturing, preclinical, and clinical issues in the development of cellular therapies for repair and regeneration of articular joint surfaces (US FDA, 2006c). In the preclinical session of this meeting the committee focused on addressing specific issues raised by the FDA with respect to the range of the subset of products (device, cellular, and tissue engineered (TE) products) that have been proposed for the repair and regeneration of articular cartilage however, the issues can be readily generalized to a broader range of regenerative medicine products. The major issues discussed were:

- The limitations and capabilities of available animal models for predicting safety and clinical activity.
- Pivotal animal toxicology studies designed to support a clinical trial of a cellular cartilage repair product.
- Additional safety concerns for allogeneic cellular products (versus autologous products) that should be addressed in an *in vivo* study prior to clinical trials.

The committee arrived at a consensus that animal studies were needed to evaluate potential products prior to human administration; however, it also was evident that there was no single animal model that is adequate to test all of the hypotheses involved in development of these complex products. For example, the committee acknowledged small animals might be useful in assessing novel biomaterials or mechanisms of action in early stages of product development. Unfortunately, due primarily, to anatomic and biomechanical considerations, large animals are needed to assess the potential clinical activity, therapeutic durability, and safety of a final product prior to initiation of clinical trials ("pivotal" toxicology studies). The committee also discussed the necessary study design features for pivotal toxicology studies and reached a general agreement that the design of such studies should incorporate:

- Assessment of the mechanism of action of the product by either interim sacrifice of subgroups of animals or by non-lethal methods of assessment such as imaging or arthroscopy (particularly if this use in animals can provide data to support use of these modalities in clinical trials by allowing comparison between in-life and necropsy assessments).
- Adequate duration prior to terminal sacrifice to demonstrate integration of the product with the native tissue.
- Histopathological examination of the site of implantation and surrounding regions, draining lymph nodes and remote organs with gross pathology at time of necropsy and histopathology of other organs as guided by data from preliminary studies.

The committee discussed the potential need to use analogous animal cells *in lieu* of human cells in testing the final product construct and the limitations posed by this approach in terms of extrapolating the data to human trials. In recognition of this concern the committee highlighted the need to understand the physiology of the animal cells relative to their human analogs. The committee also discussed the potential for additional safety concerns posed by use of allogeneic cells in these products.

Clinical Development Plan

The goal of the clinical development program is to establish product safety and efficacy. In the field of regenerative medicine, variability in the product, as well as the patient, poses unique challenges in clinical trial design and conduct. An additional challenge is the need, with many of these products, to observe their integration into the host over a prolonged period. Specific feedback regarding the adequacy of certain proposed studies and predictors of clinical benefit can be provided to Sponsors through the use of meetings with the FDA at various development time points, and the use of a Special Protocol Assessment (for products regulated as a biologic) or an Agreement Meeting (for products regulated as a device) prior to initiation of their Phase 3 studies (US FDA, 2001e, 2002e).

Cell–Scaffold Wound Healing Skin Constructs

Some of the earliest attempts at cell–scaffold combination products are skin constructs and those products are worth a special mention in this chapter because there are a number of approved products on the market. Some of these products are composed of keratinocyte and/or fibroblast cells on a scaffold (e.g. animal collagen, glucosaminoglycans, or gauze); cell–scaffold wound dressings that function primarily as physical wound coverings are generally regulated as medical devices by CDRH. In contrast, regulatory oversight of cellular constructs formulated without a scaffold intended specifically to promote wound healing are assigned to CBER.

Today, several cell–scaffold designs are commercially available. Dermagraft (a single cell construct of allogeneic neonatal fibroblasts seeded onto a bioresorbable mesh) was approved in 2001 as a medical device for treating full-thickness diabetic foot ulcers (US FDA, 2001c). In addition, in 2003 an HDE was approved for Dermagraft treatment of wounds associated with dystrophic epidermolysis bullosa (DEB).

Apligraft and Orcel are two examples of commercially available bilayered co-culture constructs comprised of allogeneic neonatal keratinocyte and fibroblast cells on bovine collagen scaffolds. Apligraft is indicated for treatment of venous insufficiency and diabetic foot ulcers (US FDA, 2000). Orcel is approved for treating split thickness donor site wounds on burn patients (US FDA, 2001d). An HDE for Orcel use in recessive DEB patient (as an adjunct in covering wounds and donor sites after the surgical release of hand contracture and deformities) was approved in 2001.

Information about the clinical performance of each product is available in the published literature and product labeling (Green and Rheinwald, 1975; Cazalet et al., 1988; Boyce et al., 1989; Hansbrough et al., 1992; Haeseker et al., 1993; Baird et al., 1998; US FDA, 2000, 2001c, d, 2005c; Currie et al., 2002). Information about the clinical studies, as well as other studies supporting approval, is available on the FDA website (US FDA, 2005c).

FDA's Standards Development Program

Since its inception, the development and use of standards has been critical to the mission of FDA. The use of standards in FDA medical product regulation began with the 1906 Federal Food and Drugs Act. Drugs, defined in accordance with the standards of strength, quality, and purity in the United States Pharmacopoeia and the National Formulary, could not be sold in any other condition unless the specific variations from the applicable standards were plainly stated on the label (Federal Food and Drugs Act, 1906). In current times, Federal government agencies including the FDA are encouraged to use voluntary consensus standards, whether domestic or international, when performing regulatory activities in lieu of government-unique standards which are developed by the government for its own uses, when practical. Standard-setting activities include the development of performance characteristics, testing methodology, manufacturing practices, product standards, scientific protocols, compliance criteria, ingredient specifications, labeling, or other technical or policy criteria.

As with guidance document development, in which Good Guidance Practices describes FDA procedures for developing and using guidance documents, there are specific regulations that describe FDA participation in outside standard-setting activities. Regulations governing this participation can be found in 21 CFR 10.95. Constructive FDA participation in organizations responsible for developing standards applicable to the products regulated by the agency is considered essential.

The FDAMA of 1997 provides for the recognition of national and international standards in medical device reviews for IDEs, HDEs, PMAs, PDPs, and 510(k)s (Marlowe and Phillips, 1998). A "recognized consensus standard" is a consensus standard that FDA has evaluated and recognized for use in satisfying a regulatory requirement and that FDA has published in a Federal Register notice. A "consensus standard" is a standard developed by a private sector standards body using an open and transparent consensus process. Conformance with recognized consensus standards is strictly voluntary for a medical device manufacturer. A manufacturer may choose to conform to applicable recognized standards or may choose to address relevant issues in another manner. A complete listing of CDRH recognized consensus standards and guidance documents can be found on the CDRH website (US FDA, 2006e).

American Society for Testing and Materials International (ASTM International) and International Standards Organization (ISO) exemplify two organizations FDA works with in standard development. ASTM FO4 Division IV is actively engaged in development of standards for tissue engineered medical products (TEMPS). F04 Division IV consists of five subcommittees: (1) Classification and Terminology, (2) Biomaterials and Biomolecules, (3) Cells and Tissue Engineered Constructs, (4) Assessment, and (5) Adventitious Agent Safety. Currently, the ASTM TEMPs group has developed 19 published standards and standard guides, and has approximately 25 draft standards under preparation. The first standards were for substrates, biomaterials, such as collagen, alginate, and chitosan, and for terminology, cells and cell processing, bone morphogenetic protein, and test methods. A standard guide for *in vivo* repair of articular cartilage has also been approved.

FDA is actively engaged in standards development. In the long term, standards development will benefit the regenerative medicine community, and by extension, the public.

ADVISORY COMMITTEE MEETINGS

As mentioned in Section "Meetings with Industry, Professional Groups, and Sponsors" above, because of the diversity of innovative technology evaluated by FDA review staff, FDA makes use of expert scientific Advisory Committees or Panels (for medical devices) to complement its internal review process. These advisors provide outside advice to contribute to scientific regulatory decision making. Outside experts can be asked to review data, or make recommendations about study designs across a product or clinical area; outside advisors can also be helpful at earlier stages of product development. Expertise on the advisory committee often includes scientific, statistical, and clinical experts, as well as consumer representation, patient advocates, and industry participation. Most meetings are public and there is an opportunity for public participation in the form of public comment.

There are 30 Advisory Committees and 18 Advisory Panels for medical devices (as well as a number of subcommittees and one Department of Health and Human Services (DHHS) Committee administered by CBER). The areas of responsibility for the panels and committees are divided along product lines.

The Advisory Committee for cellular, tissue, and gene therapy products, as mentioned earlier, is the CTGTAC. This committee has discussed a number of areas in recent years that are of potential interest to product developers in the regenerative medicine area, including:

- Hematopoietic stem cells for hematopoietic reconstitution (February 2003).
- Allogeneic islet cell therapy for diabetes (October 2003).
- Somatic cell cardiac therapies (March 2004).
- Somatic cell therapies for joint surfaces (March 2005).
- Potency measures for cell, tissue, and gene therapies (February 2006).

The presentations for each topic, as well as a transcript of the discussion, are available on the FDA website referenced at the end of this chapter (US FDA, 2006c).

FDA AND CRITICAL PATH SCIENCE

FDA recognizes the complexity of the scientific issues related to these products. FDA has introduced the Critical Path Initiative as a way to identify priority research areas that are expected to advance innovation in medical products. The Critical Path Opportunities List was recently published on March 16, 2006, and is available on FDA's website (US Department of Health and Human Services, 2006). This list "presents specific opportunities that, if implemented, can help speed the development and approval of medical products" (US Department of Health and Human Services, 2006). "Moving Manufacturing into the 21st Century – Manufacturing, Scale-up and Quality Management" was specifically identified as a critical path element (US Department of Health and Human Services, 2006). Tissue engineering was specifically addressed as an area in need of additional critical path scientific development: "A key hurdle holding back innovation in tissue engineering is the difficulty in sufficiently characterizing a finished product to enable development of meaningful quality controls and release specifications. Often, conventional techniques, such as simple cell morphology, used to evaluate cell characteristics cannot be applied to these products because, for example, the engineered product may also include non-biological materials (e.g. support matrix). Consensus on how to assess these products and ensure manufacturing consistency would give product Sponsors the predictability they need to unlock innovation in tissue engineering" (US Department of Health and Human Services, 2006).

FDA critical path research is aimed at bridging some of the challenges between product development and product evaluation. For example, following the observation of unexpected toxicity of adenoviral vector gene therapy in a clinical trial, CBER research provided insight into how adenovirus vectors cause toxicity, and developed an animal model for gene therapy in the context of preexisting liver disease (Raper et al., 2003; Smith et al., 2004). CBER researcher/regulators also worked with a consortium from industry and academia to develop reference material for adenoviral vector particles (Simek et al., 2002).

An additional example of FDA critical path research is the active intramural and collaborative research program in the area of heart valves. The challenges that overlie the development, manufacturing, and characterization of a cell–scaffold tissue engineered heart valve (TEHV) center about the complexity of native heart valve biology and valvular tissue remodeling in response to a dynamic hemodynamic environment. A TEHV must be functional and durable at the time of implantation. The identification of factors that modulate the *in vivo* remodeling of a TE construct may be difficult to predict with certainty using *in vitro* methods (e.g. cell phenotype characterization); identification of cell products such as cytokines and growth factors; and physiological preconditioning in a bioreactor. Numerous biomarkers are available to describe heart valve biology; however, there are no validated surrogate markers of *in vivo* long-term TEHV performance and durability. New regulatory approaches, based on both sound engineering and biological principles, will have to be developed to assess the preclinical safety of these novel viable tissue products.

CDRH has an integrated intramural and collaborative heart valve research program that to date has been primarily focused on the effects of preimplantation processing on heart valve tissue-derived biomaterials, cryopreservation and decellularization of allograft heart valves, and the identification of valve-related pathology and potential clinical failure modes. Extensive experience has been gained in the evaluation of replacement heart valves (mechanical, bioprosthetic, and polymeric), cryopreserved allograft, and TEHVs (Jones et al., 1982, 1989; Hilbert et al., 1987, 1990, 1992, 1994, 1999, 2004; Crescenzo et al., 1992, 1993; Hilbert and Ferrans, 1992; Schoen et al., 1992; Grehan et al., 2000). The majority of these investigations have involved preclinical *in vivo* safety and efficacy studies conducted in juvenile sheep. More recently this research program has focused on the identification of evaluative tools having potential utility for assessment of emerging TEHV and blood vessel constructs. (Hilbert et al., 2004; Xing et al., 2004a, b).

CONCLUSION

The field of regenerative medicine is an exciting field with scientific advances leading to the promise of future therapies for current unmet medical needs for patients. The FDA regulatory approach to medical products evaluation includes an ongoing assessment of how the science of those products informs regulatory policy.

FDA looks to continue ongoing dialog with the scientific community and product Sponsors to continue to develop science-based regulatory review policies that are robust and predictable in order to meet the needs of the challenging array of products that are on the horizon.

REFERENCES

Baird, L.G., Christenson, L., David, J., Du Moulin, G., Gentile, F.T., Omstead, D.R., Maxted, D.D. and Tubo, R. (1998). Voluntary guidance for the development of tissue-engineered products. *Tissue Eng. Fall* **4**(3): 239–266. (Review).

Boyce, S.T., Cooper, M.L., Foreman, T.J. and Hansbrough, J.F. (1989). Burn wound closure with cultured autologous keratinocytes and fibroblasts attached to a collagen–glycosaminoglycan substrate. *JAMA* **262**: 2125–2130.

Cazalet, C., Cherruau, B., Jaffray, P., Marien, M., Schlotterer, M., Toulon, A. and Wassermann, D. (1988). Preliminary clinical studies of a biological skin equivalent in burned patients. *Burns Incl. Therm. Inj.* **14**(4): 326–330.

Changes to an Approved Biologics Licensing Application (BLA): 21 CFR 601.12 (2007).

Crescenzo, D.G., Hilbert, S.L., Barrick, M.K., Messier Jr., R.H., Wallace, R.B. and Hopkins, R.A. (1992). Donor heart valves: electron microscopic and morphometric assessment of cellular injury induced by warm ischemia. *J. Thorac. Cardiovasc. Surg.* **103**: 253–258.

Crescenzo, D.G., Hilbert, S.L., Messier Jr., R.H., Domkowski, P.W., Barrick, M.K., Lange, P.L., Ferrans, V.J., Wallace, R.B. and Hopkins, R.A. (1993). Human cryopreserved allografts: electron microscopic analysis of cellular injury. *Ann. Thorac. Surg.* **55**: 25–31.

Currie, L., Jones, I. and Martin, R. (2002). A guide to biological skin substitutes. *Br. J. Plast. Surg.* **44**: 185–193.

Federal Food and Drugs Act, 1906, 34 Stat. 768, *repealed by* Food, Drug, and Cosmetic Act of 1938, 21 U.S.C. Sec. 329(a).

Green, H. and Rheinwald, J.G. (1975). Serial cultivation of strains of human epidermal keratinocytes: the formation of keratinizing colonies from single cells. *Cell* **6**: 331–343.

Grehan, J.F., Hilbert, S.L., Ferrans, V.J., Salerno, C.T. and Bianco, R.W. (2000). Development and evaluation of a swine model to assess the preclinical safety of mechanical heart valves. *J. Heart Valve Dis.* **9**: 710–720.

Haeseker, B., Koch, R. and Teepe, R.G. (1993). Randomized trial comparing cryopreserved cultured epidermal allografts with tulle-gras in the treatment of split-thickness skin graft donor sites. *J. Trauma* **35**(6): 850–854.

Hansbrough, J.F., Dore, C. and Hansbrough, W.B. (1992) Clinical trials of a living dermal tissue replacement placed beneath meshed, split-thickness skin grafts on excised burn wounds. *J. Burn Care Rehabil.* **13**(5): 519–529.

Hilbert, S.L. and Ferrans, V.J. (1992). Porcine aortic valve bioprostheses: morphologic and functional considerations. *J. Long-Term Eff. Med. Impl.* **2**: 99–112.

Hilbert, S.L. and Hopkins, R.A. (1994). The ultimate challenge. In: *Cardiac Reconstructions with Allograft Valves*, 2nd edn. New York: Springer-Verlag, pp. 612–620.

Hilbert, S.L. and Hopkins, R.A. Small intestinal submucosa pulmonary monocusp and mitral bi-flap valves implanted in sheep. *FDA Cooperative Research and Development Agreement* 74-03.

Hilbert, S.L., Ferrans, V.J., Tomita, Y. and Jones, M. (1987). The morphologic evaluation of explanted polyurethane trileaflet cardiac valve prostheses. *J. Thorac. Cardiovasc. Surg.* **94**: 419–429.

Hilbert, S.L., Barrick, M.K. and Ferrans, V.J. (1990). Porcine aortic valve bioprostheses: A morphologic comparison of the effects of fixation pressure. *J. Biomed. Mater. Res.* **24**, 773–787.

Hilbert, S.L., Ferrans, V.J., McAllister, H.A. and Cooley, D.A. (1992). Ionescu–Shiley bovine pericardial bioprostheses: histologic and ultrastructural studies. *Am. J. Pathol.* **140**: 1195–1204.

Hilbert, S.L., Luna, R.E., Zhang, J., Wang, Y., Hopkins, R.A., Yu, Z.X. and Ferrans, V.J. (1999). Allograft heart valves: the role of apoptosis-mediated cell loss. *J. Thorac. Cardiovasc. Surg.* **117**, 454–462.

Hilbert, S.L., Schoen, F.J., Jones, M. and Ferrans, V.J. Allograft heart valves: morphologic, biomechanical and explant pathology studies. (1994) In: *Cardiac Reconstructions with Allograft Valves*, 2nd edn. New York: Springer-Verlag, pp. 193–231.

Hilbert, S.L., Boerboom, L.E., Livesey, S.A. and Ferrans, V.J. (2004a). An explant pathology study of decellularized carotid artery vascular grafts. *J. Biomed. Mater. Res.* **69A**: 197–204.

Hilbert, S.L., Yanagida, R., Souza, J., Wolfinbarger, L., Linthurst-Jones, A., Krueger, P., Stearns, G., Bert, A. and Hopkins, R.A. (2004b). Prototype anionic detergent technique used to decellularize allograft valve conduits evaluated in the right ventricular outflow tract in sheep. *J. Heart Valve Dis.* **13**: 831–840.

Jones, M., Barnhart, G.R., Chavez, A.M., Jett, G.K., Rose, D.M., Ishihara, T. and Ferrans, V.J. (1982). Experimental evaluation of bioprosthetic valve implanted in sheep. In: Cohn, L.H. and Gallucci, V. (eds.), *Cardiac Bioprostheses*. New York: Yorke Medical Books, pp. 275–292.

Jones, M., Eidbo, E.E., Hilbert, S.L., Ferrans, V.J. and Clark, R.E. (1989). Anticalcification treatments of bioprosthetic heart valves: *in vivo* studies in sheep. *J. Cardiac Surg.* **4**: 69–73.

Marlowe, D.E. and Phillips, P.J. (1998). FDA recognition of consensus standards in the premarket notification program. In: *Biomedical Instrumentation and Technology*. Philadelphia: Hanley & Belfus, Inc., pp. 301–304.

PMA supplements: 21 CFR 814.39.(2007).

Postmarketing Reports for Applications for FDA Approval to Market a New Drug (NDA): 21 CFR 314.80 and 314.81.(2007).

Raper, S.E., Chirmule, N., Lee, F.S., Wivel, N.A., Bagg, A., Gao, G., Wilson, J.M. and Batshaw, M.L. (2003). Fatal systemic inflammatory response syndrome in a ornithine transcarbamylase deficient patient following adenoviral gene transfer. *Mol. Genet. Metab.* **80**: 148–158.

Reporting for Biological Products: 21 CFR 600.14, 600.80, 600.81, 601.28, 601.70, and 601.93.(2007).

Schoen, F.J., Levy, R.J., Hilbert, S.L. and Bianco, R.W. (1992). Antimineralization treatments for bioprosthetic heart valves: assessment of efficacy and safety. *J. Thorac. Cardiovasc. Surg.* **104**: 1285–1288.

Simek, S., Byrnes, A. and Bauer, S. (2002). FDA perspectives on the use of the adenovirus reference material. *Bioprocessing* **1**: 40–42.

Smith, J.S., Tian, J., Lozier, J.N. and Byrnes, A.P. (2004). Severe pulmonary pathology after intravenous administration of vectors in cirrhotic rats. *Mol. Ther.* **9**: 932–941.

Supplements and Changes to an Approved NDA: 21 CFR 314.70, 314.71, 314.72.(2007).

US Department of Health and Human Services, Food and Drug Administration: The Critical Path to New Medical Products (June 2006). http://www.fda.gov/oc/initiatives/criticalpath/.

US Food and Drug Administration, Center for Biological Evaluation and Research: Tissue (April 2006a). http://www.fda.gov/cber/tiss.htm

US Food and Drug Administration, Center for Biological Evaluation and Research: Xenotransplantation Action Plan, FDA Approach to the Regulation of Xenotransplantation (April 2006b). http://www.fda.gov/cber/xap/xap.htm.

US Food and Drug Administration, Center for Biologics Evaluation and Research: Cellular, Tissue and Gene Therapies Advisory Committee, (Formerly Biological Response Modifiers Advisory Committee) (March 2006c). http://www.fda.gov/cber/advisory/ctgt/ctgtmain.htm.

US Food and Drug Administration, Center for Biologics Evaluation and Research: Draft Guidance for FDA Review Staff and Sponsors: Content and Review of Chemistry, Manufacturing, and Control (CMC) Information for Human Gene Therapy Investigational New Drug Applications (INDs) (November 2004a). http://www.fda.gov/cber/gdlns/gtindcmc.htm.

US Food and Drug Administration, Center for Biologics Evaluation and Research: Draft Guidance for Industry: Eligibility Determination for Donors of Human Cells, Tissues, and Cellular and Tissue-Based Products (HCT/Ps) (May 2004b). http://www.fda.gov/cber/gdlns/tissdonor.htm.

US Food and Drug Administration, Center for Biologics Evaluation and Research: Draft Guidance for Industry: Gene Therapy Clinical Trails – Observing Participants for Delayed Adverse Events (August 2005a). http://www.fda.gov/cber/gdlns/gtclin.htm.

US Food and Drug Administration, Center for Biologics Evaluation and Research: Draft Guidance for Industry, Preventive Measures to Reduce the Possible Risk of Transmission of Creutzfeldt–Jakob Disease (CJD) and Variant Creutzfeldt–Jakob Disease (vCJD) by Human Cells, Tissues, and Cellular and Tissue-Based Products (HCT/Ps) (June 2002a). http://www.fda.gov/cber/gdlns/cjdvcjd0602.htm.

US Food and Drug Administration, Center for Biologics Evaluation and Research: Draft Guidance for Reviewers: Instructions and Template for Chemistry, Manufacturing, and Control (CMC) Reviewers of Human Somatic Cell Therapy Investigational New Drug Applications (IND's) (August 2003a). http://www.fda.gov/cber/gdlns/cmcsomcell.htm.

US Food and Drug Administration, Center for Biologics Evaluation and Research: Good Clinical Practice Program: Guidances and Information Sheets on Good Clinical Practice in FDA-Regulated Clinical (April 2001a). http://www.fda.gov/oc/gcp/guidance.html.

US Food and Drug Administration, Center for Biologics Evaluation and Research: Guidance for Industry: Acceptance of Foreign Clinical Studies (March 2001b). http://www.fda.gov/cder/guidance/fstud.htm.

US Food and Drug Administration, Center for Biologics Evaluation and Research: Guidance for Industry, Formal Meetings with Sponsors and Applicants for PDUFA Products (April 2002b). http://www.fda.gov/cber/gdlns/mtpdufa.htm.

US Food and Drug Administration, Center for Biologics Evaluation and Research: Guidance for Industry: Guidance for Human Somatic Cell Therapy and Gene Therapy (March 1998a). http://www.fda.gov/cber/gdlns/somgene.htm.

US Food and Drug Administration, Center for Biologics Evaluation and Research: Guidance for Industry: Providing Clinical Evidence of Effectiveness for Human Drugs and Biologic Products (May 1998b). http://www.fda.gov/cber/gdlns/clineff.pdf.

US Food and Drug Administration, Center for Biologics Evaluation and Research: Guidance for Industry: Source Animal, Product, Preclinical, and Clinical Issues Concerning the Use of Xenotransplantation Products in Humans (April 2003b). http://www.fda.gov/cber/gdlns/clinxeno.htm.

US Food and Drug Administration, Center for Biologics Evaluation and Research: Guidances for Submission of IND's (January 2006d). http://www.fda.gov/cber/ind/indpubs.htm.

US Food and Drug Administration, Center for Biologics Evaluation and Research: ICH Guidance on Quality of Biotechnological/Biological Products: Derivation and Characterization of Cell Substrates Used for Production of Biotechnological/Biological Products (September 1998c). www.fda.gov/cber/gdlns/qualbiot.pdf.

US Food and Drug Administration, Center for Biologics Evaluation and Research: Information on Submitting and Investigational New Drug Application for a Biological Product (June 2005b). http://www.fda.gov/cber/ind/ind.htm.

US Food and Drug Administration, Center for Biologics Evaluation and Research: Points to Consider in the Characterization of Cell Lines Used to Produce Biologicals (July 1993). http://www.fda.gov/cber/gdlns/ptccell.pdf.

US Food and Drug Administration, Center for Biologics Evaluation and Research: Points to Consider in the Manufacture and Testing of Monoclonal Antibody Products for Human Use (February 1997a). http://www.fda.gov/cber/gdlns/ptc_mab.txt.

US Food and Drug Administration, Center for Devices and Radiological Health: Apligraft (Graftskin) Product Label (September 2000). http://www.fda.gov/cdrh/pdf/p950032s016.html.

US Food and Drug Administration, Center for Devices and Radiological Health: CDRH Databases (December 2005c). http://www.fda.gov/cdrh/databases.html.

US Food and Drug Administration, Center for Devices and Radiological Health: Dermagraft Product Label (September 2001c). http://www.fda.gov/cdrh/pdf/p000036.html

US Food and Drug Administration, Center for Devices and Radiological Health: Device Advice: Investigational Device Exemptions (IDE) (July 2003c). http://www.fda.gov/cdrh/devadvice/ide/index.shtml.

US Food and Drug Administration, Center for Devices and Radiological Health: Device Advice: Premarket Approval (PMA) (November 2002c). http://www.fda.gov/cdrh/devadvice/pma/.

US Food and Drug Administration, Center for Devices and Radiological Health: Device Advice: Premarket Approval: Postapproval Requirements: General Requirements (November 2002d). http://www.fda.gov/cdrh/devadvice/pma/postapproval.html#general.

US Food and Drug Administration, Center for Devices and Radiological Health: Device Advice: Premarket Notification [510(k)]: Is a New 510(k) Required for a Modification to the Device? (February 2005d). http://www.fda.gov/cdrh/devadvice/3146.html.

US Food and Drug Administration, Center for Devices and Radiological Health: Humanitarian Use Devices (November 2003d). http://www.fda.gov/cdrh/ode/hdeinfo.html.

US Food and Drug Administration, Center for Devices and Radiological Health: Orcel Product Label (September 2001d). http://www.fda.gov/cdrh/pdf/p010016.html

US Food and Drug Administration, Center for Devices and Radiological Health: Search STANDARDS Database (March 2006e). http://www.accessdata.fda.gov/scripts/cdrh/cfdocs/cfStandards/search.cfm

US Food and Drug Administration, Draft Guidance for Industry and FDA: Current Good Manufacturing Practice (cGMPs) for Combination Products (September 2004c). http://www.fda.gov/oc/combination/OCLove1dft.html.

US Food and Drug Administration, Early Collaboration Meetings under the FDA Modernization Act (FDAMA), Final Guidance for Industry and for CDRH Staff (February 2001e). http://www.fda.gov/cdrh/ode/guidance/310.pdf.

US Food and Drug Administration, Guidance for Industry: Chronic Cutaneous Ulcer and Burn Wounds – Developing Products for Treatment (June 2006f). http://www.fda.gov/cber/gdlns/ulcburn.htm.

US Food and Drug Administration, Guidance for Industry: Special Protocol Assessment (May 2002e). http://www.fda.gov/cber/gdlns/protocol.htm.

US Food and Drug Administration: Guidance for Industry, Sterile Drug Products Produced by Aseptic Processing – Current Good Manufacturing Practice (September 2004d). http://www.fda.gov/cber/gdlns/steraseptic.htm.

US Food and Drug Administration, Proposed Approach to the Regulation of Cellular and Tissue-Based Products (February 28, 1997b). http://www.fda.gov/cber/gdlns/CELLTISSUE.txt.

US National Institutes of Health, Office of Biotechnology Activities: Recombinant DNA and Gene Transfer (September 2005). http://www4.od.nih.gov/oba/Rdna.htm.

US Public Health Service, PHS Guideline on Infectious Disease Issues in Xenotransplantation (January 2001). http://www.fda.gov/cber/gdlns/xenophs0101.htm.

Xing, Y., He, Z., Warnock, J.N., Hilbert, S.L. and Yoganathan, A.P. (2004a). Effects of constant static pressure on the biological properties of porcine aortic valve leaflets. *Ann. Biomed. Eng.* **32**: 555–562.

Xing, Y., Warnock, J.N., He, Z., Hilbert, S.L. and Yoganathan, A.J. (2004b). Cyclic pressure affects the biological properties of porcine aortic valve leaflets in a magnitude and frequency dependent manner. *Ann. Biomed. Eng.* **32**: 1461–1470.

42

Current Issues in US Patent Law

Patrea L. Pabst

Intellectual property rights provide a means for the owners of technology to recover their investment in the technology and, in some cases, to make a profit. More importantly, intellectual property rights provide a means for financing the incredibly expensive research and development and testing required for commercialization of new products and processes in the medical and biotechnology field. When the intellectual property rights have been lost, many times it is not possible to obtain the money required to see a product or process reach the clinic and benefit those for whom it is intended. It is only by protecting the technology that it can be used to help those who need it the most.

WHAT ARE INTELLECTUAL PROPERTY RIGHTS?

Intellectual property is intangible. It is embodied in patents, trademarks, service marks or trade names, copyrights, and trade secrets. Companies may have rights in such intellectual property by virtue of various licensing agreements, joint venture agreements, or by virtue of employment agreements.

The Constitution of the United States gives Congress the power to enact laws relating to patents, in Article I, Section 8, which reads "Congress shall have power . . . to promote the progress of science and useful arts, by securing for limited times to authors and inventors the exclusive right to their respective writings and discoveries." Trademarks and trade secrets are also typically protected under state law. Further protection can also be obtained using employment non-compete agreements, confidentiality agreements, and material transfer agreements. Extensions of patent rights and exclusive rights may also be obtained for delays in obtaining regulatory approval by the Food and Drug Administration (FDA), for Orphan Drug Act and under the Waxman-Hatch act. These provisions are not dealt with in detail in this chapter.

Patents

Patents are grants by a government entity that gives the patent owner the right to exclude competitors from making or using that which is defined by the claims of the patent. Patents have basically the same requirements throughout the world. In the United States, the requirements for obtaining and asserting a patent are defined by Chapter 35 of the United States Code ("USC"). Patents are governed exclusively by federal law. US Patents are granted exclusively by the US Patent and Trademark Office (PTO). Patents are granted on applications for new, useful, and non-obvious inventions. A patent is issued by the US PTO after it has examined the application for the patent and determined that it meets various criteria. Most other countries have similar laws governing granting of patents. In some cases, patents are obtained for a "region" such as the European Patent Convention countries, through a centralized patent office. The granted patents are then registered in the individual member countries in which they are enforceable.

Copyright © 2008 Elsevier Inc. All rights reserved

A patent for an invention is the grant of a property right to the inventor, issued by the PTO. Currently the term of a new patent is 20 years from the date on which the utility application for the patent was filed in the United States or, in special cases, from the date an earlier related application was filed, subject to the payment of maintenance fees and disclaimers to earlier expiring patents. Effective May 29, 2000, the patent term for subsequently filed utility applications, was extended to compensate for delays in prosecution due to the Patent Office. Design patents are granted for new, ornamental, and non-obvious designs, and have a maximum term of 14 years from the date of grant. US patent grants are effective only within the US, US territories, and US possessions. The right conferred by the patent grant is, in the language of the statute and of the grant itself, the right to exclude others from making, using, offering for sale, or selling the invention in the United States or importing the invention into the United States.[1] What is granted is not the right to make, use, offer for sale, sell, or import, but the right to exclude others from making, using, offering for sale, selling, or importing the invention.

In general, technology relating to tissue engineering is subject to the same rules as are other compositions and methods of use and manufacture thereof. As long as the "thing itself" is new, non-obvious, and subject to written description, it is patentable.

In the language of the statute, any person who "invents or discovers any new and useful process, machine, manufacture, or composition of matter, or any new and useful improvement thereof, may obtain a patent,"[2] subject to the conditions and requirements of the law. The word "process" is defined by law as a process, act or method, and primarily includes industrial or technical processes. The term "machine" used in the statute needs no explanation. The term "manufacture" refers to articles which are made, and includes all manufactured articles. The term "composition of matter" relates to chemical compositions and may include mixtures of ingredients as well as new chemical compounds. These classes of subject matter taken together include practically everything which is made by man and the processes for making the products. The patent law specifies that the subject matter must be "useful." The term "useful" in this connection refers to the condition that the subject matter has a useful purpose and also includes operativeness, that is, a machine which will not operate to perform the intended purpose would not be called useful, and therefore would not be granted a patent.

In general, patentable subject matter includes compositions, methods of manufacture, and methods of use.[2] Compositions may include, for example, biodegradable polymeric matrices seeded with cells, or isolated cells for implantation in the body. Methods of manufacture may be directed to, for example, a process for creating a unique prosthetic device. Methods of use may entail methods for administration of therapeutic compositions, or surgical implantation of, for example, a synthetic tissue matrix containing implanted isolated cells that secrete insulin.

The most significant problems we observe in patenting in pharmacology and biotechnology are with the issue of written description and enablement. The PTO has provided guidelines for the written description requirement.[3] The fundamental factual inquiry is "whether the specification conveys with reasonable clarity to those skilled in the art that, as of the filing date sought, applicant was in possession of the invention as now claimed."[3] Answering this question is not a single, simple determination, but rather is a factual determination reached by considering a number of factors. The PTO guidelines list the following factors which should be considered by an examiner in his written description analysis: (1) the level of skill and knowledge in the art; (2) partial structure, physical, and/or chemical properties disclosed in the specification; (3) functional characteristics disclosed in the specification, alone or coupled with a known or disclosed correlation between structure and function; and (4) the method of making the claimed invention disclosed in the application.

---

[1] 35 USC §271 (a) (2005).

[2] 35 USC §101 (2005).

[3] Manual of Patent Examining Procedure (MPEP) §2163 (8th Edition, August 2001) (revised August 2005).

to the publication, they conceived and diligently reduced to practice what they are claiming. In the United States, one can also get into a proceeding between multiple parties who have filed patent applications on the same invention, called an interference, to determine who was first to invent, regardless of who was first to file for a patent, which in turn determines who is entitled to the patent. Outside of the United States, rights are awarded to the first to file, and patent rights are lost immediately upon publication of the invention if made prior to filing for the patent.

What constitutes a publication? Generally, a publication is any oral, written, or physical description that conveys to the public that which an applicant would like to claim. It may be a talk at the proceedings of a society (including any slides presented), an article in a scientific journal, a grant application that is awarded (effective as prior art at the time of award, not of application), a thesis (effective at the time of cataloging), or even an offer for sale or a press release. A critical requirement is that the publication must be enabling, that is, it must convey to one of the ordinary skill in the art how to make and use that which is being claimed. Public use means more than using the composition or method in one's laboratory. However, it can include even a one patient study that is reported during clinical rounds or at a presentation at which a drug company or surgical supply representative is present. The courts, in many cases, have had to interpret what it means to be publicly available. A frequent question is when is a student's thesis available as prior art. Courts have now held that once the thesis is cataloged, it is publicly available, because it has been entered into a computer database that one searching the database will be able to access.[9] Accordingly, the publication date of a thesis is the date on which the thesis is cataloged, not the date on which it is defended or signed by the thesis committee. Slides that are not distributed, but that are shown at an oral presentation, are considered to be publications, particularly if the meeting is attended by those skilled in the art who would be able to understand and use the information in the slides. Additionally, a poster presented at a scientific meeting constitutes a "printed publication" under 35 USC §102.[10]

Disclosures to another party under the terms of a confidentiality agreement are not publications. Uses that are strictly experimental may not be public disclosures, if, among other aspects, they are designed to determine if that which is to be claimed will work, and if any other parties who are involved are clearly informed that the studies are experimental in nature. If an announcement is made publicly, which does not enable one of the ordinary skill in the art to use or make that which is later claimed, then the announcement is not a publication. For example, an announcement could be a statement made to the press that researchers X and Y have discovered a cure for cancer. Since the announcement does not tell one of the ordinary skill in the art how to cure cancer, it is not enabling. However, enablement can be difficult to prove and standards may change over time. A 1995 case involved the question of whether a publication was enabling for the development of a transdermal patch for delivery of nicotine.[11] The court found that a prior publication referring to transdermal patches for drug delivery mentioned that the drug in the transdermal patch for treatment of heart disease could be replaced with nicotine for assisting patients in quitting smoking. The court held that the article disclosed or made obvious the transdermal patch for delivery of nicotine claimed by the applicant, because the applicant merely took the transdermal patch described in the article, put nicotine in it, and demonstrate that the nicotine was delivered and would work exactly as predicted based on the delivery of the drug for treatment of heart diseases. Even though there was no information relating how the drug was to be incorporated into the transdermal patch, the publication was enabling because those of ordinary skill in the art would have been able to determine how to put the nicotine in the transdermal patch without undue experimentation.

[9] Philips Elec. & Pharmaceutical Indus. Corp. v. Thermal & Elec. Indus., Inc., 450 F.2d 1164, 1169–72, 171 U.S.P.Q. 641 (3d Cir. 1971); Gulliksen v. Halberg, 75 U.S.P.Q. 252 (Pat. Off. Bd. Int'f. 1937).

[10] In re Klopfenstein, 380 F.3d 1345, 72 U.S.P.Q.2d 1117 (Fed. Cir. 2004).

[11] Ciba-Geigy Corp. v. Alza Corp., 1995 US App. LEXIS 28214, 37 U.S.P.Q.2d 1337 (Fed. Cir. 1995).

If the invention has been described in a printed publication anywhere in the world, or if it has been in public use or on sale in this country before the date that the applicant made his/her invention or more than 1 year before the date on which an application for patent is filed in this country, a patent cannot be obtained.[7] In this connection, it is immaterial when the invention was made, or whether the printed publication or public use was by the inventor himself/herself or by someone else. If the inventor describes the invention in a printed publication or uses the invention publicly, or places it on sale, he/she must apply for a patent before 1 year has gone by, otherwise any right to a patent will be lost.

Obviousness

Even if the subject matter sought to be patented is not exactly shown by the prior art, and involves one or more differences over the most nearly similar thing already known, a patent may still be refused if the differences would be obvious. The subject matter sought to be patented must be sufficiently different from what has been used or described before that it may be said to be non-obvious to a person having ordinary skill in the area of technology related to the invention. The claimed method or composition must be non-obvious to those of ordinary skill in the art from what is publicly known.[12] This is usefully referred to outside of the United States as a requirement for an "inventive step." In 1960s, the United States Supreme Court carefully analyzed non-obviousness and those factors that are to be considered in determining whether that which is claimed is obvious from the prior art.[13] This analysis is a fact-based determination, involving not only the elements which are claimed, but also the level of skill in the art and the expectation or predictability that the claimed method or composition would perform as predicted, actual success in the marketplace, long felt need, and whether there are unexpected results. If one has no better than a 50–50 chance that a particular method may work, and the method works, it is arguably not obvious, although it may be obvious to try. If one tries something and the results are vastly different from what was expected, then the results are not obvious. For example, if one administered two drugs each in the dosage known to yield a particular effect and the combination results in a substantially greater effect than the sum of the individual effects of the two drugs, resulting in the ability to use a much lower dosage of each drug than expected, then one would have unexpected results or "synergy." If the prior art teaches away from what the applicant has done, this result would support a finding of non-obviousness. For example, if the prior art states that one cannot proliferate implanted hepatocytes, then it may be non-obvious if the applicant for a patent finds that he can administer particular growth factors with his hepatocytes and see proliferation. Many other considerations factor into whether a claimed composition or method is obvious in view of the prior art.

Enablement, Written Description, and Best Mode

Additional requirements for patentability are defined by 35 USC §112, first and second paragraphs. These include enabling one of the ordinary skill in the art to make and use that which is claimed; a written description that clearly describes that which is claimed; and, in the United States only, disclosure of the best mode for practicing the invention.[14]

The applicant must describe that which is claimed in sufficient detail, and with appropriate methods and sources of reagents or other materials or equipment, to enable one of the ordinary skill in the art to make and use that which is claimed. This sounds far simpler than it actually is in practice. In many cases, particularly when coming out of a university study or a start-up company, the invention that applicants would like to

[12] 35 USC §103 (2005).
[13] Graham v. John Deere Co., 383 US 1, 148 U.S.P.Q. 459 (1966).
[14] 35 USC §112 (2005).

claim is that which the applicant intends to develop over the next several years, based on a limited amount of data available at the time of filing. Particularly in the case of universities, where the applicant must publish or has submitted grant applications (which in and of themselves constitute prior art once they are awarded), the difficulty is in describing that which has not yet been done. The application must not only describe a specific limited example, but must describe the various ways in which one intends to practice that which is claimed.

"Invention" usually consists of two steps: "conception" and "reduction to practice." There are two kinds of reduction to practice: actual and constructive. "Constructive reduction to practice" means that the applicant has described in the application for patent **how** to make and use that which is claimed, but has not actually made and used what is claimed. This may be as simple as stating that although a biodegradable polymer such as polylactic acid–co-glycolic acid is preferred for making a matrix for culturing cells, other biodegradable polymers such as polyorthoesters or polyanhydrides could also be used. It may be less obvious that other cell types or shapes or methods of processing may be used when only one example showing reduction to practice is available. The rule of thumb in this case is the level of predictability. Therefore, in stating what kind of drugs one could deliver using the claimed technology, one might list a wide variety of drugs based on the data available with one type of drug. However, one may not be able to include in a list of drugs to be delivered peptides or very hydrophobic compounds, which usually are viewed as difficult to deliver, based on data obtained with a drug that is "easy" to deliver, such as a sugar or small molecular weight dye. Being too predictive (i.e. engaging in extensive constructive reduction to practice), which includes "non-enabling" or non-enabled technology, may in some cases be a detriment during prosecution of subsequently filed applications, because the examiner may cite the earlier work as making obvious the applicant's subsequent work. Patent attorneys frequently must play a balancing game in determining how far to go with constructive reduction to practice in order to exclude competitors while not eliminating the applicant's own ability to obtain additional, subsequent patent protection.

In the United States, there is a requirement to disclose the best mode for practicing that which is claimed at the time of filing the application. No similar requirement exists outside of the United States. Because most applicants file the same application in the United States as outside of the United States, US applicants frequently disclose their best mode in foreign-filed applications.

Patent Rights

The purpose of a patent is to exclude the competition from making and using that which is claimed, not to "protect" a product – a frequent misconception of patents. In order to exclude competition, one must describe and claim not only that which one intends to practice, but that which another party could practice in competition with the patentee. What does this mean in real terms? It means that the applicant for a patent must describe his preferred method, which is known as of the date of filing, the preferred embodiments that he or his company intends to market, as well as any embodiments that a competitor could make and use in competition with the applicant's product.

Patent Term and Patent Term Extension

Under the revised US patent law that was enacted as a result of implementation of the General Agreement for Trade and Tariffs ("GATT"), the term of a patent is 20 years from the original date of filing or the filing date of the earliest utility application to which priority is claimed. Applicants therefore have more incentive to prosecute all claims in a single application in order to minimize costs for prosecuting and maintaining the patent. Under the law in effect prior to June 8, 1995, the patent term was 17 years from the date of issue in the United States. Divisional applications were a commonly used method to extend patent protection to encompass different aspects of the technology over a period of time much greater than 17 years. For example, an application would be filed in 1990, and a single inventive concept (e.g. the composition) would be prosecuted in the first application. Three years later, when those claims were allowable and a patent was to issue, a divisional application would be filed with another set of the claims that had been restricted out of the original application. This divisional application would be prosecuted for another 2 to 3 years, the claims would be determined to be allowable, the second patent would issue with a 17-year term, and a third divisional application would be filed. The result is that patents on related technology would issue sequentially over several years, increasing the effective term of patent protection beyond 20 years. Under the new law, this mechanism to extend patent protection is not possible.

The GATT was signed into law in the United States on December 7, 1994, and the initial provisions affecting US patent practice were implemented on June 8, 1995. The most significant changes arising from enactment of that agreement, now called Uruguay Round Act, were changes in the patent term in the United States, the implementation of provisional patent applications, and the broadening of what constitutes infringement in the United States. The change in patent term has been discussed above. For those applications filed before June 8, 1995, the term of any issuing patent is 17 years from the date of issue or 20 years from the filing date, whichever is longer. The term of any patent issued on an application filed June 8, 1995 or later is 20 years from the earliest claimed non-provisional priority date (i.e. the filing date of the first utility application to which priority is claimed or the filing date of a Patent Cooperative Treaty (PCT) application designating the United States). Extensions of terms are available upon delays in issuance arising from appeals or interferences. Additional extensions of terms are available for delays in obtaining regulatory approval by the Food and Drug Administration (FDA) for a device or a drug. Recent legislation has provided for an increased patent term due to unreasonable delays by the US Patent Office. Under changes implemented in late 2000, the US Patent Office recognized that losses in patent term were resulting from delays in prosecution of applications through no fault of the applicants. Diversion of user fees to other government agencies, in particular, as well as the initial "up time" associated with contracting out the publication and other services at the US Patent Office resulted in some cases in delays of months to even years. Excessive delays may now be the basis for extensions of patent term.

Another result of the GATT was the implementation of provisional patent applications. These provide a means for applicants to make a preliminary filing, with limited claims or even no claims, and less stringent requirements as to form (but exactly the same requirements as to enablement, best mode, and written description), to preserve their initial filing date, while collecting additional data or financing or evaluating the technology. As long as a utility application is filed within 1 year claiming priority to the provisional application, one can benefit from the provisional application priority date. Provisional applications are not themselves examined and expire exactly 1 year after filing. They constitute a convenient way to preserve priority dates and provide a particular advantage in the medical and biotechnology fields, since patent term runs from the date of the utility application, not the provisional application, and therefore the effective patent term of an application claiming priority to a provisional application can be 21 years from the earliest priority date, not 20 years.

Failure in the provisional application to completely disclose and enable that which is subsequently claimed in an utility application can result in a loss of the claim to priority to the provisional application, if that which is claimed is not enabled. Merely filing as a provisional application an article that will be published or presented in order to avoid loss of foreign rights usually will not comply with the enablement requirements, and therefore will not serve as an adequate basis for priority. It is essential that applicants who file provisional applications based on an article amplify the description to encompass other embodiments and to provide the basis by which one of the ordinary skills in the art can practice that which is claimed. Application sections that are not required for enablement, which are typically included in a utility application, include the background of the invention, the problems that the claimed invention addresses, and the claims. These sections can be omitted from the provisional application, thus saving time and money in preparing the application. In many cases, fairly standard language can be used to expand or broaden the description in an article in order to meet the enablement requirements, providing a means for those with limited amounts of time or money to protect that which they are disclosing with minimum risk and expenditure.

The World Intellectual Property Organization ("WIPO"), which implements the provisions of the "PCT" and the European Patent Office, has confirmed that US provisional applications serve as an adequate basis for a claim to priority in corresponding foreign file applications. However, under the Patent Convention, all foreign applications that claim priority from an earlier filed application must still be filed within 1 year of the US filing date or the filing date of the country in which the first application is originally filed.

Because biotechnology is a complex field, especially in the areas of patentable subject matter and enablement, many of the general observations relating to patents may not be as directly applicable to more "conventional" patentable subject matter. For example, the US Patent Office consistently maintained that the change from a 17-year term from the date of issue to a 20-year term from the earliest priority date would not result in a significant loss of patent term. However, because the Patent Office applies such a stringent examination proceeding under §112 (written description and enablement) in the biotechnology area, issue time in these cases typically has been considerably longer, not uncommonly taking as many as 5–7 years from the original priority date. The result is that these complex biotechnology patents have a substantially shortened term as compared to many other types of patents. This is the case even if one obtains some extension due to delays by the US Patent Office. Because a patent extension can still be obtained for delays due to regulatory issues involving the FDA, as well as for appeals to the Board of Patent Appeals and Interferences, those in the United States who believe that their patent rights will be limited in term due to delays in prosecution should avail themselves of the Patent Extension Act, if at all possible. One must bear in mind, however, that an extension for regulatory delays can only be obtained on **one** patent for any particular product or process; thus, the inventor or licensee with multiple, related patents clearly should choose the most important patent or the patent subject to the greatest increase in patent term, when facing such a situation. The patent that is to be extended must be brought to the attention of the FDA. Following FDA approval of the claimed product or process, the extension must be applied for in a timely fashion.

Trademarks, Service Marks, and Trade Names

A trademark is a word, design, shape, number or slogan, or a combination of those elements, that identifies and distinguishes one company's products from others. It is a symbol of goodwill (e.g. Coca-Cola®, PC Paintbrush® and Budweiser®). A service mark is to services what a trademark is to products (e.g. SAKS® (retail department store services), Holiday Inn® (hotel services), and Orkin® (pest control services)). A trade name is a name used to identify a business, (i.e. a "commercial name"). A trade name can also be used as a trademark to identify products and as a service mark to identify the services (e.g. IBM®).

Trademarks and service marks can be registered on the Federal Register with the US PTO if they are used in interstate commerce, or they can be registered in a state if they are only used within that state. Trade names cannot be registered federally.

A trademark is created by use, not by filing of an application. Lack of registration is not a bar to enforcing the mark, but when a trademark owner has obtained a registration, it obtains significant procedural and substantive advantages. For example, federal registration provides constructive nationwide notice to all users of a trademark. This eliminates a second user's defense of "innocent adoption" which might otherwise permit such a second user to continue to use the mark. Federal registration can also protect an internet domain name or email address from a disruptive challenge.

"Intent to use" trademark applications allow a trademark owner to obtain "pre-approval" prior to investing money in the use of the mark. The date of filing of the intent to use application is the priority date for the mark and actual use must be shown within a period of not greater than 36 months following the approval of the mark by the US PTO.

Copyrights

A copyright protects original expression. It is totally a creature of statute, 17 USC, the US Copyright Act. The author of a copyrighted work has the exclusive right to use, distribute, modify, display, reproduce, and perform the work. The copyright protects its owner from actual copying or unauthorized use of the protected expression in the copyrighted work. It does not prohibit the independent creation of a similar work by a third party if that third party does not have access to, or knowledge of the work.

A copyright also does not grant the copyright owner a monopoly on the ideas contained within the work. Copyright protection does not extend to any "idea, procedure, process, system, method of operation, concept, principal, or discovery" embodied in the work.[15] Such concepts may be protected only through patents or trade secrets.

Trade Secrets

"Trade secret" means information including, but not limited to, technical or non-technical data, a formula, a pattern, a compilation, a program, a device, a method, a technique, a drawing, a process, financial data, financial plans, product plans, or a list of actual or potential customers or suppliers which derives economic value, actual or potential, from not being generally known to, and not being readily ascertainable by proper means by, other persons who can obtain economic value from its disclosure or use; and is the subject of efforts that are reasonable under the circumstances to maintain its secrecy.

This definition suggests three main characteristics of a trade secret. First, a trade secret generally relates to some sort of data, formula, device, method, or other similar type of information. Second, a trade secret must be valuable. This quality is generally self-evident. If a trade secret were not valuable, then a company would not consider it to be worth keeping secret.

Third, and most importantly, the information must be the subject of efforts that are reasonable under the circumstances to maintain its secrecy. An owner need not take every steps possible to maintain a secret but must take steps that are reasonable under the circumstances to maintain the secrecy. This typically means that employees will have employee agreements with non-compete provisions that the trade secrets are kept in a restricted area or under conditions wherein the documents cannot be downloaded by others and that parties accessing the information are aware that it is both valuable and confidential.

Trade secrets are becoming increasingly valuable in the biotechnology and pharmaceutical areas, with the advent of databases containing information that allows one to accumulate and collate information on a regular basis. An example of a trade secret may be a database including information regarding drug design or what amino acids can be substituted or deleted, to achieve a particular activity. It may also be what kinds of genes are in a microarrary that is being marketed or what kind of controls are required to generate useful data.

Trade secrets are undervalued assets by many investors who do not always understand that the trade off for a patent is that one must disclose who to make and use the claimed invention. To enforce rights in a patent, one must be prepared to enter into costly and time-consuming litigation. In contrast, trade secrets do not need to be enforced since the competitor never learns how to practice the trade secret. Of course, this only works if the information that is being maintained in secrecy is not subject to "reverse engineering," that is, one cannot figure out the secret from the product that is being sold. In the case of a database, it is the product that is designed using the information that is being sold, not the information *per se*. Trade secrets can last much longer than a patent, and do not cost hundreds of thousands of dollars to obtain and maintain.

---

[15] 17 USC §102(b) (2005).

WHY ARE INTELLECTUAL PROPERTY RIGHTS IMPORTANT?

Advancements in the medical and biotechnology fields often are achieved only through substantial investment of industrial, academic, and governmental resources. Patenting of these technological advancements frequently is employed to recoup that investment, to create profits which are used in part to develop new or improved products, and to enhance a competitive commercial edge. Other forms of intellectual property protection, such as trade secrets, copyrights, and trademarks, may also be used to further protect and exploit drug delivery processes, products, and services. In the rapidly expanding fields of microarray screening and genomics, where databases are revised on a daily basis and much of the information does not need to be distributed, protection of only the results obtained using the data (i.e. trade secret) is far more effective than obtaining patent protection.

One of the most frequently asked questions is why do we need to go to the trouble and expense of patenting a composition or method. The most common reason is that protecting a new composition or method of manufacture or use provides a means for obtaining the revenue required to develop a new drug or medical treatment. With the cost of developing and obtaining regulatory approval for a new drug approaching $200 million dollars in the United States, patent rights are essential to recovering expenses. For small companies that spend more time raising money than selling products, patents and patent applications represent the company's only tangible assets which it can show to potential investors. For universities and other non-profit research institutions, patents and associated know-how and, in some limited cases, trade secrets can be used to obtain royalties from license agreements, sponsored research funding in many cases, and equity in new companies started for the purpose of exploiting the technology.

Patents and other intellectual properties are valued in many different ways. For example, a process for manufacture typically would be licensed for 2–3% of the gross selling price of a product of the process. This price would be decreased if multiple licenses had to be obtained to use the process. Patents claiming compositions tend to have a greater market value, for example, between 5% and 10% of the gross selling price, due to the perception that these patents are easier to enforce than process patents. These numbers of course are affected by the stage in development, the market cap, and the number of competing technologies in the same field.

Enforcement is a risky business. A good patent strategy is to obtain patents that claim a product, methods of manufacture, and methods of use, broadly and specifically, so that a patentee is able to assert multiple patents against an alleged infringer. Patents with broad claims generally will be easier to invalidate than more specific patents. Faced with the prospect of fighting several patents, most parties will opt for settlement. The alternative, litigation, is extraordinarily expensive for both parties and can result in the patents being invalidated, or the infringer being liable not only for damages for infringement, but also for attorney's fees and punitive damages.

Patents give the patent owner the right to exclude competition. This is accomplished by asserting the patent against third parties who are marketing a product or service which falls within the scope of the claims. Referred to as "infringement," the criteria are totally different from the criteria for obtaining a patent, referred to as "patentability." In simple terms, a patent claim consists of "elements" in a defined relationship. Certain phrases expand or limit the scope of the claim. For example, the term "comprising" can be translated as "including at least," while "consisting" means "including only." If a claim reads:

Composition comprising:

- A
- B
- C

then the claim would cover any composition including A, B, C, and any other component. Use of the term "consisting" in place of "comprising" would restrict the claim to a composition including **only** A, B, and C.

In determining infringement, one must look to the claims of the patent. Claims may be clear on their face, or require reference to the specification, or description, of the patent. Claims also may be limited by agreements made during prosecution, a doctrine referred to as "file wrapper estoppel." For example, if the prosecuting attorney argues that the claims distinguish over the prior art on the basis that the prior art does not disclose a particular feature that the attorney argues is essential to the claims in the patent, then the claims will be construed to require that limitation, even if not explicitly recited in the claims as issued.

Asserting a patent allows the alleged infringer to file an action for declaratory judgment in a federal district court, asking the judge to declare the patent invalid or non-infringed. Litigation is very expensive and especially detrimental to small companies, thus providing a great deal of incentive to license the technology on terms favorable to both parties.

Since patents are now published as applications, 18 months after the original filing date, the public gets the benefit of this knowledge prior to the period of exclusivity, since at least in the United States, patent rights are not enforceable until the patent is granted, and then only to the extent that the alleged infringing method or composition falls within the issued claims, although it is now possible even in the United States to backdate damages if the claims as published are essentially the same as the issued claims. A major limitation on the ability of a patent owner to enforce his patents lies in his ability to pay for the cost of litigation. In my experience, the cost of patent infringement litigation in the United States has gone from approximately $1 million in the mid-1980s, to $3 million by 1990, to five times or more that amount by 2000. This effectively prevents small companies and individuals from enforcing their patent rights, even with legislation allowing recovery of treble damages and attorneys' fees in the event of willful infringement.

Further, as discussed briefly above, in some cases, the rights to enforce a patent have been limited by legislative or other political action. In the United States, these rights are defined by 35 USC §271. 35 USC §287, which defines the limitations on damages and other remedies, was amended on September 30, 1996, in an apparent effort to prevent issuance of patents directed to new medical treatments. The amendment, which applies only to patents issued after September 29, 1996, deprives patentees of remedies for direct infringement and induced infringement of patents to surgical or medical procedures that do not involve patented drugs or devices.[16] Although the exact scope of the exclusion is subject to interpretation of numerous terms in the amendment, it appears that the exclusion applies only to the performance by a medical practitioner (defined as a person licensed to provide medical activity or a person under the direction of such a person) of a medical or surgical procedure not involving the use of a patented machine, manufacture, or composition of matter (e.g. medical devices, implants, and drugs) or the patented use of a composition of matter. Thus, it appears that the use of patented medical devices and patented drugs will still be subjected to infringement remedies.

The amendment does not apply to activities of persons engaged in the commercial development, manufacture, sale, importation, or distribution of a machine, or composition of matter or the provision of pharmacy or clinical laboratory services where such activities are (1) directly related to the commercial development, manufacture, sale, importation, or distribution of a machine, manufacture, or composition of matter or the provision of pharmacy or clinical laboratory services, and (2) regulated under the Federal Food, Drug, and Cosmetic Act, the Public Health Service Act, or the Clinical Laboratories Improvement Act. Although it is not clear how this exclusion will be interpreted, and as of 2005, this clause has not been discussed in any published legal opinion, it appears that infringement remedies will not be limited for performance of a medical or surgical procedure that is related to the commercial development or exploitation of a product or service where the medical or surgical procedure is subject to Federal regulation.

[16] 35 USC §287 (2005).

From a practical perspective, patents are usually only enforced against manufacturers or distributors, not doctors, since it would be unduly burdensome, and typically not cost effective to pursue actions against many individuals rather than a single large entity. It is therefore doubtful that the change in law will have any detrimental impact on the health care industry in the United States.

The absence of literal infringement does not necessarily mean that a process or device does not infringe a patent. According to the judicially created "doctrine of equivalents," even though the language of a claim cannot be read literally upon a process or device, a claim can be infringed if the process or device "performs substantially the same function in substantially the same way to obtain the same result."[17] In 2002, the Supreme Court rendered its decision in *Festo Corp. v. Shoketsu Kinzoku Koygo Kabushiki Co.*, addressing the relationship between prosecution history estoppel and the doctrine of equivalents.[18] In an unanimous opinion, the Court held that prosecution history estoppel may apply to any claim amendment made to satisfy the Patent Act's requirements (e.g. novelty, obviousness, enablement, written description, etc.); it is not limited to amendments made to avoid the prior art. The Court also held that prosecution history estoppel creates a flexible bar to the doctrine of equivalents. The Court explained that amending a claim to satisfy the Patent Act's requirements need not bar suit against every equivalent to the amended claim element. By amending a claim, the inventor is deemed to concede that the patent does not extend as far as the original claim. A rebuttable presumption is made that the amendment to the claim limitation bars the patentee from asserting infringement against an equivalent to the narrowed limitation. The patentee can rebut this presumption and show that the amendment does not surrender the equivalent in question. For example, the patentee could establish that the equivalent was unforeseeable at the time the claim was drafted, the amendment did not surrender the particular equivalent in question or there was some reason why the patentee could not have recited the equivalent in the claim.

In some cases it is possible to pursue alternative remedies, such as third party reexamination proceedings in the US patent office to have a party's patent invalidated over art not considered during prosecution, or in Europe or Japan, by third party opposition proceedings. The latter are considerably less expensive than litigation in the United States and decisions are reached typically in less than 2 years, which is also quite quick compared to litigation in the United States. The results of litigation are also quite unpredictable but very risky in the medical and biotechnology fields. Especially in the biotechnology area, the late 1990s were known for the swift and drastic decisions by the Court of Appeals for the Federal Circuit, invalidating patents on the grounds the claims were not enabled by the specifications.

HOW DO YOU VALUE INTELLECTUAL PROPERTY?

A common misconception is that one must have a patent in order to license the technology or for investors to be willing to invest in the technology. This is just not the case. In most cases where the technology arises in a university and many times with individuals, licenses are entered into with only a single application having been filed, and no indication of whether or not a patent will actually issue. Numerous start-up companies have been formed based on technology which is not the subject of US patents, and may in fact be better protected solely by trade secret and collaborative agreements.

What kind of products is most attractive to investors? Certainly, in the medical area, drugs that are for known indications are attractive because the issues are simpler; one does not need PhD investors in order for the investor to understand the technology or the market. The regulatory issues are straightforward, unlike

[17] *Graver Tank & Mfg. Co., Inc. v. Linde Air Products Co.*, 339 US 605, 608, 85 U.S.P.Q. 328, 330 (1950).

[18] Festo Corp. v. Shoketsu Kinzoku Kogyo Kabushiki Co., 535 US 722, 122 S. Ct. 1831 (US 2002).

with technologies such as gene therapy and stem cells where the political and ethical issues are in a constant state of flux. The development costs can be calculated with a higher degree of certainty.

A single product usually is not enough to start a company with venture or angle funding. A single technology may be sufficient if it is basic, broad, and exclusive. There is a tremendous difference. A single product may be sufficient to start a company if the product is ready to go to market in a relatively short time frame and does not require huge amounts of funding. Certainly there are numerous examples in the e-commerce and software fields. In the medical areas, this is more difficult.

An example of a company that is typical of the models that are being used successfully uses the combination of a very broad technology, at least one product in or close to clinical trials within a few months of funding, funding, and experienced investors who evaluate the technology, the scientists and the market, and who then put together the deal with the university or other source of the technology, form the scientific advisory board, and hire the business people to run the company. Absent this essential element, experienced business people who can put together the market plan, the people, and the money, even the best technology will fail.

In summary, there are three things one must have to start a successful company: a business plan which describes a technology base, specific products, cost of development, marketing plans, and means of generating income; the individuals who can develop the technology; and the business people who can run the company and obtain funding. Failure to have anyone of the three will cause the company to fail.

A frequent question is what are typical licensing terms. Generally speaking, I suggest the following guidelines.

For a new drug, which is patented or patentable as a compound *per se*, percentages range from 5% to 15%, with 8% not being unreasonable. Variables include whether or not there is animal testing data, clinical data, formulation issues, regulatory issues, a crowded market. For example, a new Hepatitis C drug commands a high price – there are few if any good drugs on the market and the market is huge. A new HIV drug that is another variation of known drugs commands less of a price – market is crowded and it does not represent a radically different solution to the problem.

For a diagnostic assay, antibody with desirable specificity to difficult antigen, medical device, screening assay with novel reagents, percentages typically range from 3% to 6% with 4–5% being common.

For a method of manufacture or screening, the ranges are typically from 1/2% up to about 3–4%. Major variables limiting price include the need for multiple licenses to practice method, particularly if they have to be obtained from multiple third parties.

Up front fees and milestones are highly variable. In the pharmaceutical and medical areas, minimum royalties are primarily designed to insure the technology comes back to the owner if it is not developed, but are low enough not to divert the needed resources from development. Highest milestones are on completion of regulatory steps, such as filing of a new drug application or completion of Phase I clinical studies. Initial fees are usually set at a number to help offset or recover patent expenses. These fees are usually the most contentious; if the product is a success, then both licensor and licensee win. In the case of advances, the licensor has put up money for patenting expenses and development which it wants to recover; the licensee wants to defer as much as possible while it evaluates the technology, so it loses as little as possible if the technology does not pan out.

Advantages of licensing versus forming a company include:

- Less risk to the technology owner.
- Greater likelihood of having experienced partner developing, testing, and obtaining regulatory approval. More likely to have sales and distribution team in place.
- No need to provide separate manufacturing and packaging facilities.
- Partner responsible for insurance and indemnification.

- Partner assumes liability and cost of defending technology against competitors.
- Partner assumes responsibility for obtaining licenses to practice the technology, if any.
- Partner assumes cost of patenting, which can be in the tens to hundreds of thousands of dollars, especially if foreign rights are pursued.

Advantages of forming a company include:

- The party developing the technology retains control over the development.
- Some technology is too early stage to license. The company may be able to fund sufficient development of the product to license or partner the technology and both profit from the technology and see technology developed that would not otherwise be developed.

Potential for Higher Profit

There are many successful business models. Some of those that I have seen failed because they did not have sufficient initial investment, did not immediately identify products, or focused on products that took too long to get to market. These were not cases where the technology was not good; rather, it was a case of where the business people failed.

In contrast, several years after my experiences with "boom and bust" startups, I was privileged to watch a different model developed by Terry McQuire of Polaris Ventures in Cambridge, Massachusetts. Terry helped form a company called AIR (Advanced Inhalation Research), licensing in some very basic technology developed by Bob Langer of MIT and David Edwards of Penn State University. They had realized that pulmonary drug delivery was the next big target for drug delivery – potentially a means of delivering drugs currently deliverable only by injection, such as insulin. A number of other companies were working on other means for delivery such as transdermal drug delivery or ultrasound mediated drug delivery, but technical problems were delaying progress. The company was funded with very little investment money, and Terry, Bob, and David immediately set out to establish corporate deals as a source of revenue. In an extremely short time, they had deals with half a dozen companies to take their drugs and turn them into pulmonary formulations. The attraction of the technology was that it was quick to market – reformulating already approved drugs, using simple technology based on a single discovery – that "aerodynamic diameter" not "actual diameter" was determinative of whether or not the drug particles were delivered by inhalation into the "deep lung." The funding for all of the development came from the corporate partners, the time to clinicals was less than 1 year, several products were being developed simultaneously, extremely broad, dominating patents were obtained, and the company was positioned for either sale or an IPO in a little over a year. Approximately 1.5 years after formation of the company, AIR was sold to Alkermes for an excellent return on their investment.

A different model that I have seen implemented by David Scheer of Connecticut, over the last few years, has also been extremely successful. David identified markets with good growth potential and high market caps, companies that had done extensive work in developing products for that market but had made a business decision to move out of the market, then acquired the technology, intellectual property, and people to form new companies. The two companies I have had the most experience with are Esperion and Orapharma, both of which went public less than 2 years after being funded, and were then acquired shortly thereafter by Pfizer and Johnson & Johnson, respectively.

Esperion initially acquired technology based on Apolipoprotein A1 Milano from Pharmacia Upjohn, including an extensive patent portfolio, and the people involved in manufacturing, regulation, and development. Esperion then expanded to obtain rights to some small molecule technology for treatment of atherosclerosis, then rights to liposomal formulations for treatment of atherosclerosis. Phase I data had been obtained by Pharmacia so that at least one product was positioned for Phase II clinical trials within a year of funding.

Orapharma was set up similarly, with an initial product including extensive patent portfolio and people experienced with the science and regulatory sides of the technology, for treatment of periodontal disease using polymer-microsphere-based technology for administration of minocycline into the periodontal pockets. Phase I clinical trials had already been completed by American Cyanamid when it decided not to pursue the technology. Orapharma acquired all rights to the technology and used this as a basis for its IPO about 1.5 years after formation. The money from the IPO allowed the company to enter into phase II/III trials almost immediately, with regulatory approval and product on sale about 1 year after completion of its IPO. Simultaneously, it acquired rights to university technology which was much earlier stage for recruitment and regrowth of the bone in the periodontal pockets.

No matter how the property is protected, or the technology valued, the important issue is that it be developed. The amount of innovation in the medical field has been truly extraordinary, even as the cost of developing it has escalated exponentially. Indeed, we now see companies that appear to conduct their entire research and development based on identifying products such as Eli Lilly's Prozac with huge market shares and profits, reviewing the relevant patents to identify weaknesses, then spending millions of dollars to invalidate the patents, thereby not only achieving the right to market the product generically, but having a period of 180 days exclusivity under the Waxman-Hatch Act in which to recover the litigation costs. One can only hope that the current spate of litigation by generic companies to knock out patents will not discourage further innovation and development of new technologies.

Bcl-2, 410
Beta-III-tubulin, 161
Bilayer model, 271
Bile acid, 506, 511
Bioartificial kidney, 590
Bioartificial liver (BAL), 504
Bioartificial tissues, 109–110, 112–113
Biobrane™, 689
Biocompatibility, 337, 349, 372, 378, 606
  of aliphatic polyester, 313, 345
  fibrosis, and fibrous encapsulation, 387–388
  and host response, 564–566
  immunotoxicity, 388–394
  inflammation
    and wound healing, 379–387
Biodegradable polymeric scaffolds, 314, 579, 589, 603, 615, 625
Biodegradable synthetic polymers, 337, 345, 578, 624
  aliphatic polyesters, 313–314
  amino acid-derived polymers, 349
  block copolymers of polyesters with PEG, 350–351
  crosslinked polymer networks, 352–353
    polyesters, 353–357
  peptides, 349–350
  poly(amino acids), 349
  polyanhydrides, 314, 351
  polyesters, 345
    poly(α-hydroxy acids), 345–347
    polycarbonates, 349
    polylactones, 348
    polyorthoesters, 348–349
  polyphosphazenes, 351–352
  polyurethanes, 350
Biological product (42 USC 262(i)), 737
Biological scaffolds, 312, 314, 327, 330
  macroporous biodegradable scaffolds, design principles of
    fabrication, 314–316
    surface immobilization, 317–319
Biologics Control Act, 735
Biologics License Application (BLA), 701, 738
Bioluminescence imaging (BLI), 192
Biomarkers, 425, 440, 741, 748
Biomaterials, 317, 624
  biocompatibility and bioresponse, 378
    fibrosis and fibrous encapsulation, 387–388
    immunotoxicity, 388–394
    inflammation, 379–387
    wound healing, 379–387
  biodegradable synthetic polymers
    aliphatic polyesters, 313–314
    polyanhydrides, 314
  control, 306–307
  digit reconstruction, 624
  evolution, 113
  goal, 112, 337
  history, 336
  role of, 307, 584–585
  scaffolds, 486
    design principles of, 314–322
  selection, 313
  surface modification
    biological modification, 374–376
    chemical modification, 369–372
    overcoating technologies, 372–374
    surface chemical patterning, 376
    topographical modifications, 372
  sustained release, from macroporous scaffolds, 320–321
  three-dimensional, 36, 312
Biomimetic biomaterials, scaffolds of, 246, 253–254
Bioreactors, 38, 546–549
  blood vessel bioreactors, 276–278
  bone bioreactors, 241–242, 275–276
  for cardiac tissue engineering, 242–243, 566
  cartilage bioreactors, 240, 273–275
  cultures
    differentiation, 567–568
    oxygen supply, 566–567
  key function, 238
  modeling, 239
  types, 239–240
Bioscaffolds, 646
Bladder
  cell transplantation, 587
  matrices, 587
  seromuscular grafts, 586–587
  tissue expansion, 586
Bladder-derived acellular collagen matrix, 585
Blastema, 104, 106, 107–108
Block copolymers, of polyesters with PEG, 350–351
Blood cells, regeneration of, 105
Blood–material interactions, 378, 379–380
Blood urea nitrogen (BUN), 590
Blood vessel bioreactors, 276–278
Blood vessel substitutes, 39, 40
Bone, 487, 489–490
  bioreactors, 275–276
  formation of, 184, 268–269
Bone-central third of the patellar tendon-bone (BPTB), 637, 644
Bone graft, 247, 458, 466, 471
  for bone repair, 460
  carriers, 467, 472
Bone induction, 247–248
Bone–ligament–bone complex, 632–633

Bone marrow (BM), 21, 397
centrifugation, 466
compartments, 182, 185
derived MSCs, 185, 186, 577
derived progenitor cells, 426
derived stem cells, 21, 166, 168
HSCs, 21, 169
MSCs, 21, 171, 556
and stem cell niches, interaction of, 169–170
transplantation of, 21, 22, 41, 172, 176, 556
see also Bone marrow stem cells
Bone marrow aspiration technique, 463, 473
Bone marrow-derived cells (BMDCs), 641
Bone marrow mononuclear cells (BMNC), 420, 422, 425–426, 430–432, 439–440, 556
EPCs, 431–432
HSCs, 431
MSCs, 431
umbilical cord blood cells, 432
Bone marrow stem cells (BMSCs), 21, 111, 166, 438–439, 519
and bone regeneration, 172
clinical applications, for cardiac regeneration, 176
clinical trial, 424–425
endothelial progenitor cells, 169
factors regulating homing and differentiation
G-CSF, 174
IL-8, 175
morphogenic proteins, 175
SCF, 175
SDF-1, 175
TGF-β, 175
VEGF, 173–174
and heart regeneration, 170–172
hematopoietic stem cells, 167–168
in vivo differentiation, 173
and liver regeneration, 172–173
MSCs, 168–169
and nerve cell regeneration, 173
and skeletal muscle regeneration, 172
and tissue regeneration, 170
Bone marrow stromal cells, 175, 252, 253, 431
Bone morphogenetic protein (BMP), 100, 127, 128–129, 141, 142, 175, 187–189, 246, 247–251, 253–254, 255, 462, 479, 491
BMP-2, 187–189, 250, 253, 468, 489, 495
BMP-4, 249, 250
BMP-7, 250
BMP-9, 187, 189
Bone regeneration, 105, 246, 356, 496, 625
biological ingredients
osteogenic cells, delivery of, 460–462
and BMSCs, 172
carriers and growth factors, 462–463
clinical needs, for therapeutic solutions, 458–460
tissue engineering
MSCs, 186–189
Bone repair, 182, 187, 421, 460, 461, 463, 464, 472
Bone–tissue engineering, 487, 489–490
Bone-to-bone healing, 637
BOOST trial, 176, 438
Brachyury, 190, 192
Brain-derived neurotrophic factor (BDNF), 678, 681, 682
Bridge technique, 506, 507

C-kit, 175, 556
C-reactive protein (CRP), 424
C3b, complement-activated fragment, 383, 386, 388
C3H10T1/2, cell lines, 187, 188, 190, 191
Cachexia, 492
Cadaveric allograft, 459, 682
Cadavers, 682
Cadherin, 51–52
Cadherin switching, 51
Calcineurin, 273
inhibitor, 406–407
Calcium precipitation, 160
Calcium signaling, 557
CAMP, 272
Canada, 403
Canaliculi, 241, 479
Carbon nanotubes, 304–305, 306
Cardiac-derived stem cells (CSCs), 432–437
Cardiac differentiation. *See* Cardiomyocytes
Cardiac muscle, 106, 567
Cardiac patch implantation, 559–560, 568–569
Cardiac stem cells, 7, 106
Cardiac tissue, engineering of, 171, 240, 554
bioreactors
differentiation, 567–568
oxygen supply, 566–567
cell therapy, 555–557
clinical applications, of BMSCs, 176
gene therapy, 557–558
implantation, of cardiac patches, 568–569
organ function, of engineered tissue, 562
electrical conduction, 563
endothelialization, 563–564
host response and biocompatibility, 564–566
mechanical elasticity, 563
strength development, 563
thrombogenecity, 563–564
tissue architecture, 563
vascularization, 564
scaffold-based approaches
cell-based cardiac patches, 559–560

Cardiac tissue, engineering of (*Cont'd*)
cell-free cardiac patches, 559
fibrous scaffolds, 560–562
Cardiomyocytes, 7, 115, 171–172, 194–195, 432, 442, 555–557, 559, 567
Cardiovascular applications, in regenerative medicine, 39–40
Cardiovascular disease (CVD), 166, 418
cell-based repair, 420–421
cell therapy, 419–420
Carmustine, 351
Carticel, 3, 18, 40, 490
Cartilage, 491
cell response and transduction mechanisms, 242
loading conditions, 242
morphogenesis, 254
by MSCs, 189–190
regeneration, 490, 491
repair techniques, 40, 329
tissue engineering, requirements, 319
types of, 479
Cartilage-derived morphogenetic proteins (CDMPs), 254
Catagen, 103
Cauchy stress, 265
*cbfa*1, 160
Cbfa1/Runx2, 479
CD13, 169
CD31, 184
CD34, 184
CD36, 70
CD44, 69–70, 79
CD105, 159, 182, 183, 184
CD133, 169
Cell, as signal receiver and processor
growth, differentiation, and apoptosis, 237–238
mechanochemical transduction, 235–237
receptors and sensors, 231
cell–matrix adhesions, 232–234
tensegrity model, 234–235
Cell-associated mineralization, 160
Cell-based cardiac patches, 559–560
Cell-based drug delivery
cell encapsulation
challenges, 522, 524
parameters, 520–522
cell-based protein factory, 524–525
clinical applications, 526
companies working on, 526–527
definition, 518
drug-loaded tumor cell system, 525–527
engineered cells, 519
primary cells, 518–519
Cell-based protein factory, 524–525
Cell-based repair
autologous SKMBs, 426–430
BMNCs, 430–432
cardiac-derived stem cells, 432, 437
cardiovascular regeneration, 418, 420–421
neovascularization, 424
requirements, 440–445
vascular integrity
injury versus repair, 421–426
Cell-based therapies, 8, 16
adult stem cells, 34–35
allogeneic, 20–22
autologous, 17–20
commercialization, 22–23
ESCs, 34
and EGCs, 34
immunosuppressive properties, 22
production and best quality, 23
rationale for, 16
Cell-based tissue repair technique, 598
Cell biology, in musculoskeletal tissues
chondrocytes, 479
muscle cells, 479–480
osteoblasts, 478–479
osteocytes, 478–479
Cell–cell adhesion, 51–52, 55, 57, 69, 292
Cell–ECM adhesion, 52–53, 232–234
Cell–ECM interactions, 66, 484
composition and diversity, of ECM, 66–68
receptors for, ECM molecules, 68–70
during regeneration, 78–83
signal transduction events, 70–73
during skin wound healing, 73–78
Cell encapsulation
cells and materials, usage, 521
parameters, 520–522
for testosterone therapy, 609–610
therapeutic delivery system, 519
Cell-enriched grafts, bone regeneration, 469–471, 473
Cell fate potential, 126
growth, differentiation, and apoptosis, 237–238
Cell-free cardiac patches, 559
Cell isolation techniques, for bone regeneration
autologous bone marrow, on optimized matrices, 465
bone marrow centrifugation, 466
culture expansion, 466
enzymatic tissue digestion, 465
point-of-care osteogenic cell enrichment, 465
selective cell retention, 466
Cell-mediated delayed hypersensitivity, 390, 391
"Cell-painting" technologies, 474

Cell receptors and sensors, 231–232
Cell response and transduction mechanisms
  in bone tissues, 241
  in cartilage, 242
  in vasculature, 239–240
Cell–scaffold wound dressing, 746
Cell-seeded tissue-engineered bladder, 587, 588
Cell seeding techniques, 483, 546–549
Cell source, 4–10, 519, 556, 742
  for cardiac tissue engineering, 556
  tissue engineered vascular grafts, 546
Cell-support matrices, 578, 624
Cell-surface receptors, 68, 70, 388
Cell therapy, 741–743
  for cardiac tissue, 555–557
  cell isolation techniques, for bone regeneration
    autologous bone marrow, on optimized matrices, 465
    bone marrow centrifugation, 466
    culture expansion, 466
    enzymatic tissue digestion, 465
    point-of-care osteogenic cell enrichment, 465
    selective cell retention, 466
  clinical trials, 437–440, 443
  clinical studies, for bone regeneration
    autologous bone marrow, with optimized matrices, 472–473
    cell-enriched grafts, 473
  commercialization, 22–23
  future developments, for bone regeneration, 473–474
  G-CSF, 439
  for ligament and tendon healing, 641–642
  osteogenic cells, source of
    allogeneic bone marrow, 463
    autologous bone marrow, 462–463
    gene therapy, 464–465
    novel tissue sources, 463–464
  peripheral arterial disease, 439–440
  preclinical studies, for bone regeneration, 466
    autologous bone marrow, with optimized matrices, 467–469
    cell-enriched grafts, 469–471
  reparative potential
    autologous SKMBs, 426–430
    bone marrow mononuclear cells, 430–432
    cardiac-derived stem cells, 432–437
Cell therapy, reparative potential of
  autologous SKMBs, 426–430
  bone marrow mononuclear cells, 430–432
  cardiac-derived stem cells, 432–437
Cell transplant, 503, 506, 508
  adult stem cells, 110–111
  ESCs, 111–112
  fetal cells, 109–110
Cell types
  in adaptive immune system, 380, 389
Cellular microenvironments, engineering
  dynamic changes
    adhesive environment, 296–297
    soluble environment, regulation, 296
  microengineered tools
    micropatterened screening arrays, 287–289
    spatial patterning, 289–290
  microfluidics, control soluble cues, 285–287
  multicellular constructs, organization
    heterotypic interactions, 294
    homotypic interactions, 292–294
    three-dimensional patterning, 294–296
    two-dimensional patterning, 292
  substrate mechanics engineering
    MEMS devices, 290–291
    patterning stiffness, 291–292
Cellular, Tissue and Gene Therapies Advisory Committee (CTGTAC), 742, 743, 745, 748
Center for Biologics Evaluation and Research (CBER), 699, 700, 737
Center for Devices and Radiological Health (CDRH), 699, 737, 746, 747, 749
Center for Drug Evaluation and Research (CDER), 699, 737
Centers for Disease Control and Prevention (CDC), 478, 743
Ceramics, 489
Chang medium, 159, 161
Chemokine, 67, 74, 83, 173, 285–286
Chemotaxis, 247–248, 250
Chitosan, 327–329
Chondrocytes, 112, 272, 275, 479, 606–607
  bioreactor design, 242–243
  cell response and transduction mechanisms, 242
  cell therapy, 255
  transplantation, 19
Chondrogenesis, 159, 250, 479, 485
Chorionic villus sampling (CVS), 154, 159
Chronic inflammation, 349, 381, 383–384
Circulating progenitor cells, 438
ckit^{neg} cells, 160, 161
ckit^{pos} cells, 159–161
Claudin, 57
Collagen, 326–327, 493, 600
  scaffolds, 677–678
  sponge, 562, 563, 565, 568
  types, 631–632
Collagen Meniscal Implant (CMI), 327
Combination product (21 CFR 3.2(e)), 738–739
Combined ACL/MCL injuries, 638
Common lymphoid progenitors (CLP), 105

Company formation
advantages, 768
versus licensing, 767–768
Compensatory hyperplasia, 101–102
Complement system, 388
Composite scaffolds, 354, 626
Composite tissue structures, creation of, 625–627
"Composition of matter", 755, 765
Computational fluid dynamic (CFD) software, 239
Congestive heart failure, 2, 39, 40
Connected cellular network (CCN), 241
Connexin-32, 505
Connexin-43, 562
Constitutive equation, 265–267
Contigen®, 327
Copyrights, 763
Coral, 489
CorCap, 559
Corporal smooth muscle, 588
Corpus cavernosum reconstruction
acellular collagen matrix preparation, 603–604
human endothelial cell culture, from foreskin, 604–605
smooth muscle cell culture, 602–603
Corticosteroids, 406, 661
Covalent coating method, 373
COX enzymes. *See* Cyclooxygenase enzymes
Craniofacial MSCs, 186
Craniofacial tissue, 494–495
Crigler–Najjar (CN), 509–510
Critical Path Opportunities List, 748
Critical-sized defects (CSD), 356
Cross-linked polyesters
fumarate-based polymers, 353–356
polymers containing acrylate, methacrylate, or vinylsulfone functionalities, 356–357
Cross-sectional area (CSA), 108, 265, 632
Crystallinity, 346
Culture-expanded cells
in bone regeneration, 471–472
Cultured autologous articular chondrocytes. *See* Carticel
Cultured autologous epidermal keratinocytes. *See* Epicel
Current Good Manufacturing Practice (cGMP), 211, 739
Cyclin-dependent kinases, 101
Cyclooxygenase enzymes, 660
Cox-2, 114, 273, 660
Cytofluorimetric analysis, 159
Cytokines, 22, 67, 141, 169, 173, 390, 425, 439, 657–658, 659, 678
Cytoplasmic calcium, 271
Cytotoxic hypersensitivity, 391, 392
Daclizumab, 406
Decorin, 68, 76, 80, 640, 665
Dedifferentiation, 17, 106–108
Definition of
MSCs, 183–184, 478
regeneration, 101, 246
regenerative medicine, 35, 38–39
Definitive endoderm, 213, 215
Degeneration phase, in skeletal muscle healing, 657–658
Degenerative disk disease, 494
Delta, 108, 214
Delta-crystallin enhancer-binding factor 1 (δEF1), 57
Demineralized bone matrix (DBM), 246–248, 349, 489
Department of Health and Human Services (DHHS) Committee, 747
Dermagraft, 3, 689, 746
Dermal fibroblasts, 494, 569
Dermal wound model, 76
Desaminotyrosyl-tyrosine ethyl ester (DTE), 349
Design patent, 755
Desmin, 161
Desomosomes, 51
Developmental mechanisms, of regeneration and strategies, 100
regeneration mechanisms, 101
adult stem cells, activation of, 102–106
aging effects, 108
compensatory hyperplasia, 101–102
dedifferentiation, 106–108
regenerative medicine, strategies of, 109
bioartificial tissues, 112–113
cell transplants, 109–112
chemical/physical induction, of repair and regeneration, 113–117
Device (21 USC 321(h)), 737–738
Dexamethasone, 320
Dextran, 305
Diabetes, 2, 7, 28, 41, 195, 208, 218, 400, 404, 699
islet transplantation, 407–409
Diabetes Control and Complications Trial (DCCT), 400
Dialysis vascular access, 532–534
Differentiated cell, 8, 19, 31–32, 33, 39, 131, 155, 162
Differentiation assays, 23
Differentiation potential
of MSC, 184, 482
stem cells, 30, 31, 33, 34
Digit, 623, 625–627
Diisocyanate, 343–344
Dipeptidylpeptidase (DPPIV), 505
Direct perfusion bioreactor, 238–239
Discoidin domain receptors (DDR), 70
Distraction osteogenesis, 487
Divisional applications, role in patent protection, 760–761

DNA methylation, 132
Donor-specific tolerance, 411–412
Dopaminergic neurons, 7, 39, 109, 110
Dor procedure, 559
Double-quantum filtered magnetic resonance (MR) imaging, 191
Doxorubicin, 525
Drosophila melanogaster, 57, 58
Drug (21 USC 321(g)(1)), 737
Drug-loaded tumor cell (DLTC)system, 525
Dubin–Johnson syndrome, 506, 510
Duchenne muscular dystrophy (DMD), 110, 661
Dulbecco modified Eagle medium (DMEM), 155, 184
Dynamic cardiomyoplasty, 426
Dynamic compression, 269, 274
Dynamic loading, 275
"Dynamic reciprocity", 66
Dynamic regulation, of soluble environment, 296
Dystrophic epidermolysis bullosa (DEB), 746
Dystrophin, 493, 661

E-cadherin, 51–52, 55, 56, 57–58
E2A, 51, 57
Ectogenesis, 722–723
Edmonton Protocol, 41, 402–404, 406–407, 412
EEG, 729
Effector T-cells, 389
EKG, 729
Elastin, 632
Elastin-laminin receptor (ELR), 70
Elastography, 231
Electrical conduction, of cardiac tissue, 563
Electrical stimulation, to cardiac tissue, 567–568
Electropatterning methods, 295
Electrospinning, 4, 239, 316–317, 348, 492, 544, 561, 562
Electrospun scaffolds, 544, 562
Embryogenesis, 29, 558, 722
Embryoid bodies (EBs), 6, 37, 41, 126, 143
Embryonal carcinoma (EC) cells, 126, 127, 132, 138–139, 158, 585
Embryonic blastomeres, 480
Embryonic germ cells (EGCs), 29, 32, 126
  versus ESCs, 34
Embryonic stem cells (ESCs), 30–32, 100, 111–112, 212, 480–481, 624, 717, 718, 720
  culture of, 140–143
  derivation of, 20–21, 29, 138, 139–140
  differentiation, 143–144
  pluripotency, 144
  see also Human embryonic stem cells (hESCs)
Encapsulation system, 375
End-stage renal disease (ESRD), 210, 400, 401, 532
Endochondral ossification, 479
Endocrine specification, 214
Endoderm, 30
Endoderm/gut endothelium generation, 213–214
  pancreatic differentiation, 214
Endogenous tissue repair, 421
Endomysium, 656
Endostatin, 76, 519, 525
Endothelial cells, 37, 105, 169, 170, 603
Endothelial–leukocyte adhesion molecules (ELAMs), 383
Endothelial nitric oxide synthase (eNOS), 272, 273, 277
Endothelial phenotype, induction of, 37, 161
Endothelial progenitor cells (EPCs), 33, 169, 431–432
Endothelial stem cells (EnSCs), 104, 105–106
Endothelial tube formation, 75
Endothelialization, 563–564
Engelbreth-Holm-Swarm (EHS) tumors, 75
Engineered cardiac tissue, function of
  electrical conduction, 563
  endothelialization, 563–564
  host response, and biocompatibility, 564–566
  mechanical elasticity, 563
  strength development, 563
  thrombogenecity, 563–564
  tissue architecture, 563
  vascularization, 564
Engineered penile prostheses, 588–589
Engineering strain, 262, 264, 265
Engineering substrate mechanics
  MEMS devices, 290–291
  patterning substrate stiffness, 291–292
Enhanced green fluorescent protein (eGFP), 172
Enzymatic tissue digestion, 465
Enzyme-linked immunosorbent assay (ELISA), 508
Ependymal cells, 104
Epiblast, 213
Epicel, 3, 17, 18, 690
Epidermal growth factor (EGF), 67–68
Epidosembryo, 718–719, 722
Epigenetic and environmental regulation, 131–132
Epimysium, 656
Epithelial cells, 50, 51, 53, 67, 235
Epithelial-to-mesenchymal transition (EMT), 50, 104
  molecular control of, 53
    ECM signaling, 56
    signaling molecules, 54–56
    transcriptional program, 57–59
  molecular mechanisms of, 50
    basal lamina, invasion of, 53
    cell–cell adhesion, changes in, 51–52
    cell–ECM adhesion changes, 52–53
    cell motility, stimulation of, 53
eRas, 130
ErbB2/HER-2/Neu receptor, 55

ERK mitogen-activated protein kinase (MAPK), 141
Erythrocytes, 482
Erythropoietin, 193, 576
Esperion, 768
Ethical considerations, 714
  arguments do not work, 714–718
  arguments that work, 718–723
Ethical constraints, 722–723
  ectogenesis, 722–723
  humanitarian, embryo use for, 722
  oocyte donation, 723
  patent infringement claims, 723
European Medical Evaluation Agency (EMEA), 733
European Patent Office, 757, 762
Ex vivo gene therapy, 464–465, 495
  for bone regeneration, 463
Ex vivo measurements, of mechanical properties, 231
Expanded poly(tetrafluroethylene) (ePTFE), 386
Expressed sequence tags (ESTs), 756
Extracellular matrix (ECM), 50, 66, 239, 284, 302, 585
  composition and diversity, 66–68
  receptors for, 68–70
  as scaffold material, 329–330
  signaling, 56, 70
Extracellular signaling factors, 127–130

F04 Division IV, 747
Fabrication methods
  for biodegradable polymer scaffolds, 314–317
"Fact-based reasoning", 716–717
FDA Biological Response Modifiers Advisory Committee, 742
FDA Modernization Act (FDAMA), 740
  of 1997, 747
FDA regulatory process
  approval mechanism and clinical studies, 738–739
  center of, 737
  clinical development plan, 746
  and critical path science, 748–749
  gene therapy, 744
  guidance, 735–737, 740
  history of, 734
  human cells and tissues, for transplantation, 740–741
  human cellular therapies, 741–743
  laws, 735–737
  meetings, 739–740, 747–748
  organization and jurisdictional issue, 737–738
  preclinical development plan, 745–746
  regulations, 735–737, 740
  skin constructs, 746
  standards development program, 746
  xenotransplantation, 743–744
Federal Food and Drug Act, 735, 746
Federal Register, 762
Femoral bone defects, 187
Femur–graft–tibia complex (FGTC), 637
Femur–MCL–tibia complex (FMTC), 630, 638
Fetal bovine serum (FBS), 185
Fetal calf serum (FCS), 183, 184
Fetal fibroblasts, 78–79
Fetal origin, of stem cells, 212
Fetal tissues, 39, 482, 590–591
  regenerative medical application
    genitourinary system, 584, 585, 590–591
Fetal wound healing
  adhesion and migration, 78–79
  differentiation, 79–80
  proliferation, 79
Fibrin, 76, 114, 380, 524, 676, 678, 679
Fibrin–fibronectin meshwok, 73
Fibroblast alignment, 645
Fibroblast feeder layers, 141
Fibroblast growth factor (FGF), 55, 76, 128, 130, 141–142, 214, 525
Fibroblast/myofibroblast apoptosis, 77
Fibroblastic colony-forming units (CFU-Fs), 186
Fibrocartilage, 492
Fibromodulin, 80, 83
Fibronectin (FN), 57, 74, 79, 81, 235, 247, 578
Fibrosis, 81, 386, 659
  anti-fibrotic therapy, 665
  and fibrous encapsulation, 387–388
  prevention, 664–665
  and TGF-β1, 665
Fibrous scaffolds, in cardiac tissue engineering, 560–562
Ficoll, 466, 482
File wrapper estoppel, 765
Finite strain theory, 634
First Piola–Kirchoff stress, 265
Fletcher, Joseph, 720
Fluid-induced shear stress, 277
  cell response and transduction mechanisms, 239–240
Fluorescence activated cell sorter (FACS), 482
Fluorescenct *in situ* hybridization (FISH) analyses, 170
Fluorine nuclei (^{19}F), 305
^{19}FMRI, 305
Focal adhesion kinase (FAK), 232, 233, 234
Follistatin, 663
Food and Drug Administration (FDA), 211, 462, 721.734
Food, Drug, and Cosmetic Act (FD&C Act), 700, 735
Foreign body giant cell (FBGC), 385
  formation and interactions, 386–387
Freeze-gelation method, 328
Fumarate-based polymers, 353–356

Fumaric acid, 353
Functional renal tissue *in vivo*, regeneration of, 579–581
Functional tissue engineering (FTE), 487, 630
applications of
cell therapy, 642–643
future directions, 646
gene therapy, 641–642
growth factors, 639–641
mechanical factors, 645–646
scaffolding, 643–645

G-protein, 55, 83, 240, 271
G-protein-coupled receptor, 271, 272
G418 selection, 215
Gal epitope, 393
Galactose
modified PLGA, 318
Ganciclovir, 5
Gas-foaming/salt-leaching method, 315–316, 319, 320
Gastrointestinal (GI) tract, 173, 611
Gastrulation, 55
in amniotes, 50
definitive endoderm, 213
Drosophila, 57, 59
morphogenetic movements of, 100
sea urchin, 52
Gelatin, 127, 159
Gene expression
Myf5, 161, 480
Myf6 (MRF4), 161
Gene ontology (GO), 328
Gene therapy
for bone regeneration, 464–465, 489
in cardiac tissue, 557–558
in cartilage regeneration, 491
in hepatocyte transplantation, 508
in ligament and tendon healing, 640–641
MSC, role of, 494
with MSC-based vehicle, 885
in musculoskeletal repair, 487
regulations, 744
General Agreement for Trade and Tariffs (GATT), 760, 761
Genital tissues, 587
corporal smooth muscle, 588
engineered penile prostheses, 588–589
female genital tissues, 589
Genitourinary system
biomaterials, role, 584–585
fetal tissue engineering, 590–591
injectable therapies, 591–592
reconstitution strategies, 584
testicular hormone replacement, 592
urologic structures
tissue engineering, progress of, 585–590
vascularization, 585
Genzyme Biosurgery, 3
Carcitel™, 490
Epicel™, 690
Germ line intervention, 721
Glial derived neurotrophic factor (GDNF), 110, 115
Glutaraldehyde, 327, 330
Glutathione, 576
Glycosaminoglycan (GAG), 66, 68, 328, 681
GOC, 617
Goretex®, 339
Graft versus host disease (GVHD), 22, 156
Granulation tissue, 77, 329, 384
development of, 381
formation of, 74, 76
Granulocyte colony-stimulating factor (G-CSF), 173
cell therapy and administration of, 439
Green fluorescent protein (GFP), 191, 480
Green strain, 264
Growth differentiation factor 8 (Gdf-8), 663
Growth/differentiation factors (GDFs), 128–129, 249, 491
Growth factors, 67, 113, 384
role in ligament and tendon healing, 638
in vitro studies, 639
in vivo studies, 639–640
therapeutic use, 2
GSK3β inhibition, 130
GTPase proteins, 233

Hair follicle stem cells (HFSC), 103, 127
Hare, Richard, 714
HCT/Ps, 701, 702, 703, 741
Healos®, 467, 473
Health Resources and Service Administration (HRSA), 743
Heart regeneration
and BMSCs, 170–172
see also Cardiac tissue, engineering of
Heat shock protein (HSP), 558
Hedgehog, 54, 251
Helicobacter pylori, 55
Hematocrit (Hct), 193
Hematopoietic stem cells (HSCs), 21, 33, 105, 127, 156, 167–168, 431, 748
and autoimmune applications, 41
compartment, of bone marrow, 185
ex vivo expansion, 484
HSC niche, 170
and tissue regeneration, 160–161

Hematopoietic system, 167, 170
Hemodialysis, 533, 534
Hemorrhage, 635–636
Heparan sulfate proteoglycans, 81
Heparin, 67, 76–77, 155, 321, 322, 374, 468, 678
Hepatocyte allografts, 508
Hepatocyte apoptosis, 82
Hepatocyte growth factor (HGF), 55, 67, 81, 102, 106, 558, 658
Hepatocytes, 80, 101, 161, 172, 196, 213, 217, 284, 288, 294
 phenotype induction
  of pluripotent stem cells, 161
 transplantation
  in acute liver failure, 507–508
  aspects of, 511
  bridge technique, 506–507
  choice of sites for, 504
  integration of, 504–506
  for metabolic liver disease, 508–511
  for non-organ transplant candidates, 511–512
  preventing factors, 512
  shunting of, 507
Hepatocyte-like cells, 196
Herceptin®, 55
Herpes simplex virus (HSV), 5
Heterodimers, 68, 71, 232, 271
Heterotypic interactions, of multicellular constructs, 294
HIF2™α, 132
Holiday Inn®, 762
Homeobox genes (HOX), 255
"Homologous use", 701
Homotypic interactions, of multicellular constructs, 292–294
Host response
 and biocompatibility, 564–566
 to engineered cardiac tissue, 564–566
Human amniotic epithelial cells (HAEC), 157, 158
Human amniotic mesenchymal cells (hAMC), 157
Human autologous chondrocytes (HACs), 18
Human bone marrow-derived stem cells (hBMSCs), 195
Human cells, tissue, and cellular and tissue-based products (HCT/Ps), 701, 741
Human chorionic gonadotropin (hCG), 610
Human chromosome 12p13, 132
Human embryonic stem cells (hESCs), 6–7, 8, 30, 126, 480, 623
 cardiomyocyte derivation, 7
 culture of, 141–143
 condition, 31
 derivation of, 140
 developmental potential
  differentiation of, 143–144
  pluripotency of, 144–145
 differentiation potential, 31
 in vitro model of, 31, 140
 manipulation, genetic approaches to
 phenotypes, 31
 see also Embryonic stem cells (ESCs)
Human epidermal keratinocytes (HEKs), 17, 18
Human leukocyte antigen (HLA), 193, 408, 508
Human MSCs (hMSCs), 169, 182, 290, 463, 491
Human nerve growth factor (hNGF), 519
Human tissue engineering, ethical issues in, 726–733
 arguments for, 727–728
 brokenness nature and human duty, 728–729
 case study, 726
 debates of, 729–730
 deconstruction, perils of, 730
 financial backing, 732–733
 intellectual property, 732
 ownership, 730–731
 pre-market gating, 732
 principle reasons, 727–728
Humanitarian Use Device (HUD), 700, 738
Humanitarinan Device Exemption (HDE), 738
Hurler syndrome, 192
Hyaline cartilage, 18, 356, 479
Hyaluronan, 56, 191, 678
Hyaluronic acid (HA), 78, 313, 319, 678
Hybrid, 505
Hydrogels, 113, 295, 337–338, 339, 340, 341, 352, 355, 357, 559–560, 677
 based cardiac tissue engineering, 559
Hydrolytically stable polyurethanes, 343–344
Hydrophobic polymers, 351, 352
Hydrophobic polyphosphazenes, 352
Hydroxyapatite (HAP), 253, 459, 468, 469, 624
Hyperbaric oxygen (HBO), 664
Hyperglycemic liver environment, 211
Hyperplasia, 101–102
Hypersensitivity, 390, 391, 393
HYPO score, 404
Hypoglycemic unawareness, 401, 402, 404, 412
Hypoxia-inducible factor-1α (HIF-1α), 175
Hysteresis, 634

Id (inhibitor of differentiation) proteins, 100, 141
Ideal scaffold
 for ligament and tendon healing, 642
IgE antibodies, 391, 393
IgG, 383, 393
IgM, 391, 393
Iliac crest, 186, 459, 473
 autograft harvesting, 459

Imminent death, 714, 719
Immune compatibility, 8–10, 579
Immune complex hypersensitivity, 390, 391
Immunocytochemistry (ICC), 159, 163, 580
Immunofluorescence, 286, 288
Immunogenicity, 9, 393
Immunohistochemical assay, 109, 567, 579, 663
Immunomodulation
  and islet cell transplantation, 411–412
Immunomodulatory effects, of MSCs, 192–194
Immunostimulation, 392
Immunosuppression, 8, 209, 392, 402, 405, 406, 485, 568
  in islet cell transplantation, 406–407
Immunotoxicity, 378, 388–394
  cytokines
    and effects, 390
  effector T lymphocytes, 389
  hypersensitivity responses, 390–391
  persistent chronic inflammation, 391–392
  representative tests, 392
  tissue-engineered constructs, 391
Implantable protein factory (ImPACT™), 525
In silico modeling, 230
In vitro differentiation
  of MSCs, 184, 190, 196
In vitro fertilization (IVF), 140, 480
In vitro model, of human ESCs, 31, 140
In vivo differentiation
  of bone marrow stem cells, 173
  of unrestricted somatic stem cells, 158
In vivo gene therapy, for bone regeneration, 464
In vivo measurements
  mechanical properties, of tissue, 231
In vivo model, of human ESCs, 31
Indium^{111}, 510
Indomethacin, prostaglandin inhibitor, 193
Inflammation, 211, 378, 379, 381, 382, 383, 423, 424, 658, 659, 661
  and wound healing
    acute inflammation, 382–383
    blood–material interactions, 379
    chronic inflammation, 383–384
    FBGC formation and interactions, 386–387
    granulation tissue, 384
    inflammatory response, 379–380
    macrophage interactions, 384–386
    provisional matrix formation, 380–381
    temporal sequence of, 381–382
Inflammatory cell emigration, 383
Inflammatory mediators, 383
Injury-induced regeneration, 103, 115
Injury versus repair, 421–426
Inkjet printing, 559
Innate immune system, 564–565
Inner cell mass (ICM), 126, 138, 139, 144
Instant blood-mediated inflammatory reaction (IBMIR), 410, 412
Insulin-like growth factor (IGF), 344
  IGF-1, 105, 113, 480, 491, 662, 678
Insulin-producing cells
  alternative sources, 212–213
Intact extracellular matrix, as scaffold material, 329–330
Integra®, 114, 688
Integrin–ECM binding, 72
Integrin-linked kinase (ILK), 56, 73
Intellectual property rights
  copyrights, 763
  importance, 764–766
  patents, 754–757
    novelty, 757–759
    obviousness, 759
    patent rights, 760
    patent term and patent term extension, 760–762
    requirements, 759–760
  service marks, 762
  trade marks, 762–763
  trade names, 762
  trade secrets, 763
  valuation, 766–769
Interferon-gamma (IFNγ), 665
Interleukin
  IL-1, 390
  IL-2, 193, 390
  IL-4, 169, 382, 387
  IL-6, 425
  IL-8, 74, 75, 83, 175
  IL-10, 193
  IL-13, 387
International Conference on Harmonization (ICH), 737
International Society for Cellular Therapy (ISCT), 483
International Standards Organization (ISO), 747
Interstitial collagens, 77
Interstitial fluid flow (IFF), 241
Intervertebral disk (IVD)
  degenerative disk disease, 494
  regeneration, 191–192
Intestinal stem cells (ISC), 127
Intima, 276, 534
Intracellular activators, 232
Intracellular adhesion molecule (ICAM), 383
Intracellular mechanics, 236–237
Intracellular multi-molecular proteins, 232

Intracorporeal kidney support
  end stage renal failure, 576
  functional renal tissue *in vivo*, regeneration, 579–581
  kidney tissue regeneration, principles, 576–578
  renal structure *in vivo*, creation, 578–579
Intramyocardial injection, 22, 558
Investigational Device Exemption (IDE), 700, 739
Investigational New Drug (IND), 211, 739
Ischemia–reperfusion injury, 82
Ischemic cardiomyopathy trial, 437
Islet after kidney (IAK), 210
Islet cell therapy
  β-cell replacement therapy, 209–212
    benefits, 209–210
    limitations to, 211–212
  in future, 218–219
  insulin-producing cells
    alternative sources of, 212–213
  islet neogenesis, 214–217
    from adult stem cells, 215–217
    from ES cells, 214–215
  pancreatic development, 213–214
    β-cell differentiation, 214
    endocrine specification, 214
    endoderm/gut endothelium generation, 213–214
  and pancreatic stem cells, 208
  transdifferentiation, 213, 217–218
Islet cell transplantation. *See* Islet transplantation
Islet engraftment, in post-transplant period, 409–411
Islet neogenesis, 214–217
  from adult stem cells, 215–217
  from ES cells, 214–215
Islet transplantation, 209–210, 400
  clinical aspects
    patient assessment and selection, 404
    procedure, 404–406
  Edmonton Protocol, 402–404
  future challenges
    donor-specific tolerance, 411–412
    engraftment post-transplant, improvement of, 409–411
    living donor islet transplantation, 407–409
    stem cell transplantation, 409
    xenotransplantation, 409
  history of, 401–402
  risks
    immunospressive therapy and complications, 406–407
    surgical complications, 406
Islet xenotransplantation, 409
Islets of Langerhans, 208, 357
Isobutyl-1-methylxanthine, 485
Isotropic nerve grafts, 674
  agarose gels, 677
  collagen gels, 677–678
  ECM molecules, 678–679
  natural materials, as scaffolds, 674
  neuronal support cells, seeding, 679–680
  neurotrophic factors, 678
  synthetic scaffolds, for nerve repair, 674–677

Johnson & Johnson, 768
Joint loading, 268.274
Joint motion, 634–635
Jungle, The, 735

Kefauver–Harris Amendments, 735
Keloid scars, 77
Keratinocyte grafting, 691
Keratins, 17, 234
Kernicterus, 510
Kidney, 524
  bone marrow to, 173
  tissue creation, 591
  tissue generation, principles, 576–578
Knee motion, 634–635

Lability index, 404
LacZ, 189, 640
Laminin, 36, 52, 67, 75, 76, 80, 81, 559, 562, 578, 678–679, 681
Langmuir–Blodgett deposition method, 373
Large diameter vessels engineering
  tissue-engineered blood vessels (TEBVs), 534
    ECs, 535–536
    smooth muscle cells, 536–538
  tissue-engineered vascular grafts
    cell seeding and preconditioning, 546–549
    cell source, 546
    large diameter TEBV, 539
    process, schematic depiction of, 544
    scaffolds, 539, 544–546
    small diameter TEBV, 538–539
  vascular disease
    dialysis vascular access, need for, 533–534
    prevalence and impact, 532–533
Lateral inhibition process
  of endocrine specification, 214
Lef-1, 58
LEF/TCF transcription factor, 52, 58
Lentiviral vector, 487
Leukemia inhibitory factor (LIF), 127–128, 140, 141, 169
  LIF receptor (LIFR), 128
Leukocyte–endothelial cell interactions, 383

Lewis, C.S., 727, 728
Leydig cells
  encapsulation, 609–610
Licensing
  of biologic products, 700–701
  versus company formation, 767–768
LIGA, 372
Ligament and tendon injuries, 630
  biology, 631–632
  biomechanics
    joint function, 634–635
    tensile testing, 632–634
  FTE applications
    cell therapy, 641–642
    gene therapy, 640–641
    growth factors, 638–640
    mechanical factors, 644–645
    scaffolding, 642–644
  healing of, 635–636
    future directions, 645–646
    hemorrhage phase, 635–636
    inflammatory phase, 635
    remodeling phase, 636
    reparative phase, 636
  MCL, of knee, 636–638
    ACL reconstruction, 637–638
    combined ligament injuries, 638
Light microscopy, 160
Limb prosthetics, 623
Limbal stem cells, 103
Limbs regeneration, 106
Lim, 432
Lin^{neg} HSCs, 167
Lineage conversion assays, 111
Lineage-restricted stem cells, 7
Lineage-specific markers, 6
Liver regeneration, 80
  adhesion and migration, 80–81
  apoptosis, 82–83
  and BMSCs, 172–173
  bone marrow stem cells and, 172
  differentiation, 81–82
  proliferation, 81
Liver stem cells, 103
Living donor islet transplantation, 407–409
Lower critical solution temperature (LCST), 340, 341
Lymphocyte/monocyte predominant chronic
    inflammatory response, 387
Lysozyme, 328

M-cadherin, 480
Macrophage interactions, 384–386
Macrophage mannose receptor (MMR), 387
Macroporous biodegradable scaffolds
  bioactive molecule
    surface immobilization in, 317–319
    sustained release of, 320–322
  fabrication of, 314–317
Macroporous hydroxyapatite scaffolds, 187
Madin–Darby canine kidney (MDCK), 51, 57
Magnetic-activated cell sorting (MACS), 482
Magnetic resonance imaging (MRI), 231, 304, 305, 443
Major histocompatibility complex (MHC), 5, 9, 193
Mammalian stem cells, 5
Mammography Quality Standards Act (MQSA), 737
Mandibular distraction osteogenesis, 495
Mannitol, 376
Matrigel, 567, 679
Matrix metalloproteinases (MMPs), 52, 68, 356, 490–491
Maxillofacial BM, 186
McKinney, Gerry, 728
Mean arterial pressure (MAP), 484–485
Mechanical determinant, of tissue development
  bioreactors, overview of
    modeling, 239
    types, 238–239
  cell
    fates, 237–238
    mechanochemical transduction, 235–237
    receptors and sensors, 231–232
    as signal receiver and processor, 231
    tensegrity model of, 234–235
  ex vivo measurements, 231
  forces, in biological tissue, 230
  in vivo measurement, 231
  mechanical properties, of tissues, 230
  mechanics of materials, 229
    force and stress, 229
    material properties, 229–230
  practical examples of
    bone, 240–242
    cartilage, 242–243
    vasculature, 239–240
Mechanical forces, in tissue development
  biological tissues, forces, 230
  ex vivo measurements, 231
  in vivo measurements, 231
  mechanical properties, of tissues, 230
  mechanics of materials, 229–230
Mechanochemical transduction
  hard-wired nucleus, 237
  intracellular mechanics, alteration, 236–237
  ligands and binding sites, 235
  mechanosensitive ion channels, 236, 270

Mechanosensitive ion channels, 236, 270
Mechanosensory neurons, 236
Mechanotransduction
  mechanism, 269–273
    cellular response, 272–273
    force transmission, 269–270
    signal propagation, 271–272
    see also Mechanochemical transduction
Medial collateral ligament (MCL), 630
  of knee, 636
    ACL reconstruction, 637–638
    combined ligament injuries, 638
  SIS treatment, effect of, 642
Medical Device Amendments, 735
Membrane attack complex (MAC), 388
Membrane permeability, 522
MEMS devices
  to measure cellular forces, 290–291
Meniscus, 491–492
Mesangioblasts stem cells, 493
Mesenchymal cells, 50, 67, 479
Mesenchymal stem cells (MSCs), 19, 33, 105, 110, 155, 168–169, 182, 284, 431, 481
  based EC, 195
  and collagen gels, 641
  culture expansion, 466, 471
  definition of, 183–184
  and ECM interactions, 484
  gene therapy using, 494
  identification, 483–484
  immunomodulatory effects, 192–194
  immunoregulation, 485–486
  immunosuppressive properties, 22
  implantation of, 489
  in vitro behavior, 483
  isolation, 482–483
  markers, 483
  multidifferential potential of, 481, 482
  nature of, 184–185
  niche, 484
  nonskeletal tissue regeneration, 194–196
  regulation, 484–485
  schematic representation of, 461
  self-renewal capacity and multipotentiality of, 487
  skeletal tissue regeneration
    bone, 186–189
    cartilage, 189–190
    intervertebral disk, 191–192
    tendon, 190–191
  source and isolation techniques, 185–186
  tissues containing MSCs, 186
  umbilical cord blood-derived, 483
  see also Bone marrow stromal cells
Mesenchymal-to-epithelial transition (MET), 50, 51, 53, 104
Mesendoderm, 213
Mesoangioblasts, 127
Mesoderm cells, 30, 52, 57, 100
Metabolic and secretory applications, in regenerative medicine, 41
Metabolic liver disease
  clinical transplants for, 509
  hepatocyte transplantation for, 508–511
Metachromatic leukodystrophy (MLD), 192
Metastais-associated protein 3 (MTA3), 58
Micro-computed tomography (micro-CT), 189, 276
Microcontact printing, 289, 376, 562
Microencapsulation, 519–520, 610
Microengineered tools, adhesive environment
  micropatterned screening arrays, 287–289
  spatial patterning, 289–290
Microenvironment, cues in, 35–38
  3D environment, 36–37
  bioreactors, 37–38
Microfabricated post-array detector (mPAD), 291
Microfluidics, 285–287, 292
  soluble cues control, 285–286
Micropatterned screening arrays, 287–289, 293, 294, 296
Microthalamia-associated transcription factor (MITF), 58
Mill, John Stuart, 716
Minimal manipulation, 701, 703
Minocyline, 769
Mixed lymphocyte reaction (MLR), 22
Molecular organization, of cells, 50
  molecular control, of EMT, 53–59
    basal lamina invasion, 53
    cell–cell adhesion changes, 51–52
    cell–ECM adhesion changes, 52–53
    cell motility simulation, 53
  molecular mechanisms, of EMT, 50–53
    ECM signaling, 54
    signaling molecule, 54
  transcriptional program, of EMT, 57–59
Molecular weight cutoff (MWCO), 522, 523
Mononuclear cells (MNCs), 185
Mononuclear phagocytic system (MPS), 385
Morphogen, 127, 246, 247
Morphogenesis, 246
  biomimetic biomaterials, scaffolds of, 253–254
  bone marphogenetic proteins, 247–251
  cartilage-derived morphogenetic proteins, 254
  effects of stress on, 267
  regenerative medicine and surgery, 254–256
  stem cells, 251–253
  steps in, 248

Morphometric analysis, 77
Mouse embryonic stem cells (mESCs), 128, 139
  BMPs effects on, 129
  culture of, 140–141
  differentiation of, 143
  pluripotency, molecular control of, 144
  Wnt signaling in, 130
Multicellular constructs, organization of
  heterotypic interactions, 292–294
  homotypic interaction, 292–294
  three-dimensional patterning, 294–296
  two-dimensional patterning, 292–294
Multidrug resistance protein2 (MRP2), 506
Multipotent adult male germline stem cells, 126
Multipotent adult progenitor cells (MAPCs), 8, 126, 169, 195, 212
Multipotent cell, 30, 127, 166
Multi-wall carbon nanotube (MWNT), 305
Muschler's selective retention technology, 466, 469, 473
Muscle regeneration, 172, 656, 662
Muscular dystrophy, 478, 492–493
Musculoskeletal repair
  articular cartilage, 490–491
  bone, 487–489
  cell-based therapies for, 478, 486–495
    craniofacial tissue, 480–481
    strategy of, 486
  cell biology, 478
    chondrocytes, 479
    muscle cells, 479–480
    osteoblasts, 478–479
    osteocytes, 478–479
    tendons and ligament cells, 480
  connective tissue cell, types, 522
  craniofacial tissue, 494–495
  ESCs, 480–484
  gene therapy, 487
  in vivo cell-based, reports, 488
  invertebral disk, 494
  meniscus, 491–492
  MSCs, 481
    identification, 483–484
    immunoregulation, 485–486
    in vitro behavior, 483
    isolation, 482–483
    niche, 484
    regulation, 484–485
  osteochondral tissue, 492
  skeletal muscle, 492–493
  tendon and ligament, 493–494
  tissue engineering, 486–487
Myoblasts, 427–428, 493, 657
  skeletal, 555, 656
Myocardial infarction, 554, 555, 558, 569
Myocardial repair, 39, 430, 441, 555, 556
MyoD, 58, 161, 557
Myofiber, 108, 426, 656, 657, 658, 660, 664
Myofibroblast, 384
  apoptosis, 77, 80, 82
  differentiation, 73, 76–77, 79–80, 81–82, 83
Myogenic phenotype induction
  of pluripotent stem cells, 161
Myostatin (Mstn), 663

N-acetylglucosamine, 327, 328
N-cadherin, 51
N-vinyl pyrrolidone (NVP), 354
Nanocarrier components, 302, 303, 305
Nanodelivery vehicles, 305
Nanofibrous biodegradable scaffolds, 316
Nanog, 100, 131, 132, 144
Nanoparticulate imaging probes, 304
Nanotechnology, applications of, 302
  as multifunctional tool
    for biomaterial control, 306
    for cell-based therapies, 304–306
  impact, on regenerative medicine, 303–304
National Institutes of Health (NIH), 717, 744
Natural scaffolds
  chitosan, 327–329
  collagen, 326–327
  ECM, 329–330
Necrosis, 421, 715
Neo-dermis, 17
Neovascularization, 169, 384, 418, 625
  cell-based repair for
    cardiovascular regeneration and, 418
Nerve cell regeneration
  and BMSCs, 173
Nerve growth factor (NGF), 524, 662, 678
Nerve guidance conduits/channels (NGCs), 672, 674–677
Nestin-positive cells, 216
Neural applications, in regenerative medicine, 39
Neural cell adhesion molecule (NCAM), 480
Neural Ceroid Lipofuscinosis, 9
Neural stem cells (NSCs), 7, 104, 115
Neurogenic phenotype induction
  of pluripotent stem cells, 161
Neurospheres, 37
Neurotrophins, 678
New Drug Applications (NDA), 738
NGF fusion proteins, 524
Nitric oxide (NO), 536
Nitrogen
  ion beam implementation of, 372

Nodal signaling, 129, 213, 214
Non-classical type, of molecular regulation, 132
Noncomplicity defenses, 717
Non-covalent coating method, 373
Non-degradable synthetic polymers
  polymers with –C–C– backbone
    hydrolytically stable polyurethanes, 343–344
    poly(2-hydroxyethyl methacrylate), 340
    poly(ethylene terephthalate), 343
    poly(ethylene), 338
    poly(meth)acrylates
    poly(N-isopropylacrylamide), 340–341
    poly(propylene), 338
    poly(styrene), 338
    poly(tetrafluoroethylene), 338–339
    polyacrylamides, 339–341
    polyethers, 341–342
    polysiloxanes, 342–343
Nonhuman sialic acid, 142
Nonindividuation argument, 715–716
Non-invasive monitoring method, 231, 304
Non-linear viscoelastic models, 634
Non-organ transplant candidates
  hepatocyte transplantation for, 511–512
Nonprocreative cloning, 718, 720
Non-seeded acellular matrices, 585
Non-seeded allogeneic acellular bladder matrices, 587
Nonskeletal tissue regeneration, by MSCs, 194–196
Non-steroidal anti-inflammatory drugs (NSAIDs), 660
Non-traumatic peripheral nerve injuries, 672
Non-viral vectors, 464
Normal stress, in tissue development, 229, 265
Notch intracellular domain (NICD), 110
Notch signaling pathway, 56, 107, 214
NS-208, 660
Nuclear export sequence (NES), 59
Nuclear localization, 58, 59
Nuclear reprogramming
  process, 131
Nuclear transfer (NT), 34
Nuclear transplantation, 579, 718
  see also Somatic cell nuclear transfer (SCNT)
Nucleofection, 189
Nucleus pulposus (NP) cells, 191, 494

Oasis™, 114
Occludin, 57
Octamer-binding protein 4 (Oct-4), 157, 158
  knock-out embryos and stem cells, 131
  locus, 132
Ocusert®, 338
Office of Combination Product (OCP), 738
Oil-O-Red, 160
Olfactory nerve neurons, 104
Oligo(poly(ethylene glycol) fumarate) (OPF), 355, 356
Oligodeoxynucleotides (ODN), 640, 641
Oligomaltose, 376
Oocytes, 720, 721
  donation, 723
Operational tolerance, 411
Opsonins, 383
Orapharma, 768, 769
Orcel, 746
Orkin®, 762
Ornithine transcarbamylase (OTC), 503, 509, 510
Orphan Drug Act, 754
Orthopedic applications, in regenerative medicine, 40–41
Osteoarthritic TMJ, 494
Osteoarthritis (OA), 260, 478, 490, 492
Osteoblast (OB), 105, 170, 241, 465, 478–479, 484
  niche, 170
Osteocalcin, 160, 192
Osteochondral tissue, 492
Osteoclast, 241, 421
Osteoconduction, 487
Osteoconductive materials, 461
Osteoconductive scaffolds, 461
Osteocytes, 241, 478–479
Osteogenesis imperfecta, 40, 464
Osteogenetic proteins (OPs), 249
Osteogenic cells, 460, 461
  delivery of, 460–461
  sources for cell therapy
    allogeneic bone marrow, 463
    autologous bone marrow, 462–463
    gene therapy, 464–465
    novel tissue sources, 463–464
Osteogenic phenotype induction
  of pluripotent stem cells, 160
Osteogenin, 249
Osteoinduction, 487
Osteoinductive graft materials, 462
Osteoproduction, 487
Osteoprogenitor cells, 465, 466, 479
Osteopromotive graft materials, 461
Osterix, 479
Ovarian follicles
  in vitro culture of, 615–617
Ovary
  anatomy, 615
  in vitro culture, of ovarian follicles, 615–617

Overcoating technologies, of surface modification, 372
covalent coatings, 373–374
non-covalent coatings, 373
schematic representation of, 370
Oxygen transport, for cardiac tissue engineering, 566–567

P130CAS, 71, 236
PA6, 143–144
PAI-1, 79
Pancreatic beta cells, 7, 195
Pancreatic development, in islet cell therapy, 213
β-cell differentiation, 214
endocrine specification, 214
endoderm/gut endothelium generation, 213–214
pancreatic differentiation, 214
Pancreatic differentiation, 214
Pancreatic islet cells
encapsulation, 520
Pancreatic stem cells
in future, 208–209
islet cell therapy and, 208
Paracrine effects, of MSCs, 188
Parkinson's disease, 2, 39
Patch-clamp studies, 236
Patellar tendon (PT) fibroblast, 640
Patent, 723, 755
provisional applications role, 761
Patent claims, 757, 764, 765
Patent Extension Act, 762
Patent infringement, 723, 757, 765
claims, 723
Patent rights, 754, 760, 765
Paxillin, 71
Penile prothesis, for reconstruction, 606
Penis, tissue reconstruction of, 601
anatomy, 602
corpus cavernosum reconstruction
acellular collagen matrix preparation, 603–604
human endothelial cell culture, 604–606
smooth muscle cell culture, 602–603
penile prothesis, 606–607
cartilage tissue harvest, 606
Percoll, 482
Perfusion bioreactors, 37, 238, 277
Peripheral nerve regeneration, 672
animal models, 682–683
anisotropic nerve grafts, 680
aligned anisotropic scaffolds, 680–681
cell-seeded longitudinally aligned NGCs, 681
ECM molecules, 681
neurotrophic factors, 681
current strategies, 673
ECM molecules, 678–679
isotrophic nerve grafts, 674
agarose gels, 677
collagen gels, 677–678
hydrogels, as scaffolds, 677
natural materials, as scaffolds, 674
synthetic scaffolds, for nerve repair, 674–677
natural nerve grafts, 681–682
neuronal support cells for, 679–680
neurotrophic factors, 678
Peripheral nervous system (PNS), 672, 677, 679
Permacol®, 114
Peroxisome proliferation-activated receptor 2 (ppar(2)), 160
Pfizer, 768
Phagocytosis, 383
Phalanges and small joints
composite tissue structures, creation, 625–627
future perspectives, 627
principles of, 623–625
Phenobarbital, 507
Phenotype, cell change, 31
Phorbol 12-myristate 13-acetate (PMA), 193
Phosphacan, 68
Phosphate-buffered saline (PBS), 155, 525
Phosphatidylinositol bis-phosphate (PIP2), 72
Phosphatidylserine, 305
Phospholipase C (PLC), 271
PLC_{γ}, 72
Photografting, 373
Photolithography, 372, 376
Photopatterning method, 294, 295
Photopolymerized (meth)acrylated polymer networks, 356, 357
Phronesis, 729
Physical entrapment method, 375
Physical stress, 260, 265
effects on repair, and remodeling, 267–268
effects on tissue growth, 267
in vitro mechanical conditioning
blood vessel bioreactors, 276–278
bone bioreactors, 275–276
cartilage bioreactors, 273–275
mechanotransduction mechanism, 269–273
and regenerative medicine, 268–269
structural hierarchy and continuum concept, 261
Physiochemical surface modification
chemical modifications, 369–372
topographical modifications, 372

Phytohemagglutinin (PHA), 193
PI3K, 54, 55
Pipecolic acid, 511
Placenta and amniotic fluid
  differentiated cells, 155
  isolation and characterization, 159–162
  mesenchymal cells, 155–158
  pluripotent stem cells, 158–162
Placental growth factor (PLGF), 104
Plaque activation, 424
Plasma deposition, 373
Plasma fibronectin, 81
Plasmin, 81
Plasminogen activator, 53
Plasticity, 5
Platelet-derived growth factor (PDGF), 67, 142, 380, 384, 491, 636, 640, 663
Platelet-derived growth factor-BB (PDGF-BB), 169, 639
Platelet factor 4, 380, 381
Platelet-rich plasma (PRP), 461, 495
Platelets, 380
Pluripotency, 146
  of BMSCs, 166
  of human ES cells, 143
    molecular control of, 144
  molecular basis, 126–133
    BMP/GDF, 128–129
    epigenetic and environmental regulation, 131–132
    extracellular signaling factors, 127–131
    extracellular signaling transduction, 127–131
    fibroblast growth factor (FGF), 130
    leukemia inhibitory factor (LIF), 127–128
    TGF-β/activin/nodal, 129
    transcriptional networks, 131
    Wnt, 130
Pluripotent cells, 30
  in vitro, 127
  in vivo, 127
Pluripotent embryonic carcinomas, 132
Pluripotent stem cells, 130, 158, 194, 722
  isolation and characterization, 159–162
  from placenta and amniotic fluid, 158–162
Pluronics®, 341, 350
Point-of-care osteogenic cell enrichment, 465
Poisson's ratio, 262, 266
Poloxamers, 341–342
Poly tetrafluoroethylene (PTFE), 338–339, 533
Poly(2-hydroxyethyl methacrylate) (PHEMA), 339, 340, 376
Poly(amino acids), as synthetic polymers, 349
Poly(D, L-lactic acid-co-glycolic acid) (PLGA), 4, 313, 315, 321, 345, 394, 471, 480, 585, 677
Poly(dimethyl siloxane) (PDMS), 285, 289, 342, 559
Poly(DTE carbonate), 349
Poly(ethylene glycol) (PEG), 113, 315, 341, 490, 520
  and diacrylate, 355, 356, 357
  and dimethacrylate, 356, 357
  vinylsulfones, 356
Poly(ethylene glycol) diacrylate (PEGDA), 294
Poly(ethylene oxide) (PEO), 341, 347
Poly(ethylene terephthalate) (PET), 343
Poly(ethylene-*co*-vinyl acetate) (PEVAc), 338
Poly(ethyleneimine) (PEI), 320
Poly(glycolic acid) (PGA), 314, 345, 346, 350, 490, 538, 544, 585, 596, 624, 677
  scaffolds, 560, 589
Poly(iminocarbonates), 349
Poly(L-lactic acid) (PLLA), 313, 346, 607
Poly(L-lactide), 320
Poly(L-lysine), 609–610
Poly(lactic acid) (PLA), 345, 346, 490, 544, 562, 585, 677
Poly(lactic acid-co-lysine), 314
Poly(lactide-co-glycolide) (PLG), 564, 625
Poly(meth)acrylates, 340–341
Poly(methyl methacrylate) (PMMA), 336, 339–340, 386
Poly(N-isopropyl acrylamide), 340–341, 342, 352
Poly(*p*-dioxanone), 348
Poly(propylene fumarate) (PPF), 353–354, 355
Poly(propylene fumarate-co-ethylene glycol) (P(PF-*co*-EG)), 354, 355
Poly(propylene oxide) (PPO), 341
Poly(propylene) (PP), 338–339
Poly(SA–CPP), 351
Poly(styrene) (PS), 338–339
Poly(α-hydroxy acids), 345–347
Poly(α-hydroxy esters), 347
Poly(ε-caprolactone) (PCL), 348, 491, 596, 624
Poly-3-hydroxybutyrate, 677
Poly-4-hydroxybutyrate, 539
Polyacrylamide, 340–341
Polyanhydrides, 313, 314, 351
Polycarbonates, 349
Polycarprolactone, 313
Polyesters, synthetic polymers
  poly(α-hydroxy acids), 345–347
  polycarbonates, 349
  polylactones, 348
  polyorthoesters, 348–349
Polyethers, 341–342
Polyethylene, 338
  derivatives
    poly(propylene), 338
    poly(styrene), 338
    poly(tetrafluoroethylene), 338–339
Polylactones, 348
Polymerase chain reaction (PCR), 508, 625

Polymeric encapsulation systems, 520
Polyorthoesters (POEs), 348–349
Polyphosphazenes, 351–352
Polysiloxanes, 342–343, 370
Polyurethanes (PUs), 337, 350
Polyurethaneurea, 344, 350
Polyvinyl alcohol (PVA), 356
Porcine dermal matrix, 114
Porcine small intestine submucosa, 114, 329, 642
Porous scaffolds, 328
  in cardiac tissue engineering, 560
Portal vein thrombosis, 401, 405
Posterior definitive endoderm, 213
Potency, of cells, 126–127
pp125FAK, 236
Precursor cells. *See* Progenitor cells
Pre-implantation mammalian embryo, 131
Pre-market approval (PMA), 342, 699
"Pressure relief valve" mechanism, 236
Primary cells, 518–519
Primary mesenchyme cells (PMCs), 51, 52
Primatrix™, 114
Primitive endoderm, 131
Principles, of regenerative medicine
  current and future perspectives, 2–10
Processed lipoaspirate (PLA) cells, 186
Procreative cloning, 718
Progenitor cells, 29, 130, 704
Progestasert®, 338
Progressive familial intrahepatic cholestasis (PFIC), 509, 510
Pro-HGF, 102
Prostaglandin (PG), 241, 609, 660–661
  PGE, 183, 276
  PGE-2, 183, 276
Protein kinase C (PKC), 72, 272
Protein–protein cross-linking agents, 330
Protein matrices, 142
Proteoglycan, 66, 81, 632, 682
Provisional matrix formation, 379, 380, 387–388
Public Health Service (PHS) Act, 735, 765
Publication, and patentability, 757–758
Pulmonary radiotracer uptake, 510

Quantum dots (QD), 304
Quasi-linear viscoelastic theory, 634

Radiation grafting, 373
Radio-frequency glow discharge (RFGD), 370, 373
Rap1, 52–53
Ras, 52, 272
Rat amniotic epithelial (RAE) cells, 157
Rat pancreatic extract (RPE), 195
Reactive oxygen species (ROS), 53
Receptor–ligand interactions, 78–79
Receptor tyrosine kinase (RTK) ligands, 55–56
Recombinant human bone morphogenic protein-2 (rhBMP-2), 356
Recombinant human growth hormone (rhGH), 394
Recurrent autoimmunity, 411
Re-epithelialization, 74, 76, 79
Regeneration index (RI), 107
Regeneration mechanisms, 101
  adult stem cells, activation of, 102–106
  aging effects, 108
  compensatory hyperplasia, 101–102
  dedifferentiation, 106–108
Regeneration phase, in skeletal muscle healing, 658–659
  exercise and muscle regeneration, 664
  growth factors, 662–663
  hyperbaric oxygen, 664
  myostatin, neutralization of, 663
  therapeutic ultrasound, 663–664
Regeneration templates, 114
Relaxin, 665
Renal cell culture, 576
Renal cell proliferation, 578
Renal structures
  ex vivo, 590
  in vivo, 590
    creation, 578–579
Renal transplantation, 581
Reproductive system, tissue engineering of, 596
  in female
    ovary, 615–617
    uterus, 613–615
    vagina, 610–612
  in male
    penis, 601–607
    testes, 607–610
    urethra, 597–601
Request for Designation (RFD), 738
Retinoic acid, 117, 480
Reverse transcriptase polymerase chain reaction (RT-PCR), 162, 254, 579, 641
RHAMM (receptor for hyaluronate-mediated motility), 70
Rheumatoid arthritis, 478
Rosuvastatin, 425
Rotating-wall bioreactor, 238, 240
Runx2, 105, 479, 489

Salt-leaching technique
  macroporous scaffolds fabrication, 315
Sarcomeres, 567

Satellite cells, 106, 426, 480, 658, 664
Sca-1, 437, 556
Scaffolds, 642–644
as biomimetic biomaterials, 253–254
for cardiac tissue engineering
cell-based cardiac patches, 559–560
cell-free cardiac patches, 559
combination approach, 562
fibrous scaffold approach, 560–562
hydrogel approach, 559–560
porous scaffold approach, 560
thin films, 562
for nerve repair, 674, 680–681
hydrogels, 677–678
natural materials, 674
synthetic, 674–677
for small diameter tissue-engineered vessels
biologic scaffolds, 314–322
Scanning electron microscopy (SEM), 189, 291, 316, 317, 321, 580, 680
Scar tissue formation, 691–692
Schwann cells, 106, 115, 674
Screening arrays, 287–289
SeaPrep® agarose hydrogel, 677
Sebacic acid (SA), 314, 351
Secrete human leukocyte antigen-I (sHLA-I), 508
Seeded tubularized collagen matrices, 585
Self-assembled monolayers (SAMs), 297, 344, 373
"Self" tissue, 327
Serine/threonine kinases, 54, 72, 252, 272
Serum replacement (SR), 141
Service mark, 762
Severe combined immunodeficiency (SCID), 189, 569, 744
Shear strain
in tissue development, 229
Shear stress, 238, 239, 260, 265, 276
in tissue development, 229
Side population (SP) cells, 33, 167, 462–463
Signal transduction events, during cell–ECM interactions
adhesion and migration, 70–72
apoptosis, 73
differentiation, 72–73
proliferation and survival, 72
Signal transduction pathways, 73
Signaling molecules, 54, 232, 233
Notch pathway, 56
RTK ligands, 55–56
TGF-β pathway, 54
Wnt pathway, 55
Silicones, 337, 342
Simultaneous islet kidney (SIK), 210
Single cell suspension, 17
Single-wall carbon nanotube (SWNT), 305
Sirolimus, 405, 406
Skeletal muscle, 106, 492–493
regeneration and BMSCs, 172
regenerative medicine implications, 656
degeneration phase, 657–658
fibrosis phase, 659, 664–665
inflammation phase, 658, 659–662
regeneration phase, 658–659, 662–664
Skeletal muscle cells, 479–480
Skeletal muscle healing, inflammation role in, 659–662
cyclooxygenase pathway, 660–661
inflammation and muscular dystrophy, 661–662
inflammatory cells, 659–660
prostaglandins, 660–661
Skeletal myoblasts, 555
Skeletal tissue regeneration, by MSCs
bone, 186–189
cartilage, 189–190
intervertebral disk, 191–192
tendon, 190–191
Skin cell-based therapy, for burn victims, 688
clinical studies, 693–695
conventional therapy and current commercial products, 688–691
regulatory issues, 699–700
21 CFR 1271.10(a), 701
21 CFR parts 1270 and 1271, 701
"351" HCT/P, 703–704
"361" HCT/P, 703–704
biologics, licensing of, 700–701
HCT/Ps, 701
skin progenitor cells, 697–699
spray transplantation method, 692–693, 695–697
wound healing, problems of, 691–692
Skin repair, topical agents for, 113–114
Skin wound healing, cell–ECM interactions
adhesion and migration, 73–75
apoptosis, 77–78
differentiation, 76–77
proliferation, 76
Slug, 57, 58
Smad, 54, 129, 250–251
Smad interacting protein (SIP), 251
Sip1, 57
Small intestinal submucosa (SIS), 329, 393, 538, 585, 601, 624, 642
and ECM, 330
Small molecules, for chemical induction, 117
Smooth muscle cells (SMCs), 239, 424, 532, 534, 536, 538
Snail, 57
Sodium alginate, 609

Soft lithography, 289, 376
Soluble factors, 114–115
Solvent casting/salt-leaching technique, 315
Somatic cell nuclear transfer (SCNT), 111, 140
Somatic cells, 20
Sonic hedgehog (Shh), 108, 214, 558
Sox-2, 6, 100, 112, 131, 144
L-Sox-5, 479
Sox-6, 479
Sox-9, 105, 479, 494
SPARC (secreted protein acidic and rich in cysteine), 67, 76
Spatial patterning, of cellular adhesive environment, 289–290, 291
Spemann, Hans, 250
Spinal fusion, 187
Spinner flasks, 242, 489, 560, 562
Split-skin transplantation method, 688, 696
Sponsors, 739
Spontaneous differentiation model, 31
Spray transplantation method, 695–697
Stabilizing pressure input orthosis (SPIO), 305
Stage-specific embryonic antigens (SSEAs), 100
STAT 3, 59, 100, 128, 141
Static flask, bioreactor, 238
Statins, 443
Stellate cells, 82
Stem cell-derived hepatocytes, 503, 512
Stem cell factor (SCF), 170, 175
Stem cells, 20–21, 29, 166, 173, 184, 251–253, 519
  adult stem cells, 32–34
  based gene therapy
    versus BMP-2 protein delivery, 187, 188
  behavioral model, 129
  from bone marrow, 217
  cell-based therapies, issues in, 34–35
  cloning, 159
  derived from placenta and amniotic fluid, 154–163
  differentiation, hierarchical model of, 29–30
  EGCs, 29, 32
  ESCs, 29, 30–32
  of fetal origin, 212
  insulin-producing cells, source of, 208–209
  microenvironment, cues in, 35–38
  niches, 169–170
  plasticity, 30
  in regenerative medicine, 38
    cardiovascular applications, 39–40
    hematopoietic and autoimmune applications, 41
    metabolic and secretory applications, 41
    neural applications, 39
    orthopedic applications, 40–41
  of skeletal and cardiac muscle, 106
  transplantation, 409
Stemness genes, 127, 132
Stethoscopes, 729
Stirred-flask bioreactor, 238–239
Strain, 261–265, 633
  in tissue development, 229
Strategies, of regenerative medicine
  bioartificial tissues, 112–113
  cell transplants, 109–112
  chemical/physical induction, of repair and regeneration, 113–117
Strength development, of cardiac tissue, 563
Stress, 260, 265
  effects on morphogenesis, 267
  effects on repair and remodeling, 267–268
  effects on tissue growth, 267
  in tissue development, 229
Stress fibers, 233
Stress urinary incontinence (SUI), 592
Strictures, 585, 600
Stro-1, 169, 172, 183, 185, 483
Stroke, 161–163
Stroma-supportive system, 185
Stromal cell derived factor 1 (SDF-1), 169, 175
Stromal cells, 194, 252, 459
Stromal stem cells, 185
Stromal vascular fraction, 186
Structural molecules, of cytoskeleton, 234–235
Sulfated glycosaminoglycan (sGAG), 491
Suprathel®, 696
Suramin, 665
Surface analyses techniques, 369
Surface immobilization, of biomaterials
  on macroporous biodegradable scaffolds, 317–319
Surface mechanisms
  signal receiver and processor, 232, 240
Surface modification, of biomaterials
  interfaces, in regenerative medicine, 368
  overcoating technologies, 370, 372
    covalent coatings, 373–374
    non-covalent coatings, 373
  physiochemical modification
    chemical modifications, 369–372
    topographical modifications, 372
  roughness and topography, 372
  strategies, overview of, 368–369
  surface chemical patterning, 376
  surfaces, biological modification, 374–376
Surface modification, of scaffolds, 317, 318
Surface stability, 369
Surgery, regenerative medicine, 254–256
Surgical peripheral nerve injuries, 672
Surgisis®, 114

Syndecan, 68
Synthetic grafts, properties of, 458, 600
Synthetic materials, 336, 578, 624
Synthetic peptides, 349, 557
Synthetic polymers, 336, 596
applications, 357
biodegradable
aliphatic polyesters, 313–314
amino acid-derived polymers, 349–350
block copolymers of polyesters with PEG, 350–351
cross-linked polymer networks, 352–357
peptides, 350
poly(amino acids), 349
polyanhydrides, 314, 351
polyesters, 345–349
polyphosphazenes, 351–352
polyurethanes, 350
non-degradable, 338
Hydrolytically Stable Pus, 343–344
poly(2-hydroxyethyl methacrylate), 340
poly(ethylene), 338
poly(ethylene terephthalate), 343
poly(meth)acrylates, 339
poly(methyl methacrylate), 339–340
poly(N-isopropylacrylamide), 340–341
poly(propylene), 338
poly(styrene), 338
poly(tetrafluoroethylene), 338–339
polyacrylamides, 339
polyethers, 341–342
polysiloxanes, 342–343
Synthetic scaffolds, 326
for ligament and tendon healing, 642
for nerve repair, 674

T cells, 9, 22, 389, 391, 412
see also Somatic cells
T lymphocytes. *See* T cells
Tacrolimus, 403, 404, 406, 410, 411
Taurine, 376
Teflon®, 336, 337, 338, 589
Telogen, 103
Telomerase, 100
Telomerase repeat amplification protocol (TRAP) assay, 160
Temporomandibular joints (TMJ), 494–495
Tenascin C, 67–68, 79, 645
Tendon, 190–191
and ligament, 480, 493–494
injuries, 630
Tensegrity model, of cell, 234–235
Tensile testing, 632–634
Tensin, 71
Tensional integrity. *See* Tensegrity model, of cell
Teratocarcinomas, 138–139
Teratomas, 21, 39, 100, 519, 557
Testes
anatomy, 607
testosterone delivery systems, 608–609
testosterone therapy, cell encapsulation for, 609–610
transplantation, 607–608
Testicular hormone replacement, genitourinary system, 592
Testoderm, 609
Tet-off system, 188
Tether arm, 376
Tethered model
for voltage-gated ion channels, 271
TGF-β/Activin/Nodal pathway, 127, 129
Th1 and Th2 lymphocyte, 394
Th2 helper lymphocytes, 387
Therapeutic cloning, 5, 623
Therapeutic ultrasound, 663–664
Thermally induced phase separation method (TIPS), 320–321
Thin polyurethane films, 562
Three-dimensional biomaterial scaffolds. *See* Biological scaffolds
Thrombogenicity
of cardiac tissue, 563–564
Thrombopoietin (TPO), 170
Thrombospondin, 69, 76
Thrombus formation, 379
Tisseel®, 406
Tissue development, mechanical determinants, 228
bioreactors, 238–239
cell, as signal receiver and processor, 231–237
cell fates, 237–238
mechanical forces, experienced by tissues, 229–231
practical examples, 239–243
Tissue-engineered blood vessels (TEBV)
functions, 532
process, 538, 539–549
studies, 538–539
vascular physiology, 534–535
Tissue-engineered construct, 287, 383
Tissue engineered heart valve (TEHV), 748–749
Tissue engineered medical products (TEMPS), 747
Tissue-engineered vascular grafts
cell seeding and preconditioning, 546–549
cell source, 546
large diameter TEBV, 539
scaffolds, 539–546
small diameter TEBV, 538–539

Tissue engineering, 38, 284, 312, 478–479, 585
of bone, 487
cartilage production, 490–491
and cell-based therapies, 478
of fetus, 590–592
in humans, ethical issues, 726
of ligament and tendon injuries, 630
of reproductive system, 596
in female, 610–617
in male, 597–610
strategies, 486
of urologic structures, 585–590
Tissue growth, 267
effects of stress on, 260
Tissue inhibitor of metalloproteinases (TIMPs), 79
Tissue regeneration, 78, 186
and BMSCs, 170
of kidney, 576–578
Tissue repair
effects of stress on, 267–268
Tissue rules, 741
Tissue therapy
for skeletal muscle injury, 656
Titanium, 370–372
Title 21 Code of Federal Regulations (CFR), 736, 739
1270 and 1271, 701, 703, 741
3.2(e), 738
800–1200, 699
820, 700
Topographical modifications, of surface modification, 372
Totipotent stem cells, 30, 126
Trade name, 762
Trade secrets, 763
Trademark, 762–763
Trans Cyte™, 689
Transcriptional networks, 131
Transdifferentiation, 30, 217–218
see also Plasticity
Transferase proteins, 132
Transforming growth factor beta (TGF-β), 51, 58, 129, 249, 321, 462, 479
and fibrosis, 665
pathway, 54
TGF-β1, 76, 81, 83, 104, 113–114, 356, 480, 639, 658, 659
TGF-β2, 81, 640
TGF-β3, 81, 116, 273, 485
Transglutaminases, 77
Transmission electron microscopy, 77, 642
Traumatic peripheral nerve injuries, 672
Triblock copolymers, 341
Tri-fluoroacetic anhydrides, 370
Tripeptide RGD, 68, 297, 318
Trophectoderm, 131, 139
Tropomyosin, 161, 560, 562
Trypsin, 350
Tubulin, 236
Tumor necrosis factor-alpha (TNFα), 77, 170, 382, 662
Twist, 57, 58
Type 1 diabetes mellitus (T1DM), 208, 218, 400, 401, 411
Type 2 diabetes mellitus (T2DM), 208, 400
Type I collagen, 631
Type V collagen, 631, 636
L-Tyrosine, 349
Tyrosine-based polycarbonates, 349
Tyrosine-derived polyacrylates, 349
Tyrosine-derived polyethers, 349
Tyrosine kinase receptors, 70, 170
Tyrosinemia Type I, 509, 510

Ulex europaeus I (UAE-I)-coated Dynabeads, 603
Ultrasonic pulser, 231
Umbilical cord blood (UCB) stem cells, 8, 34, 156, 186, 432
Umbilical cord matrix stem (UCMS) cells, 8
Unipotent cell, 127
United States, 2, 28, 418, 458, 478, 532, 533, 554, 622, 672, 717, 755, 761, 765
patent law, current issues in intellectual property rights
copyrights, 763
importance, 764–766
patents, 754–762
service marks, 762
trade marks, 762–763
trade names, 762
trade secrets, 763
valuation, 766–769
United States Codes (USC), 736, 754, 759
United States Food and Drug Administration, 16, 211, 254, 313, 443, 462, 490
Unrestricted somatic stem cells (USSCs), 8, 158
Ureter, 586, 589
Urethra, 585
anatomy, 597–598
cell growth, 598
acellular collagen matrix preparation, 601
bladder smooth muscle cell culture, 599–601
urothelial cell culture, 598–599
Urokinase plasminogen activator (uPA), 79, 102
Urologic structures, in tissue engineering
bladder
cell transplantation, 587
matrices, 587
seromuscular grafts, 586–587
tissue expansion, 586

Urologic structures, in tissue engineering (*Cont'd*)
  genital tissues
    corporal smooth muscle, 588
    engineered penile prostheses, 588–589
    female genital tissues, 589
  renal structures
    ex vivo, 590
    in vivo, 590
  ureter, 589
  urethra, 585
Urothelial cell culture, 598–599
Uruguay Round Act, 761
US Patent and Trade mark Office (US PTO), 754, 755, 756, 762
US patent law, current issues, 754
  intellectual property rights, 754
    copyrights, 763
    importance of, 764–766
    novelty, 757–758
    obviousness, 759
    patent rights, 760
    patent term extension, 760–762
    patents, 754–757
    service marks, 762
    trade names, 762
    trade secrets, 763
    trademarks, 762–763
    written description, 759–760
  patentability, 757
  patentable subject matter, 755, 756, 757
  PTO guidelines, 755–756
  valuation of, 766–769
  "real world" context of use, 756
Uterus
  anatomy, 613–614
  uterine tissue reconstruction, 614–615
Utilitarianism, 716

Vagina, 610–611
  anatomy, 611
  tissue engineering, 611–612
Valvular replacements, 39, 40
Variant Creutzfeldt–Jakob Disease (vCJD), 741
Vascular differentiation. *See* Endothelial cells
Vascular disease
  dialysis vascular access, need for, 533–534
  prevalence and impact, 532–533
Vascular endothelial growth factor (VEGF), 67, 169, 173–174, 410, 420, 525, 558, 624, 663
  VEGF-A, 104
Vascular integrity
  injury versus repair, 421–422
Vascular niche, 170
Vascular permeability factor (VPF), 505
Vascular smooth muscle cell (VSMC), 532, 564, 614
Vascularization
  of cardiac tissue, 565
  genitourinary system, 585
  endothelial cell, 239–240
  smooth muscle cell, 239–240
Ventricular remodeling, 22, 170, 555
Vesicoureteral reflux (VUR), 591
Vimentin, 57, 547, 561
Viral vectors, 489
Visceral endoderm, 213, 215
Visceral mesoderm, 195
Vitronectin, 232
Voltage-gated ion channels, 270
Von Kossa, 160

Wavy walled bioreactor, 238, 242–243
Waxman-Hatch Act, 754, 769
Western blot analysis, 162, 589, 626
Wilson's disease, 509
Wnt, 130
  inhibitor, 130, 189
  pathway, 55, 127, 142
  signaling, 55, 58, 130, 142
  signaling pathway, 484, 485
  Wnt3a, 106, 130
  Wnt4, 55
Wolff's law, 240, 260
Women's ischemia syndrome evaluation (WISE) study, 423
World Intellectual Property Organization (WIPO), 762
Wound healing, 20, 77, 421, 691
  blood–material interactions, 379
  essential elements
    fetal healing, 78–80
  inflammatory response, initiation of, 379–380
  problems of, 692–693
  provisional matrix formation, 380–381
  spraying cells, into wound, 693–694
  and temporal sequence of inflammation, 381

Xenogeneic collagen, 327
Xenogeneic components, 31, 393
Xenotransplantation, 39, 409, 743–744
XIAP (X-chromosome linked inhibitor of apoptosis), 410

Zinc-finger transcription factors, 5
Zyderm I®, 327
Zyderm II®, 327
Zygote, 126
Zyplast®, 327

I D No b25235722
UNIVERSITY OF BRADFORD
LIBRARY
14 JUL 2016
ACCESSION No 0401636670
CLASS No
LOCATION
UNIVERSITY OF BRADFORD
WITHDRAWN AND DISCARDED
LIBRARY